ST. OLAF COLLEGE
SCIENCE LIBRARY

APR 0 6 2004

D0026282

FROM MOLECULES TO NETWORKS

AN INTRODUCTION TO CELLULAR AND
MOLECULAR NEUROSCIENCE

FROM MOLECULES TO NETWORKS

AN INTRODUCTION TO CELLULAR AND MOLECULAR NEUROSCIENCE

Edited by

JOHN H. BYRNE
*University of Texas,
Houston*

JAMES L. ROBERTS
*University of Texas,
San Antonio*

ELSEVIER
ACADEMIC
PRESS

Amsterdam   Boston   Heidelberg   London   New York   Oxford   Paris
San Diego   San Francisco   Singapore   Sydney   Tokyo

GP
3Le1
.F76
2004
+CD-Rom
This book is printed on acid-free paper. ∞

Copyright 2004, Elsevier Science (USA)

All rights reserved.
No part of this publication may be reproduced or transmitted in any form or by any means, electronic or mechanical, including photocopy, recording, or any information storage and retrieval system, without permission in writing from the publisher.

Permissions may be sought directly from Elsevier's Science & Technology Rights Department in Oxford, UK: phone: (+44) 1865 843830, fax: (+44) 1865 853333, e-mail: permissions@elsevier.com.uk. You may also complete your request on-line via the Elsevier Science homepage (http://elsevier.com), by selecting "Customer Support" and then "Obtaining Permissions."

Academic Press
An imprint of Elsevier
525 B Street, Suite 1900, San Diego, California 92101-4495, USA
http://www.academicpress.com

Academic Press
An imprint of Elsevier
84 Theobald's Road, London WC1X 8RR, UK
http://www.academicpress.com

Library of Congress Catalog Card Number: 2003107472

International Standard Book Number: 0-12-148660-5
International Standard Book Number (CD): 0-12-148661-3

Printed in Hong Kong
03 04 05 06   9 8 7 6 5 4 3 2 1

546075951

This textbook is dedicated to our wives, Susan and Poco

Full Contents

Contributors   xi
Preface   xiii

1. Cellular Components of Nervous Tissue
PATRICK R. HOF, BRUCE D. TRAPP, JEAN DE VELLIS,
LUZ CLAUDIO, AND DAVID R. COLMAN

The Neuron   1
The Neuroglia   13
The Cerebral Vasculature   22

2. Subcellular Organization of the Nervous System: Organelles and Their Functions
SCOTT BRADY, DAVID R. COLMAN, AND PETER BROPHY

Axons and Dendrites: Unique Structural Components
    of Neurons   31
Protein Synthesis in Nervous Tissue   36
The Cytoskeletons of Neurons and Glial Cells   47
Molecular Motors in the Nervous System   55
Building and Maintaining Nervous System Cells   58

3. Brain Energy Metabolism
PIERRE J. MAGISTRETTI

Energy Metabolism of the Brain as a Whole Organ   67
Tight Coupling of Neuronal Activity, Blood Flow, and
    Energy Metabolism   70
Energy-Producing and Energy-Consuming Processes in
    the Brain   73
Brain Energy Metabolism at the Cellular Level   77
Glutamate and Nitrogen Metabolism: A Coordinated
    Shuttle Between Astrocytes and Neurons   84
The Astrocyte-Neuron Metabolic Unit   87

4. Electrotonic Properties of Axons and Dendrites
GORDON M. SHEPHERD

Spread of Steady-State Signals   93
Spread of Transient Signals   98
Electrotonic Properties Underlying Propagation in
    Axons   100
Electrotonic Spread in Dendrites   102
Dynamic Properties of Passive Electrotonic
    Structure   106
Relating Passive to Active Potentials   111

5. Membrane Potential and Action Potential
DAVID A. McCORMICK

The Membrane Potential   116
The Action Potential   121

6. Molecular Properties of Ion Channels
LILY YEH JAN AND YUH NUNG JAN

Families of Ion Channels   141
Channel Gating   144
Ion Permeation   149
Ion Channel Distribution   154
Summary   157

7. Dynamical Properties of Excitable Membranes
DOUGLAS A. BAXTER, CARMEN C. CANAVIER
AND JOHN H. BYRNE

The Hodgkin–Huxley Model   161
A Geometric Analysis of Excitability   179

8. Release of Neurotransmitters

ROBERT S. ZUCKER, DIMITRI M. KULLMANN, AND
THOMAS L. SCHWARZ

Organization of the Chemical Synapse   197
Excitation–Secretion Coupling   202
The Molecular Mechanisms of the Nerve Terminal   208
Quantal Analysis   221
Short-Term Synaptic Plasticity   235

9. Pharmacology and Biochemistry of Synaptic Transmission: Classic Transmitters

ARIEL Y. DEUTCH AND ROBERT H. ROTH

Diverse Modes of Neuronal Communication   245
Chemical Transmission   246
Classic Neurotransmitters   250
Summary   276

10. Nonclassic Signaling in the Brain

ARIEL Y. DEUTCH AND JAMES L. ROBERTS

Peptide Neurotransmitters   279
Neurotensin as an Example of Peptide
    Neurotransmitters   285
Unconventional Transmitters   287
Synaptic Transmitters in Perspective   295

11. Neurotransmitter Receptors

M. NEAL WAXHAM

Ionotropic Receptors   299
G Protein-Coupled Receptors   319

12. Intracellular Signaling

HOWARD SCHULMAN

Signaling Through G-Protein-Linked Receptors   335
Modulation of Neuronal Function by Protein Kinases
    and Phosphatases   353

13. Regulation of Neuronal Gene Expression and Protein Synthesis

JAMES L. ROBERTS AND JAMES R. LUNDBLAD

Intracellular Signaling Affects Nuclear Gene
    Expression   371

Role of cAMP and Ca^{2+} in the Activation Pathways of
    Transcription   380
Summary   388

14. Mathematical Modeling and Analysis of Intracellular Signaling Pathways

PAUL D. SMOLEN, DOUGLAS A. BAXTER, AND JOHN H. BYRNE

Methods for Modelling Intracellular Signaling
    Pathways   393
General Issues in the Modeling of Biochemical
    Systems   408
Specific Modeling Methods   411
Summary   426

15. Cell–Cell Communication: An Overview Emphasizing Gap Junctions

DAVID C. SPRAY, ELIANA SCEMES, RENATO ROZENTAL,
AND ROLF DERMIETZEL

Chemical and Electrical Synapses Differ in Functional
    Characteristics   435
Biophysical and Pharmacological Properties of Gap
    Junctions in the Nervous System   439
Role of Gap Junctions in Functions of Nervous
    Tissue   442
Gap Junction-Related Neuropathologies   448

16. Postsynaptic Potentials and Synaptic Integration

JOHN H. BYRNE

Ionotropic Receptors: Mediators of Fast Excitatory and
    Inhibitory Synaptic Potentials   459
Metabotropic Receptors: Mediators of Slow Synaptic
    Potentials   472
Integration of Synaptic Potentials   475

17. Information Processing in Complex Dendrites

GORDON M. SHEPHERD

Strategies for Studying Complex Dendrites   480
Summary: The Dendritic Tree as a Complex Information
    Processing System   495

18. Learning and Memory: Basic Mechanisms

THOMAS H. BROWN, JOHN H. BYRNE, KEVIN S. LABAR,
JOSEPH E. LEDOUX, DERICK H. LINDQUIST, RICHARD F.
THOMPSON, AND TIMOTHY J. TEYLER

Long-Term Synaptic Potentiation and Depression   499
Paradigms Have Been Developed To Study Associative
    and Nonassociative Learning   529
Invertebrate Studies: Key Insights From *Aplysia* Into Basic
    Mechanisms of Learning   531

Classical Conditioning in Vertebrates: Discrete Responses
    and Fear as Models of Associative Learning   543
How Does a Change in Synaptic Strength Store a
    Complex Memory?   560
Summary   562

Index   575

Contributors

Numbers in parentheses indicate the pages on which the authors' contributions begin.

Douglas A. Baxter (161, 391)   Department of Neurobiology and Anatomy, The University of Texas-Houston, Medical School Houston, TX, USA

Scott T. Brady (31)   Department of Anatomy and Cell Biology, University of Illinois at Chicago, Il, USA

Peter J. Brophy (31)   Department of Preclinical Veterinary Sciences, University of Edinburgh, Edinburgh, Scotland, UK

Thomas H. Brown (499)   Department of Psychology, Yale University, New Haven, CT, USA

John H. Byrne (161, 391, 459, 499)   Department of Neurobiology and Anatomy, University of Texas Health Science Center, Houston, TX, USA

Carmen C. Canavier (161)   Department of Psychology, University of New Orleans, New Orleans, LA, USA

Luz Claudio (1)   Department of Community and Preventive Medicine, Mt. Sinai School of Medicine, New York, NY, USA.

David R. Colman (1, 31)   Montreal Neurological Institute, McGill University, Montreal, QC, CANADA.

Jean De Vellis (1)   Department of Neurobiology, University of California Los Angeles, School of Medicine, Los Angeles, CA, USA.

Rolf Dermietzel (431)   Institut für Anatomie, Ruhr Universität Bochum, Germany.

Ariel Y. Deutch (245, 279)   Department of Psychiatry and Pharmacology, Vanderbilt University School of Medicine, Nashville, TN, USA.

Patrick R. Hof (1)   Neurobiology of Aging Laboratories, Mt. Sinai School of Medicine, New York, NY, YSA.

Yuh Nung Jan (141)   Department of Physiology, University of California, San Francisco, CA USA

Lily Yeh Jan (141)   Department of Physiology, University of California, San Francisco, CA USA

Dimitri M. Kullmann (197)   Institute of Neurology, University College London, London, UK.

Kevin S. LaBar (499)   Center for Cognitive Neuroscience, Duke University, Durham, NC, USA.

Joseph E. LeDoux (499)   Center for Neural Science, New York University, New York, NY, USA.

Derick H. Lindquist (499)   Department of Psychology, Yale University, New Haven, CT, USA.

James R. Lundblad (371)   Division of Molecular Medicine, Oregon Health Sciences University, Portland, OR, USA.

Pierre J. Magistretti (67)   Institute de Physiologie, Université of Lausanne, Lausanne, Switzerland.

David A. McCormick (115)   Section of Neurobiology, Yale University School of Medicine, New Haven, CT, USA.

James L. Roberts (279, 371)   Department of Pharmacology, University of Texas San Antonio, San Antonio, TX, USA.

Robert H. Roth (245)   Department of Pharmacology, Yale University School of Medicine, New Haven, CT, USA.

Renato Rozental (431)   Department of Neuroscience, Albert Einstein College of Medicine, Bronx, NY, USA.

Eliana Scemes (431)   Department of Neuroscience, Albert Einstein College of Medicine, Bronx, NY, USA.

Howard Schulman (335)   SurroMed, Inc., Mountain View, CA, USA.

Thomas L. Schwarz (197)   Department of Neurology, Children's Hospital, Boston, MA, USA.

Gordon M. Shepherd (91, 479)   Section of Neurobiology, Yale University School of Medicine, New Haven, CT, USA.

Paul D. Smolen (391)   Department of Neurobiology and Anatomy, University of Texas Health Science Center, Houston, TX, USA.

David C. Spray (431)   Department of Neuroscience, Albert Einstein College of Medicine, Bronx, NY, USA.

Timothy J. Teyler (499)   Medical Education Program, University of Idaho, Moscow, ID, USA.

Richard F. Thompson (499)   Department of Psychology, The University of Southern California, Los Angeles, CA, USA.

Bruce D. Trapp (1)   Department of Neuroscience, Cleveland Clinic Foundation, Cleveland, OH, USA.

M. Neal Waxham (299)   Department of Neurobiology and Anatomy, University of Texas Health Science Center, Houston, TX, USA.

Robert S. Zucker (197)   Neurobiology Division, Molecular and Cell Biology Department, University of California, Berkeley, CA, USA.

Preface

The past twenty years have witnessed an exponential increase in the understanding of the nervous system at all levels of analyses. Perhaps the most striking developments have been in the understanding of the cell and molecular biology of the neuron. The field has moved from treating the neuron as a simple black box that added up impinging synaptic input to fire an action potential to one in which the function of nerve cells involves a host of biochemical and biophysical processes that act synergistically to process, transmit and store information. In this book, we have attempted to provide a comprehensive summary of current knowledge of the morphological, biochemical, and biophysical properties of nerve cells. The book is intended for graduate students, advanced undergraduate students, and professionals. The chapters are highly referenced so that readers can pursue topics of interest in greater detail. We have also included material on mathematical modeling approaches to analyze the complex synergistic processes underlying the operation and regulation of nerve cells. These modeling approaches are becoming increasingly important to facilitate the understanding of membrane excitability, synaptic transmission, as well gene and protein networks. The final chapter in the book illustrates the ways in which the great strides in understanding the biochemical and biophysical properties of nerve cells have led to fundamental insights into an important aspect of cognition, memory.

We are extremely grateful to the many authors who have contributed to the book, and the support and encouragement during the two past years of Jasna Markovac and Johannes Menzel of Academic Press. We would also like to thank Evangelos Antzoulatos, Evyatar Av-Ron, Diasinou Fioravanti, Yoshihisa Kubota, Rong-Yu Liu, Fred Lorenzetii, Riccardo Mozzachiodi, Gregg Phares, Travis Rodkey, and Fredy Reyes for help with editing the chapters.

John H Byrne

James L Roberts

1

Cellular Components of Nervous Tissue

Patrick R. Hof, Bruce D. Trapp, Jean de Vellis, Luz Claudio, and David R. Colman

Several types of cellular elements are integrated to yield normally functioning brain tissue. The neuron is the communicating cell, and a wide variety of neuronal subtypes are connected to one another via complex circuitries usually involving multiple synaptic connections. Neuronal physiology is supported and maintained by the neuroglial cells, which have highly diverse and incompletely understood functions. These include myelination, secretion of trophic factors, maintenance of the extracellular milieu, and scavenging of molecular and cellular debris from it. Neuroglial cells also participate in the formation and maintenance of the blood–brain barrier, a multicomponent structure that is interposed between the circulatory system and the brain substance and that serves as the molecular gateway to the brain parenchyma.

THE NEURON

Neurons are highly polarized cells, meaning that they develop, in the course of maturation, distinct subcellular domains that subserve different functions. Morphologically, in a typical neuron, three major regions can be defined: (1) the cell body, or perikaryon, which contains the nucleus and the major cytoplasmic organelles; (2) a variable number of dendrites, which emanate from the perikaryon and ramify over a certain volume of gray matter and which differ in size and shape, depending on the neuronal type; and (3) a single axon, which extends in most cases much farther from the cell body than does the dendritic arbor (Fig. 1.1). The dendrites may be spiny (as in pyramidal cells) or nonspiny (as in most interneurons), whereas the axon is generally smooth and emits a variable number of branches (collaterals). In vertebrates, many axons are surrounded by an insulating myelin sheath, which facilitates rapid impulse conduction. The axon terminal region, where

contacts with other cells are made, displays a wide range of morphological specializations, depending on its target area in the central or peripheral nervous system. Classically, two major morphological types of contacts, or *synapses*, may be recognized by electron microscopy: the asymmetric synapses, responsible for transmission of excitatory inputs, and the symmetric or inhibitory synapses.

The cell body and the dendrites are the two major domains of the cell that receive inputs, and the dendrites play a critically important role in providing a massive receptive area on the neuronal surface. In addition, there is a characteristic shape for each dendritic arbor, which is used to classify neurons into morphological types. Both the structure of the den-

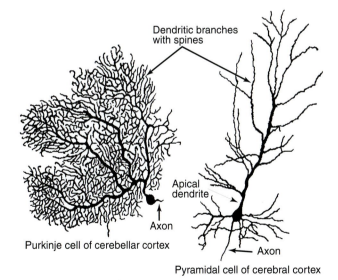

FIGURE 1.1 Typical morphology of projection neurons. On the left is a Purkinje cell of the cerebellar cortex, and on the right, a pyramidal neuron of the neocortex. These neurons are highly polarized. Each has an extensively branched, spiny apical dendrite, shorter basal dendrites, and a single axon emerging from the basal pole of the cell.

Copyright 2004, Elsevier Science (USA).
All rights reserved.

dritic arbor and the distribution of axonal terminal ramifications confer a high level of subcellular specificity in the localization of particular synaptic contacts on a given neuron. The three-dimensional distribution of the dendritic arborization is also important with respect to the type of information transferred to the neuron. A neuron with a dendritic tree restricted to a particular cortical layer may receive a very limited pool of afferents, whereas the widely expanded dendritic arborizations of a large pyramidal neuron will receive highly diversified inputs within the different cortical layers in which segments of the dendritic tree are present (Fig. 1.2) (Mountcastle, 1978; Peters and Jones, 1984; Schmitt *et al.*, 1981; Szentagothai and Arbib, 1974; Lund *et al.*, 1995; Björklund *et al.*, 1990). The structure of the dendritic tree is maintained by surface interactions between adhesion molecules and, intracellularly, by an array of cytoskeletal elements (microtubules, neurofilaments, and associated proteins), which also take part in the movement of organelles within the dendritic cytoplasm.

An important specialization of the dendritic arbor of certain neurons is the presence of large numbers of dendritic spines, which are membrane-limited organelles that project from the surface of the den-

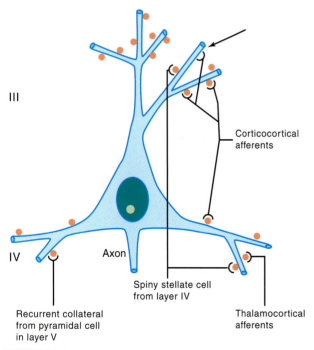

III

IV   Axon

Corticocortical
afferents

Spiny stellate cell
from layer IV

Recurrent collateral
from pyramidal cell
in layer V

Thalamocortical
afferents

FIGURE 1.2   Schematic representation of four major excitatory inputs to pyramidal neurons. A pyramidal neuron in layer III is shown as an example. Note the preferential distribution of synaptic contacts on spines. Spines are labeled in red. Arrow shows a contact directly on the dendritic shaft.

drites. They are abundant in large pyramidal neurons and are much sparser on the dendrites of interneurons. Spines are more numerous on the apical shafts of the pyramidal neurons than on the basal dendrites. As many as 30,000 to 40,000 spines are present on the largest pyramidal neurons. Spines constitute the region of the dendritic arborization that receives most of the excitatory input. Each spine generally contains one asymmetric synapse; thus, the approximate density of excitatory input on a neuron can be inferred from an estimate of its number of spines. The cytoplasm within the spines is characterized by the presence of polyribosomes and a variety of filaments, including actin and α- and β-tubulin, as well as a spine apparatus comprising cisternae, membrane vesicles, and stacks of dense lamellar material (see Box 1.1) (see (Berkley, 1896; Gray, 1959; Ramón y Cajal, 1955; Coss and Perkel, 1985; Scheibel and Scheibel, 1968; Steward and Falk, 1986; Zhang and Benson, 2000; Nimchinsky *et al.*, 2002).

The perikaryon contains the nucleus and a variety of cytoplasmic organelles. Stacks of rough endoplasmic reticulum are conspicuous in large neurons and, when interposed with arrays of free polyribosomes, are referred to as Nissl substance. Another feature of the perikaryal cytoplasm is the presence of a rich cytoskeleton composed primarily of neurofilaments and microtubules, discussed in detail in Chapter 2. These cytoskeletal elements are dispersed in "bundles" that extend into the axon and dendrites (Peters and Jones, 1984). Whereas the dendrites and the cell body can be characterized as the domains of the neuron that receive afferents, the axon, at the other pole of the neuron, is responsible for transmitting neural information. This information may be primary, in the case of a sensory receptor, or processed information that has already been modified through a series of integrative steps. The morphology of the axon and its course through the nervous system are correlated with the type of information processed by the particular neuron and by its connectivity patterns with other neurons. The axon leaves the cell body from a small swelling called the axon hillock. This structure is particularly apparent in large pyramidal neurons; in other cell types, the axon sometimes emerges from one of the main dendrites. At the axon hillock, microtubules are packed into bundles that enter the axon as parallel fascicles. The axon hillock is the part of the neuron from which the action potential is generated. The axon is generally unmyelinated in local-circuit neurons (such as inhibitory interneurons), but it is myelinated in neurons that furnish connections between different parts of the nervous system. Axons usually have larger numbers

---

BOX 1.1

SPINES

Spines are protrusions on the dendritic shafts of neurons and are the site of a large number of axonal contacts. The use of the silver impregnation techniques of Golgi or of the methylene blue used by Ehrlich in the late 19th century led to the discovery of spiny appendages on dendrites of a variety of neurons. The best known are those on pyramidal neurons and Purkinje cells, although spines occur on neuron types at all levels of the central nervous system. In 1896 Berkley observed that terminal boutons were closely apposed on spines (a fact that was later confirmed by Gray (1959) using electron microscopy) and suggested that spines may be involved in conducting impulses from neuron to neuron. In 1904, Santiago Ramón y Cajal suggested that spines could collect the electrical charge resulting from neuronal activity (Ramón y Cajal, 1955). He also noted that spines substantially increase the receptive surface of the dendritic arbor, which may represent an important factor in receiving the contacts made by the axonal terminals of other neurons. It has been calculated that the approximately 4000 spines of a pyramidal neuron account for more than 40% of its total surface area (Peters et al., 1991).

More recent analyses of spine electrical properties have demonstrated that spines are dynamic structures that can regulate many neurochemical events related to synaptic transmission and modulate synaptic efficacy (Coss et al., 1985) (see also Chapter 18). Spines are also known to undergo pathological alterations and have a reduced density in a number of experimental manipulations (such as deprivation of a sensory input) and in

many developmental, neurological, and psychiatric conditions (such as dementing illnesses, chronic alcoholism, schizophrenia, and trisomy 21) (Scheibel and Scheibel, 1968). Morphologically, spines are characterized by a narrow portion emanating from the dendritic shaft, the neck, and an ovoid bulb or head. Spines have an average length of 2 μm despite considerable variability in morphology. At the ultrastructural level (Fig. 1.3), spines are characterized by the presence of asymmetric synapses and a few vesicles and contain fine and quite indistinct filaments. These filaments most likely consist of actin and α- and β-tubulins. The microtubules and neurofilaments present in the dendritic shafts do not penetrate the spines. Mitochondria and free ribosomes are infrequent, although many spines contain polyribosomes in their head and neck. Interestingly, most polyribosomes in dendrites are located at the bases of spines, where they are associated with endoplasmic reticulum, indicating that spines possess the machinery necessary for the local synthesis of proteins (Steward and Falk, 1986).

Another classic feature of the spine is the presence in the spine head of confluent tubular cisterns that represent an extension of the dendritic smooth endoplasmic reticulum. Those cisterns are referred to as the spine apparatus. The function of the spine apparatus is not fully understood but may be related to storage of calcium ions during synaptic transmission. For additional reviews on spines, see Zhang and Benson (2000) and Nimchinsky et al. (2002).

---

of neurofilaments than do dendrites, although this distinction can be difficult to make in small elements that contain fewer neurofilaments. In addition, the axon may be extremely ramified, as in certain local-circuit neurons; it may give out a large number of recurrent collaterals, as in neurons connecting different cortical regions; or it may be relatively straight in the case of projections to subcortical centers, as in cortical motor neurons that send their very long axons to the ventral horn of the spinal cord. At the interface of axon terminals with target cells are the synapses, which represent specialized zones of contact consisting of a presynaptic (axonal) element, a narrow synaptic cleft, and a postsynaptic element on a dendrite or perikaryon. We consider the fine structure of

synapses later in this chapter. In the next section, we turn our attention to the principal morphological features of several neuronal types from the cerebral cortex, subcortical structures, and periphery as typical examples of the cellular diversity in the nervous system.

Pyramidal Cells Are the Main Excitatory Neurons in the Cerebral Cortex

All of the cortical output is mediated through pyramidal neurons, and the intrinsic activity of the neocortex can be viewed simply as a means of finely tuning their output. A pyramidal cell is a highly polarized neuron, with a major orientation axis per-

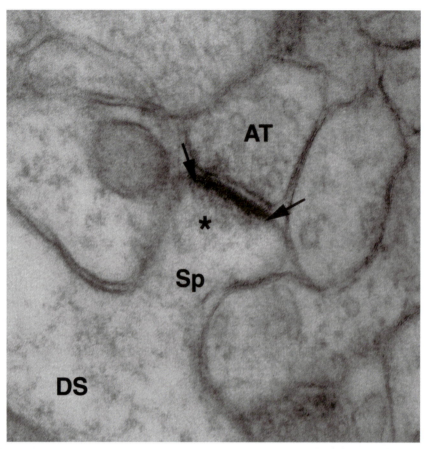

FIGURE 1.3 Ultrastructure of a single dendritic spine (Sp). Note the narrow neck emanating from the main dendritic shaft (DS) and the spine head containing filamentous material, the cisterns of the spine apparatus, and the postsynaptic density of an asymmetric synapse (arrows). AT, axon terminal.

pendicular (or orthogonal) to the pial surface of the cerebral cortex. In cross section, the cell body is roughly triangular (Fig. 1.2), although a large variety of morphological types exist with elongate, horizontal, or vertical fusiform or inverted perikaryal shapes. A pyramidal neuron typically has a large number of dendrites that emanate from the apex and form the base of the cell body. The span of the dendritic tree depends on the laminar localization of the cell body, but it may, as in giant pyramidal neurons, spread over several millimeters. The cell body and dendritic arborization may be restricted to a few layers or, in some cases, may span the entire cortical thickness (Jones, 1984).

In most cases, the axon of a large pyramidal cell extends from the base of the perikaryon and courses toward the subcortical white matter, giving off several collateral branches that are directed to cortical domains generally located within the vicinity of the cell of origin (as explained later in this section). Typically, a pyramidal cell has a large nucleus, a cytoplasmic rim that contains, particularly in large pyra-

midal cells, a collection of granular material chiefly composed of lipofuscin. The deposition of lipofuscin increases with age and is considered a benign change. Although all pyramidal cells possess these general features, they can also be subdivided into numerous classes based on their morphology, laminar location, and connectivity (Fig. 1.4) (Jones, 1975). For instance, small pyramidal neurons in layers II and III of the neocortex have restricted dendritic trees and form vast arrays of axonal collaterals with neighboring cortical domains, whereas medium-to-large pyramidal cells in deep layer III and layer V have much more extensive dendritic trees and furnish long corticocortical connections. Layer V also contains very large pyramidal neurons arranged in clusters or as isolated, somewhat regularly spaced elements. These neurons project to subcortical centers such as the basal ganglia, brainstem, and spinal cord. Finally, layer VI pyramidal cells exhibit a greater morphological variability than do pyramidal cells in other layers and are involved in certain corticocortical as well as corticothalamic projections (Feldman, 1984; Hof *et al.*, 1995a,b).

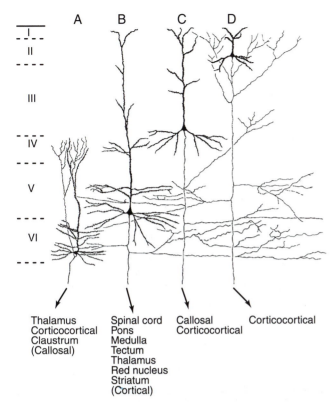

Thalamus
Corticocortical
Claustrum
(Callosal)

Spinal cord
Pons
Medulla
Tectum
Thalamus
Red nucleus
Striatum
(Cortical)

Callosal
Corticocortical

Corticocortical

FIGURE 1.4 Morphology and distribution of neocortical pyramidal neurons. Note the variability in cell size and dendritic arborization as well as the presence of axon collaterals, depending on the laminar localization (I–VI) of the neuron. Also, different types of pyramidal neurons with a precise laminar distribution project to different regions of the brain. Adapted with permission, from Jones (1984).

The excitatory inputs to pyramidal neurons can be divided into intrinsic afferents, such as recurrent collaterals from other pyramidal cells and excitatory interneurons, and extrinsic afferents of thalamic and cortical origin. The neurotransmitters in these excitatory inputs are thought to be glutamate and possibly aspartate. Although this division may appear relatively simplistic, the complexity and heterogeneity of excitatory transmission in the neocortex may not be derived from the presynaptic side, but rather from the postsynaptic side of the synapse. In other words, at the molecular level, a variety of glutamate receptor subunit combinations may confer different functional capacities on a given glutamatergic synapse (see Chapter 11).

Pyramidal cells not only furnish the major excitatory output of the neocortex, but also act as a major intrinsic excitatory input through axonal collaterals. The collaterals of the main axonal branch that exits from the cortex are referred to as recurrent collaterals because they ascend back to superficial layers; thus, the collateral branches of a pyramidal cell synapse in

layers superficial to their origin, although a deep or local system of branches is also present (see Fig. 1.4). Although many of these branches ascend in a radial, vertical pattern of arborization, there are a separate set of projections that travel horizontally over long distances (in some instances as much as 7–8 mm). One of the major functions of the vertically oriented component of the recurrent collaterals may be to interconnect layers III and V, the two major output layers of the neocortex. In layer III pyramidal cells, 95% of the synaptic targets of the recurrent cells are other pyramidal cells. This is true of both the vertical and the distant horizontal recurrent projections. In addition, the majority of these synapses are on dendritic spines and, to some degree, on dendritic shafts. It is possible that there are regional and laminar specificities to these synaptic arrangements, although such fine patterns are not yet fully elucidated (Schmitt *et al.*, 1981; Szentagothai and Arbib, 1974; Lund *et al.*, 1995; Kisvarday *et al.*, 1986a,b). These recurrent projections function to set up local excitatory patterns and coordinate multineuronal assemblies into an excitatory output.

Spiny Stellate Cells Are Excitatory Interneurons

The other major excitatory input to pyramidal cells of cortical origin is provided by the interneuron class referred to as spiny stellate cells, small multipolar neurons with local dendritic and axonal arborizations. These neurons resemble pyramidal cells in that they are the only other cortical neurons with large numbers of dendritic spines, but they differ from pyramidal neurons in that they lack an apical dendrite. Although the dendritic arbor of these neurons tends to be local, it can vary from a primarily radial orientation to one that is more horizontal. The relatively restricted dendritic arbor of these neurons is presumably a manifestation of the fact that they are high-resolution neurons that gather afferents to a very restricted region of cortex. The dendrites rarely leave the layer in which the cell body resides. The spiny stellate cell also resembles the pyramidal cell in that it provides asymmetric synapses that are presumed to be excitatory, and, like pyramidal cells, these neurons are thought to use either glutamate or aspartate as their neurotransmitter.

Spiny stellate cells exhibit extensive regional and laminar specificities in their distribution. Spiny stellate cells are found in highest concentration in layers IVC and IVA of the primary visual cortex, where they constitute the predominant neuronal type. They are also found in large numbers in layer IV of other

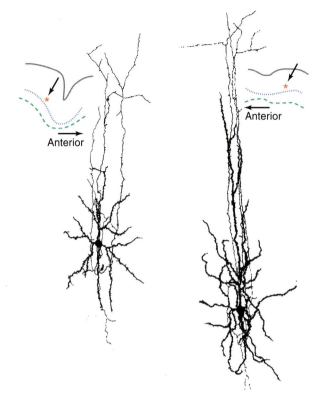

FIGURE 1.5 Drawing of Golgi-impregnated spiny stellate neurons in layer IV of the primary somatosensory cortex. The insets show the cortical localization of each neuron. Coarse branches represent the dendrites and fine branches represent the axonal plexus. Note that the axon is organized vertically. Adapted with permission, from Jones (1975).

primary sensory areas. However, several cortical regions have relatively few of these neurons, and even in areas in which these neurons are well represented, they are vastly outnumbered by aspiny interneurons (Peters and Jones, 1984; Lund *et al.*, 1995).

The axons of spiny stellate neurons are primarily intrinsic in their targets and radial in orientation and appear to play an important role in forming links between layer IV, the major thalamorecipient layer, and layers III, V, and VI, the major projection layers (Fig. 1.5). In some respects, the axonal arbor of spiny stellate cells mirrors the vertical plexuses of recurrent collaterals; however, they are more restricted than recurrent collaterals. Given its axonal distribution, the spiny stellate neuron appears to function as a high-fidelity translator of thalamic inputs, maintaining strict topographic organization and setting up initial vertical links of information transfer within sensory areas. Presumably, both pyramidal cells and aspiny nonpyramidal cells receive these radially limited inputs of the spiny stellate neuron, suggesting that this interneuron plays a key role in setting up the

excitatory component of a functional cortical domain (Peters and Jones, 1984).

Basket, Chandelier, and Double Bouquet Cells Are Inhibitory Interneurons

A large variety of inhibitory interneuron types are present in the cerebral cortex and in subcortical structures. These neurons contain the inhibitory neurotransmitter γ-aminobutyric acid (GABA) and exert strong local inhibitory effects. Three major subtypes of cortical interneurons are discussed in this section as examples. In all three cases, the dendritic and axonal arborizations offer important clues to their role in the regulation of pyramidal cell function (Cobb *et al.*, 1995; Sik *et al.*, 1995). In addition, for several GABAergic interneurons, a subtype of a given morphological class can be further defined by a particular set of neurochemical characteristics (Somogyi *et al.*, 1984). Although the following examples are taken from neurons prevalent in the neocortex and hippocampus of primates, inhibitory interneurons are present throughout the cerebral gray matter and exhibit a rich variety of morphologies, depending on the brain region as well as on the species studied (see Freund and Buzsaki, 1996).

Basket Cells

This class of GABAergic interneurons takes its name from the fact that its axonal endings form a basket of terminals surrounding a pyramidal cell soma (see Fig. 1.6) (Somogyi *et al.*, 1983). Basket cells can be divided into large and small cells. This cell class provides most of the inhibitory GABAergic synapses to the somas and proximal dendrites of pyramidal cells, although the basket cells also synapse on the shaft of the apical dendrite. One basket cell may contact numerous pyramidal cells, and, in turn, several basket cells can contribute to the pericellular basket of one pyramidal cell. The basket cells have relatively large somas and multipolar morphology, with dendrites extending in all directions for several hundred micrometers such that the vertically oriented dendrites cross several layers. The axonal pattern is the defining characteristic of this cell. The axon rises vertically, quickly bifurcates, and travels long distances (1–2 mm), forming multiple pericellular arrays as it spreads horizontally. The basket cells predominate in layers III and V in the neocortex and preferentially innervate the pyramidal cells within these layers, although they do not synapse exclusively on pyramidal cells. They are also numerous amid pyramidal neurons in the hippocampus. Thus, the basket cell is the primary source of horizontally

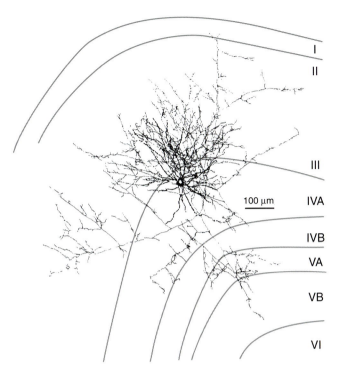

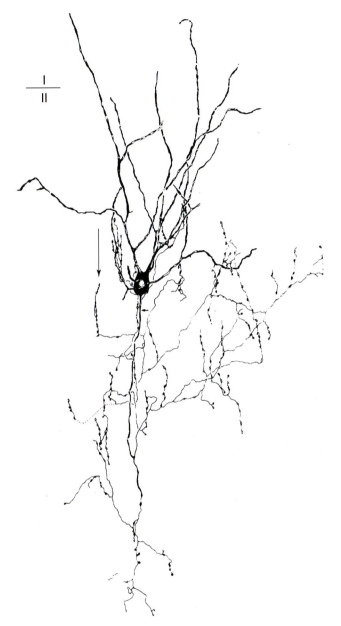

FIGURE 1.6 Drawing of a Golgi-impregnated basket cell from layer IVA of the primary visual cortex. Note the widely ramified dendritic tree and the wide horizontal spread of the axon that makes contact with many local neuronal perikarya. Cortical layers are indicated by Roman numerals. Adapted with permission, from Somogyi *et al.* (1983).

directed inhibitory inputs to the soma, proximal dendrites, and apical shaft of a pyramidal neuron. Interestingly, these cells are also characterized by certain biochemical features in that the majority of them contain the calcium-binding protein parvalbumin, and cholecystokinin appears to be the most likely neuropeptide in the large basket cells.

Chandelier Cells

The chandelier cell generally has a bitufted or multipolar dendritic tree, but the dendritic tree of this neuron is quite variable (Fig. 1.7) (Freund *et al.*, 1983). The defining characteristic of this cell class is the very striking appearance of its axonal endings. In Golgi or immunohistochemical preparations, the axon terminals appear as vertically oriented "cartridges," each consisting of a series of axonal boutons, or swellings, linked together by thin connecting pieces. These axonal specializations look like old-style chandeliers, which explains why this cell type is so named. The most salient characteristic of the chandelier cell is the extraordinary specificity of its synaptic target. These

FIGURE 1.7 Drawing of an axoaxonic chandelier neuron from layer II of the primary visual cortex. The dendritic spread of this neuron is quite limited. Note the typical axon terminal specializations (arrow). Adapted, from Freund *et al.* (1983).

neurons synapse exclusively on the axon initial segment of pyramidal cells. This characteristic is responsible for their alternate name, axoaxonic cells (Somogyi *et al.*, 1982). Most of the chandelier cells are located in layer III, and their primary target appears to be layer III pyramidal cells, although they also synapse to a lesser extent on pyramidal cells in the deep layers. One pyramidal cell may receive inputs from multiple chandelier cells, and one chandelier cell may innervate more than one pyramidal cell. Because of the high density of chandelier cell axon endings in layer III, this particular neuron may be highly involved in controlling corticocortical circuits. In addition, because the strength of the synaptic input is correlated directly with its proximity to the axon initial segment, there can be no more powerful inhibitory input to a pyramidal cell than that of the chandelier cell. Presumably, this interneuron is in a position that enables it to completely shut down the firing of a pyramidal cell (Hof *et al.*, 1995a; Somogyi *et al.*, 1982, 1983; Freund *et al.*, 1983; DeFelipe and Jones, 1992).

Double Bouquet Cells

The cell bodies of double bouquet cells are most prevalent in layers II and III, and are present in layer V of the neocortex as well. These interneurons are characterized by a vertical bitufted dendritic tree and a tight bundle of vertically oriented varicose axon collaterals that traverse layers II through V (Fig. 1.8) (Somogyi and Cowey, 1981) and are therefore entirely different from those of chandelier and basket cells.

Of the inhibitory interneurons, the double bouquet cell serves as perhaps the best example of the emerging concept of cell typology in which connectivity, location, morphology, and neurochemical phenotype are all features that are considered in "typing" a given cell (DeFelipe and Jones, 1992). It is clear that neurochemical phenotype subdivides the double bouquet cell into multiple classes. For example, a GABA–calbindin–somatostatin double bouquet cell appears to be localized primarily in layers II and III and has 40% of its synapses on spines and the remaining synapses primarily on distal shafts of pyramidal and nonpyramidal cells. Large numbers of this particular subtype of double bouquet cell are present in association cortices, with fewer in primary sensory cortices. Its regional, laminar, and synaptic organization suggests that it plays a crucial role in the regulation of pyramidal cells that furnish corticocortical projections. A different subclass of double bouquet cell contains calbindin and tachykinins as peptide neuromodulators. This subclass appears to have similar synaptic targets but is present primarily in layer V and, thus, presumably regulates the activity of a different group of pyramidal cells.

Other Interneuron Subtypes

Several other subtypes of interneurons can be distinguished on the basis of their morphology and neu-

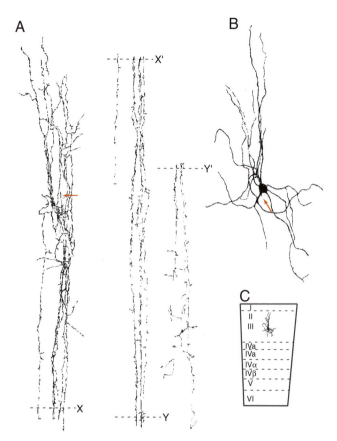

FIGURE 1.8 Drawing of a double bouquet cell in layer III of the primary visual cortex. The axonal tree (A) has been broken into three segments contiguous at *X–X'* and *Y–Y'*, to display its entire radial extent. The arrow in (B) corresponds to the arrow in (A). This neuron has very long radial axonal extensions, but very limited horizontal spread. Its location is shown in the inset (C). Adapted from Somogyi and Cowey (1981).

rochemical characteristics. A particularly interesting neuron is the poorly understood "clutch cell," which is driven primarily by thalamocortical inputs and, in turn, targets the spiny stellate cell of layer IV. Thus, this inhibitory interneuron is in essence situated so that it can regulate the firing rate of the spiny stellate cell in a fashion similar to how the three GABAergic neurons (basket, chandelier, and double bouquet) regulate the firing rate of pyramidal cells. Another important interneuron type is the bipolar neuron, which is characterized by elongated apical and basal dendrites and a locally ramifying axonal plexus, presumably making contacts with the apical dendrites of neighboring pyramidal cells. This cell is highly prevalent in the neocortex of rodents and has a modulatory role in the integration of cortical activity with noradrenergic projections from the brainstem. In the rat brain, some of these bipolar neurons may also contain the calcium-binding protein calretinin, but their homolog, if any, in the primate cortex remains to be determined.

Noncortical Neurons Have Distinct Morphological Characteristics

This section reviews characteristics of four neuronal types found in subcortical structures: the medium-sized spiny cells of the basal ganglia, the dopaminergic neurons of the pars compacta of the substantia nigra, the Purkinje cell of the cerebellum, and the alpha motor neuron of the ventral horn of the spinal cord. The rationale for choosing these particular neurons as representative is that each plays a determinant role in the pathogenic mechanisms of severe neurological disorders that affect humans. Thus, degeneration of the medium-sized spiny neurons is a central feature of Huntington's disease; the death of dopaminergic neurons is the neuropathological signature of Parkinson's disease; the loss of Purkinje cells is seen in familial cerebellar cortical degeneration; and degeneration of spinal cord motor neurons is the hallmark of lower motor neuron disease, a form of amyotrophic lateral sclerosis.

Medium-Sized Spiny Cells

These neurons are unique to the striatum, a part of the basal ganglia that comprises the caudate nucleus and putamen, where they are present in large numbers (as many as 10^8 in humans). Medium-sized spiny cells are scattered throughout the caudate nucleus and putamen and are recognized by their relatively large size, compared with other cellular elements of the basal ganglia, and by the fact that they are generally isolated neurons (Braak and Braak, 1982). These neurons differ from all others in the striatum in that they have a highly ramified dendritic arborization radiating in all directions and densely covered with spines (Fig. 1.9) (Carpenter and Sutin, 1983). Medium-sized spiny neurons are central to the function of the basal ganglia because they furnish a major output from the caudate nucleus and putamen and receive a highly diverse input from, among other sources, the cerebral cortex, thalamus, and certain dopaminergic neurons of the substantia nigra. They have long axons that leave the basal ganglia and also form a large array of recurrent collaterals that innervate neighboring medium-sized spiny cells. These neurons are neurochemically quite heterogeneous, contain GABA, and may contain several neuropeptides such as enkephalin, dynorphin, substance P, and the calcium-binding protein calbindin. In Huntington's disease, a neurodegenerative disorder of the striatum characterized by involuntary move-

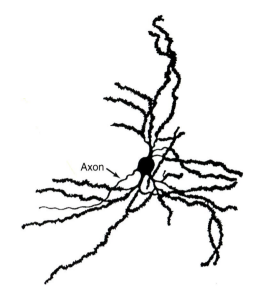

FIGURE 1.9 Drawing of a medium-size spiny neuron from the striatum. Note the highly ramified dendritic arborization radiating in all directions and the very high density of spines. Adapted, from Carpenter and Sutin (1983).

ments and progressive dementia, an early and dramatic loss of medium-sized spiny cells occurs. Interestingly, medium-sized spiny neurons that contain somatostatin appear to be relatively resistant to the degenerative process.

Dopaminergic Neurons of the Substantia Nigra

The substantia nigra is characterized by a rich diversity of neuronal types that exhibit differential distributions among the various functional compartments. Of these neurons, the most conspicuous are the large dopaminergic neurons that reside mostly within the pars compacta of the substantia nigra and in the ventral tegmental area. A distinctive feature of these cells is the presence of a pigment, neuromelanin, in compact granules in the cytoplasm. These neurons are medium-sized to large, fusiform, and frequently elongated; they have several large radiating dendrites. The axon emerges from the cell body or from one of the dendrites and projects to large expanses of cerebral cortex and to the basal ganglia. These neurons contain the catecholamine-synthesizing enzyme tyrosine hydroxylase, as well as the monoamine dopamine as their neurotransmitters; some of them colocalize calbindin and calretinin. These neurons are severely and selectively affected in Parkinson's disease, a movement disorder different from Huntington's disease and characterized by resting tremor and rigidity, and their specific loss is the neuropathological hallmark of this disorder (van Domburg and ten Donkelaar, 1991).

Purkinje Cells

The structure of the cerebellar cortex, in contrast with that of the cerebral cortex, is basically identical all over; it is composed of three layers that contain very distinct neuronal types. One of these layers contains the Purkinje cells, which are the most salient cellular elements of the cerebellar cortex. They are arranged in a single row throughout the entire cerebellar cortex between the molecular (outer) layer and the granular (inner) layer. They are the largest cerebellar neurons and have a round perikaryon with a highly branched dendritic tree shaped like a candelabra and extending into the molecular layer where they are contacted by incoming systems of afferent, parallel fibers from the granule neurons as well as other afferents from the brainstem. The apical dendrites of Purkinje cells have an enormous number of spines (more than 80,000 per cell). A particular feature of the dendritic tree of the Purkinje cell is that it is distributed in one plane, perpendicular to the longitudinal axes of the cerebellar folds, and each dendritic arbor determines a separate domain of cerebellar cortex (Fig. 1.1). The axons of Purkinje neurons course through the cerebellar white matter and contact the deep cerebellar nuclei or the vestibular nuclei. They also furnish recurrent collaterals, mostly within the granular layer. Humans have approximately 15 million Purkinje cells (Palay and Chan-Palay, 1974). These neurons contain the inhibitory neurotransmitter GABA and the calcium-binding protein calbindin. A severe disorder combining ataxic gait and impairment of fine hand movements, accompanied by dysarthria and tremor, has been documented in some families and is related directly to Purkinje cell degeneration.

Spinal Motor Neurons

The motor cells of the ventral horns of the spinal cord, also called alpha motor neurons, have their cell bodies within the spinal cord and send their axons outside the central nervous system to innervate the muscles. Different types of motor neurons are distinguished by their targets. The alpha motor neurons innervate skeletal muscles, but smaller motor neurons (the gamma motor neurons, forming about 30% of the motor neurons) innervate the spindle organs of the muscles. The alpha motor neurons are some of the largest neurons in the entire central nervous system and are characterized by a multipolar perikaryon and a very rich cytoplasm that renders them very conspicuous on histological preparations. They have a large number of spiny dendrites that arborize locally within the ventral horn. The alpha motor neuron axon leaves the central nervous system through the ventral root of the peripheral nerves. The cell bodies are arranged in a nonrandom fashion in the ventral horn so that they are grouped in functional vertical columns that span a certain number of spinal segments. This disposition corresponds to a somatotopic representation of the muscle groups of the limbs and axial musculature (Brodal, 1981). The spinal motor neurons use acetylcholine as their neurotransmitter. Large motor neurons are severely affected in lower motor neuron disease (a form of amyotrophic lateral sclerosis), a neurodegenerative disorder characterized by progressive muscular weakness that affects, at first, one or two limbs and that can be initially asymmetric. As the disease progresses, it becomes symmetric and affects more and more of the body musculature, which shows signs of wasting as a result of denervation. Neuropathologically, a massive loss of ventral horn motor neurons occurs, and the remaining motor neurons appear shrunken and pyknotic.

Retinal Photoreceptors and Cochlear Hair Cells Are Examples of Specialized Sensory Receptors

Retinal photoreceptors and cochlear hair cells are modified neuroepithelial cells that are specialized in the initial transduction of visual and acoustic stimuli respectively. Comparable specialized neuronal types exist for other sensory modalities, that is, olfactory, gustatory, and vestibular inputs. In contrast, somatosensory inputs are transmitted by peripheral nerve cells whose endings are associated with a variety of sensory structures in the peripheral tissues. Receptor neurons are extremely polarized cells, with one uniquely diversified end that is responsible for the reception of the sensory stimulus. This morphology is particularly well demonstrated in retinal photoreceptors. Photoreceptor cells are of two types, the rod and the cone, which are specialized for scotopic (light/dark) and color vision, respectively. The rods are slender cells, with an elongated cylindrical outer portion, whereas the cones are smaller elements, with shorter, conical outer portions (Fig. 1.10) (Krebs and Krebs, 1991). Each cell type consists of an outer segment and an inner segment. The inner segments of both rods and cones contain the metabolic machinery necessary for protein and lipid synthesis and oxidative metabolism. In rods, the outer segment is composed of a very large number of parallel lamellae stacked perpendicularly to the main axis of the cylinder. These lamellae are closed, flattened membranous disks that appear in thin-section electron microscopy

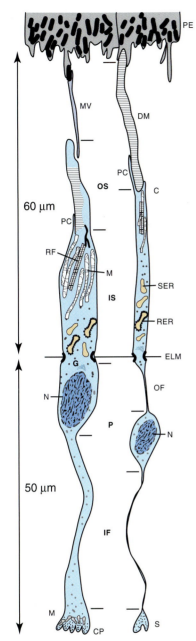

FIGURE 1.10 Drawing of a cone (left) and a rod (right) from the monkey retina. Note the differences in shape and size of these cells. They are composed of an outer segment (OS), an inner segment (IS), a perikaryon (P), and the inner fiber (IF). The outer segment is connected to the inner segment by a thin connecting cilium (C). The outer fiber (OF) is thin and well visible in rods, whereas the perikarya of both cell types are comparable in appearance. In cones the inner fiber is thicker and ends as a large cone pedicle, whereas in rods the inner fiber is rather thin and terminates in a unique spherule. Cone pedicles (CP) and rod spherules (S) are specialized synaptic endings where the photoreceptors make contact with specific subtypes of retinal relay neurons. DM, membranous disks (lamellae); ELM, external limiting membrane; G, Golgi apparatus; M, mitochondria, MV, microvilli of pigment epithelium; N, nucleus; PC, calycoid process; PE, pigment epithelium; RER (SER), rough (smooth) endoplasmic reticulum; RF, rootlet fibers. Adapted with permission, from Krebs and Krebs (1991).

as pairs of parallel membranes. In cones, these lamellar stacks are less numerous. These structures are responsible for the mechanisms of phototransduction and contain several visual pigments, located inside the membranous disks, that are necessary for the absorption of light.

Cochlear (and vestibular) hair cells also are highly polarized and present striking apical differentiation specialized in the detection of endolymphatic movements in the inner ear. In the cochlea, receptor hair cells that detect stimuli produced by sound are short, goblet-like cells embedded in supporting cells (the phalangeal cells of Deiters). Their apical domain contains a U-shaped row of stereocilia (hairs) that are in contact with the tectorial membrane of the organ of Corti. The vibrations of this membrane, generated by sound waves in the endolymph, displace the hairs and initiate the transduction of the acoustic stimulus. The other pole of the hair cell contains the nucleus and a dense population of mitochondria and receives synaptic contacts from afferent and efferent fibers from the cochlear nerve, which spreads around the lower third of the receptor cell (Fig. 1.11) (Hudspeth, 1983).

Enteric Motor Neurons Form an Independent Neural Plexus in the Gut Wall

The enteric nervous system constitutes a part of the autonomic nervous system that innervates the gastrointestinal tract, as do the sympathetic and parasympathetic systems. Although all three systems take part in the regulation of intestinal function, the enteric system is by far the most important and has the unique feature that it can function relatively independently of the control of higher centers (Furness and Costa, 1980). The enteric system consists of an extremely rich plexus of nerve fibers and neurons disseminated among all of the layers that form the wall of the intestinal tract (Fig. 1.12). It contains nerve cells arranged in ganglia interconnected by complex bundles of fibers that extend from the lower third of the esophagus to the internal anal sphincter. In humans, this system contains about 10^7 to 10^8 neurons.

The principal enteric plexus is located between the circular and longitudinal layers of the muscularis and is known as the myenteric plexus of Auerbach. It is composed of ganglia, each containing from 3 to 50 neurons linked by unmyelinated fibers and forming a continuous network. The cells in this plexus are of two main morphological types. One is a large multipolar neuron with short dendrites in direct contact with similar nearby cells and a long axon that contacts

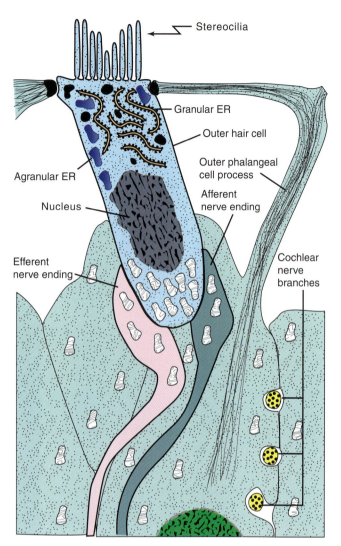

FIGURE 1.11   Schematic drawing of an electron micrograph of an outer hair cell and its relationships to supporting (outer phalangeal) cell and cochlear nerve endings. Note the apical domain containing stereocilia. The other pole of the hair cell contains the nucleus and a dense population of mitochondria and receives synaptic contacts from the cochlear nerve, which spread around the lower third of the receptor cell.

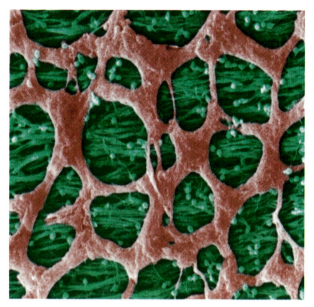

FIGURE 1.12   Scanning electron micrograph of the myenteric plexus in the intestine. Note the dense axonal bundles and synaptic boutons (pseudocolored green) and the network of large multipolar neurons with relatively short extensions contacting neighboring cells (pseudocolored red).

acetylcholine, GABA, and noradrenaline (see also Chapter 9), have been identified in enteric nerve fibers and ganglionic neurons. In addition, enteric neurons in both the myenteric and the submucosal plexuses contain a variety of neuropeptides. Neurons in the submucosal plexus are enriched in somatostatin, substance P, and vasoactive intestinal peptide but do not seem to contain Leu-enkephalin, which is observed in the myenteric nerve cells (see also Chapter 10). It is not clear, however, whether these neuropeptides are the principal transmitters of subclasses of enteric neurons or are colocalized compounds that act as local neuromodulators (Furness and Costa, 1980).

Summary

The neuron is one of the more highly specialized cell types and is the critical cellular element in the brain. All neurological processes are dependent on complex cell–cell interactions between single neurons and/or groups of related neurons. Neurons can be described according to their size, shape, neurochemical characteristics, location, and connectivity.

A neuron's size, shape, and neurochemistry are important determinants of that neuron's particular functional role in the brain. In this respect, there are three general classes of neurons: the inhibitory GABAergic interneurons that make local contacts, the local excitatory spiny stellate cells in the cerebral

different cell types in neighboring ganglia. These cells are thought to be association interneurons. The other cell type, considered an enteric motor neuron, is by far more dominant and demonstrates more variable morphology. These cells make extensive contacts with neurons of either type within the same ganglia or with distant cells. Other ganglia are found within the submucosal plexus of Meissner. Their relatively large multipolar neurons form a network interconnecting the outer nerve bundles with the submucosal tissue.

The neurochemistry of the enteric system is extremely complex and still poorly understood. A large number of classic neurotransmitters, such as cortex, and the excitatory glutamatergic efferent neurons, exemplified by the cortical pyramidal neurons. Within these general classes, the structural variation of neurons is systematic, and careful analyses of the anatomic features of neurons have led to various categorizations and to the development of the concept of cell type. The grouping of neurons into descriptive cell types (such as chandelier, double bouquet, and bipolar cells) allows the analysis of populations of neurons and the linking of specified cellular characteristics with certain functional roles. The relevant characteristics may include morphology, location, connectivity, and biochemistry.

Also, neurons form circuits, and these circuits constitute the structural basis for brain function. Macrocircuits involve a population of neurons projecting from one brain region to a distant region, and microcircuits reflect the local cell–cell interactions within a brain region. The detailed analysis of these macro- and microcircuits is an essential step in understanding the neuronal basis of a given cortical function in the healthy and the diseased brain. Thus, these cellular characteristics allow us to appreciate the special structural and biochemical qualities of that neuron in relation to its neighbors and to place it in the context of a specific neuronal subset, circuit, or function.

THE NEUROGLIA

The term *neuroglia*, or "nerve glue," was coined in 1859 by Rudolph Virchow, who conceived of the neuroglia as an inactive "connective tissue" holding neurons together in the central nervous system. The metallic staining techniques developed by Santiago Ramón y Cajal and Pio del Rio-Hortega allowed these two great pioneers to distinguish, in addition to the ependyma lining the ventricles and central canal, three types of supporting cells in the central nervous system (CNS): oligodendrocytes, astrocytes, and microglia. In the peripheral nervous system, the Schwann cell is the major neuroglial component.

Oligodendrocytes and Schwann Cells Synthesize Myelin

The more complex the brain, the more interconnections must be formed and maintained. As shown in depth later, there is a practical limit to how fast an individual bare axon can conduct an action potential (Chapters 4, 5). Thus, neurons and their associated processes cannot communicate with each other extremely rapidly through the action potential without some help. Organisms have developed two kinds of solutions for enhancing rapid communication between neurons and their effector organs. In invertebrates, the diameters of individual axons that must conduct rapidly are enlarged. In vertebrates, the myelin sheath (Fig. 1.13) has evolved to permit rapid nerve conduction.

Axon enlargement greatly accelerates the rate of conduction of the action potential, which increases with axonal diameter (Chapter 4). The net effect, therefore, is that small axons conduct at a much slower rate than larger ones. The largest axon in the invertebrate kingdom is the squid giant axon, which is about the thickness of a mechanical pencil lead. It conducts the action potential extremely rapidly, and the axon itself mediates an escape reflex, which must be rapid if the animal is to survive. An obvious trade-off in a nervous system with 10 billion neurons, as in the human brain, is that all axons cannot be as thick as pencil lead, or each human head would be very large indeed.

Thus, along the invertebrate evolutionary line, there is a natural, insurmountable limit—a constraint imposed by axonal size—to increasing the processing capacity of the nervous system beyond a certain point. Vertebrates, however, devised a way to get around this problem through the evolution of the myelin sheath, allowing the tremendous evolutionary advantage of increased rapidity of conduction of the nerve impulse along axons with fairly minute diameters (Morell and Norton, 1980). As we know, neurons interact in complex ways with the other cell types that exist within the nervous system. Virtually all axons, for example, are wrapped or ensheathed by cells that subserve what is vaguely termed a "supportive" or "trophic" function. This is true in invertebrate as well as vertebrate nervous systems. Along the vertebrate lineage, however, certain ensheathing cells have become highly specialized to generate vast quantities of plasma membrane that is compacted to form the myelin sheath, which supports rapid nerve conduction.

Not all axons in central or peripheral nervous systems are myelinated, and one of the puzzles is to determine why some are selected for myelination and others remain unmyelinated. It is believed that early in the nervous system development of an organism, signals relayed between the axon and the myelinating cell determine whether the "myelination program" is triggered in that cell. These signals have not yet been identified.

In the central nervous system, the myelin sheath (Fig. 1.14) is elaborated by oligodendrocytes, nonneuronal glial cells that, during brain development, send out a few cytoplasmic processes that engage adjacent

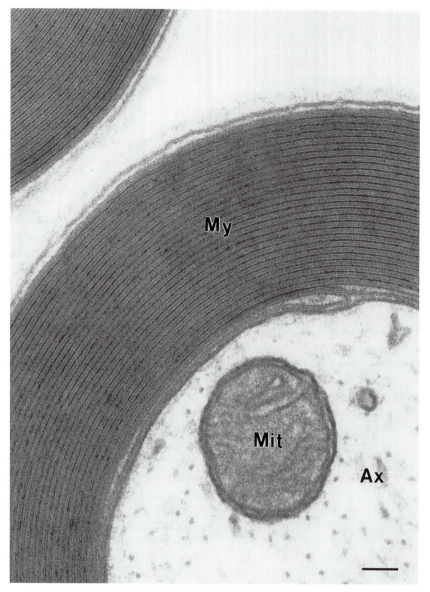

FIGURE 1.13 Electron micrograph of a transverse section through part of a myelinated axon from the sciatic nerve of a rat. The tightly compacted multilayer myelin sheath (My) surrounds and insulates the axon (Ax). Mit, mitochondria. Bar = 75 nm.

axons, some of which go on to become myelinated (Bunge, 1968). Myelin itself consists of a single sheet of oligodendrocyte plasma membrane, which is wrapped tightly around an axonal segment (Bunge *et al.*, 1962). Each myelinated segment of an axon is termed an internode because, at the end of each segment, there is a bare portion of the axon, the node of Ranvier, that is flanked by another internode. Physiologically, myelin has insulating properties such that the action potential can "leap" from node to node and therefore does not have to be regenerated continually along the axonal segment that is covered by the myelin membrane sheath. This leaping of the action

potential from node to node allows axons with fairly small diameters to conduct extremely rapidly (Ritchie, 1984). The jumping of the action potential from node to node along a given axon is called saltatory conduction (see Chapters 4 and 5 for additional details on the propagation of nerve impulses).

The evolution of a system in which a single oligodendrocyte cell body is responsible for the construction and maintenance of several myelin sheaths (Fig. 1.14) and the removal of the cytoplasm between each turn of the myelin lamellae so that only the thinnest layer of plasma membrane is left has resulted in saving a huge amount of intracranial space. Brain

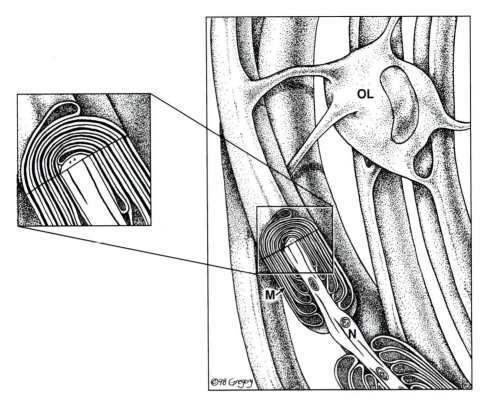

FIGURE 1.14 An oligodendrocyte (OL) in the central nervous system is depicted myelinating several axon segments. A cutaway view of the myelin sheath is shown (M). Note that the internode of myelin terminates in paranodal loops that flank the node of Ranvier (N). Inset: Enlargement of compact myelin with alternating dark and light electron-dense lines that represent the intracellular (major dense lines) and extracellular (intraperiod line) plasma membrane appositions, respectively.

volume is thus reserved for further expansion of neuronal populations (Colman *et al.*, 1996).

Conservation of space in the peripheral nervous system (PNS) does not seem to have presented such a pressing problem. Myelin in the PNS is generated by Schwann cells (Fig. 1.15), each of which wraps only a single axonal segment. The biochemical composition of the myelin derived in the CNS and the composition of that derived in the PNS differ somewhat, although there are proteins common to the two nervous system subdivisions. Myelin has a high lipid-to-protein ratio, and the lipids are specialized. The myelin sheath has become an excellent model system for studying the generation or formation of specialized plasma membrane and membrane adhesion, because each layer of the multilayered myelin sheath must adhere to the adjacent layers. This adhesion is largely accomplished by protein–protein interactions, which have been best studied in the PNS.

The major integral membrane protein of peripheral nerve myelin is protein zero (P_0), a member of a very large family of proteins termed the immunoglobulin gene superfamily. These proteins have in common

recognition or adhesion functions or both, and, although the primary amino acid sequences differ among the members of this family, all members are related to one another by certain common structural motifs. Members of the immunoglobulin (Ig) gene superfamily have one or more Ig-like domains that contain cysteines placed about 100 amino acids or so apart. These cysteines are linked to one another by disulfide bridges. Most of these Ig domains are displayed on the extracellular surfaces of cells, where they can act as ligands or receptors. Protein zero is relatively simple in primary structure, consisting of a single Ig-like domain, a transmembrane segment, and a highly charged (basic) cytoplasmic domain (Lemke and Axel, 1985). This protein makes up about 80% of the protein complement of peripheral nerve myelin. The interactions between the extracellular domains of P_0 molecules expressed on one layer of the myelin sheath and those of the apposing layer yield a characteristic regular periodicity that can be seen by thin-section electron microscopy (Fig. 1.13). This zone, called the intraperiod line, represents the extracellular apposition of the myelin bilayer as it wraps

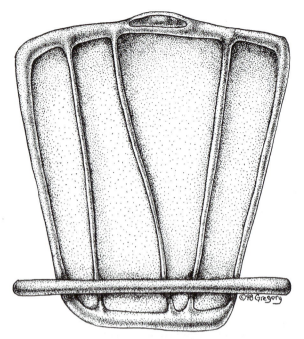

FIGURE 1.15 An "unrolled" Schwann cell in the PNS is illustrated in relation to the single axon segment that it myelinates. The broad stippled region is compact myelin surrounded by cytoplasmic channels that remain open even after compact myelin has formed, allowing exchange of materials among the myelin sheath, the Schwann cell cytoplasm, and perhaps the axon as well.

around itself. On the other side of the bilayer, the cytoplasmic side, the highly charged P_0 cytoplasmic domain probably functions to neutralize the negative charges on the polar head groups of the phospholipids that make up the plasma membrane itself, allowing the membranes of the myelin sheath to come into close apposition with one another. In electron microscopy, this cytoplasmic apposition is a bit darker than the intraperiod line and is termed the major dense line. In peripheral nerves, although other molecules are present in small quantities in compact myelin and may have important functions, compaction (i.e., the close apposition of membrane surfaces without intevening cytoplasm) is accomplished solely by P_0–P_0 interactions at both extracellular and intracellular (cytoplasmic) surfaces (Giese *et al.*, 1992).

Protein zero is a "perfect" plasma membrane compactor (Shapiro *et al.*, 1995), allowing the close apposition of adjacent bilayers such that the space between them effectively prevents the passage of anything but small ions and water along the compacted bilayer surfaces. It is in effect a "streamlined" Ig superfamily molecule that probably arose *de novo* with the development of the myelin sheath (Colman *et al.*, 1996). Curiously, P_0 is not present in all myelin sheaths in

the central nervous system of every species, an evolutionary paradox that has attracted much attention. In fish, P_0 is present in both the central and peripheral nervous systems, where it performs its compaction function, as the major integral membrane protein. However, in terrestrial vertebrates (reptiles, birds, and mammals), P_0 is limited to the PNS, and so is not found in the CNS. Instead, the compaction function is probably subserved by totally unrelated molecules, the DM20 protein and its insertion isoform, the myelin proteolipid protein (PLP) (Folch-Pi and Lees, 1951). These two proteins are generated from the same gene and are identical to each other with the exception that the proteolipid protein has, in addition, a positively charged segment exposed on the cytoplasmic aspect of the bilayer (Milner *et al.*, 1985; Nave *et al.*, 1987). Both PLP and DM20 are extremely hydrophobic and traverse the bilayer four times, and so have hydrophilic segments exposed on both cytoplasmic and extracellular surfaces of the bilayer. In this respect, the topology of these molecules is very similar to that of connexins and other polypeptides that are known to function in channel or pore formation (e.g., Chapter 15).

A large number of naturally occurring neurological mutations can affect the proteins specific to the myelin sheath. These mutations have been named according to the phenotype that is produced: the *shiverer* mouse, the *shaking* pup, the *rumpshaker* mouse, the *jimpy* mouse, the *myelin-deficient* rat, the *quaking* mouse, and so forth. Many of these mutations have been well characterized, and their analyses have allowed us to begin to understand at a molecular level what the proteins affected by each mutation actually do in the formation and maintenance of the myelin sheath (see Box 1.2) (Nave, 1994).

The first neurological mutation that was studied in this respect was the *shiverer* mouse, in which the gene that encodes a major set of peripheral membrane proteins, the myelin basic proteins (MBPs), is functionally deleted. Normally, these proteins serve to seal the cytoplasmic aspects of the myelin bilayer, possibly by charge neutralization similar in function to the cytoplasmic tail of P_0. When gene expression of these proteins is completely compromised, as in the *shiverer*, the cytoplasmic aspects fail to appose and do not fuse, and the mouse exhibits tremors and convulsions ("shivers") as it walks. This is a naturally occurring mutation and was the first neurological mutation whose effects were cured by gene transfer. This was accomplished by the introduction of an intact MBP gene into the *shiverer* genome (Readhead *et al.*, 1987). The *shiverer* mutation is an autosomal recessive, but even a single allele of the gene (i.e., the heterozygote

BOX 1.2

INHERITED PERIPHERAL NEUROPATHIES

The peripheral myelin protein-22 (PMP22) is a very hydrophobic glycoprotein and is highly expressed in compact PNS myelin. It has been mapped to the previously defined *Tr* locus on mouse chromosome 11. Comparison of marker genes on mouse chromosome 11 and human chromosome 17 revealed that PMP22 was also a candidate gene for the most common form of autosomal-dominant demyelinating hereditary peripheral neuropathy in humans, Charcot–Marie–Tooth disease type 1A (CMT1A). Indeed, the entire PMP22 gene is contained within a 1.5-Mb intrachromosomal duplication on chromosome 17p11.2, a genetic abnormality that had been linked to CMT1A by human molecular genetics. Consistent with these results, PMP22 is overexpressed in CMT1A patients who carry the characteristic duplication. The crucial role of PMP22 in the etiology of CMT1A was confirmed by generating transgenic mice and rats with increased PMP22 gene dosage, which resulted in severe PNS myelin deficits.

CMT is one of the more frequent hereditary diseases of the nervous system, with an overall prevalence of approximately 1 in 4000, and the CMT1A duplication accounts for around 70% of all cases. Why is this chromosomal abnormality so common? Detailed analysis of the CMT1A locus suggests that the duplication is due to crossing over involving repetitive sequences that flank the monomeric region. If correct, such a mechanism should also generate an allele carrying the reciprocal deletion of the same region. Indeed, the expected deletion is associated with the relatively mild recurrent neuropathy with liability to pressure palsy (HNPP). Thus, overexpression and underexpression of the myelin protein PMP22 are associated with myelin deficiencies in distinct human diseases. Although one might speculate from these data that correct stochiometry of myelin

protein expression is crucial for a myelinating Schwann cell, the exact disease mechanism remains to be clarified.

Interestingly, the finding that a myelin protein was responsible for CMT1A led to the discovery that two other components of PNS myelin are mutated in rare forms of CMT1. The adhesion protein P_0, which is largely responsible for PNS myelin compaction, is affected in CMT1B, and an X-linked form of CMT (CMTX) has been linked to mutations in the gap junction protein connexin-32. In contrast to PMP22 and P_0, connexin-32 is located in uncompacted lamellae of PNS myelin, where it is thought to facilitate the exchange of small molecules via reflexive gap junctions between adaxonal and abaxonal aspects of myelinating Schwann cells.

Finally, there is a striking correlation between the role of PMP22 in the PNS and that of PLP/DM20 in the CNS with respect to biology and involvement in disease; both genes can be affected by various genetic mechanisms, including gene duplication and gene deletion. However, despite our vast knowledge derived from human molecular genetics, the molecular functions of both proteins are largely unknown. Given the recent findings that PMP22 and PLP/DM20 are members of extended gene families and may be involved in the control of cell proliferation and cell death, these proteins may have broader functions than simply being stabilizing building blocks of compact myelin.

In summary, the combination of basic and clinical sciences has led to substantial progress in the current understanding of common hereditary neuropathies. Using clinical, genetic, and cell biology approaches in concert, we will continue to learn more about disease mechanisms involved in neuropathies to the benefit of the clinic as much as to the understanding of myelin biology.

Ueli Suter

MBP⁺/MBP⁻) produces sufficient myelin to phenotypically at least "cure" the *shiverer* of its overtly abnormal shivering behavior. These heterozygotes myelinate to a somewhat lesser extent than normal, but the fact that the *shiverer* phenotype is eliminated in the heterozygote even though the number of myelin sheaths around each axon is reduced indicates that there is a built-in safety factor in the normal situation.

Astrocytes Play Important Roles in CNS Homeostasis

As the name suggests, astrocytes are star-shaped process-bearing cells distributed throughout the central nervous system. They constitute from 20 to 50% of the volume of most brain areas. Astrocytes come in many shapes and forms. The two main forms, protoplasmic and fibrous astrocytes, predominate in

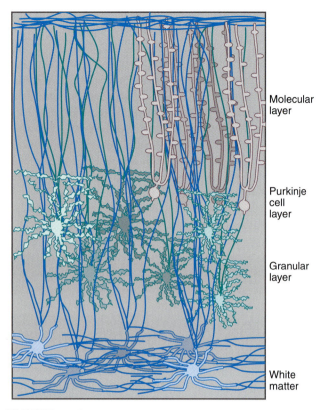

Molecular
layer

Purkinje
cell
layer

Granular
layer

White
matter

FIGURE 1.16 Arrangement of astrocytes in human cerebellar cortex. The Bergmann glial cells are in red, the protoplasmic astrocytes are in green, and the fibrous astrocytes are in blue.

gray and white matter, respectively (Fig. 1.16). Embryonically, astrocytes develop from radial glial cells, which transversely compartmentalize the neural tube. Radial glial cells serve as scaffolding for the migration of neurons and play a critical role in defining the cytoarchitecture of the CNS (Fig. 1.17). As the CNS matures, radial glia retract their processes and serve as progenitors of astrocytes. However, some specialized astrocytes of a radial nature are still found in the adult cerebellum and the retina and are known as Bergmann glial cells and Muller cells, respectively.

Astrocytes "fence in" neurons and oligodendrocytes (Arenander and de Vellis, 1983). The astrocytes achieve this isolation of the brain parenchyma by extending long processes projecting to the pia mater and the ependyma to form the glia limitans, by covering the surface of capillaries, and by making a cuff around the nodes of Ranvier. They also ensheath synapses and dendrites and project processes to cell somas (Fig. 1.18). Astrocytes are connected to each other by gap junctions, forming a syncytium that allows ions and small molecules to diffuse across the brain parenchyma. Astrocytes have in common unique cytological and immunological properties that make them easy to identify, including their star shape, the glial end feet on capillaries, and a unique population of large bundles of intermediate filaments. These filaments are composed of an astroglia-specific

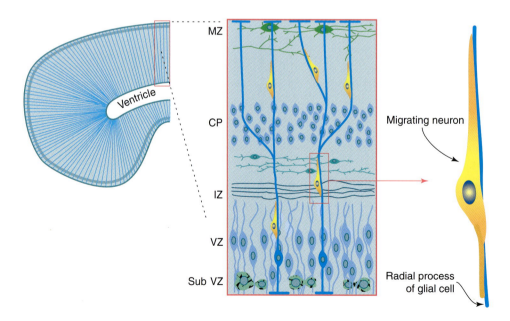

FIGURE 1.17 Radial glia perform support and guidance functions for migrating neurons. In early development, the radial glia span the thickness of the expanding brain parenchyma. Inset: Defined layers of the neural tube from the ventricular to the outer surface: VZ, ventricular zone; IZ, intermediate zone; CP, cortical plate; MZ, marginal zone. The radial process of the glial cell is indicated in blue, and a single attached migrating neuron is depicted at the right.

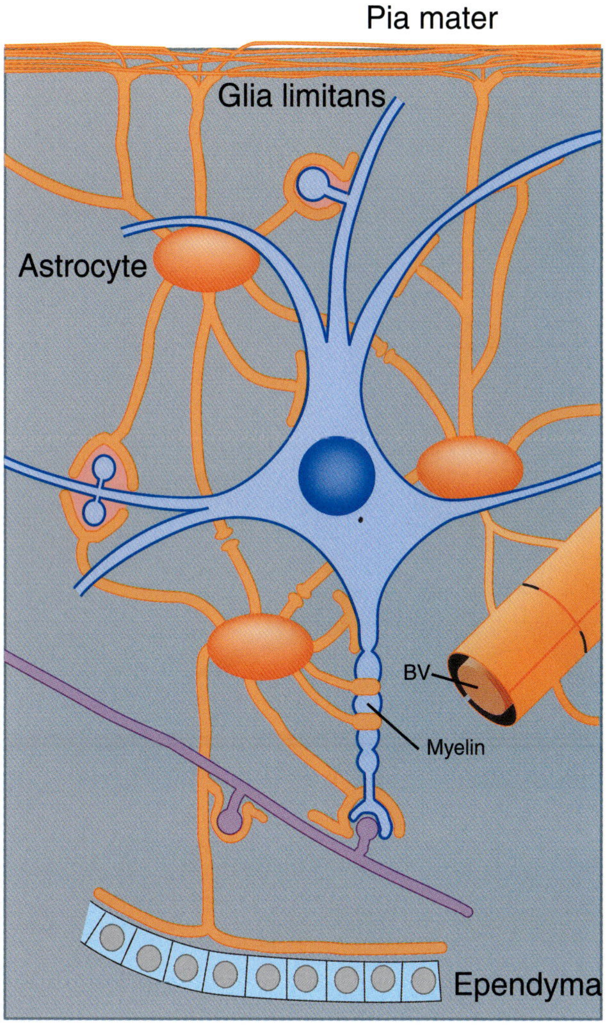

Pia mater

Glia limitans

Astrocyte

BV

Myelin

Ependyma

FIGURE 1.18 Astrocytes (in orange) are depicted *in situ* in schematic relationship with other cell types with which they are known to interact. Astrocytes send processes that surround neurons and synapses, blood vessels, and the region of the node of Ranvier and extend to the ependyma as well as to the pia mater, where they form the glia limitans.

sible for inducing and maintaining the tight junctions in endothelial cells that effectively form the barrier (Goldstein, 1988; Raub *et al.*, 1992). Astrocytes also take part in angiogenesis, which may be important in the development and repair of the CNS (Holash and Stewart, 1993). However, their role in this important process is still poorly understood.

Astrocytes Have a Wide Range of Functions

There is strong evidence for the role of radial glia and astrocytes in the migration and guidance of neurons in early development. Astrocytes are a major source of extracellular matrix proteins and adhesion molecules in the CNS; examples are nerve cell–nerve cell adhesion molecule (N-CAM), laminin, fibronectin, cytotactin, and the J–1 family members janusin and tenascin. These molecules participate not only in the migration of neurons, but also in the formation of neuronal aggregates, so-called nuclei, as well as networks.

Astrocytes produce, *in vivo* and *in vitro*, a very large number of growth factors. These factors act singly or in combination to selectively regulate the morphology, proliferation, differentiation, or survival, or all four, of distinct neuronal subpopulations. Most of the growth factors also act in a specific manner on the development and functions of astrocytes and oligodendrocytes. The production of growth factors and cytokines by astrocytes and their responsiveness to these factors are a major mechanism underlying the developmental function and regenerative capacity of the CNS. During neurotransmission, neurotransmitters and ions are released at high concentrations in the synaptic cleft. The rapid removal of these substances is important so that they do not interfere with future synaptic activity. The presence of astrocyte processes around synapses positions them well to regulate neurotransmitter uptake and inactivation (Kettenman and Ransom, 1995). These possibilities are consistent with the presence in astrocytes of transport systems for many neurotransmitters. For instance, glutamate reuptake is performed mostly by astrocytes, which convert glutamate into glutamine and then release it into the extracellular space. Glutamine is taken up by neurons, which use it to generate glutamate and GABA, potent excitatory and inhibitory neurotransmitters, respectively (Fig. 1.19). Astrocytes contain ion channels for K^+, Na^+, Cl^-, HCO^-_3, and Ca^{2+}, as well as displaying a wide range of neurotransmitter receptors. K^+ ions released from neurons during neurotransmission are soaked up by astrocytes and moved away from the area through astrocyte gap junctions. This is known

protein commonly referred to as GFAP (glial fibrillary acidic protein). S–100, a calcium-binding protein, and glutamine synthetase also are astrocyte markers. Ultrastructurally, gap junctions (connexins), desmosomes, glycogen granules, and membrane orthogonal arrays are distinct features used by morphologists to identify astrocytic cellular processes in the complex cytoarchitecture of the nervous system.

For a long time, astrocytes were thought to physically form the blood–brain barrier (considered later in this chapter), which prevents the entry of cells and diffusion of molecules into the CNS. In fact, astrocytes are indeed the blood–brain barrier in lower species. However, in higher species, the astrocytes are respon-

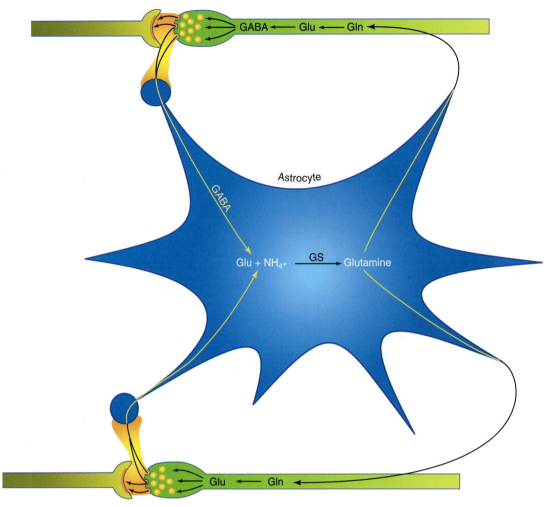

FIGURE 1.19 The glutamate–glutamine cycle is an example of a complex mechanism that involves an active coupling of neurotransmitter metabolism between neurons and astrocytes. The systems of exchange of glutamine, glutamate, GABA, and ammonia between neurons and astrocytes are highly integrated. The postulated detoxification of ammonia and inactivation of glutamate and GABA by astrocytes are consistent with the exclusive localization of glutamine synthetase in the astroglial compartment. Gln, glutamine.

as "spatial buffering." The astrocytes play a major role in detoxification of the CNS by sequestering metals and a variety of neuroactive substances of endogenous and xenobiotic origin.

In response to stimuli, intracellular calcium waves are generated in astrocytes. The propagation of the Ca^{2+} wave can be visually observed as it moves across the cell soma and from astrocyte to astrocyte. The generation of Ca^{2+} waves from cell to cell is thought to be mediated by second messengers, diffusing through gap junctions (see Chapter 15). Because they develop postnatally in rodents, gap junctions may not play an important role in development. In the adult brain, gap junctions are present in all astrocytes. Some gap junctions have also been detected between astrocytes and neurons. Thus, they may participate, along with

astroglial neurotransmitter receptors, in the coupling of astrocyte and neuron physiology.

In a variety of CNS disorders—neurotoxicity, viral infections, neurodegenerative disorders, HIV, AIDS, dementia, multiple sclerosis, inflammation, and trauma—astrocytes react by becoming hypertrophic and, in a few cases, hyperplastic. A rapid and huge upregulation of GFAP expression and filament formation is associated with astrogliosis. The formation of reactive astrocytes can spread very far from the site of origin. For instance, a localized trauma can recruit astrocytes from as far as the contralateral side, suggesting the existence of soluble factors in the mediation process. Tumor necrosis factor (TNF) and ciliary neurotrophic factors (CNTFs) have been identified as key factors in astrogliosis.

Microglia Are Mediators of Immune Responses in Nervous Tissue

The brain has traditionally been considered an "immunologically privileged site," mainly because the blood–brain barrier (see below) normally restricts the access of immune cells from the blood. However, it is now known that immunological reactions do take place in the central nervous system, particularly during cerebral inflammation. Microglial cells have been termed the tissue macrophages of the CNS, and they function as the resident representatives of the immune system in the brain. These cells are perhaps the least understood of the CNS cells. Although the function of microglia in the normal adult CNS remains to be clarified, a rapidly expanding literature describes microglia as major players in CNS development and in the pathogenesis of CNS disease. The notion that the CNS is an immune-privileged organ is no longer valid. A hallmark of microglial cells is their ability to become reactive and to respond to pathological challenges in a variety of ways.

The first description of microglial cells can be traced to Franz Nissl (1899) who used the term *rod cell* to describe a population of glial cells that reacted to brain pathology. He postulated that rod cell function was similar to that of leukocytes in other organs. Ramón y Cajal (1913) described microglia as part of his "third element" of the CNS—cells that he considered to be of mesodermal origin and distinct from neurons and astrocytes.

Del Rio-Hortega (1932) divided Ramón y Cajal's third element into oligodendrocytes and microglia, two cell types with different morphology, function, and origin. He used silver impregnation methods to visualize the ramified appearance of microglia in the adult brain, and he concluded that ramified microglia could transform into cells that were migratory, amoeboid, and phagocytic. A fundamental question raised by Del Rio-Hortega's studies was the origin of microglial cells. Although he provided evidence that microglia originated from cells that migrate into the brain from the pial surface, he also raised the possibility that microglia originate from blood "mononuclears." Controversy over the lineage of microglia still exists today.

Microglia Have Diverse Functions in Developing and Mature Nervous Tissue

Four different sources of microglia have been proposed (Dolman, 1991): (1) bone marrow-derived monocytes, (2) mesodermal pial elements, (3) neural epidermal cells, and (4) capillary-associated pericytes.

On the basis of current knowledge, it appears that most ramified microglial cells are derived from bone marrow-derived monocytes, which enter the brain parenchyma during early stages of brain development. These cells help phagocytose degenerating cells that undergo programmed cell death as part of normal development. They retain the ability to divide and have the immunophenotypic properties of monocytes and macrophages. In addition to their role in remodeling the CNS during early development, microglia may secrete cytokines or growth factors that are important in fiber tract development, gliogenesis, and angiogenesis. After the early stages of development, amoeboid microglial cells transform into the ramified microglial cells that persist throughout adulthood (Altman, 1994).

Little is known about microglial function in the normal adult vertebrate CNS. Microglia constitute a formidable percentage (5–20%) of the total cells in the mouse brain. Microglia are found in all regions of the brain, and there are more in gray than in white matter. The phylogenetically newer regions of the CNS (cerebral cortex, hippocampus) have more microglia than do older regions (brainstem, cerebellum) (Lawson *et al.*, 1990). Species variations also have been noted, as human white matter has three times more microglia than does rodent white matter.

Microglia usually have small rod-shaped somata from which numerous processes extend in a rather symmetrical fashion. Processes from different microglia rarely overlap or touch, and specialized contacts between microglia and other cells have not been described in the normal brain. Although each microglial cell occupies its own territory, microglia collectively form a network that covers much of the CNS parenchyma. Because of the numerous processes, microglia present extensive surface membrane to the CNS environment. Regional variation in the number and shape of microglia in the adult brain suggests that local environmental cues can affect microglial distribution and morphology. On the basis of these morphological observations, it is likely that microglia play a role in tissue homeostasis. The nature of this homeostasis remains to be elucidated. It is, however, clear that microglia can respond quickly and dramatically to alterations in the CNS microenvironment.

Microglia Become Activated in Pathological States

"Reactive" microglia can be distinguished from resting microglia by two criteria: (1) change in morphology and (2) upregulation of monocyte–

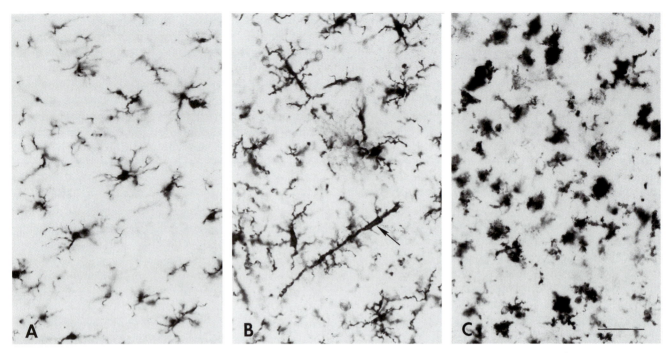

FIGURE 1.20 Activation of microglial cells in a tissue section from human brain. Resting microglia in normal brain (A). Activated microglia in diseased cerebral cortex (B) have thicker processes and larger cell bodies. In regions of frank pathology (C) microglia transform into phagocytic macrophages, which can also develop from circulating monocytes that enter the brain. Arrow in B indicates rod cell. Sections stained with antibody to ferritin. Bar = 40 μm.

macrophage molecules (Fig. 1.20). Although the two phenomena generally occur together, reactive responses of microglia can be diverse and restricted to subpopulations of cells within a microenvironment. Microglia not only respond to pathological conditions involving immune activation, but also become activated in neurodegenerative conditions that are not considered immune-mediated (Banati and Graeber, 1994). This latter response is indicative of the phagocytic role of microglia. Microglia change their morphology and antigen expression in response to almost any form of CNS injury.

Summary

Neuroglia are a set of cell types that together subserve supportive and trophic roles critical for the normal functioning of nervous tissue. Certain glial cells—the myelinating cells, for example—have clearly shaped nervous system evolution and development in that they evolved to facilitate rapid conduction of the action potential along small-caliber axons. The coordinated integrative functions of the vertebrate brain therefore depend on a normal complement of myelinated axons. Astrocytes and microglial cells also have major and extremely important functions in development and in tissue injury, but

these roles are not yet well understood. In pathological states of all kinds (autoimmune, toxic insult, trauma), these cells react to contain and limit tissue damage. They also contribute in a major way to repair mechanisms.

THE CEREBRAL VASCULATURE

Blood vessels form an extremely rich network in the central nervous system, particularly in the cerebral cortex and subcortical gray masses, whereas the white matter is less densely vascularized (Fig. 1.21) (Duvernoy et al., 1981). The vascular bed is supplied by perforating arteries that arise from a relatively small number of large, peripheral arterial trunks. The main trunks give off smaller cerebral arteries whose branches penetrate the subarachnoidal space, where they divide into many subbranches before penetrating the brain tissue. Within the cerebral gray matter, these penetrating arteries divide into a large number of small arterioles that eventually form an extremely rich, highly anastomotic capillary bed. At the other end of the capillary network are venules, draining into larger cerebral veins, which are the tributaries of large venous sinuses responsible for returning blood to the general circulation. The brain vascular system

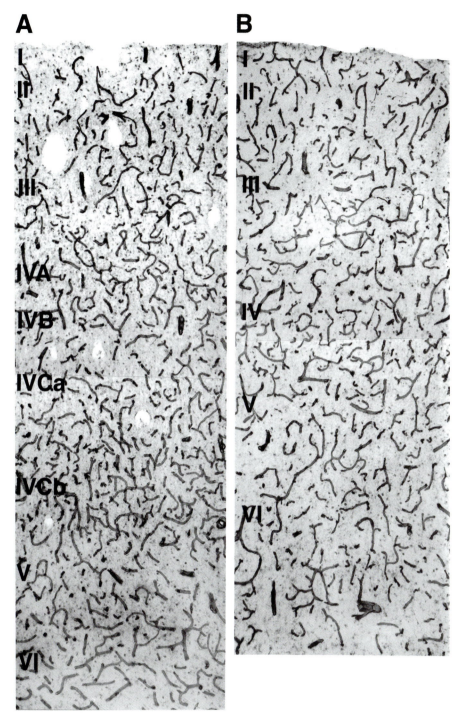

FIGURE 1.21   Microvasculature of the human neocortex. (A) Primary visual cortex (area 17). Note the presence of segments of deep penetrating arteries that have a larger diameter than the microvessels and run from the pial surface to the deep cortical layers, as well as the high density of microvessels in the middle layer (layers IVB and IVCb). (B) Prefrontal cortex (area 9). Cortical layers are indicated by Roman numerals. The microvessels are stained using an antibody against heparan sulfate proteoglycan core protein, a component of the extracellular matrix.

has no end arteries, and there is a relatively free circulation throughout the central nervous system. There are, however, distinct regional patterns of microvessel distribution in the brain. These patterns are particularly clear in certain subcortical structures that constitute discrete vascular territories and in the cerebral cortex, where regional and laminar patterns are striking. For example, layer IV of the primary visual cortex possesses an extremely rich capillary network, in comparison with other layers and adjacent regions (Fig. 1.21). Interestingly, most of the inputs from the visual thalamus terminate in this particular layer. Whether similar functional correlations may be derived from comparable vascular patterns in other brain regions remains to be determined. Nonetheless, capillary densities are higher in regions containing large numbers of neurons and where synaptic density is high. Penetrating arteries and draining veins have well-defined, tree-shaped branching patterns (Duvernoy *et al.*, 1981). With regard to cortical vessels, some arteries divide in the upper cortical layers, whereas others penetrate to the lower layers before dividing. The branches of penetrating arteries define local vascular fields of approximately similar size and

shape around the vessel of origin, which cover the entire cortical mantle in a continuous network.

Pathological factors that affect the patency of brain microvessels may result in the development of an ischemic injury localized to a variable amount of tissue, depending on the size and location of the affected arteries. For instance, the progressive occlusion of a large arterial trunk, as seen in stroke, induces an ischemic injury that may eventually lead to necrosis of the brain tissue. The size of the resulting infarction is determined in part by the worsening of the blood circulation through the cerebral microvessels. In fact, occlusion of a large arterial trunk results in rapid swelling of the capillary endothelium and surrounding astrocytes, which may reduce the capillary lumen to about one-third of its normal diameter, preventing red blood cell circulation and oxygen delivery to the tissue. The severity of these changes subsequently determines the time course of neuronal necrosis, as well as the possible recovery of the surrounding tissue and the neurological outcome of the patient. In addition, the presence of multiple microinfarcts caused by occlusive lesions of small cerebral arterioles may lead to a progressively dementing illness,

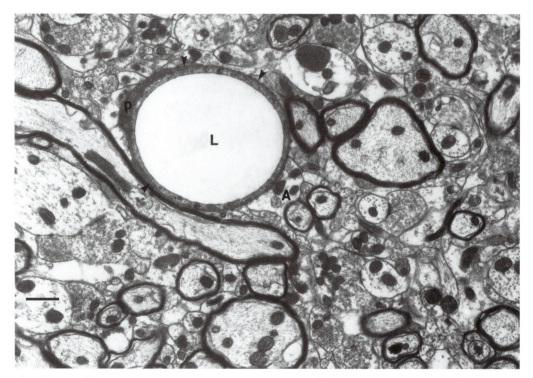

FIGURE 1.22 Electron micrograph of a blood–brain barrier (BBB) capillary. Endothelial cells joined by tight junctions form continuous capillaries with no fenestrations and restrict the passage of solutes between blood and brain. Pericytes (P) are present within the basement membrane (arrowheads) of these capillaries, serve to control vascular tone, and can also be phagocytic in the brain. Astrocyte foot processes (A) surround the basement membrane and are responsible for the induction of BBB properties on endothelial cells. Bar = 2 μm.

referred to as vascular dementia, affecting elderly humans.

The Blood–Brain Barrier Maintains the Intracerebral Milieu

Capillaries of the central nervous system form a protective barrier that restricts the exchange of solutes between blood and brain. This distinct function of brain capillaries is called the blood–brain barrier (Figs. 1.22 and 1.23) (Bradbury, 1979). The capillaries of the retina have similar properties and are termed the blood–retina barrier. It is thought that the blood–brain and blood–retina barriers function to maintain a constant intracerebral milieu, critical for neuronal function. This function is important because

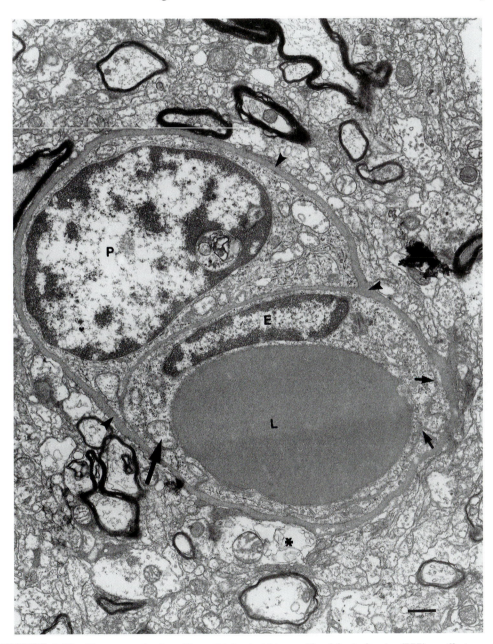

FIGURE 1.23 Human cerebral capillary obtained at biopsy. Blood–brain barrier (BBB) capillaries are characterized by the paucity of transcytotic vesicles in endothelial cells (E), a high mitochondrial content (large arrow), and the formation of tight junctions (small arrows) between endothelial cells that restrict the transport of solutes through the interendothelial space. The capillary endothelium is encased within a basement membrane (arrowheads), which also houses pericytes (P). Outside the basement membrane are astrocyte foot processes (asterisk), which may be responsible for induction of BBB characteristics on the endothelial cells. L, lumen of the capillary. Bar = 1 μm. From Claudio et al. (1995).

of the nature of intercellular communication in the CNS, which includes chemical signals across intercellular spaces. Without a blood–brain barrier, circulating factors in the blood such as certain hormones, which can also act as neurotransmitters, would interfere with synaptic communication.

When the blood–brain barrier is disrupted, edema fluid accumulates in the brain, leading to neurological impairments. Increased permeability of the blood–brain barrier plays a central role in many neuropathological conditions, including multiple sclerosis, AIDS, and childhood lead poisoning, and may also play a role in Alzheimer's disease (Claudio *et al.*, 1995; Buée *et al.*, 1994). The blood–brain barrier is composed of three cellular components—endothelial cells, pericytes, and astrocytes—and one noncellular component—the basement membrane. These components interact with each other to produce a highly selective and dynamic barrier system.

In general, the cerebral capillaries are comparable to those seen in other tissues (Peters *et al.*, 1991). The capillary wall is composed of an endothelial cell surrounded by a very thin (about 30 nm) basal lamina, similar to that seen in capillaries in peripheral tissues. End feet of perivascular astrocytes are apposed against this continuous basal lamina. Around the capillary lies a virtual perivascular space occupied by another cell type, the pericyte, which surrounds the capillary walls. The endothelial cell forms a thin monolayer around the capillary lumen, and a single endothelial cell can completely surround the lumen of the capillary (Fig. 1.23). The cytoplasm is rich in actin filaments and contains an extensive Golgi apparatus and a large number of mitochondria.

A fundamental difference between brain endothelial cells and those of the systemic circulation is the presence in brain of interendothelial tight junctions, also known as zonula occludens (Peters *et al.*, 1991). In the systemic circulation, the interendothelial space serves as a diffusion pathway that offers little resistance to most blood solutes entering the surrounding tissues. In contrast, blood–brain barrier tight junctions effectively restrict the intercellular route of solute transfer. The blood–brain barrier interendothelial junctions are not static seals; rather they are a series of active gates that can allow certain small molecules to penetrate. One such molecule is the lithium ion, used in the control of manic depression.

Another characteristic of endothelial cells of the brain is their low transcytotic activity. This is illustrated by the paucity of transcytotic vesicles in the cytoplasm, compared with endothelial cells of the systemic circulation. The frequency of transcytotic vesicles tends to increase with increasing permeability of

an endothelium. Brain endothelium, therefore, is by this index not very permeable.

It is of interest that certain regions of the brain, such as the area postrema and periventricular organs, lack a blood–brain barrier. In these regions, the perivascular space is in direct contact with the nervous tissue, and the endothelial cells are fenestrated and show many pinocytotic vesicles. In these brain regions, neurons are known to secrete hormones and other factors that require rapid and uninhibited access to the systemic circulation.

Because of the high metabolic requirements of the brain, blood–brain barrier endothelial cells must have transport mechanisms for the specific nutrients needed for proper brain function. One such mechanism is glucose transporter isoform 1 (GLUT–1), which is asymmetrically expressed on the surface of blood–brain barrier endothelial cells. In Alzheimer's disease, the expression of GLUT–1 on brain endothelial cells is reduced. This reduction may be due to a lower metabolic requirement of the brain after extensive neuronal loss. Other specific transport mechanisms on the cerebral endothelium include the large neutral amino acid carrier-mediated system that transports, among other amino acids, L–3,4-dihydroxyphenylalanine (L-dopa), used as a therapeutic agent in Parkinson's disease. Also on the surface of blood–brain barrier endothelial cells are transferrin receptors that allow transport of iron into specific areas of the brain. The amount of iron that is transported into the various areas of the brain appears to depend on the concentration of transferrin receptors on the surface of endothelial cells of that region. Thus, the transport of specific nutrients into the brain is regulated during physiological and pathological conditions by blood–brain barrier transport proteins distributed according to the regional and metabolic requirements of brain tissue.

The Basement Membrane, Pericytes, and Astrocytes Are Also Blood–Brain Barrier Components

The basement membranes are not true membranes but are extracellular matrices with a width varying from 20 to 300 nm, composed mainly of collagens, glycoproteins, laminin, proteoglycans, and other proteins. In the cerebral microvasculature, the basement membrane surrounds the endothelium and the adjacent pericytes. The nature of the basement membrane surrounding the blood vessels varies with the type of vasculature and during pathological conditions. The composition and structure of the basement membrane affect the permeability of the vessel. For example, *in vitro* studies of endothelial monolayers in which the underlying matrix was composed primarily of collagen type I restricted passage of albumin, suggesting a role for the basement membrane in blood–brain barrier permeability (Rubin *et al.*, 1991). Replacement of collagen type I with fibronectin resulted in increased permeability to albumin. This finding correlates with the lack of fibronectin around blood–brain barrier vessels.

Pericytes (Fig. 1.22) are present within the basement membrane of all vessels in the body, including the nervous system (Peters *et al.*, 1991). The functions of pericytes include the secretion of basement membrane components, the regulation of revascularization and repair, and the regulation of vascular tone in capillaries. In the central nervous system, pericytes may act as part of the vascular barrier by increasing phagocytosis after blood–brain barrier injury.

The processes of astrocytes form a sheath around blood–brain barrier microvessels. These processes are termed astrocyte foot processes or end feet, and they assist in inducing the blood–brain barrier properties of brain endothelia. (Arthur *et al.*, 1987; Janzer and Raff, 1987).

Disruption of the Blood–Brain Barrier Causes Edema

In general, disruption of the blood–brain barrier causes perivascular or vasogenic edema, which is the accumulation of fluids from the blood around the blood vessels of the brain. This is one of the main features of multiple sclerosis. In multiple sclerosis, inflammatory cells, primarily T cells and macrophages, invade the brain by migrating through the blood–brain barrier and attack cerebral elements as if these elements were foreign antigens (Claudio *et al.*, 1995). It has been observed by many investigators that it is the degree of edema accumulation that causes the neurological symptoms experienced by people with multiple sclerosis.

Studying the regulation of blood–brain barrier permeability is important for several reasons. Therapeutic treatments for neurological diseases need to be able to cross the barrier. Attempts to design drug delivery systems that take therapeutic drugs directly into the brain have been made by using chemically engineered carrier molecules that take advantage of receptors such as that for transferrin, which normally transports iron into the brain. Development of an *in vitro* test system of the blood–brain barrier is of importance in the creation of new neurotropic drugs that are targeted to the brain. This could be especially useful in the treatment of neurodegenerative diseases and the AIDS dementia complex.

Summary

The hallmark of the brain vasculature is the blood–brain barrier, a multicomponent gateway between brain tissue and other organ systems. The blood–brain barrier can be considered the gateway to the brain because it restricts access to macromolecules present in the blood. It also serves as the interface between the immune and nervous systems, acting as the "meeting site" for communication between the two.

References

Altman, J. (1994). Microglia emerge from the fog. *Trends Neurosci.* **17**, 47–49.

Arenander, A. T., and de Vellis, J. (1983). Frontiers of glial physiology. *In* "The Clinical Neurosciences" (R. Rosenberg, Ed.), Sect. 5, pp. 53–91. Churchill–Livingstone, New York.

Arthur, F. E., Shivers, R. R., and Bowman, P. D. (1987). Astrocyte-mediated induction of tight junctions in brain capillary endothelium: An efficient *in vitro* model. *Dev. Brain Res.* **36**, 155–159.

Banati, R. B., and Graeber, M. B. (1994). Surveillance, intervention and cytotoxicity: Is there a protective role of microglia? *Dev. Neurosci.* **16**, 114–127.

Berkley, H. J. (1896). The psychical nerve cell in health and disease. *Bull. Johns Hopkins Hosp.* **7**, 162–164.

Björklund, A., Hökfelt, T., Wouterlood, F. G., and Van den Pol, A. N. (Eds.) (1990). "Handbook of Chemical Neuroanatomy", Vol. 8. Elsevier, Amsterdam.

Braak, H., and Braak, E. (1982). Neuronal types in the striatum of man. *Cell Tissue Res.* **227**, 319–342.

Bradbury, M. W. B. (1979). "The Concept of a Blood–Brain Barrier", pp. 381–407. Wiley, Chichester.

Brightman, M. W., and Reese, T. S. (1969). Junctions between intimately apposed cell membranes in the vertebrate brain. *J. Cell Biol.* **40**, 648–677.

Broadwell, R. D., and Salcman, M. (1981). Expanding the definition of the BBB to protein. *Proc. Natl. Acad. Sci. USA* **78**, 7820–7824.

Brodal, A. (1981). "Neurological Anatomy in Relation to Clinical Medicine", 3rd ed. Oxford Univ. Press, New York.

Buée, L., Hof, P. R., Bouras, C., Delacourte, A., Perl, D. P., Morrison, J. H., and Fillit, H. M. (1994). Pathological alterations of the cerebral microvasculature in Alzheimer's disease and related dementing disorders. *Acta Neuropathol.* **87**, 469–480.

Bunge, M. B., and Bunge, R. P., and Pappas, G. D. (1962). Electron microscopic demonstration of connections between glia and myelin sheaths in the developing central nervous system. *J. Cell Biol.* **12**, 448–456.

Bunge, R. P. (1968). Glial cells and the central myelin sheath. *Physiol. Rev.* **48**, 197–251.

Carpenter, M. B., and Sutin, J. (1983). "Human Neuroanatomy". Williams & Wilkins, Baltimore, MD.

Claudio, L., Raine, C. S., and Brosnan, C. F. (1995). Evidence of persistent blood–brain barrier abnormalities in chronic-progressive multiple sclerosis. *Acta Neuropathol.* **90**, 228–238.

Cobb, S. R., Buhl, E. H., Halasy, K., Paulsen, O., and Somogyi, P. (1995). Synchronization of neuronal activity in hippocampus by individual GABAergic interneurons. *Nature (London)* **378**, 76–79.

Colman, D. R., Doyle, J. P., D'Urso, D., Kitagawa, K., Pedraza, L., Yoshida, M., and Fannon, A. M. (1996). Speculations on myelin sheath evolution. *In* "Glial Cell Development" (K. R. Jessen and W. D. Richardson, Eds.), pp. 85–100. Bios Scientific, Oxford.

Coss, R. G., and Perkel, D. H. (1985). The function of dendritic spines: A review of theoretical issues. *Behav. Neural Biol.* **44**, 151–185.

DeFelipe, J., and Jones, E. G. (1992). High-resolution light and electron microscopic immunocytochemistry of colocalized GABA and calbindin D-28k in somata and double bouquet cell axons of monkey somatosensory cortex. *Eur. J. Neurosci.* **4**, 46–60.

DeFelipe, J., Hendry, S. H. C., and Jones, E. G. (1989). Visualization of chandelier cell axons by parvalbumin immunoreactivity in monkey cerebral cortex. *Proc. Natl. Acad. Sci. USA* **86**, 2093–2097.

Del Rio-Hortega, P. (1932). Microglia. *In* "Cytology and Cellular Pathology of the Nervous System" (W. Penfield, Ed.), Vol. 2, pp. 481–534. Harper (Hoeber), New York.

Dolman, C. L. (1991). Microglia. *In* "Textbook of Neuropathology" (R. L. Davis and D. M. Robertson, Eds.), pp. 141–163. Williams & Wilkins, Baltimore, MD.

Duvernoy, H. M., Delon, S., and Vannson, J. L. (1981). Cortical blood vessels of the human brain. *Brain Res. Bull.* **7**, 519–579.

Feldman, M. L. (1984). Morphology of the neocortical pyramidal neuron. *In* "Cellular Components of the Cerebral Cortex" (A. Peters and E. G. Jones, Eds.), Vol. 1, pp. 123–200. Plenum, New York.

Fernandez-Moran, H. (1950). EM observations on the structure of the myelinated nerve sheath. *Exp. Cell Res.* **1**, 143–162.

Filbin, M., Walsh, F., Trapp, B., Pizzey, J., and Tennekoon, G. (1990). Role of myelin P_0 protein as a homophilic adhesion molecule. *Nature (London)* **344**, 871–872.

Folch-Pi, J., and Lees, M. B. (1951). Proteolipids, a new kind of tissue lipoproteins. *J. Biol. Chem.* **191**, 807–813.

Freund, T. F., and Buzsaki, G. (1996) Interneurons of the hippocampus. *Hippocampus* **6**, 347–470.

Freund, T. F., Martin, K. A. C., Smith, A. D., and Somogyi, P. (1983). Glutamate decarboxylase-immunoreactive terminals of Golgi-impregnated axoaxonic cells and of presumed basket cells in synaptic contact with pyramidal neurons of the cat's visual cortex. *J. Comp. Neurol.* **221**, 263–278.

Furness, J. B., and Costa, M. (1980). Types of nerves in the enteric nervous system. *Neuroscience* **5**, 1–20.

Gehrmann, J., Matsumoto, Y., and Kreutzberg, G. W. (1995). Microglia: Intrinsic immuneffector cell of the brain. *Brain Res. Rev.* **20**, 269–287.

Giese, K. P., Martini, R., Lemke, G., Soriano, P., and Schachner, M. (1992). Mouse P_0 gene disruption leads to hypomyelination, abnormal expression of recognition molecules, and degeneration of myelin and axons. *Cell (Cambridge, Mass.)* **71**, 565–576.

Goldstein, G. W. (1988). Endothelial cell–astrocyte interactions: A cellular model of the blood–brain barrier. *Ann. N.Y. Acad. Sci.* **529**, 31–39.

Gray, E. G. (1959). Axo-somatic and axo-dendritic synapses of the cerebral cortex: An electron microscope study. *J. Anat.* **93**, 420–433.

Hof, P. R., Mufson, E. J., and Morrison, J. H. (1995a). The human orbitofrontal cortex: Cytoarchitecture and quantitative immuno-histochemical parcellation. *J. Comp. Neurol.* **359**, 48–68.

Hof, P. R., Nimchinsky, E. A., and Morrison, J. H. (1995b). Neurochemical phenotype of corticocortical connections in the macaque monkey: Quantitative analysis of a subset of neurofilament protein-immunoreactive projection neurons in frontal, parietal, temporal, and cingulate cortices. *J. Comp. Neurol.* **362**, 109–133.

Holash, J. A., and Stewart, P. A. (1993). The relationship of astrocyte-like cells to the vessels that contribute to the blood–ocular barriers. *Brain Res.* **629**, 218–224.

Hudspeth, A. J. (1983). Transduction and tuning by vertebrate hair cells. *Trends Neurosci.* **6**, 366–369.

Ikenaka, K., Furuichi, T., Iwasaki, Y., Moriguchi, A., Okano, H., and Mikoshiba, K. (1988). Myelin proteolipid protein gene structure and its regulation of expression in normal and jimpy mutant mice. *J. Mol. Biol.* 1991, 587–596.

Janzer, R. C., and Raff, M. C. (1987). Astrocytes induce blood–brain barrier properties in endothelial cells. *Nature (London)* **325**, 235–256.

Jones, E. G. (1975). Varieties and distribution of non-pyramidal cells in the somatic sensory cortex of the squirrel monkey. *J. Comp. Neurol.* **160**, 205–267.

Jones, E. G. (1984). Laminar distribution of cortical efferent cells. *In* "Cellular Components of the Cerebral Cortex" (A. Peters and E. G. Jones, Eds.), Vol. 1, pp. 521–553. Plenum, New York.

Kettenman, H., and Ransom, B. R. (Eds.) (1995). "Neuroglia". Oxford Univ. Press, Oxford.

Kimbelberg, H., and Norenberg, M. D. (1989). Astrocytes. *Sci. Am.* **26**, 66–76.

Kirschner, D. A., Ganser, A. L., and Caspar, D. W. (1984). Diffraction studies of molecular organization and membrane interactions in myelin. *In* "Myelin" (P. Morell, Ed.), pp. 51–96. Plenum, New York.

Kisvarday, Z. F., Cowey, A., and Somogyi, P. (1986a). Synaptic relationships of a type of GABA-immunoreactive neuron (clutch cell), spiny stellate cells and lateral geniculate nucleus afferents in layer IVC of the monkey striate cortex. *Neuroscience* **19**, 741–761.

Kisvarday, Z. F., Martin, K. A. C., Freund, T. F., Magloczky, Z., Whitteridge, D., and Somogyi, P. (1986b). Synaptic targets of HRP-filled layer III pyramidal cells in the cat striate cortex. *Exp. Brain Res.* **64**, 541–552.

Krebs, W., and Krebs, I. (1991). "Primate Retina and Choroid: Atlas of Fine Structure in Man and Monkey". Springer-Verlag, New York.

Lawson, L. J., Perry, V. H., Dri, P., and Gordon, S. (1990). Heterogeneity in the distribution and morphology of microglia in the normal adult mouse brain. *Neuroscience* **39**, 151–170.

Lemke, G., and Axel, R. (1985). Isolation and sequence of the gene encoding the major structural protein of peripheral myelin. *Cell (Cambridge, Mass.)* **40**, 501–513.

Lum, H., and Malik, A. B. (1994). Regulation of vascular endothelial barrier function. *Am. J. Physiol.* **267**, L223–L241.

Lund, J. S., Wu, Q., Hadingham, P. T., and Levitt, J. B. (1995). Cells and circuits contributing to functional properties in area I of macaque monkey cerebral cortex: Bases for neuroanatomically realistic models. *J. Anat.* **187**, 563–581.

Milner, R., Lai, C., Nave, K.-A., Lenoir, D., Ogata, J., and Sutcliffe, J. (1985). Nucleotide sequences of two mRNAs for rat brain myelin proteolipid protein. *Cell (Cambridge, Mass.)* **42**, 931–942.

Morell, P., and Norton, W. T. (1980). *Myelin. Sci. Am.* **242**, 88–118.

Mountcastle, V. B. (1978). An organizing principle for cerebral function: The unit module and the distributed system. *In* "The Mindful Brain: Cortical Organization and the Group-Selective Theory of Higher Brain Function" (V. B. Mountcastle and G. Eddman, Eds.), pp. 7–50. MIT Press, Cambridge, MA.

Nave, K.-A. (1994). Neurological mouse mutants and the genes of myelin. *J. Neurosci. Res.* **38**, 607–612.

Nave, K.-A., Lai, C., Bloom, F. E., and Milner, R. J. (1987). Splice site selection in the proteolipid protein (PLP) gene transcript and primary structure of the DM20 protein of central nervous system myelin. *Proc. Natl. Acad. Sci. USA* **84**, 5665–5669.

Nimchinsky, E. A., Sabatini, B. L., and Svoboda, K. (2002). Structure and function of dendritic spines. *Annu. Rev. Physiol.* **64**, 313–353.

Nissl, F. (1899). U..eber einige Beziehungen zwischen Nervenzellenerkra..nkungen und glio..sen Erscheinungen bei verschiedenen Psychosen. *Arch. Psychol.* **32**, 1–21.

Palay, S. L., and Chan-Palay, V. (1974). "Cerebellar Cortex: Cytology and Organization". Springer-Verlag, Berlin.

Peters, A., and Jones, E. G. (Eds.) (1984). "Cellular Components of the Cerebral Cortex", Vol. 1. Plenum, New York.

Peters, A., Palay, S. L., and Webster, H. de F. (1991). "The Fine Structure of the Nervous System: Neurons and Their Supporting Cells", 3rd ed. Oxford Univ. Press, New York.

Ramón y Cajal, S. (1913). Contribucion al conocimiento de la neuroglia del cerebro humano. *Trab. Lab. Invest. Biol.* **11**, 255–315.

Ramón y Cajal, S. (1955). "Histologie du systeme nerveux de l'homme et des vertebres". CSIC, Instituto Cajal, Madrid.

Raub, T. J., Kuentzel, S., and Sawada, G. A. (1992). Permeability of bovine brain microvessel endothelial cells *in vitro*: Barrier tightening by a factor released from astroglioma cells. *Exp. Cell Res.* **199**, 330–340.

Readhead, C., Popko, B., Takahashi, N., Shine, H. D., Saavedra, R. A., Sidman, R. L., and Hood, L. (1987). Expression of a myelin basic protein gene in transgenic shiverer mice: Correction of the dysmyelinating phenotype. *Cell (Cambridge, Mass.)* **48**, 703–712.

Remahl, S., and Hildebrand, C. (1990). Relation between axons and oligodendroglial cells during initial myelination. Part 2. The individual axon. *J. Neurocytol.* **19**, 883–898.

Ritchie, J. M. (1984). Physiological basis of conduction in myelinated nerve fibers. *In* "Myelin" (P. Morell, Ed.), pp. 117–146. Plenum, New York.

Rosenbluth, J. (1980). Central myelin in the mouse mutant shiverer. *J. Comp. Neurol.* **194**, 639–728.

Rosenbluth, J. (1980). Peripheral myelin in the mouse mutant shiverer. *J. Comp. Neurol.* **194**, 729–753.

Rubin, L. L., Hall, D. E., Porter, S., Barbu, K., Cannon, C., Horner, H. C., Janatpour, M., Liaw, C. W., Manning, K., Morales, J., Tanner, L. I., Tomaselli, K. J., and Bard, F. (1991). A cell culture model of the blood–brain barrier. *J. Cell Biol.* **115**, 1725–1735.

Scheibel, M. E., and Scheibel, A. B. (1968). On the nature of dendritic spines: Report of a workshop. *Commun. Behav. Biol. A* **1**, 231–265.

Schmitt, O. F., Worden, F. G., Adelman, G., and Dennis, S. G. (1981). "The Organization of the Cerebral Cortex". MIT Press, Cambridge, MA.

Shapiro, L., Fannon, A. M., Kwong, P. D., Thompson, A., Lehmann, M. S., Gru..bel, G., Legrand, J.-F., Als-Nielson, J., Colman, D. R., and Hendrickson, W. A. (1995). Structural basis of cell–cell adhesion by cadherins. *Nature (London)* **374**, 327–337.

Sik, A., Penttonen, M., Ylinen, A., and Buzsaki, G. (1995). Hippocampal CA1 interneurons: An *in vivo* intracellular labeling study. *J. Neurosci.* **15**, 6651–6665.

Somogyi, P., and Cowey, A. (1981). Combined Golgi and electron microscopic study on the synapses formed by double bouquet cells in the visual cortex of the cat and monkey. *J. Comp. Neurol.* **195**, 547–566.

Somogyi, P., Freund, T. F., and Cowey, A. (1982). The axo-axonic interneuron in the cerebral cortex of the rat, cat and monkey. *Neuroscience* **7**, 2577–2607.

Somogyi, P., Hodgson, A. J., Smith, A. D., Nunzi, M. G., Gorio, A., and Wu, J. Y. (1984). Different populations of GABAergic neurons in the visual cortex and hippocampus of cat contain somatostatin- or cholecystokinin-immunoreactive material. *J. Neurosci.* **4**, 2590–2603.

Somogyi, P., Kisvarday, Z. F., Martin, K. A. C., and Whitteridge, D. (1983). Synaptic connections of morphologically identified and physiologically characterized large basket cells in the striate cortex of cat. *Neuroscience* **10**, 261–294.

Steward, O., and Falk, P. M. (1986). Protein-synthetic machinery at postsynaptic sites during synaptogenesis: A quantitative study of the association between polyribosomes and developing synapses. *J. Neurosci.* **6**, 412–423.

Szentagothai, J., and Arbib, M. A. (1974). Conceptual models of neural organization. *Neurosci. Res. Program Bull.* **12**, 306–510.

van Domburg, P. H. M. F., and ten Donkelaar, H. J. (1991). The human substantia nigra and ventral tegmental area. *Adv. Anat. Embryol. Cell Biol.* **121**, 1–132.

Williams, A. F. (1987). A year in the life of the immunoglobulin superfamily. *Immunol. Today* **8**, 298–303.

Zhang, W., and Benson, D. L. (2000). Development and molecular organization of dendritic spines and their synapses. *Hippocampus* **10**, 512–526.

2

Subcellular Organization of the Nervous System: Organelles and Their Functions

Scott Brady, David R. Colman, and Peter Brophy

Cells have many features in common, but each cell type also possesses a functional architecture related to its unique physiology. In fact, cells may become so specialized in fulfilling a particular function that virtually all cellular components may be devoted to it. For example, the machinery inside mammalian erythrocytes is completely dedicated to the delivery of oxygen to the tissues and the removal of carbon dioxide. Toward this end, this cell has evolved a specialized plasma membrane, an underlying cytoskeletal matrix that molds the cell into a biconcave disk, and a cytoplasm rich in hemoglobin. Modification of the cell machinery extends even to the discarding of structures such as the nucleus and the protein synthetic apparatus, which are not needed after the red blood cell matures. In many respects, the terminally differentiated, highly specialized cells of the nervous system exhibit comparable commitment: the extensive development of subcellular components reflects the roles that each plays.

The neuron serves as the cellular correlate of information processing and, in aggregate, all neurons act together to integrate responses of the entire organism to the external world. It is therefore not surprising that the specializations found in neurons are more diverse and complex than those found in any other cell type. Single neurons commonly interact in specific ways with hundreds of other cells—other neurons, astrocytes, oligodendrocytes, immune cells, muscle cells, and glandular cells. In this chapter, the major functional domains of the neuron are defined, the subcellular elements that compose the building blocks of these domains are described, and the processes that create and maintain neuronal functional architecture are examined.

AXONS AND DENDRITES: UNIQUE STRUCTURAL COMPONENTS OF NEURONS

Neurons and glial cells are remarkable for their size and complexity, but they do share many features with other eukaryotic cells (Peters *et al.*, 1991). As discussed in Chapter 1, the perikaryon, or cell body, contains a nucleus and its associated protein synthetic machinery. Most neuronal nuclei are large, and they typically contain a preponderance of euchromatin. This is consistent with the need to create and maintain a large cellular volume. Because protein synthesis must be kept at a high level just to maintain the neuronal extensions, transcription levels in neurons are generally high. In turn, the wide variety of different polypeptide constituents associated with the cellular domains in a neuron requires that a large number of different genes be constantly transcribed.

When specific mRNAs have been synthesized and processed, they move from the nucleus into a subcellular region that can be termed the *translational cytoplasm* (Lasek and Brady, 1982) comprising cytoplasmic ("free") and membrane-associated polysomes, the intermediate compartment of the smooth endoplasmic reticulum, and the Golgi complex. The constituents of translational cytoplasm are thus associated with the synthesis and processing of proteins. Neurons in particular have relatively large amounts of translational cytoplasm to accommodate a high level of protein synthesis. This protein synthetic machinery is arranged in discrete intracellular "granules" termed *Nissl substance* (Box 2.1) after the histologist who first discovered these structures in the 19th century. Nissl substance is actually a combination of stacks of rough endoplasmic reticulum (RER), interposed with rosettes of free polysomes

Copyright 2004, Elsevier Science (USA).
All rights reserved.

BOX 2.1

NISSL SUBSTANCE

It is interesting that Nissl recognized the composite nature of the substance named for him, although he could not have resolved either of its components.

The preceding quote and the following discussion of the Nissl substance appeared in L. Sanford, M. D. Palay, and G. E. Palade (1955). The fine structure of neurons. J. Biophys. Biochem. Cytol. 88, 69–88.

As imaged in the electron microscope, the crowded cytoplasm of the neuron contrasts sharply with the relatively open cytoplasm of many other cell types. As this compact appearance stems largely from the extensive meshwork of Nissl substance, the neuron resembles, even at the electron microscope level, certain protein-secreting glandular cells, such as those of the pancreatic acini and the salivary glands . . .

Like the basophilic substance or ergastoplasm of glandular cells, the Nissl substance is a composite material constructed of endoplasmic reticulum and fine granules, both of which have been revealed by electron microscopy. The first component of the Nissl substance appears to be part of the general endoplasmic reticulum of the neuron. The reticulum extends throughout the entire cytoplasm, but is considerably more condensed within the area of the Nissl bodies than in the rest of the cell. These condensations not only determine the size and shape of each Nissl body, but also constitute a membranous framework upon which the other components are arranged. In many types of neurons, the meshes of the endoplasmic reticulum are distributed at random in three dimensions but in all neurons, and especially in the large motor neurons, the endoplasmic reticulum may display a distinctly orderly arrangement within the Nissl bodies. This orientation consists of a layering of reticular sheets, at more or less regular intervals. Each sheet is a reticulum developed predominantly in two dimensions and comprising tubules, strings of vesicles, and numerous large and flat cisternae. Even in such highly ordered forms as the Nissl bodies of motor neurons, the continuity of the reticulum persists, as indicated by frequent branches and anastomoses between layers.

The second component of the Nissl substance is represented by small granules disposed in patterned arrays either in close contact with the outer membranes of the endoplasmic reticulum or scattered in the intervening matrix. This matrix may be considered as a third component of the Nissl substance. It is evident, therefore, that although the Nissl bodies are differentiated parts of the cytoplasm, they are continuous with it by virtue of the continuity of the endoplasmic reticulum and the matrix. No interface or membrane separates them from the rest of the cytoplasm. In this respect, as well as in general architecture, they are comparable to the ergastoplasm of glandular cells. The only differences lie in (a) a different intracellular distribution, (b) a lesser degree of preferred orientation of the endoplasmic reticulum, and (c) an apparently greater concentration of fine granules within the areas of the Nissl bodies.

We now recognize that the electron-dense "granules" are ribosomes, arranged in cytoplasmic "free" polysomal rosettes (see box in Fig. 2.1), or attached to the surface of the endoplasmic reticulum membrane. The lumen of the ER compartment (arrow) contains a "fuzz" that probably is formed by the numerous nascent chains being cotranslationally inserted into the ER lumen, some resident proteins that take part in the translation process, and certain structural components of the ER membrane.

(Box 2.1, Fig. 2.1). This arrangement is unique to neurons, and its functional significance is unknown. Most, but by no means all, proteins used throughout the neuron are synthesized in the perikaryon. During or after synthesis and processing, proteins are packaged into membrane-limited organelles, are incorporated into cytoskeletal elements, or remain as soluble constituents of the cytoplasm. After proteins have been packaged appropriately, they are transported to their sites of function.

With a few exceptions, vertebrate neurons have two discrete functional domains or compartments, the

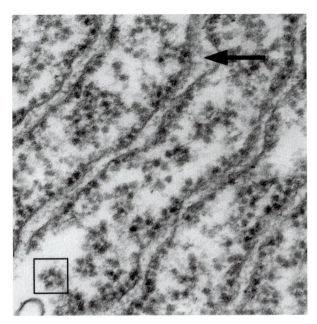

FIGURE 2.1 The "Nissl body" in neurons is an array of cytoplasmic free polysomal rosettes (boxed) interspersed between rows of rough endoplasmic reticulum (RER) studded with membrane-bound ribosomes. Nascent polypeptide chains emerging from the ribosomal tunnel on the RER are inserted into the lumen (arrow), where they may be processed before transport out of the RER. The relationship between the polypeptide products of these "free" and "bound" polysome populations in the Nissl body, an arrangement that is unique to neurons, is unknown.

axonal and the somatodendritic compartments, each of which encompasses a number of sub- or microdomains (Fig. 2.2). The axon is perhaps the most familiar functional domain of a neuron (Lasek and Brady, 1982) and is classically defined as the cellular process by which a neuron makes contact with a target cell to transmit information, providing a conducting structure for transmitting the action potential to a synapse, a specialized subdomain for transmission of a signal from neuron to target cell (neuron, muscle, etc.), most often by release of appropriate neurotransmitters. Consequently, most axons end in a presynaptic terminal, although a single axon may have many (hundreds or even thousands in some cases) presynaptic specializations known as *en passant* synapses along its length. Characteristics of presynaptic terminals are presented in greater detail later in this chapter.

The axon is the first neuronal process to differentiate during development. A typical neuron has only a single axon that proceeds some distance from the cell body before branching extensively. Usually the longest process of a neuron, axons come in many sizes. In a human adult, axons range in length from a few micrometers for small interneurons to a meter or more for large motor neurons, and they may be even

longer in large animals (such as giraffes, elephants, and whales). In mammals and other vertebrates, the longest axons generally extend approximately half the body length.

Axonal diameters also are quite variable, ranging from 0.1 to 20 μm for large myelinated fibers in vertebrates. Invertebrate axons grow to even larger diameters, with the giant axons of some squid species achieving diameters in the millimeter range (see Fig. 5.1A). Invertebrate axons reach such large diameters because they lack the myelinating glia that speed conduction of the action potential. As a result, axonal caliber must be large to sustain the high rate of conduction needed for the reflexes that permit escape from predators and capture of prey. Although axonal caliber is closely regulated in both myelinated and nonmyelinated fibers, this parameter is critical for those organisms that are unable to produce myelin.

The region of the neuronal cell body where the axon originates has several specialized features. This domain, called the axon hillock, is most readily distinguished by a deficiency of Nissl substance. Therefore, protein synthesis cannot take place to any appreciable degree in this region. Cytoplasm in the vicinity of the axon hillock may have a few polysomes but is dominated by the cytoskeletal and membranous organelles that are being delivered to the axon. Microtubules and neurofilaments begin to align roughly parallel to each other, helping to organize membrane-limited organelles destined for the axon. The hillock is a region where materials either are committed to the axon (cytoskeletal elements, synaptic vesicle precursors, mitochondria, etc.) or are excluded from the axon (RER and free polysomes, dendritic microtubule-associated proteins). The molecular basis for this sorting is not understood. Cytoplasm in the axon hillock does not appear to contain a physical "sizing" barrier (like a filter) because large organelles such as mitochondria readily enter the axon, whereas only a small number of essentially excluded structures such as polysomes are occasionally seen only in the initial segment of the axon and not in the axon proper. An exception to this general rule is during development when local protein synthesis does take place at the axon terminus or growth cone. In the mature neuron, the physiological significance of this barrier must be considerable, because axonal structures are found to accumulate in this region in many neuropathologies, including those due to degenerative diseases (such as amyotrophic lateral sclerosis) and to exposure to neurotoxic compounds (such as acrylamide).

The initial segment of the axon is the region of the axon adjacent to the axon hillock. Microtubules gener-

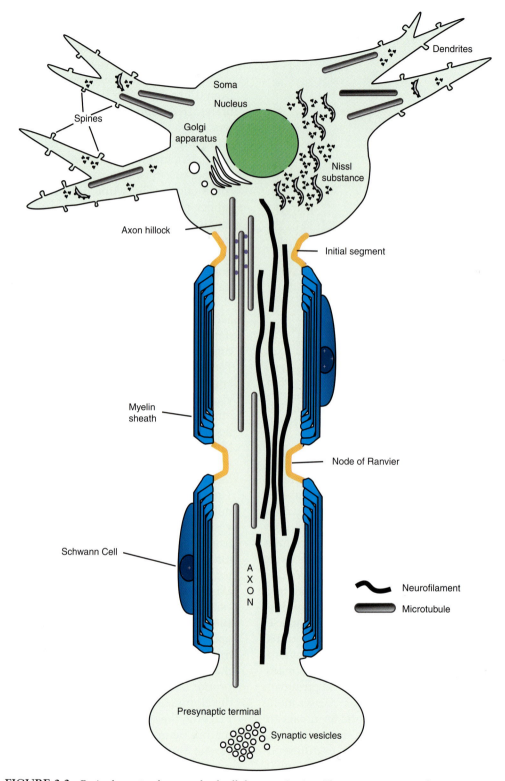

FIGURE 2.2 Basic elements of neuronal subcellular organization. The neuron consists of a soma, or cell body, in which the nucleus, multiple cytoplasm-filled processes termed dendrites, and the (usually single) axon are located. The neuron is highly extended in space; a neuron with a cell body of the size shown here could easily maintain an axon several miles in length! The unique shape of each neuron is the result of a cooperative interplay between plasma membrane components (the lipid matrix and associated proteins) and cytoskeletal elements. Most large neurons in vertebrates are myelinated by oligodendrocytes in the CNS and Schwann cells in the PNS. The compact wraps of myelin encasing the axon distal to the initial segment permit the rapid conduction of the action potential by a process termed "saltatory conduction" (see Chapters 1, 4, and 5).

ally form characteristic fascicles, or bundles, in the initial segment of the axon. These fascicles are not seen elsewhere. The initial segment and, to some extent, the axon hillock also have a distinctive specialized plasma membrane. Initially, the plasmalemma was thought to have a thick electron-dense coating actually attached to the inner surface of the membrane, but this dense undercoating is in reality separated by 5 to 10 nm from the plasma membrane inner surface and has a complex ultrastructure. Neither the composition nor the function of this undercoating is known. Curiously, the undercoating is present in the same regions of the initial segment as the distinctive fasciculation of microtubules, although the relationship is not understood. The plasma membrane is specialized in the initial segment and axon hillock in that it contains voltage-sensitive ion channels in large numbers, and most action potentials originate in this domain.

Ultimately, axonal structure is geared toward the efficient conduction of action potentials at a rate appropriate to the function of that neuron. This can be seen from both the ultrastructure and the composition of axons. Axons are roughly cylindrical in cross section with little or no taper. As discussed later in this chapter, this diameter is maintained by regulation of the cytoskeleton. Even at branch points, daughter axons are comparable in diameter to the parent axon. This constant caliber helps ensure a consistent rate of conduction. Similarly, the organization of membrane components is regulated to this end. Voltage-gated ion channels are distributed to maximize conduction. Sodium channels are distributed more or less uniformly in small nonmyelinated axons, but are concentrated at high density in the regularly spaced unmyelinated gaps, known as nodes of Ranvier (see Figs. 1.14, 2.2, and 5.8). An axon so organized will conduct an action potential or train of spikes long distances with high fidelity at a defined speed. These characteristics are essential for maintaining the precise timing and coordination seen in neuronal circuits.

Most vertebrate neurons have multiple dendrites arising from their perikarya. Unlike axons, dendrites continuously branch and taper extensively, with a reduction in caliber in daughter processes at each branching. In addition, the surface of dendrites is covered with small protrusions, or spines, which are postsynaptic specializations (see also Chapters 1, 4, 17, and 18). Although the surface area of a dendritic arbor may be quite extensive, dendrites in general remain in the relative vicinity of the perikaryon. A dendritic arbor may be contacted by the axons of many different and distant neurons or innervated by a single axon making multiple synaptic contacts.

The base of a dendrite is continuous with the cytoplasm of the cell body. In contrast to the axon, Nissl substance extends into dendrites, and certain proteins are synthesized predominantly in dendrites. There is evidence for the selective placement of some mRNAs in dendrites as well (Steward, 1995). For example, whereas RER and polysomes extend well into the dendrites, the mRNAs that are transported and translated in dendrites are a subset of the total neuronal mRNA, deficient in some mRNA species (such as neurofilament mRNAs) and enriched in mRNAs with dendritic functions (such as microtubule-associated protein mRNAs, microtubule-associated protein 2). Also, certain proteins appear to be targeted, postsynthesis, to the dendritic compartment as well.

The shapes and complexity of dendritic arborizations may be remarkably plastic. Dendrites appear relatively late in development and initially have only limited numbers of branches and spines. As development and maturation of the nervous system proceed, the size and number of branches increase. The number of spines increases dramatically, and their distribution may change. This remodeling of synaptic connectivity may continue into adulthood, and environmental effects can alter this pattern significantly. Eventually, in the aging brain, there is a reduction in complexity and size of dendritic arbors, with fewer spines and thinner dendritic shafts. These changes correlate with changes in neuronal function during development and aging.

As defined by classic physiology, axons are the structural correlates for neuronal output, and dendrites constitute the domain for receiving information. A neuron without an axon or one without dendrites might therefore seem paradoxical, but such neurons do exist. Certain amacrine and horizontal cells in the vertebrate retina have no identifiable axons, although they do have dendritic processes that are morphologically distinct from axons. Such processes may have both pre- and postsynaptic specializations or they may have gap junctions (Chapter 15) that act as direct electrical connections between two cells. Similarly, the pseudounipolar sensory neurons of dorsal root ganglia (DRG) have no dendrites. In their mature form, these DRG sensory neurons give rise to a single axon that extends a few hundred micrometers before branching. One long branch extends to the periphery, where it may form a sensory nerve ending in muscle spindles or skin. Large DRG peripheral branches are myelinated and have the morphological characteristics of an axon, but they contain neither pre- nor postsynaptic specializations. The other branch extends into the central nervous system, where it forms synaptic contacts. In

TABLE 2.1   Functional and Morphological Hallmarks of Axons and Dendrites[a]

Axons	Dendrites
With rare exceptions, each neuron has a single axon.	Most neurons have multiple dendrites arising from their cell bodies.
Axons appear first during neuronal differentiation.	Dendrites begin to differentiate only after the axon has formed.
Axon initial segments are distinguished by a specialized plasma membrane containing a high density of ion channels and distinctive cytoskeletal organization.	Dendrites are continuous with the perikaryal cytoplasm, and the transition point cannot be readily distinguished.
Axons typically are cylindrical in form with a round or elliptical cross section.	Dendrites usually have a significant taper and small spinous processes that give them an irregular cross section.
Large axons are myelinated in vertebrates, and the thickness of the myelin sheath is proportional to the axonal caliber.	Dendrites are not myelinated, although a few wraps of myelin may occur rarely.
Axon caliber is a function of neurofilament and microtubule numbers, with neurofilaments predominating in large axons.	The dendritic cytoskeleton may appear less organized, and microtubules dominate even in large dendrites.
Microtubules in axons have a uniform polarity with plus ends distal from the cell body.	Microtubules in proximal dendrites have mixed polarity, with both plus and minus ends oriented distal to the cell body.
Axonal microtubules are enriched in tau protein with a characteristic phosphorylation pattern.	Dendritic microtubules may contain some tau protein, but MAP2 is not present in axonal compartments and is highly enriched in dendrites.
Ribosomes are excluded from mature axons, although a few may be detectable in initial segments.	Both rough endoplasmic reticulum and cytoplasmic polysomes are present in dendrites, with specific mRNAs being enriched in dendrites.
Axonal branches tend to be distal from the cell body.	Dendrites begin to branch extensively near the perikaryon and form extensive arbors in the vicinity of the perikaryon.
Axonal branches form obtuse angles and have diameters similar to the parent stem.	Dendritic branches form acute angles and are smaller than the parent stem.
Most axons have presynaptic specializations that may be en passant or at the ends of axonal branches.	Dendrites are rich in postsynaptic specializations, particularly on the spinous processes that project from the dendritic shaft.
Action potentials are usually generated at the axon hillock and conducted away from the cell body.	Some dendrites can generate action potentials, but more commonly they modulate the electrical state of the perikaryon and initial segment.
Traditionally, axons are specialized for conduction and synaptic transmission, i.e., neuronal output.	Dendritic architecture is most suitable for integrating synaptic responses from a variety of inputs, i.e., neuronal input.

[a] Neurons typically have two classes of cytoplasmic extensions that may be distinguished using electrophysiological, morphological, and biochemical criteria. Although some neuronal processes may lack one or more of these features, enough parameters can generally be defined to allow unambiguous identification.

DRG neurons, the action potential is generated at distal sensory nerve endings and then transmitted along the peripheral branch to the central branch and the appropriate CNS targets, bypassing the cell body. The functional and morphological hallmarks of axons and dendrites are listed in Table 2.1.

Summary

Neurons are polarized cells that are specialized for membrane and protein synthesis as well as for conduction of the nerve impulse. In general, neurons have a cell body, a dendritic arborization that is usually located near the cell body, and an extended axon that may branch considerably before terminating to form synapses with other neurons.

PROTEIN SYNTHESIS IN NERVOUS TISSUE

Both neurons and glial cells have strikingly extended morphologies. This cytoarchitecture is ideal for a tissue whose functions depend on multiple intercellular contacts locally and at great distances. Protein and lipid components are synthesized and assembled into the membranes of these cell extensions through pathways of membrane biogenesis that have been elucidated primarily in other cell types, including the yeast *Saccharomyces cerevisiae*. However, some adaptations of these general mechanisms have been necessary, owing to the specific requirements of cells in the nervous system. Neurons, for example, have devised mechanisms for ensuring that the specific components of the axonal and dendritic plasma membranes are selectively delivered (targeted) to each plasma membrane subdomain.

The distribution to specific loci of organelles, receptors, and ion channels is critical to normal neuronal function. In turn, these loci must be "matched" appropriately to the local microenvironment and specific cell–cell interactions. Similarly, in myelinating glial cells during the narrow developmental window when the myelin sheath is being formed, these cells synthesize vast sheets of insulating plasma membrane at an unbelievably high rate. To understand how the plasma membrane of neurons and glia might be modeled to fit individual functional requirements, it is necessary to review the progress that has been made so far in the understanding of how membrane components and organelles are generated in eukaryotic cells.

There are two major categories of membrane proteins, integral and peripheral. Integral membrane proteins, which include the receptors for neurotransmitters (e.g., the acetylcholine receptor subunits) and polypeptide growth factors (e.g., the dimeric insulin receptor), have segments that are either embedded in the lipid bilayer or covalently bound to molecules that insert into the membrane, such as those proteins linked to glycosyl phosphatidylinositol at their C termini (e.g., Thy-1). A protein with a single membrane-embedded segment and an N terminus exposed at the extracellular surface is said to be of type I, whereas type II proteins retain their N termini on the cytoplasmic side of the plasma membrane. Peripheral membrane proteins are localized on the cytoplasmic surface of the membrane and do not cross any membrane during their biogenesis. They interact with the membrane either by means of their associations with membrane lipids or the cytoplasmic tails of integral proteins or by means of their affinity for other peripheral proteins (e.g., platelet-derived growth factor receptor–Grb2–Sos–Ras complex). In some cases, they may bind directly to the polar head groups of the lipid bilayer (e.g., myelin basic protein).

Integral Membrane and Secretory Polypeptides Are Synthesized *de Novo* in the Rough Endoplasmic Reticulum

The subcellular destinations of integral and peripheral membrane proteins are determined by their sites of synthesis (Blobel, 1980; Sabatini *et al.*, 1982). In the secretory pathway, integral membrane proteins, like secretory proteins, are synthesized in the rough endoplasmic reticulum (RER), whereas the mRNAs encoding peripheral proteins are translated on cytoplasmic "free"

polysomes (FP), which are not membrane-associated but which may interact with cytoskeletal structures.

The pathway by which secretory proteins are synthesized and exported was first postulated through the elegant ultrastructural studies on the pancreas by George Palade and colleagues (Palade, 1975; Jamieson and Palade, 1967a,b). Pancreatic acinar cells were an excellent choice for this work because they are extremely active in secretion, as revealed by the abun-

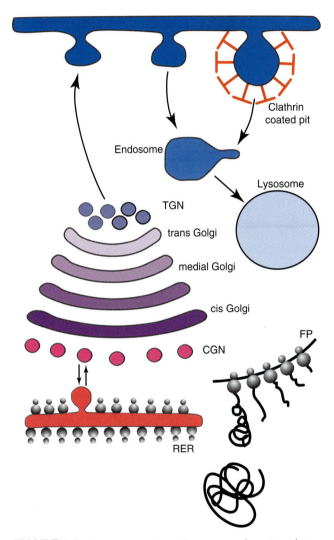

FIGURE 2.3 Secretory pathway. Transport and sorting of proteins in the secretory pathway occur as they pass through the Golgi complex before reaching the plasma membrane. Sorting occurs in the cis-Golgi network (CGN), also known as the intermediate compartment, and in the trans-Golgi network (TGN). Proteins exit from the Golgi complex at the TGN. The default pathway is the direct route to the plasma membrane. Proteins bound for regulated secretion or for transport to endosomes and from there to lysosomes are diverted from the default path by means of specific signals. In endocytosis, one population of vesicles is surrounded by a clathrin cage and is destined for late endosomes. Another population appears to be coated in a lacelike structure the composition of which is yet to be defined.

dance of their RER network, a property they share with neurons. In the 19th century, Nissl deduced that pancreatic cells and neurons would be found to have common secretory properties because of similarities in the distribution of the Nissl substance (Fig. 2.3).

Pulse–chase autoradiography has revealed that newly synthesized secretory proteins move from the RER to the Golgi apparatus, where the proteins are packaged into secretory granules and transported to the plasma membrane from which they are released by exocytosis (Palade, 1975; Jamieson and Palade 1967a,b). Pulse–chase studies in neurons reveal a similar sequence of events for proteins transported into the axon (Droz and Leblond, 1963). The unraveling of the detailed molecular mechanisms of the pathway began with the successful reconstitution of secretory protein biosynthesis *in vitro* and the direct demonstration that, very early during synthesis, secretory proteins are translocated into the lumen of RER vesicles, termed microsomes, prepared by cell fractionation, (Blobel and Dobberstein, 1975; Rothman and Lodish, 1977). A key observation here was that the fate of the protein was sealed as a result of encapsulation in the lumen of the RER at the site of synthesis. This cotranslational insertion model provided a logical framework for understanding the synthesis of integral membrane proteins with a transmembrane orientation (Sabatini *et al.*, 1982).

The process by which integral membrane proteins are synthesized closely follows the secretory pathway, except integral proteins are of course not released from the cell, but instead remain within cellular membranes. Synthesis of integral proteins begins with synthesis of the nascent chain on a polysome that is not yet bound to the RER membrane (Fig. 2.4). The emergence of the N terminus of the nascent protein from the protein-synthesizing machinery allows a ribonucleoprotein, a signal recognition particle (SRP), to bind to an emergent hydrophobic signal sequence and prevent further translation (Walter and Blobel, 1981; Walter and Johnson, 1994; Gilmore *et al.*, 1982a,b). Translation arrest is relieved when SRP docks with its cognate receptor in the RER and dissociates from the signal sequence in a process that requires GTP. Synthesis of transmembrane proteins on RER is an extremely energy-efficient process. The passage of a fully formed and folded protein through a membrane is, thermodynamically, formidably expensive (von Heijne, 1985); it is infinitely "cheaper" for cells to thread amino acids, in tandem, through a membrane during initial protein synthesis.

Protein synthesis then resumes, and the emerging polypeptide chain is translocated into the RER membrane through a conceptualized "aqueous pore"

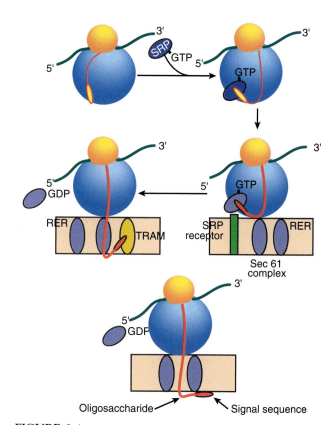

FIGURE 2.4 Translocation of proteins across the rough endoplasmic reticulum (RER). Integral membrane and secretory protein synthesis begins with partial synthesis on a free polysome not yet bound to the RER. The N terminus of the nascent protein emerges and allows a ribonucleoprotein, signal recognition particle (SRP), to bind to the hydrophobic signal sequence and prevent further translation. Translation arrest is relieved once the SRP docks with its receptor at the RER and dissociates from the signal sequence in a process that requires GTP. Once protein synthesis resumes, translocation occurs through an aqueous pore termed the translocon, which includes translocating chain-associating membrane protein (TRAM) and translocon-associated protein a (TRAP). The signal sequence is removed by a signal peptidase located in the lumen of the RER.

termed the *translocon*. Several proteins have been identified in crosslinking experiments as possible components of the translocon, including TRAM (translocating chain-associating membrane protein) and TRAP (translocon-associated protein a), which is a component of the mammalian homolog of the yeast sec 61 complex. Many others are strikingly similar to proteins that were originally discovered in yeast, revealing the common conserved nature of this process in organisms as diverse as yeasts and humans.

A few polypeptides deviate from the common pathway for secretion. For example, certain peptide growth factors, such as basic fibroblast growth factor and ciliary neurotrophic factor, are synthesized without signal peptide sequences but are potent biological modulators of cell survival and differentiation. These growth factors appear to be released under certain conditions, although the mechanisms for such release are still controversial. One possibility is that release of these factors may be associated primarily with cellular injury.

Two cotranslational modifications are commonly associated with the emergence of the polypeptide on the luminal face of the RER. First, an N-terminal hydrophobic signal sequence that is used for insertion into the RER is usually removed by a signal peptidase. Second, oligosaccharides rich in mannose sugars are transferred from a lipid carrier, dolichol phosphate, to the side chains of asparagine residues (Kornfeld and Kornfeld, 1985). The asparagines must be in the sequence N X T (or S), and they are linked to the mannose sugars by two molecules of N-acetylglucosamine. Although the prevention of glycosylation of some proteins causes their aggregation and accumulation in the RER and Golgi apparatus, for most glycoproteins, the significance of glycosylation is not apparent. Neither is it a universal feature of integral membrane proteins: some proteins, such as the proteolipid proteins of CNS myelin, neither lose their signal sequence nor become glycosylated. One clear case of a proved function for a carbohydrate moiety is the targeting of proteins to lysosomes in the trans-Golgi by means of the mannose 6-phosphate receptor. Intercellular adhesion molecules such as the selectins, which effect the sticking of lymphocytes to blood vessel walls, appear to interact with lectinlike proteins through their oligosaccharide chains. Similarly, the sialic acid side chains of NCAM (neural cell adhesion molecule) are essential for modulation of cell–cell adhesion mediated by NCAMs. Thus, for the vast majority of polypeptides destined for release from the cell (secretory polypeptides), an N-terminal "signal sequence" first mediates the passage of the protein into the RER and is immediately cleaved from the polypeptide by a signal peptidase residing on the luminal side of the RER. For proteins destined to remain as permanent residents of cellular membranes (and these form a particularly important and diverse category of plasma membrane proteins in neurons and myelinating glial cells), however, many variations on this basic theme have been observed. Simply stated:

1. Signal sequences for membrane insertions need not be only N-terminal; those that lie within a polypeptide sequence are not cleaved.

2. A second type of signal, a "halt" or "stop" transfer signal, functions to arrest translocation through the membrane bilayer. The halt transfer signal is also hydrophobic and is usually flanked by positive charges. This arrangement effectively stabilizes a polypeptide segment in the RER membrane bilayer.

3. The sequential display in tandem of insertion and halt transfer signals in a polypeptide as it is being synthesized ultimately determines its disposition with respect to the phospholipid bilayer and, thus, its final topology in its target membrane. By synthesizing transmembrane polypeptides in this way, virtually any topology may be generated.

Newly Synthesized Polypeptides Exit from the RER and Are Moved through the Golgi Apparatus

When the newly synthesized protein has established its correct transmembrane orientation in the RER, it is incorporated into vesicles and must pass through the Golgi complex before reaching the plasma membrane (Fig. 2.3). For membrane proteins, the Golgi serves two major functions: first, it sorts and targets proteins and, second, it performs further posttranslational modifications, particularly on the oligosaccharide chains that were added in the RER (Pelham, 1995). Sorting takes place in the cis-Golgi network (CGN), also known as the intermediate compartment, and in the trans-Golgi network (TGN), whereas sculpting of the oligosaccharides is primarily the responsibility of the cis-, medial-, and trans-Golgi stacks. The TGN is a tubulovesicular network wherein proteins are targeted to the plasma membrane or to organelles (Hammond and Helenius, 1994; Balch et al., 1994).

In addition to the processing of carbohydrates in the Golgi, posttranslational modifications can take place in other subcellular compartments. Some protein glycosylations are modified further post-Golgi in components of the smooth endoplasmic reticulum or transport vesicles, as described later in this section. Finally, some neuropeptides (adrenocorticotropic hormone, enkephalins, etc.) are synthesized as sequence domains in large precursor proteins that must be cleaved in transit by specific proteases to form the biologically active form.

The CGN serves an important sorting function for proteins entering the Golgi from the RER. Because most proteins that move from the RER through the secretory pathway do so by default, any resident endoplasmic reticulum proteins must be restrained from exiting or promptly returned to the RER from the CGN should they escape. Although no retention signal has been demonstrated for the endoplasmic reticulum, two retrieval signals have been identified, a Lys–Asp–Glu–Leu or KDEL sequence in type I proteins and the Arg–Arg or RR motif in the first five amino acids of proteins with a type II orientation in the membrane. The KDEL tetrapeptide binds to a receptor called *Erd 2* in the CGN, and the receptor–ligand complex is returned to the RER. There may also be a receptor for the *N*-arginine dipeptide; alternatively, this sequence may interact with other components of the retrograde transport machinery, such as microtubules.

Movement of proteins between Golgi stacks proceeds by means of vesicular budding and fusion (Rothman, 1994). Through the use of a cell-free assay containing Golgi-derived vesicles, the essential mechanisms for budding and fusion have been shown to require coat proteins (COPs) in a manner that is analogous to the role of clathrin in endocytosis. Currently, two main types of COP complex, COPI and COPII, have been distinguished. Although both have been shown to coat vesicles that bud from the endoplasmic reticulum, they may have different roles in membrane trafficking. Coat proteins provide the external framework into which a region of a flattened Golgi cisterna can bud and vesiculate (Mallabiabarrena and Malhotra, 1995; Bednarek *et al.*, 1995). A complex of these COPs forms the coatomer (coat protomer) together with a p200 protein, AP–1 adaptins, and a family of GTP-binding proteins called ADP ribosylation factors (ARFs), originally named for their role in the action of cholera toxin. Immunolocalization of one of the coatomer proteins, β-COP, predominantly to the CGN and *cis*-Golgi indicates that these proteins may also take part in vesicle transport into the Golgi (Fig. 2.5). The function of ARF is to drive the assembly of the coatomer and therefore vesicle budding in a GTP-dependent fashion. Dissociation of the coat is triggered when hydrolysis of the GTP bound to ARF is stimulated by a GTPase-activating protein (GAP) in the Golgi membrane. The cycle of coat assembly and disassembly can continue when the replacement of GDP on ARF by GTP is catalyzed by a guanine nucleotide exchange factor (GEF). The importance of this GDP–GTP exchange to normal vesicular traffic is dramatically illustrated by the effects of brefeldin A, a fungal metabolite that specifically inhibits GTP exchange and disperses the Golgi complex by preventing the return of Golgi components from the intermediate compartment.

Fusion of vesicles with their target membrane in the Golgi apparatus is believed to be regulated by a series of proteins, *N*-ethylmaleimide-sensitive factor (NSF), soluble NSF attachment proteins (SNAPs), and SNAP receptors (SNAREs), which together assist the vesicle in docking with its target membrane. In addition, the Rabs, a family of membrane-bound GTPases,

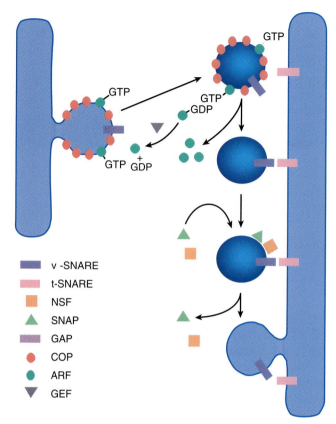

FIGURE 2.5 General mechanisms of vesicle targeting and docking in the ER and Golgi. The assembly of coat proteins (COPs) around budding vesicles is driven by ADP-ribosylation factors (ARFs) in a GTP-dependent fashion. Dissociation of the coat is triggered when hydrolysis of the GTP bound to ARF is stimulated by a GTPase-activating protein (GAP) in the Golgi membrane. The cycle of coat assembly and disassembly can continue when the replacement of GDP on ARF by GTP is catalyzed by a guanine nucleotide exchange factor (GEF). Fusion of the vesicles with their target membrane in the Golgi is regulated by a series of proteins, *N*-ethylmaleimide-sensitive factor (NSF), soluble NSF attachment proteins (SNAPs), and SNAP receptors (SNAREs), which together assist the vesicle in docking with its target membrane. SNAREs on the vesicle (v-SNAREs) are believed to associate with corresponding t-SNAREs on the target membrane.

act in concert with their own GAPs, GEFs, and a cytosolic protein that dissociates Rab–GDP from membranes after fusion called guanine nucleotide dissociation inhibitor. Rabs are believed to regulate the action of SNAREs, the proteins directly engaged in membrane–membrane contact prior to fusion. The tight control necessary for this process and the importance of ensuring that vesicle fusion takes place only at the appropriate target membrane may explain why eukaryotic cells contain so many Rabs, some of which are known to specifically take part in the internalization of endocytic vesicles at the plasma membrane (Fig. 2.3) (Orci *et al.*, 1986).

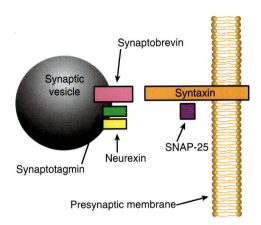

FIGURE 2.6 Mechanisms of vesicle targeting and docking in the synaptic terminal. The synaptic counterpart of v-SNARE is synaptobrevin (also known as vesicle-associated membrane protein), and syntaxin corresponds to the t-SNARE. SNAP–25 is an accessory protein that binds to syntaxin. Synaptotagmin is believed to be the Ca^{2+}-sensitive regulatory protein in the complex that binds to syntaxin. The neurexins appear to have a role in conferring Ca^{2+} sensitivity to these interactions.

Exocytosis of neurotransmitter at the synapse must occur in an even more finely regulated manner than endocytosis. Although a detailed description of the mechanisms underlying the exocytosis of neurotransmitter is provided in Chapter 8, they are described briefly here. The proteins first identified in vesicular fusion events in the secretory pathway (viz., NSF, SNAPs, and SNAREs or closely related homologs) appear to play a part in the fusion of synaptic vesicles with the active zones of the presynaptic neuronal membrane (Fig. 2.6).

The synaptic counterpart of v-SNARE is synaptobrevin (also known as vesicle-associated membrane protein, or VAMP), and syntaxin corresponds to t-SNARE. SNAP–25 is an accessory protein that binds to syntaxin. In the constitutive pathway such as between the RER and Golgi apparatus, assembly of the complex at the target membrane promotes fusion. However, at the presynaptic membrane, Ca^{2+} influx is required to stimulate membrane fusion at the presynaptic membrane. Synaptotagmin is believed to be the Ca^{2+}-sensitive regulatory protein in the complex that binds syntaxin. The neurexins appear to have a role in regulation as well, because, in addition to interacting with synaptotagmin, they are the targets of black widow spider venom α-latrotoxin, which deregulates Ca^{2+}-dependent exocytosis of neurotransmitter.

The comparison between secretion in slow-releasing cells, such as the pancreatic β cell, and neurotransmitter release at the neuromuscular junction is much like the comparison between a hand-held pocket calculator for balancing a checkbook and a state-of-the-art desktop computer. Two differences stand out. First, the speed of neurotransmitter release is much greater both in release from a single vesicle and in total release in response to a specific signal. Releasing the contents of a single synaptic vesicle at a mouse neuromuscular junction takes from 1 to 2 ms, and the response to an action potential involving the release of many synaptic vesicles is complete in approximately 5 ms. In contrast, release of the insulin in a single secretory granule by a pancreatic β cell takes from 1 to 5 s, and the full release response may take from 1 to 5 min. A 103- to 105-fold difference in rate is an extraordinary range, making neurotransmitter release one of the fastest biological events routinely encountered, but this speed is critical for a properly functioning nervous system.

A second major difference between slow secretion and fast secretion is seen in the recycling of vesicles. In the pancreas, secretory vesicles carrying insulin are used only once, and so new secretory vesicles must be assembled *de novo* and released from the TGN to meet future requirements. In the neuron, the problem is that the synapse may be at a distance of 1 m or more from the protein synthetic machinery of the perikaryon, and so newly assembled vesicles even traveling at rapid axonal transport rates (see below) may take more than a day to arrive. Now, the number of synaptic vesicles released in 15 min of constant stimulation at a single frog neuromuscular junction has been calculated to be on the order of 10^5 vesicles, but a single terminal may have only a few hundred vesicles at any one time. These measurements would make no sense if synaptic vesicles had to be constantly replaced through new synthesis in the perikaryon, as is the case with insulin-carrying vesicles. The reason that these numbers are possible is that synaptic vesicles are taken up locally by endocytosis, refilled with neurotransmitter, and reused at a rate fast enough to keep up with normal physiological stimulation levels. This takes place within the presynaptic terminal, and there is evidence that these recycled synaptic vesicles are used preferentially (Heuser and Reese, 1973). Such recycling does not require protein synthesis, because the classic neurotransmitters are small molecules, such as acetylcholine, or amino acids, such as glutamate, that can be synthesized or obtained locally.

Significantly, neurons have fast and slow secretory pathways operating in parallel in the presynaptic terminal (Sudhof, 1995). Synapses that release classic neurotransmitters (acetylcholine, glutamate, etc.) with these fast kinetics also contain dense core granules containing neuropeptides (calcitonin gene-related peptide, substance P, etc.) that are comparable to the secretory granules of the pancreatic β cell. These are used only once, because neuropeptides are produced from large polypeptide precursors that must be made by protein synthesis in the cell body. Release of neuropeptides is relatively slow; as is the case in endocrine release, the neuropeptides serve primarily as modulators of synaptic function. The small clear synaptic vesicles containing the classic neurotransmitters can in fact be pharmacologically depleted from the presynaptic terminal, while the dense core granules remain. These observations indicate that even though fast and slow secretory mechanisms have many similarities and may even have common components, in neurons they can operate independent of one another.

Proteins Exit the Golgi Complex at the trans-Golgi Network

Most of the N-linked oligosaccharide chains acquired at the RER are remodeled in the Golgi cisternae, and while the proteins are in transit, another type of glycosyl linkage to serine or threonine residues through *N*-acetylgalactosamine also can be made. Modification of existing sugar chains by a series of glycosidases and the addition of further sugars by glycosyl transferases occur from the *cis* to the *trans* stacks. Some of these enzymes have been localized to particular cisternae. For example, the enzymes β-1,4-galactosyltransferase and α-2,6-sialyltransferase are concentrated in the *trans*-Golgi. How they are retained there is a matter of some debate. One idea is that these proteins are anchored by oligomerization. Another view is that the progressively rising concentration of cholesterol in membranes more distal to the ER in the secretory pathway increases membrane thickness, which in turn anchors certain proteins and causes an arrest in their flow along the default route (Bretscher and Munro, 1993).

The default or constitutive pathway seems to be the direct route to the plasma membrane taken by vesicles that bud from the TGN (Fig. 2.3). This is how, in general, integral plasma membrane proteins reach the cell surface. Proteins bound for regulated secretion or for transport to endosomes and from there to lysosomes are diverted from the default path by means of specific signals (Kornfeld, 1987). It has been assumed that sorting of proteins for their eventual destination takes place at the TGN itself (Griffiths *et al.*, 1985). However, recent analyses of the three-dimensional structure of the TGN have provoked a revision of this view. These studies have shown that the TGN is tubular, with two major types of vesicles

that bud from distinct populations of tubules. The implication is that sorting may already have occurred in the *trans*-Golgi prior to the proteins' arrival at the TGN. One population of vesicles consists of those surrounded by the familiar clathrin cage, which are destined for late endosomes. The other population appears to be coated in a lacelike structure, which may prove to be made from the elusive coat protein required for vesicular transport to the plasma membrane. The β-COP protein and related coatomer proteins active in more proximal regions of the secretory pathway are absent from the TGN.

Endocytosis and Membrane Cycling Occur in the trans-Golgi Network

Two types of membrane invagination occur at the surface of mammalian cells and are clearly distinguishable by electron microscopy. The first type is a caveola, which has a threadlike structure on its surface made of the protein caveolin. Caveolae mediate the uptake of small molecules such as the vitamin folic acid by a process called potocytosis. They may also have a role in concentrating proteins linked to the plasma membrane by the glycosylphosphatidylinositol anchor. Recent demonstration of the targeting of protein tyrosine kinases to caveolae by the tripeptide signal MGC (Met–Gly–Cys) also suggests that caveolae may function in signal transduction cascades (Anderson *et al.*, 1992; Shenoyscaria *et al.*, 1994).

The other type of endocytic vesicle at the cell surface is that coated with the distinctive meshwork of clathrin triskelions. The triskelion (Pley and Parham, 1993) comprises three copies of a clathrin heavy chain and three copies of a clathrin light chain. The ease with which these triskelions can assemble into a cage structure demonstrates how they promote the budding of a vesicle from a membrane invagination. Clathrin binds selectively to regions of the cytoplasmic surface of membranes that are selected by adaptins. The AP–2 complex, which is active primarily at the plasma membrane, consists of 100-kDa α and β subunits and two subunits of 50 and 17 kDa each. AP–1 complexes localize to the TGN and have γ and β subunits of 100 kDa together with smaller polypeptides of 46 and 19 kDa. Adaptins bind to the cytoplasmic tails of membrane proteins, thus recruiting clathrin for budding at these sites.

A further component of the endocytic complex at the plasma membrane is the GTPase dynamin, which seems to be required for normal budding of coated vesicles during endocytosis. The dynamins (de Camilli *et al.*, 1995) are a family of 100-kDa GTPases found in both neuronal and nonneuronal cells and may interact with the AP–2 component of a clathrin-coated pit. Dynamin I is found primarily in neurons, whereas dynamin II has a widespread distribution. Oligomers of dynamin form a ring at the neck of a budding clathrin-coated vesicle, and GTP hydrolysis appears to be necessary for the coated vesicle to pinch off from the plasma membrane. The existence of a specific neuronal form of dynamin may be a manifestation of the unusually rapid rate of synaptic vesicle recycling.

The primary function of clathrin-coated vesicles at the plasma membrane is to deliver membrane proteins together with any ligands bound to them to the early endosomal apparatus. The other major site of action of clathrin is in the vesicles that bud from the TGN carrying lysosomal enzymes en route to late endosomes. Early endosomes have a tubulovesicular morphology. Receptors that will be recycled back to the plasma membrane partition into the tubules; those endocytosed proteins destined for lysosomes concentrate in the vesicular regions. Recycling seems to be the default pathway, whereas proteins must be actively targeted to lysosomes. However, the precise signals for this are unknown.

Regulation of membrane cycling in the endosomal compartment is likely to include the Rab family of small GTP-binding proteins. Indeed, each stage of the endocytic pathway may have its own Rab protein to ensure efficient targeting of the vesicle to the appropriate membrane. Rab6 is believed to have a role in transport from the TGN to endosomes, whereas Rab9 may regulate vesicular flow in the reverse direction. In neurons, Rab5a has a role in regulating fusion of endocytic vesicles and early endosomes and appears to function in endocytosis from both somatodendritic domains and the axon. The association of the protein with synaptic vesicles in nerve terminals, attached presumably by means of its isoprenoid tail, also suggests that early endosomal compartments may have a role in the packaging and recycling of synaptic vesicles.

The Lysosome Is the Target Organelle in Several Inherited Diseases That Affect the Nervous System

Lysosomes were first isolated and characterized as a distinct organelle fraction bounded by a single membrane and separable from mitochondria by differential and sucrose gradient centrifugation (de Duve, 1975, 1983). Because of their high content of acid hydrolases, the classic view is that lysosomes are organelles of terminal degradation. Indeed, the latency of hydrolase activity before membrane perme-

abilization by agents such as nonionic detergents has been used biochemically as a measure of the purity and intactness of lysosomal preparations. However, in addition to their well-established function in lipid and protein breakdown, tubulovesicular lysosomes may overlap in sorting functions with early endosomes, particularly during antigen processing in macrophages.

Inherited deficiencies in lipid metabolism in the lysosome often have particularly devastating consequences on the nervous system because of the abundance of the lipid-rich membrane myelin. Metachromatic leukodystrophy is an autosomal recessive disease caused by a deficiency in arylsulfatase A activity, which is also responsible for degrading the myelin lipid cerebroside sulfate (sulfatide). Oligodendrocytes accumulate sulfatide in metachromatic granules, causing severe disruption of myelination. Peripheral nerve myelination also is affected, as are other organs that normally contain much lower amounts of sulfatide, such as the kidney, the liver, and the endocrine system. Krabbe disease, or globoid cell leukodystrophy, also is a dysmyelinating disease in which there is an almost complete lack of oligodendrocytes and therefore myelin, caused by a deficiency in the β-galactosidase responsible for hydrolyzing galactocerebroside to ceramide and galactose. Galactocerebroside is particularly abundant in myelin, constituting about 25% of myelin lipid. Mice that lack galactocerebroside have the ability to assemble multilamellar myelin; however, this myelin does not support adequate nerve conduction; neither is it stable (Coetzee *et al.*, 1996). Unlike metachromatic leukodystrophy, Krabbe disease is limited to the CNS and PNS. However, why a buildup of galactocerebroside should prove particularly toxic to oligodendrocytes is not entirely clear. One hypothesis is that a metabolite of galactocerebroside, galactosphingosine (psychosine), is the primary culprit. Because there is an authentic mouse model for Krabbe disease, *twitcher*, gene therapy provides some hope of correcting the disease.

How are proteins destined to operate in lysosomes targeted to these organelles? Soluble lysosomal hydrolase enzymes acquire a phosphorylated mannose on their oligosaccharide chains by a two-step process in the Golgi apparatus. This mannose 6-phosphate label is recognized by specific mannose 6-phosphate receptors, which carry the proteins to late endosomes. (Ludwig *et al.*, 1995). In contrast, lysosomal membrane proteins are targeted by means of cytoplasmic tail signals that contain either leucine or tyrosine of type LJ or type YXXJ or NXXY, where J is any hydrophobic amino acid. The LJ signal seems to be essential for efficient delivery directly to endosomes, whereas the second type of signal seems to be more important in the recovery of proteins destined for lysosomes from the plasma membrane. The majority of lysosomal membrane proteins have the YXXJ but do not have the LJ signal. The implication of these observations is that many of the lysosomal membrane proteins make their way to lysosomes from the TGN endosomes through the plasma membrane.

What are the receptors for the type LJ or type YXXJ or NXXY motifs at the TGN? Because transport from the TGN to the endosomes occurs in clathrin-coated vesicles, the proteins that link such vesicles to membranes, the adaptins, may play a role. The weight of the evidence suggests that AP–1 recognizes the LJ sequence, whereas AP–2 identifies the YXXJ and NXXY motifs. Once the ligands are bound, these adaptins would direct transport of their respective ligands to the endosomes from the TGN or through the plasma membrane, respectively (Sandoval and Bakke, 1994).

At present, it is not clear how proteins in the late endosome, such as mannose 6-phosphate receptors, that cycle back to the Golgi are sorted from those whose ultimate destination is a lysosome.

How Are Peripheral Membrane Proteins Targeted to Their Appropriate Destinations?

Peripheral membrane proteins are synthesized in the same type of free polysome in which the bulk of the cytosolic proteins are made. However, the cell must ensure that these membrane proteins are sent to the plasma membrane rather than allowed to attach in a haphazard way to other intracellular organelles. The fact that a complex machinery has evolved to ensure the correct delivery of integral membrane proteins suggests that some equivalent targeting mechanism must exist for proteins that attach to the cytoplasmic surface of the plasma membrane. Such proteins are translated on "free" polysomes, but these polysomes are associated with cytoskeletal structures and are not uniformly distributed throughout the cell body. In a number of cases, mRNAs that encode soluble cytosolic proteins are concentrated in discrete regions of the cell, resulting in a local accumulation of the translated protein close to the site of action. For some peripheral membrane proteins, this is the plasma membrane.

Evidence that this mechanism might operate in peripheral membrane protein synthesis came from studies showing biochemically and by *in situ* hybridization that mRNAs encoding the myelin basic proteins are concentrated in the myelinating processes that extend from the cell body of oligodendrocytes

(Colman *et al.*, 1982). Myelin basic protein may be a special case because of its very strong positive charge and consequent propensity for binding promiscuously to the negatively charged polar head groups of membrane lipids. Nevertheless, the fact that actin mRNAs are localized to the leading edge of cultured myocytes and the mRNA for the microtubule-associated protein MAP2b is concentrated in the dendrites of neurons suggests that targeting by local synthesis is more common than originally thought. (Tucker *et al.*, 1989; Kislaukis *et al.*, 1994; Steward, 1995). This mechanism is probably less important for peripheral membrane proteins that associate with the cytoplasmic surface of the plasma membrane by means of strong specific associations with proteins already located at the membrane, because such proteins would act as specific receptors. Because only selected cytoplasmic mRNAs are localized to the periphery, the process is specific. However, no mRNAs are localized exclusively to the periphery, and a significant fraction are typically localized proximal to the nucleus in a region rich with the translational and protein-processing machinery of the cell (the Nissl substance or translational cytoplasm).

Special Mechanisms Are Used to Target Proteins to Mitochondria and Peroxisomes

The inner membrane of the mitochondrion is the site of oxidative phosphorylation in which the step-by-step transfer of electrons from oxygen intermediary metabolites to molecular oxidation is coupled to proton transport and ATP synthesis. Thus, this organelle has an essential role in providing the large amount of ATP required for the electrical activity of neurons. The fact that, in a resting adult, about 40% of the total energy consumption is required for ion pumping in the CNS accounts for the exquisite sensitivity of the brain to damage from oxygen deprivation. The sensitivity of neurons to interruptions in the provision of ATP by the mitochondrion is also seen in cases of uremia, where a buildup of ammonium ions depletes the Krebs cycle of α-oxoglutaric acid by converting it into glutamate.

Although the mitochondrion has its own circular DNA that encodes some proteins, most mitochondrial proteins are synthesized in the nucleocytoplasmic system (Zwizinski and Neupert, 1983; Hartl *et al.*, 1987). This poses the problem of how these proteins once made in the cytoplasm gain entry into the mitochondrion. Furthermore, because the mitochondrion has an inner membrane and an outer membrane, some proteins must cross two membranes to gain access to the inner matrix (Schatz and Butow, 1983;

Schleyer and Neupert, 1985). This group of proteins includes the enzymes of the Krebs cycle and the fatty acid β-oxidation pathway.

Unlike proteins inserted into the RER, mitochondrial proteins can be imported either posttranslationally or cotranslationally with the use of a cleavable amphipathic helical signal sequence usually at the N terminus (von Heijne, 1986; Roise *et al.*, 1988). At the RER, the signal sequence of a nascent polypeptide chain can be translocated across the membrane, because the polypeptide remains small and unfolded owing to the arrest of translation caused by a signal-recognition particle. However, mitochondrial proteins are typically synthesized on cytoplasmic or free polysomes and must be folded at least partially to prevent degradation. For posttranslational import, mitochondria rely on a group of molecular chaperones to prevent complete folding of the polypeptides. These hsp70 and hsp60 proteins were originally identified because they are upregulated during heat shock. Their role in binding to proteins and maintaining them in specific conformations helps to explain why these proteins have an important function in protecting proteins against the stress of elevated temperatures as well as facilitating the proper folding of newly synthesized polypeptides. In yeast, a second protein, Ydj1p, whose bacterial homolog DnaJ regulates chaperone function, has been identified. Ydj1p possesses an isoprenoid tail linked to its C-terminal amino acid and this may serve to anchor the protein to the outer membrane. The third factor that has been implicated is the mitochondrial stimulation factor, which is a heterodimer possessing an ATP-dependent protein "unfoldase" activity. This factor may be more important in cotranslational import where polysomes are known to be associated with the mitochondrial outer membrane (Hachiya *et al.*, 1993; Lill and Neupert, 1996).

Most of the current understanding of protein translocation from the mitochondrial outer membrane inward has come from studies on either the fungus *Neurospora crassa* or the yeast *Saccharomyces cerevisiae*. In both yeast and higher eukaryotes, the partially folded polypeptide targeted for the mitochondrion may be stabilized by a cytoplasmic chaperone that is a member of the hsp70 family, but this interaction is not required for import. However, several proteins in the outer membrane form an essential complex that acts as a receptor and pore for protein translocation. This complex can in turn interact with an inner membrane complex at specialized contact sites that minimize the distance across the two membranes, thereby facilitating the movement of proteins to the inner matrix. Although both pores can function independently, they

contact and cooperate when there is a transmembrane potential across the inner membrane. This accounts for early observations showing that importation of subunits of the F_1-ATPase, an inner membrane protein, required an active electron transport chain but did not need ATP synthesis. The mitochondrial import sequence extends through the pore into the inner matrix, where a second member of the hsp70 family binds and facilitates movement into the inner matrix. After proteins have crossed into the inner matrix, they must dissociate from hsp70 to fold properly, a process that requires another kind of molecular chaperone, hsp60.

Peroxisomes are so named because they contain oxidases that generate H_2O_2 and the enzyme catalase, which is responsible for detoxifying it. In addition, these organelles contain many other enzymes that take part in lipid, purine, and amino acid metabolism. Peroxisomes are of interest because of the number of inherited diseases associated with defects either in certain enzymes or indeed in the assembly of the organelle itself (Moser, 1987). Some of these diseases manifest as particularly damaging to the nervous system and include adrenoleukodystrophy (accumulation of very long chain fatty acids due to insufficient lignoceryl-CoA ligase activity caused by inefficient import of the protein) and Refsum's disease (buildup of phytanic acid due to defective α-oxidation), both of which cause demyelination.

Like many mitochondrial proteins, peroxisomal proteins are imported posttranslationally (Lazarow and Fujiki, 1985). Although cytosolic factors are implicated in peroxisomal biogenesis, no peroxisomal chaperones analogous to the hsp70 family have yet been shown to function in protein import. Therefore, unfolding and refolding are assumed not to play a role in the accumulation of proteins inside the peroxisome. Among these cytosolic proteins is presumed to be the receptor for the tripeptide C-terminal import signal SKL (Ser–Lys–Leu) known as peroxisomal targeting signal 1 (PTS1). In addition to the C-terminal PTS1, some peroxisomal proteins have a cleavable N-terminal sequence called PTS2, which signals their import. A quite distinct translocation machinery appears to operate for PTS1 and PTS2 proteins. Two possible receptor proteins in the peroxisomal membrane, one of which is the adrenoleukodystrophy protein (ALDP), have been identified. ALDP is a member of a larger family known as the ABC ATP-dependent membrane transporters (Rachubinski and Subramani, 1995; Dodt *et al.*, 1995).

A characteristic feature of peroxisomal biogenesis is that it is stimulated by drugs whose detoxification requires peroxisomal activity. It is possible that mature peroxisomes are recruited from a pool of precursor organelles, and there is some evidence for the existence of such a population in rat liver. Although the mature organelle appears to be spherical, electron microscopic evidence suggests a peroxisomal reticulum at which synthesis and protein import may take place. Mature peroxisomes might then arise from this reticulum by a process of budding.

Cytoplasmic Proteins Are Also Compartmentalized

Membrane-bound organelles are the most familiar form of compartmentation in cells, but cytoplasmic regions of the cell containing metabolic compartments exist as well. Regions of the neuronal or glial cytoplasm may have highly specialized polypeptide compositions that are important for function. For example, the neuronal phosphoprotein synapsin is highly enriched in presynaptic terminals, where it participates in localization and targeting of synaptic vesicles. Similarly, calmodulin and the glycolytic enzyme aldolase have been localized in muscle cells to the region of the I-band, where they are thought to facilitate coupling of ATP production to contractility.

As mentioned earlier, cytoplasmic proteins are synthesized on cytoplasmic polysomes, termed "free" polysomes to reflect an absence of underlying ER membrane, even though they may be restricted to specific domains of the cell cytoplasm. This restriction is particularly obvious in the neuronal perikaryon, where both cytoplasmic polysomes and membrane-associated polysomes are concentrated in areas near the nucleus and Golgi complex. In addition, cytoplasmic polysomes containing specific mRNAs may be localized to certain regions of the cell such as the proximal dendrite (those encoding the microtubule-associated protein 2 (MAP–2) and the processes of oligodendrocytes (those encoding myelin basic protein). In contrast, the protein synthetic machinery of the polysome appears to be effectively excluded from the mature axon. Therefore, cytoplasmic polysomes are representative of cytoplasmic compartmentation for proteins and nucleic acids.

In most cases, localized cytoplasmic proteins interact with cytoskeletal structures in the cytoplasm (see next section), but macromolecular complexes that form to make a cellular process more efficient or free from error have been described. Evidence exists that glycolytic enzymes of neurons and muscle cells may be organized in a labile complex that facilitates energy metabolism, but the existence of such complexes remains controversial.

Perhaps the best characterized cytoplasmic macromolecular complex is the proteasome, which is a large protein complex (2×10^6 Da, sedimenting as a 20S particle) that contains several distinct enzymatic activities, including catalytic sites for both ubiquitin-dependent and ubiquitin-independent proteolysis (Hochstrasser, 1995; Jentsch and Schlenker, 1995). Ubiquitin is a small, highly conserved polypeptide that is covalently added to cytoplasmic proteins targeted for degradation. The catalytic core of the proteasome is a barrel-shaped structure formed by four heptameric stacked rings, but additional proteins (about 16 polypeptides) may interact with the 20S core to form a larger 26S particle. Because proteasomes constitute the primary cytoplasmic pathway for protein degradation (i.e., nonlysosomal pathways), they serve a number of important physiological functions, including regulation of cell proliferation and processing of antigens for presentation. In the nervous system, however, proteasomes are likely to be most important for homeostasis, allowing turnover of cytoplasmic polypeptides at specific sites so that the elaborate cellular extensions of neurons and glia may be maintained. Recent work also indicates that the induction of some forms of memory involves the regulation of ubiquitin levels (see Chapter 18).

Cytoplasmic proteins may also be effectively compartmentalized by posttranslational modification. Two types of modification may be particularly important for this kind of compartmentalization. Local activation of kinases can lead to phosphorylation of proteins in specific domains of the neuron. For example, the reversible phosphorylation of synapsin in the presynaptic terminal appears to be responsible for the targeting of synaptic vesicles to the terminal and for the mobilization of vesicles during prolonged stimulation. An impressive variety of cytoplasmic protein kinases that may be selectively activated to modify serines or threonines presented in distinctive consensus sequences have been described. Distinct from these serine or threonine kinases, a number of other kinases that specifically modify tyrosines can be found in the brain. In some cases, the tyrosine kinase is linked directly to a membrane-spanning receptor and phosphorylates cytoplasmic proteins in the vicinity of the receptor after activation. Completing the cycle of phosphorylation and dephosphorylation are a number of phosphatases with varying specificities. The properties and physiological roles for kinases and phosphatases are discussed in greater detail in Chapter 12.

A second common posttranslational modification of cytoplasmic proteins is the addition of carbohydrate moieties. Whereas modification of membrane-associated proteins in the Golgi complex proceeds by the addition of complex carbohydrates through N linkages on selected asparagines, glycosylated cytoplasmic proteins have simpler carbohydrates added through O linkages to serine or threonine hydroxyls. This modification was first recognized as a feature of many nuclear proteins and components of the nuclear membrane, but subsequent studies showed that a number of cytoplasmic proteins also have O-linked carbohydrates. Unlike phosphorylation, relatively little is known about the functional significance of cytoplasmic glycosylation. Remarkably, however, serines and threonines subject to O-linked glycosylation would also be good sites for phosphorylation by various kinases as well. This congruence raises the possibility that glycosylation and phosphorylation of some cytoplasmic proteins may serve complementary functions.

Summary

Membrane biogenesis and protein synthesis in neurons and glial cells are accomplished by the same mechanisms that have been worked out in great detail in other cell types. Integral membrane proteins are synthesized in the RER, and peripheral membrane proteins are products of cytoplasmic free ribosomes. For transmembrane proteins and secretory polypeptides, synthesis in the RER is followed by transport to the Golgi apparatus, where membranes and proteins are sorted and targeted for delivery to precise intracellular locations. It is likely that the neuron and glial cell have evolved additional highly specialized mechanisms for membrane and protein sorting and targeting because these cells are so greatly extended in space, although these additional mechanisms have yet to be fully described. The basic features of the process of secretion, which includes neurotransmitter delivery to presynaptic terminals, are beginning to be understood as well. The key features of this process are apparently common to all cells, including yeast, although the neuron has developed certain specializations and modifications of the secretory pathway that reflect its unique properties as an excitable cell.

THE CYTOSKELETONS OF NEURONS AND GLIAL CELLS

The cytoskeleton of eukaryotic cells is an aggregate structure formed by three classes of cytoplasmic structural proteins: microtubules (tubulins), microfilaments (actins), and intermediate filaments (Fig. 2.7).

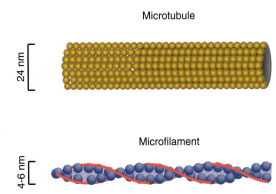

Microtubule

24 nm

Microfilament

4–6 nm

FIGURE 2.7 Two major classes of cytoskeletal structures found in all cellular components of the nervous system are microtubules and microfilaments. These structures constitute the substrates for the various motor proteins of cells. In electron micrographs, microtubules appear as hollow tubes with walls formed by 12–14 protofilaments. Each protofilament consists of a series of α- and β-tubulin dimers organized in a polar fashion, giving the microtubule a plus (fast-growing) end and a minus (slow-growing) end. In axons, the microtubules have their plus ends distal from the cell body, whereas dendritic microtubules may have either polarity. Microtubules are approximately 24 nm in diameter and may be more than 100 μm in length. Various polypeptides called microtubule-associated proteins (MAPs) are typically associated with the surface of the microtubule. These MAPs may help regulate the assembly and organization of the microtubules. In contrast, actin microfilaments form from two twisted strands of actin subunits that form filaments only 4–6 nm in diameter. The length of microfilaments is quite variable, but most neuronal filaments are short, in the range 20–50 nm. Many different proteins have been shown to interact with microfilaments in cells, including crosslinking, bundling, severing, and capping proteins (see Table 2.3).

Each of these elements exists concurrently and independently in overlapping cellular domains. Most cell types contain one or more examples of each class of cytoskeletal structure, but there are exceptions. For example, mature mammalian erythrocytes contain neither microtubules nor intermediate filaments, but they do have elaborate and highly specialized actin cytoskeletons. Among cells of the nervous system, the oligodendrocyte is unusual in that it contains no cytoplasmic intermediate filaments. Typically, each cell type in the nervous system has a unique complement of cytoskeletal proteins that are important for the differentiated function of that cell type.

Although the three classes of cytoskeletal elements interact with each other and with other cellular structures, all three are dynamic structures rather than passive structural elements. Their aggregate properties form the basis of cell morphologies and plasticity in the nervous tissue. In many cases, biochemical specialization in the cytoskeleton is characteristic of a particular cell type, function, and developmental stage. Each type of cytoskeletal element has unique functions that are essential for a working nervous system.

Microtubules Are an Important Determinant of Cell Architecture

Microtubules (Fig. 2.7) are nearly ubiquitous components of the cytoskeleton in eukaryotes (Hyams and Lloyd, 1994). They play key roles in intracellular transport, are a primary determinant of cell morphology, are the structural correlate of the mitotic spindle, and form the functional core of cilia and flagella. Microtubules are very abundant in the nervous system, and the tubulin subunits of microtubules may constitute more than 10% of total brain protein. As a result, many fundamental properties of microtubules have been defined by using microtubule protein prepared from brain extracts. At the same time, the microtubule cytoskeleton of the neuron has a variety of biochemical specializations that meet the unique demands imposed by the size and shape of the neuron.

Of the various functions defined for microtubules, intracellular transport and the generation of cellular morphology are the most important roles played by microtubules in cells of the nervous system. In part, this comes from their ability to organize cytoplasmic polarity. Microtubules *in vitro* are dynamic, polar structures with plus and minus ends that correspond to the fast- and slow-growing ends, respectively. In contrast, both stable and labile microtubules can be identified *in vivo*, where they help define both microscopic and macroscopic aspects of intracellular organization in cells. Microtubule organization, stability, and composition in nervous tissue are all highly regulated in the nervous system.

By electron microscopy, microtubules appear as hollow tubes 25 nm in diameter and in axons can be up to hundreds of micrometers in length. High-resolution electron micrographs also reveal that the walls of microtubules typically comprise 13 protofilaments formed by a linear arrangement of globular subunits, although microtubules with 12 to 14 protofilaments exist in some tissues and organisms. The globular subunits in the walls of a microtubule are heterodimers of α- and β-tubulin, whereas a variety of microtubule-associated proteins bind to the surface of microtubules.

Neuronal microtubules are remarkable for their genetic and biochemical diversity. Multiple genes exist for both α- and β-tubulins. These genes are differentially expressed according to cell type and developmental stage. Some of these genetic isotypes are expressed ubiquitously, whereas others are turned on only at specific times in development or in specific cell types or both. Most tubulin genes are expressed in nervous tissue, and some appear to be enriched or specific to neurons. When specific isotypes are prepared in a pure form, they show variability in assembly kinetics and ability to bind ligands. However, when more than one isotype is expressed in a single cell such as a neuron, they coassemble into microtubules with mixed composition.

A variety of posttranslational modifications of the tubulins have been described, the most common of which are tyrosination–detyrosination, acetylation–deacetylation, and phosphorylation. The first two of these pathways are intimately associated with assembly of microtubules, but relatively little is known about physiological functions for any of these modified tubulins. Most α-tubulin isotypes are synthesized with a Glu–Tyr dipeptide at the C terminus (Tyr-tubulin), but the tyrosine can be removed by a specific tubulin carboxypeptidase after assembly into a microtubule, leaving a terminal glutamate (Glu-tubulin). When microtubules containing detyrosinated α-tubulins are disassembled, the liberated α-tubulins are rapidly retyrosinated by a specific tubulin tyrosine ligase. The result is that microtubules that have been assembled for an extended period tend to be rich in Glu-tubulin. The tyrosination state of α-tubulin does not affect its assembly–disassembly kinetics *in vitro*, but recent evidence suggests that detyrosination may affect the interaction of microtubules with other cellular structures. In parallel with detyrosination, α-tubulins are also substrates for a specific acetylation reaction. Acetylation of tubulin was initially described for flagellar tubulins, but subsequent work demonstrated that this modification was widespread in neurons and many other cell types. Because the acetylase acts preferentially on α-tubulin assembled into microtubules, long-lived or stable microtubules tend to be rich in acetylated α-tubulin. However, the distribution of microtubules rich in acetylated tubulin may not be identical to that of Glu-tubulin. Acetylated a-tubulin also is rapidly deacetylated on disassembly of microtubules, although acetylation does not alter the stability of microtubules *in vitro*.

In contrast with these modifications, tubulin phosphorylation involves a β-tubulin and appears to be restricted to an isotype preferentially expressed in neurons and neuron-like cells. A variety of kinases have been shown to phosphorylate tubulin *in vitro*, but the endogenous kinase has not been identified. The effect of phosphorylation on assembly is unknown, but phosphorylation is upregulated during neurite outgrowth. As with the α-tubulin modifications, the physiological role of phosphorylation on neuronal β-tubulin has yet to be determined. A variety of additional posttranslational modifications have been reported, but their significance and distribution in the nervous system are not well documented.

The biochemical diversity of microtubules is increased through the association of different MAPs with different populations of microtubules (Table 2.2). The significance of microtubule diversity is not completely understood, but it may include functional differences as well as variations in assembly and stability. In particular, MAP composition may be used to define specific neuronal domains. For example, one type of MAP, MAP–2, appears to be restricted to dendritic regions of the neuron, whereas another class of MAPs, tau proteins, are differentially modified in axons. A recently identified isoform of MAP–2 is similar to MAP–2c but includes an additional repeat within the microtubule binding site; hence it is known as 4-repeat MAP–2c, or MAP–2d. Oligodendrocyte progenitors transiently express this novel isoform of MAP–2c in their cell bodies but not in their processes, suggesting that MAP–2d might have a role separate from its known capacity to bundle microtubules (Doll *et al.*, 1993; Ferhat *et al.*, 1994; Vouyiouklis and Brophy, 1995).

MAPs in nervous tissue fall into two heterogeneous groups: tau proteins and high-molecular-weight MAPs. Tau proteins have recently been the subject of intense interest, because posttranslationally modified tau proteins are the primary polypeptide constituents of neurofibrillary tangles from the brains of patients with Alzheimer's disease. Tau proteins appear to be neuronal MAPs, although reports of tau immunoreactivity outside neurons have appeared. Tau proteins bind to microtubules during assembly–disassembly cycles with a constant stoichiometry and can promote microtubule assembly and stabilization. Tau exists in a number of molecular-weight isoforms that vary with region of the nervous system and developmental stage. For example, tau proteins in the adult CNS are typically from 60 to 75 kDa, whereas PNS axons contain a higher-molecular-weight tau of approximately 100 kDa. The different isoforms of tau protein are generated from a single mRNA by alternative splicing, and additional heterogeneity is produced by phosphorylation.

In contrast with tau MAPs, high-molecular-weight MAPs are a diverse group of largely unrelated proteins found in a variety of tissues, although some are brain specific. All have apparent molecular weights

TABLE 2.2 Major Microtubule Proteins and Microtubule Motors in Mammalian Brain

Microtubule-associated proteins	Location and Function
Tubulins	
α-Tubulins and β-tubulins	Neurons, glia, and nonneuronal cells except mature mammalian erythrocytes. Multigene family with some genes expressed preferentially in brain, while others are ubiquitous. Primary structural polypeptides of microtubules.
γ-Tubulin	Present in all microtubule-containing cells, but restricted to region of microtubule-organizing center. Needed for nucleation of microtubules.
Microtubule-associated proteins (MAPs)	
MAP–1a/1b	Widely expressed in neurons and glia, including both axons and dendrites. Forms are developmentally regulated phosphoproteins.
MAP–2a/2b MAP–2c	Dendrite-specific MAPs. The smaller MAP–2c is developmentally regulated, becoming restricted to spines in adults, while MAP–2a and MAP–2b are major phosphoproteins in adult brain.
LMW tau HMW tau	Tau proteins are enriched in axons and have a distinctive phosphorylation pattern in the axon, but may be found in other compartments. A single gene with multiple forms due to alternative splicing. The HMW tau is found in adult peripheral axons.
Motor proteins	
Kinesin Neuron-specific kinesin	Present in all microtubule-containing cells. Associated with membrane-bound organelles and serves to move them along microtubules in fast axonal transport. The neuron-specific form is the product of a specific gene expressed in nervous tissue.
Kinesin-related proteins	A diverse set of motor proteins with a kinesin-related motor domain and varied tails. Some are developmentally regulated and some are restricted to dividing cells, where they act as mitotic motors.
Axonemal dynein Cytoplasmic dynein (MAP–1c)	A set of minus end-directed microtubule motors. The axonemal forms are associated with cilia and flagella. In nervous tissue, these may be associated with the ependyma. The cytoplasmic forms may be involved in the transport of either organelles or cytoskeletal elements.

of 1300 kDa and form side-arms protruding from the walls of microtubules. Many of them may participate in microtubule assembly and cytoskeletal organization. Traditionally, the high-molecular-weight MAPs comprise five polypeptides: MAPs 1a, 1b, 1c, 2a, and 2b. MAP–2 proteins are closely related and are located primarily in dendrites. In contrast, the three polypeptides known as MAP–1 are unique polypeptides with little sequence homology. MAP–1c is a cytoplasmic form of dynein (see Molecular Motors in the Nervous System later in this chapter). MAPs 1a and 1b are widespread and appear to be developmentally regulated. MAPs 1a, 1b, and 2 are all thought to play important roles in stabilizing and organizing the microtubule cytoskeleton.

In most cell types, cytoplasmic microtubules appear to be relatively dynamic structures, although stable microtubules or microtubule segments are found in all cells. In nonneuronal cells such as astrocytes and other glial cells, microtubules are typically anchored in centrosomal regions that serve as microtubule-organizing centers. As a result, their cytoplasmic microtubules are oriented so that plus ends are distal to the cell center. The biochemistry of microtubule-organizing centers is not fully understood, but they contain a novel tubulin subunit, γ-tubulin, which is thought to function as a nucleating site for microtubules.

In contrast, the dendritic and axonal microtubules of neurons are not continuous with the microtubule-organizing center, so alternate mechanisms must exist for stabilization and organization of these microtubules. The situation is complicated further by the fact that dendritic and axonal microtubules differ in both composition and organization. Recent studies show that both axonal and dendritic microtubules are nucleated at the microtubule-organizing center but are subsequently released for delivery to the appropriate compartment. Surprisingly, axonal and dendritic compartments are not equivalent. There are two striking differences. First, the MAPs of dendritic and axonal microtubules are different in both identity and phosphorylation state. Second, microtubule orientation in axons is similar to that seen in other cell types with plus end distal, but microtubules in dendrites may exhibit both polarities. Perhaps owing to these differences, dendritic microtubules are less likely to be aligned with one another and appear less regular in their spacing.

Stabilization of axonal and dendritic microtubules is essential because of the volume of cytoplasm and distance from sites of protein synthesis for tubulin. Because microtubules play critical roles in both dendritic and axonal function, mechanisms to ensure their proper extent and organization must exist. A common side effect of one class of antineoplastic drugs, the vinca alkaloids, underscores the importance of microtubule stability in axons. Vincristine and other vinca alkaloids act by destabilizing spindle microtubules, but dosage must be carefully monitored to prevent development of peripheral neuropathies due to loss of axonal microtubules.

Axonal microtubules contain a particularly stable subset of microtubule segments that are resistant to depolymerization by antimitotic drugs, cold, and calcium. The stable microtubule segments are biochemically distinct and may constitute more than half of the axonal tubulin. Stable domains in microtubules may serve to regulate the axonal cytoskeleton by nucleating and organizing microtubules as well as stabilizing them. The biochemical basis of microtubule stability is not completely understood but probably includes posttranslational modification of the tubulin or the presence of stabilizing proteins or both. There are indications that levels of cold-insoluble tubulin correlate with axonal plasticity. In contrast, relatively little is known about regulation of dendritic microtubules, but local synthesis of MAP–2 in dendrites may play a role in regulating their stability.

Microfilaments and the Actin-Based Cytoskeleton Are Involved in Intracellular Transport and Cell Movement

The actin cytoskeleton is universally present in eukaryotes, although actin microfilaments are most familiar as the thin filaments of skeletal muscle. Microfilaments (Table 2.3) play a critical role in contractility for both muscle and nonmuscle cells. Actin and its contractile partner myosin are particularly abundant in nervous tissue relative to other nonmuscle tissues. In fact, one of the earliest descriptions of nonmuscle actin and myosin was in brain (Berl *et al.*, 1973). In neurons, actin microfilaments are most abundant in presynaptic terminals, dendritic spines, growth cones, and the subplasmalemmal cortex. Although concentrated in these regions, microfilaments are also present throughout the cytoplasm of both neurons and glia in the form of short filaments from 4 to 6 nm in diameter and from 400 to 800 nm in length.

As with tubulin, multiple actin genes exist in both vertebrates and invertebrates (Sheterline and Sparrow, 1994). Four α-actin human genes have been cloned. Each of these α-actin genes is expressed specifically in a different muscle cell type (skeletal, cardiac, vascular smooth, and enteric smooth muscle). In addition to the α-actins, two nonmuscle actin genes (β- and γ-actin) are present in humans. β-Actin and γ-actin

TABLE 2.3   Selected Proteins of the Microfilament
Cytoskeleton in Brain Actins

α-Actin (smooth muscle)

β-Actin and γ-actin (neuronal and nonneuronal cells)

Actin monomer binding proteins

    Profilin

    Thymosin β4 and β10

Capping proteins

    Ezrin/radixin/moesin

    Schwannomin/merlin

Gelsolin family

    Gelsolin

    Villin

    Scinderin

Crosslinking and bundling proteins

    Spectrin (fodrin)

    Dystrophin, utrophin, and related proteins

    α-Actinin

Tropomyosin

Proteins with nonmicrofilament functions

    MAP–2

    Tau

Myosins

    Myosin Iβ

    Myosin II

    Myosin V

    Myosin VI

    Myosin VII

---

genes are expressed ubiquitously, and both are abundant in nervous tissue. The functional significance of these different genetic isotypes is not clear, because the actins are highly conserved proteins. Across the range of known actin sequences, the amino acids are identical at approximately two of three positions. Even the positions of introns within different actin genes are highly conserved across many species and genes. Despite this high degree of conservation, differences in the distribution of specific isotypes within a single neuron have been reported. For example, β-actin may be enriched in growth cones. The prominent actin bundles seen in fibroblasts and some other nonneuronal cells in culture are not characteristic of neurons, and most neuronal actin microfilaments are less than 1 μm in length.

Many microfilament-associated proteins have been described in nervous tissue (myosin, tropomyosin, spectrin, α-actinin, etc.), but less is known about their distribution and normal function in neurons and glia.

The myosins and myosin-associated proteins are considered later in this chapter under Molecular Motors in the Nervous System, but several categories of actin-binding proteins can be defined (Table 2.3). Monomer actin-binding proteins such as profilin and thymosin β4 or β10 are abundant in the developing brain and are thought to help regulate the amount of actin assembled into microfilaments by sequestering actin monomers. These monomers can be rapidly mobilized in response to appropriate signals. For example, phosphatidylinositol 4,5-bisphosphate causes the actin–profilin complex to dissociate, freeing the monomer for microfilament assembly. Such regulation may play a key role in growth cone motility, where actin assembly is an important mechanism for filopodial extension.

Several proteins that can cap actin microfilaments, serving to anchor them to other structures or to regulate microfilament length, have been identified. The ezrin–radixin–moesin gene family encodes barbed-end capping proteins that are concentrated at sites where the microfilaments meet the plasma membrane, suggesting a role in anchoring microfilaments or linking them to extracellular components through membrane proteins. A mutation in a member of this family expressed in Schwann cells, *merlin* or *schwannomin*, is thought to be responsible for the human disease neurofibromatosis type 2. Development of numerous tumors with a Schwann cell lineage in neurofibromatosis type 2 suggests that this microfilament-binding protein acts normally as a tumor suppressor.

Whereas some membrane proteins can interact directly with the actin microfilaments of the membrane cytoskeleton, others interact with the actin cytoskeleton through intermediaries such as spectrin. Proteins such as spectrin (fodrin), α-actinin, and dystrophins crosslink, or bundle, microfilaments, giving rise to higher-order complexes. Spectrin is enriched in the cortical membrane cytoskeleton and is thought to have a role in the localization of integral membrane proteins such as ion channels and receptors. Dystrophin is the best known member of a family of related proteins that all appear to be essential for the clustering of receptors in muscle and nervous tissue. A mutation in dystrophin is responsible for Duchenne muscular dystrophy. Positioning of integral membrane proteins on the cell surface is likely to be an essential function of the actin-rich membrane cytoskeleton, acting in concert with a new class of proteins that contain the protein-binding module, the PDZ domain.

Members of the gelsolin family have multiple activities. They not only can cap the barbed end of a microfilament, but also can sever microfilaments and nucleate microfilament assembly in some circumstances. These severing–capping proteins may be essential for reorganizing the actin cytoskeleton. Because gelsolin-severing activity is Ca^{2+} activated, it may provide a mechanism for altering the membrane cytoskeleton in response to Ca^{2+} transients. Other second messengers, such as phosphatidylinositol 4,5-bisphosphate, may also serve as regulators of gelsolin function, suggesting an interplay between different classes of actin-binding proteins such as gelsolin and profilin. Oligodendrocytes are the only neural cells in the CNS that express significant amounts of the actin-binding and microfilament-severing protein gelsolin (Tanaka and Sobue, 1994).

Proteins with other functions may also interact directly with actin or actin microfilaments. For example, the enzyme DNase I binds actin tightly, inhibiting both DNase activity and actin assembly. The physiological function of this interaction is unclear, but it has proved a useful tool for probing actin structure and function. Some membrane proteins, such as the epidermal growth factor receptor, bind actin microfilaments directly, which may be important in anchoring these membrane components at a particular location on the cell surface. Other cytoskeletal structures may have specific interactions. Both MAP–2 and tau microtubule-associated proteins have been shown to interact with actin microfilament *in vitro* and have the potential to mediate interactions between microtubules and microfilaments. Finally, the synaptic vesicle-associated phosphoprotein synapsin I has a phosphorylation-sensitive interaction with microfilaments that appears to be important for the targeting and storage of synaptic vesicles in the presynaptic terminal (de Camilli *et al.*, 1990). Many of these interactions have been defined by *in vitro* binding studies, and their physiological significance is not always clearly established. However, there is little doubt that interactions occur between the actin cytoskeleton and a variety of other cellular structures.

The presence of actin as a major component of both pre- and postsynaptic specializations as well as in the growth cone gives the actin cytoskeleton special significance in the nervous system. The enrichment of microfilaments and associated proteins in the membrane cytoskeleton means that they are the cytoskeletal components most subject and most responsive to changes in the local external environment of the neuron. Microfilaments also play a critical role in positioning the various receptors and ion channels at specific locations on the neuronal surface. Although many studies have emphasized the enrichment of the microfilament cytoskeleton at the plasma membrane, microfilaments are also abundant in the deep cytoplasm. In many respects, the microfilaments may be best regarded as a uniquely plastic component of the neuronal cytoskeleton that plays a critical role in local trafficking of both cytoskeletal and membrane components.

Intermediate Filaments Are Prominent Constituents of Nervous Tissue

Intermediate filaments of the nervous system appear as solid, ropelike fibrils from 8 to 12 nm in diameter that may be many micrometers long (Lee and Cleveland, 1996). Intermediate filament proteins constitute a superfamily of five classes, which have distinctive patterns of expression specific to cell type and developmental stage (Table 2.4).

Type I and type II intermediate filament proteins are the keratins, which are hallmarks of epithelial cells. The keratins are not associated with nervous tissue and are not considered further here. In contrast, all nucleated cells contain type V intermediate filament proteins, the nuclear lamins. The lamins are encoded by the most evolutionarily divergent of the intermediate filament genes, with a distinctive pattern of introns and exons, as well as a different polypeptide domain structure. Intermediate filaments in the nervous system are all produced by either type III or type IV intermediate filament proteins.

Type III intermediate filaments are a diverse family that includes, among others, vimentin (characteristic of fibroblasts and many embryonic tissues such as embryonic neurons) and glial fibrillary acidic protein (GFAP, a marker for astrocytes and Schwann cells). Type III intermediate filament subunits typically have a molecular weight between 45 and 60 kDa and consist of a conserved rod domain and relatively small gene-specific amino- and carboxy-terminal sequences. As a result, intermediate filaments formed from type III subunits form smooth filaments without side-arms. Type III polypeptides can form homopolymers but may also coassemble with other type III intermediate filament subunits.

A recently described type III intermediate filament protein, peripherin, is unique to neurons and may be coexpressed with the neurofilament triplet proteins. Peripherin has a characteristic expression during development and regeneration in specific neuronal populations. It has been shown to coassemble with neurofilament triplet proteins both *in vitro* and *in vivo*, where presumably it can substitute for the low-molecular-weight neurofilament (NFL). However, whether coassembly is generally the case is not known. Physiological roles for neuron-specific type III interme-

TABLE 2.4   Intermediate Filament Proteins of the Nervous System

Class and name	Cell type
Types I and II	
Acidic and basic keratins	Epithelial and endothelial cells
Type III	
Glial fibrillary acidic protein	Astrocytes and nonmyelinating Schwann cells
Vimentin	Neuroblasts, glioblasts, fibroblasts, etc.
Desmin	Smooth muscle
Peripherin	A subset of peripheral and central neurons
Type IV	
NF triplet (NFH, NFM, NFL)	Most neurons, expressed at highest level in large myelinated fibers
α-Internexin	Developing neurons, parallel fibers of cerebellum
Nestin	Early neuroectodermal cells. The most divergent member of this class; some have classified it as a sixth type
Type V	
Nuclear lamins	Nuclear membranes

diate filament polypeptides are uncertain. Unlike type IV intermediate filaments, intermediate filaments made from type III subunits tend to disassemble more readily under physiological conditions. Thus, the presence of type III intermediate filament subunit proteins may produce more dynamic structures, which could be important during development or regeneration.

Although other type III intermediate filament proteins are found in the nervous system, they are generally restricted to glia or to neurons at early stages of differentiation. Vimentin is abundant in a wide variety of cells during early development, including both glioblasts and neuroblasts. Some Schwann cells and astrocytes contain vimentin. Curiously, mature oligodendrocytes do not appear to have any intermediate filaments, an exception to the general rule that most metazoan cells contain all three classes of cytoskeletal structures. Oligodendrocyte precursors do, however, express vimentin and may transiently express GFAP.

In neurons, intermediate filaments typically have side-arms that limit packing density, whereas glial intermediate filaments lack side arms and may be very tightly packed. Neuronal intermediate filaments have an unusual degree of metabolic stability, which makes them well suited to the role of stabilizing and maintaining neuronal morphology. The existence of neurofilaments was established for many years before much was known about their biochemistry or function. Neurofilaments could be seen in early electron micrographs, and many traditional histological procedures visualize neurons as a result of a specific interaction of metals with neurofilaments. Recent work on the biochemistry and molecular genetics of intermediate filaments has illuminated many aspects of their function in the nervous system.

The primary type of intermediate filament in neurons is formed from three subunits, the neurofilament triplet, each encoded by a separate gene. The neurofilament triplet proteins are from type IV intermediate filament genes, which are generally expressed only in neurons and have a characteristic domain structure that can be recognized in both primary sequence and gene structure. The polypeptides were initially identified from axonal transport studies. The apparent molecular weights for the neurofilament subunits vary widely across species, but mammalian forms typically range from 180 to 200 kDa for the high-molecular-weight subunit (NFH), from 130 to 170 kDa for the medium subunit (NFM), and from 60 to 70 kDa for the low-molecular-weight subunit (NFL). Interestingly, Schwann cells in damaged peripheral nerves also transiently express NFM and NFL. Neurofilament subunits are phosphorylated in axons, with NFM and NFH having unusually high levels of phosphorylation. In some species, NFH has 50 or more repeats of a consensus phosphorylation site at its carboxy terminus, and levels of NFH phosphorylation indicate that most of these sites are phosphorylated *in vivo*. This high level of phosphorylation in neurofilament subunit tail domains is a distinctive characteristic of neurofilaments.

A second motif characteristic of neurofilaments is the presence of a glutamate-rich region in the tail adjacent to the core rod domain. This glutamate region has particular significance for neuroscientists because it appears to be the basis for the reaction of the classic neurofibrillary silver stains for neurons. These stains were first introduced in the late 19th century and have been used extensively by neurohistologists and neuroanatomists from Ramón y Cajal's time to the present day. However, the molecular basis of these neurofibrillary stains was not known until 1968, when F. O. Schmitt showed that neurofibrils were formed by the 10-nm-diameter neurofilaments. Remarkably, the ability of isolated neurofilament subunits to react with silver histological stains is retained even after separation in gel electrophoresis for neurofilaments from organisms as diverse as humans, squid, and the marine fanworm, *Myxicola*. Conservation of the glutamate-rich domain suggests both an important functional role for this motif and the early divergence of neurofilaments from the other intermediate filament families.

The neurofilaments formed from the neurofilament triplet proteins play a critical role in determining axonal caliber. As mentioned earlier, neurofilaments have characteristic side-arms, unique among intermediate filaments; these side arms are formed by NFM and NFH carboxy-terminal regions. Although all three neurofilament subunits contribute to the neurofilament central core, the side-arms are formed only by the NFM and NFH subunits. Phosphorylation of NFH and NFM side-arms alters charge density on the neurofilament surface, repelling adjacent neurofilaments with similar charge. The high density of surface charge due to phosphate groups on neurofilaments makes it difficult to imagine a stable interaction between neurofilaments and other structures of like charge. Although many reports refer to cross bridges between neurofilaments, direct studies of interactions between neurofilaments provide little evidence of stable crosslinks between neurofilaments or between neurofilaments and other cytoskeletal structures. However, dynamic interactions between neurofilaments and cellular structures or proteins may be critical for many aspects of neurofilament function and metabolism.

Alteration in expression levels for neurofilament subunits or mutations in neurofilament genes can lead to specific neuropathologies. Overexpression of genes encoding normal NFH or expression of some mutant NFL genes in transgenic mouse models leads to the accumulation of neurofilaments in the cell body and proximal axon of spinal motor neurons. These accumulations are similar to those seen in amyotrophic lateral sclerosis and related motor neuron diseases, leading to the hypothesis that disruption of normal neurofilament function is a common intermediate in the pathogenesis of motor neuron disease. Similarly, an early indicator of neuropathies caused by neurotoxins such as acrylamide and hexanedione is the accumulation of neurofilaments in either proximal or distal regions of the axon. Disruption of neurofilament organization is a hallmark of pathology for

many degenerative diseases of the nervous system, particularly those affecting large myelinated axons such as those of spinal motor neurons. Although pathology can be produced by altering neurofilament organization, the question whether neurofilament defects are a primary event in pathogenesis or a manifestation of an underlying metabolic pathology remains controversial in most cases.

Another member of the type IV intermediate filament family, α-internexin, also has been identified. Like neurofilament triplet proteins, α-internexin is expressed only in neurons. Unlike the triplet proteins, α-internexin is preferentially expressed early in development of the nervous system and then disappears from most neurons during maturation. Intermediate filaments containing α-internexin do persist in portions of the adult nervous system, such as the branched axons of granule cells in the cerebellar cortex. There is evidence that α-internexin can coassemble with members of the neurofilament triplet, but it also forms homopolymeric filaments. The primary sequence of α-internexin has features in common with both NFL and NFM that are thought to form the basis for assembly properties distinct from those of other type IV intermediate filaments.

The final type of intermediate protein present in the nervous system is nestin, which is expressed transiently during early development. Nestin is also expressed in Schwann cells and in the progenitors of oligodendrocytes, which appear late in the development of the embryonic nervous system (Gallo and Armstrong, 1995). Remarkably, nestin appears to be expressed almost exclusively in ectodermal cells after commitment to the neuroglial lineage, but prior to terminal differentiation. At 1250 kDa, nestin is the largest intermediate filament subunit and is the most divergent in sequence with several distinctive features, leading some to classify nestin as a sixth type of intermediate filament protein, whereas others group it with type IV intermediate filaments. Relatively little is known about the assembly properties of nestin *in vivo* or the physiological function of nestin intermediate filaments in neuroectodermal cells.

How Do the Various Cytoskeletal Systems Interact?

Each class of cytoskeletal structures may be found without the others in some cellular domains, but all three classes—microtubules, microfilaments, and intermediate filaments—coexist in many domains. As a result, they inevitably interact. This is not to say that individual cytoskeletal elements are necessarily crosslinked to one another. As mentioned earlier, microtubules and neurofilaments have highly phosphorylated side-arms that project from their surfaces. The high density of negative charge on the surface tends to repel structures with a like charge such as other microtubules and neurofilaments. This does not mean that microtubules and neurofilaments do not interact with each other, but it does suggest that such interactions may be transient.

One location containing longer microfilaments and more elaborate organization is the growth cone, which contains bundles of microfilaments in the filopodia as well as a more dispersed actin network. Neurofilaments are largely excluded from the growth cone, typically extending no further than the neck of the growth cone. In contrast, microtubules and microfilaments play complementary roles in the growth cone itself. Microfilaments are critical in sprouting but appear less critical for elongation, at least over short distances. Disruption of microtubules in the distal neurite does not affect sprouting but does inhibit neurite elongation.

Summary

The intracellular framework that gives shape to the neuron and glial cell is the cytoskeleton, a complicated set of filaments and tubules and their associated proteins. These organelles are responsible as well for intracellular movement of materials and, during development, for cell migration and plasma membrane extension within nervous tissue.

MOLECULAR MOTORS IN THE NERVOUS SYSTEM

Until 1985, knowledge of molecular motors in vertebrate cells of any type was restricted to myosins and flagellar dyneins. Myosins had been identified in nervous tissue, but their functions were uncertain. Because the preponderance of evidence indicated that fast axonal transport was microtubule-based, there was considerable interest in dyneins in cell cytoplasm. Despite a number of studies, no evidence for a functional cytoplasmic dynein emerged. Worse yet, the characteristic properties of fast organelle movements appeared inconsistent with both myosins and dyneins. Over the past decade, however, we have developed a good but still incomplete understanding of how these motors may work inside cells (Brady, 1991; Brady and Sperry, 1995).

Myosins and dyneins can be distinguished pharmacologically by their differential susceptibility to inhibitors of ATPase activity, but the spectrum of inhibitors active against fast axonal transport fails to match the properties of either myosin or dynein. The most striking difference between inhibitor effects on axonal transport and on myosin or dynein motors was seen in the effect of a nonhydrolyzable analog of ATP. Adenylyl-imidodiphosphate (AMP-PNP) is a weak competitive inhibitor of both myosin and dynein, requiring a 10- to 100-fold excess of analog. In contrast, within minutes of AMP-PNP perfusion into isolated axoplasm, both anterograde and retrograde axonal transport stops. Inhibition by AMP-PNP occurs even in the presence of stoichiometric concentrations of ATP. Organelles moving in both directions freeze in place and remain attached to microtubules. AMP-PNP weakens the interaction of myosin with microfilaments and of dynein with microtubules, but stabilizes the binding of membrane-bound organelles to microtubules. Thus, the effects of AMP-PNP indicate that movement of membrane-bound organelles in fast axonal transport must require another type of motor, distinct from the myosins and dyneins.

The effects of AMP-PNP both demonstrated the existence of a new type of mechanochemical ATPase and provided a basis for identifying its constituent polypeptides. Binding of the ATPase to microtubules should be increased by AMP-PNP and decreased by ATP. Polypeptides meeting this criterion were soon identified. The new ATPase was named *kinesin*, based initially on an ability to move microtubules across glass coverslips as first described in axoplasmic extracts. Studies soon established that kinesin was a microtubule-activated ATPase with minimal basal activity. This combination of ATPase activity and motility *in vitro* confirmed that kinesin was a new class of microtubule-based motor (Brady and Sperry, 1995; Bloom and Endow, 1995).

Kinesin has now been identified in a variety of organisms and tissues, leading to an extensive characterization of many biochemical, pharmacological, immunochemical, and molecular properties. Electron microscopic and biophysical analyses reveal kinesin as a long, rod-shaped protein, approximately 80 nm in length. Neuronal kinesin is a heterotetramer with two heavy chains (115–130 kDa) and two light chains (62–70 kDa). Localization of antibodies specific for kinesin subunits by high-resolution electron microscopy of bovine brain kinesin indicates that the two heavy chains are arranged in parallel, forming the heads and much of the shaft, whereas light chains are localized to the fan-shaped tail region (Fig. 2.8).

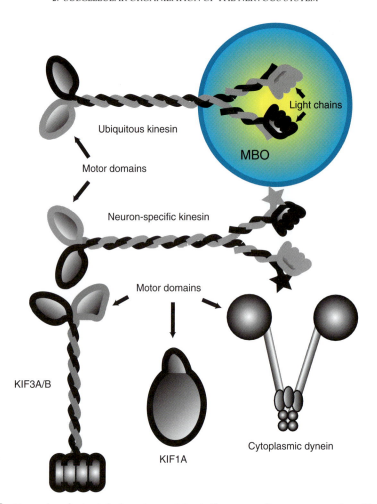

FIGURE 2.8 Examples of microtubule motor proteins in the mammalian nervous system. The first microtubule motor identified in nervous tissue was the ubiquitous form of kinesin, but subsequent studies showed that a neuron-specific form of kinesin is present in mammalian brain. The motor domains are well conserved by the tail domains and appear to be specialized for interaction with various targets, such as different membrane-bound organelles (MBO). After the sequence of the kinesin heavy chain was established, the presence of additional genes that contained sequences homologous to that of the motor domain of kinesin was soon recognized. The molecular organization of these various motor proteins is quite diverse and includes monomers (KIF1A), trimers (KIF3A/3B), and tetramers (ubiquitous and neuron-specific kinesins). Many kinesin-related proteins have been implicated in the processes of cell division, but a number can also be found in postmitotic cells such as neurons.

A variety of approaches have demonstrated that the ATP binding and microtubule binding domains of kinesin are in the head regions of the heavy chains, whereas the light chains in the tail region of kinesin appear to bind to membranes. When *in vitro* motility assays are employed for analysis of brain kinesins, movements are directed toward the plus ends of microtubules. Because axonal microtubules are oriented with their plus ends distal from the cell body, this movement would be appropriate for a motor that moves organelles in the anterograde direction.

Neuronal kinesin appears associated with a variety of membrane-bound organelles, including synaptic vesicles, mitochondria, coated vesicles, and lysosomes. The interaction of kinesin and other molecular motors with membrane surfaces is not well understood. In the case of kinesin, the interaction is thought to involve the light chains of kinesin along with the carboxy termini of the heavy chains.

The kinesins have now been shown to be a family of related proteins with a highly conserved domain that includes the ATP and microtubule binding domains. Many of these kinesin-related polypeptides appear associated with cell division, and kinesin-related proteins in vertebrate tissues are not well characterized. However, multiple members of the kinesin superfamily are expressed in both adult and developing brains. This proliferation of motor proteins has dramatically altered the questions being asked about motor function in the brain. The discovery that ncd, a kinesin-related protein from *Drosophila*, can move structures toward the minus end of microtubules increases the number of potential functions that kinesin family members might serve in nervous tissue still further, perhaps including a role in retrograde transport. Further study is needed to establish specific functions for each member of the kinesin superfamily expressed in neurons or glial cells.

As an indirect result of the discovery of kinesin, one of the high-molecular-weight microtubule-associated proteins of brain, MAP–1c, was found to be the long sought cytoplasmic form of dynein. Both MAP–1c dynein and kinesin can be isolated from bovine brain by incubation of microtubules with nucleotide-free soluble extracts. Both are bound to microtubules under these conditions and released by ATP. MAP–1c dynein moved microtubules *in vitro* with a polarity opposite that seen with kinesin and was identified as a two-headed cytoplasmic dynein by using both structural and biochemical criteria. Concurrently, a similar protein was identified in nematodes.

MAP–1c dyneins form a 40-nm-long complex of molecular mass 1.6×106 Da, which includes two heavy chains and a number of light chains (Figs. 2.8, 2.9) (Brady and Sperry, 1995; Tanaka *et al.*, 1995; Vallee, 1993). Less information is available about the distribution and properties of MAP–1c dyneins than about kinesin. Immunocytochemical studies in nonneuronal cells showed immunoreactivity on mitotic spindles. In addition, a punctate pattern of immunoreactivity also present in interphase cells was

thought to be due to dynein bound to membrane-bound organelles. Dyneins are widely thought to be the motor for fast retrograde axonal transport but are also a candidate for a motor in slow axonal transport.

Myosins from muscle were the first molecular motors identified, but in recent years interest in nonmuscle myosins has increased (Hammer, 1994; Mooseker and Cheney, 1995; Hasson and Mooseker, 1995). Nonmuscle myosins may be categorized as belonging to one of eight classes, but only a subset has been clearly demonstrated in the nervous system. The nonmuscle myosins in this subset share considerable homology in their motor domains but diverge widely in other domains.

The most familiar of the myosins are the myosin II proteins (Fig. 2.10), which are found in the thick filaments of smooth and skeletal muscle but are also present in nonmuscle cells. Two heavy chains of myosin II form a dimer that may interact with other

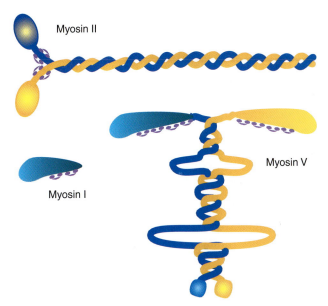

FIGURE 2.10 Examples of myosin motor proteins found in the mammalian nervous system. Myosin heavy chains contain the motor domain, whereas the light chains serve to regulate motor function. Myosin II was the first molecular motor characterized biochemically from skeletal muscle and brain. Biochemical and genetic approaches have now defined 11 classes of myosin, many of which can be found in brain. Myosin II is a classic two-headed myosin that forms thick filaments in nonmuscle cells. Myosin I motors have single motor domains, but may interact with actin microfilaments or membranes. Myosin V motors were initially identified as a mouse mutation that affected coat color and produced seizures. Myosin V has multiple binding sites for calmodulin that act as light chains. Mutations in other classes of myosin have been linked to deafness. Other myosins, including myosins I, II, and V, have been detected in growth cones as well as mature neurons. The specific roles of these various myosins in the nervous system remain to be established.

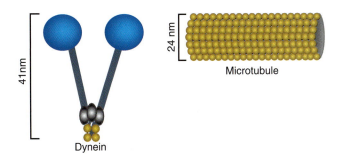

FIGURE 2.9 Biochemical studies on kinesin in brain led to the description of a cytoplasmic form of dynein that was distinct from axonemal dyneins. Cytoplasmic dynein may interact with membrane-bound organelles and cytoskeletal structures. Genetic methods have established that there may be as many as 30 kinesin-related proteins and 15 dynein heavy chains in a single organism. The diversity of microtubule-based motors is consistent with the extent of the microtubule cytoskeleton in the nervous system.

myosin II dimers to form bipolar filaments. Under tissue culture conditions, many cells contain bundles of actin microfilaments, known as stress fibers, that exhibit a characteristic distribution of myosin II into distinct patterns that may be sarcomeric equivalents, but stress fibers are not apparent in neurons and other cells of the nervous system *in situ*. However, bipolar thick filaments assembled from myosin II dimers can be isolated from nervous tissues. Many of the cellular contractile events described in nonneuronal cells, such as the contractile ring in mitosis, are thought to include myosin II. Although brain myosin II was one of the first nonmuscle myosins to be described, relatively little is known about the function of myosin II in neurons. Myosin II has been localized in the neurites of neurons in primary culture.

Myosin I proteins have a single, smaller heavy chain that does not form filaments but possesses a homologous actin-activated ATPase domain. One exciting aspect of myosin I is its ability to interact directly with membrane surfaces, which may generate movements of plasma membrane components or intracellular organelles. Myosin I has been purified from neural and neuroendocrine tissues. At least three genes in this family have been found in mammals, and multiple forms are present in brain. An interesting aspect of myosin IB is its expression in the stereocilia of hair cells in the cochlea and vestibular system, where it may play a role in mechanotransduction. Both myosin I and myosin II molecules have been proposed to have a role in the motility of lamellipodia at the leading edge of growth cones, but they are also expressed at substantial levels in adult nervous tissue in which growth cones are rare.

The mouse mutation *dilute*, which affects coat color, was shown to result from a mutation in a gene that encodes a novel myosin heavy chain distinct from both myosins I and II. Similar myosin molecules have been identified in other cell types and organisms and are classified as myosin V. The change in coat color seen in the *dilute* mouse is due to an inability of skin dendritic pigment cells to deliver the pigment to developing hairs. There are complex neurological deficits in dilute mutants, including seizures that eventually lead to death in severely altered alleles of the *dilute* mutation. Myosin V was also discovered independently in extracts from chicken and mammalian brain. The specific cellular localization and function of myosin V in the nervous system remain unclear, although it has been reported in growth cones. However, neurons in *dilute* mice without one allele of myosin V clearly develop axons and make connections. The seizures do not begin until early adulthood.

Recently, representatives from two more classes of myosin have been identified in nervous tissue. Genes for a myosin VI and a myosin VIIA have been identified in brain as well as in other tissues. Both have been implicated in forms of congenital deafness. Myosin VI appears to be the gene responsible for Snell's Waltzer deafness, and myosin VIIA has been identified as the gene responsible for a human disease involving both deafness and blindness, Usher syndrome type 1B. Both of these myosins are expressed in the mechanosensory hair cells of the cochlea and vestibular apparatus, and they exhibit a different localization from each other and from myosin IB.

The diversity of brain myosins and their distinctive localization suggest that the various myosins may have narrowly defined functions. However, relatively little is known about specific neuronal functions for the myosins despite intensive study of myosins in the nervous system. The axonal transport of myosin II-like proteins has been described, but little further progress has been made on the functions of myosin II in the mature nervous system. Even less is known about myosin I in the nervous system. However, myosins are likely to play roles in growth cone motility, synaptic plasticity, and even neurotransmitter release.

There are few instances in our knowledge of neuronal function in which we fully understand the role played by specific molecular motors, but members of all three classes are abundant in nervous tissue. This proliferation of different motor molecules and their isoforms suggests that some physiological activities may require multiple classes of motor molecules.

Summary

The concept is now firmly in place that neurons and glial cells, like other cells, contain certain molecular motors responsible for moving discrete populations of molecules, particles, and organelles through intracellular compartments.

BUILDING AND MAINTAINING NERVOUS SYSTEM CELLS

The functional architecture of neurons comprises many specializations in cytoskeletal and membranous components. Each of these specializations is dynamic, constantly changing and being renewed at a rate determined by the local environment and cellular metabolism. The processes of axonal transport represent a key to understanding neuronal dynamics and provide a basis for exploring neuronal development, regeneration, and neuropathology. Recent advances are important sources of insight into the molecular mechanisms underlying axonal transport, although many questions remain.

Slow Axonal Transport Moves Soluble Components and Cytoskeletal Structure

Slow axonal transport has two major components, both representing movement of cytoplasmic constituents (Fig. 2.11). The cytoplasmic and cytoskeletal elements of the axon in axonal transport move at rates at least two orders of magnitude more slowly than fast transport. Slow component a is composed largely of cytoskeletal proteins, neurofilaments, and microtubule protein. Slow component b is a complex and heterogeneous rate component, including hundreds of distinct polypeptides ranging from cytoskeletal proteins such as actin (and tubulin in some nerves) to soluble enzymes of intermediary metabolism (such as the glycolytic enzymes). Many characteristics of axonal transport have been described, and these characteristics provide the foundation for our understanding of mechanisms.

Neurofilaments and microtubules move as discrete cytological structures (Brady, 1992). Recent studies on transport of neurofilament protein indicate that little degradation or metabolism occurs until neurofilaments reach nerve terminals, where they are rapidly degraded. Comparable results have been obtained in studies labeling microtubule protein by radioactivity or fluorescence. Under favorable conditions, movement of individual microtubules can be detected in neurites or growth cones. Both radiolabeling studies and direct observations of individual microtubules indicate that all microtubules and neurofilaments move down the axon, but the motor protein involved is uncertain. Differential metabolism appears to be a key to the targeting of cytoplasmic and cytoskeletal proteins. Proteins with slow degradative rates accumulate and reach higher steady-state concentrations. Alteration of degradation rates changes the steady-state concentration of a protein. Concentration of actin in presynaptic terminals is explained by slower turnover in terminals relative to neurofilament proteins and tubulin, and inhibition of calpain causes neurofilament rings to appear in presynaptic terminals. Differential turnover may be accomplished by specific proteases or posttranslational modifications that affect susceptibility to degradation.

The coherent movement of neurofilaments and microtubule proteins provides strong evidence for the "structural hypothesis." For example, pulse-labeling experiments show that radiolabeled neurofilament proteins move as a bell-shaped wave with little or no trailing of neurofilament protein. This fits with the observed stability of neurofilaments under physiological conditions, which suggests that any soluble pool of neurofilament subunits is negligible. Similarly, the coherent transport of tubulin and MAPs makes sense only if microtubules are moved, because MAPs do not interact with unpolymerized tubulin.

A striking demonstration of microtubule movement can be seen with fluorescent analogs of tubulin (Tanaka and Kirschner, 1991; Reinsch et al, 1991). Tubulin labeled with caged fluorescein can be injected into one cell of a fertilized *Xenopus* oocyte at the two-cell stage. Such tubulins are fluorescent only after photoactivation. The injected oocyte is then allowed to develop into an embryo. The injected tubulin equilibrates with endogenous tubulin and is incorporated into the microtubules of all cells derived from the original injected cell. Because protein synthesis is minimal in early cell divisions of embryonic development, the labeled tubulin is reused by daughter cells until diluted out by newly synthesized tubulin. When early embryonic neurons are cultured, the caged fluorescent tubulin may be photoactivated and visualized.

When local segments of an axon are photoactivated, patches of fluorescent tubulin can be seen to move down the growing neurite. The fluorescent patches remain discrete during movements in the anterograde direction at slow transport rates. Observations of full microtubules can also be made in embryonic *Xenopus* neurons by using rhodamine-labeled tubulin. In favorable areas of axons and growth cones, individual fluorescent microtubules can be visualized. Such microtubules can be seen to move down axons and into growth cones. The forces are sufficient to bend these microtubules in conjunction with growth cone movements. When studies of axonal transport using radiolabels are combined with direct observations of individual microtubules by video microscopy, there is little doubt that microtubules and neurofilaments can and do move in the axon as intact, individual cytoskeletal elements.

Fast Axonal Transport Is the Means by Which Membrane Vesicles and Their Contents Are Rapidly Moved Long Distances within a Neuron

Early biochemical and morphological studies established that the material moving in fast axonal transport was associated with membrane-bound organelles (Fig. 2.12) (Lasek and Brady, 1982; Brady,

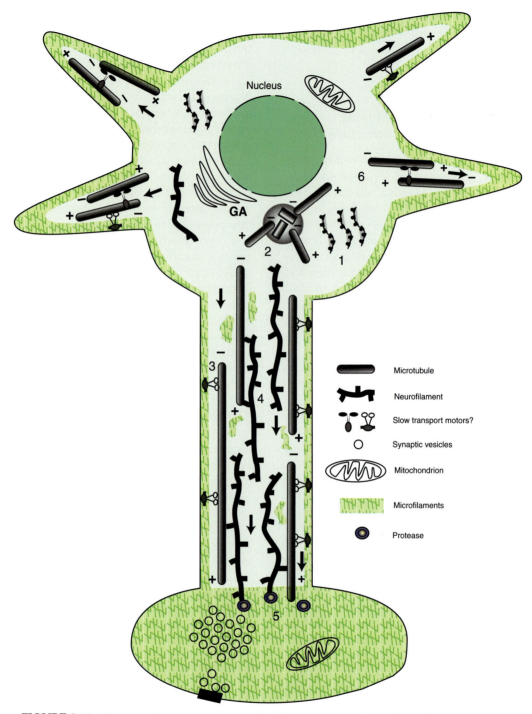

FIGURE 2.11 Slow axonal transport represents the delivery of cytoskeletal and cytoplasmic constituents to the periphery. Cytoplasmic proteins are synthesized on free polysomes and organized for transport as cytoskeletal elements or macromolecular complexes (1). The microtubules are formed by nucleation at the microtubule-organizing center near the centriolar complex (2) and then released for migration into the axon or dendrites. The molecular mechanisms are not as well understood as those for fast axonal transport, but slow transport appears to be unidirectional with no retrograde component. Recent studies suggest that motors like cytoplasmic dynein may interact with the axonal membrane cytoskeleton to move the microtubules with their plus ends leading (3). Neurofilaments do not appear able to move on their own, but may hitchhike on the microtubules (4). Other cytoplasmic proteins may do the same or they may be moved by other motors. Once cytoplasmic structures reach their destinations, they are degraded by local proteases (5) at a rate that allows either growth (in the case of growth cones) or maintenance of steady-state levels. The different composition and organization of the cytoplasmic elements in dendrites suggest that different pathways may be involved in delivery of cytoskeletal and cytoplasmic materials to the dendrite (6). In addition, some mRNAs are transported into the dendrites, but not into axons.

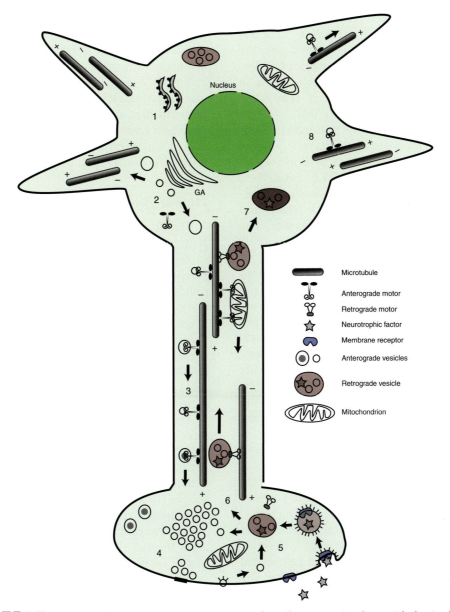

FIGURE 2.12 Fast axonal transport represents transport of membrane-associated materials, having both anterograde and retrograde components. For anterograde transport, most polypeptides are synthesized on membrane-bound polysomes, also known as rough endoplasmic reticulum **(1)**, and then transferred to the Golgi apparatus for processing and packaging into specific classes of membrane-bound organelles **(2)**. Proteins following this pathway include both integral membrane proteins and secretory polypeptides in the lumen of vesicles. Cytoplasmic peripheral membrane proteins like the kinesins are synthesized on the cytoplasmic or free polysomes. Once vesicles have been assembled and the appropriate motors associate with them, they are moved down the axon at a rate of 100–400 mm per day **(3)**. Different membrane structures are delivered to different compartments and may be regulated independently. For example, dense core vesicles and synaptic vesicles are both targeted for the presynaptic terminal **(4)**, but release of vesicle contents involves distinct pathways. After vesicles merge with the plasma membrane, their protein constituents are taken up by coated pits and vesicles via the receptor-mediated endocytic pathway and delivered to a sorting compartment **(5)**. After proper sorting into appropriate compartments, membrane proteins are either committed to retrograde axonal transport or recycled **(6)**. Retrograde moving organelles are morphologically and biochemically distinct from anterograde vesicles. These larger vesicles have an average velocity about half that of anterograde transport. The retrograde pathway is an important mechanism for delivery of neurotrophic factors to the cell body. Material delivered by retrograde transport typically fuses with cell body compartments to form mature lysosomes **(7)**, where most constituents are recycled. However, neurotrophic factors and neurotrophic viruses can act at the level of the cell body. Although there is evidence that vesicle transport also occurs into dendrites **(8)**, less is known about this process. Dendritic vesicle transport is complicated by the fact that dendritic microtubules may have mixed polarity.

1992). A variety of materials could be shown to move in fast transport. In anterograde transport, materials being moved include membrane-associated enzyme activities, neurotransmitters, and neuropeptides. Many of the materials moving down the axon in anterograde transport are returned in retrograde transport (Lasek, 1967; LaVail and La Vail, 1972), in some cases after modification in the terminal. In addition, a number of exogenous materials taken up in the distal regions of the axon are moved back to the cell body by retrograde transport (Fig. 2.12). Exogenous materials in retrograde transport include neurotrophic factors, such as nerve growth factor, and viral particles invading the nervous system. The uptake of neurotrophic factors may play a critical role in the process of regeneration.

Electron microscopic analysis of materials accumulated at a ligation or crush demonstrated that the organelles moving in the anterograde direction were morphologically distinct from those moving in the retrograde direction (Smith, 1980; Tsukita and Ishikawa, 1980). Consistent with the ultrastructural differences, radiolabel and immunocytochemical studies indicate that there are both quantitative and qualitative differences between anterograde and retrograde moving material. The differences between anterograde and retrograde transport indicate that some processing or repackaging events must occur as part of turnaround in axonal transport. Turnaround processing appears to require a proteolytic event, because certain protease inhibitors inhibit turnaround without affecting anterograde and retrograde transport.

Biochemical and morphological approaches resulted in considerable progress toward a description of the materials being transported in fast axonal transport but were not suitable for identifying the molecular motors used in translocation. A different technology that permitted direct observation of organelle movements and precise control of experimental conditions was required. Such experiments became possible with the advent of video microscopic techniques (Brady, 1991).

An early use of video-enhanced contrast (VEC) microscopy was to characterize the bidirectional movement of membrane-bound organelles in giant axons from the squid *Loligo pealeii*. Years before, studies had shown that axoplasm could be extruded from the giant axon as an intact cylinder. Properties of the isolated axoplasm had been characterized in some detail, making VEC microscopic analysis of axoplasm a natural choice. Remarkably, fast axonal transport continued unabated in isolated axoplasm for hours. Isolated axoplasm from the giant axon has no plasma membrane or other permeability barriers but can be readily maintained in an active state. Combining VEC microscopy with isolated axoplasm permitted rigorous dissection of the mechanisms for fast axonal transport with the use of biochemical and pharmacological approaches. A number of insights into axonal transport mechanisms have resulted from these studies. Most importantly, studies in isolated axoplasm have led to the identification of several families of molecular motors that may take part in axonal transport (Brady, 1991; Brady and Sperry, 1995).

How Is Axonal Transport Regulated?

The diversity of polypeptides in each axonal transport rate component and the coherent movement of proteins having many different molecular weights produce a conundrum. How can so many different polypeptides move down the axon as a group? In theory, one could propose that each protein has a motor of its own. In that case, each rate component might represent movements due to a specific class of motors or to variable affinities for a smaller number of motors. However, the relatively small numbers of motor molecules and the logistical difficulties associated with such a model effectively preclude this possibility.

The structural hypothesis mentioned earlier was formulated in response to the observation that rate components of axonal transport move as discrete waves, each with a characteristic rate and a distinctive composition (Figs. 2.11, 2.12). The hypothesis is deceptively simple (Brady, 1992): axonal transport represents the movement of discrete cytological structures. Proteins in axonal transport do not move as individual polypeptides. Instead, they move as part of a cytological structure or in association with a cytological structure. The only assumption made is that a limited number of elements can interact directly with transport motors, so transported material must be packaged appropriately to be moved. The different rate components result from packaging of transported material into different cytologically identifiable structures. In other words, membrane-associated proteins move as membrane-bound organelles (vesicles, mitochondria, etc.), whereas tubulin and MAPs move as microtubules and neurofilament protein moves as neurofilaments.

The structural hypothesis does not require movement of a crosslinked microtubule–neurofilament complex in the form of a solid axoplasmic column. The evidence for existence of a crosslinked complex of microtubules and neurofilaments is not compelling in any case. Instead, the hypothesis specifically predicts that individual microtubules and neurofilaments move rather than tubulin dimers or neurofilament monomers. No assumptions are made about higher-order interactions between cytoskeletal structures or membranous structures. Indeed, one variant of the structural hypothesis proposes that cytoskeletal proteins move in the form of small oligomers rather than polymers, although there is no evidence for the presence of such oligomers *in vivo*. For example, to the limits of detection, neurofilaments exist only as the polymer under *in vivo* conditions. Similarly, tubulin interchanges between dimer and polymer with no known oligomeric intermediate. Because a number of experiments indicate that microtubules and neurofilaments can move *in vivo* as intact polymers, there is no compelling reason to postulate additional oligomeric forms.

Because synthesis of proteins takes place at some distance from many functional domains of a neuron, transport to distal regions of the neuron is necessary, but not sufficient, for proper function. Specific materials must also be delivered to their proper site of utilization and should not be left in inappropriate locations. For example, a synaptic vesicle has no known function in axons or the cell body, so it must be delivered to a presynaptic terminal along with other components necessary for regulated neurotransmitter release. The traditional picture places the presynaptic terminal at the end of an axonal process. Such images imply that a synaptic vesicle need only move along the axonal microtubules until it reaches microtubule ends in the presynaptic terminal. However, many CNS synapses are not at the end of an axon. Numerous terminals are located sequentially along a single axon, making *en passant* contacts with multiple target cells along the way. Targeting of synaptic vesicles then becomes a more complex problem and targeting ion channels or neurotransmitter receptors to nodes of Ranvier or other appropriate sites on the neuronal surface is equally challenging.

Although the specific details of this targeting are not well understood, a simple model for the targeting of synaptic vesicles serves to illustrate how such targeting may occur (Fig. 2.13). Synapsin I (de Camilli *et al.*, 1995) is a cytoplasmic protein enriched in presynaptic terminals that can bind reversibly to both actin microfilaments and synaptic vesicles. Both binding activities are regulated by phosphorylation of synapsin, and this phosphorylation is known to be increased during stimulation. If nonphosphorylated synapsin were present at the border between the axon and the presynaptic terminal, then a fraction of the synaptic vesicles (or their precursors) traveling down the axon in fast axonal transport would be bound and crosslinked to the actin cytoskeleton that is enriched in the presynaptic terminal. These bound synaptic vesicles would become part of the reserve pool in the terminal at some distance from active zones. As the terminal was stimulated, local calmodulin-activated kinases would phosphorylate synapsin and allow these reserve synaptic vesicles to be mobilized. Consistent with this model, dephosphorylated synapsin has been shown to inhibit fast axonal transport when introduced directly into the axoplasm at concentrations comparable to those seen in the presynaptic terminal.

Although such a model is speculative at present, it does satisfy several criteria that any mechanism for targeting to specific neuronal subdomains must address. Specifically, the mechanism must be local, because distances to the cell body can be quite large. There must be some means to connect the targeting signal to an external microenvironment, such as an appropriate target cell. Finally, there must be a way of distinguishing subdomains so that synaptic vesicles are not delivered to nodes of Ranvier and voltage-gated sodium channels are not all targeted to the presynaptic terminal. The careful segregation of different organelles and polypeptides to different regions within a neuron suggests that highly efficient targeting mechanisms do exist.

Summary

A well-studied feature of the neuron is the phenomenon of axonal transport, which has been described in both the anterograde and the retrograde directions. The axonal transport system is responsible for delivery of materials from the cell body to distant parts of the neuron, for membrane retrieval and circulation, and for uptake of materials from presynaptic terminals and dendrites and their delivery to the cell soma. The precise molecular mechanisms by which anterograde and retrograde transport can be targeted within an individual dendrite and/or axon are not understood at present.

Neurons and glial cells may have unusually large cell volumes enclosed within extensive plasma membrane surfaces. Nature has evolved a number of "universal" mechanisms in other systems and adapted them for the special needs of nervous tissue cells. The synthesis and delivery of components, and in particular proteins, to cytoplasmic organelles and cell surface subdomains engage general and evolutionarily conserved molecular mechanisms and pathways that are employed in single-cell yeasts as well as in the cells in complex nervous tissue.

Once synthesized and sorted, most intracellular organelles (vesicles destined for axonal or dendritic domains, mitochondria, cytoskeletal components)

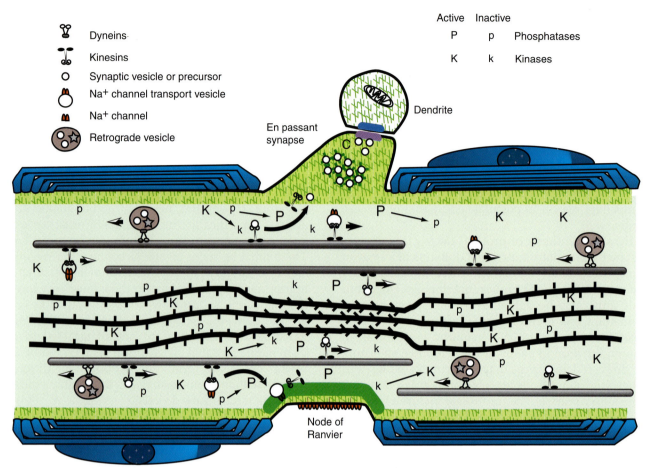

FIGURE 2.13 Axonal dynamics in a myelinated axon from the peripheral nervous system (PNS). Axons are in constant flux with many concurrent dynamic processes. This diagram illustrates a few of the many dynamic events occurring at a node of Ranvier in a myelinated axon from the PNS. Axonal transport moves cytoskeletal structures, cytoplasmic proteins, and membrane-bound organelles from the cell body toward the periphery (from right to left). At the same time, other vesicles return to the cell body by retrograde transport (retrograde vesicle). Membrane-bound organelles are moved along microtubules by motor proteins such as the kinesins and cytoplasmic dyneins. Each class of organelles must be directed to the correct functional domain of the neuron. Synaptic vesicles must be delivered to a presynaptic terminal to maintain synaptic transmission. In contrast, organelles containing sodium channels must be targeted specifically to nodes of Ranvier for saltatory conduction to occur. Cytoskeletal transport is illustrated by microtubules (rods in the upper half of the axon) and neurofilaments (bundle of ropelike rods in the lower half of the axon) representing the cytoskeleton. They move in the anterograde direction as discrete elements and are degraded in the distal regions. Microtubules and neurofilaments interact with each other transiently during transport, but their distribution in axonal cross sections suggests that they are not stably crosslinked. In axonal segments without compact myelin, such as the node of Ranvier or following focal demyelination, a net dephosphorylation of neurofilament side-arms allows the neurofilaments to pack more densely. Myelination is thought to alter the balance between kinase (K indicates an active kinase, k is an inactive kinase) and phosphatase (P indicates an active phophatase, p is an inactive phosphatase) activity in the axon. Most kinases and phosphatases have multiple substrates, suggesting a mechanism for targeting vesicle proteins to specific axonal domains. Local changes in the phosphoryation of axonal proteins may alter binding properties of proteins. The action of synapsin I in squid axoplasm suggests that dephosphorylated synapsin crosslinks synaptic vesicles to microfilaments. When a synaptic vesicle encounters the dephosphorylated synapsin and actin-rich matrix of a presynaptic terminal, the vesicle is trapped at the terminal by inhibition of further axonal transport, effectively targeting the synaptic vesicle to a presynaptic terminal. Similarly, a sodium channel-binding protein may be present at nodes of Ranvier in a high-affinity state (i.e., dephosphorylated). Transport vesicles for nodal sodium channels (Na channel vesicle) would be captured on encountering this domain, effectively targeting sodium channels to the nodal membrane. Interactions between cells could in this manner establish the functional architecture of the neuron.

must be distributed, and targeted, to precise intracellular locations. Neurons and glial cells, because they are so extended in space, have adapted and developed to a high degree common mechanisms that operate to distribute components within all

cells. In the neuron, movement of materials within the axon has been the central focus of most studies. The phenomenon of axoplasmic flow or transport is now in some measure understood at the molecular level.

References

Anderson, R. G. W., Kamen, B. A., Rothberg, K. G., and Lacey, S. W. (1992). Potocytosis: Sequestration and transport of small molecules by caveolae. *Science* **255**, 410–411.

Balch, W. E., McCaffrey, J. M., Plutner, H., and Farquhar, M. G. (1994). Vesicular stomatitis virus glycoprotein is sorted and concentrated during export from the endoplasmic reticulum. *Cell (Cambridge, Mass.)* **76**, 841–852.

Bednarek, S. Y., Ravazzola, M., Hosobuchi, M., Amherdt, M., Perrelet, A., Schekman, R., and Orci, L. (1995). COPI- and COPII-coated vesicles bud directly from the endoplasmic reticulum in yeast. *Cell (Cambridge, Mass.)* **83**, 1183–1196.

Berl, S., Puszkin, S., and Nicklas, W. J. K. (1973). Actomyosin-like protein in brain. *Science* **179**, 441–446.

Blobel, G. (1980). Intracellular protein topogenesis. *Proc. Natl. Acad. Sci. USA* **77**, 1496.

Blobel, G., and Dobberstein, B. (1975). Transfer of proteins across membranes. *J. Cell Biol.* **67**, 835–851.

Bloom, G. S., and Endow, S. A. (1995). Motor proteins. 1. Kinesins. *Protein Profile* **2**, 1109–1171.

Brady, S. T. (1991). Molecular motors in the nervous system. *Neuron* **7**, 521–533.

Brady, S. T. (1992). Axonal dynamics and regeneration. In "Neuroregeneration" (A. Gorio, Ed.), pp. 7–36. Raven Press, New York.

Brady, S. T., and Sperry, A. O. (1995). Biochemical and functional diversity of microtubule motors in the nervous system. *Curr. Opin. Neurobiol.* **5**, 551–558.

Bretscher, M. S., and Munro, S. (1993). Cholesterol and the Golgi apparatus. *Science* **261**, 1280–1281.

Coetzee, T., Fujita, N., Dupree, J., Shi, R., Blight, A., Suzuki, K., Suzuki, K., and Popko, B. (1996). Myelination in the absence of galactocerebroside and sulfatide: Normal structure with abnormal function and regional instability. *Cell (Cambridge, Mass.)* **86**, 209–219.

Colman, D. R., Kreibich, G., Frey, A. B., and Sabatini, D. D. (1982). Synthesis and incorporation of myelin polypeptides into CNS myelin. *J. Cell Biol.* **95**, 598–608.

de Camilli, P., Benfenati, F., Valtorta, F., and Greengard, P. (1990). The synapsins. *Annu. Rev. Cell Biol.* **6**, 433–460.

de Camilli, P., Takei, K., and McPherson, P. S. (1995). The function of dynamin in endocytosis. *Curr. Opin. Neurobiol.* **5**, 559–565.

de Duve, C. (1975). Exploring cells with a centrifuge. *Science* **189**, 186–194.

de Duve, C. (1983). Lysosomes revisited. *Eur. J. Biochem.* **137**, 391.

Dodt, G., Braverman, N., Wong, C., Moser, A., Moser, H. W., Watkins, P., Valle, D., and Gould, S. J. (1995). Mutations in the PTS1 receptor gene, PXR1, define complementation group-2 of the peroxisome biogenesis disorders. *Nat. Genet.* **9**, 115–125.

Doll, T., Meichsner, M., Riederer, B. M., Honegger, P., and Matus, M. (1993). An isoform of microtubule-associated protein 2 (MAP2) containing four repeats of the tubulin-binding motif. *J. Cell Sci.* **106**, 633–640.

Droz, B., and Leblond, C. P. (1963). Axonal migration of proteins in the central nervous system and peripheral nerves as shown by radioautography. *J. Comp. Neurol.* **121**, 325–346.

Ferhat, L., Ben-Ari, Y., and Khrestchatisky, M. (1994). Complete sequence of rat MAP2d, a novel MAP2 isoform. *C. R. Seanc. Acad. Sci.* **317**, 304–309.

Gallo, V., and Armstrong, R. C. (1995). Developmental and growth factor-induced regulation of nestin in oligodendrocyte lineage cells. *J. Neurosci.* **15**, 394–406.

Gilmore, R., Blobel, G., and Walter, P. (1982a). Protein translocation across the endoplasmic reticulum. I. Detection in the microsomal membrane of a receptor for a signal recognition particle. *J. Cell Biol.* **95**, 463.

Gilmore, R., Walter, P., and Blobel, G. (1982b). Protein translocation across the endoplasmic reticulum. II. Isolation and characterization of the signal recognition particle receptor. *J. Cell Biol.* **95**, 477.

Griffiths, G., Pfeifer, S., Simons, K., and Matlin, K. (1985). Exit of newly synthesized membrane proteins from trans cisternae of the Golgi complex to the plasma membrane. *J. Cell Biol.* **101**, 949.

Hachiya, N., Alam, R., Sakasegawa, Y., Sakaguchi, M., Mihara, K., and Omura, T. (1993). A mitochondrial import factor purified from rat liver cytosol is an ATP-dependent conformational modulator for precursor proteins. *EMBO J.* **12**, 1579–1586.

Hammer, J. A. (1994). The structure and function of unconventional myosins: A review. *J. Muscle Res. Cell Motil.* **15**, 1–10.

Hammond, G., and Helenius, A. (1994). Quality control in the secretory pathway: Retention of a misfolded viral membrane glycoprotein involves cycling between the ER, intermediate compartment, and Golgi apparatus. *J. Cell Biol.* **126**, 41–52.

Hartl, F.-U., Ostermann, J., Guiard, B., and Neupert, W. (1987). Successive translocation into and out of the mitochondrial matrix: Targeting of proteins to the intermembrane space by a bipartite signal peptide. *Cell (Cambridge, Mass.)* **51**, 1027.

Hasson, T., and Mooseker, M. S. (1995). Molecular motors, membrane movements and physiology: Emerging roles for myosins. *Curr. Opin. Cell Biol.* **7**, 587–594.

Heuser, J. E., and Reese, T. S. (1973). Evidence for recycling of synaptic vesicle membrane during transmitter release at the frog neuromuscular junction. *J. Cell Biol.* **57**, 315–344.

Hochstrasser, M. (1995). Ubiquitin, proteasomes, and the regulation of intracellular protein degradation. *Curr. Opin. Cell Biol.* **7**, 215–223.

Hyams, J. S., and Lloyd, C. W. (1994). Microtubules. In "Modern Cell Biology" (J. B. Harford, Ed.), p. 439. Wiley–Liss, New York.

Jamieson, J. D., and Palade, G. E. (1967a). Intracellular transport of secretory proteins in the pancreatic exocrine cell. I. Role of the peripheral elements of the Golgi complex. *J. Cell Biol.* **34**, 577.

Jamieson, J. D., and Palade, G. E. (1967b). Intracellular transport of secretory proteins in the pancreatic exocrine cell. II. Transport to condensing vacuoles and zymogen granules. *J. Cell Biol.* **34**, 597.

Jentsch, S., and Schlenker, S. (1995). Selective protein degradation: A journey's end within the proteasome. *Cell (Cambridge, Mass.)* **82**, 881–884.

Kislaukis, E. H., Zhu, X. C., and Singer, R. H. (1994). Sequences responsible for intracellular-localization of β-actin messenger-RNA also affect cell phenotype. *J. Cell Biol.* **127**, 441–451.

Kornfeld, R., and Kornfeld, S. (1985). Assembly of asparagine-linked oligosaccharides. *Annu. Rev. Biochem.* **54**, 631–664.

Kornfeld, S. (1987). Trafficking of lysosomal enzymes. *FASEB J.* **1**, 462.

Lasek, R. J. (1967). Bidirectional transport of radioactively labeled axoplasmic components. *Nature (London)* **216**, 1212–1214.

Lasek, R. J., and Brady, S. T. (1982). The axon: A prototype for studying expressional cytoplasm. In "Organization of the Cytoplasm," pp. 113–124. Cold Spring Harbor Lab. Press, Cold Spring Harbor, NY.

LaVail, J. H., and LaVail, M. M. (1972). Retrograde axonal transport in the central nervous system. *Science* **176**, 1416–1417.

Lazarow, P. B., and Fujiki, Y. (1985). Biogenesis of peroxisomes. *Annu. Rev. Cell Biol.* **1**, 489.

Lee, M. K., and Cleveland, D. W. (1996). Neuronal intermediate filaments. *Annu. Rev. Neurosci.* **19**, 187–217.

Lill, R., and Neupert, W. (1996). Mechanisms of protein import across the mitochondrial outer membrane. *Trends Cell Biol.* **6**, 56–61.

Ludwig, T., Leborgne, R., and Hoflack, B. (1995). Roles for mannose-6-phosphate receptors in lysosomal-enzyme sorting, IGF-II binding and clathrin-coat assembly. *Trends Cell Biol.* **5**, 202–206.

Mallabiabarrena, A., and Malhotra, V. (1995). Vesicle biogenesis: The coat connection. *Cell (Cambridge, Mass.)* **83**, 667–669.

Mooseker, M. S., and Cheney, R. E. (1995). Unconventional myosins. *Annu. Rev. Cell Dev. Biol.* **11**, 633–675.

Moser, H. W. (1987). New approaches in peroxisomal disorders. *Dev. Neurosci.* **9**, 1.

Orci, L., Glick, B. S., and Rothman, J. E. (1986). A new type of coated vesicular carrier that appears not to contain clathrin: Its possible role in protein-transport within the Golgi stack. *Cell (Cambridge, Mass.)* **46**, 171–184.

Palade, G. E. (1975). Intracellular aspects of the process of protein synthesis. *Science* **189**, 347–358.

Pelham, H. R. B. (1995). Sorting and retrieval between the endoplasmic-reticulum and Golgi-apparatus. *Curr. Opin. Cell Biol.* **7**, 530–535.

Peters, A., Palay, S. L., and Webster, H. D. (1991). "The Fine Structure of the Nervous System: Neurons and Their Supporting Cells", 3rd ed. Oxford Univ. Press, New York.

Pley, U., and Parham, P. (1993). Clathrin: Its role in receptor-mediated vesicular transport and specialized functions in neurons. *Crit. Rev. Biochem. Mol. Biol.* **28**, 431–464.

Rachubinski, R. A., and Subramani, S. (1995). How proteins penetrate peroxisomes. *Cell (Cambridge, Mass.)* **83**, 525–528.

Reinsch, S. S., T. J. Mitchison, and M. Kirschner (1991). Microtubule polymer assembly and transport during axonal elongation. *J. Cell Biol.* **115**, 365–380.

Roise, D., Theiler, F., Horvath, S. J., Tomich, J. M., Richards, J. H., Allison, D. S., and Schatz, G. (1988). Amphiphilicity is essential for mitochondrial presequence function. *EMBO J.* **7**, 649–653.

Rothman, J. E. (1994). Mechanism of intracellular protein transport. *Nature (London)* **372**, 55–63.

Rothman, J. E., and Lodish, H. F. (1977). Synchronised transmembrane insertion and glycosylation of a nascent membrane protein. *Nature (London)* **269**, 775–780.

Sabatini, D. D., Kreibich, G., Morimoto, T., and Adesnik, M. (1982). Mechanisms for the incorporation of proteins into membranes and organelles. *J. Cell Biol.* **92**, 1.

Sandoval, I. V., and Bakke, O. (1994). Targeting of membrane proteins to endosomes and lysosomes. *Trends Cell Biol.* **4**, 292–297.

Schatz, G., and Butow, R. A. (1983). How are proteins imported into mitochondria? *Cell (Cambridge, Mass.)* **32**, 316.

Schleyer, M., and Neupert, W. (1985). Transport of proteins into mitochondria: Translocational intermediates spanning contact sites between outer and inner membranes. *Cell (Cambridge, Mass.)* **43**, 339.

Shenoyscaria, A. M., Dietzen, D. J., Kwong, J., Link, D. C., and Lublin, D. M. (1994). Cysteine(3) of src family protein-tyrosine kinases determines palmitoylation and localization in caveolae. *J. Cell Biol.* **126**, 353–363.

Sheterline, P., and Sparrow, J. C. (1994). *Actin. Protein Profile* **1**, 1–121.

Smith, R. S. (1980). The short term accumulation of axonally transported organelles in the region of localized lesions of single myelinated axons. *J. Neurocytol.* **9**, 39–65.

Steward, O. (1995). Targeting of mRNAs to subsynaptic microdomains in dendrites [Review]. *Curr. Opin. Neurobiol.* **5**, 55–61.

Sudhof, T. C. (1995). The synaptic vesicle cycle: A cascade of protein–protein interactions. *Nature (London)* **375**, 645–653.

Tanaka, E. M., and Kirschner, M. (1991). Microtubule behavior in the growth cones of living neurons during axon elongation. *J. Cell Biol.* **115**, 345–364.

Tanaka, J., and Sobue, K. (1994). Localization and characterization of gelsolin in nervous tissues: Gelsolin is specifically enriched in myelin-forming cells. *J. Neurosci.* **14**, 1038–1052.

Tanaka, Y., Zhang, Z., and Hirokawa, N. (1995). Identification and molecular evolution of new dynein-like protein sequences in rat brain. *J. Cell Sci.* **108**, 1883–1893.

Tsukita, S., and Ishikawa, H. (1980). The movement of membranous organelles in axons: Electron microscopic identification of anterogradely and retrogradely transported organelles. *J. Cell Biol.* **84**, 513–530.

Tucker, R. P., Garner, C. C., and Matus, A. (1989). In-situ localization of microtubule-associated protein messenger-RNA in the developing and adult-rat brain. *Neuron* **2**, 1245–1256.

Vallee, R. B. (1993). Molecular analysis of the microtubule motor dynein. *Proc. Natl. Acad. Sci. USA* **90**, 8769–8772.

von Heijne, G. (1985). Structural and thermodynamic aspects of the transfer of proteins into and across membranes. *Curr. Top. Membr. Transp.* **24**, 151.

von Heijne, G. (1986). Mitochondrial targeting sequences may form amphiphilic helices. *EMBO J.* **5**, 1335.

Vouyiouklis, D. A., and Brophy, P. J. (1995). Microtubule-associated proteins in developing oligodendrocytes: Transient expression of a MAP2c isoform in oligodendrocyte precursors. *J. Neurosci. Res.* **42**, 803–817.

Walter, P., and Blobel, G. (1981). Translocation of proteins across the endoplasmic reticulum. III. Signal recognition protein causes signal sequence-dependent and site-specific arrest of chain elongation that is released by microsomal membranes. *J. Cell Biol.* **91**, 557–561.

Walter, P., and Johnson, A. E. (1994). Signal sequence recognition and protein targeting to the endoplasmic-reticulum membrane. *Annu. Rev. Cell Biol.* **10**, 87–119.

Zwizinski, C., and Neupert, W. (1983). Precursor proteins are transported into mitochondria in the absence of proteolytic cleavage of the additional sequences. *J. Biol. Chem.* **258**, 13340.

Brain Energy Metabolism

Pierre J. Magistretti

All the processes described in this textbook require energy. Ample clinical evidence indicates that the brain is exquisitely sensitive to perturbations of energy metabolism. This chapter covers the topics of energy delivery, production, and utilization by the brain. Careful consideration of the basic mechanisms of brain energy metabolism is an essential prerequisite to a full understanding of the physiology and pathophysiology of brain function. The chapter reviews the features of brain energy metabolism at the global, regional, and cellular levels and, at the cellular level, extensively describes recent advances in the understanding of neuron–glia metabolic exchanges. A particular focus is the cellular and molecular mechanisms that tightly couple neuronal activity to energy consumption. This tight coupling is at the basis of functional brain imaging techniques, such as positron emission tomography (PET) and functional magnetic resonance imaging (fMRI).

ENERGY METABOLISM OF THE BRAIN AS A WHOLE ORGAN

Glucose Is the Main Energy Substrate for the Brain

The human brain constitutes only 2% of the body weight, yet the energy-consuming processes that ensure proper brain function account for approximately 25% of total body glucose utilization. With a few exceptions that will be reviewed later, glucose is the obligatory energy substrate of the brain. In any tissue, glucose can follow various metabolic pathways; in the brain, glucose is almost entirely oxidized to CO_2 and water through its sequential processing by glycolysis (Fig. 3.1), the tricarboxylic acid (TCA) cycle (Fig. 3.2), and the associated oxidative phosphorylation, which yield, on a molar basis, 38 ATP per

glucose. Indeed, the oxygen consumption of the brain, which accounts for almost 20% of the oxygen consumption of the whole organism, is 160 µmol per 100 g of brain weight per minute and roughly corresponds to the value determined for CO_2 production. This O_2/CO_2 relation corresponds to what is known in metabolic physiology as a respiratory quotient of nearly 1 and demonstrates that carbohydrates, and glucose in particular, are the exclusive substrates for oxidative metabolism. This rather detailed information of whole-brain energy metabolism was obtained by using an experimental approach in which the concentration of a given substrate in the arterial blood entering the brain through the carotid artery is compared with that present in the venous blood draining the brain through the jugular vein (Kety and Schmidt, 1948). If the substrate is used by the brain, the arteriovenous (A–V) difference is positive; in certain cases, the A–V difference may be negative, indicating that metabolic pathways resulting in the production of the substrate predominate. In addition, when the rate of cerebral blood flow (CBF) is known, the steady-state rate of utilization of the substrate can be determined per unit time and normalized per unit brain weight according to the following relation: *CMR = CBF (A–V)*, where *CMR* is the cerebral metabolic rate of a given substrate. This approach was pioneered by Seymour Kety and C. F. Schmidt in the late 1940s and further developed in the 1950s and 1960s. In normal adults, CBF is approximately 57 ml per 100 g of brain weight per minute, and the calculated glucose utilization by the brain is 31 µmol per 100 g of brain weight per minute, as determined with the A–V difference method (Kety and Schmidt, 1948). This value is slightly higher than that predicted from the rate of oxygen consumption of the brain. Thus, in an organ such as the brain with a respiratory quotient of 1, the stoichiometry would predict that 6 µmol of oxygen is needed to fully oxidize 1 µmol of the six-carbon mole-

Copyright 2004, Elsevier Science (USA).
All rights reserved.

cule of glucose; given an oxygen consumption rate of 160 μmol per 100 g of brain weight per minute, the predicted glucose utilization would be 26.6 μmol per 100 g of brain weight per minute (160 divided by 6), yet the actual measured rate is 31 μmol. What then is the fate of the excess 4.4 μmol? First, glucose metabolism may proceed, to a very limited extent, only through glycolysis, resulting in the production of lactate without oxygen consumption (see Fig. 3.1); glucose can also be incorporated into glycogen (Fig. 3.1). Second, glucose is an essential constituent of macromolecules such as glycolipids and glycoproteins present in neural cells. Finally, glucose enters the metabolic pathways that result in the synthesis of three key neurotransmitters of the brain: glutamate, GABA, and acetylcholine (see Chapter 9).

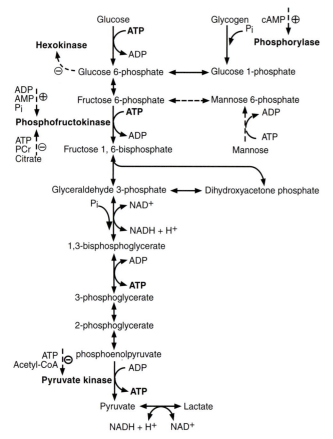

FIGURE 3.1   Glycolysis (Embden–Meyerhof pathway). Glucose phosphorylation is regulated by hexokinase, an enzyme inhibited by glucose 6-phosphate. Glucose must be phosphorylated to glucose 6-phosphate to enter glycolysis or to be stored as glycogen. Two other important steps in the regulation of glycolysis are catalyzed by phosphofructokinase and pyruvate kinase. Their activity is controlled by the levels of high-energy phosphates as well as of citrate and acetyl-CoA. Pyruvate, through lactate dehydrogenase, is in dynamic equilibrium with lactate. This reaction is essential to regenerate NAD+ residues necessary to sustain glycolysis downstream of glyceraldehyde 3-phosphate. PCr, phosphocreatine.

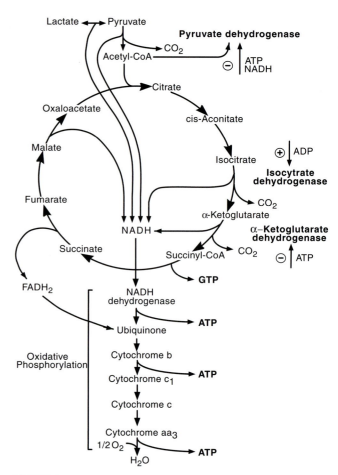

FIGURE 3.2   Tricarboxylic acid cycle (Krebs cycle) and oxidative phosphorylation. Pyruvate entry into the cycle is controlled by pyruvate dehydrogenase activity that is inhibited by ATP and NADH. Two other regulatory steps in the cycle are controlled by isocitrate and α-ketoglutarate dehydrogenases, the activity of which is controlled by the levels of high-energy phosphates.

Ketone Bodies Become Adequate Energy Substrates for the Brain in Particular Circumstances

In particular circumstances, substrates other than glucose can be used by the brain. For example, breast-fed neonates have the capacity to use the ketone bodies acetoacetate (AcAc) and D-3-hydroxybutyrate (3-HB), in addition to glucose, as energy substrates for the brain (Girard *et al.*, 1992). This capacity is an interesting example of a developmentally regulated adaptive mechanism, because maternal milk is highly enriched in lipids, resulting in a lipid-to-carbohydrate ratio much higher than that present in postweaning nutrients. Indeed, lipids account for approximately 55% of the total calories contained in human milk, in contrast to 30–35% for a balanced postweaning diet. In addition to the ketone bodies AcAc and 3-HB,

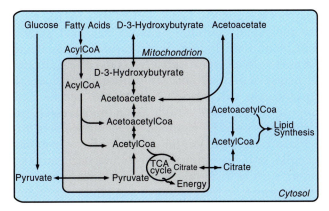

FIGURE 3.3 Relationship between lipid metabolism and the TCA cycle. Under particular dietary conditions, such as lactation in newborns and fasting in adults, the ketone bodies acetoacetate and D-3-hydroxybutyrate and circulating fatty acids can provide substrates to the TCA cycle after conversion into acetyl-CoA. Carbon atoms for lipid synthesis can be provided by glucose through citrate produced in the TCA cycle, a particularly relevant process for the developing brain.

other products of lipid metabolism relevant to brain metabolic processes are free fatty acids. Acetoacetate, 3-HB, and free fatty acids can all be processed to acetyl-CoA, thus providing ATP through the TCA cycle (Fig. 3.3). As will be described later, brain energy metabolism is highly compartmentalized, with certain metabolic pathways specifically localized in a given cell type. It is therefore not surprising that, whereas ketone bodies can be oxidized by neurons, oligodendrocytes, and astrocytes, the β-oxidation of free fatty acids is localized exclusively in astrocytes (Auestad *et al.*, 1991). Another consideration regarding the lipid-rich diet provided during the suckling period relates to its contribution to the process of myelination. The question is whether the polar lipids and cholesterol that make up myelin are derived from dietary sources or are synthesized within the brain. There is evidence that brain lipids can be synthesized from bloodborne precursors such as ketone bodies (Nehlig and Pereira de Vasconcelos, 1993). In addition, when suckling rats are fed a diet low in ketones, carbon atoms for lipogenesis can also be provided by glucose. To summarize, ketone bodies and AcAc are adequate energy substrates as well as precursors for lipogenesis during the suckling period; however, the developing brain appears to be metabolically quite flexible because glucose, in addition to its energetic function, can be metabolized to generate substrates for lipid synthesis.

Starvation and diabetes are two situations in which the availability of glucose to tissues is inadequate and in which plasma ketone bodies are elevated because of enhanced lipid catabolism. Under these conditions, the adaptive mechanisms described for breastfed neonates become operative in the brain, allowing it to use AcAc or 3-HB as energy substrates (Owen *et al.*, 1967).

Mannose, Lactate, and Pyruvate Serve as Instructive Cases

A number of metabolic intermediates have been tested as alternative substrates to glucose for brain energy metabolism. Among the numerous molecules tested, mannose is the only one that can sustain normal brain function in the absence of glucose. Mannose readily crosses the blood–brain barrier and, in two enzymatic steps, is converted into fructose 6-phosphate, an intermediate of the glycolytic pathway (see Fig. 3.1). However, mannose is not normally present in the blood and, therefore, is not considered a physiological substrate for brain energy metabolism.

Lactate and pyruvate can be sources of insight into the intrinsic properties of isolated brain tissue versus those of the brain as an organ receiving substrates from the circulation. Lactate and pyruvate can adequately sustain the synaptic activity of isolated brain samples, usually thin slices, maintained *in vitro* in a physiological medium lacking glucose (Schurr *et al.*, 1999). *In vivo*, until recently, it was thought that their permeability across the blood–brain barrier was limited, hence preventing circulating lactate or pyruvate from substituting for glucose to adequately maintain brain function. However, recent evidence using magnetic resonance spectroscopy (MRS) indicates that the permeability of circulating lactate across the blood–brain barrier may actually be higher than previously thought (Hassel and Bratte, 2000). In addition, monocarboxylate transporters are found on intraparenchymal brain capillaries (Pierre *et al.*, 2000). Thus, there is a need for the reappraisal of the use by the brain of monocarboxylates. However, if formed within the brain parenchyma from glucose that has crossed the blood–brain barrier, lactate and pyruvate may in fact become the preferential energy substrates for activated neurons (see below).

Summary

Glucose is the obligatory energy substrate for brain, and it is almost entirely oxidized to CO_2 and H_2O. This simple statement summarizes, with few exceptions, more than four decades of careful studies of brain energy metabolism at the organ and regional levels. Under ketogenic conditions, such as starvation and diabetes and during breastfeeding, ketone bodies may provide an adequate energy source for the brain. Lactate and pyruvate, formed from glucose within the brain parenchyma, are adequate energy substrates as well.

TIGHT COUPLING OF NEURONAL ACTIVITY, BLOOD FLOW, AND ENERGY METABOLISM

A striking characteristic of the brain is its high degree of structural and functional specialization. Thus, when we move an arm, motor areas and their related pathways are selectively activated; intuitively, one can predict that as "brain work" increases locally (e.g., in motor areas), the energy requirements of the activated regions will increase in a temporally and spatially coordinated manner. Because energy substrates are provided through the circulation, blood flow should increase in the modality-specific activated area. More than a century ago, the British neurophysiologist Charles Sherrington showed, in experimental animals, increases in blood flow localized to the parietal cortex in response to sensory stimulation (Roy and Sherrington, 1890). He postulated that "the brain possesses intrinsic mechanisms by which its vascular supply can be varied locally in correspondence with local variations of functional activity." With remarkable insight, he also proposed that "chemical products of cerebral metabolism" produced in the course of neuronal activation could provide the mechanism to couple activity with increased blood flow.

Which Mechanisms Couple Neuronal Activity to Blood Flow?

Since Sherrington's seminal work, the search for the identification of chemical mediators that can couple neuronal activity with local increases in blood flow has been intense. These signals can be broadly grouped into two categories: (1) molecules or ions that transiently accumulate in the extracellular space after neuronal activity and (2) specific neurotransmitters that mediate the coupling in anticipation or at least in parallel with local activation (neurogenic mechanisms). The increases in extracellular K^+, adenosine, and lactate and the related changes in pH are all a consequence of increased neuronal activity, and all have been considered mediators of neurovascular coupling because of their vasoactive effects (Villringer and Dirnagl, 1995). However, the spatial and temporal resolution achieved by these mediators

may not be sufficient to account entirely for the activity-dependent coupling between neuronal activity and blood flow. Indeed, these vasoactive agents are formed with a certain delay (seconds) after initiation of neuronal activity and can diffuse at considerable distance. In this respect, neurogenic mechanisms appear to be better fitted. Brain microvessels are richly innervated by neuronal fibers. These fibers may have an extrinsic origin (e.g., in the autonomic ganglia) or be part of neuronal circuits intrinsic to the brain, such as local interneurons or long projections that originate in the brainstem (e.g., those containing monoaminergic neurotransmitters) (see Chapter 9). In addition, functional receptors coupled to signal transduction pathways have been identified for several neurotransmitters on intraparenchymal microvessels. The neurotransmitters with potential roles in coupling neuronal activity with blood flow include the amines noradrenaline, serotonin, and acetylcholine and the peptides vasoactive intestinal peptide, neuropeptide Y (NPY), calcitonin gene-related peptide (CGRP), and substance P (SP). The neurogenic mode of neurovascular coupling implies that vasoactive neurotransmitters are released from perivascular fibers as excitatory afferent volleys activate a discrete and functionally defined brain volume.

A recent and very attractive addition to the list of potential mediators for coupling neuronal activity to blood flow is nitric oxide (NO) (see also Chapter 10). Indeed, NO is an ideal candidate; it is formed locally by neurons and glial cells under the action of a variety of neurotransmitters likely to be released by depolarized afferents to an activated brain area (Iadecola *et al.*, 1994). Nitric oxide is a diffusible and potent vasodilator whose short half-life spatially and temporally restricts its domain of action. However, in several experimental models in which the activity of NO synthase, the enzyme responsible for NO synthesis, was inhibited, a certain degree of coupling was still observed, indicating that NO is probably only one of the regulators of local blood flow acting in synergy with others (Iadecola *et al.*, 1994).

In summary, several products of activity-dependent neuronal and glial metabolism such as lactate, H^+, adenosine, and K^+ have vasoactive effects and are therefore putative mediators of coupling, although the kinetics and spatial resolution of this mode do not account for all the observed phenomena. As attractive as it is, an exclusively neurogenic mode of coupling neuronal activity to blood flow is unlikely and, moreover, still awaits firm functional confirmation *in vivo*. Nitric oxide is undoubtedly a key element in coupling, particularly in view of the fact that glutamate, the principal excitatory neurotransmitter, triggers receptor-mediated NO formation in neurons and glia; this is consistent with the view that whenever a functionally defined brain area is activated and glutamate is released by the depolarized afferents, NO may be formed, thus providing a direct mechanism contributing to the coupling between activity and local increases in blood flow.

Through the activity-linked increase in blood flow, more substrates, namely glucose and oxygen, necessary to meet the additional energy demands are delivered to the activated area per unit time. The cellular and molecular mechanisms involved in oxygen consumption and glucose utilization are treated in a later section.

Blood Flow and Energy Metabolism Can Be Visualized in Humans

Modern functional brain imaging techniques enable the *in vivo* monitoring of human blood flow and the two indices of energy metabolism, namely glucose utilization and oxygen consumption (Box 3.1). For instance, with the use of positron emission tomography (PET) and appropriate positron-emitting isotopes such as ^{18}F and ^{15}O, basal rates as well as activity-related changes in local blood flow or oxygen consumption can be studied by using ^{15}O-labeled water or ^{15}O, respectively (Frackowiak *et al.*, 1980). Local rates of glucose utilization (also defined as local

---

BOX 3.1

PET AND FMRI

We have seen that neuronal activity is tightly coupled to blood flow and metabolism. With the advent of sophisticated imaging procedures such as positron emission tomography (PET) and functional magnetic resonance imaging (fMRI), it is now possible to detect the signals generated by the metabolic processes associated with neuronal activity, thus providing a unique opportunity to see the "brain at work." Indeed, local changes in blood flow, glucose utilization, and oxygen consumption can be noninvasively monitored under basal and activated conditions in human subjects. How is this possible?

For PET, a solution containing slightly radioactive molecules is injected into the circulation, and its sites of brain uptake can be visualized. The molecule is radioactively labeled with an unstable radionuclide possessing an excess number of protons; as a consequence of normal radioactive decay, the excess proton is converted into a neutron. In this process, a positron (a positively charged electron) is emitted and collides with an electron, releasing energy in the form of two photons with opposite trajectories. The photons are sensed by specialized detectors placed around the head; when two photons simultaneously reach two detectors positioned at 180° of each other, the origin of the positron–electron collision can be localized with a resolution of 5 to 10 mm. Commonly used positron-emitting radionuclides are oxygen-15 (^{15}O), carbon-11 (^{11}C), and fluorine-18 (^{18}F). Blood flow is monitored with ^{15}O-labeled water and glucose utilization with ^{18}F-2-deoxyglucose. Oxygen consumption is visualized directly with ^{15}O. With the use of sophisticated algorithms to process the data, the localization of activity

to specific brain areas during a given task (sensory, motor, cognitive) can be achieved. Thus, the activity of neuronal ensembles, and of the associated glia, coupled to increased blood flow and glucose utilization results in a localized signal due to the augmented concentration of ^{15}O-labeled water (monitoring blood flow) and ^{18}F-2-deoxyglucose (assessing glucose utilization) in the activated area.

PET is one of the ways in which brain work can be visualized. An increasingly popular technique, fMRI relies on the magnetic signals detected in an activated brain region in relation to its degree of oxygenation. Depending on the degree to which it is saturated by oxygen, hemoglobin (by acting as a paramagnetic contrast agent) can alter the magnetic signal detected in a tissue exposed to the magnetic fields used for structural MRI. In other words, different MRI signals can be obtained depending on the oxyhemoglobin/deoxyhemoglobin ratio in a given brain area. Local activation of a brain area results in increased blood flow. Although the precise mechanisms are still being discussed, it is currently thought that this phenomenon, by leading to a localized enrichment in oxyhemoglobin, alters the oxyhemoglobin/deoxyhemoglobin ratio, providing the signal for fMRI. Functional MRI is a remarkably convenient and powerful technique: the signal acquisition time is extremely rapid (on the order of seconds) and the resolution equals that of PET (i.e., a few millimeters). In addition, fMRI is totally noninvasive and can thus be frequently repeated on the same subject, who can then serve as its own control.

cerebral metabolic rates for glucose, or LCMRglu) can be determined with ^{18}F-labeled 2-deoxyglucose (2-DG) (Phelps *et al.*, 1979). The use of 2-DG as a marker of LCMRglu was pioneered by Louis Sokoloff and his associates at the National Institutes of Health, first in laboratory animals (Sokoloff, 1981). The method is based on the fact that 2-DG crosses the blood–brain barrier, is taken up by brain cells, and is phosphorylated by hexokinase with kinetics similar to those for glucose; however, unlike glucose 6-phosphate, 2-deoxyglucose 6-phosphate cannot be metabolized further and therefore accumulates intracellularly (Fig. 3.4).

For studies in laboratory animals, tracer amounts of radioactive 2-DG are injected intravenously; the animal is subjected to the behavioral paradigms of interest and sacrificed at the end of the experiment. Serial thin sections of the brain are prepared and

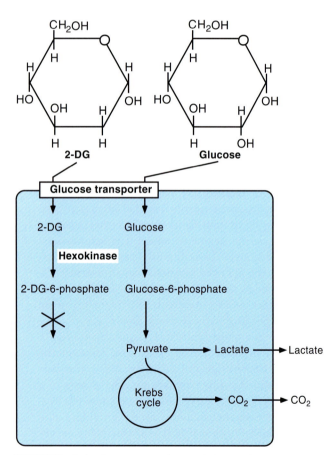

FIGURE 3.4 Structure and metabolism of glucose and 2-deoxyglucose. 2-Deoxyglucose (2-DG) is transported into cells through glucose transporters and phosphorylated by hexokinase to glucose 6-phosphate without significant further processing or dephosphorylation back to glucose. Therefore, when radioactively labeled, 2-DG used in tracer concentrations is a valuable marker of glucose uptake and phosphorylation, which directly indicates glucose utilization.

processed for autoradiography. This autoradiographic method provides, after appropriate corrections, an accurate measurement of LCMRglu with a spatial resolution of approximately 50–100 μm (Sokoloff *et al.*, 1977). Using this method, researchers have determined LCMRglu in virtually all structurally and functionally defined brain structures in various physiological and pathological states, including sleep, seizures, and dehydration, and after a variety of pharmacological treatments (Sokoloff, 1981). Furthermore, glucose utilization increases in the pertinent brain areas during motor tasks (Roland, 1985) or activation of pathways subserving specific modalities, such as visual, auditory, olfactory, or somatosensory stimulation (Sokoloff, 1981). For example, in mice, sustained stimulation of the whiskers results in marked increases in LCMRglu in discrete areas of the primary sensory cortex called the barrel fields, where each whisker is represented with an extreme degree of topographical specificity (Melzer *et al.*, 1985). Basal glucose utilization of the gray matter as determined by 2-DG autoradiography varies, depending on the brain structure, between 50 and 150 μmol per 100 g wet weight per minute in the rat.

In humans, LCMRglu determined by PET with the use of ^{18}F–2-DG is approximately 50% lower than that in rodents, and physiological activation of specific modalities increases LCMRglu in discrete areas of the brain that can be visualized with a spatial resolution of a few millimeters. For example, visual stimulations presented to subjects as checkerboard patterns reversing at frequencies ranging from 2 to 10 Hz selectively increase LCMRglu in the primary visual cortex and a few connected cortical areas. With the use of this stimulation paradigm, the combined PET analysis of local cerebral blood flow (LCBF) and local oxygen consumption (LCMRO$_2$), in addition to LCMRglu, has revealed a unique and unexpected feature of the regulation of human brain energy metabolism. The canonical view was that the three metabolic parameters were tightly coupled, implying that, if, for example, CBF increased locally during physiological activation, LCMRglu and LCMRO$_2$ would increase in parallel. In what is now referred to as the phenomenon of "uncoupling," physiological stimulation of the visual system increases LCBF and LCMRglu (both by 30–40%) in the primary visual cortex without a commensurate increase in LCMRO$_2$ (which increases only 6%) (Fox *et al.*, 1988), indicating that the additional glucose used during neuronal activation can be processed through glycolysis rather than through the TCA cycle and oxidative phosphorylation (Gusnard and Raichle, 2001). The phenomenon of uncoupling has been confirmed in other cortical areas, although its magnitude may differ depending on the modality, and may actually be absent in certain cases (Vafaee *et al.*, 1999). A glance at the metabolic pathways reveals that if glucose does not enter the TCA cycle to be oxidized, then lactate will be produced (see Figs. 3.1 and 3.2). Lactate, like several other metabolically relevant molecules, can be determined with the technique of magnetic resonance imaging (MRI) spectroscopy for ^{1}H, which provides a means of unequivocally identifying in living tissues the presence of molecules that bear the naturally occurring isotope ^{1}H. Consistent with the prediction that if during activation glucose is processed predominantly glycolytically, then lactate should be produced locally in the activated region, a transient increase in the lactate signal is detected with ^{1}H MRI spectroscopy in the human primary visual cortex during appropriate visual stimulation (Prichard *et al.*, 1991). These observations support the view that to face the local increases in energy demands linked to neuronal activation, the brain transiently resorts to glycolysis followed by at least oxidative phosphorylation (Magistretti and Pellerin, 1999). This transient uncoupling may vary in amplitude depending on the modalities of activation and is likely to occur in different cellular compartments, i.e., astrocytes versus neurons (Frackowiak *et al.*, 2001).

Summary

Studies at the whole-organ level, based on the A–V differences of metabolic substrates, have revealed a great deal about the global energy metabolism of the brain. They have indicated that, under normal conditions, glucose is virtually the sole energy substrate for the brain and that it is entirely oxidized. New techniques that allow imaging of the three fundamental parameters of brain energy metabolism, namely, blood flow, oxygen consumption, and glucose utilization, provide a more refined level of spatial resolution and demonstrate that brain energy metabolism is regionally heterogeneous and is tightly coupled to the functional activation of specific neuronal pathways (Magistretti *et al.*, 1999).

ENERGY-PRODUCING AND ENERGY-CONSUMING PROCESSES IN THE BRAIN

What are the cellular and molecular mechanisms that underlie the regulation of brain energy metabolism revealed by the foregoing studies at the global and regional levels? In particular, what are the meta-

bolic events taking place in the cell types that make up the brain parenchyma? How is it possible to reconcile whole-organ studies indicating complete oxidation of glucose with transient activation-induced glycolysis at the regional level? These and other related questions are addressed here and in the next sections.

Glucose Metabolism Produces Energy

Before describing an analysis of the cell-specific mechanisms of brain energy metabolism, it seems appropriate to briefly review some basic aspects of the brain's energy balance. Because glucose, in normal circumstances, is the main energy substrate of the brain, the overview is restricted to its metabolic pathways. Glucose metabolism in the brain is similar to that in other tissues and includes three principal metabolic pathways: glycolysis, the tricarboxylic acid cycle, and the pentose phosphate pathway. Because of the global similarities with other tissues, these pathways are simply summarized in Figs. 3.1, 3.2, and 3.5, and only a few aspects specific to the nervous tissue are discussed.

Glycolysis

Glycolysis (Embden–Meyerhof pathway) is the metabolism of glucose to pyruvate (see Fig. 3.1). It results in the net production of only two molecules of ATP per glucose molecule; indeed, four ATP are formed in the processing of glucose to pyruvate, whereas two ATP are consumed to phosphorylate glucose to glucose 6-phosphate and fructose 6-phosphate to fructose 1,6-bisphosphate, respectively (see Fig. 3.1). Under anaerobic conditions, pyruvate is converted into lactate, allowing the regeneration of nicotinamide adenine dinucleotide (NAD$^+$), which is essential to maintain a continued glycolytic flux. Indeed, if NAD$^+$ were not regenerated, glycolysis could not proceed beyond glyceraldehyde 3-phosphate (see Fig. 3.1). Another situation in which the end product of glycolysis is lactate rather than pyruvate is when oxygen consumption does not match glucose utilization, implying that the rate of pyruvate production through glycolysis exceeds pyruvate oxidation by the TCA cycle (see Fig. 3.2). This condition has been well described in skeletal muscle during intense exercise and appears to share similarities with the transient uncoupling observed between glucose utilization and oxygen consumption that has been described in the human cerebral cortex during activation with the use of PET (Fox *et al.*, 1988).

The Tricarboxylic Acid Cycle

Under aerobic conditions, pyruvate is oxidatively decarboxylated to yield acetyl-CoA, in a reaction cat-

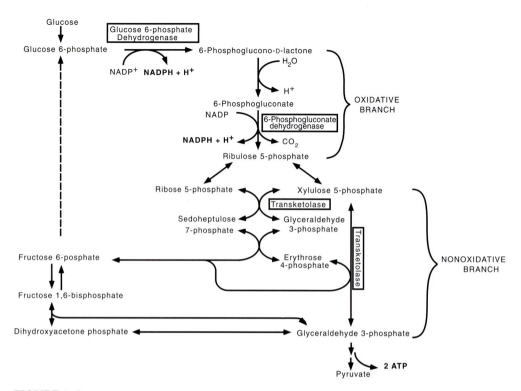

FIGURE 3.5 Pentose phosphate pathway. In the oxidative branch of the pentose phosphate pathway, two NADPH are generated per glucose 6-phosphate. The first, rate-limiting reaction of the pathway is catalyzed by glucose-6-phosphate dehydrogenase; the second NADPH is generated through the oxidative decarboxylation of 6-phosphogluconate, a reaction catalyzed by glucose-6-phosphogluconate dehydrogenase. The nonoxidative branch of the pentose phosphate pathway provides a reversible link with glycolysis, by regenerating the two glycolytic intermediates glyceraldehyde 3-phosphate and fructose 6-phosphate. This regeneration is achieved through three sequential reactions. In the first, catalyzed by transketolase, xylulose 5-phosphate and ribose 5-phosphate (which originate from ribulose 5-phosphate, the end product of the oxidative branch) yield glyceraldehyde 3-phosphate and sedoheptulose 7-phosphate. Under the action of transaldolase, these two intermediates yield fructose 6-phosphate and erythrose 4-phosphate. The latter intermediate combines with glyceraldehyde 3-phosphate, in a reaction catalyzed by transketolase, to yield fructose 6-phosphate and glyceraldehyde 3-phosphate. Thus, through the nonoxidative branch of the pentose phosphate pathway, two hexoses (fructose 6-phosphate) and one triose (glyceraldehyde 3-phosphate) of the glycolytic pathway are regenerated from three pentoses (ribulose 5-phosphate).

alyzed by the enzyme pyruvate dehydrogenase (PDH). Acetyl-coenzyme A condenses with oxaloacetate to produce citrate (see Fig. 3.2). This is the first step of the tricarboxylic acid cycle, in which three pairs of electrons are transferred from NAD$^+$ to NADH—and one pair from flavin adenine dinucleotide (FAD) to its reduced form (FADH$_2$)—through four oxidation–reduction steps (see Fig. 3.2). NADH and FADH$_2$ transfer their electrons to molecular O$_2$ through the mitochondrial electron transfer chain to produce ATP in the process of oxidative phosphorylation. Thus, under aerobic conditions (i.e., when glucose is fully oxidized through the TCA cycle to CO$_2$ and H$_2$O), NAD$^+$ is regenerated, and glycolysis proceeds to pyruvate, not lactate. However, as soon as a mismatch, even a transient one, occurs between glucose utilization and oxygen consumption, lactate is produced. As discussed earlier, such transient production of lactate appears to occur in the human brain during activation. Experiments performed in freely moving rats also have demonstrated a transient increase in lactate content in the extracellular space of discrete brain regions during physiological sensory stimulation (Fellows *et al.*, 1993). In these experiments, lactate was determined in the extracellular fluid collected by microdialysis.

The Pentose Phosphate Pathway

Although glycolysis, the TCA cycle, and oxidative phosphorylation are coordinated pathways that produce ATP, using glucose as a fuel, ATP is not the only form of metabolic energy. Indeed, for several biosynthetic reactions in which the precursors are in a more oxidized state than the products, metabolic energy in the form of reducing power is needed in addition to ATP. This is the case for the reductive synthesis of free fatty acids from acetyl-CoA, which are components of myelin and of other structural elements of neural cells, such as the plasma membrane. In cells of the brain, as in other organs, the reducing power is provided by the reduced form of nicotinamide adenine dinucleotide phosphate (NADPH). The processing of glucose through the pentose phosphate pathway produces NADPH. The first reaction in the pentose phosphate pathway is the conversion of glucose 6-phosphate into ribulose 5-phosphate (Fig. 3.5). This dehydrogenation, in which two molecules of NADPH are generated per molecule of glucose 6-phosphate, is the rate-limiting step of the pentose phosphate pathway. The NADP/NADPH ratio is the single most important factor regulating the entry of glucose 6-phosphate into the pentose phosphate pathway. Thus, if high reducing power is needed, NADPH levels decrease and the pentose phosphate pathway is activated to generate new reducing equivalents. In addition to reductive biosynthesis, NADPH is needed for the scavenging of reactive oxygen species (ROS). The superoxide anion, hydrogen peroxide, and the hydroxy radical are three ROS, generated by the transfer of single electrons to molecular oxygen as by-products of certain physiological cellular processes (Halliwell, 1992). Examples of such processes are the electron transfer chain associated with oxidative phosphorylation and the activities of monoamine oxidase, tyrosine hydroxylase, nitric oxide synthase, and the eicosanoid-forming enzymes lipoxygenases and cyclooxygenases. Reactive oxygen species are highly damaging to cells because they can cause DNA disruption and mutations, as well as activation of enzymatic cascades including proteases and lipases that can eventually lead to cell death (Greenlund *et al.*, 1995).

Scavenging of ROS is ensured by the sequential action of superoxide dismutase (SOD) and glutathione peroxidase (Ben-Yoseph *et al.*, 1994) (Fig. 3.6). Thus, two superoxide anions formed by the aforementioned cellular processes are converted by SOD into H_2O_2, still a ROS. Glutathione peroxidase converts H_2O_2 into H_2O and O_2 at the expense of reduced glutathione, which is regenerated by glutathione reductase in the presence of NADPH. In addition to the scavenging mechanisms for ROS, the pentose phosphate pathway is also tightly connected to glycolysis through two enzymes, transketolase and transaldolase, which recycle ribulose 5-phosphate to fructose 6-phosphate and glyceraldehyde 3-phosphate, two intermediates of glycolysis (see Fig. 3.5).

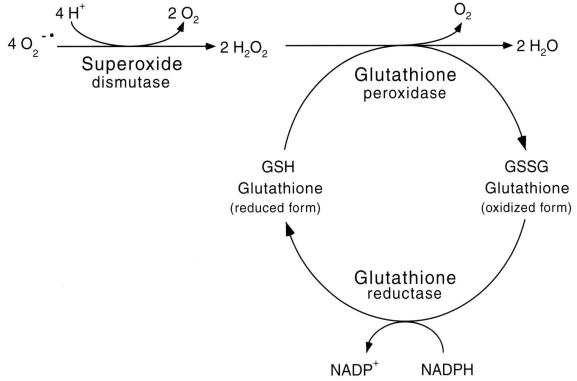

FIGURE 3.6 Enzymatic reactions for scavenging reactive oxygen species (ROS). The toxic superoxide anion (O_2^-) formed by a variety of physiological reactions, including oxidative phosphorylation, is scavenged by superoxide dismutase, which converts the superoxide anion into hydrogen peroxide (H_2O_2) and molecular oxygen. Glutathione peroxidase converts the still toxic hydrogen peroxide into water; reduced glutathione (GSH) is required for this reaction, in which it is converted into its oxidized form (GSSG). GSH is regenerated through the action of glutathione reductase, a reaction requiring NADPH.

The Wernicke–Korsakoff Syndrome: A Neuropsychiatric Disorder Caused by a Dysfunction of Energy Metabolism

A well-characterized neuropsychiatric disorder, the Wernicke–Korsakoff syndrome, is caused by transketolase hypoactivity (Blass and Gibson, 1977). The Wernicke–Korsakoff syndrome is characterized by a severe impairment of memory and of other cognitive processes accompanied by balance and gait dysfunction and by paralysis of oculomotor muscles. The syndrome is due to a lack of thiamine (vitamin B_1) in the diet; it affects only susceptible persons who are also alcoholics or chronically undernourished. Thiamine pyrophosphate is a thiamine-containing cofactor essential for the activity of transketolase. In patients with the Wernicke–Korsakoff syndrome, thiamine pyrophosphate binds 10 times less avidly to transketolase than to the enzyme of normal persons. This enzymatic dysfunction renders patients with the Wernicke–Korsakoff syndrome much more vulnerable to thiamine deficiency. This syndrome illustrates the ways in which an anomaly in a discrete metabolic pathway of energy metabolism may result in severe alterations in behavior and motor function.

The Processes Linked to Neuronal Function Consume Energy

The main energy-consuming process of the brain is the maintenance of ionic gradients across the plasma membrane, a condition that is crucial for excitability. Maintenance of these gradients is achieved predominantly through the activity of ionic pumps fueled by ATP, particularly Na^+,K^+-ATPase, localized in neurons as well as in other cell types such as glia. Activity of these pumps accounts for approximately 50% of basal glucose oxidation in the nervous system (Erecinska and Dagani, 1990). Very recently elegant theoretical calculations of the cost of synaptic transmission have been provided by Attwell and Laughlin (2001). They estimated the energy budget of an average glutamatergic pyramidal neuron firing at 4 Hz, with the assumption that > 80% of cortical neurons are pyramidal cells and that > 90% of the synapses release glutamate. First, the cost of the recycling of released glutamate via reuptake and metabolism in astrocytes and the restoration of the postsynaptic ion gradient has been estimated. Glutamate recycling requires 2.67 ATP/glutamate molecule; as one vesicle contains 4×10^3 molecules of glutamate, the cost of transmitter recycling is ~ 1.1×10^4 ATP/vesicle. The restoration of postsynaptic ionic gradients disrupted by activity of NMDA and non-NMDA receptors is ~ 1.4×10^5 ATP/vesicle, giving a total of

1.51×10^5 ATP/vesicle. Estimating the total number of synapses formed by a single pyramidal neuron at 8×10^3 (Braitenberg and Schüz, 1998) and a firing rate of 4 Hz (implying a 1:4 chance that an active potential releases one vesicle) yields the figure of 3.2×10^8 ATP/action potential/ neuron.

Contrary to previous estimates based on measurement of heat production in peripheral *unmyelinated* nerves (Creutzfeld, 1975), the cost of action potential propagation is rather elevated. Thus, if an action potential is considered to actively depolarize the cell body and axons by 100 mV and passively depolarize the dendrites by 50 mV, the calculation yields a value of 3.8×10^8 ATP/neuron. This calculation is based on the estimate of the minimal Na^+ influx required to depolarize the cell (Attwell and Laughlin, 2001). If calculations also include Ca^{2+}-mediated depolarization of dendrites, the cost is increased by 7%. Remember that these energetic costs are due to the activation of ATPases needed to restore ion gradients. Thus, the overall cost of synaptic transmission plus action potential propagation for a pyramidal neuron firing at 4 Hz would be 2.8×10^9 ATP/neuron/s. The basal energy consumption for maintenance of the resting potential based on the estimates of input resistance, reversal potential, and membrane conductance yields values of 3.4×10^8 ATP/cell/s for neurons and 1×10^8 ATP/cell/s for glia and, thus, a combined consumption of 3.4×10^9 ATP/cell/s assuming a 1:1 ratio between neurons and glia (Haug, 1987). On the basis of this calculation one can conclude that ~ 87% of total energy consumed reflects activity of glutamate-mediated neurotransmission and 13% reflects the energy requirements of resting potential maintenance (Fig. 3.7). This value is in remarkable agreement with the estimates made *in vivo* using magnetic resonance spectroscopy (Sibson *et al.*, 1998). If the total energy consumption per neuron and the associated glia is compounded per gram of tissue per minute (the conventional form for expressing glucose utilization) the figure obtained is 30 μmol ATP/g/min, a value that is very close to that determined *in vivo* for brain glucose utilization, i.e., 30–50 μmol ATP/g/min (Clarke and Sokoloff, 1999).

In addition to the maintenance of ionic gradients that are disrupted during activity, other energy-consuming processes exist in neurons. Thus, the permanent synthesis of molecules needed for communications, such as neurotransmitters, or for general cellular purposes consumes energy (Siesjo, 1978). Axonal transport of molecules synthesized in the nucleus to their final destination along the axon or at the axon terminal is yet another process fueled by cellular energy metabolism (Siesjo, 1978).

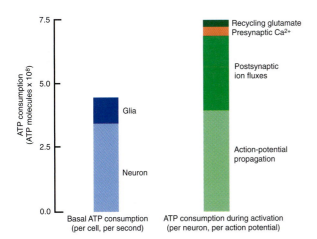

FIGURE 3.7 Energy budget for the rodent central cortex (Attwell and Laughlin, 2001). Relative rates of ATP consumption by resting neurons and glia (left). Relative cost of the various processes associated with a firing rate of 4 Hz for a glutamatergic pyramidal neuron. Modified from Laughlin and Attwell in Frackowiak *et al.* (2001).

Summary

Exactly as in other tissues, the metabolism of glucose, the main energy substrate of the brain, produces two forms of energy: ATP and NADPH. Glycolysis and the TCA cycle produce ATP, whereas energy in the form of reducing equivalents stored in the NADPH molecule is predominantly produced through the pentose phosphate pathway. Maintenance of the electrochemical gradients, particularly for Na^+ and K^+, needed for electrical signaling via the action potential and for chemical signaling through synaptic transmission, is the main energy-consuming process of neural cells.

BRAIN ENERGY METABOLISM AT THE CELLULAR LEVEL

Glia and Vascular Endothelial Cells, in Addition to Neurons, Contribute to Brain Energy Metabolism

Neurons exist in a variety of sizes and shapes and express a large spectrum of firing properties (see Chapters 5 and 7). These differences are likely to imply specific energy demands. For example, large pyramidal cells in the primary motor cortex, which must maintain energy-consuming processes such as ion pumping over a large membrane surface or axonal transport along several centimeters, have considerably larger energy requirements than do local interneurons. However, it is now clear that the other cell types of the nervous system, namely glia and vascular endothelial cells, not only consume energy but also play a crucial role in the flux of energy substrates to neurons. The arguments for such an active role for nonneuronal cells, in particular, glia, are both quantitative and qualitative. Glial cells make up approximately half of the brain volume (O'Kusky and Colonnier, 1982). A conservative figure is a 1:1 ratio between the number of astrocytes, one of the predominant glial cell types (see Chapter 1), and the number of neurons (Haug, 1987). Higher ratios have been described, depending on the region, developmental age, or species. Indeed, the astrocyte-to-neuron ratio increases with the size of the brain and is thus high in humans. It is therefore clear that glucose reaching the brain parenchyma provides energy substrates to a variety of cell types, only some of which are neurons.

Even more compelling for the realization of the key role that astrocytes play in providing energy substrates to active neurons are the cytological relations that exist between brain capillaries, astrocytes, and neurons. These relations, which are illustrated in Fig. 3.8, are as follows. First, through specialized processes, called end feet, astrocytes surround brain capillaries (Kacem *et al.*, 1998). This implies that astrocytes form the first cellular barrier that glucose entering the brain parenchyma encounters and makes them a likely site of prevalent glucose uptake and energy substrate distribution. More than a century ago, the Italian histologist Camillo Golgi and his pupil Luigi Sala sketched such a principle. A lucid formulation of it was presented by the British neuropathologist W. L. Andriezen in an article describing the features of the perivascular glia (Andriezen, 1893): "The development of a felted sheath of neuroglia fibers in the ground-substance immediately surrounding the blood vessels of the Brain seems therefore . . . to allow the free passage of lymph and metabolic products which enter into the fluid and general metabolism of the nerve cells." In addition to perivascular end feet, astrocytes bear processes that ensheathe synaptic contacts. Astrocytes also express receptors and uptake sites with which neurotransmitters released during synaptic activity can interact (Barres, 1991). These features endow astrocytes with an exquisite sensitivity to detect increases in synaptic activity. In summary, because of the foregoing structural and functional characteristics, astrocytes are ideally suited to couple local changes in neuronal activity with coordinated adaptations in energy metabolism (see Fig. 3.8).

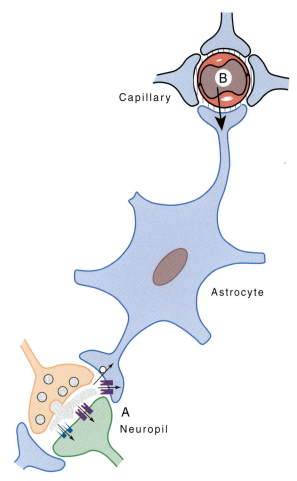

Capillary

Astrocyte

A
Neuropil

FIGURE 3.8 Schematic representation of the cytological relations existing between intraparenchymal capillaries, astrocytes, and the neuropil. Astrocyte processes surround capillaries (endfeet) and ensheathe synapses; in addition, receptors and uptake sites for neurotransmitters are present on astrocytes. These features make astrocytes ideally suited to sense synaptic activity (A) and to couple it with uptake and metabolism of energy substrates originating from the circulation (B).

A Tightly Regulated Glucose Metabolism Occurs in All Cell Types of the Brain, Neuronal and Nonneuronal

Given the high degree of cellular heterogeneity of the brain, understanding the relative role played by each cell type in the flux of energy substrates has depended largely on the availability of purified preparations such as primary cultures enriched in neurons, astrocytes, or vascular endothelial cells. Such preparations have some drawbacks, because they may not necessarily express all the properties of the cells *in situ*. In addition, one of the parameters of energy metabolism *in vivo*, namely blood flow, cannot be examined in cultures. Despite these limitations, *in*

vitro studies in primary cultures have proved very useful in identifying the cellular sites of glucose uptake and its subsequent metabolic fate—in particular, glycolysis and oxidative phosphorylation—thus providing illuminating correlations of two parameters of brain energy metabolism that are monitored *in vivo*: (1) glucose utilization and (2) oxygen consumption.

Glucose Transporters in the Brain

Glucose is a highly hydrophilic molecule that enters cells through facilitated transport mediated by specific transporters. Eleven genes, encoding glucose transporter proteins have been identified and cloned so far; these are designated GLUT1 to GLUT11 (Maher *et al.*, 1994; Doege *et al.*, 2001). Glucose transporters belong to a family rather homologous to glycosylated membrane proteins with 12 transmembrane-spanning domains, and both the amino and the carboxyl terminals are exposed to the cytoplasmic surface of the membrane. In the brain, three transporters are predominantly expressed in a cell-specific manner: GLUT1, GLUT3, and GLUT5 (Maher *et al.*, 1994)

Two forms of GLUT1 with molecular masses of 55 and 45 kDa, respectively, are detected in the brain, depending on their degree of glycosylation. The 55-kDa form of GLUT1 is essentially localized in brain microvessels, choroid plexus, and ependymal cells. In microvessels, the distribution of GLUT1 is asymmetric, with a higher density on the ablumenal (parenchymal) side than on the vascular side (Maher *et al.*, 1994). An intracellular pool of GLUT1 also has been identified in vascular endothelial cells. In the brain *in situ*, the 45-kDa form of GLUT1 is localized predominantly in astrocytes (Morgello *et al.*, 1995). Under culture conditions, all neural cells, including neurons and other glial cells, express GLUT1; however, this phenomenon appears to be due to the capacity of GLUT1 to be induced by cellular stress (Maher *et al.*, 1994).

The glucose transporter specific to neurons is GLUT3 (Maher *et al.*, 1994). Its cellular distribution appears to predominate in the cell bodies rather than in the axon terminal compartment.

GLUT5 is localized to microglial cells, the resident macrophages of the brain, taking part in the immune and inflammatory responses of the nervous system (Maher *et al.*, 1994). In peripheral tissues, particularly in the small intestine (from which it was cloned), GLUT5 functions as a transporter for fructose, whose concentrations are very low in the brain. In the nervous system, therefore, GLUT5 may have diverse transport functions.

Another glucose transporter, GLUT2, has been localized selectively in astrocytes of discrete brain

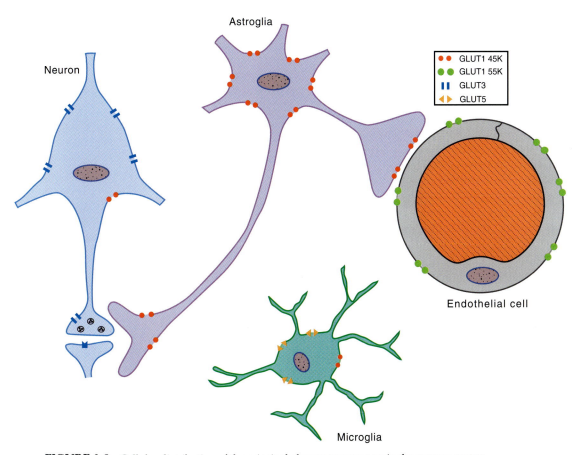

FIGURE 3.9 Cellular distribution of the principal glucose transporters in the nervous system.

areas, such as certain hypothalamic and brainstem nuclei, which participate in the regulation of feeding behavior and in the central control of insulin release. The insulin-sensitive glucose transporter GLUT4 has been localized in brain vascular endothelium (Maher *et al.*, 1994).

It is clear that glucose uptake into the brain parenchyma is a highly specified process regulated in a cell-specific manner by glucose transporter subtypes. Figure 3.9 summarizes this process: Glucose enters the brain through 55-kDa GLUT1 transporters localized on endothelial cells of the blood–brain barrier. Uptake into astrocytes is mediated by 45-kDa GLUT1 transporters, whereas GLUT3 transporters mediate this process in neurons. GLUT2 transporters on astrocytes may "sense" glucose, a function of this glucose transporter subtype in pancreatic β cells. Finally, GLUT5 mediates the uptake of an unidentified substrate into microglial cells. Other glucose transporters recently identified on neurons are GLUTx1 (or GLUT8) and GLUT9 (Ibberson *et al.*, 2000; Doege *et al.*, 2000).

Cell-Specific Glucose Uptake and Metabolism

As described above, glucose utilization can be assessed with radioactively labeled 2-DG. To determine the cellular site of basal and activity-related glucose utilization, this technique has been applied to homogeneous cultures of astrocytes or neurons. For quantitative purposes and to allow comparisons with *in vivo* studies, these *in vitro* experiments, in which radioactive 2-DG is used as a tracer, must be conducted in a medium containing a concentration of glucose near that measured *in vivo* in the extracellular space of the brain, for which values ranging between 0.5 and 2.0 mM have been reported (Fellows *et al.*, 1992). The basal rate of glucose utilization is higher in astrocytes than in neurons, with values of about 20 and 6 nmol per milligram of protein per minute, respectively (Magistretti and Pellerin, 1996). These values are of the same order as those determined *in vivo* for cortical gray matter (10–20 nmol mg^{-1} min^{-1}), with the 2-DG autoradiographic technique (Sokoloff *et al.*, 1977). In view of this difference and of the quantitative preponderance of astrocytes compared with neurons in the gray matter, these data reveal a significant contribution by astrocytes to basal glucose utilization as determined by 2-DG autoradiography or PET *in vivo*.

The contribution of astrocytes to glucose utilization during activation is even more striking. *In vitro*, activation can be mimicked by exposure of the cells to glutamate, the principal excitatory neurotransmitter (see Chapters 9, 11, 16), because, during activation of a given cortical area, the concentration of glutamate in the extracellular space increases considerably owing to its release from the axon terminals of activated pathways (Braitenberg and Schüz, 1998). As shown in Fig. 3.10A, L-glutamate stimulates 2-DG uptake and phosphorylation by astrocytes in a concentration-dependent manner, with an EC_{50} of 60 to 80 μM (Pellerin and Magistretti, 1994; Takahashi *et al.*, 1995). Unlike other actions of glutamate, stimulation of glucose utilization in astrocytes is mediated not by specific glutamate receptors but by glutamate transporters. Indeed, in addition to the maintenance of extracellular K^+ homeostasis, one of the well-established functions of astrocytes is to ensure the reuptake of certain neurotransmitters, in particular, that of glutamate at excitatory synapses (Danbolt, 2001) (see also Chapter 9). At least three glutamate transporter subtypes have been cloned in various species, including humans (Danbolt, 2001). The GLT–1 subtype is localized exclusively in astrocytes, whereas the EAAC1 subtype is exclusively neuronal; GLAST, the third subtype, is expressed both in glia and in neurons, with a predominant distribution in astrocytes. The density of GLT–1 and GLAST is particularly high on astrocytes that surround nerve terminals and dendritic spines, consistent with the prominent role of these transporters in the reuptake of synaptically released glutamate (Danbolt, 2001). The driving force for glutamate uptake through the specific transporters is the transmembrane Na^+ gradient (Bergles and Jahr, 1997); indeed, glutamate is cotransported with Na^+ in a ratio of one glutamate for every two or three Na^+ ions. The selective loss of GLT–1, the astrocyte-selective glutamate transporter, has been demonstrated in the motor cortex and spinal cord of patients who died of amyotrophic lateral sclerosis, a neurodegenerative disease affecting motor neurons.

Glutamate-Stimulated Uptake of Glucose by Astrocytes Is a Source of Insight into the Cellular Bases of ^{18}F-2-DG PET *in Vivo*

The glutamate-stimulated uptake of glucose by astrocytes is a source of insight into the cellular bases of the activation-induced local increase in glucose utilization visualized with ^{18}F-2-DG PET *in vivo*. As

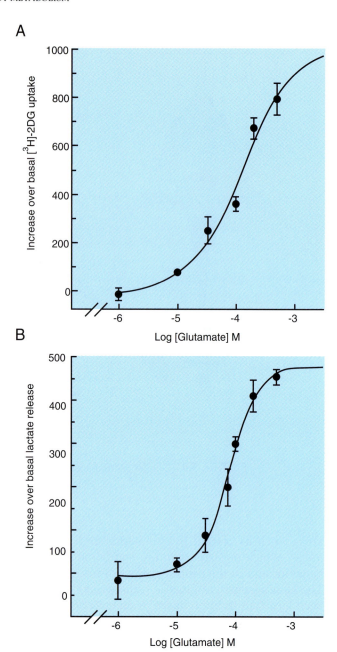

FIGURE 3.10 Stimulation by glutamate of glycolysis in astrocytes. Glutamate stimulates glucose uptake and phosphorylation (A) and lactate production (B) in astrocytes. This effect is concentration dependent with an EC_{50} of ~ 60 μM.

described previously, focal physiological activation of specific brain areas is accompanied by increases in glucose utilization; because glutamate is released from excitatory synapses when neuronal pathways subserving specific modalities are activated, the stimulation by glutamate of glucose utilization in astrocytes provides a direct mechanism for coupling neuronal activity to glucose utilization in the brain (Fig. 3.11). The intracellular molecular mechanism of

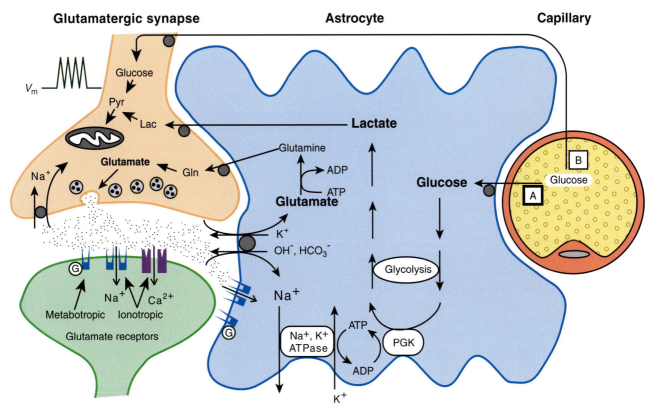

FIGURE 3.11 Schematic representation of the mechanism for glutamate-induced glycolysis in astrocytes during physiological activation. At glutamatergic synapses, presynaptically released glutamate depolarizes postsynaptic neurons by acting at specific receptor subtypes. The action of glutamate is terminated by an efficient glutamate uptake system located primarily in astrocytes. Glutamate is cotransported with Na^+, resulting in an increase in the intra-astrocytic concentration of Na^+, leading to activation of the astrocyte Na^+,K^+-ATPase. Activation of the Na^+,K^+-ATPase stimulates glycolysis (i.e., glucose utilization and lactate production). The stoichiometry of this process is such that for one glutamate molecule taken up by two to three Na^+ ions, one glucose molecule enters astrocytes, two ATP molecules are produced through glycolysis, and two lactate molecules are released. Within the astrocyte, one ATP fuels one "turn of the pump," while the other provides the energy needed to convert glutamate to glutamine by glutamine synthase (see Fig. 3.13). Lactate, once released by astrocytes, can be taken up by neurons and serve as an adequate energy substrate. (For graphic clarity only, lactate uptake into presynaptic terminals is indicated. However, this process could also take place at the postsynaptic neuron.) In accord with recent evidence, glutamate receptors are also shown on astrocytes. This model, which summarizes *in vitro* experimental evidence indicating glutamate-induced glycolysis, is taken to show cellular and molecular events occurring during activation of a given cortical area (arrow labeled A, activation). Direct glucose uptake into neurons under basal conditions is also shown (arrow labeled B, basal conditions). Pyr, pyruvate; Lac, lactate; Gln, glutamine; G, G protein; PGK, phosphoglucokinase. Modified from Pellerin and Magistretti (1994).

this coupling requires Na^+, K^+-ATPase, because ouabain completely inhibits the glutamate-evoked 2-DG uptake by astrocytes (Pellerin and Magistretti, 1994). The astrocytic Na^+, K^+-ATPase responds predominantly to increases in intracellular Na^+ (Na^+_i) for which it shows a K_m of about 10 mM (Erecinska, 1989). In astrocytes, the Na^+_i concentration ranges between 10 and 20 mM (Kimelberg *et al.*, 1993), and so Na^+, K^+-ATPase is set to be readily activated when Na^+_i rises concomitantly with glutamate uptake (Chatton *et al.*, 2000). These observations indicate that a major determinant of glucose utilization is the activity of Na^+, K^+-ATPase. In this context, we should note that, *in vivo*, the main mechanism that accounts for the activation-induced 2-DG uptake is the activity of Na^+, K^+-ATPase.

It is important here to briefly consider the relative participation of the neuronal and astrocytic Na^+, K^+-ATPases in glucose utilization. When glutamate is released from depolarized neuronal terminals, it is taken up predominantly into astrocytes. The stoichiometry of glutamate reuptake being one molecule of glutamate cotransported with two or three Na^+ ions, the increase in intracellular astrocytic Na^+ concentration associated with glutamate reuptake massively activates the pump. Thus, although the tonic activity of the Na^+, K^+-ATPase is needed to maintain the transmembrane neuronal and glial ionic gradients and accounts for basal glucose utilization, on a short-term temporal scale (from milliseconds to seconds), when glutamate is released from depolarized axon terminals of modality-specific afferents, the astrocytic Na+, K+-ATPase is briskly activated, owing to the massive increase (by at least 10 mM) in intracellular Na+ associated with glutamate reuptake, providing the signal for activation-dependent glucose utilization. Increases in glutamate as small as 10 μM are sufficient to double the activity of Na+, K+-ATPase (Chatton *et al.*, 2000).

How does activation of Na+, K+-ATPase cause increased glucose utilization? The mechanism was explained in pioneering studies on erythrocytes by Joseph Hoffmann and his colleagues at Yale University (Proverbio and Hoffman, 1977), which have been confirmed in a number of other cell systems including brain (Erecinska, 1989) and vascular smooth muscle. The increase in pump activity consumes ATP, which is a negative modulator of phosphofructokinase, the principal rate-limiting enzyme of glycolysis (see Fig. 3.1). Thus, when ATP concentration is low, phosphofructokinase activity is stimulated, resulting in increased glucose utilization. The activity of hexokinase, the enzyme responsible for glucose and 2-DG phosphorylation (see Fig. 3.4), also is increased under these conditions. This explains why the increase in glucose utilization, associated with the stimulation of Na+, K+-ATPase, can be monitored with 2-DG, which is not processed beyond the hexokinase step.

Compartmentalization of glucose uptake during activation has also been unequivocably found by Tsacopoulos and his colleagues (1988) in the honey-bee drone retina. In this highly organized, crystallike, nervous tissue preparation, photoreceptor cells form rosette-like structures that are surrounded by glial cells. In addition, mitochondria are exclusively present in the photoreceptor neurons. Light activation reveals an increase in radioactive 2-DG uptake in the glial cells surrounding the rosettes but not in the photoreceptor neurons (Tsacopoulos *et al.*, 1988). An increase in O_2 consumption is nevertheless measured in photoreceptor neurons. After activation of photoreceptors by light, glucose is probably taken up predominantly by glial cells, which then release a metabolic substrate to be oxidized by photoreceptor neurons.

In summary, as indicated in the operational model described in Fig. 3.11, on activation of a particular brain area, glutamate released from excitatory terminals is taken up by a Na+-dependent transporter located on astrocytes. The ensuing local increase in intracellular Na+ concentration activates the Na+, K+-ATPase, which in turn stimulates glucose uptake by astrocytes. This model delineates a simple mechanism for coupling synaptic activity to glucose utilization; in addition, it is consistent with the notion that the

signals detected during physiological activation in humans with ^{18}F-2-DG PET and in laboratory animals with autoradiography may reflect predominantly uptake of the tracer into astrocytes. This conclusion does not question the validity of the 2-DG-based techniques; rather, it provides a cellular and molecular basis for these functional brain imaging techniques (Magistretti *et al.*, 1999).

The fact that the increase in glucose uptake during activation can be ascribed predominantly, if not exclusively, to astrocytes indicates that energy substrates must be released by astrocytes to meet the energy demands of neurons. As indicated earlier, lactate and pyruvate are adequate substrates for brain tissue *in vitro* (Schurr *et al.*, 1999). In fact, synaptic activity can be maintained *in vitro* in cerebral cortical slices with only lactate or pyruvate as a substrate (Schurr *et al.*, 1999). Lactate is quantitatively the main metabolic intermediate released by cultured astrocytes at a rate of 15 to 30 nmol per milligram of protein per minute. Other, quantitatively less important intermediates released by astrocytes are pyruvate (approximately 10 times less than lactate) and α-ketoglutarate, citrate, and malate, which are released in marginal amounts (Westergaard *et al.*, 1995). For lactate (or pyruvate) to be an adequate metabolic substrate for neurons, particularly during activation, two additional conditions must be fulfilled: (1) during activation lactate release by astrocytes increases and (2) lactate uptake by neurons must be demonstrated. Both mechanisms have been demonstrated. Mimicking activation *in vitro* by exposing cultured astrocytes to glutamate results in a marked release of lactate and, to a lesser degree, pyruvate (Pellerin and Magistretti, 1994) (see Fig. 3.10B). This glutamate-evoked lactate release shows the same pharmacology and time course as does the glutamate-evoked glucose utilization and indicates that glutamate stimulates the processing of glucose through glycolysis. As noted earlier, *in vivo* [1]H MRI studies in humans that show a transient lactate peak in the primary visual cortex during physiological stimulation are consistent with the notion of activation-induced glycolysis (Prichard *et al.*, 1991). In addition, lactate levels in the rat somatosensory cortex transiently increase subsequent to forepaw stimulation (Fellows *et al.*, 1993). Finally, monocarboxylate transporters have been demonstrated on neurons and astrocytes in addition to capillaries (Pierre *et al.*, 2000).

Thus, a metabolic compartmentation whereby glucose taken up by astrocytes and metabolized glycolytically to lactate is then released in the extracellular space to be used by neurons is consistent with biochemical and electrophysiological observations (Magistretti *et al.*, 1999; Tsacopoulos and Magistretti,

1996). This array of *in vitro* and *in vivo* experimental evidence is summarized in the model of cell-specific metabolic regulation illustrated in Fig. 3.11.

Studies of the well-compartmentalized honeybee drone retina and of isolated preparations of guinea pig retina containing photoreceptors attached to Müller (glial) cells corroborate the existence of such metabolic fluxes between glia and neurons (Poitry-Yamate *et al.*, 1995). In addition to the glial localization of glucose uptake during activation, glycolytic products have been shown to be released. In particular, during activation, glial cells in the honeybee drone retina release alanine produced from pyruvate by transamination; the released alanine is taken up by photoreceptor neurons and, after reconversion into pyruvate, can enter the TCA cycle to yield ATP through oxidative phosphorylation (see Fig. 3.2). In the guinea pig retina, lactate, formed glycolytically from glucose, is released by the Müller cells to fuel photoreceptor neurons (Poitry-Yamate *et al.*, 1995).

Although plasma lactate is unlikely to be a substitute for glucose as a metabolic substrate for the brain, lactate formed within the brain parenchyma (e.g., through glutamate-activated glycolysis in astrocytes) can fulfill the energetic needs of neurons. Lactate, after conversion into pyruvate by a reaction catalyzed by lactate dehydrogenase (LDH), can provide, on a molar basis, 18 ATP through oxidative phosphorylation. Conversion of lactate into pyruvate does not require ATP, and, in this regard, lactate is energetically more favorable than the first, obligatory step of glycolysis, in which glucose is phosphorylated to glucose 6-phosphate at the expense of one molecule of ATP (see Fig. 3.1). Recently, another metabolic fate for lactate has been shown *in vitro* and *in vivo* by magnetic resonance spectroscopy (Hassel and Brathe, 2000). Thus, once converted to pyruvate, lactate may enzymatically yield glutamate and hence be a substrate for the replenishment of the neuronal pool of glutamate. As this reaction is not associated with oxygen consumption, part of the uncoupling between glucose utilization and oxygen consumption described in certain paradigms of activation may be explained by the processing of glucose-derived lactate into the glutamate neuronal pool.

Glycogen, the Storage Form of Glucose, Is Localized in Astrocytes

Glycogen is the single largest energy reserve of the brain (Magistretti *et al.*, 1993); it is localized mainly in astrocytes, although ependymal and choroid plexus cells, as well as certain large neurons in the brainstem, contain glycogen. When compared with the contents in liver and muscle, the glycogen content of the brain is exceedingly small, about 100 and 10 times inferior, respectively. Thus, the brain can hardly be considered a glycogen storage organ, and here the function of glycogen should be viewed as that of providing a metabolic buffer during physiological activity.

Glycogen Metabolism Is Coupled to Neuronal Activity

Glycogen turnover in the brain is extremely rapid, and glycogen levels are finely coordinated with synaptic activity (Magistretti *et al.*, 1993). For example, during general anesthesia, a condition in which synaptic activity is markedly attenuated, glycogen levels rise sharply. Interestingly, however, the glycogen content of cultures containing exclusively astrocytes is not increased by general anesthetics; this observation indicates that the *in vivo* action of general anesthetics on astrocyte glycogen is due to the inhibition of neuronal activity, stressing the existence of a tight coupling between synaptic activity and astrocyte glycogen. Accordingly, reactive astrocytes, which develop in areas where neuronal activity is decreased or absent as a consequence of injury, contain large amounts of glycogen (Watanabe and Passonneau, 1974).

In addition to glycogen, glucose is incorporated into other macromolecules such as proteins (glycoproteins) and lipids (glycolipids) at rates specific for the turnover of each macromolecule, which can span from a few minutes to a few days.

Certain Neurotransmitters Regulate Glycogen Metabolism in Astrocytes

Glycogen levels in astrocytes are tightly regulated by various neurotransmitters. Several monoamine neurotransmitters, namely noradrenaline, serotonin, and histamine, are glycogenolytic in the brain, in addition to certain peptides, such as vasoactive intestinal peptide (VIP) and pituitary adenylate cyclase activating peptide (PACAP), and adenosine and ATP (Magistretti *et al.*, 1981, 1993). The effects of all these neurotransmitters are mediated by their cogent specific receptors coupled to second-messenger pathways that are under the control of adenylate cyclase or phospholipase C. The initial rate of glycogenolysis activated by VIP and noradrenaline is between 5 and 10 nmol per milligram of protein per minute, a value that is remarkably close to glucose utilization of the gray matter, as determined by the 2-DG autoradiographic method. This correlation indicates that the glycosyl units mobilized in response to the glycogenolytic neurotransmitters can provide quantitatively adequate substrates for the energy demands of the brain parenchyma. At present, whether the glycosyl units mobilized through glycogenolysis are used by astrocytes to meet their energy demands during activation or are metabolized to a substrate such as lactate, which is then released for the use of neurons, is not clear. It appears, however, that glucose is not released by astrocytes after glycogenolysis, supporting the view that the activity of glucose-6-phosphatase (see Fig. 3.1) in astrocytes is very low (Dringen *et al.*, 1993). *In vitro* evidence suggests that lactate may be the metabolic intermediate produced through glycogenolysis and exported from astrocytes (Dringen *et al.*, 1993).

These observations show that neuronal signals (e.g., certain neurotransmitters) can exert receptor-mediated metabolic effects on astrocytes in a manner similar to that of peripheral hormones on their target cells. However, the action of this type of neurotransmitter is temporally specified and spatially restricted to activated areas. Indeed, brain glycogenolysis visualized by autoradiography in laboratory animals has also been demonstrated *in vivo* after physiological activation of a modality-specific pathway (Swanson *et al.*, 1992). Repeated stimulation of whiskers resulted in a marked decrease in the density of glycogen-associated autoradiographic grains in the somatosensory cortex of rats (barrel fields), as well as in the relevant thalamic nuclei (Swanson *et al.*, 1992). These observations indicate that the physiological activation of specific neuronal circuits results in the mobilization of glial glycogen stores.

Summary

Under basal conditions, glucose uptake and metabolism occur in every brain cell type. Glucose uptake is mediated by specific transporters that are distributed in a cell-specific manner. Astrocytes play a critical role in the utilization of glucose coupled to excitatory synaptic transmission. The molecular mechanisms of this coupling are stoichiometrically directed: for each synaptically released glutamate molecule taken up with two to three Na^+ ions by an astrocyte, one glucose molecule enters the same astrocyte, two ATP molecules are produced through glycolysis, and two lactate molecules are released and consumed by neurons to yield 18 ATP through oxidative phosphorylation. Neuronal signals, for example, certain neurotransmitters, can exert receptor-mediated glycogenolysis in astrocytes in a manner similar to peripheral hormones on their target cells. However, this type of effect by neurotransmitters is temporally specified

and spatially restricted within activated areas, possibly to provide additional energy substrates in register with local increases in neuronal activity.

GLUTAMATE AND NITROGEN METABOLISM: A COORDINATED SHUTTLE BETWEEN ASTROCYTES AND NEURONS

As discussed above, synaptically released glutamate is rapidly removed from the extracellular space by a transporter-mediated reuptake system that is particularly efficient in astrocytes (Danbolt, 2001). This mechanism contributes in a crucial manner to the fidelity of glutamate-mediated neurotransmission. Indeed, glutamate levels in the extracellular space are low (< 3 μM), allowing for optimal glutamate-mediated signaling after depolarization while preventing overactivation of glutamate receptors, which could eventually result in excitotoxic neuronal damage.

One may wonder how astrocytes dispose of the glutamate that they take up, because, unlike carbohydrates or lipids, amino acids cannot be stored. The predominant pathway in peripheral tissues for disposing of amino acids is the transfer of their α-amino group to a corresponding α-keto acid; this reaction is catalyzed by aminotransferases (Fig. 3.12). In astrocytes, the α-amino group of glutamate can be transferred to oxaloacetate to yield α-ketoglutarate (α-KG) and aspartate in a reaction catalyzed by aspartate amino transferase (AAT) (Yudkoff *et al.*, 1986). The α-KG generated is an intermediate of the TCA cycle and is therefore further oxidized. Another transamination reaction catalyzed by

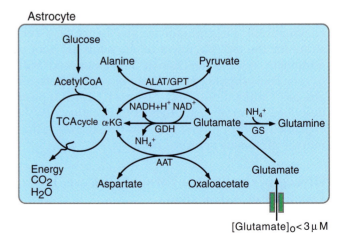

FIGURE 3.12 Metabolic fate of glutamate taken up by astrocytes. ALAT, alanine aminotransferase; GDH, glutamate dehydrogenase; GS, glutamine synthase; AAT, aspartate aminotransferase; GPT, glutamate dehydrogenase; α-KG, α-ketoglutarate.

alanine amino transferase (ALAT) transfers the α-amino group of glutamate to pyruvate, resulting in the formation of alanine and α-KG.

Two other pathways exist in astrocytes to metabolize glutamate. First, glutamate can be directly converted into α-KG through an NAD-requiring oxidative deamination catalyzed by glutamate dehydrogenase (GDH) (Yudkoff *et al.*, 1986) (see Fig. 3.12). Glutamate, by indirectly (through AAT or ALAT) or directly (through GDH) entering the TCA cycle, is an energy substrate for astrocytes. Second, the quantitatively predominant metabolic pathway of glutamate in astrocytes is its amidation to glutamine, an ATP-requiring reaction in which an ammonium ion is fixed on glutamate (see Fig. 3.12) (Van den Berg and Garfinkel, 1971). This reaction is catalyzed by glutamine synthase (GS), an enzyme almost exclusively localized in astrocytes (Norenberg and Martinez-Hernandez, 1979), and provides an efficient means of disposing not only of glutamate but also of ammonium (Box 3.2). Glutamine is released by astrocytes and taken up by neurons, where it is hydrolyzed back

to glutamate by the phosphate-dependent mitochondrial enzyme glutaminase (Erecinska and Silver, 1990). This metabolic pathway, often referred to as the glutamate–glutamine shuttle, is a clear example of cooperation between astrocytes and neurons (Fig. 3.13). It allows the removal of potentially toxic excess glutamate from the extracellular space, while returning to the neuron a synaptically inert (glutamine does not affect neurotransmission) precursor with which to regenerate the neuronal pool of glutamate.

Not all glutamate, however, is regenerated through the glutamate–glutamine shuttle, because some of the glutamate released by neurons enters at the α-KG, the TCA cycle in astrocytes; therefore, *de novo* synthesis is required to maintain the neuronal glutamate pool. Glutamate can be synthesized through NADPH-dependent reductive amination of α-KG catalyzed by GDH (note that here the cofactor is NADPH, whereas, for the opposite reaction also catalyzed by GDH, the oxidant is NAD; see Figs. 3.12 and 3.13) (Erecinska and Silver, 1990). For the synthesis of glutamate, glucose provides the carbon backbone as α-KG

BOX 3.2

HEPATIC ENCEPHALOPATHY IS A DISORDER OF ASTROCYTE FUNCTION RESULTING IN A NEUROPSYCHIATRIC SYNDROME

Hepatic encephalopathy is observed in patients with severe liver failure. The disease can be in one of two forms: an acute form, called fulminant hepatic failure, and (2) a chronic form, portosystemic encephalopathy (Plum and Hindfeld, 1976). The neuropsychiatric symptoms of fulminant hepatic failure are delirium, coma, and seizures associated with acute toxic or viral hepatic failure. Patients having portosystemic encephalopathy may present personality changes, episodic confusion, or stupor, and, in the most severe cases, coma. The current view on the pathophysiology of hepatic encephalopathy is that, owing to liver failure, "toxic" substances that affect brain function accumulate in the circulation (Norenberg *et al.*, 1992). One of the substances thought to be responsible for the neuropsychiatric "toxicity" is ammonia. The neuropathological findings are rather striking: astrocytes are the brain cells that appear principally affected. In the acute form, astrocyte swelling is prominent and likely to be the cause of the observed acute brain edema. In portosystemic encephalopathy, astrocytes adopt morphological features characteristic of what is defined as an Alzheimer type II astrocyte: in these cells, the nucleus is pale and enlarged, chromatin is mar-

ginated, and a prominent nucleolus is often observed. Lipofuscin deposits may be present, and the amount of the astrocyte-specific protein glial fibrillary acidic protein (see Chapter 2) is decreased. Neurons appear structurally normal. All the foregoing histopathological changes have been reproduced *in vitro* by acutely or chronically applying ammonium chloride to primary astrocyte cultures. As mentioned earlier, detoxification of ammonium is an ATP-requiring, astrocyte-specific reaction catalyzed by glutamine synthase (see Fig. 3.12). It is therefore not surprising that excess ammonia perturbs energy metabolism; indeed, ammonia stimulates glycolysis (McKhann and Tower, 1961) whereas it inhibits TCA cycle activity (Muntz and Hurwitz, 1951). In addition, ammonia markedly decreases the glycogen content of astrocytes.

In summary, although the precise pathophysiological mechanisms of the neuropsychiatric syndrome in hepatic encephalopathy are still unknown, this clinical condition provides a striking illustration of the fundamental importance of neuron–astrocyte metabolic interactions, because structural and functional alterations apparently restricted to astrocytes result in severe behavioral perturbations.

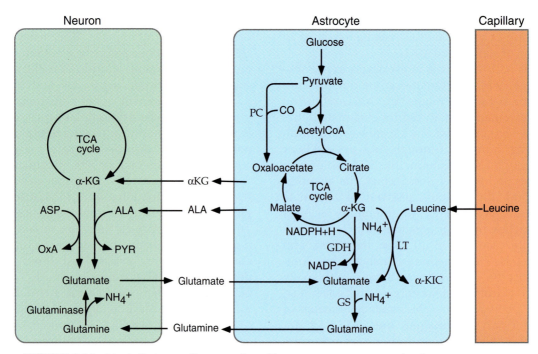

FIGURE 3.13 Metabolic intermediates are released by astrocytes to regenerate the glutamate neurotransmitter pool in neurons. Glutamine, formed from glutamate in a reaction catalyzed by glutamine synthase (GS), is released by astrocytes and taken up by neurons, which convert it into glutamate under the action of glutaminase. GS is an enzyme selectively localized in astrocytes. This metabolic cycle is referred to as the glutamate–glutamine shuttle. Other, quantitatively less important sources of neuronal glutamate are lactate, alanine, and α-ketoglutarate (α-KG). In astrocytes, glutamate is synthesized *de novo* from α-KG in a reaction catalyzed by glutamate dehydrogenase (GDH). The carbon backbone of glutamate is exported by astrocytes after conversion into glutamine under the action of GS; the conversion of leucine into α-ketoisocaproate (α-KIC), catalyzed by leucine transaminase (LT), provides the amino group for the synthesis of glutamine from glutamate. The carbons "lost" from the TCA cycle as α-KG is converted into glutamate are replenished by oxaloacetate (OxA) formed from pyruvate in a reaction catalyzed by pyruvate carboxylase (PC), another astrocyte-specific enzyme.

through the TCA cycle, whereas an exogenous source of nitrogen is necessary (see Fig. 3.13). Convincing evidence, obtained by using ^{15}N-labeled amino acids whose metabolic fate was determined by gas chromatography and mass spectrometry, indicates that plasma leucine provides the nitrogen required for net glutamate synthesis from α-KG (Yudkoff *et al.*, 1994). Thus, leucine taken up from the circulation at astrocytic end feet provides the amino group to α-KG in a reaction catalyzed by leucine transaminase (LT), resulting in the formation of glutamate and α-ketoisocaproate (α-KIC) (see Fig. 3.13). Because this reaction takes place in astrocytes, to replenish the neuronal glutamate pool, the astrocytes export glutamate as glutamine. As noted earlier, the neuronal glutamate pool could also be replenished by lactate released by astrocytes (Hassel and Brathe, 2000).

Finally, another potential pathway described by Arne Schousboe and colleagues exists for the *de novo* synthesis of glutamate in neurons from substrates provided by astrocytes. With the use of uniformly[13]C-

labeled compounds in combination with magnetic resonance spectroscopy, astrocytes have been shown to release significant amounts of alanine and α-KG (Westergaard *et al.*, 1995). Both metabolic intermediates are taken up by neurons and can be converted into glutamate and pyruvate in a transamination reaction catalyzed by ALAT (see Fig. 3.12). In this case, as for the glutamate–glutamine shuttle (Fig. 3.13), astrocytes provide the substrate(s) necessary for glutamate synthesis in neurons.

Note that because α-KG is used for glutamate synthesis, metabolic intermediates downstream of α-KG must be available to maintain a sustained flux through the TCA cycle in astrocytes (see Fig. 3.13). This need is met by the activity of the enzyme pyruvate carboxylase (PC), which fixes CO_2 on pyruvate to generate oxaloacetate, which, by condensing with acetyl-CoA, maintains the flux through the TCA cycle. The carboxylation of pyruvate to oxaloacetate is referred to as an *anaplerotic* (Greek for "fill up") reaction. Interestingly, like glutamine synthase, PC is selectively localized in astrocytes (Shank *et al.*, 1985). The fact that these two enzymes are localized in astrocytes in conjunction with the existence of a glutamate–glutamine shuttle stresses that astrocytes are essential for maintaining the neuronal glutamate pool used for neurotransmission (see Fig. 3.13).

As noted earlier, the metabolic intermediate α-KG lies at the branching point of glucose and glutamate metabolism (see Fig. 3.12). Any change in the activities of the enzymes that convert α-KG into glutamate or into succinyl-CoA, the next intermediate in the TCA cycle, may affect the efficacy of the TCA cycle or glutamate levels. Interestingly, a marked decrease in the activity of α-ketoglutarate dehydrogenase (α-KGDH), the enzyme catalyzing the conversion of α-KG into succinyl-CoA, was observed in a very large proportion of postmortem brains from patients with Alzheimer's disease (Sheu *et al.*, 1994); in addition, a similar decrease in α-ketoglutarate dehydrogenase activity has been demonstrated in the fibroblasts of patients affected by the familial form of Alzheimer's disease (Sheu *et al.*, 1994).

Summary

A key function of astrocytes is to remove synaptically released glutamate. A large proportion of glutamate is transformed to glutamine through an energy-requiring process that also allows for the detoxification of ammonium. Glutamine released by astrocytes regenerates the neuronal glutamate pool. Some of the glutamate is also regenerated from lactate and through the fixation of the amino group of leucine onto the TCA intermediate α-KG, providing another indication of the tight link existing between glutamate and nitrogen metabolism and of the crucial function that astrocytes play in maintaining the neuronal glutamate pool at levels that ensure the maintenance of synaptic transmission.

THE ASTROCYTE–NEURON METABOLIC UNIT

From a strictly energetic viewpoint, the brain can be seen as an almost exclusive glucose-processing machine producing H_2O and CO_2. However, the metabolism of glucose in the brain is temporally, spatially, and functionally specified. Thus, glucose metabolism increases with exquisite spatiotemporal precision in register with neuronal activity. The site of this increase is not the neuronal cell body; rather, it is the neuropil, where presynaptic terminals, postsynaptic elements, and astrocytes

ensheathing synaptic contacts are localized (Sokoloff, 1981). This cytological relation between astrocytes and neurons is also manifested by a functional metabolic partnership: in response to a neuronal signal (glutamate), astrocytes release a glucose-derived metabolic substrate for neurons (lactate). Glucose also provides the carbon backbone for regeneration of the neuronal pool of glutamate. This process results from a close astrocyte–neuron cooperation. Indeed, the selective localization of pyruvate carboxylase in astrocytes, indicating the need to replenish the TCA cycle with carbon backbones, strongly suggests that glucose-derived metabolic intermediates are used for glutamate (and other amino acid) synthesis. The newly synthesized glutamate is not provided as such by astrocytes to neurons; rather, it is converted into glutamine by glutamine synthase, another enzyme selectively localized in astrocytes. Glutamate, taken up by astrocytes during synaptic activity, undergoes the same metabolic process, also being released as glutamine (the glutamate–glutamine shuttle).

Summary

In conclusion, the axon terminal of glutamatergic neurons, which are the main communication lines in the nervous system, and the astrocytic processes that surround them should be viewed as a metabolic unit, in which the neuron furnishes the activation signal (glutamate) to the astrocyte and the astrocyte provides not only the precursors needed to maintain the neurotransmitter pool (glutamine and, in part, lactate and alanine) but also the energy substrate (lactate) (Fig. 3.14). The efficacy of the predominant excitatory

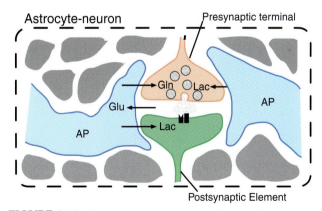

FIGURE 3.14 The astrocyte–neuron metabolic unit. Glutamatergic terminals and the astrocytic processes that surround them can be viewed as a highly specialized metabolic unit in which the activation signal (glutamate) is furnished by the neuron to the astrocyte, whereas the astrocyte provides the precursors needed to maintain the neurotransmitter pool (glutamine, lactate, alanine), as well as the energy substrate (lactate). AP, astrocyte process.

synapse in the brain, the glutamatergic synapse, cannot be maintained without a close astrocyte–neuron interaction.

References

Andriezen, W. L. (1893). On a system of fibre-like cells surrounding the blood vessels of the brain of man and mammals, and its physiological significance. *Int. Monatsschr. Anat. Physiol.* **10**, 532–540.

Attwell, D., and Laughlin, S. B. (2001). An energy budget for signalling in the grey matter of the brain. *J. Cereb. Blood Flow Metab.* **21**, 1133–1145.

Auestad, N., Korsak, R. A., Morrow, J. W., and Edmond, F. (1991). Fatty acid oxidation and ketogenesis by astrocytes in primary culture. J. Neurochem. **56**, 1376–1386.

Barres, B. A. (1991). New roles for glia. *J. Neurosci.* **11**, 3685–3694.

Ben-Yoseph, O., Boxer, P. A., and Ross, B. D. (1994). Oxidative stress in the central nervous system: Monitoring the metabolic response using the pentose phosphate pathway. *Dev. Neurosci.* **16**, 328–336.

Bergles, D. E., and Jahr, C. E. (1997). Synaptic activation of glutamate transporters in hippocampal astrocytes. *Neuron* **19**, 1297–1308.

Blass, J. P., and Gibson, G. E. (1977). Deleterious aberrations of a thiamine-requiring enzyme in four patients with Wernicke–Korsakoff syndrome. *N. Engl. J. Med.* **297**, 1367–1370.

Braitenberg, V., and Schüz, A. (1998). "Cortex: Statistics and Geometry of Neuronal Connectivity," 2nd ed. Springer-Verlag, New York.

Chatton, J. Y., Marquet, P., and Magistretti, P. J. (2000). A quantitative analysis of L-glutamate-regulated Na^+ dynamics in mouse cortical astrocytes: Implications for cellular bioenergetics. *Eur. J. Neurosci.* **12**, 3843–3853.

Clarke, D. D., and Sokoloff, L. (1999). Circulation and energy metabolism of the brain. *In* "Basic Neurochemistry: Molecular, Cellular and Medical Aspects" (G. Siegel, B. Agranoff, R. W. Albers, S. K. Fisher, and M. D. Uhler, Eds.), pp. 637–669. Lippincott–Raven, Philadelphia.

Creutzfeldt, O. D. (1975). Neurophysiological correlates of different functional states of the brain. *In* "Brain Work, the Coupling of Function, Metabolism and Blood Flow in the Brain," Proceedings, Alfred Benzon Symp VII (D. H. Ingvar and N. A. Lassen, Eds.), pp. 21–46. Academic Press, Copenhagen/Munksgaard/New York.

Danbolt, N. C. (2001). Glutamate uptake. *Prog. Neurobiol.* **65**, 1–105.

Doege, H., Bocianski, A., Joost, H.-G., and Schürmann, A. (2000) Activity and genomic organization of human glucose transporter 9 (GLUT9), a novel member of the family of sugar-transport facilitators predominantly expressed in brain and leucocytes. *Biochem. J.* **350**, 771–776.

Dringen, R., Gebhardt, R., and Hamprecht, B. (1993). Glycogen in astrocytes: Possible function as lactate supply for neighboring cells. *Brain Res.* **623**, 208–214.

Erecinska, M. (1989). Stimulation of the Na^+/K^+ pump activity during electrogenic uptake of acidic amino acid transmitters by rat brain synaptosomes. *J. Neurochem.* **52**, 135–139.

Erecinska, M., and Dagani, F. (1990). Relationships between the neuronal sodium/potassium pump and energy metabolism: effects of K^+, Na^+, and adenosine triphosphate in isolated brain synaptosomes. *J. Gen. Physiol.* **95**, 591–616.

Erecinska, M., and Silver, I. A. (1990). Metabolism and role of glutamate in mammalian brain. *Prog. Neurobiol.* **35**, 245–296.

Fellows, L. K., Boutelle, M. G., and Fillenz, M. (1992). Extracellular brain glucose levels reflect local neuronal activity: A microdialysis study in awake, freely moving rats. *J. Neurochem.* **59**, 2141–2147.

Fellows, L. K., Boutelle, M. G., and Fillenz, M. (1993). Physiological stimulation increases nonoxidative glucose metabolism in the brain of the freely moving rat. *J. Neurochem.* **60**, 1258–1263.

Fox, P. T., Raichle, M. E., Mintun, M. A., and Dence, C. (1988). Nonoxidative glucose consumption during focal physiologic neural activity. *Science* **241**, 462–464.

Frackowiak, R. S. J., Lenzi, G. L., Jones, T., and Heather, J. D. (1980). Quantitative measurement of regional cerebral blood flow and oxygen metabolism in man using 15O and positron emission tomography: Theory, procedure and normal values. *J. Comput. Assist. Tomogr.* **4**, 727–736.

Frackowiak, R. S. J., Magistretti, P. J., Shulman, R. G. and Adams, M. (2001). "Neuroenergetics: Relevance for Functional Brain Imaging." HFSP, Strasbourg.

Girard, J., Ferre, P., Pegorier, J.-P., and Duee, P. H. (1992). Adaptations of glucose and fatty acid metabolism during the perinatal period and suckling–weaning transition. *Physiol. Rev.* **72**, 507–562.

Greenlund, L. J. S., Deckwreth, T. L., and Johnson, E. M., Jr. (1995). Superoxide dismutase delays neuronal apoptosis: A role for reactive oxygen species in programmed neuronal death. *Neuron* **14**, 303–315.

Gusnard, A., and Raichle, M. E. (2001). Searching for a baseline: Functional imaging and the resting human brain. *Nat. Rev. Neurosci.* **2**, 685–694.

Halliwell, B. (1992). Reactive oxygen species and the central nervous system. *J. Neurochem.* **59**, 1609–1623.

Hassel, B., and Brathe, A. (2000). Cerebral metabolism of lactate *in vivo*: Evidence for neuronal pyruvate carboxylation. *J. Cereb. Blood Flow Metab.* **20**, 327–336.

Haug, H. (1987). Brain sizes, surfaces, and neuronal sizes of the cortex cerebri: A stereological investigation of man and his variability and a comparison with some mammals (primates, whales, marsupials, insectivores, and one elephant). *Am. J. Anat.* **180**, 126–142.

Iadecola, C., Pelligrino, D. A., Moskowitz, M. A., and Lassen, N. A. (1994). Nitric oxide synthase inhibition and cerebrovascular regulation. *J. Cereb. Blood Flow Metab.* **14**, 175–192.

Ibberson, M., Uldry, M., and Thorens, B. (2000). GLUTX1, a novel mammalian glucose transporter expressed in the central nervous system and insulin-sensitive tissues. *J. Biol. Chem.* **275**, 4607–4612.

Kacem, K., Lacombe, P., Seylaz, J., and Bonvento, G. (1998). Structural organization of the perivascular astrocyte endfeet and their relationship with the endothelial glucose transporter: A confocal microscopy study. Glia **23**, 1–10.

Kety, S. S., and Schmidt, C. F. (1948). The nitrous oxide method for the quantitative determination of cerebral blood flow in man: Theory, procedure, and normal values. *J. Clin. Invest.* **27**, 476–483.

Kimelberg, H. K., Jalonen, T., and Walz, W. (1993). Regulation of brain microenvironment: Transmitters and ions. *In* "Astrocytes: Pharmacology and Function" (S. Murphy, Ed.), pp. 193–228. Academic Press, San Diego, CA.

Magistretti, P. J., and Pellerin, L. (1996). Cellular bases of brain energy metabolism and their relevance to functional brain imaging: Evidence for a prominent role of astrocytes. *Cereb. Cortex* **6**, 50–61.

Magistretti, P., and Pellerin, L. (1999). Cellular mechanisms of brain energy metabolism and their relevance to functional brain imaging. *Phil. Trans. R. Soc. London Ser. B* **354**, 1155–1163.

Magistretti, P. J., Morrison, J. H., Shoemaker, W. J., Sapin, V., and Bloom, F. E. (1981). Vasoactive intestinal polypeptide induces glycogenolysis in mouse cortical slices: A possible regulatory mechanism for the local control of energy metabolism. *Proc. Natl. Acad. Sci. USA* **78**, 6535–6539.

Magistretti, P. J., Pellerin, L., Rothman, D. L., and Shulman, R. G. (1999). Energy on demand. *Science* **283**, 496–497.

Magistretti, P. J., Sorg, O., and Martin, J. L. (1993). Regulation of glycogen metabolism in astrocytes: Physiological, pharmacological, and pathological aspects. *In* "Astrocytes: Pharmacology and Function" (S. Murphy, Ed.), pp. 243–265. Academic Press, San Diego, CA.

Maher, F., Vannucci, S. J., and Simpson, I. A. (1994). Glucose transporter proteins in brain. *FASEB J.* **8**, 1003–1011.

McKhann, G. M., and Tower, D. B. (1961). Ammonia toxicity and cerebral oxidative metabolism. *Am. J. Physiol.* **200**, 420–424.

Melzer, P., Van der Loos, H., Dörfl, J., Welker, E., Robert, P., Emery, D., and Berrini, J. C. (1985). A magnetic device to stimulate selected whiskers of freely moving or restrained small rodents: Its application in a deoxyglucose study. *Brain Res.* **348**, 229–240.

Morgello, S., Uson, R. R., Schwartz, E. J., and Haber, R. S. (1995). The human blood–brain barrier glucose transporter (GLUT1) is a glucose transporter of gray matter astrocytes. *Glia* **14**, 43–54.

Muntz, J. A., and Hurwitz, J. (1951). The effect of ammonia ions upon isolated reactions of the glycolytic scheme. *Arch. Biochem. Biophys.* **32**, 137–149.

Nehlig, A., and Pereira de Vasconcelos, A. (1993). Glucose and ketone body utilization by the brain of neonatal rats. *Prog. Neurobiol.* **40**, 163–221.

Norenberg, M. D., and Martinez-Hernandez, A. (1979). Fine structural localization of glutamine synthetase in astrocytes of rat brain. *Brain Res.* **161**, 303–310.

Norenberg, M. D., Neary, J. T., Bender, A. S., and Dombro, R. S. (1992). Hepatic encephalopathy: A disorder in glial–neuronal communication. *Prog. Brain Res.* **94**, 261–269.

Ogawa, S., Tank, D. W., Menon, R., Ellermann, J. M., Kim, S.-G., Merkle, H., and Ugurbil, K. (1992) Intrinsic signal changes accompanying sensory stimulation: Functional brain mapping with magnetic resonance imaging. *Proc. Natl. Acad. Sci. USA* **89**, 5951–5955.

O'Kusky, J., and Colonnier, M. (1982). A laminar analysis of the number of neurons, glia and synapses in the visual cortex (area 17) of the adult macaque monkey. *J. Comp. Neurol.* **210**, 278–290.

Owen, O. E., Morgan, A. P., Kemp, H. G., Sullivan, J. M., Herrera, M. G., and Cahill, G. F. J. (1967). Brain metabolism during fasting. *J. Clin. Invest.* **46**, 1589–1595.

Pellerin, L., and Magistretti, P. J. (1994). Glutamate uptake into astrocytes stimulates aerobic glycolysis: A mechanism coupling neuronal activity to glucose utilization. *Proc. Natl. Acad. Sci. USA* **91**, 10625–10629.

Phelps, C. H. (1972). Barbiturate-induced glycogen accumulation in brain: An electron microscopic study. *Brain Res.* **39**, 225–234.

Phelps, M. E., Huang, S. C., Hoffman, E. J., Selin, C., Sokoloff, L., and Kuhl, D. E. (1979). Tomographic measurement of local cerebral glucose metabolic rate in humans with (F–18)2-fluoro-2-deoxy-D-glucose: Validation of method. *Ann. Neurol.* **6**, 371–388.

Pierre, K., Pellerin, L., Debernardi, R., Riederer, B. M., and Magistretti, P. J. (2000). Cell-specific localization of monocarboxylate transporters, MCT1 and MCT2 in the adult mouse brain revealed by double immunohistochemical labeling and confocal microscopy. *Neuroscience* **100**, 617–727.

Plum, F., and Hindfeld, B. (1976). Handbook of clinical neurology. *In* "Handbook of Clinical Neurology" (P. J. Vinken and G. W. Bruyn, Eds.), Vol. 27, pp. 349–377. North-Holland, Amsterdam.

Poitry-Yamate, C. L., Poitry, S., and Tsacopoulos, M. (1995). Lactate released by Müller glial cells is metabolized by photoreceptors from mammalian retina. *J. Neurosci.* **15**, 5179–5191.

Prichard, J., Rothman, D., Novotny, E., Petroff, O., Kuwabara, T., Avison, M., Howseman, A., Hanstock, C., and Shulman, R. (1991). Lactate rise detected by ^{1}H NMR in human visual cortex during physiologic stimulation. *Proc. Natl. Acad. Sci. USA* **88**, 5829–5831.

Proverbio, F., and Hoffman, J. F. (1977). Membrane compartmentalized ATP and its preferential use by the Na$^+$–K$^+$ ATPase of human red cell ghosts. *J. Gen. Physiol.* **69**, 605–632.

Roland, P. E. (1985). Cortical organization of voluntary behavior in man. *Hum. Neurobiol.* **4**, 155–167.

Roy, C. S., and Sherrington, C. S. (1890). On the regulation of the blood supply of the brain. *J. Physiol. (London)* **11**, 85–108.

Schurr, A., Miller, J. J., Payne, R. S., and Rigor, B. M. (1999). An increase in lactate output by brain tissue serves to meet the energy needs of glutamate-activated neurons. *J. Neurosci.* **19**, 34–39.

Shank, R. P., Bennet, G. S., Freytag, S. O., and Campbell, G. L. (1985). Pyruvate carboxylase: An astrocyte-specific enzyme implicated in the replenishment of amino acid neurotransmitter pools. *Brain Res.* **329**, 364–367.

Sheu, K. F., Cooper, A. J., Koike, M., Lindsay, J. G., and Blass, J. P. (1994). Abnormality of the alpha-ketoglutarate dehydrogenase complex in fibroblasts from familial Alzheimer's disease. *Ann. Neurol.* **35**, 312–318.

Sibson, N. R., Dhankhar, A., Mason, G. F., Rothman, D. L., Behar, K. L., and Shulman, R. G. (1998). Stoichiometric coupling of brain glucose metabolism and glutamatergic neuronal activity. *Proc. Natl. Acad. Sci. USA* **95**, 316–321.

Siesjo, B. K. (1978) "Brain Energy Metabolism." Wiley, New York.

Sokoloff, L. (1981). Localization of functional activity in the central nervous system by measurement of glucose utilization with radioactive deoxyglucose. *J. Cereb. Blood Flow Metab.* **1**, 7–36.

Sokoloff, L., Reivich, M., Kennedy, C., Des Rosiers, M. H., Patlak, C. S., Pettigrew, K. D., Sakurada, O., and Shinohara, M. (1977). The [^{14}C]deoxyglucose method for the measurement of local cerebral glucose utilization: Theory, procedure, and normal values in the conscious and anesthetized albino rat. *J. Neurochem.* **28**, 897–916.

Swanson, R. A., Morton, M. M., Sagar, S. M., and Sharp, F. R. (1992). Sensory stimulation induces local cerebral glycogenolysis: Demonstration by autoradiography. *Neuroscience* **51**, 451–461.

Takahashi, S., Driscoll, B. F., Law, M. J., and Sokoloff, L. (1995). Role of sodium and potassium ions in regulation of glucose metabolism in cultured astroglia. *Proc. Natl. Acad. Sci. USA* **92**, 4616–4620.

Tsacopoulos, M., and Magistretti, P. J. (1996). Metabolic coupling between glia and neurons. *J. Neurosci.* **16**, 877–885.

Tsacopoulos, M., Evequoz-Mercier, V., Perrottet, P., and Buchner, E. (1988). Honeybee retinal glial cells transform glucose and supply the neurons with metabolic substrates. *Proc. Natl. Acad. Sci. USA* **85**, 8727–8731.

Vafaee, M. S., Meyer, E., Marrett, S., Paus, T., Evans, A. C., and Gjedde, A. (1999). Frequency-dependent changes in cerebral metabolic rate of oxygen during activation of human visual cortex. *J. Cereb. Blood Flow Metab.* **19**, 272–277.

Van den Berg, C. J., and Garfinkel, D. (1971). A simulation study of brain compartments: Metabolism of glutamate and related substances in mouse brain. *Biochem. J.* **123**, 211–218.

Villringer, A., and Dirnagl, U. (1995). Coupling of brain activity and cerebral blood flow: Basis of functional neuroimaging. *Cerebrovasc. Brain Metab. Rev.* **7**, 240–276.

Watanabe, H., and Passonneau, J. V. (1974). The effect of trauma on cerebral glycogen and related metabolites and enzymes. *Brain Res.* **66**, 147–159.

Westergaard, N., Sonnewald, U., and Schousboe, A. (1995). Metabolic trafficking between neurons and astrocytes: The glutamate/glutamine cycle revisited. *Dev. Neurosci.* **17**, 203–211.

Yudkoff, M., Daikhin, Y., Lin, Z.-P., Nissim, I., Stern, J., and Pleasure, D. (1994). Inter-relationships of leucine and glutamate metabolism in cultured astrocytes. *J. Neurochem.* **62**, 1953–1964.

Yudkoff, M., Nissim, I., Hummeler, K., Medow, M., and Pleasure, D. (1986). Utilization of [^{15}N]glutamate by cultured astrocytes. *Biochem. J.* **234**, 185–192.

4

Electrotonic Properties of Axons and Dendrites

Gordon M. Shepherd

Neurons characteristically have elaborate dendritic trees arising from their cell bodies and single axons with their own terminal branching patterns (see Chapters 1 and 2). With this structural apparatus, neurons carry out five basic functions (Fig. 4.1):

1. *Generate intrinsic activity* (at any given site in the neuron through voltage-dependent membrane properties and internal second-messenger mechanisms; see Chapters 5–7 and 12 for details).

2. *Receive synaptic inputs* (mostly in dendrites, to some extent in cell bodies, and in some cases in axon terminals; see also Chapters 15–18).

3. *Integrate signals* (combine synaptic responses with intrinsic membrane activity).

4. *Encode output patterns* (in graded potentials or action potentials).

5. *Distribute synaptic outputs* (from axon terminals and, in some cases, from cell bodies and dendrites; see also Chapters 8 and 16–18).

In addition to synaptic inputs and outputs, neurons may receive and send nonsynaptic signals in the form of electric fields, volume conduction of neurotransmitters and gases, and release hormones into the bloodstream.

Toward a Theory of Neuronal Information Processing

A **fundamental goal of neuroscience** is to develop quantitative descriptions of these functional operations and their coordination within the neuron that enable it to function as an integrated **information-processing unit**. This is the necessary basis for testing experiment-driven hypotheses that can lead to **realistic empirical computational models** of neurons (see

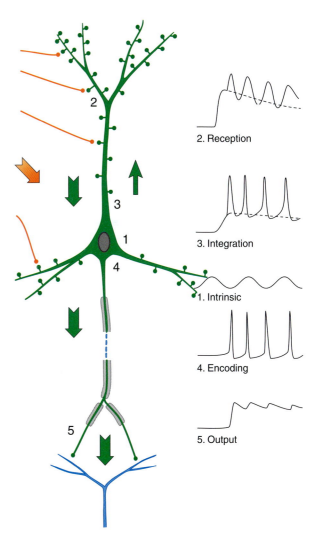

FIGURE 4.1 Nerve cells have four main regions and five main functions. Electrotonic potential spread is fundamental for coordinating the regions and their functions.

Copyright 2004, Elsevier Science (USA).
All rights reserved.

Chapter 7), neural systems, and networks and their roles in information processing and behavior (see Chapters 16 and 18).

Toward these ends, the first task is to understand how activity spreads within and between different parts of the neuron. To do this for the single axon is difficult enough; for the branching dendrites it becomes extremely challenging. It is no exaggeration to say that the task of understanding how intrinsic activity, synaptic potentials, and active potentials spread through and are integrated within the complex geometry of dendritic trees is one of the last frontiers of molecular and cellular neuroscience.

We begin in this chapter with the passive properties of the membrane that underlie the spread of most types of neuronal activity. In Chapter 17 we consider the active membrane properties that contribute to more complex types of information processing, particularly the types that take place in dendrites. *Together, these chapters provide an integrated theoretical framework for understanding the neuron as a complex information processing unit.* Both draw on other chapters for the specific properties—membrane receptors (Chapter 11), internal receptors (Chapter 12), synaptically gated membrane channels (Chapter 16), intrinsic voltage-gated channels (Chapters 5–7), and second-messenger systems (Chapter 12)—that mediate the neuron's operations.

The Basic Tools: Cable Theory and Compartmental Models

Slow spread of neuronal activity is by ionic or chemical diffusion or active transport. Our main interest in this chapter is in rapid spread by electric current. What are the factors that determine this spread? The most basic are electrotonic properties.

The origins of the study of electrotonic properties are summarized in Box 4.1, which highlights the interesting fact that the understanding of electrotonus arose from a merging of the study of current spread in nerve cells and muscle with the development of cable theory for transmission through electrical cables on the ocean floor. The electrotonic properties of neurons are therefore often referred to as *cable properties*. As indicated in Box 4.1, electrotonic theory was first applied mathematically to the nervous system in the late 19th century for the spread of electric current through nerve fibers. By the 1930s and 1940s, it was applied to simple invertebrate (crab and squid) axons—the first steps toward the development of the Hodgkin– Huxley equations (Chapters 5,7) for the action potential in the axon.

Mathematically, the analytical approach is impractical for the complexity of dendritic systems, but the development of computational compartmental methods for dendritic trees by Rall (1964, 1967, 1977; Rall and Shepherd, 1968; Segev *et al.*, 1995), beginning in the 1960s, opened the way to *compartmental models*. Together with the analytical methods of cable theory, these models have provided a sound basis for a theory of dendritic function (Segev *et al.*, 1995; Jack *et al.*, 1975). Combined with mathematical models for generation of synaptic potentials and action potentials (see Chapters 7 and 16), they provide the basis for a complete theoretical description of neuronal activity.

BOX 4.1

THE ORIGINS OF ELECTROTONUS

The mathematical treatment of axonal electrotonus began in the 1870s with the work of Hermann (1872, 1879), supported by Weber's (1873) mathematical analysis of the external field in the surrounding volume conductor. Hermann recognized the mathematical analogy of this problem with the problems in heat conduction, but the analogy with Kelvin's (1855) treatment of the submarine telegraph cable in the 1850s was first recognized by Hoorweg in 1898. This cable analogy was developed independently by Cremer (1899, 1909) and by Hermann (1905) early in the 20th century and has been widely used since that time. These mathematical analo-

gies are important because of the extensive literature devoted to both general mathematical methods and special solutions applicable to problems of this kind (Carslaw and Jaeger, 1959). Important papers on the steady-state distributions of axonal electrotonus are those of Rushton (1927, 1934) and Cole and Hodgkin (1939) published in the 1920s and 1930s. The two most useful mathematical presentations of axonal electrotonus (including consideration of transients) are those provided by Hodgkin and Rushton (1946) and Davis and Lorente de No (1947) in the 1940s. (From Rall, 1958.)

Wilfrid Rall

BOX 4.2

WEB SITES FOR NEURONAL MODELING

Components and Tools for Accessible Computer Modeling in Biology (Catacomb)
http://www.compneuro.org/catacomb/

Electrophysiology
http://pb010.anes.ucla.edu/

General Monte Carlo Simulator of Cellular Microphysiology (M-Cell)
http://www.mcell.cnl.salk.edu/

GEneral NEural SImulation System (Genesis)
http://www.genesis-sim.org/GENESIS/

Neural Open Simulation
http://www.neosim.org/

Neural Simulation Technology (NEST)
http://www.synod.uni-freiburg.de/

Neuron
http://www.neuron.yale.edu/

Simulator for Neural Networks and Action Potentials (SNNAP)
http://snnap.uth.tmc.edu

Further orientation to neuron modeling software
http://senselab.med.yale.edu/senselab/ModelDB/mdbre-sources.asp

A variety of software packages now make it possible for even a beginning student to explore functional properties and construct realistic neuron models (Shepherd and Brayton, 1979; Hines, 1984; Bower and Beeman, 1995; Ziv *et al.*, 1994) (see Chapter 7). Box 4.2 lists some web sites where tools are available to support the building of models. We therefore present modern electrotonic theory within the context of constructing these compartmental models. Exploration of these models will greatly aid the student in understanding the complexities that are present in even the simplest types of passive spread of current in axons and dendrites.

SPREAD OF STEADY-STATE SIGNALS

Modern Electrotonic Theory Depends on Simplifying Assumptions

The successful application of cable theory to nerve cells requires that it be based as closely as possible on the structural and functional properties of neuronal processes. The problem confronting the neuroscientist is that most of these processes are quite complicated. As discussed in Chapters 1 and 2, a segment of axon or dendrite may be filled with various organelles, bounded by a plasma membrane with its own complex structure and irregular outline, and surrounded by myriad neighboring processes (see Fig. 4.2A).

Constructing a model of the spread of electric current through such a segment therefore requires some carefully chosen simplifying assumptions,

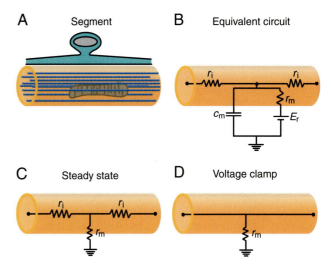

FIGURE 4.2 Construction of a compartmental model of the passive electrical properties of a nerve cell process begins with (A) identification of a segment of the process and its organelles, followed by (B) abstraction of an equivalent electrical circuit based on the membrane capacitance (c_m), membrane resistance (r_m), resting membrane potential (E_r), and internal resistance (r_i); and (C) abstraction of the circuit for steady-state electrotonus, in which c_m and E_r can be ignored. (D) The space clamp used in voltage-clamp analysis reduces the equivalent circuit even further to only the membrane resistance (r_m), usually depicted as the membrane conductances (g) for different ions. In a compartmental modeling program, the equivalent circuit parameters are scaled to the size of each segment.

which allow for the construction of an *equivalent circuit* of the electrical properties of such a segment. These are summarized in Box 4.3. Students should work their way through this series of simple equations; for the mathematically disinclined, a grasp of the principles is important. The main message of this box is that even the simplest representation of the passive spread of electrotonic potential during a steady-state input to a neuron requires a number of assumptions. Understanding them is essential for describing electrotonic spread under the different conditions that the nervous system presents. It is important for the student to gain at least an intuitive feel for the way that the complexity of the neuron is reduced by these assumptions to extract principles that give insight into function.

Electrotonic Spread Depends on the Characteristic Length

To describe the passive spread of electrotonic potential, we use the assumptions described in Box 4.3 to represent a segment of a process as an internal resistance r_i connected both to the r_i of the next segment and, through the membrane resistance r_m, to ground (see Fig. 4.2B). Let us first consider the spread of electrotonic potential under steady-state conditions (Fig. 4.2C). In standard cable theory, this is described by

$$V = \frac{r_m}{r_i} \cdot \frac{d^2V}{dx^2}. \tag{4.1}$$

This equation states that if there is a steady-state current input at point $x = 0$, the electrotonic potential (V) spreading along the cable is proportional to the product of the second derivative of the potential with respect to distance and the ratio of the membrane resistance (r_m) to the internal resistance (r_i) over that distance. The steady-state solution of this equation for a cable of infinite extension for positive values of x is

$$V = V_0 e^{-x/\lambda} \tag{4.2}$$

where λ is defined as the square root of r_m/r_i (in centimeters) and V_0 is the value of V at $x = 0$.

Inspection of this equation shows that when $x = \lambda$, the ratio of V to V_0 is $e^{-1} = 1/e = 0.37$. Thus, λ is a critical parameter defining the length over which the electrotonic potential spreading along an infinite cable, with given values for internal and membrane resistance, decays (is attenuated) to a value of 0.37 of the value at the site of the input. It is accordingly referred to as the *characteristic length* (space constant, length constant) of the cable. The higher the value of the

specific membrane resistance (R_m), the higher the value of r_m for that segment, the larger the value for λ, and the greater the spread (the less the attenuation) of electrotonic potential through that segment (Fig. 4.3). Specific membrane resistance (R_m) is thus an important variable in determining the spread of activity in a neuron. Most of the passive electrotonic current may be carried by K^+ "leak" channels, which are open at "rest" and are largely responsible for holding the cell at its resting potential. However, as mentioned earlier, many cells or regions within a cell are seldom at "rest" but are constantly active, in which case electrotonic current is carried by a variety of open channels. Thus, the effective R_m can vary from values of less than 1000 Ω cm^2 to more than 100 000 Ω cm^2 in different neurons and in different parts of a neuron. Note that λ varies with the square root of R_m, so a 100-fold difference in R_m translates into only a 10-fold difference in λ.

Conversely, the higher the value of the specific internal resistance (R_i), the higher the value of r_i for that segment, the smaller the value of λ, and the less

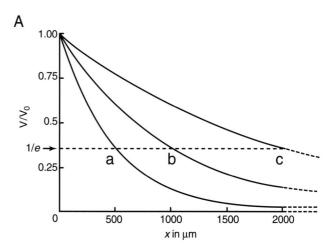

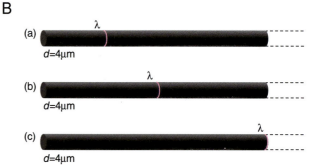

FIGURE 4.3 The space constant governing the spread of electrotonic potential through a nerve cell process depends on the square root of the ratio between the specific membrane resistance (R_m) and the specific internal resistance (R_i). (A) Potential profiles for processes with three different values of λ. (B) Red lines represent the location of λ on each of the three processes.

BOX 4.3

BASIC ASSUMPTIONS UNDERLYING CABLE THEORY

1. *Segments are cylinders.* A segment is assumed to be a cylinder with constant radius. This is the simplest assumption; however, compartmental simulations can readily incorporate different geometrical shapes with differing radii if needed (Fig. 4.2B).

2. *The electrotonic potential is due to a change in the membrane potential.* At any instant of time, the "resting" membrane potential (E_r) at any point on the neuron can be changed by several means: injection of current into the cell, extracellular currents that cross the membrane, and changes in membrane conductance (caused by a driving force different from that responsible for the membrane potential). Electric current then begins to spread between that point and the rest of the neuron, in accord with the equation

$$V = V_m - E_r,$$

where V is the electrotonic potential and V_m is the changed membrane potential. Modern neurobiologists recognize that the membrane potential is rarely at rest. In practice, "resting" potential means the membrane potential at any given instant of time other than during an action potential or rapid synaptic potential.

3. *Electrotonic current is ohmic.* Passive electrotonic current flow is usually assumed to be ohmic, that is, in accord with the simple linear equation

$$E = IR,$$

where E is the potential, I is the current, and R is the resistance. This relation is largely inferred from macroscopic measurements of the conductance of solutions having the composition of the intracellular medium, but is rarely measured directly for a given nerve process. Also largely untested is the likelihood that at the smallest dimensions (0.1 μm diameter or less), the processes and their internal organelles may acquire submicroscopic electrochemical properties that deviate significantly from macroscopic fluid conductance values; compartmental models permit the incorporation of estimates of these properties.

4. *In the steady state, membrane capacitance is ignored.* The simplest case of electrotonic spread occurs from the point on the membrane of a steady-state change (e.g., due to injected current, a change in synaptic conductance, or a change in voltage-gated conductance), so that time-varying properties (transient charging or discharg-

ing of the membrane) due to the membrane capacitance can be ignored (Fig. 4.2C).

5. *The resting membrane potential can usually be ignored.* In the simplest case, we consider the spread of electrotonic potential (V) relative to a uniform resting potential (E_r), so that the value of the resting potential can be ignored. Where the resting membrane potential may vary spatially, V must be defined for each segment as

$$V = V_m - F_r.$$

6. *Electrotonic current divides between internal and membrane resistances.* In the steady state, at any point on a process, current divides into two local resistance paths: further within the process through an internal (axial) resistance (r_i) or across the membrane through a membrane resistance (r_m) (see Fig. 4.2C).

7. *Axial current is inversely proportional to diameter.* Within the volume of the process, current is assumed to be distributed equally (in other words, the resistance across the process, in the Y and Z axes, is essentially zero). Because resistances in parallel sum to decrease the overall resistance, axial current (I) is inversely proportional to cross-sectional area ($I \propto 1/A \propto 1/\pi r^2$); thus, a thicker process has a lower overall axial resistance than does a thinner process. Because the axial resistance (r_i) is assumed to be uniform throughout the process, the total cross-sectional axial resistance of a segment is represented by a single resistance,

$$r_i = R_i/A,$$

where r_i is the internal resistance per unit length of cylinder (in ohms per centimeter of axial length), R_i is the specific internal resistance (in ohmcentimeter, or Ω cm), and A ($= \pi r^2$) is the cross-sectional area.

The internal structure of a process may contain membranous or filamentous organelles that can raise the effective internal resistance or provide high-conductance submicroscopic pathways that can lower it. In voltage-clamp experiments, the space clamp eliminates current through r_i, so that the only current remaining is through r_m, thereby permitting isolation and analysis of different ionic membrane conductances, as in the original experiments of Hodgkin and Huxley (Fig. 4.2D) (see also Chapters 5 and 7).

8. *Membrane current is inversely proportional to membrane surface area.* For a unit length of cylinder, the

BOX 4.3 *Continued*

membrane current (i_m) and the membrane resistance (r_m) are assumed to be uniform over the entire surface. Thus, by the same rule of the summing of parallel resistances, the membrane current is inversely proportional to the membrane area of the segment, so that a thicker process has a lower overall membrane resistance. Thus,

$$r_m = R_m/c,$$

where r_m is the membrane resistance for unit length of cylinder (in Ω cm of axial length), R_m is the specific membrane resistance (in Ω cm^2), and c (= $2\pi r$) is the circumference. For a segment, the entire membrane resistance is regarded as concentrated at one point; that is, there is no axial current flow within a segment but only between segments (see Fig. 4.2C).

Membrane current passes through ion channels in the membrane. The density and types of these channels vary in different processes and indeed may vary locally in different segments and branches. These differences are readily incorporated into compartmental representations of the processes.

9. *The external medium along the process is assumed to have zero resistivity.* In contrast with the internal axial resistivity (r_i), which is relatively high because of the small dimensions of most nerve processes, the external medium has a relatively low resistivity for current because of its relatively large volume. For this reason, the resistivity of the paths either along a process or to ground is generally regarded as negligible, and the potential outside the membrane is assumed to be everywhere equivalent to ground (see Fig. 4.2C). This greatly simplifies the equations that describe the spread of electrotonic potentials inside and along the membrane.

Compartmental models can simulate any arbitrary distribution of properties, including significant values

for extracellular resistance where relevant. Particular cases in which external resistivity may be large, such as the special membrane caps around synapses on the cell body or axon hillock of a neuron, can be addressed by suitable representation in the simulations. However, for most simulations, the assumption of negligible external resistance is a useful simplifying first approximation.

10. *Driving forces on membrane conductances are assumed to be constant.* It is usually assumed that ion concentrations across the membrane are constant during activity. Changes in ion concentrations with activity may occur, particularly in constricted extracellular or intracellular compartments; these changes may cause deviations from the assumptions of constant driving forces for the membrane currents, as well as the assumption of uniform E_r. For example, accumulations of extracellular K$^+$ may change local E_r (Pongracz *et al.*, 1992), and intracellular accumulations of ions within the tiny volumes of spine heads may change the driving force on synaptic currents (Qian and Sejnowski, 1989). These special properties are easily included in most compartmental models.

11. *Cables have different boundary conditions.* In classic electrotonic theory, a cable such as that used for long-distance telecommunication is very long and can be considered of infinite length (one customarily assumes a semi-infinite cable with $V = 0$ at $x = 0$ and only positive values of length x). This assumption carries over to the application of cable theory to long axons, but most dendrites are relatively short. This imposes boundary conditions on the solutions of the cable equations, which have very important effects on electrotonic spread.

In highly branched dendritic trees, boundary conditions are difficult to deal with analytically but are readily represented in compartmental models.

---

the spread of electrotonic potential through that segment (see Fig. 4.3). Traditionally, the value of R_i has been believed to be in the range of approximately 50–100 Ω cm based on muscle cells and the squid axon. In mammalian neurons, estimates now tend toward a value of 100–250 Ω cm. This limited range may suggest that R_i is less important than R_m in controlling passive current spread in a neuron. The square-root relation further reduces the sensitivity of λ to R_i. However, as noted in assumption 7 in Box 4.3, the membranous and filamentous organelles in the cytoplasm may alter the effective R_i. The presence of these organelles in very thin processes, such as distal

dendritic branches, spine stems, and axon preterminals, may thus have potentially significant effects on the spread of electrotonic current through them. Furthermore, the relative significance of R_i and R_m greatly depends on the length of a given process, as will be seen shortly.

Electrotonic Spread Depends on the Diameter of a Process

The space constant (λ) depends not only on the internal and membrane resistance, but also on the diameter of a process (Fig. 4.4). Thus, from the rela-

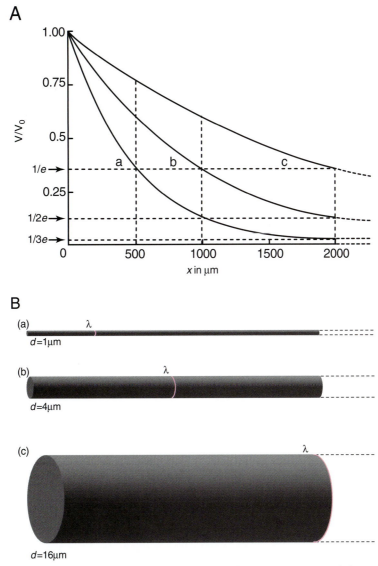

FIGURE 4.4 The space constant governing the spread of electrotonic potential through a nerve cell process also depends on the square root of the diameter of the process. (A) Potential profiles for processes with three different diameters but fixed values of R_i and R_m. (B) The three axon profiles in (A). Note that to double λ, the diameter must be quadrupled.

tions between r_m and R_m, and r_i and R_i, discussed in the preceding section,

$$\lambda = \sqrt{\frac{r_m}{r_i}} = \sqrt{\frac{R_m}{R_i} \cdot \frac{d}{4}}. \tag{4.3}$$

Neuronal processes vary widely in diameter. In the mammalian nervous system, the thinnest processes are the distal branches of dendrites, the necks of some dendritic spines, and the cilia of some sensory cells; these processes may have diameters of only 0.1 μm or less (the thinnest processes in the nervous system are approximately 0.02 μm). In contrast, the thickest processes in the mammal are the largest myelinated axons and the largest dendritic trunks, which may have diameters as large as 20 to 25 μm. This means that the range of diameters is approximately three orders of magnitude (1000-fold). Note, again, that λ varies with the square root of d; thus, for a 10-fold difference in diameter, the difference in λ is only about 3-fold (Fig. 4.4).

Electrotonic Properties Must Be Assessed in Relation to the Lengths of Neuronal Processes

Application of classic cable theory to neuronal processes assumes that the processes are infinitely long (assumption 11 in Box 4.3). However, because neuronal processes have finite lengths, the length of a given process must be compared with λ to assess the extent to which λ accurately describes the actual electrotonic spread in that process. One of the largest processes in any nervous system, the squid giant axon, has a diameter of approximately 1 mm. R_m for this axon has been estimated as 600 Ω cm^2 (a very low value compared with most values of R_m in mammals), and R_i as approximately 80 Ω cm, the value of Ringer solution. Entering these values into Eq. (4.3) gives a λ of approximately 4.5 mm (note that the very large diameter is counterbalanced by the very low R_m). The real length of the giant axon is several centimeters; to relate real length to characteristic length, we define electrotonic length (L) as

$$L = \frac{x}{\lambda}. \tag{4.4}$$

Thus, if $x = 30$ mm, then $L = 30$ mm/4.5 mm = 7; that is, the real length of the giant axon is 7 λ. The electrotonic potential decays to only a small percentage of the original value by only three characteristic lengths (see Fig. 4.4), so for this case the assumption of an infinite length is justified. A reason often given for why the nervous system needs action potentials is that they overcome the severe attenuation of passively spreading potentials that occurs over the considerable lengths required for transmission of signals by axons.

In contrast to axons, dendritic branches have lengths that are usually much shorter than three characteristic lengths. In dendrites, therefore, the branching patterns come to dominate the extent of potential spread. We discuss the methods for dealing with these branching patterns later in this chapter.

The relative importance of R_i and R_m in controlling current spread depends on the length of a segment relative to its characteristic length λ. Consider, for example, a neuronal process (large axon or dendrite) 10 μm in diameter, with $R_i = 100$ Ω cm and $R_m = 10\,000$ Ω cm^2. By the preceding definitions, the longitudinal resistance r_i per unit length (1 cm) would equal 130 MΩ, whereas the membrane resistance r_m would be only 3 MΩ. Thus, the relatively low specific internal resistance has a relatively large effect over the unit distance (cm) because of the small diameter of the nerve process. Such a process would have a characteristic length of 1500 μm, where, by definition (see Eq. 4.4), $r_m = r_i$, and the current would be equally

divided between the two. At shorter distances, more current tends to flow through r_i as the membrane area is reduced (and r_m thereby increases); this becomes an important factor in shaping current flow through small dendritic branches and dendritic spines (as will be seen shortly and in Chapter 17).

Summary

Passive spread of electrical potential along the cell membrane underlies all types of electrical signaling in the neuron. It is thus the foundation for understanding how the diverse functions of the neuron are coordinated within the neuron, so that the neuron can generate, receive, integrate, encode, and send signals in interacting with its neighboring neurons and glial cells.

Electrotonic spread shares properties with electrical transmission through electrical cables; the study of cable transmission has put these properties on a sound quantitative basis. The theoretical basis for extension of cable theory to complex dendritic trees has been developed in parallel with compartmental modeling methods for simulating dendritic signal processing.

Cable theory depends on a number of reasonable simplifying assumptions about the geometry of neuronal processes and current flow within them. Steadystate electrotonus in dendrites depends, to begin with, on passive resistance of the membrane and of the internal cytoplasm and on the diameter and length of a nerve process.

SPREAD OF TRANSIENT SIGNALS

Electrotonic Spread of Transient Signals Depends on Membrane Capacitance

Until now, we have considered only the passive spread of steady-state inputs. However, the essence of many neural signals is that they change rapidly. In mammals, fast action potentials characteristically last from 1 to 5 ms, and fast synaptic potentials last from 5 to 30 ms. How do the electrotonic properties affect spread of these rapid signals?

Rapid signal spread depends not only on all of the factors discussed thus far, but also on the membrane capacitance (c_m), which is due to the lipid moiety of the plasma membrane. Classically, the value of the specific membrane capacitance (C_m) has been considered to be 1 microfarad per square centimeter (1 μF cm^{-2}). However, values of 0.6–0.75 μF cm^{-2} are now preferred for the lipid moiety itself, the remainder being due to gating charges on membrane proteins (Jack *et al.*, 1975).

The simplest case demonstrating the effect of the membrane capacitance on transient signals is that of a single segment or a cell body with no processes (a very unrealistic assumption, but a simple starting point). In the equivalent electrical circuit for a neural process, the membrane capacitance is placed in parallel with the ohmic components of the membrane conductance and the driving potentials for ion flows through those conductances (see Fig. 4.2B). Again neglecting the resting membrane potential, we take as an example the injection of a current step into a soma; in this case, the time course of the current spread to ground is described by the sum of the capacitative and resistive currents (plus the input current, I_{pulse}):

$$C\frac{dV_m}{dt} + \frac{V_m}{R} = I_{pulse} \cdot \quad (4.5)$$

Rearranging,

$$RC\frac{dV_m}{dt} + V_m = I_{pulse} \cdot R, \quad (4.6)$$

where $RC = \tau$ (τ is the time constant of the membrane). The solution of this equation for the response to a step change in current (I_{pulse}) is

$$V_m(T) = I_{pulse}R(1 - e^{-T}), \quad (4.7)$$

where $T = t/\tau$. When the pulse is terminated, the decay of the steady-state potential (V_∞) to rest is given by

$$V_m(T) = V_\infty e^{-T}. \quad (4.8)$$

These "on" and "off" transients are shown in Fig. 4.5. The significance of τ is shown in the diagram; it is the time required for the voltage change across the membrane to reach $1/e = 0.37$ of its final value (i.e., the change from V_∞ back to the resting potential). This time constant of the membrane defines the transient voltage response of a patch of membrane to a current step in terms of the electrotonic properties of the patch, analogous to the way that the length constant defines the spread of voltage change over distance in terms of the electrotonic properties of a segment.

A Two-Compartment Model Defines the Basic Properties of Signal Spread

These spatial and temporal cable properties can be combined in a two-compartment model (Shepherd, 1994) that can be applied to the generation and spread of any arbitrary transient signal (Fig. 4.6).

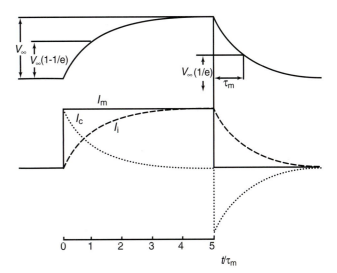

FIGURE 4.5 The equivalent circuit of a single isolated compartment responds to an injected current step by charging and discharging along a time course determined by the time constant τ. In actuality, nerve cell segments are parts of longer processes (axonal or dendritic) or larger branching trees, so the actual time courses of charging or discharging are modified. V_∞, steady-state voltage; I_m (I_{pulse}), injected current applied to membrane; I_c, current to charge the capacitance; I_i current through the ionic leak conductance; τ_m, membrane time constant.

In the simplest case, current is injected into one of the compartments. For a positive current pulse, the positive charge injected into compartment A attempts to flow outward across the membrane, partially opposing the negative charge on the inside of the lipid membrane (the charge responsible for the negative resting potential), thereby depolarizing the membrane capacitance (C_m) at that site. At the same time, the charge begins to flow as current across the membrane through the resistance of the ionic membrane channels (R_m) that are open at that site. The proportion of charge divided between C_m and R_m determines the rate of charge of the membrane, that is, the membrane time constant, τ. However, charge also starts to flow through the internal resistance (R_i) into compartment B, where the same events take place. The charging (and discharging) transient in compartment A departs from the time constant of a single isolated compartment, being faster because of the conductance load (e.g., current sink) of the rest of the cable (represented by compartment B). Thus, *the time constant of the system no longer describes the charging transient in the system because of the conductance load of one compartment on another* (this is analogous to the way that the conductance load makes the electrotonic potential more attenuated than the space constant). The system is entirely passive and invariant; the response to a second current pulse sums linearly with that of the first.

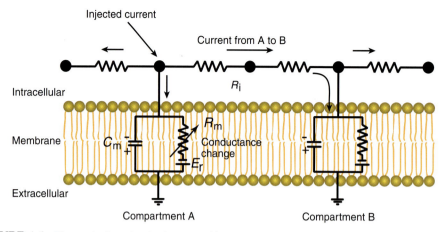

FIGURE 4.6 The equivalent circuit of two neighboring compartments or segments (A and B) of an axon or dendrite shows the pathways for current spread in response to an input (injected current or increase in membrane conductance) at segment A.

This case is a useful starting point because an experimenter often uses electrical currents as stimuli in analyzing nerve function. However, a neuron normally generates current spread by means of a localized conductance change across the membrane. Consider such a change in the example of compartment A in Fig. 4.6. Assume a change in the ionic conductance for Na$^+$, as in the initiation of an action potential or an excitatory postsynaptic potential (see Chapter 16), producing an inward positive current in compartment A. The charge transferred to the interior surface of the membrane attempts to follow the same paths followed by the injected current just described, by opposing the negativity inside the membrane capacitance, crossing the membrane through the open membrane channels to ground, and spreading through the internal resistance to the next compartment, where the charge flows are similar.

Thus, the two cases start with different means of transferring positive charge within the cell, but from that point the current paths and the associated spread of the electrotonic potential are similar. The electrotonic current that spreads between the two segments is referred to as the *local current*. Note again that the charging transient in compartment A is faster than the time constant of the resting membrane; this difference is due both to the conductance load of compartment B and to the fact that the imposed conductance increase in compartment A reduces the time constant of compartment A (by reducing effective R_m). This illustrates a critical point first emphasized by Wilfrid Rall: *changes in membrane conductance alter the system so that it is no longer a linear system, even though it is a passive system* (Rall, 1964). Thus, passive electrotonic spread is not so simple as most people think! Nonlinear sum-

mation of synaptic responses is further discussed later in this chapter.

Summary

In addition to the properties underlying steady-state electrotonus, passive spread of *transient* potentials depends on the membrane capacitance. Initiation of electrotonic spread by intracellular injection of a transient electrical current pulse produces an electrotonic potential that spreads by passive local currents from point to point. It is more attenuated in amplitude than the steady-state case as it spreads along an axon or dendrite, due to the low-pass filtering action of the membrane capacitance. Simultaneous current pulses at that site or other sites produce potentials that add linearly, because the passive properties are invariant. However, transient conductance changes, as in synaptic responses, generate electrotonic potentials that do not sum linearly because of the nonlinear interactions of the conductances.

ELECTROTONIC PROPERTIES UNDERLYING PROPAGATION IN AXONS

Impulses Propagate in Unmyelinated Axons by Means of Local Electrotonic Currents

Let us apply the knowledge of electrotonic current properties to the propagation of an action potential in an unmyelinated axon (i.e., one that is not surrounded by myelin or other membranes that restrict the spread of extracellular current). Details on the ionic mechanisms of the nerve impulse can be found in Chapters 5, 6, and 7. The local current spreading through the internal resistance to the neighboring compartment enables the action potential to spread along the membrane of the axon. The rate of spread is determined by both the passive cable properties and the kinetics of the action potential mechanism.

Each of the cable properties is relevant in specific ways. For brief signals such as the action potential, C_m is critical in controlling the rate of change of the membrane potential. For long processes such as axons, R_i increasingly opposes electrotonic current flow as the value of r_i increases beyond the characteristic length, whereas the effect of r_m decreases, due to the increased membrane area for parallel current paths (see earlier). This effect is greater in thinner axons, which have shorter characteristic lengths. Finally, R_m is a parameter that can vary widely. Thus, each of these parameters must be assessed to understand the exquisite effects of passive variables on the rates of impulse spread in axons.

A high value of R_m, for example, forces current further along the membrane, increasing the characteristic length and consequently the spread of electrotonic potential, as we have seen; however, at the same time, it increases the membrane time constant, thus slowing the response of a neighboring compartment to a rapid change. Increasing the diameter of the axon (d) lowers the effective internal resistance of a compartment, thereby also increasing the characteristic length, but without a concomitant effect on the time constant. Thus, changing the diameter is a direct way of affecting the rate of impulse propagation through changes in passive electrotonic properties. The conduction rate of any given axon depends on the particular combination of these properties (Rushton, 1951; Ritchie, 1995). For example, in the squid giant axon, the very large diameter (as large as 1 mm) promotes rapid impulse propagation; the very low value of R_m (600 Ω cm^2) lowers the time constant (promoting rapid current spread) but also decreases the length constant (limiting the spatial extent of current spread).

The effects of these passive properties on impulse velocity also depend on other factors. For example, on the basis of the cable equations, it can be shown that the conduction velocity should be related to the square root of the diameter (Rushton, 1951). However, the density of Na$^+$ channels in fibers of different diameters is not constant; thus, the binding of saxitoxin molecules, for example, to Na$^+$ channels varies greatly with diameter, from almost 300 μm^{-2} in the squid axon to only 35 μm^{-2} in the garfish olfactory nerve (Ritchie, 1995). Thus, both active and passive

properties must be assessed to understand a particular functional property.

Myelinated Axons Have Membrane Wrappings and Booster Sites for Faster Conduction

The evolution of larger brains to control larger bodies and more complex behavior required communication over longer distances within the brain and body. This requirement placed a premium on the ability of axons to conduct impulses as rapidly as possible. As noted in the preceding section and Chapter 5, a direct way of increasing the rate of conduction is by increasing the diameter, but larger diameters mean fewer axons within a given space, and complex behavior must be mediated by many axons. Another way of increasing the rate of conduction is to make the kinetics of the impulse mechanism faster; that is, make the rate of increase in Na$^+$ conductance with increasing membrane depolarization faster. The Hodgkin–Huxley equations (Chapter 7) for the action potential in mammalian nerves in fact have this faster rate.

As we have seen, rapid spread of local currents is promoted by an increase in R_m but opposed by an associated increase in the time constant. What is needed is an increase in R_m with a concomitant decrease in C_m. This is brought about by putting more resistances in series with the membrane resistance (because resistances in series add) while putting more capacitances in series with the membrane capacitance (capacitances in series add as the reciprocals, much like resistances in parallel, as noted earlier). The way the nervous system does this is through a special satellite cell called a Schwann cell, a type of glial cell. As described in Chapters 1 and 2, Schwann cells wrap many layers of their plasma membranes around an axon. The membranes contain special constituents and together are called myelin. Myelinated nerves contain the fastest conducting axons in the nervous system. A general empirical finding known as the Hursh factor (Hursh, 1939) states that *the rate of propagation of an impulse along a myelinated axon in meters per second is six times the diameter of the axon in micrometers.* Thus, the largest axons in the mammalian nervous system are approximately 20 μm in diameter, and their conduction rate is approximately 120 m s^{-1}, whereas the thin myelinated axons of about 1 μm in diameter have conduction rates of approximately 5 to 10 m s^{-1}.

As discussed in Chapter 5, myelinated axons are not myelinated along their entire length; at regular intervals (approximately 1 mm in peripheral nerves), the myelin covering is interrupted by a node of

Ranvier. The node has a complex structure. The density of voltage-sensitive Na⁺ channels at the node is high (10,000 μm^{-2}), whereas it is very low (20 μm^{-2}) in the internodal membrane. This difference in density means that the impulse is actively generated only at the node; the impulse jumps, so to speak, from node to node, and the process is therefore called *saltatory conduction. A myelinated axon therefore resembles a passive cable with active booster stations.*

In rapidly conducting axons the impulse may extend over considerable lengths; for example, in a 20-μm-diameter axon conducting at 120 m s^{-1}, at any instant of time an impulse of 1-ms duration extends over a 120-mm length of axon, which includes more than 100 nodes of Ranvier. It is therefore more appropriate to conceive that *the impulse is generated simultaneously by many nodes, with their summed local currents spreading to the next adjacent nodes to activate them.*

The specific membrane resistance (R_m) at the node is estimated to be only 50 Ω cm^2, owing to a large number of open ionic channels at rest. This value of R_m reduces the time constant of the nodal membrane to approximately 50 microseconds, which enables the nodal membrane to charge and discharge quickly, greatly aiding rapid impulse generation. For axons of equal cross-sectional area, myelination is estimated to increase the impulse conduction rate 100-fold.

In all axons, a critical relationship exists between the amount of local current spreading down an adjacent axon and the threshold for opening Na⁺ channels in the membrane of the adjacent axon so that propagation of the impulse can continue. This introduces the notion of a *safety factor*, that is, the amount by which the electrotonic potential exceeds the threshold for activating the impulse. The safety factor must protect against a wide range of operating conditions, including adaptation (during high-frequency firing), fatigue, injury, infection, degeneration, and aging.

Normally, an excess of local current ensures an adequate margin of safety against these factors. In the squid axon, the safety factor ranges from 4 to 5. In myelinated axons, an exquisite matching between the internodal electrotonic properties and the nodal active properties ensures that the electrotonic potential reaching a node has an adequate amplitude and the node has sufficient Na⁺ channels to generate an action potential that will spread to the next node. The safety factors for myelinated axons range from 5 to 10. Thus, the interaction of passive and active properties underlies the safety factors for impulse propagation in axons. Similar considerations apply to the orthodromic spread of signals in dendritic branches and the backpropagation of action potentials from the axon hillock into the soma and dendrites.

Theoretically, the conduction velocity, space constant, and impulse wavelength of myelinated fibers scale linearly with fiber diameter (Rushton, 1951; Ritchie, 1995), as indeed is indicated in the aforementioned Hursh factor. This difference between myelinated and unmyelinated fibers in their dependence on diameter is thus related to the scaling of the internodal length. At approximately 1 μm in diameter, the Hursh factor breaks down; at less than 1 μm in diameter, there is an advantage, all other factors being equal, for an axon to be unmyelinated. However, myelinated axons are found down to a diameter of 0.2 μm, and this has been correlated with shorter internodal distances (Waxman and Bennett, 1972). Thus, conduction velocity in myelinated nerve depends on a complex interplay between passive and active properties.

Summary: Passive Spread and Active Propagation

A frequent source of confusion in describing nerve activity is use of the terms *spread* and *propagation*. Rall has used these terms to make the distinction between electrotonus and action potential conduction. Thus, electrotonic potentials are said to *spread* whereas action potentials (impulses) are said to *propagate*. These and related terms are discussed further in Chapter 17.

Impulses propagate continuously through unmyelinated fibers because the local currents spread directly to neighboring sites on the membrane. The rate of propagation is directly determined by the electrotonic properties of the fiber. In myelinated axons, the impulse propagates discontinuously from node to node. The electrotonic properties of both the nodal and internodal regions determine not only the rate of impulse propagation but also the safety factor for impulse transmission.

ELECTROTONIC SPREAD IN DENDRITES

Dendrites are the main neuronal compartment for reception of synaptic inputs. Spread of synaptic responses through the dendritic tree depends critically on the electrotonic properties of the dendrites. Because dendrites are branching structures, understanding the rules governing dendritic electrotonus and the resulting integration of synaptic responses in dendrites is much more difficult than understanding the rules of simple spread in a single axon.

Dendritic Electrotonic Spread Depends on Boundary Conditions of Dendritic Termination and Branching

As noted earlier, compared with axons, dendrites are relatively short, and their length becomes an important factor in assessing their electrotonic properties. Consider, in the mammalian nervous system, a moderately thin dendrite of 1 μm (three orders of magnitude smaller than the squid axon) that has a typical R_m of 60,000 Ω cm^2 (two orders of magnitude larger than that of the squid axon) and an R_i of 240 Ω cm (three times the squid value). Inserting these values into the equation for characteristic length (Eq. 4.3) gives a λ of approximately 790 μm. This illustrates that λ tends to be relatively long in comparison with the actual lengths of the dendrites; in other words, because of the relatively high membrane resistance, *electrotonic spread of potentials is relatively effective within a dendritic branching tree.*

This essential property underlies the integration of signals in dendrites. The effective spread immediately leads to a second property. The assumption of infinite length no longer holds; dendritic branches are bounded by their terminations, on the one hand, and the nature of their branching, on the other. These are termed boundary conditions. *The spread of electrotonic potentials is therefore exquisitely sensitive to the boundary conditions of the dendrites.*

This problem is approached most easily by considering two extreme types of termination of a dendritic branch. First, consider that at $x = a$ the branch ends in a sealed end with infinite resistance. In this case, the axial component of the current can spread no further and must therefore seek the only path to ground, which is across the membrane of the cylinder. This current is added to the current already crossing the membrane; in the equation for Ohm's law ($E = IR$), I is increased, giving a larger E. The membrane will thus be more depolarized up to the terminal point a; in fact, near point a, axial current is negligible and almost all the current is across the membrane, which amounts to a virtual space clamp near point a (Fig. 4.7). If at point a the infinite resistance is replaced by the more realistic assumption of an end that is sealed with surface membrane, only a small amount of current crosses this membrane and attenuation of electrotonic potential is only slightly greater. Infinite resistance is therefore a useful approximation for assessing the effects of a sealed end on electrotonic spread in a terminal dendritic branch.

At the other extreme, consider that at point a a small dendritic branch opens out into a very large conductance. Examples are, in the extreme, a hole in

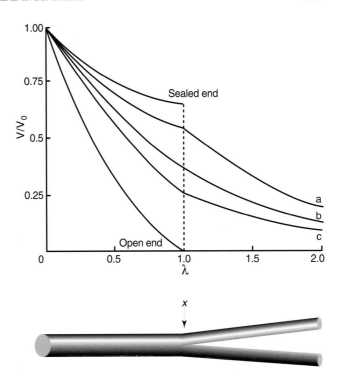

FIGURE 4.7 The spread of electrotonic potential through a short nerve cell process such as a dendritic branch is governed by the space constant and by the size of the branches; the latter imposes a boundary condition at the branch point. Curves a–c represent a range of realistic assumptions about the sizes of the branches relative to the size of the stem, together with the limiting conditions of an open circuit (corresponding to an infinite conductance load) and a closed circuit (corresponding to a sealed tip).

the membrane; less extreme are a very small dendritic branch on a large soma and a small twig or spine on a large dendritic branch. Recall that large processes sum their resistances in parallel, which gives low current density and small voltage changes. Therefore, a current spreading through the high resistance of a small branch into a large branch encounters a very low resistance. For steady-state current spread, this situation is referred to as a large conductance load; for a transient current, we refer to it as a low impedance (which includes the effect of the membrane capacitance). This introduces the key principle of *impedance matching between interacting compartments*, an important principle generally in biological systems. In our example, an *impedance mismatch* exists between the high-impedance thin branch and the lower-impedance thick branch. This mismatch reduces any voltage change due to the current and, in the extreme, effectively clamps the membrane to the resting potential (E_r) at that point. The electrotonic potential is thus attenuated through the branch much more rapidly than would be predicted by the characteristic length (see Fig. 4.7). This does not invalidate λ as a measure of electrotonic properties; rather, it means that, as with the time constant, *each cable property must be assessed within the context of the size and branching of the dendrites.*

All the different types of branching found in neuronal dendrites lie between these two extremes with a corresponding range of boundary conditions at $x = a$. Consider a segment of dendrite that divides into two branches at $x = a$. We can appreciate intuitively that the amount of spread of electrotonic potential into the two branches will be governed by the factors just considered. One possibility is that the two branches have very small diameters, so their input impedance is higher than that of the segment; in this case, the situation will tend toward the sealed-end case (Fig. 4.7, top trace). In contrast, the segment may give rise to two very fat branches, so the situation will tend toward the large-conductance-load case (Fig. 4.7, bottom trace).

For many cases of dendritic branching, the input impedance of the branches is between the two extremes (see Fig. 4.7, traces a–c), providing for a reasonable degree of impedance matching between the stem branch and its two daughter branches. This situation thus approximates the infinite-cylinder case, in which by definition the input impedance at one site matches that at its neighboring site along the cylinder. The general rules for impedance matching at branch points were worked out by Rall and colleagues (Rall, 1964, 1977; Segev *et al.*, 1995), who showed that the input conductance of a dendritic segment varies with the diameter raised to the 3/2 power. There is electrotonic continuity at a branch point equivalent to the infinitely extended cylinder if the diameter of the segment raised to the 3/2 power equals the sum of the diameters raised to the 3/2 power of all the daughter branches. An idealized branching pattern that satisfies this rule is shown in Fig. 4.8. When the branching tree reduces to a single chain of compartments, as in this case, it is called an "equivalent cylinder." When the branching pattern departs from the $d^{3/2}$ rule, the compartment chain is referred to as an "equivalent dendrite" (cf. Rall and Shepherd, 1968).

Dendritic Synaptic Potentials Are Delayed and Attenuated by Electrotonic Spread

We are now in a position to assess the effects of cable properties on the time course of the spread of synaptic potentials through dendritic branches and trees. Consider in Fig. 4.8 the case of recording from a soma while delivering a brief excitatory synaptic con-

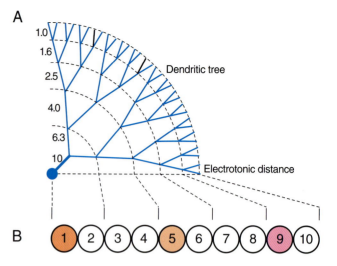

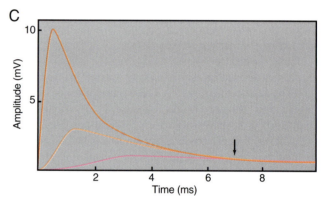

FIGURE 4.8 The spread of electrotonic potentials is accompanied by a delay and an attenuation of amplitude. (A) Dendritic diameters (left) satisfy the $d^{3/2}$ rule, so that the tree can be portrayed by an equivalent cylinder. An excitatory postsynaptic potential (EPSP) is generated in compartment 1, 5, or 9 (B) while recordings are made from compartment 1. The graph (C) shows the short latency, large amplitude, and rapid transient response in compartment 1 at the site of input, as well as the later, smaller, and slower responses recorded in compartment 1 for the same input to compartments 5 and 9. Despite the initial differences in time course, the responses converge at the arrow to decay together. Based on Rall (1964); computer simulation in (C) by K. L. Marton.

ductance change to different locations in the dendritic tree. The response to the nearest site is a rapidly rising synaptic potential that peaks near the end of the conductance change and then rapidly decays toward baseline. When the input is delivered to the middle of the chain of compartments, the response in the soma begins only after a delay, rises more slowly, reaches a much lower peak (which is reached after the end of the conductance change in the soma), and decays slowly toward baseline. For input to the terminal compartment, the voltage delay at the soma is so long that the response has scarcely started by the end of the conductance change in the distal dendrite; the response rises slowly to a delayed (several milliseconds) and prolonged plateau that subsides very slowly (see Fig. 4.8).

Although the synaptic potentials thus decrease in amplitude as they spread, *the rate of electrotonic spread can be calculated in terms of the half-amplitude at any point*. If distance is expressed in units of λ and time in units of τ, then for spread through a semi-infinite cable, we have the simple equation (Jack *et al.*, 1975).

$$\text{Velocity} = 2\frac{\lambda}{\tau}. \qquad (4.9)$$

Thus, if boundary effects are ignored, for the 10-mm process mentioned earlier in which λ = 1500 μm and τ = 10 ms, the velocity of spread would be 0.3 m s^{-1}, or 300 μm ms^{-1}. It can be seen that electrotonic spread can be relatively fast over short distances within a dendritic tree but is very slow in comparison with impulse transmission for an axon of this diameter (60 m s^{-1}). Thus, *both the severe decrement and the slow velocity make passive spread by itself ineffective for transmission over long distances.*

These general rules of delay and attenuation govern the passive spread of all transient potentials in dendritic branches and trees. As a rule of thumb, spread within one space constant (see the decrement between compartments 1 and 5 in Fig. 4.8) mediates relatively effective linkage for rapid signal integration, whereas spread over one or two space constants (see the decrement between compartments 1 and 9 in Fig. 4.8) is limited to slower background modulation. In real dendrites, these limitations are often overcome through boosting of the signals at intermediate sites by voltage-gated properties (see Chapter 17).

The spread of electrotonic potential from a point of input involves the *equalization of charge* on the membrane throughout the system. After cessation of the input, a time is reached when charge has become equalized and the entire system is equipotential; from this time on, the remaining electrotonic potential decays equally at every point in the system. This time is indicated by the vertical arrow in Fig. 4.8C. Before this time, the decaying transients are governed by equalizing time constants, indicating electrotonic spread, which can be identified by "peeling" on semilogarithmic plots of the potentials (Rall, 1977). After this time, the decay of electrotonic potential is governed solely by the membrane time constant, τ. In experimental recordings of synaptic potentials, the overall electrotonic length of the dendritic system,

considered as an "equivalent cylinder" or "equivalent dendrite" (see above), can be estimated from measurements of the membrane time constant and the equalizing time constants. The electrotonic lengths of the dendritic trees of many neuron types lie between 0.3 and 1.5.

What is the spread of the postsynaptic potential throughout the system when a synaptic input is delivered to *a single terminal dendritic branch* (Fig. 4.9) (Rall and Rinzel, 1973; Rinzel and Rall, 1974)? First consider a steady-state potential. Two main factors are involved. First, in the terminal branch, both the effective membrane resistance and the internal resistance are very large; hence, the branch has a very high input resistance, which produces a very large voltage change for any given synaptic conductance change. Balanced against this high input resistance is a second factor: the small branch has a very large conductance load on it because of the rest of the dendritic tree. As a result, there is a steep decrement in the electrotonic potential spreading from the branch through the tree to the cell body (see Fig. 4.9A). For comparison, a direct input to the soma produces only a small potential change there because of the relatively very low input resistance at that site.

Now for a transient synaptic input, a third factor— the membrane capacitance—must be taken into account. The small surface area of a terminal branch has little capacitance, according to our earlier calculations, so the amplitude of a transient response differs little from a steady-state response in the branch. However, in spreading out from a small process (such as a distal dendritic twig or spine), the transient synaptic potential is attenuated by the impedance mismatch between the process and the rest of the dendritic tree. Spread of the transient through the dendritic tree is further attenuated by the need to charge the capacitance of the dendritic membrane and is slowed by the time taken for the charging. The amount of slowing is so precise that the relative distance of a synapse in the dendritic tree from the soma can be calculated from experimental measurements in the soma of the time to peak of the recorded synaptic potential (Rall, 1977; Johnston and Wu, 1995). For these reasons, the peak of a synaptic potential transient spreading from the distal dendrites toward the soma may be severely attenuated, severalfold more than for the case of steady-state attenuation. This is often referred to as the *filtering effect* of the cable properties. However, the integrated response (the area under the transient voltage) is approximately equivalent to the steady-state amplitude, indicating that there is only a small loss of total charge (see Fig. 4.9B).

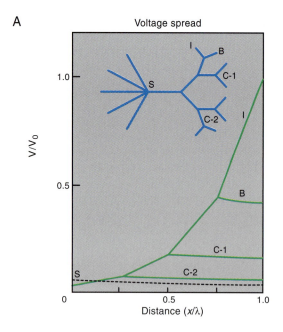

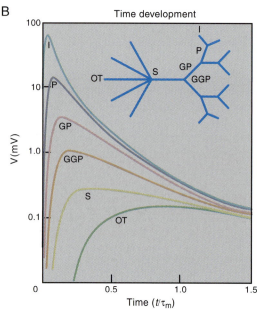

DYNAMIC PROPERTIES OF PASSIVE ELECTROTONIC STRUCTURE

The Electrotonic Structure of the Neuron Changes Dynamically

These considerations show that, compared with the anatomical structure of a dendritic system, which is relatively fixed over short periods, the electrotonic properties have complex, shifting effects on signal integration. The effects depend on multiple factors, including the directions of signal spread, inhomogeneities in passive properties, rates of signal transfer, and interactions between synaptic or active conductances, to name a few. The effects can be illustrated in a graphic fashion for the entire soma – dendritic system by taking a stained neuron and replacing it with a representation based on its electrotonic properties. This is termed a *morpho-electrotonic transform* (MET) (Zador *et al.*, 1995), or *neuromorphic transform* (Carnevale *et al.*, 1997).

The method is illustrated in Fig. 4.10 for a CA1 hippocampal pyramidal cell. The problem is to compare the spread of a signal from the soma to the dendrites (voltage out, V_{out}) with spread from the dendrites to

FIGURE 4.9 Electrotonic spread from a single small dendritic branch. (A) For a steady-state input (*I*) the electrotonic potential (*V*), relative to the initial potential (V_0) at the site of input, spreads from the distal branch through the dendritic tree, with large decrement into the parent branch (due to the large conductance load) but small decrement into neighboring branches B, C–1, and C–2 (due to the small conductance loads). The resulting potential in the soma (S) is much reduced, as is the response to the same input delivered directly to the soma (because of the low input resistance at the soma and the large conductance load of the dendritic tree). The dashed line indicates the response when the same amount of current is injected into the soma. (B) For transient input (*I*) to a distal branch, the transient electrotonic potentials decrease sharply in amplitude and are delayed and slower as they spread toward the soma through the parent (P), grandparent (GP), and great-grandparent (GGP) branches, eventually reaching the soma (S) and output trunk (OT). Modified from Segev (1995) based on Rall and Rinzel (1973, 1974).

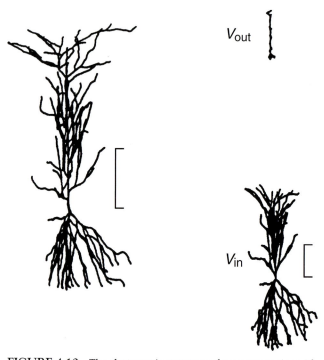

FIGURE 4.10 The electrotonic structure of a neuron varies with the direction of spread of signals. (Left) Stained CA1 pyramidal neuron. (Right) Electrotonic transform of the stained morphology for the case of a voltage spreading toward the cell body (below, V_{in}) and away from the cell body (above, V_{out}). Calibration bar-1 electrotonic length. See text. From Carnevale *et al.* (1997).

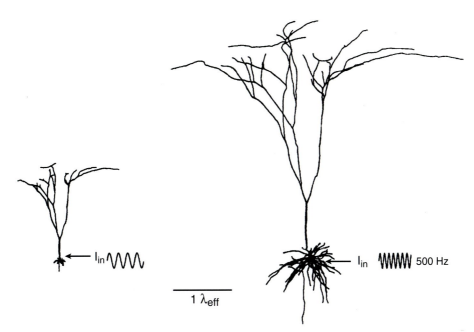

FIGURE 4.11 The electrotonic structure of a neuron varies with the rapidity of signals. (Left) Electrotonic transform of a pyramidal neuron in response to a sinusoidal current of 100 Hz injected into the soma (i.e., this is an example of V_{out}). (Right) Electrotonic transform of same cell in response to 500 Hz. Calibration bar-1 electrotonic length. See text. From Zador *et al.* (1995).

the soma (voltage in, V_{in}). On the left is the stained neuron, giving rise to a long apical dendrite with many branches and shorter basal dendrites and their branches. On the lower right of the diagram is an electrotonic representation of the neuron for signals spreading from the distal dendrites toward the soma. There is severe decrement from each distal branch (cf. Fig. 4.9), so that apical and basal dendritic trees have electrotonic lengths of approximately 3 and 2, respectively. By comparison, on the upper right of the diagram, is an electrotonic representation of this neuron for a signal spreading from the soma to the dendrites. The basal dendrites have shrunk to almost nothing, indicating that they are nearly isopotential. This is because they are relatively short compared with their electrotonic lengths and because the sealed-end boundary condition greatly reduces the decrement in electrotonic potential through them (cf. Fig. 4.7). The apical dendrite has shrunk to an electrotonic length of approximately 1. Thus, *distal synaptic responses decay considerably in spreading all the way to the soma, which active properties help to overcome*, as is described in Chapter 17. On the other hand, signals at the soma "see" a relatively compact dendritic tree.

The analysis in Fig. 4.10 applies to spread of steady-state or very slowly changing signals. What about spread of rapid signals? As discussed previously, the membrane capacitance makes the den-

drites act as a low-pass filter, further reducing rapid signals. The electrotonic transforms can include this effect, as shown in Fig. 4.11. On the left, the electrotonic representation of a pyramidal neuron is shown for a slow (100 Hz) current injected in the soma. The form is similar to that of the cell in Fig. 4.10, with tiny, virtually isopotential basal dendrites, and a longer apical dendritic tree of electrotonic length of approximately 1.5. By comparison, a rapid (500 Hz) signal is severely attenuated in spreading into the dendrites, as shown by the basal dendrites with L of approximately 1 and the apical dendritic tree electrotonic lengths of 4–5. Thus, a somatic action potential could backpropagate into the basal dendrites rather effectively, but would require active properties to invade very far into the apical dendrites. There is direct evidence for these properties underlying backpropagating action potentials in apical dendrites (Chapter 17).

The electrotonic structure of a neuron is not necessarily fixed, but may vary under synaptic control. An example is shown in Fig. 4.12 for the case of a medium spiny cell in the basal ganglia. During low levels of resting excitatory synaptic input, the electrotonic transform of this cell type is relatively large (left) because of the action of a specific K^+ current (known as I_h) in the dendrites that holds them relatively hyperpolarized (see arrow at −90 mV). When synaptic excitation increases, the K^+ current is deactivated, reducing the membrane conductance and thereby

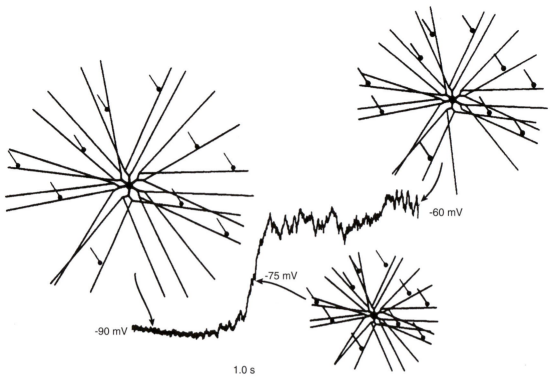

FIGURE 4.12 The electrotonic structure of a neuron can vary with shifts in the resting membrane potential. In this medium spiny cell, the electrotonic transform varies with the resting membrane potential, which in turn reflects the combination of resting voltage-gated K⁺ currents and excitatory synaptic currents. See text. From Wilson (1998).

increasing the input resistance of the cell; the dendritic tree becomes more compact electrotonically (middle) so that synaptic inputs are more effective in activating the cell. As the cell responds to the synaptic excitation, the resulting depolarization activates other K⁺ currents, which expand the electrotonic structure again (right). This example illustrates the ways in which *cable properties and voltage-gated properties interact to control the integrative actions of the neuron.*

Synaptic Conductances in Dendrites Tend to Interact Nonlinearly

Dynamic interactions also occur between synaptic conductances. It is often assumed that synaptic responses sum linearly, but as has already been noted this assumption is not generally true. In an electrical cable, responses to simultaneous current inputs sum linearly (they show "superposition") because the cable properties remain invariant at all times. However, as noted, synaptic responses in real neurons generate current by means of changes in the membrane conductance at the synapse. In addition to generating current, the change in synaptic conductance

alters the overall membrane resistance of that segment and with it the input resistance, thereby changing the electrotonic properties of the whole system. As pointed out by Rall, excitatory and inhibitory conductance changes involve *"a change in a conductance which is an element of the system; the system itself is perturbed; the value of a constant coefficient in the linear differential equation is changed; hence the simple superposition rules do not hold."* (Rall, 1964).

This effect is easily illustrated by the two-compartment model of Fig. 4.6. Consider a synaptic input to compartment A, which decreases the membrane resistance of that compartment. Now consider a simultaneous synaptic input to compartment B, which has the same effect on the membrane resistance of that compartment. The internal current flowing between the two compartments encounters a much lower impedance and hence has much less effect on the membrane potential than would have been the case for current injection. The integration of these two responses therefore gives a smaller summed potential than the summation of the two responses taken individually. This effect is referred to as *occlusion*. In essence, each compartment partially short-circuits the

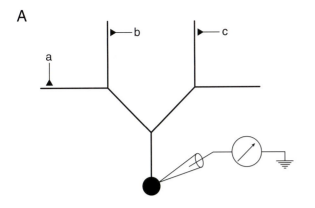

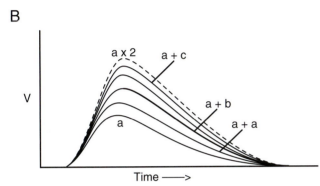

FIGURE 4.13 Schematic diagram of a dendritic tree to illustrate graded effects of nonlinear interactions between synaptic conductances. (A) Three sites of synaptic input (*a–c*) are shown, with recording site in the soma. (B) The voltage response (*V*) is shown for the response to a single input at *a*, the theoretical linear summation for two inputs at *a* (*a* × 2), and the gradual reduction in summation from *c* due to increasing shunting between the conductances. See text. From Shepherd and Koch (1991).

other through a larger conductance load, thus reducing the combined response.

These properties mean that, as noted earlier, *synaptic integration in dendrites in general is not linear even for purely passive electrotonic properties*. The further apart the synaptic sites, the fewer the interactions between the conductances, and the more linear the summation becomes (see Fig. 4.13). These nonlinear properties of passive dendrites combined with the nonlinear properties of voltage-gated channels at local sites on the membrane contribute to the complexity of signal processing that takes place in dendrites, as is discussed in Chapter 17.

The Significance of Active Conductances in Dendrites Depends on Their Relation to Cable Properties

In electrophysiological recordings from the cell body, dendritic synaptic responses often appear small

and slow (cf. Fig. 4.8). However, at their sites of origin in the dendrites, the responses tend to have a large amplitude (because of the high input resistances of the thin distal dendrites) and a rapid time course (because of the small membrane capacitance) (cf. Fig. 4.9). These properties have important implications for the signal processing that takes place in dendrites. In particular, the fact that the distal dendrites contain sites of voltage-gated channels means that local integration, local boosting, and local threshold operations can take place. *These most distal responses need spread no further than to neighboring local active sites to be boosted by these sites*; thus, a rapid integrative sequence of these actions ultimately produces significant effects on signal integration at the cell body. These properties are considered further in Chapter 17.

In addition to their role in local signal processing, the cable properties of the neuron are also important for controlling the spread of synaptic potentials from the dendrites through the soma to the site of action potential initiation in the axon hillock initial segment, and for backpropagation of an action potential into the soma – dendritic compartments, where it can activate dendritic outputs and interact with the active properties involved in signal processing. These properties are also discussed further in Chapter 17.

Dendritic Spines Form Electrotonic and Biochemical Compartments

The rules governing electrotonic interactions within a dendritic tree also apply at the level of the smallest process of a nerve cell, called a spine. This may vary from a bump on a dendritic branch to a twig to a lollipop-shaped process several micrometers long (Fig. 4.14). A dendritic spine usually receives a single excitatory synapse; an axonal initial segment spine characteristically receives an inhibitory synapse.

Dendritic spines receive most of the excitatory inputs to pyramidal neurons in the cerebral cortex and to Purkinje cells in the cerebellum, as well as to a variety of other neuron types, so an understanding of their properties is critical for understanding brain function (Shepherd, 1996; Zador and Koch, 1994; Harris *et al.*, 1994; Yuste and Tank, 1996) (see Chapter 18). As with the whole dendritic tree, one begins with their electrotonic properties. Given the rules we have built earlier in this chapter, by simple inspection of spine morphology as shown in Fig. 4.14, it is possible to postulate several distinctive features that may have important functional implications (see Box 4.4).

In addition to its electrotonic properties, the spine may have interesting biochemical properties. The same cable equations that govern electrotonic

A   B   C

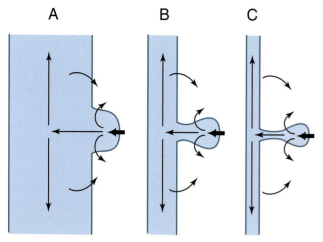

FIGURE 4.14 Diagrams illustrating different types of spines and the current flows generated by a synaptic input. (A) Stubby spine arising from a thick process. (B) Moderately elongated spine from a medium-diameter branch. (C) Spine with a long stem originating from a thin branch. Parallel considerations apply to diffusion between the spine head and dendritic branch. Modified from Shepherd (1974).

properties also have their counterparts in describing the diffusion of substances (as well as the flow of heat). Thus, as already noted, accumulations of only small numbers of ions are needed within the tiny volumes of spine heads to change the driving force on an ion species or to effect significant changes in the concentrations of subsequent second messengers. This interest is intensifying, as the ability to image ion fluxes, such as for Ca^{2+}, and measure other molecular properties of individual spines, increases with new technology such as two-photon microscopy (reviewed in Matus and Shepherd (2000); see also Chapters 17 and 18). The interpretation of those results for the integrative properties of the neuron will require considerations in the biochemical domain that parallel those we have discussed in the electrotonic domain. The range of properties and possible functions of spines are discussed further in Chapters 17 and 18.

Summary

In addition to being dependent on membrane properties, spread of electrotonic potentials in branching dendritic trees is dependent on the boundary conditions set by the modes of branching and termination within the tree. In general, other parts of the dendritic tree constitute a conductance load on activity at a given site; the spread of activity from that site is determined by the impedance match or mismatch between that site and the neigh-

---

BOX 4.4

SOME BASIC ELECTROTONIC PROPERTIES OF DENDRITIC SPINES

High input resistance. The smaller the size and the narrower the stem, the higher the input resistance; this gives a large-amplitude synaptic potential for a given synaptic conductance. Such a large depolarizing excitatory postsynaptic potential (EPSP) can have powerful effects on the local environment within the spine.

Low spine membrane capacitance. The small size also means a small single spine membrane capacitance, implying that synaptic (and any active) potentials may be rapid; this means that spines on dendrites can be involved in rapid information transmission.

Increases in total dendritic membrane capacitance. Although the membrane capacitance of an individual spine is small, the combined spine population increases the total capacitance of its parent dendrite. This increases the filtering effect of the dendrite on transmission of signals through it.

Decrement of potentials spreading from the spine. There is an impedance mismatch between the spine head and its parent dendrite; this means that potentials spreading from the spine to the dendrite will suffer considerable decrement, unless there are active properties of the dendrite or of neighboring spines to boost the signal.

Ease of potential spread into the spine. The other side of the impedance mismatch is that membrane potential changes within the dendrite spread into the spine with little decrement; thus, the spine tends to follow the potential of its dendrite, except for the transient large-amplitude responses to its own synaptic input. This means that a spine can serve as a coincidence detector for nearby synaptic responses or for an action potential backpropagating into the dendritic tree.

boring sites. Rules governing these impedance relations have been worked out relative to the case in which the sum of the daughter branch diameters raised to the 3/2 power is equal to that of the parent branch, in which case the system of branches is an "equivalent cylinder," resembling a single continuous cable. This provides a starting point in analyzing synaptic integration which can be adapted for different types of branching patterns in terms of "equivalent dendrites."

Synchronous synaptic potentials in several branches spread relatively effectively through most dendritic trees. Responses in individual branches may be relatively isolated because of the decrement of passive spread and require local active boosting for effective communication with the rest of the tree. Passive spread can be characterized in terms of several measures, including characteristic length of the equivalent cylinder. There is scaling within individual branches, such that in finer branches electrotonic spread is relatively effective over their shorter lengths. Integration of synaptic potentials in passive dendrites is fundamentally nonlinear, because of interactions between the synaptic conductances. The rules for electrotonic spread in dendrites are the basis for understanding the contributions of active properties of dendrites (see Chapter 17).

RELATING PASSIVE TO ACTIVE POTENTIALS

It is now possible to begin to gain insight into the relationship between passive and active potentials in a neuron by applying the principles of this chapter to a model, the olfactory mitral cell. A basic problem is to understand the factors that decide where the action potential will be initiated with different levels of excitatory or inhibitory inputs. The possible sites are anywhere from the axon through the soma to the most distal dendrites. The mitral cell is advantageous for this analysis, because all the excitatory synaptic input is through olfactory nerve terminals that make their synapses on the distal dendritic tuft, and because the primary dendrite that connects the tuft to the cell body is an unbranched cylinder (see Fig. 4.15A). Applying depolarizing current to distal dendrite or soma, the experimental findings were

counterintuitive: with weak distal inputs the action potential initiation site is far away, in the axon, but with increasing excitation it shifts to the distal dendrite, as illustrated in Fig. 4.15B (Chen *et al.*, 1997). How can the weak response spread so far passively, and why does it not excite the active dendrites along the way?

This is much too complex a problem to solve in one's head or with "back of the envelope" calculations. The only effective method is a computational simulation (cf. Mainen *et al.*, 1995 and Chapter 7). A compartmental model of the mitral cell was therefore constructed, with Na^+ and K^+ conductances scaled to the structure of the mitral cell. Fitting of computed experimental responses was carried out under stringent constraints, with minimization of eight simultaneous simulations (distal and soma recording sites, distal and soma sites of excitatory current input; strong and weak levels of excitation) (Shen *et al.*, 1999).

The active properties are analyzed in Chapter 17. The focus here is on fitting the passive properties. Two steps were essential. First, each experimental recording began with a period of passive charging of the mitral cell membrane (c in Fig. 14.15A). Figure 4.15A shows that the model gave a very accurate simulation, even when the charging was long-lasting (left, weak stimulation). This was a critical fit for giving the correct latency of action potential initiation. Second, the longitudinal spread of passive current between the axon and the distal dendrite was calculated. This showed that with weak distal excitation (Fig. 4.15B), the electrotonic current spread with a shallow gradient from the site of injection along the dendrite to the axon (bottom traces); the action potential arose first in the axon because of the much higher density of Na^+ channels there compared with the dendrite. However, with strong distal excitation (C), the direct depolarization of the less excitable distal dendrite led the weaker electrotonic depolarization of the more excitable axon, and dendritic action potential initiation occurred first. Interested readers can explore this model online at *senselab.med.yale.edu/senselab/modeldb/*.

The computational simulations thus show precisely how the interactions of passive and active potentials control the sites of action potential initiation in the neuron. This is a model for the complex integrative properties of the neuron, which we explore further in Chapter 17.

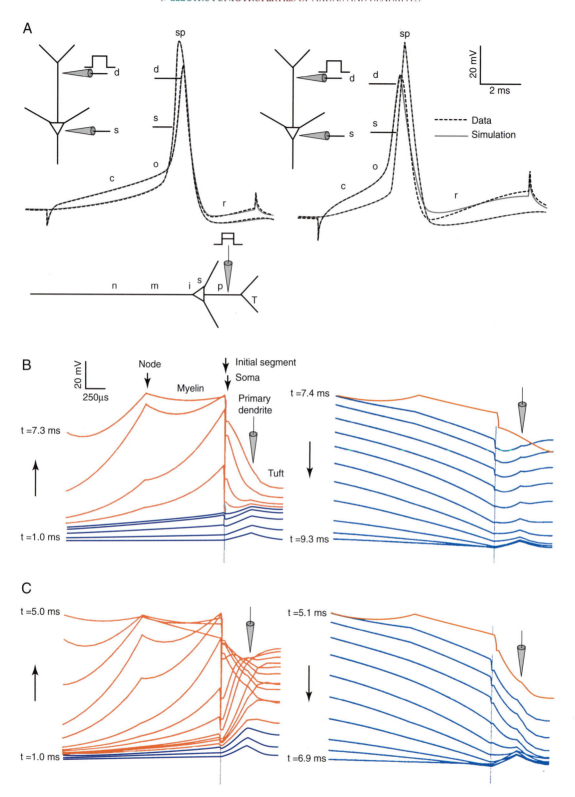

FIGURE 4.15 Interactions of passive and active potentials in the olfactory mitral cell. (A). Insets show diagrams of a mitral cell with recording sites at soma and distal dendrite. Curves show fitting of experimental and computed responses to weak and strong depolarizing currents injected into the distal primary dendrite. Note the nearly exact superposition of experimental (solid lines) and computed (dashed lines) responses. (B). Longitudinal distribution of membrane potential changes during the responses to weak distal dendritic excitation. (C). Same to strong distal dendritic excitation. Blue lines, predominantly passively generated potentials; red lines, predominantly actively generated potentials. d, dendrite; s, soma; c, passive charging; o, onset of action potential; sp, spike peak; r, recovery. See text. Adapted from Shen *et al.* (1999).

References

Bower, J. M., and Beeman, D. (Eds.) (1995). *"The Book of Genesis."* Springer-Verlag (Telos), New York.

Carnevale, N. T., Tsai, K. Y., Claiborne, B. J., and Brown, T. H. (1997). Comparative electrotonic analysis of 3 classes of rat hippocampal neurons. *J. Neurophysiol.* **78,** 703–720.

Carslaw, H. S., and Jaeger, J. C. (1959). *"Conduction of Heat in Solids."* Oxford Univ. Press, London.

Chen, W. R., Midtgaard, J., and Shepherd, G. M. (1997). Forward and backward propagation of dendritic impulses and their synaptic control in mitral cells. *Science* **278,** 463–467.

Cole, K. C., and Hodgkin, A. L. (1939). Membrane and protoplasm resistance in the squid giant axon. *J. Gen. Physiol.* **22,** 671–687.

Cremer, M. (1899). Zum kernleiterproblem. *Z. Biol.* **37,** 550–553.

Cremer, M. (1909). Die allgemeine physiologie ner nerven. *In "Handbuch der Physiologie des Menschen,"* p. 793. Vieweg, Braunschweig.

Davis, L., and Lorente de No, R. (1947). Contribution to the mathematical theory of electrotonus. *Stud. Rockefeller Inst. Med. Res.* **131,** 50–62.

Harris, K. M., and Kater, S. B. (1994). Dendritic spines: Cellular specializations imparting both stability and flexibility to synaptic function. *Annu. Rev. Neurosci.* **17,** 341–371.

Hermann, L. (1872). *Arch. Ges. Physiol. Menschen Tiere* **6,** 312.

Hermann, L. (1905). Beitrage zur physiologie und physik des nerven. *Arch. Ges. Physiol. Menschen Tiere* **109,** 95.

Hermann, L. (Ed.) (1879). *"Handbuch der Physiologie."* Vogel, Leipzig.

Hines, M. (1984). Efficient computation of branched nerve equations. *Int. J. Bio-Med. Comput.* **15,** 69–76.

Hodgkin, A. L., and Rushton, W. A. H. (1946). The electrical constants of a crustacean nerve fibre. *Proc. R. Soc. London Ser. B* **133,** 444–447.

Hoorweg, J. L. (1898). Ueber die elektrischen Eigenschaften der Nerven. *Arch. Ges. Physiol. Menschen Tiere* **71,** 128.

Hursh, J. B. (1939). Conduction velocity and diameter of nerve fibers. *Am. J. Physiol.* **127,** 131–139.

Jack, J. J. B., Noble, D., and Tsien, R. W. (1975). *"Electrical Current Flow in Excitable Cells."* Oxford Univ. Press (Clarendon), London.

Johnston, D., and Wu, S. M.-S. (1995). *"Foundations of Cellular Neurophysiology."* MIT Press, Cambridge, MA.

Kelvin, W. T. (1855). On the theory of the electric telegraph. *Proc. R. Soc. London* **7,** 382–399.

Kelvin, W. T. (1856). On the theory of the electric telegraph. *Philos. Mag. [4]* **11,** 146–160.

Mainen, Z. F., Joerges, J., Huguenard, J. R., and Sejnowski, T. J. (1995) A model of spike initiation in neocortical pyramidal neurons. *Neuron* **6,** 1427–1439.

Matus, A. and Shepherd, G. M. (2000). The millenium of the dendrite? *Neuron* **27,** 431–434.

Pongracz, F., Poolos, N. P., Kocsis, J. D., and Shepherd, G. M. (1992). A model of NMDA receptor-mediated activity in dendrites of hippocampal CA1 pyramidal neurons. *J. Neurophysiol.* **6,** 2248–2259.

Qian, N., and Sejnowski, T. J. (1989). An electro-diffusion model for computing membrane potentials and ionic concentrations in branching dendrites, spines and axons. *Biol. Cybernet.* **62,** 1–15.

Rall, W. (1958). Dendritic current distribution and whole neuron properties. *Nav. Med. Res. Inst. Res. Rep.* **NM 0105.01.02,** 479–525.

Rall, W. (1959). Branching dendritic trees and motoneuron membrane resistivity. *Exp. Neurol.* **1,** 491–527.

Rall, W. (1964). Theoretical significance of dendritic trees for neuronal input-output relations. *In "Neural Theory and Modelling"* (R. F. Reiss, Ed.), pp. 73–97. Stanford Univ. Press, Stanford, CA.

Rall, W. (1967). Distinguishing theoretical synaptic potentials computed for different soma–dendritic distributions of synaptic input. *J. Neurophysiol.* **30,** 1138–1168.

Rall, W. (1977). Core conductor theory and cable properties of neurons. *In "The Nervous System: Cellular Biology of Neurons"* (E. R. Kandel, Ed.), Vol. 1, pp. 39–97. Am. Physiol. Soc., Bethesda, MD.

Rall, W., and Rinzel, J. (1973). Branch input resistance and steady attenuation for input to one branch of a dendritic neuron model. *Biophys. J.* **13,** 648–688.

Rall, W., and Shepherd, G. M. (1968). Theoretical reconstruction of field potentials and dendrodendritic synaptic interactions in olfactory bulb. *J. Neurophysiol.* **3,** 884–915.

Rinzel, J., and Rall, W. (1974). Transient response in a dendritic neuron model for current injected at one branch. *Biophys. J.* **14,** 759–790.

Ritchie, J. M. (1995). Physiology of axons. *In "The Axon: Structure, Function, and Pathophysiology"* (S. G. Waxman, J. D. Kocsis, and P. K. Stys, Eds.), pp. 68–69. Oxford Univ. Press, New York.

Rushton, W. A. H. (1927). The effect upon the threshold for nervous excitation of the length of nerve exposed and the angle between current and nerve. *J. Physiol. (London)* **63,** 357.

Rushton, W. A. H. (1934). A physical analysis of the relation between threshold and interpolar length in the electric excitation of medullated nerve. *J. Physiol. (London)* **82,** 332–352.

Rushton, W. A. H. (1951). A theory of the effects of fibre size in medullated nerve. *J. Physiol. (London)* **115,** 101–122.

Segev, I. (1995). Cable and compartmental models of dendritic trees. *In "The Book of Genesis"* (J. M. Bower and D. Beeman, Eds.), pp. 53–82. Springer-Verlag (Telos), New York.

Segev, I., Rinzel, J., and Shepherd, G. M., eds. (1995). *"The Theoretical Foundation of Dendritic Function."* MIT Press, Cambridge, MA.

Shen, G., Chen, W. R., Midtgaard, J., Shepherd, G. M., and Hines, M. L. (1999) Computational analysis of action potential initiation in mitral cell soma and dendrites based on dual patch recordings. *J. Neurophysiol.* **82,** 3006–3020.

Shepherd, G. M., ed. (2004). *The Synaptic Organization of the Brain.* Oxford Univ. Press, New York.

Shepherd, G. M. (1994). *Neurobiology.* New York: Oxford University Press.

Shepherd, G. M. (1996). The dendritic spine: A multifunctional integrative unit. *J. Neurophysiol.* **75,** 2197–2210.

Shepherd, G. M., and Brayton, R. K. (1979). Computer simulation of a dendrodendritic synaptic circuit for self- and lateral-inhibition in the olfactory bulb. *Brain Res.* **175,** 377–382.

Waxman, S. G., and Bennett, M. V. L. (1972). Relative conduction velocities of small myelinated and nonmyelinated fibres in the central nervous system. *Nat. New Biol.* **238,** 217.

Weber, H. (1873). Uber die stationaren stromungen der elektricitat in cylindren. *J. Reine Angewandte Math.* **7,** 1–20.

Wilson, C. J. (1998). Basal ganglia. *In "The Synaptic Organization of the Brain"* (G. M. Shepherd, Ed.), 4th ed., pp. 329–376. Oxford Univ. Press, New York.

Yuste, R., and Tank, D. (1996). Dendritic integration in mammalian neurons, a century after Cajal. *Neuron* **13,** 23–43.

Zador, A. M., Agmon-Snir, H., and Segev, I. (1995). The morphoelectrotonic transform: A graphical approach to dendritic function. *J. Neurosci.* **15,** 1169–1682.

Zador, A., and Koch, C. (1994). Linearized models of calcium dynamics: Formal equivalence to the cable equation. *J. Neurosci.* **14,** 4705–4715.

Ziv, I., Baxter, D. A., and Byrne, J. H. (1994). Simulator for neural networks and action potentials: Description and application. *J. Neurophysiol.* **71,** 294–308.

5

Membrane Potential and Action Potential

David A. McCormick

The communication of information between neurons and between neurons and muscles or peripheral organs requires that signals travel over considerable distances. A number of notable scientists have contemplated the nature of this communication through the ages. In the second century AD, the great Greek physician Claudius Galen proposed that "humors" flowed from the brain to the muscles along hollow nerves. A true electrophysiological understanding of nerve and muscle, however, depended on the discovery and understanding of electricity itself. The precise nature of nerve and muscle action became clearer with the advent of new experimental techniques by a number of European scientists, including Luigi Galvani, Emil Du Bois-Reymond, Carlo Matteucci, and Hermann von Helmholtz, to name a few (Brazier, 1959, 1988). Through the application of electrical stimulation to nerves and muscles, these early electrophysiologists demonstrated that the conduction of commands from the brain to muscle for the generation of movement was mediated by the flow of electricity along nerve fibers.

With the advancement of electrophysiological techniques, electrical activity recorded from nerves revealed that the conduction of information along the axon was mediated by the active generation of an electrical potential, called the action potential. But what precisely was the nature of these action potentials? To know this in detail required not only a preparation from which to obtain intracellular recordings but also one that could survive *in vitro*. The squid giant axon provided precisely such a preparation, as was first demonstrated by J. Z. Young in 1936 (Young, 1936). Many invertebrates contain unusually large axons for the generation of escape reflexes; large axons conduct more quickly than small ones and so the response time for escape is reduced (see Chapter 4). The squid possesses an axon approximately 0.5 mm in diameter, large enough to be impaled by

even a coarse micropipet (Fig. 5.1). By inserting a glass micropipet filled with a salt solution into the squid giant axon, Alan Hodgkin and Andrew Huxley demonstrated in 1939 that axons at rest are electrically polarized, exhibiting a resting membrane potential of approximately –60 mV inside versus outside (Hodgkin and Huxley, 1939; Hodgkin, 1976). In the generation of an action potential, the polarization of the membrane is removed (referred to as depolarization) and exhibits a rapid swing toward, and even past, 0 mV (Fig. 5.1B). This depolarization is followed by a rapid swing in the membrane potential to more negative values, a process referred to as *hyperpolarization*. The membrane potential following an action potential typically becomes even more negative than the original value of approximately –60 mV. This period of increased polarization is referred to as the *afterhyperpolarization* or the undershoot.

The development of electrophysiological techniques to the point that intracellular recordings could be obtained from the small cells of the mammalian nervous system revealed that action potentials in these neurons are generated through mechanisms similar to that of the squid giant axon (Brock *et al.*, 1952; Buser and Albe–Fessard, 1953; Tasaki *et al.*, 1954; Phillips, 1956).

It is now known that action potential generation in nearly all types of neurons and muscle cells is accomplished through mechanisms similar to those first detailed in the squid giant axon by Hodgkin and Huxley. In this chapter, we consider the cellular mechanisms by which neurons and axons generate a resting membrane potential and how this membrane potential is briefly disrupted for the purpose of propagation of an electrical signal, the action potential. The following chapters describe in more detail the molecular properties of ion channels (Chapter 6) and the quantitative analysis and dynamic properties of action potentials (Chapter 7).

Copyright 2004, Elsevier Science (USA).
All rights reserved.

A

B

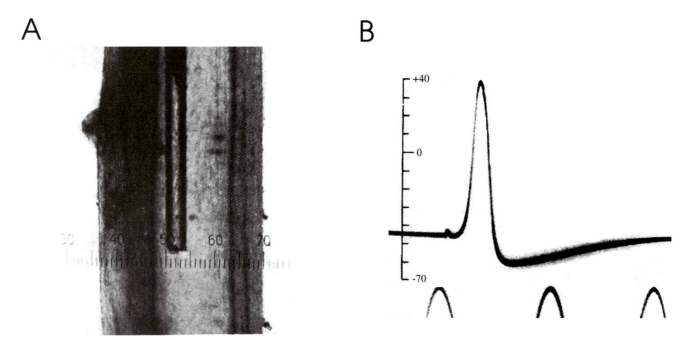

FIGURE 5.1 Intracellular recording of the membrane potential and action potential generation in the squid giant axon. (A) A glass micropipette, about 100 μm in diameter, was filled with seawater and lowered into the giant axon of the squid after it had been dissected free. The axon is about 1 mm in diameter and is transilluminated from behind. (B) One action potential recorded between the inside and the outside of the axon. Peaks of a sine wave at the bottom provided a scale for timing, with 2 ms between peaks. From Hodgkin and Huxley (1939).

THE MEMBRANE POTENTIAL

The Membrane Potential Is Generated by the Differential Distribution of Ions

Through the operation of ionic pumps and special ionic buffering mechanisms, neurons actively maintain precise internal concentrations of several important ions, including Na^+, K^+, Cl^-, and Ca^{2+}. The mechanisms by which they do so are illustrated in Figs. 5.2 and 5.3. The intracellular and extracellular concentrations of Na^+, K^+, Cl^-, and Ca^{2+} differ markedly (see Fig. 5.2 and Table 5.1); K^+ is actively concentrated inside the cell, and Na^+, Cl^-, and Ca^{2+} are actively extruded to the extracellular space. However, this does not mean that the cell is filled only with positive charge; anions (denoted A^-) to which the plasma membrane is impermeant are also present inside the cell and almost balance the high concentration of K^+. The osmolarity inside the cell is approximately equal to that outside the cell.

Electrical and Thermodynamic Forces Determine the Passive Distribution of Ions

Ions tend to move down their concentration gradients through specialized ionic pores, known as ionic channels, in the plasma membrane. Through simple laws of thermodynamics, the high concentration of K^+ inside glial cells, neurons, and axons results in a tendency for K^+ ions to diffuse down their concentration gradient and leave the cell or cell process (see Fig. 5.3). However, the movement of ions across the membrane also results in a redistribution of electrical charge. As K^+ ions move down their concentration gradient, the intracellular voltage becomes more negative, and this increased negativity results in an electrical attraction between the negative potential inside the cell and the positively charged K^+ ions, thus offsetting the outward flow of these ions. The membrane is selectively permeable; that is, it is impermeable to the large anions inside the cell, which cannot follow the potassium ions across the membrane. At some membrane potential, the "force" of the electrostatic attraction between the negative membrane potential inside the cell and the positively charged K^+ ions exactly balances the thermal "forces" by which K^+ ions tend to flow down their concentration gradient (see Fig. 5.3). In this circumstance, it is equally likely that a K^+ ion exits the cell by movement down the concentration gradient as it is that a K^+ ion enters the cell owing to the attraction between the negative membrane potential and the positive charge of this ion. At this membrane potential, there is no net flow

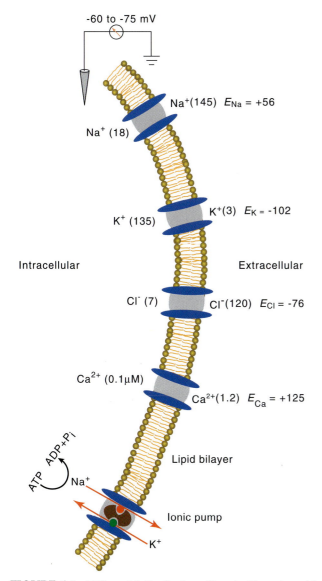

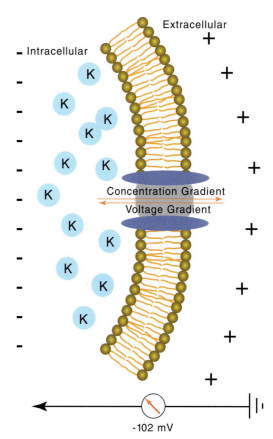

FIGURE 5.2 Differential distribution of ions inside and outside plasma membrane of neurons and neuronal processes, showing ionic channels for Na$^+$, K$^+$, Cl$^-$, and Ca^{2+}, as well as an electrogenic Na$^+$–K$^+$ ionic pump (also known as Na$^+$, K$^+$-ATPase). Concentrations (in millimoles except that for intracellular Ca^{2+}) of the ions are given in parentheses; their equilibrium potentials (E) for a typical mammalian neuron are indicated.

FIGURE 5.3 The equilibrium potential is influenced by the concentration gradient and the voltage difference across the membrane. Neurons actively concentrate K$^+$ inside the cell. These K$^+$ ions tend to flow down their concentration gradient from inside to outside the cell. However, the negative membrane potential inside the cell provides an attraction for K$^+$ ions to enter or remain within the cell. These two factors balance one another at the equilibrium potential, which in a typical mammalian neuron is –102 mV for K$^+$.

of K$^+$, and these ions are said to be in equilibrium. The membrane potential at which this occurs is known as the equilibrium potential. (See Box 5.1 for the calculation of the equilibrium potential.)

To illustrate, let us consider the passive distribution of K$^+$ ions in the squid giant axon as studied by Hodgkin and Huxley. The K$^+$ concentration inside the squid giant axon [K$^+$]$_i$ is about 400 mM, whereas that outside the axon [K$^+$]$_o$ is about 20 mM (Table 5.1). Because [K$^+$]$_i$ is greater than [K$^+$]$_o$, potassium ions

TABLE 5.1 Ion Concentrations and Equilibrium Potentials

	Inside (mM)	Outside (mM)	Equilibrium potential (mV)
	Squid giant axon		
Na$^+$	50	440	+55
K$^+$	400	20	–76
Cl$^-$	40	560	–66
Ca^{2+}	0.4μM	10	+145
	Mammalian neuron		
Na$^+$	18	145	+56
K$^+$	140	3	–102
Cl$^-$	7	120	–76
Ca^{2+}	100nM	1.2	+125

BOX 5.1

THE NERNST EQUATION

The equilibrium potential is determined by (1) the concentration of the ion inside and outside the cell, (2) the temperature of the solution, (3) the valence of the ion, and (4) the amount of work required to separate a given quantity of charge. The equation that describes the equilibrium potential was formulated by a German physical chemist named Walter Nernst in 1888:

$$E_{ion} = RT/zF \, \ln[ion]_o/[ion]_i.$$

Here, E_{ion} is the membrane potential at which the ionic species is at equilibrium, R is the gas constant (8.315 joules per Kelvin per mole), T is the temperature in Kelvin ($T_{Kelvin} = 273.16 + T_{Celsius}$), F is Faraday's con-

stant (96,485 coulombs per mole), z is the valence of the ion, and $[ion]_o$ and $[ion]_i$ are the concentrations of the ion outside and inside the cell, respectively. For a monovalent, positively charged ion (cation) at room temperature (20°C), substituting the appropriate numbers and converting natural log (ln) into log base 10 (log) results in the equation

$$E_{ion} = 58.2 \log[ion]_o/[ion]_i;$$

at a body temperature of 37°C, the Nernst equation is

$$E_{ion} = 61.5 \log[ion]_o/[ion]_i.$$

---

tend to flow down their concentration gradient, taking positive charge with them. The equilibrium potential (at which the tendency for K^+ ions to flow down their concentration gradient is exactly offset by the attraction for K^+ ions to enter the cell because of the negative charge inside the cell) at a room temperature of 20°C can be determined by the Nernst equation as:

$$E_K = 58.2 \log(20/400) = -76 \text{ mV}.$$

Therefore, at a membrane potential of –76 mV, K^+ ions have an equal tendency to flow either into or out of the axon. The concentrations of K^+ in mammalian neurons and glial cells differ considerably from that in the squid giant axon (see Table 5.1). By substituting 3.1 mM for $[K^+]_o$ and 140 mM for $[K^+]_i$ in the Nernst equation, with $T = 37$°C, we obtain

$$E_K = 61.5 \log(3.1/140) = -102 \text{ mV}.$$

Movements of Ions Can Cause Either Hyperpolarization or Depolarization

In mammalian cells, at membrane potentials positive to –102 mV, K^+ ions tend to flow out of the cell. Increasing the ability of K^+ ions to flow across the membrane, that is, increasing the conductance of the membrane to K^+ (g_K), causes the membrane potential to become more negative, or hyperpolarized, owing to the exit of positively charged ions from inside the cell (Fig. 5.4).

At membrane potentials negative to –102 mV, K^+ ions tend to flow into the cell; increasing the mem-

brane conductance to K^+ causes the membrane potential to become more positive, or depolarized, owing to the flow of positive charge into the cell. The membrane potential at which the net current "flips" direction is referred to as the *reversal potential*. If the channels conduct only one type of ion (e.g., K^+ ions), then the reversal potential and the Nernst equilibrium potential for that ion coincide (see Fig. 5.4A). Increasing the membrane conductance to K^+ ions while the membrane potential is at the equilibrium potential for K^+ (E_K) does not change the membrane potential, because no net driving force causes K^+ ions to either exit or enter the cell. However, this increase in membrane conductance to K^+ decreases the ability of other species of ions to change the membrane potential, because any deviation of the potential from E_K increases the drive for K^+ ions either to exit or to enter the cell, thereby drawing the membrane potential back toward E_K (see Fig. 5.4B).

The exit from and entry to the cell of K^+ ions during generation of the membrane potential give rise to a curious problem. When K^+ ions leave the cell to generate a membrane potential, the concentration of K^+ changes both inside and outside the cell. Why does this change in concentration not alter the equilibrium potential, thus changing the tendency for K^+ ions to flow down their concentration gradient? The reason is that the number of K^+ ions required to leave the cell to achieve the equilibrium potential is quite small. For example, if a cell were at 0 mV and the membrane suddenly became permeable to K^+ ions, only about 10–12 mol of K^+ ions per square centimeter of mem-

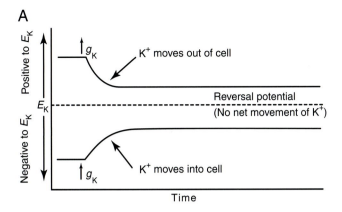

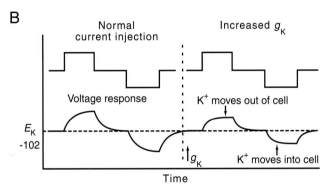

FIGURE 5.4 Increases in K$^+$ conductance can result in hyperpolarization, depolarization, or no change in membrane potential. (A) Opening K$^+$ channels increases the conductance of the membrane to K$^+$, denoted g_K. If the membrane potential is positive to the equilibrium potential (also known as the reversal potential) for K$^+$, then increasing g_K will cause some K$^+$ ions to leave the cell, and the cell becomes hyperpolarized. If the membrane potential is negative to E_K when g_K is increased, then K$^+$ ions enter the cell, making the inside more positive (more depolarized). If the membrane potential is exactly E_K when g_K is increased, then there is no net movement of K$^+$ ions. (B) Opening K$^+$ channels when the membrane potential is at E_K does not change the membrane potential; however, it reduces the ability of other ionic currents to move the membrane potential away from E_K. For example, a comparison of the ability of the injection of two pulses of current, one depolarizing and one hyperpolarizing, to change the membrane potential before and after opening K$^+$ channels reveals that increases in g_K noticeably decrease the responses of the cell.

brane would move from inside to outside the cell in bringing the membrane potential to the equilibrium potential for K$^+$. In a spherical cell of 25-μm diameter, this would amount to an average decrease in intracellular K$^+$ of only about 4 μM (e.g., from 140 to 139.996 mM). However, there are instances when significant changes in the concentrations of K$^+$ may occur, particularly during the generation of pronounced activity, such as that related to an epileptic seizure. During the occurrence of a tonic–clonic generalized (grand mal) seizure, large numbers of neurons discharge throughout the cerebral cortex in a

synchronized manner. This synchronous discharge of large numbers of neurons significantly increases the extracellular K$^+$ concentration, by as much as a couple of millimoles, resulting in a commensurate positive shift in the equilibrium potential for K$^+$ (Hotson *et al.*, 1973; Prince *et al.*, 1973). This shift in the equilibrium potential can increase the excitability of affected neurons and neuronal processes and thus promote the spread of the seizure activity. Fortunately, the extracellular concentration of K$^+$ is tightly regulated and kept at normal levels through uptake by glial cells as well as by diffusion through the fluid of the extracellular space (Kuffler and Nicholls, 1966).

As is true for K$^+$ ions, each of the membrane-permeable species of ions possesses an equilibrium potential that depends on the concentration of the ions inside and outside the cell. Thus, equilibrium potentials may vary between different cell types, such as those found in animals adapted to live in salt water versus mammalian neurons (see Table 5.1). In mammalian neurons, the equilibrium potential is approximately +56 mV for Na$^+$, approximately –76 mV for Cl$^-$, and about +125 mV for Ca^{2+} (see Table 5.1 and Fig. 5.2). Thus, increasing the membrane conductance to Na$^+$ (g_{Na}) through the opening of Na$^+$ channels depolarizes the membrane potential toward +56 mV; increasing the membrane conductance to Cl$^-$ brings the membrane potential closer to –76 mV; and finally increasing the membrane conductance to Ca^{2+} depolarizes the cell toward +125 mV.

Na$^+$, K$^+$, and Cl$^-$ Contribute to the Determination of the Resting Membrane Potential

If a membrane is permeable to only one ion and no electrogenic ionic pumps are operating (see next section), then the membrane potential is necessarily at the equilibrium potential for that ion. At rest, the plasma membrane of most cell types is not at the equilibrium potential for K$^+$ ions, indicating that the membrane is also permeable to other types of ions. For example, the resting membrane of the squid giant axon is permeable to Cl$^-$ and Na$^+$, as well as K$^+$, owing to the presence of ionic channels that not only allow these ions to pass but also are open at the resting membrane potential. Because the membrane is permeable to K$^+$, Cl$^-$, and Na$^+$, the resting potential of the squid giant axon is not equal to E_K, E_{Na}, or E_{Cl}, but is somewhere within these three. A membrane permeable to more than one ion has a steady-state membrane potential whose value is between those of the equilibrium potentials for each of the permeant ions (Box 5.2) (Goldman, 1943; Hodgkin and Katz, 1949).

BOX 5.2

THE GOLDMAN–HODGKIN–KATZ EQUATION

An equation developed by Goldman (1943) and later used by Hodgkin and Katz (1949) describes the steady-state membrane potential for a given set of ionic concentrations inside and outside the cell and the relative permeabilities of the membrane to each of those ions:

$$V_{\mathrm{m}} = RT/F \ln \{(pK[K^+]_o + pNa[Na^+]_o + pCl[Cl^-]_i)/$$
$$(pK[K^+]_i + pNa[Na^+]_i + pCl[Cl^-]_o)\}.$$

The relative contribution of each ion is determined by its concentration differences across the membrane and the relative permeability (p_K, p_{Na}, p_{Cl}) of the membrane to each type of ion. If a membrane is permeable to only one ion, then the Goldman–Hodgkin–Katz equation reduces to the Nernst equation. In the squid giant axon, at resting membrane potential, the permeability ratios are:

$$pK{:}pNa{:}pCl = 1.00{:}0.04{:}0.45.$$

The membrane of the squid giant axon, at rest, is most permeable to K^+ ions, less so to Cl^-, and least permeable to Na^+. (Chloride appears to contribute considerably less to the determination of the resting potential of mammalian neurons.) These results indicate that the resting membrane potential is determined by the resting permeability of the membrane to K^+, Na^+, and Cl^-. In theory, this resting membrane potential may be anywhere between E_K (e.g., –76 mV) and E_{Na} (+55 mV). For the three ions at 20°C, the equation is

$$V_{\mathrm{m}} = 58.2 \log\{(1.20 + 0.04 \cdot 440 + 0.45 \cdot 40)/$$
$$(1.400 + 0.04 \cdot 50 + 0.45 \cdot 560)\} = -62 \text{ mV}.$$

This suggests that the squid giant axon should have a resting membrane potential of –62 mV. In fact, the resting membrane potential may be a few millivolts hyperpolarized to this value through the operation of the electrogenic Na^+–K^+ pump.

Different Types of Neurons Have Different Resting Potentials

Intracellular recordings from neurons in the mammalian CNS reveal that different types of neurons exhibit different resting membrane potentials. Indeed, some types of neurons do not even exhibit a true "resting" membrane potential; they spontaneously and continuously generate action potentials even in the total lack of synaptic input. In the visual system, intracellular recordings have shown that the photoreceptor cells of the retina—the rods and cones—have a membrane potential of approximately –40 mV at rest and are hyperpolarized when activated by light (Tomita, 1965). Cells in the dorsal lateral geniculate nucleus, which receive axonal input from the retina and project to the visual cortex, have a resting membrane potential of approximately –70 mV during sleep and –55 mV during waking (Hirsch et al., 1983; Jahnsen and Llinas, 1984a,b), whereas pyramidal neurons of the visual cortex have a resting membrane potential of about –75 mV (McCormick et al., 1985). Presumably, the resting membrane potentials of different cell types in the central and peripheral nervous system are highly regulated and are functionally important. For example, the depolarized membrane potential of photoreceptors presumably allows the

membrane potential to move in both negative and positive directions in response to changes in light intensity. The hyperpolarized membrane potential of thalamic neurons during sleep (–70 mV) dramatically decreases the flow of information from the sensory periphery to the cerebral cortex (Livingstone and Hubel, 1981; Steriade and McCarley, 1990), presumably to allow the cortex to be relatively undisturbed during sleep, and the 20-mV membrane potential between the resting potential and the action potential threshold in cortical pyramidal cells may permit the subthreshold computation and integration of multiple neuronal inputs in single neurons (see Chapters 4, 16, and 17).

Ionic Pumps Actively Maintain Ionic Gradients

Because the resting membrane potential of a neuron is not at the equilibrium potential for any particular ion, ions constantly flow down their concentration gradients. This flux becomes considerably larger with the generation of electrical and synaptic potentials, because ionic channels are opened by these events. Although the absolute number of ions traversing the plasma membrane during each action potential or synaptic potential may be small in individual cells, the collective influence of a large neural network of cells, such as in the brain, and the presence of ion fluxes even at rest can substantially change the distribution of ions inside and outside neurons. As described in Chapter 3, cells have solved this problem with the use of active transport of ions against their concentration gradients. The proteins that actively transport ions are referred to as ionic pumps, of which the Na^+–K^+ pump is perhaps the most thoroughly understood (Hodgkin and Keynes, 1955; Skou, 1957, 1988; Thomas, 1972). The Na^+–K^+ pump is stimulated by increases in the intracellular concentration of Na^+ and moves Na^+ out of the cell while moving K^+ into it, achieving this task through the hydrolysis of ATP (see Fig. 5.2). Three Na^+ ions are extruded for every two K^+ ions transported into the cell. Owing to the unequal transport of ions, the operation of this pump generates a hyperpolarizing electrical potential and is said to be electrogenic. The Na^+–K^+ pump typically results in the membrane potential of the cell being a few millivolts more negative than it would be otherwise.

The Na^+–K^+ pump consists of two subunits, α and β, arranged in a tetramer $(\alpha\beta)_2$. The α subunit has a molecular mass of about 100 kDa and six hydrophobic regions capable of forming transmembrane helices (Mercer, 1993; Horisberger et al., 1991). The β subunit is smaller (about 38 kDa) and has only one hydrophobic membrane-spanning region. The Na^+–K^+ pump is believed to operate through conformational changes that alternatively expose a Na^+ binding site to the interior of the cell (followed by the release of Na^+) and a K^+ binding site to the extracellular fluid (see Fig. 5.2). Such a conformation change may be due to the phosphorylation and dephosphorylation of the protein.

The membranes of neurons and glia contain multiple types of ionic pumps, used to maintain the proper distribution of each ionic species important for cellular signaling (Pedersen and Carafoli, 1987; Läuger, 1991). Many of these pumps are operated by the Na^+ gradient across the cell, whereas others operate through a mechanism similar to that of the Na^+–K^+ pump (i.e., the hydrolysis of ATP). For example, the calcium concentration inside neurons is kept at very low levels (typically 50–100 nM) through the operation of both types of ionic pumps as well as special intracellular Ca^{2+} buffering mechanisms. Ca^{2+} is extruded from neurons through both a Ca^{2+},Mg^{2+}-ATPase and a Na^+–Ca^{2+} exchanger. The Na^+–Ca^{2+} exchanger is driven by the Na^+ gradient across the membrane and extrudes one Ca^{2+} ion for each Na^+ ion allowed to enter the cell.

The Cl^- concentration in neurons is actively maintained at a low level through the operation of a chloride–bicarbonate exchanger, which brings in one ion of Na^+ and one ion of HCO_3^- for each ion of Cl^- extruded (Reithmeier, 1994; Thompson et al., 1988). Intracellular pH also can markedly affect neuronal excitability and is therefore tightly regulated, in part by a Na^+–H^+ exchanger that extrudes one proton for each Na^+ allowed to enter the cell.

Summary

The membrane potential is generated by the unequal distribution of ions, particularly K^+, Na^+, and Cl^-, across the plasma membrane. This unequal distribution of ions is maintained by ionic pumps and exchangers. K^+ ions are concentrated inside the neuron and tend to flow down their concentration gradient, leading to a hyperpolarization of the cell. At the equilibrium potential, the tendency of K^+ ions to flow out of the cell is exactly offset by the tendency of K^+ ions to enter the cell owing to the attraction of the negative potential inside the cell. The resting membrane is also permeable to Na^+ and Cl^- and therefore the resting membrane potential is approximately –75 to –40 mV, in other words, substantially positive to E_K.

THE ACTION POTENTIAL

An Increase in Na^+ and K^+ Conductance Generates Action Potentials

Hodgkin and Huxley not only recorded the action potential with an intracellular microelectrode (see Fig. 5.1), but also went on to perform a remarkable series of experiments that qualitatively and quantitatively explained the ionic mechanisms by which the action potential is generated (Hodgkin and Huxley, 1952a–d; Hodgkin et al., 1952). As mentioned earlier, these investigators found that during the action potential, the membrane potential of the cell rapidly overshoots 0 mV and approaches the equilibrium potential for Na^+. After generation of the action potential, the membrane potential repolarizes and becomes more negative than before, generating an afterhyperpolarization. Cole and Curtis (1939) had previously shown that these changes in membrane potential during the generation of the action potential are associated with a large increase in conductance of the plasma membrane. But to what does the membrane become conductive to generate the action potential? The prevailing hypothesis was that there was a nonselective increase in conductance causing the negative resting potential to increase toward 0 mV. Since publication

BOX 5.3

THE VOLTAGE-CLAMP TECHNIQUE

In the voltage-clamp technique, two independent electrodes are inserted into the squid giant axon: one for recording the voltage difference across the membrane and the other for intracellularly injecting the current (Fig. 5.5). These electrodes are then connected to a feedback circuit that compares the measured voltage across the membrane with the voltage desired by the experimenter. If these two values differ, then current is injected into the axon to compensate for this difference. This continuous feedback cycle, in which the voltage is measured and current is injected, effectively "clamps" the membrane at a particular voltage. If ionic channels were to open, then the resultant flow of ions into or out of the axon would be compensated for by the injection of positive or negative current into the axon through the current-injection electrode. The current injected through this electrode is necessarily equal to the current flowing through the ionic channels. It is this injected current that is measured by the experimenter. The benefits of the voltage-clamp technique are twofold. First, the current injected into the axon to keep the membrane potential "clamped" is necessarily equal to the current flowing through the ionic channels in the membrane, thereby giving a direct measurement of this current. Second, ionic currents are both voltage and time dependent; they become active at certain membrane potentials and do so at a particular rate. Keeping the voltage constant in the voltage clamp allows these two variables to be separated; the voltage dependence and the kinetics of the ionic currents flowing through the plasma membrane can be directly measured.

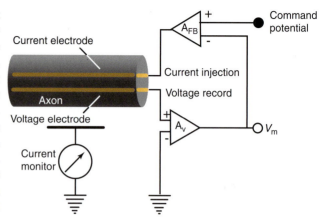

FIGURE 5.5 Voltage-clamp technique. The voltage-clamp technique keeps the voltage across the membrane constant so that the amplitude and time course of ionic currents can be measured. In the two-electrode voltage-clamp technique, one electrode measures the voltage across the membrane while the other injects current into the cell to keep the voltage constant. The experimenter sets a voltage to which the axon or neuron is to be stepped (the command potential). Current is then injected into the cell in proportion to the difference between the present membrane potential and the command potential. This feedback cycle occurs continuously, thereby clamping the membrane potential to the command potential. By measuring the amount of current injected, the experimenter can determine the amplitude and time course of the ionic currents flowing across the membrane.

---

of the experiments of Overton in 1902, the action potential had been known to depend on the presence of extracellular Na^+. Reducing the concentration of Na^+ in the artificial seawater bathing the axon resulted in a marked reduction in the amplitude of the action potential. On the basis of these and other data, Hodgkin and Katz proposed that the action potential is generated through a rapid increase in the conductance of the membrane to Na^+ ions. Quantitative proof of this theory was lacking, however, because ionic currents could not be observed directly. The development of the voltage-clamp technique by Cole (1949) at the Marine Biological Laboratory in Massachusetts resolved this problem and allowed quantitative measurement of the Na^+ and K^+ currents underlying the action potential (Box 5.3).

Hodgkin and Huxley used the voltage-clamp technique to investigate the mechanisms of generation of the action potential in the squid giant axon. Neurons have a threshold for the initialization of an action potential of about −45 to −55 mV. Increasing the voltage from −60 to 0 mV produces a large, but transient, flow of positive charge into the cell (known as inward current). This transient inward current is followed by a sustained flow of positive charge out of the cell (the outward current). By voltage-clamping the cell and substituting different ions inside or outside the axon or both, Hodgkin, Huxley, and colleagues demonstrated that the transient inward current is carried by Na^+ ions flowing into the cell and the sustained outward current is mediated by a sustained flux of K^+ ions moving out of the cell (Fig. 5.6) (Hodgkin and Huxley, 1952a–d; Hodgkin et al., 1952; Hille, 1977).

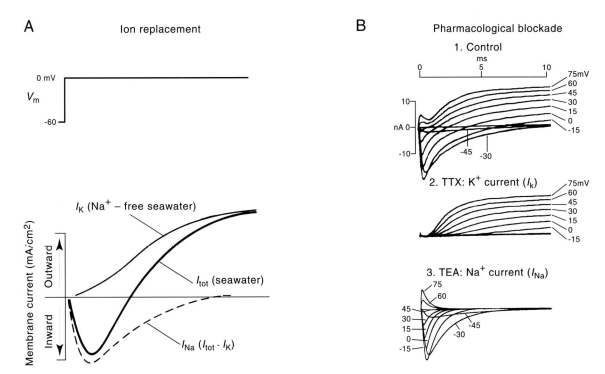

FIGURE 5.6 Voltage-clamp analysis reveals the ionic currents underlying action potential generation. (A) Increasing the potential from –60 to 0 mV across the membrane of the squid giant axon activates an inward current followed by an outward current. If the Na^+ in seawater is replaced by choline (which does not pass through Na^+ channels), then increasing the membrane potential from –60 to 0 mV results in only the outward current, which corresponds to I_K. Subtracting I_K from the recording in normal seawater illustrates the amplitude–time course of the inward Na^+ current, I_{Na}. Note that I_K activates more slowly than I_{Na} and that I_{Na} inactivates with time. (B) These two ionic currents can also be isolated from one another through the use of pharmacological blockers. (1) Increasing the membrane potential from –45 to +75 mV in 15-mV steps reveals the amplitude–time course of the inward Na^+ and outward K^+ currents. (2) After the block of I_{Na} with the poison tetrodotoxin (TTX), increasing the membrane potential to positive levels activates I_K only. (3) After the block of I_K with tetraethylammonium (TEA), increasing the membrane potential to positive levels activates I_{Na} only. (A) from Hodgkin and Huxley (1952); part (B) from Hille (1977).

The Na^+ and K^+ currents (I_{Na} and I_K, respectively) can be blocked, allowing each current to be examined in isolation (see Fig. 5.6B) (see also Chapter 7). Tetrodotoxin (TTX), a powerful poison found in the puffer fish *Spheroides rubripes* (Kao, 1966), selectively blocks voltage-dependent Na^+ currents (the puffer fish remains a delicacy in Japan and must be prepared with the utmost care by the chef). Using TTX, one can selectively isolate I_K and examine its voltage dependence and time course (see Fig. 5.6B).

Armstrong and Hille (1972) and others demonstrated that tetraethylammonium (TEA) is a useful pharmacological tool for selectively blocking I_K (see Fig. 5.6B). The use of TEA to examine the voltage dependence and time course of the Na^+ current underlying action-potential generation (see Fig. 5.6B) reveals some fundamental differences between Na^+ and the K^+ currents. First, the inward Na^+ current activates, or "turns on," much more rapidly than

does the K^+ current (giving rise to the name "delayed rectifier" for this K^+ current). Second, the Na^+ current is transient; it inactivates, even if the membrane potential is maintained at 0 mV (see Fig. 5.6A). In contrast, the outward K^+ current, once activated, remains "on" as long as the membrane potential is clamped to positive levels; that is, the K^+ current does not inactivate; it is sustained. Remarkably, from one experiment, we see that the Na^+ current both rapidly activates and inactivates, whereas the K^+ current only slowly activates. These fundamental properties of the underlying Na^+ and K^+ channels allow the generation of action potentials.

Hodgkin and co-workers (Hodgkin and Huxley, 1952a–d; Hodgkin *et al.*, 1952) proposed that the K^+ channels possess a voltage-sensitive "gate" that opens by the depolarization and closes by the subsequent repolarization of the membrane potential. This process of "turning on" and "turning off" the

K$^+$ current came to be known as *activation* and *deactivation*. The Na$^+$ current also exhibits voltage-dependent activation and deactivation (see Fig. 5.6), but the Na$^+$ channels also become inactive despite maintained depolarization. Thus, the Na$^+$ current not only activates and deactivates, but also exhibits a separate process known as *inactivation*, whereby the channels become blocked even though they are activated. The removal of this inactivation is achieved by removal of the depolarization and is a process known as *deinactivation*. Thus, the Na$^+$ channels possess two voltage-sensitive processes: activation–deactivation and inactivation–deinactivation. The kinetics of these two properties of Na$^+$ channels are different: inactivation takes place at a slower rate than activation.

The functional consequence of the two mechanisms is that Na$^+$ ions are allowed to flow across the membrane only when the channel is activated but not inactivated. Accordingly, Na$^+$ ions do not flow at resting membrane potentials, because the activation gate is closed (even though the inactivation gate is not). On depolarization, the activation gate opens, allowing Na$^+$ ions to flow into the cell. However, this depolarization also results in closure (at a slower rate) of the inactivation gate, which then blocks the flow of Na$^+$ ions. On repolarization of the membrane potential, the activation gate once again closes and the inactivation gate once again opens, preparing the axon for generation of the next action potential (Fig. 5.7) (see also Chapter 7, Fig. 7.6). Depolarization allows ionic current to flow by virtue of activation of the channel. The rush of Na$^+$ ions into the cell further depolarizes the membrane potential and more Na$^+$ channels become activated, forming a positive feedback loop that rapidly (within 100 μs or so) brings the membrane potential toward E_{Na}. However, the depolarization associated with the generation of the action potential also inactivates Na$^+$ channels, and, as a larger and larger percentage of Na$^+$ channels become inactivated, the rush of Na$^+$ into the cell diminishes. This inactivation of the Na$^+$ channels and activation of K$^+$ channels result in repolarization of the action potential. This repolarization deactivates the Na$^+$ channels. Then, the inactivation of the channel is slowly removed, and the channels are ready, once again, for generation of another action potential (see Fig. 5.7).

By measuring the voltage sensitivity and kinetics of these two processes, activation–deactivation and inactivation–deinactivation of the Na$^+$ current, as well as activation–deactivation of the delayed rectifier K$^+$ current, Hodgkin and Huxley generated a series of mathematical equations (see Chapter 7 for details)

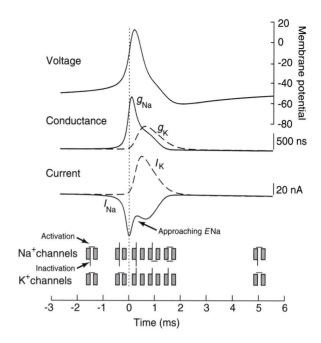

FIGURE 5.7 Generation of the action potential is associated with an increase in membrane Na$^+$ conductance and Na$^+$ current followed by an increase in K$^+$ conductance and K$^+$ current. Before action potential generation, Na$^+$ channels are neither activated nor inactivated (illustrated at the bottom of the figure). Activation of the Na$^+$ channels allows Na$^+$ ions to enter the cell, depolarizing the membrane potential. This depolarization also activates K$^+$ channels. After activation and depolarization, the inactivation particle on the Na$^+$ channels closes and the membrane potential repolarizes. The persistence of the activation of K$^+$ channels (and other membrane properties) generates an afterhyperpolarization. During this period, the inactivation particle of the Na$^+$ channel is removed and the K$^+$ channels close. From Huguenard and McCormick (1994).

that quantitatively describe the generation of the action potential (calculation of the propagation of a single action potential required an entire week of cranking a mechanical calculator). According to these early experimental and computational neuroscientists, the action potential is generated as follows. Depolarization of the membrane potential increases the probability of Na$^+$ channels being in the activated, but not yet inactivated, state. At a particular membrane potential, the resulting inflow of Na$^+$ ions tips the balance of the net ionic current from outward to inward (remember that depolarization also increases K$^+$ and Cl$^-$ currents by moving the membrane potential away from E_K and E_{Cl}). At this membrane potential, known as the action potential threshold (typically about –55 mV), the movement of Na$^+$ ions into the cell depolarizes the axon and opens more Na$^+$ channels, causing yet more depolarization of the membrane;

repetition of this process yields a rapid, positive feedback loop that brings the axon close to E_{Na}. However, even as more and more Na⁺ channels are becoming activated, some of these channels are also inactivating and, therefore, no longer conducting Na⁺ ions. In addition, the delayed rectifier K⁺ channels also are opening, owing to the depolarization of the membrane potential, and allowing positive charge to exit the cell. At some point, close to the peak of the action potential, the inward movement of Na⁺ ions into the cell is exactly offset by the outward movement of K⁺ ions out of the cell. After this point, the outward movement of K⁺ ions dominates, and the membrane potential is repolarized, corresponding to the fall of the action potential. The persistence of the K⁺ current for a few milliseconds following generation of the action potential generates the afterhyperpolarization. During this afterhyperpolarization, which is lengthened by the membrane time constant, the inactivation of the Na⁺ channels is removed, preparing the axon for generation of the next action potential (see Fig. 5.7).

The occurrence of an action potential is *not* associated with substantial changes in the intracellular or extracellular concentrations of Na⁺ or K⁺, as we saw earlier for the generation of the resting membrane potential. For example, the generation of a single action potential in a 25-μm-diameter hypothetical spherical cell should increase the intracellular concentration of Na⁺ by only approximately 6 μM (from about 18 to 18.006 mM). Thus, the action potential is an electrical event generated by a change in the distribution of charge across the membrane and not by a marked change in the intracellular or extracellular concentration of Na⁺ or K⁺.

Refractory Periods Prevent "Reverberation"

The ability of depolarization to activate an action potential varies as a function of the time since the last generation of an action potential, owing to the inactivation of Na⁺ channels and the activation of K⁺ channels. Immediately after generation of an action potential, another action potential usually cannot be generated regardless of the amount of current injected into the axon. This period corresponds to the absolute refractory period and is mediated largely by the inactivation of Na⁺ channels. The relative refractory period occurs during the action potential afterhyperpolarization and follows the absolute refractory period. The relative refractory period is characterized by a requirement for the increased injection of ionic current into the cell to generate another action potential and results from the persistence of the outward K⁺ current. The

practical implication of refractory periods is that action potentials are not allowed to "reverberate" between the soma and the axon terminals.

The Speed of Action Potential Propagation Is Affected by Myelination

Axons may be either myelinated or unmyelinated. Invertebrate axons or small vertebrate axons are typically unmyelinated, whereas larger vertebrate axons are often myelinated. As described in Chapters 1 and 4, sensory and motor axons of the peripheral nervous system are myelinated by specialized cells (Schwann cells) that form a spiral wrapping of multiple layers of myelin around the axon (Fig. 5.8). Several Schwann cells wrap around an axon along its length; between the ends of successive Schwann cells are small gaps (nodes of Ranvier). In the central nervous system, a single oligodendrocyte, a special type of glial cell, typically ensheathes several axonal processes (Bunge, 1968).

In unmyelinated axons, the Na⁺ and K⁺ channels taking part in action potential generation are distributed along the axon, and the action potential propagates along the length of the axon through local depolarization of each neighboring patch of membrane (see Chapter 4), causing that patch of membrane also to generate an action potential (Fig. 5.8). In myelinated axons, on the other hand, the Na⁺ channels are concentrated at the nodes of Ranvier (Ritchie and Rogart, 1977). The generation of an action potential at each node results in the depolarization of the next node and subsequently the generation of an action potential with an internode delay of only about 20 μs, referred to as *saltatory conduction* (from the Latin *saltare*, "to dance"). Growing evidence indicates that, between the nodes of Ranvier and underneath the myelin covering, K⁺ channels may play a role in determining the resting membrane potential and the repolarization of the action potential. A cause of some neurological disorders, such as multiple sclerosis and Guillain–Barré syndrome, is the demyelination of axons, resulting in a block of conduction of the action potentials (see Chapter 1).

Ion Channels Are Membrane-Spanning Proteins with Water-Filled Pores

The generation of ionic currents useful for the propagation of action potentials requires the movement of significant numbers of ions across the membrane in a relatively short time. The rate of ionic flow during the generation of an action potential is far too

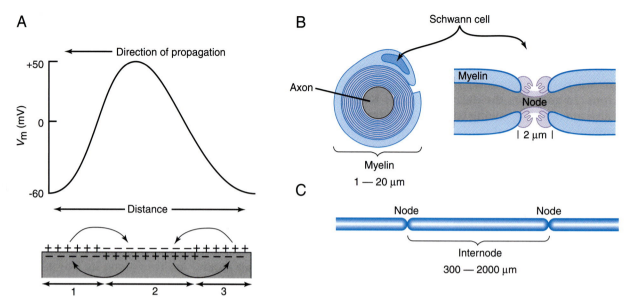

FIGURE 5.8 Propagation of the action potential in unmyelinated and myelinated axons. (A) Action potentials propagate in unmyelinated axons through the depolarization of adjacent regions of membrane. In the illustrated axon, region 2 is undergoing depolarization during the generation of the action potential, while region 3 has already generated the action potential and is now hyperpolarized. The action potential will propagate further by depolarizing region 1. (B) Vertebrate myelinated axons have a specialized Schwann cell that wraps around them in many spiral turns. The axon is exposed to the external medium at the nodes of Ranvier (Node). (C) Action potentials in myelinated fibers are regenerated at the nodes of Ranvier, where there is a high density of Na^+ channels. Action potentials are induced at each node through the depolarizing influence of the generation of an action potential at an adjacent node, thereby increasing the conduction velocity.

high to be achieved by an active transport mechanism and results instead from the opening of ion channels. Although the existence of ionic channels in the membrane has been postulated for decades, their properties and structure have only recently become known in detail. The powerful combination of electrophysiological and molecular techniques, and, most recently, X-ray crystallography, has greatly enhanced the knowledge of the structure–function relations of ionic channels (Jiang *et al.*, 2002a,b; Catterall, 1995, 2000a,b; Yellen, 2002) (Box 5.4).

Various neural toxins were particularly useful in the initial isolation of ionic channels. For example, three subunits (α, $\beta 1$, $\beta 2$) of the voltage-dependent Na^+ channel were isolated with the use of a derivative of a scorpion toxin (Catterall, 2000b; Beneski and Catterall, 1980). The α-subunit of the Na^+ channel is a large glycoprotein with a molecular mass of 270 kDa, whereas the $\beta 1$ and $\beta 2$ subunits are smaller polypeptides of molecular masses 39 and 37 kDa, respectively (Fig. 5.9). The α subunit, of which there are at least nine different isoforms, is the building block of the water-filled pore of the ionic channel, whereas the β subunits have some other role, such as in the regulation or structure of the native channel.

The α subunit of the Na^+ channel contains four internal repetitions (see Fig. 5.9B). Hydrophobicity analysis of these four components reveals that each contains six hydrophobic domains that may span the membrane as an α-helix. Of these six membrane-spanning components, the fourth (S4) has been proposed to be critical to the voltage sensitivity of the Na^+ channels. Voltage-sensitive gating of Na^+ channels is accomplished by the redistribution of ionic charge ("gating charge") in the Na^+ channel (Armstrong, 1992). Positive charges in the S4 region may act as voltage sensors such that an increase in the positivity of the inside of the cell results in a conformational change of the ionic channel. In support of this hypothesis, site-directed mutagenesis of the S4 region of the Na^+ channel to reduce the positive charge of this portion of the pore also reduces the voltage sensitivity of activation of the ionic channel. (Catterall, 2000b).

The mechanisms of inactivation of ionic channels have been analyzed with a combination of molecular and electrophysiological techniques. The most convincing hypothesis is that inactivation is achieved by a block of the inner mouth of the aqueous pore. Ionic channels are inactivated without detectable move-

BOX 5.4

ION CHANNELS AND DISEASE

Cells cannot survive without functional ion channels. It is therefore not surprising that an ever-increasing number of diseases have been found to be associated with defective ion channel function. There are a number of different mechanisms by which this may occur.

1. Mutations in the coding region of ion channel genes may lead to gain or loss of channel function, either of which may have deleterious consequences. For example, mutations producing enhanced activity of the epithelial Na^+ channel are responsible for Liddle syndrome, an inherited form of hypertension, whereas other mutations in the same protein that cause reduced channel activity give rise to hypotension. The most common inherited disease in Caucasians is also an ion channel mutation. This disease is cystic fibrosis (CF), which results from mutations in the epithelial chloride channel, known as CFTR. The most common mutation, the deletion of a phenylalanine at position 508, results in defective processing of the protein and prevents it from reaching the surface membrane. CFTR regulates chloride fluxes across epithelial cell membranes, and this loss of CFTR activity leads to reduced fluid secretion in the lung, resulting in potentially fatal lung infections.

2. Mutations in the promoter region of the gene may cause under- or overexpression of a given ion channel.

3. Other diseases result from defective regulation of channel activity by cellular constituents or extracellular ligands. This defective regulation may be caused by mutations in the genes encoding the regulatory molecules themselves or defects in the pathways leading to their production. Some forms of maturity-onset diabetes of the young (MODY) may be attributed to such a mechanism. ATP-sensitive potassium (K-ATP) channels play a key role in the glucose-induced insulin secretion from pancreatic β cells, and their defective regulation is responsible for two forms of MODY.

4. Autoantibodies to channel proteins may cause disease by downregulating channel function—often by causing internalization of the channel protein itself. Well-known examples are myasthenia gravis, which results from antibodies to skeletal muscle acetylcholine channels, and Eaton–Lambert myasthenic syndrome, in which patients produce antibodies against presynaptic Ca channels.

5. Finally, a number of ion channels are secreted by cells as toxic agents. They insert into the membrane of the target cell and form large nonselective pores, leading to cell lysis and death. The hemolytic toxin produced by the bacterium *Staphylococcus aureus* and the toxin secreted by the protozoan *Entamoeba histolytica*, which causes amebic dysentery, are examples.

Natural mutations in ion channels have been invaluable in studying the relationship between channel structure and function. In many cases, genetic analysis of a disease has led to cloning of the relevant ion channel. The first K channel to be identified (Shaker), for example, came from the cloning of the gene that caused *Drosophila* to shake when exposed to ether. Likewise, the gene encoding the primary subunit of a cardiac potassium channel (KVLQT1) was identified by positional cloning in families carrying mutations that caused a cardiac disorder known as long QT syndrome (see below). Conversely, the large number of studies on the relationship between Na channel structure and function have greatly assisted our understanding of how mutations in Na channels produce their clinical phenotypes.

Many diseases are genetically heterogeneous, and the same clinical phenotype may be caused by mutations in different genes. Long QT syndrome is a relatively rare inherited cardiac disorder that causes abrupt loss of consciousness, seizures, and sudden death from ventricular arrhythmia in young people. Mutations in three different genes, two types of cardiac muscle K channels (HERG and KVLQT1) and the cardiac muscle sodium channel (SCN1A), give rise to long QT syndrome. The disorder is characterized by a long QT interval in the electrocardiogram, which reflects the delayed repolarization of the cardiac action potential. As might therefore be expected, the mutations in the cardiac Na channel gene that cause long QT syndrome enhance the Na current (by reducing Na channel inactivation), while those in the potassium channel genes cause loss of function and reduce the K current.

Mutations in many different types of ion channels have been shown to cause human diseases. In addition to the examples listed above, mutations in water channels cause nephrogenic diabetes insipidus; mutations in gap junction channels cause Charcot–Marie–Tooth disease (a form of peripheral neuropathy) and hereditary deafness; mutations in the skeletal muscle Na channel cause a range of disorders known as the periodic paralyses; mutations in intracellular Ca-release channels cause malignant hyperthermia (a disease in which inhalation anesthetics trigger a potentially fatal rise in body

BOX 5.4 *Continued*

temperature); and mutations in neuronal voltage-gated Ca channels cause migraine and episodic ataxia. The list increases daily. As is the case with all single-gene disorders, the frequency of these diseases in the general population is very low. However, the insight they have provided into the relationship between ion channel structure and function, and into the physiological role of the different ion channels, has been invaluable. As William Harvey said in 1657, "nor is there any better way to advance the proper practice of medicine than to give our minds to the discovery of the usual form of nature, by careful investigation of the rarer forms of disease."

Frances M. Ashcroft

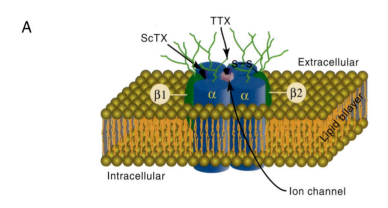

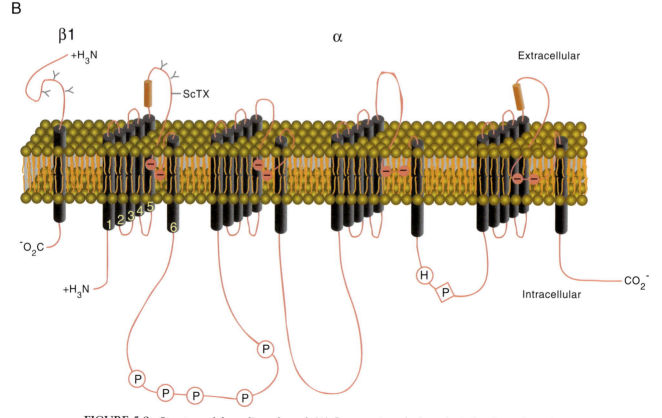

FIGURE 5.9 Structure of the sodium channel. (A) Cross section of a hypothetical sodium channel consisting of a single transmembrane α subunit in association with a β1 subunit and a β2 subunit. The α subunit has receptor sites for α-scorpion toxin (ScTX) and tetrodotoxin (TTX). (B) Primary structures of α and β1 subunits of sodium channel illustrated as transmembrane folding diagrams. Cylinders represent probable transmembrane α-helices.

ment of ionic current through the membrane; thus inactivation is probably not directly gated by changes in the membrane potential alone. Rather, inactivation may be triggered or facilitated as a secondary consequence of activation. Site-directed mutagenesis or the use of antibodies has shown that the part of the molecule between regions III and IV may be allowed to move to block the cytoplasmic side of the ionic pore after the conformational change associated with activation (Vassilev *et al.*, 1988, 1989; Stuhmer *et al.*, 1989). Additional information on the molecular properties of voltage-gated ion channels is provided in Chapter 6.

Neurons of the Central Nervous System Exhibit a Wide Variety of Electrophysiological Properties

The first intracellular recordings of action potentials in mammalian neurons by Sir John Eccles and colleagues revealed a remarkable similarity to those of the squid giant axon and gave rise to the assumption that the electrophysiology of neurons in the CNS was really rather simple: when synaptic potentials brought the membrane potential positive to action potential threshold, action potentials were produced through an increase in Na^+ conductance followed by an increase in K^+ conductance, as in the squid giant axon. The assumption, therefore, was that the complicated patterns of activity generated by the brain during the resting, sleeping, or active states were brought about as an interaction of the very large numbers of neurons present in the mammalian CNS (Brock *et al.*, 1952; Eccles, 1957). However, intracellular recordings of invertebrate neurons revealed that different cell types exhibit a wide variety of different electrophysiological behaviors, indicating that neurons may be significantly more complicated than the squid giant axon (Alving, 1968; Arvanitaki and Chalazonitis, 1961; Jackelet, 1989). Elucidation of the basic electrophysiology and synaptic physiology of different types of neurons and neuronal pathways within the mammalian CNS was facilitated by the *in vitro* slice technique, in which thin (~0.5 mm) slices of brain can be maintained for several hours. Intracellular recordings from identified cells revealed that neurons of the mammalian nervous system, similar to those of invertebrate networks, can generate complex patterns of action potentials entirely through intrinsic ionic mechanisms and without synaptic interaction with other cell types. For example, Rodolfo Llinás and colleagues discovered that Purkinje cells of the cerebellum can generate high-frequency trains (> 200 Hz) of Na^+- and K^+-mediated action potentials interrupted by Ca^{2+} spikes

in the dendrites, (Llinás and Sugimori, 1980a,b), whereas a major afferent to these neurons, the inferior olivary cell, can generate rhythmic sequences of broad action potentials only at low frequencies (< 15 Hz) through an interaction between various Ca^{2+}, Na^+, and K^+ condutances (Llinás and Yarom, 1981a,b) (Fig. 5.10). These *in vitro* recordings confirmed a major finding obtained with earlier intracellular recordings *in vivo*: each morphologically distinct class of neuron in the brain exhibits distinct electrophysiological features (Llinás, 1988). Just as cortical pyramidal cells are morphologically distinct from cerebellar Purkinje cells, which are distinct from thalamic relay cells, the electrophysiological properties of each of these different cell types also are markedly distinct.

Although no uniform classification scheme has been formulated in which all the different types of neurons of the brain can be classified, a few characteristic patterns of activity seem to recur. The first general class of action potential generation is characterized by those cells that generate trains of action potentials one spike at a time. The more prolonged the depolarization of these cells, the more prolonged their discharge. The more intensely these cells are depolarized, the higher the frequency of action potential generation. This type of relatively linear behavior is typical of brainstem and spinal cord motor neurons functioning in muscle contraction. A modification of this basic pattern of "regular firing" is characterized by the generation of trains of action potentials that exhibit a marked tendency to slow down in frequency with time, a process known as spike frequency adaptation. Examples of cells that discharge in this manner are cortical and hippocampal pyramidal cells (McCormick *et al.*, 1985; Madison and Nicoll, 1984; Pennefather *et al.*, 1985).

In addition to these regular firing cells, many neurons in the central nervous system exhibit the intrinsic propensity to generate rhythmic bursts of action potentials (Fig. 5.10) (see also Chapter 7). Examples of such neurons are thalamic relay neurons, inferior olivary neurons, and some types of cortical and hippocampal pyramidal cells (Jahnsen and Llinás, 1984a,b; Llinás and Yarom, 1981a,b; Wang and McCormick, 1993). In these cells, clusters of action potentials can occur together when the membrane is brought above firing threshold. These clusters of action potentials are typically generated through the activation of specialized Ca^{2+} currents that, through their slower kinetics, allow the membrane potential to be depolarized for a sufficient period to result in the generation of a burst of regular, Na^+- and K^+-dependent action potentials (discussed in the next section).

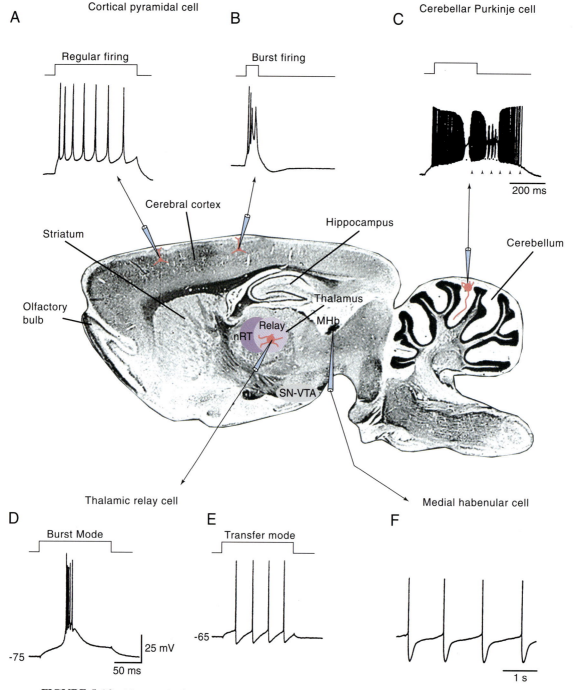

FIGURE 5.10 Neurons in the mammalian brain exhibit widely varying electrophysiological properties. (A) Intracellular injection of a depolarizing current pulse in a cortical pyramidal cell results in a train of action potentials that slow down in frequency. This pattern of activity is known as "regular firing." (B) Some cortical cells generated bursts of three or more action potentials, even when depolarized only for a short period. (C) Cerebellar Purkinje cells generate high-frequency trains of action potentials in their cell bodies that are disrupted by the generation of Ca^{2+} spikes in their dendrites. These cells can also generate "plateau potentials" from the persistent activation of Na^+ conductances (arrowheads). (D) Thalamic relay cells may generate action potentials either as bursts (D) or as tonic trains (E) of action potentials owing to the presence of a large low-threshold Ca^{2+} current. (F) Medial habenular cells generate action potentials at a steady and slow rate, in a "pacemaker" fashion.

Yet another general category of neurons in the brain comprises cells that generate relatively short duration (< 1 ms) action potentials and can discharge at relatively high frequencies (> 300 Hz). Such electrophysiological properties are often found in neurons that release the inhibitory amino acid γ-aminobutyric acid (Llinás and Sugimori, 1980a,b) (see Fig. 5.10) and some types of interneurons in the cerebral cortex, thalamus, and hippocampus (McCormick *et al.*, 1985; Schwartzkroin and Mathers, 1978; Pape and McCormick, 1995). Finally, the last general category of neurons consists of those that spontaneously generate action potentials at relatively low frequencies (e.g., 1–10 Hz). This type of electrophysiological behavior is often associated with neurons that release neuromodulatory transmitters, such as acetylcholine, norepinephrine, serotonin, and histamine (Vandermaelen and Aghajanian, 1983; Williams *et al.*, 1984; Reiner and McGeer, 1987). Neurons that release these neuromodulatory substances often innervate wide regions of the brain and appear to set the "state" of the different neural networks of the CNS, in a manner similar to the modulation of the different organs of the body by the sympathetic and parasympathetic nervous systems (Steriade and McCarley, 1990; McCormick, 1992).

Each of these unique intrinsic patterns of activity in the nervous system is due to the presence of a distinct mixture and distribution of different ionic currents in the cells. As in the classic studies of the squid giant axon, these different ionic currents have been characterized, at least in part, with voltage-clamp and pharmacological techniques, and the basic electrophysiological properties have been replicated with computational simulations (Belluzzi and Sacchi, 1991; McCormick and Huguenard, 1992; Huguenard and McCormick, 1994) (see Figs. 5.7 and 5.12).

Neurons Have Multiple Active Conductances

The search for the electrophysiological basis of the varying intrinsic properties of different types of neurons of vertebrates and invertebrates revealed a wide variety of ionic currents. Each type of ionic current is characterized by several features: (1) the type of ions conducted by the underlying ionic chan-

TABLE 5.2   Neuronal Ionic Currents

Current	Description	Function
Na$^+$ currents		
$I_{Na,t}$	Transient; rapidly activating and inactivating	Action potentials
$I_{Na,p}$	Persistent; noninactivating	Enhances depolarization; contributes to steady-state firing
Ca^{2+} currents		
I_T, low threshold	Transient; rapidly inactivating; threshold negative to –65 mV	Underlies rhythmic burst firing
I_L, high threshold	Long-lasting; slowly inactivating; threshold around –20 mV	Underlies Ca^{2+} spikes that are prominent in dendrites; involved in synaptic transmission
I_N	Neither; rapidly inactivating; threshold around –20 mV	Underlies Ca^{2+} spikes that are prominent in dendrites; involved in synaptic transmission
I_P	Purkinje; threshold around –50 mV	
K$^+$ currents		
I_K	Activated by strong depolarization	Repolarization of action potential
I_C	Activated by increases in [Ca^{2+}]$_i$	Action potential repolarization and interspike interval
I_{AHP}	Slow afterhyperpolarization; sensitive to increases in [Ca^{2+}]$_i$	Slow adaptation of action potential discharge; the block of this current by neuromodulators enhances neuronal excitability
I_A	Transient; inactivating	Delayed onset of firing; lengthens interspike interval; action potential repolarization
I_M	Muscarine sensitive; activated by depolarization; noninactivating	Contributes to spike frequency adaptation; the block of this current by neuromodulators enhances neuronal excitability
I_h	Depolarizing (mixed cation) current that is activated by hyperpolarization	Contributes to rhythmic burst firing and other rhythmic activities
$I_{K,leak}$	Contributes to neuronal resting membrane potential	Block of this current by neuromodulators can result in a sustained change in membrane potential

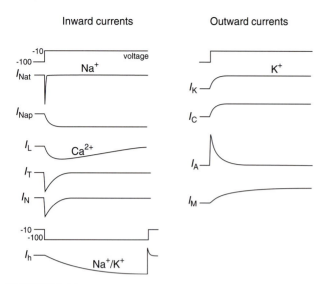

FIGURE 5.11 Voltage dependence and kinetics of different ionic currents in the mammalian brain. Depolarization of the membrane potential from –100 to –10 mV results in the activation of currents entering or leaving neurons.

nels (e.g., Na^+, K^+, Ca^{2+}, Cl^-, or mixed cations), (2) their voltage and time dependence, and (3) their sensitivity to second messengers. In vertebrate neurons, two distinct Na^+ currents have been identified and six distinct Ca^{2+} currents and more than seven distinct K^+ currents are known (Table 5.2 and Fig. 5.11). This is a minimal number, as these currents are formed from a much greater pool of channel subunits. The following sections briefly review these classes of ionic currents and their ionic channels, relating them to the different patterns of behavior mentioned earlier for neurons in the mammalian CNS.

Na$^+$ Currents Are Both Transient and Persistent

Depolarization of many different types of vertebrate neurons results not only in the activation of the rapidly activating and inactivating Na^+ current (I_{Nat}) underlying action potential generation but also in the rapid activation of a Na^+ current that does not inactivate and is therefore known as the "persistent" Na^+ current (I_{Nap}) (Llinás, 1988; Hotson *et al.*, 1979; Stafstrom *et al.*, 1982; Alzheimer *et al.*, 1993). The threshold for activation of the persistent Na^+ current is typically about –65 mV, that is, below the threshold for the generation of action potentials. This property gives this current the interesting ability to enhance or facilitate the response of the neuron to depolarizing, yet subthreshold, inputs. For example, synaptic events that depolarize the cell activate I_{Nap}, resulting

in an extra influx of positive charge and therefore a larger depolarization than otherwise would occur. Likewise, hyperpolarizations may result in deactivation of I_{Nap}, again resulting in larger hyperpolarizations than would otherwise occur. In this manner, the persistent Na^+ current may play an important regulatory function in the control of the functional responsiveness of the neuron to synaptic inputs and may contribute to the dynamic coupling of the dendrites to the soma.

Persistent activation of I_{Nap} may also contribute to another electrophysiological feature of neurons, namely, the generation of plateau potentials (Llinás, 1988). A plateau potential refers to the ability of many different types of neurons to generate, through intrinsic ionic mechanisms, a prolonged (from tens of milliseconds to seconds) depolarization and action potential discharge in response to a short-lasting depolarization (see Fig. 5.10C). One can wonder whether such plateau potentials contribute to persistent firing in neurons during the performance of visual memory tasks, as has been found in some types of neurons in the frontal neocortex and superior colliculus of behaving primates (Goldman-Rakic, 1995).

K$^+$ Currents Vary in Their Voltage Sensitivity and Kinetics

Potassium currents that contribute to the electrophysiological properties of neurons are numerous and exhibit a wide range of voltage-dependent and kinetic properties (Yellen, 2002; Jan and Jan, 1990; Storm, 1990; Johnston and Wu, 1995) (see also Chapter 7 for additional details). Perhaps the simplest K^+ current is that characterized by Hodgkin and Huxley: this K^+ current, I_K, rapidly activates on depolarization and does not inactivate (see Fig. 5.11). Other K^+ currents activate with depolarization but also inactivate with time. For example, the rapid activation and inactivation of I_A give this current a transient appearance (see Fig. 5.11), and I_A is believed to be important in controlling the rate of action potential generation, particularly at low frequencies (Connor and Stevens, 1971a,b) (Fig. 5.12). Like the Na^+ channel, I_A channels are inactivated by the plugging of the inner mouth of the pore through the movement of an inactivation particle (Yellen, 2002; Hoshi *et al.*, 1990; Zagotta *et al.*, 1990).

Another broad class of K^+ channels consists of those that are sensitive to changes in the intracellular concentration of Ca^{2+} (Blatz and Magleby, 1987; Latorre *et al.*, 1989). These K^+ currents are collectively referred to as I_{KCa}, with two examples being the K^+ currents that underlie slow afterhyperpolarizations following repetitive action potential discharge, I_{AHP}

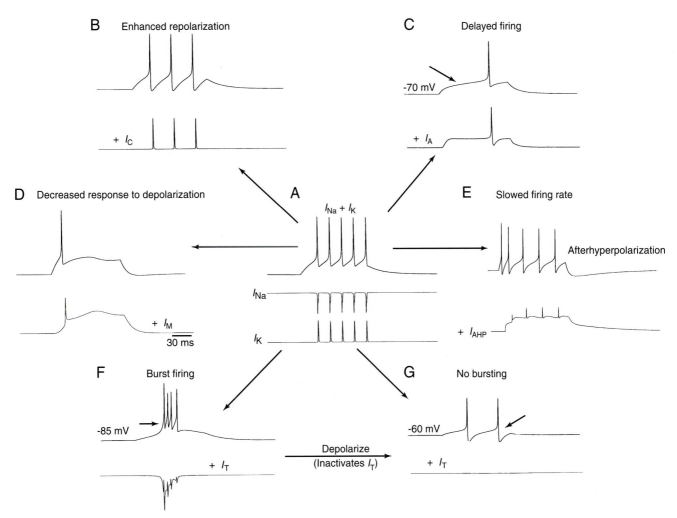

FIGURE 5.12 Simulation of the effects of the addition of various ionic currents to the pattern of activity generated by neurons in the mammalian CNS. (A) The repetitive impulse response of the classic Hodgkin–Huxley model (voltage recordings above, current traces below). With only I_{Na} and I_K, the neuron generates a train of five action potentials in response to depolarization. Addition of I_C (B) enhances action potential repolarization. Addition of I_A (C) delays the onset of action potential generation. Addition of I_M (D) decreases the ability of the cell to generate a train of action potentials. Addition of I_{AHP} (E) slows the firing rate and generates a slow afterhyperpolarization. Finally, addition of the transient Ca^{2+} current I_T results in two states of action potential firing: (F) burst firing at –85 mV and (G) tonic firing at –60 mV. From Huguenard and McCormick (1994).

and a fast K$^+$ current that helps repodarize action potentials, I_C (see Fig. 5.11). Still other K$^+$ channels are not only activated by voltage but also modulated by activation of various modulatory neurotransmitter receptors, such as the M current (see Fig. 5.11). By investigating the ionic mechanisms by which the release of acetylcholine from preganglionic neurons in the brain results in prolonged changes in the excitability of neurons of the sympathetic ganglia, Brown and Adams (1980) discovered a unique K$^+$ current (I_M) that slowly (over tens of milliseconds) turns on with depolarization of the neuron (see Fig. 5.12D). The slow activation of this K$^+$ current results in a decrease in the responsiveness of the cell to depolarization and,

therefore, regulates how the cell responds to excitation. This K$^+$ current, like I_{AHP}, is reduced by the activation of a wide variety of receptors, including muscarinic receptors, for which it is named. Reduction of I_M results in a marked increase in responsiveness of the affected cell to depolarizing inputs and again may contribute to the mechanisms by which neuromodulatory systems control the state of activity in cortical and hippocampal networks (McCormick, 1992; Nicoll, 1988; Nicoll *et al.*, 1990).

Between these classic examples of K$^+$ currents are a variety of other types that have not been fully characterized, including K$^+$ currents that vary from one another in their voltage sensitivity, kinetics, and response to various second messengers. Molecular biological studies of voltage-sensitive K^+ channels, first done in *Drosophila* and later in mammals, have revealed a large number of genes that generate K^+ channels. They consist of four distinct families: Kv1, Kv2, Kv3, and Kv4 (Yellen, 2002; Chandy and Gutman, 1995). These genes generate a wide variety of different K^+ channels due not only to the large number of genes involved, but also to alternative RNA splicing, gene duplication, and other posttranslational mechanisms. Functional expression of different K^+ channels reveals remarkable variation in the rate of inactivation, such that some are rapidly inactivating (A-current-like), whereas others inactivate more slowly. Finally, some K^+ channels do not inactivate, such as I_K. One of the largest subfamilies of K^+ channels comprises those that give rise to the resting membrane potential, so-called "leak channels." Interestingly, these channels appear to be opened by gaseous anesthetics, indicating that the hyperpolarization of central neurons is a major component of general anesthesia. Recently, MacKinnon and colleagues have succeeded in crystallizing different types of K^+ channels, leading to a great leap in our knowledge of their structure and how they function, including the mechanisms by which channels are opened by voltage and ligands (Jiang *et al.*, 2002a,b). It is now clear that each type of neuron in the nervous system is likely to contain a unique set of functional voltage-sensitive K^+ channels, perhaps selected, modified, and placed in particular spatial locations in the cell in a manner to facilitate the unique role of that cell type in neuronal processing.

Ca²⁺ Currents Control Electrophysiological Properties and Ca²⁺-Dependent Second-Messenger Systems

Ionic channels that conduct Ca^{2+} are present in all neurons. These channels are special in that they serve two important functions. First, Ca^{2+} channels are present throughout the different parts of the neuron (dendrites, soma, synaptic terminals) and contribute greatly to the electrophysiological properties of these processes (Llinás, 1988; Regehr and Tank, 1994; Johnston *et al.*, 1996). Second, Ca^{2+} channels are unique in that Ca^{2+} is an important second messenger in neurons, and entry of Ca^{2+} into the cell can affect numerous physiological functions, including neurotransmitter release, synaptic plasticity, neurite outgrowth during development, and even gene expression.

On the bases of their voltage sensitivity, their kinetics of activation and inactivation, and their ability to be blocked by various pharmacological agents, Ca^{2+} currents can be separated into at least four distinct categories, three of which are I_T ("transient"), I_L ("long lasting"), and I_N ("neither") (Nowycky *et al.*, 1985; Carbonne and Lux, 1984), illustrated in Fig. 5.11A. A fourth, I_P, is found in the Purkinje cells of the cerebellum, as well as in many different cell types of the CNS (Llinás *et al.*, 1992) Calcium channels are formed from at least 10 different α subunits as well as a variety of β and γ subunits, indicating that an even greater number of Ca^{2+} currents are yet to be characterized (Tsien *et al.*, 1991; Ertel *et al.*, 2000).

Neurons Possess Multiple Subtypes of High-Threshold Ca²⁺ Currents

High-voltage-activated Ca^{2+} channels are activated at membrane potentials more positive than approximately -40 mV and include the currents I_L, I_N, and I_P. The L-type calcium currents exhibit a high threshold for activation (about -10 mV) and give rise to rather persistent, or long-lasting, ionic currents (see Fig. 5.11A). Dihydropyridines, Ca^{2+} channel antagonists, are clinically useful for their effects on the heart and vascular smooth muscle (e.g., for the treatment of arrhythmias, angina, and migraine headaches) and selectively block L-type Ca^{2+} channels (Bean, 1989; Stea *et al.*, 1985). In contrast with I_L, I_N is not blocked by dihydropyridines; rather it is selectively blocked by a toxin found in Pacific cone shells (ω-conotoxin-GVIA). The N-type Ca^{2+} channels have a threshold for activation of about -20 mV, inactivate with maintained depolarization, and are modulated by a variety of neurotransmitters. In some cell types, I_N has a role in the Ca^{2+}-dependent release of neurotransmitters at presynaptic terminals (Wheeler *et al.*, 1994). The P-type calcium channel is distinct from N and L types in that it is not blocked by either dihydropyridines or ω-conotoxin-GVIA but is blocked by a toxin (ω-agatoxin-IVA) present in the venom of the funnel web spider (Llinás *et al.*, 1992; Stea *et al.*, 1995). This type of calcium channel activates at relatively high thresholds and does not inactivate. Prevalent in Purkinje cells as well as other cell types, as mentioned earlier, the P-type Ca^{2+} channel participates in the generation of dendritic Ca^{2+} spikes, which can strongly modulate the firing pattern of the neuron in which it resides (see Fig. 5.10C).

Collectively, the high-threshold-activated Ca^{2+} channels contribute to the generation of action potentials in mammalian neurons. The activation of Ca^{2+} currents adds somewhat to the depolarizing part of the action potential, but, more importantly, these channels allow Ca^{2+} to enter the cell and this has the secondary consequence of activation of various Ca^{2+}-

activated K⁺ currents (Latorre *et al.*, 1989) and protein kinases (see Chapters 12 and 18). As mentioned earlier, the activation of these K⁺ currents modifies the pattern of action potentials generated in the cell (see Figs. 5.10 and 5.12). High-threshold Ca^{2+} channels are similar to the Na⁺ channel in that they are composed of a central $\alpha 1$ subunit that forms the aqueous pore and several regulatory or auxiliary subunits. As in the Na⁺ channel, the primary structure of the $\alpha 1$ subunit of the Ca^{2+} channel consists of four homologous domains (I–IV), each containing six regions (S1–S6) that may generate transmembrane α-helices. Genes for at least 10 different Ca^{2+} channel α subunits have been cloned and are separated into three families (CaV1, CaV2, CaV3). The properties of the products of these genes indicate that I_L is likely to correspond to the CaV1 subfamily, whereas I_N corresponds to CaV2.2 and I_T is formed by the CaV3 subfamily (Catterall, 2000a; Bean, 1989).

Low-Threshold Ca^{2+} Currents Generate Bursts of Action Potentials

Low-threshold Ca^{2+} currents (see Fig. 5.11A) often take part in the generation of rhythmic bursts of action potentials (see Figs. 5.10 and 5.12). The low-threshold Ca^{2+} current is characterized by a threshold for activation of about –65 mV, which is below the threshold for generation of typical Na⁺–K⁺-dependent action potentials (–55 mV). This current inactivates with maintained depolarization. Owing to these properties, the role of low-threshold Ca^{2+} currents differs markedly from that of the high-threshold Ca^{2+} currents. Through activation and inactivation of the low-threshold Ca^{2+} current, neurons can generate slow (about 100 ms) Ca^{2+} spikes, which can result, owing to their prolonged duration, in the generation of a high-frequency "burst" of short-duration Na⁺–K⁺ action

---

BOX 5.5

JELLYFISH—WHAT A NERVE!

Research on jellyfish provides intriguing insight into how the properties and distribution of ion channels within a nerve membrane can affect the behavior of the whole animal. *Aglantha digitale* can swim slowly when feeding or quickly if escaping predators just through the "behavior" of a single muscle sheet coupled to a simply organized nervous system.

The jellyfish does this through an unusual form of signaling. Each "giant" motor nerve axon not only has voltage-dependent sodium channels and three types of potassium channels, but also crucial T-type calcium channels. *Aglantha* motor axons are unusual because they develop two entirely different propagating action potentials (Mackie and Meech, 1985). The T-type calcium channels contribute to a low-amplitude calcium-dependent spike that propagates along the motor axon without gaining amplitude or decrementing in the way that electrotonic potentials do (Meech and Mackie, 1995). The motor axon makes direct synaptic contact with the muscle epithelium that makes up the bell of the jellyfish and so the propagating calcium spike induces the weak contractions responsible for propulsion during the regular slow swimming the animal performs when feeding.

Aglantha lives in the colder waters of the world at a depth of about 100 m. Studied in their natural habitat, they are seen to avoid predators by generating an altogether stronger form of swimming. In the laboratory this "escape" swimming can be reproduced by stimulating vibration-sensitive receptors at the base of the bell of the animal (Arkett *et al.*, 1988). The strong synaptic depolarization that this stimulus induces in each of the eight giant motor axons drives its membrane potential beyond the peak of the calcium spike and induces a full-sized sodium action potential. As the sodium spike propagates more rapidly than the slow swim calcium spike, there is a coordinated contraction of the body wall that drives the animal forward.

Sodium and calcium spikes like those seen in *Aglantha* have been recorded from a variety of sites in the mammalian CNS (Llinás and Yarom, 1981b; Llinás and Jahnsen, 1982). However, unlike in *Aglantha*, the peak of the calcium spike always exceeds the threshold of the sodium spike and the two impulses form a single complex signal. Patch-clamp analysis of *Aglantha* axons has revealed a family of potassium channels that are responsible for setting thresholds and repolarizing each of the two different impulses. Each potassium channel class has an identical unitary conductance and appears to be organized in a mosaic fashion over the surface of the axon (Meech and Mackie, 1993). Sodium and T-type calcium channels are clustered together into well-defined "hot spots." George Mackie and I have suggested that the mosaic organization facilitates the turnover of ion channels; channels inserted into the membrane in clusters age together and are eliminated together.

Robert W. Meech

potentials (see Fig. 5.10 and Box 5.5) (Llinás and Jahnsen, 1982).

In the mammalian brain, this pattern is especially well exemplified by the activity of thalamic relay neurons; in the visual system, these neurons receive direct input from the retina and transmit this information to the visual cortex. During periods of slow wave sleep, the membrane potential of these relay neurons is relatively hyperpolarized, resulting in the removal of inactivation (deinactivation) of the low-threshold Ca^{2+} current. This deinactivation allows these cells to spontaneously generate low-threshold Ca^{2+} spikes and bursts of from two to five action potentials (Fig. 5.13) (McCormick and Pape, 1990). The large number of thalamic relay cells bursting during sleep in part gives rise to the spontaneous synchronized activity that early investigators were so surprised to find during recordings from the brains of sleeping animals (Steriade et al., 1993). It has even proved possible to maintain one of the sleep-related brain rhythms (spindle waves) intact in slices of thalamic tissue maintained in vitro, owing to the generation of

this rhythm by the interaction of a local network of thalamic cells and their electrophysiological properties (von Krosigk et al., 1993).

The transition to waking or the period of sleep when dreams are prevalent (rapid eye movement sleep) is associated with a maintained depolarization of thalamic relay cells to membrane potentials ranging from about –60 to –55 mV. The low-threshold Ca^{2+} current is inactivated and therefore the burst discharges are abolished. In this way, the properties of a single ionic current (I_T) help to explain in part the remarkable changes in brain activity taking place in the transition from sleep to waking (Fig. 5.13).

Low-threshold Ca^{2+} channels were recently cloned and shown to have some similarities to other Ca^{2+} channels (Perez-Reyes et al., 1998). Evidence suggests that some antiepileptic drugs may exert their therapeutic actions through a reduction in I_T. This is especially true of the drugs useful in the treatment of generalized absence (petit mal) seizures, which are known to rely on the thalamus for their generation (Coulter et al., 1990).

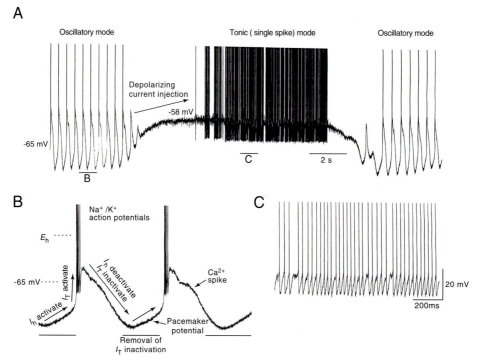

FIGURE 5.13   Two different patterns of activity generated in the same neuron, depending on membrane potential. (A) The thalamic neuron spontaneously generates rhythmic bursts of action potentials owing to the interaction of the Ca^{2+} current I_T and the inward "pacemaker" current I_h. Depolarization of the neuron changes the firing mode from rhythmic burst firing to tonic action potential generation in which spikes are generated one at a time. Removal of this depolarization reinstates the rhythmic burst firing. This transition from rhythmic burst firing to tonic activity is similar to that which occurs in the transition from sleep to waking. (B) Expansion of detail of rhythmic burst firing. (C) Expansion of detail of tonic firing. From McCormick and Pape (1990).

Hyperpolarization-Activated Ionic Currents Are Involved in Rhythmic Activity

In most types of neurons, hyperpolarization negative to approximately -50 mV activates an ionic current, known as I_h, that conducts both Na^+ and K^+ ions (see Fig. 5.11A). This current typically has very slow kinetics, turning on with a time constant on the order of tens of milliseconds to seconds. Because the channels underlying this current allow the passage of both Na^+ and K^+ ions, the reversal potential of I_h is typically about -35 mV—between E_{Na} and E_K. Because this current is activated by hyperpolarization below approximately -60 mV, it is typically dominated by the inward movement of Na^+ ions and is therefore depolarizing. For what purpose could neurons use a depolarizing current that activates when the cell is hyperpolarized? A clue comes from cardiac cells in which this current, known as I_f for "funny," is important for determining heart rate (DiFrancesco, 1985). Activation of I_f results in a slow depolarization of the membrane potential between adjacent cardiac action potentials. The more that I_f is activated, the faster the membrane depolarizes between beats and, therefore, the sooner the threshold for the next action potential is reached and the next beat is generated. In this manner, the amplitude, or sensitivity to voltage, of I_f can modify heart rate. Interestingly, the sensitivity of I_f to voltage is adjusted by the release of noradrenaline and acetylcholine; the activation of β-adrenoceptors by noradrenaline increases I_f and therefore increases the heart rate, whereas the activation of muscarinic receptors decreases I_f, thereby decreasing the heart rate (DiFrancesco, 1993; DiFrancesco et al., 1989). This continual adjustment of I_f results from a "push–pull" arrangement between adrenergic and muscarinic cholinergic receptors and is mediated by the adjustment of intracellular levels of cyclic AMP. Indeed, the recent cloning of H channels reveals that their structure is similar to that of cyclic nucleotide-gated channels (Ludwig et al., 1998).

Could I_h play a role in neurons similar to that of I_f in the heart? Possibly. Synchronized rhythmic oscillations in the membrane potential of large numbers of neurons, in some respects similar to those of the heart, are characteristic of the mammalian brain. Oscillations of this type are particularly prevalent in thalamic relay neurons during some periods of sleep, as mentioned earlier. Intracellular recordings from these thalamic neurons reveal that they often generate rhythmic "bursts" of action potentials mediated by the activation of a slow spike that is generated through the activation of the low-threshold, or transient, Ca^{2+} current, I_T (McCormick and Huguenard, 1992; McCormick and Pape, 1990) (see Fig. 5.13). Between the low-threshold Ca^{2+} spikes is a slowly depolarizing membrane potential generated by activation of the mixed Na^+–K^+ current I_h, as with I_f in the heart. The amplitude, or voltage sensitivity, of I_h adjusts the rate at which the thalamic cells oscillate, and, as with the heart, this sensitivity is adjusted by the release of modulatory neurotransmitters (see Fig. 5.13). In a sense, the thalamic neurons are "beating" in a manner similar to that of the heart.

Summary

An action potential is generated by the rapid influx of Na^+ ions followed by a slightly slower efflux of K^+ ions. Although the generation of an action potential does not disrupt the concentration gradients of these ions across the membrane, the movement of charge is sufficient to generate a large and brief deviation in the membrane potential. Propagation of the action potential along the axon allows communication of the output of the cell to its synapses. Neurons possess many different types of ionic channels in their membranes, allowing complex patterns of action potentials to be generated and complex synaptic computations to occur within single neurons.

References

Alving, B. O. (1968). Spontaneous activity in isolated somata of *Aplysia* pacemaker neurons. *J. Gen. Physiol.* **51**, 29–45.

Alzheimer, C., Schwindt, P. C., and Crill, W. E. (1993). Modal gating of Na^+ channels as a mechanism of persistent Na^+ current in pyramidal neurons from rat and cat sensorimotor cortex. *J. Neurosci.* **13**, 660–673.

Arkett, S., Mackie, G. O., and Meech, R. W. (1988). Hair-cell mechanoreception in the jellyfish *Aglantha digitale. J. Exp. Biol.* **135**, 329–342.

Armstrong, C. M. (1992). Voltage-dependent ionic channels and their gating. *Physiol. Rev.* **72**(Suppl.), 5–13.

Armstrong, C. M., and Hille, B. (1972). The inner quaternary ammonium ion receptor in potassium channels of the node of Ranvier. *J. Gen. Physiol.* **59**, 388–400.

Arvanitaki, A., and Chalazonitis, N. (1961). Slow waves and associated spiking in nerve cells of *Aplysia. Bull. Inst. Oceanogr. Monaco* **58**, 1–15.

Bean, B. P. (1989). Classes of calcium channels in vertebrate cells. *Annu. Rev. Physiol.* **51**, 367–384.

Belluzzi, O., and Sacchi, O. (1991). A five-conductance model of the action potential in the rat sympathetic neurone. *Prog. Biophys. Mol. Biol.* **55**, 1–30.

Beneski, D. A., and Catterall, W. A. (1980). Covalent labeling of protein components of the sodium channel with a photoactivable derivative of scorpion toxin. *Proc. Natl. Acad. Sci. USA* **77**, 639–643.

Blatz, A. L., and Magleby, K. L. (1987). Calcium-activated potassium channels. *Trends Neurosci.* **11**, 463–467.

Brazier, M. A. B. (1959). The historical development of neurophysiology. In "Handbook of Physiology" (J. Field, Ed.), Sect. 1, Vol. 1, pp. 1–58. Am. Physiol. Soc., Washington, DC.

Brazier, M. A. B. (1988). "A History of Neurophysiology in the 19th Century." Raven Press, New York.

Brock, L. G., Coombs, J. S., and Eccles, J. C. (1952). The recording of potentials from motoneurones with an intracellular electrode. *J. Physiol. (London)* **117**, 431–460.

Brown, D. A., and Adams, P. R. (1980). Muscarinic suppression of a novel voltage sensitive K^+ current in a vertebrate neurone. *Nature (London)* **283**, 673–676.

Bunge, R. P. (1968). Glial cells and the central myelin sheath. *Physiol. Rev.* **48**, 197–251.

Buser, P., and Albe-Fessard, D. (1953). Premiers resultats d'une analyse l'activite electrique du cortex cerebral du Chat par microelectrodes intracellulaires. *C. R. Hebd. Seances Acad. Sci.* **236**, 1197–1199.

Carbonne, E., and Lux, H. D. (1984). A low voltage-activated, fully inactivating Ca channel in vertebrate sensory neurones. *Nature (London)* **310**, 501–502.

Catterall, W. A. (1995). Structure and function of voltage-gated ion channels. *Annu. Rev. Biochem.* **64**, 493–531.

Catterall, W. A. (2000a). Structure and regulation of voltage-gated Ca^{2+} channels. *Annu. Rev. Cell Dev. Biol.* **16**, 521–555.

Catterall, W. A. (2000b). From ionic currents to molecular mechanisms: The structure and function of voltage-gated sodium channels. *Neuron* **26**, 13–25.

Chandy, K. G., and Gutman, G. A. (1995). Voltage-gated potassium channel genes. *In* "Ligand- and Voltage-Gated Channels" (A. North, Ed.), pp. 1–72. CRC Press, Boca Raton, FL.

Cole, K. S. (1949). Dynamic electrical characteristics of the squid axon membrane. *Arch. Sci. Physiol.* **3**, 253–258.

Cole, K. S., and Curtis, H. J. (1939). Electric impedance of the squid giant axon during activity. *J. Gen. Physiol.* **22**, 649–670.

Connor, J. A., and Stevens, C. F. (1971a). Voltage clamp studies of a transient outward membrane current in gastropod neural somata. *J. Physiol. (London)* **213**, 21–30.

Connor, J. A., and Stevens, C. F. (1971b). Prediction of repetitive firing behaviour from voltage clamp data on an isolated neurone soma. *J. Physiol. (London)* **213**, 31–53.

Coulter, D. A., Huguenard, J. R., and Prince, D. A. (1990). Differential effects of petit mal anticonvulsants and convulsants on thalamic neurones: Calcium current reduction. *Br. J. Pharmacol.* **100**, 800–806.

DiFrancesco, D. (1985). The cardiac hyperpolarizing-activated current. II. Origins and developments. *Prog. Biophys. Mol. Biol.* **46**, 163–183.

DiFrancesco, D. (1993). Pacemaker mechanisms in cardiac tissue. *Annu. Rev. Physiol.* **55**, 455–472.

DiFrancesco, D., Ducouret, P., and Robinson, R. B. (1989). Muscarinic modulation of cardiac rate at low acetylcholine concentrations. *Science* **243**, 669–671.

Eccles, J. C. (1957). "The Physiology of Nerve Cells." Johns Hopkins Univ. Press, Baltimore, MD.

Ertel, E. A., Campbell, K. P., Harpold, M. M., Hofmann, F., Mori, Y., Perez-Reyes, E., Schwartz, A., Snutch, T. P., Tanabe, T., Birnbaumer, L., Tsien, R. W., and Catterall, W. A. (2000). Nomenclature of voltage-gated calcium channels. *Neuron* **25**, 533–535.

Goldman, D. F. (1943). Potential, impedance, and rectification in membranes. *J. Gen. Physiol.* **27**, 37–60.

Goldman-Rakic, P. S. (1995). Cellular basis of working memory. *Neuron* **14**, 477–485.

Hille, B. (1977). Ionic basis of resting potentials and action potentials. *In* "Handbook of Physiology" (E. R. Kandel, Ed.), Sect. 1, Vol. 1, pp. 99–136. Am. Physiol. Soc., Bethesda, MD.

Hirsch, J. C., Fourment, A., and Marc, M. E. (1983). Sleep-related variations of membrane potential in the lateral geniculate body relay neurons of the cat. *Brain Res.* **259**, 308–312.

Hodgkin, A. L. (1976). Chance and design in electrophysiology: An informal account of certain experiments on nerve carried out between 1934 and 1952. *J. Physiol. (London)* **263**, 1–21.

Hodgkin, A. L., and Huxley, A. F. (1939). Action potentials recorded from inside a nerve fiber. *Nature (London)* **144**, 710–711.

Hodgkin, A. L., and Huxley, A. F. (1952a). Currents carried by sodium and potassium ions through the membrane of the giant axon of *Loligo*. *J. Physiol. (London)* **116**, 449–472.

Hodgkin, A. L., and Huxley, A. F. (1952b). The components of membrane conductance in the giant axon of *Loligo*. *J. Physiol. (London)* **116**, 473–496.

Hodgkin, A. L., and Huxley, A. F. (1952c). The dual effect of membrane potential on sodium conductance in the giant axon of *Loligo*. *J. Physiol. (London)* **116**, 497–506.

Hodgkin, A. L., and Huxley, A. F. (1952d). A quantitative description of membrane current and its application to conduction and excitation in nerve. *J. Physiol. (London)* **117**, 500–544.

Hodgkin, A. L., and Katz, B. (1949). The effect of sodium ions on the electrical activity of the giant axon of the squid. *J. Physiol. (London)* **108**, 37–77.

Hodgkin, A. L., and Keynes, D. (1955). Active transport of cations in giant axons from *Sepia* and *Loligo*. *J. Physiol. (London)* **128**, 28–60.

Hodgkin, A. L., Huxley, A. F., and Katz, B. (1952). Measurement of current–voltage relations in the membrane of the giant axon of *Loligo*. *J. Physiol. (London)* **116**, 424–448.

Horisberger, J.-D., Lemas, V., Kraehenbu..hl, J.-P., and Rossier, B. C. (1991). Structure–function relationship of Na,K-ATPase. *Annu. Rev. Physiol.* **53**, 565–584.

Hoshi, T., Zagotta, W. N., and Aldrich, R. W. (1990). Biophysical and molecular mechanisms of Shaker potassium channel inactivation. *Science* **250**, 533–538.

Hotson, J. R., Prince, D. A., and Schwartzkroin, P. A. (1979). Anomalous inward rectification in hippocampal neurons. *J. Neurophysiol.* **42**, 889–895.

Hotson, J. R., Sypert, G. W., and Ward, A. A. (1973). Extracellular potassium concentration changes during propagated seizures in neocortex. *Exp. Neurol.* **38**, 20–26.

Huguenard, J., and McCormick, D. A. (1994). "Electrophysiology of the Neuron." Oxford Univ. Press, New York.

Jackelet, J. W. (1989). "Neuronal and Cellular Oscillators." Dekker, New York.

Jahnsen, H., and Llinás, R. (1984a). Electrophysiological properties of guinea-pig thalamic neurons: an *in vitro* study. *J. Physiol. (London)* **349**, 205–226.

Jahnsen, H., and Llinás, R. (1984b). Ionic basis for the electroresponsiveness and oscillatory properties of guinea-pig thalamic neurons *in vitro*. *J. Physiol. (London)* **349**, 227–247.

Jan, L. Y., and Jan, Y. N. (1990). How might the diversity of potassium channels be generated? *Trends Neurosci.* **13**, 415–419.

Jiang, Y., Lee, A., Cadene, M., Chalt, B. T., and MacKinnon, R. (2002b) The open pore conformation of potassium channels. *Nature* **417**, 523–526.

Jiang, Y., Lee, A., Chen, J., Cadene, M., Chait, B. T., and MacKinnon, R. (2002a) Crystal structure and mechanism of calcium-gated potassium channel. *Nature* **417**, 515–522.

Johnston, D., and Wu, S. M.-S. (1995). "Foundations of Cellular Neurophysiology." MIT Press, Cambridge, MA.

Johnston, D., Magee, J. C., Colbert, C. M., Cristie, B. R. (1996) Active properties of neuronal dendrites. *Annu. Rev. Neurosci.* **19**, 165–186.

Kao, C. T. (1966). Tetrodotoxin, saxotoxin and their significance in the study of excitation phenomena. *Pharmacol. Rev.* **18**, 997–1049.

Kuffler, S. W., and Nicholls, J. G. (1966). The physiology of neuroglia cells. *Ergeb. Physiol.* **57**, 1–90.

Läuger, P. (1991). Electrogenic Ion Pumps. Sinauer, Sunderland, MA.

Latorre, R., Oberhauser, A., Labarca, P., and Alvarez, O. (1989). Varieties of calcium-activated potassium channels. *Annu. Rev. Physiol.* **51**, 385–399.

Livingstone, M. S., and Hubel, D. H. (1981). Effects of sleep and arousal on the processing of visual information in the cat. *Nature (London)* **291**, 554–561.

Llinás, R. R. (1988). The intrinsic electrophysiological properties of mammalian neurons: Insights into central nervous system function. *Science* **242**, 1654–1664.

Llinás, R., and Jahnsen, H. (1982). Electrophysiology of mammalian thalamic neurones *in vitro*. *Nature (London)* **297**, 406–408.

Llinás, R., and Sugimori, M. (1980a). Electrophysiological properties of *in vitro* Purkinje cell somata in mammalian cerebellar slices. *J. Physiol. (London)* **305**, 171–195.

Llinás, R., and Sugimori, M. (1980b). Electrophysiological properties of *in vitro* Purkinje cell dendrites in mammalian cerebellar slices. *J. Physiol. (London)* **305**, 197–213.

Llinás, R., and Yarom, Y. (1981a). Electrophysiology of mammalian inferior olivary neurones *in vitro*: Different types of voltage-dependent ionic conductances. *J. Physiol. (London)* **315**, 569–584.

Llinás, R., and Yarom, Y. (1981b). Properties and distribution of ionic conductances generating electroresponsiveness of mammalian inferior olivary neurones *in vitro*. *J. Physiol. (London)* **315**, 569–584.

Llinás, R., Sugimori, M., Hillman, D. E., and Cherksey, B. (1992). Distribution and functional significance of the P-type, voltage-dependent Ca^{2+} channels in the mammalian nervous system. *Trends Neurosci.* **15**, 351–355.

Ludwig, A., Zong, X., Jeglitsch, M., Hofmann, F., and Biel, M. (1998). A family of hyperpolarization-activated mammalian cation channels. *Nature* **393**, 587–591.

Mackie, G. O., and Meech, R. W. (1985). Separate sodium and calcium spikes in the same axon. *Nature (London)* **313**, 791–793.

Madison, D. V., and Nicoll, R. A. (1984). Control of repetitive discharge of rat CA1 pyramidal neurons *in vitro*. *J. Physiol. (London)* **354**, 319–331.

McCormick, D. A. (1992). Neurotransmitter actions in the thalamus and cerebral cortex and their role in neuromodulation of thalamocortical activity. *Prog. Neurobiol.* **39**, 337–388.

McCormick, D. A., and Huguenard, D. A. (1992). A model of the electrophysiological properties of thalamocortical relay neurons. *J. Neurophysiol.* **68**, 1384–1400.

McCormick, D. A., and Pape, H.-C. (1990). Properties of a hyperpolarization-activated cation current and its role in rhythmic oscillation in thalamic relay neurones. *J. Physiol. (London)* **431**, 291–318.

McCormick, D. A., Connors, B. W., Lighthall, J. W., and Prince, D. A. (1985). Comparative electrophysiology of pyramidal and sparsely spiny neurons of the neocortex. *J. Neurophysiol.* **54**, 782–806.

Meech, R. W., and Mackie, G. O. (1993). Potassium channel family in giant motor axons of *Aglantha digitale*. *J. Neurophysiology* **69**, 894–901.

Meech, R. W., and Mackie, G. O. (1995). Synaptic events underlying the production of calcium and sodium spikes in motor giant axons of *Aglantha digitale*. *J. Neurophysiol.* **74**, 1662–1669.

Mercer, R. W. (1993). Structure of the Na,K-ATPase. Int. Rev. Cytol. C **137**, 139–168.

Nernst, W. (1888). On the kinetics of substances in solution. Translated from *Z. Phys. Chem.* **2**, 613–622, 634–637. *In* "Cell Membrane Permeability and Transport" (G. R. Kepner, Ed.), pp. 174–183. Dowden, Hutchinson & Ross, Stroudsburg, PA, 1979.

Nicoll, R. A. (1988). The coupling of neurotransmitter receptors to ion channels in the brain. *Science* **241**, 545–551.

Nicoll, R. A., Malenka, R. C., and Kauer, J. A. (1990). Functional comparison of neurotransmitter receptor subtypes in mammalian central nervous system. *Physiol. Rev.* **70**, 513–565.

Nowycky, M. C., Fox, A. P., and Tsien, R. W. (1985). Three types of neuronal calcium channel with different calcium agonist sensitivity. *Nature (London)* **316**, 440–443.

Overton, E. (1902). Beitra..ge zur allgemeinen Muskelund Nerven physiologie. II. Ueber die Urentbehrlichkeit von Natrium- (oder Lithium-) Ionen fu..r den Contractsionact des Muskel. *Pfluegers Arch. Ges. Physiol. Menschen Tiere* **92**, 346–386.

Pape, H.-C., and McCormick, D. A. (1995). Electrophysiological and pharmacological properties of interneurons in the cat dorsal lateral geniculate nucleus. *Neuroscience* **68**, 1105–1125.

Pedersen, P. L., and Carafoli, E. (1987). Ion motive ATPases. I. Ubiquity, properties, and significance to cell function. *Trends Biochem. Sci.* **12**, 146–150.

Pennefather, P., Lancaster, B., Adams, P. R., and Nicoll, R. A. (1985). Two distinct Ca-dependent K currents in bullfrog sympathetic ganglion cells. *Proc. Natl. Acad. Sci. USA* **82**, 3040–3044.

Perez-Reyes, E., Cribbs, L. L., Daud, A., Lacerda, A. E., Barclay, J., Williamson, M. P., Fox, M., Rees, M. and Lee, J.-H. (1998). Molecular characterization of a neuronal low-voltage-activated T-type calcium channel. *Nature* **391**, 896–900.

Phillips, C. G. (1956). Intracellular records from betz cells in the cat. *Q. J. Exp. Physiol.* **41**, 58–69.

Prince, D. A., Lux, H. D., and Neher, E. (1973). Measurements of extracellular potassium activity in cat cortex. *Brain Res.* **50**, 489–495.

Regehr, W. G., and Tank, D. W. (1994). Dendritic calcium dynamics. *Curr. Opin. Neurobiol.* **4**, 373–382.

Reiner, P. B., and McGeer, E. G. (1987). Electrophysiological properties of cortically projecting histamine neurons of the rat hypothalamus. *Neurosci. Lett.* **73**, 43–47.

Reithmeier, R. A. F. (1994). Mammalian exchangers and co-transporters. *Curr. Opin. Cell Biol.* **6**, 583–594.

Ritchie, J. M., and Rogart, R. B. (1977). Density of sodium channels in mammalian myelinated nerve fibers and nature of the axonal membrane under the myelin sheath. *Proc. Natl. Acad. Sci. USA* **74**, 211–215.

Schwartzkroin, P. A., and Mathers, L. H. (1978). Physiological and morphological identification of a nonpyramidal hippocampal cell type. *Brain Res.* **157**, 1–10.

Skou, J. C. (1957). The influence of some cations on an adenosine triphosphatase from peripheral nerves. *Biochim. Biophys. Acta* **23**, 394–401.

Skou, J. C. (1988). Overview: The Na,K pump. *In* "Methods in Enzymology" (S. Fleischer and B. Fleischer, Eds.), Vol. 156, pp. 1–25. Academic Press, Orlando, FL.

Stafstrom, C. E., Schwindt, P. C., and Crill, W. E. (1982). Negative slope conductance due to a persistent subthreshold sodium current in cat neocortical neurons *in vitro*. *Brain Res.* **236**, 221–226.

Stea, A. Soong, T. W., and Snutch, T. P. (1995). Voltage-gated calcium channels. *In* "Ligand and Voltage-Gated Ion Channels" (A. North, Ed.), pp. 113–152. CRC Press, Boca Raton, FL.

Steriade, M., and McCarley, R. W. (1990). "Brainstem Control of Wakefulness and Sleep." Plenum, New York.

Steriade, M., McCormick, D. A., and Sejnowski, T. (1993). Thalamocortical oscillations in the sleep and aroused brain. *Science* **262**, 679–685.

Storm, J. F. (1990). Potassium currents in hippocampal pyramidal cells. *Prog. Brain Res.* **83**, 161–187.

Stuhmer, W., Conti, F., Suzuki, H., Wang, X., Noda, M., Yahadi, N., Kobu, H., and Numa, S. (1989). Structural parts involved in activation and inactivation of the sodium channel. *Nature (London)* **339**, 597–603.

Tasaki, I., Polley, E. H., and Orrego, F. (1954). Action potentials from individual elements in cat geniculate and striate cortex. *J. Neurophysiol.* **17**, 454–474.

Thomas, R. C. (1972). Electrogenic sodium pump in nerve and muscle cells. *Physiol. Rev.* **52**, 563–594.

Thompson, S. M., Deisz, R. A., and Prince, D. A. (1988). Relative contributions of passive equilibrium and active transport to the distribution of chloride in mammalian cortical neurons. *J. Neurophysiol.* **60**, 105–124.

Tomita, T. (1965). Electrophysiological study of the mechanisms subserving color coding in the fish retina. *Cold Spring Harbor Symp. Quant. Biol.* **30**, 559–566.

Tsien, R. W., Ellinor, P. T., and Horne, W. A. (1991). Molecular diversity of voltage-dependent Ca^{2+} channels. *Trends Pharmacol. Sci.* **12**, 349–354.

Vandermaelen, C. P., and Aghajanian, G. K. (1983). Electrophysiological and pharmacological characterization of serotonergic dorsal raphe neurons recorded extracellularly and intracellularly in rat brain slices. *Brain Res.* **289**, 109–119.

Vassilev, P. M., Scheuer, T., and Catterall, W. A. (1988). Identification of an intracellular peptide segment involved in sodium channel inactivation. *Science* **241**, 1658–1661.

Vassilev, P., Scheuer, T., and Catterall, W. A. (1989). Inhibition of inactivation of single sodium channels by a site-directed antibody. *Proc. Natl. Acad. Sci. USA* **86**, 8147–8151.

von Krosigk, M., Bal, T., and McCormick, D. A. (1993). Cellular mechanisms of a synchronized oscillation in the thalamus. *Science* **261**, 361–364.

Wang, Z., and McCormick, D. A. (1993). Control of firing mode of corticotectal and corticopontine layer V burst-generating neurons by norepinephrine, acetylcholine, and 1S,3R-ACPD. *J. Neurosci.* **13**, 2199–2216.

Wheeler, D. B., Randall, A., and Tsien, R. W. (1994). Roles of N-type and Q-type Ca^{2+} channels in supporting hippocampal synaptic transmission. *Science* **264**, 107–111.

Williams, J. T., North, R. A., Shefner, S. A., Nishi, S., and Egan, T. M. (1984). Membrane properties of rat locus coeruleus neurones. *Neuroscience* **13**, 137–156.

Yellen, G. (2002). The voltage-gated potassium channels and their relatives. *Nature* **419**, 35–42.

Young, J. Z. (1936). The giant nerve fibers and epistellar body of cephalopods. *Q. J. Microsc. Sci.* **78**, 367.

Zagotta, W. N., Hoshi, T., and Aldrich, R. W. (1990). Restoration of inactivation in mutants of Shaker potassium channels by a peptide derived from ShB. *Science* **250**, 568–571.

6

Molecular Properties of Ion Channels

Lily Yeh Jan and Yuh Nung Jan

Ion channels are present in most if not all cells. In the nervous system, they set the resting membrane potential of neurons, and control the firing pattern and waveform of action potentials. Ion channels alter their activities in response to the actions of transmitters and the metabolic state of the cell, so as to modulate neuronal excitability (Hille, 1992). To fulfill these physiological functions, each type of ion channel allows only a certain kind of ions to pass through. To maintain levels of channel activity appropriate for the various physiological conditions, it is important to have just the right number of channels at the right locations on the cell membrane, as well as proper regulation of these channels by second messengers.

How does an ion channel selectively allow certain ions to pass through? How does a channel determine when to open and when to close? How does a cell control the number of channels on its cell surface? How are the different channels distributed on the dendrite, soma, axon, and nerve terminals of a neuron? These key questions to the understanding of ion channel functions have been the focus of recent molecular studies. We discuss some of these issues in more detail, following a brief overview of how ion channels are grouped into different families based on their molecular properties.

The first issue to be examined in this chapter is channel gating (i.e., the opening and closing of channels):

How does voltage open channels?
How does calcium affect channel opening?
How do transmitters and second messengers affect the opening and closing of channels?
How could the metabolic state of a cell influence channel opening?

The second issue is ion selectivity:

Why do potassium channels let the larger potassium ions rather than the smaller sodium ions go through?
The narrowest part of a sodium channel pore is actually larger than that of a potassium channel pore (Hille, 1992). How does a sodium channel allow sodium rather than potassium ions to pass through?
How does a calcium channel allow only calcium ions to pass through under physiological conditions where calcium ions are far outnumbered by sodium ions of nearly the same size?
Typically millions of ions stream through a channel in single file in a second, so that there is less than one microsecond of interaction between the channel and the permeant ion. How does the channel manage to distinguish between different types of ions?

The last issue to be dealt with in this chapter is the distribution of different types of ion channels in a neuron and the physiological significance of controlling the ion channel type and number in various compartments of the cell membrane.

FAMILIES OF ION CHANNELS

Ion channels are grouped into several families. Channels in the same family typically share the same membrane topology for their pore-lining α subunits, and display significant sequence similarity (Fig. 6.1). Interestingly, some of these families bear weak though recognizable resemblance to one another, indicating that they are likely to be evolutionarily related. The major families of ion channels are:

• *Voltage-gated ion channels and other family members (the 6-TM family).* The pore of channels in this family

Copyright 2004, Elsevier Science (USA).
All rights reserved.

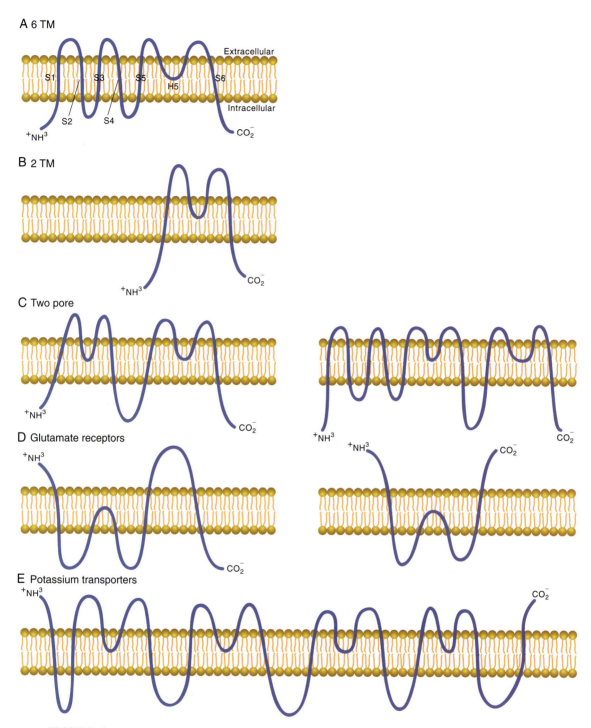

FIGURE 6.1 Several major families use similar designs. (A) The 6-TM family includes voltage-gated sodium, calcium, and potassium channels, hyperpolarization-activated cation channels, and cyclic nucleotide-gated cation channels. (B) The 2-TM family includes inwardly rectifying potassium channels and the bacterial potassium channel KcsA. The amiloride-sensitive epithelial sodium channel, the P_{2X} ATP receptor, and a bacterial mechanosensitive cation channel also contain two transmembrane segments per subunit. (C) Two-pore potassium channels in the animal kingdom contain four transmembrane segments (left) whereas a yeast two-pore channel contains eight transmembrane segments (right). (D) Ionotropic glutamate receptors contain pore-lining domains that appear to be upside down relative to other channels in this figure. Glutamate receptors in the animal kingdom (left) are cation channels whereas a prokaryotic glutamate receptor (right) is selective for potassium ions. (E) Potassium transporters in bacteria and yeast contain four repeats of the 2-TM pore-lining structure.

is lined by four subunits, or pseudo-subunits, linked together in a large α subunit as in the case of voltage-gated sodium channels illustrated in Figure 6.1. Each contains six transmembrane (TM) segments (S1-S6) and an H5 region, or P loop, between the last two transmembrane segments (Fig. 6.1A). The founding members of this family are voltage-gated sodium channels, calcium channels, and potassium channels (Catterall, 1998, 2000; Jan and Jan, 1997; Plummer and Meisler, 1999). Other members include calcium-activated potassium channels (Meera *et al.*, 1997; Vergara *et al.*, 1998), hyperpolarization-activated cation channels involved in rhythmic activities (Luthi and McCormick, 1998), plant potassium channels that appear to be activated by hyperpolarization (Gaymard *et al.*, 1996; Marten *et al.*, 1999; Schachtman *et al.*, 1992; Tang *et al.*, 2000), and cyclic nucleotide-gated cation channels and vanilloid receptors that are important for sensory transduction (Caterina *et al.*, 2000; Davis *et al.*, 2000; Zufall *et al.*, 1997). This last group of cation channels, like calcium-activated potassium channels of small or intermediate conductance, exhibits little voltage sensitivity. Whereas voltage-gated calcium channels contain four pseudo-subunits in one α subunit, analogous to voltage-gated sodium channels, calcium channels that are activated on emptying of internal calcium stores (also known as Icrac, see Figure 12.10) and allow calcium entry into the cell are composed of subunits with a design similar that of potassium channels and vanilloid receptors (Yue *et al.*, 2001).

• *Inwardly rectifying potassium channels and other channels with two transmembrane segments in each pore-lining subunit (the 2-TM family).* Named for their ability to allow much larger potassium influx than efflux, inwardly rectifying potassium (Kir) channels also have four α subunits lining the pore, and their transmembrane domains resemble the second half of the transmembrane domains of the voltage-gated potassium (Kv) channel α subunit (Fig. 6.1B) (Jan and Jan, 1997). Other channels that also have two transmembrane segments per subunit, bearing little sequence similarity to Kir, include the amiloride-sensitive epithelial sodium channel (Sheng *et al.*, 2000; Snyder *et al.*, 1999), a bacterial mechanosensitive cation channel (Rees *et al.*, 2000), the P_{2X} ATP receptor (Brake *et al.*, 1994; Valera *et al.*, 1994), and the Phe–Met–Arg–Phe-amide-activated sodium channel (Coscoy *et al.*, 1998). The last two are ligand-gated ion channels activated by the purinergic transmitter ATP and the peptide transmitter Phe–Met–Arg–Phe-amide.

• *"Two-pore" potassium channels (4-TM or 8-TM).* Each pore-lining α subunit of these channels appears to be a tandem dimer of two Kir-like, or one Kir-like and one Kv-like, α subunits (Fig. 6.1C) (Lesage and Lazdunski, 2000). These channels are thought to be "leak" potassium channels that are active at rest, contributing to the determination of the resting membrane potential. Some channels of this family may be modulated by volatile anesthetics (Patel *et al.*, 1999).

• *Ionotropic glutamate receptors.* The pore-lining domain of these ligand-gated ion channels is topologically equivalent to an "upside-down" Kir α subunit (Wo and Oswald, 1995), as illustrated in Figs. 6.1A and 11.12. A prokaryotic potassium-selective glutamate receptor with this membrane topology (Chen *et al.*, 1999) has been identified as a missing link between potassium channels and eukaryotic glutamate receptors, which are permeable to cations and contain one additional transmembrane segment at the C terminus of the pore-lining domain.

• *Nicotinic acetylcholine receptors and related ionotropic transmitter receptors.* These receptors contain five subunits, each with four transmembrane segments. The second (M2) transmembrane segment, the pore-lining helix, of the acetylcholine receptor rotates on acetylcholine binding so as to open the channel (Unwin, 1995), as illustrated in Fig. 11.7. Like acetylcholine receptors, the $5HT_3$ serotonin receptor is permeable to cations and mediates fast excitatory synaptic transmission (Maricq *et al.*, 1991). Other family members such as glycine receptors and $GABA_A$ receptors are permeable to anions and mediate fast inhibitory synaptic transmission (Galzi *et al.*, 1992). These transmitter receptors differ in structure from ionotropic glutamate receptors. Instead, they may be structurally related to certain bacterial endotoxins (Unwin, 1995).

• *Intracellular calcium channels.* Calcium channels of likely six transmembrane segments and a large cytoplasmic domain in each of their four subunits are responsible for releasing calcium from internal stores such as the endoplasmic reticulum (ER) (Mikoshiba, 1997). Family members include the inositol triphosphate (IP_3) receptors that are activated by binding to the second messenger IP_3, and ryanodine receptors that can be activated by direct interaction with voltage-gated calcium channels on the cell membrane.

• *Chloride channels.* Chloride channels of 10–12 transmembrane segments per subunit of the ClC family are widely distributed (Jentsch, 1996). The cystic fibrosis transmembrane regulator (CFTR) protein of the ATP-binding cassette (ABC) superfamily also forms chloride channels in the heart, in the airway epithelium, and in exocrine tissue (Sheppard and Welsh, 1999). The membrane topology of chloride channels resembles that of transporters.

Different families of ion channels may share common functional modules. There is some resemblance between the basic pore-lining structures of different ion channels (e.g., voltage-gated ion channels of the 6-TM family, Kir channels of the 2-TM family, two-pore potassium channels, and ionotropic glutamate receptors). Likewise, there is weak but discernible similarity between potassium transporters in bacteria and yeast and the pore-lining domain of potassium channels (Durell et al., 1999; Jan and Jan, 1994); a transporter appears to contain four Kir-like pseudo-subunits linked in tandem (Fig. 6.1E). Apparently, different ion channels and transporters may adopt the same basic structural motif for transporting ions across the membrane although the specific designs vary due to divergence of the physiological requirements.

Different members of the same family may have divergent functions. Whereas ion channels from the same family often have similar functional modules for ion permeation and/or channel gating, their functions can diverge to a remarkable degree. For example, voltage-gated ion channels are permeable to cations in general, but individual members are selective to sodium, potassium, or calcium permeation, or exhibit little selectivity among cations. Also, not all voltage-gated ion channels are activated by depolarization; some appear to be activated by hyperpolarization, whereas still others show very weak, if any, voltage sensitivity. Not only could the functional modules have evolved to take on different functional characteristics, divergent channel functions may also arise due to differences in the temporal and spatial expression patterns of channels. Such divergence may be further encouraged following channel gene duplication.

Transmitter receptors that are ligand-gated ion channels, primarily members of the fourth and fifth families of ion channels, are the topic of Chapter 11. In this chapter we focus primarily on voltage-gated ion channels that mediate action potentials and transmitter release, and inwardly rectifying potassium channels that mediate slow synaptic potentials and possibly provide protection of neurons under metabolic stress.

CHANNEL GATING

How Does Voltage Open Channels?

Controlling channel activity by voltage is key to neuronal excitability and signaling. Voltage-gated sodium channels and potassium channels mediate the generation of action potentials, which allow signals to be propagated from one end of a neuron to the other (Hodgkin and Huxley, 1952). Calcium entry due to activation of voltage-gated calcium channels in the nerve terminal then triggers transmitter release. The ability of these channels to be gated by voltage across the membrane is therefore fundamental to signaling in the nervous system. Intensive biophysical and molecular studies have yielded a framework for the mechanism for channel gating by voltage across the membrane (Fig. 6.2).

Voltage-gated ion channels contain intrinsic voltage sensors. Voltage-gated ion channels typically are closed at the resting membrane potential but open on membrane depolarization. These channels are intrinsically sensitive to membrane potential; they contain intrinsic voltage sensors that can detect changes in membrane potential and trigger conformational changes of the channel. This leads to movement of charges intrinsic to the channel protein; the resulting gating currents and gating charge have been measured in biophysical experiments (Keynes, 1994; Sigworth, 1994).

The S4 Segment Corresponds to the Voltage Sensor

Each of the four subunits or pseudo-subunits of a voltage-gated ion channel contains an intrinsic voltage sensor, corresponding primarily to the fourth transmembrane segment S4 which bears basic residues at every third position. Depolarization of the cell membrane causes the (S4) segment to move outward relative to the "membrane" that is inaccessible to water, as shown for the voltage-gated sodium channel and potassium channel.

S4 movements in the voltage-gated sodium channel. Movement of the S4 segment in the fourth pseudo-subunit of the voltage-gated sodium channel has been detected experimentally (Horn, 2000). By replacing an arginine of this S4 segment with a cysteine, Horn and colleagues tested whether this cysteine is exposed to water and can therefore react with thiol reagents such as MTSET. This reaction some-

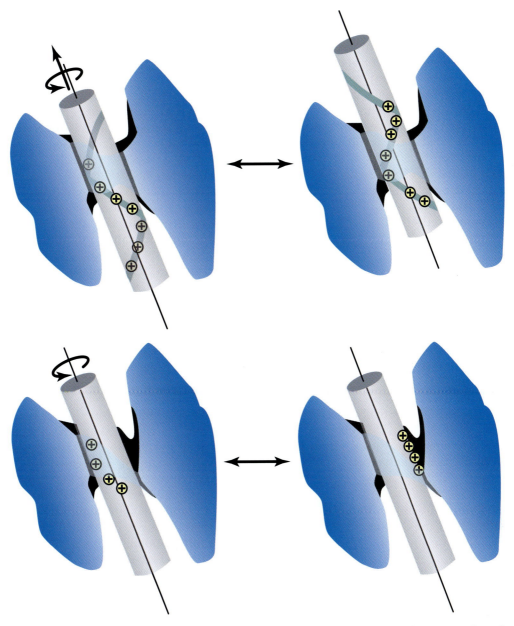

FIGURE 6.2 Activation motion of the voltage sensor S4 segment of voltage-gated potassium channels can be accounted for by a helical twist of 180°. Top: Helical screw model of S4. A 180° twist yields an axial translation of nine residues. Bottom: Model of S4 twist without axial translation. Adapted from Glauner *et al.* (1999).

times results in an alteration in the kinetic properties of the channel, thereby providing an electrophysiological readout of the covalent modification. Remarkably, a cysteine at the position of the second or third arginine of this S4 segment is accessible from the cytoplasmic side of the membrane when the channel is closed, but becomes accessible from the extracellular side when the channel is activated by depolarization (Yang et al., 1996).

S4 movements in the voltage-gated potassium channel. In the homotetrameric Shaker potassium channel, each S4 segment contains seven basic residues, evenly spaced at every third position. As shown by Isacoff and colleagues, only the second arginine of the S4 segment is buried in the closed channel; a cysteine at that position cannot react with thiol reagents from either side of the membrane. As the channel opens on membrane depolarization, the second arginine becomes exposed to the extracellular side of the membrane, whereas the third, fourth and fifth basic residues move from the cytoplasmic side to sites buried in the membrane (Larsson et al., 1996). A twist or rotation of the S4 segment probably takes place in this voltage-induced movement of the intrinsic voltage sensor (Cha et al., 1999; Glauner et al., 1999). The S4 movement can also be monitored by a rhodamine fluorophore that is covalently attached to the S4 segment and reports changes in its environment by changing its fluorescence. The time course of the fluorescence change parallels the time course of the gating current (Mannuzzu et al., 1996), verifying that the S4 movement reflects channel gating that leads to channel opening.

The four S4 segments of four identical subunits undergo first independent, then concerted movements. The presence of four S4 segments that function as intrinsic voltage sensors of a voltage-gated ion channel accounts for the steep voltage dependence of channel activation. On depolarization, the S4 segments of the four identical subunits in a Shaker potassium channel move outward in two discernible steps, initially independently of one another, then cooperatively in a concerted step, leading to channel opening (Mannuzzu and Isacoff, 2000). This cooperativity between subunits is an important factor in determining the voltage sensitivity of the channel.

How Can a Voltage-Gated Ion Channel Not Stay Open Indefinitely on Prolonged Depolarization?

Channels Inactivate after They Activate and Open

Whereas some voltage-gated ion channels such as the M-type voltage-gated potassium channel stay open as long as the membrane potential is above the threshold for channel activation, most voltage-gated ion channels *inactivate*. In other words, the channel stops conducting ions even though the membrane potential is maintained at a depolarized level. Channels can be inactivated in different ways. N-Type inactivation takes place near the cytoplasmic end of the pore, whereas P-type and C-type inactivation involves the extracellular end of the pore.

N-Type inactivation takes place near the cytoplasmic end of the pore. N-Type inactivation is also known as the "ball and chain" mechanism for inactivation (Armstrong and Bezanilla, 1977). The ball peptide at the N terminus of the α or β subunit (Rettig et al., 1994; Wallner et al., 1999; Zagotta et al., 1990), a

part of the channel's cytoplasmic domain, appears to bind to a "receptor" that becomes accessible when the channel opens, resulting in blockade of ion permeation. The ball peptide resides at the N terminus of the channel. Residues implicated in either electrostatic or physical interaction with the inactivation ball have been found in the sequences just preceding the first transmembrane segment S1 (Gulbis et al., 2000), and the cytoplasmic loop connecting the S4 and the S5 segments, the S4–S5 loop (Isacoff et al., 1991; Yellen, 1998). Whereas the inactivation ball at the N terminus of the α or β subunit is thought to physically plug the pore at the cytoplasmic end, thereby blocking ion permeation, there is surprisingly little room that allows access of the inactivation ball to the pore (Fig. 6.3). It is also conceivable that inactivation could result from conformational changes triggered by channel interaction with the ball, which functions even in isolation.

N-Type inactivation may couple to voltage gating in two ways. First, some channels can inactivate only after some or all of their voltage sensors have undergone movements in response to membrane depolarization to cause channel activation. Second, the voltage sensors may be "immobilized" in their activated configuration once the channel has entered the N-type inactivated state. These types of coupling would tend to simplify the "state diagram" of possible transitions among different states of the channel. In the extreme case, bringing the membrane potential from a hyperpolarized to a depolarized level would cause a channel to shift from deactivated to activated state, then to N-type inactivated state. On reversal of the membrane potential to a hyperpolarized level, the channel would have to reverse course, going from N-type inactivated to activated state, then to the deactivated state. This way the channel would open briefly after the membrane potential is brought back to a hyperpolarized level. Such reopening of voltage-gated calcium channels could cause "delayed release" of transmitters following an action potential (Slesinger and Lansman, 1991).

P-Type inactivation takes place near the extracellular end of the pore. P-Type inactivation involves movements of pore-lining structures near the extracellular end of the pore (Loots and Isacoff, 1998; Olcese et al., 1997). Following channel opening, ion permeation stops due to movements of the P region that forms the narrowest part of the pore. The P region in the bacterial KcsA potassium channel forms a pore–helix and a pore–loop, with carbonyl groups of the pore–loop surrounding the permeant ion (Doyle et al., 1998). Sequences that connect the pore–helix with the pre-

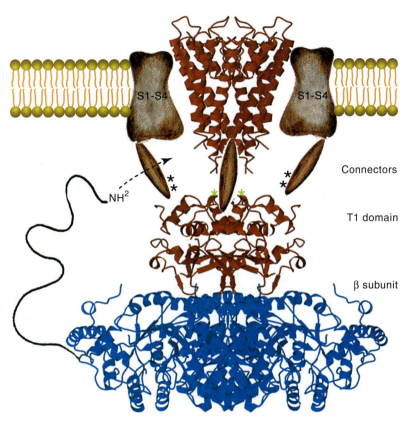

FIGURE 6.3 Composite model of a voltage-dependent potassium channel. The α subunit is shown in red and the β subunit is in blue. The model of the pore region is based on the KcsA potassium channel. The structure of the voltage-sensing region and connectors is unknown (depicted schematically). An NH$_2$-terminal inactivation peptide is shown entering a lateral opening to gain access to the pore. Asterisks indicate locations of mutations that affect inactivation kinetics. Adapted from Gulbis *et al.* (2000).

ceding transmembrane segment (S5 in voltage-gated ion channels) form a "turret" on the extracellular side of the membrane. Movements of this turret accompany the P-type inactivation, as indicated by fluorescence changes of fluorophores attached to the turret and by state-dependent formation of disulfide bridges in this region (Gandhi et al., 2000).

C-Type inactivation follows P-type inactivation and further prevents reactivation of the channel. Subsequent to P-type inactivation, more global movements of the channel take place and stabilize the S4 segment in the C-type inactivated state (Yellen, 1998; Loots and Isacoff, 1998; Olcese et al., 1997; Larsson and Elinder, 2000). Once the channel has entered the C-type inactivated state, a greater amount of hyperpolarization is necessary to revert the channel to the deactivated state so that the channel can once again be induced by depolarization to activate and open.

Channel inactivation may be caused by permeant ions. There are other forms of inactivation. For

example, channels can be inactivated by their permeant ions, as in the case of calcium-induced calcium channel inactivation described below (Lee et al., 1999; Zühlke et al., 1999).

How Does Calcium Affect Channel Activity?

Neuronal Activities May Regulate Channel Activities via Calcium

Besides changing the membrane potential, neuronal activities often cause increases in cytosolic calcium concentration, as a result of calcium entry through voltage-gated calcium channels or certain transmitter-gated ion channels or of release of calcium from internal stores by second messengers. Thus, calcium modulation of ion channels represents one way for channels to be regulated by neuronal activities. There are at least two different ways for calcium to modulate channel functions. Calcium may either directly interact with the channel protein or indirectly modulate channel activities via calcium-binding proteins such as calmodulin.

Direct calcium action is likely to underlie the calcium activation of the large-conductance calcium-activated potassium (BK) channels. These channels are sensitive to both voltage and calcium; increasing intracellular calcium concentrations over six orders of magnitude causes the voltage dependence curve of channel activation to shift progressively to the left (Meera et al., 1997; Barrett et al., 1982; Cui et al., 1997; Marty, 1981). How is this amazing feat accomplished? Recent molecular studies have provided valuable clues. The BK channels have a large cytoplasmic domain C-terminal to the transmembrane domain containing seven transmembrane segments. Preceding the six transmembrane segments that are commonly found in voltage-gated potassium channels is another transmembrane segment and an extracellular N-terminal domain (Meera et al., 1997). The large C-terminal domain contains multiple calcium binding sites and an inhibitory region (Schreiber et al., 1999). This part of the C-terminal domain appears to stabilize the closed state(s), thereby inhibiting the channel in the absence of calcium. Calcium interaction with multiple calcium binding sites in the C-terminal domain destabilizes the closed state(s), causing a shift in the voltage dependence curve of the channel to the left.

Calmodulin mediates modulation of many different ion channels. Calmodulin has four calcium binding sites; calcium binding causes calmodulin to undergo conformation changes (Meador et al., 1993). Thus, channel activities may be modulated either by their binding to calmodulin in a calcium-dependent manner or by their sensing the conformation changes of calmodulin molecules that are bound to the channel constitutively, in the absence as well as in the presence of calcium. Such interactions underlie activation of calcium-activated potassium channels of small and intermediate conductance (SK and IK) (Xia et al., 1998), calcium-induced calcium channel inactivation, frequency-dependent facilitation of calcium channels (Lee et al., 1999; Zühlke et al., 1999), and calcium modulation of cyclic nucleotide-gated cation channels (Chen and Yau, 1994)

How Do Transmitters Cause Channels to Open or Close?

Metabotropic Transmitter Receptors Are Coupled to the Trimeric G Protein

As described in detail in Chapter 12, transmitter activation of G protein-coupled receptors facilitates GTP exchange for GDP bound to the α subunit of the trimeric G protein, leading to dissociation of the α-GTP subunit from the βγ subunit (see Figs. 6.4 and 12.3). Either or both of these G-protein subunits may bind to channel proteins and modulate channel activities. These G-protein subunits may also modulate the activities of other effectors, thereby liberating second messengers such as calcium, cyclic nucleotides, phosphoinositides, and kinases. Different isoforms of G protein coupled receptors couple to different G proteins, which may activate adenylyl cyclase (Gs), inhibit adenylyl cyclase (Gi), or activate phospholipase C (PLC) (Gq) or other signaling pathways. The various downstream second messengers, in turn, may alter channel activities.

Some channels are modulated directly by interaction with Gβγ subunits. Direct interaction between the βγ subunits of the G protein and G protein-activated inwardly rectifying potassium channels (GIRK channels of the 2-TM Kir family) causes channel acti-

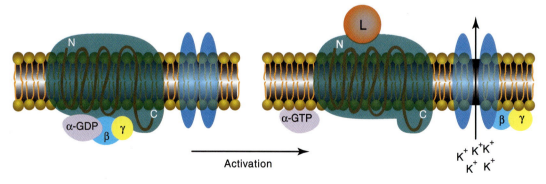

FIGURE 6.4 G protein-activated inwardly rectifying potassium channels (Kir3) are activated by direct interaction with the βγ subunits of G protein. L represents the ligand for the G protein-coupled receptor with seven transmembrane segments, e.g., the parasympathetic transmitter acetylcholine for slowing the heart rate or the inhibitory transmitter GABA for generating the slow inhibitory postsynaptic potential in the central nervous system.

vation (Fig. 6.4). This allows inhibitory transmitters such as GABA to generate slow inhibitory postsynaptic potentials (IPSPs) in the brain, and the parasympathetic transmitter acetylcholine to slow the heart rate (Luscher et al., 1997; Wickman et al., 1998). The βγ subunits also bind to calcium channels to cause channel inhibition (Herlitze et al., 1996; Ikeda, 1996), a likely mechanism for presynaptic inhibition given that calcium channels are found in complexes with synaptic proteins such as syntaxin and synaptotagmin (Catterall, 1998).

Multiple second messengers may converge on the same channel, resulting in integration of signaling processes. For example, protein kinase C (PKC) phosphorylation not only increases calcium channel activities but also prevents channel inhibition by the βγ subunits of the G protein (Zamponi et al., 1997). Likewise, the sensitivity of GIRK channels to the βγ subunits may be modulated by PIP$_2$ (Huang et al., 1998). In the heart, acetylcholine first activates GIRK channels via m2 muscarinic acetylcholine receptors due to direct action of the Gβγ subunits on the channel. The same transmitter then causes GIRK channel desensitization by activating m3 muscarinic acetylcholine receptors, which in turn activate PLC and reduce PIP$_2$ levels (Kobrinsky et al., 2000). In these examples multiple second messengers converge on the α subunit of the channel. In many other cases second messengers impinging on α subunits, as well as the regulatory β subunits, modulate channel activities.

How Could Metabolic State of a Cell Influence Channel Activity?

One Well-Known Example of Metabolic Regulation of Channel Activities is the ATP-Sensitive Potassium Channel

In Chapter 11 we see purinergic receptors that respond to extracellular ATP; the A-type and most of the P-type purinergic receptors are G protein-coupled receptors except for the P$_{2X}$ receptors, which are ligand-gated ion channels. Intracellular ATP may also alter the activity of channels. For example, intracellular ATP inhibits ATP-sensitive potassium channels. ATP-sensitive potassium channels open in response to increases in the blood sugar level and intracellular metabolic state of the pancreatic β cell to trigger insulin release. Similar or identical ATP-sensitive potassium channels are present in the brain, the heart, and skeletal and smooth muscles (Ashcroft and Gribble, 1998; Ashcroft, 2000; Babenko et al., 1998;

Nichols and Lopatin, 1997; Quayle et al., 1997). ATP-sensitive potassium channels in arterial smooth muscle may open during ischemia so as to cause dilation of blood vessels. Metabolic stress in central neurons may also activate these channels, thereby protecting the neurons from death. Besides metabolic regulation, transmitters and their G protein-coupled receptors also modulate ATP-sensitive potassium channel activity. For example, vasodilators activate these channels in the smooth muscle via protein kinase A (PKA), whereas vasoconstrictors inhibit these channels via protein kinase C (PKC).

ATP and Mg-ADP mediate metabolic regulation of ATP-sensitive potassium channels. ATP-sensitive potassium channels are inhibited by ATP but stimulated by Mg-ADP. These channels contain not only four pore-lining subunits (Kir6.2) but also four regulatory subunits of the ATP-binding cassette family (SUR1 or SUR2A/B) (Fig. 6.5). ATP can act on the Kir6.2 subunits and cause channel inhibition though the ATP sensitivity is greatly enhanced by interaction between Kir6.2 and the regulatory SUR subunits. In contrast, Mg-ADP acts on the SUR subunits to cause channel activation in a manner that requires the nucleotide binding domains of the SUR, suggesting that ATP hydrolysis by the SUR subunits is necessary for channel regulation. The SUR subunits also mediate channel inhibition by sulfonylurea drugs such as glibenclamide and tolbutamide, used to treat type II diabetes. ATP-sensitive potassium channel openers (KCOs) such as diazoxide act on the SUR subunits and stimulate ATP hydrolysis, a feature reminiscent of transporters in the ABC family. Whereas close relatives of SUR such as MRP transport hydrophobic substrates across the membrane, SUR interacts with the first transmembrane segment of Kir6.2 via its transmembrane domain. This interaction may potentially play a role in allowing the SUR subunit to regulate channel activity.

ION PERMEATION

Proper physiological functions of ion channels depend on their exquisite ion selectivity, a remarkable feat. How does a calcium channel allow only calcium ions to pass through even though sodium ions are of nearly the same size and are much more abundant under physiological conditions? How does a potassium channel select for the larger potassium ions over the smaller sodium ions, given that millions of potassium ions stream through the channel in single file in a

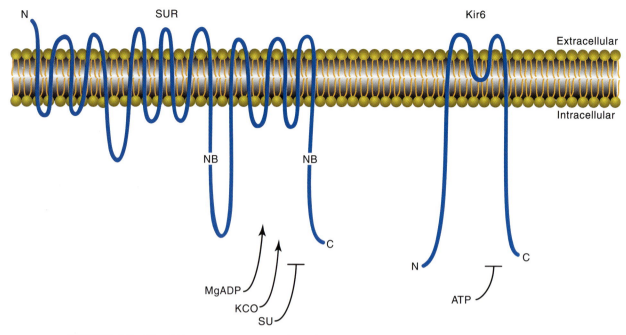

FIGURE 6.5 The ATP-sensitive potassium channels contain four pore-lining α subunits (Kir6) and four regulatory β subunits (SUR). SUR is a member of the ATP-binding cassette (ABC) family and contains two nucleotide binding (NB) domains. ATP acts on Kir6 to inhibit the channel whereas Mg-ADP acts on SUR to activate the channel. Sulfonylurea (SU) drugs that inhibit the channel and KCO compounds that activate the channel also act on SUR.

second so that there is less than one microsecond of interaction between the channel and a potassium ion? It is known that the narrowest part of a sodium channel pore is actually larger than that of a potassium channel pore. So how does a sodium channel allow sodium rather than potassium ions to pass through? These questions have attracted much scrutiny in biophysical and molecular studies, leading to the following model (Hille, 1992).

To achieve high selectivity without holding onto the ion for too long, ion channels have multiple binding sites for the permeant ion. The ion selectivity could be achieved if the channel contains a binding site with much higher affinity for calcium, or sodium, or potassium than for other ions. But then, why does the preferred ion not get stuck in the channel? One way to attain both exquisite ion selectivity and large ion flux through the channel would be to have more than one permeant ion in the channel pore, perhaps each interacting with a separate binding site. Electrostatic or other long-range interactions between these ions in the pore could facilitate their dissociation from their binding sites, thereby allowing rapid flow of permeant ions across the membrane.

Evidence for Multiple Ions Residing in a Channel Pore

Evidence for Ion Binding Site(s)

A dependence of ion permeation on the composition and concentrations of ions implicates ion binding sites in the channel. Early indications for binding sites arose from examination of the amount of current flowing through a channel as a function of ion concentration; saturation of current level at high ion concentrations indicates the presence of at least one binding site. Another indication for ion binding sites came from the "test of independence." If one assumes that permeant ions move through the channel, from one side of the membrane to the other, independent of one another, the permeability ratio for the two types of permeant ions being tested should vary with ion concentration in a certain way. This prediction does not fit the experimental data, indicating that there are ions in the channel pore, resulting in interdependence between ions as they permeate the channel (Hille, 1992).

Evidence for More Than One Binding Site

The presence of more than one ion in a channel pore is revealed by several different experiments. In the 1950s, Hodgkin and Keynes (1955) showed that the

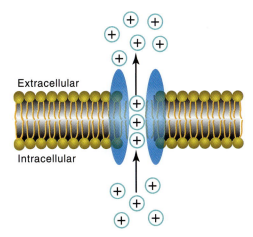

FIGURE 6.6 Movement of multiple (three in this example) ions in a channel pore in concert causes the ratio of potassium efflux to influx to be greater than one (2.5) even though each potassium ion carries one unit charge.

ratio of potassium efflux to influx is 2.5, indicating that more than one of the monovalent potassium ions move in concert through the potassium channel (Fig. 6.6). Similarly, the voltage dependence of blocking an inwardly rectifying potassium channel by the monovalent cesium ion indicates that more than one cesium ion can reside in the channel pore at once (Hille, 1992; Hagiwara et al., 1976).

Another indication is the so-called anomalous mole-fraction effect. First, let us suppose that a channel is permeable to two types of ions (Fig. 6.7). Compared with permeant ion B, permeant ion A binds to the channel's external binding site more tightly. Permeant ion A also moves through the channel at a slower rate. If the extracellular solution contains predominantly B-type permeant ions, mixed with a small amount of A-type permeant ions, occupation of the channel's external binding site by permeant ion A would block the binding and passage of permeant B through the channel. Because of the difference in affinity, it would be more difficult for permeant ion B than permeant ion A to displace permeant ion A from the external binding site to the next binding site. In this way, the rate of ion flow may turn out to be smaller than when either permeant ion species alone is present. This apparently paradoxical, or anomalous, phenomenon arises when a channel contains more than one ion binding site (Hille, 1992; Hagiwara *et al.*, 1977).

Crystallographic Evidence for Multiple Ions in a Pore

Direct evidence for the presence of multiple permeant ions in the channel pore is provided by visualization of the bacterial KcsA potassium channel in the crystal form (Doyle *et al.*, 1998) (Fig. 6.8). Two potas-

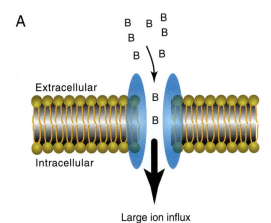

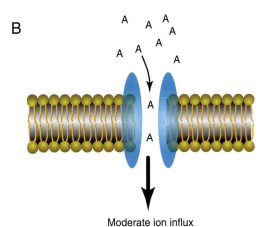

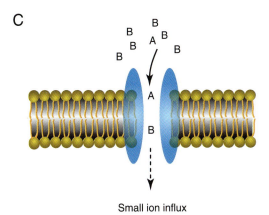

FIGURE 6.7 Anomalous mole-fraction effect. (A) In the presence of only permeant ion B the ion flux is large because permeant ion B moves through the channel at a fast rate. (B) In the presence of only permeant ion A the ion flux is moderate because permeant ion A binds more tightly to the channel's external binding site and moves through the channel more slowly. (C) In the presence of predominantly permeant ion B mixed with a small fraction of permeant ion A, the ion flux is smaller than in either of the above cases, because permeant ion A binds tightly to the external binding site in the channel and blocks passage of permeant ion B.

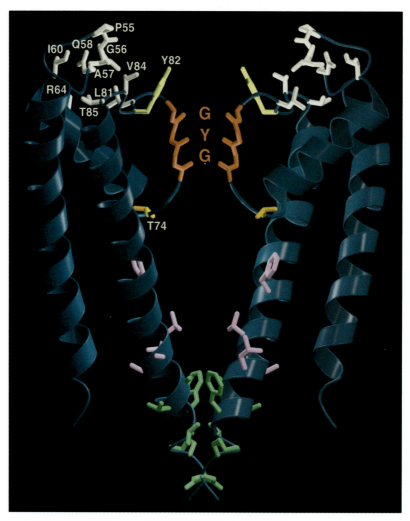

FIGURE 6.8 KcsA structure. Ribbon representation of two subunits is shown. The pore loops containing the GYG (glycine–tyrosine–glycine) sequence form the narrowest part (the selectivity filter) of the channel pore. Mutations in the Shaker voltage-gated potassium channel mapped onto their equivalent positions. Mutation of any of the white side chains significantly alters the affinity of agitoxin 2 or charybdotoxin for the Shaker potassium channel. Changing the yellow side chain affects both agitoxin 2 and TEZ binding from the extracellular solution. The mustard-colored side chain at the base of the selectivity filter affects TEA binding from the intracellular solution. The side chains colored green, when mutated to cysteine, are modified by cysteine-reactive agents whether or not the channel gate is open, whereas those colored pink react only when the channel is open. Adapted from Doyle *et al.* (1998).

sium ions are located within the upper half of the pore adjacent to the extracellular surface (Fig. 6.9B). These two potassium ions are coordinated by the backbone carbonyl groups of the pore–loop, which is part of the H5 or P region that connects the two transmembrane segments. The last transmembrane segment lines the inner half of the pore, which contains one discernible ion.

The known structure is for KcsA at pH 7.5. KcsA channels are closed at pH 7 but can be activated at pH 4, due to conformational changes that include movements of the last transmembrane segment (Perozo *et al.*, 1999). The exact form of the channel in its open conformation remains to be determined. A gating model based on random mutagenesis and yeast screen for fully functional Kir channels (Yi *et al.*, 2001) indicates that the membrane-spanning helices of each Kir3.2 subunit rotate clockwise as the channel opens (Fig. 6.9).

The Ion Selectivity of Voltage-Gated Sodium Channels and Calcium Channels

Sodium channels and calcium channels have wider pores than potassium channels and yet have to favor the passage of ions smaller than potassium. It is of physiological importance to have voltage-gated sodium channels for the generation of action potentials in neurons, and to have voltage-gated calcium

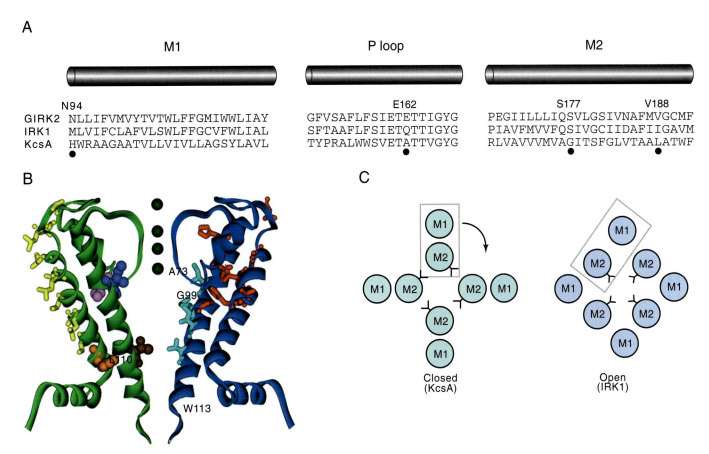

FIGURE 6.9 Gating model for G protein-activated inwardly rectifying potassium channel Kir3.2 (GIRK2) deduced from yeast mutant screens. (A) Sequence alignment of the first transmembrane segment M1, the P loop and the second transmembrane segment M2 of GIRK2, IRK1 with an open probability close to one, and KcsA. Yeast screens of randomly mutagenized GIRK2 channels reveal that mutations of the four residues marked above the GIRK2 sequence cause the channel to be constitutively open. (B) Model of Kir channels based on the KirBac1.1 structure, showing ten M1 residues per subunit predicted to face lipid (yellow), four M2 residues predicted to line the pore (cyan), and eleven M1 and M2 residues predicted to be buried within the channel protein (red), based on analyses of IRK1 (Kir2.1) mutant channels that rescue potassium-transport-deficient yeast for growth in low-potassium medium. The upper pair (pink and blue) and lower pair (orange and brown) yielding GIRK2 (Kir3.2) gating mutants isolated from unbiased yeast screens are shown on the green subunit on the left. The amino acids in IRK1 (and corresponding residues given in parenthesis for KirBac1.1) are: I87(S66), L90(A69), A91(L70), L94(V73), L97(T76), F98(L77), C101(L80), W104(Q83), L105(L84) and L108(A87) for lipid-facing, S165(I131), C169(M135), D172(I138) and I176(T142) for pore-lining, and F92(F71), S95(N74), W96(N75), F99(F78), G100(A79), A107(D86), A157(A123), V158(H124), V161(A127), Q164(E130) and G168(G134) for buried residues. The outer pair and inner pair of GIRK2 residues important for holding the channel in the closed conformation (and corresponding residues given in parenthesis for KirBac1.1) are: E152(L108) in pink and S177(I131) in blue; N94(F63) in orange and V188(T142) in brown. (C) Models of the transmembrane helical arrangement in the open form and the closed form. V188 (drawn) in the closed GIRK2 channel aligns with a residue in KcsA that is involved in M2–M2 interactions. In the open IRK1 channel, the residue corresponding to V188 of GIRK2 faces the pore. During gating, M1 and M2 as a unit (boxed) may rotate clockwise to bring V188 from a buried position in the closed channel to a pore-lining position in the open channel.

From Minor D. L., Jr., Masseling, S. J., Jan, Y. N., and Jan, L. Y. 1999. *Cell* 96:879–891; Yi, B. A., Lin, Y. F., Jan, Y. N., and Jan, L. Y. 2001. *Neuron* 29:657–667; and Kuo, A., Gulbis, J. M., Antcliff, J. F., Rahman, T., Lowe, E. D., Zimmer, J., Cuthbertson, J., Ashcroft, F. M., Ezaki, T., Doyle, D. A. Crystal structure of the potassium channel KirBac1.1 in the closed state. Science. 2003 Jun 20; 300(5627):1922–6.

channels for triggering transmitter release and for allowing entry of calcium which can function as a second messenger. Although sodium and calcium ions carry different amounts of charge, they are about the same size. How are channels designed to be selectively permeable to sodium or to calcium?

Using organic ions of different size to gauge the pore dimension, Hille showed in the seventies that the narrowest part of sodium channels and calcium channels is actually wider than potassium channels, though not quite large enough to allow permeation of fully hydrated ions (Hille, 1992). Thus, ion channels cannot simply discriminate different ions based on their size. Rather, dehydration energy appears to be an important factor; the pore of a channel probably approximates the hydration shell of its permeant ion so as to ease the process of losing a significant fraction of the water molecules in the hydration shell as the ion moves through the channel. This consideration, combined with the features afforded by a multi-ion pore, may account for the different ion selectivity of these channels.

One remarkable feature of voltage-gated calcium channels is the dependence of its ion selectivity on ion concentration. Under physiological conditions, with millimolar calcium ions in the extracellular medium, these channels are selectively permeable to calcium even though the extracellular sodium concentration is much higher, typically greater than 100 mM. If the extracellular calcium concentration is reduced to submicromolar levels, calcium channels conduct sodium ions. At intermediate calcium concentrations, neither calcium ions nor sodium ions conduct currents through calcium channels (Almers and McCleskey, 1984; Hess and Tsien, 1984) (Fig. 6.10A).

This behavior of the calcium channel could be accounted for if the calcium channel has more than one binding site with higher affinity for calcium than sodium. In the absence of calcium ions, sodium ions occupy the binding sites and go through the channel. As the calcium concentration is raised, one of the binding sites becomes occupied by calcium. The higher affinity of the binding site for calcium makes it difficult for a sodium ion to displace the calcium ion. Thus, the bound calcium ion in effect blocks sodium permeation through the channel. Having more than one calcium ion in the same channel would effectively reduce the affinity of these binding sites for calcium, because of the electrostatic repulsion between these doubly charged calcium ions. This explains why at low calcium concentrations, calcium is much more likely to occupy only one of the binding sites. At sufficiently high calcium concentrations, under physiological conditions, it becomes more likely for a calcium ion in the external solution to displace the calcium ion at the external binding site and cause it to move to the next binding site, leading to calcium permeation.

Key molecular differences between voltage-gated calcium and sodium channels. In each of the four repeats of the voltage-gated calcium channel there is a highly conserved glutamate in the H5 or P region. These negatively charged residues are crucial for the channel's affinity for calcium, and probably form two binding sites for calcium (Yang et al., 1993) (Figs. 6.10B–D). At equivalent positions in the voltage-gated sodium channel there are two glutamate residues, one alanine residue, and one lysine residue (Fig. 6.10E). Glutamate substitutions of these latter residues cause the mutant sodium channel to behave similarly to calcium channels (Heinemann et al., 1992). In the absence of calcium, the mutant channels also exhibit altered selectivity among monovalent cations. Taken together, these studies suggest that residues at the position of the highly conserved glutamate in the H5 or P region of voltage-gated calcium channels and sodium channels line the pore and interact with more than one permeant ion.

ION CHANNEL DISTRIBUTION

As described in detail in Chapter 4, neurons are highly polarized. Their dendrites receive synaptic inputs, thereby generating fast and slow synaptic potentials. Integration of these synaptic inputs sometimes leads to the generation, in or near the soma, of action potentials, which propagate from the soma along the axon to the nerve terminals. Transmitter release from the nerve terminals is triggered by the arrival of action potentials, but may also be regulated by transmitters acting on receptors located at the nerve terminal. Proper function of these neuronal activities depends on adequate placement of ion channels of the appropriate number and type in each domain of the neuronal membrane. We use a few examples to illustrate this point, in preparation for the consideration of dendritic information processing to be discussed in Chapter 17.

Different Ion Channels Are Localized to Different Parts of the Neuron

Whereas it has long been recognized that transmitter receptors are located near the sites of transmitter release, it is now evident that many other types of ion channels are also targeted to discrete regions of the neuronal membrane. Once the molecular entity of ion channels becomes known, specific probes can be generated to determine their expression patterns. These studies have revealed an intricate mosaic-like distribution of different ion channels. Presumably, ion channels are targeted to specific locations because their channel properties are most suited for the physiological functions at those sites. The possibility of different channel isoforms

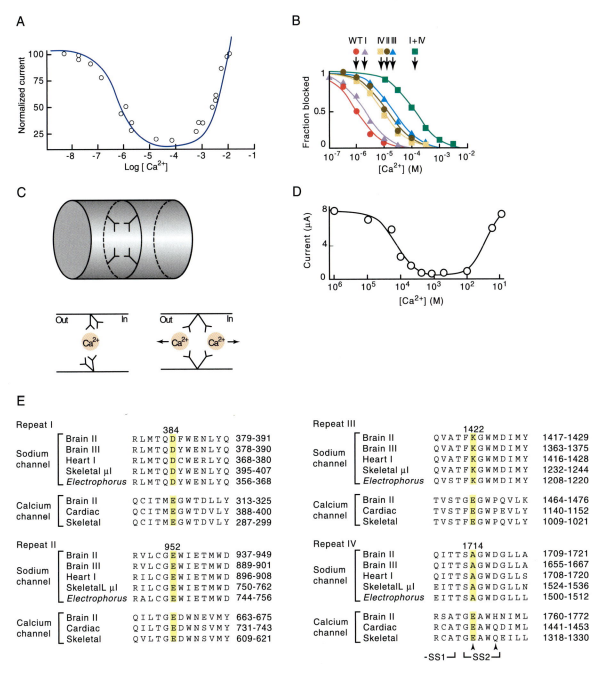

FIGURE 6.10 Dependence of voltage-gated calcium channel ion selectivity on calcium concentration. (A) In the presence of sodium ions and varying concentration of calcium ions, calcium channels are permeable to sodium ions at submicromolar calcium concentration. At submillimolar calcium concentration, calcium ion occupies one binding site in the channel and blocks sodium permeation. At still higher calcium concentration, calcium may occupy multiple binding sites; the presence of multiple calcium ions in the same channel pore allows them to dissociate from the binding site more readily and pass through the channel. Adapted from Almers and McCleskey (1984). (B) The affinity of the calcium binding site as indicated by the blocking action of calcium on lithium permeation is reduced by substituting a glutamate in the P loop with glutamine. WT, wild-type calcium channel. I, II, III, and IV indicate glutamine substitution in the first, second, third, and fourth repeats of the channel. I + IV indicates double mutations in the first and fourth repeats. (C) How the ring of four glutamate residues in the calcium channel pore might bind one or two calcium ions. (B) and (C) are adapted from Yang *et al.* (1993). (D) Glutamate substitution of lysine 1422 of the P loop in the third repeat of voltage-gated sodium channels causes the mutant channel to behave like a calcium channel. (E) Alignment of the P loop sequences for each of the four repeats of the voltage-gated sodium channels and calcium channels. (D) and (E) are adapted from Heinemann *et al.* (1992).

being regulated differently allows further dynamic modulation of channel properties by neuronal activities.

Different potassium channels are targeted to axons and dendrites. Myelinated axons have voltage-gated sodium channels confined to the node of Ranvier (Salzer, 1997), flanked by certain isoforms of voltage-gated potassium channels (Kv1.1 and Kv1.2) forming two rings in the juxtaparanodal regions (Rasband et al., 1998; Wang et al., 1993; Zhou et al., 1999). Another member of the same family, Kv1.4, is found in patches along the axon and near the nerve terminals (Cooper et al., 1998), whereas members of a closely related family such as Kv4.2 are located on the dendrite and on the postsynaptic membrane (Alonso and Widmer, 1997; Sheng et al., 1993; Tkatch et al., 2000). Large conductance (BK) calcium-activated potassium channels, on the other hand, are present on the presynaptic membrane (Wanner et al., 1999). Heteromeric channels formed by different α and β subunits may be localized to different domains of the neuronal membrane, thereby further increasing the diversity of these channels (Rhodes et al., 1995; Veh et al., 1995).

Different calcium channels have different distributions. Pharmacological and electrophysiological studies have shown that N-type and P/Q-type voltage-gated calcium channels mediate the calcium entry that triggers transmitter release from the nerve terminal, whereas L-type voltage-gated calcium channels are found in the soma. An even more refined picture has emerged from immunocytochemical studies (Catterall, 1998). L-Type calcium channels may contain α_{1C} or α_{1D} subunits, which are localized predominantly in cell bodies and proximal dendrites. However, α_{1D} subunits appear to be smoothly distributed on the cell membrane, whereas α_{1C} subunits form clusters extending far into the dendrites and to postsynaptic membranes of glutamatergic synapses. The α_{1B} subunits of N-type calcium channels and α_{1A} subunits of P/Q-type subunits are present in low density in dendrites but at high density in presynaptic nerve terminals, where they colocalize with the SNARE proteins (see Chapter 8) involved in synaptic vesicle docking and exocytosis.

Action Potentials May "Backpropagate" from the Soma to the Dendrites or Even Be Generated in the Dendrite

Whereas it was thought earlier that action potentials are generated in the axon hillock region near the soma and propagate in one direction, down the axon, recent molecular and electrophysiological studies reveal many other possibilities. Propagation of action potential past the axonal branch point may be controlled by voltage-gated potassium channels (Debanne et al., 1997; Obaid and Salzberg, 1996). Not only can action potentials generated in or near the soma propagate "backward" into the dendrites via activation of dendritic voltage-gated sodium channels, action potentials may also be initiated within the dendrites as a result of primarily activation of voltage-gated calcium channels (Schiller et al., 1997; Spruston et al., 1995). For example, backpropagation of action potentials into dendrites has been observed not only in brain slice preparations but also in awake animals (Buzsaki and Kandel, 1998; Stuart et al., 1997). And sensory stimulation of the whiskers sometimes generates calcium action potentials in pyramidal neurons in the somatosensory cortex of rodents (Helmchen et al., 1999).

The extent of action potential back propagation varies with the types of neuron. In dopamine neurons in the midbrain and mitral cells in the olfactory bulb (see Fig. 17.11), neurons known to release transmitters from their dendrites, there is hardly any attenuation of action potentials as they propagate to dendrites. In contrast, the amplitude of dendritic action potentials decreases sharply within 0.1 mm of the soma of Purkinje cells in the cerebellum (see Fig. 17.14). In many other central neurons, including cortical neurons in layer 5 (see Fig. 17.10), hippocampal pyramidal neurons in the CA1 region, and spinal neurons, dendritic action potentials may be attenuated by less than 50% over a distance of half a millimeter (Stuart et al., 1997). These observations indicate that prior neuronal activities may exert influence over subsequent synaptic inputs in dendrites via backpropagating action potentials.

Action potential back propagation can be regulated. Repetitive firing results in a rapid decline in dendritic action potentials as a result of cumulative sodium channel inactivation, which is more prominent in apical dendrites (Spruston et al., 1995; Colbert et al., 1997; Mickus et al., 1999). The lower safety factor is also evident from the tendency of backpropagating action potentials to fail at a branch point of dendrites. Moreover, in certain central neurons, not only is the density of the rapidly inactivating A-type voltage-gated potassium channels higher in apical dendrites than in proximal dendrites, those channels at apical dendrites are more active due to a left shift in the voltage dependence of channel activation (Hoffman et al., 1997). These spatial gradients present another factor that influences the extent of action potential back propagation. Both sodium channel inactivation and potassium channel activities can be further modulated by protein kinases and transmitters such as acetylcholine and dopamine (Johnston et al., 1999). The extent to which synaptic potentials may be altered by backpropagating action potentials, therefore, may vary depending on the pattern of prior neuronal activities as well as the action of transmitters that stimulate G protein-coupled metabotropic transmitter receptors.

Spatial Gradients of Ion Channels Allow Synaptic Potentials Generated over Large Distance on the Dendrites to Reach the Cell Soma with Similar Size and Duration

As in the case of voltage-gated ion channels, the dendritic distribution is not uniform for channels that contribute to the input resistance as well as ionotropic transmitter receptors. These spatial gradients play an important role in keeping the amplitude and waveform of the excitatory postsynaptic potential (EPSP) constant in some of the central neurons, even though they may be generated at synaptic sites that are up to 1 mm apart. If the channel distributions and membrane properties were uniform throughout the dendrite, the more distal on the dendrite an EPSP was generated the smaller and longer lasting it would be when it reached the soma, as illustrated in Fig. 4.8. This is prevented from occurring in certain central neurons by the presence of a greater number of hyperpolarization-activated cation channels (I_h), which affect the input resistance and the membrane time constant, and a larger quantal size of the transmitter response at the more distal dendrite (Magee, 1998; Magee and Cook, 2000; Williams and Stuart, 2000). Presumably, a greater number of glutamate receptors can be activated by glutamate released onto apical dendrites of these central neurons.

The distance a synaptic potential travels and the extent to which multiple synaptic potentials summate can be regulated. In addition to the spatial gradient of I_h channel density, there is a left shift in the voltage dependence of channel activation at more distal dendrites of certain central neurons (Magee, 1998), possibly due to a spatial gradient of second messengers. These channels are sensitive to modulation by transmitters and second messengers such as cyclic AMP (Luthi and McCormick, 1998). Thus, modulation of channel activities due to neuronal or synaptic activities could profoundly alter the size and duration of subsequent synaptic potentials, as well as the extent of temporal and spatial summation of synaptic inputs.

SUMMARY

Signal processing in the nervous system is mediated by a wide variety of ion channels localized to different compartments of the highly polarized neuron. This can be better appreciated now that it is possible to follow the activity of single channels by patch recording from the dendrites, as well as the soma, and to examine channel distribution using molecular probes specific for individual channel types. Molecular analyses of the mechanisms for channel permeation and gating, as well as how these processes may be modulated by transmitters and second messengers mobilized by neuronal activities, provide further insight into the plasticity of these signaling processes.

References

Almers, W., and McCleskey, E. W. (1984). Non-selective conductance in calcium channels of frog muscle: Calcium selectivity in a single-file pore. *J. Physiol. (London)* **353**, 585–608.

Alonso, G., and Widmer, H. (1997). Clustering of KV4.2 potassium channels in postsynaptic membrane of rat supraoptic neurons: An ultrastructural study. *Neuroscience* **77**, 617–621.

Armstrong, C. M., and Bezanilla, F. (1977). Inactivation of the sodium channel. II. Gating current experiments. *J. Gen. Physiol.* **70**, 567–590.

Ashcroft, S. J. H. (2000). The beta-cell KATP channel. *J Membr. Biol.* **176**.

Ashcroft, F. M., and Gribble, F. M. (1998). Correlating structure and function in ATP-sensitive K+ channels. *Trends Neurosci.* **21**, 288–294.

Babenko, A. P., Aguilar-Bryan, L., and Bryan, J. (1998). A view of sur/KIR6.X, KATP channels. *Annu Rev. Physiol.* **60**, 667–687.

Barrett, J. N., Magleby, K. L., and Pallotta, B. S. (1982). Properties of single calcium-activated potassium channels in cultured rat muscle. *J. Physiol. (London)* **331**, 211–230.

Brake, A. J., Wagenbach, M. J., and Julius, D. (1994). New structural motif for ligand-gated ion channels defined by an ionotropic ATP receptor. *Nature* **371**, 519–523.

Buzsaki, G., and Kandel, A. (1998). Somadendritic backpropagation of action potentials in cortical pyramidal cells of the awake rat. *J. Neurophysiol.* **79**, 1587–1591.

Catterall, W. A. (1998). Structure and function of neuronal Ca2+ channels and their role in neurotransmitter release. *Cell Calcium* **24**, 307–323.

Catterall, W. A. (2000). From ionic currents to molecular mechanisms: The structure and function of voltage-gated sodium channels. *Neuron* **26**, 13–25.

Caterina, M. J., Leffler, A., Malmberg, A. B., Martin, W. J., Trafton, J., Petersen-Zeitz, K. R., Koltzenburg, M., Basbaum, A. I., and Julius, D. (2000). Impaired nociception and pain sensation in mice lacking the capsaicin receptor. *Science* **288**, pp. 306–313.

Cha, A., Snyder, G. E., Selvin, P. R., and Bezanilla, F. (1999). Atomic scale movement of the voltage-sensing region in a potassium channel measured via spectroscopy. *Nature* **402**, 809–813.

Chen, T. Y., and Yau, K. W. (1994). Direct modulation by Ca2+-calmodulin of cyclic nucleotide-activated channel of rat olfactory receptor neurons. *Nature* **368**, 545–548.

Chen, G.-Q., Cui, C., Mayer, M. L., and Gouaux, E. (1999). Functional characterization of a potassium-selective prokaryotic glutamate receptor. *Nature* **402**, 817–821.

Colbert, C. M., Magee, J. C., Hoffman, D. A., and Johnston, D. (1997). Slow recovery from inactivation of Na⁺ channels underlies the activity-dependent attenuation of dendritic action potentials in hippocampal CA1 pyramidal neurons. *J. Neurosci.* **17**, 6512–6521.

Cooper, E. C., Milroy, A., Jan, Y. N., Jan, L. Y., and Lowenstein, D. H. (1998). Presynaptic localization of Kv1.4-containing A-type potassium channels near excitatory synapses in the hippocampus. *J. Neurosci.* **18**, 965–974.

Coscoy, S., Lingueglia, E., Lazdunski, M., and Barbry, P. (1998). The Phe–Met–Arg–Phe-amide-activated sodium channel is a tetramer. *J. Biol. Chem.* **273**, 8317–8322.

Cui, J., Cox, D. H., and Aldrich, R. W. (1997). Intrinsic voltage dependence and Ca²⁺ regulation of mslo large conductance Ca-activated K⁺ channels. *J. Gen. Physiol.* **109**, 647–673.

Davis, J. B., Gray, J., Gunthorpe, M. J., Hatcher, J. P., Davey, P. T., Overend, P., Harries, M. H., Latcham, J., Clapham, C., Atkinson, K., Hughes, S. A., Rance, K., Grau, E., Harper, A. J., Pugh, P. L., Rogers, D. C., Bingham, S., Randall, A., and Sheardown, S. A. (2000). Vanilloid receptor–1 is essential for inflammatory thermal hyperalgesia. *Nature* **405**, 183–187.

Debanne, D., Guerineau, N. C., Gahwiler, B. H., and Thompson, S. M. (1997). Action-potential propagation gated by an axonal I(A)-like K⁺ conductance in hippocampus [published erratum appears in Nature 1997; 390: 536]. *Nature* **389**, 286–289.

Doyle, D. A., Morais Cabral, J. H., Pfuetzner, R. A., Kuo, A., Gulbis, J. M., Cohen, S. L., Chait, B. T., and MacKinnon, R. (1998). The structure of the potassium channel: Molecular basis of K⁺ conduction and selectivity. *Science* **280**, 69–77.

Durell, S. R., Hao, Y., Nakamura, T., Bakker, E. P., and Guy, H. R. (1999). Evolutionary relationship between K⁺ channels and symporters. *Biophys. J.* **77**, 775–788.

Galzi, J. L., Devillers-Thiery, A., Hussy, N., Bertrand, S., Changeux, J. P., and Bertrand, D. (1992). Mutations in the channel domain of a neuronal nicotinic receptor convert ion selectivity from cationic to anionic. *Nature* **359**, 500–505.

Gandhi, C. S., Loots, E., and Isacoff, E. Y. (2000). Reconstructing voltage sensor–pore interaction from a fluorescence scan of a voltage-gated K⁺ channel. *Neuron* **27**, 585–595.

Gaymard, F., Cerutti, M., Horeau, C., Lemaillet, G., Urbach, S., Ravallec, M., Devauchelle, G., Sentenac, H., and Thibaud, J. B. (1996). The baculovirus/insect cell system as an alternative to *Xenopus* oocytes: First characterization of the AKT1 K⁺ channel from *Arabidopsis thaliana*. *J. Biol. Chem.* **271**, 22863–22870.

Glauner, K. S., Mannuzzu, L. M., Gandhl, C. S., and Isacoff, E. Y. (1999). Spectroscopic mapping of voltage sensor movement in the *Shaker* potassium channel. *Nature* **402**, 813–817.

Gulbis, J. M., Zhou, M., Mann, S., and MacKinnon, R. (2000). Structure of the cytoplasmic beta subunit–T1 assembly of voltage-dependent K⁺ channels. *Science* **289**, 123–127.

Hagiwara, S., Miyazaki, S., and Rosenthal, N. P. (1976). Potassium current and the effect of cesium on this current during anomalous rectification of the egg cell membrane of a starfish. *J. Gen. Physiol.* **67**, 621–638.

Hagiwara, S., Miyazaki, S., Krasne, S., and Ciani, S. (1977). Anomalous permeabilities of the egg cell membrane of a starfish in K⁺-Tl⁺ mixtures. *J. Gen. Physiol.* **70**, 269–281.

Heinemann, S. H., Terlau, H., Stuhmer, W., Imoto, K., and Numa, S. (1992). Calcium channel characteristics conferred on the sodium channel by single mutations. *Nature* **356**, 441–443.

Helmchen, F., Svoboda, K., Denk, W., and Tank, D. W. (1999). *In vivo.* dendritic calcium dynamics in deep-layer cortical pyramidal neurons. *Nat. Neurosci.* **2**, 989–996.

Herlitze, S., Garcia, D. E., Mackie, K., Hille, B., Scheuer, T., and Catterall, W. A. (1996). Modulation of Ca²⁺ channels by G-protein beta gamma subunits. *Nature* **380**, 258–262.

Hess, P., and Tsien, R. W. (1984). Mechanism of ion permeation through calcium channels. *Nature* **309**, 453–456.

Hille, B. (1992). "Ionic Channels of Excitable Membranes". Sinauer Associates, Inc., Sunderland, MA.

Hodgkin, A. L., and Huxley, A. F. (1952). A quantitative description of membrane current and its application to conduction and excitation in nerve. *J. Physiol.* **117**, 500–544.

Hoffman, D. A., Magee, J. C., Colbert, C. M., and Johnston, D. (1997). K⁺ channel regulation of signal propagation in dendrites of hippocampal pyramidal neurons [see comments]. *Nature* **387**, 869–875.

Horn, R. (2000). A new twist in the saga of charge movement in voltage-dependent ion channels. *Neuron* **25**, 511–514.

Huang, C. L., Feng, S., and Hilgemann, D. W. (1998). Direct activation of inward rectifier potassium channels by PIP₂ and its stabilization by Gbetagamma. *Nature* **391**, 803–806.

Ikeda, S. R. (1996). Voltage-dependent modulation of N-type calcium channels by G-protein beta gamma subunits. *Nature* **380**, 255–258.

Isacoff, E. Y., Jan, Y. N., and Jan, L. Y. (1991). Putative receptor for the cytoplasmic inactivation gate in the Shaker K⁺ channel. *Nature* **353**, 86–90.

Jan, L. Y., and Jan, Y. N. (1994). Potassium channels and their evolving gates. *Nature* **371**, 119–122.

Jan, L. Y., and Jan, Y. N. (1997). Cloned potassium channels from eukaryotes and prokaryotes. *Annu. Rev. Neurosci.* **20**, 91–123.

Jentsch, T. J. (1996). Chloride channels: A molecular perspective. *Curr. Opin. Neurobiol.* **6**, 303–310.

Johnston, D., Hoffman, D. A., Colbert, C. M., and Magee, J. C. (1999). Regulation of back propagating action potentials in hippocampal neurons. *Curr. Opin. Neurobiol.* **9**, 288–292.

Keynes, R. D. (1994). The kinetics of voltage-gated ion channels. *Quarterly Rev Biophys* **27**, 339–434.

Kobrinsky, E., Mirshahi, T., Zhang, H., Jin, T. and Logothetis, D. E. (2000). Receptor-mediated hydrolysis of plasma membrane messenger PIP2 leads to K⁺-current desensitization. *Nat. Cell. Biol.* **2**, 507–514.

Larsson, H. P., and Elinder, F. (2000). A conserved glutamate is important for slow inactivation in K⁺ channels. *Neuron* **27**, 573–583.

Larsson, H. P., Baker, O. S., Dhillon, D. S., and Isacoff, E. Y. (1996). Transmembrane movement of the *Shaker* K⁺ channel S4. *Neuron* **16**, 387–397.

Lee, A., Wong, S. T., Gallagher, D., Li, B., Storm, D. R., Scheuer, T., and Catterall, W. A. (1999). Ca²⁺/calmodulin binds to and modulates P/Q-type calcium channels. *Nature* **399**, 155–159.

Lesage, F., and Lazdunski, M. (2000). Molecular and functional properties of two-pore-domain potassium channels. *Am. J. Physiol. Renal Physiol.* **279**, 793–801.

Loots, E., and Isacoff, E. Y. (1998). Protein rearrangements underlying slow inactivation of the Shaker K⁺ channel. *J. Gen. Physiol.* **112**, 377–389.

Luscher, C., Jan, L. Y., Stoffel, M., Malenka, R. C., and Nicoll, R. A. (1997). G protein-coupled inwardly rectifying K⁺ channels (GIRKs) mediate postsynaptic but not presynaptic transmitter actions in hippocampal neurons [published erratum appears in Neuron 1997; 19: following 945]. *Neuron* **19**, 687–695.

Luthi, A., and McCormick, D. A. (1998). H-current: Properties of a neuronal and network pacemaker. *Neuron* **21**, 9–12.

Magee, J. C. (1998). Dendritic hyperpolarization-activated currents modify the integrative properties of hippocampal CA1 pyramidal neurons. *J. Neurosci.* **18**, 7613–7624.

Magee, J. C., and Cook, E. P. (2000). Somatic EPSP amplitude is independent of synapse location in hippocampal pyramidal neurons [see comments]. *Nat. Neurosci.* **3**, 895–903.

Mannuzzu, L. M., and Isacoff, E. Y. (2000). Independence and cooperativity in rearrangements of a potassium channel voltage sensor revealed by single subunit fluorescence. *J. Gen. Physiol.* **115**, 257–268.

Mannuzzu, L. M., Moronne, M. M., and Isacoff, E. Y. (1996). Direct physical measure of conformational rearrangement underlying potassium channel gating. *Science* **271**, 213–216.

Maricq, A. V., Peterson, A. S., Brake, A. J., Myers, R. M., and Julius, D. (1991). Primary structure and functional expression of the 5HT3 receptor, a serotonin-gated ion channel. *Science* **254**, 432–437.

Marten, I., Hoth, S., Deeken, R., Ache, P., Ketchum, K. A., Hoshi, T., and Hedrich, R. (1999). AKT3, a phloem-localized K^+ channel, is blocked by protons. *Proc. Natl. Acad. Sci. USA* **96**, 7581–7586.

Marty, A. (1981). Ca-dependent K channels with large unitary conductance in chromaffin cell membranes. *Nature* **291**, 497–500.

Meador, W. E., Means, A. R., and Quiocho, F. A. (1993). Modulation of calmodulin plasticity in molecular recognition on the basis of x-ray structures. *Science* **262**, 1718–1721.

Meera, P., Wallner, M., Song, M., and Toro, L. (1997). Large conductance voltage- and calcium-dependent K^+ channel, a distinct member of voltage-dependent ion channels with seven N-terminal transmembrane segments (S0–S6), an extracellular N terminus, and an intracellular (S9–S10) C terminus. *Proc. Natl. Acad. Sci. USA* **94**, 14066–14071.

Mickus, T., Jung, H., and Spruston, N. (1999). Properties of slow, cumulative sodium channel inactivation in rat hippocampal CA1 pyramidal neurons. *Biophys. J.* **76**, 846–860.

Mikoshiba, K. (1997). The InsP3 receptor and intracellular Ca^{2+} signaling. *Curr. Opin. Neurobiol.* **7**, 339–345.

Nichols, C. G., and Lopatin, A. N. (1997). Inward rectifier potassium channels. *Annu. Rev. Physiol.* **59**, 171–191.

Obaid, A. L., and Salzberg, B. M. (1996). Micromolar 4-aminopyridine enhances invasion of a vertebrate neurosecretory terminal arborization: Optical recording of action potential propagation using an ultrafast photodiode-MOSFET camera and a photodiode array. *J. Gen. Physiol.* **107**, 353–368.

Olcese, R., Latorre, R., Toro, L., Bezanilla, F., and Stefani, E. (1997). Correlation between charge movement and ionic current during slow inactivation in Shaker K^+ channels. *J. Gen. Physiol.* **110**, 579–589.

Patel, A. J., Honore, E., Lesage, F., Fink, M., Romey, G., and Lazdunski, M. (1999). Inhalational anesthetics activate two-pore-domain background K^+ channels. *Nat. Neurosci.* **2**, 422–426.

Perozo, E., Cortes, D. M., and Cuello, L. G. (1999). Structural rearrangements underlying K^+-channel activation gating. *Science* **285**, 73–78.

Plummer, N. W., and Meisler, M. H. (1999). Evolution and diversity of mammalian sodium channel genes. *Genomics* **57**, 323–331.

Quayle, J. M., Nelson, M. T., and Standen, N. B. (1997). ATP-sensitive and inwardly rectifying potassium channels in smooth muscle. *Physiol. Rev.* **77**, 1165–1232.

Rasband, M. N., Trimmer, J. S., Schwarz, T. L., Levinson, S. R., Ellisman, M. H., Schachner, M., and Shrager, P. (1998). Potassium channel distribution, clustering, and function in remyelinating rat axons. *J. Neurosci.* **18**, 36–47.

Rees, D. C., Chang, G., and Spencer, R. H. (2000). Crystallographic analyses of ion channels: Lessons and challenges. *J. Biol. Chem.* **275**, 713–716.

Rettig, J., Heinemann, S. H., Wunder, F., Lorra, C., Parcej, D. N., Dolly, J. O., and Pongs, O. (1994). Inactivation properties of voltage-gated K^+ channels altered by presence of beta-subunit. *Nature* **369**, 289–294.

Rhodes, K. J., Keilbaugh, S. A., Barrezueta, N. X., Lopez, K. L., and Trimmer, J. S. (1995). Association and colocalization of K^+ channel alpha- and beta-subunit polypeptides in rat brain. *J. Neurosci.* **15**, 5360–5371.

Salzer, J. L. (1997). Clustering sodium channels at the node of Ranvier: close encounters of the axon–glia kind. *Neuron* **18**, 843–846.

Schachtman, D. P., Schroeder, J., Lucas, W. J., Anderson, J. A., and Gaber, R. F. (1992). *Science* **258**, 1654–1658.

Schiller, J., Schiller, Y., Stuart, G., and Sakmann, B. (1997). Calcium action potentials restricted to distal apical dendrites of rat neocortical pyramidal neurons. *J. Physiol. (London)* **505**, 605–616.

Schreiber, M., Yuan, A., and Salkoff, L. (1999). Transplantable sites confer calcium sensitivity to BK channels. *Nat. Neurosci.* **2**, 416–421.

Sheng, S., Li, J., McNulty, K. A., Avery, D., and Kleyman, T. R. (2000). Characterization of the selectivity filter of the epithelial sodium channel. *J. Biol. Chem.* **275**, 8572–8581.

Sheng, M., Liao, Y. J., Jan, Y. N., and Jan, L. Y. (1993). Presynaptic A-current based on heteromultimeric K^+ channels detected *in vivo*. *Nature* **365**, 72–75.

Sheppard, D. N., and Welsh, M. J. (1999). Structure and function of the CFTR chloride channel. *Physiol. Rev.* **79**, S23–S45.

Sigworth, F. J. (1994). Voltage gating of ion channels. *Q. Rev. Biophys.* **27**, 1–40.

Slesinger, P. A., and Lansman, J. B. (1991). Reopening of Ca^{2+} channels in mouse cerebellar neurons at resting membrane potentials during recovery from inactivation. *Neuron* **7**, 755–762.

Snyder, P. M., Olson, D. R., and Bucher, D. B. (1999). A pore segment in DEG/ENaC Na(+) channels. *J. Biol. Chem.* **274**, 28484–28490.

Spruston, N., Schiller, Y., Stuart, G., and Sakmann, B. (1995). Activity-dependent action potential invasion and calcium influx into hippocampal CA1 dendrites [see comments]. *Science* **268**, 297–300.

Stuart, G., Spruston, N., Sakmann, B., and Hausser, M. (1997). Action potential initiation and backpropagation in neurons of the mammalian CNS. *Trends Neurosci.* **20**, 125–131.

Tang, X. D., Marten, I., Dietrich, P., Ivashikina, N., Hedrich, R., and Hoshi, T. (2000). Histidine(118) in the S2–S3 linker specifically controls activation of the KAT1 channel expressed in *Xenopus* oocytes. *Biophys. J.* **78**, 1255–1269.

Tkatch, T., Baranauskas, G., and Surmeier, D. J. (2000). Kv4.2 mRNA abundance and A-type K(+) current amplitude are linearly related in basal ganglia and basal forebrain neurons. *J. Neurosci.* **20**, 579–588.

Unwin, N. (1995). Acetylcholine receptor channel imaged in the open state. *Nature* **373**, 37–43.

Valera, S., Hussy, N., Evans, R. J., Adami, N., North, R. A., Surprenant, A., and Buell, G. (1994). A new class of ligand-gated ion channel defined by P2x receptor for extracellular ATP [see comments]. *Nature* **371**, 516–519.

Veh, R. W., Lichtinghagen, R., Sewing, S., Wunder, F., Grumbach, I. M., and Pongs, O. (1995). Immunohistochemical localization of five members of the Kv1 channel subunits: Contrasting subcellular locations and neuron-specific co-localizations in rat brain. *Eur. J. Neurosci.* **7**, 2189–2205.

Vergara, C., Latorre, R., Marrion, N. V., and Adelman, J. P. (1998). Calcium-activated potassium channels. *Curr. Opin. Neurobiol.* **8**, 321–329.

Wallner, M., Meera, P., and Toro, L. (1999). Molecular basis of fast inactivation in voltage and Ca^{2+}-activated K^+ channels: A transmembrane beta-subunit homolog. *Proc. Natl. Acad. Sci. USA* **96**, 4137–4142.

Wang, H., Kunkel, D. D., Martin, T. M., Schwartzkroin, P. A., and Tempel, B. L. (1993). Heteromultimeric K^+ channels in terminal and juxtaparanodal regions of neurons. *Nature* **365**, 75–79.

Wanner, S. G., Koch, R. O., Koschak, A., Trieb, M., Garcia, M. L., Kaczorowski, G. J., and Knaus, H. G. (1999). High-conductance calcium-activated potassium channels in rat brain: Pharmacology, distribution, and subunit composition. *Biochemistry* **38**, 5392–5400.

Wickman, K., Nemec, J., Gendler, S. J., and Clapham, D. E. (1998). Abnormal heart rate regulation in GIRK4 knockout mice. *Neuron* **20**, 103–114.

Williams, S. R., and Stuart, G. J. (2000). Site independence of EPSP time course is mediated by dendritic $I(h)$ in neocortical pyramidal neurons. *J. Neurophysiol.* **83**, 3177–3182.

Wo, Z. G., and Oswald, R. E. (1995). Unraveling the modular design of glutamate-gated ion channels. *Trends Neurosci.* **18**, 161–168.

Xia, X. M., Fakler, B., Rivard, A., Wayman, G., Johson-Pais, T., Keen, J. E., Ishii, T., Hirschberg, B., Bond, C. T., Lutsenko, S., Maylie, J., and Adelman, J. P. (1998). Mechanism of calcium gating in small-conductance calcium-activated potassium channels. *Nature* **395**, 503–507.

Yang, J., Ellinor, P. T., Sather, W. A., Zhang, J. F., and Tsien, R. W. (1993). Molecular determinants of Ca^{2+} selectivity and ion permeation in L-type Ca^{2+} channels [see comments]. *Nature* **366**, 158–161.

Yang, N., George, A. L. J., and Horn, R. (1996). Molecular basis of charge movement in voltage-gated sodium channels. *Neuron* **16**, 113–122.

Yellen, G. (1998). The moving parts of voltage-gated ion channels. *Q. Rev. Biophys.* **31**, 239–295.

Yi, B. A., Lin, Y., Jan, Y. N., and Jan, L. Y. (2001). Yeast screen for constitutively active mutant G protein-activated potassium channels. *Neuron* **29**, 657–667.

Yue, L., Peng, J. B., Hediger, M. A., and Clapham, D. E. (2001). CaT1 manifests the pore properties of the calcium-release-activated calcium channel. *Nature* **410**, pp. 705–709.

Zagotta, W. N., Hoshi, T., and Aldrich, R. W. (1990). Restoration of inactivation in mutants of Shaker potassium channels by a peptide derived from ShB [see comments]. *Science* **250**, 568–571.

Zamponi, G. W., Bourinet, E., Nelson, D., Nargeot, J. and Snutch, T. P. (1997). Crosstalk between G proteins and protein kinase C mediated by the calcium channel alpha1 subunit [see comments]. *Nature* **385**, 442–446.

Zhou, L., Messing, A., and Chiu, S. Y. (1999). Determinants of excitability at transition zones in Kv1.1-deficient myelinated nerves. *J. Neurosci* **19**, 5768–5781.

Zufall, F., Shepherd, G. M., and Barnstable, C. J. (1997). Cyclic nucleotide gated channels as regulators of CNS development and plasticity. *Curr. Opin. Neurobiol.* **7**, 404–412.

Zühlke, R. D., Pitt, G. S., Deisseroth, K., Tsien, R. W., and Reuter, H. (1999). Calmodulin supports both inactivation and facilitation of L-type calcium channels. *Nature* **399**, 159–162.

Dynamical Properties of Excitable Membranes

Douglas A. Baxter, Carmen C. Canavier and John H. Byrne

The nervous system functions to encode, process, store, and transmit information. To perform these tasks, neurons have evolved sophisticated means of generating electrical and chemical signals. Chapters 12 and 14 describe several aspects of intracellular chemical signaling mechanisms (e.g., second-messenger cascades and genetic-regulatory networks) and Chapters 5, 8, and 16 describe several aspects of electrical signaling in neurons (e.g., the membrane potential, the action potential, and synaptic transmission). The goals of this chapter are to provide a more detailed mathematical description of neuronal excitability and to introduce several mathematical tools based on the theory of nonlinear dynamical systems that can be used to analyze neuronal excitability. These mathematical tools (e.g., phase plane analysis, bifurcation theory) provide graphical or geometric representations of the system dynamics and can be used to understand, predict, and interpret biophysical features such as threshold phenomena and oscillatory and bursting behavior, as well as the mechanisms of bistability and hysteresis. Such analyses can provide novel insights into the capabilities of individual neurons to process and store information.

THE HODGKIN–HUXLEY MODEL

Since 1952, the understanding and methods for studying the biophysics of excitable membranes have been profoundly influenced by the pioneering work of Alan Hodgkin and Andrew Huxley. Using the techniques of voltage clamping, which had recently been developed by Cole (1949) and Marmont (1949) and improved on by Hodgkin *et al.* (1952) (see also Box 5.3), Hodgkin and Huxley characterized the time and voltage dependency of the ionic conductances that underlie an action potential in the squid giant axon (Hodgkin *et al.*, 1952; Hodgkin and Huxley, 1952a–c). In addition, they developed a cogent mathematical model that accurately predicted the waveform of the action potential and several other physiological properties such as refractory period, propagation of the action potential along an axon, anode break excitation, and accommodation (Hodgkin and Huxley, 1952d). In biology, quantitatively predictive theories are rare, and this work stands out as one of the most successful combinations of experimental and computational approaches to understanding a fundamentally important issue in neuroscience. For their seminal work on neuronal excitability, Hodgkin and Huxley received the Nobel Prize in Physiology or Medicine in 1963. [To view biographical information about Hodgkin and Huxley and other Nobel laureates, you may visit the Nobel Prize web site at *http://www.nobel.se/nobel*. The lectures that Hodgkin (1964) and Huxley (1964) delivered when they received the Nobel Prize were published in 1964. In addition, Hodgkin (1976, 1977) and Huxley (2000, 2002) have published informal narratives describing events surrounding these studies, and others have reviewed the work of Hodgkin and Huxley from a historical perspective (Cole, 1968; Hille, 2001; Nelson and Rinzel, 1998; Rinzel, 1990).] Their empirical studies and quantitative model remain influential to the present day and provide an analytical framework for investigating and modeling a large class of diverse membrane phenomena. In the first half of this chapter we review the steps taken by Hodgkin and Huxley to derive a mathematical description of neuronal excitability. In the second half of this chapter we discuss the application of techniques from the mathematical field of nonlinear dynamics to the analysis of neuronal excitability.

Perhaps the best way to begin a discussion of the Hodgkin–Huxley model is to analyze the *equivalent electrical circuit* employed by Hodgkin and Huxley to

Copyright 2004, Elsevier Science (USA).
All rights reserved.

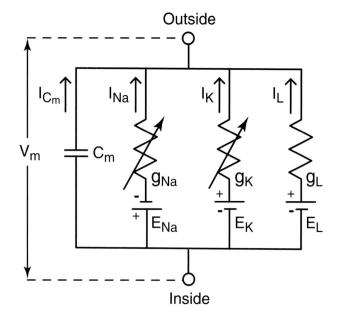

FIGURE 7.1 Equivalent electrical circuit proposed by Hodgkin and Huxley for a patch of squid giant axon. The electrical circuit has four parallel branches. The capacitive branch (C_m) represents the dielectric properties of the membrane. The variable resistors (resistors with arrows) represent voltage- and time-dependent conductances of Na^+ (g_{Na}) and K^+ (g_K), whereas the leakage conductance (g_L) is constant. Each conductance element is in series with a battery (E_{Na}, E_K, E_L) that represents its electromotive force. Adapted from Hodgkin and Huxley (1952d).

represent a patch of membrane (Fig. 7.1). In this approach, the membrane is considered to be an electrical circuit composed of a capacitive element (C_m) in parallel with conductances (g), which are in series with a battery. The capacitive element represents the dielectric properties of the lipid bilayer of biological membranes. The conductances represent channels in the membrane through which ions can pass (see Chapter 6), and the batteries represent the electrochemical potential gradient that is associated with a given species of ion (see Chapter 5). In the equivalent circuit, the current that flows across the membrane (I_m) has two major components, one associated with charging the membrane capacitance (I_{Cm}) and one associated with the movement of ions across the membrane (I_{ionic}). Thus,

$$I_m = I_{Cm} + I_{ionic} = C_m \frac{dV_m}{dt} + I_{ionic}, \qquad (7.1)$$

where V_m is potential across the membrane, and t is time.

Separating the Membrane Current into its Major Components

As a first step in their analysis, Hodgkin and Huxley separated the membrane current into its

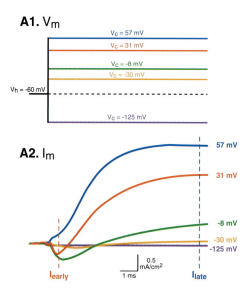

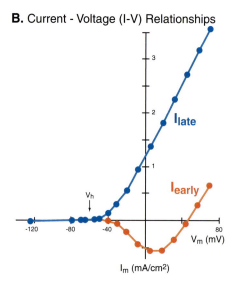

FIGURE 7.2 Membrane currents (I_m) recorded with voltage clamp of squid axon. Data are from an experiment by Hodgkin, *et al.* (1952) (i.e., axon 41 at 3.8°C). (**A1**) In this simple voltage-clamp protocol, the membrane potential (V_m) was held near its rest potential, which is taken to be –60 mV (i.e., $V_h = V_m = -60$ mV). Voltage-clamp commands (V_c) were used to step the membrane potential to various hyperpolarized and depolarized potentials, which are indicated above each trace. (Hodgkin and Huxley used a different convention for labeling membrane potentials and displaying currents; see Box 7.1 for an explanation. In this and all subsequent figures, the data have been modified to reflect modern conventions.) (**A2**) Responses elicited by a series of voltage-clamp steps. Successive current traces have been superimposed. Inward current is indicated by a downward deflection and outward current is indicated by an upward deflection. (Only the response immediately prior to and during the initial phase of the voltage-clamp step is shown. The response at the end of the voltage-clamp step is not illustrated.) The time course, direction, and magnitude of I_m vary with V_m. (**B**) Current–voltage relationship (i.e., *I–V plot*) from voltage-clamp experiment illustrated in (A). [Additional data points are included in the *I–V* plot that were not illustrated in (A).] The magnitude of the currents at 0.5 and 8 ms (I_{early} and I_{late}, respectively) are plotted as functions of V_m. Adapted from Hodgkin *et al.* (1952).

major components (i.e., capacitative, I_{Cm}, and ionic, I_{ionic}, currents). It was the voltage-clamp technique that provided the method for this separation. By using a feedback circuit (see Chapter 5), the voltage-clamp technique maintains (i.e., "clamps") the membrane potential at a designated voltage (Hodgkin *et al.*, 1952). In a typical experiment (Fig. 7.2), the membrane potential was held at voltage (i.e., the *holding voltage*, V_h), which was often near the resting potential of the neuron, and voltage-clamp commands (i.e., the *command voltage*, V_c) stepped the membrane potential from V_h to various depolarized or hyperpolarized levels for a few milliseconds and, then, back to V_h. The command voltage is so named because it determines the membrane potential that will be maintained by the voltage-clamp amplifier during the brief voltage step. The stepwise depolarizations (or hyperpolarizations) of the membrane have two advantages for measuring ionic current. First, except for the brief moment of transition between V_h and V_c, the membrane potential is constant (i.e., $dV_m/dt = 0$). Thus, the capacitive current is eliminated (i.e., $I_{Cm} = C_m \times 0$ in Eq. 7.1) and the ionic current can be measured in isolation. Second, with the voltage constant, the time dependency of the ionic currents can be measured.

The voltage-clamp technique provides a quantitative measure of the ionic currents that flow through an excitable membrane such as the squid giant axon. An example of the types of membrane currents that were recorded by Hodgkin and Huxley is presented in Fig. 7.2 (see also Chapter 5). When the membrane potential was depolarized, a transient inward current (i.e., the flow of positive charge into the cell) was observed. This transient inward current (i.e., I_{early}) was followed by a sustained outward current (i.e., I_{late}), which represents the flow of positive charge out of the cell. The time course and magnitude of these ionic currents depended markedly on the magnitude of the depolarizing step (i.e., V_c). For example, the amplitude of I_{early} increased and then decreased as V_c became more positive, whereas I_{late} increased monotonically with V_c (Fig. 7.2B). Because of the voltage dependency of some membrane currents, the underlying channels, through which the ions flow, are often referred to as *voltage-gated channels*.

The second step in the analysis was to separate the complex ionic current into its components. By altering the ionic composition of the external solutions (e.g., substituting choline for Na^+) (Hodgkin and Huxley, 1952a), the inward current (i.e., I_{early}) was eliminated (Fig. 7.3), leaving only outward current (i.e., I_{late}), which was assumed to be carried by K^+ (Huxley, 1951). Thus, the ionic current was subdivided into two primary components: one carried by Na^+ (I_{Na}) and another carried by K^+ (I_K). In addition,

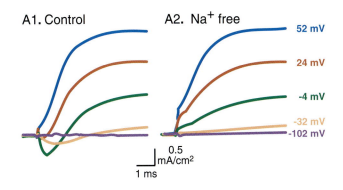

A1. Control   **A2. Na⁺ free**

52 mV
24 mV
-4 mV
-32 mV
-102 mV

0.5 mA/cm²
1 ms

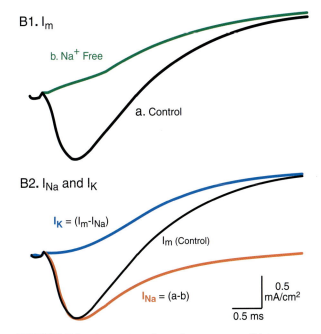

B1. I_m

b. Na⁺ Free

a. Control

B2. I_{Na} and I_K

$I_K = (I_m - I_{Na})$

I_m (Control)

$I_{Na} = (a-b)$

0.5 mA/cm²
0.5 ms

FIGURE 7.3 Separation of membrane current (I_m) into components carried by Na^+ (I_{Na}) and K^+ (I_K). Data are from an experiment by Hodgkin and Huxley (1952a) (i.e., axon 21 at 8.5°C). (**A**) Total membrane currents (I_m) were measured in control (**A1**) and Na^+-free (**A2**) solutions. The membrane potential was held at –60 mV, and was stepped to the various potentials indicated to the right of each trace. (Only the response immediately prior to and during the voltage-clamp step is illustrated.) In control solutions, command voltages that depolarized the membrane potential elicited an inward current followed by an outward current. After the Na^+ in the solution was replaced with impermeant choline ions, the inward component of I_m was abolished and only the outward component remained. (**B1**) Enlarged view of the membrane currents elicited by the voltage-clamp step to –4 mV, in (A). Membrane currents were first elicited in control solution (i.e., trace **a**) and again after removal of Na^+ (trace **b**), which blocked the inward component of I_m. (**B2**) Subtracting the response in the Na^+-free solution from the response in control solution (i.e., trace a – trace b) revealed the amplitude and time course of the Na^+-dependent component of I_m (i.e., I_{Na}). Similarly, I_K could be isolated by subtracting I_{Na} from I_m. Adapted from Hodgkin and Huxley (1952a).

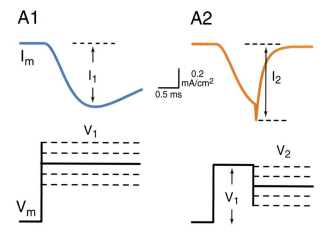

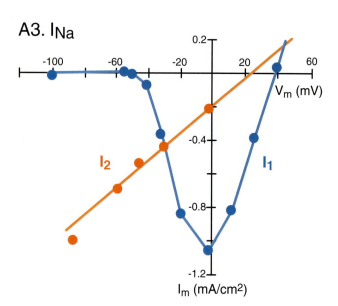

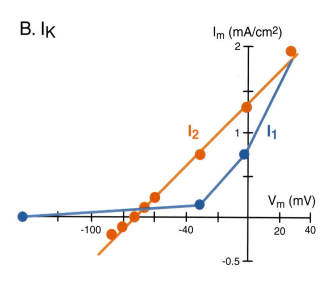

FIGURE 7.4 Instantaneous current–voltage (I–V) relationships for the early and late components of I_m (i.e., I_{Na} and I_K, respectively). **(A1)** In a simple voltage-clamp protocol, the membrane potential is stepped from a holding potential to various depolarized (or hyperpolarized) command voltages (V_1), which activate ion channels and elicit the flow of membrane currents (I_1). **(A2)** To measure the instantaneous I–V relationship, two command-voltage steps are used. The first voltage-clamp step (V_1) is to a fixed potential. In studies of Na^+ channels, $V_1 = -31$ mV, and in studies of K^+ channels, $V_1 = 24$ mV. V_1 was very brief, and had a duration of ~ 1 ms. The second voltage-clamp step (V_2), which was relatively long, varied between –100 mV and 50 mV. The instantaneous current (I_2) is measured immediately (i.e., within ~30 μs) of the second voltage-clamp step. **(A3)** Current–voltage relationships for the early inward component of I_m (i.e., I_{Na}). The current–voltage relationship of the maximum inward current during a single voltage-clamp step [i.e., I_1 in (A1)] is extremely nonlinear (blue curve). In contrast, the current–voltage relationship of the instantaneous current at the beginning of the second voltage-clamp step [i.e., I_2 in (A2)] is approximately linear (red curve). Data are from an experiment by Hodgkin and Huxley (1952b) (i.e., axon 31 at 4°C). **(B)** Current–voltage relationships for the late outward component of I_m (i.e., I_K), which were measured while the axon was bathed in Na^+-free saline. The current–voltage relationship of current measured ~0.6 ms after the beginning of the first voltage-clamp step (i.e., I_1) versus V_1 is nonlinear (blue trace), whereas the current–voltage relationship of the instantaneous current (i.e., I_2) is approximately linear (red trace). Data are from an experiment by Hodgkin and Huxley (1952b) (i.e., axon 26 at 20°C). The linear I–V relationships of the instantaneous currents (i.e., I_2) were in striking contrast to the extremely nonlinear I–V relationships that were obtained when the currents were measured at later intervals (i.e., I_1). The curvature of the I_1 I–V relationships [i.e., blue traces in (A3) and (B)] reflects the voltage- and time-dependent opening of Na^+ and K^+ channels. In contrast, the linear nature of the instantaneous current–voltage relationships [i.e., red traces in (A3) and (B)] indicate that the flow of ionic currents in open channels obeys Ohm's law. Adapted from Hodgkin and Huxley (1952b).

there is a small component that is referred to as the *leakage current* (I_l), which represents ions flowing through non-voltage-gated channels. Thus, the total ionic current is expressed as

$$I_{ionic} = I_{Na} + I_K + I_l. \tag{7.2}$$

In addition to separating the ionic current into its components, it was necessary to determine the relationship between ionic current and membrane potential at a constant permeability. To examine this issue, Hodgkin and Huxley developed a voltage-clamp protocol that measured what they referred to as the *instantaneous current–voltage (I–V) relationship* (Hodgkin and Huxley, 1952b). In this protocol, two voltage-clamp commands were applied to the axon (Fig. 7.4). The first voltage-clamp command (V_1) was brief and had a fixed amplitude, whereas the amplitude of the second voltage-clamp step (V_2) was varied. V_1 served to activate (i.e., open) the Na^+ channels (i.e., the channels that mediate I_{early} in Fig. 7.2).

After the Na^+ channels were activated by V_1, the membrane potential was suddenly stepped to V_2 and the current (I_2) was measured within the first few microseconds of V_2 (i.e., at the instant the membrane potential changed from V_1 to V_2). (Because I_2 is measured at the "tail end" of V_1, such currents are also referred to as *tail currents*.) The purpose of this protocol was to measure the current without time-dependent influences. Immediately after the step from V_1 to V_2, the level of activation (and inactivation) is constant because it has not yet had time to change. Thus, only the driving force $\Delta V = (V_m - E_{ion})$, where $V_m = V_2$, differs for different values of V_2. The results indicated that the I–V relationship of I_2 was approximately linear. The instantaneous I–V relationship of K^+ channels was also studied, and a similar linear relationship was observed. These results indicated that under normal ionic conditions the flow of ionic current in open Na^+ and K^+ channels obeys *Ohm's law* (i.e, $\Delta V = I \times R$, where $G = 1/R$, and R stands for resistance and G for its inverse, conductance). It therefore follows, that by using their empirical measurements of I_{Na} and I_K, Hodgkin and Huxley were able to determine g_{Na} and g_K (see below).

Ohm's law implies that the individual ionic currents in Eq. 7.2 were proportional to the conductance (i.e., g) times the *driving force* (i.e., the difference between the membrane potential, V_m, and the Nernst potential, E_{ion}, for a given ion species), resulting in a general equation for an ionic current of the form

$$I_{ion}(V_m, t) = g_{ion}(V_m, t)\,(V_m - E_{ion}), \tag{7.3}$$

where $I_{ion}(V, t)$ is the current created by the movement of a given species of ion across the membrane, $g_{ion}(V, t)$ represents the voltage- and time-dependent conductance (i.e., permeability) of the membrane to that ionic species, and E_{ion} is the Nernst potential of the ion (see Box 5.1). Thus, I_{Na}, I_K and I_l were described by

$$I_{Na}(V_m, t) = g_{Na}(V_m, t)\,(V_m - E_{Na}), \tag{7.4}$$

$$I_K(V_m, t) = g_K(V_m, t)\,(V_m - E_K), \tag{7.5}$$

$$I_l = g_l\,(V_m - E_l). \tag{7.6}$$

After substitution of Eq. 7.2 into Eq. 7.1, the description of membrane current becomes

$$I_m = C_m \frac{dV_m}{dt} + I_{Na} + I_K + I_l. \tag{7.7}$$

Finally, Eq. 7.7 can be expanded by including Eqs. 7.4, 7.5, and 7.6, and becomes

$$I_m = C_m \frac{dV_m}{dt} + g_{Na}(V_m, t)(V_m - E_{Na}) +$$

$$g_K(V_m, t)(V_m - E_K) + g_l(V_m - E_l). \tag{7.8}$$

The final stage of the analysis was to characterize the active conductances, i.e., $g_{Na}(V, t)$ and $g_K(V, t)$.

Analyzing the Time and Voltage Dependency of Ionic Conductances

To analyze the ionic conductances, it was first necessary to devise a method for obtaining measures of g_{Na} and g_K. The linear nature of the instantaneous I–V relationship for I_{Na} and I_K (see above) indicated that these ionic currents and their underlying ionic conductances were related by Ohm's law. This provided Hodgkin and Huxley with the means of calculating g_{Na} and g_K. From Eqs. 7.4 and 7.5, it is possible to define ionic conductances as

$$g_{Na}(V_m, t) = \frac{I_{Na}(V_m, t)}{V_m - E_{Na}}, \tag{7.9}$$

$$g_K(V_m, t) = \frac{I_K(V_m, t)}{V_m - E_K}. \tag{7.10}$$

Thus, changes in the conductances g_{Na} and g_K during a voltage-clamp step could be calculated by applying Eqs. 7.9 and 7.10 to separated ionic currents. Figure 7.5 illustrates data obtained by Hodgkin and Huxley in which the magnitude and time course of g_K and g_{Na} were calculated for two different voltage-clamp steps. As discussed in Chapter 5, these conductances have several striking properties. First, the magnitude of the conductances increases with more positive values of V_c. Second, the rising phase of the conductances become more rapid with increasing V_c. Third, there is a delay in the onset of the change in the conductances, particularly for g_K. Finally, the increase in g_{Na} is transient, whereas the increase in g_K is not.

To explain these experimental data, Hodgkin and Huxley suggested a model that could account for the voltage- and time-dependent properties of g_K and g_{Na}. This model is often referred to as the *gate model* (Fig. 7.6; see also Chapter 5). The model assumes that the macroscopic conductances as measured with the voltage-clamp procedure arise from the combined effects of many individual ion channels, each with a microscopic conductance to a specific species of ion (i.e., K^+ or Na^+). Each individual channel has one or more "gates" that regulate the flow of ions through the channel. Each gate can be in one of two states: open or closed. When all the gates for a particular channel are open, ions can pass through the channel

A

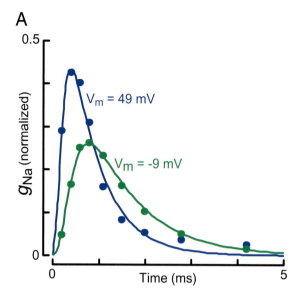

B

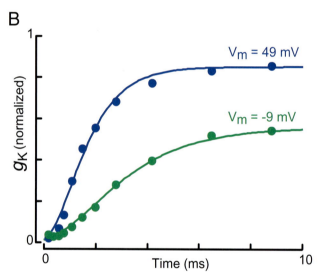

FIGURE 7.5 Experimental voltage-clamp data illustrating the voltage- and time-dependent properties of the g_{Na} (**A**) and g_K (**B**). By recording membrane currents in solutions of different ionic composition (e.g., normal seawater vs Na$^+$-free seawater), Hodgkin and Huxley (1952a) were able to isolate I_{Na} and I_K (see also Fig. 7.3 and Chapter 5). The conductances were calculated using Eqs. 7.9 and 7.10. Although Hodgkin and Huxley examined a wide range of membrane potentials, the results from only two voltage-clamp steps are illustrated. The blue traces are responses elicited by a command voltage (V_c) to 49 mV, and the green traces are responses elicited during a voltage-clamp step to –9 mV. The voltage-clamp steps are not illustrated, but they begin at $t = 0$ and extend beyond the end of the illustrations. The filled circles are data from an experiment by Hodgkin and Huxley (i.e., axon 17) and the solid lines are best-fit curves for Eqs. 7.24 and 7.32. Adapted from Hodgkin and Huxley (1952d).

(i.e., the single-channel conductance is > 0). If any of the gates are closed, ions cannot pass through the channel (i.e., the single channel conductance is 0). The

status of a gate (i.e., open vs closed) was assumed to be controlled by distribution of one or more charged "particles" within the membrane (i.e., *gating particles* or *gating charges*). These gating particles act as "molecular voltmeters," and as the electrical field across the membrane changes, the distribution of these gating particles is altered such that the gates transition between states. A gate is open only when all of the particles that are associated with the gate are in a permissive state. The possible molecular structures responsible for gating particles are described in Chapters 5 and 6.

Mathematically, the voltage dependence of channel opening (and closing) can be derived using the *Boltzmann equation* of statistical mechanics, which describes the equilibrium distribution of independent particles in force fields. From the Boltzmann principle, the proportion of gating particles that are at a location associated with a *permissive state* (P_i) is related to the proportion of gating particles in *nonpermissive* locales (P_0) by the function

$$\frac{P_i}{P_o} = \exp\left(\frac{w + zeV}{kT}\right),\qquad(7.11)$$

where w is the work required to move the gating particle from the nonpermissive state to the permissive state, z is the valence of the particle, e is the elementary charge of the particle, V is the potential difference between the inside and outside of the membrane, k is Boltzmann's constant, and T is the absolute temperature. Since $P_i + P_0 = 1$, Eq. 7.11 can be rearranged to give the proportion of gating particles in the permissive state as a function of voltage:

$$P_i = \frac{1}{1 + \exp\left(\frac{-(w + zeV)}{kT}\right)}.\qquad(7.12)$$

Equation 7.12 quantifies the voltage dependence of gating in the system and is sometimes referred to as the *activation function of a channel*. As V increases in Eq. 7.12 (i.e., the membrane potential is depolarized), the proportion of gating particles in the permissive state approaches unity (i.e., $P_i \rightarrow 1$). With an increasing proportion of gating particles in the permissive state, a greater number of channels are likely to be open, and thus, the macroscopic membrane conductance increases. Figure 7.7 illustrates semilogarithmic plots of Eq. 7.12 that are superimposed on the maximum values of g_{Na} and g_K that were measured during a voltage-clamp experiment. The two curves are very similar in shape. One of the most striking properties of the curves is the extreme steepness of the relation between ionic conductance and mem-

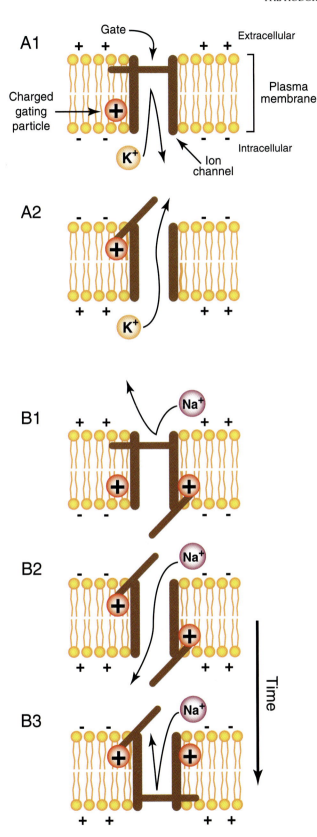

FIGURE 7.6 Schematic representation of the gate model that was proposed by Hodgkin and Huxley (1952d). The gate model assumes that many individual ion channels, each with a small ionic conductance, determine the behavior of the macroscopic membrane conductance. The ion channels have "gates" that are controlled by voltage-sensitive gating charges or particles. If the gating particles are in the permissive state, then the gates are open and ions can pass through the channel. Otherwise, the gates are closed and passage of ions through the channel is blocked. In this example, each gate is regulated by a single gating particle and the gating particles are assumed to be positively charged. Other scenarios are possible, however. (**A**) K+ channels are regulated by a single activation gate, which is embodied in the Hodgkin–Huxley model as n (see Eq. 7.33). At the resting potential (as indicated by the minus signs on the intracellular surface and the plus signs on the extracellular surface), the gating particles are distributed primarily at locations within the membrane that are nonpermissive; i.e., the gate is closed (**A1**). As the membrane potential is depolarized (**A2**) (note translocation of – and + signs), the probability increases that a gating particle will be located in a position that is permissive; i.e., the gate is open (i.e., activation). Potassium ions (K+) can flow through the open channel. (**B**) Na+ channels are regulated by two gates: an activation gate and an inactivation gate. These two gates are embodied in the Hodgkin–Huxley model as m and h, respectively (see Eq. 7.25). Unlike the activation gate, the inactivation gate is normally open at the resting potential (**B1**). On depolarization (**B2**), the probability increases that the activation gate will open, whereas the probability that the inactivation gate will remain open decreases. While both gates are open, Na+ passes through the channel. As the depolarization continues, however (**B3**), the inactivation gate closes and ions can no longer pass through the channel (i.e., inactivation).

brane potential. At low depolarizations (i.e., near the resting potential), the curves approach straight lines. Since the ordinate is plotted on a logarithmic scale, this means that the peak conductances increase exponentially with membrane depolarization, until at high depolarizations, the curves reach a maximal value and flatten (i.e., saturate). These maximal conductances are denoted $\bar{g}_{Na}$ and $\bar{g}_{K}$. To describe the time and voltage dependency of $\bar{g}_{Na}$ and $\bar{g}_{K}$ (see Fig. 7.5), $\bar{g}_{Na}$ and $\bar{g}_{K}$ must be multiplied by coefficients that represent the fractions of the maximum conductances expressed at any given time and at any given membrane potential. Thus, g_{Na} and g_{K} can be written in general form as

$$g_{Na}(V_m, t) = y_{Na}(V_m, t)\bar{g}_{Na}, \tag{7.13}$$

$$g_{K}(V_m, t) = y_{K}(V_m, t)\bar{g}_{K}, \tag{7.14}$$

where y_{Na} and y_{K} are functions of one or more gating variables (y_i) that vary between zero and one.

The excellent agreement between the observed voltage dependency of g_{Na} and g_{K} and the Boltzmann equation (i.e., Fig. 7.7) lent support to Hodgkin and Huxley's proposition that changes in ionic permeability depended on the movement of some component of the membrane that behaved as though it were a

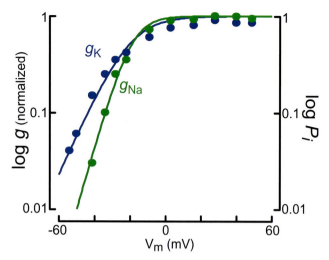

FIGURE 7.7 Voltage dependence of ionic conductances. The maximum g_{Na} and g_K were measured at several different membrane potentials under voltage clamp. The filled circles represent data from an experiment by Hodgkin and Huxley (i.e., axon 17) and the solid lines represent best-fit curves of the form given in Eq. 7.12. (The calculations assumed $w = 0$.) The most striking feature of the conductances is the extreme steepness of the relation between ionic conductance and the membrane potential. A depolarization of only ~4 mV can increase g_{Na} by e-fold ($e \approx 2.72$), while the corresponding figure for g_K is ~5 mV. Adapted from Hodgkin and Huxley (1952a).

charged particle. In addition to providing a possible mechanism for the voltage dependency (i.e., the non-linear I–V relations in Fig. 7.2) of membrane conductance, the gate model also offered a possible explanation for the time dependency of ionic conductances (Fig. 7.8). The gate model assumed that the rate of change in an ionic conductance following a step depolarization was governed by the rate of redistribution of the gating particles within the membrane and that the transitions between *permissive* (i.e., open or activated) and *nonpermissive* (i.e., closed or deactivated) *states* can be described by the first-order kinetic model

$$(1-y) \underset{\beta_y(V_m)}{\overset{\alpha_y(V_m)}{\rightleftharpoons}} y, \tag{7.15}$$

where y is the probability of finding a single gating particle in the permissive state, $(1 - y)$ is the probability of finding the particle in the non permissive state, and $\alpha_y(V_m)$ and $\beta_y(V_m)$ are voltage-dependent rate constants describing the rate at which a particle moves from the nonpermissive to the permissive (α_y) state and from the permissive to the non permissive (β_y) state. If a gate is regulated by a single particle, then the probability that the gate will open over a short interval of time is proportional to the probability of finding the gate closed multiplied by the opening rate constant, i.e., $\alpha_y(V_m)(1 - y)$. Conversely,

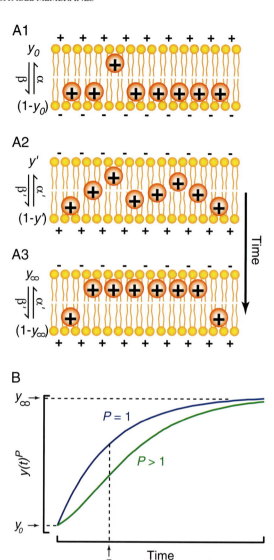

FIGURE 7.8 Kinetics of the increasing probability of channel activation during a voltage-clamp step. (**A**) Changes in the membrane potential alter the distribution of gating particles within the membrane. In this example, gating particles are assumed to have a positive charge and the permissive site for opening a gate is assumed to be at the outer surface of the membrane (see Fig. 7.6). (**A1**) At the resting potential, the probability of a gating particle being in a permissive state (y_0) is low, whereas the probability of a gating particle being in a nonpermissive state ($1 - y_0$) is high. (**A2**) Immediately following a depolarization (note the translocation of the minus and plus signs), the gating particles begin to redistribute. The rate of movement of the gating particles within the membrane is described by Eq. 7.16. The voltage-dependent rate constant α represents the rate at which particles move from the inner to the outer surface (i.e., from nonpermissive to permissive states), and β is the rate of reverse movement. (**A3**) Eventually, the distribution reaches a steady state in which the probability of a gating particle being in a permissive state has increased (y_∞). (**B**) The kinetics of the redistribution of gating particles is described by Eq. 7.18. If a channel is controlled by a single gating particle (i.e., $P = 1$ in Eq. 7.20), then the solution is a simple exponential. If a channel is controlled by several identical and independent gating particles (i.e., $P > 1$), then a delay is noted in the change in conductance.

the probability that the gate will close over a short interval of time is proportional to the probability of finding the gate open multiplied by the closing rate constant, i.e., $\beta_y(V_m)y$. The rate at which the open probability for a single gating particle changes following a change in membrane potential is the difference of these two terms:

$$\frac{dy}{dt} = \alpha_y(V_m)(1-y) - \beta_y(V_m)y. \qquad (7.16)$$

The first term in Eq. 7.16 [$\alpha_y(V_m)(1-y)$] describes the opening (i.e., *activation*) of the gate, and the second term [$\beta_y(V_m)y$], describes the closing (i.e., *deactivation*) of the gate.

Although y is usually taken to represent the probability of finding a single gate in the open state, it can also be interpreted as the fraction of open gates in a large population of gates, and $1-y$ would be the fraction of gates in the closed state. If the membrane potential is voltage clamped to some fixed value, then the fraction of gates in the open state will eventually reach a steady-state value (i.e., $dy/dt = 0$) as $t \to \infty$. Solving Eq. 7.16 for the steady-state value [$y_\infty(V_m)$] yields

$$y_\infty(V_m) = \frac{\alpha_y(V_m)}{\alpha_y(V_m) + \beta_y(V_m)}. \qquad (7.17)$$

During a voltage-clamp step, the time course for approaching this steady state (i.e., the solution of a first-order kinetic expression like Eq. 7.16) is described by a simple exponential function,

$$y(t) = y_\infty(V_m) - (y_\infty(V_m) - y_0)\exp(-t/\tau_y(V_m)), \qquad (7.18)$$

where y_0 is the initial value of y (i.e., the value of y at the holding potential, V_h) and the time constant, $\tau_y(V_m)$, is given by

$$\tau_y(V_m) = \frac{1}{\alpha_y(V_m) + \beta_y(V_m)}. \qquad (7.19)$$

If P independent and identical gating particles are involved in gating a channel, then the probability that all of the particles will simultaneously be in the permissive state is the product of their individual probabilities, $y(t)^P$. Thus, substituting Eq. 7.19 and including the possibility of more than one gating particle changes the time course for approaching the steady state to

$$y(t)^P = \left(y_\infty(V_m) - (y_\infty(V_m) - y_0) \right.$$
$$\left. \exp\left(\frac{-(\alpha_y(V_m) + \beta_y(V_m))}{t}\right) \right)^P. \qquad (7.20)$$

As the number of gating particles increases (i.e., $P > 1$), a delay and a sigmoidal rising phase are introduced to the time course of $y(t)^P$ (Fig. 7.8B).

Instead of using the rate constants α_y and β_y, Eq. 7.16 can be written in terms of the steady-state value $y_\infty(V_m)$ (i.e., Eq. 7.17) and the voltage-dependent time constant (i.e., Eq. 7.19). Thus, Eq. 7.16 becomes

$$\frac{dy}{dt} = \frac{y_\infty(V_m) - y_0}{\tau_y(V_m)}. \qquad (7.21)$$

Equation 7.21 indicates that for a fixed voltage (V_m), the gating particle (y) approaches the steady-state value [$y_\infty(V_m)$] exponentially with the time constant $\tau_y(V_m)$. Although Eqs. 7.16 and 7.21 are equivalent, Eq. 7.21 is simpler to interpret and is more conveniently fit to experimental data. In addition, from Eqs. 7.17 and 7.19, it is possible to calculate α_y and β_y from experimental data:

$$\alpha_y(V_m) = \frac{y_\infty(V_m)}{\tau_y(V_m)}, \qquad (7.22)$$

$$\beta_y(V_m) = \frac{1 - y_\infty(V_m)}{\tau_y(V_m)}, \qquad (7.23)$$

where y_∞ and τ_y are measured empirically (see below).

In the discussion provided above, the descriptions of gating particles (i.e., Eqs. 7.13–7.21) have been presented using generalized notation that can be applied to a wide variety of conductances. The key remaining tasks in the development of the Hodgkin–Huxley model are determining the number and type of gating particles that regulate g_{Na} and g_K and quantitatively describing the voltage dependency of the rate constants that govern these gating particles. In brief, this was done by measuring the $\tau_y(V_m)$ and $y_\infty(V_m)$ from the time records of g_{Na} and g_K (e.g., Fig. 7.5) and then calculating the rate constants using Eqs. 7.22 and 7.23. The values for the rate constants were plotted as functions of membrane potential and the data were fit with empirically derived exponential functions (see below).

Characterizing the Na⁺ Conductance

As illustrated in Fig. 7.5A, the dynamics of g_{Na} are complex. This led Hodgkin and Huxley to postulate that g_{Na} was regulated by two types of gates. One gate regulated the activation of g_{Na} and was termed m, and the other gate regulated inactivation and was termed h. To analyze the empirical voltage-clamp data and extract values for g_{Na}, τ_m, τ_h, and P (see Eqs. 7.13, 7.20,

A

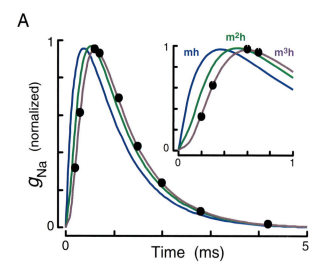

B

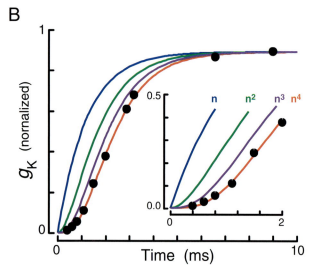

FIGURE 7.9 Estimating the number of gating particles that regulate each Na$^+$ and K$^+$ channel. The filled circles represent data from an experiment by Hodgkin and Huxley (i.e., axon 17). The insets illustrate an enlargement of the first few milliseconds of the response. The ionic conductances, g_{Na} (**A**) and g_K (**B**), increase with a delay during a voltage-clamp step. This observation suggests that $P > 1$ in Eq. 7.20 (see also Fig. 7.8). To estimate appropriate values for P, empirical data are fit (solid lines) with equations similar to Eq. 7.20 (i.e., Eqs. 7.24 and 7.32) where the value for P is increased from 1 until an adequate fit of the data is achieved. The initial delay in g_{Na} is well fit with $P = 3$, while the corresponding value for g_K is $P = 4$. Adapted from Hodgkin and Huxley (1952d).

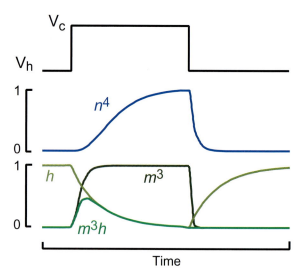

FIGURE 7.10 Temporal dynamics of n^4, h, m^3, and m^3h during a hypothetical voltage-clamp step. Initially, the membrane potential is voltage clamped at a holding potential (V_h) near the resting potential. A command voltage (V_c) briefly steps the membrane potential to a depolarized level. During the voltage step, n and m are being activated, and thus, the curves for n^4 and m^3 followed the $(1 - \exp(-t/\tau))^p$ time course. Conversely, h is inactivated by depolarization, and thus, the curve for h follows the $\exp(-t/\tau)$ time course. In this schematic, the values for the time constants τ_m, τ_n, and τ_h were adjusted to the ratio 1:4:5, and the duration of the voltage-clamp step was the equivalent of $20 \times \tau_m$. The curves illustrate how n^4 and m^3h closely imitate the time courses of g_{Na} and g_K that are observed empirically (compare with Fig. 7.9).

where $g'_{Na}(V_c)$ is the value that g_{Na} would attain during the voltage-clamp step if h remained at its resting value (i.e., $h = 1$). This extrapolation is illustrated in Fig. 7.10. During the voltage-clamp step to V_c, m changes from its resting value, which in this example is 0, to a new steady-state level, which in this example is 1 (the curve labeled m^3). Conversely, the resting level of h in this example is 1, and during the voltage-clamp step to V_c, h approaches a new steady-state value, which in this example is 0 (curve labeled h). In the absence of h (i.e., the curve labeled m^3) the activation of g_{Na} attains a higher level and is maintained as compared with when h is included in the calculation (i.e., the curve labeled m^3h). The multiplicative interaction of the activation and inactivation gating variables (i.e., m and h) produces the transient response of g_{Na} during a voltage-clamp step and reduces the apparent magnitude of activation g_{Na}. The best fit of Eq. 7.24 to voltage-clamp data provided Hodgkin and Huxley with measurements of g_{Na}, τ_m, and τ_h at each value of V_c. In addition, the best fit of Eq. 7.24 to the voltage-clamp data also indicated that $P = 3$ (see Eq. 7.20), which suggested that three gating particles regulated the activation gate of g_{Na} (see Fig. 7.8) and which produced a

7.22, and 7.23), an exponential function was fit to the time course of g_{Na} during voltage-clamp steps to various membrane potentials. Hodgkin and Huxley used an exponential function to fit the time course of g_{Na} during a voltage-clamp step (Fig. 7.9A),

$$g_{Na}(t) = g'_{Na}(V_c)\left(1 - e^{-t/\tau_m}\right)^3 \exp\left(-t/\tau_h\right), \qquad (7.24)$$

sigmoidal time course of $m(t)$ (see inset in Fig. 7.9). Thus, g_{Na} was described by

$$g_{Na} = \bar{g}_{Na} m^3 h, \tag{7.25}$$

where

$$\frac{dm}{dt} = \alpha_m(V_m)(1-m) - \beta_m(V_m)m, \tag{7.26a}$$

$$\frac{dh}{dt} = \alpha_h(V_m)(1-h) - \beta_h(V_m)h, \tag{7.26b}$$

or the equivalent expressions

$$\frac{dm}{dt} = \frac{m_\infty(V_m) - m}{\tau_m(V_m)}, \tag{7.27a}$$

$$\frac{dh}{dt} = \frac{h_\infty(V_m) - h}{\tau_h(V_m)}, \tag{7.27b}$$

where

$$m_\infty(V_m) = \frac{\alpha_m(V_m)}{\alpha_m(V_m) + \beta_m(V_m)} \tag{7.28a}$$

$$\tau_m(V_m) = \frac{1}{\alpha_m(V_m) + \beta_m(V_m)}, \tag{7.28b}$$

$$h_\infty(V_m) = \frac{\alpha_h(V_m)}{\alpha_h(V_m) + \beta_h(V_m)}, \tag{7.29a}$$

$$\tau_h(V_m) = \frac{1}{\alpha_h(V_m) + \beta_h(V_m)}. \tag{7.29b}$$

The time constants [$\tau_m(V_m)$ and $\tau_h(V_m)$] and the steady-state values of m and h [$m_\infty(V_m)$ and $h_\infty(V_m)$] were measured by fitting Eq. 7.24 to the voltage-clamp records of I_{Na} (Fig. 7.11A), and from these data the rate constants [$\alpha_m(V_m)$, $\beta_m(V_m)$, $\alpha_h(V_m)$, and $\beta_h(V_m)$] were calculated using Eqs. 7.22 and 7.23. The values for the rate constants were plotted as functions of voltage, and expressions for the voltage dependency of the rate constants were derived. Empirically, Hodgkin and Huxley derived for the rate constants the equations

$$\alpha_m(V_m) = \frac{0.1(V_r - V_m + 25)}{\exp\left(\dfrac{V_r - V_m + 25}{10}\right) - 1} \tag{7.30a}$$

$$\beta_m(V_m) = 4 \exp\left(\frac{V_r - V_m}{18}\right), \tag{7.30b}$$

$$\alpha_h(V_m) = 0.07 \exp\left(\frac{V_r - V_m}{20}\right), \tag{7.31a}$$

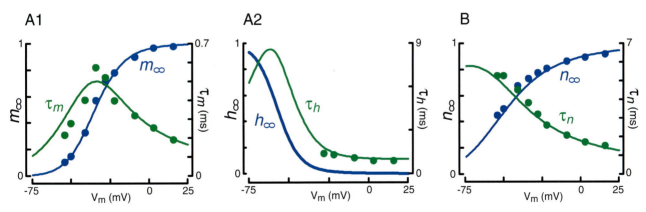

FIGURE 7.11 Voltage dependence of gating variables and their respective time constants. The time constants (τ_m, τ_h, and τ_n) and the steady-state activation (m_∞ and n_∞) and inactivation (h_∞) gating variables are plotted as functions of membrane potential (V_m) for the squid giant axon at 6.3°C. The filled circles represent data from an experiment by Hodgkin and Huxley (i.e., axon 17). The smooth curves were calculated using Eqs. 7.28 (A1), 7.29 (A2), and 7.36 (B). (**A1**) Activation gating variable (m_∞) and its time constant (τ_m) for Na⁺ conductance. (**A2**) inactivation gating variable (h_∞) and its time constant (τ_h) for Na⁺ conductance. (**B**) Activation gating variable (n_∞) and its time constant (τ_n) for K⁺ conductance. The steady-state inactivation h_∞ is a monotonically decreasing function of voltage, whereas the steady-state activation variables m_∞ and n_∞ increase with membrane depolarization. Note that in Eqs. 7.25 and 7.33, the activation gating variables for Na⁺ (i.e., m) and K⁺ (i.e., n) are raised to the third and fourth powers, respectively. Thus, the functional voltage-dependent activation of I_{Na} and I_K is much steeper than represented in this illustration. Adapted from Hodgkin and Huxley (1952d).

$$\beta_h(V_m) = \frac{1}{\exp\left(\dfrac{V_r - V_m + 30}{10}\right) + 1}. \qquad (7.31b)$$

where V_r is the resting potential (which is usually taken to be either –60 or –65 mV), V_m is the membrane potential in units of millivolts, and the rate constants (i.e., α_i and β_i) are expressed per millisecond. (It should be noted that the conventions used by Hodgkin and Huxley to describe voltage were different from those in use today. Box 7.1 explains this difference. Also note that the resting potential V_r is sometimes referred to as E_r.)

Although not indicated in Eqs. 7.26 and 7.27, the kinetics of the ionic conductances are influenced by temperature. Hodgkin and Huxley performed most of their voltage-clamp experiments with the preparations cooled to ~6°C. The cooler temperature slowed the kinetics of the membrane currents, which, in turn, made them easier to record and analyze. The standard parameters of the Hodgkin–Huxley equations reflect a temperature of 6.3°C. The kinetics of the Hodgkin–Huxley equations can be adjusted to reflect some other temperature (T) by multiplying the kinetic equations by a temperature coefficient, $\Phi = Q_{10}^{\Delta T/10}$ (see below) (Hodgkin and Huxley, 1952d; FitzHugh, 1966).

Characterizing the K+ Conductance

Unlike I_{Na}, I_K did not inactivate during voltage-clamp steps (e.g., Fig. 7.3B2). Thus, Hodgkin and Huxley postulated that g_K was governed by a single type of gate, which governed the activation of g_K and

---

BOX 7.1

UPDATING THE PARAMETERS IN THE HODGKIN–HUXLEY EQUATIONS

Although the original papers of Hodgkin *et al.*, 1952; Hodgkin and Huxley, 1952a–d were published more than 50 years ago, this series of papers remains influential to the present day. The papers have been reprinted on several occasions (Cooke and Lipkin, 1952; Hodgkin and Huxley, 1990; Moore and Stuart, 2000), and the detailed descriptions of what became known as the Hodgkin–Huxley equations (Cole *et al.*, 1955) are often included in modern textbooks about cellular neurophysiology and computational neuroscience (Hille, 2001; Koch, 1999; Bower and Beeman, 1998; Byrne and Schultz, 1994; Cronin, 1987; Dayan and Abbott, 2001; DeSchutter, 2001; Johnston and Wu, 1997; MacGregor, 1987; Tuckwell, 1988; Ventriglia, 1994; Weiss, 1997). Although these modern descriptions of the Hodgkin–Huxley equations are similar to those in the original publications, the present day values for the parameters often do not appear to agree with those used by Hodgkin and Huxley in 1952.

This apparent discrepancy arises from the conventions used by Hodgkin and Huxley to represent voltage. In their original series of papers, Hodgkin and Huxley chose to regard the resting potential as a positive quantity and the action potential as a negative (i.e., downward) deflection (Hodgkin *et al.*, 1952). In addition, the variable V (i.e., voltage) in the Hodgkin–Huxley equations denoted the *displacement* of the membrane potential from its resting value. Thus, Hodgkin and Huxley defined V as $V = E - E_r$, where E was the absolute value of the membrane potential and E_r was the absolute value of the resting potential. With their choice of conventions, depolarizations of the membrane potential were negative values (i.e., downward deflections) and inward membrane currents had a positive sign (i.e., upward deflections). [For additional explanation of the conventions used by Hodgkin and Huxley, see Rinzel (1998).]

This aspect of the papers by Hodgkin and Huxley differs from the current practices in which the intracellular electrode measures the membrane potential with respect to an external ground (i.e., the resting potential is a negative quantity), action potentials and depolarizations are positive (i.e., upward) deflections, inward membrane currents have a negative sign (i.e., downward deflections), and the voltage variable (i.e., V_m) in modern representations of the Hodgkin–Huxley equations denotes the absolute membrane potential. The Hodgkin–Huxley equations can be recast into modern conventions for polarity and using absolute membrane potential by simple subtraction and multiplication (Palti, 1971a,b). In this chapter, the Hodgkin–Huxley equations for a spaced-clamped patch of membrane, as well as the data that are plotted in the figures, are expressed using modern conventions for V_m, where V_m is the absolute membrane potential and V_r is the absolute value for the resting potential, which is usually taken to be –60 mV. At a conceptual level, the choice of conventions for membrane currents and voltage is inconsequential. It does matter a great deal, however, when one wishes to implement and simulate the Hodgkin–Huxley model.

which they termed n. To determine values for g_K, τ_n, and P, the time course of g_K during voltage clamps to various membrane potentials was best described by

$$g_K(t) = \left(g_\infty^{\frac{1}{4}}(V_m) - \left(g_\infty^{\frac{1}{4}}(V_m) - g_0^{\frac{1}{4}} \right) \exp\left(\frac{-t}{\tau_{n(V_m)}} \right) \right)^4, \quad (7.32)$$

which indicated that four particles regulated the activation gate of g_K (Fig. 7.9B). The fourth power produces a sigmoidal time course of $n(t)$ (Fig. 7.10). Thus, g_K was described by:

$$g_K = \bar{g}_K n^4, \quad (7.33)$$

where

$$\frac{dn}{dt} = \alpha_n(V_m)(1-n) - \beta_n(V_m)n, \quad (7.34)$$

or the equivalent expression

$$\frac{dn}{dt} = \frac{n_\infty(V_m) - n}{\tau_n(V_m)}, \quad (7.35)$$

where

$$n_\infty(V_m) = \frac{\alpha_n(V_m)}{\alpha_n(V_m) + \beta_n(V_m)}, \quad (7.36a)$$

$$\tau_n(V_m) = \frac{1}{\alpha_n(V_m) + \beta_n(V_m)}. \quad (7.36b)$$

The time constant [$\tau_n(V_m)$] and steady values of n [$n_\infty(V_m)$] were measured (Fig. 7.11B), and from these

data the rate constants [$\alpha_n(V_m)$ and $\beta_n(V_m)$] were calculated (Fig. 7.12B). The empirically determined expressions for the voltage dependency of the rate constants are

$$\alpha_n(V_m) = \frac{0.01(V_r - V_m + 10)}{\exp\left(\frac{V_r - V_m + 10}{10} \right) - 1}, \quad (7.37a)$$

$$\beta_n(V_m) = 0.125 \exp\left(\frac{V_r - V_m}{80} \right), \quad (7.37b)$$

where V_r is the resting potential of the cell in millivolts, V_m is the membrane potential in millivolts, and $\alpha_n(V_m)$ and $\beta_n(V_m)$ are given per millisecond.

Simulations of the Hodgkin–Huxley Equations

By incorporating Eqs. 7.25 and 7.33 into Eq. 7.8, it is possible to write a single equation that describes the total membrane current (I_m):

$$I_m = C_m \frac{dV}{dt} + \bar{g}_{Na} m^3 h(V_m - E_{Na}) +$$

$$\cdot \bar{g}_K n^4 (V_m - E_K) + \bar{g}_l(V_m - E_l). \quad (7.38)$$

This nonlinear differential equation, in addition to the three linear differential equations that describe the temporal evolution of the rate constants (i.e., Eqs. 7.26a, 7.26b, and 7.34), constitutes the four-dimensional Hodgkin and Huxley model for a space-

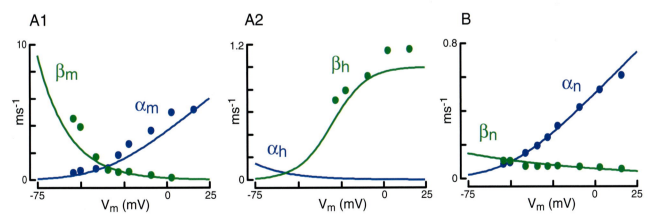

FIGURE 7.12 Voltage dependence of the rate coefficients for the Hodgkin–Huxley model of a squid giant axon at 6.3°C. The filled circles represent data from an experiment by Hodgkin and Huxley (i.e., axon 17). The smooth curves were calculated from Eqs. 7.30 (A1), 7.31 (A2), and 7.37 (B). (**A1**) Forward (α_m) and backward (β_m) rate coefficients for activation of Na$^+$ conductance. (**A2**) Forward (α_h) and backward (β_h) rate coefficients for inactivation of Na$^+$ conductance (**B**) Forward (α_n) and backward (β_n) rate coefficients for activation of K$^+$ conductance. The forward rate coefficients for activation of Na$^+$ and K$^+$ (α_m and α_n, respectively) increase with membrane depolarization because m and n gating particles move into a permissive state in response to membrane depolarization (see Fig. 7.6). Conversely, the forward rate coefficient for inactivation of Na$^+$ (α_h) decreases with membrane depolarization because the h gating particle moves into a nonpermissive state in response to membrane depolarization. Adapted from Hodgkin and Huxley (1952d).

BOX 7.2

COMPUTING SOLUTIONS TO THE HODGKIN–HUXLEY EQUATIONS

Hodgkin and Huxley conducted their voltage-clamp experiments on the squid giant axon at the Marine Biological Laboratory in Plymouth, England. Although the Plymouth laboratory was badly damaged during the great air raids of 1941, it was partially rebuilt by the time Hodgkin and Huxley arrived in July of 1949. With the help of Bernard Katz and their improved voltage-clamp apparatus, it took them only a month to obtain all of the voltage-clamp records that were used in the five papers published in 1952 (Hodgkin et al., 1952; Hodgkin and Huxley, 1952a–d). On returning to the University of Cambridge, they spent the next 2 years analyzing the data and preparing the manuscripts. By March 1951, they had settled on a set of the equations and parameters that adequately described the time course, magnitude, and voltage dependency of the membrane currents that were observed during voltage-clamp steps. It was by no means a foregone conclusion, however, that these same equations would describe the behavior of the membrane under its normal operating conditions. Thus, the final stage of their analysis was to calculate the response of their mathematical representation of the nerve to the equivalent of an electrical stimulus. If the calculations produced an action potential that agreed favorably with experimental data, this would help validate their model.

Hodgkin and Huxley planned to solve their equations on the first electronic computer at the University of Cambridge, EDSAC (Electronic Delay Storage Automatic Calculator). Construction of EDSAC was completed in 1949, and at that time, it was the state-of-the-art for electronic digital computing (Wheeler, 1992a,b). EDSAC was approximately 12 ft × 12 ft in size, had a power consumption of 12 kW, contained some 3000 vacuum tubes, and had about 2000 bytes of nonrandom access memory, which was constructed from a series of 5-ft-long tubes filled with mercury. EDSAC operated at 500 kHz and could perform 650 instructions per second. Division of two numbers took the EDSAC about 200 ms. Unfortunately for Hodgkin and Huxley, EDSAC was undergoing major modifications in March of 1951 and would be unavailable for 6 months.

Hodgkin and Huxley overcame this setback by solving the differential equations numerically with a mechanical Brunsviga desk calculator. This was a laborious task. A space-clamped action potential (e.g., Fig. 7.14A) took a matter of days to compute, and a propagated action potential [e.g., Fig. 15 of Hodgkin and Huxley (1952d)] took a matter of weeks. In addition to the space-clamped and propagated action potentials, Hodgkin and Huxley's computations included the impedance changes and the total movements of Na^+ and K^+ ions into and out of the axon during an action potential; recovery during the relative refractory period; anode break excitation; and the oscillatory response of the membrane to a rectangular pulse of current. These results were published in 1952 (Hodgkin and Huxley, 1952d) and showed surprisingly good agreement with the available empirical data from the giant axon. This agreement suggested that the formulations developed by Hodgkin and Huxley were substantially correct.

The scope of Hodgkin and Huxley's computations was limited, however, by the fact that an automatic computer was unavailable. The manual methods of solving the Hodgkin–Huxley equations were so laborious as to discourage more detailed and broader investigations of the ability of the Hodgkin–Huxley equations to predict and interpret the well-established and fundamental characteristics of nerve behavior. For example, the all-or-none nature of initiating an action potential was considered to be a key feature of neuronal excitability, but it was unknown whether the Hodgkin–Huxley equations manifested this key feature.

To address these issues, Kenneth Cole and his colleagues (1955) wrote the machine language program necessary to run simulations of the Hodgkin–Huxley equations on the first fully operational stored-program electronic computer in the United States, SEAC (Standards Eastern Automatic Computer). Construction of SEAC was completed in 1950. Its design and capabilities were comparable to those of EDSAC (Kirsch, 1998). Although it was not their goal to cross-check the original calculations of Hodgkin and Huxley, the first computer simulation of the Hodgkin–Huxley equations by Cole et al. in 1955 was of a space-clamped action potential (see Fig. 1 of Cole et al., 1955). The results from Huxley's hand calculations were indistinguishable from those of computer simulation. Moreover, the SEAC calculations provided evidence that the Hodgkin–Huxley equations manifest an all-or-none response to current stimulation. Although some [including Huxley (1959)] doubted the threshold phenomena observed by Cole et al., the independent replication of the action potential calculation served to increase confidence in the Hodgkin–Huxley equations.

BOX 7.2 (continued)

COMPUTING SOLUTIONS TO THE HODGKIN–HUXLEY EQUATIONS

In theory, using a computer should increase the speed and accuracy of solving the Hodgkin–Huxley equations. In practice, however, this was not always the case in the early days of computers. Although SEAC could calculate the space-clamped action potential in about 30 min, accessing the computer proved very slow (FitzHugh, 1960). Solutions took a week or more, including the time for relaying instructions to the programming and operating technicians, scheduling time on the computer, and receiving the results. In addition, a flaw was detected in the machine language program for the first computer simulation of the Hodgkin–Huxley equations (FitzHugh and Antosiewicz, 1959). The program contained division by zero in the calculations of α_m and α_n. The major effect of these errors was to produce a spurious saddle point (see Box 7.5). It was an unfortunate accident that the spurious saddle point appeared near the membrane potential at which the threshold was believed to occur. In 1959, FitzHugh and Antosiewicz reprogrammed the Hodgkin–Huxley equations in FORTRAN, avoiding the division by zero, and, using an IBM 704, reexamined the issue of all-or-none responses in the Hodgkin–Huxley equations. On recalculation, the all-or-none threshold was lost, and the Hodgkin–Huxley equations were found to manifest a *"quasi-threshold"* phenomenon (FitzHugh and Antosiewicz, 1959); i.e., over a sufficiently small range of stimulus intensities the amplitude of the action potential decreases continuously from an "all" to a "none" [see Fig. 2 of FitzHugh and Antosiewicz (1959)]. The sharpness of threshold phenomenon is determined by the steepness of the stimulus–response curve for the peak amplitude versus stimulus intensity. The Hodgkin–Huxley equations manifest a very sharp threshold. An increase in the stimulus intensity of only one part in 10^8 was sufficient to distinguish between a very small graded response and a complete action potential. Thus, the lack of a mathematically correct threshold (i.e., saddle point) seemed entirely academic, and faith in the Hodgkin–Huxley equations was not shaken. Indeed, the computational prediction of quasi-threshold behavior was subsequently confirmed with experimental studies (Cole *et al.*, 1970), which further increased confidence in the Hodgkin–Huxley equations.

At present, access to computers is no longer a limiting factor to individuals who wish to simulate the Hodgkin–Huxley model. Commonly available personal computers are more than adequate to calculate solutions to the Hodgkin–Huxley equations. In addition, simulating the Hodgkin–Huxley model is no longer limited to individuals with skills necessary to develop the computer programs required to solve the Hodgkin–Huxley equations. It is possible to solve the Hodgkin–Huxley equations using commonly available spreadsheet programs (Brown, 1999, 2000), and many freely available and user-friendly software packages have been developed to simulate Hodgkin–Huxley-type models of neurons. Software packages that are specifically designed to build and simulate models of Hodgkin–Huxley-type neurons and neural circuits are commonly referred to as *neurosimulators*. Some examples of neurosimulators include GENESIS (Bower and Beeman, 1998), NEURON (Hines and Carnevale, 1997), and SNNAP (Ziv *et al.*, 1994). Several reviews describe the features and availability of neurosimulators (DeSchutter, 1989, 1992; Hayes *et al.*, 2002).

---

clamped patch of membrane (i.e., Fig. 7.1). These equations and their associated algebraic functions and parameters were derived to mathematically describe the magnitude and time course of I_{Na} and I_K produced by a series of voltage-clamp step depolarizations. As a first step toward validating their model, Hodgkin and Huxley tested the ability of the model to correctly calculate the total membrane current during a series of voltage-clamp steps. At a constant voltage, $dV/dt = 0$ and the steady-state values of the rate constants [$\alpha(V_m)$ and $\beta(V_m)$] are constant. The solution is then obtained directly in terms of the expressions given for $m(V_m,t)$, $h(V_m,t)$, $n(V_m,t)$ (i.e., Eqs. 7.26a, 7.26b, and 7.34). Using only a mechanical desk calculator (see

Box 7.2), Andrew Huxley computed I_m for a number of different voltages and compared these computations with similar empirical data. This comparison is illustrated in Fig. 7.13. There is excellent agreement between the calculated membrane currents and empirical voltage-clamp records, which lent credence to the model.

The overriding goal of Hodgkin and Huxley's quantitative analysis of voltage-clamp currents, however, was to explain neuronal excitability, and the ultimate test of the model was to determine whether it could quantitatively describe the action potential. Thus, Hodgkin and Huxley concluded their studies with calculations of the membrane potential changes

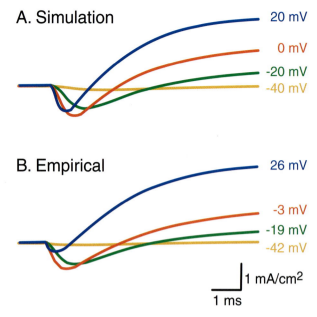

A. Simulation

20 mV
0 mV
-20 mV
-40 mV

B. Empirical

26 mV
-3 mV
-19 mV
-42 mV

1 mA/cm²

1 ms

FIGURE 7.13 Comparison of simulated membrane current and empirical observations. (**A**) As a first test of their model, Hodgkin and Huxley used Eq. 7.38 to compute the membrane current that would be elicited by a series of voltage-clamp steps. Four of the calculated traces are illustrated. They represent the predicted total membrane current during voltage-clamp steps to –40, –20, 0, and 20 mV. (**B**) Empirical data from one of Hodgkin and Huxley's voltage-clamp experiments (i.e., axon 31 at 4°C). The empirical data were collected from a series of voltage-clamp steps to –42, –19, –3, and 26 mV. There was excellent agreement between the simulated and empirical data, which suggests that the Hodgkin–Huxley equations captured the salient features of the experimental data. Adapted from Hodgkin and Huxley (1952d).

predicted by their equations. In the absence of the feedback amplifier (i.e., without the voltage clamp), dV/dt was no longer constant, and there is no explicit solution to Eq. 7.38. Using numerical methods, Hodgkin and Huxley solved the equations and computed an action potential waveform that duplicated with remarkable accuracy the naturally occurring action potential (Fig. 7.14). In addition to describing the space-clamped action potential, Hodgkin and Huxley demonstrated the considerable power of their model to predict many other properties of neuronal excitability, including subthreshold responses, a sharp threshold for firing, membrane conductance changes during an action potential, the effects of temperature on the action potential waveform, propagated action potentials, ionic fluxes, absolute and relative refractory periods, anode break excitation, and accommodation. Although the model has some limitations (see Box 7.3), the remarkable success of the Hodgkin–Huxley model in accurately describing such a wide array of phenomena remains to this day a triumph of classic biophysics in understanding a fundamental neuronal property (i.e., excitability).

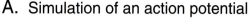

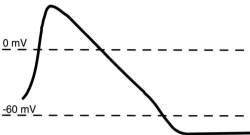

A. Simulation of an action potential

0 mV

-60 mV

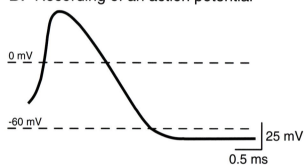

B. Recording of an action potential

0 mV

-60 mV

25 mV

0.5 ms

FIGURE 7.14 Comparison of simulated space-clamped action potential and empirical observations. If a giant axon is stimulated simultaneously over a substantial portion of its length (e.g., by applying a shock to a long internal electrode), all points within that length will undergo an action potential simultaneously (see Chapter 5). There will be no difference of potential along the axis of the nerve fiber and, therefore, no longitudinal current. This type of action potential is referred to as a "space-clamped" (as opposed to a propagated) action potential. (**A**) Simulation of a space-clamped action potential. The solution of Eq. 7.38 describes the membrane potential of a space-clamped patch of membrane. If the equivalent of a suprathreshold electrical shock is incorporated into Eq. 7.38, an action potential is produced. This example was produced by Hodgkin and Huxley, who used a mechanical desk calculator to solve the set of differential equations that have come to be known as the Hodgkin–Huxley equations (i.e., Eqs. 7.39–7.42). (**B**) An action potential recorded under experimental conditions that matched those simulated in (A) (i.e., spaced-clamp patch of membrane, temperature of 6°C). There is good agreement between the theoretical and empirical action potentials. This agreement suggests that the formulations developed by Hodgkin and Huxley accurately represented the underlying molecular processes. Adapted from Hodgkin and Huxley (1952d).

One of the great advantages of having a mathematical model of a complex process (e.g., neuronal excitability) is that it provides an opportunity to examine the component processes in ways that may not be experimentally possible. For example, it is possible to calculate the time courses and magnitudes of the different ionic currents, conductances, and gating variables during an action potential (see Chapter 5). Similarly, it is possible to calculate the individual ionic currents during voltage-clamp steps (Fig. 7.15). Given that the tools (i.e., computers and software)

BOX 7.3

LIMITATIONS OF THE HODGKIN–HUXLEY EQUATIONS

The Hodgkin–Huxley model has been so widely accepted as a paradigm for excitable membranes that its appropriateness for the squid giant axon itself has generally not been questioned. The model fails, however, to provide a good description for some electrophysiological properties of the axon. For example, in response to a relatively long duration, suprathreshold current pulse, the axon generally produces a single action potential (i.e., accommodation). In contrast, the Hodgkin–Huxley model predicts sustained spiking activity throughout the stimulus [see Fig. 3 of Clay (1998)]. This discrepancy is attributed to the assumption by Hodgkin and Huxley that the activation (m) and inactivation (h) of the Na$^+$ conductance were independent processes, whereas empirical evidence indicates the two processes are coupled (Bezanilla and Armstrong, 1977). A revised version of the Hodgkin–Huxley model (Clay, 1998; Vandenberg and Bezanilla, 1991), which incorporates coupling between activation and inactivation, provides a better description of the squid axon.

Other assumptions inherent in the Hodgkin–Huxley model have been examined and found to be only approximately valid. In the model, temperature is assumed to affect only the kinetics of the ionic conductances (see Eqs. 7.40–7.42). The conductance of an ionic channel is also altered by temperature, albeit to a relatively small degree (Hille, 2001). If the Hodgkin–Huxley equations are modified to incorporate an effect of temperature on the conductances, a better fit of empirical data is achieved (FitzHugh, 1966). Another assumption within the model is that the flow of ionic current through open channels obeys Ohm's law. Current data, however, suggest that the linearity is only approximate and holds neither under all ionic conditions nor in Na$^+$ and K$^+$

channels of all organisms (Hille, 2001). For example, Na$^+$ channels are not ohmic in nodes of Ranvier (Dodge and Frankenhaeuser, 1959).

Rather than detracting from the Hodgkin–Huxley model, these experimental results highlight some of the advantages of formulating a detailed, quantitative model. First, the formulation of the model forces one to clearly and quantitatively state the assumptions that underlie the model and to evaluate the impact of these assumptions on the behavior of the model. This procedure, in turn, provides guidelines for future experimental studies that can directly test the enumerated assumptions. Second, the model provides a modifiable framework with which new data and concepts can be incorporated and evaluated. Hodgkin and Huxley were well aware that their model had limitations. In their discussion of the model (Hodgkin and Huxley, 1952d), they acknowledged the shortcomings of the model and pointed out some discrepancies between the calculated and observed behavior of the squid giant axon. For example, the waveform of the calculated action potential had a sharper peak and small "hump" in the falling phase that was not present in the recorded action potential (closely compare the two action potentials in Fig. 7.14). As Huxley stated in 1964 *"I would not like to leave you with the impression that the particular equations we produced in 1952 are definitive … Hodgkin and I felt that these equations should be regarded as a first approximation which needs to be refined and extended in many ways in the search for the actual mechanisms of the permeability changes on the molecular scale."* Even if its details cannot be taken literally, the Hodgkin–Huxley model continues to have important general properties with mechanistic implications that are helping to direct future studies, both empirical and computational.

necessary to simulate the Hodgkin–Huxley model are readily available (see Box 7.2), anyone who wishes to explore the rich dynamical properties of these equations can easily do so.

Summary

The Hodgkin–Huxley model describes neuronal excitability in terms of four variables: the membrane potential, $V_m(t)$, and three gating variables, $m(V,t)$, $h(V,t)$, and $n(V,t)$, which describe the permeability

(i.e., conductance) of the membrane to Na$^+$ and K$^+$ (i.e., g_{Na} and g_K). The magnitude of the activation gating variables [$m(V,t)$ and $n(V,t)$] increases with increasing depolarization, whereas the magnitude of the inactivation gating variable [$h(V,t)$] decreases with depolarization. The gating variables were described by first-order differential equations with two voltage-dependent terms: the steady-state activation or inactivation [$m_\infty(V_m)$, $h_\infty(V_m)$, $n_\infty(V_m)$] and the time constant [$\tau_m(V_m)$, $\tau_h(V_m)$, $\tau_n(V_m)$]. Thus, the Hodgkin–Huxley model (Hodgkin and Huxley, 1952d) for a spaced-

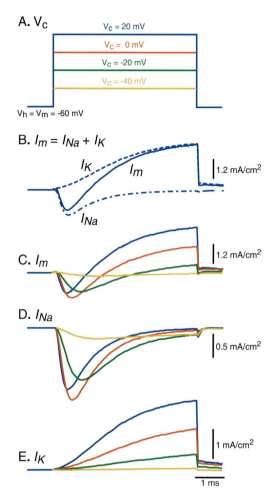

A. V_c

V_c = 20 mV

V_c = 0 mV

V_c = -20 mV

V_c = -40 mV

$V_h = V_m = -60$ mV

B. $I_m = I_{Na} + I_K$

I_K

I_m

I_{Na}

1.2 mA/cm²

C. I_m

1.2 mA/cm²

D. I_{Na}

0.5 mA/cm²

E. I_K

1 mA/cm²

1 ms

FIGURE 7.15 Simulation of a voltage-clamp experiment. Solving the Hodgkin–Huxley equations is much easier today than it was for Hodgkin and Huxley in 1952 (see Box 7.2). Digital computers are widely available and many software packages are available that have been designed to simulate neuronal properties. Such simulations can provide a useful tool for gaining insights into the complex and nonlinear processes that underlie neuronal excitability. (**A**) In these simulations, which were produced using the simulation package SNNAP (Ziv et al., 1994), the holding potential (V_h) was equal to the resting membrane potential (V_m), which was taken to be –60 mV. The command voltage (V_c) briefly steps the membrane potential to various depolarized values. The values of V_c are indicated above each trace. The colors of each voltage-clamp step in (A) correspond to the membrane currents in (B)–(E). (**B**) The simulated membrane current (I_m) is composed primarily of I_{Na} and I_K. (**C**) The total membrane current is computed by solving Eq. 7.38. I_m has an early inward component that is followed by a large, sustained outward component. The overall appearance of these simulated membrane currents agrees well with empirical data (e.g., Fig. 13B) and with the original calculations of Hodgkin and Huxley (e.g., Fig. 13A). (**D**) I_{Na} in isolation. Note that the largest I_{Na} trace (i.e., the red trace) does not occur during the largest voltage-clamp step (i.e., the blue trace). As V_c approaches E_{Na}, the driving force for I_{Na} approaches zero and the magnitude of the current decreases (see Eq. 7.3). (**E**) I_K in isolation. Unlike I_{Na}, I_K increases monotonically as a function of increasing V_c. In addition, because the description of g_K (i.e., Eq. 7.33) has only an activation gating variable, I_K is sustained throughout the duration of the voltage-clamp step.

clamped patch of membrane (i.e., an isopotential compartment) is a system of four ordinary differential equations:

$$I_m = C_m \frac{dV}{dt} + \bar{g}_{Na} m^3 h (V_m - E_{Na}) +$$

$$+ \bar{g}_K n^4 (V_m - E_K) + \bar{g}_l (V_m - E_l), \qquad (7.39)$$

$$\frac{dm}{dt} = \Phi(T) \frac{m_\infty(V_m) - m}{\tau_m(V_m)}$$

$$= \Phi(T) \left[\alpha_m(V_m)(1-m) - \beta_m(V_m)m \right], \quad (7.40)$$

$$\frac{dh}{dt} = \Phi(T) \frac{h_\infty(V_m) - h}{\tau_h(V_m)}$$

$$= \Phi(T) \left[\alpha_h(V_m)(1-h) - \beta_h(V_m)h \right], \qquad (7.41)$$

$$\frac{dn}{dt} = \Phi(T) \frac{n_\infty(V_m) - n}{\tau_n(V_m)}$$

$$= \Phi(T) \left[\alpha_n(V_m)(1-n) - \beta_n(V_m)n \right], \qquad (7.42)$$

where the coefficient **Φ** (T) describes the effects of temperature on the three gating variables (see below), and the voltage and time dependency of the gating variables is given by

$$m_\infty(V_m) = \frac{\alpha_m(V_m)}{\alpha_m(V_m) + \beta_m(V_m)}, \qquad (7.43a)$$

$$\tau_m(V_m) = \frac{1}{\alpha_m(V_m) + \beta_m(V_m)}, \qquad (7.43b)$$

$$h_\infty(V_m) = \frac{\alpha_h(V_m)}{\alpha_h(V_m) + \beta_h(V_m)}, \qquad (7.44a)$$

$$\tau_h(V) = \frac{1}{\alpha_h(V_m) + \beta_h(V_m)}, \qquad (7.44b)$$

$$n_\infty(V_m) = \frac{\alpha_n(V_m)}{\alpha_n(V_m) + \beta_n(V_m)}, \qquad (7.45a)$$

$$\tau_n(V_m) = \frac{1}{\alpha_n(V_m) + \beta_n(V_m)}. \qquad (7.45b)$$

These equations relate the membrane potential (i.e., V_m) to the permeability of the membrane to Na⁺, K⁺, and the nonspecific leakage of ions. The equations contain several auxiliary parameters representing equilibrium potentials of the ions (i.e., E_{Na}, E_K, and E_l);

maximum ionic conductances ($\bar{g}_{Na}$, $\bar{g}_K$, and $\bar{g}_l$), a temperature coefficient (Φ), and the capacitance of the membrane (C_m). Values for these parameters were obtained from analyses of experimental data (Hodgkin et al., 1952; Hodgkin and Huxley, 1952a–c). The values used by Hodgkin and Huxley (1952d) to describe the squid giant axon were

$$E_{Na} = V_r + 115 \text{ mV}, \qquad \bar{g}_{Na} = 120 \text{ mS/cm}^2,$$

$$E_K = V_r - 12 \text{ mV}, \qquad \bar{g}_K = 36 \text{ mS/cm}^2,$$

$$E_l = V_r + 10.613 \text{ mV}, \qquad \bar{g}_l = 0.3 \text{ mS/cm}^2,$$

$$\Phi(T) = 3^{(T-6.3)/10}, \qquad C_m = 1 \text{ } \mu\text{F/cm}^2,$$

where V_r is resting membrane potential of the cell, and T is temperature in degrees centigrade. The coefficient Φ provides the three gating variables with a Q_{10} of 3, and equals 1 at Hodgkin and Huxley's standard temperature of 6.3°C. The rate constants (α_i and β_i) were estimated by fitting empirically derived exponential functions of voltage to the experimental data. For the squid giant axon at 6.3°C, these functions were (Hodgkin and Huxley, 1952d):

$$\alpha_m(V_m) = \frac{0.1(V_r - V_m + 25)}{\exp\left(\dfrac{V_r - V_m + 25}{10}\right) - 1}, \tag{7.46a}$$

$$\beta_m(V_m) = 4 \exp\left(\frac{V_r - V_m}{18}\right), \tag{7.46b}$$

$$\alpha_h(V_m) = 0.07 \exp\left(\frac{V_r - V_m}{20}\right), \tag{7.47a}$$

$$\beta_h(V_m) = \frac{1}{\exp\left(\dfrac{V_r - V_m + 30}{10}\right) + 1}, \tag{7.47b}$$

$$\alpha_n(V_m) = \frac{0.01(V_r - V_m + 10)}{\exp\left(\dfrac{V_r - V_m + 10}{10}\right) - 1}, \tag{7.48a}$$

$$\beta_n(V_m) = 0.125 \exp\left(\frac{V_r - V_m}{80}\right). \tag{7.48b}$$

The work of Hodgkin and Huxley was a landmark in the field of biophysical research. Their protocol involved voltage-clamp analysis of membrane currents, separation of the membrane current into its components, development and fitting of kinetic schemes for the time and voltage dependence of the ionic conductances, and, finally, reconstruction of the action potential. This work established a precedent for combining experimental and computational techniques to explore excitable membrane systems. This interdisciplinary approach is still commonly used, and Hodgkin–Huxley formalisms remain a cornerstone of quantitative models of neuronal excitability.

A GEOMETRIC ANALYSIS OF EXCITABILITY

The Hodgkin–Huxley equations constitute a remarkably successful quantitative model. With reasonable accuracy, the model describes the membrane currents and the action potential of the squid giant axon, as well as a number of other dynamical properties of neuronal excitability, such as anode break excitation, accommodation, and refractory period (see Chapter 5). Although this quantitative, conductance-based model has been enormously fruitful in terms of providing a mathematical framework for modeling neuronal excitability, it has a serious drawback, i.e., its numerical complexity. The Hodgkin–Huxley equations are highly nonlinear (i.e., m is raised to the third power and n is raised to the fourth power) and complex (i.e., the model consists of four coupled differential equations, a large number of algebraic equations, and a host of parameters). This complexity makes it difficult to intuitively understand the workings of the model. Indeed, Andrew Huxley may have said it best when he stated (Huxley, 1964): "Very often my expectations turned out to be wrong, and an important lesson I learned from these manual computations was the complete inadequacy of one's intuition in trying to deal with a system of this degree of complexity."

To gain intuitive insight into the dynamical properties of neurons, it is helpful to examine less complex models that manifest the salient features of neuronal excitability. By exploiting a low-dimensional (i.e., two or three differential equations) model of an excitable membrane and by applying techniques from the mathematical field of nonlinear dynamics, many dynamical properties of neurons (e.g., threshold behavior, excitability, repetitive firing, autonomous bursting) can be understood, predicted, and interpreted. Others have used this approach and have found that it is considerably easier—from a numerical and a conceptual point of view—to study the dynamical properties of neurons described by reduced models rather than simulate the behavior of biophysically complex neurons (Abbott, 1994; Bertram, 1994; Av-Ron et al., 1991; Alexander and Cai, 1991; Bertram et al., 1995; Butera et al., 1996; Butera, 1998; Canavier et al., 2002; Ermentrout, 1996; FitzHugh, 1960; Gall and Zhou, 2000; Hoppensteadt and Izhikevich, 2001;

Izhikevich, 2000; Kepler *et al.*, 1992; Koch, 1999; Krinskii and Kokoz, 1973; Rinzel, 1985; Rinzel and Ermentrout, 1998) (see also Chapter 14). In the second half of this chapter we illustrate how to develop reduced models of excitability and how to analyze their dynamical properties.

Two-Dimensional Reduction of the Hodgkin–Huxley Model

The Hodgkin–Huxley model can be reduced to a two-variable model by identifying and combining variables with similar time scales and biophysical roles and by allowing relatively fast variables to be instantaneous. For example, the time constant for m (τ_m) is an order of magnitude faster than that of h or n (Fig. 7.11). Thus, it is reasonable to approximate m by m_∞ (V_m), which eliminates one of the differential equations (i.e., Eq. 7.26a). In addition, the variables n and h evolve on

similar time scales and with an approximately constant relationship between their values; i.e., the sum of $n(V_m, t)$ and $-h$ (V_m, t) is approximately constant. Thus, it is reasonable to combine h and n into a single "recovery" variable, which has been termed w (Rinzel, 1985; FitzHugh, 1961). Using a single recovery variable replaces two of the differential equations (i.e., Eqs. 7.26b and 7.34) with a single expression for dw/dt (see below). By incorporating these approximations and some additional simplifications in the algebraic equations, it is possible to produce a tractable, two-variable model (i.e., a model with only dV_m/dt and dw/dt) that is versatile, that manifests many of the salient features of neuronal excitability, and that manifest a rich and diverse array of dynamical properties.

The reduced model (Av-Ron *et al.*, 1991) includes a linear leakage current (I_l) and an externally applied stimulus current (I_s), as well as a time- and voltage-dependent inward current (I_{Na}) and outward current

FIGURE 7.16 Two-dimensional model of neuronal excitability. The reduced model (Eqs. 7.49 and 7.50) is excitable and produces an action potential in response to a brief, suprathreshold stimulus. (**A1**) Time-series plot of the computed membrane potential, $V_m(t)$. The resting membrane potential is –60 mV, and in response to a brief depolarizing current (i.e., $I_s = 5\ \mu A\ cm^{-2}$ for 3 ms, lower trace), the model produces an action potential. The parameters of the model (see below) were adjusted to closely approximate the squid giant axon. (**A2**) Time-series plot of recovery variable, $w(t)$. At the resting potential, $w(t)$ has a value of ~0.4. During the action potential, the magnitude of $w(t)$ increases and this increase helps to restore the rest state. This variable acts much like the n and h gating variables of the Hodgkin and Huxley model. (**A3**) The phase plane of the reduced model has coordinates V_m and w. The evolution of the variables is a point moving through this phase plane (black line). The direction of the trajectory (i.e., time) is indicated by the arrow. The values of w and V_m at the resting potential are indicated by the dashed lines. In the absence of any external stimulus, the trajectory (i.e., solution path) remains at the resting values of V_m and w (i.e., the intersection of the dashed lines). The brief increase in I_s [see (A1)] displaces the trajectory away from the resting point. If the displacement crosses an apparent threshold along the V_m axis, the trajectory follows a large amplitude pseudo-orbit through the phase plane and back to the resting point. The large excursion of the trajectory represents the projection of the action potential in the phase plane. (**B**) Nullclines of the reduced model. Nullclines in the phase plane are curves along which derivatives of a given variable are constant and equal to zero (i.e., $dV_m/dt = 0$, $dw/dt = 0$). In addition, nullclines divide the phase plane into regions where the derivatives have a constant sign. (**B1**) w nullcline (blue line). The w nullcline specified by the steady-state function of w (Eq. 7.54). (**B2**) V_m nullcline (green line). The V_m nullcline represents the pairs of values of $V_m(t)$ and $w(t)$ for which the net current is equal to zero (see Eq. 7.49). E_K and E_{Na} indicate the equilibrium potentials of I_{Na} and I_K. (**C**) The trajectory of an action potential (yellow line) and the nullclines for V_m (green line) and w (blue line) are superimposed in the V_m–w plane. The intersection of the nullclines represents the resting point (i.e., quiescent state) in the phase plane. An instantaneous displacement of system (arrowhead) that crosses the threshold for generating an action potential and the system produces an action potential. Although time is not explicitly plotted in the phase plane, the red dots along the action potential trajectory are separated by 50 μs. Thus, it is possible to visualize the speed of the solution path through the phase plane. The action potential trajectory passes through four regions in the phase plane. The initial up stroke of the spike (i.e., V_m is increasing) occurs in the region where both dV_m/dt and dw/dt are greater than zero. The trajectory crosses the V_m nullcline (i.e., the peak of the action potential) and enters a region where $dV_m/dt < 0$ (i.e., the falling phase of the action potential) and $dw/dt > 0$. As the trajectory crosses the w nullcline, it enters a region where both dV_m/dt and dw/dt are less than zero (i.e., the absolute refractory period; see Chapter 5). Finally, the trajectory crosses the V_m nullcline a second time (i.e., the minimum of the afterhyperpolarization) and enters a region where $dV_m/dt > 0$ and $dw/dt < 0$ (i.e., the relative refractory period; see Chapter 5). **Inset:** The time-series plot of $V_m(t)$ that is equivalent to the trajectory in the phase plane. For these simulations $C_m = 1\ \mu F/cm^2$, $\bar{g}_{Na} = 120\ mS/cm^2$, $\bar{g}_K = 36\ mS/cm^2$, $\bar{g}_l = 0.3\ mS/cm^2$, $E_{Na} = 55\ mV$, $E_K = -72\ mV$, $E_l = -49.4\ mV$, $V_{1/2}^{(m)} = -33\ mV$, $V_{1/2}^{(w)} = -55\ mV$, $a^{(m)} = 0.055$, $a^{(w)} = 0.045$, $mp = 3$, $wp = 4$, $s = 1.3$, and $\bar{\lambda} = 0.2$. With these parameters, the reduced model simulates the properties of the squid giant axon. Adapted from Av-Ron *et al.* (1991).

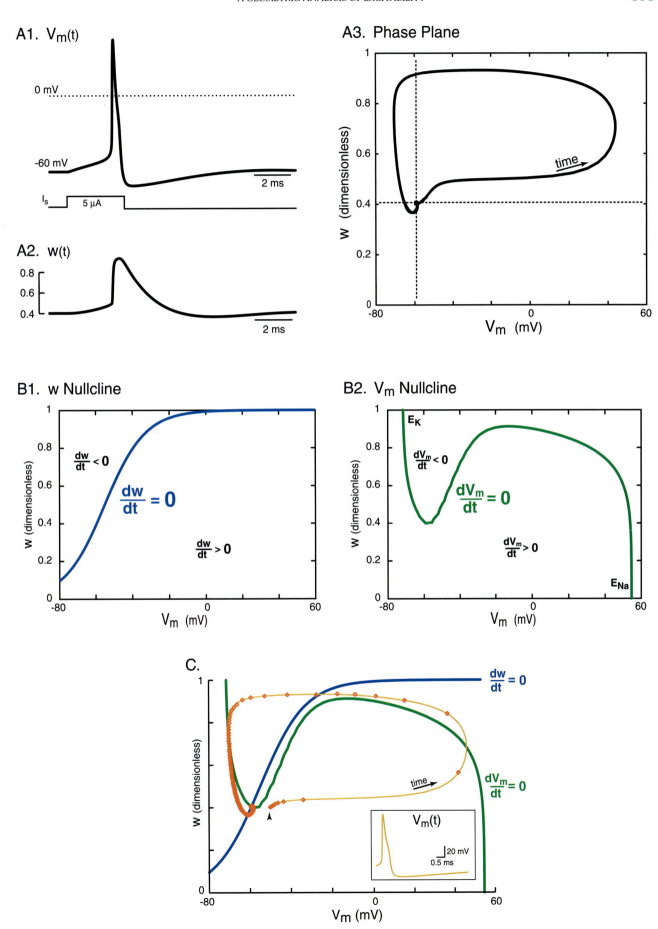

A1. V_m(t)

0 mV

-60 mV

2 ms

I_s   5 µA

A2. w(t)

0.8
0.6
0.4

2 ms

A3. Phase Plane

w (dimensionless)

time

V_m (mV)

B1. w Nullcline

w (dimensionless)

$\frac{dw}{dt} < 0$

$\frac{dw}{dt} = 0$

$\frac{dw}{dt} > 0$

V_m (mV)

B2. V_m Nullcline

w (dimensionless)

E_K

$\frac{dV_m}{dt} < 0$

$\frac{dV_m}{dt} = 0$

$\frac{dV_m}{dt} > 0$

E_Na

V_m (mV)

C.

w (dimensionless)

$\frac{dw}{dt} = 0$

$\frac{dV_m}{dt} = 0$

time

V_m(t)

20 mV

0.5 ms

V_m (mV)

(I_K). The inward current is rapidly activated by depolarization, and contributes to its own further activation by producing further depolarization (i.e., autocatalytic or positive feedback). This positive feedback process has an instantaneous dependence on membrane potential. There is also a slower, negative feedback process. With depolarization, w increases, leading to activation of I_K and inactivation of I_{Na}. The dynamics of this two-dimensional system are described by the differential equations:

$$\frac{dV_m}{dt} = \frac{-\left(I_{Na} + I_K + I_l - I_s\right)}{C_m}, \qquad (7.49)$$

$$\frac{dw}{dt} = \Phi(T)\frac{w_\infty(V_m) - w}{\tau_w(V_m)}, \qquad (7.50)$$

where V_m is the absolute membrane potential, C_m is the membrane capacitance, and $\Phi(T)$ is similar to the temperature coefficient in the Hodgkin–Huxley equations. The currents are described by

$$I_{Na} = \bar{g}_{Na}m_\infty^{mp}(V_m)(1 - w(V_m,t))(V_m - E_{Na}), \qquad (7.51)$$

$$I_K = \bar{g}_K\left(\frac{w(V_m,t)}{s}\right)^{wp}(V_m - E_K), \qquad (7.52)$$

$$I_l = \bar{g}_l(V_m - E_l) \qquad (7.53)$$

where $\bar{g}_i$ represent the maximum conductances of a given ionic current, E_i represent the equilibrium potential for a given ionic current, mp and wp are parameters, and s is a scaling factor that allows the magnitude of recovery variable (i.e., w) to be different for I_{Na} and I_K. The following equations give the voltage dependencies of the steady state of the recovery variable (i.e., w_∞), the steady state of the Na$^+$ activation variable (i.e., m_∞), and the relaxation function of the recovery variable (i.e., τ_w):

$$w_\infty(V_m) = \frac{1}{1 + \exp\left(-2a^{(w)}\left(V_m - V_{1/2}^{(w)}\right)\right)}, \qquad (7.54)$$

$$m_\infty(V_m) = \frac{1}{1 + \exp\left(-2a^{(m)}\left(V_m - V_{1/2}^{(m)}\right)\right)}, \qquad (7.55)$$

$$\tau_w(V_m) = \frac{1}{\bar{\lambda}\exp\left(a^{(w)}\left(V_m - V_{1/2}^{(w)}\right)\right) + \bar{\lambda}\exp\left(-a^{(w)}\left(V_m - V_{1/2}^{(w)}\right)\right)}. \qquad (7.56)$$

Here, Eqs. 7.54 and 7.55 are sigmoid curves and $V_{1/2}^{(i)}$ is the voltage for the half-maximal value and parameter $a^{(i)}$ controls the slope of the curve at this midpoint (i.e., inflection point). Equation 7.56 is a

bimodal sigmoid and the parameter $\bar{\lambda}$ represents the effects of $\Phi(T)$ in Eq. 7.40. [For additional details of this reduced model, see Av-Ron et al. (1991).]

This reduced model (Eqs. 7.49 and 7.50) can exhibit two major types of qualitative behavior. For some parameter regimes, the model is normally quiescent, but excitable in that it can fire one or more action potentials in response to transient stimuli (Fig. 7.16A). For different parameter regimes, the model fires action potentials in a repetitive fashion (i.e., limit cycle oscillations) (Fig. 7.17b). In addition, for some parameter regimes, the model can manifest at least one type of bistability (Fig. 7.19B), which underlies some types of burst firing. The same nonlinear properties that allow the model to fire a single action potential also enable it to function as a nonlinear oscillator, generating sustained rhythmic activity. How can such distinct behaviors emerge from such a simple system, and how can the behavior of the system be predicted? With parameter variations, a series of examples can be generated in a way that explains how the nonlinear properties of this simple two-dimensional model endow the system with such rich dynamical behaviors. Several methods (e.g., phase planes, nullclines, stability, and bifurcations) also can be used to develop an intuitive understanding of the workings of this model (see Box 7.4). This approach closely follows that used by Rinzel and Ermentrout (1998) to analyze a different model (the Morris–Lecar model) and has been used by Av-Ron et al. (1991, 1993; Av-Ron, 1994) and by Canavier et al. (2002) to analyze the two-variable model described above (i.e., Eqs. 7.49 and 7.50).

The Action Potential Trajectory in the Phase Plane

Several methods can be used to visualize and analyze the dynamical behavior of the reduced model. For example, after numerically integrating the differential equations, the dependent variables [$V_m(t)$ and $w(t)$] can be plotted as a function of time (Figs. 7.16A1 and A2). Although this method of visualizing dynamical behavior may be familiar and is useful for comparing temporal relationships between the variables, it is not analytically powerful. A more valuable way to view the response of multiple variables is by *phase plane* profiles (i.e., curves of one dependent variable against another; Fig. 7.16A3). The solution path in the phase plane is referred to as a *trajectory*. The phase plane is completely filled with trajectories, since each point can serve as an initial condition for the model. For example, the simulation illustrated in Figs. 7.16A1 and A2 began with the initial conditions

BOX 7.4

FIXED POINTS AND BIFURCATIONS IN NONLINEAR DYNAMICAL SYSTEMS

Systems that change, that evolve in time, are referred to as *dynamical systems*. Mathematical representations of dynamical systems fall into two general categories: *difference equations* (or *iterative maps*) and *differential equations*. Iterative maps are used to represent systems in which time is taken to be discrete, whereas differential equations are used to represent systems in which time is taken to be continuous. Differential equations are widely used in science and engineering. The two most common types of differential equations are *ordinary differential equations* (*ODEs*), in which there is only one independent variable such as time *t*, and partial differential equations, in which there are two or more independent variables such as time *t* and space *x*. Both types of differential equations were used by Hodgkin and Huxley to model action potentials (i.e., the spaced-clamped versus the propagated action potential). The solution and analysis of partial differential equations are complex, however. Thus, ODEs are more commonly used, and many tools are available to investigate and understand the dynamical behavior of ODEs (Ermentrout, 1996; Izhikevich, 2000; Strogatz, 1994; Ermentrout, 2002; Abraham and Shaw, 1992; FitzHugh, 1969; Jackson, 1991; Pavlidis, 1973; Tufillaro *et al.*, 1992).

Dynamical systems are described in terms of how many differential equations are included in the system (i.e., an *"nth-order"* or *"n-dimensional"* system) and whether they are *linear* or *nonlinear*. For example, consider the general system

$$\frac{dx_1}{dt} = k_1 x_1 + k_2 x_2, \tag{7.57}$$

$$\frac{dx_2}{dt} = k_3 x_1 + k_4 x_2, \tag{7.58}$$

where k_i are constants and x_i are variables. This is a second-order or two-dimensional system, because there are two ODEs. In addition, these ODEs are referred to as coupled because x_1 is defined in terms of x_2, and vice versa. This is also a linear system, because all of the x_i on the right-hand side appear to the first power only (see also Chapter 14). Otherwise, the system would be nonlinear. Typical nonlinear terms are products (e.g., $x_i x_j$), powers (e.g., x_i^3), and functions (e.g., $\sin x_i$).

Unlike linear systems (e.g., Eqs. 7.57 and 7.58), most nonlinear systems are impossible to solve analytically, and must be studied by using techniques such as numerical integration of the equations, geometric methods such as phase plane analysis, and stability theory. For example, consider a general form for a nonlinear, two-dimensional system,

$$\frac{1}{\mu} \frac{dx_1}{dt} = f_1(x_1, x_2), \tag{7.59}$$

$$\frac{dx_2}{dt} = f_2(x_1, x_2), \tag{7.60}$$

where the functions f_1 and f_2 are determined by the problem at hand and x_1 and x_2 are variables. For example, f_i may describe processes underlying neuronal excitability, and the variables x_1 and x_2 might represent membrane potential, gating variables for membrane conductances, or the intracellular concentration of a second messenger. The parameter μ can be thought of as a rate constant that scales the relative rates of the two functions. When $0 < \mu << 1$, then Eq. 7.59 is referred to as the slow subsystem and Eq. 7.60 is referred to as the fast subsystem. Numerical integration of Eqs. 7.59 and 7.60 will produce a set of ordered pairs of real numbers $x_1(t)$ and $x_2(t)$; where $x_1(0)$ and $x_2(0)$ represent the initial values (i.e., *initial conditions*) of the two variables (i.e., at time $t = 0$), and $x_1(t)$ and $x_2(t)$ represent the values of the two variables at time t (i.e., the *temporal evolution* of the variables). A common method for visualizing these solutions is to plot $x_1(t)$ and/or $x_2(t)$ versus time (e.g., Figs. 7.16A1 and A2). Such a plot is referred to as a *time series* plot of the variables. Alternatively, an abstract space with coordinates (x_1, x_2) can be constructed. In this space, the solution $[x_1(t), x_2(t)]$ corresponds to a point moving along a curve (e.g., Fig. 7.16A3). This curve (i.e., the solution to the system of differential equations) is referred to as a *trajectory*, and the direction of motion along a trajectory is often indicated by an arrowhead. The abstract space is called the *phase space* for the system, and the *phase portrait* of the system shows the overall picture of trajectories in phase space. Because Eqs. 7.59 and 7.60 constitute a two-dimensional system, the phase space of the system is a plane (i.e., a *phase plane*).

BOX 7.4 (continued)

FIXED POINTS AND BIFURCATIONS IN NONLINEAR DYNAMICAL SYSTEMS

If a trajectory asymptotically approaches a constant, time-independent solution, then this point in the phase space is referred to as a *stable fixed point* (e.g., Fig. 7.17A2). (*Note*: dynamical systems theory is rife with conflicting terminology and different terms are often used for the same thing. For example, fixed points are also referred to as points of equilibrium, or singularities, or critical points. This lack of a standard terminology can be a source of great frustration and confusion for individuals not intimately involved in the field of dynamical systems theory.) In the phase plane, stable fixed points often are indicated by solid black dots. Alternatively, if a trajectory of a nonlinear system asymptotically approaches a time-dependent solution that precisely returns to itself in a time T (i.e., the period), then this periodic solution is referred to as a *stable limit cycle*, or simply a limit cycle (e.g., Fig. 7.17B2). A limit cycle is represented as a closed curve on a phase plane. These two asymptotically stable trajectories (i.e., stable fixed point and limit cycle) are examples of *attractors*, because trajectories approach and coalesce on them. If a phase plane has only one attractor, then all trajectories ultimately lead to that solution, which is referred to as *globally attracting* (e.g., Fig. 7.16C). Alternatively, if a phase plane has more than one attractor, then the system can manifest more than one stable steady state and, thus, is referred to as *multistable* (e.g., Fig. 7.19B; see also Chapter 14).

Because of the special topological properties of the plane, phase plane analyses can provide fundamental insights into the dynamical properties of a two-dimensional system such as Eqs. 7.59 and 7.60. For example, the Jordan Curve Theorem implies that the only attractors in the phase plane are limit cycles and fixed points. The fixed points in a phase plane can be identified by plotting the *nullclines* (e.g., Fig. 7.16B). A nullcline is a curve in the phase plane along which the rate of change of a particular variable is zero (i.e., $dx_i/dt = 0$). Nullclines are useful because they divide the phase plane into regions in which the derivative of each variable has a constant sign and because any place the two nullclines intersect is a fixed point (i.e., $dx_1/dt = dx_2/dt = 0$).

Fixed points have several features that can be used to classify them. For example, a fixed point can be either *stable* (e.g., Fig. 7.17A2) or *unstable* (e.g., Fig. 7.17B2). A fixed point is stable if all sufficiently small perturbations away from it dampen out with time (i.e., the solution returns to the fixed point). Alternatively, if the perturbation grows with time, the fixed point is unstable. The

stability of fixed points can be defined more rigorously in mathematical terms, and readers who want a more detailed discussion of this matter should consult one of the many excellent textbooks that deal with nonlinear systems (e.g., Strogatz, 1994).

In addition to stability, fixed points can be classified on the basis of how trajectories behave in the neighborhood of the fixed point. Such behavior is often referred to as the *flow* or *motion*. For example, trajectories approach *stable nodes* and leave *unstable nodes*. Trajectories in the neighborhood of the *saddle point* are hyperbolic (i.e., they do not approach the saddle point, but pass by it, looking somewhat like members of a family of hyperbolas near their common center). Saddle points organize boundaries (i.e., a *threshold separatrix*) between classes of trajectories with qualitatively different properties. For example, on one side of the separatrix may reside a stable node that represents a quiescence or rest state (e.g., the resting membrane potential of a neuron), and on the other side may reside a large-amplitude trajectory that starts and ends near the equilibrium (e.g., a spike). Thus, small perturbations of the solution decay if they do not lead beyond the separatrix, whereas those crossing the separatrix grow away exponentially. A system with a saddle point has well-defined threshold and all-or-none behavior. Because the solution eventually returns to the stable node (i.e., rest state), however, the system is not oscillatory, but rather it is excitable. Thus, a saddle point can be used to describe threshold phenomenon mathematically.

Fixed points and closed orbits can be created or destroyed or destabilized as parameters are varied. If the phase portrait changes its topological structure as a parameter is varied, this is termed a *bifurcation*. Examples include changes in the number or stability of fixed points, closed orbits, or saddle connections as a parameter is varied. Bifurcations are most clearly illustrated in what is termed a bifurcation diagram (e.g., Fig. 7.18B). A *bifurcation diagram* plots a system parameter (e.g., μ in Eq. 7.59), which is referred to as the *bifurcation parameter*, on the horizontal axis and a representation of an attractor (e.g., x_i in Eqs. 7.59 and 60) on the vertical axis (see also Chapter 14). As the value for the parameter is systematically varied the stability of the fixed points (and closed orbits) will change, which is usually indicated by a branch in the bifurcation diagram. Bifurcations are important, because they provide insights into when transitions and instabilities may occur as some control parameter is varied.

BOX 7.4 (*continued*)

FIXED POINTS AND BIFURCATIONS IN NONLINEAR DYNAMICAL SYSTEMS

For those who wish to explore the behavior of dynamical systems, dynamical systems software has recently become available for personal computers. In general, with these software packages all one has to do is type in the equations and the parameters; the program solves the equations numerically and provides analytical tools. For example, the software package XPPAUT can plot variables in two- and three-dimensional phase space, calculate and plot nullclines, analyze the stability of fixed points, and perform bifurcation analyses (Ermentrout, 2002). The features and availability of several of these software packages were recently reviewed (Ermentrout, 2002; Hayes *et al.*, 2002; Hubbard and West, 1992; Kocak, 1989).

$V_m(0) = -59.407$ mV and $w(0) = 0.402$. This point is marked in the V_m–w plane by the dashed lines in Fig. 7.16A3. With these initial conditions the solution of the model does not change over time; i.e., the system is quiescent. Since there is no tendency for the variables to change, this point is referred to as a stable *fixed point* or *steady state*. If the system is subjected to a sufficiently large perturbation (e.g., a brief depolarizing current is injected), then the solution path rapidly evolves along a curve in the phase plane (i.e., the black line in Fig. 7.16A3). This trajectory represents the evolution of an action potential in the V_m–w phase plane. Although time is not explicitly plotted in the phase plane, the direction of the solution is usually indicated by an arrow. The trajectory of the action potential begins and ends at the fixed point. Thus, this fixed point is an example of an *attractor*. Indeed, for the set of parameter values used in Fig. 7.16 and $I_s = 0$ μA/cm², all trajectories will ultimately return to this fixed point, which is said to be *globally attracting*.

To gain a more complete understanding of the dynamical behavior of the reduced model, it is useful to combine the phase plane profiles with a nullcline analysis. A *nullcline* for a given variable is a curve in the phase plane along which one of the derivatives is constant and is equal to zero. In addition, nullclines divide the phase plane into regions where the derivatives have a constant sign (see below). Nullclines for the reduced model are illustrated in Fig. 7.16B. The nullcline associated with the slow variable (i.e., w) is specified by the steady-state w curve (i.e., Eq. 7.54). If the system is currently located on the w nullcline, then its imminent trajectory must be horizontal, because only V_m can change (i.e., $dw/dt = 0$). Horizontal movements to the right of the w nullcline represent depolarizations, which, in turn, would cause w to increase. Conversely, horizontal movements to the left of the w nullcline represent hyperpolarization, which,

in turn, would cause w to decrease. Thus, w is decreasing ($dw/dt < 0$) in the region to the left of the w nullcline and w is increasing ($dw/dt > 0$) in regions to the right of the w nullcline. The nullcline associated with the fast variable (i.e., V_m) is a cubic function (Fig. 7.16B2), and is composed of pairs of values of $V_m(t)$ and $w(t)$ for which the net current is equal to zero (see Eq. 7.49). If the evolution of the system brings it onto the V_m nullcline, its imminent trajectory must be vertical, because only w can change (i.e., $dV_m/dt = 0$). Downward vertical movements represent a decrease in w, which, in turn, would cause V_m to increase. Conversely, upward vertical movements represent an increase in w, which, in turn, would cause V_m to decrease. Thus, V_m is decreasing ($dV_m/dt < 0$) in regions above the V_m nullcline, and V_m is increasing ($dV_m/dt > 0$) in regions below the V_m nullcline. Given that the rate of change of V_m is fast compared with that of w, the nullclines define two very important features of the phase plane. First, the fixed points of the system can be predicted from the nullclines. At intersections of the nullclines (i.e., $dV_m/dt = dw/dt = 0$), there is no tendency for any variable to change, so these intersections represent fixed points or steady states. A fixed point does not guarantee a quiescent membrane potential, however. Fixed points may not be stable to small perturbations (see below). Second, the qualitative dynamics of the system can be predicted from the nullclines. The qualitative prediction is possible because the system will quickly relax to near the potential nullcline, then move in the direction dictated by whether w, the slow variable, is increasing or decreasing. Furthermore, the stability of a fixed point can be predicted based on which branch of the V_m nullcline contains it (under the assumption that voltage changes much more rapidly than w). In the middle branch, positive feedback dominates so a fixed point in the branch is unstable with respect to perturbations. The other branches are stable. For the system to be excitable or oscillatory, w must be decreasing above the w nullcline (left stable branch) and increasing below it (right stable branch). This will cause a loop in the trajectory as shown in Fig. 7.16C.

The configuration of the nullclines can be changed by altering values of the parameters. For example (Fig. 7.17), the nullclines can be altered by varying the parameter I_s (i.e., the externally applied stimulus current). If the value of I_s is set to -2 μA cm^{-2}, the system is quiescent (Fig. 7.17A1). In the quiescent

case, the single fixed point (filled circle in Fig. 7.17A2) falls on the left branch of the V_m nullcline and thus is stable. Although the system is quiescent, it remains excitable, and a sufficiently large perturbation will elicit an action potential (e.g., Fig. 7.16A1). If the value of I_s is set to 60 μA cm^{-2}, the dynamical behavior of the system is qualitatively different. The system continuously generates action potentials with a constant interspike interval (Fig. 7.17B1). This type of activity is often referred to as *beating* or *pacemaker activity*. In the pacemaker case, the fixed point is moved onto the

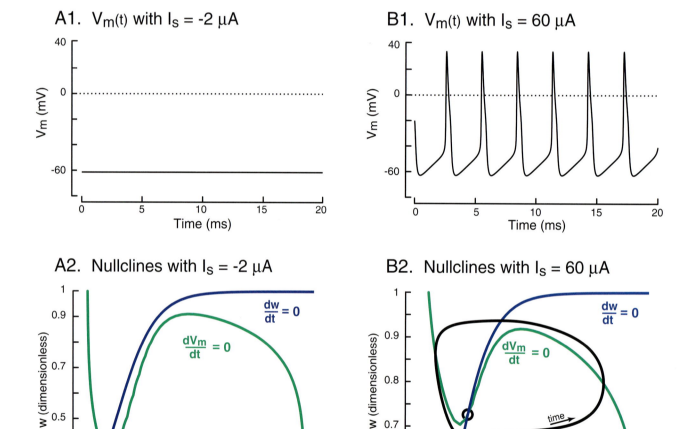

FIGURE 7.17 Oscillations emerge as fixed points lose stability. In a phase plane, the intersection of nullclines defines a fixed point (i.e., $dV_m/dt = dw/dt = 0$). Fixed points can be either stable or unstable. (**A1**) When the value of $I_s = -2$ μA cm^{-2}, the computed membrane potential is quiescent. Although sufficiently large perturbation elicits an action potential (not shown), the system ultimately returns to the stable resting state. (**A2**) The fixed point is indicated with a filled black circle and it occurs on a stable branch of the V_m nullcline (green line). Thus, the fixed point is stable. (**B1**) When the value of $I_s = 60$ μA cm^{-2}, the computed membrane potential is oscillating (i.e., repetitive spiking). Although perturbations may transiently alter this pattern of firing, the system ultimately returns to this exact pattern of repetitive spiking. (**B2**) The fixed point (open circle at the intersection of the nullclines) is now located on the unstable branch of the V_m nullcline, and thus, the resting state is not stable. The closed curve (black line) represents the trajectory of the system in the phase plane and corresponds to a stable limit cycle. With the exception of I_s, all values for parameters were as in Fig. 7.16.

unstable middle branch (open circle in Fig. 7.17B2). Thus, rather than returning to a single point in the phase plane, the solution path continually travels along a closed curve in the phase plane (black line in Fig. 7.17B2), which is referred to as a *limit cycle*. This limit cycle is stable, and if the system is momentarily perturbed, the trajectory will only transiently leave the limit cycle. Regardless of the magnitude of the perturbation, the solution path will ultimately return to the limit cycle.

Bifurcations and Bistability

A nonlinear system can make a transition from a stable fixed point to a limit cycle in several ways [for review see Izhikevich (2000)]. Such transitions are termed *bifurcations*. A bifurcation occurs any time the phase portrait (i.e., the overall picture of trajectories in the phase space) is changed to a topologically non-equivalent portrait by a change in the value of a control or *bifurcation parameter*. For example, increasing I_s from 0 to 60 μA causes the reduced model to transition (i.e., a bifurcation) from a resting state to an oscillatory state (Fig. 7.18A). In the phase portrait, this transition would represent the loss of a stable fixed point (e.g., Fig. 7.17A2) and the emergence of a stable limit cycle (e.g. Fig. 7.17B2). Conversely, if I_s is decreased from 60 to 0 μA cm^{-2}, there is a bifurcation of the oscillatory state, which is represented in the phase portrait as a loss of the stable limit cycle and the emergence of a stable fixed point.

The bifurcation is most clearly illustrated in what is termed a *bifurcation diagram* (Fig. 7.18B; see also Chapter 14). A bifurcation diagram plots a system parameter (I_s) on the horizontal axis and a system variable (V_m) on the vertical axis. For example, in Fig. 7.18B, the value of I_s was systematically increased from 0 to 400 μA cm^{-2}. At each new value of I_s, the steady-state value of V_m (i.e., the value of V_m at the fixed point) was determined and the stability of the fixed point was determined. For values of I_s between 0 and 16.31 μA cm^{-2} the fixed point is stable, which is indicated by the solid black line. At I_s = 16.31 μA cm^{-2}, the fixed point loses stability and a periodic solution emerges (i.e., a bifurcation). The transition illustrated in Fig. 7.18 is a type of bifurcation known as a Hopf bifurcation (filled circle labeled HB). For values of I_s between 16.31 and 336.8 μA cm^{-2}, the fixed point remains unstable, as indicated by the dashed black line. At I_s = 336.8 μA cm^{-2}, the periodic solution loses stability and a stable fixed point emerges, again via a Hopf bifurcation. The fixed point remains stable for all further increases in I_s.

Figure 7.18B also illustrates the maximum and minimum values of V_m for the oscillatory response

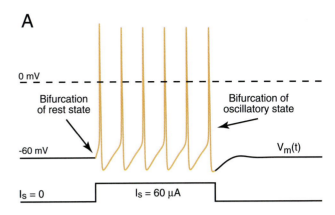

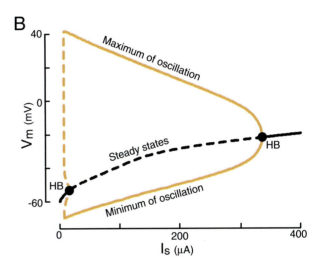

FIGURE 7.18 Bifurcations between quiescent and oscillatory states. As parameters are changed, nonlinear systems can make transitions between a stable steady state (i.e., a quiescent state) and a limit cycle (i.e., oscillations). These transitions are referred to as bifurcations. (**A**) When I_s = 0 μA cm^{-2}, the system is in a stable steady state. When I_s = 60 μA cm^{-2}, the resting state loses stability and stable oscillations emerge. If I_s is returned to 0 μA cm^{-2}, the periodic solution loses stability, and the system returns to a stable steady state. With the exception of I_s, all values for parameters were as in Fig. 7.16. (**B**) A bifurcation diagram plots the steady states (black line) of a system as the value of a parameter is systematically varied. Stable steady states are indicated by the solid line, and unstable steady states are indicated by the dashed line. For values of I_s between 0 and 16.31 μA cm^{-2}, the steady state is stable and the system is quiescent. At I_s = 16.31 μA cm^{-2}, however, the steady state loses stability and oscillations emerge. The mathematical characteristics of this bifurcation classify it as a Hopf bifurcation (point HB). [For a review of the mathematical criteria used to classify various types of bifurcations see Strogatz (1994).] The periodic solution loses stability as I_s > 336.8 μA cm^{-2}, and once again the system is quiescent (i.e., the steady state is stable). The second transition is also a Hopf bifurcation. In addition to the steady states, the bifurcation diagram plots the maximum and minimum values of V_m during periodic solutions (yellow line). Periodic solutions can be either unstable (dashed line) or stable (solid line). With the exception of I_s, all values for parameters were as in Fig. 7.16. In this and subsequent figures, the bifurcation analyses were performed using the AUTO (Doedel, 1981) algorithm, which is incorporated into the XPPAUT (Ermentrout, 2002) software package.

(the yellow lines). Just as a fixed point can be stable or unstable, a periodic solution can also be stable or unstable. The unstable periodic solutions are indicated by the dashed, yellow line, and the stable limit cycles are indicated by the solid, yellow line. When a Hopf bifurcation leads to *unstable* periodic solutions (i.e., the Hopf bifurcation that occurs at $I_s = 16.31 \, \mu\text{A/cm}^2$), then the bifurcation is termed *subcritical* (Fig. 7.19A1). When a Hopf bifurcation leads to only *stable* periodic solutions (i.e., the Hopf bifurcation that occurs at $I_s = 336.8 \, \mu\text{A cm}^{-2}$), it is termed *supercritical* (Fig. 7.19A2). The two Hopf bifurcations are illustrated in greater detail in Fig. 7.19. The supercritical Hopf bifurcation is the simpler type and, because of its appearance in a bifurcation diagram, is also referred to as a pitchfork bifurcation. At the supercritical Hopf bifurcation, the amplitude of the periodic solution slowly decreases in amplitude until a stable steady state is reached. This lethargic decay is called "critical slowing down" in the physics literature (Strogatz, 1994). In contrast to the supercritical Hopf bifurcation, the subcritical Hopf bifurcation has a relatively complex configuration (Fig. 7.19A1). The emergent branches of the unstable periodic solutions (dashed yellow lines) fold back into the parameter region where the steady state is stable. This backward folding has several important implications. First, because of this backward fold, there is a parameter region where the stable fixed points (i.e., solid black line) and the stable limit cycle (i.e., solid yellow lines) overlap. In this parameter region, the model is *bistable*, i.e., both a stable quiescent and a stable oscillatory solutions coexist (see also Chapter 14). In the phase plane (not shown), this type of bistability would be represented by a stable fixed point at the intersection of the nullclines, and surrounding the stable fixed point would first be an unstable period orbit and then a stable limit cycle. [*Note:* Bistability can result from other types of bifurcation schemata which in turn would have different properties and different phase plane portraits. For some examples of different mechanisms for achieving bistability, see Bertram *et al.* (1995), Canavier *et al.* (2002), Izhikevich (2000), or Rinzel and Ermentrout (1998).] Second, the existence of different stable states allows for the possibility of "jumps" between the two stable states and of hysteresis (i.e., the lagging of an effect behind its cause). Finally, as the solution jumps from the stable steady state to the stable periodic solution, large-amplitude oscillations appear dramatically from the resting state. In the physics literature, this dramatic jumping between states is referred to an "explosive instability" or a "blowup" (Strogatz, 1994).

Bistability can endow a neuron with some very interesting dynamical properties. For example, in the

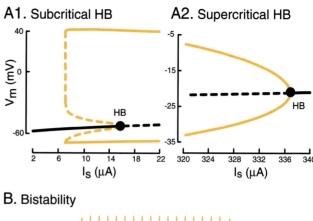

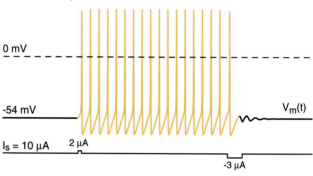

FIGURE 7.19   Bistability. In a nonlinear system, it is possible for more than one stable solution to coexist in a given parameter regime. Thus, the system is said to be bistable (i.e., two stable solutions) or multistable (more than two stable solutions). For an example of a multistable system see Canavier *et al.* (1993) and Butera (1998) (**A**) The regions of the two Hopf bifurcations in Fig. 7.18 are expanded and replotted. (**A1**) The first Hopf bifurcation (i.e., $I_s = 16.31 \, \mu\text{A cm}^{-2}$) is a subcritical Hopf bifurcation. The unstable branch of the periodic solution folds back over a region in which the steady state is stable. This results in a parameter region where the stable steady state and stable periodic solution overlap. Thus, the system can reside on either of the two stable attractors and perturbations can switch the system between the two stable states (see below). (**A2**) The second Hopf bifurcation is a supercritical Hopf bifurcation. The amplitude of the periodic solution decreases in amplitude through exponentially damped oscillations. (**B**) At the beginning of the simulation, the system resides in a stable resting state. A brief perturbation (see trace labeled I_s), however, induces a switch from the stable resting state to the stable oscillatory state. A second perturbation can switch the system back to the stable resting state. At a subcritical Hopf bifurcation, the trajectory must "jump" between distant attractors. Thus, large-amplitude oscillations appear dramatically from the resting state. With the exception of I_s, all values for parameters were as in Fig. 7.16. At the beginning of the simulation (i.e., at $t = 0$), the values for the two state variables [i.e., $V_m(t)$ and $w(t)$] were initially set to $V_m(0) = -54.314 \, \text{mV}$ and $w(0) = 0.515$. The values of state variables at the beginning of a simulation are referred to as the *initial conditions*.

bistable regime, sufficiently large perturbations can induce jump or switch from a resting state (i.e., stable steady solution) to a spiking state (i.e., stable oscillatory state) and vice versa. Figure 7.19B illustrates such a case. At the beginning of the simulation (i.e., at

$t = 0$), the values for the two-state variables [$V_m(t)$ and $w(t)$] were initially set to $V_m(0) = -54.314$ mV and $w(0) = 0.515$. The values of state variables at the beginning of a simulation are referred to as the *initial conditions*. With these initial conditions and parameter values, the system resides on the stable fixed point and is quiescent [black line in the $V_m(t)$ trace of Fig. 7.19A]. In the absence of any large perturbations, the system will remain in this resting state indefinitely. However, a sufficiently large perturbation (e.g., a 2 μA cm⁻², 3 ms injection of current) can send the solution path off the stable fixed point and onto the stable limit cycle. In the absence of any large perturbations, the system will remain in the oscillatory state indefinitely. However, a sufficiently large perturbation (e.g., a –3 μA cm⁻², 12 ms injection of current) can send the solution path off the stable limit cycle and back onto the stable fixed point. Such behavior has been observed empirically, for example, in the squid giant axon (Guttman *et al.*, 1980) and in the bursting neuron R15 of *Aplysia* (Lechner *et al.*, 1996). The switch from the resting state to the oscillatory state is often referred to as *hard excitation*, and the switch from the oscillatory state to the resting state is often referred to as *annihilation*. Moreover, this bistable behavior is critical for the occurrence of bursting when a slow conductance is added to the system (see below).

Dynamical Underpinnings of Bursting Activity

In contrast to pacemaker activity (i.e., continuous spiking activity with an approximately constant interspike interval), *bursting activity* is characterized by the clustering of spikes into groups that are separated by quiescent periods. Bursting cells can be classified as conditional or endogenous bursters. *Endogenous bursters* fire in a bursting pattern in the absence of any input, whereas *conditional bursters* can fire in bursts if they receive appropriate input from other neurons within a network. Endogenous bursting activity relies on bistability in the system.

In the example of bistability described above (Fig. 7.19), the bifurcation parameter was an externally applied current. Other, more physiologically relevant, parameters can achieve similar results. For example, adjusting the magnitude of an ionic conductance such as $\bar{g}_K$ (i.e., the maximal conductance of I_K; see Eq. 7.52) can alter the dynamical properties of the system (Av-Ron *et al.*, 1991). Figure 7.20 illustrates a bifurcation diagram in which $\bar{g}_K$ was used as the bifurcation parameter. As the value of $\bar{g}_K$ is decreased, the steady state loses stability and an oscillatory state emerges via a subcritical Hopf bifurcation. This oscillatory state represents the pacemaker activity, which is similar to that illustrated in Fig. 7.17B1. With

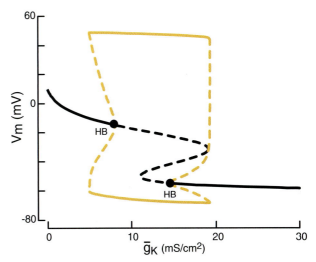

FIGURE 7.20 Bifurcation diagram as a function of $\bar{g}_K$. As in previous bifurcation diagrams stable solutions are indicated by solid lines and unstable solutions are indicated by dashed lines. The steady states are indicated by the black line, and the periodic solutions are indicated by the yellow line. Moving from right to left along the X axis, the maximum conductance for the K⁺ current (i.e., $\bar{g}_K$) decreases. As $\bar{g}_K$ is decreased, the steady state loses stability and oscillations emerge via a subcritical Hopf bifurcation (point HB). As $\bar{g}_K$ continues to decrease, the periodic solution loses stability and a stable steady state emerges via a second subcritical Hopf bifurcation. With the exception of $\bar{g}_K$, all values for parameters were as in Fig. 7.16. Adapted from Av-Ron *et al.* (1991).

further reductions in $\bar{g}_K$, the oscillatory state loses stability and a stable steady state emerges via a subcritical Hopf bifurcation. Thus, there are two parameter regions where bistability exists, and in principle, the system could undergo cyclical transitions between quiescence and oscillatory states (i.e., bursting) if the value of $\bar{g}_K$ was forced to traverse back and forth between parameter regions associated with a stable steady state and a stable limit cycle.

Generally, bursting cannot occur in a two-variable model (Rinzel and Ermentrout, 1998). Bursting can be realized, however, by incorporating an additional process (i.e., a relatively slow negative feedback process). To illustrate the general principles underlying burst firing, the reduced model presented above (i.e., Eqs. 7.49 and 7.50) is extended to include two more currents (I_{Ca} and $I_{K,Ca}$; see also Chapter 5) and an additional variable ([Ca^{2+}]), which describes the intracellular concentration of Ca^{2+}. Although several schemata are possible, the two additional currents considered here are an inward Ca^{2+} current (I_{Ca}) that activates with increasing depolarization and an outward K⁺ current that is activated by intracellular Ca^{2+} ($I_{K,Ca}$). The two new currents are described by

$$I_{K,Ca} = \bar{g}_{K,Ca} z([Ca^{2+}])(V_m - E_K), \qquad (7.61)$$

$$I_{Ca} = \bar{g}_{Ca} m_{\infty}^{mp}(V_m)(1 - w(V_m, t))(V_m - E_{Ca}), \qquad (7.62)$$

where $\bar{g}_i$ represent the maximal conductances of the ionic currents, E_i represent their respective equilibrium potentials, and z is a function that describes the Ca^{2+} dependency of the $I_{K,Ca}$, which is given by

$$z([Ca^{2+}]) = \frac{[Ca^{2+}]}{[Ca^{2+}] + K_d}, \qquad (7.63)$$

where $[Ca^{2+}]$ represents the concentration of intracellular Ca^{2+} and K_d is the concentration at which $I_{K,Ca}$ is half-activated. The dynamics of intracellular Ca^{2+} are described by

$$\frac{d[Ca^{2+}]}{dt} = K_p(-I_{Ca}) - R[Ca^{2+}], \qquad (7.64)$$

where K_p is a conversion factor from current to concentration and R is the removal rate constant. (See Chapter 14 for more detailed models of Ca^{2+} regulation.) Note that by extending Eq. 7.49 to include $I_{K,Ca}$, the total K^+ conductance of the reduced model becomes

$$g_K^{(total)} = \bar{g}_K \left(\frac{w(V_m, t)}{s}\right)^{wp} + \bar{g}_{K,Ca} z([Ca^{2+}]). \qquad (7.65)$$

The conductances $\bar{g}_K$ and $\bar{g}_{K,Ca}$ are determined such that the system traverses a region of bistability (see Fig. 7.20). [For additional details see Av-Ron *et al.* (1991, 1993).]

Figure 7.21 illustrates a bursting solution to the three-variable model (Eqs. 7.49, 7.50, and 7.64). During the quiescent phase, V_m (Fig. 7.21A) ramps up and fast subthreshold oscillations give rise to a burst of action potentials. During the burst of action potentials, $[Ca^{2+}]$ increases (Fig. 7.21B), which in turn increases the magnitude of $I_{K,Ca}$ (Fig. 7.21C). Once $g_K^{(total)}$ becomes sufficiently large, the oscillation is halted. Calcium is removed during the quiescent period, which in turn reduces $I_{K,Ca}$, and oscillations return. This pattern of activity does not require an external stimulus and will continue indefinitely. Thus, this type of bursting behavior is that of an *endogenous burster*. This type of bursting activity has been observed in mesencephalic trigeminal sensory neurons (Pedroarena *et al.*, 1999).

Although phase plane analysis cannot provide a full description for higher-order systems such as the three-variable model (i.e., Eqs. 7.49, 7.50, and 7.64), judicious two-dimensional projections can yield useful insights into the dynamical behavior of higher-order systems. [Canavier *et al.* (1993, 1994) provide an

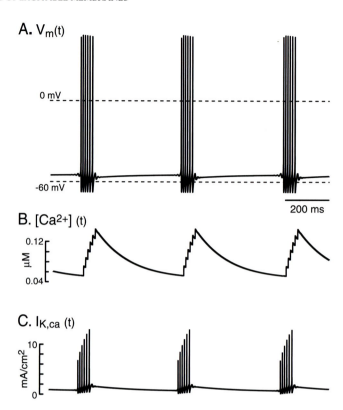

FIGURE 7.21 Elliptic (Type III) bursting in a three-dimensional model. In the absence of any external stimulus, the three-variable model generates sustained bursting activity. (**A**) Time-series plot of the computed membrane potential [$V_m(t)$]. Brief bursts of action potentials are separated by quiescent periods. Small subthreshold oscillations can be seen waxing and waning before and after each burst of spikes. (**B**) Time-series plot of the computed intracellular concentration of Ca^{2+} ([Ca^{2+}](t)). Calcium enters the cell during the spikes and slowly accumulates during the burst. During the intervening quiescent period, the levels of Ca^{2+} fall. Note the slow time scale with which levels of Ca^{2+} vary. (**C**) Time-series plot of $I_{K,Ca}(t)$ during the bursting activity. As Ca^{2+} accumulates during the burst [see (B)], the magnitude of $I_{K,Ca}$ increases. Once the total K^+ conductance (see Eq. 7.65) is sufficiently large, the spiking activity is terminated. As the level of Ca^{2+} slowly decreases [see (B)] during the quiescent period, $I_{K,Ca}$ begins to decrease. Once the total K^+ conductance has decreased sufficiently, spiking resumes. This cycle of events repeats indefinitely and does not require an external stimulus. Thus, this bursting activity is intrinsic (or endogenous) to the system. With the exception of $\bar{g}_{Ca} = 5$ mS/cm², $E_{Ca} = 124$ mV, $\bar{g}_K = 12$ mS/cm², $E_l = -50$ mV, $\bar{g}_{K,Ca} = 0.5$ mS/cm², $K_d = 0.5$ mM, $K_p = 0.00052$, and $R = 0.0045$, all values for parameters were as in Fig. 7.16. Adapted from Av-Ron et al. (1993).

example of how a phase plane can be used to analyze an 11-order model of a bursting neuron.] Figure 7.22A illustrates a bifurcation diagram for the three-variable model. To construct this diagram, the observation was made that the intracellular concentration of Ca^{2+} (i.e., Eq. 7.64) changed very slowly relative to the other state variables, i.e., $V_m(t)$ and $w(t)$. Thus, the function z, which is directly related to the intracellular concen-

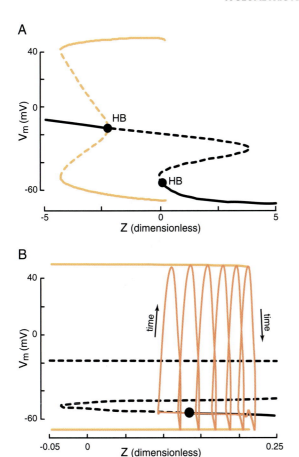

tration of Ca^{2+} (see Eq. 7.63), can be treated as a *para-meter* (i.e., a bifurcation parameter) rather than as a *function* when analyzing the dynamics of V_m and w. Because V_m and w are relatively fast as compared with $[Ca^{2+}]$ (and z), Eqs. 7.49 and 7.50 (i.e., the equations that describe V_m and w) are termed the fast subsystem and Eq. 7.64 (i.e., the equation that describes $[Ca^{2+}]$) is termed the slow subsystem. For each value of z, the dynamics of V_m and w can be analyzed. For some values of z, V_m and w may oscillate (i.e., spiking activity), whereas for other values of z, V_m, and w may settle to a fixed point (i.e., quiescent membrane potential). As the magnitude of z is varied, the fast subsystem may undergo bifurcations where V_m and w are first at a fixed point and then begin to oscillate, or conversely, where V_m and w are oscillating and then become quiescent (see Fig. 7.22).

In Fig. 7.22A, the branch of the steady states (black line with stable steady states represented by a sold line and unstable steady states represented by a dashed line) forms the S-shaped curve and the oscillatory solutions are represented by the forked curve (yellow line with stable oscillatory states represented by a solid line and unstable oscillation represented by a dashed line). As the magnitude of z decreases (i.e., moving from right to left along the X axis), the stable steady state destabilizes at a turning point (point HB at $z = 0.13$), which is the subcritical Hopf bifurcation. The unstable oscillations born at this Hopf bifurcation, which are not illustrated, terminate in a saddle loop. A stable period solution emerges from this saddle loop bifurcation. As the magnitude of z continues to decrease, the stable periodic solutions destabilize, again via a subcritical Hopf bifurcation (point HB at $z = -2.3$), and a stable steady state is reestablished. Although negative values of z are physiologically irrelevant, the bifurcation diagram serves to illustrate that the stable steady states of the three-variable model destabilize via subcritical Hopf bifurcations, and that for physiologically relevant values of z (i.e., $z \geq 0$), the stable periodic solutions disappear abruptly when they reach the saddle node of periodics (SNP). [For more detailed explanations of this bifurcation diagram, see Bertram *et al.* (1995), Canavier *et al.* (2002), Rinzel and Ermentrout, (1998), and Izhikevich (2001).]

The transitions between resting and oscillatory states can be visualized by projecting the burst solution in the $z–V_m$ plane onto the bifurcation diagram (Fig. 7.22B). The resolution of the bifurcation diagram has been increased to focus attention on the ranges of values for z that are observed during the bursting activity (see Fig. 7.21). The trajectory (red line) tracks the stable periodic solution (i.e., the fast subsystem)

FIGURE 7.22 Bifurcation diagram of the three-dimensional system with z as the bifurcation parameter. Although z is a variable, it is sufficiently slow relative to changes in either $V_m(t)$ or $w(t)$, which constitute a fast subsystem, that it can be treated as a parameter for the purpose of generating this bifurcation analysis. (**A**) Bifurcation diagram using z as the control parameter. As before, solid lines indicate stable solutions and dashed lines indicated unstable solutions. The black line indicates the steady states, and the yellow line indicates periodic solutions. From right to left (i.e., as the magnitude of z is decreased), the steady state loses stability via a subcritical Hopf bifurcation (point HB at $z = 0.13$). The unstable oscillations that are born at the subcritical Hopf bifurcation (see Fig. 7.20) terminate in a saddle loop (not shown) and a stable periodic solution emerges beyond the saddle-loop bifurcation. As the magnitude of z continues to decrease, the stable periodic solution destabilizes, and a stable steady state is reestablished, again via a Hopf bifurcation (point HB at $z = -2.3$). (**B**) Projection of the bursting solution in the $z – V_m$ phase plane (red line) onto the bifurcation diagram for positive values of z. Note the resolution of the bifurcation has been substantially increased. The direction of the solution path (i.e., time) is indicated by the arrows. The trajectory tracks the fast subsystem (i.e. spiking in the periodic solution) until it reaches the unstable regime, at which point, the trajectory "falls" onto the stable branch of the steady state. As intracellular levels of Ca^{2+} decrease (see Fig. 7.21B) and z decreases, the trajectory is forced past the point where the stable steady state disappears (black, filled circle). Once in this unstable regime, the trajectory "jumps" onto the stable branch of the periodic solution. Thus, bursting emerges as the slow subsystem [i.e., $z(t)$] forces the trajectory to slowly drift through bifurcations in the fast subsystem. With the exception of z, all values for parameters were as in Fig. 7.21.

BOX 7.5

ALTERNATIVE MODELS OF EXCITABILITY

Over the past 50 years, the formalisms developed by Hodgkin and Huxley have been used as a starting point for developing conductance-based models. These models were usually developed to simulate specific systems, such as the endogenous bursting cell R15 in *Aplysia* (Canavier *et al.*, 1991) or the thalamocortical relay neurons (McCormick and Huguenard, 1992). These models often incorporate descriptions of 3 to 10 ionic conductances and may include additional variables that describe processes such as the dynamics of intracellular levels of Ca^{2+} or the dynamics of a second messenger such as cAMP. Thus, these models are high dimensional and difficult to analyze. Others have taken an alternative approach and developed low-dimensional models. Although these reduced models often lack biophysical detail, they still manifest salient features of neuronal excitability (e.g., threshold behavior and excitability) and have provided some fundamental insights into excitability. Three of these reduced models are described briefly.

The FitzHugh–Nagumo Model

Perhaps the best known and most widely used low-dimensional model of an excitable system is the FitzHugh–Nagumo model (FitzHugh, 1961; Nagumo *et al.*, 1962) (for reviews see Rinzel, 1990; Koch, 1999; Murray, 1993). This model qualitatively describes the events occurring in an excitable membrane. The FitzHugh–Nagumo equations are often written, assuming dimensionless variables, in the general form

$$\frac{dv}{dt} = f(v) - w + I_s, \qquad (7.66)$$

$$\frac{dw}{dt} = \phi(v - \gamma w), \qquad (7.67)$$

where $f(v)$ is the cubic:

$$f(v) = v(1-v)(v-a). \qquad (7.68)$$

Here v corresponds to membrane potential, w is a recovery variable (i.e., combines the effects of h and n), which activates slowly (i.e., the rate constant, ϕ, is less than one), and γ is a positive constant. The cubic expression represents a nonlinear I–V relationship, which follows from the assumption that inward current activates instantaneously; a is the voltage threshold for this current. The qualitative behavior of the FitzHugh–Nagumo model is very similar to that of the Hodgkin–Huxley system in that for some parameter regimes the model has a globally stable resting state yet is excitable in that a sufficiently large perturba-

tion will elicit a spike. The system does not have a saddle node and, thus, manifests a pseudo-threshold, similar to the Hodgkin–Huxley system (see Box 7.2). In addition, with sustained current injection, a limit cycle appears and the system oscillates. The simplicity of this model has made it a model of choice for nonlinear dynamical analyses of excitability and for simulations of large-scale networks of excitable elements.

The Morris–Lecar Model

The Morris–Lecar model (Morris and Lecar, 1981) was formulated and studied in the context of investigating electrical activity in barnacle muscle fibers. The model incorporates a voltage-gated Ca^{2+} current and a voltage-gated, delay-rectifier K^+ current; neither current inactivates. The Morris–Lecar equations are often written, assuming dimensionless variables, in the general form

$$C\frac{dv}{dt} = I_s - \bar{g}_{Ca}m_\infty(v)(v - v_{ca}) - \qquad (7.69)$$

$$\bar{g}_K w(v)(v - v_K) - \bar{g}_l(v - v_l),$$

$$\frac{dw}{dt} = \Phi\frac{w_\infty(v) - w}{\tau_w(v)}, \qquad (7.70)$$

where

$$m_\infty(v) = 0.5\left(1 + \tanh\frac{v - v_1}{v_2}\right), \qquad (7.71)$$

$$w_\infty(v) = 0.5\left(1 + \tanh\frac{v - v_3}{v_4}\right), \qquad (7.72)$$

$$\tau_w(v) = \frac{1}{\cosh\dfrac{v - v_3}{v_4}}, \qquad (7.73)$$

where v_i are parameters, and w is the fraction of K^+ channels open.

The qualitative behavior of the Morris–Lecar model is very similar to that of the Hodgkin–Huxley system in that for some parameter regimes the model has a globally stable resting state yet is excitable. The system manifests a pseudo-threshold and oscillates with sustained current injection. The Morris–Lecar model has been studied extensively by Rinzel and Ermentrout (1998). They demonstrated an intriguing feature of this system. For some parameter regimes the oscillations emerge via a subcritical Hopf bifurcation. Thus, this two-dimensional system manifests bistability. Indeed, with different parameter regimes the two-variable system manifests tristability (i.e.,

---

BOX 7.5 *(continued)*

ALTERNATIVE MODELS OF EXCITABILITY

two distinct stable steady states and a stable limit cycle coexist for a given parameter regime). The greatest advantage of the model has come from an expanded version, which includes a third variable that describes the dynamics of intracellular Ca^{2+}, and by adding a Ca^{2+}-dependent K^+ current. This three-variable system can simulate a variety of bursting behaviors. The simplicity of this model and its rich dynamical repertoire make it an excellent choice for nonlinear dynamical analyses of complex neuronal firing patterns such as bursting.

The Hindmarsh–Rose Model

Although similar to the FitzHugh–Nagumo model, the Hindmarsh–Rose model (Hindmarsh and Rose, 1982) was developed from first principles with the assumptions that the rate of change of membrane potential (dx/dt) depends linearly on z (an externally applied current) and y (an intrinsic current). The forms of the functions and values for the parameters were selected to fit data from a large neuron in the visceral ganglion of the pond snail *Lymnaea stagnalis*. The equations for this reduced model are

$$\frac{dx}{dt} = -a\big(f(x) - y - z\big), \qquad (7.74)$$

$$\frac{dy}{dt} = b\big(f(x) - q\,\exp(rx) + s - y\big), \qquad (7.75)$$

where $f(x)$ is given by the cubic

$$f(x) = cx^3 + dx^2 + ex + h, \qquad (7.76)$$

where a, b, c, d, e, f, h, q, r, and s are constants. This two-variable system offers some advantages, such as a better fit of the frequency–current relationship and a better fit to aspects of the spike waveform. As with the Morris–Lecar model, however, the greatest advantages of the Hindmarsh–Rose model have come from extended versions. A three-dimensional version (Rose and Hindmarsh, 1984, 1985, 1989a,b; Wang, 1993) manifests bursting activity and multistability, including multirhythmicity (i.e., multiple forms of bursting activity coexisting for a single-parameter regime). Additional extensions of the model (Rose and Hindmarsh, 1989c) have been used to investigate the dynamical properties of thalamic neurons. Of the reduced models of excitability, the Hindmarsh–Rose model appears to exhibit the richest repertoire of dynamical properties.

---

until it loses stability (i.e., at the SNP). During spiking activity, the levels of Ca^{2+} increase, and hence the magnitude of z also increases (see Eq. 7, Fig. 7.21). Thus, the general movement of the trajectory along the stable branch of the periodic solution is from left to right. At the termination of the stable branch of the periodic solution (i.e., at the SNP), the trajectory then "falls" onto the stable branch of the steady state (i.e., a quiescent membrane potential). During the quiescent phase, the level of Ca^{2+} decreases, and hence the magnitude of z also decreases (see Fig. 7.21). Thus, general movement of the trajectory along the stable branch of the steady state solution is from right to left. As z decreases (see Fig. 7.21B), the trajectory slowly moves through the Hopf bifurcation, and as the steady state loses stability, the trajectory "jumps" onto the stable branches of the periodic solution and the general movement of the trajectory once again is from left to right. Thus, the oscillations in the intracellular levels of Ca^{2+}, which in turn induce oscillations in z and

$I_{K,Ca}$, drive the trajectory back and forth through the bifurcation between spiking and nonspiking.

The envelope of $V_m(t)$ during the burst activity of Fig. 21A is "elliptical" as the subthreshold oscillations gradually increase (i.e., wax) in amplitude before a burst and gradually decrease (i.e., wane) at the end of the burst. Thus, this type of bursting activity is often referred to as *elliptic bursting*. Other qualitative types of bursting have been observed, such as *parabolic bursting* (i.e., the frequency of spiking activity gradually increases and then decreases during the burst) and *square-wave bursting* (i.e., the burst of spikes rides atop a plateau-like depolarization). Bursting activity can also be classified on the basis of the types of bifurcations that give rise to the oscillatory states. The history of formally classifying bursting began with Rinzel and Lee (1987), who contrasted the bifurcation mechanisms inherent in the elliptic, parabolic, and square-wave bursters. Bertram *et al.* (1995) suggested referring to the types of bursting using Roman numerals (i.e., elliptic bursting is Type I, parabolic bursting is Type II, and square-wave bursting is Type III). There are many ways that a nonlinear system can make a transition from a stable resting state to a stable periodic solution and there are many possible combinations. Izhikevich (2000) has described all possible combinations and reviewed examples of each.

Summary

Some neurons are normally quiescent, but excitable in that they can fire one or more action potentials in response to transient stimuli. Others are capable of firing action potentials in a repetitive fashion in the absence of external stimuli. The same nonlinear properties of the cell membrane that allow a neuron to fire a single action potential can enable it to function as a nonlinear oscillator, generating sustained rhythmic activity. The simplest type of repetitive action potential firing is pacemaker activity in which single, apparently identical action potentials are generated at relatively fixed intervals. Another type of rhythmic firing, bursting, is characterized by action potentials clustered into bursts that are separated by quiescent periods. To generate either an action potential or a sustained oscillation of any kind, two opposing processes are required, one relatively rapid autocatalytic process and one somewhat slower, restorative process. For rhythmic activity to be sustained, the opposing processes must have the appropriate steady-state characteristics and kinetics to continue to alternately dominate the system dynamics. In addition to the two processes required for action potential generation, a bursting oscillation requires at least one additional, slower process to modulate action potential firing so as to group them in bursts.

References

Abbott, L. F. (1994). Single neuron dynamics: An introduction. In "Neural Modeling and Neural Networks" (F. Ventriglia, ed.), pp. 57–78. Pergamon Press, New York.

Abraham, R. H., and Shaw, C. D. (1992). "Dynamics: The Geometry of Behavior." Addison–Wesley, Redwood City, CA.

Alexander, J. C., and Cai, D.-Y. (1991). On the dynamics of bursting systems. J. Math. Biol. **29**, 405–423.

Av-Ron, E. (1994). The role of a transient potassium current in a bursting neuron model. J. Math. Biol. **33**, 71–87.

Av-Ron, E., Parnas, H., and Segel, L. (1991). A minimal biophysical model for an excitable and oscillatory neuron. Biol. Cybern. **65**, 487–500.

Av-Ron, E., Parnas, H., and Segel, L. (1993). A basic biophysical model for bursting neurons. Biol. Cybern. **69**, 87–95.

Bertram, R. (1994). Reduced-system analysis of the effects of serotonin on a molluscan burster neuron. Biol. Cybern. **70**, 359–368.

Bertram, R., Butte, M. J., Kiemel, T., and Sherman, A. (1995). Topological and phenomenological classification of bursting oscillations. Bull. Math. Biol. **57**, 413–439.

Bezanilla, F., and Armstrong, C. M. (1977). Inactivation of the sodium channel. I. Sodium current experiments. J. Gen. Physiol. **70**, 549–566.

Bower, J. M., and Beeman, D. (1998). "The Book of GENESIS." Springer-Verlag, Santa Clara.

Brown, A. M. (1999). A methodology for simulating biological systems using Microsoft EXCEL. Comput. Methods Programs Biomed. **58**, 181–190.

Brown, A. M. (2000). Simulation of axonal excitability using a spreadsheet template created in Microsoft EXCEL. Comput. Methods Programs Biomed. **63**, 47–54.

Butera, R. J. (1998). Multirhythmic bursting. Chaos **8**, 274–284.

Butera, R. J., Clark, J. W., and Byrne, J. H. (1996). Dissection and reduction of a modeled bursting neuron. J. Comput. Neurosci. **3**, 199–223.

Byrne, J. H., and Schultz, S. G. (1994). "An Introduction to Membrane Transport and Bioelectricty." Raven Press, New York.

Canavier, C. C., Baxter, D. A., and Byrne, J. H. (2002). Repetitive action potential firing. In "Encyclopedia of Life Sciences" (N.P. Group, Ed.), on line (http://www.els.net). Grove's Dictionaries, New York.

Canavier, C. C., Baxter, D. A., Clark, J. W., and Byrne, J. H. (1993). Nonlinear dynamics in a model neuron provide a novel mechanism for transient synaptic inputs to produce long-term alterations of postsynaptic activity. J. Neurophysiol. **69**, 2252–2257.

Canavier, C. C., Baxter, D. A., Clark, J. W., and Byrne, J. H. (1994). Multiple modes of activity in a model neuron suggest a novel mechanism for the effects of neuromodulators. J. Neurophysiol. **72**, 872–882.

Canavier, C. C., Clark, J. W., and Byrne, J. H. (1991). Simulation of the bursting activity of neuron R15 in Aplysia: Role of ionic currents, calcium balance, and modulatory transmitters. J. Neurophysiol. **66**, 2107–2124.

Clay, J. R. (1998). Excitability of squid giant axon revisited. J. Neurophysiol. **80**, 903–913.

Cole, K. S. (1949). Dynamic electrical characteristics of the squid axon membrane. Arch. Sci. Physiol. **22**, 253–258.

Cole, K. S. (1968). "Membranes, Ions, and Impulses." Univ. of California Press, Berkeley.

Cole, K. S., Antosiewicz, H. A., and Rabinowitz, P. (1955). Automatic computation of nerve excitation. J. Soc. Ind. Appl. Math. **3**, 153–172.

Cole, K. S., Guttman, R., and Bezanilla, F. (1970). Nerve excitation without threshold. Proc. Natl. Acad. Sci. USA **65**, 884–891.

Cooke, I., and Lipkin, M. (Eds.) (1972). "Cellular Neurophysiology: A Source Book." Holt, Rinehart and Winston, New York.

Cronin, J. (1987). "Mathematical Aspects of Hodgkin–Huxley Neural Theory." Cambridge Univ. Press, New York.

Dayan, P., and Abbott, L. F. (2001). "Theoretical Neuroscience: Computational and Mathematical Modeling of Neural Systems." MIT Press, Cambridge, MA.

DeSchutter, E. (1989). Computer software for development and simulation of compartmental models of neurons. Comput. Biol. Med. **19**, 71–81.

DeSchutter, E. (1992). A consumer guide to neuronal modeling software. Trends Neurosci. **15**, 462–464.

DeSchutter, E. (Ed.) (2001). "Computational Neuroscience: Realistic Modeling for Experimentalists." CRC Press, New York.

Dodge, F., and Frankenhaeuser, B. (1959). Sodium currents in the myelinated nerve fibre of Xenopus laevis investigated with the voltage clamp technique. J. Physiol. (London) **148**, 188–200.

Doedel, E. J. (1981). AUTO: A program for the automatic bifurcation and analysis of autonomous systems. Congr. Num. **30**, 265–284.

Ermentrout, G. B. (1996). Type I membranes, phase resetting curves, and synchrony. *Neural Comp.* **8**, 979–1001.

Ermentrout, G. B. (2002). "Simulating, Analyzing and Animating Dynamical Systems: A Guide to XPPAUT for Researchers and Students." SIAM, Philadelphia.

FitzHugh, R. (1960). Thresholds and plateaus in the Hodgkin–Huxley equations. *J. Gen. Physiol.* **43**, 867–896.

FitzHugh, R. (1961). Impulses and physiological states in theoretical models of nerve membrane. *Biophys. J.* **1**, 445–466.

FitzHugh, R. (1966). Theoretical effects of temperature on threshold in the Hodgkin–Huxley nerve model. *J. Gen. Physiol.* **49**, 989–1005.

FitzHugh, R. (1969). Mathematical models of excitation and propagation in nerve. *In* "Biological Engineering" (H.P. Schwan, Ed.), pp. 1–85. McGraw–Hill, New York.

FitzHugh, R., and Antosiewicz, H. A. (1959). Automatic computation of nerve excitation: detailed correction and addition. *J. Soc. Ind. Appl. Math.* **7**, 447–458.

Gall, W. G., and Zhou, Y. (2000). An organizing center for planar neural excitability. *Neurocomp.* **32/33**, 757–765.

Guttman, R., Lewis, S., and Rinzel, J. (1980). Control of repetitive firing in squid axon membrane as a model for a neuroneoscillator. *J. Physiol. (London)* **305**, 377–395.

Hayes, R., Byrne, J. H., and Baxter, D. A. (2002). Neurosimulation: Tools and resources. *In* "The Handbook of Brain Theory and Neural Networks" (M.A. Arbib, Ed.), MIT Press, Cambridge, MA.

Hille, B. (2001). "Ionic Channels of Excitable Membranes." Sinauer Associates, Sunderland, MA.

Hindmarsh, J. L., and Rose, R. M. (1982). A model of the nerve impulse using two first-order differential equations. *Science* **296**, 162–164.

Hindmarsh, J. L., and Rose, R. M. (1984). A model of neuronal bursting using three coupled first order differential equations. *Proc. R. Soc. London Ser. B* **221**, 87–102.

Hines, M. L., and Carnevale, N. T. (1997). The NEURON simulation environment. *Neural Comput.* **9**, 1179–1209.

Hodgkin, A. L. (1964). The ionic basis of nervous conduction. *Science* **145**, 1148–1154.

Hodgkin, A. L. (1976). Chance and design in electrophysiology: An informal account of certain experiments on nerve carried out between 1934 and 1952. *J. Physiol. (London)* **263**, 1–21.

Hodgkin, A. L. (1977). "The Pursuit of Nature: Informal Essays on the History of Physiology." Cambridge Univ. Press, Cambridge.

Hodgkin, A. L., and Huxley, A. F. (1952a). Currents carried by sodium and potassium ions through the membrane of the giant axon of *Loligo. J. Physiol. (London)* **116**, 449–472.

Hodgkin, A. L., and Huxley, A. F. (1952b). The components of membrane conductance in the giant axon of *Loligo. J. Physiol. (London)* **116**, 473–496.

Hodgkin, A. L., and Huxley, A. F. (1952c). The dual effect of membrane potential on sodium conductance in the giant axon of *Loligo. J. Physiol. (London)* **116**, 497–506.

Hodgkin, A. L., and Huxley, A. F. (1952d). A quantitative description of membrane current and its application to conduction and excitation in nerve. *J. Physiol. (London)* **117**, 500–544.

Hodgkin, A. L., and Huxley, A. F. (1990). A quantitative description of membrane current and its application to conduction and excitation in nerve: 1952. *Bull. Math. Biol.* **52**, 5–23.

Hodgkin, A. L., Huxley, A. F., and Katz, B. (1952). Measurements of current–voltage relations in the membrane of the giant axon of Loligo. *J. Physiol. (London)* **116**, 424–448.

Hoppensteadt, F., and Izhikevich, E. (2001). Canonical neural models. *In* "Brain Theory and Neural Networks" (M. A. Arbib, Ed.), pp. 1–7. MIT Press, Cambridge, MA.

Hubbard, J. H., and West, B. H. (1992). "MacMath: A Dynamical Systems Software Package for the Macintosh." Springer-Verlag, New York.

Huxley, A. F. (1951). The ionic basis of electrical activity in nerve and muscle. *Biol. Rev.* **26**, 339–409.

Huxley, A. F. (1959). Can a nerve propagate a subthreshold disturbance? *J. Physiol. (London)* **148**, 80–81P.

Huxley, A. F. (1964). Excitation and conduction in nerve: Quantitative analysis. *Science* **145**, 1154–1159.

Huxley, A. F. (2000). Reminiscences: Working with Alan, 1939–1952. *J. Physiol. (London)* **527P**, 13S.

Huxley, A. F. (2002). Hodgkin and the action potential 1939–1952. *J. Physiol. (London)* **538**, 2.

Izhikevich, E. (2001). Synchronization of elliptic bursters. *SIAM Rev.* **43**, 315–344.

Izhikevich, E. M. (2000). Neural excitability, spiking and bursting. *Int. J. Bifurcation Chaos* **10**, 1171–1266.

Jackson, E. A. (1991). "Perspectives of Nonlinear Dynamics." Cambridge Univ. Press, Cambridge, MA.

Johnston, D., and Wu, S. M. (1997). "Foundations of Cellular Neurophysiology." MIT Press, Cambridge, MA.

Kepler, T. B., Abbott, L. F., and Marder, E. (1992). Reduction of conductance-based neuron models. *Biol. Cybern.* **66**, 381–387.

Kirsch, R. A. (1998). SEAC and start of image processing at the National Bureau of Standards. *IEEE Ann. Hist. Comput.* **20**, 7–13.

Kocak, H. (1989). "Differential and Difference Equations through Computer Experiments, 2nd ed. Springer-Verlag, New York.

Koch, C. (1999). "Biophysics of Computation: Information Processing in Single Neurons". Oxford Univ. Press, New York.

Krinskii, V. I., and Kokoz, Y. M. (1973). Analysis of equations of excitable membranes. I. Reduction of the Hodgkin–Huxley equations to a second order system. *Biofizika* **18**, 506–511.

Lechner, H. A., Baxter, D. A., Clark, J. W., and Byrne, J. H. (1996). Bistability and its regulation by serotonin in the endogenously bursting neuron R15 in *Aplysia. J. Neurophysiol.* **75**, 957–962.

MacGregor, R. J. (1987). "Neural and Brain Modeling." Academic Press, New York.

Marmont, G. (1949). Studies on the axon membrane. I. A new method. *J. Cell. Comp. Physiol.* **34**, 351–382.

McCormick, D. A., and Huguenard, J. (1992). A model of the electrophysiological properties of thalamocortical relay neurons. *J. Neurophysiol.* **68**, 1384–1440.

Moore, J. W., and Stuart, A. E. (2000). "Neurons in Action: Computer Simulations with NeuroLab." Sinauer Associates, Sunderland, MA.

Morris, C., and Lecar, H. (1981). Voltage oscillations in the barnacle giant muscle fiber. *Biophys. J.* **35**, 193–213.

Murray, J. D. (1993). "Mathematical Biology." Springer-Verlag, New York.

Nagumo, J. S., Arimato, S., and Yoshizawa, S. (1962). An active pulse transmission line simulating a nerve axon. *Proc. IRE* **50**, 2061–2070.

Nelson, M., and Rinzel, J. (1998). The Hodgkin–Huxley model. *In* "The Book of Genesis: Exploring Realistic Neural Models with the GEeneral NEural SImulation System" (J. M. Bower and D. Beeman, Eds.), pp. 29–49. Spring-Verlag, New York.

Palti, Y. (1971a). Description of axon membrane ionic conductances and currents. *In* "Biophysics and Physiology of Excitable Membrane" (W. J. Adelman, Ed.), pp. 168–182. Van Nostrand Reinhold, New York.

Palti, Y. (1971b). Digital computer solutions of membrane currents in the voltage clamped giant axon. *In* Biophysics and Physiology of Excitable Membranes (W. J. Adelman, Ed.), pp. 183–193. Van Nostrand Reinhold, New York.

Pavlidis, T. (1973). "Biological Oscillators: Their Mathematical Analysis." Academic Press, New York.

Pedroarena, C. M., Pose, I. E., Yamuy, J., Chase, M. H., and Morales, F. R. (1999). Oscillatory membrane potential activity in the soma of a primary afferent neuron. *J. Neurophysiol.* **82,** 1465–1476.

Rinzel, J. (1985). Excitation dynamics: Insights from simplified membrane models. *Fed. Proc.* **44,** 2944–2946.

Rinzel, J. (1990). Electrical excitability of cells, theory and experiment: Review of the Hodgkin–Huxley foundation and an update. *Bull. Math. Biol.* **52,** 5–23.

Rinzel, J. (1998). The Hodgkin–Huxley model. *In* "The Book of Genesis: Exploring Realistic Neural Networks with the GEneral NEural SImulation System" (J.M. Bower and D. Beeman, Eds.). Springer-Verlag, New York.

Rinzel, J., and Ermentrout, G. B. (1998). Analysis of neural excitability and oscillations. *In* "Methods in Neuronal Modeling: From Ions to Networks" (C. Koch and I. Segev, Eds.), 2nd ed., pp. 251–292. MIT Press, Cambridge, MA.

Rinzel, J., and Lee, Y. S. (1987). Dissection of a model for neuronal parabolic bursting. *J. Math. Biol.* **25,** 653–675.

Rose, R. M., and Hindmarsh, J. L. (1985). A model of a thalamic neuron. *Proc. R. Soc. London Ser. B.* **225,** 161–193.

Rose, R. M., and Hindmarsh, J. L. (1989a). The assembly of ionic currents in a thalamic neuron I. The three-dimensional model. *Proc. R. Soc. London Ser. B* **237,** 267–288.

Rose, R. M., and Hindmarsh, J. L. (1989b). The assembly of ionic currents in a thalamic neuron II. The stability and state diagrams. *Proc. R. Soc. London Ser. B* **237,** 289–312.

Rose, R. M., and Hindmarsh, J. L. (1989c). The assembly of ionic current in a thalamic neuron III. The seven-dimensional model. *Proc. R. Soc. London Ser. B* **237,** 313–334.

Strogatz, S. H. (1994). "Nonlinear Dynamics and Chaos: With Applications to Physics, Biology, Chemistry and Engineering." Perseus Books, Reading, MA.

Tuckwell, H. C. (1988). "Introduction to Technical Neurobiology", Vol. 2: "Nonlinear and Stochastic Theories." Cambridge Univ. Press, Cambridge, MA.

Tufillaro, N. B., Abbott, T., and Reilly, J. (1992). "An Experimental Approach to Nonlinear Dynamics and Chaos." Addison–Wesley, Redwood City, CA.

Vandenberg, C. A., and Bezanilla, F. (1991). A sodium gating model based on single channel, macroscopic ionic, and gating currents in the squid giant axon. *Biophys. J.* **60,** 1511–1533.

Ventriglia, F. (Ed.) (1994). "Neural Modeling and Neural Networks." Pergamon Press, New York.

Wang, X. -J. (1993). Genesis of bursting oscillations in the Hindmarsh–Rose model and homoclinicity to a chaotic saddle. *Physica D* **62,** 263–274.

Weiss, T. F. (1997). "Cellular Biophysics," Vol. 2: "Electrical Properties." MIT Press, Cambridge, MA.

Wheeler, D. J. (1992a). The EDSAC programming systems. *IEEE Ann. Hist. Comput.* **14,** 34–40.

Wheeler, J. (1992b). Applications of the EDSAC. *IEEE Ann. Hist. Comput.* **14,** 27–33.

Ziv, I., Baxter, D. A., and Byrne, J. H. (1994). Simulator for neural networks and action potentials: Description and application. *J. Neurophysiol.* **71,** 294–308.

8

Release of Neurotransmitters

Robert S. Zucker, Dimitri M. Kullmann, and Thomas L. Schwarz

The synapse is the point of functional contact between one neuron and another. It is the primary place at which information is transmitted from neuron to neuron in the central nervous system or from neuron to target (gland or muscle) in the periphery. The simplest way for one cell to inform another of its activity is by direct electrical interaction, in which the current generated extracellularly from the action potential in the first cell passes through neighboring cells. Owing to the shunting of current by the highly conductive extracellular fluid, a 100-mV action potential may generate only 10–100 μV in a neighboring neuron. This coupling can be improved if neighboring cells are joined by a specialized conductive pathway through gap junctions (see Chapter 15); even then, a presynaptic spike is not likely to generate more than about 1 mV postsynaptically, unless the presynaptic process is nearly as large or larger than the postsynaptic process. This biophysical constraint limits the number of presynaptic cells that can converge on and influence a postsynaptic cell, and such electrical connections can normally only be excitatory and short-lasting, are bidirectional in transmission, and show little plasticity or modifiability. They have limited potential for complex computation, but can be useful when a postsynaptic neuron must be activated with high reliability and speed or when concurrent activity in a large number of presynaptic afferents must be signaled.

ORGANIZATION OF THE CHEMICAL SYNAPSE

Most interneuronal communication relies on the use of a chemical intermediary, or *transmitter*, secreted subsequent to action potentials by presynaptic cells to influence the activity of postsynaptic cells. In chemical transmission, a single action potential in a small presynaptic terminal can generate a large *postsynaptic potential* (PSP) (as large as tens of millivolts). This is accomplished by the release of thousands to hundreds of thousands of molecules of transmitter that can bind to postsynaptic *receptor molecules* and open (or close) as many ion channels in about 1 ms. There is room for many afferents (often thousands) to interact and influence a postsynaptic neuron, and the effect can be either excitatory or inhibitory, depending on the ions that permeate the channels operated by the receptor. The resulting responses are either *excitatory postsynaptic potentials* (EPSPs) or *inhibitory postsynaptic potentials* (IPSPs), depending on whether they drive the cell toward a point above or below its firing threshold. Different afferents can have different effects, with different strengths and kinetics, on each other as well as on postsynaptic cells. These differences depend on the identity of the transmitter(s) released and the receptors present (see Chapters 9, 11, and 16). Chemical synapses are often modified by prior activity in the presynaptic neuron. Chemical synapses are also particularly subject to modulation of presynaptic ion channels by substances released by the postsynaptic or neighboring neurons. This flexibility is essential for the complex processing of information that neural circuits must accomplish, and it provides an important locus for modifiability of neural circuits underlying adaptive processes such as learning (Chapter 18).

Transmitter Release Is Quantal

One of the first applications of the microelectrode was the discovery that transmitter release is *quantal* in nature (Katz, 1969). Transmitter is released spontaneously in multimolecular packets called *quanta* in the absence of presynaptic electrical activity. Each packet generates a small postsynaptic signal—either a *miniature excitatory* or a *miniature inhibitory postsynaptic*

Copyright 2004, Elsevier Science (USA).
All rights reserved.

potential (MEPSP or MIPSP, respectively, or just MPSP); under voltage clamp, a *miniature excitatory* or a *miniature inhibitory postsynaptic current* (mEPSC or mIPSC, respectively, or just mPSC) is generated. An action potential tremendously, but very briefly, accelerates the rate of secretion of quanta and synchronizes them to evoke a PSP. Vertebrate skeletal neuromuscular junctions are frequently used as model synapses, because both receptors and nerve terminals are relatively accessible for anatomical, electrophysiological, and biochemical studies. At the neuromuscular junction, the motor nerve forms a cluster of small unmyelinated processes that lie in shallow gutters in the muscle to form a structure called an *end plate*, and PSPs, PSCs, mPSPs, and mPSCs are called *end-plate potentials* (EPPs), *end-plate currents* (EPCs), *miniature end-plate potentials* (mEPPs), and *miniature end-plate currents* (mEPCs), respectively.

Why is transmission quantized? Neural circuits must process complex and quickly changing information fast enough to generate timely appropriate responses. This requires rapid transmission across synapses. Fast-acting chemical synapses accomplish this by concentrating transmitter in membrane-bound structures, ~50 nm in diameter, called *synaptic vesicles* and docking these vesicles at specialized sites called *active zones* along the presynaptic membrane (Fig. 8.1A). Vesicles not docked at the membrane are clustered behind it and associated with cytoskeletal elements (Heuser, 1977). Action potentials release transmitter by depolarizing the presynaptic membrane and opening Ca^{2+} channels that are strategically colocalized with the synaptic vesicles in the active zone (Robitaille *et al.*, 1990). The local intense rise in Ca^{2+} concentration triggers the fusion of docked vesicles with the plasma membrane (called *exocytosis*; Figs. 8.1B and C) (Heuser and Reese, 1981) and the release of their contents into the narrow *synaptic cleft* (about 100 nm wide) separating the presynaptic terminal from high concentrations of postsynaptic receptors. The fusion of one vesicle releases about 5000 transmitter molecules within a millisecond (Kuffler and Yoshikami, 1975; Fletcher and Forrester, 1975; Whittaker, 1988) and generates the quantal response recorded postsynaptically. No membrane carrier can release so much transmitter this fast, nor can a pore or channel unless some mechanism exists to concentrate the transmitter behind the pore, which may be regarded as the function of synaptic vesicles. Evidence that transmitter is released from vesicles and that one quantum is due to exocytosis of a vesicle is summarized in Box 8.1 (Eccles, 1964; Edwards, 1992; Torri-Tarelli *et al.*, 1985, 1992; von Wedel *et al.*, 1981; Van der Kloot, 1988, 1991; Searl *et al.*, 1991; Prior,

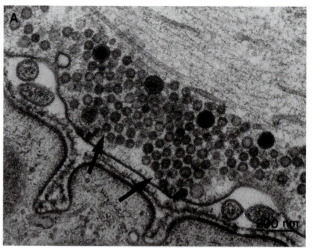

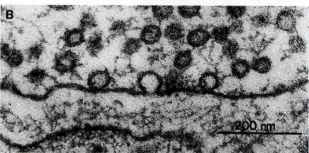

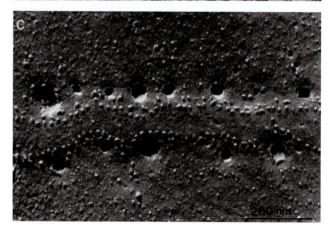

FIGURE 8.1 Ultrastructural images of synaptic vesicle exocytosis. Synapses from frog sartorius neuromuscular junctions were quick-frozen milliseconds after stimulation in 4-aminopyridine to broaden action potentials and enhance transmission. (A) A thin section from which water was replaced with organic solvents (freeze substitution) and fixed in osmium tetroxide, showing vesicles clustered in the active zone, some docked at the membrane (arrows). (B) Shortly (5 ms) after stimulation, vesicles were seen to fuse with the plasma membrane. (C) After freezing, presynaptic membranes were freeze-fractured and a platinum replica was made of the external face of the cytoplasmic membrane leaflet. Vesicles fuse about 50 nm from rows of intramembranous particles thought to include Ca^{2+} channels. (A) and (B) from Heuser (1977); (C) from Heuser and Reese (1981).

BOX 8.1

EVIDENCE THAT A QUANTUM IS A VESICLE

Transmitter Is Released from Vesicles

1. All chemically transmitting synaptic terminals contain presynaptic vesicles (Eccles, 1964).

2. Synaptic vesicles concentrate and store transmitter (Edwards, 1992).

3. Rapid freezing of neuromuscular junctions during stimulation shows vesicle exocytosis occurring at the moment of transmitter release (Torri-Tarelli *et al.*, 1985).

4. Intravesicular proteins appear on the external terminal surface after secretion (von Wedel *et al.*, 1981; Torri-Tarelli *et al.*, 1992).

5. Retarding the filling of vesicles by using transport inhibitors (e.g., vesamicol for acetylcholine) or by reducing the transvesicular pH gradient generates a class of small mEPSPs that probably represent partially filled vesicles; drugs that enhance vesicle loading increase mEPSP size (Van der Kloot, 1991; Searl *et al.*, 1991; Prior, 1994).

6. Quantal size is independent of membrane potential or cytoplasmic acetylcholine concentration altered osmotically (Van der Kloot, 1988).

7. Synaptic vesicles formed by endocytosis load with extracellular electron-dense and fluorescent dyes (horseradish peroxidase and FM1–43, respectively) after nerve stimulation; the dye is released by subsequent stimulation (Heuser and Reese, 1973; Betz and Bewick, 1993).

8. False transmitters synthesized from choline derivatives load slowly into cholinergic vesicles; they are co-released with acetylcholine in proportion to their concentrations in vesicles (Large and Rang, 1978).

9. Clostridial toxins that interfere with the synaptic vesicle–plasma membrane interaction block neurosecretion (Schiavo *et al.*, 1994).

One Quantum Is One Vesicle

1. The number of acetylcholine molecules in isolated vesicles corresponds to the number of molecules released in a quantum (Kuffler and Yoshikami, 1975; Fletcher and Forrester, 1975; Whittaker, 1988).

2. When release is enhanced and the collapse of vesicle fusion images is prolonged by treatment with the potassium channel blocker 4-aminopyridine to broaden action potentials, the number of vesicle fusions observed corresponds to the number of quanta released by an action potential (Heuser *et al.*, 1979). In these special circumstances, several vesicles are released at each active zone (Fig. 8.1C).

3. The number of vesicles present in nerve terminals corresponds to the total store of releasable quanta. When endocytosis is blocked by the temperature-sensitive *Drosophila* mutant *shibire* (Van der Kloot and Molgó, 1994) or pharmacologically (Hurlbut *et al.*, 1990) and the motor nerve is stimulated to exhaustion, the number of quanta released corresponds to the original number of presynaptic vesicles.

4. The statistical variations in quantal release match the statistical variations in vesicle release and recovery measured with staining and destaining of lipophilic dyes (see Box 8.3 and section on Quantal Analysis) (Ryan *et al.*, 1997; Murthy and Stevens, 1988).

1994; Heuser and Reese, 1973; Betz and Bewick, 1993; Large and Rang, 1978; Schiavo *et al.*, 1994; Heuser *et al.*, 1979; Hurlbut *et al.*, 1990; Van der Kloot and Molgó, 1994; Ryan *et al.*, 1997; Murthy and Stevens, 1998).

At neuromuscular junctions, transmitter from one vesicle diffuses across the synaptic cleft in 2 µs and reaches a concentration of about 1 mM at the postsynaptic receptors (Matthews-Bellinger and Salpeter, 1973). These receptors bind transmitter rapidly, opening from 1000 to 2000 postsynaptic ion channels (Van der Kloot *et al.*, 1994) (two molecules of transmitter must bind simultaneously to receptors to open each channel; see Chapters 11 and 16). Each channel

has a 25-pS conductance and remains open for about 1.5 ms, admitting a net inflow of 35,000 positive ions. A single action potential in a motor neuron can release 300 quanta within about 1.5 ms along a junction that contains about 1000 active zones. The resulting postsynaptic depolarization, which begins after a *synaptic delay* of about 0.5 ms and reaches a peak of tens of millivolts, is typically sufficient to generate an action potential in the muscle fiber.

At fast central synapses, postsynaptic cells make contact with presynaptic axon swellings called *varicosities* when they occur along fine axons and *boutons* when they are located at the tips of terminals. Each varicosity or bouton contains one active zone or a few of them. The postsynaptic process is often on a fine dendritic branch or tiny spine with a length of a few micrometers, having a very high input resistance and capable of generating active propagating responses (Chapter 17). At inhibitory GABAergic synapses and excitatory glutamatergic synapses (Edwards *et al.*, 1990; Jonas *et al.*, 1993), each action potential releases from 5 to 10 quanta, and each quantum released elevates the transmitter concentration (Clements *et al.*, 1992; Tang *et al.*, 1994; Tong and Jahr, 1994) in the cleft to about 1 mM and activates about 30 ion channels. At excitatory synapses, this release may be sufficient to generate EPSPs of 1 mV or less in amplitude, clearly subthreshold for generating action potentials. But central neurons often receive thousands of inputs, each of which has a "vote" on how the cell should respond (see Chapter 16). No input has absolute, or even majority, control over postsynaptic cell activity, but the matching of quantal size to input resistance ensures that inputs are reasonably effective. Consequently, at synapses onto larger central neurons with lower input resistances, quanta open between 100 and 1000 postsynaptic channels.

Synaptic Vesicles Are Recycled

A constant supply of vesicles filled with transmitter must be available for release from the nerve terminal at all times. Maintaining this supply requires the efficient recycling of synaptic vesicles. For this purpose, two partly overlapping cycles are used: one for the components of the synaptic vesicle membrane and another for the vesicle contents (transmitter substances). The cycles overlap from the time of transmitter packaging into vesicles until exocytosis. The cycles are distinct during the stages in which vesicle membrane and transmitter are recovered for reuse. The various steps of these cycles are common to all chemical synapses and are summarized in Fig. 8.2.

Vesicle Membrane Cycle

The components of the synaptic vesicle membrane are initially synthesized in the cell body before being transported to nerve terminals by fast axoplasmic transport (Bennett and Scheller, 1994; Jahn and Südhof, 1994) (see Chapter 2). Within the nerve terminal, the synaptic vesicles are loaded with transmitter and either anchored to each other and actin filaments (McGuiness *et al.*, 1989) or targeted to plasma membrane docking sites at active zones. These docking sites are also rich in clusters of high-voltage-activated Ca^{2+} channels (Robitaille *et al.*, 1990; Haydon *et al.*, 1994) (mainly N- and P/Q-type Ca^{2+} channels, depending on the synapse (Wheeler *et al.*, 1994;

Dunlap *et al.*, 1995), see Chapters 5 and 6). Depolarization of the plasma membrane by an invading action potential opens these voltage-dependent Ca^{2+} channels to admit Ca^{2+} ions in the neighborhood of docked vesicles. The local high concentration of Ca^{2+} resulting from the opening of multiple Ca^{2+} channels triggers exocytosis. After exocytosis, some vesicles may rapidly reclose, whereas others fuse fully with the plasma membrane (Murthy and Stevens, 1998; Zimmerman, 1979; Ceccarelli and Hurlbut, 1980; Koenig and Ikeda, 1996; Klingauf *et al.*, 1998). The latter are recovered by *endocytosis*, a budding off of the vesicular membrane to form a new "coated" vesicle covered by the protein *clathrin*. Endocytosis may also be regulated by presynaptic [Ca^{2+}] (Thomas *et al.*, 1994; Von Gersdorff and Matthews, 1994). Recovered vesicular membrane often fuses to form large membranous sacs, called *endosomes* or *cisternae*, from which new synaptic vesicles are formed. The molecular mechanisms of the vesicle cycle of exo- and endocytosis are discussed later in this chapter.

Transmitter Cycle

The steps of the transmitter cycle vary with the type of transmitter (for additional details see Chapters 9 and 10). Some transmitters are synthesized from precursors in the cytoplasm before transport into synaptic vesicles, whereas other transmitters are synthesized in synaptic vesicles from transported precursors. Peptide transmitters are synthesized exclusively in the cell body and are not locally recycled. At most synapses, a transporter that harnesses the energy in the proton gradient across the vesicular membrane functions to concentrate transmitter (or transmitter precursors) in vesicles (Edwards, 1992). The pH gradient arises from the action of a vacuolar proton ATPase that uses the energy of ATP hydrolysis to transport protons into vesicles. After exocytosis, released transmitter diffuses across the synaptic cleft and rapidly binds to receptors. As transmitter falls off receptors, it is typically recovered from the synaptic cleft by sodium-dependent uptake transporters (see Chapter 9). At cholinergic synapses, acetylcholine is hydrolyzed to acetate and choline by the enzyme acetylcholinesterase present in the synaptic cleft. This enzyme is saturated by the initial gush of transmitter following exocytosis but can keep up with its subsequent slower release from receptors. The choline so produced is recovered by a presynaptic choline transporter and made available for the synthesis of new transmitter. Much of the evidence for the steps outlined in Fig. 8.2 comes from ultrastructural and pharmacological experiments. Some of this evidence is outlined in Box 8.2 (Elmgvist and Quastel, 1965;

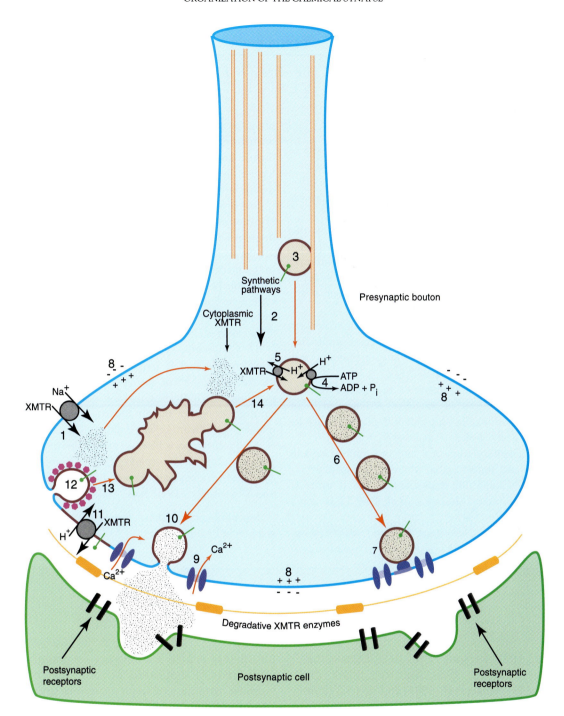

FIGURE 8.2 Steps in the life cycle of synaptic vesicles: (1) Na$^+$-dependent uptake of transmitter (XMTR) or XMTR precursors into the cytoplasm, (2) synthesis of XMTR, (3) delivery of vesicle membrane containing specialized transmembrane proteins by axoplasmic transport on microtubules, (4) production of transvesicular H$^+$ gradient by vacuolar ATPase, (5) concentration of XMTR in vesicles by H$^+$/XMTR antiporter, (6) synapsin I-dependent anchoring of vesicles to actin filaments near active zones, (7) releasable vesicles docked in active zones near Ca^{2+} channels, (8) depolarization of nerve terminal and presynaptic bouton by action potential, (9) opening of Ca^{2+} channels and formation of regions of local high [Ca^{2+}] ("Ca^{2+} microdomains") in active zones, (10) triggering of exocytosis of docked vesicles comprising quantal units of XMTR released by overlapping Ca^{2+} microdomains, (11) nonquantal leakage of XMTR through vesicle membrane fused with plasma membrane and exposure of vesicle proteins to synaptic cleft, (12) recovery of vesicle membrane by dynamin-dependent endocytosis of clathrin-coated vesicles, (13) fusion of coated vesicles with endosomal cisternae, (14) formation of synaptic vesicles from endosomes. Also shown are postsynaptic receptors with multiple XMTR binding sites and extracellular XMTR-degradative enzymes in synaptic cleft.

BOX 8.2

EVIDENCE FOR SOME OF THE EVENTS IN THE LIFE HISTORY OF VESICLES

Numbers refer to the steps in Fig. 8.2.

1. Uptake of transmitter or transmitter precursors is prevented by specific inhibitors, such as *hemicholinium-3* block of choline uptake at cholinergic synapses, ultimately leading to failure of synaptic transmission (Elmgvist and Quastel, 1965).

2. Cholinergic synapses can be identified by the presence of the synthetic enzyme choline acetyltransferase, GABAergic synapses by the enzyme glutamic acid decarboxylase, adrenergic synapses by the enzyme dopamine β-hydroxylase, and so forth (Cooper *et al.*, 1991).

3. Vesicular transport into nerve terminals is blocked by inhibitors of axoplasmic transport such as antibodies to the microtubule motor protein *kinesin* (Bennett and Scheller, 1994; Jahn and Südhof, 1994).

5. The storage of transmitter in vesicles can be blocked by inhibitors of vacuolar ATPase, such as *bafilomycin A_1*, or of an H^+-dependent transporter, such as *vesamicol* for acetylcholine (Parsons *et al.*, 1993).

6. Dephosphorylation of synapsin I inhibits vesicle movements and transmission, whereas its phosphorylation by Ca^{2+}-calmodulin-dependent kinase II protects against this inhibition (McGuiness *et al.*, 1989; Llinás *et al.*, 1991).

7. Toxins from *Clostridium* bacteria, which proteolyze the vesicular protein *synaptobrevin* or plasma membrane proteins *SNAP-25* and *syntaxin*, block exocytosis, whereas mutants deficient in the vesicle protein synaptotagmin and injection of peptides derived from *synaptotagmin* show defects in evoked transmitter release (more details later in this chapter).

8. Block of action potential propagation by local application of *tetrodotoxin* prevents transmission, and depolarization by elevating potassium in the bath accelerates mPSP frequency, as long as Ca^{2+} is present in the medium (Katz, 1969).

9, 10. N- and P/Q-type calcium channel antagonists, such as ω-conotoxin and ω-agatoxin IVA, prevent Ca^{2+} influx and block transmission at many synapses (Wheeler *et al.*, 1994; Dunlap *et al.*, 1995).

11. Cholinergic synapses show a nonquantal leak of acetylcholine that is enhanced after stimulation; it is blocked by vesamicol, the vesicular acetylcholine transport inhibitor, indicating that the leak is due to transport through vesicular membrane fused with the plasma membrane (Edwards *et al.*, 1985).

12. Endocytosis is blocked at high temperature in the *shibire* mutant of *Drosophila*, which affects the protein *dynamin* in endocytosis of coated vesicles (Van der Bliek and Meyerowitz, 1991; Chan *et al.*, 1991).

Morphological evidence for steps 13 and 14 in Fig. 8.2 is given in Box 8.3.

Cooper *et al.*, 1991; Parsons *et al.*, 1993; Llinás *et al.*, 1991; Edwards *et al.*, 1985; Van der Bliek and Meyerowitz, 1991; Chan *et al.*, 1991) and Box 8.3 (Ryan *et al.*, 1993).

Summary

Chemical synapses are ideally suited to permit one neuron to rapidly and effectively excite or inhibit the activity of another cell. A diversity of transmitters and receptors guarantees a multiplicity of postsynaptic responses. The opportunity for presynaptic and postsynaptic interactions between inputs provides for marvelously complex computational capabilities. The packaging of transmitter into vesicles and its release in quanta enable a single action potential to secrete hundreds of thousands of molecules of transmitter almost instantaneously at a synapse onto another cell. Neurochemical and ultrastructural studies have provided a rich picture of the life cycle of synaptic vesicles from their exocytosis at active zones to their recovery by endocytosis, their refilling with transmitter, and redocking at release sites.

EXCITATION–SECRETION COUPLING

Shortly after an action potential invades presynaptic terminals at fast synapses, the synchronous release of many quanta of transmitter generates the postsy-

---

BOX 8.3

HISTOLOGICAL TRACERS CAN BE USED TO FOLLOW VESICLE RECYCLING

An elegant picture of the life history of synaptic vesicles comes from studies using electron-dense or fluorescent markers of intracellular regions that have been in contact with the extracellular space. *Horseradish peroxidase* (HRP) is an enzyme that catalyzes the oxidation of diaminobenzidine, forming an electron-dense product that can easily be identified in tissues fixed with osmium tetroxide for electron microscopy; *FM1–43* is an amphipathic styryl dye that becomes highly fluorescent on partitioning into cell membranes. When frog muscles were soaked in HRP and the motor neurons were stimulated at 10 Hz for 1 min, the enzyme appeared in coated vesicles in nerve terminals in regions outside active zones. After more prolonged stimulation, most of the HRP collected in endosomal cisternae, owing to the fusion of endocytotic vesicles with these organelles. When the HRP was washed out and the neurons were rested for an hour before fixation, HRP appeared in small clear synaptic vesicles in active zones. When rested neurons were stimulated again before fixation, this time in the absence of HRP, the filled vesicles gradually disappeared owing to their release by exocytosis (Heuser and Reese, 1973).

Another study traced the uptake of FM1–43 into living motor nerve terminals with the use of confocal fluorescence microscopy. High-frequency stimulation for just 15 s in FM1–43 was marked by uptake of dye into nerve terminals. More prolonged stimulation followed by a period of rest without the dye in the bath resulted in the persistent staining of synaptic vesicles in active zones. Subsequent stimulation at 10 Hz gradually destained the terminals in minutes; destaining required the presence of Ca^{2+} in the medium and represented exocytosis of stained vesicles. After about 1 min, the rate of destaining decreased as the vesicle pool began to be diluted with unstained vesicles newly recovered by endocytosis (Betz and Bewick, 1993). Exposing dissociated hippocampal neurons to FM1–43 at various times after stimulation showed that endocytosis proceeded for about 1 min after exocytosis. Cells loaded with dye and then restimulated began to destain about 30 s after endocytosis, which is a measure of the time needed for recycling of recovered vesicles into the pool of releasable vesicles (Ryan *et al.*, 1993). These experiments provide a dynamic view of the life cycle of synaptic vesicles.

---

naptic potential. Since the work of Locke in 1894, the presence of calcium in the external medium has been known to be a requirement for transmission. What is the central role of Ca^{2+} in triggering neurosecretion?

Calcium Triggers Release of Transmitters at Internal Sites

Calcium was originally believed to act at an external site to enable neurons to release transmitter. The pioneering work of Bernard Katz (1969) and his co-workers showed that Ca^{2+} acts intracellularly. This conclusion is based on many lines of evidence:

1. Calcium must be present only at the moment of invasion of the nerve terminal by an action potential for transmitter to be released.

2. Calcium entry is retarded by a large presynaptic depolarization, and transmitter release is delayed until the voltage gradient is reversed at the end of the pulse, whereupon Ca^{2+} enters and release occurs as an off-EPSP until Ca^{2+} channels close. Sodium influx is not necessary for secretion, and K^+ ions also play no role.

3. Elevation of intracellular $[Ca^{2+}]$ accelerates the spontaneous release of quanta of transmitter (Steinbach and Stevens, 1976; Rahamimoff *et al.*, 1980). Stimulation in a $[Ca^{2+}]$-free medium reduces intracellular $[Ca^{2+}]$ and MEPSP frequency.

4. The presence of Ca^{2+} channels in presynaptic terminals is shown by the ability to stimulate local action potentials that trigger release in a high-$[Ca^{2+}]$ medium when Na^+ action potentials are blocked with tetrodotoxin and K^+ channels are blocked with tetraethylammonium.

5. Divalent cations that permeate Ca^{2+} channels, such as Ba^{2+} and Sr^{2+}, support transmitter release, although only weakly. Cations that block Ca^{2+} channels, such as Co^{2+} and Mn^{2+}, block transmission (Augustine *et al.*, 1987); Mg^{2+} reduces transmission, perhaps by screening fixed surface charge and effectively hyperpolarizing the nerve (Muller and Finkelstein, 1974).

6. Transmission depends nonlinearly on $[Ca^{2+}]$ in the bath, varying with the fourth power of $[Ca^{2+}]$, whereas Ca^{2+} influx remains a linear function of $[Ca^{2+}]$, indicating a high degree of Ca^{2+} cooperativity in triggering exocytosis (Llinas *et al.*, 1981).

7. At giant synapses in the stellate ganglion of squid, voltage-clamp recording of the presynaptic Ca^{2+} current reveals a close correspondence between Ca^{2+} influx and transmitter release, including an association between the off-EPSP and a delay in Ca^{2+} current until the end of large pulses (called a tail current) (Llinás et al., 1981).

8. Action potentials trigger no phasic release of transmitter when Ca^{2+} influx is blocked, even when presynaptic Ca^{2+} is tonically elevated by photolysis of photosensitive Ca^{2+} chelators; however, the elevated presynaptic $[Ca^{2+}]$ accelerates the frequency of MEPSPs (Mulkey and Zucker, 1991).

Vesicles Are Released by Calcium Microdomains

Single action potentials generate a Ca^{2+} rise of about 10 nM, which lasts a few seconds (Charlton et al., 1982; Zucker et al., 1991). This increment in $[Ca^{2+}]$ is a small fraction of the typical resting $[Ca^{2+}]$ of 100 nM. How can such a tiny change in $[Ca^{2+}]$ trigger a massive synchronous release of quanta, and why is secretion so brief compared with the duration of the $[Ca^{2+}]$ change? As mentioned earlier, postsynaptic responses begin only 0.5 ms after an action potential invades nerve terminals. This synaptic delay includes the time taken for Ca^{2+} channels to begin to open after the peak of the action potential (300 μs) (Llinás et al., 1981), leaving only about 200 μs after that for transmitter secretion and the start of a postsynaptic response. At this time, Ca^{2+} has barely begun to diffuse away from Ca^{2+} channel mouths. In an aqueous solution, $[Ca^{2+}]$ would be confined mainly to within 1 μm of channel mouths estimated roughly from the solution of the diffusion equation for a brief influx of M moles of Ca^{2+},

$$[Ca^{2+}] = \frac{M}{8(\pi Dt)^{3/2}} e^{-r^2/4Dt}$$

where t is time after the influx, r is distance from the channel mouth, and D is the diffusion constant for Ca^{2+}, ~6 × 10⁻⁶ cm² s⁻¹. In the cytoplasm, Ca^{2+} diffusion is retarded by intracellular organelles and the presence of millimolar concentrations of fast-acting protein-associated Ca^{2+} binding sites with an average dissociation constant of a few micromolar. Together, these effects restrict Ca^{2+} microdomains to about 50 nm around channel mouths.

Furthermore, the 200 μs preceding the postsynaptic response must include not only the time required for Ca^{2+} to reach its target but also the time required for Ca^{2+} to bind and initiate exocytosis and for transmit-

ter to diffuse across the synaptic cleft, bind to receptors, and begin to open channels. Thus, the presynaptic Ca^{2+} targets must be located within a few tens of nanometers of Ca^{2+} channel mouths. Neuromuscular junctions that are fast frozen during the act of secretion show vesicle fusion images in freeze-fracture planes of the presynaptic membrane about 50 nm from intramembranous particles thought to be Ca^{2+} channels (see Fig. 8.1C). Solution of the diffusion equation for a steady point source of Ca^{2+} influx in the presence of a nearly immobile fast-binding Ca^{2+} buffer reveals that approximately 100 μs after a Ca^{2+} channel opens, $[Ca^{2+}]$ increases to more than 10 μM at 50 nm from its source and to more than 100 μM at a distance of 10 nm (Llinás et al., 1981).

This calculation considers only what happens in the neighborhood of a single open Ca^{2+} channel. However, when individual Ca^{2+} channels are labeled with biotinylated ω-conotoxin tagged with colloidal gold particles, more than 100 channels per active zone are seen in terminals of chick parasympathetic ganglia (Haydon et al., 1994) Any vesicle docked at such an active zone is likely to be surrounded by as many as 10 Ca^{2+} channels within a 50-nm distance. Even though not all these channels will open during each action potential, more than one channel is likely to open, so a vesicle will be influenced by Ca^{2+} entering through several nearby channels. At the squid giant synapse, more than 50 channels open in each ~0.6 μm^2 active zone, whereas 10 channels open within the more compact active zones of frog saccular hair cells (Roberts, 1994; Yamada and Zucker, 1992). The Ca^{2+} microdomains of these channels overlap at single vesicles, and they cooperate in triggering secretion of a vesicle. Calculations of diffusion of Ca^{2+} ions from arrays of Ca^{2+} channels in the presence of a saturable buffer indicate that the $[Ca^{2+}]$ at sites where neurotransmitter release is triggered may reach as high as 100–200 μM (Fig. 8.3).

Three indications that $[Ca^{2+}]$ in fact reaches very high levels in active zones during action potentials are:

1. $[Ca^{2+}]$ levels greater than 100 μM have been measured in presynaptic submembrane regions of squid giant synapses likely to be active zones by using the low-affinity Ca^{2+}-sensitive photoprotein n-aequorin-J (Llinás et al., 1992).

2. Estimates of $[Ca^{2+}]$ based on the activity of Ca^{2+}-activated K^+ channels in active zones of mechanosensory hair cells are similar (Roberts et al., 1991).

3. Transmitter release is blocked only by presynaptic injection of at least millimolar concentrations of fast high-affinity Ca^{2+} chelators, indicating that

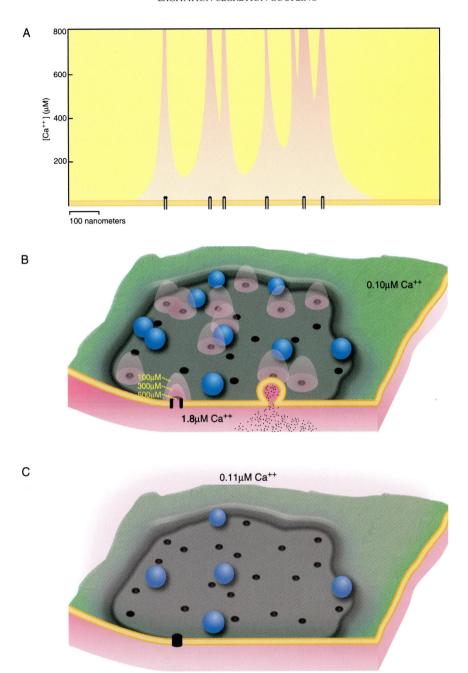

FIGURE 8.3 Microdomains with high Ca^{2+} concentrations form in the cytosol near open Ca^{2+} channels and trigger the exocytosis of synaptic vesicles. (A) In this adaptation of a model of Ca^{2+} dynamics in the terminal, a set of Ca^{2+} channels is spaced along the x axis, as if in a cross section of a terminal. The channels have opened and, while they are open, the cytosolic Ca^{2+} concentration (y axis) is spatially inhomogeneous. Near the mouth of the channel, the influx of Ca^{2+} drives the local concentration to as high as 800 μM, but within just 50 nm of the channel, the concentration drops off to below 100 μM. The channels are irregularly spaced but are often sufficiently close to one another that their clouds of Ca^{2+} can overlap and sum. (B). In the active zone (gray), an action potential has opened a fraction of the Ca^{2+} channels and microdomains of high cytosolic Ca^{2+} (pink) arise around these open channels as Ca^{2+} flows into the cell. In the rest of the cytoplasm, the Ca^{2+} concentration is at resting levels (0.10 μM), but within these microdomains, and particularly near the channel mouth, Ca^{2+} concentrations are much higher, as in (A). Synaptic vesicles docked and primed at the active zone may come under the influence of one or more of these microdomains and thereby be triggered to fuse with the membrane. (C) A few milliseconds after the action potential, the channels have closed and the microdomains have dispersed. The overall Ca^{2+} concentration in the terminal is now slightly higher (0.11 μM) than before the action potential. If no other action potentials occur, the cell will pump the extra Ca^{2+} out across the plasma membrane and restore the initial condition after several hundred milliseconds. (A) is adapted from Roberts *et al.* (1991).

release is triggered locally by high concentrations of Ca^{2+} (Adler *et al.*, 1991).

Vesicle Exocytosis Is Normally Triggered by Overlapping Ca^{2+} Channel Microdomains of High $[Ca^{2+}]$

Although release of a quantum of transmitter subsequent to the opening of a single presynaptic Ca^{2+} channel has been observed (Stanley, 1993), exocytosis may normally be due to Ca^{2+} entering through clusters of Ca^{2+} channels in active zones and contributing to local high $[Ca^{2+}]$ at docked vesicles:

1. When transmitter release is increased under voltage clamp with pulses of increasing amplitude, a third-order power law relationship exists between presynaptic Ca^{2+} current and postsynaptic response (Augustine and Charlton, 1986). If each vesicle were released by Ca^{2+} entering through a single Ca^{2+} channel, then increasing depolarizations should recruit additional channel openings and proportionally more vesicle releases (Simon and Llinás, 1985). However, if Ca^{2+} channel microdomains from neighboring clustered Ca^{2+} channels overlap at docked vesicles, the $[Ca^{2+}]$ at each vesicle will rise with increasing depolarization as more channels are recruited, and some cooperativity of Ca^{2+} action in triggering secretion will be expressed (Zucker and Fogelson, 1986).

2. In some neurons, more than one Ca^{2+} channel type contributes to secretion (Wheeler *et al.*, 1994; Dunlap *et al.*, 1995). When contributions of each channel type are isolated pharmacologically, their combined effects add nonlinearly, much as would be predicted by a fourth-order cooperativity, indicating that the Ca^{2+} microdomains of different channels overlap and summate at the Ca^{2+} sensors at vesicles docked within individual active zones.

3. At some synapses, (Borst and Sakmann, 1996), presynaptic injection of the slow-acting Ca^{2+} buffer ethylene glycol bis (β-aminoethyl ether)-*N,N,N′,N′*-tetraacetic acid (EGTA) reduces transmission, indicating that the target for Ca^{2+} action is not particularly close to any one Ca^{2+} channel, but rather is affected by Ca^{2+} ions entering through many channels.

4. When transmitter release is increased by prolonging presynaptic depolarizations (e.g., by broadening action potentials with K^+ channel blockers), more channels are not likely to be opened simultaneously. Rather, as channels that open early in the action potential close, others open; so the pattern of presynaptic Ca^{2+} microdomains is not so much intensified as prolonged, leading to a more nearly linear relationship between increases in Ca^{2+} influx and transmitter release (Zucker *et al.*, 1991).

5. Large depolarizations admit little Ca^{2+} as they approach the Ca^{2+} equilibrium potential; they are therefore accompanied by a reduced Ca^{2+} current and reduced transmitter release during a pulse. However, large depolarizations can release more transmitter than can small depolarizations evoking a given macroscopic Ca^{2+} current (Llinás *et al.*, 1981; Augustine *et al.*, 1985). This apparent voltage dependence of transmitter release may be due to the different spatial profiles of $[Ca^{2+}]$ in the active zone, with greater overlap of $[Ca^{2+}]$ from the larger number of more closely apposed open Ca^{2+} channels during large depolarizations (Zucker and Fogelson, 1986).

The Exocytosis Trigger Must Have Fast, Low-Affinity, Cooperative Ca^{2+} Binding

The brevity of the synaptic delay implies not only that Ca^{2+} acts near Ca^{2+} channels to evoke exocytosis but also that Ca^{2+} must bind to its receptor extremely rapidly. This is confirmed by the finding that presynaptic injection of relatively slow Ca^{2+} buffers such as EGTA have almost no effect on transmitter release to single action potentials. Only millimolar concentrations of fast Ca^{2+} buffers such as 1,2-bis(2-aminophenoxy)ethane-*N,N,N′,N′*-tetraacetic acid (BAPTA), with on-rates of about 5×10^8 M^{-1} s^{-1}, can capture Ca^{2+} ions before they bind to the secretory trigger (Adler *et al.*, 1991), indicating that the on-rate of Ca^{2+} binding to this trigger is similarly fast. At a rate of 5×10^8 M^{-1} s^{-1}, 100 μM $[Ca^{2+}]$ reaches equilibrium with its target in about 50 μs.

From the dependence of transmitter release on external $[Ca^{2+}]$, it is known that at least four Ca^{2+} ions cooperate in the release of a vesicle. The off-rate of Ca^{2+} dissociation from these sites also must be fast, at least 10^3 s^{-1}, to account for the rapid termination of transmitter release (0.25-ms time constant) after Ca^{2+} channels close and Ca^{2+} microdomains collapse. The high temperature sensitivity of the time course of transmitter release ($Q_{10} \approx 3$) indicates that exocytosis is rate limited by a step with a high energy barrier (Yamada and Zucker, 1992). This step is likely to be the process of exocytosis itself. If Ca^{2+} binding is not rate limiting, its dissociation rate must be substantially faster than 10^3 s^{-1}. This means that the affinity of the secretory trigger for Ca^{2+} is low, with a dissociation constant (K_D) above 10 μM.

The Ca^{2+}-binding trigger is not saturated under normal conditions, because increasing $[Ca^{2+}]$ in the bath increases release. Furthermore, because of the speed with which Ca^{2+} binds to its sites, this reaction nearly equilibrates during the typical 0.5–1.0 ms that $[Ca^{2+}]$ remains high before Ca^{2+} channels close at the end of an action potential. If $[Ca^{2+}]$ reaches 100 μM or more in equilibrium with unsaturated release sites, the affinity of at least some of those sites binding Ca^{2+} must be similar to 100 μM or lower.

These predictions are consistent with experiments in which neurosecretion is triggered by photolysis of caged Ca^{2+} chelators such as DM-nitrophen. Partial flash photolysis of partially Ca^{2+}-loaded DM-nitrophen generates a $[Ca^{2+}]$ "spike" of a duration similar to the lifetime of Ca^{2+} microdomains around Ca^{2+} channels opened by an action potential (Landò and Zucker, 1994). This spike results in a postsynaptic response that closely resembles the normal EPSC at crayfish neuromuscular junctions, confirming that no presynaptic depolarization is necessary to obtain high levels of phasic transmitter release. Secretion depended on the fourth power of peak $[Ca^{2+}]$, and about 50 μm Ca^{2+} activated release at the same rate as an action potential.

In similar experiments on retinal bipolar neurons from fish (Heidelberger et al., 1994), fully loaded DM-nitrophen was photolyzed to produce a stepped increase in $[Ca^{2+}]$ while secretion was monitored as an increase in membrane capacitance, a measure of cell membrane area increased by fusion of vesicles. The Ca^{2+} concentration had to be raised by more than 20 μM before a fast phase of secretion developed. The sharp Ca^{2+} dependence of release and short synaptic delays were fitted by a model with a high degree of positive Ca^{2+} cooperativity, in which four successive Ca^{2+} ions bind with affinities increasing (or K_D decreasing) from 140 to 9 μM, followed by a Ca^{2+}-independent rate-limiting step. In contrast, at the calyx of Held synapse, a $[Ca^{2+}]$ level of about 10 μM was sufficient to activate release at a rate similar to that in an EPSP (Schneggenburger and Neher, 2000), possibly without positive cooperativity (Bollmann et al., 2000). These experiments differentiate Ca^{2+} receptors triggering release at various synapses.

Calcium Ions Must Mobilize Vesicles to Docking Sites at Slowly Transmitting Synapses

Most peptidergic synapses and some synapses releasing biogenic amines display kinetics remarkably different from those of fast synapses. In these slower synapses, single action potentials often have no discernible postsynaptic effect. During repetitive stimulation, postsynaptic responses rise slowly, often with a delay of seconds from the beginning of stimulation,

and persist just as long after stimulation ceases. Such slow responses are due to many factors: the postsynaptic receptors may have intrinsically sluggish second messengers or G proteins (Chapters 11 and 16); the postsynaptic receptors are often distant from release sites, so extracellular diffusion takes significant time; and release starts after the beginning of stimulation and continues after stimulation stops. Given these limitations, it is not surprising that single quanta are never discernible, either as spontaneous PSPs or as components of evoked responses.

The ultrastructural anatomy of presynaptic terminals of slowly transmitting synapses also is different from that of fast synapses. Transmitter is stored in large, dense core vesicles scattered randomly throughout the cytoplasm; vesicles do not tend to cluster at active zones or to line up at the membrane, docked and ready for release (De Camilli and John, 1990; Leenders et al., 1999). Nevertheless, there is no doubt that transmitter is released from vesicles, because it is both stored and often synthesized in them, and they can be seen to undergo exocytosis during high-frequency stimulation causing high rates of release (Verhage et al., 1994).

A High-Affinity Calcium Binding Step Controls Secretion of Slow Transmitters

Calcium ions are required for excitation–secretion coupling in slow synapses, but the dependence of release on $[Ca^{2+}]$ is linear, in contrast with fast synapses (Sakaguchi et al., 1991). Furthermore, because few vesicles are predocked at active zones, most of those released by repetitive activity are not exposed to the local high $[Ca^{2+}]$ near Ca^{2+} channels. Thus, an important event triggered by Ca^{2+} influx in action potentials is likely to be the translocation of dense core vesicles to plasma membrane release sites, followed by exocytosis. This process has a very different dependence on $[Ca^{2+}]$ than does the release of docked vesicles. Measurements of $[Ca^{2+}]$ during stimulation indicate that release correlates well with $[Ca^{2+}]$ levels in the low micromolar range above a minimum, or threshold, level of a few hundred nanomolar (Lindau et al., 1992; Peng and Zucker, 1993).

A striking difference between the release of fast transmitters, such as γ-aminobutyric acid (GABA) and glutamate, and peptide transmitters, such as cholecystokinin, was found in studies of *synaptosomes*, isolated nerve terminals prepared from homogenized brain tissue by differential centrifugation (Verhage et al., 1991). When terminals were depolarized to admit Ca^{2+} through Ca^{2+} channels, the amino acid transmitters GABA and glutamate were released at much lower levels of bulk cytoplasmic [Ca^{2+}] than when Ca^{2+} was admitted more uniformly and gradually across the membrane by use of the Ca^{2+}-transporting ionophore ionomycin. Peptides were released at the same low levels of [Ca^{2+}] no matter which method was used to elevate [Ca^{2+}]. Thus, only amino acids were sensitive to the difference in [Ca^{2+}] gradients imposed by the two methods and were preferentially released by local high submembrane [Ca^{2+}] caused by depolarization. Apparently, peptides are released by a high-affinity rate-limiting step not especially sensitive to submembrane [Ca^{2+}] levels.

Slow and Fast Transmitters May Be Co-released from the Same Neuron Terminal

Some neurons have both small synaptic vesicles containing acetylcholine or glutamate and large, dense core vesicles containing neuropeptides (Lundberg and Hökfelt, 1986). Often, the two transmitters act on different targets. Single action potentials release only the fast transmitter, so different patterns of activity can have very different relative effects on the targets. For example, postganglionic parasympathetic nerves to the salivary gland release acetylcholine, which stimulates salivation, and vasoactive intestinal peptide, which stimulates vasodilation. Many examples of the co-release of multiple transmitters have been described.

Summary

Ca^{2+} acts as an intracellular messenger tying the electrical signal of presynaptic depolarization to the act of neurosecretion. At fast synapses, Ca^{2+} enters through clusters of channels near docked synaptic vesicles in active zones. It acts at extremely short distances (tens of nanometers) in remarkably little time (200 μs) and at very high local concentrations ($\geq$100 μM), in calcium microdomains, by binding cooperatively to a low-affinity receptor with fast kinetics to trigger exocytosis. Some transmitters, such as peptides and some biogenic amines, are stored in larger, dense core vesicles not docked at the plasma membrane in active zones. Release of these transmitters, as well as their diffusion to postsynaptic targets and their postsynaptic actions, is much slower than that of transmitters such as acetylcholine and amino acids at fast synapses. Release of slow transmitters depends linearly on [Ca^{2+}] and may be governed by a Ca^{2+}-sensitive rate-limiting step different from that triggering exocytosis of docked vesicles at fast synapses.

THE MOLECULAR MECHANISMS OF THE NERVE TERMINAL

To release neurotransmitter in response to an action potential, a synaptic vesicle must fuse with the plasma membrane with great rapidity and fidelity and thus the synapse requires an effective and well-regulated molecular machine. The mechanisms of the terminal must also include the means to load the vesicle with transmitter, to dock the vesicle near the membrane so that it can fuse with a short latency, to define a release site on the plasma membrane, and to restrict fusion to the active zone rather than other points on the surface of the terminal or axon. Additionally, a reserve of synaptic vesicles must be held near the active zone and those vesicles must be recruited to the plasma membrane as needed. The number of vesicles that are ready and waiting to fuse must be strictly determined and the protein and lipid components of the vesicle must be recycled to form a new vesicle after fusion has occurred. For each of these processes, a molecular understanding remains incomplete, but a decade of rapid scientific progress in this field has made considerable headway.

A Cycle of Membrane Trafficking

Active neurons need to secrete transmitter in a constant, ongoing fashion. A bouton in the CNS, for example, may contain a store of 200 vesicles, but if it releases even one of these with each action potential and if the cell is firing at an unexceptional rate such as 5 Hz, the store of vesicles would be consumed within less than a minute. This calculation exemplifies the need for an efficient mechanism to recycle and reload vesicles within the terminal. Transport of newly synthesized vesicles from the cell body would be far too slow to support such a demand and the axonal traffic needed to supply an entire arbor of nerve terminals would be staggering. As a consequence, only peptide neurotransmitters are supplied in this fashion, because they must be synthesized in the endoplasmic-reticulum and then sent down the axon. Not surprisingly, therefore, vesicles with peptide transmitters are released at very low rates. For most neurotransmitters, however, the exocytosis of a synaptic vesicle is rapidly followed by its endocytosis and within approximately 30 s the vesicle is again available for release (Ryan and Smith, 1995). This pathway is sometimes referred to as the exo–endocytic cycle (Fig. 8.2). Moreover, each step in this cycle represents a potential control point for modulating the efficacy of the synapse. Modulation of the strength or fidelity of synaptic signaling, commonly known as *synaptic plasticity*, plays an important role both in the development of synaptic connections and in the functioning of the mature nervous system (see also Chapter 18). Indeed, regulation by Ca^{2+} and other second messenger systems is known to affect the docking, fusing, and recycling of vesicles at some synapses and is also likely to regulate the balance between reserve stores and those vesicles actively engaged in the exo–endocytic cycle (Beutner *et al.*, 2001; Dinkelacker *et al.*, 2000; Smith *et al.*, 1998; Neher and Zucker, 1993). Understanding the mechanisms of this modulation is an important goal and will certainly require a detailed understanding of the fundamental machinery itself. The question has been approached through the combined use of biochemical, genetic, and biophysical techniques.

Transmitter Release Is Rapid

As discussed earlier in this chapter, the delay between the arrival of an action potential at a terminal and the secretion of the transmitter can be less than 200 μs. This places some severe constraints on the fusion mechanism. Vesicles must already be present at the release sites, as there is no time to mobilize them from a distance. A catalytic cascade during fusion, such as that involved in phototransduction or in excitation–contraction coupling in smooth muscle, would also be far too slow for excitation–secretion coupling at the nerve terminal. Indeed, even a single bimolecular catalytic step might be too slow for such short latencies. Models therefore favor the idea that a fusion-ready complex of the vesicle and plasma membrane is preassembled at release sites and that Ca^{2+} binding need only trigger a simple conformation change in this complex to open a pathway for the transmitter to exit the vesicle. Because the volume of the synaptic vesicle is small, the diffusion of transmitter from the vesicle proceeds almost instantaneously as soon as a pore has opened up between the vesicle lumen and the extracellular space. This structure is referred to as the fusion pore, but its biochemical nature is unknown. Thus, the time-critical step comes between the influx of Ca^{2+} and the formation of the fusion pore. The complete merging of the vesicle and plasma membrane, if it occurs at all, can occur on a slower time course. The movement of the vesicle to the release site and any biochemical events that need to occur to reach the fusion-ready state can also be slower. These largely theoretical steps are sometimes referred to as "docking" and "priming." Docking and priming a vesicle cannot be too slow, however; a CNS synapse is estimated to have 2 to 20 vesicles in this fusion-ready state (Harris and Sultan, 1995; Stevens and Tsujimoto, 1995) and, therefore, if a synapse is to respond faithfully to a sustained train of action potentials, it must be able to replace the fusion-ready vesicles with a time course of seconds. The rate at which this occurs may determine some of the dynamic properties of the synapse.

The short latency of transmission would seem to preclude the involvement of ATP hydrolysis at the fusion step. This is born out by numerous physiological studies in which exocytosis persists after Mg^{2+}-ATP has been dialyzed from the cell and in which all the Mg^{2+} has been chelated (which would render any residual ATP inert to most enzymes) (Ahnert-Hilger *et al.*, 1985; Holz *et al.*, 1989; Hay and Martin, 1992; Parsons *et al*, 1995). Thus, if energy is needed to fuse the membranes, it should be stored in the fusion-ready state of the vesicle–membrane complex and released on addition of Ca^{2+}.

Transmitter Release Is a Cell Biological Question

Exocytosis and endocytosis are not unique to neurons; these processes go on in every eukaryotic cell. Moreover, exocytosis itself is only one representative of a general class of membrane trafficking steps in which one membrane-bound compartment must fuse with another. Other examples would include the fusion of recycling vesicles with endosomes, transport from the endoplasmic reticulum (ER) to the Golgi, or transport from endosomal compartments to lysosomes. In each case, the same biophysical problem must be overcome and the mechanisms for all these membrane fusion steps appear to have much in common.

For a vesicle to fuse with the plasma membrane, or any other target, there is a large energy barrier to surmount. To bring the lipid bilayers within a few nanometers of one another so that they can fuse, the hydration shell around the polar lipid head groups must be disrupted. Simply to split open each membrane so that the bilayer of one could be connected to the bilayer of the other would require exposing the hydrophobic core of each membrane to the aqueous milieu of the cytoplasm and this barrier is sufficiently great that it does not occur under normal conditions. Thus a specialized mechanism is required to bring the membranes close together and then drive fusion. At present, it appears that all the membrane trafficking steps within the cell use a similar set of proteins to accomplish this task (Fig. 8.4). As is discussed below, homologues of proteins found at the synapse and known to be essential for exocytosis have also been

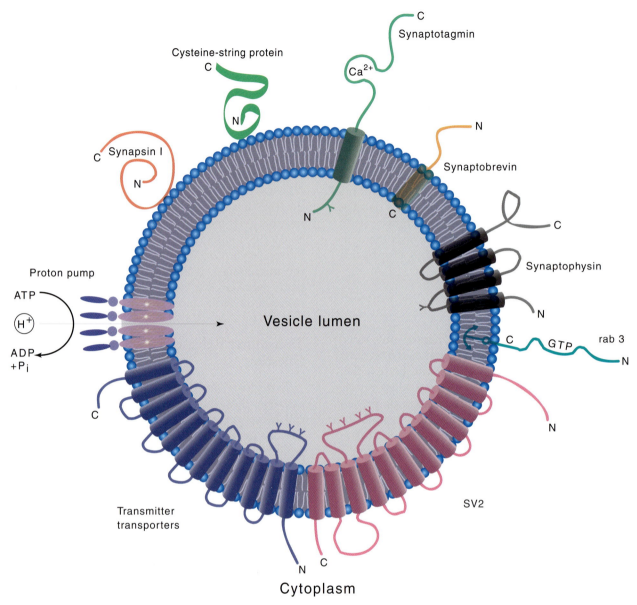

FIGURE 8.4 Schematic representation of the structure and topology of the major synaptic vesicle membrane proteins (see also Table 8.1). N, amino terminal; C, carboxy terminal.

shown to function in ER-to-Golgi transport in mammalian cells, in endosomal fusion, and in exocytosis in yeast. Indeed, it appears that representatives of this core set of proteins are present on every trafficking vesicle or target membrane and are required for every fusion step (Chen and Scheller, 2001). This discovery, which grew from the conjunction of independent studies of different model systems (Bennett and Scheller, 1993; Fischer von Mollard *et al.*, 1991), has led to the exciting hypothesis that all intracellular membrane fusions will be united by a single and universal mechanism. To the extent that this proves true, it will be a great boon to cell biology: experiments in one

model system, for example, the highly developed genetic analysis of membrane fusion in yeast, can be absorbed into neuroscience. Similarly, the abundance of synaptic vesicles for biochemical analysis and the unparalleled precision of electrophysiological assays of single vesicle fusions can deepen the understanding of other cellular events. Will these disparate membrane fusion events be truly identical in their mechanisms? The jury is still out. Most likely, however, different membrane fusions will present variations on a common theme; the fundamental processes will be adapted to the specific requirements of each physiological step. Perhaps an instructive analogy may be drawn with muscle contraction (Leavis and Gergely, 1984): actin and myosin offer a fundamentally conserved mechanism for producing force and movement and every muscle contains isoforms of actin, heavy and light myosin chains, troponins, and tropomyosins. But skeletal muscle and smooth muscle differ enormously in the details of how Ca^{2+} interacts with these proteins and regulates their activity. Regulation via troponins or via myosin light chain kinase, differences in the intrinsic ATPase rates of the myosins, and other critical features have adapted these related sets of proteins to their particular purpose in the individual type of muscle. Similarly, the need for extremely tight control of fusion in the nerve terminal—for rapid rates of fusion with very short latencies to occur in brief bursts at precise points on the plasma membrane—may cause some profound differences in how the synaptic isoforms of these proteins function compared with the isoforms involved in general cellular traffic.

Identifying the Synaptic Proteins: Vesicle Purification

The most important starting point in the elucidation of the machinery of the nerve terminal was the observation that synaptic vesicles can be purified. Synaptic vesicles are abundant in nervous tissue (Fig. 8.5) and, owing to their unique physical properties (uniform small diameter and low buoyant density), can be purified to homogeneity by simple subcellular fractionation techniques (Nagy *et al.*, 1976; Carlson *et al.*, 1978). As a result, at a biochemical level, synaptic vesicles are among the most thoroughly characterized organelles. One of the first sources for

the purification of synaptic vesicles was the electric organ of marine elasmobranchs. This structure, a specialized adaptation of the neuromuscular junction, is highly enriched in synaptic vesicles. It has also proven simple to isolate synaptic vesicles from mammalian brain. The protein compositions of synaptic vesicles from these sources are remarkably similar, demonstrating the evolutionary conservation of synaptic vesicle function. This similarity also points to the fact that many of the proteins present on the synaptic vesicle membrane perform general functions that are not restricted to a single class of transmitter.

Our knowledge of many of the important proteins in vesicle fusion commenced with the purification and characterization of synaptic vesicles. These vesicles contained a discrete set of abundant proteins and individual proteins could be isolated and subsequently cloned. The major constituents of the synaptic vesicle are shown in Fig. 8.4. For some of these proteins there are well-established functions, but for others the functions remain uncertain (Table 8.1). The proton transporter, for example, is an ATPase that acidifies the lumen of the vesicle. The resulting proton gradient provides the energy by which transmitter is moved into the vesicle. This task is carried out by another vesicular protein, a vesicular transporter that allows protons to move down their electrochemical gradient (leaving the vesicle) in exchange for moving the transmitter into the lumen. Several vesicular transporters, all structurally related, are known, and by their substrate selectivity they help to specialize a terminal for release of the appropriate transmitter (see below). Two other large proteins with multiple transmembrane domains are abundant in vesicles: SV2 (synaptic vesicle protein 2) and synaptophysin. The

TABLE 8.1 Function of Synaptic Vesicle Proteins

Protein	Function
Proton pump	Generation of electrochemical gradient of protons
Vesicular transmitter transporter	Transmitter uptake into vesicle
VAMP/synaptobrevin	Component of SNARE complex; acts in a late, essential step in vesicle fusion
Synaptotagmin	Ca^{2+} binding; possible trigger for fusion and component of vesicle docking at release sites via interactions with SNARE complex and lipid; promotes clathrin-mediated endocytosis by binding AP-2 complex
Rab3	Possible role in regulating vesicle targeting and availability
Synapsin	Likely to tether vesicle to actin cytoskeleton
Cysteine string protein	Promotes reliable coupling of action potential to exocytosis
SV2	Unknown
Synaptophysin	Unknown

functional significance of these proteins remains elusive despite extensive biochemical and genetic characterization (Feany et al., 1992; Bajjalieh et al., 1992; Janz et al., 1999a, b; Crowder et al., 1999; Brose and Rosenmund, 1999; Buckley et al., 1997; Südhof et al., 1987; McMahon et al., 1996). Other vesicular proteins are discussed later in this chapter.

Transmitter release depends on more than just vesicular proteins; proteins of the plasma membrane and cytoplasm are also important. In many cases, the identification of these additional components or an appreciation of their importance to the synapse derived from investigations of vesicular proteins. Two examples serve to illustrate this point. The first example begins with synaptotagmin, an integral membrane protein of the synaptic vesicle that was purified from synaptic vesicles and cloned (Bixby and Reichardt, 1985; Perin et al., 1990). The portion of synaptotagmin that extends into the cytoplasm (the majority of the protein, see Fig. 8.4) was subsequently used for affinity column chromatography. In this manner, a protein called syntaxin was identified as a synaptotagmin-binding protein (Bennett et al., 1992). This protein resides in the plasma membrane and is now appreciated as one of the critical players in vesicle fusion (see below). More recently, yeast two-hybrid screens have been conducted and used to identify proteins that bind to syntaxin. One such protein, syntaphilin, may serve as a regulator or modulator of syntaxin function (Lao et al., 2000). Thus, subsequent to the isolation of an abundant vesicular protein, biochemical assays have led to a fuller picture. A second and similar example began with the realization that an abundant small GTP-binding protein, rab3, was present on the vesicle surface (Südhof et al., 1987). This protein, discussed further below, may regulate vesicle availability or docking to release sites. Subsequently, rabphilin was identified on the basis of its affinity for the GTP-bound form of rab3 (Shirataki et al., 1993). Rabphilin lacks a transmembrane domain but is recruited to the surface of the synaptic vesicle by binding to rab3. Rabphilin may have a role in modulating transmission, particularly in mossy fiber terminals of the hippocampus (Lonart and Südhof, 1998). Subsequently, many additional rab3-binding proteins have been identified, including RIM (rab3-interacting molecule), a component of the active zone (Dresbach et al., 2001; Wang et al., 1997). Rab3 can exist in either GTP- or GDP-bound states and additional factors that regulate these states were also identified: a GDP dissociation inhibitor (GDI), GDP/GTP exchange protein (GEP), and GTPase-activating protein (GAP) are found in the synaptic cytosol (Südhof, 1997). Thus, from the identification of a synaptic vesicle protein, an array of additional factors have come to light, all of which are likely to figure in the exo–endocytic cycle.

Genetics Identifies Synaptic Proteins

Genetic screens have provided an independent method for identifying the machinery of transmitter release. One of the most fertile screens was carried out not in the nervous system per se, but rather in yeast. Because membrane trafficking in yeast is closely parallel to vesicle fusion at the terminal, mutations that alter the secretion of enzymes from yeast can be a springboard for the identification of synaptic proteins. In the early 1980s a series of such screens were carried out (Novick et al., 1980, 1981) and a collection of more than 50 mutants was obtained. In many of these mutants, post-Golgi vesicles accumulated in the cytoplasm and thus the mutation appeared to block a late stage of transport, such as the targeting or fusion of these vesicles at the plasma membrane. Screens for suppressors and enhancers of these secretion mutations uncovered further components (Aalto et al., 1993). Subsequently, excellent in vitro assays have been established in which to study the fusion of vesicles derived from yeast with their target organelles (Conradt et al., 1994; Mayer et al., 1996; Wickner and Haas, 2000). Among the secretion mutants and their interacting genes were homologues of some of the proteins discussed above: sec4 encodes a small GTP-binding protein like rab3, and Sso1 and Sso2 encode plasma membrane proteins that are homologues of syntaxin (Aalto et al., 1993; Salminen and Novick, 1987). The sec1 gene encodes a soluble protein with a very high affinity for Sso1, and the mammalian homologue of this protein, n-sec1, is tightly bound to syntaxin in nerve terminals and has an essential function in transmission (Aalto et al., 1991; Pevsner et al., 1994). Yeast is not the only organism in which a genetic screen uncovered an important protein for the synapse: the unc–13 mutation of Caenorhabditis elegans, for example, identified a component of the active zone membrane that may be important in priming vesicles for fusion and in the modulation of the synapse (Maruyama and Brenner, 1991; Richmond et al., 1999).

Genetics has further contributed to the understanding of synaptic proteins by allowing tests of the significance of an identified protein for synaptic transmission. Such studies have been carried out in C. elegans, Drosophila, and mice and can reveal either an absolute requirement for the protein, as in the case of syntaxin mutants in Drosophila (Schulze et al., 1995; Burgess et al., 1997), or relatively subtle effects, as in

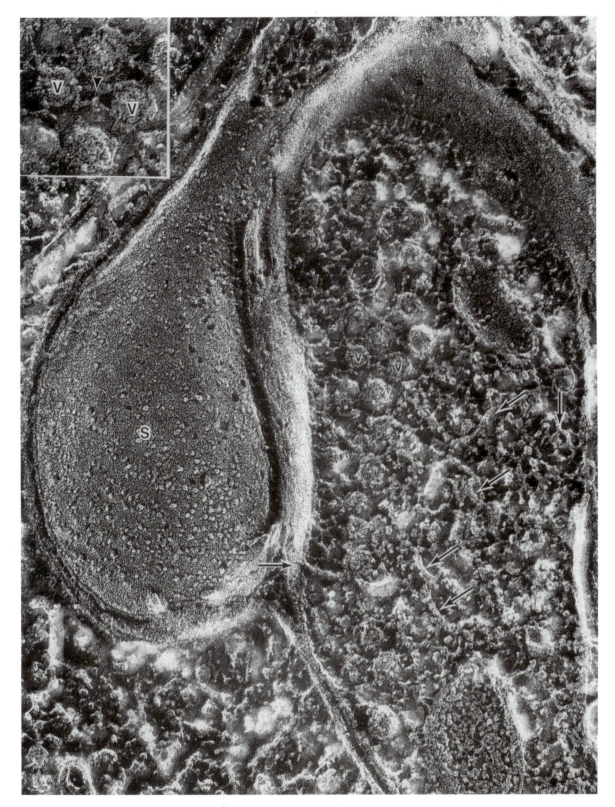

FIGURE 8.5 Structure of a synapse between a parallel fiber and a Purkinje cell spines (S) in the cerebellum. The sample was rapidly frozen and then freeze-fractured, shallow-etched, and rotary-shadowed to reveal the details of the synaptic architecture. Inset shows synaptic vesicles (V), and arrows indicate actin attachments. From Landis *et al.* (1988) Neuron 1: 201–209. Copyright © by Cell Press.

TABLE 8.2   Additional Proteins Implicated in Transmitter Release

Protein	Function
Syntaxin	SNARE protein present on plasma membrane (and on synaptic vesicles to a lesser extent); forms core complex with SNAP-25 and VAMP/synaptobrevin; essential for late step in fusion
SNAP-25	SNARE protein present on plasma membrane (and on synaptic vesicles to a lesser extent); forms core complex with syntaxin and VAMP/synaptobrevin; essential for late step in fusion
Nsec-1/munc-18	Syntaxin-binding protein required for all membrane traffic to the cell surface; likely bound to syntaxin before and after formation of SNARE complex
Synaphin/complexin	Syntaxin-binding protein; may oligomerize core complexes.
Syntaphilin	Binds syntaxin; prevents formation of SNARE complex (?)
Snapin	Binds SNAP-25; associated with synaptic vesicles; unknown function
NSF	ATPase that can disassemble the SNARE complex; likely to disrupt complexes after exocytosis
α-SNAP	Cofactor for NSF in SNARE complex disassembly
unc-13/munc-13	Active zone protein; vesicle priming for release(?); modulation of transmission(?)
Rabphilin	C2 domain protein; Ca^{2+}-binding protein; binds rab3 and associates with synaptic vesicle; modulation of transmission (?)
DOC2	C2 domain protein; Ca^{2+}-binding protein; binds munc-18; unknown function
RIM1 and related proteins	Active zone proteins; bind rab3; modulation of transmission (?)
Piccolo	Likely scaffolding protein to tether vesicles near active zone
Bassoon	Likely scaffolding protein to tether vesicles near active zone
Exocyst (sec6/8 complex)	Marks plasma membrane sites of vesicle fusion in yeast; synaptic role uncertain

the case of rab3 mutations in mice (Castillo *et al.*, 1997).

From biochemical purifications, *in vitro* assays, genetic screens, and fortuitous discoveries, an ever-growing list of nerve terminal proteins has been assembled (Tables 8.1 and 8.2). The manner in which these proteins coordinate the release of transmitter as well as all the other cell biological functions of the exo–endocytic cycle remains uncertain, but a consensus has emerged in recent years that puts one set of proteins at the core of the vesicle fusion.

The Mechanism of Membrane Fusion: SNAREs and the Core Complex

Three synaptic proteins, VAMP (vesicle-associated membrane protein)/synaptobrevin, syntaxin, and SNAP-25 (synaptosome-associated protein of 25 kDa), are capable of forming an exceptionally tight complex that is generally referred to as either the *core complex* or *SNARE complex* (Söllner *et al.*, 1993; Sutton *et al.*, 1998). The interaction of these three is essential for synaptic transmission and is likely to lie very close to or indeed

at the final fusion step of exocytosis (Figs. 8.6 and 8.7) What are these proteins? VAMP (also called synaptobrevin) was among the first synaptic vesicle proteins to be cloned (Trimble *et al.*, 1988). It is anchored to the synaptic vesicle by a single transmembrane domain and has a cytoplasmic domain that contributes a coiled-coil strand to the core complex. Syntaxin has a very similar structure but is located primarily in the plasma membrane (though some is present on vesicles as well) (Bennett *et al.*, 1992). SNAP-25 is also a protein primarily of the plasma membrane but, unlike the others, lacks a transmembrane domain and is instead anchored in its central region by acylations (Chapman *et al.*, 1994). SNAP-25 contributes two strands to the SNARE complex. Thus the interactions of these proteins can be envisioned as bringing closely together the two membranes that are meant to fuse. The proteins of this complex are archetypes of a class of membrane trafficking proteins collectively called the SNAREs. The vesicle-associated proteins, such as VAMP, are referred to as v-SNAREs, and those of the target membrane, such as SNAP–25 and syntaxin, are referred to as t-SNAREs.

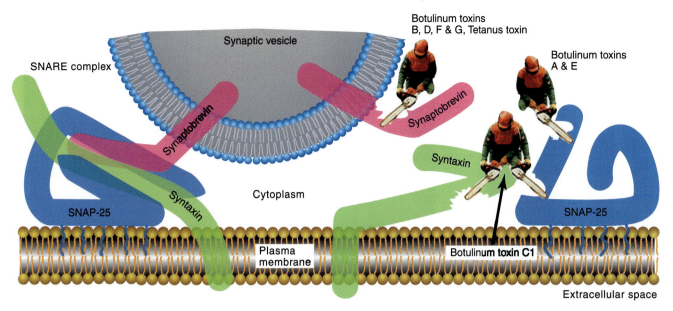

FIGURE 8.6 SNARE proteins and the action of the clostridial neurotoxins. The SNARE complex shown on the left brings the vesicle and plasma membranes into close proximity and likely represents one of the last steps in vesicle fusion. Vesicular VAMP, also called synaptobrevin, binds with the syntaxin and SNAP-25 that are anchored to the plasma membrane. Transmitter release can be blocked by tetanus toxin and the botulinum toxins—proteases that cleave specific SNARE proteins.

A SNARE complex is found at each membrane trafficking step within a eukaryotic cell (Fig. 8.7). Through a combination of SNARE proteins on the opposing membranes, a four-stranded coiled coil is formed. Alongside the example of the synapse in Fig. 8.7A, three analogous cases are shown from exocytosis in yeast (Fig. 8.7B), the fusion of late endosomes with one another (Fig. 8.7C) (Antonin *et al.*, 2000), and the fusion of vesicles that form the yeast vacuole (Fig. 8.7D) (Wickner and Haas, 2000; Sato *et al.*, 2000). Abundant genetic and biochemical data argue for an essential role for SNAREs, and the combination of yeast genetics, *in vitro* assays, synaptic biochemistry, and synaptic physiology has had a synergistic effect in advancing the field. It was the discovery that secretory mutations in yeast were homologues to the synaptic SNAREs that provided some of the first functional data on these proteins (Bennett and Scheller, 1993), while data on synapses first placed the proteins on vesicles and the plasma membrane (Bennett *et al.*, 1992; Trimble *et al.*, 1988; Chapman *et al.*, 1994). Exploring the mechanism of the SNAREs similarly brought together neurons and yeast. The crystal structure, for example, was solved first for the synaptic proteins (Sutton *et al.*, 1998). For crystallization, however, the transmembrane domains and of course the membranes themselves were absent. Thus it is not possible to conclude from the structure alone that the SNAREs formed a bridge between the membranes rather than complexes within the same membrane. In an *in vitro* assay with purified yeast vesicles, however, it was possible to establish this point (Nichols *et al.*, 1997).

Some of the strongest evidence for an essential role for SNAREs at the synapse has come from the study of a potent set of eight neurotoxins produced by clostridial bacteria. These toxins (tetanus toxin and the family of related botulinum toxins) have long been known to block the release of neurotransmitter from the terminal. The discovery that they do so by proteolytically cleaving individual members of the SNARE complex (Schiavo *et al.*, 1992; Link *et al.*, 1992) provided neurobiologists with a set of tools with which to probe SNARE function (Fig. 8.6). Each of the toxins comprises a heavy chain and a light chain that are linked by disulfide bonds (Pellizzari *et al.*, 1999). The heavy chain binds the toxin to surface receptors on neurons and thereby enables the toxin to be endocytosed. Once inside the cell, the disulfide bond is reduced and the free light chain enters the cytoplasm of the cell. This light chain is the active portion of the toxin and is a member of the Zn^{2+}-dependent family of proteases. The catalytic nature of the toxin accounts for its astonishing potency; a few tetanus toxin light chains, for example, at a synapse can suffice to proteolyze all the VAMP/synaptobrevin and thereby shut down transmitter release. The toxins are highly specific, recognizing unique sequences within an individual SNARE protein as summarized in Fig. 8.6. VAMP/synaptobrevin can be cleaved not only by

A. Transmitter release

B. Exocytosis in yeast

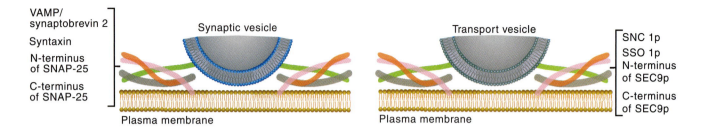

VAMP/
synaptobrevin 2

Syntaxin

N-terminus
of SNAP-25

C-terminus
of SNAP-25

Synaptic vesicle

Plasma membrane

Transport vesicle

Plasma membrane

SNC 1p

SSO 1p

N-terminus
of SEC9p

C-terminus
of SEC9p

C. Fusion of mammalian late endosomes

D. Yeast vacuole assembly

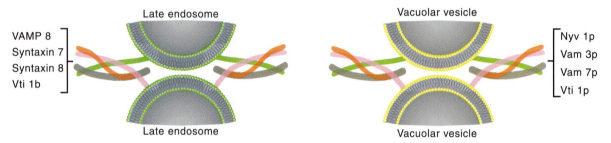

VAMP 8

Syntaxin 7

Syntaxin 8

Vti 1b

Late endosome

Late endosome

Nyv 1p

Vam 3p

Vam 7p

Vti 1p

Vacuolar vesicle

Vacuolar vesicle

FIGURE 8.7 Neurotransmitter release probably shares a core mechanism with all membrane fusion events within eukaryotic cells. The fusion of synaptic vesicles (A) is driven by a particular complex of four coiled-coil domains contributed by three different proteins. Exocytosis in yeast (B), the fusion of late endosomes in mammalian cells (C), and the fusion of vacuolar vesicles in yeast (D) exemplify the closely related four-stranded coiled-coil complexes that are required to drive fusion in other membrane trafficking steps.

tetanus toxin but by the botulinum toxin types B,D,F, and G and each toxin cleaves at a different peptide bond within the structure (Schiavo *et al.*, 1992; Yamasaki *et al.*, 1994). SNAP-25 is cleaved by botulinum toxins A and E, but again at different sites (Blasi *et al.*, 1993a; Schiavo *et al.*, 1993). Botulinum toxin C1 cleaves both syntaxin (Blasi *et al.*, 1993b) and, less efficiently, SNAP-25 (Foran *et al.*, 1996).

What precisely is the function of the SNARE proteins in promoting transmitter release? How does the assembly of this complex relate to membrane fusion? Studies with botulinum toxins, yeast mutants, mutants of *Drosophila* and *C. elegans*, and permeabilized mammalian cells all place the SNAREs late in the process. Synapses that lack an individual SNARE, for example, synapses whose VAMP has been mutated or cleaved by toxin, have the expected population of synaptic vesicles and these vesicles accumulate at active zones in the expected manner. Indeed, as judged by electron microscopy, there may even be an excess of "docked" vesicles in close proximity to the

plasma membrane at the active zone (Hunt *et al.*, 1994; Broadie *et al.*, 1995). Yet these synapses are incapable of secreting transmitter. Thus the SNAREs appear to be essential for a step that comes after the docking of the vesicle at the release site, but before the fusion pore opens and transmitter can diffuse into the cleft. The details of how the complex functions, however, are less certain. One attractive model (Chen and Scheller, 2001) is that the energy released by the formation of this very high affinity complex is used to drive together the two membranes. A loose complex of the SNARES would form and then "zipper up" and pull the membranes together. This may correspond to the actual fusion of the membranes or, alternatively, to a priming step that requires a subsequent rearrangement of the lipids to open the pore that will connect the vesicle lumen to the extracellular space. Evidence from the fusion of yeast vacuolar vesicles (Peters *et al.*, 2001; Peters and Mayer, 1998) implicates a distinct downstream step, regulated by Ca^{2+}/calmodulin and involving subunits of the proton pump, but whether or not this is true of neurons as well remains unknown. Vesicles consisting of only lipid bilayers and SNAREs can fuse *in vitro*, suggesting that no other proteins will be essential for the fusion step (Nickel *et al.*, 1999).

One additional function may reside with the SNARE proteins: the identification of an appropriate target membrane (Rothman and Warren, 1994; McNew *et al.*, 2000). Within a cell, there are myriad membrane compartments with which a transport vesicle can fuse: How then is specificity achieved? The great diversity of SNAREs (Fig. 8.7) may account for some of this specificity because not all combinations of v- and t-SNAREs form functional complexes. This potential mechanism, however, is likely to be only a part of the story. Particularly in the nerve terminal, it appears that synaptic vesicles can find the active zone even in the absence of the relevant SNAREs (Hunt *et al.*, 1994; Broadie *et al.*, 1995). In addition, the t-SNAREs syntaxin and SNAP-25 can be present along the entire axon and thus are inadequate in explaining the selective release of transmitter at synapses and active zones (Sesack and Snyder, 1995; Garcia *et al.*, 1995).

NSF: An ATPase for Membrane Trafficking

At some point as vesicles move through their exo–endocytic cycle, energy must be added to the system. *N*-Ethylmaleimide-sensitive factor (NSF), an ATPase involved in membrane trafficking, is one likely source. NSF was first identified as a required cytosolic factor in an *in vitro* trafficking assay (Block *et al.*, 1988). The importance of NSF was confirmed when it was found to correspond to the yeast sec18 gene, an essential gene for secretion (Eakle *et al.*, 1988). NSF hexamers bind a cofactor called a-SNAP (soluble NSF attachment protein) or sec17 and this complex, in turn, can bind to the SNARE complex. When Mg-ATP is hydrolyzed, the SNARE complex is disrupted into its component proteins (Söllner *et al.*, 1993). Originally it was speculated that this disassembly of the complex might correspond to fusion—that the action of NSF might catalytically wrench the SNAREs so that the membranes were brought together. As discussed above, however, a late role for ATP is unlikely in transmitter release and *in vitro* SNAREs can stimulate fusion without NSF present. More recent models put NSF action well before or after the fusion step (Mayer *et al.*, 1996; Banerjee *et al.*, 1996). If SNARE complexes form between VAMP, syntaxin, and SNAP-25 all in the same membrane, these futile complexes can be split apart by NSF so that productive complexes bridging the membrane compart-

ments can be formed. After fusion, the tight SNARE complex needs to be disrupted so that the VAMP can be recycled to synaptic vesicles while the other SNAREs remain on the plasma membrane. If, indeed, the energy of forming a tight SNARE complex is part of the energy that drives fusion, NSF, by restoring the SNAREs to their dissociated, high-energy state, will be an important part of the energetics of membrane fusion.

Docking and Priming the Vesicles for Fusion

Although SNAREs are thought to act late in the fusion reaction and fusion itself is an extremely rapid state, many preparatory and regulatory steps may precede the action of the SNAREs. These steps must tether the vesicle at an appropriate release site in the active zone and hold the vesicle in a fusion-ready state. These mechanisms remain among the most obscure at the synapse, but the list of proteins that may participate is growing. One example is the protein n-sec1 (also called munc18), the neuronal homologue of the product of the yeast sec1 gene. This protein binds to syntaxin with a very high affinity and, when so bound, prevents syntaxin from binding to SNAP-25 or VAMP (Pevsner *et al.*, 1994; Garcia *et al.*, 1994). Mutations of this protein prevent trafficking in both yeast and higher organisms. It appears likely that n-sec1 serves two functions. It may promote membrane fusion by priming syntaxin so that, once it has dissociated, it can participate correctly in fusion. But it may also be a negative regulator, keeping syntaxin inert until an appropriate vesicle or signal displaces n-sec1 and allows a SNARE complex to form.

Rab3 is another protein for which a priming or regulatory role is often invoked at the synapse. This small GTP-binding protein, mentioned above, is the homologue of the yeast sec4 gene product. In yeast, and at other membrane trafficking steps within mammalian cells, rab proteins have essential roles (Salminen and Novick, 1987; Gorvel *et al.*, 1991). They appear to help a vesicle to recognize its appropriate target and begin the process of SNARE complex formation. At the synapse, however, the significance of rab3 is still uncertain. Though it is clearly associated with synaptic vesicles, genetic disruption of rab3 in mice has a surprisingly slight phenotype and causes only subtle alterations in synaptic properties (Castillo *et al.*, 1997). Whether this is due to additional, redundant rab proteins or whether the rab family has been relegated to a more minor role at the synapse remains to be determined.

Because synaptic vesicles dock and fuse specifically at the active zone, this region of the nerve terminal membrane must have unique properties that promote docking and priming. The special nature of this domain is easily discernible in electron micrographs: a "fuzz" of electron-dense material can be observed opposite the postsynaptic density. In some synapses, such as photoreceptors, hair cells, and many insect synapses, the structures are more elaborate and include ribbons, dense bodies, and T-bars that extend into the cytoplasm and appear to have a special relationship with the nearby pool of vesicles. Recent advances in electron microscopy have allowed a more detailed look at the association of vesicles and plasma membrane at the active zone (Harlow *et al.*, 2001). At the neuromuscular junction of the frog (Fig. 8.1), it has been revealed that the electron-dense "fuzz" adjacent to the presynaptic plasma membrane is actually a highly ordered structure—a lattice of proteins that connect the vesicles to a cytoskeleton, to one another, and to the plasma membrane (Fig. 8.8). The molecules that correspond to these structures are not yet known. A few proteins, however, are known to be concentrated in the active zone or in the cloud of vesicles near the active zone. These proteins, piccolo, bassoon, RIM1, and unc-13, may be a part of the machinery that defines the active zone as the appropriate target for synaptic vesicle fusion (Garner *et al.*, 2000).

In addition to the specializations of the active zone, additional machinery must be present to preserve a dense cluster of synaptic vesicles extending approximately 200 nm back from the active zone (Fig. 8.1). The vesicles in this domain are not likely to be releasable within microseconds of the arrival of an action potential but are instead likely to represent a reserve pool from which vesicles can be mobilized to release sites on the plasma membrane. The equilibrium between this pool and the vesicles actually at the membrane may be an important determinant of the number of vesicles released per impulse, but remains poorly understood. One protein that is likely to play a role in the maintenance of the reserve pool is synapsin, a family of peripheral membrane proteins on synaptic vesicles (De Camilli *et al.*, 1983; Südhof *et al.*, 1989). Synapsins can also bind actin filaments and thus may provide a linker that tethers the vesicles in the cluster to the synaptic cytoskeleton (Fig. 8.5). Disruption of this link can cause the vesicle cluster to be diminished (Pieribone *et al.*, 1995) and reduce the number of vesicles in the releasable pool of the terminal (Ryan *et al.*, 1996). Synapsin has attracted considerable interest because it is the substrate for phosphorylation by both cAMP and Ca^{2+}-dependent protein kinases (Greengard *et al.*, 1993). These phosphorylations may influence the availability of reserve vesicles for recruitment to release sites.

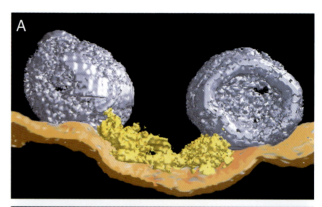

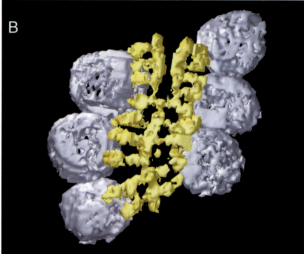

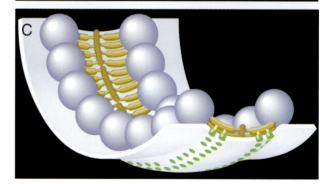

FIGURE 8.8   Fine structure of the active zone at a neuromuscular junction. (A) As revealed by electron tomography, synaptic vesicles (silver) are seen docked adjacent to the plasma membrane (tan) and associated with the proteins that make up the active zone (gold). (B) Viewed from the cytoplasmic side, the proteins are seen to extend from the vesicles and connect in the center. (C) Schematic rendering of an active zone based on the tomographic analysis. An ordered structure of ribs, pegs, and beams aligns the vesicles and connects them to the plasma membrane and to one another. After Harlow *et al.* (2001) Nature 409:479–484.

Coupling the Action Potential to Vesicle Fusion

The most striking difference between synaptic transmission and traffic between other cellular compartments is the rapid triggering of fusion by action potentials. As already discussed, the opening of Ca^{2+} channels and the focal rise of intracellular Ca^{2+} activate the fusion machinery. How does the terminal sense the rise in Ca^{2+}? What is the Ca^{2+} trigger and how does it open the fusion pore? Is it a single Ca^{2+}-binding protein or do several components respond to the altered Ca^{2+} concentration? Does Ca^{2+} remove a brake that normally prevents a docked, primed vesicle from fusing or does Ca^{2+} induce a conformational change that is actively required to promote fusion? Whence does the steep, exponential relationship of release to intracellular Ca^{2+} arise? These questions are an active area of investigation and debate.

Some clues may come from other systems: Ca^{2+}-dependent membrane fusion is not unique to the synapse. Ca^{2+} can trigger both exocrine and endocrine secretion. Furthermore, Ca^{2+} released from intracellular stores now appears to be essential in trafficking steps in yeast that previously had been viewed as constitutive and unregulated. In yeast vacuolar fusion, for example, calmodulin senses a local increase in Ca^{2+} and triggers fusion (Peters and Mayer, 1998). Calmodulin, however, has not been the leading candidate for the synaptic trigger. The affinity of calmodulin for Ca^{2+} has generally been taken to be too high to explain the relatively high levels of Ca^{2+} (at least 10 μM) that are needed to evoke transmitter release. However, the apparent affinity of calmodulin for Ca^{2+} is very dependent on the proteins to which calmodulin binds, and the four Ca^{2+} binding sites of free calmodulin probably have an average affinity of about 11 μM (Jurado et al., 1999). Evidence for involvement of calmodulin in synaptic transmission continues to arise (Chen and Scheller, 2001; Arredondo et al., 1998; Quetglas et al., 2000; Chamberlain et al., 1995), suggesting that yeast vacuolar traffic and synaptic transmission may not be as divergent as one might think. At present, however, a modulatory role for calmodulin is favored over a requirement in the final triggering step at the synapse.

The leading candidate for synaptic Ca^{2+} sensor is synaptotagmin, an integral membrane protein of the synaptic vesicle (Fig. 8.4) (Perrin et al., 1991; Brose et al., 1992; Geppert et al., 1994; Li et al., 1995). Synaptotagmin has a large cytoplasmic portion that comprises two Ca^{2+} binding C2 domains, called C2A and C2B. These domains can also interact with the SNARE complex proteins and with phospholipids in a Ca^{2+}-dependent manner. It has been hypothesized that one or more of these interactions is the molecular correlate of the triggering event for fusion. Consistent with this hypothesis, mutations that remove synaptotagmin profoundly reduce synaptic transmission in flies, worms, and mice (Geppert et al., 1994; Di Antonio et al., 1993; DiAntonio and Schwarz, 1994; Broadie et al., 1994; Littleton et al., 1993; Nonet et al., 1993) while having little effect on or enhancing the rate of spontaneous release of transmitter. Whether or not this reduction in evoked release is due specifically to loss of the Ca^{2+} trigger, however, is harder to demonstrate. Synaptotagmin is likely to be involved in endocytosis and potentially in vesicle docking as well (Jorgensen et al., 1995; Reist et al., 1998; Zhang et al., 1994; Haucke and De Camilli, 1999), which has complicated the analysis. These processes, as mentioned above, are also likely to be regulated by Ca^{2+}, and the Ca^{2+} binding sites on synaptotagmin may be relevant for this regulation as well.

Packaging Transmitter into the Vesicle

A central requirement of quantal synaptic transmission is the synchronous release of thousands of molecules of transmitter from the presynaptic nerve terminal. This requirement is partly met by the capacity of synaptic vesicles to accumulate and store high concentrations of transmitter. In cholinergic neurons, the concentration of acetylcholine within the synaptic vesicle can reach 0.6 M, more than 1000-fold greater than that in the cytoplasm (Nagy et al., 1976; Carlson et al., 1978). Two synaptic vesicle proteins mediate the uptake of transmitter: the vacuolar proton pump and a family of transmitter transporters. The vacuolar proton pump is a multisubunit ATPase that catalyzes the translocation of protons from the cytoplasm into the lumen of a variety of intracellular organelles, including synaptic vesicles (Nelson, 1992). The resulting transmembrane electrochemical proton gradient is used as the energy source for the active uptake of transmitter by transmitter transporters. Transmitter uptake has been characterized in isolated synaptic vesicle preparations in which at least four types of distinct transporters have been identified: one for acetylcholine, another for biogenic amines (catecholamines and serotonin), a third for the excitatory amino acid glutamate, and the fourth for inhibitory amino acids (GABA and glycine) (Edwards, 1992). At least one gene for each of these classes of transmitter transporter has now been cloned (Takamori et al., 2000; Bellocchio et al., 2000; Reimer et al., 1998). As expected, these distinct transporters are differentially expressed by neurons. The type of transporter expressed in a cell dictates the type of transmitter stored in the synaptic vesicles of a particular neuron, and when investigators drive the expression of a glutamate transporter, for example, in a GABA-releasing neuron, they can trick the cell into now releasing glutamate.

The vesicular transporters are integral membrane proteins with 12 membrane-spanning domains that display sequence similarity with bacterial drug resistance transporters. The synaptic vesicle transporters are clearly distinct from the plasma membrane transmitter transporters that remove transmitter from the synaptic cleft and thereby contribute to the termination of synaptic signaling (see Fig. 8.2 and Chapter 9). The distinguishing characteristics include their transport topology, energy source, pharmacology, and structure (Reimer *et al.*, 1998).

In contrast to small chemical transmitters, proteinaceous signaling molecules, including neuropeptides and hormones, are typically stored in granules that are larger and have a higher electron density than synaptic vesicles. The contents of these granules are not recycled at the release sites; as a result, their replenishment requires new protein synthesis followed by packaging into secretory vesicles in the cell body. Because of the slow kinetics of their release, the slow responsiveness of their postsynaptic receptors, and their inability to be locally recycled, proteinaceous signaling molecules typically mediate regulatory functions.

Endocytosis Recovers Synaptic Vesicle Components

After exocytosis, the components of the synaptic vesicle membrane must be recovered from the presynaptic plasma membrane, as discussed above. Vesicle recycling is accomplished by either of two mechanisms. The first is simply a reversal of the fusion process. In this case, a fusion pore opens to allow transmitter release and then rapidly closes to re-form a vesicle. Often nicknamed "kiss and run," it has the theoretical advantage that it would allow all the vesicular components to remain together on a single vesicle that would be immediately available for reloading with transmitter. The mechanism is potentially quick and energetically efficient. This mechanism is employed in some systems (Monck and Fernandez, 1992), but its relevance for the synapse is an unresolved issue (Stevens and Williams, 2000; Ales *et al.*, 1999; Klingauf *et al.*, 1998; Sankaranarayaran and Ryan, 2001).

The predominant pathway for synaptic vesicle recycling is more likely to be endocytosis. Endocytosis

of synaptic vesicle components, like receptor-mediated endocytosis in other cell types, is mediated by vesicles coated with the protein clathrin (Maycox *et al.*, 1992). Accessory proteins select the cargo incorporated into these vesicles as they assemble. One class of accessory proteins known as AP–2 displays a high affinity for synaptotagmin (Zhang *et al.*, 1994), which may be important, therefore, in recruiting the clathrin coat to the vesicle. The final pinching off of the clathrin-coated vesicle requires the protein dynamin, which can form a ringlike collar around the neck of an endocytosing vesicle (De Camilli *et al.*, 1995; Warnock and Schmid, 1996). A crucial role for dynamin in synaptic vesicle recycling is most clearly demonstrated in a temperature-sensitive *Drosophila* mutant known as *shibire*. The *shibire* gene encodes the *Drosophila* homologue of dynamin (Van der Bliek and Meyerowitz, 1991; Chen *et al.*, 1991). At the nonpermissive temperature, the *shibire* mutant flies rapidly become paralyzed owing to nearly complete depletion of synaptic vesicles from their nerve terminals. Dynamin is a GTPase whose activity is modulated by calcium-regulated phosphorylation and dephosphorylation (Robinson *et al.*, 1994). Thus, the endocytic recycling of synaptic vesicles provides another site at which calcium regulates the synaptic vesicle exo–endocytic cycle. When the components of the synaptic vesicle membrane are recovered in clathrin-coated vesicles, recycling is completed by vesicle uncoating and, perhaps, passage through an endosomal compartment in the nerve terminal (see Fig. 8.2).

Summary

The life of the synaptic vesicle involves much more than just the Ca^{2+}-dependent fusion of a vesicle with the plasma membrane. It is a cyclical progression that must include endocytosis, transmitter loading, docking, and priming steps as well. In many regards, this cell biological process shares mechanistic similarities with membrane trafficking in other parts of the cell and with simpler organisms such as yeast. The interaction of the vesicular and plasma membrane proteins of the SNARE complex—VAMP/synaptobrevin, syntaxin, and SNAP–25—is an essential, late step in fusion. Many other proteins have been identified that are likely to precede the action of the SNAREs, regulate the SNAREs, and recycle them. Components of the synaptic vesicle membrane, the presynaptic plasma membrane, and the cytoplasm all contribute to the regulation of synaptic vesicle function. Together, these proteins build on the fundamental core apparatus to create an astonishingly accurate, fast, and reliable means of delivering transmitter to the synaptic cleft.

QUANTAL ANALYSIS

A quantitative description of the signal passing across a synapse is of utmost importance in understanding the function of the nervous system. This signal is the final output of all the integrative processes taking place in the presynaptic cell, and a complete statistical description should be able to capture the flow of information between neurons, as well as between neurons and effector cells. At many chemical synapses, a quantitative description is also a source of unique insight into the biophysics of transmission. The postsynaptic signal often fluctuates from trial to trial in a *quantal* manner; that is, it adopts preferred levels, which arise from the summation of various numbers of discrete events, thought to result from the release of individual vesicles of neurotransmitter (Katz, 1969, Martin, 1977). Examination of the trial-to-trial amplitude fluctuation of the synaptic signal allows the size of the *quantum*, as well as the average number of quanta released for a given presynaptic action potential, to be estimated. This approach is also a source of insight into the probabilistic processes underlying transmitter release from the presynaptic terminal and into the mechanisms by which transmission can be modified by physiological, pharmacological, and pathological phenomena.

Transmission at the Frog Neuromuscular Junction Is Quantized

The quantal nature of transmission was first demonstrated in the early 1950s by Bernard Katz and his colleagues, who studied the frog motor end plate. By recording from a muscle fiber immediately under a branch of the motor axon, they measured the postsynaptic potential both at rest and in response to stimulation of the axon. Spontaneous signals (MEPPs) were observed to occur at random intervals, measuring between 0.1 and 2 mV in amplitude (Fatt and Katz, 1952). Stimulating the presynaptic axon produced a postsynaptic signal approximately 100 times larger. At any one site, the *spontaneous MEPPs* were of roughly the same amplitude, with a coefficient of variation of 30%, compatible with the intermittent release of multimolecular packets of transmitter from the presynaptic terminal. When the presynaptic axon was stimulated under conditions designed to depress transmitter release to very low levels, the evoked EPP

fluctuated from trial to trial between preferred amplitudes, which coincided with integral multiples of the MEPP amplitude (del Castillo and Katz, 1954; Liley, 1956; Boyd and Martin, 1956) (see Fig. 8.9). This implied that the MEPP was a *quantal building* block, variable numbers of which were released to make up the evoked signal. The relative numbers of trials resulting in 0, 1, 2, … , quanta were well described by a Poisson distribution, a statistical distribution that arises in many instances where a random process operates (Box 8.4), implying that the process governing quantal release may also depend on a simple underlying mechanism.

On the basis of this evidence, Katz and colleagues proposed the following model of transmission, which has gained wide acceptance and is referred to as the *standard Katz model*.

- Arrival of an action potential at the presynaptic terminal briefly raises the probability of release of quanta of transmitter.
- Several quanta are available to be released, and every quantum gives roughly the same electrical signal in the postsynaptic cell. This is the quantal amplitude, Q, which sums linearly with all other quanta released.
- The average number of quanta released, m, is given by the product of n, the number of available quanta, and p, the average release probability: $m = np$. The relative probability of observing 0, 1, 2, . . ., n quanta released is then given by a binomial distribution, with parameters n and p (see Box 8.4).
- Under conditions of depressed transmission, p is low, and the system approximates a Poisson process. This is the limiting case of the binomial distribution where p tends to 0 and n tends to ∞ and is determined by the unique parameter m (see Box 8.4).

The standard Katz model has been supported by similar experiments in other preparations, which have shown that evoked EPSPs, IPSPs, EPSCs, or IPSCs tend to cluster near preferred values corresponding to integral multiples of a unit. In many cases, this quantal amplitude corresponds closely to the amplitude of spontaneous miniature postsynaptic signals (potentials or currents) occurring in the absence of presynaptic action potentials (Fig. 8.10) (Paulsen and Heggelund, 1994).

A major impetus for accurate measurement of quantal parameters is that it may help determine the locus of modulatory effects on synaptic transmission. An increase or decrease in the probability of presynaptic transmitter release should be detected as a change in quantal content, m, whereas an alteration in the

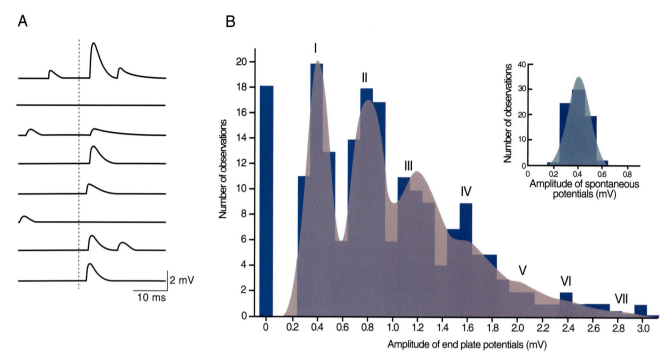

FIGURE 8.9   Quantal transmission at the neuromuscular junction. (A) Intracellular recordings from a rat muscle fiber in response to repeated presynaptic stimulation of the motor axon. Extracellular $[Ca^{2+}]$ and $[Mg^{2+}]$ were kept low and high, respectively, to depress transmission to a very low level. The size of the postsynaptic response, seen after the stimulus (not shown) that occurs at the dashed vertical line, fluctuated from trial to trial, with some trials giving failures of transmission. Spontaneous mEPPs, occurring in the background, had approximately the same amplitude as the smallest evoked EPPs, implying that they arose from the release of single quanta of acetylcholine. (B) The peak amplitudes of 200 evoked EPPs from a similar experiment, plotted as an amplitude histogram. Eighteen trials resulted in failures of transmission (indicated by the bar at 0 mV), and the rest gave EPPs whose amplitude tended to cluster at integral multiples of 0.4 mV. This coincides with the mean amplitude of the spontaneous mEPPs, whose amplitude distribution is shown in the inset together with a Gaussian fit. The shading through the EPP histogram is a fit obtained by assuming a Poisson model of quantal release (see Box 8.4). The parameters describing this model were the average number of quanta released, m, obtained by dividing the mean EPP amplitude by the mean mEPP amplitude, and the quantal amplitude, Q, and its variance, Var_Q, obtained from the mEPP amplitude distribution. There is a good agreement with the observed amplitude distribution. The Poisson model predicts 19 failures. Roman numerals indicate the number of quanta corresponding to each component in the distribution. (A) Liley (1956); (B) from Boyd and Martin (1956).

postsynaptic density or efficacy of receptors should be detected as a change in quantal amplitude, Q.

Biophysical Phenomena Underlie the Quantal Parameters

Considerable effort has been directed at establishing the physical correlates of the quantal parameters, Q, n, and p.

Quantal Parameter Q

A MEPP was thought to be too large to be accounted for by the release of an individual molecule of acetylcholine, because low concentrations of exogenous acetylcholine generated responses much smaller than a MEPP (Fatt and Katz, 1952). Vesicles were sub-

sequently observed in electron micrographs of presynaptic terminals (see Chapter 1). Because these vesicles could act as packaging devices for transmitter, it was proposed that the quantal amplitude, Q, represents the discharge of one vesicle into the synaptic cleft. A large body of evidence (Box 8.1) has since led to almost universal agreement that Q is the postsynaptic response to exocytosis of a single vesicle of transmitter.

Figure 8.11 (Bartol *et al.*, 1991) shows a computer simulation of the diffusion of molecules of acetylcholine and binding to receptors at the neuromuscular junction: acetylcholine spreads out in a disk in the synaptic cleft and binds most of the underlying postsynaptic receptors, although many more spare receptors beyond the edge of the disk remain unoccupied by transmitter (Matthews-Bellinger and Salpeter, 1973).

BOX 8.4

BINOMIAL AND POISSON MODELS

Binomial Model

According to the binomial description, n quanta can be released in response to a presynaptic action potential, each of which has a probability, p, of being discharged. For convenience, let us define q as the probability that a given quantum is not discharged in a given trial: $q = 1 - p$. In any given trial, the number of quanta observed is between 0 and n. Imagine that only three quanta are available ($n = 3$), each of which has a 40% chance of being discharged in response to the action potential ($p = 0.4$, $q = 0.6$). The average number of quanta released is $m = np = 1.2$.

The probability that no quanta are released is $q^3 = 0.216$.

The probability that only one quantum is released is the sum of the probabilities that only the first is released, only the second is released, and only the third is released: $3pq^2 = 0.432$.

The probability that two of three are released is, conversely, $3p^2q = 0.288$.

Finally, the probability that all three are released is $p^3 = 0.064$.

The relative probabilities of observing 0, 1, . . ., n quanta are the coefficients of the expansion of $G = (p + qz)^n = 0.216 + 0.432z + 0.288z^2 + 0.064z^3$. This is known as the generating function for the binomial distribution, and z is simply a "dummy variable"; that is, it serves no function except to allow the binomial expansion. These coefficients can be obtained from the binomial distribution, which gives the probability of observing 0, 1, . . ., n quanta released for any n and p. Writing P_x for the probability of observing x quanta, we obtain

$$P_x = \frac{n!}{(n-x)!\,x!}\,p^x q^{n-x}.$$

Poisson Model

As n becomes very large and p very small, the probability of observing 0, 1, . . ., quanta is equally well described by a Poisson distribution. This is a limiting case of the binomial distribution when n tends to ∞ and p tends to 0, and, instead of two parameters (n and p), it is described by the sole parameter m. Again, m is the average value for P_x. The relative probability of observing 0, 1, . . ., quanta is now given by

$$P_x = \frac{m^x e^{-m}}{x!}.$$

For the same average quantal content m as in the binomial model, the relative probabilities of observing 0, 1, . . ., quanta are now approximately

$$P_0 \cong 0.30,$$
$$P_1 \cong 0.36,$$
$$P_2 \cong 0.21,$$
$$P_3 \cong 0.08,$$
$$P_4 \cong 0.02,$$
$$\vdots$$

Compound or Nonuniform Binomial Model

In the binomial model, what happens if different release sites have different, but still independent, probabilities? Let us take the following example: $p_1 = 0.1$, $p_2 = 0.2$, $p_3 = 0.9$ (again defining $q = 1 - p$, $q_2 = 1 - p_2$, $q_3 = 1 - p_3$). The average quantal content is again $p_1 + p_2 + p_3 = 1.2$.

$$P_0 = q_1 q_2 q_3 = 0.07,$$
$$P_1 = p_1 q_2 q_3 + q_1 p_2 q_3 + q_1 q_2 p_3 = 0.67,$$
$$P_2 = q_1 p_2 p_3 + p_1 q_2 p_3 + p_1 p_2 q_3 = 0.23,$$
$$P_3 = p_1 p_2 p_3 = 0.01.$$

Generally, P_x are again obtained from the coefficients of the polynomial expansion of the generating function:

$$G = (p_1 + q_1 z)(p_2 + p_2 z) \text{ K } (p_n + q_n z)$$
$$= \Pi_k (p_k + q_k z), \, k = 1, \text{ K }, n.$$

Quantal Parameter n

The physical correlate of n has been more elusive. In the original description of quantal transmission, n represented the number of releasable quanta. As the ultrastructure of the presynaptic terminal was eluci-

dated, however, vesicles were found to be clustered near specializations, or active zones, in the presynaptic membrane. Each active zone is composed of a dense bar on the cytoplasmic face of the terminal membrane, bordered by rows of intramembranous

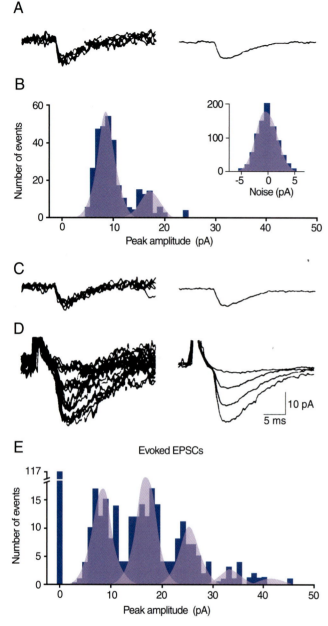

FIGURE 8.10 Quantal transmission in a thalamic neuron in a guinea pig brain slice. (A) Several spontaneous EPSCs superimposed (left), and the average time course (right). (B) An amplitude histogram showing that these events are clustered principally near 8.3 pA, with a smaller peak near 17 pA possibly representing the synchronous release of two quanta. (C) When presynaptic action potentials were abolished by tetrodotoxin, mean amplitudes of mEPSCs and spontaneous EPSCs were similar, implying that most were uniquantal. (D) EPSCs evoked by presynaptic stimulation in the optic tract. (E) Amplitude histogram of (D), showing clear clustering at integral multiples of approximately 8.3 pA. The superimposed Gaussian curves in (B) and (E) have approximately the same variance as the background noise, implying that quantal variability was negligible. From Paulsen and Heggelund (1994).

particles, thought to be voltage-sensitive calcium channels (Heuser and Reese, 1973). As stated earlier, entry of calcium ions through these channels raises the intracellular calcium concentration in a small volume immediately adjacent to the channels, triggering the exocytosis. Can n then be the number of active zones, if they are release sites? At the frog neuromuscular junction, simultaneously evoked quanta seem to summate linearly; however, when acetylcholine is added to the bath the relationship between depolarization and acetylcholine concentration is nonlinear (Hartzell *et al.*, 1975). The discrepancy between the linear summation of quanta and the nonlinear dose dependency of the response to acetylcholine can be explained by proposing that the multiple quanta making up an EPP activate separate populations of receptors. This could occur if each active zone normally releases only one quantum, and is consistent with the view that n is indeed the number of active zones.

Quantal Parameter p

Parameter p is the probability of exocytosis in response to a presynaptic action potential. Because this interval is of finite duration, it is more strictly a time integral of the probability during this transient event (Katz and Miledi, 1965). Moreover, because discharge of a quantum may leave a release site empty, p should be treated as a product of two probabilities: (1) that a release site is occupied by a quantum (p_1) and (2) that a presynaptic action potential evokes release (p_2) (Zucker, 1973).

With improved ultrastructural resolution of presynaptic terminals, it has become apparent that some vesicles are especially intimately related to the active zone. The number of such "docked" vesicles may represent a readily releasable pool (Schikorski and Stevens, 1997), which corresponds to the product of n and p_1.

The Standard Katz Model Does Not Always Apply

Before accepting the standard Katz model in all its details, we more closely examine some of its implications to see how they tally with our knowledge of the underlying molecular mechanisms.

Quantal Uniformity

At the vertebrate neuromuscular junction, the amount of acetylcholine released into the synaptic cleft determines the quantal size (Kuffler and Yoshikami, 1975; Fletcher and Forrester, 1975; Whittaker, 1988). Thus, for the quantal amplitude to

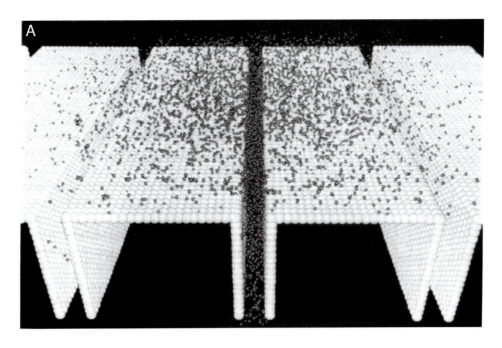

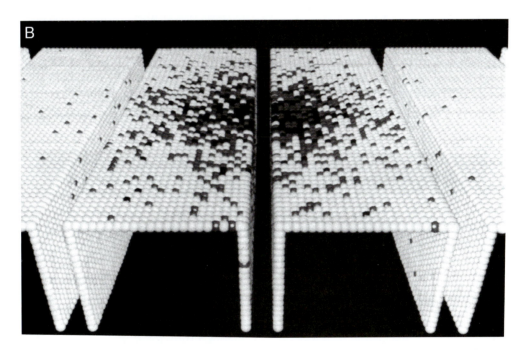

FIGURE 8.11 Monte Carlo simulation of quantal release of acetylcholine at the neuromuscular junction. The postsynaptic receptors are represented as a sheet of spheres—white if unliganded, gray if singly liganded, and black if doubly liganded. Presynaptic structures and the ends of the junctional folds are not shown, and the effect of acetylcholinesterase is not modeled. (A) Thirty microseconds after synchronous release of 9500 molecules of acetylcholine (small gray spheres) from a point source opposite the central fold. (B) Postsynaptic response, showing an effectively saturated area at the center, opposite the release site, surrounded by singly bound and unbound receptors. From Bartol *et al.* (1991).

be constant at different release sites, a uniform population of vesicles must be available to be released, as must receptors with identical properties opposite each release site. Electron microscopic images of vesicles in the presynaptic terminal indicate that their diameters are indeed remarkably uniform, although whether the neurotransmitter content of the vesicles is unvarying is not known. Similarly, although postsynaptic receptors are clustered opposite the active zone, their density and properties may not be uniform between different sites.

Uniform and Independent Release Probabilities

The release sites must be identical, with a uniform probability of exocytosis. If this condition were not satisfied, evoked signals would still cluster at integral multiples of the quantal amplitude, but the relative proportion of trials resulting in 0, 1, ... , n quanta would no longer be described by a simple binomial (or Poisson) distribution. As a limiting case, if p at some sites is effectively 0, then the meaning of n is questionable.

Rapid and Synchronous Transmitter Release

All-or-none exocytosis is clearly necessary for quantization and is supported by freeze-fracture images of terminals taken during intense evoked release (Heuser *et al.*, 1979). However, all-or-none exocytosis may not be the only mode of transmitter release, because secretory vesicles in mast cells can release some of their contents through a fusion pore that opens reversibly without necessarily leading to full exocytosis (Alvarez de Toledo *et al.*, 1993). Whether this mode of release also occurs in synapses remains to be determined. Quantization in the size of the evoked response can also be concealed by asynchrony of transmitter release from individual sites. (Isaacson and Walmsley, 1995).

Ion Channel Noise

Stochastic properties of postsynaptic ligand-gated ion channels must not add excessive variability to the size of the postsynaptic signal. Again, if this condition were not satisfied, it would be difficult to identify the quantal amplitude, and clustering of amplitudes at integral multiples of the quantum would be concealed. If we assume that individual ionophores act independently of one another, the variance of the quantal current arising from their stochastic opening is described by the binomial formula

$$\text{Var} = i^2 \, kp_o \, (1 - p_o),$$

where i, k, and p_o refer to the single channel current, the number of channels, and their probability of opening in response to transmitter release, respectively. Because the average quantal current amplitude is ikp_o, the coefficient of variation of the quantal amplitude is $\sqrt{(1 - p_o) / kp_o}$. A low quantal variability, which is required for quantal behavior to be detected, therefore implies either a large number of ionophores, k, or a high probability of opening, p_o, in response to transmitter release.

Postsynaptic Summation and Distortion of Signals

Postsynaptic currents or potentials arising from different release sites must sum linearly. If this is not satisfied, clustering of evoked postsynaptic signals may not occur at integral multiples of a quantal amplitude. If the postsynaptic membrane becomes appreciably depolarized as a result of the activation of many receptors, quanta may no longer summate linearly, either because the driving force for ion fluxes decreases or because voltage-gated channels open to cause regenerative currents to flow.

Stationarity

The state of the synapse must be relatively stable with time. A drift in the release probability with time could preclude a binomial or Poisson model, and changes in the quantal amplitude could prevent clear clustering in the distribution of evoked signals.

Thus, many of the requirements for the standard Katz model cannot realistically be expected to hold in all cases. In the presence of nonuniformity of release probability, the trial-to-trial amplitude fluctuation of the postsynaptic signal is unlikely to be described by a binomial or Poisson model. Indeed, the evidence that these simple probabilistic models are correct is far from compelling. On the other hand, the fact that evoked synaptic signals are often found clustered at integral multiples of an underlying unit strongly argues that vesicle filling and the postsynaptic phenomena determining the quantal amplitude are sufficiently uniform to ensure that a more general quantal description of transmission applies.

Central Nervous System Synapses Behave Differently from the Frog End Plate

A number of differences have emerged between quantal transmission at the neuromuscular junction and in central synapses in vertebrates.

One-Quantum Release

A correlation of histological and electrophysiological evidence obtained in the same preparation led to the proposal that many individual terminals in the CNS have only one release site (Korn and Faber, 1987, 1991; Redman, 1990; Walmsley, 1991). There are some notable exceptions to this rule. Calyceal synapses in the brainstem auditory pathway, for instance, have multiple active zones, and glutamate released from one release site can interact with transmitter released from neighboring sites (Trussell *et al.*, 1993).

Nonuniform Release Probabilities

In the mammalian spinal cord, release probabilities may vary between individual sites supplied by an individual muscle afferent. Postsynaptic signals have been shown to fluctuate in a manner that cannot be described by a binomial model, unless the individual release probabilities are allowed to vary (Redman, 1990; Walmsley, 1991; Jack *et al.*, 1981). Because the release sites are often segregated in different terminals, they may be subject to differing amounts of tonic presynaptic inhibition mediated by axo-axonic synapses.

Relatively Few Receptors Are Available to Detect Presynaptic Transmitter Release

A major difference between vertebrate CNS synapses and the neuromuscular junction is that the quantal amplitude is often determined not only by the vesicle contents but also by the number of available receptors (Edwards *et al.*, 1990; Jonas *et al.*, 1993; Korn and Faber, 1987, 1991; Redman, 1990; Walmsley, 1991). At some excitatory synapses, the glutamate content of a quantum appears to be sufficient to bind a large proportion of the available postsynaptic receptors (Clements *et al.*, 1992; Tang *et al.*, 1994; Tong and Jahr, 1994). Fewer than 100 receptors open, compared with 1000–2000 at the neuromuscular junction (Edwards *et al.*, 1990; Jonas *et al.*, 1993).

Different Receptors May Sample Different Quantal Contents

In contrast to the neuromuscular junction, CNS synapses frequently have several pharmacologically distinct postsynaptic receptors. Although little is known of quantal signaling via metabotropic receptors, glutamatergic synapses contain different combinations of AMPA, kainate, and *N*-methyl-D-aspartate (NMDA) receptors, all of which can open in response to glutamate release (see Chapter 11). The quantal content sampled by NMDA receptors is often larger than that sampled by AMPA receptors (Kullmann and Siegelbaum, 1995). Because NMDA receptors are unable to open at hyperpolarized membrane potentials, such synapses are functionally silent in the absence of postsynaptic depolarization. This discrepancy in signaling is partly explained by differential expression of receptors at synapses, but, in addition, differences in the affinity of AMPA and NMDA receptors for glutamate may also play a role.

Spontaneous Miniature Postsynaptic Signals

In the central nervous system, spontaneous mEPSCs and mIPSCs vary widely in amplitude (Edwards *et al.*, 1990; Jonas *et al.*, 1993; Manabe *et al.*, 1992) implying a quantal coefficient of variation considerably greater than that at the neuromuscular junction: between 40 and 80% instead of 30%. At first sight, this would preclude unambiguous peaks in histograms of evoked signals. However, quantal variability must be divided into variability from trial to trial at an individual release site (intrasite) and variability among sites (intersite). Spontaneous mEPSCs and mIPSCs arise from a large number of different sites, so their amplitude range includes both sources of quantal variability. If intrasite quantal variability were very large, then we would not expect to be able to detect quantal clustering in the amplitudes of evoked synaptic signals. The fact that such clustering is sometimes seen (see Fig. 8.10) implies that intrasite variability can be modest, and the wide range of amplitudes of miniature events principally indicates a large intersite variability. Thus, the mean amplitude of spontaneous miniature events cannot be used as a guide to the quantal amplitude underlying an evoked synaptic signal.

Quantal Parameters Can Be Estimated from Evoked and Spontaneous Signals

The goal is to establish, with a reasonable degree of precision, the quantal amplitude, Q, the average quantal content, m, and, if appropriate, the number of release sites, n, and the average release probability, p. The realization that release sites in the CNS may not always be uniform (Redman, 1990; Walmsley, 1991) makes a complete statistical description of transmission much more difficult to achieve. The parameters that need estimation must then include the release probability and the quantal amplitude and variability at each site. In principle, it should be possible to estimate the quantal parameters from the probability density of a statistic measured from the evoked postsynaptic signal. Most workers have measured the peak amplitude of the postsynaptic voltage or current on a large number of trials and displayed the results in the form of a histogram. Considerable information can also be obtained from the amplitude distribution of spontaneous miniature events. Two major obstacles, *sampling artifact* and *noise*, immediately arise.

Because the data sample is finite, the true probability density of a desired statistic is not known; only an approximation can be obtained from the recordings. This problem can be mitigated by obtaining a larger sample, but the sample is generally limited by nonstationarity in the recording and time constraints imposed by the experiment. Noise also conceals the true amplitude of spontaneous or evoked signals. If the noise amplitude is comparable to the quantal amplitude, then not only can spontaneous miniature events be missed, but features of the probability density of evoked signals also can be concealed. Relying entirely on visual inspection to determine whether the peaks and troughs in an amplitude histogram are "genuine" (i.e., that they do not arise from sampling artifact and noise) is misleading. A number of different computational approaches have been developed to overcome this obstacle. These approaches differ in the degree to which they rely on assumptions about the underlying probabilistic process. Clearly, if the assumptions are incorrect, then nothing has been achieved, because the parameters will have been estimated incorrectly.

Spontaneous Miniature Signals

If Q can be obtained from the amplitude distribution of mEPSCs or mIPSCs, then the average quantal content, m, can be obtained by dividing the average evoked signal amplitude by Q. This can be done only when the amplitude distribution of spontaneous signals is narrow, and when there is no a priori reason that the quanta underlying the evoked signals should be different.

Spontaneous miniature signals can also be used to detect changes in quantal parameters caused by a conditioning treatment affecting a large number of synapses; if the average amplitude of mEPSCs or mIPSCs becomes larger after an experimental perturbation, the implication is that a widespread increase in

quantal amplitude has been distributed among the synapses that give rise to the miniature currents. Changes in the frequency of spontaneous miniature events, on the other hand, usually imply an alteration in the average release probability (Van der Kloot, 1991; Van der Kloot and Molgó, 1994). An important difficulty with analysis of spontaneous miniature signals in CNS neurons is that their amplitude distribution is generally skewed, with a long tail toward larger values. At the other end of the distribution, small events often fall at the threshold for detection. To compare mEPSCs or mIPSCs obtained before and after a manipulation, cumulative distributions are generally easier to interpret than raw histograms (Van der Kloot, 1991). A genuine widespread change in amplitude is then seen as a shift in the position of the cumulative distribution, whereas a change in frequency should have no effect on the position of the line, other than that which can be accounted for by sampling artifact (Fig. 8.12). The Kolmogorov–Smirnov test can then be applied to test the hypothesis that any difference between the two curves arose by chance.

Multimodal amplitude distributions have occasionally been described for spontaneous miniature signals (Edwards *et al.*, 1990; Jonas *et al.*, 1993). When the modes are at equal intervals on the amplitude axis, multiquantal release, possibly arising from regenerative processes in the presynaptic terminal, is implied, although sampling error as a source of spurious peaks must be ruled out.

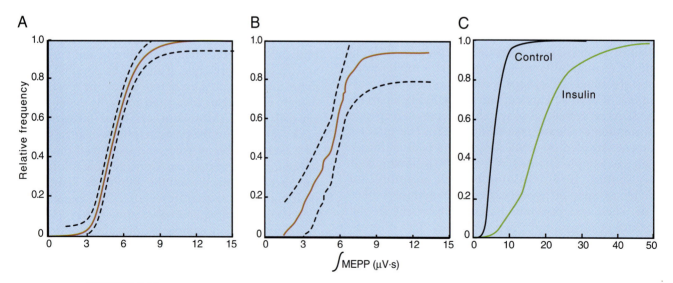

FIGURE 8.12   Cumulative distribution of MEPPs. (A) One thousand consecutive MEPPs recorded at the frog neuromuscular junction are plotted cumulatively. The dashed lines are the 95% confidence limits, calculated by applying Kolmogorov–Smirnov statistics. (B) The first 100 MEPPs are plotted in the same way, showing that, as the sample size is reduced, the confidence limits broaden. (C) Effect of insulin on the cumulative distribution. The curve is shifted to the right, indicating an increase in quantal amplitude. From Van der Kloot (1991).

The principal shortcoming of miniature spontaneous signal analysis is that, generally, the synapses giving rise to the spontaneous signals cannot be identified unambiguously, although localized application of hypertonic sucrose or barium can selectively increase the frequency of discharge in relatively restricted areas (Fatt and Katz, 1952; Bekkers and Stevens, 1989).

Quantal Amplitude Estimation

When Q cannot be estimated from the amplitude distribution of spontaneous MEPSCs or MIPSCs, it can often be obtained from the positions of the peaks in histograms of evoked signals. However, plotting data in the form of histograms can result in misleading estimates of Q in the presence of noise and finite sampling; spurious peaks and troughs can emerge, depending on the position and width of the bins. An approach for attacking this problem is to convolve the data with a *kernel*, generally a gaussian function (Silverman, 1986). *Convolution* describes the mathematical equivalent of "smearing" one function across another. Gaussian kernel convolution has the effect of producing a smooth function with inflections that are no longer susceptible to *binning artifact*. The main pitfall of this approach is that the optimal width of the smoothing kernel cannot be known a priori, and, if it is too narrow, artifactual inflections, arising from noise and finite sampling, will still appear in the convolved function.

Whatever method is used to display the data, the question remains whether all the inflections in the resulting function could have arisen by chance because of finite sampling. A testable null hypothesis is that the underlying function is in fact continuous; that is, transmission is not quantal. This hypothesis can be modeled by choosing a unimodal distribution with the same overall shape as that of the data distribution, but without any peaks or troughs. If this hypothesis can be rejected at a given degree of confidence, the implication is that the clustering in the data sample does indeed indicate a genuine underlying quantized process. An easily implemented general method is to draw random samples repeatedly from the smooth function, with a sample size equal to that of the data. If the peaks and troughs in these random samples are never or only rarely as prominent as those in the data sample, then the null hypothesis can be rejected. This is an example of a *Monte Carlo* test (Horn, 1987).

Poisson Model

If the Poisson model holds, m can be estimated by counting the quanta released on each of a large

number of trials (Isaacson and Walmsley, 1995). Alternatively, since the variance of a Poisson distribution is equal to its mean, m can be estimated from the trial-to-trial variability in the number of quanta released. However, it is rarely possible to count quanta unambiguously, principally because recordings are affected by noise and because the quantal amplitude has some intrinsic variability, and so indirect methods frequently have to be used to estimate m. The first of these is to count the proportion of trials that result in a failure of transmission (N_0 of N trials). The first term of the Poisson expansion (Box 8.4) gives the probability of observing zero quanta released and is equal to e^{-m}. Therefore, m is given by taking the natural logarithm of the inverse of this ratio: $m = \log_e(N/N_0)$. The second indirect approach to estimating m is to measure the coefficient of variation of the postsynaptic response, CV_R. CV_R must be corrected for two other sources of variability in the postsynaptic response: (1) variability in quantal size, expressed as the quantal coefficient of variation, CV_Q; and (2) background noise variance, Var_I, which can be measured separately by collecting data in the absence of evoked activity. If we assume that the variances arising from the Poisson process, quantal variability, and noise add linearly, m is then given by the formula

$$m = \frac{1 + CV_Q^2}{CV_R^2 - Var_I / \overline{R}^2},$$

where $\overline{R}$ is the average evoked response amplitude (McLachlan, 1978). The variance method of estimation is often used incorrectly when the necessary corrections for quantal size and noise fluctuations are ignored.

Agreement among the estimates of m obtained with all these methods constitutes circumstantial evidence in favor of the model.

Binomial Models

The binomial model has more parameters than the Poisson (n and p replace m). If the number of trials resulting in 0, 1, ..., n trials is known unambiguously, then n and p can be estimated as follows. From the binomial theorem, the variance of the number of quanta, Var_m is equal to $np(1 - p)$. It follows that $p = 1 - Var_m/m$ and $n = m/p$. If the number of quanta released cannot be estimated unambiguously, then p can be obtained from the proportion of failures of transmission. Because $N_0 = N(1 - p)^n$, it follows that $p = 1 - (N_0/N)^{1/n}$. The usefulness of this method is limited by the requirements that there be an appreciable proportion of failures and that n be known. Alternatively, the variance method may be used, again with an appropriate correction for the quantal variability and background noise (McLachlan, 1978):

$$p = 1 + CV_Q{}^2 - \frac{\left(\overline{R} \cdot CV_R{}^2 - Var_I / \overline{R}\right)}{Q}.$$

This method requires that estimates be made of both the average quantal size, $\overline{Q}$ and its coefficient of variation, CV_Q, severely limiting its usefulness.

If we relax the assumption of uniform release probabilities while continuing to assume that different sites are independent of one another, then a nonuniform or compound binomial model must be applied (Jack *et al.*, 1981). In this case, the desired parameters include the individual release probabilities: $p_1, p_2, ..., p_n$. If the sample was perfect, they could be obtained by treating the observed proportions of trials resulting in 0, 1, ..., n quanta as the polynomial expansion of $\Pi_k (p_k + q_K z)$ (see Box 8.4). Solving the polynomial would then yield $p_1, p_2, ..., p_n$ (Jack *et al.*, 1981). In practice, however, the sample is incomplete; that is, some rare events may never have been observed, and others may be spuriously overrepresented. Root-finding algorithms generally yield complex roots in this situation. An alternative approach is to use a numerical optimization, that is, to find the release probabilities that give the best agreement with the data, taking into account the fact that the data sample is incomplete (Kullmann, 1989).

Noise Deconvolution

In many cases, the proportion of trials resulting in 0, 1, ..., trials cannot be determined unambiguously because of excessive noise, and the assumption that Poisson or simple binomial statistics apply is untenable. How then can one resolve the underlying quantal process at the synapses under investigation? The method of noise deconvolution (Edwards *et al.*, 1976) again relies on the assumption that noise adds linearly to the synaptic signal, which means that the sampled probability density function is a convolution of the underlying quantal density function with the noise density function. Because the noise can be measured independently, by recording the background signal in the absence of evoked synaptic activity, it should be possible to undo the convolution to reconstruct the probability density function that describes the underlying signal.

This operation is not trivial, because the evoked signal and noise samples are finite: their true probability density functions are not known, and only an approximation can be obtained from the measured signals. The underlying noise-free probability density function must therefore be estimated by applying an optimization method. The underlying function is generally assumed to comprise a number of discrete components, representing different numbers of quanta released. The task is then formally equivalent to solving a *mixture problem*, in which the data are sampled from a mixture of overlapping distributions, or components, each having a membership (probability), mean amplitude, and variance that need to be estimated. Optimization algorithms work as follows: (1) the data distribution is compared with an initial solution reconvolved with the noise function; (2) the solution is then adjusted to improve the goodness of fit; and (3) the cycle is repeated until no further improvement is detectable. The best results are obtained by maximizing likelihood, and a robust and versatile algorithm to use for this purpose is known as the expectation–maximization algorithm (Kullmann, 1989; Stricker and Redman, 1994). A number of constraints can be imposed on the solution to accommodate physiological assumptions. As a rule, as more constraints are imposed, the quantal parameters are more accurately estimated, but only as long as the underlying assumptions are justified.

A major obstacle is that the number of components in the solution, which cannot generally be known a priori, is a critical parameter. As the number of parameters available to fit the data is increased, the maximum likelihood value increases, because finer details of the data distribution, many of which are due to sampling error and noise, can be accounted for. An alternative approach, which avoids the problem of overfitting, is to treat the underlying probability density function not as a mixture of discrete components but as a continuous function. The solution is biased toward the flattest, most featureless function that is just compatible with the data. This method, known as maximum entropy noise deconvolution, can give an estimate of quantal amplitude if there are periodic inflections in the solution (Kullmann and Nicoll, 1992).

Figure 8.13 shows the results of applying several deconvolution methods to an amplitude histogram.

Model Discrimination

An important goal of parameter estimation is to choose between different models of transmission. The simplest approach is to ask if a given model is able to fit the data, by applying a conventional goodness-of-fit test, such as the χ^2 test. If the fit is unsatisfactory, the model can tentatively be rejected with the corresponding degree of confidence. However, the model's being in good agreement with the data does not necessarily mean that the assumptions underlying the solution are correct, because many alternative models also may give adequate fits.

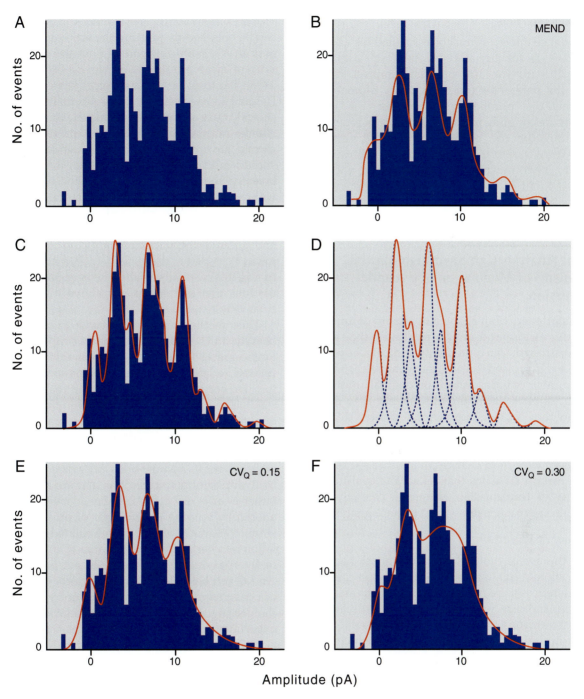

FIGURE 8.13 Noise deconvolution. (A) Amplitude histogram for 400 EPSCs recorded in a CA1 cell in a hippocampal slice in response to repeated stimulation of afferent fibers. EPSCs appear to cluster at integral multiples of approximately 3.6 pA. The continuous line in (B) is the maximum entropy noise deconvolution solution. This function, convolved with the noise, just fits the data at the 5% level of confidence (i.e., a curve any smoother and more featureless would have to be rejected at the 5% level). The periodic inflections seen in this function imply that clustering results not simply from noise, sampling, and binning artifact but from an underlying quantal process. (C) The result of maximum likelihood deconvolution, with nine underlying components, each with the same variance as the background noise but with no constraint on their amplitudes and probabilities. The continuous line represents the solution reconvolved with the noise, showing a very good fit to the data. The underlying components (dashed lines) are plotted in (D), together with their sum (continuous line). Although the agreement of this solution with the data is excellent, some of its features may arise from sampling artifact and noise. (E, F) A quantal model has been fitted to the data; that is, the components have been constrained to occur at equal intervals, with the first component at 0, and with a quantal coefficient of variation (CV_Q) of 0.15 (E) or 0.3 (F). The maximum likelihood solution in (F), but not in (E), can be rejected at the 5% level, implying that $CV_Q < 0.3$.

Confidence Intervals

Confidence intervals must be estimated for quantal parameters, as for any statistic. Such an estimation can be difficult for any but the simplest model because the parameter space has many dimensions, and even the number of dimensions is often unknown. Resampling methods that rely on repeating the optimization on a large number of random samples drawn from the original data set (Efron and Tibshirani, 1993) must be used with caution, and it is important in all cases to be aware of the limitations and biases of optimization algorithms by testing them extensively with Monte Carlo simulations.

Quantal Analysis Can Shed Light on the Mechanisms of Modulation of Synaptic Transmission

As mentioned above, a comparison of quantal parameters before and after a treatment that alters synaptic strength can potentially indicate how this alteration is expressed. Ideally, the mean release probability, number of release sites, and quantal amplitude could be estimated at different time points to determine how they change. In practice, it is often difficult to establish all of these parameters unambiguously. This difficulty does not preclude separating changes in one parameter from changes in another parameter, because certain statistics that reflect the trial-to-trial variability of transmission change in characteristic ways. This approach is known as variance analysis.

Coefficient of Variation

With the use of the coefficient of variation of the evoked signal, the binomial and Poisson models allow inferences to be made about the site of modulation of transmission without the need to estimate the quantal parameters (McLachlan, 1978). To correct for the background noise variance, Var_I, the coefficient of variation of the underlying signal, CV_S, is given by

$$CV_s = \frac{\sqrt{Var_R - Var_I}}{\bar{R}}.$$

CV_S is determined by probabilistic quantal release as well as by the quantal variability and is a useful statistic because it is dimensionless. It can be used to distinguish between changes in quantal amplitude and changes in quantal content. Briefly, if CV_S changes with a conditioning treatment that alters the average amplitude of the postsynaptic signal, the implication is a change in quantal content. If, conversely, CV_S is unaffected, the implication is that the conditioning treatment altered quantal amplitude.

If a Poisson model is assumed, further information can be obtained by plotting the ratio of $1/CV_S^2$ before and after a manipulation against the corresponding ratio of mean amplitude $\bar{R}$ (Fig. 8.14) (Manabe et al., 1993). Because the variance of a Poisson distribution is equal to its mean, a change in quantal content, m, should cause an excursion along the line of identity. A change in quantal amplitude, Q, on the other hand, should have no effect on $1/CV_S^2$, so the data points should fall on the horizontal line. If the points fall between the line of identity and the horizontal line, we can conclude that both quantal content and quantal amplitude changed. If a simple binomial model is assumed, then the results are slightly different, because the variance $np(1-p)$ is less than the mean np. A plot of the ratio of $1/CV_S^2$ against the ratio of $\bar{R}$ in this case falls on the line of identity for manipulations that increase n and above it for manipulations that increase p.

This method, although easy to use, depends heavily on the assumption that a simple binomial or Poisson model applies. As soon as this assumption is relaxed, a wide range of explanations can be put forward for virtually any outcome (Faber and Korn, 1992). The method is also sensitive to changes in the quantal coefficient of variation, so it must be assumed that changes in Q are accompanied by proportional changes in $\sqrt{Var_Q}$. It is, however, often possible to test whether these assumptions are correct, by deliberately applying manipulations that are known to alter either n, or p, or Q. For instance, n can sometimes be altered by varying the number of presynaptic axons stimulated, while p can be manipulated in relative isolation by altering the extracellular $[Ca^{2+}]/[Mg^{2+}]$ ratio or by applying drugs known to act presynaptically. And Q can be scaled by applying low concentrations of postsynaptic receptor blockers or by manipulating the driving force for the synaptic current. This set of experiments establishes characteristic trajectories for a plot of ratios of $1/CV_S^2$ against ratios of $\bar{R}$, which may or may not coincide with those expected of binomial or Poisson models. These trajectories can then be compared with the effect of the new manipulation under investigation. If the observed ratio of $1/CV_S^2$ plotted against the ratio $\bar{R}$ runs along one of the trajectories established for manipulations of either n, p or Q, then it can be inferred that an alteration in the corresponding parameter has occurred. Although this is a potentially powerful approach to verify the validity of variance analysis, it is still potentially flawed if the synaptic plasticity under investigation is not reproduced by any of the experimental alterations of n, p or Q. A notable example is the controversy over the site of expression of long-term potentiation (LTP):

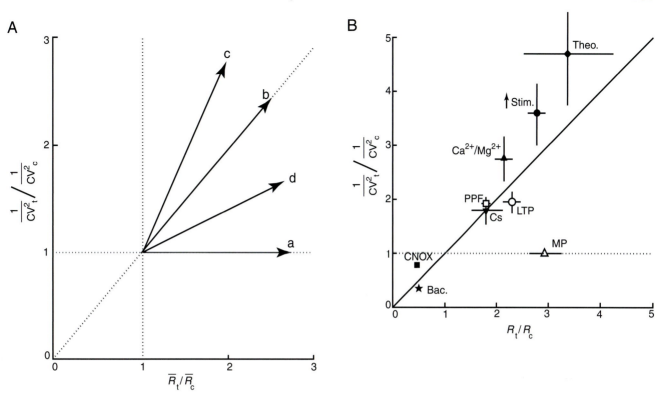

FIGURE 8.14 Coefficient of variation method for determining the site of modulation of synaptic transmission. (A) The possible excursions in the ratio of $1/CV_s^2$ plotted against the ratio of the mean response amplitude $\bar{R}$ for a manipulation that increases synaptic strength. Subscripts "c" and "t" refer to the control and test conditions, respectively. A horizontal excursion (a) implies an increase in quantal amplitude, Q, for instance, through a postsynaptic increase in the number of receptors. If a Poisson model applies, an excursion along the 45° line (b) implies an increase in m. If a simple binomial model holds, the ratio should fall on the 45° line for an increase in n and above it (c) for an increase in p. Ratios falling below the 45° line and above the horizontal (d) imply an increase in both quantal content and quantal amplitude. Conversely, modulations that decrease synaptic strength should cause the ratios to fall to the left of the point (1, 1) and below it. These rules apply only if a Poisson or simple binomial model holds. (B) Experimental results obtained by recording from CA1 hippocampal pyramidal cells with various modulations. Each point shows the mean of several cells (with standard errors). Increasing the driving force for the synaptic current by changing the postsynaptic membrane potential (MP) produces no change in $1/CV_s$, as expected from a purely postsynaptic modification. Extracellular theophylline or Cs^+, an increase in the extracellular $[Ca^{2+}]/[Mg^{2+}]$ ratio, and facilitation by a conditioning prepulse (PPF) cause the ratios to fall in region (c) of (A), as expected from an increase in binomial p. Increasing the stimulus strength also causes the points to fall in region (c), although a simple binomial model predicts that an increase in n should cause the ratio to fall on the 45° line. Long-term potentiation of transmission causes the points to fall in region (d), implying an increase in both quantal content and quantal amplitude. Conversely, baclofen decreases transmitter release, and the glutamate receptor antagonist CNQX decreases quantal amplitude. (B) from Manabe *et al.* (1992).

although variance analysis consistently shows an increase in quantal content, implying a presynaptic alteration in n and/or p, an alternative explanation that has received substantial support from alternative methods is that postsynaptic receptor clusters are uncovered at previously silent sites (see Chapter 18). That is, a postsynaptic modification, with Q at some sites switching to nonzero values, mimics a presynaptic alteration.

Manipulating the Release Probability Experimentally Can Yield an Estimate of Quantal Parameters

The previous section described how experimental manipulation of the quantal parameters can be used to test the validity of the variance method. Although this method can shed light on how quantal parameters change with an alteration in synaptic strength, it is not designed to determine the absolute values of these parameters. Paradoxically, manipulating the release probability can actually allow an estimate to be obtained of the quantal parameters under conditions where they cannot be ascertained from the other methods described above (Clements and Silver, 2000). This approach is analogous to nonstationary variance analysis, a powerful method used to estimate single-channel conductance from membrane currents. For a simple binomial model, the trial-to-trial variance of the postsynaptic response is given by $\mathrm{Var}_R = npQ^2 (1 + \mathrm{CV}_Q^2) - n p^2 Q^2$. [The term $(1 + \mathrm{CV}_Q^2)$ takes into account the contribution of quantal variability within release sites, assuming that the quantal coefficient of variation CV_Q is uniform across different sites.] This relationship can be rewritten as a function of the mean postsynaptic response amplitude $\bar{R}$: $\mathrm{Var}_R = A\bar{R} - B\bar{R}^2$. The quantal parameter estimates are then given by $Q = A/(1 + \mathrm{CV}_Q^2)$, $p = B \cdot \bar{R}(1 + \mathrm{CV}_Q^2)/A$, and $n = 1/B$. In practice the method works as follows. The transmitter release probability is manipulated by varying the extracellular $[\mathrm{Ca}^{2+}]/[\mathrm{Mg}^{2+}]$ ratio or by applying drugs that act presynaptically, to construct a *variance–mean plot* (Fig. 8.15). If the binomial model holds, this has a parabolic shape. Fitting the formula $\mathrm{Var}_R = A\bar{R} - B\bar{R}^2$ then yields estimates for Q, n, and p. If the simple binomial assumption is relaxed, the parabolic shape is often preserved or can become skewed. Under these conditions, the estimates for Q and p can still be informative, although they should be more correctly interpreted as weighted means of the underlying parameters, because release sites with low probabilities and/or low quantal amplitudes contribute less to the postsynaptic response variability. This method can be useful even if it is not possible to increase the mean release probability sufficiently to obtain a clear parabolic variance–mean relationship: the initial slope can be fitted with $\mathrm{Var}_R = A\bar{R}$, yielding an estimate of Q. In this situation, however, it is not possible to estimate n and p.

Estimation of Transmitter Release Probability Does Not Rely Exclusively on Quantal Analysis of Evoked Postsynaptic Signals

With the development of fluorescent dyes that label presynaptic vesicles as they are recycled (Box 8.3), it has become possible to estimate the probability with which vesicles are released following presynaptic action potentials. This approach has given an elegant confirmation of the quantal hypothesis (Ryan *et al.*, 1997).

An alternative method for estimating the release probability at glutamatergic synapses makes use of the pharmacological properties of N-methyl-D-aspar-

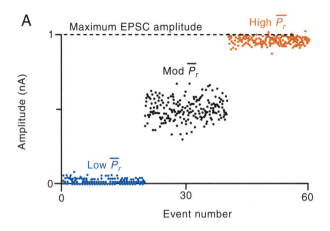

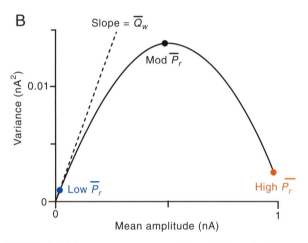

FIGURE 8.15   Variance–mean analysis. (A) Simulated synaptic response amplitudes collected under conditions of low, moderate, and high release probability. The variance is maximal with an intermediate release probability. (B) The variance–mean relationship for a binomial model is parabolic. The initial slope of the plot (dashed line) yields an estimate of the mean quantal amplitude. The number of release sites and the mean release probability can be estimated from the shape of the parabola. From Clements and Silver (2000).

tate (NMDA) receptors (Rosenmund *et al.*, 1993). After application of MK-801, an irreversible open channel blocker, the size of the population synaptic signal mediated by NMDA receptors gradually decays with repeated stimulation of presynaptic fibers. Because the rate at which the postsynaptic signal decays is related to the probability that postsynaptic receptors are activated by released glutamate, this method gives an indirect estimate of the probability of transmitter release. Interestingly, the time course of the decay cannot be described by a single exponential, implying that the release probability must vary considerably across the population of synapses contributing to the signal.

Another indirect method for monitoring changes in transmitter release probability is to examine the short-term facilitation or depression when two or more presynaptic stimuli are delivered in rapid succession, and then to repeat this measurement after a manipulation that alters the strength of the synapse. In general, either increasing or reducing the probability of transmitter release has a relatively greater effect on the size of the response to the first pulse than on that to the second. Therefore, a change in the ratio of the responses to the two pulses implies that at least part of the effect of the manipulation is mediated presynaptically.

Summary

Quantal analysis has greatly improved our understanding of biophysical and pharmacological mechanisms of transmission. Numerical methods can be used to estimate the probability of transmitter release and the size of the postsynaptic effect of an individual quantum of neurotransmitter. Although these methods must be applied with caution, they yield a unique insight into the mechanisms of synaptic plasticity, both at the neuromuscular junction and in the central nervous system.

SHORT-TERM SYNAPTIC PLASTICITY

Chemical synapses are not static transmitters of information. Their effectiveness waxes and wanes, depending on frequency of stimulation and history of prior activity (Zucker and Regehr, 2002). At most synapses, repetitive high-frequency stimulation (called a tetanus) is initially dominated by a growth in successive PSP amplitudes, called *synaptic facilitation*. This process builds to a steady state within about 1 s and decays equally rapidly when stimulation stops. Decay is measured by single test stimuli given at various intervals after a conditioning train. Facilitation can often be divided into two exponential phases, called its first and second components, and may reach appreciable levels (e.g., a doubling of PSP size) after a single action potential. At most synapses, a slower phase of increase in efficacy, which has a characteristic time constant of several seconds and is called *augmentation*, succeeds facilitation. Finally, with prolonged stimulation, some synapses display a third phase of growth in PSP amplitude that lasts minutes and is called *potentiation*.

Often, a phase of decreasing transmission, called *synaptic depression*, is superimposed on these processes. Synaptic depression leads to a dip in transmission during repetitive stimulation, which often tends to overlap and obscure the augmentation and potentiation phases. When stimulation ceases, recovery from the various processes occurs in the same order as their development during the tetanus, with facilitation decaying first, then depression and augmentation, and finally potentiation (Fig. 8.16). Thus, potentiation is often visible in isolation only long after a tetanus and is thus called *posttetanic potentiation* (PTP). At some synapses, even longer-lasting effects (persisting for hours), named *long-term potentiation* (LTP) (Baxter *et al.*, 1985; Minota *et al.*, 1991), have been observed. This long-term potentiation should not be confused with the form of synaptic plasticity bearing the same name and prominent at mammalian cortical synapses (Chapter 18). In almost all synapses in which a quantal analysis has been done, all these forms of synaptic plasticity (except some forms of cortical LTP) are due to changes in the number of quanta released by action potentials. When binomial parameters were estimated, correlated changes in p and n were usually observed. This result is expected if release sites with nonuniform p become more effective while "silent" sites are recruited during enhanced transmission, and vice versa during depression.

Depression May Arise from Depletion of Readily Releasable Transmitter or from Autoinhibition

In contrast with the growth phases in synaptic plasticity, the rate at which depression develops

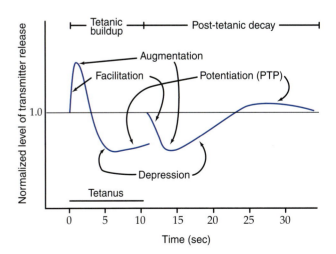

FIGURE 8.16 Accumulation of the effects of facilitation, augmentation, depression, and potentiation on transmitter release by each action potential in a tetanus, and the posttetanic decay of these phases of synaptic plasticity measured by single stimuli after the tetanus.

usually depends on stimulation frequency, whereas recovery from depression proceeds with a single time constant of seconds to minutes in different preparations. At many synapses, depression is relieved when transmission is reduced by lowering $[Ca^{2+}]$ or raising $[Mg^{2+}]$ in the medium. These characteristics are consistent with depression being due to the depletion of a readily releasable store of docked or nearly docked vesicles and recovery being due to their replenishment from a supply store (Fig. 8.17) (Zucker, 1989). In fact, styryl dyes appear to label distinct reserve and readily releasable pools of vesicles at *Drosophila* neuromuscular junctions (Kuromi and Kidokoro, 1998). The parameters of such a depletion model can be estimated from the rate of recovery from depression, which gives the rate of refilling the releasable store, and the fractional drop in PSPs given at short intervals, which gives the fraction of the releasable store liberated by each action potential.

Recent data suggest that the classic depletion model is too simple. As the readily releasable store is depleted, both the probability of it releasing a vesicle (Wu and Borst, 1999; Dobrunz and Stevens, 1997) and its maximum capacity (Dobrunz and Stevens, 1997) appear to drop. More realistic models take account of the finite capacity of the readily releasable vesicle pool (Stevens and Wesseling, 1999; Wu and Betz, 1998). At the frog neuromuscular junction depression also involves a reduction in the rates of endocytotic vesicle recovery and of their transport into the readily releasable store (Wu and Betz, 1998). At other synapses a Ca^{2+}-dependent process speeds mobilization of vesicles into the readily releasable pool to partially compensate for its depletion (Dittman and Regehr, 1998; Wang and Kaczmarek, 1998; Stevens and Wesseling, 1998).

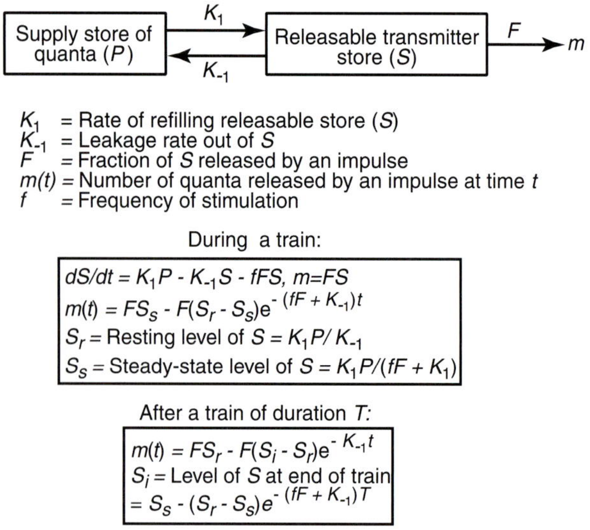

It is now clear that vesicle depletion cannot account for all forms of synaptic depression. At some synapses, depression is due to an inhibitory action of released transmitter on presynaptic receptors called *autoreceptors*. For example, in rat hippocampal cortex, depression of GABA responses is blocked by antagonists of presynaptic $GABA_B$ receptors (Davies *et al.*, 1990). At synapses made by dorsal root ganglion neurons and by some brainstem neurons, depression appears to be due to inactivation of Ca^{2+} channels during repetitive activity (Jia and Nelson, 1986; Forsythe *et al.*, 1998); this mechanism, however, has been specifically rejected at other synapses (Charlton *et al.*, 1982). Moreover, in other brainstem neurons and at some central synapses in the sea slug *Aplysia*, depression is due in part to desensitization of postsynaptic cholinergic and AMPA-type glutamatergic receptors (Wachtel and Kandel, 1971; Otis *et al.*, 1996; Antzoulatos et al, 2003). Depression can also be caused by a reduction in sensitivity of NMDA-type glutamatergic receptors arising from postsynaptic Ca^{2+} accumulation (Mennerick and Zorumski, 1996).

Facilitation, Augmentation, and Potentiation Are Due to Effects of Residual Ca^{2+}

With few exceptions (Wojtowicz and Atwood, 1988), all the phases of increased short-term plasticity are Ca^{2+} dependent in the sense that little or no facilitation, augmentation, or potentiation is generated by stimulation in Ca^{2+}-free medium. Originally, these phases of increased transmission were thought to be due to the effect of residual Ca^{2+} remaining in active zones after presynaptic activity and summating with Ca^{2+} influx during subsequent action potentials to generate slightly higher peaks of $[Ca^{2+}]$ (Zucker, 1989; Katz and Miledi, 1968). Owing to the highly nonlinear dependence of transmitter release on $[Ca^{2+}]$, a small residual $[Ca^{2+}]$ could activate a substantial increase in phasic transmitter release during an action potential while simultaneously increasing MPSP frequency. Temporal correlations between increased spike-evoked transmission after single action potentials or tetani and increases in mPSP frequency supported this idea. But the tetanic accumulation of facilitation, augmentation, and potentiation did not accord quantitatively with predictions of a model of the accumulation of Ca^{2+} acting at one site (Magleby and Zengel, 1982). Simulations of expected levels of peak and residual $[Ca^{2+}]$ levels also were unable to account for the full magnitude of facilitation (Yamada and Zucker, 1992). Presynaptic $[Ca^{2+}]$ measurements using fura-2

K_1 = Rate of refilling releasable store (S)
K_{-1} = Leakage rate out of S
F = Fraction of S released by an impulse
$m(t)$ = Number of quanta released by an impulse at time t
f = Frequency of stimulation

During a train:

$$dS/dt = K_1P - K_{-1}S - fFS, \quad m = FS$$
$$m(t) = FS_s - F(S_r - S_s)e^{-(fF + K_{-1})t}$$
$$S_r = \text{Resting level of } S = K_1P/K_{-1}$$
$$S_s = \text{Steady-state level of } S = K_1P/(fF + K_1)$$

After a train of duration T:

$$m(t) = FS_r - F(S_i - S_r)e^{-K_{-1}t}$$
$$S_i = \text{Level of } S \text{ at end of train}$$
$$= S_s - (S_r - S_s)e^{-(fF + K_{-1})T}$$

FIGURE 8.17 The classic depletion model of synaptic depression.

confirmed the persistence of residual $[Ca^{2+}]$ after repetitive stimulation during facilitation, augmentation, PTP, and LTP, but showed that it was too weak to explain augmentation or potentiation by simply summating with peak $[Ca^{2+}]$ from Ca^{2+} channels at release sites (Delaney et al., 1991; Delaney and Tank, 1994; Atluri and Regehr, 1996). These findings led to the proposal that, in addition to summating with peak $[Ca^{2+}]$ transients at release sites, Ca^{2+} acts to increase transmission at one or more targets distinct from the sites triggering exocytosis.

Two possibilities exist. Residual free Ca^{2+} could continue to act in equilibrium with such sites to increase transmission, or Ca^{2+} could bind to these targets and activate processes that increase release after residual Ca^{2+} has dissipated. The latter idea is suggested by experiments in which facilitation, augmentation, or potentiation persists when residual Ca^{2+} should be absorbed by presynaptic introduction of exogenous chelators; however, results of this sort of experiment have not been consistent (Zucker, 1994). The former idea is supported by experiments in which photolabile BAPTA derivatives were injected presynaptically at crayfish neuromuscular junctions (Kamiya and Zucker, 1994). Posttetanic photolysis to produce enough chelator to suddenly remove residual Ca^{2+} after conditioning stimulation sharply reduced facilitation within a few milliseconds, and it reduced augmentation and potentiation within about 1 s. These results suggest that Ca^{2+} can "prime" subsequent phasic release by action potentials by acting at two additional targets distinct from the exocytosis trigger: a fast site responsible for facilitation and a slow one for augmentation and potentiation.

Several additional indications of separate sites of Ca^{2+} action in synaptic plasticity are as follows:

1. Facilitation grows while secretion decays after one action potential at very low temperature at frog neuromuscular junctions (Van der Kloot, 1994).

2. Sr^{2+} and Ba^{2+} selectively enhance facilitation and augmentation, respectively, of both evoked transmitter release and mEPP frequency at frog neuromuscular junctions (Zengel and Magleby, 1980, 1981).

3. Drosophila mutants defective in enzymes affecting cAMP-dependent phosphorylation show reduced facilitation and potentiation (Zhong and Wu, 1991).

4. Other transformed Drosophila carrying an inhibitor of Ca^{2+}-calmodulin-dependent protein kinase II have impaired facilitation, augmentation, and potentiation (Wang et al., 1994).

5. Mice lacking synapsin I show a specific defect in facilitation but not potentiation in hippocampal pyramidal cells (Rosahl et al., 1993).

6. Facilitation is approximately linearly related to residual $[Ca^{2+}]_i$, while secretion involves a high degree of Ca^{2+} cooperativity (Wright et al., 1996; Vyshedskiy and Lin, 1997).

A particular site of Ca^{2+} action proposed to contribute to potentiation is the Ca^{2+}-calmodulin-dependent kinase II phosphorylation of synapsin I and the mobilization of vesicles to docking sites (Greengard et al., 1993). However, inhibitors of calmodulin and inhibitors or mutants with defects in the kinase failed to affect transmission, facilitation, augmentation, or potentiation at crayfish neuromuscular junctions and mammalian cortical synapses (Zucker, 1994). But short-term synaptic enhancement in Drosophila (see above) and LTP in bull frog sympathetic ganglia may depend on this enzymatic pathway (Baxter et al., 1985; Minota et al., 1991).

Potentiation lasts for a long period after a strong tetanus because residual Ca^{2+} is present for the duration of PTP, owing to overloading of the processes responsible for removing excess Ca^{2+} from neurons. These processes include Ca^{2+} extrusion pumps, such as the plasma membrane ATPase and Na^+–Ca^{2+} exchange, and Ca^{2+} uptake into organelles such as endoplasmic reticulum and mitochondria (Fossier et al., 1992; David et al., 1998; Tang and Zucker, 1997). The accumulation of Ca^{2+} in mitochondria during a tetanus and its gradual posttetanic release into cytoplasm produce a prolonged elevation in residual $[Ca^{2+}]_i$ and lead to PTP (Tang and Zucker, 1997). In addition, Na^+ accumulation during tetanic activity prolongs residual Ca^{2+} by reducing its extrusion by Na^+–Ca^{2+} exchange (Zhong et al., 2001).

Summary

Short-term synaptic plasticity allows synaptic strength to be modulated as a function of prior activity. Synapses may show a decline in transmission (depression) or an increase in synaptic efficacy, with time constants ranging from seconds (facilitation and augmentation) to minutes (potentiation or PTP) to hours (LTP); many synapses show a mixture of several of these phases. Depression may be due to depletion of a readily releasable supply of vesicles or to the inhibitory action of transmitter or enzymatically produced transmitter products on presynaptic autoreceptors. Depression makes synapses selectively responsive to brief stimuli or to changes in level of activity. Frequency-dependent increases in synaptic efficacy are due to the effects of residual presynaptic Ca^{2+} acting to modulate the release process. At least part of Ca^{2+} action is mediated by separate targets for facilitation and for augmentation and potentiation. PTP is prolonged after a long tetanus because residual Ca^{2+} remains in synaptic terminals for minutes after such stimulation. These frequency-dependent increases in synaptic efficacy allow synapses to distinguish significant signals from noise and respond to selected patterns of activity.

References

Aalto, M. K., Ronne, H., and Keranen, S. (1993). Yeast syntaxins Sso1p and Sso2p belong to a family of related membrane proteins that function in vesicular transport. *Embo J.* **12**, 4095–4104.

Aalto, M. K., Ruohonen, L., Hosono, K., and Keranen, S. (1991). Cloning and sequencing of the yeast *Saccharomyces cerevisiae* SEC1 gene localized on chromosome IV. *Yeast* **7**, 643–650.

Adler, E. M., Augustine, G. J., Duffy, S. N., and Charlton, M. P. (1991). Alien intracellular chelators attenuate neurotransmitter release at the squid giant synapse. *J. Neurosci.* **11**, 1496–1507.

Ahnert-Hilger, G., Bhakdi, S., and Gratzl, M. (1985). Minimal requirements for exocytosis: A study using PC 12 cells permeabilized with staphylococcal alpha-toxin. *J. Biol. Chem.* **260**, 12730–12734.

Ales, E., Tabares, L., Poyato, J. M., Valero, V., Lindau, M., and Alvarez de Toledo, G. (1999). High calcium concentrations shift the mode of exocytosis to the kiss-and-run mechanism. *Nat. Cell. Biol.* **1**, 40–44.

Alvarez de Toledo, G., Fernandez-Chacon, R., and Fernandez, J. M. (1993). Release of secretory products during transient vesicle fusion. *Nature (London)* **363**, 554–558.

Antonin, W., Holroyd, C., Fasshauer, D., Pabst, S., Von Mollard, G. F., and Jahn, R. (2000). A SNARE complex mediating fusion of late endosomes defines conserved properties of SNARE structure and function. *EMBO J.* **19**, 6453–6464.

Antzoulatos, E. G., Cleary, L. J., Eskin, A., Baxter, D. A. and Byrne, J. H. (2003). Desensitization of postsynaptic glutamate receptors contributes to high-frequency homosynaptic depression of *Aplysia* sensorimotor connections. *LearnMem.*, in press.

Arredondo, L., Nelson, H. B., Beckingham, K., and Stern, M. (1998). Increased transmitter release and aberrant synapse morphology in a Drosophila *calmodulin* mutant. *Genetics* **150**, 265–274.

Atluri, P. P., and Regehr, W. G. (1996) Determinants of the time course of facilitation at the granule cell to Purkinje cell synapse. *J. Neurosci.* **16**, 5661–5671.

Augustine, G. J., and Charlton, M. P. (1986). Calcium-dependence of presynaptic calcium current and post-synaptic response at the squid giant synapse. *J. Physiol. (London)* **381**, 619–640.

Augustine, G. J., Charlton, M. P., and Smith, S. J. (1985). Calcium entry and transmitter release at voltage-clamped nerve terminals of squid. *J. Physiol. (London)* **367**, 163–181.

Augustine, G. J., Charlton, M. P., and Smith, S. J. (1987). Calcium action in synaptic transmitter release. *Annu. Rev. Neurosci.* **10**, 633–693.

Bajjalieh, S. M., Peterson, K., Shinghal, R., and Scheller, R. H. (1992). SV2, a brain synaptic vesicle protein homologous to bacterial transporters. *Science* **257**, 1271–1273.

Banerjee, A., Barry, V. A., DasGupta, B. R., and Martin, T. F. (1996). N-Ethylmaleimide-sensitive factor acts at a prefusion ATP-dependent step in Ca^{2+}-activated exocytosis. *J. Biol. Chem.* **271**, 20223–20226.

Bartol, T. M., Land, B. R., Salpeter, E. E., and Salpeter, M. M. (1991). Monte Carlo simulation of miniature endplate current generation in the vertebrate neuromuscular junction. *Biophys. J.* **59**, 1290–1307.

Baxter, D. A., Bittner, G. D., and Brown, T. H. (1985). Quantal mechanism of long-term synaptic potentiation. *Proc. Natl. Acad. Sci. USA* **82**, 5978–5982.

Bekkers, J. M., and Stevens, C. F. (1989). NMDA and non-NMDA receptors are co-localized at individual excitatory synapses in cultured rat hippocampus. *Nature (London)* **341**, 230–233.

Bellocchio, E. E., Reimer, R. J., Fremeau, R. T., Jr., and Edwards, R. H. (2000). Uptake of glutamate into synaptic vesicles by an inorganic phosphate transporter. *Science* **289**, 957–960.

Bennett, M. K., Calakos, N., and Scheller, R. H. (1992). Syntaxin: A synaptic protein implicated in docking of synaptic vesicles at presynaptic active zones. *Science* **257**, 255–259.

Bennett, M. K., and Scheller, R. H. (1993). The molecular machinery for secretion is conserved from yeast to neurons. *Proc. Natl. Acad. Sci. USA* **90**, 2559–2563.

Bennett, M. K., and Scheller, R. H. (1994). A molecular description of synaptic vesicle membrane trafficking. *Annu. Rev. Biochem.* **63**, 63–100.

Betz, W. J., and Bewick, G. S. (1993). Optical monitoring of transmitter release and synaptic vesicle recycling at the frog neuromuscular junction. *J. Physiol. (London)* **460**, 287–309.

Beutner, D., Voets, T., Neher, E., and Moser, T. (2001). Calcium dependence of exocytosis and endocytosis at the cochlear inner hair cell afferent synapse. *Neuron* **29**, 681–690.

Bixby, J. L., and Reichardt, L. F. (1985). The expression and localization of synaptic vesicle antigens at neuromuscular junctions *in vitro*. *J. Neurosci.* **5**, 3070–3080.

Blasi, J., Chapman, E. R., Link, E., Binz, T., Yamasaki, S., De Camilli, P., Südhof, T. C., Niemann, H., and Jahn, R. (1993a). Botulinum neurotoxin A selectively cleaves the synaptic protein SNAP-25. *Nature* **365**, 160–163.

Blasi, J., Chapman, E. R., Yamasaki, S., Binz, T., Niemann, H., and Jahn, R. (1993b). Botulinum neurotoxin C1 blocks neurotransmitter release by means of cleaving HPC-1/syntaxin. *EMBO J.* **12**, 4821–4828.

Block, M. R., Glick, B. S., Wilcox, C. A., Wieland, F. T., and Rothman, J. E. (1988). Purification of an N-ethylmaleimide-sensitive protein catalyzing vesicular transport. *Proc. Natl. Acad. Sci. USA* **85**, 7852–7856.

Bollmann, J. J., Sakmann, B., and Borst, J. G. G. (2000). Calcium sensitivity of glutamate release in a calyx-type terminal. *Science* **289**, 953–956.

Borst, J. G. G. and Sakmann, B. (1996) Calcium influx and transmitter release in a fast CNS synapse. *Nature* **383**, 431–434.

Boyd, I. A., and Martin, A. R. (1956). Spontaneous subthreshold activity at mammalian neuromuscular junctions. *J. Physiol. (London)* **132**, 74–91.

Broadie, K., Bellen, H. J., DiAntonio, A., Littleton, J. T., and Schwarz, T. L. (1994). Absence of synaptotagmin disrupts excitation–secretion coupling during synaptic transmission. *Proc. Natl. Acad. Sci. USA* **91**, 10727–10731.

Broadie, K., Prokop, A., Bellen, H. J., O'Kane, C. J., Schulze, K. L., and Sweeney, S. T. (1995). Syntaxin and synaptobrevin function downstream of vesicle docking in *Drosophila*. *Neuron* **15**, 663–673.

Brose, N., Petrenko, A. G., Südhof, T. C., and Jahn, R. (1992). Synaptotagmin: A calcium sensor on the synaptic vesicle surface. *Science* **256**, 1021–1025.

Brose, N., and Rosenmund, C. (1999). SV2: SVeeping up excess Ca^{2+} or tranSVorming presynaptic Ca^{2+} sensors? *Neuron* **24**, 766–768.

Buckley, K. M., Floor, E., and Kelly, R. B. (1987). Cloning and sequence analysis of cDNA encoding p38, a major synaptic vesicle protein. *J. Cell Biol.* **105**, 2447–2456.

Burgess, R. W., Deitcher, D. L., and Schwarz, T. L. (1997). The synaptic protein syntaxin1 is required for cellularization of *Drosophila* embryos. *J. Cell Biol.* **138**, 861–875.

Carlson, S. S., Wagner, J. A., and Kelly, R. B. (1978). Purification of synaptic vesicles from elasmobranch electric organ and the use of biophysical criteria to demonstrate purity. *Biochemistry* **17**, 1188–1199.

Castillo, P. E., Janz, R., Südhof, T. C., Tzounopoulos, T., Malenka, R. C., and Nicoll, R. A. (1997). Rab3A is essential for mossy fibre long-term potentiation in the hippocampus. *Nature* **388**, 590–593.

Ceccarelli, B., and Hurlbut, W. P. (1980). Ca^{2+}-dependent recycling of synaptic vesicles at the frog neuromuscular junction. *J. Cell Biol.* **87**, 297–303.

Chamberlain, L. H., Roth, D., Morgan, A., and Burgoyne, R. D. (1995). Distinct effects of alpha-SNAP, 14–3–3 proteins, and calmodulin on priming and triggering of regulated exocytosis. *J. Cell Biol.* **130**, 1063–1070.

Chan, M. S., Obar, R. A., Schroeder, C., Austin, T. W., Poodry, C. A., Wadsworth, S. A., and Vallee, R. B. (1991). Multiple forms of dynamin are encoded by *shibire*, a *Drosophila* gene involved in endocytosis. *Nature (London)* **351**, 583–586.

Chapman, E. R., An, S., Barton, N., and Jahn, R. (1994). SNAP-25, a t-SNARE which binds to both syntaxin and synaptobrevin via domains that may form coiled coils. *J. Biol. Chem.* **269**, 27427–27432.

Charlton, M. P., Smith, S. J., and Zucker, R. S. (1982). Role of presynaptic calcium ions and channels in synaptic facilitation and depression at the squid giant synapse. *J. Physiol. (London)* **323**, 173–193.

Chen, M. S., Obar, R. A., Schroeder, C. C., Austin, T. W., Poodry, C. A., Wadsworth, S. C., and Vallee, R. B. (1991). Multiple forms of dynamin are encoded by *shibire*, a *Drosophila* gene involved in endocytosis. *Nature* **351**, 583–586.

Chen, Y. A., and Scheller, R. H. (2001). SNARE-mediated membrane fusion. *Nat. Rev. Mol. Cell. Biol.* **2**, 98–106.

Clements, J. D., Lester, R. A. J., Tong, G., Jahr, C. E., and Westbrook, G. L. (1992). The time course of glutamate in the synaptic cleft. *Science* **258**, 1498–1501.

Clements, J. D., and Silver, R. A. (2000). Unveiling synaptic plasticity: A new graphical and analytical approach. *Trends Neurosci.* **23**, 105–113.

Conradt, B., Haas, A., and Wickner, W. (1994). Determination of four biochemically distinct, sequential stages during vacuole inheritance *in vitro*. *J. Cell Biol.* **126**, 99–110.

Cooper, J. R., Bloom, F. E., and Roth, R. H. (1991). "The Biochemical Basis of Neuropharmacology," 6th ed. Oxford Univ. Press, New York.

Crowder, K. M., Gunther, J. M., Jones, T. A., Hale, B. D., Zhang, H. Z., Peterson, M. R., Scheller, R. H., Chavkin, C., and Bajjalieh, S. M. (1999). Abnormal neurotransmission in mice lacking synaptic vesicle protein 2A (SV2A). *Proc. Natl. Acad. Sci. USA* **96**, 15268–15273.

David, G., Barrett, J. N., and Barrett, E. F. (1998). Evidence that mitochondria buffer physiological Ca^{2+} loads in lizard motor nerve terminals. *J. Physiol.* **509**, 59–65.

Davies, C. H., Davies, S. N., and Collingridge, G. L. (1990). Paired-pulse depression of monosynaptic GABA-mediated inhibitory postsynaptic responses in rat hippocampus. *J. Physiol. (London)* **424**, 513–531.

De Camilli, P., Harris, S. M., Jr., Huttner, W. B., and Greengard, P. (1983). Synapsin I (Protein I), a nerve terminal-specific phosphoprotein. II. Its specific association with synaptic vesicles demonstrated by immunocytochemistry in agarose-embedded synaptosomes. *J. Cell Biol.* **96**, 1355–1373.

De Camilli, P., and Jahn, R. (1990). Pathways to regulated exocytosis in neurons. *Annu. Rev. Physiol.* **52**, 625–645.

De Camilli, P., Takei, K., and McPherson, P. S. (1995). The function of dynamin in endocytosis. *Curr. Opin. Neurobiol.* **5**, 559–565.

del Castillo, J., and Katz, B. (1954). Quantal components of the end-plate potential. *J. Physiol. (London)* **124**, 560–573.

Delaney, K. R., and Tank, D. W. (1994). A quantitative measurement of the dependence of short-term synaptic enhancement on presynaptic residual calcium. *J. Neurosci.* **14**, 5885–5902.

Delaney, K. R., Zucker, R. S., and Tank, D. W. (1991). Presynaptic calcium in motor nerve terminals associated with posttetanic potentiation. *J. Neurosci.* **9**, 3558–3567.

DiAntonio, A., Parfitt, K. D., and Schwarz, T. L. (1993). Synaptic transmission persists in synaptotagmin mutants of *Drosophila*. *Cell* **73**, 1281–1290.

DiAntonio, A., and Schwarz, T. L. (1994). The effect on synaptic physiology of synaptotagmin mutations in *Drosophila*. *Neuron* **12**, 909–920.

Dinkelacker, V., Voets, T., Neher, E., and Moser, T. (2000). The readily releasable pool of vesicles in chromaffin cells is replenished in a temperature-dependent manner and transiently overfills at 37°C. *J. Neurosci.* **20**, 8377–8383.

Dittman, J. S., and Regehr, W. G. (1998). Calcium dependence and recovery kinetics of presynaptic depression at the climbing fiber to Purkinje cell synapse. *J. Neurosci.* **18**, 6147–6162.

Dobrunz, L. E., and Stevens, C. F. (1997). Heterogeneity of release probability, facilitation, and depletion at central synapses. *Neuron* **18**, 995–1008.

Dresbach, T., Qualmann, B., Kessels, M. M., Garner, C. C., and Gundelfinger, E. D. (2001). The presynaptic cytomatrix of brain synapses. *Cell. Mol. Life Sci.* **58**, 94–116.

Dunlap, K., Luebke, J. I., and Turner, T. J. (1995). Exocytotic Ca^{2+} channels in mammalian central neurons. *Trends Neurosci.* **18**, 89–98.

Eakle, K. A., Bernstein, M., and Emr, S. D. (1988). Characterization of a component of the yeast secretion machinery: Identification of the SEC18 gene product. *Mol. Cell. Biol.* **8**, 4098–4109.

Eccles, J. C. (1964). "The Physiology of Synapses." Springer-Verlag, Berlin. (A comprehensive summary of early work.)

Edwards, C., Doležal, V., Tuček, S., Zemková, H., and Vyskočil, F. (1985). Is an acetylcholine system transport system responsible for nonquantal release of acetylcholine at the rodent myoneural junction? *Proc. Natl. Acad. Sci. USA.* **82**, 3514–3518.

Edwards, F. A., Konnerth, A., and Sakmann, B. (1990). Quantal analysis of inhibitory synaptic transmission in the dentate gyrus of rat hippocampal slices: A patch-clamp study. *J. Physiol. (London)* **430**, 213–249.

Edwards, F. R., Redman, S. J., and Walmsley, B. (1976). Statistical fluctuation in charge transfer at Ia synapses on spinal motoneurones. *J. Physiol. (London)* **259**, 665–688.

Edwards, R. H. (1992). The transport of neurotransmitters into synaptic vesicles. *Curr. Opin. Neurobiol.* **2**, 586–594.

Edwards, R. H. (1992). The transport of neurotransmitters into synaptic vesicles. *Curr. Opin. Neurobiol.* **2**, 586–594.

Efron, B., and Tibshirani, R. (1993). "An Introduction to the Bootstrap". Chapman & Hall, New York.

Elmqvist, D., and Quastel, D. M. J. (1965). Presynaptic action of hemicholinium at the neuromuscular junction. *J. Physiol. (London)* **177**, 463–482.

Faber, D. S., and Korn, H. (1992). Application of the coefficient of variation method for analyzing synaptic plasticity. *Biophys. J.* **60**, 1288–1294.

Fatt, P., and Katz, B. (1952). Spontaneous sub-threshold activity at motor nerve endings. *J. Physiol. (London)* **119**, 109–128.

Feany, M. B., Lee, S., Edwards, R. H., and Buckley, K. M. (1992). The synaptic vesicle protein SV2 is a novel type of transmembrane transporter. *Cell* **70**, 861–867.

Fischer von Mollard, G., Südhof, T. C., and Jahn, R. (1991). A small GTP-binding protein dissociates from synaptic vesicles during exocytosis. *Nature* **349**, 79–81.

Fletcher, P., and Forrester, T. (1975). The effect of curare on the release of acetylcholine from mammalian motor nerve terminals and an estimate of quantum content. *J. Physiol. (London)* **251**, 131–144.

Foran, P., Lawrence, G. W., Shone, C. C., Foster, K. A., and Dolly, J. O. (1996). Botulinum neurotoxin C1 cleaves both syntaxin and SNAP-25 in intact and permeabilized chromaffin cells: Correlation with its blockade of catecholamine release. *Biochemistry* **35**, 2630–2636.

Forsythe, I. D., Tsujimoto, T., Barnes-Davies, M., Cuttle, M. F., and Takahashi, T. (1998). Inactivation of presynaptic calcium current contributes to synaptic depression at a fast central synapse. *Neuron* **20**, 797–807.

Fossier, P., Baux, G., Trudeau, L. -E., and Tauc, L. (1992). Involvement of Ca^{2+} uptake by a reticulum-like store in the control of transmitter release. *Neuroscience* **50**, 427–434.

Garcia, E. P., Gatti, E., Butler, M., Burton, J., and De Camilli, P. (1994). A rat brain Sec1 homologue related to Rop and UNC18 interacts with syntaxin. *Proc. Natl. Acad. Sci. USA* **91**, 2003–2007.

Garcia, E. P., McPherson, P. S., Chilcote, T. J., Takei, K., and De Camilli, P. (1995). rbSec1A and B colocalize with syntaxin 1 and SNAP-25 throughout the axon, but are not in a stable complex with syntaxin. *J. Cell Biol.* **129**, 105–120.

Garner, C. C., Kindler, S., and Gundelfinger, E. D. (2000). Molecular determinants of presynaptic active zones. *Curr. Opin. Neurobiol.* **10**, 321–327.

Geppert, M., Goda, Y., Hammer, R. E., Li, C., Rosahl, T. W., Stevens, C. F., and Südhof, T. C. (1994). Synaptotagmin I: A major Ca^{2+} sensor for transmitter release at a central synapse. *Cell* **79**, 717–727.

Gorvel, J. P., Chavrier, P., Zerial, M., and Gruenberg, J. (1991). rab5 controls early endosome fusion *in vitro. Cell* **64**, 915–925.

Greengard, P., Valtorta, F., Czernik, A. J., and Benfenati, F. (1993). Synaptic vesicle phosphoproteins and regulation of synaptic function. *Science* **259**, 780–785.

Greengard, P., Valtorta, F., Czernik, A. J., and Benfenati, F. (1993). Synaptic vesicle phosphoproteins and regulation of synaptic function. *Science* **259**, 780–785.

Harlow, M. L., Ress, D., Stoschek, A., Marshall, R. M., and McMahan, U. J. (2001). The architecture of active zone material at the frog's neuromuscular junction. *Nature* **409**, 479–484.

Harris, K. M., and Sultan, P. (1995). Variation in the number, location and size of synaptic vesicles provides an anatomical basis for the nonuniform probability of release at hippocampal CA1 synapses. *Neuropharmacology* **34**, 1387–1395.

Hartzell, H. C., Kuffler, S. W., and Yoshikami, D. (1975). Postsynaptic potentiation: Interaction between quanta of acetylcholine at the skeletal neuromuscular synapse. *J. Physiol. (London)* **251**, 427–463.

Haucke, V., and De Camilli, P. (1999). AP-2 recruitment to synaptotagmin stimulated by tyrosine-based endocytic motifs. *Science* **285**, 1268–1271.

Hay, J. C., and Martin, T. F. (1992). Resolution of regulated secretion into sequential MgATP-dependent and calcium-dependent stages mediated by distinct cytosolic proteins. *J. Cell. Biol.* **119**, 139–151.

Haydon, P. C., Henderson, E., and Stanley, E. F. (1994). Localization of individual calcium channels at the release face of a presynaptic nerve terminal. *Neuron* **13**, 1275–1280.

Heidelberger, R., Heinemann, C., Neher, E., and Matthews, G. (1994). Calcium dependence of the rate of exocytosis in a synaptic terminal. *Nature (London)* **371**, 513–515.

Heuser, J. E. (1977). Synaptic vesicle exocytosis revealed in quick-frozen frog neuromuscular junctions treated with 4-aminopyridine and given a single electrical shock. *Soc. Neurosci. Symp.* **2**, 215–239. (A fine anatomical analysis of vesicle recycling.)

Heuser, J. E., and Reese, T. S. (1973). Evidence for recycling of synaptic vesicle membrane during transmitter release at the frog neuromuscular junction. *J. Cell Biol.* **57**, 315–344.

Heuser, J. E., and Reese, T. S. (1981). Structural changes after transmitter release at the frog neuromuscular junction. *J. Cell Biol.* **88**, 564–580.

Heuser, J. E., Reese, T. S., Dennis, M. J., Jan, Y., Jan, L., and Evans, L. (1979). Synaptic vesicle exocytosis captured by quick freezing and correlated with quantal transmitter release. *J. Cell Biol.* **81**, 275–300.

Holz, R. W., Bittner, M. A., Peppers, S. C., Senter, R. A., and Eberhard, D. A. (1989). MgATP-independent and MgATP-dependent exocytosis: Evidence that MgATP primes adrenal chromaffin cells to undergo exocytosis. *J. Biol. Chem.* **264**, 5412–5419.

Horn, R. (1987). Statistical methods for model discrimination: Applications to gating kinetics and permeation of the acetylcholine receptor channel. *Biophys. J.* **51**, 255–263.

Hunt, J. M., Bommert, K., Charlton, M. P., Kistner, A., Habermann, E., Augustine, G. J., and Betz, H. (1994). A postdocking role for synaptobrevin in synaptic vesicle fusion. *Neuron* **12**, 1269–1279.

Hurlbut, W. P., Iezzi, N., Fesce, R., and Ceccarelli, B. (1990). Correlation between quantal secretion and vesicle loss at the frog neuromuscular junction. *J. Physiol. (London)* **425**, 501–526.

Isaacson, J. S., and Walmsley, B. (1995). Counting quanta: Direct measurements of transmitter release at a central synapse. *Neuron* **15**, 875–884.

Jack, J. J. B., Redman, S. J., and Wong, K. (1981). The components of synaptic potentials evoked in cat spinal motoneurones by impulses in single group Ia afferents. *J. Physiol. (London)* **321**, 65–96.

Jahn, R., and Südhof, T. C. (1994). Synaptic vesicles and exocytosis. *Annu. Rev. Neurosci.* **17**, 219–246.

Janz, R., Goda, Y., Geppert, M., Missler, M., and Südhof, T. C. (1999a). SV2A and SV2B function as redundant Ca^{2+} regulators in neurotransmitter release. *Neuron* **24**, 1003–1016.

Janz, R., Südhof, T. C., Hammer, R. E., Unni, V., Siegelbaum, S. A., and Bolshakov, V. Y. (1999b). Essential roles in synaptic plasticity for synaptogyrin I and synaptophysin I. *Neuron* **24**, 687–700.

Jia, M., and Nelson, P. G. (1986). Calcium currents and transmitter output in cultured spinal cord and dorsal root ganglion neurons. *J. Neurophysiol.* **56**, 1257–1267.

Jonas, P., Major, G., and Sakmann, B. (1993). Quantal components of unitary EPSCs at the mossy fibre synapse on CA3 pyramidal cells of rat hippocampus. *J. Physiol. (London)* **472**, 615–663.

Jorgensen, E. M., Hartwieg, E., Schuske, K., Nonet, M. L., Jin, Y., and Horvitz, H. R. (1995). Defective recycling of synaptic vesicles in synaptotagmin mutants of *Caenorhabditis elegans. Nature* **378**, 196–199.

Jurado, L. A., Chockalingam, P. S., and Jarrett, H. W. (1999). Apocalmodulin. *Physiol. Rev.* **79**, 661–682.

Kamiya, H., and Zucker, R. S. (1994). Residual Ca^{2+} and short-term synaptic plasticity. *Nature (London)* **371**, 603–606.

Katz, B. (1969). "The Release of Neural Transmitter Substances." Thomas, Springfield, IL. (Describes the classic experiments on transmitter release.)

Katz, B., and Miledi, R. (1965). The release of acetylcholine from nerve endings by graded electric pulses. *Proc. R. Soc. London Ser. B* **167**, 28–38.

Katz, B., and Miledi, R. (1968). The role of calcium in neuromuscular facilitation. *J. Physiol. (London)* **195**, 481–492.

Klingauf, J., Kavalali, E. T. and Tsien, R. W. (1998) Kinetics and regulation of fast endocytosis at hippocampal synapses. *Nature* **394**, 581–585.

Koenig, J., and Ikeda, I. (1996) Synaptic vesicles have two distinct recycling pathways. *J. Cell Biol.* **135**, 797–808.

Korn, H., and Faber, D. S. (1987). Regulation and significance of probabilistic release mechanisms at central synapses. *In* "Synaptic Function" (G. Edelman, W. E. Gall, and W. M. Cowan, Eds.), pp. 57–108. Wiley, New York.

Korn, H., and Faber, D. S. (1991). Quantal analysis and synaptic efficacy in the CNS. *Trends Neurosci.* **14**, 439–445.

Kuffler, S. W., and Yoshikami, D. (1975). The number of transmitter molecules in a quantum: An estimate from iontophoretic application of acetylcholine at the neuromuscular junction. *J. Physiol. (London)* **251**, 465–482.

Kullmann, D. M. (1989). Applications of the expectation–maximization algorithm to quantal analysis of postsynaptic potentials. *J. Neurosci. Methods* **30**, 231–245.

Kullmann, D. M., and Nicoll, R. A. (1992). Long-term potentiation is associated with increases in quantal content and quantal amplitude. *Nature (London)* **357**, 240–244.

Kullmann, D. M., and Siegelbaum, S. A. (1995). The site of expression of NMDA receptor-dependent LTP: New fuel for an old fire. *Neuron* **15**, 997–1002.

Kuromi, H., and Kidokoro, Y. (1998). Two distinct pools of synaptic vesicles in single presynaptic boutons in a temperature-sensitive *Drosophila* mutant, *shibire*. *Neuron* **20**, 917–925.

Landò, L., and Zucker, R. S. (1994). Ca²⁺ cooperativity in neurosecretion measured using photolabile Ca²⁺ chelators. *J. Neurophysiol.* **72**, 825–830.

Lao, G., Scheuss, V., Gerwin, C. M., Su, Q., Mochida, S., Rettig, J., and Sheng, Z. H. (2000). Syntaphilin: A syntaxin-1 clamp that controls SNARE assembly. *Neuron* **25**, 191–201.

Large, W. A., and Rang, H. P. (1978). Variability of transmitter quanta released during incorporation of a false transmitter into cholinergic nerve terminals. *J. Physiol. (London)* **285**, 25–34.

Leavis, P. C., and Gergely, J. (1984). Thin filament proteins and thin filament-linked regulation of vertebrate muscle contraction. *CRC Crit. Rev. Biochem.* **16**, 235–305.

Leenders, A. G., Scholten, G., Wiegant, V. M., Da Silva, F. H., and Ghijsen, W. E. (1999). Activity-dependent neurotransmitter release kinetics: Correlation with changes in morphological distributions of small and large vesicles in central nerve terminals. *Eur. J. Neurosci.* **11**, 4269–4277.

Li, C., Ullrich, B., Zhang, J. Z., Anderson, R. G., Brose, N., and Südhof, T. C. (1995). Ca²⁺-dependent and -independent activities of neural and non-neural synaptotagmins. *Nature* **375**, 594–599.

Liley, A. W. (1956). The quantal components of the mammalian endplate potential. *J. Physiol. (London)* **133**, 571–587.

Lindau, M., Stuenkel, E. L., and Nordmann, J. J. (1992). Depolarization, intracellular calcium and exocytosis in single vertebrate nerve endings. *Biophys. J.* **61**, 19–30.

Link, E., *et al.* (1992). Tetanus toxin action: Inhibition of neurotransmitter release linked to synaptobrevin proteolysis. *Biochem. Biophys. Res. Commun.* **189**, 1017–1023.

Littleton, J. T., Stern, M., Schulze, K., Perin, M., and Bellen, H. J. (1993). Mutational analysis of *Drosophila* synaptotagmin demonstrates its essential role in Ca²⁺-activated neurotransmitter release. *Cell* **74**, 1125–1134.

Llinás, R., Gruner, J. A., Sugimori, M., McGuiness, T. L., and Greengard, P. (1991). Regulation by synapsin I and Ca²⁺-calmodulin-dependent protein kinase II of the transmitter release at the squid giant synapse. *J. Physiol. (London)* **436**, 257–282.

Llinás, R., Steinberg, I. Z., and Walton, K. (1981). Relationship between presynaptic calcium current and postsynaptic potential in squid giant synapse. *Biophys. J.* **33**, 323–351.

Llinás, R., Sugimori, M., and Silver, R. B. (1992). Microdomains of high calcium concentration in a presynaptic terminal. *Science* **256**, 677–679.

Locke, F. S. (1894). Notiz über den Einfluβ physiologischer Kochsalzlösung auf die elektrische Erregbarkeit von Muskel und Nerv. *Zentralbl. Physiol.* **8**, 166–167.

Lonart, G., and Südhof, T. C. (1998). Region-specific phosphorylation of rabphilin in mossy fiber nerve terminals of the hippocampus. *J. Neurosci.* **18**, 634–640.

Lundberg, J. M., and Hökfelt, T. (1986). Multiple co-existence of peptides and classical transmitters in peripheral autonomic and sensory neurons: Functional and pharmacological implications. *Prog. Brain Res.* **68**, 241–262.

Magleby, K. L., and Zengel, J. E. (1982). A quantitative description of stimulation-induced changes in transmitter release at the frog neuromuscular junction. *J. Gen. Physiol.* **30**, 613–638.

Manabe, T., Renner, P., and Nicoll, R. (1992). Postsynaptic contribution to long-term potentiation revealed by the analysis of miniature synaptic currents. *Nature (London)* **355**, 50–55.

Manabe, T., Wyllie, D. J. A., Perkel, D. J., and Nicoll, R. A. (1993). Modulation of synaptic transmission and long-term potentiation: Effects on paired pulse facilitation and EPSC variance in the CA1 region of the hippocampus. *J. Neurophysiol.* **70**, 1451–1459.

Martin, A. R. (1977). Junctional transmission. II. Presynaptic mechanisms. *In* "Handbook of Physiology" (E. Kandel, Ed.), Sect. 1, pp. 329–355. Am. Physiol. Soc., Bethesda, MD.

Maruyama, I. N., and Brenner, S. (1991). A phorbol ester/diacylglycerol-binding protein encoded by the *unc-13* gene of *Caenorhabditis elegans*. *Proc. Natl. Acad. Sci. USA* **88**, 5729–5733.

Matthews-Bellinger, J., and Salpeter, M. M. (1973). Distribution of acetylcholine receptors at frog neuromuscular junctions with a discussion of some physiological implications. *J. Physiol. (London)* **279**, 197–213.

Maycox, P. R., Link, E., Reetz, A., Morris, S. A., and Jahn, R. (1992). Clathrin-coated vesicles in nervous tissue are involved primarily in synaptic vesicle recycling. *J. Cell Biol.* **118**, 1379–1388.

Mayer, A., Wickner, W., and Haas, A. (1996). Sec18p (NSF)-driven release of Sec17p (alpha-SNAP) can precede docking and fusion of yeast vacuoles. *Cell* **85**, 83–94.

McGuiness, T. L., Brady, S. T., Gruner, J. A., Sugimori, M., Llinás, R., and Greengard, P. (1989). Phosphorylation-dependent inhibition by synapsin I of organelle movement in squid axoplasm. *J. Neurosci.* **9**, 4138–4149.

McLachlan, E. M. (1978). The statistics of transmitter release at chemical synapses. *Int. Rev. Physiol.* **17**, 49–117.

McMahon, H. T., Bolshakov, V. Y., Janz, R., Hammer, R. E., Siegelbaum, S. A., and Südhof, T. C. (1996). Synaptophysin, a major synaptic vesicle protein, is not essential for neurotransmitter release. *Proc. Natl. Acad. Sci. USA* **93**, 4760–4764.

McNew, J. A., Parlati, F., Fukuda, R., Johnston, R. J., Paz, K., Paumet, F., Söllner, T. H., and Rothman, J. E. (2000). Compartmental specificity of cellular membrane fusion encoded in SNARE proteins. *Nature* **407**, 153–159.

Mennerick, S., and Zorumski, C. F. (1996) Postsynaptic modulation of NMDA synaptic currents in rat hippocampal microcultures by paired-pulse stimulation. *J. Physiol.* **490**, 405–417.

Minota, S., Kumamoto, E., Kitakoga, O., and Kuba, K. (1991). Long-term potentiation induced by a sustained rise in the intraterminal Ca^{2+} in bull-frog sympathetic ganglia. *J. Physiol. (London)* **435**, 421–438.

Monck, J. R. and Fernandez, J. M. (1992). The exocytotic fusion pore. *J. Cell Biol.* **119**, 1395–1404.

Mulkey, R. M., and Zucker, R. S. (1991). Action potentials must admit calcium to evoke transmitter release. *Nature (London)* **350**, 153–155.

Muller, R. U., and Finkelstein, A. (1974). The electrostatic basis of Mg^{2+} inhibition of transmitter release. *Proc. Natl. Acad. Sci. USA* **71**, 923–926.

Murthy, V. N., and Stevens, C. F. (1998). Synaptic vesicles retain their identity through the endocytic cycle. *Nature* **392**, 497–501.

Nagy, A., Baker, R. R., Morris, S. J., and Whittaker, V. P. (1976). The preparation and characterization of synaptic vesicles of high purity. *Brain Res.* **109**, 285–309.

Neher, E., and Zucker, R. S. (1993). Multiple calcium-dependent processes related to secretion in bovine chromaffin cells. *Neuron* **10**, 21–30.

Nelson, N. (1992). The vacuolar H^+-ATPase: One of the most fundamental ion pumps in nature. *J. Exp. Biol.* **172**, 19–27.

Nichols, B. J., Ungermann, C., Pelham, H. R., Wickner, W. T., and Haas, A. (1997). Homotypic vacuolar fusion mediated by t- and v-SNAREs. *Nature* **387**, 199–202.

Nickel, W., Weber, T., McNew, J. A., Parlati, F., Söllner, T. H., and Rothman, J. E. (1999). Content mixing and membrane integrity during membrane fusion driven by pairing of isolated v-SNAREs and t-SNAREs. *Proc. Natl. Acad. Sci. USA* **96**, 12571–12576.

Nonet, M. L., Grundahl, K., Meyer, B. J., and Rand, J. B. (1993). Synaptic function is impaired but not eliminated in *C. elegans* mutants lacking synaptotagmin. *Cell* **73**, 1291–1305.

Novick, P., Ferro, S., and Schekman, R. (1981). Order of events in the yeast secretory pathway. *Cell* **25**, 461–469.

Novick, P., Field, C., and Schekman, R. (1980). Identification of 23 complementation groups required for post-translational events in the yeast secretory pathway. *Cell* **21**, 205–215.

Otis, T., Zhang, S., and Trussell, L. O. (1996) Direct measurement of AMPA receptor desensitization induced by glutamatergic synaptic transmission. *J. Neurosci.* **16**, 7496–7504.

Parsons, S. M., Prior, C., and Marshall, I. G. (1993). Acetylcholine transport, storage, and release. *Int. Rev. Neurobiol.* **35**, 279–390.

Parsons, T. D., Coorssen, J. R., Horstmann, H., and Almers, W. (1995). Docked granules, the exocytic burst, and the need for ATP hydrolysis in endocrine cells. *Neuron* **15**, 1085–1096.

Paulsen, O., and Heggelund, P. (1994). The quantal size at retinogeniculate synapses determined from spontaneous and evoked EPSCs in guinea-pig thalamic slices. *J. Physiol. (London)* **480**, 505–511.

Pellizzari, R., Rossetto, O., Schiavo, G., and Montecucco, C. (1999). Tetanus and botulinum neurotoxins: Mechanism of action and therapeutic uses. *Philos. Trans. R. Soc. London Ser. B* **354**, 259–268.

Peng, Y. -Y., and Zucker, R. S. (1993). Release of LHRH is linearly related to the time integral of presynaptic Ca^{2+} elevation above a threshold level in bullfrog sympathetic ganglia. *Neuron* **10**, 465–473.

Perin, M. S., Brose, N., Jahn, R., and Südhof, T. C. (1991). Domain structure of synaptotagmin (p65). *J. Biol. Chem.* **266**, 623–629.

Perin, M. S., Fried, V. A., Mignery, G. A., Jahn, R., and Südhof, T. C. (1990). Phospholipid binding by a synaptic vesicle protein homologous to the regulatory region of protein kinase C. *Nature* **345**, 260–263.

Peters, C., and Mayer, A. (1998). Ca^{2+}/calmodulin signals the completion of docking and triggers a late step of vacuole fusion. *Nature* **396**, 575–580.

Peters, C., Bayer, M. J., Buhler, S., Andersen, J. S., Mann, M., and Mayer, A. (2001). Trans-complex formation by proteolipid channels in the terminal phase of membrane fusion. *Nature* **409**, 581–588.

Pevsner, J., Hsu, S. C., and Scheller, R. H. (1994). n-Sec1: a neural-specific syntaxin-binding protein. *Proc. Natl. Acad. Sci. USA* **91**, 1445–1449.

Pieribone, V. A., Shupliakov, O., Brodin, L., Hilfiker-Rothenfluh, S., Czernik, A. J., and Greengard, P. (1995). Distinct pools of synaptic vesicles in neurotransmitter release. *Nature* **375**, 493–497.

Prior, C. (1994). Factors governing the appearance of small-mode miniature endplate currents at the snake neuromuscular junction. *Brain Res.* **664**, 61–68.

Quetglas, S., Leveque, C., Miquelis, R., Sato, K., and Seagar, M. (2000). Ca^{2+}-dependent regulation of synaptic SNARE complex assembly via a calmodulin- and phospholipid-binding domain of synaptobrevin. *Proc. Natl. Acad. Sci. USA* **97**, 9695–9700.

Rahamimoff, R., Lev-Tov, A., and Meiri, H. (1980). Primary and secondary regulation of quantal transmitter release: Calcium and sodium. *J. Exp. Biol.* **89**, 5–18.

Redman, S. (1990). Quantal analysis of synaptic potentials in neurons of the central nervous system. *Physiol. Rev.* **70**, 165–198.

Reimer, R. J., Fon, E. A., and Edwards, R. H. (1998). Vesicular neurotransmitter transport and the presynaptic regulation of quantal size. *Curr. Opin. Neurobiol.* **8**, 405–412.

Reist, N. E., Buchanan, J., Li, J., DiAntonio, A., Buxton, E. M., and Schwarz, T. L. (1998). Morphologically docked synaptic vesicles are reduced in *synaptotagmin* mutants of *Drosophila*. *J. Neurosci.* **18**, 7662–7673.

Richmond, J. E., Davis, W. S., and Jorgensen, E. M. (1999). UNC-13 is required for synaptic vesicle fusion in *C. elegans*. *Nat. Neurosci.* **2**, 959–964.

Roberts, W. M. (1994). Localization of calcium signals by a mobile calcium buffer in frog saccular hair cells. *J. Neurosci.* **14**, 3246–3262.

Roberts, W. M., Jacobs, R. A., and Hudspeth, A. J. (1991). Colocalization of ion channels involved in frequency selectivity and synaptic transmission at presynaptic active zones of hair cells. *J. Neurosci.* **11**, 1496–1507.

Robinson, P. J., Liu, J. P., Powell, K. A., Fykse, E. M., and Südhof, T. C. (1994). Phosphorylation of dynamin I and synaptic-vesicle recycling. *Trends Neurosci* **17**, 348–353.

Robitaille, R., Adler, E. M., and Charlton, M. P. (1990). Strategic location of calcium channels at transmitter release sites of frog neuromuscular synapses. *Neuron* **5**, 773–779.

Rosahl, T. W., Geppert, M., Spillane, D., Herz, J., Hammer, R. E., Malenka, R. C., and Südhof, T. C. (1993). Short-term synaptic plasticity is altered in mice lacking synapsin I. *Cell* **75**, 661–670.

Rosenmund, C., Clements, J. D., and Westbrook, G. L. (1993). Nonuniform probability of glutamate release at a hippocampal synapse. *Science* **262**, 754–757.

Rothman, J. E., and Warren, G. (1994). Implications of the SNARE hypothesis for intracellular membrane topology and dynamics. *Curr. Biol.* **4**, 220–233.

Ryan, T. A. and Smith, S. J. (1995). Vesicle pool mobilization during action potential firing at hippocampal synapses. *Neuron* **14**, 983–989.

Ryan, T. A., Li, L., Chin, L. S., Greengard, P., and Smith, S. J. (1996). Synaptic vesicle recycling in synapsin I knock-out mice. *J. Cell Biol.* **134**, 1219–1227.

Ryan, T. A., Reuter, H., Wendland, B., Schweizer, F. E., Tsien, R. W., and Smith, S. J. (1993). The kinetics of synaptic vesicle recycling measured at single presynaptic boutons. *Neuron* **11**, 713–724.

Ryan, T. A., Reuter, H., and Smith, S. J. (1997). Optical detection of a quantal presynaptic membrane turnover. *Nature* **388**, 478–482.

Sakaguchi, M., Inaishi, Y., Kashihara, Y., and Kuno, M. (1991). Release of calcitonin gene-related peptide from nerve terminals in rat skeletal muscle. *J. Physiol. (London)* **434**, 257–270.

Salminen, A., and Novick, P. J. (1987). A ras-like protein is required for a post-Golgi event in yeast secretion. *Cell* **49**, 527–538.

Sankaranarayanan, S., and Ryan, T. A. (2001). Calcium accelerates endocytosis of vSNAREs at hippocampal synapses. *Nat. Neurosci.* **4**, 129–136.

Sato, T. K., Rehling, P., Peterson, M. R., and Emr, S. D. (2000). Class C Vps protein complex regulates vacuolar SNARE pairing and is required for vesicle docking/fusion. *Mol. Cell* **6**, 661–671.

Schiavo, G., Benfenati, F., Poulain, B., Rossetto, O., Polverino de Laureto, P., DasGupta, B. R., and Montecucco, C. (1992). Tetanus and botulinum-B neurotoxins block neurotransmitter release by proteolytic cleavage of synaptobrevin. *Nature* **359**, 832–835.

Schiavo, G., Rossetto, O., Catsicas, S., Polverino de Laureto, P., DasGupta, B. R., Benfenati, F., and Montecucco, C. (1993). Identification of the nerve terminal targets of botulinum neurotoxin serotypes A, D, and E. *J. Biol. Chem.* **268**, 23784–23787.

Schiavo, G., Rossetto, O., and Montecucco, C. (1994). Clostridial neurotoxins as tools to investigate the molecular events of transmitter release. *Semin. Cell Biol.* **5**, 221–229.

Schikorski, T., and Stevens, C. F. (1997) Quantitative ultrastructural analysis of hippocampal excitatory synapses. *J. Neurosci.* **17**, 5858–5867.

Schneggenburger, R., and Neher, E. (2000). Intracellular calcium dependence of transmitter release rates at a fast central synapse. *Nature* **406**, 889–893.

Schulze, K. L., Broadie, K., Perin, M. S., and Bellen, H. J. (1995). Genetic and electrophysiological studies of *Drosophila* syntaxin-1A demonstrate its role in nonneuronal secretion and neurotransmission. *Cell* **80**, 311–320.

Searl, T., Prior, C., and Marshall, I. G. (1991). Acetylcholine recycling and release at rat motor nerve terminals studied using (–)-vesamicol and troxpyrrolium. *J. Physiol. (London)* **444**, 99–116.

Sesack, S. R., and Snyder, C. L. (1995). Cellular and subcellular localization of syntaxin-like immunoreactivity in the rat striatum and cortex. *Neuroscience* **67**, 993–1007.

Shirataki, H., Kaibuchi, K., Sakoda, T., Kishida, S., Yamaguchi, T., Wada, K., Miyazaki, M., and Takai, Y. (1993). Rabphilin-3A, a putative target protein for smg p25A/rab3A p25 small GTP-binding protein related to synaptotagmin. *Mol. Cell. Biol.* **13**, 2061–2068.

Silverman, B. W. (1986). "*Density Estimation for Statistics and Data Analysis.*" Chapman & Hall, London.

Simon, S. M., and Llinás, R. R. (1985). Compartmentalization of the submembrane calcium activity during calcium influx and its significance in transmitter release. *Biophys. J.* **48**, 485–498.

Smith, C., Moser, T., Xu, T., and Neher, E. (1998). Cytosolic Ca²⁺ acts by two separate pathways to modulate the supply of release-competent vesicles in chromaffin cells. *Neuron* **20**, 1243–1253.

Söllner, T., Bennett, M. K., Whiteheart, S. W., Scheller, R. H., and Rothman, J. E. (1993). A protein assembly–disassembly pathway *in vitro* that may correspond to sequential steps of synaptic vesicle docking, activation, and fusion. *Cell* **75**, 409–418.

Stanley, E. F. (1993). Single Ca²⁺ channels and acetylcholine release at a presynaptic nerve terminal. *Neuron* **11**, 1007–1011.

Steinbach, J. H., and Stevens, C. F. (1976). Neuromuscular transmission. *In* "Frog Neurobiology" (R. Llinás and W. Precht, Eds.), pp. 33–92. Springer-Verlag, Berlin.

Stevens, C. F., and Tsujimoto, T. (1995). Estimates for the pool size of releasable quanta at a single central synapse and for the time required to refill the pool. *Proc. Natl. Acad. Sci. USA* **92**, 846–849.

Stevens, C. F., and Wesseling, J. F. (1998). Activity-dependent modulation of the rate at which synaptic vesicles become available to undergo exocytosis. *Neuron* **21**, 415–424.

Stevens, C. F., and Wesseling, J. F. (1999). Identification of a novel process limiting the rate of synaptic vesicle cycling at hippocampal synapses. *Neuron* **24**, 1017–1028.

Stevens, C. F., and Williams, J. H. (2000). "Kiss and run" exocytosis at hippocampal synapses. *Proc. Natl. Acad. Sci. USA* **97**, 12828–12833.

Stricker, C., and Redman, S. (1994). Statistical models of synaptic transmission evaluated using the expectation–optimization algorithm. *Biophys. J.* **67**, 656–670.

Südhof, T. C., Lottspeich, F., Greengard, P., Mehl, E., and Jahn, R. (1987). A synaptic vesicle protein with a novel cytoplasmic domain and four transmembrane regions. *Science* **238**, 1142–1144.

Südhof, T. C., Czernik, A. J., Kao, H. -T., Takei, K., Johnston, P. A., Horiuchi, A., Kanazir, S. D., Wagner, M. A., Perin, M. S., de Camilli, P., and Greengard, P. (1989). Synapsins: Mosaics of shared and individual domains in a family of synaptic vesicle phosphoproteins. *Science* **245**, 1474–1480.

Südhof, T. C. (1997). Function of Rab3 GDP–GTP exchange. *Neuron* **18**, 519–522.

Sutton, R. B., Fasshauer, D., Jahn, R., and Brunger, A. T. (1998). Crystal structure of a SNARE complex involved in synaptic exocytosis at 2.4 A resolution. *Nature* **395**, 347–353.

Takamori, S., Rhee, J. S., Rosenmund, C., and Jahn, R. (2000). Identification of a vesicular glutamate transporter that defines a glutamatergic phenotype in neurons. *Nature* **407**, 189–194.

Tang, C.-M., Margulis, M., Shi, Q.-Y., and Fielding, A. (1994). Saturation of postsynaptic glutamate receptors after quantal release of transmitter. *Neuron* **13**, 1385–1393.

Tang, Y. -G., and Zucker, R. S. (1997). Mitochondrial involvement in post-tetanic potentiation of synaptic transmission. *Neuron* **18**, 483–491.

Thomas, P., Lee, A. K., Wong, J. G., and Almers, W. (1994). A triggered mechanism retrieves membrane in seconds after Ca²⁺-stimulated exocytosis in single pituitary cells. *J. Cell Biol.* **124**, 667–675.

Tong, G., and Jahr, C. E. (1994). Multivesicular release from excitatory synapses of cultured hippocampal neurons. *Neuron* **12**, 51–59.

Torri Tarelli, F., Bossi, M., Fesce, R., Greengard, P., and Valtorta, F. (1992). Synapsin I partially dissociates from synaptic vesicles during exocytosis induced by electrical stimulation. *Neuron* **9**, 1143–1153.

Torri-Tarelli, F., Grohovaz, F., Fesce, R., and Ceccarelli, B. (1985). Temporal coincidence between synaptic vesicle fusion and quantal secretion of acetylcholine. *J. Cell Biol.* **101**, 1386–1399.

Trimble, W. S., Cowan, D. M., and Scheller, R. H. (1988). VAMP-1: A synaptic vesicle-associated integral membrane protein. *Proc. Natl. Acad. Sci. USA* **85**, 4538–4542.

Trussell, L. O., Zhang, S., and Raman, I. M. (1993). Desensitization of AMPA receptors upon multiquantal neurotransmitter release. *Neuron* **10**, 1185–1196.

van der Bliek, A. M. and Meyerowitz, E. M. (1991). Dynamin-like protein encoded by the *Drosophila shibire* gene associated with vesicular traffic. *Nature* **351**, 411–414.

Van der Kloot, W. (1988). Acetylcholine quanta are released from vesicles by exocytosis (and why some think not). *Neuroscience* **24**, 1–7.

Van der Kloot, W. (1991). The regulation of quantal size. *Prog. Neurobiol.* **36**, 93–130.

Van der Kloot, W. (1994). Facilitation at the frog neuromuscular junction at 0°C is not maximal at time zero. *J. Neurosci.* **14**, 5722–5724.

Van der Kloot, W., and Molgó, J. (1994). Quantal acetylcholine release at the vertebrate neuromuscular junction. *Physiol. Rev.* **74**, 899–991. (An extremely comprehensive modern review.)

Van der Kloot, W., Balezina, O. P., Molgó, J., and Naves, L. A. (1994). The timing of channel opening during miniature endplate currents at the frog and mouse neuromuscular junctions: Effects of fasciculin-2, other anti-cholinesterases and vesamicol. *Pfluegers Arch.* **428**, 114–126.

Verhage, M., Ghijsen, W. E. J. M., and Lopes da Silva, F. H. (1994). Presynaptic plasticity: The regulation of Ca^{2+}-dependent transmitter release. *Prog. Neurobiol.* **42**, 539–574.

Verhage, M., McMahon, H. T., Ghijsen, W. E. J. M., Boomsma, F., Scholten, G., Wiegant, V. M., and Nicholls, D. G. (1991). Differential release of amino acids, neuropeptides, and catecholamines from isolated nerve terminals. *Neuron* **6**, 517–524.

Von Gersdorff, H., and Matthews, G. (1994). Inhibition of endocytosis by elevated internal calcium in a synaptic terminal. *Nature (London)* **370**, 652–655.

von Wedel, R. J., Carlson, S. S., and Kelly, R. B. (1981). Transfer of synaptic vesicle antigens to the presynaptic plasma membrane during exocytosis. *Proc. Natl. Acad. Sci. USA* **78**, 1014–1018.

Vyshedskiy, A. and Lin, J. -W. (1997). Activation and detection of facilitation as studied by presynaptic voltage control at the inhibitor of the crayfish opener muscle. *J. Neurophysiol.* **77**, 2300–2315.

Wachtel, H., and Kandel, E. R. (1971). Conversion of synaptic excitation to inhibition at a dual chemical synapse. *J. Neurophysiol.* **34**, 56–68.

Walmsley, B. (1991). Central synaptic transmission: Studies at the connection between primary muscle afferents and dorsal spinocerebellar tract (DSCT) neurones in Clarke's column of the spinal cord. *Prog. Neurobiol.* **36**, 391–423.

Wang, J., Renger, J. J., Griffith, L. C., Greenspan, R. J., and Wu, C. -F. (1994). Concomitant alterations of physiological and developmental plasticity in Drosophila CaM kinase II-inhibited synapses. *Neuron* **13**, 1373–1384.

Wang, L. -Y., and Kaczmarek, L. K. (1998). High-frequency firing helps replenish the readily releasable pool of synaptic vesicles. *Nature* **394**, 384–388.

Wang, Y., Okamoto, M., Schmitz, F., Hofmann, K., and Südhof, T. C. (1997). Rim is a putative Rab3 effector in regulating synaptic–vesicle fusion. *Nature* **388**, 593–598.

Warnock, D. E., and Schmid, S. L. (1996). Dynamin GTPase, a force-generating molecular switch. *Bioessays* **18**, 885–893.

Wheeler, D. B., Randall, A., and Tsien, R. W. (1994). Roles of N-type and Q-type Ca^{2+} channels in supporting hippocampal synaptic transmission. *Science* **264**, 107–111.

Whittaker, V. P. (1988). Model cholinergic systems: An overview. *Handb. Exp. Pharmacol.* **86**, 3–22.

Wickner, W., and Haas, A. (2000). Yeast homotypic vacuole fusion: A window on organelle trafficking mechanisms. *Annu. Rev. Biochem.* **69**, 247–275.

Wojtowicz, J. M., and Atwood, H. L. (1988). Presynaptic long-term facilitation at the crayfish neuromuscular junction: Voltage-dependent and ion-dependent phases. *J. Neurosci.* **8**, 4667–4674.

Wright, S. N., Brodwick, M. S., and Bittner, G. D. (1996). Calcium currents, transmitter release and facilitation of release at voltage-clamped crayfish nerve terminals. *J. Physiol.* **496**, 363–378.

Wu, L. -G., and Betz, W. J. (1998). Kinetics of synaptic depression and vesicle recycling after tetanic stimulation of frog motor nerve terminals. *Biophys. J.* **74**, 3003–3009.

Wu, L. -G., and Borst, J. G. G. (1999). The reduced release probability of releasable vesicles during recovery from short-term synaptic depression. *Neuron* **23**, 821–832.

Yamada, W. M., and Zucker, R. S. (1992). Time course of transmitter release calculated from simulations of a calcium diffusion model. *Biophys. J.* **61**, 671–682.

Yamasaki, S., *et al.* (1994). Cleavage of members of the synaptobrevin/VAMP family by types D and F botulinal neurotoxins and tetanus toxin. *J. Biol. Chem.* **269**, 12764–12772.

Zengel, J. E., and Magleby, K. L. (1980). Differential effects of Ba^{2+}, and Sr^{2+}, and Ca^{2+} on stimulation-induced changes in transmitter release at the frog neuromuscular junction. *J. Gen. Physiol.* **76**, 175–211.

Zengel, J. E., and Magleby, K. L. (1981). Changes in miniature endplate potential frequency during repetitive nerve stimulation in the presence of Ca^{2+}, Ba^{2+}, and Sr^{2+} at the frog neuromuscular junction. *J. Gen. Physiol.* **77**, 503–529.

Zhang, J. Z., Davletov, B. A., Südhof, T. C., and Anderson, R. G. (1994). Synaptotagmin I is a high affinity receptor for clathrin AP-2: Implications for membrane recycling. *Cell* **78**, 751–760.

Zhong, N., Beaumont, V., and Zucker, R. S. (2001). Roles for mitochondrial and reverse mode Na^+/Ca^{2+} exchange and the plasmalemma Ca^{2+} ATPase in post-tetanic potentiation at crayfish neuromuscular junctions. *J. Neurosci.* **21**, 9598–9607.

Zhong, Y., and Wu, C. -F. (1991). Altered synaptic plasticity in *Drosophila* memory mutants with a defective cyclic AMP cascade. *Science* **251**, 198–201.

Zimmermann, H. (1979). Vesicle recycling and transmitter release. *Neuroscience* **4**, 1773–1804.

Zucker, R. S. (1973). Changes in the statistics of transmitter release during facilitation. *J. Physiol. (London)* **229**, 787–810.

Zucker, R. S. (1989). Short-term synaptic plasticity. *Annu. Rev. Neurosci.* **12**, 13–31.

Zucker, R. S. (1994). Calcium and short-term synaptic plasticity. *Neth. J. Zool.* **44**, 495–512.

Zucker, R. S., and Fogelson, A. L. (1986). Relationship between transmitter release and presynaptic calcium influx when calcium enters through discrete channels. *Proc. Natl. Acad. Sci. USA* **83**, 3032–3036.

Zucker, R. S., Delaney, K. R., Mulkey, R., and Tank, D. W. (1991). Presynaptic calcium in transmitter release and posttetanic potentiation. *Ann. N. Y. Acad. Sci.* **635**, 191–207.

Zucker, R. S., and Regehr, W. G. (2002). Short-term synaptic plasticity. *Annu. Rev. Physiol.* **64**, 355–405.

Pharmacology and Biochemistry of Synaptic Transmission: Classic Transmitters

Ariel Y. Deutch and Robert H. Roth

The study of the nervous system 100 years ago was a period of claim and counterclaim, confusion, and recrimination—not unlike politics today or, for that matter, science. The reason for the tempestuous entry into the 20th century was the radical overthrow of the idea that the brain is one large continuous melded network (a syncytium), with each cell in physical contact. In contrast to this view, the pioneering studies of Santiago Ramón y Cajal revealed a very different picture, in which neurons, the units of the brain, are independent structures (see Shepherd, 1991, and Chapter 1). Although final confirmation of this view would await the development and application of electron microscopy, there soon became a generalized acceptance of neurons as the independent units of the nervous system. In turn, this acceptance brought about a new debate: What is the mode of communication between neurons? The answer is not static, but is evolving continuously. In this chapter we discuss briefly several means through which cells communicate with each other and then discuss in considerable detail one such mechanism—chemical synaptic transmission. Another means of communication, electrical transmission through gap junctions, is discussed in Chapter 15.

As discussed in Chapter 1, neurons, although varying widely in both morphology and function, share certain structural characteristics. A cell body (soma) from which processes emanate is present, with the processes (axons and dendrites) representing polarized compartments of the cell. Axons can be short or long, and remain local or alternatively project to distant areas. In contrast, dendrites are local. The general concept arose that axons transmit information, which is conveyed to the dendrites or soma of follower cells. The critical gap between the transmit-

ting element of the neurons (axon) and the recipient zone of the follower cell (e.g., the dendrite) is the area across which transmission of information occurs; this area was termed the *synapse* by Charles Sherrington (Shepherd, 1991). Thus, there were presynaptic and postsynaptic neurons. This general conceptual framework remains in place today, although there are many exceptions, including dendrites that release neuroactive substances and axons that receive inputs from other neurons. One other characteristic proposed by Sherrington that is central to the concept of chemical communication between neurons is that synaptic transmission does not follow all-or-none rules, but is graded in strength and flexible (Sherrington, 1906).

DIVERSE MODES OF NEURONAL COMMUNICATION

The controversy surrounding the nature of neuronal communication—chemical or electrical—was in full force for the first half of the 20th century, even though evidence was marshaled in support of the chemical mode of communication midway through the 19th century. In 1849, Claude Bernard noted that curare, which is the active constituent of a poison applied to arrows in South America, blocks nerve-to-muscle neurotransmission (Bernard, 1849). This effect was subsequently shown to be due to binding of curare to postsynaptic receptors for the neurotransmitter acetylcholine (ACh), thus resulting in antagonism of neuromuscular transmission. Half a century later Thomas Elliott (1905) observed that adrenaline resulted in contraction of smooth muscle that had been deprived of its neural innervation, thereby indi-

Copyright 2004, Elsevier Science (USA).
All rights reserved.

cating that muscle contraction depended on the action of molecules liberated from nerves. In a seminal series of studies using the isolated frog heart Otto Loewi provided firm evidence of chemical neurotransmission by showing that ACh was released on nerve stimulation and activated a target muscle (Loewi, 1921).

Although it is now clear that the major means of interneuronal communication is chemical in nature, neurons also use several other processes for intercellular communication. These include electrical synaptic transmission (see Chapter 15), ephaptic interactions, and autocrine, paracrine, and long-range signaling, to which molecules produced by both neural and nonneural cells contribute. The nonsynaptic mode of intercellular communication with the longest range (distance) is humoral or hormonal signaling. For example, some peripheral hormones can enter the central nervous system to drive, inhibit, or modulate neuronal activity. The neuronal targets of these hormones have specific receptors for these hormones. The actions of hormones may be either short term (to acutely change neuronal activity) or more often long term, such as long-lasting changes in gene expression by targeting nuclear hormone receptors.

Molecules produced by neurons can also be used in intercellular communication that does not require synaptic specializations. Water-soluble factors that are secreted by neurons or diffuse from the cells in which they are generated include classic neurotransmitters, neuropeptides, and neurosteroids, as well as the gases nitric oxide and carbon monoxide. These factors may act through autocrine mechanisms (activating receptors on the same cell that releases them; see Fig. 9.1) or through paracrine pathways to nearby cells (Fig. 9.1), and the gases nitric oxide and carbon monoxide in particular may act as retrograde neurotransmitters, signaling a presynaptic element from its postsynaptic localization, i.e., working backward. Soluble factors can act on high- or low-affinity receptors and can act either locally or over some distance. The role of such molecules is thought to be primarily in modulating neural activity, although they may also provide trophic support ("guidance cues") for neurons in the process of outgrowth and for the establishment and maintenance of synaptic connections. A more extended discussion of the "unconventional" transmitters as well as peptide transmitters and growth factors can be found in the next chapter.

Cells may also signal one another through structural components and adhesion molecules that are bound to cell surfaces. Adhesion molecules are typically transmembrane proteins, comprised on an extracellular segment that mediates adhesion interactions across the intercellular space, and a cytoplasmic domain that is linked to signal transduction pathways or the cytoskeleton. The intercellular domains of adhesion molecules may be "activated" by the binding of the extracellular segment to a cognate membrane-bound "ligand" protruding from another cell. Only in the context of diffusible transmitters is this a strange idea, with molecules reaching across the synapse to "touch and kiss" and thereby drive intracellular signaling cascades. In many cases cell-specific adhesion molecules and extracellular matrix components are critical signals during development of the organism, and provide cues for cell differentiation.

The most intimate nonsynaptic mode of intercellular communication is the ephapse, in which electrical impulses in one cell or extracellular ion accumulation in the vicinity of one cell can directly affect the activity of an adjacent cell. In this case (see Fig. 9.1), the only morphological requirement is closely apposed membranes. Ephaptic interactions are either transient or sustained, and can be induced by various treatments.

The two categories of intercellular communication considered to be true synaptic transmission are electrical (see Chapter 15) and chemical synapses (Fig. 9.1). We focus on chemical synaptic transmission, which is thought to be the dominant mode of interneuronal communication.

CHEMICAL TRANSMISSION

Chemically mediated transmission is the major means by which a signal is communicated from one nerve cell to another, and is the mode of neuronal communication on which this and the next chapter focus. The general acceptance of chemical communication between neurons as the dominant mode of signaling required that certain criteria be met for a compound to be accepted as a neurotransmitter. These "classic" criteria for transmitters have resulted in a relatively small number of compounds being recognized as "classic" transmitters. However, over the past generation it has become apparent that there are an even larger number of chemical messengers that broadly qualify as intercellular transmitters, although these often do not meet (and indeed serve as exceptions) the classic criteria.

The Criteria for Definition as a Neurotransmitter

Neurotransmitters are endogenous substances that are released from neurons, act on receptor sites that are

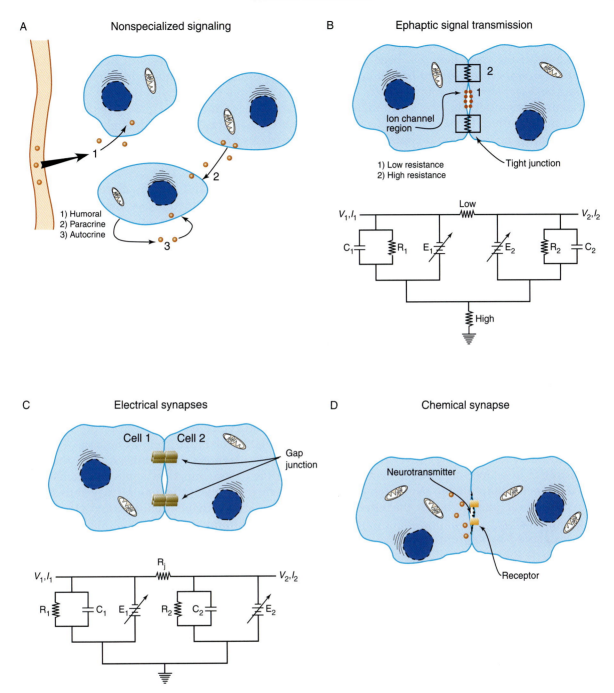

FIGURE 9.1 The multiple modes of intercellular signaling. (A) Substances (red) produced outside the nervous system **(1)** or by cells within the CNS **(2, 3)** can affect neuronal activity, acting through **(1)** humoral, **(2)** paracrine, and **(3)** autocrine mechanisms. (B) Ephaptic transmission. Two apposed cell membranes showing regions of low **(1,** red label indicating channels) and high **(2,** sawtooths, representing tight junctions) resistances. The electrical equivalent circuit is shown below. Communication between these cells is permitted through membrane regions of low resistance. The presence of tight junctions **(2)** between apposed cells favors an increase in current density by preventing current flow into the bulk extracellular space. Charges accumulate in the narrow intercellular space and affect capacitance and resistive components of the cell membranes. (C) Electrical synapses. Gap junction channels provide low-resistance pathways between adjacent cells, allowing direct communication between the cytoplasms of both cells. In contrast with the ephaptic mode of transmission, current flows directly from cell to cell and not through the extracellular space. The electrical equivalent circuit is shown below, differing from that of the ephapse in (B) primarily in the absence of a resistance to ground. (D) Chemical synapses. Neurotransmitters released from the presynaptic terminal (on the left) diffuse across the cleft to bind to postsynaptic receptors, thereby opening ion channels and increasing conductance of the postsynaptic membrane, producing currents that excite or inhibit the cell.

typically present on membranes of postsynaptic cells, and produce a functional change in the properties of the target cell. Over the years there has been general agreement that several criteria should be met before a substance can be designated a neurotransmitter.

First, a neurotransmitter must be synthesized by and released from neurons. In many cases, this means that the presynaptic neuron should contain a transmitter and the appropriate enzymes required for synthesis of that neurotransmitter. However, synthesis in the nerve terminal is not an absolute requirement. For example, peptide transmitters are synthesized in the cell body and transported to distant sites, where they are released (see Chapter 10).

Second, the substance should be released from nerve terminals in a chemically or pharmacologically identifiable form. Thus, one should be able to isolate the transmitter and characterize its structure using biochemical or other techniques.

Third, a neurotransmitter should reproduce at the postsynaptic cell the specific events (such as changes in membrane properties) that are seen after stimulation of the presynaptic neuron; the concentrations that approximate those seen after release of the neurotransmitter by nerve stimulation should mimic the effects of presynaptic stimulation.

Fourth, the effects of a putative neurotransmitter should be blocked by known competitive antagonists of the transmitter in a dose-dependent manner. In addition, treatments that inhibit synthesis of the candidate transmitter should block the effects of presynaptic stimulation.

Fifth, there should be appropriate active mechanisms to terminate the action of the putative neurotransmitter. Such mechanisms can include enzymatic degradation and reuptake of the substance into the presynaptic neuron or glial cells through specific transporter molecules.

The Five Steps of Chemical Neurotransmission: Synthesis, Storage, Release, Receptor Binding, and Inactivation

The general mechanisms of chemical synaptic transmission are depicted in Fig. 9.2. Synaptic transmission consists of a number of steps, and each of these steps is a potential site of drug action.

1. *Biosynthesis of the neurotransmitter in the presynaptic neuron.* For the transmitter to be synthesized, precursors should be present in the appropriate places within the neurons. The enzymes taking part in the conversion of the precursor(s) into the transmitter should be present in an active form and localized to the appropriate compartment in the neuron, and any necessary cofactors for enzyme activity should be present. The biosynthesis of neurotransmitters has long been an important site for clinically useful drugs. An example is α-methyl-*p*-tyrosine, a drug used in the treatment of an adrenal gland tumor (pheochromocytoma) that causes strikingly high blood pressure and increases the risk of strokes. This tumor releases massive amounts of norepinephrine, a neurotransmitter that acts to constrict blood vessels and increase cardiac output and thereby elevates blood pressure. α-Methyl-*p*-tyrosine blocks the synthesis of catecholamines such as norepinephrine and thus prevents the actions of these amine transmitters on target cells, lowering blood pressure.

2. *Storage of the neurotransmitter or its precursor or both in the presynaptic nerve terminal.* Classic and peptide transmitters are stored in synaptic vesicles, where they are sequestered and protected from enzymatic degradation, and are readily available for release. In the case of so-called classic neurotransmitters (acetylcholine, biogenic amines, and amino acids), the synaptic vesicles are small (~ 50 nm in diameter). In contrast, neuropeptide transmitters are stored in large dense-core vesicles (~ 100 nm in diameter) and typically released in response to repetitive stimulation or burst firing of neurons. Because most neurotransmitters are synthesized in the cytosol of neurons, there must be some mechanism through which the transmitter gains entry into the vesicle. Recent studies have identified and cloned vesicular transporter proteins.

3. *Release of the neurotransmitter into the synaptic cleft.* The transmitter-containing vesicle fuses with the cellular membrane and release (exocytosis) of the transmitter occurs. Neurons use two pathways to secrete proteins. The release of most neurotransmitters occurs by a regulated pathway controlled by extracellular signals. The neurotransmitter release process is discussed more fully in Chapter 8. There is a second (constitutive) pathway that is not triggered by extracellular stimulation and is used to secrete membrane components, viral proteins, and extracellular matrix molecules; the degree to which a few unconventional transmitters (e.g., growth factors) are synthesized and released in the constitutive pathway remains unclear.

4. *Binding and recognition of the neurotransmitter by target receptors.* Released transmitters interact with receptors located on the target (postsynaptic) cell. These receptors fall into two broad classes. The first are membrane proteins called metabotropic receptors, which are coupled to intracellular G proteins as effectors (see Chapters 11, 12, and 16). The other group of receptors, termed ionotropic receptors, form ion chan-

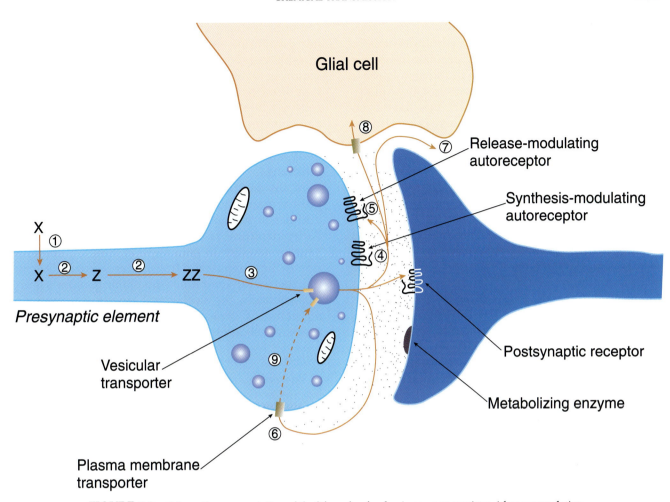

Glial cell

Release-modulating autoreceptor

Synthesis-modulating autoreceptor

X

①

X → Z → ZZ ③

② ②

Presynaptic element

Postsynaptic receptor

Vesicular transporter

Metabolizing enzyme

Plasma membrane transporter

FIGURE 9.2 Schematic representation of the life cycle of a classic neurotransmitter. After accumulation of a precursor amino acid (X) into the neuron (step 1), the amino acid precursor is sequentially metabolized (step 2) to yield the mature transmitter (ZZ). The transmitter is then accumulated into vesicles by the vesicular transporter (step 3), where it is poised for release and protected from degradation. The released transmitter can interact with postsynaptic receptors (step 4) or autoreceptors (step 5) that regulate transmitter release, synthesis, or firing rate. Transmitter actions are terminated by means of a high-affinity membrane transporter (step 6) that is usually associated with the neuron that released the transmitter. Alternatively, the actions of the transmitter may be terminated by means of diffusion (step 7) or by accumulation into glia through a membrane transporter (step 8). When the transmitter is taken up by the neuron, it is subject to metabolic inactivation (step 9).

nels that are either ligand-gated or voltage-gated and through which ions such as sodium and calcium enter the cell. We typically think of receptors to neurotransmitters as being localized to the postsynaptic neuron. However, there are also transmitter receptors on presynaptic neurons that respond to the transmitter from that cell. These autoreceptors modulate transmitter release or synthesis, or impulse flow, and can be considered a homeostatic feedback mechanism. A more thorough discussion of neurotransmitter receptors may be found in Chapter 11.

5. *Inactivation and termination of the action of the released transmitter.* If there are no mechanisms to terminate the actions of neurotransmitters, the conse-

quences for the neuron may be dire. Adverse sustained activation of postsynaptic targets can result in tetanus in muscles or seizure discharges in neurons. There are multiple processes to terminate the action of neurotransmitters, both active and passive. Among the active mechanisms are reuptake of the neurotransmitter through specific neuronal transporter proteins on the presynaptic neuron, enzymatic inactivation to an inactive substance, and a combination of these processes. In addition, in certain cases glial cells can accumulate released transmitters (see also Chapter 3). Diffusion away from the synaptic region is a passive mechanism that may contribute to inactivation of released transmitters.

The five steps described above form a logical scaffold for understanding chemical neurotransmission. However, there are particular and peculiar intricacies for each of the steps and for each of the many neurotransmitters. Catecholamines constitute a structurally defined group of neurotransmitters that have been extensively studied and for which there is a relatively complete understanding of the nature of chemical neurotransmission. This chapter therefore first considers the catecholamine neurotransmitters to illustrate the various steps of chemical neurotransmission. This discussion is then followed by an examination of the particulars of chemical neurotransmission for other classic neurotransmitters. These include the indoleamine serotonin (5-hydroxytryptamine), acetylcholine, and the amino acids GABA (γ-aminobutyric acid) and glutamate. (*Note*: The catecholamines and the indoleamines are grouped together as biogenic amines.) We also note the key differences between classic and other (nonclassic) neurotransmitters or chemical messengers, among which are peptide neurotransmitters and unconventional transmitters such as nitric oxide and growth factors, which are described in detail in the next chapter.

CLASSIC NEUROTRANSMITTERS

The term *classic* is used to differentiate acetylcholine, the biogenic amines, and the amino acid transmitters from other transmitters. Although the designation is somewhat arbitrary, the two groups of transmitters can be differentiated on several grounds. As discussed above and in Chapter 8, storage vesicles, when present, are smaller for classic transmitters. In addition, classic transmitters or their metabolic products are subject to reuptake by the presynaptic cell and can be viewed as homoeostatically conserved, but there is no energy-dependent high-affinity reuptake process for nonclassic transmitters. Finally, most classic transmitters are synthesized in the nerve terminal via enzymatic action; in contrast, peptides and some unconventional transmitters are synthesized in the soma from a precursor protein and then transported to the nerve terminal.

The Catecholamine Neurotransmitters

The term *catecholamine* refers generically to organic compounds that contain a catechol nucleus (a benzene ring with two adjacent hydroxyl substitutions) and an amine group. In practice, however, the term is usually used to describe the endogenous

compounds dopamine (dihydroxyphenylethylamine), norepinephrine, and epinephrine. These three neurotransmitters are formed by successive enzymatic steps requiring distinct enzymes (see Fig. 9.3). The localization of particular synthesizing enzymes to different cells results in distinct dopamine-, norepinephrine-, and epinephrine-containing neurons in the brain. The catecholamines also have transmitter roles in the peripheral nervous system and have certain hormonal functions.

In the peripheral nervous system, dopamine is present mainly as a precursor for norepinephrine but also has important biological activity in the kidney. Norepinephrine is the postganglionic sympathetic neurotransmitter in mammals; in contrast, epinephrine is the sympathetic transmitter in frogs. Despite this species difference in sympathetic nervous system characteristics, the biochemical aspects of neurotransmission as a general rule are remarkably constant across vertebrate species and, indeed, invertebrates.

Biosynthesis of Catecholamines

The amino acids phenylalanine and tyrosine are precursors for catecholamines. These amino acids are present in the plasma and brain in high concentrations. In mammals, tyrosine can be derived from dietary phenylalanine by an enzyme (phenylalanine hydroxylase) that is found primarily in liver. Phenylketonuria, a disorder caused by insufficient amounts of phenylalanine hydroxylase, results in very high plasma and brain levels of phenylalanine. Unless dietary phenylalanine intake is restricted, this can result in intellectual impairment (see Box 9.1). Catecholamines are formed in the brain, adrenal chromaffin cells, and sympathetic nerves. The processes regulating catecholamine synthesis are generally the same in the different tissues.

Catecholamine synthesis is usually considered to begin with tyrosine, which represents a branch point for many important biosynthetic processes in animal tissues. The sequence of enzymatic steps in the synthesis of catecholamines from tyrosine was first postulated by Blaschko in 1939 and finally confirmed by Nagatsu and co-workers in 1964, when they demonstrated that the enzyme tyrosine hydroxylase (TH) converts the amino acid L-tyrosine into 3,4-dihydroxyphenylalanine (L-DOPA). All of the component enzymes in the catecholamine biosynthetic pathway have been purified to homogeneity, which allowed detailed analyses of the kinetics, substrate specificity, and cofactor requirements of these enzymes and aided in the development of useful inhibitors of the enzymes. Moreover, the development of antibodies against the purified enzymes has permitted the

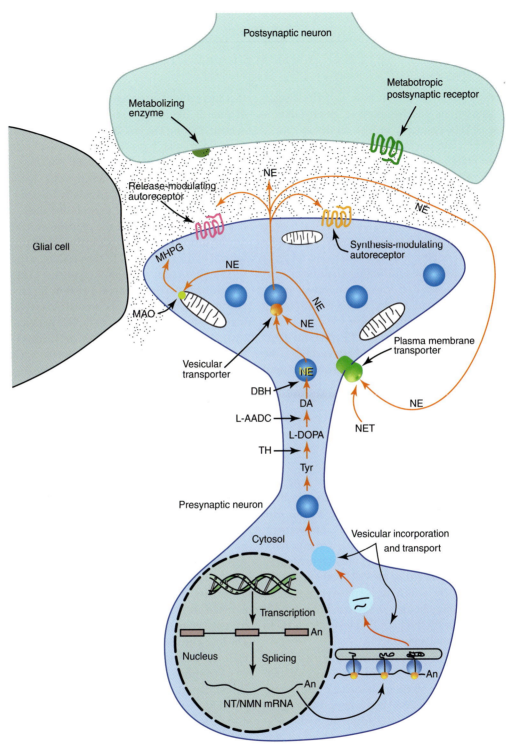

FIGURE 9.3 Characteristics of a norepinephrine (NE)-containing catecholamine neuron. On accumulation of tyrosine (Tyr) by the neuron, tyrosine is sequentially metabolized by tyrosine hydroxylase (TH) and L-aromatic amino acid decarboxylase (L-AADC) to dopamine (DA). The DA is then accumulated by the vesicular monoamine transporter. In dopaminergic neurons this is the final step. However, in this noradrenergic neuron, the DA is metabolized to NE by dopamine-β-hydroxylase (DBH), which is found in the vesicle. Once NE is released, it can interact with postsynaptic noradrenergic receptors or different types of presynaptic noradrenergic autoreceptors. The accumulation of NE by the high-affinity membrane norepinephrine transporter (NET) terminates the extracellular actions of NE. Once accumulated by the neuron, the NE can be metabolized to inactive species (for example, MHPG) by key degradative enzymes, such as monoamine oxidase (MAO), or taken back up by the vesicular transporter.

BOX 9.1

PKU AND METABOLISM

Classic phenylketonuria (PKU) is a genetic disease caused by mutations in the enzyme phenylalanine hydroxylase (PAH), resulting in loss of the enzyme's ability to hydroxylate phenylalanine (Phe) to tyrosine. PAH plays an integrated dual role in the metabolism of humans and other mammals. First, it provides an endogenous supply of tyrosine, thereby making consumption of this amino acid unnecessary for normal growth. Second, the reaction catalyzed by PAH is an essential step in the complete oxidation of Phe. When PAH levels are low or absent, as in PKU, blood levels of Phe are typically 20- to 50-fold higher than normal. A tiny fraction of the elevated Phe is converted to the phenylketone phenylpyruvic acid, which is excreted in the urine, hence the name of the disease.

The increased concentration of Phe seen in PKU spares the body, but spoils—indeed, devastates—the developing brain, and typically leads to severe mental retardation unless steps are taken to limit dietary Phe intake. The majority of untreated PKU patients suffer severe intellectual impairment (IQ < 20). They also have a somewhat higher incidence of seizures and tend to have fair skin and hair. The latter effect is due to the inhibition of melanin formation by the excess Phe.

PKU is inherited as an autosomal recessive trait. The vast majority of PKU babies are conceived when both parents are heterozygotes, each one harboring one normal gene and one "PKU" gene. Thus, on average, one-fourth of the children born to such parents have PKU, one-fourth are normal, and half are heterozygotes. The incidence of heterozygosity for PKU is about 1 in 55.

The mechanism by which hyperphenylalaninemia damages the developing brain is unknown, but probably involves competition by the high Phe levels with brain uptake of other essential amino acids.

Phenylalanine hydroxylase functions *in vivo* as part of a complex multicomponent system consisting of two other enzymes, dihydropteridine reductase and pterin 4α-carbinolamine dehydratase, and a nonprotein coenzyme, tetrahydrobiopterin (BH4). During the hydroxylation reaction, BH4 is stoichiometically oxidized; i.e., for every molecule of Phe converted to tyrosine by PAH, a molecule of BH4 is oxidized. The function of the two ancillary enzymes is to regenerate BH4.

With the realization that the hydroxylating system consists of four essential components, it was predicted that there might be variant forms of PKU caused by lack of one of the other components of the hydroxylating system. During the last 20 years, patients with these predicted variants have been described. Defects in the reductase or in one of the several enzymes essential for the synthesis of BH4 (but not the one due to defects in the dehydratase) were originally called "lethal" or "malignant" PKU. In all probability, these variants are deadly because BH4 and the reductase are also essential for the functions of tyrosine hydroxylase (TH) and tryptophan hydroxylase (TPH), thus leading to deficits in catecholamine and serotonin systems. These patients therefore suffer from three different metabolic lesions. Fortunately, these variants are extremely rare, accounting for between 1 and 2% of all PKU patients, and can be treated (with varying degrees of success) by feeding them the compounds beyond the metabolic blocks in TH and TPH (i.e., L-dopa and 5-hydroxytryptophan) and, when needed, large doses of BH4.

If one can dare say that there is anything fortunate about this dreadful disease, it is that it is extremely rare, with an average incidence of about 1/12,000. The frequency, however, varies widely among different ethnic groups, being only 1/200,000 in Japan but as high as 1/5000 in Ireland.

One auspicious feature of the disease is that the affected infants are essentially normal at birth, which raised the hope that some way might be found to prevent the intellectual deterioration of PKU. This hope was realized about 50 years ago with the introduction of a low-Phe diet (not a no-Phe diet!), which has proven to be largely if not totally effective in preventing brain damage, at least as reflected by the normal IQ of PKU patients who are started on the diet shortly after birth. The low-Phe diet is a heavy burden for both patients and their families, and it was once hoped that the diet could be discontinued after 6 or 7 years. It now appears that a longer period is beneficial. The goal of the diet is to keep blood Phe levels from rising no more than five- to sixfold normal levels.

Women with PKU who are contemplating having children raises dietary issues as well. If women with PKU went off the diet at some earlier time, they must resume it before they become pregnant, or the fetus risks *in utero* damage caused by the mother's high levels of Phe, a condition called "maternal PKU."

BOX 9.1 (*continues*)

PKU AND METABOLISM

Since the *sine qua non* of the successful dietary treatment of PKU is to start the diet as soon as possible after birth, its success was closely tied to the development of a cheap and rapid test for PKU. The Guthrie test, which was introduced in 1961, gives a semiquantitative measure of Phe levels in a drop of blood and has been widely used for screening newborns; the blood drop can be collected on a piece of filter paper and then mailed to a suitable laboratory.

PKU is noteworthy in that the brain is the only organ that is damaged by mutations in an enzyme that is found only in the liver in humans. This provides a valuable lesson: metabolically, no organ in the body is an island unto itself.

Seymour Kaufman

Box 9.1 is a U.S. government work in the public domain.

---

precise localization of the enzymes by immunohistochemical techniques.

The hydroxylation of L-tyrosine by TH results in the formation of the dopamine precursor L-DOPA. The L-DOPA is almost immediately metabolized to dopamine by L-aromatic amino acid decarboxylase (AADC); this step is so rapid that it is very difficult to measure L-DOPA in brain without first inhibiting AADC. In dopamine-containing neurons of the brain the decarboxylation of L-DOPA to dopamine is the final step in transmitter synthesis. However, in neurons using norepinephrine (also known as noradrenaline) or epinephrine (adrenaline) as transmitters, the enzyme dopamine β-hydroxylase (DBH) is present; this enzyme oxidizes dopamine to yield norepinephrine. Finally, in neurons in which epinephrine serves as the transmitter, a third enzyme, phenylethanolamine *N*-methyltransferase (PNMT), is present and converts norepinephrine into epinephrine. Thus, an adrenergic neuron (which uses epinephrine as its transmitter) contains four enzymes (TH, AADC, DBH, PNMT) that sequentially metabolize tyrosine to epinephrine. Noradrenergic neurons express only the enzymes TH, AADC, and DBH, and thus norepinephrine cannot be further metabolized to epinephrine. Similarly, because dopamine neurons lack DBH and PNMT, the catecholamine end product is the transmitter dopamine. The enzymes and cofactors taking part in the synthesis of the catecholamines are illustrated in Fig. 9.4.

Tyrosine hydroxylase. In human beings, a single tyrosine hydroxylase gene gives rise to four TH mRNA species through alternative splicing, resulting in four distinct TH isoforms (Lewis *et al.*, 1993). In contrast, in most primates two TH isoforms are present. Still different is the rat, which possesses but a single form of TH. It has been speculated that the different forms of TH in human beings are associated with differences in activity of the enzyme, but conclusive data addressing this point are lacking.

Tyrosine hydroxylase function is determined by two factors: changes in enzyme activity (the rate at which the enzyme converts the precursor into its product) and changes in the amount of enzyme protein present. One determinant of TH activity is phosphorylation of the enzyme (Fig. 9.4), which takes place at four different serine sites at the N terminus of the TH protein (Haycock and Haycock, 1991). These four serine residues are differently phosphorylated by various kinases. A second means of regulating the activity of the enzyme is through end-product inhibition: catecholamines can inhibit the activity of TH through competition for a required pterin cofactor for the enzyme (see Cooper *et al.*, 1996).

An increased demand for catecholamine synthesis can be met by inducing TH protein or by activating (by phosphorylation) the enzyme. The degree to which increases in catecholamine synthesis depend on *de novo* synthesis of new enzyme protein or changes in enzymatic activity differs across brain regions. For example, increased synthetic demand in noradrenergic neurons of the brainstem nucleus locus coeruleus appears to be accomplished primarily by increasing TH gene expression. In contrast, the same conditions and treatments that increase TH gene expression in brainstem noradrenergic neurons fail to increase TH mRNA levels in dopamine-containing

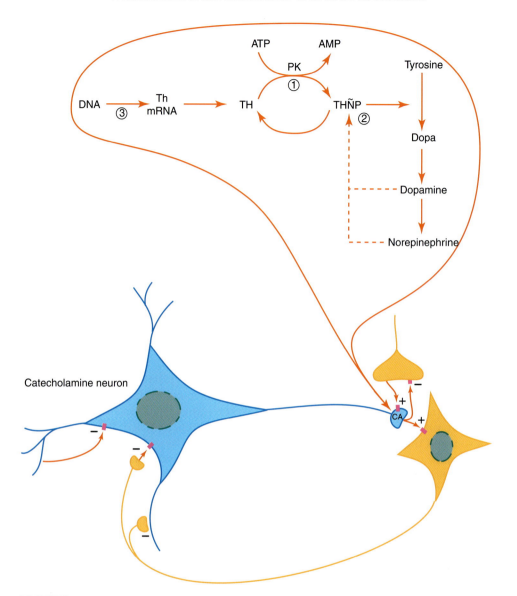

FIGURE 9.4   Schematic for the regulation of TH enzymatic activity. The numbered sites depict the three major types of regulation from TH phosphorylation **(1)**, accomplished by the action of specific protein kinases (PK), to end-product inhibition **(2)** to changes in TH gene transcription **(3)** and the subsequent increase in protein.

neurons of the midbrain. In these dopaminergic neurons it appears that synthesis is regulated primarily by altering the activity of TH (i.e., by posttranslational events).

The synthesis of catecholamines starts with the entry of tyrosine into the brain. This process is an energy-dependent one at which tyrosine competes with large neutral amino acids as a substrate for the transporter. Because brain levels of tyrosine are high enough to saturate TH, under basal conditions synthesis of catecholamines cannot be increased by administration of tyrosine. An exception is synthesis in catecholamine neurons that have a high basal firing

rate, such as the DA neurons that innervate the prefrontal cortex, particularly under pathological conditions (Tam *et al.*, 1990).

Since TH is saturated by tyrosine under basal conditions, TH is the rate-limiting step for catecholamine synthesis under basal conditions. However, under conditions of neuronal activation, the enzyme responsible for norepinephrine synthesis, DBH, becomes rate limiting (Scatton *et al.*, 1984), and thus under certain conditions tyrosine availability may regulate catecholamine synthesis.

Tyrosine hydroxylase is a mixed-function oxidase that has moderate substrate specificity, hydroxylating phenylalanine as well as tyrosine. The major substrate in brain is tyrosine, with molecular oxygen required. The actions of TH require a biopterin cofactor and iron (Fe^{2+}). Tetrahydrobiopterin (BH4) is an essential cofactor not only for tyrosine hydroxylase but also for phenylalanine hydroxylase, tryptophan hydroxylase, and NO synthase (Thorny *et al.* 2000). Because the levels of the cofactor (reduced BH4) are not saturated under basal conditions, endogenous levels of the pterin cofactor are significant in regulating TH activity. The formation and intracellular concentration of this essential cofactor are regulated by GTP cyclohydrolase, the rate-limiting enzyme in the synthesis of the biopterin cofactor (Ichinose *et al.*, 1994). Thus, the activity of this important enzyme plays a critical role in the regulation of tyrosine hydroxylase and other pteridine-dependent enzymes. The pterin cofactor is of clinical significance. For example, the genetic defect in a movement disorder called DOPA-responsive dystonia has now been shown to be due to mutations in the gene encoding GTP cyclohydrolase 1.

L-Aromatic amino acid decarboxylase. The hydroxylation of tyrosine by TH generates L-DOPA, which is then decarboxylated by L-aromatic amino acid carboxylase (also known in the brain as "DOPA decarboxylase") to the neurotransmitter dopamine. AADC has low substrate specificity and decarboxylates tryptophan as well as tyrosine. Since this enzyme is present in both catecholaminergic and serotonergic neurons, it plays an important role in the biosynthesis of both groups of transmitters. In dopaminergic neurons, AADC is the final enzyme of the synthetic pathway.

The transmitter dopamine does not cross the blood–brain barrier. In contrast, L-DOPA readily enters the brain. Accordingly, the dopamine precursor has achieved fame as a means of treating Parkinson's disease, which is due to loss of dopamine in the striatum: L-DOPA administration to patients with Parkinson's increases brain dopamine concentrations and thus provides symptomatic relief. Although L-DOPA readily enters the brain, decarboxylating enzymes in the liver and capillary endothelial cells readily degrade the dopamine precursor; L-DOPA is therefore administered in combination with a peripheral decarboxylase inhibitor that does not readily enter the brain. The administration of the peripheral decarboxylase inhibitor protects L-DOPA from metabolism before it enters the brain, and thereby sharply increases central dopamine concentrations.

The AADC gene has been cloned. The AADC gene of the fruit fly *Drosophila* undergoes alternative splicing, but in mammalian organisms it appears that different transcripts are generated through different promoter sequences, leading to one transcript being expressed in the brain and another in peripheral tissues. AADC mRNA is expressed in all catecholamine- and indoleamine-containing neurons in the CNS.

Levels of L-DOPA are virtually unmeasurable in the CNS under basal conditions. This is because the activity of AADC is so high that L-DOPA is converted into dopamine almost instantaneously. AADC requires pyridoxal 5-phosphate as a cofactor. The regulation of AADC has not been as intensively studied as that of TH, but converging data suggest that AADC is regulated primarily through induction of new protein rather than changes in activity. It is interesting to note that although L-DOPA is used extensively in Parkinson's disease, recent studies suggest that chronic administration of L-DOPA or other dopamine agonists decreases the activity of endogenous AADC. Thus, as is often the case with therapeutic administration of pharmacological agents, these drugs can have side effects or even be counter-therapeutic.

Dopamine β-hydroxylase. Noradrenergic and adrenergic neurons contain the enzyme dopamine-β-hydroxylase (DBH), a mixed-function oxidase that converts dopamine into norepinephrine. In noradrenergic neurons, this conversion is the final step in catecholamine synthesis. Humans appear to possess two different DBH mRNAs that are generated from a single gene. DBH mRNAs in the nervous system are restricted to noradrenergic and adrenergic neurons.

DBH is a copper-containing glycoprotein that requires ascorbic acid as the electron source during the hydroxylation of dopamine. Dicarboxylic acids such as fusaric acid are not absolute requirements but stimulate the enzymatic conversion of dopamine into norepinephrine. DBH does not have a high degree of substrate specificity and, *in vitro*, oxidizes almost any phenylethylamine to its corresponding phenylethanolamine. For example, in addition to forming norepinephrine from dopamine, DBH converts tyramine into octopamine and α-methyldopamine into α-methylnorepinephrine. Interestingly, many of the resultant structurally similar metabolites can replace norepinephrine at the noradrenergic nerve ending, acting as false neurotransmitters.

The regulation of DBH, like that of AADC, is less completely understood than that of TH. It appears that under conditions that increase the activity of locus coeruleus noradrenergic neurons and increase TH activity, DBH (rather than TH) becomes saturated and is the rate-limiting step in catecholamine synthesis. This results in the accumulation of dopamine and acidic dopamine metabolites in noradrenergic neurons when these neurons are firing faster, which results in the release of dopamine and its acidic metabolites from noradrenergic neurons.

Phenylethanolamine **N**-*methyltransferase.* Phenylethanolamine N-methyltransferase (PNMT) is present at high levels in the adrenal medulla, where it methylates norepinephrine to form epinephrine, the major adrenal catecholamine. A single PNMT gene with three exons has been cloned. The transcript is present in the adrenal medulla and in the brainstem, and thus PNMT is also formed in two nuclei of the brainstem as well as the adrenal gland.

PNMT requires *S*-adenosylmethionine as the methyl donor for methylation of the amine nitrogen of norepinephrine. The enzyme has modest substrate specificity and will transfer methyl groups to the nitrogen atom on a variety of β-hydroxylated amines. However, adrenal PNMT is distinct from the non-specific N-methyltransferases present in lung, which methylate indoleamines (such as serotonin).

The regulation of PNMT activity in the brain has not been extensively studied. In the adrenal gland, glucocorticoids regulate activity of the enzyme, and activity is increased in response to nerve growth factor.

Storage of Catecholamines and Their Enzymes

Vesicular storage. It has been known since the 1960s that much of the norepinephrine in sympathetic nerve endings and in adrenal chromaffin cells is present in highly specialized subcellular particles termed granules. Similarly, most of the norepinephrine and other catecholamines in the brain is stored in similar vesicles. These granules contain ATP in a catecholamine:ATP molar ratio of 4:1. The anionic phosphate groups of ATP are thought to form a salt link with norepinephrine, which exists as a cation at physiological pH and thereby binds the amines within the vesicles.

Vesicular storage of transmitters serves as a depot of the transmitter that can be released by appropriate physiological stimuli. Catecholamine transmitters are stored in small vesicles located near the synapse, where they are ready for fusion with the cellular membrane and subsequent exocytosis. Catecholamine storage in vesicles has two other key functions. First, the ability to sequester catecholamines in vesicles retards their diffusion out of the neuron. Second, vesicular storage offers protection from metabolic inactivation by intraneuronal enzymes or attack by toxins that have gained entry into the neuron.

The norepinephrine-synthesizing enzyme DBH differs from the other catecholamine-synthesizing enzymes by being present in the vesicles rather than the cytosol. This means that only after dopamine is accumulated into the vesicles by the vesicular monoamine transporter (VMAT) is it metabolized to norepinephrine. The vesicular storage of DBH has one other interesting consequence: DBH is actually released from cells when its product (norepinephrine) is released.

Vesicular monoamine transporters. The ability of vesicles to take up dopamine or other catecholamines depends on the presence of the VMAT (Weihe and Eiden, 2000). The VMAT is distinct from the neuronal membrane transporter (which is discussed in detail below) in terms of both substrate affinity and localization. Two vesicular monoamine transporter genes have been cloned. One is found in the adrenal medulla, in the adrenal chromaffin cells that synthesize and release monoamines. The other is present in catecholamine and serotonin neurons in the central nervous system. VMAT2, the isoform found in the brain, shows modest substrate specificity and transports catecholamines and indoleamines, as well as histamine, into vesicles. Both VMATs are Mg^{2+}-dependent and are inhibited by reserpine, a drug that disrupts vesicular storage of monoamines (Henry *et al.*, 1998).

Reserpine has been used in India for centuries as a folk medicine to treat hypertension and psychoses. The discovery that reserpine depletes vesicular stores of monoamines was critical to the understanding of the mechanisms through which reserpine alleviates psychotic symptoms, and it shed light on the means through which certain toxins can cause a Parkinson's disease-like syndrome. The use of reserpine in the treatment of hypertension and psychoses was reported in international journals in the early 1930s, but the therapeutic actions of reserpine were not widely appreciated in Western medicine until a generation later. At that time, Bernard Brodie and co-workers discovered that reserpine depleted brain stores of serotonin. Contemporaneously, it became known that the hallucinogen LSD (lysergic acid diethylamide) is structurally related to serotonin, which resulted in the proposal that the antipsychotic actions of reserpine were due to its ability to deplete brain stores of serotonin. However, it was soon realized that reserpine depletes both serotonin and catecholamines in the brain, and thus the antipsychotic effects of reserpine might be due to either serotonin or catecholamine depletion (or both).

To determine which transmitters were more important, Arvid Carlsson and colleagues (Carlsson, 1972) administered catecholamine and serotonin precursors to reserpine-treated rats to replenish monoamine levels; they then examined locomotor activity, which is severely depressed by reserpine treatment. Motor function was restored by administration of the dopamine precursor L-DOPA but not on treatment with the serotonin precursor 5-hydroxytryptophan. Subsequent neurochemical studies revealed that despite the improvement in motor behavior, L-DOPA treatment did not restore brain concentrations of norepinephrine and epinephrine. Soon thereafter dopamine was characterized as a neurotransmitter, and L-DOPA treatment of reserpinized animals was shown to increase brain concentrations of dopamine. These data were interpreted to suggest that the primary mechanism through which reserpine exerts antipsychotic effects is through its ability to disrupt dopamine neurotransmission. These and subsequent studies led to the hypothesis that dysfunction of central dopamine systems underlies schizophrenia. This hypothesis soon became the dominant view guiding schizophrenia research. Interestingly, recent data suggest that to treat most effectively the full spectrum of psychotic symptoms in schizophrenia, drugs that are antagonists at both dopamine and serotonin receptors may be superior to drugs that are antagonists only at dopamine receptors.

One critical aspect of vesicles is that they determine quantal release of transmitter. This is perhaps best shown in a study in which adenovirus mediated transfection of VMAT2 in cultures of ventral midbrain DA neurons, which resulted in overexpression of the vesicular transporter in small synaptic vesicles, and increased both quantal size and frequency of release (Pothos *et al.*, 2000). This observation is consistent with the recruitment of vesicles that do not normally release DA.

Cloning of the VMATs revealed a significant homology of these vesicular transporters to a group of bacterial antibiotic drug resistance transporters, suggesting a role for VMAT in detoxification. This indeed is the case. VMAT allows the vesicles to sequester toxins and thereby reduce toxicity. This is best exemplified by studies in VMAT2 knockout mice and heterozygous mice bearing one copy of VMAT2 (see Edwards, 1993). In such heterozygous mice there is an increase in the toxicity of DA neurons in response to administration of the parkinsonian toxin MPTP. The loss of VMAT means less sequestration of the toxin, which can then exert its toxic actions by targeting mitochondrial respiration.

One final function of VMAT appears to be the indirect role that the transporter plays in the generation of neuromelanin, black pigmented deposits that accumulate in midbrain DA neurons as a function of age and are responsible for the term *substantia nigra* (black stuff). Neuromelanin can be induced in rat dopamine neurons of the substantia nigra treated with the dopamine precursor L-DOPA. This effect of L-DOPA is abolished in cells in which VMAT2 is overexpressed.

Release of Catecholamines

Catecholamine release typically occurs by the same Ca^{2+}-dependent process (exocytosis) that has been described for other transmitters (see Chapter 8 for a more extended discussion). This is thought to occur at the synaptic cleft, where the presynaptic axon terminal is apposed to a specialized postsynaptic density, but also at synapses present at varicosities formed as the catecholaminergic axons course to their target region, although this view has been challenged. Such *en passant* varicosities appear like beads along an axon string.

Catecholamine release has also been observed to occur through at least two other mechanisms. First, catecholamines can be released by a reversal of the catecholamine (cell membrane) transporters. This occurs in response to certain drugs (e.g., amphetamine) and has been reported to occur following the application of excitatory amino acids as well. Second, dopamine (and perhaps other catecholamines) can be released from dendrites through a process that may not always involve conventional exocytosis, since some studies have found that dendritic release is not Ca^{2+}-dependent.

Regulation of catecholamine synthesis and release by autoreceptors. The enzymes that control the synthesis of catecholamines can be regulated, as noted earlier, at the transcriptional level and by posttranslational modifications that alter enzymatic activity. In addition, the synthesis of catecholamines such as dopamine and norepinephrine can be regulated by the interaction of the catecholamine released from the nerve terminal with specific dopamine or norepinephrine autoreceptors that are located on the nerve terminal. Similarly, the release of catecholamines is regulated by autoreceptors, as is the firing rate of catecholaminergic neurons.

Dopamine autoreceptors are perhaps the best characterized of the catecholamine autoreceptors. Autoreceptors exist on most parts of the neuron, including the soma, the dendrites, and nerve terminals. They can be defined functionally in relation to the events that they regulate. Thus, synthesis-, release-, and impulse-modulating dopamine autoreceptors have been described (see Cooper *et al.*, 1996). All three types of dopamine autoreceptors belong to the D2 family of dopamine receptors, which includes three cloned receptors (D_2, D_3, and D_4). Although it is clear that there are D_2 autoreceptors, there has been considerable controversy surrounding the presence of D_3 autoreceptors. However, even assuming that there are D_3 as well as D_2 autoreceptors, there are three functionally different types of dopaminergic autoreceptors, raising the possibility that the same receptor protein may couple to different autoreceptor roles through distinct transduction mechanisms.

Dopamine that is released from the neuron interacts with an autoreceptor and dampens further release of the transmitter. The ability of the released transmitter to dampen subsequent transmitter release can be conceptualized as a homeostatic mechanism. Release-modulating autoreceptors appear to be a common regulatory feature on catecholamine neurons and other neurons that use classic transmitters. Because intracellular levels of dopamine regulate tyrosine hydroxylase activity by binding the pterin cofactor, changes in the release of dopamine may also alter synthesis of transmitter.

Dopamine autoreceptors also directly regulate the synthesis of dopamine. Again, the released transmitter acts homoeostatically at the synthesis-modulating autoreceptor to control synthesis: dopamine agonists decrease synthesis, whereas dopamine antagonists increase synthesis of the transmitter. Interestingly, synthesis-modulating autoreceptors are not found on all dopamine neurons: some midbrain dopamine neurons that project to the prefrontal cortex appear to lack synthesis-modulating autoreceptors, as do the tuberoinfundibular dopamine neurons of the hypothalamus. Because release-modulating autoreceptors may indirectly regulate synthesis, the presence of synthesis-modulating autoreceptors may not be necessary in certain neurons.

Impulse-modulating autoreceptors are located on the soma and dendrites of dopamine neurons and regulate the firing rate of dopaminergic neurons. As noted earlier, because the release of dopamine can alter dopamine synthesis, impulse-modulating autoreceptors can also be expected to change dopamine synthesis. Thus, all three types of dopamine autoreceptors may regulate synthesis. This interdependence of regulatory processes over catecholamine neurons appears to be characteristic of monoamine neurons.

Although dopamine autoreceptors have been perhaps the most intensely studied autoreceptors, there are also norepinephrine autoreceptors that regulate release. However, the direct regulation of norepinephrine synthesis by synthesis-modulating autoreceptors on noradrenergic neurons is not well established. We do know, however, that there are two norepinephrine autoreceptors. One of these, an α_2 receptor, inhibits norepinephrine release, but a second norepinephrine β receptor actually facilitates release. Little is known about the role of autoreceptors in regulating epinephrine release in the CNS.

Inactivation of Catecholamine Neurotransmission

Continuous stimulation of neuronal receptors is not a desirable condition on two grounds. The first is that the nonpathological activity of neurons is not continuous, but fluctuates, with neurons sometimes firing and sometimes not; as such, continuous stimulation does not convey information concerning the activity of the presynaptic neuron accurately to the follower cell. This can be most easily understood with respect to receptors that form ion channels, in which continued action of a neurotransmitter would lead to inappropriate ion concentrations across the membrane and thus disrupt neurotransmission. The second reason is that continuous stimulation is typically pathological, and results in damage to and overt loss of postsynaptic neurons. However, the principle for all neurotransmitters is simply that continued activation of target cell receptors does not convey appropriate information about the dynamic state of the presynaptic neuron.

There are several different mechanisms for terminating the actions of a catecholamine. Perhaps the simplest is diffusion away from the receptor sites followed by dilution in extracellular fluid or plasma to subthreshold concentrations. More critical are active modes of terminating transmitter action, including the uptake of catecholamines by a neuronal membrane-associated transporter protein. In turn, this may be followed by uptake of the catecholamines into storage vesicles, from where they can be reused, or catabolism by monoamine oxidase (MAO). A third possibility is direct catabolism by catechol-*O*-methyltransferase (COMT).

Enzymatic inactivation of catecholamines. Enzymatic inactivation, once thought to be the predominant mechanism through which catecholamines are inactivated in the CNS, plays a secondary role in the termination of action of catecholamines and, indeed, of most transmitters. Nonetheless, enzymatic inactivation remains important for two reasons. The first is that certain drug treatments for neuropsychiatric disorders are based on manipulation of the key enzymes that act on catecholamines (see Box 9.2). Second, enzymatic inactivation is the major mode of terminating the action of circulating catecholamines in the bloodstream.

Two major enzymes take part in catecholamine catabolism: MAO and COMT. Both enzymes can act independently or on the products of the other, leading to catecholamine metabolites that are deaminated, *O*-methylated, or both. COMT is a relatively nonspecific enzyme that transfers methyl groups from the donor *S*-adenosylmethionine (SAM) to the *m*-hydroxy group of catechols. COMT is found in both peripheral tissues and central nervous system, and is the major means of inactivating circulating catecholamines that are released from the adrenal gland.

Two forms of MAO have been identified on the basis of substrate specificities and selective enzyme inhibitors. MAO$_A$ has a high affinity for norepinephrine and serotonin and is selectively inhibited by clorgyline. In contrast, MAO$_B$ has a higher affinity for *o*-phenylethylamines and is selectively inhibited by the monoamine oxidase inhibitor (MAOI) deprenyl. Both MAO$_A$ and MAO$_B$ are associated with the outer mitochondrial membrane. The MAOs oxidatively deaminate catecholamines and their *O*-methylated derivatives to form inactive and unstable aldehyde derivatives. These aldehydes can be further catabolized by dehydrogenases and reductases to form corresponding acids and alcohols.

The enzymatic inactivation of catecholamines appears to be the primary mode of terminating the action of circulating catecholamines, while in the CNS reuptake mechanisms are thought to be more important. Several drugs that target the enzymatic inactivation of catecholamines have proved useful in the treatment of diverse neuropsychiatric disorders, ranging from depression to Parkinson's disease (see Box 9.2).

Neuronal catecholamine transporters. The reuptake of a transmitter released by a neuron is the major mode of inactivation of the released transmitter in the brain. In addition, the accumulation of the transmitter also allows intracellular enzymes that degrade the transmitter to act, thus bolstering the actions of extracellular enzymes.

Neuronal reuptake of catecholamines, and indeed of all transmitters for which a reuptake process has been identified, has several characteristics (Clark and Amara, 1993). The reuptake process is energy-dependent and saturable, and depends on Na$^+$ co-transport as well as requiring extracellular Cl$^-$. Because reuptake depends on coupling to the Na$^+$ gradient across the neuronal membrane, toxins that inhibit Na$^+$, K$^+$-

ATPase inhibit reuptake. However, under certain conditions, the coupling of transporter function to Na$^+$ flow may lead to local changes in the membrane Na$^+$ gradient and thereby paradoxically extrude ("release") the transmitter.

The membrane catecholamine transporters are not Mg^{2+}-dependent and are not inhibited by reserpine. These characteristics distinguish the neuronal membrane transporters from the monoamine transporters localized to neuronal vesicles (see below). The catecholamine transporters are localized to neurons; although there appears to be a reuptake process that accumulates catecholamines in glial cells, the process is not a high-affinity one and the functional significance of the glial reuptake of catecholamines remains unknown.

Two distinct mammalian catecholamine transporter proteins, the dopamine transporter (DAT) and norepinephrine transporter (NET), have been cloned and characterized pharmacologically. The two transporters share significant sequence homology and are members of a class of transporter proteins (including serotonin and amino acid transmitter transporters) with 12 transmembrane domains. Neither transporter is very specific, with each accumulating both dopamine and norepinephrine. In fact, the NET has a higher affinity for dopamine than for norepinephrine. A specific transporter for epinephrine-containing neurons has been identified in the frog but not in mammalian species.

The regional distribution of DAT and NET largely follows the expected localization to distinct dopamine and norepinephrine neurons, respectively. However, DAT does not appear to be expressed in all dopamine cells. The tuberoinfundibular dopamine neurons, which are hypothalamic cells that release dopamine into the pituitary portal blood system, lack demonstrable DAT mRNA and protein. Because dopamine released from tuberoinfundibular neurons is carried away in the vasculature, the existence of a transporter protein on these dopamine neurons would be superfluous.

Although studies defining the cellular localization of the two catecholamine transporters did not uncover any major surprises, immunohistochemical studies of the subcellular localization of the transporters did yield an unexpected finding. The use of antibodies generated against DAT revealed that the transporter is typically expressed outside of the synapse, in the extrasynaptic region of the axon terminal. This finding suggests that the transporter may be used to inactivate (accumulate) dopamine that has escaped from the synaptic cleft and, thus, that diffusion is the initial process by which dopamine is

BOX 9.2

MAO AND COMT INHIBITORS IN THE TREATMENT OF NEUROPSYCHIATRIC DISORDERS

Depression

One hypothesis concerning the pathophysiology of depression posits a decrease in noradrenergic tone in the brain. MAO_A inhibitors such as tranylcypromine effectively increase norepinephrine levels (as well as dopamine and serotonin concentrations) and were once a mainstay of the treatment of depression. More recently the use of MAO inhibitors in depression has been largely supplanted by the introduction of drugs that increase extracellular norepinephrine levels by blocking reuptake of the transmitter (tricyclic antidepressants) and other agents to increase serotonin or dopamine levels by blocking SERT or DAT [*fluoxetine* (Prozac) and *bupropion* (Welbutrin), respectively].

The treatment of depression with MAO_A inhibitors, although still useful for certain patients who do not respond to other antidepressants, is marred by a large number of side effects. Among the most serious side effects is hypertensive crisis. Patients who are treated with MAO_A inhibitors and eat foods that contain large amounts of tyramine (such as aged cheeses) cannot metabolize the ingested tyramine. Because tyramine releases catecholamines from nerve endings and relatively small amounts of tyramine increase blood pressure significantly, a marked increase in blood pressure and a high risk for stroke may develop.

Parkinson's Disease

Deprenyl, a specific inhibitor of MAO_B, has been used as an initial treatment for Parkinson's disease (PD). The use of deprenyl in the treatment of PD and the rationale for its use were based on data from studies of a neurotoxin, 1-methyl-4-phenyl-1,2,3,6-tetrahydropyridine (MPTP). The systemic administration of MPTP to humans and other primates results in a relatively specific degeneration of midbrain dopamine neurons and a marked parkinsonian syndrome. MPTP toxicity was first noted in a group of opiate addicts. In an attempt to synthesize a designer drug that was a meperidine (Demerol) derivative, the structurally related MPTP was inadvertently produced. Addicts who injected this drug developed a severe parkinsonian syndrome. Subsequent animal studies showed that MPTP itself is not toxic, but that the active metabolite of MPTP, MPP^+, is highly toxic. The formation of MPP^+ from MPTP is catalyzed by MAO_B, and animal studies soon revealed that treatment with MAO inhibition by deprenyl could prevent MPTP toxicity.

The realization that MPTP administration rather faithfully reproduces the cardinal signs and symptoms of Parkinson's disease reawakened interest in environmental toxins as a cause of PD. This interest led to the idea that treatment with deprenyl might be useful in slowing the progression of PD, putatively caused by an environmental toxin. The first evaluation of clinical trials of newly diagnosed PD patients indicated that daily administration of deprenyl increased the amount of time required before patients needed other drugs for the relief of symptoms; however, when the MAO_B inhibitor was withdrawn, patients treated with deprenyl regressed and appeared no better than untreated subjects. It now appears that the actions of deprenyl are due at least in part to the symptomatic improvement that results from increasing dopamine levels by inhibiting degradation of the transmitter rather than to slowing of the progression of PD. Moreover, low levels of methamphetamine are generated by the metabolism of deprenyl; because methamphetamine potently releases dopamine from nerve terminals, this would result in a symptomatic improvement.

Catechol *O*-methyltransferase, which together with MAO is responsible for the enzymatic degradation of cacholamines, is also a target in the treatment of PD. Two inhibitors of COMT, one that is active peripherally and another that is active both peripherally and centrally, are used to prevent the enzymatic inactivation of L-DOPA. By inhibiting COMT these drugs prolong the therapeutic action of L-DOPA and smooth out the characteristic fluctuations in therapeutic response to DOPA.

Schizophrenia

Changes in catecholamine function have been a subject of intense scrutiny in schizophrenia, with much attention focusing on a loss of dopaminergic tone in the prefrontal cortex. One allelic variant of the COMT gene substitutes a single methionine for a valine, and results in a much reduced activity of the enzyme. Recent data have examined COMT alleles for full versus low COMT activity in normal subjects and schizophrenics. Individuals bearing the allele that confers lower COMT activity have improved performance on cognitive tasks that involve the prefrontal cortex; the performance of schizophrenic persons on these tasks is impaired. There is a significant increase in transmission of the COMT allele conferring high enyme activity to schizophrenic subjects, and it has been proposed that COMT activity may confer increased risk of developing schizophrenia.

removed from the synapse. This observation is consistent with recent studies indicating that perisynaptic concentrations of dopamine can reach 1.0 mM or more, a value roughly comparable to the affinity of the cloned DAT for DA. Receptors for dopamine and many other transmitters are also found extrasynaptically (indeed, along the length of axons); this observation, coupled with the presence of catecholamine transporters to extrasynaptic regions, suggests that extrasynaptic ("paracrine" or volume) neurotransmission (Zoli et al., 1999) may be of considerable importance for catecholaminergic signaling.

How neurotransmitter transporter proteins are regulated has only recently been studied. Chronic treatment with inhibitors of catecholamine reuptake alters the number of transporter sites, but the precise regulatory mechanisms remain unclear. There are phosphorylation sites on DAT and NET, and thus changes in neurotransmitter release may alter function through interaction with autoreceptors and subsequent activation of serine–threonine kinases.

The DAT knockout has been particularly useful in clarifying the role of DAT in dopaminergic neurons and, by extension, the role of other monoamine transmitter transporters. The constitutive loss of the DA transporter results in a remarkably wide array of deficits in dopaminergic function, ranging from an increase in extracellular DA levels and delayed clearance of released DA to a striking decrease in tissue concentrations of DA in the face of increased DA synthesis (Gainetdinov et al., 1998). In addition, there is a striking loss of autoreceptor-mediated tone, including deficits in release-, synthesis-, and impulse-modulating autoreceptor function (Jones et al., 1999). The alterations in DA knockout mice have been suggested to reflect a disinhibition of tyrosine hydroxylase due to a lack of intraneuronal DA to provide feedback inhibition of the enzyme and a markedly increased rate of DA turnover, such that synthesis and release of the neurons are accelerated. Interestingly, the widespread characteristic deficits in the DAT knockout mice are similar to the normal function of the DA neurons that innervate the prefrontal cortex (see Roth and Elsworth, 1995).

Psychostimulants, such as cocaine and amphetamine, exert their effects on arousal by increasing extracellular levels of catecholamines; cocaine also increases extracellular serotonin levels. The mechanism through which psychostimulants increase catecholamine levels is by blocking DAT and NET. In particular, cocaine shows a very high affinity for the dopamine transporter; amphetamine is a less potent inhibitor of reuptake but also induces release of catecholamines from the cytoplasm. Studies in mice with targeted null mutations of the DAT have surprisingly revealed that these mice will still self-administer cocaine. However, mice with double knockouts of both the DA and serotonin transporters fail to self-administer cocaine (Sora et al., 2000). Cocaine administration results in sharp increases in both extracellular dopamine and serotonin, and the failure of DAT–serotonin transporter (SERT) knockout mice to sustain cocaine self-administration suggests that both are critical for the effects of cocaine. Studies in animals with transient suppression of DAT and SERT expression are needed to ensure that the SERT is not compensating developmentally for the loss of DAT.

The NET is also a target of clinically important drugs. The tricyclic antidepressants potently inhibit norepinephrine reuptake, with significantly weaker effects on the dopamine and serotonin transporters. In addition, new agents that inhibit NET selectively without significantly decreasing serotonin or dopamine reuptake are now being used in the treatment of depression in Europe. Mice with targeted null mutations of NET act like antidepressant-treated wild-type mice; interestingly, NET knockout mice are hyperresponsive to psychostimulant-elicited locomotor stimulation (Xu et al., 2000).

Serotonin

Scientists have been aware of a substance found in the blood that induces powerful contractions of smooth muscle organs for more than 150 years. However, more than a century passed until Page and his collaborators succeeded in isolating the compound (which they proposed to be a possible cause of high blood pressure) from platelets. At the same time, Italian researchers were characterizing a substance present in high concentrations in intestinal mucosa that caused contractions of gastrointestinal smooth muscle. The material isolated from blood platelets was given the name "serotonin," and the substance isolated from the intestinal tract was called "enteramine." Subsequently, both materials were purified and crystallized, and shown to be the identical substance, 5-hydroxytryptamine (5-HT), usually referred to as serotonin. The laboratory synthesis of serotonin soon allowed direct comparison of serotonin with the purified compound isolated from platelets, which conclusively demonstrated that serotonin possessed all the biological features of the natural substance.

Serotonin is found in neurons as well as several different peripheral cells. The latter include platelets, mast cells, and enterochromaffin cells. In fact, the brain accounts for only about 1% of body stores of serotonin.

Although the purification and identification of serotonin were based on studies of blood pressure regulation, the possible relation of serotonin to psychiatric disorders propelled research on the central effects of serotonin. The observation that the indole structure of serotonin was similar to that of the psychedelic agent LSD and a number of other psychotropic compounds soon led to theories linking abnormalities of serotonin function to various psychiatric disorders, including schizophrenia and depression. This linkage remains a major focus of research on central serotonergic systems.

The basic principles of the biochemical neuropharmacology of synaptic transmission as revealed by studies of catecholamines are also applicable to neurons that use serotonin as a transmitter. We therefore outline the nature of chemical transmission in serotonergic neurons, focusing on differences that are unique to serotonergic neurons.

Synthesis of Serotonin

The basic outline of serotonin biosynthesis is very similar to that of catecholamine transmitters. A peripheral amino acid gains entry into the central nervous system and is metabolized in serotonergic neurons via a series of enzymatic steps that culminate in the synthesis of serotonin.

Once tryptophan enters the serotonergic neuron it is hydroxylated by tryptophan hydroxylase, the rate-limiting step in serotonin synthesis (see Fig. 9.5). The resultant serotonin precursor, 5-hydroxytryptophan (5-HTP), is subsequently decarboxylated by aromatic amino acid decarboxylase. Thus, only two critical enzymes (tryptophan hydroxylase and AADC) are involved in the synthesis of serotonin (Cooper *et al.*, 1996; Frazer and Hensler, 1994).

Serotonin is the final product of this synthetic pathway, with no subsequent enzymes generating other transmitters in other parts of the brain, with one exception. In the pineal gland, serotonin is metabolized further to the hormone melatonin. Although serotonin is not metabolized to form other transmitters, there are enzymatic steps from tryptophan metabolism, called the kynurenic acid shunt, that result in the formation of several compounds that may be involved in neurotoxicity.

Tryptophan hydroxylase. The rate-limiting step in serotonin synthesis is enzymatic, requiring tryptophan hydroxylase. However, the availability of the precursor amino acid tryptophan plays an important role in regulating the synthesis of serotonin; this is in sharp contrast to catecholamine synthesis, which under normal conditions is not regulated by precursor availability.

Because serotonin cannot cross the blood–brain barrier, brain cells must synthesize the amine. The precursor to serotonin is the amino acid tryptophan, which is present in high levels in plasma. Changes in dietary sources of tryptophan can substantially alter brain levels of serotonin. An active uptake process facilitates entry of tryptophan into the brain. However, other large neutral aromatic amino acids compete for this transport process. Accordingly, brain levels of tryptophan are determined by plasma concentrations of competing neutral amino acids as well as the plasma levels of tryptophan itself.

The gene encoding tryptophan hydroxylase has been cloned and sequenced. Although some differences in the biochemical properties of tryptophan hydroxylase isolated from brain and peripheral tissues have been reported, these differences appear to be due to posttranslational modifications of the protein rather than to different mRNAs.

Neuronal systems need to be able to adapt to either short- or long-term demands on activity. In serotonergic neurons, the synthesis of serotonin from tryptophan is increased in a frequency-dependent manner in response to electrical stimulation of serotonergic cells. Tryptophan hydroxylase requires both molecular oxygen and a reduced pterin cofactor. The product of tryptophan hydroxylase, 5-HTP, is very rapidly converted to serotonin [5-hydroxytryptamine (5-HT)].

L-Aromatic Amino Acid Decarboxylase

Aromatic amino acid decarboxylase metabolizes the serotonin precursor 5-HTP to the transmitter serotonin. This enzyme is the same as that found in catecholaminergic neurons. Just as in catecholamine cells, in which the precursor L-DOPA is almost instantaneously converted into dopamine by AADC, the precursor 5-HTP is so rapidly decarboxylated in serotonergic cells that levels of 5-HTP under basal conditions are almost nil. Thus, because AADC is not saturated with 5-HTP under physiological conditions, it is possible to increase the content of serotonin in brain not only by increasing the dietary intake of tryptophan but also by administering 5-HTP, which readily enters the brain.

Alternative Tryptophan Metabolic Pathways. Although serotonin is generally thought of as the final product of tryptophan synthesis, in the brain and periphery serotonin can be further metabolized to yield important active products. In the pineal gland, serotonin is metabolized to 5-methoxy-*N*-acetyltryptamine (melatonin), a hormone that is thought to play an important role in both sexual behavior and sleep. The production of melatonin from serotonin requires

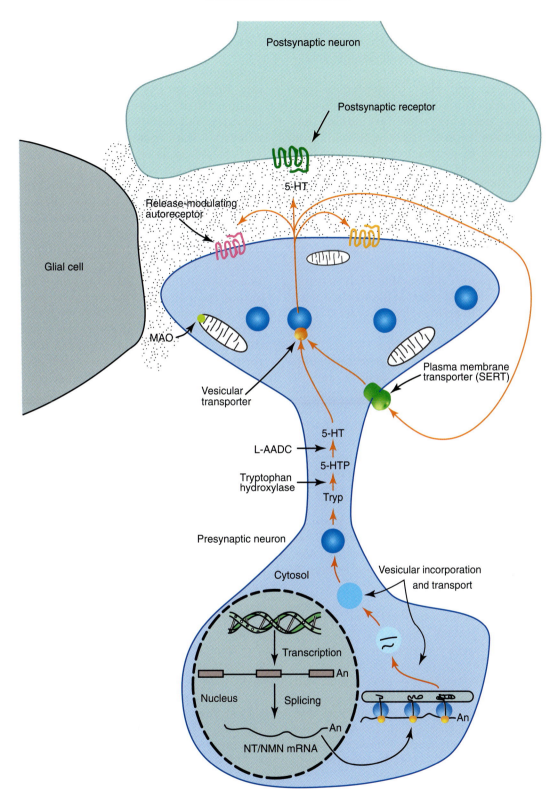

FIGURE 9.5 Depiction of a serotonergic neuron. Tryptophan (Tryp) in the neuron is sequentially metabolized by tryptophan hydroxylase and L-AADC to yield serotonin (5-HT). The serotonin is accumulated by the vesicular monoamine transporter. When released, the serotonin can interact with both postsynaptic receptors and presynaptic autoreceptors. 5-HT is taken up by the high-affinity serotonin transporter (SERT), and once inside the neuron it can be reaccumulated by vesicular transporter or metabolically inactivated by MAO and other enzymes.

two enzymatic steps: *N*-acetylation of serotonin to form *N*-acetylserotonin, which is rapidly methylated by 5-hydroxyindole-*O*-methyltransferase to melatonin.

In peripheral tissues, most tryptophan is not metabolized to serotonin but is instead metabolized in the kynurenine pathway. Recent data indicate that the kynurenine shunt is present in the CNS and leads to the accumulation of neuroactive substances that may be of clinical importance in cases of trauma and stroke (Stone, 1993). The two major tryptophan metabolites that are generated by the kynurenine shunt are quinolinic acid and kynurenic acid. Quinolinic acid is a potent agonist at *N*-methyl-D-aspartate (NMDA) receptors and causes neurotoxicity and convulsions. In contrast, kynurenine is an antagonist at NMDA receptors. Considerable effort is being directed to determining the role that these tryptophan metabolites may play in neurological disorders.

Storage and Release of Serotonin

Vesicular accumulation and storage of serotonin. Serotonin is stored primarily in vesicles and is released by an exocytotic mechanism. The vesicular amine transporter that accumulates serotonin is the same as that in catecholamine neurons (VMAT2), and, in most respects, vesicles that store serotonin resemble those that contain catecholamines. It has been suggested that serotonin-containing (but not catecholamine-containing) vesicles may express a specific high-affinity serotonin-binding protein.

Since catecholamines and serotonin share a common vesicular transporter, it is not surprising that reserpine, which depletes vesicular stores of catecholamines, also depletes serotonin from serotonin neurons.

Regulation of serotonin synthesis and release. There are some important regulatory differences between catecholaminergic and serotonergic neurons. As noted earlier, serotonin neurons are sensitive to changes in plasma levels of the precursor amino acid tryptophan, and thus dietary changes can regulate serotonin levels in brain. In addition, it appears that increases in intracellular serotonin levels do not significantly alter serotonin synthesis *in vivo*; in contrast, in catecholaminergic neurons transmitter synthesis is influenced by end-product inhibition.

Short-term requirements for increases in serotonin synthesis appear to be accomplished by a Ca^{2+}-dependent phosphorylation of tryptophan hydroxylase, which changes its kinetic properties without necessitating the synthesis of more enzyme. In contrast, serotonergic neurons respond to the need for long-term increases in serotonin availability by the induction of (new) tryptophan hydroxylase protein.

Serotonin autoreceptors regulate serotonin release and synthesis. As is the case in catecholamine neurons, there are functionally dissociable somatodendritic and terminal autoreceptors on serotonin neurons. The release- and impulse-modulating autoreceptor in serotonin neurons is a 5-HT$_{1A}$ receptor. This receptor is found on presynaptic (serotonin-containing) neurons and as a postsynaptic receptor on nonserotonergic cells. In addition, there are data indicating the 5-HT$_{1D}$ receptor is a nerve terminal release-modulating autoreceptor in humans. It has been more difficult to clearly delineate putative autoreceptor roles for 5-HT$_{1D}$ and 5-HT$_{1B}$ receptors in nonhuman brain. The affinity of many ligands is higher for the autoreceptor than for the postsynaptic site, thereby allowing some degree of selectivity in dampening serotonin function by targeting the autoreceptor. The difficulty in definitively ascribing an autoreceptor role for the 5-HT$_{1D}$ and 5-HT$_{1B}$ receptors is due in large part to the lack of drugs that act selectively at these two sites; available drugs show some degree of preference for one site over another *in vitro*, but when examined *in vivo* the difference in affinity of the various agents is smaller.

Inactivation of Released Serotonin

As in the case of the catecholamine neurotransmitters, reuptake serves as a major means of terminating the action of serotonin. Released serotonin is taken up by a plasma membrane carrier, the serotonin transporter. In addition, the same enzymatic inactivation that is operative within catecholamine neurons is functional in serotonergic neurons.

Reuptake. Serotonin that is released into the synapse is inactivated primarily by the reuptake of the transmitter by a plasma membrane serotonin transporter. The serotonin transporter (SERT) has been cloned and sequenced and belongs to the same family of 12-transmembrane-domain transporters as the catecholamine transporters. SERT has in common with other transporter family members an absolute requirement for Na^+ co-transport.

SERT is also an important clinical target for therapeutic drugs. Just as the norepinephrine transporter is the target of tricyclic antidepressant drugs, SERT is the target of the new class of antidepressant drugs termed selective serotonin reuptake inhibitors (SSRIs), which includes such drugs as fluoxetine (Prozac). The ability of antidepressant drugs to alter monoamine inactivation by disrupting serotonin and norepinephrine transporters or by disrupting enzymatic inactivation of the monoamines has led to the dominant theories of the pathogenesis of depression, which suggest a critical modulatory role for norepinephrine and serotonin (Heninger *et al.*, 1996).

As noted above, cocaine and other psychostimulants block the dopamine transporter and thereby sharply increase extracellular dopamine levels. However, cocaine also increases extracellular serotonin levels. Interestingly, even though the dopamine transporter is a major target of cocaine, DAT knockout mice continue to self-administer cocaine, as do SERT knockout mice. However, in mice bearing double DAT–SERT knockouts cocaine self-administration is reduced, suggesting that both transporters must be targeted for the rewarding effects of psychostimulants to be manifested.

Enzymatic degradation. The primary catabolic pathway for serotonin is oxidative deamination by the enzyme monoamine oxidase. The product of this reaction, 5-hydroxyindole acid aldehyde, can be further oxidized to 5-hydroxyindoleacetic acid (5-HIAA), the primary serotonin metabolite, or can be reduced to 5-hydroxytryptophol. In the brain and cerebrospinal fluid 5-HIAA is the primary metabolite of serotonin. Monoamine oxidase inhibitors increase serotonin levels and have been used extensively as antidepressants. MAOIs are effective antidepressants, but have the potential to produce serious side effects. The introduction of tricyclic antidepressants, which inhibit both the norepinephrine and serotonin transporters, and drugs such as fluoxetine, a serotonin selective reuptake blocker, has largely supplanted the use of monoamine oxidase inhibitors as antidepressant medications, although some patients who fail to respond to the newer drugs respond well to the monoamine oxidase inhibitors.

γ-Aminobutyric Acid: The Major Inhibitory Neurotransmitter

A number of amino acids fulfill most of the criteria for neurotransmitters. The three best studied of these amino acid transmitters are GABA, the major inhibitory transmitter in brain; glutamate, which is the major excitatory transmitter in brain; and glycine, another inhibitory amino acid (Paul, 1995). The broad principles outlined in the discussion of catecholamine neurotransmitters are also applicable to the amino acid transmitters, although certain aspects of the synthesis of amino acid transmitters are less completely understood compared with the catecholamines.

A major difference between the biogenic amines transmitters and the amino acid transmitters is that the latter are derived from intermediary glucose metabolism. This dual role for the amino acid transmitters means that there must be mechanisms to segregate the transmitter and general metabolic pools of the amino acid transmitters (see Delorcy and Olsen, 1994). A second difference between amino acid and biogenic amine transmitters is that amino acid transmitters that are released from neurons are readily taken up by glial cells as well as neurons. We review GABA as a prototypic amino acid transmitter, focusing on differences between the catecholamine transmitters and GABA.

GABA was discovered in 1950 by Eugene Roberts, whose subsequent study (Roberts, 1986) revealed that GABA has a neurotransmitter role. GABA is ubiquitous in the CNS, as might be expected for a transmitter derived from the metabolism of glucose. Although the presence of GABA as a transmitter in neurons is widespread, it nonetheless has a distinct distribution. Although it was originally thought that (with a few exceptions) GABA was a neurotransmitter in local circuit interneurons but not in projection neurons, it has become apparent that there are many examples of GABAergic projection neurons.

GABA Biosynthesis

Several aspects of the synthesis of GABA differ from that of the monoamines (see Fig. 9.6). These differences are due to precursors of GABA being part of cellular intermediary metabolism rather than dedicated solely to a neurotransmitter synthetic pool.

The GABA shunt and GABA transaminase. GABA is ultimately derived from glucose metabolism. α-Ketoglutarate formed by the Krebs (tricarboxylic acid) cycle is transaminated to the amino acid glutamate by the enzyme GABA α-oxoglutarate transaminase (GABA-T). In those cells in which GABA is used as a transmitter, the presence of the enzyme glutamic acid decarboxylase (GAD) permits the formation of GABA from glutamate derived from α-ketoglutarate (Paul, 1995).

One unusual feature of the GABA synthetic pathway is that intraneuronal GABA is inactivated by the actions of GABA-T, which appears to be associated with mitochondria (Fig. 9.6). Thus, GABA-T is both a key synthetic enzyme and a degradative enzyme. GABA-T metabolizes GABA to succinic semialdehyde, but only if α-ketoglutarate is present to receive the amino group that is removed from GABA. This unusual GABA shunt serves to maintain supplies of GABA.

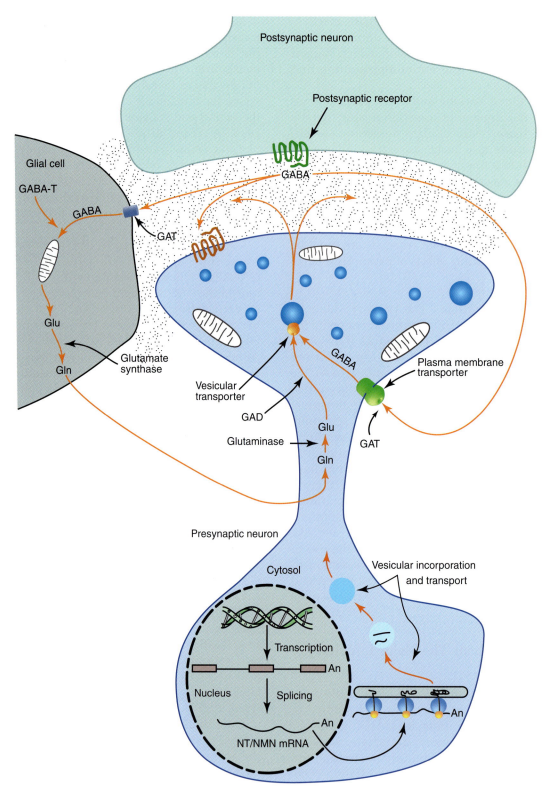

FIGURE 9.6   Schematic depiction of the life cycle of a GABAergic neuron. α-Ketoglutarate formed in the Krebs cycle is transaminated to glutamate (Glu) by GABA transaminase (GABA-T). The transmitter GABA is formed from the Glu by glutamic acid decarboxylase (GAD). GABA that is released is taken by high-affinity GABA transporters (GAT) present on neurons and glia. Gln, glutamine.

Glutamic acid decarboxylase. The critical biosynthetic enzyme for the neurotransmitter GABA is glutamic acid decarboxylase (GAD). GAD is localized exclusively in the central nervous system to neurons that use GABA as a transmitter.

There are two isoforms of GAD, which are encoded by two distinct genes (Erlander and Tobin, 1991). These two isoforms, designated GAD65 and GAD67 in accord with their molecular weights, exhibit somewhat different intracellular distributions, suggesting that the two GAD forms may be regulated in different ways. This appears to be the case. GAD requires a pyridoxal phosphate cofactor for activity. GAD65 and GAD67 differ significantly in their affinity for the pyridoxal cofactor: GAD65 shows a relatively high affinity for the cofactor, whereas the larger GAD isoform does not. The affinity of GAD65 for the cofactor results in the ability of GAD65 enzyme activity to be efficiently and quickly regulated. In contrast, the activity of GAD67 is determined through induction of new enzyme protein rather than through posttranslational mechanisms.

A major question concerning amino acid transmitters is how the transmitter pools are kept distinct from the general metabolic pools in which the amino acids serve. GAD is necessary for synthesis of the transmitter GABA, and the presence of the GAD mRNAs or proteins are markers of GABAergic neurons. GAD is a cytosolic enzyme, but GABA-T, which converts α-ketoglutarate into the GAD substrate glutamate, is present in mitochondria. Thus, the metabolic pool is present in the mitochondria, but glutamate destined for the transmitter pool must be exported from the mitochondria to the cytosolic compartment. This export process is poorly understood.

Glutamate is not only a precursor to the formation of GABA, but is also the major excitatory neurotransmitter. GAD is not present in neurons in which glutamate functions as a transmitter, and thus glutamatergic neurons do not use GABA as a transmitter. What prevents GABA neurons from using the precursor glutamate as a transmitter is not well understood but may require two different biosynthetic enzymes for glutamate as a transmitter and as a metabolic intermediary and the necessity of a vesicular transporter for glutamatergic neurons (Takamori *et al.*, 2000). A specific form of glutaminase [a phosphate-activated glutaminase (PAG)] has been proposed to be responsible for the synthesis of the transmitter pool of glutamate. PAG is localized to certain vesicles. Because both GABA and glutamate cause very rapid changes in postsynaptic neurons, one depolarizing neurons and the other hyperpolarizing them, it is not surprising (and probably fortunate) that the two amino acid transmitter pools are not generally co-localized. However, in the rat olfactory bulb and the chicken retina (which is a neural tissue), anatomical studies have suggested that there are a few isolated neurons in which GABA and glutamate are co-localized (Quaglino *et al.*, 1999). The functional significance of such an arrangement is not clear.

Storage and Release of GABA

Vesicular inhibitory amino acid transporter. A vesicular transporter in GABAergic cells accumulates GABA. The transporter was cloned on the basis of homology to *unc-47* in *Caenorhabditis elegans* (McIntire *et al.*, 1997), a strategy of moving from the worm to mammalian species that has proven to be a very useful strategy for identifying a variety of mammalian transmitter-related genes. The vesicular GABA transporter differs from the two VMATs by belonging to a different class, having 10 rather than 12 transmembrane domains.

The vesicular GABA transporter shares with the VMATs, however, a lack of substrate specificity, and will transport the inhibitory transmitter glycine as well as GABA. Consistent with this pharmacology, the vesicular GABA transporter has been found in glycine as well as GABA-containing neurons. Accordingly, it has been suggested that the transporter can be more accurately designated as a vesicular inhibitor amino acid transporter (Weihe and Eiden, 2000). Interestingly, there are some rare GABA neurons that lack the transporter, raising the specter of another (related) transporter in these neurons or, alternatively, some unique functional attribute of these cells.

Regulation of GABA release by autoreceptors. The major postsynaptic GABA receptor is the GABA$_A$ receptor, which contains the chloride ion channel (see Chapters 11 and 16). This multimeric receptor complex is formed by a number of different subunit proteins. Pharmacological studies indicate that autoreceptor-mediated regulation of GABA neurons takes place predominantly through GABA$_B$ receptors located on GABAergic nerve terminals. Immunohistochemical studies have revealed that both GABA$_B$ and GABA$_A$ receptors are present on postsynaptic non-GABAergic neurons. It is possible that these GABA$_A$ postsynaptic receptors respond to GABA released from a neuron that is presynaptic to another GABA neuron expressing the GABA$_A$ site. However, because an anatomical arrangement of one GABA neuron terminating on another GABA cell would have the same functional consequence as an autoreceptor (decreasing subsequent transmitter release), it is therefore difficult to distinguish between true autoreceptors and heteroreceptors.

Inactivation of Released GABA

Uptake of several transmitters by glial cells as well as neurons has been reported. The dual glial–neuronal reuptake is common in neurons using amino acid transmitters, probably because amino acids can play dual roles as both transmitters and metabolic intermediaries. However, the ability of glia to avidly accumulate GABA and other amino acids distinguishes amino acid transmitters from other classic transmitters.

GABA transporter proteins. Reuptake is the primary mode of inactivation of GABA that is released from neurons. At least three specific GABA transporter (GAT) proteins are expressed in the CNS, providing a diverse means of regulating GABA neurons (see Cherubini and Conti, 2001). In addition, a betaine transporter that accumulates GABA has been cloned. Two types of GABA transporters were long known as being neuronal and glial, and were defined on the basis of pharmacological specificity. However, the cloning of GABA transporters, which belong to the same family of transporter genes that includes the catecholamine transporters, revealed an unexpected finding. *In situ* hybridization and immunohistochemical studies revealed that one of the GATs found in brain, which on pharmacological grounds was defined as a "glial" transporter, is present in both neurons and glia. Moreover, the other GATs appear to be expressed in both neurons and glia.

The presence of multiple transporter proteins for the same transmitter, all localized to neurons, differs from the situation for catecholamine transmitters, in which a single membrane-associated transporter protein with relatively poor substrate specificity is found in a neuron. An obvious question arises: "Why are there multiple transporters for GABA?" GATs are expressed in both GABAergic neurons and non-GABAergic cells (presumably cells that receive a GABA innervation). However, it is not clear if multiple GATs are found in the same cell, and the precise intracellular localization of the transporter proteins is not yet known. It is possible that different transporters are targeted differently in the cell. For example, one might be present in dendrites and another expressed in axons, with corresponding different functional requirements. Another possibility is that the GATs that have been cloned may serve as cotransporters for other amino acids. For example, transporters for β-alanine and taurine have not been cloned, but these amino acids are accumulated by

GATs. Finally, it is possible that one or more of these transporters frequently works in the outward direction, serving as a paradoxical mechanism for the release of GABA.

Enzymatic inactivation of GABA. GABA-T is both a synthetic and a degradative enzyme, with both enzymatic functions acting to conserve the transmitter pool of GABA. GABA-T is a particulate enzyme that is present in high concentration in GABAergic neurons. GABA-T is found in non-GABAergic as well as GABA-containing neurons and is present in a number of peripheral tissues. Electron microscope data suggest that GABA-T is associated with mitochondria. However, pharmacological studies of various subcellular fraction preparations suggest that the activity of GABA-T associated with synaptosomes that contain mitochondria is less than that seen in synaptosomal membrane fractions (without mitochondria), suggesting that GABA may be metabolized either extraneuronally or in postsynaptic neurons.

Glutamate and Aspartate: The Excitatory Amino Acid Transmitters

Excitatory amino acid transmitters account for most of the fast synaptic transmission that occurs in the mammalian CNS. Glutamate and aspartate are the major excitatory amino acid neurotransmitters, but several related amino acids (including *N*-acetylaspartylglutamate and homocysteic acid) also appear to have neurotransmitter roles. The excitatory amino acids, like the inhibitory amino acid transmitter GABA, participate in intermediary metabolism as well as cellular communication; the problem of dissociating neurotransmitter from metabolic roles therefore holds for excitatory amino acids. The intertwining of the transmitter roles of amino acids and intermediary metabolism makes it difficult to fulfill all of the criteria that would give amino acids fully legitimate status as neurotransmitters. Despite these issues, it is now widely accepted that glutamate and aspartate function as excitatory transmitters in the CNS. We briefly consider glutamate biosynthesis and regulation, focusing on the differences between excitatory and inhibitory amino acid transmitters. Many of the general principles addressed in the section on GABA are applicable to glutamate and therefore are not discussed in detail.

Biosynthesis of Glutamate

Although glutamic acid is present in very high concentrations in the CNS, brain glutamate and aspartate levels are derived solely by local synthesis from glucose, because neither amino acid crosses the blood–brain barrier. Two processes contribute to the synthesis of glutamate in the nerve terminal. As mentioned previously (in the section on GABA), glutamate is formed from glucose through the Krebs cycle and transamination of α-ketoglutarate. In addition, glutamate can be formed directly from glutamine (see Fig. 9.7). Because glutamine is synthesized in glial cells, there is an unusual degree of interaction between glia and neurons in the determination of availability of the transmitter pool of glutamate. The glutamine that is formed in glia is transported into nerve terminals and then locally converted by glutaminase into glutamate (Dingledine and McBain, 1994). Thus, the synthesis of glutamate depends critically on the enzyme glutaminase. A phosphate-acti-

vated glutaminase (PAG) has been suggested to be the specific form of the enzyme responsible for the synthesis of the transmitter pool of glutamate. However, PAG is also found in relatively high concentrations in peripheral tissues such as the liver (Conti and Minelli, 1994). PAG is localized to mitochondria; as discussed in the section on GABA, the process by which glutamate is exported to allow vesicular storage of the transmitter remains poorly understood.

Storage and Release of Glutamate

Vesicular glutamate transporter. Glutamate is stored in synaptic vesicles from which the transmitter is released in a calcium-dependent manner on depolarization of the nerve terminal. Although the vesicular storage of glutamate was convincingly demonstrated quite some time ago and was well char-

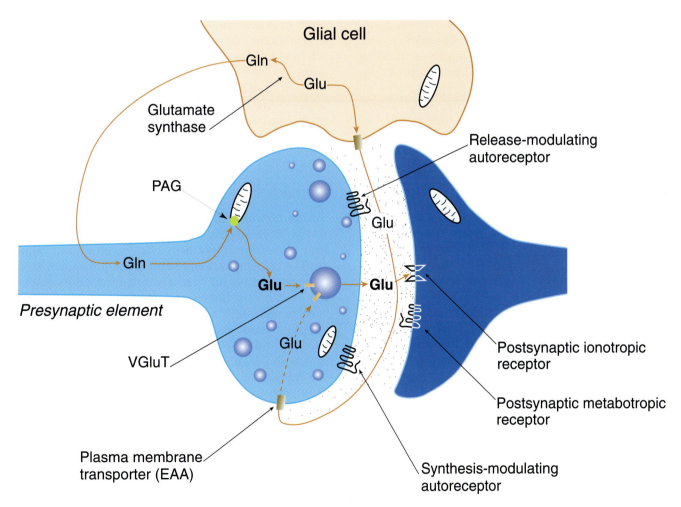

FIGURE 9.7 Depiction of an excitatory amino acid (glutamate) synapse. Glutamate, synthesized via metabolic pathways, is concentrated through a vesicular transporter into secretory granules. After release from the presynaptic terminal, glutamate can interact with postsynaptic and/or release-modulating receptors. Glutamate is then cleared from the synaptic region by the high-affinity plasma membrane transporters or by recycling through adjacent glia.

acterized biochemically, only recently has the vesicular transporter for glutamate been cloned (Bellocchio *et al.*, 2000). This is in part due to the fact that the vesicular glutamate transporter is not related to other known transmitter transporters, although it shares significant sequence homology with EAT-4, a worm protein implicated in glutamatergic transmission. The vesicular glutamate transporter was identified as a protein that was previously suggested to mediate the sodium-dependent transport of inorganic phosphate across the membrane. It is found predominantly in axon terminals, particularly those that form asymmetric (excitatory) synapses.

Regulation of glutamate release. The release of glutamate from nerve terminals is regulated by a metabotropic autoreceptor (see Chapter 11 for discussion of metabotropic receptors). Indeed, eight different receptors (and various splice variants) that constitute three distinct classes of metabotropic glutamate receptors have been identified. The release-modulating autoreceptor is a member of one class of metabotropic receptors, the class II metabotropic glutamate receptors, that are negatively coupled to adenylyl cyclase. Both members of the class II family (the $mGluR_2$ and $mGluR_3$ receptors) have been localized to presynaptic glutamatergic axon terminals, and a large number of studies have revealed that class II receptors function as release-modulating glutamate receptors. In addition, electrophysiological studies have suggested an impulse-modulating glutamate autoreceptor, which is thought to be either an $mGluR_1$ or $mGluR_5$ site. The $mGluR_4$ receptor has been localized to presynaptic nerve terminals, but is present on nonexcitatory nerve terminals and thus is likely to be a heteroreceptor rather than autoreceptor. With eight different primary metabotropic glutamatergic receptors, it is reasonable to ask why have so many? The answer may be that these receptors subserve an extremely broad array of functions, and appear to be critically involved in regulating not only glutamatergic function, but also the activity (including release) of a dizzying number of transmitters, ranging from classic transmitters such as dopamine to peptide transmitters such as substance P (Cartmell and Schoepp, 2000).

Inactivation of Glutamate

Glutamate inactivation occurs predominantly by reuptake of the amino acid by dicarboxylic acid plasma membrane transporters. In contrast to GABA and other classic transmitters, there does not appear to be a significant role for enzymatic inactivation of glutamate. The extent to which diffusion regulates synaptic and extracellular levels of glutamate is not clear.

Five glutamate transporters have been cloned, with some localized to glia and others to neurons. Electron microscopic studies suggest that glutamate transporters are heavily expressed in astrocytes but relatively weakly present in neurons. The glutamate transporters accumulate L-glutamate and D- and L-aspartate; although the affinities of the transporters are similar for glutamate they differ for other amino acids. The transporters have distinct brain distributions, and even the glial transporters exhibit regional and intracellular differences in expression (Chaudhry *et al.*, 1996), underscoring the heterogeneity of glia as well as neurons.

The presence of certain glutamate transporters on glial cells is consistent with the intricate interplay of glial and neuronal elements in the synthesis of glutamate. Because glutamate that is released from neurons is accumulated by glia and then metabolized to glutamine, there is an ultimate recycling of the released amino acid. The fate of glutamate accumulated by the neuronal glutamate transporter is unclear. It has not been established if glutamate released from a given neuron is taken up by a glutamate transporter on that particular neuron or, alternatively, by glutamate transporters on other neurons or glia.

Acetylcholine

Our concepts of chemical synaptic transmission rest on the early studies of acetylcholine (ACh), the first transmitter identified. First noted as the vagal stuff of Loewi (1921) and subsequently demonstrated to be responsible for transmission at the neuromuscular junction by Loewi and Navratil, it has been a century since ACh was first proposed as a transmitter.

A key reason ACh has assumed such a prominent role in guiding studies of neurotransmitters is the ease with which ACh can be studied. Acetylcholine is the transmitter at the neuromuscular junction, and thus both the nerve terminal and its target can be readily accessed for experimental manipulations. Subsequent investigations also focused on another peripheral site, the superior cervical ganglion, which was also easy to isolate and study. Lessons learned from experiments conducted on these peripheral tissues have shaped our current approaches to defining the characteristics of neurotransmitters.

The ability to expose the neuromuscular junction and maintain isolated preparations of the junction permitted electrophysiological and biochemical studies of synaptic transmission. Electrophysiological studies revealed fast excitatory responses of muscle fibers to stimulation of the nerve innervating the muscle. The presence of miniature end-plate potentials (MEPPs) in the muscle fiber was noted, and Fatt and Katz (1952) demonstrated that these MEPPs resulted from the slow "leakage" of ACh, with each MEPP representing the release of transmitter in one vesicle (termed a quantum) (see Chapter 8 for additional details on MEPPs). Overt depolarization generated an increase in the number of quanta released over a given period. In addition, studies of the neuromuscular junction allowed detailed analyses of the enzymatic inactivation of ACh, setting the reference for subsequent studies.

Over the past half century many of the rules that govern ACh neurotransmission have been shown to be general principles that apply to other transmitters. For example, the concept of the quantal nature of neurotransmission is central to current ideas of transmitter release. Although the discovery of different neurotransmitters has expanded our knowledge, studies of ACh continue to provide a foundation for modern concepts of chemical neurotransmission.

Acetylcholine Synthesis

The synthesis of ACh is arguably the most simple transmitter synthesis, with but a single step: the acetyl group from acetyl-coenzyme A is transferred to choline by the enzyme choline acetyltransferase (ChAT). The requirements for ACh synthesis are correspondingly few: the substrate choline, the donor acetyl-coenzyme A, and the enzyme ChAT (see Fig. 9.8).

The acetyl-CoA that serves as the donor is derived from pyruvate generated by glucose metabolism. This obligatory dependence on a metabolic intermediary is similar to the situation present in GABA synthesis, where the immediate precursor glutamate is formed from α-ketoglutarate. Acetyl-CoA is localized to mitochondria. Because the synthetic enzyme ChAT is cytoplasmic, acetyl-CoA must exit the mitochondria to gain access to ChAT; this process is poorly understood.

Choline acetyltransferase. Choline acetyltransferase is the definitive marker for cholinergic neurons (Wu and Hersch, 1994). Multiple mRNAs encode ChAT, resulting from differential use of three promoters and alternative splicing of the 5' noncoding region. In the rat the different transcripts encode the same protein, but in human give rise to multiple forms of the enzyme, including both active and inactive (truncated) forms. The functional significance of the different transcripts under normal conditions is a topic of considerable interest. Myasthenia gravis, a disease marked by decreased muscle activity, is linked to a variety of deficits in neuromuscular cholinergic function; in congenital forms of myasthenia mutations in both nicotinic receptors and AChE have been found. Recently, a particular ChAT mutation has been linked to a form of myasthenia that is characterized by often-fatal episodes of apnea (Ohno *et al.*, 2001).

Although ChAT is the sole enzyme in ACh synthesis, ChAT is not the rate-limiting step in ACh synthesis. As in the case for TH in the synthesis of catecholamines, the full enzymatic activity of ChAT is not expressed *in vivo*: when ChAT activity is measured *in vitro* it is much greater than would be expected on the basis of ACh synthesis *in vivo*. The reason for this discrepancy has been suggested to be related to the need to transport acetyl-CoA from the mitochondria to the cytoplasm, which may be rate-limiting in ACh synthesis. Alternatively, intracellular choline concentrations may ultimately determine the rate of ACh synthesis. This latter speculation has led to the use of choline precursors in attempts to enhance ACh synthesis in Alzheimer's disease, in which there is a marked decrease in ACh in the cerebral cortex. Attempts have been made to treat Alzheimer's disease with lecithin, a choline precursor; unfortunately, lecithin does not appear to diminish dementia, although it does markedly increase bad breath!

Acetylcholine Storage and Release

Vesicular cholinergic transporter. ACh is synthesized by ChAT and transported into vesicles by the vesicular cholinergic transporter (VAChT). This transporter is distinct from the membrane transporter that accumulates choline. VAChT was cloned on the basis of homology to a *C. elegans* gene (*unc-17*) that encodes a protein that is homologous with VMAT (Roghani *et al.*, 1994). VAChT is expressed in cholinergic neurons throughout the brain.

When the mammalian VAChT was initially cloned it was noted that the human VAChT is present in chromosome 10, near the gene for ChAT. It was subsequently demonstrated that the VAChT is unique in that its entire coding region is contained in the first intron of the ChAT gene (Usdin *et al.*, 1995). This suggests that both genes are coordinately regulated, a suspicion that has been confirmed (Bernard *et al.*, 1995).

Cholinergic autoreceptor function. Cholinergic release-modulating autoreceptors have been identified both in peripheral tissues and in the brain. This receptor is a muscarinic cholinergic receptor,

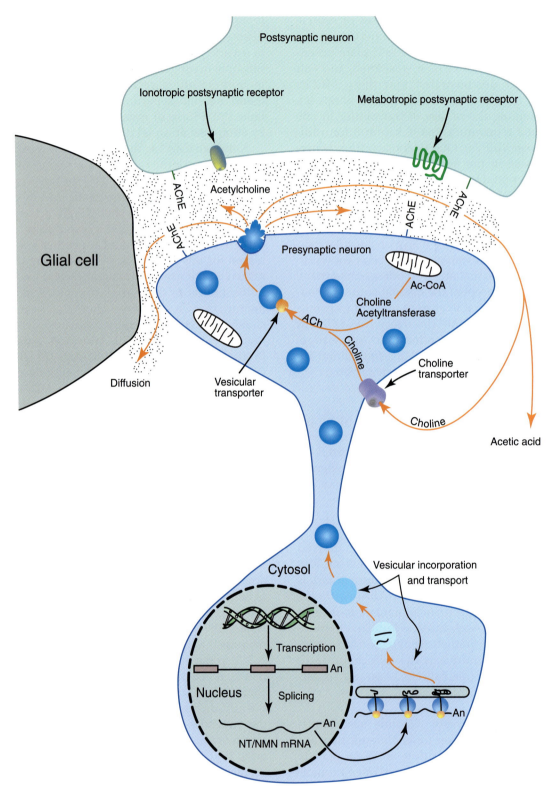

FIGURE 9.8 Acetylcholine (ACh) synthesis, release, and termination of action. A choline transporter accumulates choline. The enzyme choline acetyltransferase (ChAT) acetylates the choline using acetyl-CoA (Ac-CoA) to form the transmitter ACh, which is accumulated into vesicles by the vesicular transporter. The released ACh may interact with postsynaptic muscarinic or nicotinic cholinergic receptors, or can be taken up into the neuron by a choline transporter. Acetylcholine can be degraded after release by the enzyme acetylcholine esterase (AChE).

rather than the nicotinic cholinergic receptor found at the neuromuscular junction. A cholinergic receptor regulating release of ACh from cholinergic axons in the cortex, hippocampus, and striatum has been identified, and converging data indicate that this receptor is an M2 cholinergic receptor, one of five muscarinic receptors; the M2 site is also found on noncholinergic (cholinoreceptive) neurons, where it serves as a conventional heteroceptor. There are few direct data that point to the presence of a synthesis-modulating cholinergic autoreceptor.

Inactivation of Acetylcholine

Acetylcholinesterase. The primary mode of inactivation of ACh appears to be enzymatic. Several esterases that hydrolyze choline esters can degrade ACh, but the major esterase in the central nervous system is acetylcholinesterase (AChE; see Box 9.3).

The enzymatic inactivation of ACh is simply the hydrolysis of ACh to choline. Two groups of cholinesterases have been defined on the basis of substrate specificity, acetyl cholinesterases and butyrylcholinesterases (Taylor and Brown, 1994). The first are relatively specific for ACh and are present in high concentration in the brain, where there are multiple AChE species (Fernandez *et al.*, 1996). Butyrylcholinesterases also efficiently hydrolyze choline esters, but are found primarily in the liver, with lower levels being present in the adult brain.

AChE is present in high concentration in cholinergic neurons; however, AChE is also present in moderately high concentration in some noncholinergic neurons that receive cholinergic inputs (i.e., that are cholinoreceptive). This observation is consistent with the fact that AChE is a secreted enzyme that is associated with the cell membrane. Thus, ACh hydrolysis takes place extracellularly, and the choline generated is conserved by the high-affinity reuptake process.

In addition to its role in inactivating released acetylcholine, AChE has been proposed to function as a chemical messenger in the CNS (Greenfield, 1991). The release of AChE from neurons in the substantia nigra and cerebellum is calcium-dependent, and cerebellar release is enhanced by electrical stimulation of cerebellar afferents. Electrophysiological studies have revealed that AChE elicits changes in the threshold for Ca^{2+} spikes, and local application of AChE enhances the responses of cerebellar neurons to glutamate and aspartate, the transmitters present in the climbing and mossy fiber innervations of the cerebellum (Appleyard and Jahnsen, 1992).

There are several AChE species, all of which are encoded by a single gene that is alternatively spliced (Schumacher *et al.*, 1988), with tissue-specific expression of different transcripts (Seidman *et al.*, 1995). Among the multiple mRNAs encoding AChE is one that represents the primary form of the enzyme expressed in brain and muscle.

BOX 9.3

ACETYLCHOLINESTERASE INHIBITORS, NERVE GASES, AND PHARMACOTHERAPY

The enzymatic inactivation of acetylcholine (ACh) has been fertile ground for the development of a large number of pharmaceutical agents. Anticholinesterases such as sarin are potent neurotoxins and have been used as nerve gases since World War I. Other anticholinesterases include organophosphates (such as parathion), which are widely used insecticides. Anticholinesterases, whether the target is a human or a tomato hornworm, function in the same way: instead of the released ACh leading to discrete single depolarizations of muscle fibers, the accumulation of acetylcholine at the neuromuscular junction leads to muscle fibrillation and ultimately depolarization inactivation of the muscle; i.e., the muscle is so excited it stops!

Anticholinesterases have some less aggressive uses as well. Competitive neuromuscular blocking agents such as succinylcholine are used as an adjunct to anesthetics during surgery to increase muscle relaxation; conversely, anticholinergics can be used to reverse the muscle paralysis caused by succinylcholine. Anticholinesterases are the mainstay of treatment of myasthenia gravis, a disorder of the neuromuscular junction that is usually marked by the presence of anti-nicotinic receptor antibodies. Attempts have also been made to treat Alzheimer's disease, in which there is a sharp decrease in cortical ACh, by administering an anticholinesterase to inhibit breakdown of ACh. Unfortunately, this approach has not proven very effective.

High affinity choline transporter. Choline is found in the plasma in high concentration. A low-affinity reuptake process for choline is widely distributed in the body; however, both low-affinity and high-affinity choline uptake processes are present in brain. Cholinergic neurons in the brain express a sodium-dependent transporter that is saturated at plasma levels of choline, consistent with the high-affinity component of choline uptake.

The choline transporter, in contrast to other plasma membrane transmitter transporters, is not directly involved in termination of the action of release transmitter. Since ACh is hydrolyzed by AChE, enzymatic inactivation is the major means of terminating cholinergic transmission. In the absence of direct evidence that choline binds with high affinity to cholinergic receptors, it appears that the function of the choline transporter is conservation of transmitter stores.

The choline transporter has long been known but was not cloned until recently (Okuda *et al.*, 2000). Several years ago a putative high affinity choline transporter was cloned. However, this transporter was present in peripheral tissues as well as brain and, when expressed in cell lines, displayed kinetics different from those of CNS tissue. Moreover, this transporter was not sensitive to hemicholinium 3, which blocks the high-affinity cholinergic reuptake process. It now appears that this gene cloned encoded a creatinine transporter rather than the choline transporter (Happe and Murrin, 1995).

Paralleling the situation surrounding the identification of the vesicular glutamate transporter gene, which was cloned on the basis of a related worm gene, a cDNA encoding the rat high-affinity choline transporter was recently cloned based on similarities to *cho–1* in *C. elegans*. The vesicular cholinergic transporter is not homologous to other neurotransmitter transporters, being related instead to the sodium-dependent glucose transporter family.

Why Do Neurons Have So Many Transmitters?

We have discussed in varying amounts of detail a moderate number, but not all classic transmitters. About a dozen classic transmitters and literally dozens of neuropeptides function as transmitters. Still more molecules serve as "unconventional" transmitters, including growth factors and gases such as nitric oxide. If the role of neurotransmitters is to serve as a chemical bridge that conveys information between two spatially distinct cells, why have so many chemical messengers?

Afferent Convergence on a Common Neuron

Perhaps the simplest explanation for multiple transmitters is that many nerve terminals abut on a single neuron. A neuron must be able to distinguish between the multiple inputs that bring information to the neuron. To some degree, this can be accomplished by the site on a neuron at which an afferent (input) terminates: at the cell body, axon, or dendritic shaft or spine. However, because many afferents terminate in close proximity, another means of distinguishing the inputs and their information is necessary. One way in which this can be accomplished is by chemically coding afferent neurons. The information conveyed by distinct transmitters is then distinguished by the different receptors present on the targeted neuron and their various transduction mechanisms.

Neurotransmitter Colocalization

A major conceptual change in the neurosciences over the past generation has been the realization that a cell can use more than one neurotransmitter. The idea that a neuron is limited to one transmitter can be traced to Henry Dale, or, more properly, to an informal restatement of what is termed Dale's principle. About 60 years ago, Dale posited that a metabolic process that takes place in the cell body can reach or influence events in all the processes of the neuron. John Eccles restated Dale's view to suggest that a neuron releases the same transmitter at all its processes. Illustrating the dangers of the scientific equivalent of sound bites, this principle was soon misinterpreted to indicate that only a single transmitter can be present in a given neuron. This is clearly not the case. Neurons can co-localize two or more transmitters. For example, a neuron can use both a classic transmitter such as dopamine and a peptide transmitter such as neurotensin. Indeed, it now appears that few if any neurons contain only one transmitter, and there are many cases in which three or even four transmitters are found in a single neuron.

The presence of multiple transmitters in a single neuron may indicate that different transmitters are used by a neuron to signal different functional states to its target cell. For example, the firing rates of neurons differ considerably, and thus it may be useful for a neuron to encode fast firing by one transmitter and slower firing by another transmitter. In addition, the firing pattern of cells is of significance. For example, a neuron might show an absolute firing rate of 5 impulses per second. This frequency could result from a neuron discharging every 200 ms or discharging five times in an initial 200-ms period followed by 800 ms of silence. Recent data indicate that in cases in which there is a co-localization of a peptide and a classic transmitter, peptide transmitters are often released at higher firing rates and particularly under burst-firing patterns.

In many ways, the different biosynthetic strategies used by peptides and classic transmitters may lead to differential release. Classic transmitters can be rapidly replaced because their synthesis occurs in nerve terminals. In contrast, peptide transmitters are synthesized in the cell body and transported to the terminal. It is therefore useful to conserve peptide transmitters for situations of high demand, because they would otherwise be rapidly depleted.

Transmitter Release from Different Processes

The restatement of Dale's principle by Eccles held that a transmitter or other protein is present in all processes of a neuron. However, it now appears that a transmitter can be specifically localized to different parts of a neuron (see Deutch and Bean, 1995). For example, in the marine mollusck *Aplysia*, different transmitters are targeted to different processes of a single neuron (Sossin *et al.*, 1990). If a transmitter is restricted to a particular part of a neuron, it follows that the neuron would need multiple transmitters to account for different release sites. Considerable evidence supports distinct spatial localizations of receptors (e.g., for ionotropic glutamate receptors) on a neuron, and even indicates that there is movement and clustering of receptors to maximize information transfer from pre- to postsynaptic neurons.

Synaptic Specializations versus Nonjunctional Appositions between Neurons

In addition to the diversity in transmitters that may result from transmitters being targeted to different intraneuronal sites, the anatomical relationships between one cell and its follower may contribute to the need for different transmitters. We usually think of synaptic specializations (see Chapters 1 and 8) as the morphological substrate of communication between two neurons. However, there may also be nonsynaptic forms of communication between two neurons. These could occur across distances that are smaller (e.g., gap junctions) or much larger than the separation of pre- and postsynaptic neurons by the synaptic cleft. The requirements for transmitter action would differ from those discussed previously if the distance traversed by a transmitter is larger than that typically present at a synaptic apposition. Thus, transmitters that lack an efficient reuptake system, such as peptide transmitters, might be favored at nonsynaptic sites. Because a single neuron can form both synaptic and nonsynaptic specializations, a single neuron may require more than one neurotransmitter.

Fast versus Slow Responses of Target Neurons to Neurotransmitters

We have seen that different firing rates or patterns may be accompanied by changes in the amount of transmitter being released from a neuron. As described in detail in Chapter 16, the postsynaptic response to a transmitter occurs over different time scales. For example, transmitter activation of ionotropic receptors (i.e., those that form ion channels) leads to very rapid changes, because the ionic gradients across the cell are almost instantaneously changed. In contrast, metabotropic receptors that respond to catecholamines and peptide transmitters are coupled to intracellular events via various transduction molecules, such as G proteins, and respond to neurotransmitter stimulation on a slower time scale than is seen when ionotropic receptors are activated. This difference in temporal response characteristics is useful, because it allows the receptive neuron to respond differently to a stimulus, depending on the antecedent activity in the cell. A transmitter can change the response characteristics of a particular cell to subsequent stimuli on the order of seconds or even minutes, and thus short-term changes can occur independent of changes in gene expression.

Nontransmitter Roles of Neurotransmitters

Over the past few years several of the key proteins involved in regulating chemical neurotransmission have been identified based on homologies to proteins found in invertebrate species, such as the worm *Caenorhabditis elegans* and the fly *Drosophila melanogaster*.

It now appears that some of the molecular players in neurotransmission are found even in plants! Plant homologues of glutamate receptors have been identified and shown to be important in regulating diverse functions, ranging from calcium utilization to morphogenesis (Brenner *et al.*, 2000; Kim *et al.*, 2001), and geneological analysis has suggested that these glutamate receptors may predate the divergence of plants and animals (Chiu *et al.*, 1999). As nervous systems have become elaborate through evolution, many neurotransmitter-related proteins have roles that are not related to transmitter function or alternatively are involved in less discrete and more spatially elaborate signaling.

An example is acetylcholine. As discussed above, ChAT is a cytosolic protein that drives the synthesis of

ACh from choline and acetyl-CoA; however, one form of human ChAT is localized to the nucleus, where it seems unlikely to play a transmitter role (Resendes et al., 1999). ChAT mRNA is found in the testes, where it is translated and the protein appears in spermatozoa (Ibanez et al., 1991). Moreover, ChAT mRNAs have been reported to be present in lymphocytes, as have certain muscarinic cholinergic receptors (Kawashima et al., 1998).

In addition, AChE mRNAs appear to be present in lymphocytes, where both acetylcholinesterase and butyrylcholinesterase enzyme activity has been reported as decreased in Alzheimer's disease (Bartha et al., 1987; Inestrosa et al., 1994). AChE is present in high abundance in bone marrow cells and peripheral blood cells in certain types of leukemias (Lapidot-Lifson et al., 1989). Recent data indicate that inhibition of AChE gene expression in bone marrow cultures suppresses apoptosis (programmed cell death) and leads to progenitor cell expansion (Soreq et al., 1994), suggesting a role for AChE in the development of leukemias.

The presence of neurotransmitter-related proteins in peripheral tissues is not restricted to molecules related to ACh function. Three dopamine receptor mRNAs, encoding the D3, D4, and D5 receptors, are present in lymphocytes, and expression of the D3 receptor transcript is increased in schizophrenic subjects (Ilani et al., 2001; Kwak et al., 2001).

It is relatively easy to envision how transmitter receptors that are expressed on peripheral nonneural tissues can respond to transmitters present in the periphery, essentially functioning as hormonal signals. Thus, dopamine or other catecholamines that are circulating at low levels in the periphery may bind to DA receptors present on lymphocytes. However, another means of signaling is via axonal noradrenergic innervation of nonneural immune tissues in the periphery. Thus, sympathetic noradrenergic fibers innervate not only the vasculature, but also primary (bone marrow) and secondary (spleen, lymph nodes) immune lymphoid structures (Felton et al., 1985). This noradrenergic innervation may be the central regulator of peripheral immune and stress responses, since chemical lesions of the sympathetic nervous system markedly alter T-and B-cell proliferation and activity (Madden et al., 2000). Not only are the transmitters of neurons communicating between neural and immune system cells the same, but there are similarities between the structural substrates of communication between different nervous system cells (neurons) and immune system cells, through junctional specializations called synapses (see Trautman and Vivier, 2001).

SUMMARY

Classic neurotransmitters are small molecules that are derived from amino acids or intermediary metabolism and share several characteristics. The sequential actions of key enzymes result in the biosynthesis of these transmitters, usually in the general vicinity of where they will be released. The synthesized transmitter is stored in vesicles where it is poised for release and protected from degradation; the vesicular transporters also sequester xenobiotics and thus protect the neuron from certain toxins. Neurotransmitter release is elicited by depolarization and is calcium-dependent. The action of the released neurotransmitter is terminated by a reuptake mechanism involving plasma membrane transporters and by enzymatic means.

The criteria for designation as a classic transmitter have been based on experiments that were conducted in sites that were easily accessible (such as the neuromuscular junction). Although many of the key principles of chemical synaptic transmission have been found to be the same in other areas that are less accessible to experimental manipulation (neurons in the brain), our ideas of the defining characteristics of transmitters have evolved to account for new knowledge and the emergence of many exceptions to the rules enunciated above. The relatively high concentrations of classic transmitters permitted the easy measurement of these compounds, and thus the measurement of transmitter release became a key criterion for defining a neurotransmitter. Unfortunately, transmitter release has proven to be a difficult criterion to meet for many putative transmitters discovered over the past 30 years. Nevertheless, the increasing sensitivity of analytical techniques coupled with the ingenuity of neuroscientists led to the uncovering of a large number of peptides, growth factors, and even gases that function as transmitters. We explore in the next chapter the similarities and differences of the classic transmitters with these new kids on the block. These differences have often illuminated unknown fundamental processes of neurons and expanded our concept of information flow between neurons.

References

Appleyard, M., and Jahnsen, H. (1992). Actions of acetylcholinesterase in the guinea-pig cerebellar cortex in vitro. Neuroscience 47, 291–301.

Bartha, E., Szelenyi, J., Szilagyi, K., Venter, V., Thu Ha, N. T., Paldi-Haris, P., and Hollan, S. (1987). Altered lymphocyte acetylcholinesterase activity in patients with senile dementia. Neurosci. Lett. 79, 190–194.

Bean, A. J., and Roth, R. H. (1992). Dopamine–neurotensin interactions in mesocortical neurons: Evidence from microdialysis studies. *Ann. N.Y. Acad. Sci.* **668**, 43–53.

Bellocchio, E. E., Reimer, R. J., Fremeau, R. T., Jr. and Edwards, R. H. (2000). Uptake of glutamate into synaptic vesicles by an inorganic phosphate transporter. *Science* **289**, 957–60.

Bernard, C. (1849). Action physiologique des venins (curare). *C. R. Seances Soc. Biol. Ses Fil.* **1**, 90.

Bernard, S., Varoqui, H., Cervine, R., Israel, M., Mallet, J., and Diebler, M. F. (1995). Coregulation of two embedded gene products, choline acetyltransferase and the vesicular acetylcholine transporter. *J. Neurochem.* **65**, 939–942.

Brenner, E. D., Martinez-Barboza, N., Clark, A. P., Liang, Q. S., Stevenson, D. W., and Coruzzi, G. M. (2000). Arabidopsis mutants resistant to S(+)-beta-methyl-alpha, beta-diaminoproprionic acid, a cycad-derived glutamate receptor agonist. *Plant Physiol.* **124**, 1615–1624.

Carlsson, A. (1972). Biochemical and pharmacological aspects of parkinsonism. *Acta Neurol. Scand.* (*Suppl.*) **51**, 11–42.

Cartmell, J, and Schoepp, D. D. (2000). Regulation of neurotransmitter release by metabotropic glutamate receptors. *J. Neurochem.* **75**, 889–907.

Chaudhry, F. A., Lehre, K. P., van Lookeren Campagne, M., Otterson, O. P., Danbolt, N. C., and Storm-Mathisen, J. (1996). Glutamate transporters in glial plasma membranes: Highly differentiated localizations revealed by quantitative ultrastructural immunocytochemistry. *Neuron* **15**, 711–720.

Cherubini, E., and Conti, F. (2001). Generating diversity at GABAergic synapses. *Trends Neurosci.* **24**, 155–162.

Chiu, J., DeSalle, R., Lam, H. M., Meisel, L., and Coruzzi, G. (1999). Molecular evolution of glutamate receptors: A primitive signaling mechanism that existed before plants and animals diverged. *Mol. Biol. Evol.* **16**, 826–838.

Clark, J. A., and Amara, S. G. (1993). Amino acid neurotransmitter transporters: Structure, function, and molecular diversity. *BioEssays* **15**, 323–332.

Conti, F., and Minelli, A. (1994). Glutamate immunoreactivity in rat cerebral cortex is reversibly abolished by 6-diazo-5-oxo-L-norleucine (DON), an inhibitor of phosphate-activated glutaminase. *J. Histochem. Cytochem.* **42**, 717–726.

Cooper, J. R., Bloom, F. E., and Roth, R. H. (2003). "The Biochemical Basis of Neuropharmacology," 8th ed. Oxford Univ. Press, New York.

DeLorcy, T. N., and Olsen, R. W. (1994). GABA and glycine. *In* "Basic Neurochemistry" (G. J. Siegel, B. W. Agranoff, R. W. Albers, and P. B. Molinoff, Eds.), 5th ed., pp. 389–400. Raven Press, New York.

Deutch, A. Y., and Bean, A. J. (1995). Colocalization in dopamine neurons. *In* "Psychopharmacology: The Fourth Generation of Progress" (F. E. Bloom and D. J. Kupfer, Eds.), Raven Press, New York, pp. 197–206.

Dingledine, R., and McBain, C. J. (1994). Excitatory amino acid transmitters. *In* "Basic Neurochemistry" (G. J. Siegel, B. W. Agranoff, R. W. Albers, and P. B. Molinoff, Eds.), pp. 367–388. 5th ed. Raven Press, New York.

Edwards, R. H. (1993) Neural degeneration and the transport of neurotransmitters. *Ann. Neurol.* **34**, 638–45.

Elliott, T. R. (1905). On the action of adrenaline. *J. Physiol.* **32**, 401.

Erlander, M. G., and Tobin, A. J. (1991) The structural and functional heterogeneity of glutamic acid decarboxylase: a review. *Neurochem. Res.* **16**, 215–26.

Fatt, P., and Katz, B. (1952). Spontaneous subthreshold activities at motor nerve endings. *J. Physiol.* **117**, 109–128.

Felton, D. L., Felton, S. Y., Carlson, S. L., Olschowka, J. A., and Livnat, S. (1985). Noradrenergic and peptidergic innervation of lymphoid tissue. *J. Neuroimmunol.* **135**, (Suppl. 2), 755–765.

Fernandez, H. L., Moreno, R. D., and Inestrosa, N. C. (1996). Tetrameric (G4) acetylcholinesterase: Structure, localization, and physiological regulation. *J. Neurochem.* **66**, 1335–1346.

Frazer, A., and Hensler, J. G. (1994). Serotonin. *In* "Basic Neurochemistry" (G. J. Siegel, B. W. Agranoff, R. W. Albers, and P. B. Molinoff, Eds.), 5th ed., pp. 283–309. Raven Press, New York.

Gainetdinov, R. R., Jones, S. R., Fumagalli, F., Wightman, R. M., and Caron, M. G. (1998). Re-evaluation of the the role of the dopamine transporter in dopamine system homeostasis. *Brain Res. Rev.* **26**, 148–53.

Greenfield, S. A. (1991). A non-cholinergic role of ACHE in the substantia nigra: From neuronal secretion to the generation of movement. *Mol. Cell. Neurobiol.* **11**, 55–77.

Happe, H. K., and Murrin, L. C. (1995). In situ hybridization analysis of CHOTT, a creatine transporter, in the rat central nervous system. *J. Comp. Neurol.* **351**, 94–103.

Haycock, J. W., Haycock, D. A. (1991). Tyrosine hydroxylase in rat brain dopaminergic nerve terminals: Multiple-site phosphorylation *in vivo* and in synaptosomes. *J. Biol. Chem.* **266**, 5650–5657.

Heninger, G. R., Delgado, P. L., and Charney, D. S. (1996). The revised monoamine theory of depression: A modulatory role for monoamines, based on new findings from monoamine depletion experiments in humans. *Pharmacopsychiatry* **29**, 2–11.

Henry, J. P., Sagne, C., Bedet, C., and Gasnier, B. (1998). The vesicular monoamine transporter: From chromaffin granule to brain. *Neurochem. Int.* **32**, 227–246.

Ichinose, H., Ohye, T., Takahashi, E., Seki, N., Hori, T., Segawa, M., Nomura, Y., Endo, K., Tanaka, K., Tanaka, H., and Tsuji, S. (1994). Hereditary progressive dystonia with marked diurnal fluctuation caused by mutations in the GTP cyclohydrolase I gene. *Nat. Genet.* **8**, 236–242.

Ilani, T., Ben-Shachar, D., Strous, R. D., Mazor, M., and Sheinkman, A., Kotler, M., and Fuchs, S. (2001). A peripheral marker for schizophrenia: Increased levels of D3 dopamine receptor mRNA in blood lymphocytes. *Proc. Natl. Acad. Sci. USA* **98**, 625–628.

Inestrosa, N. C., Alarcon, R., Arriagada, J., Donoso, A., Alvarez, J., and Campos, E. O. (1994). Blood markers in Alzheimer disease: Subnormal acetylcholinesterase and butyrylcholinesterase in lymphocytes and erythrocytes. *J. Neurol. Sci.* **122**, 1–5.

Jones, S. R., Gainetdinov, R. R., Hu, X. T., Cooper, D. C., Wightman, R. M., White, F. J., and Caron, M. G. (1999). Loss of autoreceptor functions in mice lacking the dopamine transporter. *Nat. Neurosci.* **2**, 649–655.

Kawashima, K., Fujii, T., Watanabe, Y., and Misawa, H. (1998). Acetylcholine synthesis and muscarinic receptor subtype mRNA expression in T-lymphocytes. *Life Sci.* **62**, 1701–1705.

Kim, S. A., Kwak, J. M., Jae, S. K., Wang, M. H., and Nam, H. G. (2001). Overexpression of the AtGluR2 gene encoding an Arabidopsis homolog of mammalian glutamate receptors impairs calcium utilization and sensitivity to ionic stress in transgenic plants. *Plant Cell Physiol.* **42**, 74–84.

Kwak, Y. T., Koo, M. S., Choi, C. H., and Sunwoo, I. (2001). Change of dopamine receptor mRNA expression in lymphocyte of schizophrenic patients. *BMC Med. Genet.* **2**, 3.

Lapidot-Lifson, Y., Prody, C. A., Ginzberg, D., Meytes, D., Zakut, H., and Soreq, H. (1989). Coamplification of human acetylcholinesterase and butrylcholinesterase genes in blood cells: Correlation with various leukemias and abnormal megakaryacytopoiesis. *Proc. Natl. Acad. Sci. USA* **86**, 4715–4719.

Lewis, D. A., Melchitzky, D. S., and Haycock, J. W. (1993) Four isoforms of tyrosine hydroxylase are expressed in human brain. *Neuroscience* **54**, 477–492.

Loewi, O. (1921). Uber Humorale Ubertragbarkeit Herznervenwirkung. *Pfluegers Arch Ges. Physiol. Menschen Tiere* **189**, 239.

Madden, K. S., Stevens, S. Y., Felton, D. L., and Bellinger, D. L. (2000). Alterations in T lymphocyte activity following chemical sympathectomy in young and old Fisher 344 rats. *J. Neuroimmunol.* **103**, 131–145.

McIntire, S. L., Reimer, R. J., Schuske, K., Edwards, R. H., and Jorgensen, E. M. (1997). Identification and characterization of the vesicular GABA transporter. *Nature* **389**, 870–876.

Ohno, K., Tsujino, A., Brengman, J. M., Harper, C. M., Bajzer, Z., Udd, B., Beyring, R., Robb, S., Kirkham, F. J., and Engel, A. G. (2001). Choline acetyltransferase mutations cause myasthenic syndrome associated with episodic apnea in humans. *Proc. Natl. Acad. Sci. USA* **98**, 2017–2022.

Okuda, T., Haga, T., Kanai, Y., Endou, H., Ishihara, T., and Katsura, I. (2000). Identification and characterization of the high-affinity choline transporter. *Nat. Neurosci.* **3**, 120–125.

Paul, S. P. (1995). GABA and glycine. *In* "Neuropyschopharmacology: The Fourth Generation of Progress" (F. E. Bloom and D. J. Kupfer, Eds.), pp. 87–94. Raven Press, New York.

Pothos, E. N., Larsen, K. E., Krantz, D. E., Liu, Y., Haycock, J. W., Setlik, W., Gershon, M. D., Edwards, R. H., and Sulzer, D. (2000). Synaptic vesicle transporter expression regulates vesicle phenotype and quantal size. *J. Neurosci.* **20**, 7297–7306.

Resendes, M. C., Dobransky, T., Ferguson, S. S., Rylett, R. J. (1999). Nuclear localization of the 82 kDA form of human choline acetyltransferase. *J. Biol. Chem* **274**, 19417–19421.

Roberts, E. (1986). GABA: The road to neurotransmitter status. *In* "Benzodiazepine/GABA Receptors and Chloride Channels: Structural and Functional Properties" (Olsen and Venter, Eds.), pp. 1–39. Liss, New York.

Roghani, A., Feldman, J., Kohan, S. A., Shirzadi, A., Gundersen, C. B., Brecha, N., and Edwards, R. H. (1994). Molecular cloning of a putative vesicular transporter for acetylcholine. *Proc. Natl. Acad. Sci. USA* **91**, 10620–10624.

Roth, R. H., and Elsworth, J. D. (1995). Biochemical pharmacology of midbrain dopamine neurons. *In* "Psychopharmacology: The Fourth Generation of Progress" (F. E. Bloom and D. J. Kupfer, Eds.), pp. 227–243, Raven Press, New York.

Scatton, B., Dennis, T., and Curet, O. (1984). Increase in dopamine and DOPAC levels in noradrenergic nerve terminals after electrical stimulation of the ascending noradrenergic pathways. *Brain Res.* **298**, 193–196.

Schumacher, M., Maulet, Y., Camp, S., and Taylor, P. (1988). Multiple messenger RNA species give rise to the structural diversity of acetylcholinesterase. *J. Biol. Chem.* **263**, 18979–18987.

Seidman, S., Sternfeld, M., Ben Azziz-Aloya, R., Timberg, R., Kaufer-Nachum, D., and Soreq, H. (1995). Synaptic and epidermal accumulations of human acetylcholinesterase are encoded by alternative 3'-terminal exons. *Mol. Cell. Biol.* **15**, 2993–3002.

Shepherd, G. M. (1991). "Foundations of the Neuron Doctrine." Oxford Univ. Press, New York.

Sherrington, C. S. (1906). "The Integrative Action of the Nervous System." Scribner's, New York.

Soreq, H., Pantinkin, D., Lev-Lehman, E., Grifman, M., Ginzberg, D., Eckstein, F., and Zakut, H. (1994). Antisense oligonucleotide inhibiton of acetylcholinesterase gene expression induces progenitor cell expansion and suppresses hematopoeietic apoptosis ex vivo. *Proc. Natl. Acad. Sci. USA.* **91**, 7907–7911.

Sossin, W. S., Sweet-Cordero, A., and Scheller, R. H. (1990). Dale's hypothesis revisited: Different neuropeptides derived from a common prohormone are targeted to different processes. *Proc. Natl. Acad. Sci. USA* **87**, 4845–4848.

Stone, T. W. (1993). Neuropharmacology of quinolinic and kynurenic acids. *Pharmacol. Rev.* **45**, 309–379.

Takamori, S., Rhee, J. S., Rosenmund, C., and Jahn, R. (2000). Identification of a vesicular glutamate transporter that defines a glutamatergic phenotype in neurons. *Nature* **407**, 189–194.

Tam, S. Y., Elsworth, J. D., Bradberry, C. W., and Roth, R. H. (1990). Mesocortical dopamine neurons: High basal firing frequency predicts tyrosine dependence of dopamine synthesis. *J. Neural Transm. (Gen. Sect.)* **81**, 97–110.

Taylor, P., and Brown, J. H. (1994). Acetylcholine. *In* "Basic Neurochemistry" (G. J. Siegel, B. W. Agranoff, R. W. Albers, and P. B. Molinoff, Eds.), 5th ed. Raven Press, New York.

Thorny, B., Auerbach, G., and Blau, N. (2000). Tetrahydrobiopterin biosynthesis, regeneration and functions. *Biochem. J.* **347**, 1–16.

Trautmann, A., and Vivier, E. (2001). Agrin: A bridge between the nervous and immune systems. *Science* **292**, 1667–1668.

Usdin, T. B., Eiden, L. E., Bonner, T. I., and Erickson, J. D. (1995). Molecular biology of the vesicular ACh transporter. *Trends Neurosci.* **18**, 218–224.

Weihe, E., and Eiden, L. E. (2000) Chemical neuroanatomy of the vesicular amine transporters. *FASEB J* **14**, 2435–2449.

Wu, D., and Hersch, L. B. (1994). Choline acetyltransferase: Celebrating its fiftieth year. *J. Neurochem.* **62**, 1653–1663.

Xu, F., Gainetdinov, R. R., Wetsel, W. C., Jones, S. R., Bohn, L. M., Miller, G. W., Wang, Y. M., and Caron, M. G. (2000). Mice lacking the norepinephrine transporter are supersensitive to psychostimulants. *Nat. Neurosci.* **3**, 465–471.

Zoli, M., Jansson, A., Sykova, E., Agnati, L. F., and Fuxe, K. (1999). Volume transmission in the CNS and its relevance for neuropsychopharmacology. *Trends Pharmacol. Sci.*

10

Nonclassic Signaling in the Brain

Ariel Y. Deutch and James L. Roberts

As described in Chapter 9, chemically mediated transmission is the major means by which a signal is communicated from one nerve cell to another. The neuron doctrine, developed in the late 19th and early 20th centuries, put forth that neurons are not part of a continuous physical network, but are discrete, spatially distinct elements. This formulation led to the revolutionary idea that the release of some chemical substance from a nerve cell might influence another nerve cell or target. Today it is widely accepted that neuronal communication involves the release of specific neurotransmitters from one neuronal element to affect another element. However, there are a variety of forms of chemical communication that do not conform to the basic concept of synaptic signaling. These range from humoral influences, such as those of steroids and sugars, derived from peripheral sources to a cornucopia of different peptide/proteins, to readily diffusible gases produced within the brain. The goal of this chapter is to outline the different types of chemical communication, with an emphasis on describing how these signaling systems function in the brain.

In general, the nonclassic neurotransmitters share many of the same fundamental properties of the classic neurotransmitters with a few notable exceptions. They are not always locally synthesized, some being derived from other tissues in the body or regions of the brain. Nor are they always stored to await a specific release signal; in some cases the signaling mechanism for release is the same as the stimulus for synthesis, and thus these messengers are released as quickly as they can be synthesized.

PEPTIDE NEUROTRANSMITTERS

There are many more peptide transmitters than classic transmitters. There are some similarities between these two classes of transmitters, but also as many differences. Both classic and peptide transmitters are typically well conserved across species. In fact, many of the peptide transmitters, or closely related peptides, were initially isolated from amphibian species. Moreover, both classic and peptide transmitters are released in a calcium-dependent manner. However, the biosynthetic mechanisms and the modes of inactivation of peptide and classic transmitters are quite different. We discuss first the question of the significance of multiple neurotransmitters. We then describe the general principles of peptide transmitter biosynthesis and inactivation, which are illustrated by examining in detail one particular peptide transmitter, neurotensin.

Why Have So Many Transmitters?

Chemical neurotransmission, as we have loosely defined it, appears to be overwhelmingly redundant. There are about a dozen classic transmitters and literally dozens of neuropeptides that function as transmitters. If the role of neurotransmitters is to serve as a chemical bridge that conveys information between two spatially distinct cells, why have so many chemical messengers?

Several different factors, ranging from the intracellular localization of transmitters to the different firing rates and patterns of neurons, probably contribute to the need for multiple transmitters. The characteristics of neurons and neuronal communication that may require multiple transmitters are discussed below.

Afferent Convergence on a Common Neuron

Perhaps the simplest explanation for multiple transmitters is that many afferents terminate on a single neuron. It is apparent that a neuron must be able distinguish between the multiple afferent inputs

Copyright 2004, Elsevier Science (USA).
All rights reserved.

that bring information to the neuron. To some degree this can be accomplished by the site on a neuron at which an afferent terminates: at the axon, or soma, or dendritic spine, or shaft. However, since many afferents terminate in relatively close approximation, another means of distinguishing the inputs and their information is necessary. One way this can be accomplished is by chemically coding afferent neurons. The information conveyed by distinct transmitters is then distinguished by the different receptors present on a neuron and their various transduction mechanisms.

Co-localization of Neurotransmitters

A major conceptual revolution in the neurosciences over the past generation has been the realization that more than one neurotransmitter can be in the same cell. The idea that a neuron is limited to one transmitter can be traced to Sir Henry Dale or, more properly, to an informal restatement of what is termed Dale's principle. About 60 years ago Dale posited that a metabolic process that occurs in the cell body can reach or influence events in all the processes of the neuron. Sir John Eccles restated Dale's view to suggest that a neuron releases the same transmitter at all of its processes. Illustrating the dangers of the scientific equivalent of sound bites, this principle was soon misinterpreted to indicate that a single transmitter is present in a given neuron. This is clearly not the case. Neurons can co-localize two or more transmitters. For example, a neuron can use both a classic transmitter such as dopamine and a peptide transmitter such as neurotensin. Indeed, it now appears that few if any neurons contain only one transmitter, and in several cases three or even four transmitters have been found in a single neuron.

The presence of multiple transmitters in a single neuron may suggest that the information that neurons transmit to follower cells is encoded by different transmitters for different functional states. For example, the firing rates of neurons differ considerably, and thus it may be useful for neurons to encode fast firing by one transmitter and a slower firing frequency by another transmitter. In addition, the firing pattern of cells is of significance. For example, a neuron might show an absolute rate of 5 spikes every second. This frequency could result from a neuron discharging every 200 ms, or by discharging five times in an initial 200-ms period followed by 800 ms of silence. Recent data indicate that in cases of co-localization of peptides and classic transmitters, peptide transmitters are often released at higher firing rates and particularly under burst firing patterns (Bean and Roth, 1991).

In many ways, the different biosynthetic strategies used by peptides and classic transmitters may lead to differential release. Classic transmitters can be rapidly synthesized because their synthesis takes place in nerve terminals, in vesicles in which the biosynthetic enzymes are stored. In contrast, peptide transmitters must be synthesized in the cell body and transported to the terminal. Thus, it might be useful to conserve peptide transmitters for situations of high demand, since they would be rapidly depleted otherwise.

Transmitter Release from Different Processes: Axonal versus Dendritic Release

Neurons are polarized cells. This means that not only during development but also as mature cells they possess specialized regions for different functions, which may be thought of as heads and tails, or axons and dendrites. The prototypic site of release of transmitter is the region of the axon terminal that synapses onto a postsynaptic cell. However, transmitters are also released from dendrites. In addition, studies of peripheral nerves suggest that transmitters can be released from varicosities that are seen along the preterminal axons of some neurons. These different sites of transmitter release may be occupied by different transmitters.

As noted in Chapter 9, Eccles' restatement of Dale's principle states that a transmitter or other protein is present in all processes of a neuron. However, it appears that a transmitter can also be specifically localized to different parts of a neuron. For example, in the marine mollusk *Aplysia* different transmitters are targeted to different processes of a single neuron. Although differential targeting of transmitters in mammalian neurons has not been conclusively demonstrated, if a transmitter were restricted to a particular part of a neuron, it follows that the neuron would need multiple transmitters to account for different release sites. Consistent with this suggestion is the observation that a variety of neurotransmitter receptors are differentially distributed on mammalian neurons.

Synaptic Specializations versus Nonjunctional Appositions between Neurons

In addition to the diversity in transmitters that may result from transmitters being targeted to different intraneuronal sites, the anatomical relationships between one cell and its follower may possibly contribute to the need for different transmitters. We typically think of synaptic specializations (see Chapter 2) as the morphological substrate of communication between two neurons. However, there may also be nonjunctional appositions between two neurons, and it appears likely that transmitters are released at these sites.

The requirements for transmitter action at a nonjunctional apposition and a synapse would differ, since the distance traversed by the transmitter molecule would be larger than at a synaptic apposition. Thus, transmitters that lack an efficient reuptake system, such as peptide transmitters, might be favored at nonjunctional synapses. Since a single neuron can form both synaptic specializations and nonjunctional appositions, a single neuron may require more than one neurotransmitter.

Fast versus Slow Responses of Target Neurons to Neurotransmitters

We have seen that different firing rates or patterns are accompanied by changes in the transmitter being released from a neuron. The postsynaptic response to a transmitter can occur over different time scales. For example, receptors that form ion channels lead to very rapid changes on stimulation by a released transmitter, since the ionic gradients across the cell are almost instantaneously changed. In contrast, metabotropic receptors that respond to classical transmitters and peptide transmitters are coupled to intracellular events through specific transduction molecules, such as G proteins. Thus, metabotropic receptors respond to neurotransmitter stimulation on a slower time scale than do ionotropic receptors (see also Chapters 11 and 16). This difference in temporal response characteristics is useful, since it allows the receptive neuron to respond differently to a stimulus depending on the antecedent activity in the neuron. A transmitter can change the response characteristics of a particular cell to subsequent stimuli on the order of seconds or even minutes and, thus, can occur independent of changes in gene expression.

Comparison of Synthesis and Inactivation of Peptide and Classic Transmitters

There are two major differences between classic and peptide transmitters. Peptide transmitters are synthesized in the cell body, rather than at the terminal processes of neurons; this has significant functional consequences. In addition, peptide transmitters are inactivated by enzymatic actions and not by a reuptake process.

Synthesis and Storage of Peptide Transmitters

Classic transmitters are synthesized in the process (axon, dendrite) from which they are released, but peptide transmitters are not. In the majority of cases, peptide transmitter genes encode a prohormone, a larger precursor protein from which the peptide transmitter is subsequently cleaved. This synthesis from a larger precursor protein allows for an additional layer of synthetic strategy (see Box 10.1). The prohormone is incorporated into secretory granules after translation, where it can be acted on by peptidases, called prohormone convertases (Seidah and Chrétien, 1999), to form the functional neuropeptide. In contrast, classic transmitters are formed by successive small enzymatic transformations of a transmitter precursor, rather than from a larger precursor, and do not require transport to distal processes. Although the synthesis of a prohormone is the major strategy used for generation of peptide transmitters, certain small peptides can be enzymatically synthesized. An example is carnosine (N-β-alanyl-L-histidine), which is synthesized by the enzyme carnosine synthase.

Increases in the amount of a classic transmitter that is available for release occur by local synthesis. However, increasing the amount of a peptide transmitter requires an increase in gene expression of the prohormone mRNA, either by transcription of the gene or by stabilization of the mRNA (see Chapter 13); the subsequent delivery of the prohormone/peptide-containing granules to the terminal via axonal transport may take hours. Thus, classic transmitters can respond to increased demand for transmitter release quite rapidly, but peptide transmitters cannot. This difference in biosynthetic strategies contributed to the initial difficulties in localizing the sources of peptide-containing innervations of certain brain regions. Because peptides or their prohormones are transported from the soma immediately after translation from mRNA, the cell body region and proximal processes of these neurons typically contain very low concentrations of the peptide transmitters. Although immunohistochemical methods can easily localize the cell bodies of classic transmitters, to demonstrate the cell bodies in which peptide transmitters are formed it is usually necessary to disrupt microtubule-mediated axonal transport to allow the peptide to accumulate in the soma.

The methods of storage of peptide and classic transmitters also differ. Classic neurotransmitters are stored in small (~50 nm) synaptic vesicles; neuropeptide transmitters are stored in large (~100 nm) dense-core vesicles. Since peptide transmitters are typically released at a high neuronal firing frequency or burst firing pattern, it is reasonable to assume that there are different mechanisms for the exocytosis and subsequent release of peptide and classic transmitter vesicles. Recent data suggest that there are distinct but

BOX 10.1

COORDINATE SYNTHESIS OF MULTIPLE PEPTIDES IN A SINGLE PRECURSOR

The biosynthetic pathway for a neuropeptide follows that of most secreted proteins in that there is a signal peptide at the N terminus that directs the polyribosome to the rough endoplasmic reticulum (RER) for co-translational vesicular discharge of the prohormone into the lumen of the RER. This also dictates a minimal size to the prohormone of approximately 60–70 amino acids for the signal peptide to emerge from the ribosome and be recognized by the signal recognition particle. As mature neuropeptides range in size from few to tens of amino acids, nature has evolved to use the extra "spacer" peptide material in many instances. Figure 10.1 shows many of the different strategies that have evolved, from coordinate synthesis of different peptides that work at different receptors to multiple copies of the same peptide. The presence of multiple neuropeptides within one precursor protein can also allow for differential processing of the same prohormone in different cell types. For example, the pro-opiomelanocortin (POMC) prohormone is processed to adrenocorticotrophic hormone (ACTH) in the anterior pituitary, but ACTH is further processed to α-melanocyte-stimulating hormone (α-MSH) in the hypothalamus. Another interesting strategy is reflected in the synthesis of two neuroendocrine peptides, vasopressin and oxytocin, which are made in the brain and released for action into the peripheral circulation. In this case, the prohormone also contains the binding protein, neurophysin, which aids in the transport of the neuropeptide in the bloodstream. All of these different strategies give an additional level of complexity to synthesis of neuropeptides, adding to their uniqueness from the classic neurotransmitters.

FIGURE 10.1   The synthesis of neuropeptides in a larger precursor form allows for multiple different synthetic strategies.

Strategies for precursor peptide synthesis

Multiple copies of the neuropeptide

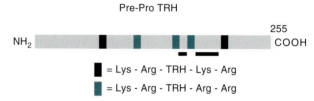

Coordinate synthesis of different peptides

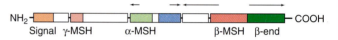

Coordinate synthesis with carrier protein

Allow for proper folding and S-S bonding

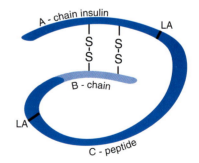

related molecular mechanisms that subserve release of small and large dense-core vesicles (Bean *et al.*, 1995). The release of peptide transmitters, like the classic transmitters, is calcium-dependent (see Chapter 8 for further discussion).

Inactivation of Peptide Transmitters

Different strategies are used to synthesize peptide and classic transmitters; these differences are paralleled by differences in inactivation of the released transmitter. Classic transmitters have high-affinity reuptake processes that remove the transmitter from the synaptic or extracellular space. In contrast, peptide transmitters appear to be inactivated enzymatically or by diffusion, but lack a high-affinity reuptake process. The enzymatic inactivation of peptide transmitters also differs from that of classic transmitters. Since peptide transmitters are short chains of amino acids, the inactivating enzymes show specificity for certain types of amino acids at the cleavage site but are not specific to any single peptide (see Box 10.2). For example, the metalloendopeptidase that inactivates enkephalins, which are small pentapeptide opioid-like transmitters, is frequently called enkephalinase, but is also critically involved in the inactivation of other neuropeptides, such as neurotensin and somatostatin. One final difference in inactivation of peptide and classic transmitters is the product. In the case of classic transmitters that are catabolized, the product is inactive at the receptor site, hence the term *inactivation*. However, in the case of peptide transmitters, certain peptide fragments derived from the enzymatic "inactivation" of the peptide are biologically active. For example, the peptide angiotensin I is metabolized to yield angiotensin II and angiotensin III, which are successively more active than the parent peptide. Hence, it is difficult to distinguish between extracellular synthetic processing (of a prohormone) and inactivation. The peptide that is stored in vesicles and then released is therefore considered the transmitter, although the actions of certain peptidases may lead to other biologically active fragments. This is an exciting area of investigation in neuroscience as more becomes understood on how the peptide transmitters are metabolized *in vivo* (see Box 10.2).

Approaches to the Study of Neuropeptide Synthesis and Release

The unusual aspects of peptide transmitter biosynthesis and inactivation have resulted in different methods being emphasized in the study of peptide transmitters and classic transmitters.

Anatomical approaches are used extensively to define the neurons in which peptides serve as neurotransmitters. Immunohistochemistry, using antibodies generated against the prohormone, can be used to identify neurons in which peptide transmitters are synthesized. Similarly, antibodies against the peptide transmitter itself are often used to localize the peptide to cell bodies, particularly when axonal transport has been blocked by prior administration of colchicine. Recent anatomical studies have emphasized the use of *in situ* hybridization histochemistry, in which cRNA probes or oligonucleotides complementary to a defined sequence of an mRNA are used to identify the cells in which the gene encoding a peptide prohormone is transcribed. The advantage of this approach is that localization of the mRNA is unaffected by axonal transport of secretory granules since peptide synthesis takes place in the cell body.

Biochemical studies of peptide synthesis and inactivation also use approaches different from those undertaken in the study of classic transmitters. The synthesis of peptide transmitters has been extensively studied by following the rapid incorporation of radiolabeled amino acids into peptides, in so-called "pulse–chase" experiments. More recent studies have emphasized molecular approaches analyzing specific mRNAs, since a prohormone from which the peptide will subsequently be cleaved is directly transcribed from mRNA.

The release of peptides has been studied in slices of brain, much as has been the case with classic transmitters. *In vivo* studies of release have not been possible until relatively recently, but can now be accomplished. The measurement of metabolites as an index of release is not a good method, since the peptide fragments generated may be similar to those of other peptides and since larger peptides are further degraded to smaller peptide fragments. This requires antisera with exquisite specificity for the parent peptide but not fragments; such antibodies are rare indeed. Nonetheless, if one knows the principal peptide fragments catabolically generated from the released peptide, one can use immunoassays coupled with chromatographic separation methods to measure levels of the parent peptide and certain peptide fragments to gain an appreciation of the release of the peptide. The introduction of *in vivo* microdialysis methods has been useful for the measurement of extracellular concentrations of neuropeptides, particularly small neuropeptides. Since the released peptide in relatively quick order diffuses across the dialysis membrane, while the peptidases are generally too large to do so, one can obtain measurable levels of

BOX 10.2

NEUROPEPTIDE-METABOLIZING ENZYMES COME IN A VARIETY OF DIFFERENT FORMS

Neuropeptide metabolism takes place after the peptide has been released into the extracellular space. Unlike protein synthesis, the cleavage of a peptide bond requires no outside energy input, such as ATP, and hence can take place readily in the extracellular environment. The neuropeptidases that mediate peptide metabolism are a more diverse group than the subtilisin-like family of enzymes that constitute the majority of the prohormone convertases (Seidah and Chrétien, 1999). Essentially all classes of peptidases have been shown to be involved in neuropeptide metabolism, from the Zn-containing endopeptidases (which cleave internal to the peptides) enkephalinase and thimet oligopeptidase, to the aspartylpeptidase, post-proline cleavage enzyme, to exopeptidases (which cleave from the outside in) like aminopeptidase N. Often, the N and C termini of neuropeptides are blocked by posttranslational modifications such as N-acetylation and C-amidation, which protect them from exopeptidase degradation. Thus, the neuropeptides are protected from degradation until they are cleaved internally by endopeptidases. The peptidases are broadly distributed in the nervous system, reflecting the broad distribution of neuropeptides.

Several of these metabolizing peptidases, like enkephalinase, are membrane anchored proteins, synthesized as classic membrane proteins with an N-terminal signal sequence and a membrane-spanning C-terminal sequence that locks the peptidase with its active site facing the lumen of the RER which subsequently becomes an extracellular plasma membrane-facing enzyme. Others, however, are synthesized as soluble cytosolic enzymes, such as thimet oligopeptidase, that are released from neurons or glia via a yet uncharacterized mechanism. These peptidases either remain soluble, become associated with extracellular matrix, or become anchored on the extracellular face of the cell and function to metabolize neuropeptides, either degrading or converting them to a different biological activity (see Fig. 10.2). An exciting area of current research deals with the factors involved in regulating the location of these metabolizing peptidases at the cell surface and their proximity to the receptors that bind the peptides they metabolize. This becomes a crucial issue since peptides are not directly released into the synapse and generally must diffuse to their site of action, providing an opportunity for peptidase action.

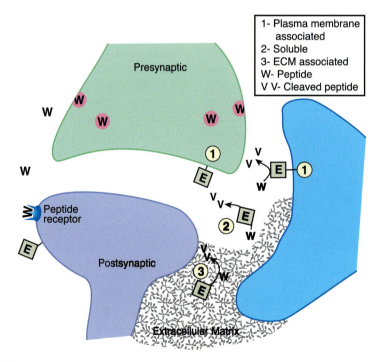

FIGURE 10.2   Localization of neuropeptidases (E) in the extracellular environment.

peptides in the dialysate. The dialysis approach, however, requires very sensitive analytic methods because of the poor recovery of peptides, and does not offer good temporal resolution. Another *in vivo* approach has been to insert electrodes coated with antibodies into a specific area of the brain for a short period. The removed probes are then exposed to a radiolabeled peptide, which will bind to receptors that are not already occupied by the peptide that was released endogenously. This method does not allow repeated measurements, and requires very precise experiments to optimize the time the probe is left in the brain.

Pharmacological studies of neuropeptide transmitters have been hampered by the lack of specific compounds that interact with peptide receptors. Many specific peptide analogues of neurotransmitter peptides have been synthesized. However, while these peptides interact specifically with appropriate receptors, peptides and proteins do not readily enter the brain and, thus, have been of very limited utility in *in vivo* studies. Recently, several nonpeptide antagonists have been developed that enter the central nervous system with relative ease. In contrast, there are very few nonpeptide agonists that can be used to study central peptide transmitters; a notable exception is morphine and related opiates used to label central opioid receptors.

NEUROTENSIN AS AN EXAMPLE OF PEPTIDE NEUROTRANSMITTERS

Neurotensin (NT) is a peptide of 13 amino acids that is expressed in the central nervous system and in peripheral tissues, particularly the small intestine. In addition, another peptide termed neuromedin N (NMN) is also transcribed from the gene that encodes NT. NMN is a structurally related hexapeptide that occurs in mammals; a nearly identical peptide called LANT-6 is found in birds. Peptides often have central and peripheral roles, one involving chemical communication between neurons and the other involving varied peripheral functions. Indeed, neurotensin and many other mammalian brain peptides are structurally similar to peptides that are found in nonmammalian organisms, particularly amphibian species, and a large number of peptides were discovered in the skin of certain toads. An example is xenopsin, a neurotensin (NT)-like octapeptide, which was isolated from the skin of *Xenopus*.

Neurotensin Synthesis

A 170-amino-acid prohormone precursor of NT is encoded by a single gene that is transcribed to yield

two mRNAs (see Fig. 10.3). The smaller transcript is the predominant form in the intestine, while both mRNA species are present in equal abundance in most brain areas (Kislauskis *et al.*, 1988). The 170-amino-acid precursor contains one copy each of NT and NMN. However, differential processing of the precursor can occur, leading to different molar ratios of the two peptides.

NT is present in very high concentrations in the small intestine and in lower amounts in the stomach and large intestine. NMN is also present in these tissues, but NT:NMN molar ratios differ across the different tissues, suggesting differential enzymatic processing of the prohormone or the generation of different transcripts. Because NT and NMN are contained in the same exon of the NT/NMN gene, differences in relative abundance of NT and NMN are due to differential processing of the precursor.

Storage and Release of Neurotensin

Neurotensin and NMN, and other peptide transmitters, are stored in large dense-core vesicles. Studies of the release of NT and NMN have been examined in the hypothalamus, an area of the brain enriched in NT and NMN. Superfusing hypothalamic tissue with high concentrations of K^+ to depolarize cells evokes release of both NT and NMN (Kitabgi *et al.*, 1992); the molar ratio of NT:NMN release is virtually identical to that in extracts of hypothalamic tissue. The release of NT and NMN is not seen in tissue perfused with a medium that lacks calcium. Thus, the release of the peptides is calcium-dependent (reflecting dependence of vesicular docking with the membrane and exocytosis on calcium) and is evoked by stimulation. These criteria are considered to be necessary for the designation of a compound as a neurotransmitter.

As noted earlier, the release of peptides such as NT is impulse-dependent. NT release is regulated by both the firing frequency and firing pattern of neurons: when neurons discharge rapidly in bursts, NT release is markedly enhanced (Bean and Roth, 1992).

Inactivation of Neurotensin

There is no high-affinity reuptake process for peptides. Accordingly, other than possible diffusion, the major means of inactivating peptide transmitters that are released from a neuron is enzymatic.

NT is enzymatically inactivated by a group of enzymes known as metallo-endopeptidases. In particular, three of these endopeptidases (known with great flair as EP24.11, EP24.15, and EP24.16) are involved in the catabolism of NT. In both brain and gut, NT is

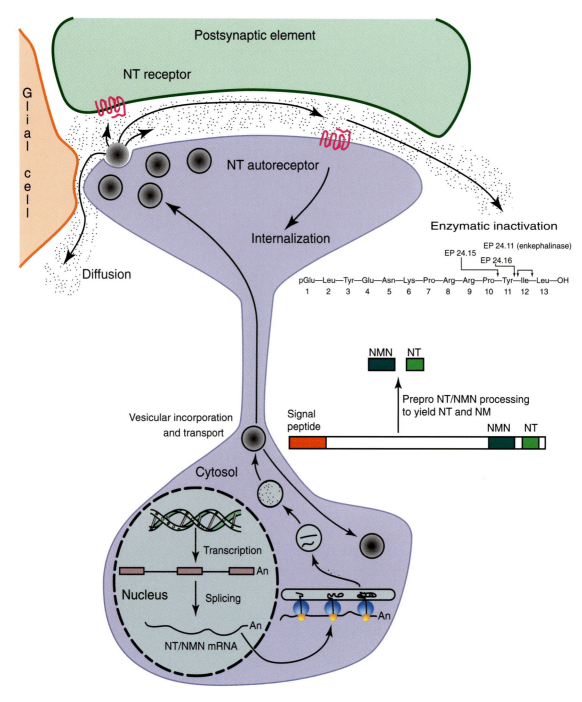

FIGURE 10.3   Schematic illustration of neurotensin (NT) and neuromedin (NMN) biosynthesis and action, from gene transcription to translation and processing in the vesicular secretory pathway to diffusion from the synaptic cleft and inactivation by a group of endopeptidases (EP) extracellularly.

hydrolyzed by combinations of the three endopeptidases (Kitabgi *et al.*, 1992). Endopeptidase 24.11 (also known as enkephalinase for its well-characterized actions on enkephalins) cleaves NT at two sites, the Pro[10]–Tyr[11] and the Tyr[11]–Ile[12] residues. Endopeptidase 24.15 acts at the Arg[8]–Arg[9] site and endopeptidase

24.16 acts at the same Pro[10]–Tyr[11] site as enkephalinase. Thus, the enzymes involved in peptide degradation, including those that metabolize NT, do not exhibit a high degree of specificity (see Box 10.2).

We noted in the previous chapter that certain proteins that have important functions in neurotransmission (such as acetylcholinesterase) may have very different functions in other (nonneural) systems. Enkephalinase, which hydrolyzes NT, somatostatin, and enkephalins, appears to be similar to acetylcholinesterase by having a role in leukemia. The common acute lymphoblastic leukemia antigen (CALLA) is a protein expressed by most acute lymphoblastic leukemias and normal lymphoid progenitors. The human CALLA cDNA encodes a protein that appears nearly identical to that of enkephalinase (Shipp et al., 1989).

As noted earlier, there is no high-affinity reuptake process for peptides. Nonetheless, certain peptides, including NT, have been shown to be accumulated by neurons. The high-affinity NT receptor is a G protein-coupled receptor; such receptors, on binding of ligand, are internalized and thus accumulate the ligand into the neuron. A series of studies have revealed that NT, when injected into the striatum, is internalized by nerve terminals that express the high-affinity NT receptor; once internalized, the peptide dissociates from the receptor and is retrogradely transported to the cell bodies of origin, which are dopaminergic neurons in the midbrain.

The functional significance of internalization of NT and its subsequent retrograde transport is not clear. It could conceivably serve as a growth factor, regulating the targeted growth and survival of afferents to the striatum. An intriguing possibility is that NT may actually function over a relatively extended period to alter the transmitter function of the cell (Deutch and Zahm, 1992). Local injections of NT into the striatum result in retrograde labeling of midbrain dopamine neurons and increase the number of midbrain cells expressing tyrosine hydroxylase mRNA.

Coexistence of Neurotensin and Classic Transmitters

Neurotensin is found in many dopaminergic neurons in mammalian species, including certain hypothalamic and midbrain dopamine cells. In particular, certain midbrain DA neurons of rodents use NT as a transmitter. Since all of the neurotensin in the prefrontal cortex of the rat is thought to be contained in DA axons, the study of the rodent prefrontal cortex has afforded a unique opportunity to investigate interactions between co-localized transmitters in vivo. Such studies have revealed that neurotensin release in the prefrontal cortex is increased when neuronal firing is increased, or when DA neurons enter into a burstlike firing pattern (Bean and Roth, 1992). In addition, there appears to be a close association between the dopaminergic autoreceptor present on DA axons

in the prefrontal cortex and NT release from these axons. In contrast to most forebrain areas innervated by DA axons, transmitter release but not synthesis is regulated by DA autoreceptors in the prefrontal cortex. Low autoreceptor selective doses of dopamine agonists, which decrease subsequent dopamine release, enhance neurotensin release from these axons; conversely, antagonists at the release-modulating autoreceptor decrease NT release but enhance DA release. Thus, NT and dopamine release are regulated in a reciprocal fashion by autoreceptors in neurons in which both transmitters are present.

Interestingly, although neurotensin-containing cells are found in many areas of the primate brain, very few midbrain dopamine cells express the NT/NMN gene in primate species, and the projection target of these neurons is not known.

UNCONVENTIONAL TRANSMITTERS

The designation of a substance as a neurotransmitter rests on the fulfillment of certain criteria, which were discussed in Chapter 9. However, these criteria were formulated early in the modern neuroscience era, and are based mainly on studies of peripheral sites, particularly the neuromuscular junction and superior cervical ganglion. Over the past several decades there have been remarkable increases in technical sophistication that allow us to measure substances in the brain that are present in minute quantities or are very unstable. The ability to measure these substances has opened the possibility that several substances that previously did not meet the requirements for designation as a neurotransmitter may indeed be transmitters.

The advances in technical wizardry, however, have also led us to unanticipated mechanisms of intercellular signaling that have required a new definition of neurotransmitters. One simple approach would be to designate a chemical mediator a neurotransmitter if it allows information to flow from one neuron to another. This definition circumvents the issues of glial contribution to the ionic milieu of the neuron, which certainly imparts information concerning the function of the glia (and parenthetically points to the increasing awareness of the active roles that glia play in the CNS—see Box 10.3). On the other hand, such a definition closely approximates the definition for a hormone, in that the temporal characteristics of transmitter action are not defined, nor is the distance of the target cell to the transmitter substance specified. Finally, this definition does not accommodate unconventional roles for transmitters, such as the regulation

BOX 10.3

ASTROCYTES: FROM SUPPORT PLAYER TO CENTER STAGE

For much of the past century neuroscience has focused on the integrative activity of neurons. Although neurons are certainly essential elements in central nervous system function, it is increasingly clear that glial cells, particularly astrocytes, are important: these cells are excitable and can send and receive signals using mechanisms similar to those employed by neurons. The diversity of signaling and contributions of astrocytes was not appreciated initially because recordings of the membrane potential of an astrocyte offered little to observe. The membrane potential is stable and maintained at about −80 mV, and regardless of the magnitude of depolarization elicited by current injection, an action potential is never elicited. Although this stable negative resting potential is critical to one function of the astrocyte, the electrogenic uptake of glutamate from the synaptic cleft, it does not reflect the diversity of regulated biochemical signaling that we are beginning to appreciate in these nonneuronal cells.

Technical developments over the past 15 years have allowed us to determine that astrocytes exhibit a form of nonelectrical excitability that depends on calcium. Neurotransmitters, which we commonly think of as the signals used at neuronal chemical synapses, also elicit calcium oscillations in astrocytes. These calcium signals arise from the release of calcium from internal stores, which is gated by metabotropic receptor-dependent activation of phospholipase C and the production of inositol trisphosphate. What, however, are the functional consequences of such calcium signals? One of the major surprises in the study of astrocytes has been the observation that even relatively small elevations of astrocytic intracellular calcium lead to the calcium-dependent release of the chemical transmitter glutamate! In addition, ATP, D-serine, and prostaglandin E_2 can also be released by astrocytes in response to metabotropic receptor activation. Thus, astrocytic calcium excitability results in the release of a variety of chemical transmitters that can regulate both local astrocytes and neurons.

The concept of tripartite synaptic transmission has emerged in recent years, with the astrocyte being considered an active participant in the control of communication between two neurons. Many synapses are associated with an astrocytic process. A series of complex studies have been performed to determine if active signaling is present between the neuronal and glial elements of the tripartite synapse. For instance, stimulation of presynaptic neuronal afferents not only evokes postsynaptic potentials but also stimulates calcium oscillations in synaptically associated astrocytes. Moreover, experimentally induced calcium increases in astrocytes can modulate neighboring synapses as a result of the release of "gliotransmsitters." These functional studies suggest that astrocytes, as part of the tripartite synapse, integrate neuronal inputs and provide feedback regulation of the synapse by way of the calcium-dependent release of chemical transmitters.

Calcium signaling within astrocytes is a complex process, and does not exhibit simple all-or-none features. Synaptic activity can evoke calcium elevations within local portions of the processes of a single astrocyte; such local increases in Ca^{2+} do not necessarily propagate throughout the entire cell. However, increased neuronal activity can switch the behavior from a local, synaptically associated process to a more global, cell-wide Ca^{2+} elevation. Moreover, cellwide Ca^{2+} elevations can propagate to neighboring astrocytes: recent studies suggest that astrocytes may be interconnected with one another in short-range circuits. Since each astrocyte contacts thousands of synapses, the distance over which a calcium signal spreads within one astrocyte and whether signals spread between interconnected astrocytes are likely to play a major role in modulating synaptic transmission.

These are early days in a new field of study. The full extent to which astrocytes are involved in the regulation of the synapse remains to be determined. Although most studies of astrocytic function have been performed using cell cultures, recent slice and *in vivo* studies have made it clear that synaptically associated glia serve a regulatory role in brain regions such as the hippocampus, as well as regions including the retina and neuromuscular junction. Whether biochemical integration in these nonneuronal cells plays roles in neuronal processes such as synaptic plasticity awaits a new level of experimental study. Regardless, the fact that astrocytes integrate neuronal inputs and exhibit calcium excitability that controls the release of chemical transmitters changes the way in which we view the roles of these cells in nervous system function.

Philip G. Haydon, Ph.D.

of neuronal development and intracellular trafficking of proteins. Since growth factors can clearly influence neuritic outgrowth and cell survival, as well as influence neuronal polarity and other various functions thought to involve changes in intracellular trafficking, the definition of transmitter changes as one considers what we have termed "unconventional" transmitters.

The discovery of peptidergic neurotransmitters over a generation ago was accompanied by considerable debate over the definition of a transmitter. Yet peptide transmitters are quite similar to classic transmitters, sharing such features as conventional storage, calcium-dependent release, and the ability to influence the activity (e.g., the firing rate) of postsynaptic neurons over a relatively brief interval. Given these basic similarities, and the broad acceptance of peptide transmitters into the language of neurotransmission, it seems necessary to contrast peptide transmitters with other substances that influence target cells in unique ways. Accordingly, we have designated these novel "transmitters" as unconventional, at least for a few years. Among these are neuroactive gases (nitric oxide and carbon monoxide), growth factors, and neurosteroids.

Nitric Oxide and Carbon Monoxide: Gases as Unconventional Transmitters

Nitrates have been extensively used in the treatment of cardiovascular disorders to dilate blood vessels of the heart and thereby relieve the symptoms of angina, but the mechanisms of action of nitroglycerine and similar nitrates were not known until recently. In 1980, an endothelium-derived relaxing factor present in cells lining blood vessels was shown to potently and rapidly dilate blood vessels. Shortly thereafter it was demonstrated that this endothelial factor was a gas, nitric oxide. In addition, it was observed that glutamate, acting at NMDA receptors, releases a factor that causes vasodilation and increases cGMP levels in brain. It soon became apparent that the endothelium-derived relaxing factor and the glutamate-induced factor that dilate blood vessels were the same factor, and that nitric oxide (NO) was present in neurons as well as vasculature. These data, coupled with the identification of neuronal as well as vascular isoforms of the NO synthetic enzyme (nitric oxide synthase), led to the concept that NO is a molecule involved in intercellular communication, including neurotransmission.

Nitric oxide is a well-known air pollutant and, thus, an unlikely candidate for a neurotransmitter. The idea that an unstable toxic gas could be a transmitter led to several questions concerning the nature of neurotransmission, the most obvious being how a gas can be stored for release in an impulse-dependent manner. Because the simple answer is it can't, the classic definition of a neurotransmitter has become blurred or untenable, depending on one's perspective. Many theories can accommodate one exception. However, subsequent studies revealed that NO is not the only gaseous neurotransmitter and that carbon monoxide plays a similar transmitter-like role (Dawson and Snyder, 1994). These findings led to the realization that neurotransmitters as classically defined may be the exception rather than the norm. The exceptions posed by NO to the dogma of traditional neurotransmitters are substantial. For example, NO is not stored in cells, is not exocytotically released, lacks an active process that terminates its action, does not interact with specific membrane receptors on target cells, and regulates the function of synaptic terminals proximal to the neuron in which NO is synthesized. It is not difficult to understand the skepticism that met the hypothesis that gases such as NO could be neurotransmitters, or to have some sympathy for the view that NO is not a transmitter but some alien event, intent on making neuroscientists question their most cherished beliefs.

Synthesis of Nitric Oxide

The synthesis of NO is not complicated, consisting of one step: the conversion of L-arginine to NO and citrulline (see Fig. 10.4). The enzyme responsible for this step is nitric oxide synthase (NOS). Three distinct isoforms of NOS have been cloned. Macrophage-inducible NOS (iNOS) is present in microglia, while endothelial NOS (eNOS) is found in the endothelial cells lining blood vessels, and the localization of neuronal NOS (nNOS) is obvious. All three forms require tetrahydrobiopterin as a cofactor and NADPH as a coenzyme. In fact, NOS is identical to the previously described enzyme NADPH diaphorase, which is expressed in high concentrations in several types of neurons in the CNS.

Regulation of Nitric Oxide Levels

NOS is among the most regulated of enzymes, and NO levels in various tissues can be modified in several ways. Although the general use of the term inducible NOS for macrophage-inducible NOS may lead one to suspect that only iNOS is regulated, all NOS isoforms are regulated. Regulatory processes include phosphorylation (which decreases NOS activity) and hormonal control. In addition, levels of NOS can be modified by direct inhalation of NO.

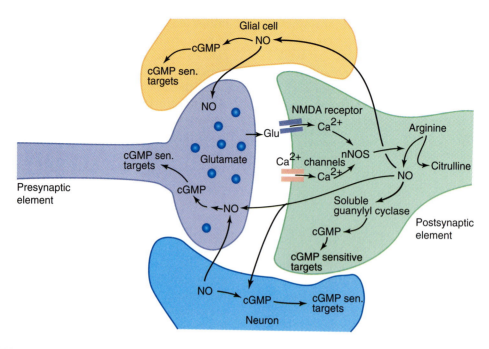

FIGURE 10.4 Schematic representation of a nitric oxide (NO)-containing neuron. NO is formed from arginine by the actions of different nitric oxide synthases (NOS). NO freely diffuses across cell membranes and can thereby influence both presynaptic neurons (such as the glutamatergic presynaptic neuron in the figure) or other cells that are not directly apposed to the NOS-containing neuron; these other cells can be neurons or glia.

Storage, Release, and Inactivation of Nitric Oxide

The criteria for classic transmitter status include intraneuronal synthesis of the substance and release from these cells by calcium-dependent exocytosis. These criteria imply an intermediary storage step. However, NO is not in stored in vesicles, since NO is an uncharged molecule that diffuses freely across cell membranes. Nor is NO released by calcium-dependent exocytosis.

Although NO is synthesized in neurons, the ability of the gas to diffuse across membranes allows it to modify the activity of other cells. The lack of storage conditions and exocytotic release suggests that NO simply modifies the activity of targets in a tonic fashion, but the tight regulation of NOS and the critical observation that neuronal stimulation elicits NO release indicate a more phasic (and, hence, more "transmitter-like") role.

NO also differs from conventional transmitters by lacking an active process to terminate its actions. Instead, inactivation of NO is essentially passive: NO decays spontaneously to nitrate with a half-life of less than 30 s. Another means of terminating the action of NO is by reaction of the gas with iron-containing compounds, including hemoglobin.

Actions of Nitric Oxide Synthase on Receptive Neurons and Functional Effects

Yet another unique aspect of NO as a neurotransmitter is that NO does not interact with specific receptor proteins on postsynaptic target cells, but instead interacts with second-messenger molecules in the target cell to initiate a cascade of intracellular processes. In particular, NO stimulates guanlylyl cyclase to increase cGMP levels.

A final aspect of NO that confounds the traditional boundaries of neurotransmitters is the cellular target of NO. Conventional neurotransmitters have effects on postsynaptic cells that are "downstream" of the neuron releasing the transmitter. In contrast, in many cases NO acts as a retrograde messenger, modifying the metabolism and release of transmitter from presynaptic terminals. For example, NO appears to be released from certain hippocampal neurons after NMDA receptor stimulation to enhance transmitter release from presynaptic elements; this role has been suggested to be of critical importance in long-term potentiation (O'Dell *et al.*, 1994). The release of NO on NMDA receptor stimulation may also underlie potential neurotoxic actions of NO, which is highly reactive. It has been suggested that an excess release of NO associated with NMDA receptor stimulation may be important in the toxic cell loss caused by strokes.

Carbon Monoxide

A large and still growing body of literature documents the function of NO in the brain and argues for NO as an unconventional neurotransmitter. Despite these data, the acceptance of NO as an unconventional transmitter would have been delayed considerably were it not for data indicating that a sibling of NO, carbon monoxide (CO), is also a neurotransmitter.

There are several striking parallels between NO and CO: both are gases that diffuse freely across cells, freely target guanlylyl cyclase rather than a membrane protein as a receptor, and lack a cellular storage compartment. Although CO has not yet received the attention devoted to NO, awareness of the role of CO as an unconventional transmitter is growing.

CO is formed by the enzyme heme oxygenase, which catalyzes the conversion of heme to biliverdin with the accompanying liberation of CO. There are two major isoforms of heme oxygenase (HO). HO-1 is an inducible enzyme that is present in glia and a few neurons, and is present in very high concentrations in the liver and spleen. In contrast, HO-2 is constitutively active rather than inducible, and is found in high concentration in central neurons, particularly in the cerebellum and hippocampus (Zakhary *et al.*, 1996). Cytochrome P450 reductase is a necessary electron donor for heme oxygenase and NOS.

Functional studies of CO are relatively recent and it is difficult to gauge the degree to which CO may parallel NO. Recent data, however, do suggest that CO (like NO) may be an endothelium-derived relaxing factor (Zakhary *et al.*, 1996). Moreover, recent studies of eNOS knockout mice indicate that cerebral circulatory responses are relatively normal, despite the absence of eNOS, and suggest that CO may function cooperatively with NO to maintain vascular tone.

Growth Factors

The discussion of nitric oxide illustrated several striking differences between conventional and unconventional transmitters. Another class of unconventional transmitters, the growth factors, share some features with the transmitter gases but have still other properties that differentiate them from classic and gas transmitters.

Growth factors are an extremely heterogeneous group of proteins that regulate the growth and differentiation of various cell types, some of which specifically target neurons (see Russell, 1995; Thoenen, 1995). One similarity between the neurotrophic growth factors and NO is that both are capable of influencing presynaptic cells. Thus, NO has been shown to diffuse out of cells and alter transmitter release from presynaptic axons in the hippocampus, and neurotrophic factors provide trophic support for developing axons innervating a region in which growth factors are expressed in cells (hence the term trophic factors). Growth factors are also unconventional in that the release of these proteins may be through both the constitutive and regulated pathways.

There are several different classes of growth factors, each class comprising many different growth factors (see Fig. 10.5). For example, the neurotrophins include nerve growth factor, brain derived-neurotrophic factor, neurotrophin 3 (NT-3), NT-4, and NT-5 (Chao, 2003). In many cases, growth factors were originally identified from functions in peripheral tissues, but have subsequently been found to have profound effects in brain, for example, fibroblast growth factor. Unfortunately, there are many gaps in our knowledge of the basic cellular biology of any single growth factor, even one known as long as nerve growth factor. Accordingly, we draw on examples from several different growth factors to illustrate what may be general characteristics of relevance.

Synthesis of Growth Factors

Growth factors have been identified historically by examining the effects of crude extracts of certain tissues or biological fluids on a bioassay of cell survival or growth. Following the determination that some serum or tissue factor is capable of sustaining cell survival or differentiation, further purification of the crude extract is performed, culminating in the isolation and purification of the protein. The protein is then sequenced. This approach is tedious, requires very large amounts of starting material, and requires a sensitive and reliable bioassay. This approach is essentially the same—and therefore confronted with the same difficulties—as the early methods used to discover and characterize pituitary and central peptides. Contemporary approaches to the identification of growth factors emphasize cloning of growth factors on the basis of sequence homology; the cloned genes are then expressed in various cell lines and tested in various bioassays. This approach has allowed us to gain some insight into potential regulatory features of growth factors and their cellular biology.

Molecular approaches have provided interesting and unexpected information on the synthesis of growth factors. For example, certain growth factors are translated from multiple mRNAs. Brain-derived growth factor (BDNF) is a neurotrophin that has a heterogeneous distribution in the CNS and several distinct functions, some of which are those classically associated with growth factors, others of which are

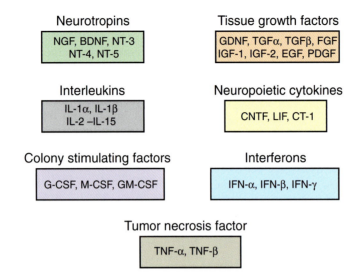

FIGURE 10.5   Outline of the different families of growth factors in the brain.

more "transmitter-like." The BDNF gene has five different exons that encode the mature BDNF protein. There is a separate promotor for each of four 5' exons, and alternative use of these promoters gives rise to eight different BDNF mRNAs. Despite this wonderful complexity in the molecular "synthesis" of BDNF, there is only one mature BDNF protein; it has been speculated that various BDNF transcripts may give rise to different amounts of protein and are differentially associated with pre- and postsynaptic elements (see Chapter 13).

Surprisingly little is known about the posttranslational processing of growth factors. In the case of the neurotrophin nerve growth factor (NGF), the mature protein is cleaved from a prohormone, in a manner similar to the synthesis of peptide transmitters. The processing of the NGF prohormone is unique, however, in that three subunits are present. One of these, the γ subunit, is a serine protease that is thought to cleave the prohormone to yield the mature protein. Other neurotrophins, including BDNF and NT-3, are also formed by cleavage of a prohormone, although the identity and degree of specificity of the responsible enzymes are only now being determined. One final aspect of growth factor biosynthesis (in its broad definition) is their unusual multimeric assembly. The neurotrophins form biologically active heteromers *in vitro*, but *in vivo* have been thought to form exclusively homodimers. However, recent data suggest that heterodimers of NGF, BDNF, and NT-3 can be formed both *in vitro* and *in vivo* (Heymach and Shooter, 1995).

Storage and Release of Growth Factors

Surprisingly little is known about the storage and release of growth factors. There have been very few studies of the storage of growth factors, particularly electron microscopic studies of intracellular localization. In part this reflects the difficulty in generating specific antibodies to growth factors, since individual growth factors (e.g., BDNF) within a family (neurotrophins) share a striking degree of homology with other members of the family (e.g., NT-3). However, the production of new specific antisera, in conjunction with studies of localization of growth factors in cells transfected with specific growth factor genes, has made it is possible to determine the intracellular localization of some growth factors.

Neurons secrete proteins by two distinct processes. The constitutive pathway for secretion is not triggered by extracellular stimulation, and is used to secrete membrane components, viral proteins, and extracellular matrix molecules. In this pathway there is continuous fusion of Golgi-derived vesicles with the plasma membrane. Growth factors have generally been considered to be secreted by the constitutive pathway. In contrast, the release of conventional neurotransmitters is controlled by extracellular signals and uses the so-called regulated pathway. For example, peptide protein precursors contain an N-terminal signal sequence, which targets the protein to the endoplasmic reticulum and then to the Golgi network ultimately to be packaged into vesicles.

Because some growth factors lack a signal sequence, the growth factors have generally been considered to be processed by the constitutive pathway, despite clear data indicating release of these proteins. Recent studies of neurotrophins suggest that BDNF and NGF are secreted through both the constitutive and regulated pathways. Using antibodies to examine the localization of BDNF in a transfected cell line, Goodman *et al.* (1996) showed that BDNF is present in chromogranin-containing secretory granules. This observation suggests that BDNF may be present in certain vesicles in neurons, but this remains to be shown.

Several published studies on BDNF and NGF suggest that under basal conditions there is constitutive release from the soma and proximal dendrites, but that depolarization induced by high potassium concentration or glutamate leads to regulated release from both the distal and proximal processes of the neuron (Goodman *et al.*, 1996; Blochl and Thoenen, 1996). After seizures, BDNF can be seen to move into the axons of hippocampal pyramidal cells, and then into perikarya, suggesting that it is poised to be released; 2 to 3 h after seizures, BDNF immunoreactivity is observed in the neuropil surrounding pyramidal cells and in the area where mossy fibers terminate, suggesting that the growth factor has been released from somatodendritic sites (Wetmore *et al.*, 1994). The activity-dependent release does not appear to depend on extracellular calcium, but is critically dependent on intracellular calcium stores (Blochl and Thoenen, 1996); this is distinct from release of conventional transmitters, which has an absolute dependence on extracellular calcium. Thus, the release of (at least certain) growth factors differs from that of conventional transmitters by occurring through both constitutive and regulated pathways and by not being dependent of extracellular calcium concentrations.

Functional Significance of Growth Factors as Neurotransmitters

Growth factors have a wide array of functions. As indicated by their name, growth factors support the development and differentiation of neurons. Thus, the survival and axonal growth of many neurons to their final target require certain factors expressed in the targeted cells. These functions are developmentally specific. Once the axon of a neuron reaches its target and survives, it usually does not require further support from growth factors under normal conditions. This ability of growth factors to determine survival, differentiation, and final target destination is certainly not part of the normal list of functions ascribed to neurotransmitters. It is not clear that the release of the growth factor in this context is a chemical signal that conveys information to neurons, as opposed to providing critical sustenance.

What information then supports the contention that growth factors may be unconventional transmitters, even under the broad definition that we have used? First, growth factors are stored and released from neurons. The release of growth factors is now thought to occur through the regulated as well as constitutive pathways, suggesting some specificity in release. Some data are consistent with the speculation that BDNF is stored in vesicles. The synthesis and release of growth factors are also under transynaptic control. For example, expression of NGF and BDNF is controlled by neuronal activity, with glutamate and acetylcholine increasing expression and GABA decreasing expression of these neurotrophic factors. Moreover, the induction of BDNF by various treatments is regionally, spatially, and temporally distinct. Thus, seizures result in patterns of increases of various BDNF exon-specific mRNAs that differ across hippocampal subregions, and the specific pattern of promotor activation depends on the stimulus (Kokaia *et al.*, 1994). These characteristics all suggest that in addition to the synthesis and storage of neurotrophins in neurons, the synthesis and release of these growth factors are regulated by neuronal activity, and thus growth factors may participate in both receiving information and conveying information across cells.

The second characteristic suggesting a transmitter role for growth factors is that they appear to regulate other neurons. The receptors for neurotrophins, the trk receptors, are expressed in neurons, are members of a class of transmembrane receptor tyrosine kinases, and have an intracellular catalytic domain that is activated on ligand binding. Among the functions subserved by growth factors are the regulation of growth, differentiation, migration, transcription, and protein synthesis in neurons. In addition, neurotrophins appear to change the functional activity of other cells. BDNF has been shown to stimulate phosphoinositide turnover in neurons (Widmer *et al.*, 1993), and thus shares with NO the ability to regulate a key intracellular transduction mechanism. BDNF has also been shown to increase NT-3 expression in the cerebellum and hippocampus (Leingartner *et al.*, 1994; Lindholm *et al.*, 1994), again suggesting that growth factors regulate the functional activity of target cells.

The above-described data suggest that growth factors can regulate the functional activity of neurons, but do not address a physiological role for growth factors independent of their effects on growth and survival. However, several recent studies have suggested that BDNF may regulate hippocampal function during the induction of long-term potentiation (LTP). Mice carrying a deletion in the coding region of the BDNF gene show significantly reduced LTP (Korte *et al.*, 1995), while BDNF administration to hippocampal slices markedly enhances LTP (Patterson *et al.*, 1992).

Steroid Hormones

Steroid hormones are classically understood as hormones derived from peripheral endocrine glands that communicate back to the brain the physiological status of the animal. It was believed that the steroids acted via their well-characterized nuclear receptors, suggesting that they would have a slow time-course of action, many minutes to days. Studies in the last decade, however, have expanded and refined this concept to include local synthesis and/or modification of steroids in the brain and new sites of steroid hormone action.

It has long been understood that the brain is capable of metabolizing peripheral steroids to modify their function in the brain. For example, in the male, the primary circulating androgen, testosterone, is converted into estrogen by the enzyme aromatase, which then acts on estrogen receptors (see Fig. 10.6). Indeed, there are more estrogen receptors than androgen receptors in male brain tissue! The brain also produces some unique steroids as discussed below.

More recently it has become clear that the brain can actually synthesize steroids *de novo* (see Fig. 10.6). All of the enzymes involved in the synthesis of steroids have been found in the brain. Cultures of brain cells have also been shown to convert labeled cholesterol

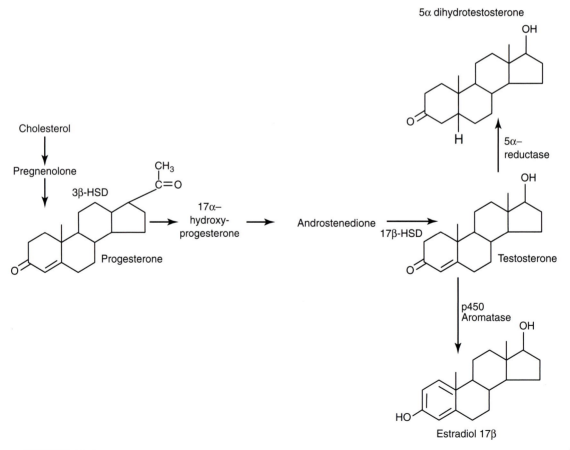

FIGURE 10.6 Neurosteroid synthesis and metabolism in the brain. While the brain can synthesize all of these steroids *de novo* from circulating cholesterol, circulating steroids such as testosterone and progesterone can also enter the brain and be metabolized further to other steroids.

into labeled mature steroids, showing that those enzymes indeed function. It is believed that much of the *de novo* synthesis takes place in glia as opposed to neurons, although this is still a controversial point.

There are several steroid metabolites (enriched in the brain and, hence, termed neurosteroids) that have a modulatory effect on GABA$_A$ receptor, sigma opiate receptor, or NMDA receptor function (Gibbs *et al.*, 1999; Monet *et al.*, 1995). The sulfated derivatives of 3β-hydroxy-Δ5-steroids, DHEA, or pregnenalone are inhibitory to GABA and hence are excitatory. On the other hand, 3α,5α-tetrahydroprogesterone enhances the effect of GABA, and is inhibitory to neuronal depolarization. Thus, depending on which enzymes are present in a given brain region for steroid modification, there can be differential effects on neurotransmitter action. It is currently unclear whether the source of these steroids is local synthesis or modification of gonad- or adrenal-derived steroids brought to the brain by the bloodstream. In any event, it is clear that these freely diffusing substances can have a great effect on brain function.

Steroids are capable of functioning as neurotransmitters in a variety of circumstances, even though they are "unconventional" in their mode of synthesis and action. In addition to their action via nuclear receptors, they can directly modulate multiple neurotransmitter receptors and also affect signal transduction systems directly. The action of steroids via nuclear receptors and direct effects on transcription is well understood (see Chapter 13). There is now an emerging body of research that points to a cytosolic site of action of steroids, which results in a more rapid effect of steroids on neuronal function. Studies on the electrical activity of neurons showed that steroids could change the electrical properties within seconds to a few minutes, much too quickly to be explained by nuclear transcription events. At first this was passed off as nonspecific membrane effects of the steroids, e.g., the effect of the lipid-soluble steroid intercalating into the plasma membrane. There was, however, a specificity to the action; for example, different glucocorticoids would work, while sex steroids would not, even though their basic chemical structures were similar. There are now multiple reports of membrane-associated steroid receptors that can mediate steroid-dependent functions (reviewed in Lösel and Wehling, 2003; Watson and Gametchu, 1999). Some studies have shown that these "membrane" steroid-associated events are dependent on expression of the well-characterized "nuclear" steroid receptor, and suggest that the nuclear steroid receptor can also associate with membrane elements and mediate steroid signaling via cytosolic pathways (see Fig. 10.7) (Kim *et al.*, 1999; Razandi *et al.*, 2002). For example, it is becoming well established that estrogen can also elicit its action via the MAP kinase pathway, independent of direct estrogen receptor action in the nucleus (Edwards and Boonyaratanakornkit, 2003). Thus, steroid hormones can have profound effects at all levels of signaling in the brain, from electrical activity to synaptic transmission to long-term changes in gene expression.

SYNAPTIC TRANSMITTERS IN PERSPECTIVE

We have discussed, in what may seem to be overwhelming detail, the biochemical aspects of chemically coded interneuronal transmission. During this discussion several things have become apparent. One is that it is difficult to discuss the biochemical nature of synaptic transmission without reference to other critical information about the structure and function of neurons. Neuroscience is multidisciplinary, requiring an appreciation of several different aspects of cellular function to come to grips with basic principles of integrated neuronal function. In addition, synaptic transmission is a dynamic process that is constantly changing; befitting this situation, the study of synaptic transmission is also dynamic, requiring frequent reevaluation and revision.

This can be most clearly seen in the changing definitions of a neurotransmitter. Over the years there has been a shift from a rather strict definition based on criteria that evolved primarily from studies of peripheral sites. The status of classic transmitters as the sole occupants of the royal family of transmitters was challenged a generation ago with the emergence of peptide transmitters, the pretenders to the throne. Sufficient data amassed to establish that the peptides were at least cousins, if not siblings or offspring, of the classic transmitters. Soon after this trauma came the unexpected finding that more than one transmitter is present in a neuron. And now we are just beginning to come to grips with the concept that there may be transmitters that are gases!

The new developments in unconventional transmitters have been paralleled by a greater appreciation of the intricacies of classic transmitters. The use of the terms conventional and unconventional transmitters reflects our current unease with the expanding definition of transmitters. However, all aspects of neuroscience are expanding, which is part of the reason for the excitement of neuroscience.

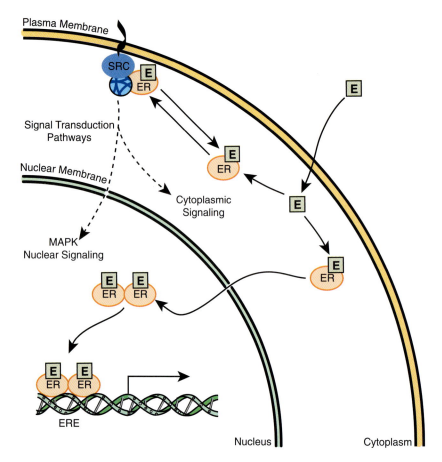

FIGURE 10.7 Estrogen acts in the cytoplasm to mediate signal transduction via the MAP kinase pathway. While the nuclear estrogen receptor appears to be capable of mediating the activation of MAP kinase, it still remains controversial as to whether a true plasma membrane-associated receptor can also mediate the estrogenic effects.

References

Baulieu, E. E. (1997). Neurosteroids: of the nervous system, by the nervous system, for the nervous system. *Recent Prog. Horm. Res.* **52**, 1–32.

Bean, A. I., Roth, R. H. (1992). Dopamine-neurotensin interactions in mesocortical neurons. Evidence from microdialysis studies. *Ann N Y Acad Sci* **668**, 43–53.

Bean, A. J., Zhang, X., and Hokfelt, T. (1995). Peptide secretion: What do we know? *FASEB J.* **8**, 630–638.

Bloch, A., Thoenen, H. (1996). Localization of cellular storage compartments and sites of constitutive and activity-dependent release of nerve growth factor (NGF) in primary cultures of hippocampal neurons. *Mol Cell Neurosci* **7**, 173–90.

Chao, M. V. (2003). Neurotrophins and their receptors: A convergence point for many signaling pathways. *Nat. Rev.* **4**, 299–309.

Chaudhry, F. A., Lehre, K. P., van Lookeren Campagne, M., Otterson, O. P., Danbolt, N. C., and Storm-Mathisen, J. (1995). Glutamate transporters in glial plasma membranes: Highly differentiated localizations revealed by quantitative ultrastructural immunocytochemistry. *Neuron* **15**, 711–720.

Clark, J. A., and Amara, S. G. (1993). Amino acid neurotransmitter transporters: Structure, function, and molecular diversity. *BioEssays* **15**, 323–332.

Cooper, J. R., Bloom, F. E., and Roth, R. H. (1996). "The Biochemical Basis of Neuropharmacology." Oxford Univ. Press, New York.

Dawson, T. M., and Snyder, S. H. (1994). Gases as biological messengers: Nitric oxide and carbon monoxide in the brain. *J. Neurosci.* **14**, 5147–5159.

Deutch, A. Y., Zahm, D. S. (1992). The current status of neurotensin-dopamine interactions. Issues and speculations. *Ann N Y Acad Sci* **668**, 232–52.

Dores, R.M., Lecaudé, S., Bauer, D., and Danielson, P.B. (2002). Analyzing the evolution of the opioid/orphanin gene family. *Mass Spectrom. Rev.* **21**, 220–243.

Edwards, D. P., and Boonyaratanakornkit, V. (2003). Rapid extranuclear signaling by the estrogen receptor (ER): MNAR couples ER and Src to the MAP kinase signaling pathway. *Mol. Interven.* **3**, 12–15.

Erickson, J. D., Varoqui, H., Schafer, M. K., Modi, W., Diebler, M. F., Weihe, E., Rand, J., Eiden, L. E., Bonner, T. I., and Usdin, R. B. (1994). Functional identification of a vesicular acetylcholine transporter and its expression from a "cholinergic" gene locus. *J. Biol. Chem.* **269**, 21929–21932.

Gibbs, T. T., Yagoubi, N., Weaver, C. E., Park-Chung, M., Russek, S. J., and Farb, D. H. (1999). Modulation of ionotropic glutamate receptors by neuroactive steroids. In (Eds.), "Neurosteroids: A New Regulatory Function in the Nervous System," (E. E. Baulieu, P. Robel, and M. Schumacher, Eds.), pp. 167–190. Humana Press, Totowa, NJ.

Goodman, L. J., Valverde, J., Lim, F., Geschwind, M. D., Federoff, H. J., Geller, A. I., Hefti, F. (1996). Regulated release and polarized localization of brain-derived neurotrophic factor in hippocampal neurons. *Mol Cell Neurosci* **7**, 222–238.

Greenfield., S. A. (1991). A non-cholinergic role of AChE in the substantia nigra: From neuronal secretion to the generation of movement. *Mol. Cell. Neurobiol.* **11**, 55–77.

Haydon, P. G. (2001). GLIA: Listening and talking to the synapse. *Nat. Rev. Neurosci.* **2**, 185–193.

Heymach, J. V., Jr. and Shooter, E. M. (1995). The biosynthesis of neurotrophin heterodimers by transfected mammalian cells. *J. Biol. Chem.* **270**, 12297–12304.

Kim, H.P., Lee, J.Y., Jeong, J.K., Bae, S.W., Lee, H.K., and Jo, I. (1999). Nongenomic stimulation of nitric oxide release by estrogen is mediated by estrogen receptor α localized in caveolae. *Biochem. Biophys. Res. Commun.* **263**, 257–262.

Kislauskis, E., Bullock, B., McNeil, S., and Dobner, P. R. (1988). The rat gene encoding neurotensin and neuromedin N: Structure, tissue-specific expression, and evolution of exon sequences. *J. Biol. Chem.* **263**, 4963–4968.

Kitabgi, P., De Nadal, F., Rovere, C., and Bidard, J. N. (1992). Biosynthesis, maturation, release, and degradation of neurotensin and neuromedin N. *Ann. N.Y. Acad. Sci.* **668**, 30–42.

Kokaia, Z., Metsis, M., Kokaia, M., Bengzon, J., Elmer, E., Smith, M. J., Timmusk, T., Siesjo, B. K., Persson, H., Lindvall, O. (1994). Brain insults in rats induce increased expression of the BDNF gene through differential use of multiple promoters. *Eur J Neurosci* **6**, 587–596.

Korte, M., Carroll, P., Wolf, E., Brem, G., Thoenen, H., and Bonhoeffer, T. (1995). Hippocampal long-term potentiation is impaired in mice lacking brain-derived neurotrophic factor. *Proc. Natl. Acad. Sci. USA* **92**, 8856–8860.

Leingartner, A., Heisenberg, C. P., Kolbeck, R., Thoenen, H., and Lindholm, D. (1994). Brain-derived neurotrophic factor increase neurotrophin-3 expression in cerebellar granule cells. *J. Biol. Chem.* **269**, 828–830.

Lindholm, D., da Penha Berzaghi, M., Cooper, J., Thoenen, H., and Castren, E. (1994). Brain-derived neurotrophic factor and neurotrophin-4 increase neurotrophin-3 expression in rat hippocampus. *Int. J. Dev. Neurosci.* **12**, 745–751.

Lindvall, O., Kokaia, Z., Bengzon, J., Elmer, E., and Kokaia, M. (1994). Neurotrophins and brain insults. *Trends Neurosci.* **17**, 490–496.

Lösel, R., and Wehling, M. (2003). Nongenomic actions of steroid hormones. *Nat. Rev. Mol. Cell Biol.* **4**, 46–55.

McAllister, A., Katz, L., and Lo, D. (1999). Neurotrophins and synaptic plasticity. *Annu. Rev. Neurosci.* **22**, 295–318.

Monet, P. P., Mahe, V., Robel, P., Baulieu, E. E. (1995). Neurosteroids via sigma receptors modulate the [³H] norepinephrine release evoked by NMDA in the rat hippocampus *Proc Natl Acad Sci USA* **92**, 3774–3777.

O'Dell, T. J., Huang, P. L., Dawson, T. M., Dinnerman, J. L., Snyder, S. H., Kandel, E. R., and Fishman, M. C. (1994). Endothelial NOS and the blockade of LTP by NOS inhibitors lacking neuronal NOS. *Science* **265**, 542–546.

Patterson, S. L., Grover, L. M., Schwartzkroin, P. A., and Bothwell, M. (1992). Neurotrophin expression in rat hippocampal slices: A stimulus paradigm inducing LTP in CA1 evokes increases in BDNF and NT-3 mRNAs. *Neuron* **9**, 1081–1088.

Plassart-Schiess, E., and Baulieu, E-E. (2001). Neurosteroids: Recent findings. *Brain Res. Rev.* **37**, 133–140.

Razandi, M., Oh, P., Pedram, A., Schnitzer, J., and Levin, E. R. (2002). ERs associate with and regulate the production of caveolin: Implications for signaling and cellular actions. *Mol. Endocrinol.* **16**, 100–115.

Russell, D. S. (1995). Neurotrophins: Mechanisms of action. *Neuroscientist* **1**, 3–6.

Seidah, N. G., and Chrétien, M. (1999). Proprotein and prohormone convertases: A family of subtilases generating diverse bioactive polypeptides. *Brain Res.* **848**, 45–62.

Seidman, S., Sternfeld, M., Ben Azziz-Aloya, R., Timberg, R., Kaufer-Nachum, D., and Soreq, H. (1995). Synaptic and epidermal accumulations of human acetylcholinesterase are encoded by alternative 3'-terminal exons. *Mol. Cell. Biol.* **15**, 2993–3002.

Shipp, M. A., Vijayaraghavan, J., Schmidt, E. V., Masteller, E. L., D'Adamio, L., Hersch, L. B., and Reinherz, E. L. (1989). Common acute lymphoblastic leukemia antigen (CALLA) is active neutral endopeptidase 24.11 ("enkaphalinase"): Direct evidence by cDNA transfection analysis. *Proc. Natl. Acad. Sci. USA* **86**, 297–301.

Siegel, G. J., Agranoff, B. W., Albers, R. W., and Molinoff, P. B. (1994). "Basic Neurochemistry," 5th ed. Raven Press, New York.

Thoenen, H. (1995). Neurotrophins and neuronal plasticity. *Science* **270**, 593–598.

Watson, C. S., and Gametchu, B. (1999). Membrane-initiated steroid actions and the proteins that mediate them. *Proc. Soc. Exp. Biol. Med.* **220**, 9–19.

Wetmore, C., Olson, L., and Bean, A. J. (1994). Regulation of brain-derived neurotrophic factor (BDNF) expression and release from hippocampal neurons is mediated by non-NMDA type glutamate receptors. *J. Neurosci.* **14**, 1688–1700.

Widmer, H. R., Ohsawa, F., Knusel, B., and Hefti, F. (1993). Downregulation of phosphatidylinositol response to BDNF and NT-3 in cultures of cortical neurons. *Brain Res.* **614**, 325–334.

Zakhary, R., Gaine, S. P., Dinerman, J. L., Ruat, M., Flavahan, N. A., and Snyder, S. H. (1996). Heme oxygenase 2: Endothelial and neuronal localization and role in endothelial-dependent relaxation. *Proc. Natl. Acad. Sci. USA* **93**, 795–798.

11

Neurotransmitter Receptors

M. Neal Waxham

Chemical synaptic transmission plays a fundamental role in the process of neuron-to-neuron and neuron-to-muscle communication. The type of receptors present in the plasma membrane determines in large part the nature of the response of a neuron or muscle cell to a neurotransmitter. The nature of the response can be either through the direct opening of an ion channel (ionotropic receptors) or through alteration of the concentration of intracellular metabolites (metabotropic receptor) (see also Chapter 16). The response magnitude is determined by receptor number, the "state" of the receptors, and the amount of transmitter released. Finally, the sign of the response can be inhibitory or excitatory. The temporal and spatial summation of information conveyed by receptor activation determines whether the postsynaptic cell will fire an action potential or the muscle will contract. As one can see, there is remarkable flexibility and diversity in molding the response to neurotransmitter by constructing a synapse with the desired receptor types.

There exist two broad classifications for receptors. An ionotropic receptor is a relatively large, multisubunit complex typically composed of five individual proteins that combine to form an ion channel through the membrane (Fig. 11.1A). These ion channels exist in a closed state in the absence of neurotransmitter and are largely impermeable to ions. Neurotransmitter binding induces rapid conformational changes that open the channel, permitting ions to flow down their electrochemical gradients. Changes in membrane current resulting from ligand binding to ionotropic receptors are generally measured on a millisecond time scale. The ion flow ceases when transmitter dissociates from the receptor or when the receptor becomes desensitized, a process discussed in more detail later in this chapter.

In contrast, a metabotropic receptor is composed of a single polypeptide (Fig. 11.1B) and exerts its effects not through the direct opening of an ion channel but rather by binding to and activating GTP-binding proteins (G proteins). Transmitters that activate metabotropic receptors typically produce responses of slower onset and longer duration (from tenths of seconds to potentially hours) owing to the series of enzymatic steps necessary to produce a response. The metabotropic receptors have more recently been named G protein coupled receptors, or GPCRs for short, to more accurately capture their properties and the latter nomenclature is adopted in this chapter.

We consider the structure of the ionotropic receptor family first and then turn to a description of the structure of GPCRs. In each section, information is presented to establish a general structural model of each receptor type. These models are then used to guide the description of other related ionotropic receptors or GPCRs. The order in which receptor types are presented is based predominantly on structural relatedness and should not be interpreted as representing their relative importance in the function of the nervous system.

IONOTROPIC RECEPTORS

All ionotropic receptors are membrane-bound protein complexes that form an ion-permeable pore in the membrane. By comparing the amino acid sequences of the cloned ionotropic receptors one can deduce that they are similar in overall structure, although two independent ancestral genes have given rise to two distinct families. One family includes the nicotinic acetylcholine receptor (nAChR), the γ-aminobutyric acid A (GABA$_A$) receptor, the glycine receptor, and one subclass of serotonin receptors (Ortell and Lunt, 1995). The other family comprises the many types of ionotropic glutamate receptors (Hollmann and Heinemann, 1994).

Copyright 2004, Elsevier Science (USA).
All rights reserved.

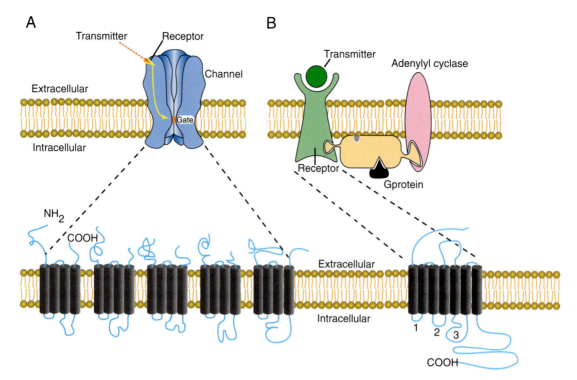

FIGURE 11.1   Structural comparison of ionotropic and metabotropic receptors. (A) Ionotropic receptors bind transmitter, and this binding directly translates into the opening of the ion channel through a series of conformational changes. Ionotropic receptors are composed of multiple subunits. Shown are the five subunits that together form the functional nAChR. Note that each nAChR subunit wraps back and forth through the membrane four times and that the mature receptor is composed of five subunits. (B) Metabotropic receptors bind transmitter and, through a series of conformational changes, bind to G proteins and activate them. G proteins then activate enzymes such as adenylyl cyclase to produce cAMP. Through the activation of cAMP-dependent protein kinase, ion channels become phosphorylated, which affects their gating properties. Metabotropic receptors are single subunits. They contain seven transmembrane-spanning segments, with the cytoplasmic loops formed between the segments providing the points of interactions for coupling to G proteins. Adapted from Kandel *et al.* (1991).

The understanding of ionotropic receptor structure and function has expanded enormously in the past 25 years. Molecular approaches have provided elegant and extensive descriptions of gene families encoding different receptors and systems for expressing cloned cDNAs have permitted detailed structure–function analysis of each receptor subtype. Expression of subunits independently and together has resulted in a detailed concept of the necessity and sufficiency of the multisubunit nature of the ionotropic receptor family. With the addition of biophysical and X-ray structural analysis, events associated with the opening of at least one ionotropic receptor, nAChR, are available at nearly atomic resolution (Unwin, 1993a, b, 1995).

nAChR Is a Model for the Structure of Ionotropic Receptors

nAChR is so named because the plant alkaloid nicotine can bind to the ACh binding site and activate the receptor. Nicotine is therefore called an agonist of ACh because it binds to the receptor and opens it. In contrast, antagonists are molecules that bind to the receptor and inhibit its function. Agonists and antagonists are powerful tools that permit characterization of the structure and function of individual receptor subtypes.

More is known about the structure of nAChR than about any other ionotropic receptor, primarily because electric organs of certain species of fish, such as the *Torpedo* ray, contain nearly crystalline arrays of this molecule (Fig. 11.2). The electric organ is a specialized form of skeletal muscle that has the potential to generate large voltages (as much as 500 V in some cases) from the simultaneous opening of arrays of ion channels activated through the binding of ACh. The majority of biochemical and structural analyses of nAChRs have been done on receptors isolated from the ray electric organ. Purification of nAChRs was aided significantly by utilization of a toxin from snake

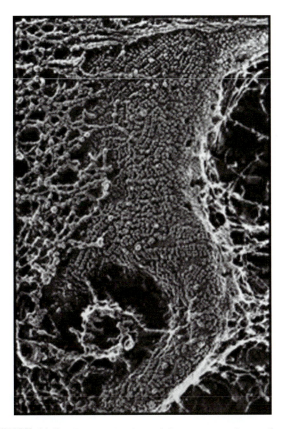

FIGURE 11.2 Panoramic view of the postsynaptic membrane of an electrocyte in the *Torpedo* electric organ, revealed by "deep-etch" electron microscopy. The vase-like structure in the center of the field is the external surface of the postsynaptic membrane, which is revealed by removal of the basal lamina. Clusters and linear arrays of 8- to 9-nm protrusions can be clearly seen. These represent the AChR oligomers. To the left of the vase-like structure, a lace-like basal lamina lies above the membrane, obscuring it from view. To the right of the vaselike structure, the postsynaptic membrane has been freeze-fractured away, thus revealing an underlying meshwork of cytoplasmic filaments that supports the postsynaptic membrane and its receptors. ×175,000. Original courtesy of J. Heuser.

venom called α-bungarotoxin. Affinity columns constructed with α-bungarotoxin bind to nAChR with high affinity and specificity, providing a means of purifying nearly homogeneous nAChRs in a single chromatographic step.

nAChR Is a Heteromeric Protein Complex with a Distinct Architecture

The structure of nAChR is typical of ionotropic receptors. nAChR purified as described from *Torpedo* is composed of five subunits (see Fig. 11.1) and has a native molecular mass of approximately 290 kDa. The subunits are designated α, β, γ, and δ, and each receptor complex contains two copies of the α subunit. The

subunits are homologous membrane-bound proteins that assemble in the bilayer to form a ring enclosing a central pore. Pioneering electron microscopic analyses by Nigel Unwin have provided the best image of the structural appearance of nAChR (Fig. 11.3). In fact, nAChR is the only membrane-bound neurotransmitter receptor for which high-resolution structural information is available. The extracellular domains of the subunits together form a funnel-shaped opening that extends approximately 100 Å outward from the outer leaflet of the plasma membrane. The funnel at the outer portion of the receptor has an inside diameter of 20–25 Å (Unwin, 1993a). The funnel shape is thought to concentrate and force ions and transmitter to interact with amino acids in the limited space of the pore without producing a major barrier to diffusion. This funnel narrows near the center of the lipid bilayer to form the domain of the receptor that determines the opened or closed state of the ion pore. The intracellular domain of the receptor forms short exits for ions traveling into the cell and an entrance for ions traveling out of the cell. The intracellular domain also establishes the association of the receptor with other intracellular proteins that determine the subcellular localization of nAChR. The arrangement of the subunits in the receptor is somewhat debatable; however, most data support a model whereby the β subunit lies between the two α subunits (Unwin, 1993a,b, 1995).

Each nAChR Subunit Has Multiple Membrane-Spanning Segments

The primary structure of each nAChR subunit was obtained by the efforts of Shosaka Numa and his colleagues (Noda *et al.*, 1982, 1983). The deduced amino acid sequence from cloned mRNAs indicates that nAChR subunits range in size from 40 to 65 kDa. A general domain structure for each subunit was derived from primary sequence data and toxin and antibody binding studies. Each subunit consists of four transmembrane-spanning segments referred to as TM1–TM4 (Fig. 11.4). Each segment is composed mainly of hydrophobic amino acids that stabilize the domain within the hydrophobic environment of the lipid membrane. The four transmembrane domains are arranged in an antiparallel fashion, wrapping back and forth through the membrane. The N terminus of each subunit extends into the extracellular space, as does the loop connecting TM2 and TM3 as well as the C terminus. The amino acids linking TM1 and TM2 and those linking TM3 and TM4 form short loops that extend into the cytoplasm.

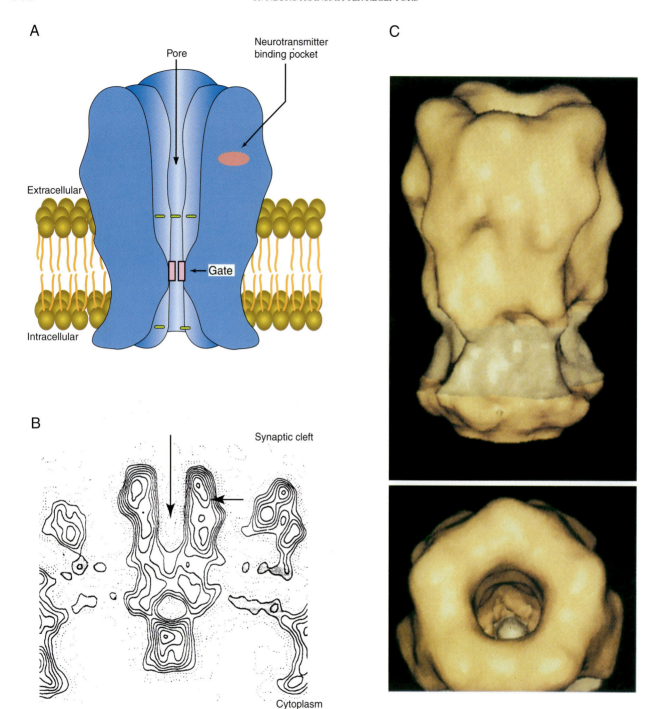

FIGURE 11.3 (A) Vertical section diagramming the structure of nAChR as it is believed to exist in the membrane. Note that the funnel-shaped structure narrows to a small central point referred to as the gate. Strategically placed rings of negatively charged amino acids on both sides of the gate form part of the selectivity filter for positively charged ions. The approximate position of the neurotransmitter binding site is shown in relation to the gate and the plasma membrane. (B) Protein density map derived from reconstructions of nAChR imaged by cryoelectron microscopy. The vertical arrow indicates the direction of ion flow from outside to inside within the funnel-shaped part of the receptor. The horizontal arrow indicates the predicted position of the neurotransmitter-binding site that resides approximately 30 Å above the bilayer. The additional protein density attached to the bottom of the receptor is suggested to be a protein that anchors the nAChR to synapses. (C) Three-dimensional computer rendering of the nAChR. (Top) Side view of nAChR similar to that in (A). The darker shaded area near the bottom of the receptor delineates the approximate location where the receptor contacts the lipid bilayer. (Bottom) A view looking down into the funnel-shaped opening of the receptor. Note that the funnel narrows forming the gate.

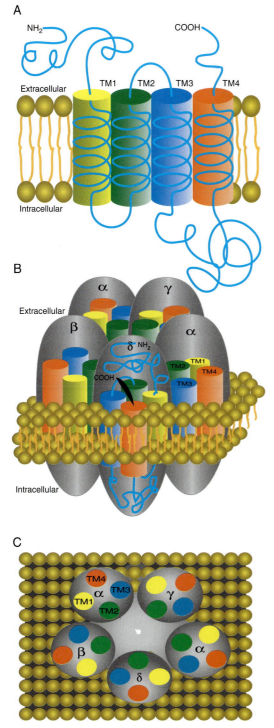

FIGURE 11.4 (A) Diagram highlighting the orientation of the membrane-spanning segments of one subunit of nAChR. The amino and carboxy termini extend in the extracellular space. The four membrane-spanning segments are designated TM1–TM4. Each forms an α helix as it traverses the membrane. (B) Side view of the five subunits in their approximate positions within the receptor complex. There are two α subunits present in each nAChR. (C) Top view of all five subunits highlighting the relative positions of their membrane-spanning segments, TM1–TM4, and the position of TM2 that lines the channel pore. Adapted from Kandel *et al.* (1991).

The Structure of the Channel Pore Determines Ion Selectivity and Current Flow

In the model shown in Fig. 11.4C, each subunit of nAChR can be seen to contribute one cylindrical component (representing a membrane-spanning segment) that presents itself to a central cavity that forms the ion channel through the center of the complex. The membrane-spanning segments that line the pore are the five TM2 regions, one contributed by each subunit. The amino acids that compose the TM2 segment are arranged in such a way that three rings of negatively charged amino acids are oriented toward the central pore of the channel (Fig. 11.5). These rings of negative charge appear to provide much of the selectivity filter so that only cations can pass through the central channel, whereas anions are largely excluded owing to charge repulsion (Karlin, 1993; Imoto *et al.*, 1988). nAChR is permeable to most cations, such as Na^+, K^+, and Ca^{2+}, although monovalent cations are preferred. This mechanism for selectivity is poor in relation to the selectivity described for the family of voltage-gated ion channels (e.g., voltage-gated Ca^{2+} channels; see Chapters 5 and 6). From analysis of the permeation of various sized cations, the dimensions of the pore forming the final barrier for ion permeation were estimated to be approximately 8.5 Å (Hille, 1992). This size is in excellent agreement with the measurements of 9–10 Å for the pore diameter from images of the open state of the receptor (Unwin, 1995). The restricted physical dimensions of the pore contribute greatly to the selectivity for particular ions. When the pore of nAChR opens, positively charged ions move down their respective electrochemical gradients, resulting in an influx of Na^+ and Ca^{2+} and an efflux of K^+. A coarse filtering that also contributes to selectivity appears to be a shielding effect produced by other negatively charged amino acids surrounding the outer channel region of the receptor.

Ions do not directly enter or exit through the central pore of the cytoplasmic end of nAChR. Two narrow openings are present on the lateral aspects of the cytoplasmic portion of the receptor through which ions must travel to exit or gain access to the central pore (Figs. 11.5A, C, and D) (Miyazawa *et al.*, 1999). α-Helical rods extending down from each subunit form an inverted pentagonal cone to produce these openings. Although too large (8 × 15 Å) to be a significant barrier to ion flow, these lateral pores could serve as an additional filtering step for the passage of certain ions.

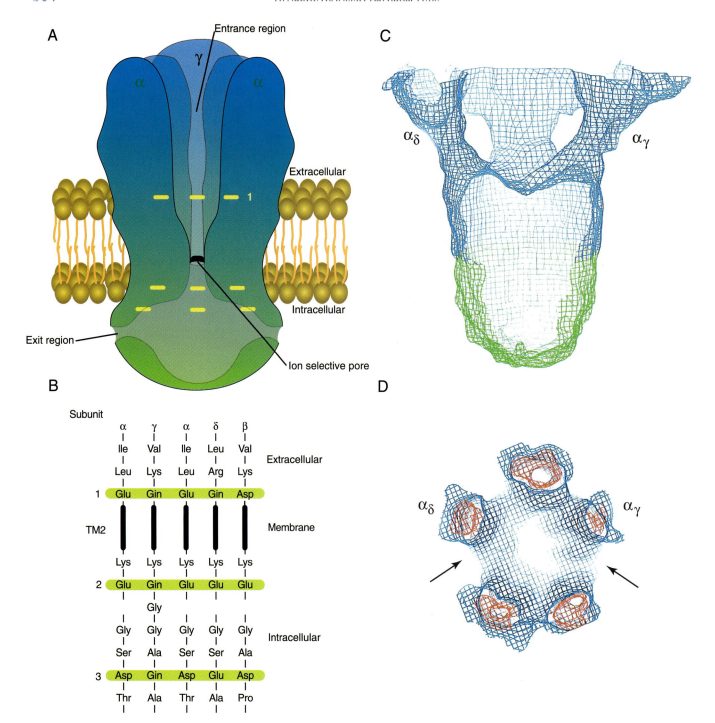

FIGURE 11.5   (A) Vertical section highlighting the relative positions of the three rings of negatively charged amino acids that help form the cation selectivity of nAChR. The regions where ions exit or enter from the intracellular side of the receptor are disposed laterally at the base of the receptor (B) Amino acid sequence of each of the TM2 membrane-spanning segments of the five nAChR subunits. Numbers 1–3 correspond to the positions of the amino acids taking part in the formation for the three rings of negatively charged amino acids that determine the cation selectivity of the pore. Aspartate (Asp) and glutamate (Glu) are negatively charged amino acids. (C) Wireframe portrayal of the protein density distribution of the intracellular portion of nAChR. The front portion of the receptor was cut away to reveal the inverted cone-shaped cavity of the intracellular domain. The green wireframe represents protein density contributed by the anchoring protein rapsyn. (D) Wireframe portrayal of the protein density distribution of the intracellular domain of nAChR looking downward from within the receptor. The arrows indicate the major gaps in the lateral walls of the receptor where ions enter and exit.

There Are Two Binding Sites for ACh on nAChR

Each receptor complex has two ACh binding sites that reside in the extracellular domain and lie approximately 30 Å from the outer leaflet of the membrane (see Fig. 11.3). The ACh binding site is formed for the most part by six amino acids in the α subunits; however, amino acids in both the γ and the δ subunits also contribute to binding (Karlin, 1993) (Fig. 11.6A). Mutations introduced at these critical amino acids in the α subunit significantly attenuate ligand binding. The two binding sites are not equivalent because of the receptor's asymmetry due to the different neighboring subunits (either γ or δ) adjacent to the two α subunits. Significant cooperativity also exists within the receptor molecule, and so binding of the first molecule of ACh enhances binding of the second (Changeux *et al.*, 1984). A higher-resolution structure of nAChR (Miyazawa *et al.*, 1999) has revealed that access to the ACh binding sites appears to be through small channels that open into the interior mouth of the pore (Figs. 11.6B and C). Thus, ACh molecules must enter the pore and traverse these channels to gain access to their binding sites. It is speculated that similar attractive forces that bring positively charged ions into the pore also attract the positively charged ACh molecules favoring entrance into the channels that lead to the ACh binding sites. Two adjacent Cys residues (Cys-192 and Cys-193) in each α subunit form a disulfide bond that also appears to contribute to the stability of the ACh binding pocket (Fig. 11.6A). These Cys residues are highly conserved in most ionotropic receptors and must form an essential bond for stabilizing high-affinity neurotransmitter binding. α-Bungarotoxin binds to the α subunit in close proximity to the two adjacent Cys residues (Karlin, 1993).

Opening of nAChR Occurs through Concerted Conformational Changes Induced by ACh Binding

When nAChR binds two molecules of ACh, the channel opens almost instantaneously [time constants for opening are approximately 20 μs (Colquhoun and Sakmann, 1985; Colquhoun and Ogden, 1988)], thus permitting the passage of ions (see Fig. 4 of Chapter 16). A model developed from electron micrographic reconstructions of the nicotine-bound form of nAChR indicates that the closed-to-open transition is associated with a rotation of the TM2 segments (Unwin, 1995) (Fig. 11.7). The TM2 segments are helical and exhibit a kink in their structure that forces a Leu residue from each segment into a tight ring that effec-

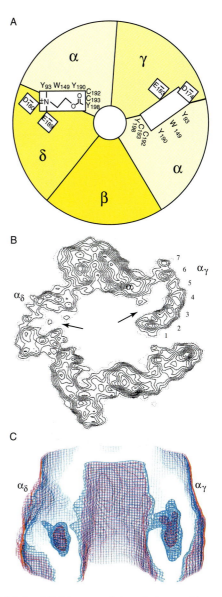

FIGURE 11.6 (A) Diagram of the relative positions of amino acids that form the Ach binding site in nAChR. The view is from above the receptor looking down into the pore. Each subunit is represented by a wedge. At the left, Ach is shown bound to its site at the interface between the α and δ subunits. The length of the binding site is shown slightly contracted relative to the site (between the α and γ subunits) without bound Ach. Critical amino acids for transmitter binding are indicated. Residues shown in boxes are amino acids predicted to make contact with the positively charged part of the Ach molecule. Note that many of the residues important for Ach binding are contributed by the α subunit. Cysteine (Cys) residues at positions 192 and 193 form a disulfide bond essential for stabilizing the Ach binding pocket. (B) A top-down view of the protein density map of nAChR sliced through the area where the Ach-binding areas reside. Note that Ach molecules must gain access to their binding sites through channels whose openings are on the inside of the funnel-shaped portion of the receptor. (C) Lateral view of a wireframe portrayal of nAChR at the level of the Ach binding sites. A portion of the receptor was cut away to highlight the fact that the cavities where Ach bind must be accessed from the pore of the receptor through short channels.

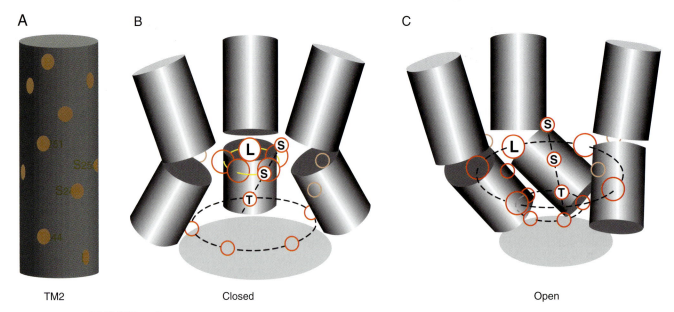

FIGURE 11.7   (A) Relative positions of amino acids in the TM2 segment of one of the nAChR α subunits modeled as an α helix. The glutamate residues (E) that form parts of the negatively charged rings for ion selectivity are shown at the top and bottom of the helix. (B) Arrangement of three of the TM2 segments of nAChR modeled with the receptor in the closed (Ach-free) configuration. In the closed configuration, leucine (L) residues form a right ring in the center of the pore that blocks ion permeation. (C) Arrangement of the three TM2 segments after Ach binds to the receptor. In the open configuration, the construction formed by the ring of leucine (L) residues opens as the helices twist about their axes. Note that the polar serine (S) and threonine (T) residues align when Ach binds, which apparently helps the water-solvated ions travel though the pore. Adapted from Unwin (1995).

tively blocks the flow of ions through the central pore of the receptor. When the TM2 segments rotate because of ACh binding, the kinks also rotate, relaxing the constriction formed by the Leu ring, and ions can then permeate the pore. The rotation also orients a series of Ser and Thr residues (amino acids with a polar character) into the central area of the pore, which facilitates the permeation of water-solvated cations. As the resolution of the structure increases, refinements in this model are likely to be forthcoming; however, the architecture of nAChR is well established and provides a structural framework with which all other ionotropic receptors can be compared.

The main features of the nAChR transition from a closed to open state are summarized in Fig 11.8. ACh gains access to its binding sites by entering the central pore of the receptor where it then enters small channels that provide access to the binding sites. Once both binding sites are occupied, the receptor rapidly opens to permit ion flow. It is the twisting of the TM2 segments induced by ACh binding that gates open the ion pore. In addition to negatively charged rings spaced within the central channel, charge screening also occurs in the lateral openings of the cytoplasmic domain of the receptor (Fig. 11.8). The ion selectivity of nAChR for positively charged ions (Na$^+$ and K$^+$

mainly) comes from the charge screening at these different levels and the physical constriction of the gate of the pore.

The Muscle Form of nAChR Is Very Similar to nAChR from *Torpedo*

nAChRs at the neuromuscular junction are a concentrated collection of homogeneous receptors having a structure similar to that of the *Torpedo* electric organ. This similarity is not surprising, because the electric organ is a specialized form of muscle tissue. The adult form of the muscle receptor has the pentameric structure $\alpha_2\beta\varepsilon\delta$. An embryonic form of the receptor has an analogous structure, except that the ε subunit is replaced by a unique γ subunit. The embryonic and adult subunits of both mouse and bovine muscle receptors have been cloned and expressed in heterologous systems, such as the *Xenopus laevis* oocyte (Box 11.1) (Mishina *et al.*, 1984), and the receptors differ in both channel kinetics and channel conductance. These differences in channel properties appear to be necessary for the proper function of nAChRs as they undergo the transition from developing to mature neuromuscular junction synapses.

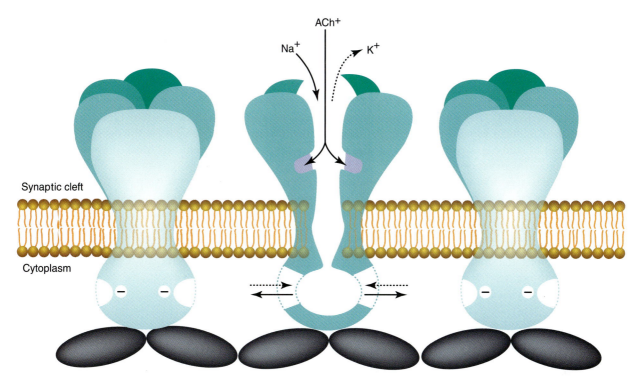

FIGURE 11.8 Summary figure highlighting the structural features of nAChR. The ACh-free form of the receptor remains closed to ion flow. ACh gains access to its binding site by entering the outer portion of the central pore of nAChR which produces a relaxation of the central pore and an expansion of the holes in the lateral walls of the intracellular portion of the receptor. The protein rapsyn, represented by the gray ovals, anchors nAChR to synapses by interacting with the intracellular domain.

nAChR Matures as a Typical Membrane-Bound Protein and Has Well-Ordered Assembly

The pathway of nAChR assembly in muscle is a tightly regulated process. For example, the five subunits of nAChR have the potential to randomly assemble into 208 different combinations. Nevertheless, in vertebrate muscle, only one of these configurations ($\alpha_2\beta\epsilon\delta$) is typically found in mature tissue, indicating a very high degree of coordinated assembly and, ultimately, little structural variability (Paulson *et al.*, 1991; Green and Claudio, 1993). The well-ordered assembly of specific intermediates is essential for this coordinated process, and the intermediates formed appear to start with a dimer between α and either ϵ (γ in mature muscle) or δ. The heterodimers then bind to β and to each other to form the final receptor (Gu *et al.*, 1991). An alternative pathway in which α, β, and γ first form a trimer has also been proposed (Green and Claudio, 1993). All of this assembly takes place within the endoplasmic reticulum. During intracellular maturation, each subunit is glycosylated, and, if glycosylation is inhibited, the production of mature nAChRs decreases. Two highly conserved disulfide bonds in the N-terminal extracellular domain are essential for

efficient assembly of the mature receptor. The first is between two adjacent Cys residues (Cys-192 and Cys-193) and, as noted, resides very close to the ACh binding site on the receptor (see Fig. 11.6). The second bond is between two Cys residues 15 amino acids apart, forming a loop in the extracellular domain.

Phosphorylation Is a Common Posttranslational Modification of Receptors

Many ionotropic receptors, such as nAChR, are phosphorylated, although the functional significance of the phosphorylation is not always evident. nAChR is phosphorylated by at least three protein kinases. cAMP-dependent protein kinase (PKA; phosphorylates the γ and δ subunits), Ca^{2+}/phospholipid-dependent protein kinase (PKC; phosphorylates the δ subunit), and an unidentified tyrosine kinase phosphorylates the β, γ, and δ subunits (Huganir and Greengard, 1990). The phosphorylation sites are all found in the intracellular loop between the TM3 and TM4 membrane-spanning segments. Phosphorylation by these three protein kinases appears to increase the rapid phase of desensitization of the receptor. Desensitization of receptors is a common observation, and this process limits the

BOX 11.1

THE *XENOPUS* OOCYTE

The *Xenopus* oocyte has been used extensively to study the properties of cDNAs encoding receptor subunits and their mutated forms. In addition, the oocyte has been used to study how combinations of different subunits interact to produce receptors with different properties. The large size and efficient translational machinery of *Xenopus* oocytes make them ideal for electrophysiological analyses of cDNAs encoding prospective receptors and channels. For example, mRNAs produced by *in vitro* transcription of cDNAs encoding each of the individual nAChR subunits were introduced into oocytes by microinjection (A). Several days later, the oocytes were voltage clamped to study the properties of the expressed channels (B). When ACh was applied through a separate pipet, a significant inward current was detected in the oocyte (C, panel 1). The response was specifically blocked by addition of an antagonist, tubocurarine (C, panel 2), and the block was reversed by a 15-min wash (C, panel 3). Details of this study indicate that all four of the nAChR subunits (α, β, γ, and δ) were required for ACh to produce an electrophysiological response (Mishina *et al.*, 1984). More recently, the patch-clamp technique has also been applied to oocyte expression of receptors to analyze the behavior of single channels.

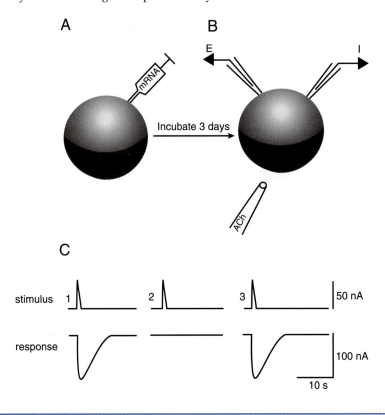

amount of ion flux through a receptor by producing transitions into a closed state (one that does not permit ion flow) in the continued presence of neurotransmitter. For nAChR, the rate of desensitization has a time constant of approximately 50–100 ms. This rate appears to be too slow to have much significance in shaping the synaptic response at the neuromuscular junction, where the response typically lasts from 5 to 10 ms. This slow desensitization is not true of the brain forms of nAChR and is discussed further in a later section of this chapter.

The Structures of Other Ionotropic Receptors Are Variations of the nAChR Structure

On the basis of similarity of structure, clear evolutionary relationships exist for the family of ionotropic receptors. Figure 11.9 shows an evolutionary tree for

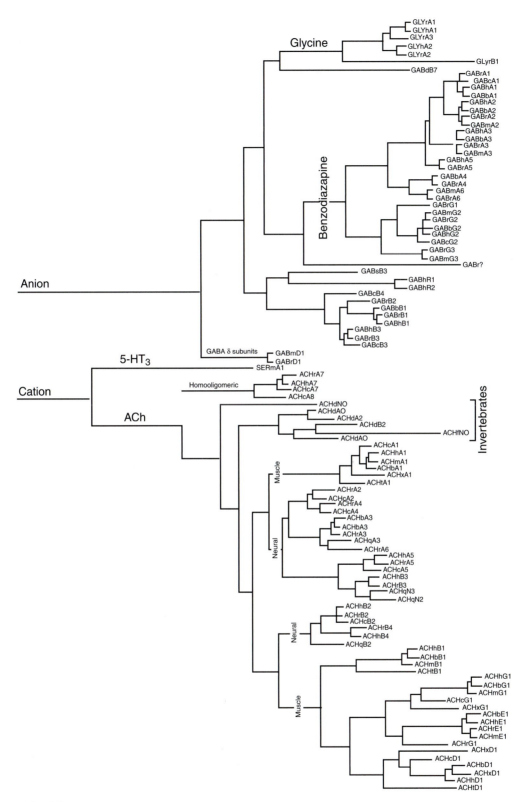

FIGURE 11.9 Evolutionary relationships of the family of cloned ionotropic receptor subunits. The nomenclature used to describe each receptor subunit is RRRsS#, where *RRR* represents the type of receptor, *s* the organism, *S* the subunit type, and # the subunit number. Type of receptor (RRR): Ach, acetylcholine; GAB, GABA; GLY, glycine; SER, serotonin. Organism (s): b, bovine; c, chicken; d, *Drosophila*; f, filaria; g, goldfish; h, human; l, locust; m, mouse; n, nematode; r, rat; s, snail; t, *Torpedo*; x, *Xenopus*. Subunit type (S): A, alpha; B, beta; G, gamma; D, delta; E, epsilon; R, rho; N, non-alpha; ?, undetermined. Adapted from Ortell and Lunt (1995) Trends Neurosci. **18**, 122.

the family of ionotropic receptors related to nAChR. An early major subdivision separates those receptors permeable to anions from those permeable to cations. The former group includes the GABA_A and glycine receptors, whereas the latter group includes the 5-hydroxytryptamine (5-HT_3, serotonin) and ACh receptors. One can begin to appreciate that structural similarities can predict a degree of functional similarity. Each of these receptor types is described in the next section.

Neuronal nAChRs Contain Two Types of Subunits

Structurally, neuronal nAChRs are similar to, yet distinct from, the *Torpedo* isoform of the receptor (Figs. 11.9 and 11.10). For example, neuronal nAChR appears to have only two types of subunits, α and β, that combine to produce the functional receptor, and the majority of these receptors do not bind to α-bungarotoxin. At least nine different α subtypes (α1 being the muscle α subunit) have been identified, and some are species specific (α8 is found only in chicken and α9 is found only in rat). Four different β subtypes (β1 being the muscle β subunit) have been identified. The neuronal β subunits are not closely related to the muscle β1 subunit and are sometimes referred to simply as non-α subunits. One structural feature that distinguishes neuronal α subunits from β subunits is the presence of particular Cys residues in the extracellular domain of the α subunit. Two of these Cys residues are adjacent to one another and form a disulfide bond. The β subunits do not have these adjacent Cys residues. Because these Cys residues are critical for ACh binding, the α subunits of neuronal AChRs, like muscle α subunits, contain the main contact points for ACh binding (Fig. 11.10). All of the α and β genes encode proteins with four transmembrane-spanning segments (TM1–TM4). Although the physical structure of this receptor family has not been well characterized, it appears that each functional receptor is a pentameric assembly.

Structural Diversity of Neuronal nAChRs Produces Channels with Unique Properties

Neuronal nAChRs have diverse functions and are the receptors presumed to be responsible for the psychophysical effects of nicotine addiction. One major function of nAChRs in the brain is to modulate excitatory synaptic transmission through a presynaptic action (McGehee *et al.*, 1995). The diversity in function can be related to the heterogeneous structure contributed by the thousands of possible combinations between the different α and β subunits. Control mechanisms for receptor assembly in neurons do not appear to be as stringent as those of nAChR in *Torpedo* and muscle. Functional neuronal nAChRs can be assembled from a single subunit (as in α7, α8, and α9), and a single type of α subunit can be assembled with multiple types of β subunits (e.g., α3 with β2 or β4 or both) and vice versa (Fig. 11.10). These additional possibilities produce a staggering array of potential receptor molecules, each with distinct properties including differences in single-channel kinetics and rates of desensitization. This type of diversity is not unique to neuronal nAChRs. For most receptor classes studied in detail, diversity is the rule and not the exception. It is intriguing to speculate that subunit composition may also play roles in targeting the receptors to different intracellular locations.

Neuronal nAChRs exhibit a range of single-channel conductances between 5 and 50 pS, depending on the tissue analyzed or the specific subunits expressed. Most, but not all, are blocked by neuronal bungarotoxin, a snake venom distinct from α-bungarotoxin. All of the neuronal nAChRs are cation-permeable channels that, in addition to permitting the influx of Na^+ and the efflux of K^+, permit an influx of Ca^{2+}. This Ca^{2+} permeability is greater than that for muscle nAChR (Vernino *et al.*, 1992) and is variable among the different neuronal receptor subtypes. Indeed, some receptors have very high Ca^{2+}/Na^+ permeability ratios; for example, α7 nAChRs exhibit a Ca^{2+}/Na^+ permeability ratio of nearly 20 (Seguela *et al.*, 1993). The Ca^{2+} permeability of the α7 nAChR can be eliminated by the mutation of a single amino acid residue in the second transmembrane domain (Glu-237 for Ala) without significantly affecting other aspects of the receptor (Bertrand *et al.*, 1993). This key Glu residue must lie within the pore of the receptor and presumably enhances the passage of Ca^{2+} ions through an interaction with its negatively charged side chain.

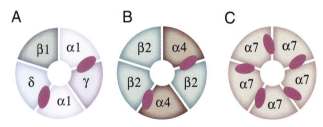

FIGURE 11.10 Diagrams of top-down views of AChR from muscle (A), one of the the neuronal AChRs composed of α2 and β4 subunits (B), and the homoligomeric form of neuronal AChR produced by assembly of α7 subunits (C). Purple ovals represent the ACh binding sites on each receptor complex. The ACh receptors from brain are diverse in both structure and properties due to the variety of different receptor complexes produced from subunit mixing.

Activation of α7 receptors through the binding of ACh could therefore produce a significant increase in the level of intracellular Ca^{2+} without the opening of voltage-gated Ca^{2+} channels. Subunits α7, α8, and α9 are also the α-bungarotoxin-binding subtypes of neuronal nAChRs. Other neuronal isoforms exhibit Ca^{2+}/Na^{+} permeability ratios of about 1.0–1.5.

Neuronal nAChRs Desensitize Rapidly

For nAChR from muscle, desensitization is minor and probably is not of physiological significance in determining the shape of the synaptic response at the neuromuscular junction. However, for some neuronal nAChRs, desensitization likely plays a major role in determining the effects of the actions of ACh. Receptors composed of α7, α8, and certain α/β combinations exhibit desensitization time constants between 100 and 500 ms, whereas others exhibit desensitization constants between 2 and 20 s. Given the diverse functions of neuronal nAChRs, the variable rates of desensitization likely play important roles whereby this inherent property of the receptor shapes the physiological response generated from binding ACh.

One Serotonin Receptor Subtype, 5-HT₃, Is Ionotropic and Is a Close Relative of nAChR

Serotonin (5-hydroxytryptamine, 5-HT) is historically thought of as a transmitter that binds to and activates only GPCRs (described in more detail later). The 5-HT₃ subclass is an exception forming an ionotropic receptor activated by binding serotonin. The 5-HT₃ receptor is permeable to Na^{+} and K^{+} ions and is similar in many ways to nAChR in that both desensitize rapidly and are blocked by tubocurarine. From expression studies of the cloned cDNA (Marica *et al.*, 1991), it appears that the 5-HT₃ receptor is a homomeric complex composed of five copies of the same subunit. The deduced amino acid sequence of the cDNA indicates that the protein is 487 amino acids long (56 kDa) and has a structure most analogous to the α7 subtype of neuronal nAChRs which also forms a homo-oligomeric receptor.

The 5-HT₃ receptor is mostly impermeable to divalent cations. For example, Ca^{2+} is largely excluded from permeation and, in fact, effectively blocks current flow through the pore, even though the predicted pore size of the channel (7.6 Å) is approximately the same as that for nAChR (8.4 Å). Apparently, other physical or electrochemical barriers limit the capacity of divalent ions to permeate the 5-HT₃ pore. Dose–response studies indicate that at least two ligand binding sites must be occupied for the channel to open; however, the binding of agonist and/or opening of the channel appears to be approximately 10 times slower than for most other ligand-gated ion channels. The functional significance or physical explanation of this slow opening is not known. The native 5-HT₃ receptor also exhibits desensitization (time constant 1–5 s), although the rate varies widely, depending on the methodology used for analysis and the source of receptor. Interestingly, this desensitization can be significantly slowed or enhanced by single amino acid substitutions at a Leu residue in the TM2 transmembrane-spanning segment of the subunit (Yakel *et al.*, 1993).

The 5-HT₃ receptors are sparsely distributed on primary sensory nerve endings in the periphery and are widely distributed at low concentrations in the mammalian CNS. The 5-HT₃ receptor is clinically significant because antagonists of 5-HT₃ receptors have important applications as antiemetics, anxiolytics, and antipsychotics.

GABA₍A₎ Receptors Are Related in Structure to nAChRs, but Exhibit an Inhibitory Function

Synaptic inhibition in the mammalian brain is mediated principally by GABA receptors. The most widespread ionotropic receptor activated by GABA is designated GABA₍A₎. The subunits composing the GABA₍A₎ receptor have sequence homology with the nAChR subunit family, and the two families have presumably diverged from a common ancestral gene. In fact, the general structures of the two receptors appear to be quite similar. The GABA₍A₎ receptor is composed of multiple subunits, probably forming a heteropentameric complex of approximately 275 kDa. Five different types of subunits are associated with GABA₍A₎ receptors and are designated α, β, γ, δ, and ε. An additional subunit, ρ, is found predominantly in the retina, whereas the other subunits are widely distributed in the brain. Each subunit group also has different subtypes; for example, six different α, four β, four γ, and two ρ subunits have been identified. The predicted amino acid sequences indicate that each of these subunits has a molecular mass ranging between 48 and 64 kDa. Like neuronal nAChR, these subunits mix in a heterogeneous fashion to produce a wide array of GABA₍A₎ receptors with different pharmacological and electrophysiological properties. The predominant GABA₍A₎ receptor in brain and spinal cord is α1, β2, and γ2, with a likely stoichiometry of two α1s, two β2s and one γ2. Expression of subunit cDNAs in oocytes indicates that the α subunit is essential for producing a functional channel. The α subunit also appears to contain the high-affinity binding site for GABA (Seighart, 1992).

The ion channel associated with the $GABA_A$ receptor is selective for anions (in particular, Cl^-), and the selectivity is provided by strategically placed positively charged amino acids near the ends of the ion channel (Barnard *et al.*, 1987). When GABA binds to and activates this receptor, Cl^- flows into the cell, producing a hyperpolarization by moving the membrane potential away from the threshold for firing an action potential (see also Chapter 16). The neuronal $GABA_A$ receptor exhibits multiple conductance levels, with the predominant conductance being 27–30 pS. Measurements and modeling of single-channel kinetics suggest that two sequential binding sites exist for anions within the pore (Bormann, 1988).

The $GABA_A$ Receptor Binds Several Compounds That Affect Its Properties

The $GABA_A$ receptor is an allosteric protein, its properties being modulated by the binding of a number of compounds. Two well-studied examples are barbiturates and benzodiazepines, both of which bind to the $GABA_A$ receptor and potentiate GABA binding. The net result is that in the presence of barbiturates or benzodiazepines or both, the same concentration of GABA will increase inhibition (see Fig. 8 of Chapter 16 for example). Benzodiazepine binding is conferred on the receptor by the γ subunit (Pritchett *et al.*, 1989), but the presence of the α and β subunits is necessary for the qualitative and quantitative aspects of benzodiazepine binding. The benzodiazepine binding site appears to lie along the interface between the α and γ subunits and only certain subtypes are sensitive to benzodiazepines. Benzodiazepine binding to $GABA_A$ receptors requires $\alpha 1$, $\alpha 2$, or $\alpha 5$ and $\gamma 2$ or $\gamma 3$; other subunit combinations are insensitive to benzodiazepines (Rudolph *et al.*, 1999).

Picrotoxin, a potent convulsant compound, appears to bind within the channel pore of the $GABA_A$ receptor and prevent ion flow (Seighart, 1992). Single-channel experiments indicate that picrotoxin either slowly blocks an open channel or prevents the GABA receptor from undergoing a transition into a long-duration open state. Apparently, barbiturates produce similar changes in channel properties, but they potentiate rather than inhibit $GABA_A$ receptor function. Bicuculline, another potent convulsant, appears to inhibit $GABA_A$ receptor channel activity by decreasing the binding of GABA to the receptor. Steroid metabolites of progesterone, corticosterone, and testosterone also appear to have potentiating effects on GABA currents that are similar in many ways to the action of barbiturates; however, the binding sites for these steroids and the barbiturates are distinct. Finally,

penicillin directly inhibits GABA receptor function, apparently by binding within the pore and thus being designated an open channel blocker.

The physiological effects of compounds such as picrotoxin, bicuculline, and penicillin are striking. Each of these compounds at a sufficiently high concentration can produce widespread and sustained seizure activity. Conversely, many, but not all, of the sedative properties associated with barbiturates and benzodiazepines can be attributed to their ability to augment inhibition in the brain by enhancing GABA's inhibitory potency.

Interestingly, ρ subunit-containing GABA receptors, found in abundance in the retina, are pharmacologically unique. They are resistant to bicuculline's inhibitory action, although they remain sensitive to blockage by picrotoxin. In addition, these retinal receptors are not sensitive to modulation by barbiturates or benzodiazepines. Thus, ρ subunit-containing receptors are distinct from $GABA_A$ receptors and are similar to receptors earlier designated $GABA_C$ (Bormann and Fiegenspan, 1995).

Several studies indicate that phosphorylation of the $GABA_A$ receptor likely modifies its functions; however, whether the receptor itself is phosphorylated *in vivo* and whether phosphorylation increases or decreases the current flowing through the channel remain debated. $GABA_A$ receptors are modulated by protein kinase A, protein kinase C, Ca^{2+}-calmodulin-dependent protein kinase, and an undefined protein kinase.

Glycine Receptor Structure Is Closely Related to $GABA_A$ Receptor Structure

Glycine receptors are the major inhibitory receptors in the spinal cord (Betz, 1991) and within the CNS, particularly in the brainstem; glycine receptors provide similar inhibitory functions. Glycine receptors are similar to $GABA_A$ receptors in that both are ion channels selectively permeable to the anion Cl^- (see Fig. 11.9). The structure of the glycine receptor is indicative of this similarity in properties. The native complex is approximately 250 kDa and is composed of two main subunits, α (48 kDa) and β (58 kDa). The receptor appears to be pentameric, most likely composed of three α and two β subunits. Apparently, three molecules of glycine must bind to the receptor to open it to ion flow (Young and Snyder, 1974), suggesting that the α subunit may contain the glycine binding site. The glycine receptor has an open-channel conductance of approximately 35–50 pS, similar to that of the $GABA_A$ receptor. A potent antagonist of the glycine receptor is the compound strychnine.

Four distinct α subunits and one β subunit of the glycine receptor have been cloned. Each exhibits the typical predicted four transmembrane segments and they are approximately 50% identical with one another at the amino acid level. Expression of a single α subunit in oocytes is sufficient to produce functional glycine receptors, indicating that the α subunit is the pore-forming unit of the native receptor. The β subunits play exclusively modulatory roles, affecting, for example, sensitivity to the inhibitory actions of picrotoxin. They are widespread in the brain, and their distribution does not specifically co-localize with glycine receptor α-subunit mRNA. The β subunits may serve other functions independent of their association with glycine receptor.

Certain Purinergic Receptors Are Also Ionotropic

Purinergic chemical transmission is distributed throughout the body and the receptor subtypes and myriad effects are considered in greater detail in the later section on GPCRs. Purinergic receptors bind to ATP (or other nucleotide analogs) or its breakdown product adenosine. ATP is released from certain synaptic terminals in a quantal manner and is often packaged within synaptic vesicles containing another neurotransmitter, the best described being acetylcholine and the catecholamines.

Although not included in Fig. 11.9, a few purinergic receptor subtypes are related to the family of ionotropic receptors. Two subtypes of ATP-binding purinergic receptors (P2x and P2z) were recently discovered to be ionotropic receptors, but data on their functions and properties are sparse. P2x receptors appear to mediate a fast depolarizing response in neurons and muscle cells to ATP by the direct opening of a nonselective cation channel. cDNAs encoding the P2x receptor indicate that its structure comprises only two transmembrane domains, with some homology in its pore-forming region with K$^+$ channels (Brake *et al.*, 1994; Valera *et al.*, 1994). The P2z receptor also is a ligand-gated channel that permits permeation of either anions or cations and even molecules as large as 900 Da. Its primary structure has not yet been defined.

Glutamate Receptors Are Derived from a Different Ancestral Gene and Are Structurally Distinct from Other Ionotropic Receptors

Glutamate receptors are widespread in the nervous system where they are responsible for mediating the vast majority of excitatory synaptic transmission in the brain and spinal cord. Early studies

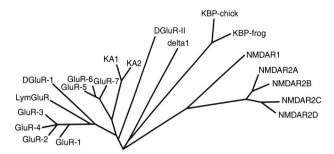

FIGURE 11.11 Evolutionary relationships of the ionotropic glutamate receptor family. Adapted from Hollmann and Heinemann (1994).

suggested that the glutamate receptor family was composed of several distinct subtypes. In the 1970s, Jeffrey Watkins and his colleagues significantly advanced this field by developing agonists that could pharmacologically distinguish between different glutamate receptor subtypes. Four of these agonists—*N*-methyl-D-aspartate (NMDA), amino-3-hydroxy-5-methylisoxazoleproprionic acid (AMPA), kainate, and quisqualate—are distinct in the type of receptors to which they bind and have been used extensively to characterize the glutamate receptor family (Hollmann and Heinemann, 1994; Watkins *et al.*, 1990). A convenient distinction for describing the ionotropic glutamate receptors has been to classify them as either NMDA or non-NMDA subtypes, depending on whether they bind the agonist NMDA. Non-NMDA receptors also bind the agonist kainate or AMPA. Both NMDA and non-NMDA receptors are ionotropic. Quisqualate is unique within this group in having the capacity to activate both ionotropic and GPCR glutamate receptor subtypes (Hollmann and Heinemann, 1994). A family tree highlighting the evolutionary relationship of the glutamate receptors is shown in Fig. 11.11.

Non-NMDA Receptors Are a Diverse Family

In 1989, Stephen Heinemann and his colleagues reported the isolation of a cDNA that produced a functional glutamate-activated channel when expressed in *Xenopus* oocytes (Hollmann *et al.*, 1989). The initial glutamate receptor was termed GluR-K1, and the cDNA encoded a protein with an estimated molecular mass of 99.8 kDa. Not long after this original report, Heinemann's group (Boulter *et al.*, 1990), Peter Seeburg's group (Keinanen *et al.*, 1990), and Richard Axel's group (Nakanishi *et al.*, 1990) independently reported the isolation of families of glutamate receptor subunits, termed GluR$_1$–GluR$_4$ by Heinemann's group and GluRA–GluRD by Seeburg's group. Each GluR subunit consists of approximately

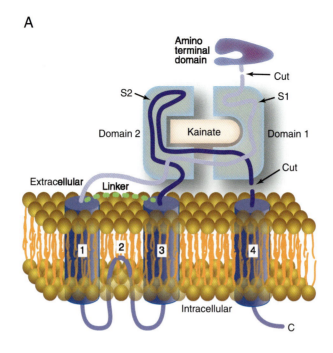

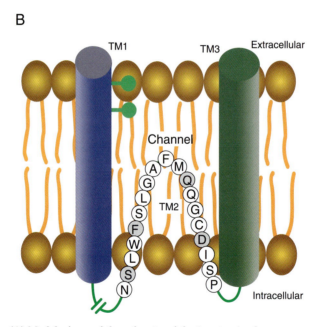

FIGURE 11.12 (A) Model of one of the subunits of the ionotropic glutamate receptor. The ionotropic glutamate receptors have four membrane-associated segments; however, unlike nAChR, only three of them completely traverse the lipid bilayer. TM2 forms a loop and re-exits into the cytoplasm. Thus, the large N-terminal region extends into the extracellular space, while the C terminus extends into the cytoplasm. Two domains in the extracellular segments associate with each other to form the binding site for transmitter, in this example kainate, a naturally occurring agonist of glutamate. (B) Enlarged area of the predicted structure and amino acid sequence of the TM2 region of the glutamate receptor, GluR$_3$. TM1 and TM3 are drawn as cylinders in the membrane flanking TM2. The residue that determines the Ca^{2+} permeability of the non-NMDA receptor is the glutamine residue (Q) highlighted in gray. In NMDA receptors, an asparagine residue at this same position is the proposed site of interaction with Mg^{2+} ions that produce the voltage-dependent channel block. The serine (S) and phenylalanine (F) also shaded in gray are highly conserved in the non-NMDA receptor family. The aspartate (D) residue is also conserved and is thought to form part of the internal cation binding site. The break in the loop between TM1 and TM2 indicates a domain that varies in length among ionotropic glutamate receptors. Adapted from Wo and Oswald (1995).

900 amino acids and has four predicted membrane-spanning segments (TM1–TM4). However, there is an important distinction in the TM2 domain making the GluRs distinct from the nAChR family. The native form of GluR subunits appears to be a tetrameric complex with an approximate molecular mass of 600 kDa (Blackstone *et al.*, 1992; Wenthold *et al.*, 1992). Thus, the size of the glutamate receptor is almost twice that of nAChR, mostly because of the large extracellular domain where glutamate binds to the receptor.

Unique Properties of Non-NMDA Receptors Are Determined by Assembly of Different Subunits

When the cDNAs encoding these receptors were expressed in either oocytes or HeK-293 cells, application of the non-NMDA receptor agonist AMPA produced substantial inward currents. In these same experiments, the agonist kainate was demonstrated to produce larger currents, mainly because of rapid and significant desensitization of the receptor when AMPA was used as the agonist. A striking observation was made from these expression studies. Specifically, when the $GluR_2$ subunit alone was expressed in the oocytes, little current was obtained when the preparation was exposed to agonist, unlike the large currents found when either $GluR_1$ or $GluR_3$ was expressed (Boulter *et al.*, 1990; Nakanishi *et al.*, 1990; Verdoorn *et al.*, 1991). $GluR_2$ subunits by themselves appear to form poorly conducting receptors. However, when $GluR_2$ is expressed with either $GluR_1$ or $GluR_3$, the behavior of the heteromeric receptor is distinctly different. Examination of the current/voltage relationships (i.e., I/V plots, see also Chapter 16) indicates that when $GluR_1$ and $GluR_3$ are expressed alone or together, they produce channels with strong inward rectification. Co-expression of $GluR_2$ with either $GluR_1$ or $GluR_3$ produces a channel with little rectification and a near-linear I/V plot. Further analyses (Hollmann and Heinemann, 1994) indicated that $GluR_1$ and $GluR_3$, either independently or when co-expressed, exhibited channels permeable to Ca^{2+}. In contrast, any combination of receptor that included the $GluR_2$ subunit produced channels impermeable to Ca^{2+}. Earlier single-channel analyses of glutamate receptors expressed in embryonic hippocampal neurons indicated that two distinct receptors are present: one relatively impermeable to Ca^{2+}, exhibiting a linear I/V plot, and another significantly permeable to Ca^{2+}, exhibiting a rectifying I/V plot (Iino *et al.*, 1990). Clearly, the properties of glutamate receptors can be quite different and can initiate unique

intracellular responses, depending on the subunit composition expressed in a particular neuron. In a series of elegant studies, the replacement of a single amino acid (Arg for Gln) in the second transmembrane region of the $GluR_2$ subunit (see Fig. 11.12B for identification of this amino acid) was shown to switch its behavior from a non-Ca^{2+}-permeable to a Ca^{2+}-permeable channel (Hume *et al.*, 1991; Burnashev *et al.*, 1992). Apparently, an Arg at this position blocks Ca^{2+} from traversing the pore formed in the center of the GluR channel.

Functional Diversity in GluRs Is Produced by mRNA Splicing and RNA Editing

Analysis of the mRNAs encoding GluR subunits indicated that each could be expressed in one of two splice variants, termed *flip* and *flop* (Sommer *et al.*, 1990). These flip and flop modules are small (38 amino acid) segments just preceding the TM4 transmembrane domain in all four GluR subunits. The receptor channel expressed from these splice variants has distinct properties, depending on which of the two modules is present. Specifically, the flop-containing receptors exhibit significantly greater magnitudes of desensitization during glutamate application. Therefore, GluRs with flop modules express smaller steady-state currents than GluRs with flip modules. Both flip- and flop-containing GluRs are widely expressed in the brain with a few exceptions. One unique cell type appears to be pyramidal CA3 cells in the rat hippocampus, where the GluRs are deficient in flop modules. In neighboring CA1 pyramidal cells and dentate granule cells, flop-containing GluRs appear to dominate. The significance of these splice variations for information processing in the brain is not known, but the physiological prediction would be that CA3 neurons exhibit larger steady-state glutamate-activated currents owing to decreased desensitization from the absence of flop modules.

Typically, one believes that there is absolute fidelity in the process of transcribing DNA into mRNA and then into protein (i.e., the nucleotides present in the DNA are accurate predictors of the ultimate amino acid sequence of the protein). However, Peter Seeburg and his colleagues discovered a novel mechanism in the neuronal nucleus that edits mRNAs posttranscriptionally, and at least three of the four GluR subunits are subjected to this editing mechanism (Sommer *et al.*, 1991). In fact, one of the sites edited is the critical Arg residue regulating Ca^{2+} permeability in the $GluR_2$ subunit. At another edited site, Gly replaces Arg-764 in the $GluR_2$ subunit, and this editing also takes place in $GluR_3$ and $GluR_4$. The Arg-to-Gly conversion at amino acid 764 produces receptors that exhibit significantly faster rates of recovery from the desensitized state (Lomeli *et al.*, 1994). The extent to which other receptors or other protein molecules undergo this form of editing is an area rich for investigation. At a minimum, this editing mechanism produces dramatic differences in the function of GluRs.

Glutamate Receptors Do Not Conform to the Typical Four Transmembrane-Spanning Segment Structures Described for nAChR

Although the field of glutamate receptors is advancing at a rapid pace, few structural data are available on the native molecule or on the topology of any single GluR subunit as it exists in the membrane. The receptor has a large extracellular domain that serves as the binding site for glutamate (Fig. 11.12A). Through the use of sophisticated genetic engineering, a crystal structure has been obtained of the glutamate-binding site for GluR in the presence of the agonist kainate. Intricate interactions between the extracellular loops of the GluR subunits forms the kainate binding sites (Fig. 11.12A).

Superficially, the remainder of the receptor was originally thought to resemble nAChR in having four TM segments that wrap back and forth through the membrane in an antiparallel fashion. However, the original model has now been proven incorrect by a number of elegant molecular and biochemical studies. The most recent information indicates that the TM2 membrane-spanning segment does not completely traverse the membrane (Fig. 11.12). Instead, it forms a kink within the membrane and enters back into the cytoplasm, similar in some ways to the pore forming domain (P segment) of voltage-activated K^+ channels (Wo and Oswald, 1995) (see Chapter 6). An enlargement of this P segment (Fig 11.12B) highlights the amino acids conserved in all the GluRs and further identifies the critical Gln (Q) residues responsible for Ca^{2+} permeability of the receptor. It also appears that glutamate receptors do not conform to the five-subunit structure of nAChR. There is both biochemical (Armstrong and Bouaux, 2000; Mano and Teichberg, 1998) and electrophysiological (Rosenmund *et al.*, 1998) evidence that functional glutamate receptors are composed of four, not five subunits. Thus, it appears glutamate receptors are rather highly divergent from the nAChR receptor family. In fact, their structure conforms more closely to the family of K^+ channels in that both appear tetrameric and both have a unique P segment that forms the selectivity filter.

Other Non-NMDA GluRs Have Poorly Characterized Functions

Three other members, $GluR_5$–$GluR_7$, now form a second non-NMDA receptor subfamily, whose contribution to producing functionally distinct receptors is less well understood. Their overall structure is similar to that of $GluR_1$–$GluR_4$, and they exhibit about 40% sequence homology. However, their agonist binding profile and their electrophysiological properties are distinct. They are expressed at lower levels in the brain than the $GluR_1$–$GluR_4$ family (Hollmann and Heinemann, 1994).

Two members of the glutamate receptor family, KA-1 and KA-2, are the high-affinity kainate-binding receptors found in brain. Clearly distinct from the glutamate receptors discussed so far, KA-1 and KA-2 are more similar to the $GluR_5$–$GluR_7$ subfamily than to the $GluR_1$–$GluR_4$ subfamily. Neither KA-1 nor KA-2 produces a functional channel when expressed in cells or oocytes, even though high-affinity kainate binding sites were detected. KA-1 does not appear to form functional receptors or channels with any of the other GluR subunits, and its physiological relevance remains obscure. It is expressed at high concentrations in only two cell types, hippocampal CA3 and dentate granule cells. KA-2 exhibits interesting properties when combined with other GluR subunits. For example, co-expression of $GluR_6$ and KA-2 produces functional receptors that respond to AMPA, although neither subunit itself responds to this agonist (Herb *et al.*, 1992). This information indicates that agonist binding sites are at least partly formed at the interfaces between subunits.

Although other kainate-binding proteins and glutamate receptors have been described, their functions and biological significance are not currently understood. These receptors include two kainate-binding proteins, one from chicken and the other from frog, several invertebrate glutamate receptors, and two "orphan" receptors termed α1 and δ2 (Hollmann and Heinemann, 1994).

The NMDA Receptors Are a Family of Ligand-Gated Ion Channels That Are Also Voltage-Dependent

NMDA receptors appear to be at least partly responsible for aspects of development, learning, and memory and neuronal damage due to brain injury. The particular significance of this receptor to neuronal function comes from two of its unique properties. First, the receptor exhibits associativity (see Chapter 18 for a more detailed discussion of associativity and the role of the NMDA receptor in memory mechanisms.) For the channel to be open the receptor must bind glutamate and the membrane must be depolarized. This behavior is due to a Mg^{2+}-dependent block of the receptor at normal membrane resting potentials (Ascher and Nowak, 1988; Mayer and Westbrook, 1987) and gives rise to the dramatic voltage dependence of the channel (see Fig. 16.9). Second, the receptor permits a significant influx of Ca^{2+}, and increases in intracellular Ca^{2+} activate a variety of processes that alter the properties of the neuron. Excess Ca^{2+} is also toxic to neurons, and the hyperactivation of NMDA receptors is thought to contribute to a variety of neurodegenerative disorders.

Many pharmacological compounds produce their effects through interactions with the NMDA receptor. For example, certain hallucinogenic compounds, such as phencyclidine (PCP) and dizocilpine (MK-801), are effective blockers of the ion channel associated with the NMDA receptor (Fig. 11.13). These potent antagonists require the receptor channel to be open to gain access to their binding sites and are therefore referred to as open-channel blockers. They also become trapped when the channel closes and are therefore difficult to wash out of the NMDA receptor's channel. Antagonists for the glutamate-binding site also have

been developed, and some of the most well known are AP-5 and AP-7. These and other antagonists specific for the glutamate binding site also produce hallucinogenic effects in both animal models and humans. NMDA remains a specific agonist for this receptor; however, it is about one order of magnitude less potent than L-glutamate for receptor activation. L-glutamate is the predominant neurotransmitter that activates the NMDA receptor; however, L-aspartate can also activate the receptor, as can an endogenous dipeptide in the brain, N-acetylaspartylglutamate (Hollmann and Heinemann, 1994).

NMDA Receptor Subunits Show Similarity to Non-NMDA Receptor Subunits

The primary structure of the NMDA receptor was revealed in 1990 when Nakanishi and his colleagues isolated the first cDNA encoding a subunit of the NMDA receptor (Moriyoshi et al., 1991). The first cloned subunit was aptly named NMDAR1, and the deduced amino acid sequence indicated a protein of approximately 97 kDa, similar to other members of the GluR family. Four potential transmembrane domains were identified, and the current assumption is that four individual subunits compose the macromolecular

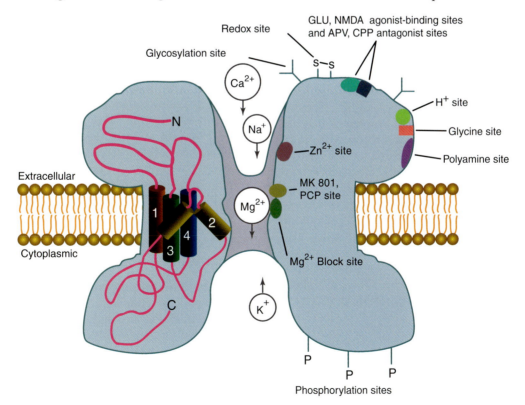

FIGURE 11.13 Diagram of a NMDA receptor highlighting binding sites for numerous agonists, antagonists, and other regulatory molecules. The location of these sites is a crude approximation for the purpose of discussion. Adapted from Hollmann and Heinemann (1994).

NMDA receptor complex. However, recall that the transmembrane organization of GluR subunits indicates that TM2 does not fully traverse the membrane. It seems likely that the NMDA receptor subunits will also follow this recent modification of the model. The TM2 segment of each subunit clearly lines the pore of the NMDA receptor channel, as does the TM2 segment of the GluR subunits. In fact, a single Asn residue, analogous to that in the $GluR_2$ subunit, regulates the Ca^{2+} permeability of the NMDA receptor (Burnashev et al., 1992; Mori et al., 1992). Mutation of this Asn residue markedly reduces Ca^{2+} permeability.

Three of the best-characterized facets of the NMDA receptor were found when the $NMDAR_1$ subunit was initially expressed by itself in oocytes, although currents were relatively small. These characteristics are (1) a Mg^{2+}-dependent voltage-sensitive ion channel block, (2) a glycine requirement for effective channel opening, and (3) Ca^{2+} permeability (Moriyoshi et al., 1991). As described below, other NMDAR subunits contribute to assembly of the receptors thought to exist in the nervous system.

Functional Diversity of NMDA Receptors Occurs through RNA Splicing

At least eight splice variants have now been identified for the $NMDAR_1$ subunit and these variants produce differences, ranging from subtle to significant, in the properties of the expressed receptor (Hollmann and Heinemann, 1994). For example, $NMDAR_1$ receptors lacking a particular N-terminal insert owing to alternative splicing exhibit enhanced blockade by protons and exhibit responses that are potentiated by Zn^{2+} in micromolar concentrations. Zn^{2+} has classically been described as an NMDA receptor antagonist that significantly blocks its activation. Clearly, the particular splice variant incorporated into the receptor complex affects the types of physiological response generated. Spermine, a polyamine found in neurons and in the extracellular space, also slightly increases the amplitude of NMDA responses, and this modulatory effect also appears to be associated with a particular splice variant. The physiological role of spermine in regulating NMDA receptors remains unclear.

Multiple NMDA Receptor Subunit Genes Also Contribute to Functional Diversity

Four other members of the NMDA receptor family have been cloned ($NMDAR_2A$–$NMDAR_2D$), and their deduced primary structures are highly related. These four NMDA receptor subunits do not form channels when expressed singly or in combination unless they are co-expressed with $NMDAR_1$ (Monyer et al., 1992; Kutsuwada et al., 1992; Meguro et al., 1992). Apparently, $NMDAR_1$ serves an essential function for the formation of a functional pore by which activation of NMDA receptors permits the flow of ions. $NMDAR_2A$–$NMDAR_2D$ play important roles in modulating the receptor activity when mixed as heteromeric forms with $NMDAR_1$. Co-expression of $NMDAR_1$ with any of the other subunits produces much larger currents (from 5- to 60-fold greater) than when $NMDAR_1$ is expressed in isolation, and NMDA receptors expressed in neurons are likely to be hetero-oligomers of $NMDAR_1$ and $NMDAR_2$ subunits. The C-terminal domains of $NMDAR_2A$–$NMDAR_2D$ are quite large relative to the $NMDAR_1$C terminus and they appear to play roles in altering channel properties and in affecting the subcellular localization of the receptors. All of the NMDAR subunits have an Asn residue at the critical point in the TM2 domain essential for producing Ca^{2+} permeability. This Asn residue also appears to form at least part of the binding site for Mg^{2+}, which suggests that the sites for Mg^{2+} binding and Ca^{2+} permeation overlap (Burnashev et al., 1992; Mori et al., 1992).

The distribution of $NMDAR_2$ subunits is generally more restricted than the homogeneous distribution of $NMDAR_1$, with the exception of $NMDAR_2A$, which is expressed throughout the nervous system. $NMDAR_2C$ is restricted mostly to cerebellar granule cells, whereas 2B and 2D exhibit broader distributions. As noted, the large size of the C terminus of the $NMDAR_2$ subunit suggests a potential role in association with other proteins, possibly to target or restrict specific NMDA receptor types to areas of the neuron. Mechanisms related to receptor targeting are now becoming understood and will clearly play major roles in determining the efficacy of synaptic transmission (Ehlers et al., 1995; Komau et al., 1995).

NMDA Receptors Exhibit Complex Channel Properties

The biophysical properties of the NMDA receptor are complex (Ascher and Nowak, 1988; Mayer and Westbrook, 1987). The single-channel conductance has a main level of 50 pS; however, subconductances are evident, and different subunit combinations produce channels with distinct single-channel properties. A binding site for the Ca^{2+}-binding protein calmodulin has also been identified on the $NMDAR_1$ subunit (Ehlers et al., 1996). Binding of Ca^{2+}-calmodulin to NMDA receptors produces a fourfold decrease in open-channel probability. Ca^{2+} influx through the

NMDA receptor could induce calmodulin binding and lead to an immediate short-term feedback inhibition, decreasing ion flow through the receptor.

Summary

A general model for ionotropic receptors has emerged mainly from analyses of nAChR. Ionotropic receptors are large membrane-bound complexes generally composed of five subunits. The subunits each have four transmembrane domains, and the amino acids in the transmembrane segment TM2 form the lining of the pore. Transmitter binding induces rapid conformational changes that are translated into an increase in the diameter of the pore, permitting ion influx. Cation or anion selectivity is obtained through the coordination of specific negatively or positively charged amino acids at strategic locations in the receptor pore. How well the details of structural information obtained for nAChR will generalize to other ionotropic receptors awaits structural analyses of these other members. However, it is already clear that this model does not adequately describe the orientation of the transmembrane domains or the subunit number of the glutamate receptor family. The TM2 domain of glutamate receptors forms a hairpin instead of traversing the membrane completely, causing the remainder of the receptor to adopt an architecture different than that described for the nAChR family. It also appears that glutamate receptors are composed of four, not five, subunits. These differences are perhaps not surprising given that the nAChR family and the glutamate receptor family appear to have arisen from two different ancestral genes.

G PROTEIN-COUPLED RECEPTORS

The number of members in the G protein-coupled receptor family is enormous, with more than 1000 identified and the number growing. The historical term *metabotropic* was used to describe the fact that intracellular metabolites are produced when these receptors bind ligand. However, there are now clearly documented cases where activation of metabotropic receptors does not produce alterations in metabolites but instead produces their effects by interacting with G proteins that alter the behavior of ion channels. Thus, these receptors are now referred to as G protein-coupled receptors or GPCRs.

When a GPCR is activated it couples to a G protein initiating the exchange of GDP for GTP, activating the G protein. Activated G proteins then couple to many downstream effectors and most alter the activity of other intracellular enzymes or ion channels. Many of the G-protein target enzymes produce diffusible second messengers (metabolites) that stimulate further downstream biochemical processes, including the activation of protein kinases (see Chapter 12). Time is required for each of these coupling events, and the effects of GPCR activation are typically slower in onset than those observed following activation of ionotropic receptors. Because there is a lifetime associated with each intermediate the effects produced by activation of GPCRs are also typically longer in duration than those produced by activation of ionotropic receptors. Most small neurotransmitters, such as ACh, glutamate, serotonin, and GABA, can bind to and activate both ionotropic and GPCRs. Thus, each of these transmitters can induce both fast responses (milliseconds), such as typical excitatory or inhibitory postsynaptic potentials, and slow-onset and longer-duration responses (from tenths of seconds to, potentially, hours). Other transmitters, like neuropeptides, produce their effects largely by binding only to GPCRs. These effects across multiple time domains provide the nervous system with a rich source for temporal information processing that is subject to constant modification. Currently, the GPCR family can be divided into three subfamilies on the basis of their structures: (1) the rhodopsin–adrenergic receptor subfamily, (2) the secretin–vasoactive intestinal peptide receptor subfamily, and (3) the metabotropic glutamate receptor subfamily (Strader *et al.*, 1995).

GPCR Structure Conforms to a General Model

A GPCR consists of a single polypeptide with a generally conserved structure. The receptor contains seven membrane-spanning helical segments that wrap back and forth through the membrane (Fig. 11.14). G protein-coupled receptors are homologous to rhodopsin from both mammalian and bacterial sources, and detailed structural information on rhodopsin has been used to provide a framework for developing a general model for GPCR structures (Palczewski *et al.*, 2000; Henderson *et al.*, 1990). Aside from rhodopsin, two of the best structurally characterized GPCRs are the β-adrenergic receptor (βAR) and the muscarinic acetylcholine receptor (mAChR), and biochemical analyses to date support the use of rhodopsin as a structural framework for the family of GPCRs (Mizobe *et al.*, 1996).

The most conserved feature of GPCRs is the seven membrane-spanning segments; however, other generalities can be made about their structure. The N terminus of the receptor extends into the extracellular

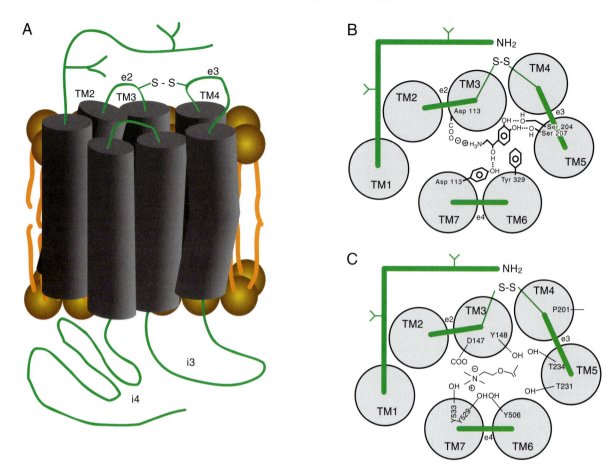

FIGURE 11.14 (A) Diagram showing the approximate position of the catecholamine binding site in βAR. The transmitter binding site is formed by amino acids whose side chains extend into the center of the ring produced by the seven transmembrane domains (TM1–TM7). Note that the binding site exists at a position that places it within the plane of the lipid bilayer. (B) A view looking down on a model of βAR identifying residues important for ligand binding. The seven transmembrane domains are represented as gray circles labeled TM1 though TM7. Amino acids composing the extracellular domains are represented as green bars labeled e1 through e4. The disulfide bond (–S–S–) that links e2 to e3 also is shown. Each of the specific residues indicated makes stabilizing contact with the transmitter. (C) A view looking down on a model of mAChR identifying residues important for ligand binding. Stabilizing contacts, mainly through hydroxyl groups (–OH), are made with the transmitter on four of the seven transmembrane domains. The chemical nature of the transmitter (i.e., epinephrine versus ACh) determines the type of amino acids necessary to produce stable interactions in the receptor binding site (compare B and C). Adapted from Strosberg (1990).

space, whereas the C terminus resides within the cytoplasm (Fig. 11.14). Each of the seven transmembrane domains between the N and C termini consists of approximately 24 mostly hydrophobic amino acids. These seven TM domains associate together to form an oblong ring within the plasma membrane (Fig. 11.14B). Between each transmembrane domain is a loop of amino acids of various sizes. The loops connecting TM1 and TM2, TM3 and TM4, and TM5 and TM6 are intracellular and are labeled i1, i2, and i3, respectively, whereas those between TM2 and TM3, TM4 and TM5, and TM6 and TM7 are extracellular and are labeled e1, e2, and e3, respectively (see Fig. 11.14A for examples).

The Neurotransmitter Binding Site Is Buried in the Core of the Receptor

The neurotransmitter binding site for many GPCRs (excluding the metabotropic glutamate, GABA_B and neuropeptide receptors) resides within a pocket formed in the center of the seven membrane-spanning segments (Fig. 11.14A). In βAR, this pocket resides ~11 Å into the hydrophobic core of the receptor, placing the ligand binding site within the plasma membrane lipid bilayer (Mizobe *et al.*, 1996; Kobilka, 1992). Strategically positioned charged and polar residues in the membrane-spanning segments point inward into a central pocket that forms the binding site for the ligand. For example, Asn residues in the second and third segments, two Ser residues in the fifth segment, and a Phe residue in the sixth segment provide major contact points in the βAR-binding site for the transmitter (Kobilka, 1992) (Fig. 11.14B). Replacing the Asp in TM3 with a Glu reduced transmitter binding by more than 100-fold, and replacement with a less conserved amino acid, such as Ser, reduces binding by more than 10,000-fold. Two Ser residues in TM5 are also essential for efficient transmitter binding and receptor activation, as are an Asp residue in TM2 and a Phe residue in TM6. In total, the two Asp, two Ser, and one Phe residue are highly conserved in all receptors that bind catecholamines. Variations in the amino acids at these five positions appear to provide the specificity between binding of different transmitters to the individual GPCRs.

The neurotransmitter binding site of mAChRs, like that of β2AR, has been investigated in great detail (Fig. 11.14C). The Asp residue in TM3 is also critical for ACh binding to mAChRs. Mutagenesis studies indicate important roles for Tyr and Thr residues in TM3, TM5, TM6, and TM7 in contributing to the ligand binding site for ACh. Interestingly, many of these mutations do not affect antagonist binding, indi-

cating that distinct sets of amino acids participate in binding agonists and antagonists. When the TM domains are examined from a side view (Fig. 11.15), all of the key amino acids implicated in agonist binding lie at about the same level within the core of the receptor structure, buried approximately 10–15 Å from the surface of the plasma membrane (yellow boxed amino acids). An additional amino acid identified as essential for agonist binding of mAChR is a Pro residue in TM4. This residue is also highly conserved among the GPCRs, and structural predictions suggest that it affects ligand binding not by interacting with agonist directly but by stabilizing a conformation essential for high-affinity binding. Structural predictions also place this Pro residue in the same plane as the Asp, Tyr, and Thr residues that form the ligand binding site of mAChR (Fig. 11.15).

Transmitter Binding Causes a Conformational Change in the Receptor and Activation of G Proteins

Proposed models for GPCR activation assume that the receptor can spontaneously isomerize between the inactive and active states (Perez *et al.*, 1996; Premont *et*

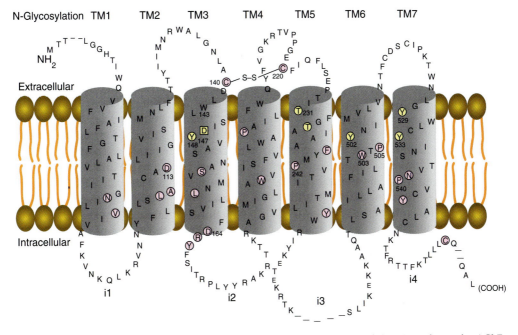

FIGURE 11.15 Amino acid sequence and predicted domain topology of the M3 isoform of mAChR. The transmembrane domains are TM1–TM7. The NH₂ terminus of the protein is at the left and extends into the extracellular space. The COOH terminus is intracellular and is at the right; i1 to i4 are the four intracellular domains. The conserved disulfide bond (–S–S–) connects extracellular loop 2 to loop 3. The dashes in the amino acid sequence represent inserts of various lengths that are not shown. Conserved amino acids for all members of the G protein-coupled receptor family of receptors are marked in purple. The amino acids taking part in ACh binding to the receptor are highlighted in yellow. Note that all amino acids associated with ligand binding lie in approximately the same horizontal plane across the receptor. Adapted from *Trends Pharmacol. Sci.*, Vol. 14, 1993.

al., 1995). Only the active state interacts with G proteins in a productive fashion. This isomerization is analogous to the spontaneous isomerization proposed for ion channels as they oscillate between open and closed states. At equilibrium, in the absence of agonist, the inactive state of GPCRs is favored, and little G-protein activation occurs. Agonist binding stabilizes the active conformation and shifts the equilibrium toward the active form, and G-protein activation ensues. Conversely, receptor antagonists block G-protein activation through two proposed mechanisms: (1) negative antagonism in which antagonists bind to the inactive state of the receptor, thus favoring an equilibrium with the inactive form; and (2) neutral antagonism in which antagonists bind to both the active and inactive forms, thus stabilizing both and preventing a complete transition into the active form. This kinetic model indicates that agonist binding is not necessary for the receptor to undergo a transition into the active state; instead, it stabilizes the activated state of the receptor. This proposed model is supported by observations of both spontaneously arising and engineered mutants of βARs and αARs. Specific amino acid replacements produce receptors that exhibit constitutive activity in the absence of agonists (Perez *et al.*, 1996; Premont *et al.*, 1995). The amino acid changes apparently stabilize the active conformation of the molecule in a state more similar to the agonist-bound form of the receptor, leading to productive interactions with G proteins in the agonist-free state.

The Third Intracellular Loop Forms a Major Determinant for G-Protein Coupling

Extensive studies using site-directed mutagenesis and the production of chimeric molecules have revealed the domains and amino acids essential for G-protein coupling to GPCRs. Receptor domains within the second (i2) and third (i3) intracellular loops (Fig. 11.15) appear largely responsible for determining the specificity and efficiency of coupling for adrenergic and muscarinic cholinergic receptors and are the likely sites for G-protein coupling of the entire GPCR family. In particular, the 12 amino acids of the N-terminal region of the third intracellular loop significantly affect the specificity of G-protein coupling. Other regions in the C terminus of the third intracellular loop and the N-terminal region of the C-terminal tail appear to be more important for determining the efficiency of G-protein coupling than for determining its specificity (Kobilka, 1992). The third intracellular loop varies enormously in size among the different G protein-coupled receptors, ranging from 29 amino acids in the substance P (a neuropep-

tide) receptor to 242 amino acids in mAChR (Strader *et al.*, 1994). The intracellular loop connecting TM5 and TM6 is the main point of receptor coupling to G proteins, and ligand binding to amino acids in TM5 and TM6 may be responsible for triggering G protein–receptor interaction by transmitting a conformational change to the third intracellular loop (i3).

Specific Amino Acids Are Involved in Transducing Transmitter Binding into G-Protein Coupling

Residues associated with transmitting the conformational change induced by ligand binding to the activation of G proteins have been investigated with the use of mAChRs. These studies revealed that an Asp residue in TM2 is important for receptor activation of G proteins, and altering the Asp by site-directed mutagenesis has a major negative effect on G protein–receptor activation (Fraser *et al.*, 1988, 1989). A Thr residue in TM5 and a Tyr residue in TM6 also are essential. Because these residues are connected by i3, they are assumed to play fundamental roles in transmitting the conformational change induced by ligand binding to the area of i3 essential for G protein coupling and activation. When mutated, a Pro residue on TM7 produces a major impairment in the ability of the TM3 segment to induce activation of phospholipase C through a G protein and, presumably, is another key element in propagating the conformational changes necessary for efficient coupling to G proteins. As informative as mutagenesis studies can be, a true molecular understanding of the conformational changes induced by agonist binding will likely require a structural approach similar to that applied by Nigel Unwin to nAChR.

As mentioned earlier, GPCRs are single polypeptides; however, they are clearly separable into distinct functional domains. For example, β2AR can be physically split, with the use of molecular techniques, into two fragments, one fragment containing TM1–TM5 and the other containing TM6 and TM7. In isolation, neither of these fragments can produce a functional receptor; however, when co-expressed in the same cell, functional β2ARs that can bind ligand and activate G proteins are produced (Fig. 11.16). This remarkable experiment indicates that physical contiguity in the primary sequence is not essential for producing functional β2ARs, but it does emphasize the contribution of domains in the separate fragments (TM1–TM5 and TM6, TM7) to both ligand binding and G-protein coupling. Like β2AR, the m2 and m3 members of the mAChR family can form functional receptors even if split into two separate domains. A

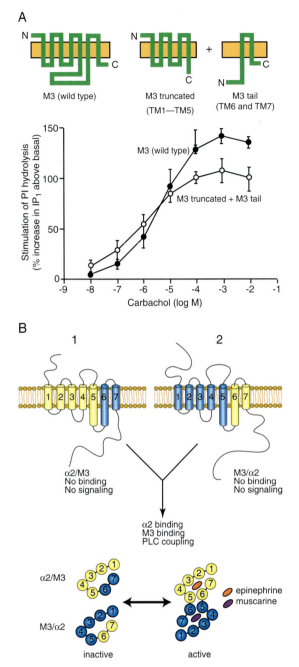

fragment containing the first five TM domains, when expressed with a fragment containing TM6 and TM7, forms a functional receptor (Strosberg, 1990) (Fig. 11.16).

GPCRs Also Exist as Homo-oligomers or Hetero-oligomers

The observation that GPCRs can be physically split through genetic engineering and when recombined produced functional channels provided the first hint that full-length GPCRs might also oligomerize with each other into functional molecules. A test of such a hypothesis was accomplished by making chimeric receptors composed of TM domains 1–5 of α2-AR and TM domains 6 and 7 of the m3 muscarinic receptors and vice versa (Fig. 11.16B) (Maggio *et al.*, 1993). When either of these chimeric molecules was expressed in isolation, neither formed a functional receptor. However, when co-expressed, receptors were formed that bind both muscarinic and adrenergic ligands and ligand binding led to functional activation of downstream effectors. Through domain swapping the ligand binding sites for both receptor ligands were reconstituted by oligomerization of the two chimeric receptors into one bifunctional chimeric dimer (Fig. 11.16B). Oligomerization of GPCRs is also supported by crosslinking and immunoprecipitation experiments and with experiments examining the direct biophysical association of the receptors in living cells (Maggio *et al.*, 1993; Salahpour *et al.*, 2000; Overton and Blumer, 2000; Lee *et al.*, 2000). While some debate remains, the evidence now seems overwhelming that oligomerization of GPCRs is adding a new layer of complexity and diversity to the study of these receptors. The functional impacts of GPCR oligomerization are just beginning to be appreciated. Important functional consequences could relate to alterations in: (1) ligand binding, (2) efficiency and specificity of coupling to downstream effectors, (3) subcellular localization, and (4) receptor desensitization. The evolving and apparently widespread nature of direct receptor–receptor interactions leads one to believe that our current understanding of neurotransmitter receptors and their biological impact will be undergoing continual modifications for many years to come.

G-Protein Coupling Increases the Affinity of the Receptor for Neurotransmitter

The affinity of GPCR for agonist increases when the receptor is coupled to the G protein. This positive feedback effectively increases the lifetime of the agonist-bound form of the receptor by decreasing the dissociation rate of the agonist. An excellent demon-

FIGURE 11.16 (A) mAChR can be split into two physically separated domains that, when added back together, retain the ability to bind transmitter and activate G proteins. (Upper left) Model of full-length mAChR; (upper right) two engineered pieces of the receptor. The graph indicates that, when coexpressed in the same cells, the two fragments can produce a functional mAChR that responds to the agonist carbachol producing activation of G protein and subsequent activation of an enzyme that hydrolyzes phosphatidylinositol (PI). Adapted from *Trends Pharmacol. Sci.*, Vol. 14, 1993. (B) Some GPCRs can function as dimers. In this example, chimeric receptors were produced between α2AR (α2) and mAChR(M3) by swapping certain TM domains through genetic engineering. When α2/M3 or M3/α2 are expressed separately, they are not active. However, if both chimeric molecules are expressed in the same cells, they form receptors that can be activated by either epinephrine or muscarine. The bottom panel shows a top-down view of how this domain swapping might occur when two molecules dimerize to produce receptors that can respond to both transmitters.

stration of this effect comes from studies using engineered βARs that are constitutively active in their ability to couple to G proteins. These mutant receptors show significantly increased affinity for agonists (Perez *et al.*, 1996). When G protein dissociates, the agonist binding affinity of the receptor returns to its original state. The changes induced by ligand binding apparently stabilize the receptor in a conformation with both higher affinity for ligand and higher affinity for coupling to G proteins.

The Specificity and Potency of G-Protein Activation Are Determined by Several Factors

GPCRs associate with G proteins to transduce ligand binding into intracellular effects. This coupling step can lead to diverse responses, depending on the type of G protein and the type of effector enzyme present. In addition, ligand binding to a single subtype of GPCR can activate multiple G protein-coupled pathways. Activated α2ARs have been shown to couple to as many as four different G proteins in the same cell (Strader *et al.*, 1994). Some of the specificity for G-protein activation can be determined by the specific conformations assumed by the receptor, and a single receptor can assume multiple conformations. For example, α2ARs can isomerize into at least two states. One state interacts with a G protein that couples to phospholipase C, and a second state interacts with G proteins that couple to both phospholipase C and phospholipase A2 (Perez *et al.*, 1996). Thus, a single GPCR can produce a diversity of responses, making it difficult to assign specific biological effects to individual receptor subtypes in all settings.

Activated GPCRs are free to couple to many G-protein molecules, permitting a significant amplification of the initial transmitter binding event (Cassel and Selinger, 1977). This catalytic mechanism is referred to as "collision coupling" (Tolkovsky *et al.*, 1982), whereby a transient association between the activated receptor and the G protein is sufficient to produce the exchange of GDP for GTP, activating the G protein. Because enzymes such as adenylyl cyclase appear to be tightly coupled to the G protein, the rate-limiting step in the production of cAMP is the number of productive collisions between the receptor and the G protein. A constant GTPase activity hydrolyzes GTP, bringing the G protein and therefore the adenylyl cyclase back to the basal state. Transmitter concentration clearly plays a role in the number of activated receptors present at any given time, and GPCRs exhibit saturable dose–response

curves. This apparent maximal rate is achieved when all of the G protein–cyclase complexes have become activated (more accurately, when the rate of formation is maximal with respect to the rate of GTP hydrolysis). A less intuitive consequence that evolves from these models is that receptor number can significantly affect the concentration of transmitter that produces a half-maximal response of cAMP accumulation. Thus, the larger the receptor number, the greater the probability that a productive collision will occur between an agonist-bound receptor and the G protein. Experimental evidence for this prediction was obtained for βAR expressed at various levels in eukaryotic cells. Increasing concentrations of βAR produced a decrease in the concentration of agonist required to produce half-maximal production of cAMP. Apparently, the cell can adjust the magnitude of its response by adjusting the number of receptors available for transmitter interaction. In addition, the important process of receptor desensitization can also regulate the number of receptors capable of productive G-protein interactions.

Receptor Desensitization Is a Built-in Mechanism for Decreasing the Cellular Response to Transmitter

Desensitization is a very important process whereby cells can decrease their sensitivity to a particular stimulus to prevent saturation of the system. Desensitization involves a complex series of events (Kobilka, 1992; Clark *et al.*, 1999). For GPCRs, desensitization is defined as an increase in the concentration of agonist required to produce half-maximal stimulation of, for example, adenylyl cyclase. In practical terms, desensitization of receptors produces less response for a constant amount of transmitter.

There are two known mechanisms for desensitization. One mechanism is a decrease in response brought about by the covalent modifications produced by receptor phosphorylation and is quite rapid (seconds to minutes). The other mechanism is the physical removal of receptors from the plasma membrane (likely through a mechanism of receptor-mediated endocytosis) and tends to require greater periods (minutes to hours). The latter process can be either reversible (sequestration) or irreversible (downregulation).

The Rapid Phase of GPCR Desensitization Is Mediated by Receptor Phosphorylation

Desensitization of βAR appears to involve at least three protein kinases: PKA, PKC, and β-adrenergic receptor kinase [βARK; also referred to as G-protein receptor kinase (GRK)]. Phosphorylation of ARs by PKA does not require that agonist be bound to the receptor and appears to be a general mechanism by which the cell can reduce the effectiveness of all receptors, independent of whether they are in the agonist-bound or unbound state (Fig. 11.17). This process is also referred to as heterologous desensitization because the receptor does not require bound agonist (for simplicity PKA is shown phosphorylating only the agonist-bound form of the receptor in Fig. 11.17). PKA and PKC phosphorylate sites on the third intracellular loop and possibly the C-terminal cytoplasmic domain. Phosphorylation of these sites functionally interferes with the receptor's ability to couple to G proteins, thus producing the desensitization (Fig. 11.17). Whether the same sites on βAR are phosphorylated by both PKA and PKC is controversial. Some researchers conclude that the effects of phosphorylation by either kinase on decreasing coupling of the receptor to G proteins are similar (suggesting that the sites phosphorylated are similar) (Huganir and Greengard, 1990). Others find that the effects are additive (Yuan *et al.*, 1994). Although the details of the role played by each of these kinases are ambiguous, phosphorylation by either enzyme desensitizes the receptor.

G Protein receptor kinases (GRKs) can also phosphorylate GPCRs and lead to receptor desensitization (Inglese *et al.*, 1993; Sterne-Marr and Benovic, 1995). Six members of the GRK family of kinases have been identified: rhodopsin kinase (GRK1), βARK (GRK2), and GRK3 through GRK6 (Premont *et al.*, 1995; Sterne-Marr and Benovic, 1995). GRK2 (originally called β-adrenergic receptor kinase or βARK) is a Ser- and Thr-specific protein kinase initially identified

by its capacity to phosphorylate βAR. GRK2 phosphorylates only the agonist-bound form of the receptor, usually when agonist concentrations reach the micromolar level, as typically found in the synaptic cleft. This process is referred to as homologous desensitization because the regulation is specific for those receptor molecules that are in the agonist-bound state. Phosphorylation of βAR by GRK2 does not substantially interfere with coupling to G proteins. Instead, an additional protein, arrestin, binds the GRK2-phosphorylated form of the receptor, thus blocking receptor–G protein coupling (Fig. 11.17). This process is analogous to the desensitization of the light-sensitive receptor molecule rhodopsin produced by GRK1 phosphorylation and the binding of arrestin. The phosphorylation sites on βAR for GRK2 reside on the C-terminal cytoplasmic domain and are distinct from those phosphorylated by PKA.

The cycle of homologous desensitization starts with the activation of a GPCR, which induces activation of G proteins and dissociation of the βγ subunit complex from α subunits. At least one role for the βγ complex appears to be to bind to GRKs, which leads to their recruitment to the membrane in the area of the locally activated G protein–receptor complex. The recruited GRK is then activated, leading to phosphorylation of the agonist-bound receptor and subsequent binding of arrestin. Arrestin binds to the same domains on the receptor necessary for coupling to G proteins, thus terminating the actions of the activated receptor (Fig. 11.18). The ensuing process of sequestration follows GPCR phosphorylation and arrestin binding.

Desensitization Can Also Be Produced by Loss of Receptors from the Cell Surface

Desensitization of GPCRs is also produced by removal of the receptor from the cell surface. This process can be either reversible (sequestration or internalization) or irreversible (downregulation). Sequestration is the term used to describe the rapid (within minutes) but reversible endocytosis of receptors from the cell surface after agonist application (Fig. 11.18). Neither G-protein coupling nor receptor phosphorylation appears to be absolutely essential for this process, but phosphorylation by GRKs clearly enhances the rate of sequestration (Ferguson *et al.*, 1995). The binding of arrestins to the phosphorylated receptor also enhances sequestration (Ferguson *et al.*, 1996). Thus, arrestin binding appears to promote not only rapid desensitization by disrupting the receptor–G protein interaction but also receptor sequestration. Because the receptor can be functionally

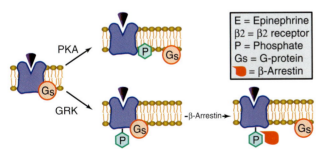

FIGURE 11.17 Different modes of desensitization of GPCRs. This diagram indicates that the epinephrine (E)-bound form of β2AR normally couples to the G protein G_s. PKA can phosphorylate the receptor, leading to an inhibition of binding to G_s. G-protein receptor kinase (GRK) also can phosphorylate the receptor; however, this phosphorylation does not directly interfere with binding to G_s. GRK phosphorylation is needed for the binding of another protein, β-arrestin, which, by its association with the receptor, prevents G_s from binding. Adapted from Ehlers *et al.* (1996).

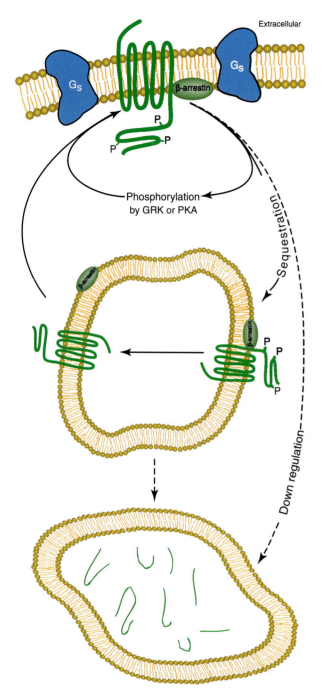

uncoupled from the G protein through the rapid phosphorylation-dependent phase of desensitization, the physiological role(s) of sequestration remains an open issue, although decreasing the number of receptor molecules on the cell surface would contribute to the overall process of desensitization to agonist. Receptor cycling through intracellular organelles is a trafficking mechanism that leads to an enhanced rate of dephosphorylation of the phosphorylated receptor, returning it to the cell surface in its basal state (Fig. 11.18) (Barak *et al.*, 1994).

Downregulation occurs more slowly than sequestration and is irreversible (Fig. 11.18). The early phase (within 4 h) may involve both a PKA-dependent and a PKA-independent process (Bouvier *et al.*, 1989; Proll *et al.*, 1992). This early phase of downregulation is apparently due to receptor degradation after endocytotic removal from the plasma membrane. The later phases (>14 h) of downregulation appear to be further mediated by a reduction in receptor biosynthesis through a decrease in the stability of the receptor mRNA (Bouvier *et al.*, 1989) and a decreased transcription rate.

Other Posttranslational Modifications Are Required for Efficient Metabotropic Receptor Function

Like many proteins expressed on the cell surface, GPCRs are glycosylated, and the N-terminal extracellular domain is the site of carbohydrate attachment. Relatively little is known about the effect of glycosylation on the function of GPCRs. Glycosylation does not appear to be essential to the production of a functional ligand binding pocket (Strader *et al.*, 1994), although prevention of glycosylation may decrease membrane insertion and alter intracellular trafficking of β2AR.

Another important structural feature of most GPCRs is the disulfide bond formed between two Cys residues present on the extracellular loops (e2 and e3; Figs. 11.14 and 11.15). Apparently, the disulfide bond stabilizes a restricted conformation of the mature receptor by covalently linking the two extracellular domains, and this conformation favors ligand binding. Disruption of this disulfide bond significantly decreases agonist binding (Kobilka, 1992).

A third Cys residue, in the C-terminal domain of GPCRs (Fig. 11.15, pink circled C in i4), appears to serve as a point for covalent attachment of a fatty acid (often palmitate). Presumably, fatty acid attachment stabilizes an interaction between the C-terminal domain of a GPCR and the membrane (Casey, 1995). The full consequences of this posttranslational

FIGURE 11.18   Additional intracellular pathways associated with desensitization of GPCRs. GPCRs are phosphorylated (noted with P) on their intracellular domains by PKA, GRK, and other protein kinases. The phosphorylated form of the receptor can be removed from the cell surface by a process called sequestration with the help of the adapter protein β-arrestin; thus fewer binding sites remain on the cell surface for transmitter interactions. In intracellular compartments, the receptor can be dephosphorylated and returned to the plasma membrane in its basal state. Alternatively, the phosphorylated receptors can be degraded (downregulated) by targeting to a lysosomal organelle. Degradation requires replenishment of the receptor pool through new protein synthesis. Adapted from Kobilka (1992).

modification are not understood, because replacing the normally palmitoylated Cys with an amino acid that cannot be acylated appears to have little effect on receptor binding; however, G-protein coupling may not be as efficient (O'Dowd *et al.*, 1989).

GPCRs can Physically Associate with Ionotropic Receptors

There is now good evidence that metabotropic and ionotropic receptors can interact directly with each other (Liu *et al.*, 2000). GABA$_A$ receptors (ionotropic) were shown to couple to dopamine (D$_5$) receptors (metabotropic) through the second intracellular loop of the γ subunit of the GABA$_A$ receptor and the C-terminal domain of the D$_5$, but not the D$_1$, receptor. Dopamine binding to D$_5$ receptors produced down-regulation of GABA$_A$ currents and pharmacologically blocking the GABA$_A$ receptor produced decreases in cAMP production when cells were stimulated with dopamine. It further appeared that ligand binding to both receptors was necessary for their stable interaction. Whether this form of receptor regulation is unique to this pair of partners or is a widespread phenomenon remains an open question ripe for further investigation.

GPCRs All Exhibit Similar Structures

The family of GPCRs exhibits structural similarities that permit the construction of "trees" describing the degree to which they are evolutionarily related (Fig. 11.19). Some remarkable relations become evident in such an analysis. For example, the D$_1$ and D$_5$ subtypes of dopamine receptors are more closely related to α2AR than to the D$_2$, D$_3$, and D$_4$ dopamine receptors. The similarities and differences among GPCR families are highlighted in the remainder of this chapter.

Muscarinic ACh Receptors

Muscarine is a naturally occurring plant alkaloid that binds to muscarinic subtypes of AChRs and activates them. mAChRs play a dominant role in mediating the actions of ACh in the brain, indirectly producing both excitation and inhibition through binding to a family of unique receptor subtypes. mAChRs are found both presynaptically and postsynaptically and, ultimately, their main neuronal effects appear to be mediated through alterations in the properties of ion channels. Presynaptic mAChRs take part in important feedback loops that regulate neurotransmitter release. ACh released from the presynaptic ter-

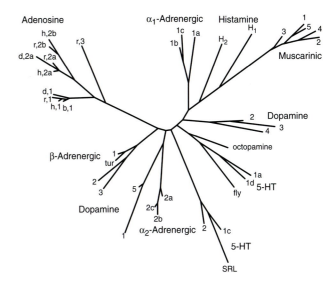

FIGURE 11.19 Evolutionary relationship of the metabotropic receptor family. To assemble this tree, sequence homologies in the transmembrane domains were compared for each receptor. Distance determines the degree of relatedness. r, rat; d, dog; h, human; tur, turkey; SRL, a putative serotonin receptor; 5-HT, 5-hydroxytryptamine (serotonin). Adapted from Linden, Chapter 21, Fig. 2, "Basic Neurochemistry", 1993. Original tree construction was by William Pearson and Kevin Lynch, University of Virginia.

minal can bind to mAChRs on the same nerve ending, thus activating enzymatic processes that modulate subsequent neurotransmitter release. This modulation is typically an inhibition; however, activation of the m5 AChR produces an enhancement in subsequent release. These autoreceptors are an important regulatory mechanism for short-term (milliseconds to seconds) modulation of neurotransmitter release.

The family of mAChRs now includes five members (m1–m5), ranging from 55 to 70 kDa, and each of the five subtypes exhibits the typical architecture of seven TM domains. Much of the diversity in this family of receptors resides in the third intracellular loop (i3) responsible for the specificity of coupling to G proteins. The m1, m3, and m5 mAChRs couple predominantly to G proteins that activate the enzyme phospholipase C. The m2 and m4 receptors couple to G proteins that inhibit adenylyl cyclase, as well as to G proteins that directly regulate K$^+$ and Ca^{2+} channels. As is the case for other GPCRs, the domain near the N terminus of the third intracellular loop is important for the specificity of G-protein coupling. This domain is conserved in m1, m3, and m5 AChRs, but is unique in m2 and m4. Several other important residues also have been identified for G-protein coupling. A particular Asp residue near the N terminus of the second intracellular loop (i2) is important for G protein coupling, as are residues residing in the C-terminal region of the i3 loop.

The major mAChRs found in the brain are m1, m3, and m4, and each is diffusely distributed. The m2 subtype is the heart isoform and is not highly expressed in other organs. The genes for m4 and m5 lack introns, whereas those encoding m1, m2, and m3 contain introns, although little is known concerning alternatively spliced products of these receptors. Atropine is the most widely used antagonist for mAChR and binds to most subtypes, as does N-methylscopolamine. The antagonist pirenzipine appears to be relatively specific for the m1 mAChR, and other antagonists such as AF-DX116 and hexahydrosiladifenidol appear to be more selective for the m2 and m3 subtypes.

Adrenergic Receptors

The catecholamines epinephrine (adrenaline) and norepinephrine (noradrenaline) produce their effects by binding to and activating adrenergic receptors. Interestingly, epinephrine and norepinephrine can both bind to the same adrenergic receptor. Adrenergic receptors are currently separated into three families, α1, α2, and β. The α1 and α2 families are further subdivided into three subclasses each. Similarly, the β family also contains three subclasses (β1, β2, and β3). The main adrenergic receptors in the brain are the α2 and β1 subtypes. α2ARs have diverse roles, but the function that is best characterized (in both central and peripheral nervous tissue) is their role as autoreceptors (i.e., presynaptic receptors that bind transmitter and alter the release apparatus so that subsequent release is modulated, usually, in an inhibitory fashion). Different adrenergic receptor subtypes bind to G proteins that can alter the activity of phospholipase C, Ca^{2+} channels, and, probably best studied, adenylyl cyclase. For example, activation of α2ARs produces inhibition of adenylyl cyclase, whereas all βARs activate the cyclase.

Only a few agonists or antagonists cleanly distinguish the adrenergic receptor subtypes. One of them, isoproterenol, is an agonist that appears to be highly specific for βARs. Propranolol is the best-known antagonist for β receptors, and phentolamine is a good antagonist for α receptors but weakly binds at β receptors. The genomic organization of the different AR subtypes is unusual. Like many GPCRs, β1ARs and β2ARs are encoded by genes lacking introns. β3ARs, which apparently have a role in lipolysis and are poorly characterized, are encoded by an intron-containing gene, as are αARs, providing an opportunity for alternative splicing as a means of introducing functional heterogeneity into the receptor.

Dopamine Receptors

Some 80% of the dopamine in the brain is localized to the corpus striatum, which receives major input from the substantia nigra and takes part in coordinating motor movements. Dopamine is also found diffusely throughout the cortex, where its specific functions remain largely undefined. However, many neuroleptic drugs appear to exert their effects by blocking dopamine binding, and imbalances in the dopaminergic system have long been associated with neuropsychiatric disorders.

Dopamine receptors are found both pre- and postsynaptically, and their structure is homologous to that of the receptors for other catecholamines (Bunzow et al., 1988; Civelli et al., 1993). Five subtypes of dopamine receptors can be grouped into two main classes, D_1-like and D_2-like receptors. D_1-like receptors include D_1 and D_5, and D_2-like receptors include D_2, D_3, and D_4 (see Fig. 11.19). The main distinction between these two classes is that D_1-like receptors activate adenylyl cyclase through interactions with G_s, whereas D_2-like receptors inhibit adenylyl cyclase and other effector molecules by interacting with G_i/G_o. D_1-like receptors are also slightly larger in molecular mass than D_2-like receptors. An additional point of interest, as noted above, is that D_5 receptors selectively bind to $GABA_A$ receptors, impacting their function and vice versa (Liu et al., 2000). The deduced amino acid sequence for the entire family ranges from 387 amino acids (D_4) to 477 amino acids (D_5). The main structural differences between D_1-like and D_2-like receptors are that the intracellular loop between the sixth and seventh TM segments is larger in the D_2-like receptors and the D_2-like receptors have smaller C-terminal intracellular segments. Two isoforms of the D_2 receptor have been isolated called D_2 long and D_2 short; alternative splicing generates these isoforms. D_2 long contains a 29-amino-acid insert in the large intracellular loop between the fifth and sixth membrane-spanning segments. Functional or anatomical differences have yet been fully resolved for the short and long forms of D_2.

D_1-like receptors, like βARs, are transcribed from intronless genes (Dohlman et al., 1987). Conversely, all D_2-like receptors contain introns, thus providing for possibilities of alternatively spliced products. Posttranslational modifications include glycosylation at one or more sites, disulfide bonding of the two Cys residues in e2 and e3, and acylation of the Cys residue in the C-terminal tail (analogous to the β2AR). The dopamine binding site includes two Ser residues in TM5 and an Asp residue in TM3, analogous to βAR.

Because of dopamine's presumed role in neuropsychiatric disorders, enormous effort has been focused on developing pharmacological tools for manipulating this system. Dopamine receptors bind amphetamines, bromocriptine, lisuride, clozapine, melperone, fluperlapine, and haloperidol. Because these drugs do not show great specificity for receptor subtypes, their usefulness for dissecting effects specifically related to binding to one or another dopamine receptor subtype is limited. However, their role in the treatment of human neuropsychiatric disorders is enormous.

Purinergic Receptors

Purinergic receptors bind to ATP (or other nucleotide analogs) or its breakdown product adenosine. Although ATP is a common constituent found within synaptic vesicles, adenosine is not and is therefore not considered a "classic" neurotransmitter. However, the multitude of receptors that bind and are activated by adenosine indicates that this molecule has important modulatory effects on the nervous system. Situations of high metabolic activity that consume ATP and situations of insufficient ATP-regenerating capacity can lead to the accumulation of adenosine. Because adenosine is permeable to membranes and can diffuse into and out of cells, a feedback loop is established in which adenosine can serve as a local diffusible signal that communicates the metabolic status of the neuron to surrounding cells and vice versa (Linden, 1994).

The original nomenclature describing purinergic receptors defined adenosine as binding to P1 receptors and ATP as binding to P2 receptors. Families of both P1 and P2 receptors have since been described, and adenosine receptors are now identified as A-type purinergic receptors, consisting of A_1, A_2a, A_2b, and A_3. ATP receptors are designated as P type and consist of P2x, P2y, P2z, P2t, and P2u. Recall that P2x and P2z subtypes are ionotropic receptors.

A-type receptors exhibit the classic arrangement of seven transmembrane-spanning segments but are typically shorter than most GPCRs, ranging in size between 35 and 46 kDa. The ligand-binding site of A-type receptors is unique in that the ligand, adenosine, has no inherent charged moieties at physiological pH. A-type receptors appear to use His residues as their points of contact with adenosine, and, in particular, a His residue in TM7 is essential, because its mutation eliminates agonist binding. Other His residues in TM6 and TM7 are conserved in all A-type receptors and may serve as other points of contact with agonists. Work with chimeric A-type receptors has further substantiated the importance of residues

in TM5, TM6, and TM7 for ligand binding. A_1 receptors are highly expressed in the brain, and their activation downregulates adenylyl cyclase and increases phospholipase C activity. A_2a and A_2b receptors are not as highly expressed in nervous tissue and are associated with the stimulation of adenylyl cyclase and phospholipase C, respectively. The A_3 subtype exhibits a unique pharmacological profile in that binding of xanthine derivatives, which blocks adenosine's action competitively, is absent. Very low levels of the A_3 receptor are found in brain and peripheral nervous tissue. The A_3 receptor appears to be coupled to the activation of phospholipase C.

The human A_1 receptor has a unique mode of receptor expression (Olah and Stiles, 1995). Introns in the 5' untranslated sequence of the mRNA, spliced in a tissue-specific manner, are capable of affecting the translational efficiency of the mRNA. Two extra start codons upstream from the start codon that initiates translation of the A_1 receptor exert a negative effect on translation. Mutating these two extra start codons can relieve the translational repression. This process is an effective way of controlling the level of receptor expression and may serve as a more general model for translational regulation for other mRNAs.

The P-type receptors, P2y, P2t, and P2u, are typical G protein-linked GPCRs, mostly localized to the periphery. However, direct effects of ATP have been detected in neurons, and often the response is biphasic; an early excitatory effect followed, with its breakdown to adenosine, by a secondary inhibitory effect. Interestingly, P-type receptors exhibit a higher degree of homology to peptide-binding receptors than they do to the A-type purinergic receptors. As in A-type receptors, P-type receptors have a His residue in the third transmembrane domain; however, other sites for ligand binding have not been specifically identified.

Serotonin Receptors

Serotonin-containing cell bodies are found in the raphe nucleus in the brainstem and in nerve endings distributed diffusely throughout the brain (Julius, 1991). Serotonin has been implicated in sleep, modulation of circadian rhythms, eating, and arousal. Serotonin also has hormone-like effects when released in the bloodstream, regulating smooth muscle contraction and affecting the platelet aggregating and immune systems.

Serotonin receptors are classified into four subtypes: $5-HT_1$ to $5-HT_4$, with a further subdivision of the $5-HT_1$ subtypes. Recall that the $5-HT_3$ receptor is ionotropic. The other 5-HT receptors exhibit the typical seven transmembrane-spanning segments, and all couple to G proteins to exert their effects. For example, 5-HT$_1$a, 5-HT$_1$b, 5HT$_1$d, and 5HT$_4$ either activate or inhibit adenylyl cyclase. 5-HT$_1$c and 5-HT$_2$ receptors preferentially stimulate activation of phospholipase C to produce increased intracellular levels of diacylglycerol and inositol 1,4,5-trisphosphate.

Serotonin receptors can also be grossly distributed into two groups on the basis of their gene structures. Both 5-HT$_1$c and 5-HT$_2$ are derived from genes that contain multiple introns. In contrast, similar to the βAR family, 5-HT$_1$ is coded by a gene lacking introns. Interestingly, 5-HT$_1$a is more closely related ancestrally to the βAR family than it is to other membranes of the serotonin receptor family and was originally isolated by using the cDNA for β2AR as a molecular probe (Kobilka *et al.*, 1987). This observation helps explain some pharmacological data suggesting that both 5-HT$_1$a and 5-HT$_1$b can bind certain adrenergic antagonists.

Glutamate GPCRs

The GPCRs that bind glutamate (metabotropic glutamate receptors or mGluRs) are similar in general structure in having seven transmembrane-spanning segments to other GPCRs; however, they are divergent enough to be considered to have originated from a separate evolutionary derived receptor family (Hollmann and Heinemann, 1994; Nakanishi, 1994). In fact, sequence homology between the mGluR family and the other GPCRs is minimal except for the GABA$_B$ receptor. The mGluR family is heterogeneous in size, ranging from 854 to 1179 amino acids. Both the N-terminal and C-terminal domains are unusually large for G protein-coupled receptors. One great difference in the structures of mGluRs is that the binding site for glutamate resides in the large N-terminal extracellular domain and is homologous to a bacterial amino acid-binding protein (Armstrong and Gouaux, 2000; O'Hara *et al.*, 1993). In most of the other families of GPCRs, the ligand binding pocket is formed by the transmembrane segments partly buried in the membrane. In addition, mGluRs exist as functional dimers in the membrane in contrast to the single-subunit forms of most GPCRs (Kunishima *et al.*, 2000). These significant structural distinctions support the idea that mGluRs evolved separately from other GPCRs. The third intracellular loop, thought to be the major determinant responsible for G-protein coupling of mGluRs is relatively small, whereas the C-terminal domain is quite large. The coupling between mGluRs and their respective G proteins may be through unique determinants that exist in the large C-terminal domain.

Currently, eight different mGluRs can be subdivided into three groups on the basis of sequence homologies and their capacity to couple to specific enzyme systems. Both mGluR1 and mGluR5 activate a G protein coupled to phospholipase C. mGluR1 activation can also lead to the production of cAMP and arachidonic acid, by coupling to G proteins that activate adenylyl cyclase and phospholipase A2 (Aramori and Nakanishi, 1992). mGluR5 seems more specific, activating predominantly the G protein-activated phospholipase C.

The other six mGluR subtypes are distinct from one another in favoring either *trans*-1-aminocyclopentane-1,3-dicarboxylate (mGluR$_2$, mGluR$_3$, and mGluR$_8$) or 1,2-amino-4-phosphonobutyrate (mGluR$_4$, mGluR$_6$, and mGluR$_7$) as agonists for activation. mGluR$_2$ and mGluR$_4$ can be further distinguished pharmacologically by using the agonist 2-(carboxycyclopropyl)glycine, which is more potent at activating mGluR$_2$ receptors (Hayashi *et al.*, 1992). Less is known about the mechanisms by which these receptors produce intracellular responses; however, one effect is to inhibit the production of cAMP by activating an inhibitory G protein.

mGluRs are widespread in the nervous system and are found both pre- and postsynaptically. Presynaptically, they serve as autoreceptors and appear to participate in the inhibition of neurotransmitter release. Their postsynaptic roles appear to be quite varied and depend on the specific G protein to which they are coupled. mGluR$_1$ activation has been implicated in long-term synaptic plasticity at many sites in the brain, including long-term potentiation in the hippocampus and long-term depression in the cerebellum (see Chapter 18).

GABA$_B$ Receptor

GABA$_B$ receptors are found throughout the nervous system, where they are sometimes co-localized with ionotropic GABA$_A$ receptors. GABA$_B$ receptors are present both pre- and postsynaptically. Presynaptically, they appear to mediate inhibition of neurotransmitter release through an autoreceptor-like mechanism by activating K$^+$ conductances and diminishing Ca^{2+} conductances. In addition, GABA$_B$ receptors may affect K$^+$ channels through a direct physical coupling to the K$^+$ channel, not mediated through a G protein intermediate. Postsynaptically, GABA$_B$ receptor activation produces a characteristic slow hyperpolarization (termed the slow inhibitory postsynaptic potential) through the activation of a K$^+$ conductance. This effect appears to be through a pertussis toxin-sensitive G protein that inhibits adenylyl cyclase.

The cloning of the $GABA_B$ receptor ($GABA_BR_1$) revealed that it has high sequence homology to the family of glutamate GPCRs, but shows little similarity to other G protein-coupled receptors. The large N-terminal extracellular domain of the $GABA_B$ receptor is the presumed site of GABA binding. With the exception of this large extracellular domain, the $GABA_B$ receptor structure is typical of the GPCR family, exhibiting seven TM domains. The initial cloning of the $GABA_B$ receptor was made possible by the development of a high-affinity, high-specificity antagonist termed CGP64213. This antagonist is several orders of magnitude more potent at inhibiting $GABA_B$ receptor function than the more widely known antagonist saclofen. Baclofen, an analog of saclofen, remains the best agonist for activating $GABA_B$ receptors.

Functional $GABA_B$ receptors appear to exist primarily as dimers in the membrane (Jones et al., 1998; White et al., 1998; Kaupmann et al., 1998). Expression of the cloned $GABA_BR_1$ isoform does not produce significant functional receptors. However, when co-expressed with the $GABA_BR_2$ isoform, receptors that are indistinguishable functionally and pharmacologically from those in brain were produced. In addition, $GABA_B$ dimers exist in neuronal membranes and all data point to the conclusion that $GABA_B$ receptors dimerize and that the dimer is the functionally important form of the receptor. As noted earlier in this chapter (Fig. 11.16B), GPCRs can interact with themselves and other receptors. It is good to keep in mind that these types of direct receptor interactions may be more widespread than currently appreciated.

Peptide Receptors

Neuropeptide receptors are an immense family. Because of their diversity, they cannot be covered in detail in this chapter. Despite their diversity, none of the receptors that bind peptides appear to be coupled directly to the opening of ion channels. Neuropeptide receptors exert their effects either through the typical pathway of activation of G proteins or through a more recently described pathway related to activation of an associated tyrosine kinase activity.

The peptide binding domain of neuropeptide receptors includes residues in both the large N-terminal extracellular domain and the transmembrane domain (Strader et al., 1995). These additional stabilizing contacts presumably provide the receptors with their remarkably high affinity for neuropeptides (in the nanomolar concentration range). For example, residues in the first and second extracellular domains, as well as those in at least four of the TM domains of the NK1 neurokinin receptor, interact with substance

P to form stabilizing contacts. Many small-molecule antagonists are known to inhibit activation of the NK1 neurokinin receptor, and these antagonists bind to some, but not all, of the same amino acids in the TM segments as does substance P. The possible mechanisms for inhibition of the peptide receptors range from complete structural overlap between agonist and antagonist binding to complete allosteric exclusion (Strader et al., 1995). Knowledge of the activated structure of the neuropeptide receptors provides remarkable opportunities for future drug design.

Summary

GPCRs are single polypeptides composed of seven transmembrane-spanning segments. In general, the binding site for neurotransmitter is located within the core of the circular structure formed by these segments. Transmitter binding produces conformational changes in the receptor that expose parts of the i3 region, among others, for binding to G proteins. G protein binding increases the affinity of the receptor for transmitter. Desensitization is common among GPCRs and leads to decreased response of the receptor to neurotransmitter by several distinct mechanisms. mGluRs are structurally distinct from other GPCRs; mGluRs have large N-terminal extracellular domains that form the binding site for glutamate. Otherwise, the basic structure of mGluRs appears to be similar to that of the rest of the GPCR family.

References

Aramori, I., and Nakanishi, S. (1992). Signal transduction and pharmacological characteristics of a metabotropic glutamate receptor, mGluR1, in transfected CHO cells. Neuron 8, 757–765.

Armstrong, N., and Gouaux, E. (2000). Mechanisms for activation and antagonism of an AMPA-sensitive glutamate receptor: Crystal structures of the GluR2 ligand binding core. Neuron 28, 165–181.

Ascher, P., and Nowak, L. (1988). The role of divalent cations in the N-methyl-D-aspartate responses of mouse central neurones in culture. J. Physiol. 399, 247–266.

Barak, L. S., Tiberi, M., Freedman, N. J., Kwatra, M. M., Lefkowitz, R. J., and Caron, M. G. (1994). A highly conserved tyrosine residue in G protein-coupled receptors is required for agonist-mediated beta 2-adrenergic receptor sequestration. J. Biol. Chem. 269, 2790–2795.

Barnard, E. A., Darlison, M. G., and Seeburg, P. (1987). Molecular biology of the GABAA receptor: The receptor/channel super-family. Trends Neurosci. 10, 502–509.

Bertrand, D., Galzi, J. L., Devillers-Thiery, A., Bertrand, S., and Changeux, J. P. (1993). Mutations at two distinct sites within the channel domain M2 alter calcium permeability of neuronal alpha 7 nicotinic receptor. Proc. Natl. Acad. Sci. USA 90, 6971–6975.

Betz, H. (1991). Glycine receptors: Heterogeneous and widespread in the mammalian brain. *Trends Neurosci.* **14**, 458–461.

Blackstone, C. D., Moss, S. J., Martin, L. J., Levey, A. I., Price, D. L., and Huganir, R. L. (1992). Biochemical characterization and localization of a non-*N*-methyl-D-aspartate glutamate receptor in rat brain. *J. Neurochem.* **58**, 1118–1126.

Bormann, J. (1988). Electrophysiology of GABA$_A$ and GABA$_B$ receptor subtypes. *Trends Neurosci.* **11**, 112–116.

Bormann, J., and Fiegenspan, A. (1995). GABA$_C$ receptors. *Trends Neurosci.* **18**, 515–519.

Boulter, J., Hollmann, M., O'Shea-Greenfield, A., Hartley, M., Deneris, E., Maron, C., and Heinemann, S. (1990). Molecular cloning and functional expression of glutamate receptor subunit genes. *Science* **249**, 1033–1037.

Bouvier, M., Collins, S., O'Dowd, B. F., Campbell, P. T., de Blasi, A., Kobilka, B. K., MacGregor, C., Irons, G. P., Caron, M. G., and Lefkowitz, R. J. (1989). Two distinct pathways for cAMP-mediated down-regulation of the beta 2-adrenergic receptor: Phosphorylation of the receptor and regulation of its mRNA level. *J. Biol. Chem.* **264**, 16786–16792.

Brake, A. J., Wagenbach, M. J., and Julius, D. (1994). New structural motif for ligand-gated ion channels defined by an ionotropic ATP receptor. *Nature* **371**, 519–523.

Bunzow, J. R., Van Tol, H. H., Grandy, D. K., Albert, P., Salon, J., Christie, M., Machida, C. A., Neve, K. A., and Civelli, O. (1988). Cloning and expression of a rat D2 dopamine receptor cDNA [see comments]. *Nature* **336**, 783–787.

Burnashev, N., Schoepfer, R., Monyer, H., Ruppersberg, J. P., Gunther, W., Seeburg, P. H., and Sakmann, B. (1992). Control by asparagine residues of calcium permeability and magnesium blockade in the NMDA receptor. *Science* **257**, 1415–1419.

Casey, P. J. (1995). Protein lipidation in cell signaling. *Science* **268**, 221–225.

Cassel, D., and Selinger, Z. (1977). Mechanism of adenylate cyclase activation by cholera toxin: Inhibition of GTP hydrolysis at the regulatory site. *Proc. Natl. Acad. Sci. USA* **74**, 3307–3311.

Changeux, J. P., Devillers-Thiery, A., and Chemouilli, P. (1984). Acetylcholine receptor: An allosteric protein. *Science* **225**, 1335–1345.

Civelli, O., Bunzow, J. R., and Grandy, D. K. (1993). Molecular diversity of the dopamine receptors. *Annu. Rev. Pharmacol. Toxicol.* **33**, 281–307.

Clark, R. B., Knoll, B. J., and Barber, R. (1999). Partial agonists and G protein-coupled receptor desensitization. *Trends Pharmacol. Sci.* **20**, 279–286.

Colquhoun, D., and Sakmann, B. (1985). Fast events in single-channel currents activated by acetylcholine and its analogues at the frog muscle end-plate. *J. Physiol. (London)* **369**, 501–557.

Colquhoun, D., and Ogden, D. C. (1988). Activation of ion channels in the frog end-plate by high concentrations of acetylcholine. *J. Physiol.* **395**, 131–159.

Dohlman, H. G., Caron, M. G., and Lefkowitz, R. J. (1987). A family of receptors coupled to guanine nucleotide regulatory proteins. *Biochemistry* **19/26**, 2657–2664.

Ehlers, M. D., Whittemore, G. T., and Huganir, R. L. (1995). Regulated subcellular distribution of the NR1 subunit of the NMDA receptor. *Science* **269**, 1734–1737.

Ehlers, M. D., Zhang, S., Bernhardt, J. P., and Huganir, R. L. (1996). Inactivation of NMDA receptors by direct interaction of calmodulin with the NR1 subunit. *Cell* **84**, 745–755.

Ferguson, S. S. G., Menard, L., Barak, L. S., Koch, W. J., Colapietro, A.-M., and Caron, M. G. (1995). Role of phosphorylation in agonist-promoted gb2-adrenergic receptor sequestration. *J. Biol. Chem.* **270**, 24782–24789.

Ferguson, S. S. G., Downey, W. E., Colapietro, A.-M., Barak, L. S., Menard, L., and Caron, M. G. (1996). Role of arrestin in mediating agonist-promoted G-protein coupled receptor internalization. *Science* **271**, 363–366.

Fraser, C. M., Chung, F. Z., Wang, C. D., and Venter, J. C. (1988). Site-directed mutagenesis of human beta-adrenergic receptors: Substitution of aspartic acid-130 by asparagine produces a receptor with high-affinity agonist binding that is uncoupled from adenylate cyclase. *Proc. Natl. Acad. Sci. USA* **85**, 5478–5482.

Fraser, C. M., Wang, C. D., Robinson, D. A., Gocayne, J. D., and Venter, J. C. (1989). Site-directed mutagenesis of m1 muscarinic acetylcholine receptors: Conserved aspartic acids play important roles in receptor function. *Mol. Pharmacol.* **36**, 840–847.

Green, W. N., and Claudio, T. (1993). Acetylcholine receptor assembly: Subunit folding and oligomerization occur sequentially. *Cell (Cambridge, Mass.)* **74**, 57–69.

Gu, Y., Forsayeth, J. R., Verrall, S., Yu, X. M., and Hall, Z. W. (1991). Assembly of the mammalian muscle acetylcholine receptor in transfected COS cells. *J. Cell Biol.* **114**, 799–807.

Hayashi, Y., Tanabe, Y., Aramori, I., Masu, M., Shimamoto, K., Ohfune, Y., and Nakanishi, S. (1992). Agonist analysis of 2-(carboxycyclopropyl)glycine isomers for cloned metabotropic glutamate receptor subtypes expressed in Chinese hamster ovary cells. *Br. J. Pharmacol.* **107**, 539–543.

Henderson, R., Baldwin, J. M., Ceska, T. A., Zemlin, F., Beckmann, E., and Downing, K. H. (1990). Model for the structure of bacteriorhodopsin based on high-resolution electron cryomicroscopy. *J. Mol. Biol.* **213**, 899–929.

Herb, A., Burnashev, N., Werner, P., Sakmann, B., Wisden, W., and Seeburg, P. H. (1992). The KA-2 subunit of excitatory amino acid receptors shows widespread. *Neuron* **8**, 775–785.

Hille, B. (1992). "Ionic Channels of Excitable Membranes." Sinauer, Sunderland, MA.

Hollmann, M., and Heinemann, S. (1994). Cloned glutamate receptors. *Annu. Rev. Neurosci.* **17**, 31–108.

Hollmann, M., O'Shea-Greenfield, A., Rogers, S. W., and Heinemann, S. (1989). Cloning by functional expression of a member of the glutamate receptor family. *Nature* **342**, 643–648.

Huganir, R. L., and Greengard, P. (1990). Regulation of neurotransmitter receptor desensitization by protein phosphorylation. *Neuron* **5**, 555–567.

Hume, R. I., Dingledine, R., and Heinemann, S. F. (1991). Identification of a site in glutamate receptor subunits that controls calcium permeability. *Science* **253**, 1028–1031.

Iino, M., Ozawa, S.. and Tsuzuki, K. (1990). Permeation of calcium through excitatory amino acid receptor channels in cultured hippocampal neurones. *J. Physiol.* **424**, 151–165.

Imoto, K., Busch, C., Sakmann, B., Mishina, M., Konno, T., Nakai, J., Bujo, H., Mori, Y., Fukuda, K., and Numa, S. (1988). Rings of negatively charged amino acids determine the acetylcholine receptor channel conductance. *Nature* **335**, 645–648.

Inglese, J., Freedman, N. J., Koch, W. J., and Lefkowitz, R. J. 1993. Structure and mechanism of the G protein-coupled receptor kinases. *J. Biol. Chem.* **268**, 23735–23738.

Jones, K. A., Borowsky, B., Tamm, J. A., *et al.* (1998). GAGA$_B$ receptors function as a heteromeric assembly of the subunits GABA$_B$R1 and GABA$_B$R2. *Nature* **396**, 674–679.

Julius, D. (1991). Molecular biology of serotonin receptors. *Annu. Rev. Neurosci.* **14**, 335–360.

Kandel, E. R., Schwartz, J. H., and Jessell, T. M. (1991). "Principles of Neural Science," 3rd ed. Elsevier, New York/Amsterdam.

Karlin, A. (1993). Structure of nicotinic acetylcholine receptors. *Curr. Opin. Neurobiol.* **3**, 299–309.

Kaupmann, K., Malitschek, B., Schuler, V., *et al.* (1998). GABA$_B$-receptor subtypes assemble into functional heteromeric complexes. *Nature* **396**, 683–687.

Keinanen, K., Wisden, W., Sommer, B., Werner, P., Herb, A., Verdoorn, T. A., Sakmann, B., and Seeburg, P. H. (1990). A family of AMPA-selective glutamate receptors. *Science* **249**, 556–560.

Kobilka, B. (1992). Adrenergic receptors as models for G protein-coupled receptors. *Annu. Rev. Neurosci.* **15**, 87–114.

Kobilka, B. K., Frielle, T., Collins, S., Yang-Feng, T., Kobilka, T. S., Francke, U., Lefkowitz, R. J., and Caron, M. G. (1987). An intronless gene encoding a potential member of the family of receptors coupled to guanine nucleotide regulatory proteins. *Nature* **329**, 75–79.

Komau, H.-C., Schenker, L. T., Kennedy, M. B., and Seeburg, P. H. (1995). Domain interaction between NMDA receptor subunits and the postsynaptic density protein PSD-95. *Science* **269**, 1737 1740.

Kunishima, N., Shimada, Y., Tsuji, Y., *et al.* (2000). Structural basis of glutamate recognition by a dimeric metabotropic glutamate receptor. *Nature* **407**, 971–977.

Kutsuwada, T., Kashiwabuchi, N., Mori, H., Sakimura, K., Kushiya, E., Araki, K., Meguro, H., Masaki, H., Kumanishi, T., Arakawa, M., and Mishina, M. (1992). Molecular diversity of the NMDA receptor channel [see comments]. *Nature* **358**, 36–41.

Lee, S. P., O'Dowd, B. F., Ng, G. Y. K., Varghese, G., Akil, H., Mansour, A., Nguyen, T., and George, S. R. (2000). Inhibition of cell surface expression by mutant receptors demonstrates that D2 dopamine receptors exist as oligomers in the cell. *Mol. Pharmacol.* **58**, 120–128.

Linden, J. (1994). *In* "Basic Neurochemistry" (G. J. Siegel, B. W. Agranoff, R. W. Albers, and P. B. Molinoff, Eds.), pp. 401–416. Raven Press, New York.

Liu, F., Wan, Q., Pristupa, Z. B., Yu, X. -M., Want, Y. T., and Niznik, H. B. (2000). Direct protein–protein coupling enables cross-talk between dopamine D5 and γ-aminobutyric acid A receptors. *Nature* **403**, 274–278.

Lomeli, H., Mosbacher, J., Melcher, T., Hoger, T., Geiger, J. R., Kuner, T., Monyer, H., Higuchi, M., Bach, A., and Seeburg, P. H. (1994). Control of kinetic properties of AMPA receptor channels by nuclear RNA editing. *Science* **266**, 1709–1713.

Maggio, R., *et al.* (1993). Coexpression studies with mutant muscarinic/adrenergic receptors provide evidence for intermolecular "cross-talk" between G-protein coupled receptors. *Proc. Natl. Acad. Sci. USA* **90**, 3103–3107.

Mano, I., and Teichberg, V.I. (1998). A tetrameric subunit stoichiometry for a glutamate receptor–channel complex. *NeuroReport* **26**, 327–331.

Maricq, A. V., Peterson, A. S., Brake, A. J., Myers, R. M., and Julius, D. (1991). Primary structure and functional expression of the 5HT3 receptor, a serotonin-gated ion channel. *Science* **254**, 432–437.

Mayer, M. L., and Westbrook, G. L. (1987). Permeation and block of N-methyl-D-aspartic acid receptor channels by divalent cations in mouse cultured central neurones. *J. Physiol.* **394**, 501–527.

McGehee, D. S., Heath, M. J., Gelber, S., Devay, P. and Role, L. W. (1995). Nicotine enhancement of fast excitatory synaptic transmission in CNS by presynaptic receptors [see comments]. *Science* **269**, 1692–1696.

Meguro, H., Mori, H., Araki, K., Kushiya, E., Kutsuwada, T., Yamazaki, M., Kumanishi, T., Arakawa, M., Sakimura, K., and Mishina, M. (1992). Functional characterization of a heteromeric NMDA receptor channel expressed from cloned cDNAs. *Nature* **357**, 70–74.

Mishina, M., Kurosaki, T., Tobimatsu, T., Morimoto, Y., Noda, M., Yamamoto, T., Terao, M., Lindstrom, J., Takahashi, T., Kuno, M., and Numa, S. (1984). Expression of functional acetylcholine receptor from cloned cDNAs. *Nature* **307**, 604–608.

Miyazawa, A., Fujiyoshi, Y., Stowell, M., and Unwin, N. (1999). Nicotinic acetylcholine receptor at 4.6A resolution: Transverse tunnels in the channel wall. *J. Mol. Biol.* **288**, 765–786.

Mizobe, T., Maze, M., Lam, V., Suryanarayana, S., and Kobilka, B. K. (1996). Arrangement of transmembrane domains in adrenergic receptors: Similarity to bacteriorhodopsin. *J. Biol. Chem.* **271**, 2387–2389.

Monyer, H., Sprengel, R., Schoepfer, R., Herb, A., Higuchi, M., Lomeli, H., Burnashev, N., Sakmann, B., and Seeburg, P. H. (1992). Heteromeric NMDA receptors: Molecular and functional distinction of subtypes. *Science* **256**, 1217–1221.

Moriyoshi, K., Masu, M., Ishii, T., Shigemoto, R., Mizuno, N., and Nakanishi, S. (1991). Molecular cloning and characterization of the rat NMDA receptor. *Nature* **354**, 31–37.

Mori, H., Masaki, H., Yamakura, T., and Mishina, M. (1992). Identification by mutagenesis of a Mg(2+)-block site of the NMDA receptor channel. *Nature* **358**, 673–675.

Nakanishi, N., Shneider, N. A., and Axel, R. (1990). A family of glutamate receptor genes: Evidence for the formation of heteromultimeric receptors with distinct channel properties. *Neuron* **5**, 569–581.

Nakanishi, S. (1994). Metabotropic glutamate receptors: Synaptic transmission, modulation, and plasticity. *Neuron* **13**, 1031–1037.

Noda, M., Takahashi, H., Tanabe, T., Toyosato, M., Furutani, Y., Hirose, T., Asai, M., Inayama, S., Miyata, T., and Numa, S. (1982). Primary structure of alpha-subunit precursor of *Torpedo californica* acetylcholine receptor deduced from cDNA sequence. *Nature* **299**, 793–797.

Noda, M., Takahashi, H., Tanabe, T., Toyosato, M., Kikyotani, S., Furtani, Y., Hirose, T., Takashima, H., Inayama, S., Miyata, T., and Numa, S. (1983). Structural homology of *Torpedo californica* acetylcholine receptor subunits. *Nature* **302**, 528–532.

O'Dowd, B. F., Hnatowich, M., Caron, M. G., Lefkowitz, R. J., and Bouvier, M. (1989). Palmitoylation of the human beta 2-adrenergic receptor: Mutation of Cys341 in the carboxyl tail leads to an uncoupled nonpalmitoylated form of the receptor. *J. Biol. Chem.* **264**, 7564–7569.

O'Hara, P. J., Sheppard, P. O., Thogersen, H., Venezia, D., Haldeman, B. A., McGrane, V., Houamed, K. M., Thomsen, C., Gilbert, T. L., and Mulvihill, E. R. (1993). The ligand-binding domain in metabotropic glutamate receptors is related to bacterial periplasmic binding proteins. *Neuron* **11**, 41–52.

Olah, M. E., and Stiles, G. L. (1995). Adenosine receptor subtypes: Characterization and therapeutic regulation. *Annu. Rev. Pharmacol. Toxicol.* **35**, 581–606.

Ortell, M. O., and Lunt, G. G. (1995). Evolutionary history of the ligand-gated ion-channel superfamily of receptors. *Trends Neurosci.* **18**, 121–128.

Overton, M. C., and Blumer, K. J. (2000). G-protein-coupled receptors function as oligomers *in vivo. Curr. Biol.* **10**, 341–344.

Palczewksi, K., Kumasaka, T., Hori, T., Behnke, C. A., Motoshima, H., Fox, B. A., Le Trong, I. L., Teller, D. C., Okada, T., Stenkamp, R. E., Yamamoto, M., and Miyano, M. (2000). Crystal structure of rhodopsin: A G protein-coupled receptor. *Science* **289**, 739–745.

Paulson, H. L., Ross, A. F., Green, W. N., and Claudio, T. (1991). Analysis of early events in acetylcholine receptor assembly. *J. Cell Biol.* **113**, 1371–1384.

Perez, D. M., Hwa, J., Gaivin, R., Manjula, M., Brown, F, and Graham, R. M. (1996). Constitutive activation of a single effector pathway: Evidence for multiple activation states of a G protein-coupled receptor. *Mol. Pharmacol.* **49**, 112–122.

Premont, R. T., Inglese, J., and Lefkowitz, R. J. (1995). Protein kinases that phosphorylate activated G protein-coupled receptors. *FASEB J.* **9**, 175–182.

Pritchett, D. B., Sontheimer, H., Shivers, B. D., Ymer, S., Kettenmann, H., Schofield, P. R., and Seeburg, P. H. (1989). Importance of a novel GABAA receptor subunit for benzodiazepine pharmacology. *Nature* **338**, 582–585.

Proll, M. A., Clark, R. B., Goka, T. J., Barber, R., and Butcher, R. W. (1992). Adrenergic receptor levels and function after growth of S49 lymphoma cells in low concentrations of epinephrine. *Mol. Pharmacol.* **42**, 116–122.

Rosenmund, C., Stern-Bach, Y., and Stevens, C.F. (1998). The tetrameric structure of a glutamate receptor channel. *Science* **280**, 1596–1599.

Rudolph, U., Crestani, F., Benke, D., Brunig, I., Benson, J. A., Fritschy, J.-M., Martin, J. R., Bluethmann, H., and Mohler, H. (1999). Benzodiazepine actions mediated by specific γ-aminobutyric acidA receptor subtypes. *Nature* **401**, 796.

Salahpour, A., Angers, S., and Bouvier, M. (2000). Functional significance of oligomerization of G-protein-coupled receptors. *Trends Endocrinol. Metab.* **11**, 163–168.

Seguela, P., Wadiche, J., Dineley-Miller, K., Dani, J. A., and Patrick, J. W. (1993). Molecular cloning, functional properties, and distribution of rat brain alpha 7: A nicotinic cation channel highly permeable to calcium. *J. Neurosci.* **13**, 596–604.

Seighart, W. (1992). GABAA receptors: Ligand-gated Cl- ion channels modulated by multiple drug-binding sites. *Trends Pharmacol. Sci.* **13**, 446–450.

Sommer, B., Keinanen, K., Verdoorn, T. A., Wisden, W., Burnashev, N., Herb, A., Kohler, M., Takagi, T., Sakmann, B., and Seeburg, P. H. (1990). Flip and flop: A cell-specific functional switch in glutamate-operated channels of the CNS. *Science* **249**, 1580–1585.

Sommer, B., Kohler, M., Sprengel, R., and Seeburg, P. H. (1991). RNA editing in brain controls a determinant of ion flow in glutamate-gated channels. *Cell* **67**, 11–19.

Sterne-Marr, R., and Benovic, J. L. (1995). Regulation of G protein-coupled receptors by receptor kinases and arrestins. *Vitam. Horm. (N.Y.)* **51**, 193–234.

Strader, C. D., Fong, T. M., Tota, M. R., Underwood, D., and Dixon, R. A. (1994). Structure and function of G protein-coupled receptors. *Annu. Rev. Biochem.* **63**, 101–132.

Strader, C. D., Fong, T. M., Graziano, M. P., and Tota, M. R. (1995). The family of G-protein-coupled receptors. *FASEB J.* **9**, 745–754.

Strosberg, A. D. (1990). Biotechnology of beta-adrenergic receptors. *Mol. Neurobiol.* **4**, 211–250.

Tolkovsky, A. M., Braun, S., and Levitzki, A. (1982). Kinetics of interaction between receptors, GTP protein, and the catalytic unit of turkey erythrocyte adenylate cyclase. *Proc. Natl. Acad. Sci. USA* **79**, 213–217.

Unwin, N. (1993a). Neurotransmitter action: Opening of ligand-gated ion channels. *Cell* **7**(Suppl.), 31–41.

Unwin, N. (1993b). Nicotinic acetylcholine receptor at 9 A resolution. *J. Mol. Biol.* **229**, 1101–1124.

Unwin, N. (1995). Acetylcholine receptor channel imaged in the open state. *Nature* **373**, 37–43.

Valera, S., Hussy, N., Evans, R. J., Adami, N., North, R. A., Surprenant, A., and Buell, G. (1994). A new class of ligand-gated ion channel defined by P2x receptor for extracellular ATP [see comments]. *Nature* **371**, 516–519.

Verdoorn, T. A., Burnashev, N., Monyer, H., Seeburg, P. H., and Sakmann, B. (1991). Structural determinants of ion flow through recombinant glutamate receptor channels. *Science* **252**, 1715–1718.

Vernino, S., Amador, M., Luetje, C. W., Patrick, J., and Dani, J. A. (1992). Calcium modulation and high calcium permeability of neuronal nicotinic acetylcholine receptors. *Neuron* **8**, 127–134.

Watkins, J. C., Krogsgaard-Larsen, P., and Honore, T. (1990). Structure–activity relationships in the development of excitatory amino acid receptor agonists and competitive antagonists. *Trends Pharmacol. Sci.* **11**, 25–33.

Wenthold, R. J., Yokotani, N., Doi, K., and Wada, K. (1992). Immunochemical characterization of the non-NMDA glutamate receptor using subunit-specific antibodies: Evidence for a hetero-oligomeric structure in rat brain. *J. Biol. Chem.* **267**, 501–507.

White, J. H., Wise, A., Main, M. J., *et al.* (1998). Heterodimerization is required for the formation of a functional GABAB receptor. *Nature* **396**, 679–682.

Wo, Z. G., and Oswald, R. E. (1995). Unraveling the modular design of glutamate-gated ion channels. *Trends Neurosci.* **18**, 161–168.

Yakel, J. L., Lagrutta, A., Adelman, J. P., and North, R. A. (1993). Single amino acid substitution affects desensitization of the 5-hydroxytryptamine type 3 receptor expressed in *Xenopus* oocytes. *Proc. Natl. Acad. Sci. USA* **90**, 5030–5033.

Young, A. B., and Snyder, S. H. (1974). The glycine synaptic receptor: Evidence that strychnine binding is associated with the ionic conductance mechanism. *Proc. Natl. Acad. Sci. USA* **71**, 4002–4005.

Yuan, N., Friedman, J., Whaley, B. S., and Clark, R. B. (1994). cAMP-dependent protein kinase and protein kinase C consensus site mutations of the adrenergic receptor. *J. Biol. Chem.* **269**, 23,032–23,038.

12

Intracellular Signaling

Howard Schulman

Almost all aspects of neuronal function, from its maturation during development, to its growth and survival, cytoskeletal organization, gene expression, neurotransmission, and use-dependent modulation, are dependent on intracellular signaling initiated at the cell surface. The response of a neuron to neurotransmitters, growth factors, and other signaling molecules is determined by its complement of receptors, pathways available for transducing signals into the neuron and transmitting these signals to its intracellular compartments, and the enzymes, ion channels, and cytoskeletal proteins that ultimately mediate the effects of the neurotransmitters. The molecules involved in signal transmission and transduction are highly represented in mammalian and invertebrate genomes. Individual neuronal responses are further determined by the concentration and localization of signal transduction components and modified by the prior history of neuronal activity. Several primary classes of signaling systems, operating on different time courses, provide great flexibility for intercellular communication. One class comprises ligand-gated ion channels, such as the nicotinic receptor considered in Chapter 11. This class of signaling systems provides fast transmission (see Chapter 16) that is activated and deactivated within about 10 ms. It forms the underlying "hard wiring" of the nervous system that makes rapid multisynaptic computations possible. A second class consists of receptor tyrosine kinases, which typically respond to growth factors and to trophic factors and produce major changes in the growth differentiation, or survival of neurons (Chapter 10). A third class uses G protein-linked signals and constitutes the largest number of receptors. This signaling system requires several steps for transduction and transmission of the signal, thus slowing the response from 100–300 ms to many minutes (see Chapter 16). The relatively slow speed is offset, however, by a richness in the diversity of its

modulation and its inherent capacity for amplification and plasticity. The initial steps in this signaling system typically generate a second messenger inside the cell, and this second messenger then activates a number of proteins, including protein kinases that modify cellular processes. Signal transduction also modulates the level of transcription of genes, which determine the differentiated and functional state of cells.

SIGNALING THROUGH G-PROTEIN-LINKED RECEPTORS

Signal transduction through G protein-linked receptors requires three membrane-bound components: (1) a cell surface receptor that determines to which signal the cell can respond; (2) a G protein on the intracellular side of the membrane that is stimulated by the activated receptor; and (3) either an effector enzyme that changes the level of a second messenger or an effector channel that changes ionic fluxes in the cell in response to the activated G protein. The human genome encodes for more than 600 receptors for catecholamines, odorants, neuropeptides, and light that couple to one or more of the 27 identified G proteins. These, in turn, regulate one or more of more than two dozen different effector channels and enzymes. The key feature of this information flow is the ability of G proteins to detect the presence of activated receptors and to amplify the signal by altering the activity of appropriate effector enzymes and channels.

G proteins are GTP-binding proteins that couple the activation of seven-helix receptors by neurotransmitters at the cell surface to changes in the activity of effector enzymes and effector channels. A common effector enzyme is adenylyl cyclase, which synthe-

335

Copyright 2004, Elsevier Science (USA).
All rights reserved.

sizes cyclic AMP—an intracellular surrogate for the neurotransmitter, the first messenger. Phospholipase C (PLC), another effector enzyme, generates diacylglycerol (DAG) and inositol trisphosphate (IP3), the latter of which releases intracellular stores of Ca^{2+}. Information from an activated receptor flows to the second messengers that typically activate protein kinases, which modify a host of cellular functions. Cyclic AMP, Ca^{2+}, and DAG have in common the ability to activate protein kinases with broad substrate specificities. They phosphorylate key intracellular proteins, ion channels, enzymes, and transcription factors taking part in diverse cellular biological processes. The activities of protein kinases and phosphatases are in balance, constituting a highly regulated process, as revealed by the phosphorylation state of these targets of the signal transduction process. In addition to regulating protein kinases, second messengers such as cAMP, cGMP, Ca^{2+}, and arachidonic acid can directly gate, or modulate, ion channels. G proteins can also couple directly to ion channels without the interception of second messengers or protein kinases. In these diverse ways, a neurotransmitter outside the cell can modulate essentially every aspect of cell physiology and encode the history of cell stimuli in the form of altered activity and expression of its cellular constituents. An overview of G-protein signaling to protein kinases is presented in Fig. 12.1.

G-Protein Signaling Operates on Common Principles

The many types of G proteins and the many types of effector enzymes have certain features in common

(Hille, 1992). First, each receptor can couple to one or only a few G proteins, thus specifying the stimulus response. Second, the second messengers are typically synthesized in one or two steps from a precursor (e.g., ATP) that is readily available in those cells at high concentration but is itself inactive as a signaling molecule. G proteins can also stimulate enzymes that eliminate second messengers. Third, the initial signal can be greatly amplified: each receptor can activate many G protein molecules; each adenylyl cyclase can synthesize many cAMP molecules; and each protein kinase can phosphorylate many copies of each of its substrates. Fourth, the process is slower in onset and persists longer than ligand-gated ion channel signaling.

A fifth feature of G-protein signaling is the ability to orchestrate a variety of effects through the same second messenger. For example, the cyclic AMP-dependent protein kinase (PKA), stimulated by serotonin exposure to sensory neurons in the mollusk *Aplysia*, phosphorylates and inhibits a K^+ channel (see Chapter 18). This inhibition results in a prolonged action potential and a greater influx of Ca^{2+} with each stimulation of the sensory neuron and in a sensitization of the withdrawal response regulated by this circuit. On a slower time course, cAMP and PKA can also modify carbohydrate metabolism to keep up with activity and the transcription of genes and translation of proteins that ultimately modify the number of synaptic contacts made by the sensory neuron after prolonged periods of activity or inactivity.

A nervous system with information flow by fast transmission alone would be capable of stereotyped or reflex responses. Modulation of this transmission and changes in other cellular functions by G protein-

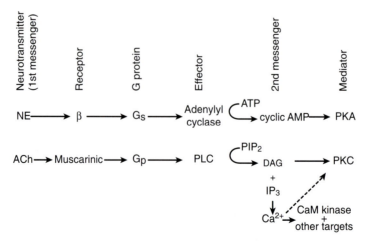

FIGURE 12.1 Overview of G-protein signaling to protein kinases. Norepinephrine (NE) and acetylcholine (ACh) can stimulate certain receptors that couple through distinct G proteins to different effectors, which results in increased synthesis of second messengers and activation of protein kinases (PKA and PKC). PLC, phospholipase C; PIP₂, phosphatidylinositol bisphosphate; DAG, diacylglycerol; CaM, Ca^{2+}-calmodulin-dependent; IP₃, inositol 1,4,5-trisphosphate.

linked systems and by receptor-tyrosine kinase-linked systems enables an orchestrated response. The large diversity of signaling molecules and their intracellular targets offer nearly unlimited flexibility of response over a broad time scale and with high amplification. This signaling is a key feature of neuronal plasticity, regulating every step of the way from neurotransmitter receptors and ion channels, signal transduction pathways, neurotransmitter synthesis, and release to the expression of genes in the nucleus that underlie synaptic changes linked to learning and memory (see Chapter 18 for exended discussion).

Receptors Catalyze the Conversion of G Proteins into the Active GTP-Bound State

G proteins undergo a molecular switch between two interconvertible states that are used to "turn on" or "turn off" downstream signaling. G proteins taking part in signal transduction use a regulatory motif that is seen in other GTPases engaged in protein synthesis and in intracellular vesicular traffic. G proteins are switched on by stimulated receptors, and they switch themselves off after a time delay. Whether a G protein is turned on or off depends on the guanine nucleotide to which it is bound. The G proteins are inactive when GDP is bound and are active when GTP is bound. The sole function of the seven-helix receptors in activating G proteins is to catalyze an exchange of GTP for GDP. This is a temporary switch because G proteins are designed with a GTPase activity that hydrolyzes the bound GTP and converts the G protein back into the GDP-bound, or inactive, state. Thus, a G protein must continuously sample the state of activation of the receptor, and it transmits downstream information only while the neuron is exposed to neurotransmitter. A fast GTPase means that the signal transduction pathway is very responsive to the presence of neurotransmitter outside, whereas a slow GTPase provides greater amplification but is less responsive to the elimination of the neurotransmitter. A cell cannot produce distinct responses to each stimulus presented at high frequency when the participating G protein has a slow GTPase. The benefit of a slow GTPase is that it allows amplification of the signal by permitting a given receptor to activate many G proteins and allows G proteins to produce changes in many effector enzymes for each G-protein cycle. The GTPase activity of G proteins serves both as a timer and as an amplifier (Fig. 12.2).

The G-Protein Cycle

G proteins are trimeric structures composed of two functional units: (1) an α subunit (39–52 kDa) that cat-

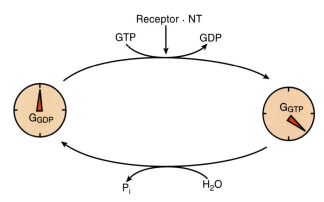

FIGURE 12.2 GTPase activity of G proteins serves as a timer and amplifier. Receptors activated by neurotransmitters (NT) initiate the GTPase timing mechanism of G proteins by displacement of GDP by GTP. Neurotransmitters thus convert G_{GDP} ("turned-off state") to G_{GTP} (time-limited "turned-on" state).

alyzes the GTPase activity and (2) a βγ dimer (35 and 8 kDa, respectively) that tightly interacts with the α subunit when bound to GDP (Stryer and Bourne, 1986; Neer, 1995). The role of the three subunits in the G-protein cycle is depicted in Fig. 12.3. In the basal state, GDP is bound tightly to the α subunit, which is associated with the βγ pair to form an inactive G protein. In addition to blocking interaction of the α subunit with its effector, the βγ pair increases the affinity of the α subunit for activated receptors. Binding of the neurotransmitter to the receptor produces a conformational change that positions previously buried residues that promote increased affinity of the receptors for the inactive G protein (Palczewski et al., 2000). A given receptor can interact with only one or a limited number of G proteins, and the α subunit produces most of this specificity. Coupling

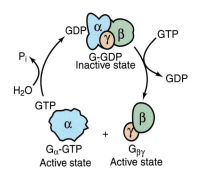

FIGURE 12.3 Interconversion, catalyzed by excited receptors, of G-protein subunits between inactive and active states. Displacement of GDP with GTP dissociates the inactive heterotrimeric G protein, generating α-GTP and βγ, both of which can interact with their respective effectors and activate them. The system converts into the inactive state after GTP has been hydrolyzed and the subunits have reassociated. From "Biochemistry," 4th ed. by Stryer. © 1995 by Lubert Stryer. Used with permission of W. H. Freeman and Company.

with the activated receptor reduces the affinity of the α subunit for GDP, facilitating its dissociation and thus leaving the nucleotide binding site empty. Either GDP or GTP can bind to the vacant site; however, because the level of GTP in the cell is much greater than the level of GDP, the dissociation of GDP is usually followed by the binding of GTP. Thus, the receptor effectively catalyzes an exchange of GTP for GDP (Cassel and Selinger, 1978). Binding of GTP has two consequences: (1) it dissociates the G protein into α-GTP and βγ (Northup et al., 1983) and (2) the α-GTP subunit has a reduced affinity for the stimulated receptor, leading to dissociation of the complex.

The GTP–GDP exchange is inherently very slow because the amount of activation energy required to open the nucleotide binding site is large. This slow exchange ensures that very little of the G protein is in the on state under basal conditions. The stimulated receptor does not provide the energy for the cycle, but it lowers the amount of free energy for the dissociation from the nucleotide binding site and greatly accelerates the exchange reaction. When the stimulated receptor increases the GTP–GDP exchange, the GTPase reaction becomes the rate-limiting step of the cycle and α-GTP accumulates. The level of G protein in the on state can increase from 1% to more than 50% of all G protein. The direction of the cycle is determined by the GTPase reaction, which uses the energy of GTP to make the reaction irreversible and to maintain a low level of α-GTP in the basal state (Stryer and Bourne, 1986).

Information Flow

One of the more tense and public debates in signal transduction has been the question of whether the α subunit alone conveys information that specifies which effector is activated or whether the βγ pair has a role. One of the contestants even paid for a vanity license plate proclaiming "α not β." The α subunit was thought to be responsible for specifying which effector enzyme was activated by a G protein, with βγ playing a nonspecific role in keeping the α subunit inactive while presenting it to receptors for activation (Cassel and Selinger, 1978). This notion was eventually changed because of the finding that βγ can directly activate certain K+ channels (Logothetis et al., 1987; Huang et al., 1995). The historic association of G-protein function with α has persisted for the purpose of nomenclature, with G_s and α_s referring to the G protein and its corresponding α subunit, which stimulates adenylyl cyclase. These names have been retained even though it is now apparent that α and βγ subunits can both modify effector enzymes and channels and that a given α subunit may combine with a

number of βγ pairs. The α subunits may act either independently or in concert with βγ (Clapham and Neer, 1993). Furthermore, β and γ subunits in a βγ pair can combine in many different ways. The terms G_i, G_p, and G_o were used for G-protein activities that inhibited adenylyl cyclase, stimulated phospholipase, or were presumed to have other effects, respectively. At this stage, cloning and genomic sequencing have outpaced functional studies. The human genome encodes 27 distinct genes for α subunits, along with five β's and 13 γ's (Neer, 1995; Venter et al., 2001). The protein composition of most G proteins is not known, so a cell may contain $\alpha_i\beta_2\gamma_3$ and $\alpha_i\beta_1\gamma_4$, each having different properties. The inherent affinities of βγ pairs for a particular α and spatial segregation of G proteins probably greatly limit the number of combinations of subunits. See Box 12.1.

Function of α Subunits

Determination of the crystal structures of different conformational states of some α subunits and βγ subunits has been a source of insight into their functional domains and the critical GTPase activity of all α subunits (Sondek et al., 1996; Wall et al., 1996). The α subunits are compact molecules that must accommodate a number of protein–protein interactions in addition to their GTP-hydrolyzing activity. The β subunit consists of repeating motifs arranged like blades of a propeller or wedges of a pie with a central tunnel (see next subsection). In the trimeric GDP-bound state, the α subunit is docked to the β subunit; the α amino terminus interacts with one side of β while its catalytic domain interacts with the tunnel region of β (Wall et al., 1996). Receptors interact with the carboxy terminus of the α subunit; this interaction, along with other contacts, specifies which receptors can activate which α subunits (Conklin et al., 1993). The guanine nucleotide binding site is highly conserved and likely altered by interaction with ligand-bound receptors. Movement of the C terminus of the α subunit may reduce contacts with GDP to facilitate dissociation of the nucleotide and exchange for GTP, thus producing an activated α-GTP conformation. The γ phosphate present in GTP, but not in GDP, interacts with a region of α that participates in the docking of α to βγ. The resulting conformational changes reduce the affinity between this surface and βγ, resulting in dissociation of the trimer to produce α and βγ capable of interacting with their respective effectors. Dissociation also releases the receptor, which can then catalyze activation of other G proteins. Thus, each functional half of the G protein inhibits the action of the other: βγ does not convey information to its effectors in the presence

BOX 12.1

WHY ARE G PROTEIN-REGULATED SYSTEMS SO COMPLEX?

Transmembrane signaling systems all contain two fundamental elements: one that recognizes an extracellular signal (a receptor) and another that generates an intracellular signal. These elements can be easily incorporated into a single molecule, for example, in receptor tyrosine kinases and guanylyl cyclases. Why then are G protein-regulated systems so complex, minimally containing five gene products in the basic module (receptor, heterotrimeric G protein, and effector)? The design of these systems permits both integration and branching at its two interfaces: between receptor and G protein and between G protein and effector. Each component of the system can thus be regulated independently—transcriptionally, posttranslationally, or by interactions with other regulatory proteins. Furthermore, hundreds of genes encode receptors for hormones and neurotransmitters, dozens of genes encode G-protein subunits (α, β, and γ), and dozens more genes encode G protein-regulated effectors. Each cell in an organism thus has the opportunity to sample the genome and construct a highly customized and sophisticated switchboard in its plasma membrane, permitting the organism to make an extraordinary variety of responses to complex situations. The choices that each cell makes include much more than just the components of the basic modules, extending as well to regulators of G protein-mediated pathways such as receptor kinases and GTPase-activating proteins. And, of course, the identity and concentrations of the components of the switchboard can be sculpted within minutes or hours to permit adaptation to developmental needs or environmental stresses.

The classic stress response of mammals to the hormone epinephrine provides an elegant example of the power of modular signal transduction systems. A single hormone is used to initiate responses of opposite polarity in very similar cells. Thus, vascular smooth muscle in skin contracts to minimize bleeding (if there is a wound)

and to maintain blood pressure, whereas vascular smooth muscle in skeletal muscle relaxes to provide increased blood flow during heightened physical activity. Smooth muscle in the intestine relaxes, whereas cardiac muscle is powerfully stimulated. In addition, hepatocytes hydrolyze glycogen to glucose, adipocytes hydrolyze triglycerides and release free fatty acids for fuel, and certain endocrine and exocrine secretions are stimulated or inhibited. Several distinct receptors for epinephrine are selectively expressed in various cell types to achieve this beneficial orchestration of stimulatory and inhibitory events. These receptors vary in their capacities to interact with G proteins from three different subfamilies (G_s, G_i, G_q) and thus to activate or inhibit several effects to achieve the desired responses. Each cell's choices of particular receptors, G proteins, and effectors permit additional choices of regulators of each of these components to adjust the magnitude and/or the kinetics of the response.

The past two decades of research in this area have witnessed identification and characterization of the molecular players involved in G protein-mediated signaling and appreciation of the basic mechanisms that underlie the protein–protein interactions that drive these systems. Current research is expanding this basic core of knowledge: on the one hand, toward greater understanding of how the individual modules contribute to the integrated networks of intact cells and, on the other, toward elucidation of the physical and structural bases of these complex cellular reactions. We can begin to construct a movie of G protein-mediated signaling at atomic resolution, and many of its most important frames are shown in Fig. 12.4.

Alfred G. Gilman
Stephen R. Sprang

BOX 12.1 (continued)

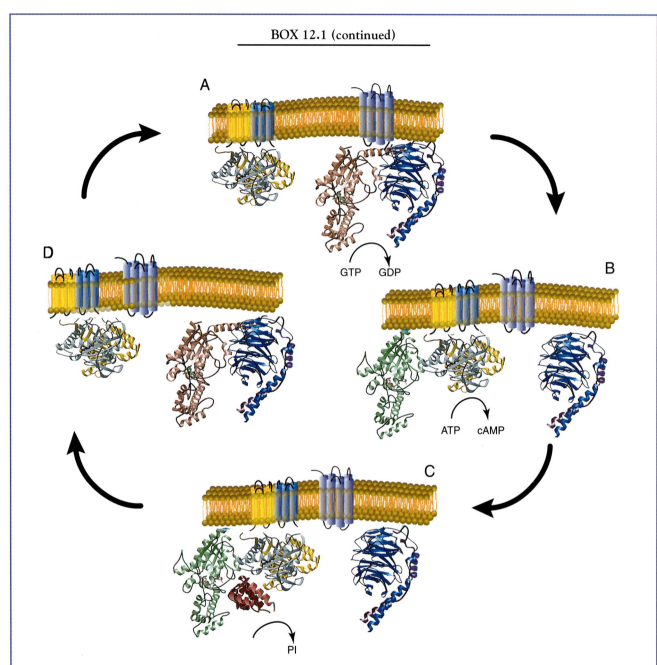

FIGURE 12.4   (A) G proteins are held in an inactive state because of very high affinity binding of GDP to their α subunits. When activated by an agonist, membrane-bound seven-helical receptors (right, glowing magenta) interact with heterotrimeric G proteins (α, amber; β, teal; γ, burgundy) and stimulate dissociation of GDP. This permits GTP to bind to and activate α, which then dissociates from the high-affinity dimer of β and γ subunits. (B) Both activated (GTP-bound) α (lime) and βγ are capable of interacting with downstream effectors. Also illustrated is the interaction of αs-GTP with adenylyl cyclase (catalytic domains are mustard and ash). Adenylyl cyclase then catalyzes the synthesis of the second messenger cyclic AMP (cAMP) from ATP. (C) Signaling is terminated when α hydrolyzes its bound GTP to GDP. In some signaling systems, GTP hydrolysis is stimulated by GTPase-activating proteins or GAPs (cranberry) that bind to α and stablize the transition state for GTP hydrolysis. (D) Hydrolysis of GTP permits α-GDP to dissociate from its effector and associate again with βγ. The heterotrimeric G protein is then ready for another signaling cycle if an activated receptor is present. The figure is based on the original work of Mark Wall and John Tesmer.

TABLE 12.1 Functions of α Subunits of G Proteins

Class	Member	Modifying toxin	Some functions
α_s	α_s, α_{olf}	Cholera	Stimulate adenylyl cyclase, regulate Ca^{2+} channels
α_i	$\alpha_{i\text{-}1}$, $\alpha_{i\text{-}2}$, $\alpha_{i\text{-}3}$, α_o, α_z	Pertussis	Inhibit adenylyl cyclase, regulate K^+ and Ca^{2+} channels
α_t	α_{gust}, $\alpha_{t\text{-}1}$, $\alpha_{t\text{-}2}$	Cholera and pertussis	Activate cGMP phosphodiesterase
α_q	α_q, α_{11}, α_{14}, α_{15}, α_{16}	—	Activate PLC
α_{12}	α_{12}, α_{13}	—	Regulate Na^+/K^+ exchange

Source: Summarized from Neer (1995).

of α-GDP, whereas α is ineffective at conveying information to its effectors when complexed with βγ as the trimeric G protein. Dissociation of these subunits exposes the residues needed for interactions with effectors. GTP hydrolysis reverses the conformational changes, exposing a region of α that can interact with βγ and creating a binding pocket for residues in the β subunit that bind α. Table 12.1 indicates the range of effectors that are regulated by G proteins through their α subunits.

Function of βγ Subunits

The crystal structure of the βγ heterodimer both in its inactive form bound to the a subunit and in its free active form has been determined (Sondek *et al.*, 1996; Wall *et al.*, 1996). The C-terminal half of the β subunit has a sevenfold repeat of a structural motif termed the WD repeat, which interacts with γ and is the likely site of interaction between β and its effector enzymes and channels. The WD repeat is a sequence of 25–40 amino acids and ends with the amino acids tryptophan (W) and aspartic acid (D). Each repeat forms a wedge of a circular disk with a central tunnel in both the free and bound states. Whether each WD repeat specifies interaction with a particular target effector of β or whether several effectors can be bound simultaneously to a single β to form a large complex is not known. The γ subunit is in an extended conformation, circling and making contact with several WD domains of the β subunit. Prenylation of the C terminus of γ anchors the βγ dimer to membranes.

The βγ subunits regulate numerous effector enzymes and channels (Table 12.2), as well as participating in the overall function of G proteins. The identified effector targets of βγ are several adenylyl cyclases, the β isoform of phospholipase C, ion channels (for K^+ and Ca^{2+}), phospholipase A2, and phosphatidylinositol 3-kinase (Clapham and Neer, 1993). The other functions of βγ are as follows. First, they keep G proteins inactive in the basal state by complexing with α and reducing its intrinsic GTP–GDP

TABLE 12.2 Effector Functions of β/γ Dimers of G Proteins

Inhibition of adenylyl cyclase

Stimulation of adenylyl cyclase types II and IV (with α)

Stimulation of phospholipase Cβ (PLCβ)

Stimulation of K^+ channel

Stimulation of Ca^{2+} channel

Stimulation of phospholipase A_2 (PLA$_2$)

Stimulation of phosphatidylinositol-3-kinase (PI-3-kinase)

exchange rate. Second, they help to target the α subunits to the membrane and increase the affinity of α-GDP for ligand-bound receptors. Third, both β and γ help to specify which receptors couple to the G protein. Fourth, the WD repeats of β serve as anchors for a protein kinase, called βARK, that terminates signaling by ligand-bound β-adrenergic receptors.

Examination and Manipulation of G-Protein-Coupled Signals

Neurotransmitters can produce their cellular effects by a variety of signal transduction pathways. A number of experimental tools and approaches must be used to delineate the pathway used in any system of interest. Table 12.3 summarizes some of these experimental tools.

TABLE 12.3 Experimental Tools for and Approaches to Testing the Role of G Proteins

Cholera toxin and pertussis toxin

GTPγS or GTPβS, NaF

Antibody to G-protein subunits

Antisense oligonucleotides or RNA

Knockouts or other forms of gene disruption

Reconstitution from purified components

Toxins

Differential sensitivities to cholera toxin and pertussis toxin can be used to implicate a G-protein-mediated pathway. Both G_s and transducin are sensitive to cholera toxin, which selectively ADP-ribosylates α_s and α_t. The α subunits of G_i, G_o, and transducin, but not of G_s, are ADP-ribosylated by pertussis toxin. These toxins transfer the ADP-ribose moiety of NAD^+ to an arginine (cholera toxin) or cysteine (pertussis toxin) on the appropriate α subunit. The toxins act at distinct steps in the GTPase cycle to lock it in either the on or the off position (see Fig. 12.3). Cholera toxin reduces the GTPase activity to insignificant levels, thereby generating a persistently on state. For example, cholera toxin treatment of G_s stimulates robust and continuous production of cAMP by adenylyl cyclase. In contrast, pertussis toxin acts on the inactive G protein, blocking its interaction with receptors so that it cannot be activated and thus remains in the GDP-bound, or off, state. Toxin sensitivity can be used to narrow the choice of possible G proteins taking part in a process or to block a known pathway.

Guanine Nucleotides and NaF

The GTPase cycle of all G proteins can also be modified for experimental purposes by GTPγS or GDPβS. GTPγS is a nonhydrolyzable analog of GTP with high affinity for the α subunit. Activation of any of the G proteins in the presence of GTPγS leads to the exchange of GTPγS for GDP to produce α-GTPγS. The G protein is thereby activated for prolonged periods because the GTPase cycle is blocked and remains so until GDP exchanges with GTPγS. A similar response is obtained by addition of aluminum fluoride. Fluoride forms a complex with trace amounts of Al^{3+} and binds to α-GDP. The aluminum fluoride moiety simulates the γ phosphate of GTP so that, like GTPγS, α-GDP-AlF_3 persistently activates the G protein. (Coleman *et al.*, 1994; Sondek *et al.*, 1994). In contrast, GDPβS can exchange with GDP, leaving the G protein in the inactive state. GDPβS has a higher affinity than GDP for the G protein and, as a result, the GTPase cycle is slowed and spends more of its time with an inactive G protein.

Effector Enzymes, Channels, and Transporters Decode Receptor-Mediated Cell Stimulation in the Cell Interior

The function of the trimeric G proteins is to decode information about the concentration of neurotransmitters bound to appropriate receptors on the cell surface and convert this information into a change in the activity of enzymes and channels that mediate the effects of the neurotransmitter. The effector can be an enzyme that synthesizes or degrades a diffusible second messenger or it can be an ion channel. The number of identified effectors of G-protein signaling has increased markedly in the past few years and now also includes membrane transport proteins.

Response Specificity in G-Protein Signaling

The modular design of G-protein signaling may appear to be incapable of providing specificity. Receptors can stimulate one or more G proteins, G proteins can couple to one or more effector enzymes or channels, and the resulting second messengers will affect many cellular processes. Signals originating from activated receptors can either converge or diverge, depending on the receptor and on the complement of G proteins and effectors in a given neuron (Fig. 12.5) (Ross, 1989).

How can a neurotransmitter produce a specific response if G-protein coupling has the potential for such a diversity of effectors? A given neuron has only a subset of receptors, G proteins, and effectors, thereby limiting possible signaling pathways. Transducin, for example, is confined to the visual system, where the predominant effector is the cGMP phosphodiesterase and not adenylyl cyclase. A

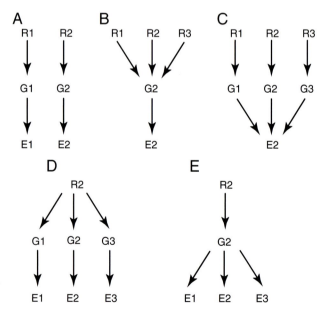

FIGURE 12.5 Signals can converge or diverge on the basis of interactions between receptors (R) and G proteins (G) and between G proteins and effectors (E). The complement of receptors, G proteins, and effectors in a given neuron determines the degree of integration of signals, as well as whether cell stimulation will produce a focused response to a neurotransmitter or a coordination of divergent responses. Adapted from Ross (1989).

number of other factors combine to increase signal specificity (Hille, 1992; Neer, 1995; Sondek *et al.*, 1994).

Specificity and choice. Receptors and G proteins have higher intrinsic affinities and efficacies for modulating the activity of the "correct" G protein(s) and effector(s), respectively. Some coupling obtained when high concentrations of these components are generated by overexpression in cell lines or in reconstitution experiments with purified proteins does not occur at concentrations found *in vivo*.

Spatial compartmentalization. Second messenger systems can be compartmentalized, thus adding specificity and localized control of signaling. The same receptor may regulate a Ca^{2+} channel through one G protein at a nerve terminal and regulate PLCβ at a distal dendrite through another G protein. Although the addition of the neurotransmitter in culture would produce both effects, synaptic inputs would be able to elicit specific effects at nerve terminals or at dendrites.

GTPase activity. The degree of amplification by the G protein (based on GTPase) and by the effector (based on its specific activity or conductance) can determine which of the possible pathways is more prominent. Furthermore, some effectors appear to act as GTPase-activating proteins (GAPs), which modify the intrinsic GTPase activity of the G protein. Such an effector terminates signaling faster when stimulated by one G protein relative to another and fine-tunes the flow of information through the various forks of the signaling system. The signaling strength (i.e., the speed and efficiency) of any branch of the pathway can be modulated. Inherent affinities, level of expression of the various components, compartmentalization, GTPase rates, and GAP activity combine to produce either a well-focused response by a single pathway or a richer and more diffuse response through several pathways.

Fine-tuning of cAMP by Adenylyl Cyclases

The level of cAMP is highly regulated owing to a balance between synthesis by adenylyl cyclases and degradation by cAMP phosphodiesterases (PDEs). Each of these enzymes can be independently regulated and manipulated. Adenylyl cyclase was the first G-protein effector to be identified, and now a group of related adenylyl cyclases are known to be differentially regulated by both α and βγ subunits. (Taussig and Gilman, 1995). G proteins can both activate and inhibit adenylyl cyclases either synergistically or antagonistically.

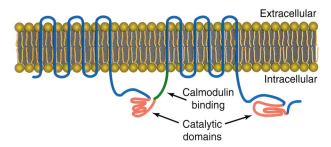

FIGURE 12.6 Domain structure of adenylyl cyclase. Two sets of transmembrane segments with two catalytic sites are characteristic of mammalian adenylyl cyclases. Some isoforms of the kinase also have a calmodulin binding domain (shown in green).

Adenylyl cyclases are large proteins of approximately 120 kDa. Their topology, shown in Fig. 12.6, has been deduced from sequence information. (Krupinski *et al.*, 1989). All the known classes of adenylyl cyclase consist of a tandem repeat of the same structural motif: a short cytoplasmic region followed by six putative transmembrane segments and then a highly conserved catalytic domain of approximately 35 kDa on the cytoplasmic side. The catalytic domains resemble each other as well as the catalytic domain of guanylyl cyclase. It is therefore likely that these two domains interact with G_s, bind ATP, and catalyze its conversion into cyclic AMP. Some isoforms are activated by calmodulin and have one calmodulin binding domain in the link between the first catalytic domain and the second set of transmembrane sequences.

Differential regulation of adenylyl cyclase isoforms. The most common and familiar pathway for modulating adenylyl cyclase is activation through α_s. As more isoforms were cloned and studied, however, it became clear that, though all were activated by α_s, they differed in the degree and type of regulation by other G proteins. The activity of adenylyl cyclase corresponds to input from receptors that are stimulatory or inhibitory and whose concurrent action can be additive or synergistic. Mammals have at least nine adenylyl cyclase isoforms, designated I–IX (not including alternative splicing), that differ in their regulatory properties and tissue distribution. (Taussig and Gilman, 1995; Hanoune and Defer, 2001; Tang and Gilman, 1991) Many isoforms are found throughout the body, but Type I is restricted largely to nerve cells and Type III to the olfactory bulb. Additional isoforms, such as Type II, also are prominent in the brain. Types V and VI may be the predominant forms of adenylyl cyclase in peripheral tissues.

All adenylyl cyclase isoforms are stimulated by G_s through its α_s subunit. The known isoforms can be

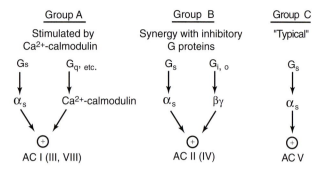

FIGURE 12.7 Isoforms of adenylyl cyclase (AC). All isoforms are stimulated by α_s but differ in the degree of interaction with Ca^{2+}-calmodulin and with $\beta\gamma$ derived from inhibitory G proteins. Not shown is the ability of excess $\beta\gamma$ to complex with and inhibit group A and group C adenylyl cyclases. Adapted from Taussig and Gilman (1995).

minimally divided into at least three groups on the basis of additional regulatory properties (Fig. 12.7). Group A (Types I, III, and VIII) possesses a calmodulin binding domain and is activated by Ca^{2+}-calmodulin. Group B (Types II and IV) is weakly responsive to direct interaction with α_s or $\beta\gamma$ but is highly activated when both are present. (Krupinski *et al.*, 1989). As described later, this synergistic effect enables this cyclase to function as a coincidence detector (see also Chapter 18). Group C is typified by Types V and VI (and IX), which differ from group A cyclases in their inhibitory regulation.

Inhibition of adenylyl cyclases. Adenylyl cyclases are also subject to several forms of inhibitory control. First, activation of all adenylyl cyclases can be antagonized to some extent by $\beta\gamma$ released from abundant G proteins, such as G_i, G_o, and G_z, which complex with α_s-GTP and shift the equilibrium toward an inactive trimer by mass action (Northup *et al.*, 1983). Second, either α or $\beta\gamma$ subunits derived from G_i, G_o, or G_z can directly inhibit group A cyclases, and the α subunit from G_i or G_z can inhibit group C cyclases (Tang and Gilman, 1991). Many G proteins can generate $\beta\gamma$ subunits capable of directly inhibiting group A and activating group B adenylyl cyclases. However, not all G proteins are sufficiently abundant to produce enough $\beta\gamma$ to bring about these effects. The level of $\beta\gamma$ required for these actions is higher than the level needed for α subunits to produce their effects. The level of G_s in particular is low; thus, the α_s derived from G_s is sufficient to activate adenylyl cyclases, but the $\beta\gamma$ derived from it is insufficient to directly inhibit or activate adenylyl cyclases. Therefore, the sources of $\beta\gamma$ for modulation of Type I and II adenylyl cyclases are likely the abundant G proteins, such as G_i and G_o. This explains the

apparent paradox that receptors that couple to G_s produce effects only through α_s, whereas receptors that couple to G_i produce effects through both α_i and $\beta\gamma$ even though they can share the same $\beta\gamma$.

Receptor coupling to adenylyl cyclase. Dozens of neurotransmitters and neuropeptides work through cyclic AMP as a second messenger and do so by activation of either G_s to stimulate adenylyl cyclase or G_i or G_o to inhibit adenylyl cyclase. Among the neurotransmitters that increase cyclic AMP are the amines norepinephrine, epinephrine, dopamine, serotonin, and histamine and the neuropeptides vasointestinal peptide (VIP) and somatostatin. In the olfactory system, a special form of G-protein α subunit, termed α_{olf}, serves the same function as α_s. Odorants are detected by several hundred seven-helix receptors that activate G_{olf}, which in turn activates the Type III adenylyl cyclase in the neuroepithelium. Many of the same neurotransmitters activate distinct receptors that couple to G_i or G_o. They include acetylcholine, dopamine, serotonin, norepinephrine, and opiate peptides.

Adenylyl cyclases as coincidence detectors. The properties of adenylyl cyclases described in Fig. 12.7 suggest an integrative capacity for adenylyl cyclase. Type I and Type II adenylyl cyclases appear to be specifically designed to detect concurrent stimulation of neurons by two or more neurotransmitters (Bourne and Nicoll, 1993).

Type I adenylyl cyclase is stimulated by neurotransmitters that couple to G_s and by neurotransmitters that elevate intracellular Ca^{2+}. This adenylyl cyclase can convert the depolarization of neurons into an increase in cAMP (Wayman *et al.*, 1994). Cells possess many mechanisms for increasing intracellular Ca^{2+}, including voltage gated Ca^{2+} channels that allow Ca^{2+} influx in response to depolarization and a G_s-coupled pathway. This class of adenylyl cyclase has been implicated in several associative forms of learning, a role that may be related to its ability to link cAMP-based and Ca^{2+}-based signals (Levin *et al.*, 1992; Wu *et al.*, 1995).

Stimulation of Type II adenylyl cyclase by α_s is conditional on the presence of $\beta\gamma$ derived from a G protein other than G_s, thus enabling the cyclase to serve as a coincidence detector (Tang and Gilman, 1991; Federman *et al.*, 1992). As indicated earlier, $\beta\gamma$ derived from G_s is not sufficient to produce synergistic activation of this enzyme. Thus, activation of a second receptor, presumably coupled to the abundant G_i and G_o, is needed to provide the $\beta\gamma$. In tissues lacking the Type II adenylyl cyclase, neurotransmitters can couple to G_i and inhibit the other adenylyl cyclases. In tissues, such as cortex and hippocampus, which contain the type II adenylyl cyclase, the same neurotransmitters can couple to G_i and potentiate increases in cyclic AMP resulting from concurrent stimulation by neurotransmitters coupled to G_s.

Sources of Second Messengers: Phospholipids

Two phospholipids, phosphatidylinositol 4,5-bisphosphate (PIP_2) and phosphatidylcholine (PC), are primary precursors for a G protein-based second messenger system. Three second messengers, diacylglycerol, arachidonic acid and its metabolites, and Ca^{2+}, are ultimately produced. A single step converts the inert phospholipid precursors into the lipid messengers (Divecha and Irvine, 1995). The conversion of Ca^{2+} from an inert into an active form is accomplished by the regulated entry of Ca^{2+} from a concentrated pool sequestered in the endoplasmic reticulum or from outside the cell into the lumen of the cytosol or nucleus, where its concentration is low. DAG action is mediated by protein kinase C (Takai *et al.*, 1979; Tanaka and Nishizuka, 1994). Ca^{2+} has many cellular targets but mediates most of its effects through calmodulin, a Ca^{2+}-binding protein that activates many enzymes after it binds Ca^{2+}. One class of calmodulin-dependent enzymes is a family of protein kinases that enable Ca^{2+} signals to modulate a large number of cellular process by phosphorylation (Braun and Schulman, 1995).

Generation of DAG and IP_3 from G_q and G_i coupled to PLCβ.
The phosphatidyl inositide-signaling pathway is just as prominent in neuronal signaling as the cyclic AMP pathway and is similar to it in overall design. Stimulation of a large number of neurotransmitters and hormones [including acetylcholine (M1, M3), serotonin ($5HT_2$, $5HT_{1C}$), norepinephrine (α_{1A}, α_{1B}), glutamate (metabotropic), neurotensin, neuropeptide Y, and substance P] is coupled to the activation of a phosphatidylinositide-specific PLC. Turnover of phosphatidylinositol—its degradation followed by resynthesis in response to a large variety of neurotransmitters and hormones—has been a curiosity for decades. Only recently has it become clear that in addition to their function as membrane phospholipids, the phosphoinositides are precursors for second messengers (Berridge, 1993; Clapham, 1995).

Phosphatidylinositol (PI) is composed of a diacylglycerol backbone with *myo*inositol attached to the *sn*-3 hydroxyl by a phosphodiester bond (Fig. 12.8). The six positions of the inositol are not equivalent: the 1-position is attached by a phosphate to the DAG

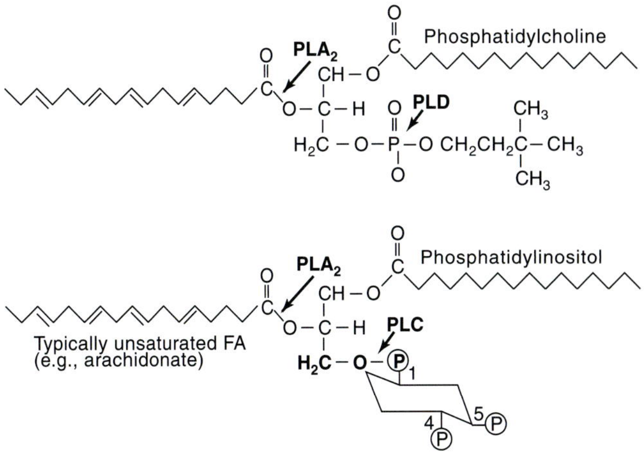

FIGURE 12.8 Structures of phosphatidylinositol and phosphatidylcholine. The sites of hydrolytic cleavage by PLC, PLD, and PLA_2 are indicated by arrows. FA, fatty acid.

moiety. PI is phosphorylated by PI kinases at the 4-position and then at the 5-position to form PIP_2. In response to the appropriate G-protein coupling, PLC hydrolyzes the bond between the *sn*-3 hydroxyl of the DAG backbone and the phosphoinositol to produce DAG, a hydrophobic molecule, and inositol 1,4,5-trisphosphate (IP_3), which is water soluble (Divecha and Irvine, 1995; Hokin and Hokin 1955) (Fig. 12.9). Three classes of PLC that hydrolyze PIP_2 with some selectivity have been cloned and characterized. Two dozen genes encode the three classes designated PLCβ, PLCγ, and PLCδ, soluble enzymes that have in common a catalytic domain structure but differ in their regulatory properties. G proteins couple to several variants of PLCβ. PLCγ is regulated by growth factor tyrosine kinases. In contrast, PLCδ in brain is primarily glial, and its mode of regulation is not well understood, although it may be activated by arachidonic acid.

PLCβ is coupled to neurotransmitters by G_i and G_q (Clapham, 1995). At least two G proteins were suspected because PLC stimulation was fully inhibited by pertussis toxin in some systems and only partially inhibited in others. The pertussis toxin-sensitive pathway is mediated by G_i. However, the $\beta\gamma$ rather than the α subunit of G_i is responsible. The pertussis toxin-insensitive pathway is mediated by a number of isoforms originally termed G_p, with the subscript "p" for PLC activation. Because the PLC-activating function resides in several cloned α subunits given the designation α_q, this activity is now more commonly referred to as G_q and α_q rather than G_p and α_p. G_q couples to PLCβ through its α subunit. There are several PLCβ isoforms, and they show distinct regulation by G proteins. One isoform is most sensitive to

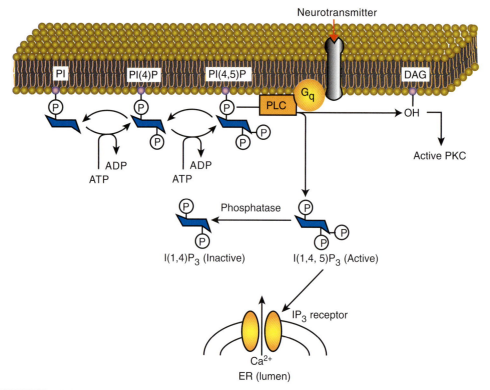

FIGURE 12.9 Schematic pathway of IP_3 and DAG synthesis and action. Stimulation of receptors coupled to G_q activates PLCβ, which leads to release of DAG and IP_3. DAG activates PKC, whereas IP_3 stimulates the IP_3 receptor in the endoplasmic reticulum (ER), leading to mobilization of intracellular Ca^{2+} stores. Adapted from Berridge (1993).

α_q; another is more sensitive to βγ (e.g., from G_i), and a third shows little activation by either G protein.

DAG is derived from activation of phospholipase D. The time course of DAG production after cell stimulation suggests a complex process. DAG is initially derived from PIP_2 as a direct consequence of G-protein coupling to PLCβ. This DAG activates protein kinase C (PKC), and its action is quickly terminated by conversion into phosphatidic acid and recycling into phospholipids. However, a second phase of DAG synthesis often follows, and this phase produces a level of DAG far in excess of that available from PIP_2. This second phase can last many minutes and may include the hydrolysis of a major cellular phospholipid, phosphatidylcholine, by phospholipase D (PLD) activity (Divecha and Irvine, 1995; Nishizuka, 1995). PLD cleaves phosphatidylcholine at the phosphodiester bond to produce phosphatidic acid and choline. Dephosphorylation of phosphatidic acid produces DAG. The PLD pathway may be used by some mitogens and growth factors and likely contains a variety of activation schemes that may include G proteins. Stimulation of the cell generates a transient cleavage of IP_3 and DAG from PIP_2 and a transient rise in Ca^{2+}

due to IP_3-mediated release of intracellular stores. This produces the early activation of PKC. Activation is sustained in the cell in which PLD is stimulated, leading to the second phase of DAG, from phosphatidylcholine. If PLA_2 is activated, arachidonic acid is released from PIP_2 and phosphatidylcholine.

Regulation of PLCγ by receptor tyrosine kinases DAG and IP_3 is also produced when certain receptor tyrosine kinases are activated. This G protein-independent pathway uses PLCγ rather than PLCβ. Stimulation of receptor tyrosine kinases such as epidermal growth factor (EGF) leads to their autophosphorylation on tyrosine residues and activation. Specific phosphotyrosine moieties on the receptor then recruit effector enzymes, such as PLCγ, that possess Src homology 2 (SH2) domains. These structural elements specifically recognize certain protein sequences with phosphotyrosine and lead to the translocation of effectors such as PLCγ to the receptor at the membrane. After binding to the receptor, PLCγ is activated by phosphorylation on tyrosine and hydrolyzes PIP_2 to DAG and IP_3.

Additional lipid messengers. PLA_2 cleaves the fatty acid at the *sn*–2 position of the DAG backbone

(Fig. 12.8). The fatty acid composition of phospholipids is quite heterogeneous; PIP_2 is composed largely of stearic acid at the *sn*–1 position and arachidonic acid at the *sn*–2 position. Thus, arachidonic acid is released in response to PLA_2 hydrolysis of PIP_2. This fatty acid is a 20-carbon *cis*-unsaturated fatty acid with four double bonds. PLA_2 can be activated by $\beta\gamma$ and α subunits of G proteins, but the identity of the endogenous G proteins that mediate this activation is not known. A cytosolic PLA_2 has been cloned and shown to be activated by mitogen-activated protein (MAP) kinases in response to growth factor signaling. Phosphorylation leads to its translocation to the membrane, where it can act on membrane phospholipids. Arachidonic acid has biological activity of its own in addition to serving as a precursor for prostaglandins and leukotrienes. Arachidonic acid and other *cis*-unsaturated fatty acids can modulate K^+ channels, $PLC\gamma$, and some forms of PKC. Other lipids also may generate signaling molecules.

A subfamily of lipid kinases that are specific for addition of a phosphate moiety on the 3-position, phosphoinositide 3-kinases (PI–3 kinases) also play a regulatory role (Rameh and Cantley, 1999). Depending on their preferred lipid substrate, they can produce PI-3-P, PI–3,4-P_2, PI–3,5-P_2, and PI–3,4,5-P_3. A number of signals, including growth factors, activate PI–3 kinases to generate these lipid messengers. In turn, these lipids then directly bind to a number of proteins and enzymes to modify vesicular traffic, protein kinases involved in survival and cell death. There is also evidence that another lipid, sphingomyelin, is a precursor for intracellular signals as well.

IP_3, a potent second messenger that produces its effects by mobilizing intracellular Ca^{2+}. The main function of IP_3 is to stimulate the release of Ca^{2+} from intracellular stores (Streb *et al.*, 1983). The concentration of cytosolic free Ca^{2+} is approximately 100 nM in unstimulated neurons, whereas its concentration in the extracellular space is 1.5–2.0 mM. This provides a tremendous driving force for movement down its concentration gradient; its reversal potential is more than 100 mV. Ca^{2+} is the most common second messenger in neurons, yet it can be neurotoxic. Neurons have therefore developed several mechanisms for maintaining a low interstimulus level of free Ca^{2+}. A Ca^{2+}-ATPase and a Na^+–Ca^{2+} exchanger in the plasma membrane catalyze the active transport of Ca^{2+} to the extracellular space, and a different Ca^{2+}-ATPase in the ER membrane sequesters Ca^{2+} in the ER network. Much of the Ca^{2+} in the ER is complexed with low-affinity-binding proteins that enable the ER to concentrate Ca^{2+} yet enable Ca^{2+} to readily flow down its concentration gradient into the cell lumen on opening of Ca^{2+} channels in the ER. The ER is the major IP_3-sensitive Ca^{2+} store in cells (Fig. 12.9).

The IP_3 receptor is a macromolecular complex that functions as an IP_3 sensor and a Ca^{2+}-release channel. It has a broad tissue distribution but is highly concentrated in the cerebellum. Purification and cloning of the IP_3 receptor show it to be a 313-kDa membrane glycoprotein with a single IP_3 binding site at its N terminal, facing the cytoplasm. The functional channel is composed of four such subunits. The C-terminal half of the molecule contains eight putative transmembrane domains; four such sets of transmembrane segments combine to form a relatively nonselective channel or pore. Ca^{2+} release by IP_3 is highly cooperative, with a Hill coefficient of 2.7. Thus, a small change in IP_3 has a large effect on Ca^{2+} release from the ER. The IP_3 receptor has low activity at either high or low levels of cytoplasmic Ca^{2+}, with peak release requiring 200–300 nM Ca^{2+}, a property that may be used in the generation of some Ca^{2+} waves (Berridge, 1993; Tsien and Tsien, 1990; Bootman *et al.*, 1997). The mouse mutants *pcd* and *nervous* have deficient levels of the IP_3 receptor and exhibit defective Ca^{2+} signaling, and a genetic knockout of the IP_3 receptor leads to motor and other deficits.

The structure of the IP_3 receptor is quite similar to that found earlier for the ryanodine receptor, which serves as a Ca^{2+}-sensitive Ca^{2+} channel in muscle and brain. The ryanodine receptor is a tetramer composed of 560-kDa subunits. In muscle, it is activated by voltage changes that are detected by the dihydropyridine receptor and conveyed to the ryanodine receptor by direct protein–protein interaction. In other cells, the ryanodine receptor is gated by Ca^{2+} or by cyclic ADP-ribose. The localizations of the IP_3 and the ryanodine receptors in the brain are distinct, suggesting that they subserve different aspects of Ca^{2+} signaling. For example, the IP_3 receptor is more enriched in cerebellar Purkinje cells and hippocampal CA1 neurons, whereas the ryanodine receptor is more enriched in the dentate gyrus and CA3 neurons in hippocampus. Electron microscopy reveals that IP_3 receptors in the hippocampus are often on dendritic shafts and cell bodies, whereas ryanodine receptors are in axons and in dendritic spines and the nearby shaft.

Termination of the IP_3 signal. IP_3 is a transient signal terminated by dephosphorylation to inositol. Inactivation is initiated either by dephosphorylation to inositol 1,4-bisphosphate (Fig. 12.9) or by an initial phosphorylation to a tetrakisphosphate form that is dephosphorylated by a different pathway. Both pathways have in common an enzyme that cleaves the phosphate on the 1-position. Complete dephosphorylation yields inositol, which is recycled in the biosynthetic pathway. Recycling is important because most tissues do not contain *de novo* biosynthetic pathways for making inositol. Thus, the phosphatases not only terminate the signal but also serve as a salvage step that may be particularly important when cells are actively undergoing PI turnover. It is intriguing that the simple salt Li$^+$ selectively inhibits the salvage of inositol by inhibiting the enzyme that dephosphorylates the 1-position and is common to the two pathways. This simple salt is the drug used to treat manic–depressive disorders. At therapeutic doses of Li$^+$, the reduced salvage of inositol in cells with high phosphoinositide signaling may lead to depletion of PIP$_2$ and a selective inhibition of this signaling pathway in active cells.

Calcium Ion

Calcium has a dual role as a carrier of electrical current and as a second messenger. Its effects are more diverse than those of other second messengers such as cyclic AMP and DAG because its actions are mediated by a much larger array of proteins, including protein kinases (Carafoli and Klee, 1999). Furthermore, many signaling pathways directly or indirectly increase cytosolic Ca^{2+} concentration from 100 nM to 0.5–1.0 μM. The source of elevated Ca^{2+} can be either the ER or the extracellular space (Fig. 12.10). As indicated earlier, mobilization of ER Ca^{2+} is mediated by IP$_3$ derived from PLCβ activation through G proteins and from PLCγ activation by receptor tyrosine kinases acting on the IP$_3$ receptor. In addition, Ca^{2+} can activate its own mobilization through the ryanodine receptor on the ER. Mechanisms for Ca^{2+} influx from outside the cell include several voltage-sensitive Ca^{2+} channels and ligand-gated cation channels that are permeable to Ca^{2+} [e.g., nicotinic receptor and *N*-methyl-D-aspartate (NMDA) receptor]. In the *Drosophila* visual system and in nonexcitable mammalian cells, depletion of Ca^{2+} from the cytosol and ER initiates an unknown signal that stimulates a low-conductance influx current called I_{CRAC} (Ca^{2+}-release-activated current) across the plasma membrane (Niemeyer *et al.*, 1996). In mammalian cells, a unique channel protein responsible for the current (Yue *et al.*, 2001) provides a slow but prolonged rise in intracellular Ca^{2+} that not only serves as a second messenger but also replenishes ER stores. Ca^{2+} is efficiently sequestered in the ER and extruded out of the cell, which can rapidly lower Ca^{2+} to baseline levels.

Dynamics of Ca^{2+} signaling revealed by fluorescent Ca^{2+} indicators. A great deal is known about the spatial and temporal regulation of Ca^{2+} signals because of the development of fluorescent Ca^{2+} indicators. A variety of fluorescent compounds related to the Ca^{2+} chelator ethylene glycol bis(β-aminoethyl ether) N,N′-tetraacetic acid selectively bind Ca^{2+} at various concentration ranges and change their fluorescent properties on binding Ca^{2+} (Minta and Tsien, 1989). They have dissociation constants in the physiological range of Ca^{2+} and provide a rapid and fairly accurate measurement of ionized Ca^{2+}. Digital fluorescence imaging can be used to detect Ca^{2+} in subcellular compartments such as dendrites and spines, the nucleus, and the cytosol and has demonstrated localized changes in free Ca^{2+} (see Chapter 18, Fig. 18.8). Some cells also undergo oscillations in free Ca^{2+} with a frequency that is increased by an increased concentration of hormone. Stimulation of cultured astrocytes with glutamate generates Ca^{2+} oscillations, which propagate as a wave that spreads across a multicellular network and may coordinate some actions of glia.

Lack of uniformity in Ca^{2+} levels. The concentration of Ca^{2+} entering the cytosol through voltage-sensitive Ca^{2+} channels in the plasma membrane or through the IP$_3$ receptor in the ER is extremely high because of the large concentration gradient across these membranes. Relatively low-affinity Ca^{2+}-dependent processes can produce effects of Ca^{2+} near the membrane, such as synaptic release and modulation of Ca^{2+} channels. However, by the time Ca^{2+} diffuses a few membrane diameters away, it is rapidly buffered by many Ca^{2+}-binding proteins, and its concentration drops from 100 to 1 μM or less. The diffusion of Ca^{2+} is greatly slowed in biological fluid because of the high concentration of binding proteins (0.2–0.3 mM). Ca^{2+} diffuses a distance of 0.1–0.5 μm, and diffusion lasts approximately 30 ms before Ca^{2+} is bound. Ca^{2+} is therefore a second messenger that acts locally, a feature that makes Ca^{2+} subdomains possible where Ca^{2+} signaling is spatially segregated. In contrast, IP$_3$ is a global intermediate with an effective range that can span a typical soma before being terminated by dephosphorylation (Allbritton *et al.*, 1992).

Calmodulin-Mediated Effects of Ca^{2+}

Ca^{2+} acts as a second messenger to modulate the activity of many mediators. The predominant mediator of Ca^{2+} action is calmodulin. This abundant and ubiquitous 17-kDa calcium-binding protein is highly conserved across phyla. Ca^{2+} binds to calmodulin in the physiological range and converts it into an activator. Calmodulin has no intrinsic enzymatic activity. It serves a central regulatory role by modulating the activity of various cellular targets (Cohen and Klee, 1988). Binding of Ca^{2+} to calmodulin produces a conformational change that greatly increases its affinity for target enzymes. Ca^{2+}-calmodulin binds and activates more than 20 eukaryotic enzymes, including cyclic nucleotide PDEs, adenylyl cyclase, nitric oxide synthase, Ca^{2+}-ATPase, calcineurin (a phosphoprotein phosphatase), and several protein kinases (Fig. 12.10). This activation of calmodulin allows neurotransmitters that change the concentration of Ca^{2+} to affect dozens of cellular proteins, presumably in an orchestrated fashion. Ca^{2+} also affects proteins and enzymes independently of calmodulin, including calcium-binding proteins and enzymes such as calpain and PKC. Ion channels (K channels and IP_3 receptors or channels) are directly modulated by Ca^{2+} (Fig. 12.10). First discovered as a protein factor necessary for Ca^{2+}-dependent activation of a cyclic nucleotide PDE, the substance was renamed calmodulin (Ca^{2+} response modulator) when it was subsequently found to modulate many enzymes in addition to PDE.

Calmodulin interacts with its targets in several ways. The "conventional" interaction with enzyme targets requires a stimulated rise in Ca^{2+} so that the four Ca^{2+} binding sites on calmodulin become occupied and it can bind. At basal levels of Ca^{2+}, however, much of calmodulin may be bound to a diverse group of proteins such as GAP–43 (neuromodulin), neurogranin, and unconventional myosins. Interactions with these targets are not well understood and may serve to localize calmodulin near other targets or to "buffer" calmodulin so that the level of free calmodulin is kept very low and is less likely to activate the conventional

targets at basal Ca^{2+} (Persechini and Cronk, 1999). A third group of enzymes, which includes the inducible form of nitric oxide synthase, bind calmodulin in a manner that makes it sensitive to basal Ca^{2+}, and the enzymes are therefore active at basal Ca^{2+}.

Actions of enzymes and proteins modulated by calmodulin. Calmodulin has four Ca^{2+} binding sites, or binding folds, described as EF hands, a recurring Ca^{2+} binding structural motif first identified in paravalbumin from carp muscle. Calmodulin is composed of a number of helical segments designated by capital letters and separated by loops. Orientation of the helix–loop–helix EF segment of 29 amino acids is similar to that of the thumb and index finger, which positions amino acids in the loop for coordination with Ca^{2+}, hence the name EF hand. Two EF hands, or calmodulin folds, stabilize each other to form two Ca^{2+} binding sites. Binding of Ca^{2+} is cooperative, with binding to the first set of high-affinity sites facilitating binding to low-affinity sites (K_d approximately 1 μM). The ability of Ca^{2+} to be accommodated in an asymmetric coordination shell with multiple and distant amino acids, including uncharged oxygens, enables it to compete with Mg^{2+} and to produce large conformation changes that are the basis for interconversion between inactive and active states of proteins.

Calmodulin recognizes a short segment of the enzymes that it regulates; however, there is no strict consensus sequence for calmodulin binding. X-Ray crystallography and NMR have provided a three-dimensional structure of Ca^{2+}-calmodulin with and without a peptide whose sequence is based on calmodulin binding sites in several kinases. In the absence of a target protein, Ca^{2+}-calmodulin is in an extended structure composed of two globular regions, each containing a set of calmodulin folds separated by a long α-helical tether. Binding of Ca^{2+} allows movement of the globular regions around the calmodulin binding site, "gripping" it as would hands around a rope (Ikura *et al.*, 1992). The two lobes of this compact structure surround the target peptide, making dozens of hydrophobic contacts as well as ionic interactions between Arg and Lys typically found in target peptides and Glu residues in calmodulin. The large number of possible interactions allows calmodulin to accommodate many target sequences, with the calmodulin grip slightly different in each case to maximize interactions. This binding likely produces the necessary displacement of the calmodulin binding domain for activation of the target enzymes.

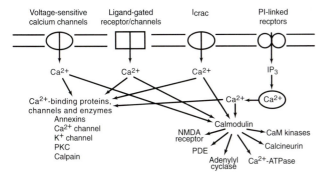

FIGURE 12.10 Multiple sources of Ca^{2+} converge on calmodulin and other Ca^{2+}-binding proteins. Cellular levels of Ca^{2+} can rise either by influx (e.g., calcium release-activated calcium current (I_{CRAC}) channels) or by redistribution from intracellular stores triggered by IP_3. Calcium modulates dozens of cellular processes by the action of the Ca^{2+}-calmodulin complex on many enzymes, and calcium has some direct effects on enzymes such as PKC and calpain. CaM kinase, Ca^{2+}-calmodulin-dependent kinase.

Regulation of guanylyl cyclase by nitric oxide. An important target of Ca^{2+}-calmodulin is the enzyme nitric oxide synthase (NOS) (see Chapter 10). This enzyme synthesizes one of the simplest known messengers, the gas NO (Schmidt and Walter, 1994; Baranano. *et al.*, 2001). Nitric oxide was first recognized as a signaling molecule that mediates the action of acetylcholine on smooth muscle relaxation (Furchgott and Zawadzki, 1980). An intercellular signal was necessary to explain how stimulation of endothelial cells by acetylcholine could produce a relaxation in tone of the underlying smooth muscle cells that surround the vessels. A labile substance operationally termed endothelium-derived relaxing factor (EDRF) was detected and has since been shown to be NO. Acetylcholine stimulates the PI signaling pathway in the endothelium to increase intracellular Ca^{2+}, which activates NOS so that more NO is made. NO then diffuses radially from the endothelial cells across two cell membranes to the smooth muscle cell, where it activates guanylyl cyclase to make cyclic GMP. This in turn activates a cyclic GMP-dependent protein kinase that phosphorylates proteins, leading to a relaxation of muscle (Schmidt and Walter, 1994). In 1998, Robert F. Furchgott, Louis J. Ignarro, and Ferid Murad received the Nobel Prize for their discoveries concerning nitric oxide as a signaling molecule and therapeutic mediator in the cardiovascular system. This intercellular signaling in which NO made in one cell increases cyclic GMP in a nearby cell is the conventional view of how NO works. In a related scheme, heme oxygenase may be stimulated to make carbon monoxide (CO), which also can act intercellularly through guanylyl cyclase and cyclic GMP.

Let us now turn to the details of the NO pathway. We see that other pathways can activate NOS, mediate the actions of NO, stimulate guanylyl cyclases, and mediate the actions of cyclic GMP.

Nitric oxide is derived from L-arginine in a reaction catalyzed by NOS, a complex enzyme that has one equivalent each of flavin adenine dinucleotide (FAD), flavin mononucleotide (FMN), tetrahydrobiopterin, and heme (iron protoporphyrin IX) per monomer. NOS converts L-arginine and O_2 into NO and L-citrulline in a five-electron oxidation reaction that requires nicotinamide adenine dinucleotide phosphate (NADPH) as a cofactor. NOS likely produces the neutral free radical NO as the active agent. As we have already seen in the generation of second messengers, stimulation of a single step converts a common inert precursor (in this case, L-arginine) into a powerful intercellular and intracellular signal (NO). Although NO is stable in oxygen-free water, it is labile and lasts only a few seconds in biological fluids because of its inactivation by superoxides and its complex formation with heme-containing proteins

such as oxyhemoglobin. Thus, no specialized processes are needed to inactivate this particular signaling molecule.

As a gas, NO is soluble in both aqueous and lipid media and can readily diffuse from its site of synthesis across the cytosol or cell membrane and affect targets in the same cell or in nearby neurons, glia, and vasculature (Baranano *et al.*, 2001). Neuronal communication by synaptic vesicles is unidirectional, from presynaptic to postsynaptic neuron. NO provides the capability for a retrograde message and thus reverses the usual flow of information; that is, information can travel from a postsynaptic site of synthesis to a presynaptic site of modulation. When synaptic activity stimulates NO production at a spine, NO signals are unlikely to be restricted to that synapse alone, the consequences of which are important for models of long-term potentiation that propose NO as a messenger.

NO produces a variety of effects, including relaxation of smooth muscle (as mentioned earlier) of the peripheral vasculature and perhaps control of cerebral blood flow, relaxation of smooth muscle of the gut in peristalsis, and killing of foreign cells by macrophages (Schmidt and Walter, 1994). It was first recognized as a neuronal messenger that couples glutamate receptor stimulation to increases in cyclic GMP. Analogs of L-arginine, such as nitroarginine and monomethyl arginine, block NOS unless there is an excess of L-arginine. Such inhibitors of NO synthase have been used to implicate NO in long-term potentiation and long-term depression in the hippocampus and cerebellum, respectively (Schuman and Madison, 1994). (see also Chapter 18).

Representative clones from three classes of NO synthase have been characterized. Two of them are constitutively expressed and activated by Ca^{2+}-calmodulin formed when intracellular Ca^{2+} is elevated. The constitutive forms are designated neuronal or endothelial on the basis of their original source, although they overlap in this tissue distribution. The third class of NOS is an inducible form; its level is markedly increased by transcription and protein synthesis in response to cell stimulation, and it is prominent in macrophages stimulated by cytokines. After translation, the inducible form is active at basal Ca^{2+} levels because it has a tightly bound calmodulin in a conformation that greatly enhances its Ca^{2+} sensitivity. Because NOS is a major flavin-containing enzyme, it possesses NADPH diaphorase activity. It reduces nitroblue tetrazolium to formazan, a reaction that can be used as a cytochemical stain for NOS because much of the diaphorase staining in the brain is due to this enzyme. This family of enzymes displays distinct regional distributions in the brain, suggesting some regional specificity in the use of NO as a signaling molecule in the brain. The constitutive neuronal isoform is concentrated in cerebellar granule cells and likely provides the NO that activates guanylyl cyclase in nearby Purkinje cells during the induction of long-term depression in the cerebellum (see Chapter 18).

The action of NO is often mediated by guanylyl cyclase and cyclic GMP. However, a number of physiological and pathological effects of NO are independent of cyclic GMP. NO stimulates ADP-ribosylation of a number of proteins in the brain and other tissues. The nature of the ADP-ribosyltransferases and their effects is not known. Another effect of NO is stimulation of release of neurotransmitters, apparently in a Ca^{2+}-independent fashion.

Activation of guanylyl cyclases. Two types of guanylyl cyclase, a soluble one regulated by NO and a membrane-bound enzyme directly regulated by neuropeptides (Garbers and Lowe, 1994) synthesize cyclic GMP from GTP in a reaction similar to the synthesis of cyclic AMP from ATP. The soluble enzyme is a heterodimer, with catalytic sites resembling those of adenylyl cyclase and a heme group. NO activates the soluble enzyme by binding to the iron atom of the heme moiety. This is the basic mechanism for regulation of soluble guanylyl cyclases. Stimulation of guanylyl cyclase is the major, but not only, effect of NO in the brain and other tissues. A number of therapeutic muscle relaxants, such as nitroglycerin and nitroprusside, are NO donors that produce their effects by stimulating cyclic GMP synthesis.

The membrane-bound guanylyl cyclases are transmembrane proteins with a binding site for neuroendocrine peptides on the extracellular side of the plasma membrane and a catalytic domain on the cytosolic side. Several isoforms of membrane-bound guanylyl cyclase, each with a binding site for a distinct neuropeptide such as atrial natriuretic peptide and brain natriuretic peptide, have been characterized. In the periphery, these peptides regulate sodium excretion and blood pressure; in the brain, their functions are less clear.

Cyclic GMP Phosphodiesterase, an Effector Enzyme in Vertebrate Vision

The versatility of G-protein signaling is illustrated in vertebrate phototransduction, in which a specialized G protein called transducin (G_t) is activated by light rather than by a hormone or neurotransmitter. Without transducin, we would not be able to see. Transducin stimulates cyclic GMP phosphodiesterase, an effector enzyme that hydrolyzes cyclic GMP and

ultimately turns off the dark current. Nature has devised an elegant mechanism for using photons of light to modify a hormone-like molecule, retinal, that activates a seven-helix receptor called rhodopsin (Baylor, 1996). This receptor has a built-in prehormone that is converted into the active form by light. Light photoisomerizes the inactive 11-*cis*-retinal to the active all-*trans*-retinal, which functions as a neurotransmitter to activate its receptor. Activated rhodopsin triggers the GTP–GDP exchange of transducin, leading to dissociation of its α_t and $\beta\gamma$ subunits. The active species in transducin is the α subunit. It activates a soluble cyclic GMP phosphodiesterase by binding to and displacing an inhibitory subunit of the enzyme. In the dark, retinal rods contain high levels of cyclic GMP, which maintains a cyclic GMP-gated channel permeable to Na^+ and Ca^{2+} in the open state and thus provides a depolarizing dark current. As the levels of cyclic GMP drop, the channel closes to hyperpolarize the cell.

Rods can detect a single photon of light because the signal-to-noise ratio of the system is very low owing to a very low spontaneous conversion of the 11-*cis*-retinal into the all-*trans*-retinal. Furthermore, the amplification factor is quite high; one rhodopsin molecule stimulated by a single photon can activate 500 transducins. Transducin remains in the "on" state long enough to activate 500 PDEs. PDE is designed for speed and can hydrolyze 10^5 cyclic GMP molecules in the second before it is deactivated by GTP hydrolysis and dissociated from transducin (Stryer, 1991). Cyclic GMP in rods regulates a cyclic GMP-gated cation channel, leading to additional amplification of the signal.

Modulation of Ion Channels by G Protein

Each neuron has a set of ion channels that it uses to integrate incoming signals, propagate action potentials, and introduce Ca^{2+} into terminals specialized for the release of neurotransmitters. Because the repertoire of ion channels gives neurons their individual response signatures, it is not surprising that several types of mechanisms regulate these channels. Second messengers derived from G protein and other pathways activate protein kinases that phosphorylate ion channels (see also Chapter 16). In addition, certain ion channels are effector proteins that are directly modulated by G proteins.

The first ion channel demonstrated to undergo regulation by G proteins was the cardiac K^+ channel that mediates slowing of the heart by acetylcholine released from the vagus nerve. When this I_{KACh} channel is examined in a membrane patch delimited by the seal of a cell-attached electrode, addition of acetylcholine within the electrode dramatically increases the frequency of channel opening, whereas addition of acetylcholine to the cell surface outside the seal does not. Although acetylcholine stimulates muscarinic M2 receptors when added either inside or outside the seal, the receptors outside do not have access to the channels being recorded in the sealed patch because this signaling pathway does not include diffusible second messengers that can affect the channel. The process is therefore described as membrane delimited, which is explained most simply by a direct interaction between the G protein and the channel. Subsequent studies have shown that the pathway is pertussis toxin sensitive and that purified G_i activated by GTPγS added to the underside of the patch will activate the channel.

Whether the active component of G_i is the α or the $\beta\gamma$ subunit was controversial. Dogmas do not die easily and, for many years, $\beta\gamma$ was not considered a direct activator or inhibitor of effector enzymes and channels. The I_{KACh} channel appears to be composed of heteromultimers of two types of subunits that can be activated either by α_i or by $\beta\gamma$. The α_i subunit is more potent in regulating the channel, but the membrane contains enough $\beta\gamma$ to enable $\beta\gamma$ to activate the channel as well (Huang et al., 1995; Wickman et al., 1994). Different combinations of channel subunits may be preferentially activated by α_i and by $\beta\gamma$ subunits.

Of the ion channels other than the K$^+$ channel, evidence is most compelling for the stimulation or inhibition of Ca^{2+} channel subtypes by G proteins. The central role played by Ca^{2+} in muscle contraction, in synaptic release, and in gene expression makes the modulation of Ca^{2+} influx a common target for regulation by neurotransmitters. In the heart, where L-type Ca^{2+} channels are critical for regulation of contractile strength, the Ca^{2+} current is enhanced by α_S formed by β-adrenergic stimulation of G_S. In contrast, N-type Ca^{2+} channels, which modulate synaptic release in nerve terminals, are often inhibited by muscarinic and α-adrenergic agents and by opiates acting at receptors coupled to G_i and G_o. In sympathetic ganglia, norepinephrine reduces synaptic release by inhibiting Ca^{2+} influx through the N channel. The G protein couples to the channel in a membrane-delimited process that shifts the temporal distribution of gating modes, favoring the time spent in a low-open probability mode, thereby effectively lowering the open time of the channel (Delcour and Tsien, 1993).

Inhibition of L-type Ca^{2+} currents can exhibit strict G-protein specificity for both the α and $\beta\gamma$ subunits. In GH3 cells, a pituitary cell line, antisense oligonucleotides that eliminate expression of α_{o1} and α_{o2} block the inhibitory effect of muscarinic agents (at M4 receptors) and of somatostatin, respectively. Microinjection of antisense oligonucleotides to β_1 and β_3 blocked inhibition by somatostatin and muscarinic agents, respectively (Kleuss et al., 1992). Finally, the γ subunit subtype also was critical. Thus, γ_3 was required for coupling to the somatostatin receptor, whereas γ_4 coupled to muscarinic receptors (Kleuss et al., 1993). Thus, in the same cell, somatostatin couples to Ca^{2+} channels through $\alpha_{o2}\beta_1\gamma_3$, whereas the muscarinic receptor couples to $\alpha_{o1}\beta_3\gamma_4$. The $\beta\gamma$ subunits may directly affect the channels, perhaps in synergy with the α subunit or as a requirement for the appropriate presentation of the α subunits to the correct receptor.

G-Protein Signaling Gives Special Advantages in Neural Transmission

The G protein-based signaling system provides several advantages over fast transmission (Hille, 1992; Neer, 1995). These advantages include amplification of the signal, modulation of cell function over a broad temporal range, diffusion of the signal to a large cellular volume, cross-talk, and coordination of diverse cell functions.

Amplification

Several thousandfold amplification can be initiated by a single neurotransmitter–receptor complex that activates numerous G proteins, each of which activates many effector enzymes and channels. Each enzyme can generate many second–messenger molecules, and each channel allows the flux of many ions. As we see in the next section, second messengers often activate protein kinases that phosphorylate many substrates before deactivation.

Temporal Range

The sacrifice in speed relative to signaling by ligand-gated ion channels is compensated by a broad range of signaling that facilitates integration of signals by the G-protein system. Transmission through membrane-delimited coupling of ion channels to G proteins is relatively fast, with only some sacrifice in speed. Signaling that includes second messengers is much slower. It can be as fast as 100–300 ms, as in olfactory signaling in which cAMP and IP3 take part, or it can take from seconds to minutes.

Spatial Range

A slower time frame means that cellular processes that are quite distant from the receptor can be modulated. Diffusion of second messengers such as IP3, Ca^{2+}, and DAG can extend neurotransmission through the cell body and to the nucleus to alter gene expression.

Cross-Talk

Both the signal transduction machinery and the ultimate mediators of their responses, such as the protein kinases, are capable of cross-talk. This is seen in coincident detection of signals from two receptors converging on Type I and Type II adenylyl cyclase.

Coordinated Modulation

Neurotransmitters acting through G proteins can elicit a coordinated response of the cell that can modulate synaptic release, resynthesis of neurotransmitter, membrane excitability, the cytoskeleton, metabolism, and gene expression.

Summary

A major class of signaling using G protein-linked signals affords the nervous system a rich diversity of modulation, amplification, and plasticity. Signals are mediated through second messengers activating proteins that modify cellular processes and gene transcription. A key feature is the ability of G proteins to detect the presence of activated receptors and to amplify the signal through effector enzymes and channels. Phosphorylation of key intracellular proteins, ion channels, and enzymes activates diverse, highly regulated cellular processes. Specificity of response is ensured through receptors reacting only with a limited number of G proteins. The response of the system is determined by the speed of activation of GTPase. The function of G-protein subunits is now being elucidated. In addition to speed of response, the spatial compartmentalization of the system enables specificity and localized control of signaling. Phospholipids and phosphoinositols provide substrates for second messenger signaling for G proteins. Stimulation of release of intracellular calcium is often the mediator of the signal. Calcium itself has a dual role as a carrier of electrical current and as a second messenger. Calmodulin is a key regulator that provides complexity and enhances specificity of the signaling system. Sensitivity of the system is imparted by an extremely robust amplification system, as seen in the visual system, which can detect single photons of light.

MODULATION OF NEURONAL FUNCTION BY PROTEIN KINASES AND PHOSPHATASES

Protein phosphorylation and dephosphorylation are key processes that regulate cellular function. They play a fundamental role in mediating signal transduc-

tion initiated by neurotransmitters, neuropeptides, growth factors, hormones, and other signaling molecules. The primary determinants of a cell's morphology and function are the protein constituents expressed in that cell. However, the functional state of many of these proteins is modified by phosphorylation–dephosphorylation, the most ubiquitous posttranslational modification in eukaryotes. More than 2% of mammalian genes encode a kinase or phosphatase, and a fifth of all proteins may serve as targets for kinases and phosphatases. Phosphorylation can rapidly modify the function of enzymes, structural and regulatory proteins, receptors, and ion channels taking part in diverse processes without a need to change the level of their expression. As is described in greater detail in the following chapter, phosphorylation and dephosphorylation can also produce long-term alterations in cellular properties by modulating transcription and translation and changing the complement of proteins expressed by cells.

Protein kinases catalyze the transfer of the terminal, or γ, phosphate of ATP to the hydroxyl moieties of Ser, Thr, or Tyr residues at specific sites on target proteins. Most protein kinases are either Ser/Thr kinases or Tyr kinases, with only a few designed to phosphorylate both categories of acceptor amino acids. Protein phosphatases catalyze the hydrolysis of the phosphoryl groups from phosphoserine–phosphothreonine, phosphotyrosine, or both types of phosphorylated amino acids on phosphoproteins.

Protein phosphatases reverse the effects of protein kinases and protein kinases reverse the effects of protein phosphatases (Fig. 12.11). This statement may

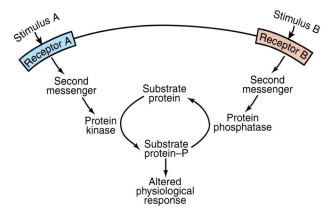

FIGURE 12.11 Regulation by protein kinases and protein phosphatases. Enzymes and other proteins serve as substrates for protein kinases and phosphoprotein phosphatases, which modify their activity and control them in a dynamic fashion. Multiple signals can be integrated at this level of protein modification. Adapted from Greengard *et al.* (1998).

seem odd only because signal transduction schemes typically depict unidirectional and not bidirectional regulation; that is, a stimulus activates a kinase that phosphorylates a substrate protein, and basal phosphatase activity dephosphorylates the substrate protein after kinase activity subsides. In fact, regulation of the phosphorylation state of proteins is bidirectional, and the phosphorylation state of proteins *in vivo* ranges widely, from minimal to almost fully phosphorylated, even in the absence of cell stimulation (Rosenmund *et al.*, 1994; Greengard *et al.*, 1998). The phosphorylation state can be dynamically altered either upward or downward from the steady state, depending on the cell's inputs and its complement of kinases and phosphatases. Although phosphatases clearly serve to reverse a stimulated phosphorylation, our lack of understanding of phosphatase regulation is largely to blame for our viewing phosphatases in this limited role.

The activity of protein kinases and protein phosphatases is typically regulated either by a second messenger (e.g., cAMP or Ca^{2+}) or by an extracellular ligand (e.g., nerve growth factor). In general, the second messenger-regulated kinases modify Ser and Thr, whereas the receptor-linked kinases modify Tyr. Among the thousands of protein kinases and protein phosphatases in neurons, a relatively small number serve as master regulators to orchestrate neuronal function.

The cAMP-dependent protein kinase is a prototype for the known regulated Ser/Thr kinases; they are similar in overall structure and regulatory design. PKA is emphasized here because the experimental strategies currently being used in the study of kinases have come from the investigation of PKA-mediated processes. As its name implies, cyclic AMP-dependent protein kinase carries out the posttranslational modification of numerous protein targets in response to signal transduction processes that act through G proteins and alter the level of cAMP in cells. PKA is the predominant mediator for signaling through cAMP, the only other being a cAMP-liganded ion channel in olfaction. In a similar fashion, the related cGMP-dependent protein kinase (PKG) mediates most of the actions of cGMP. Ca^{2+}-calmodulin-dependent protein kinase II and several other kinases mediate many of the actions of stimuli that elevate intracellular Ca^{2+}. Finally, the PI signaling system increases both DAG and Ca^{2+}, which activate any of a family of protein kinases collectively called protein kinase C. Each of these kinases has a broad substrate specificity and is therefore able to phosphorylate diverse substrates throughout the cell. The activities of protein kinases and phosphatases are balanced, as

revealed by the phosphorylation state of these targets of the signal transduction process. Here, again, although there are thousands of protein phosphatases, a relatively small number exemplified by protein phosphatase 1 (PP-1), protein phosphatase 2A (PP-2A), and protein phosphatase 2B (PP-2B, or calcineurin) are responsible for most of the dephosphorylation at Ser and Thr residues on phosphoproteins that are under the regulation of the aforementioned kinases. The Nobel Prize for Physiology or Medicine was awarded to Edwin G. Krebs and Edmund H. Fischer in 1992 for their pioneering work on regulation of cell function by protein kinases and phosphatases.

Certain Principles Are Common in Protein Phosphorylation and Dephosphorylation

Protein kinases and protein phosphatases are described either as multifunctional, if they have a broad specificity and therefore modify many protein targets, or as dedicated, if they have a very narrow substrate specificity and may modify only a single protein target. The Ser/Thr kinases and phosphatases described here are multifunctional, giving them the ability to coordinate the regulation of many cellular processes in response to cell stimulation. But how is response specificity achieved with kinases and phosphatases that are designed to recognize many substrates? These enzymes are by no means promiscuous; their substrates conform either to a consensus sequence along the primary protein sequence (for the kinases) or to general features of the three-dimensional structure of the phosphoprotein (for the phosphatases). Furthermore, spatial positioning of kinases and their substrates in the cell either increases or decreases the likelihood of phosphorylation–dephosphorylation of a given substrate.

The amplification of signal transduction described earlier is continued during the transmission of the signal by protein kinases and protein phosphatases. In some cases, the kinases are themselves subject to activation by phosphorylation in a cascade in which one activated kinase phosphorylates and activates a second, and so on, to provide amplification and a switchlike response termed *ultrasensitivity* (Ferrell and Machleder, 1998). (See Chapter 14 for additional discussion of ultrasensitivity.)

Kinases and phosphatases integrate cellular stimuli and encode the stimuli as the steady-state level of phosphorylation of a large complement of proteins in the cell (Hunter, 1995). Phosphorylation and dephosphorylation are reversible processes, and the net activity of the two processes determines the phosphorylation state of each substrate. The phosphorylation state depends on the degree of activation or inactivation of the protein kinase or protein phosphatase, the affinity of the protein target for these enzymes, and the concentration and access of the kinase, phosphatase, and target protein. Some proteins are phosphorylated largely in the basal state and are subject primarily to regulation of phosphatases. Distinct signal transduction pathways can converge on the same or different target substrates. In some cases, these substrates can be phosphorylated by several kinases at distinct sites.

Phosphorylation produces specific changes in the function of a target protein, but these changes are completely dependent on the site of phosphorylation and the nature of the target protein. Phosphorylation may increase or decrease the catalytic activity of an enzyme or its affinity for its substrate or cofactor. It can modify interactions between the phosphoprotein and other proteins, DNA, phospholipids, or other cellular constituents and thereby alter the function of the phosphoprotein in gene expression, synaptic vesicle recycling, and membrane transport. Phosphorylation can regulate desensitization of receptors, their coupling to other signaling molecules, or their localization at synaptic sites. Any of several characteristics of ion channels can be altered by phosphorylation, including voltage dependence, probability of being opened, open and close time kinetics, and conductance. The number of possible effects is almost limitless and enables the fine-tuning of numerous cellular processes over broad time scales, from milliseconds to hours. Kinases and phosphatases do this fine-tuning by regulating the presence of a highly charged and bulky phosphoryl moiety on Ser, Thr, or Tyr at a precise location on the substrate protein. The phosphate may introduce a steric constraint at the surface of the protein in interactions with other cellular constituents, or the negative charge of the phosphoryl moiety may elicit a conformational change because of attractive or repulsive ionic interactions between the phosphorylated segment and other charged amino acids on the protein.

Finally, each of the three kinases described here is capable of functioning as a cognitive kinase, that is, a kinase capable of a molecular memory. Although each is activated by its respective second messenger, it can undergo additional modification that reduces its requirement for the second messenger. As is described in greater detail in Chapter 18, this molecular memory potentiates the activity of these kinases and may enable them to participate in aspects of neuronal plasticity.

cAMP-Dependent Protein Kinase Was the First Well-Characterized Kinase

Neurotransmitters that stimulate the synthesis of cAMP exert their intracellular effects primarily by activating PKA (Nairn *et al.*, 1985). The functions (and substrates) regulated by PKA include gene expression (cAMP response element-binding protein, or CREB), catecholamine synthesis (tyrosine hydroxylase), carbohydrate metabolism (phosphorylase kinase), cell morphology (microtubule-associated protein 2, or MAP-2), postsynaptic sensitivity (AMPA receptor), and membrane conductance (K^+ channel). Paul Greengard and Eric Kandel received the Nobel Prize for Physiology or Medicine in 2000 (along with Arvid Carlsson) for their discoveries concerning signal transduction via PKA and phosphoprotein phosphatases in the nervous system.

PKA is a tetrameric protein composed of two types of subunits: (1) a dimer of regulatory (R) subunits (either two RI subunits for Type I PKA or two RII subunits for Type II PKA) and (2) two catalytic subunits (C subunit) (Scott, 1991). Two or more isoforms of the RI, RII, and C subunits have distinct tissue and developmental patterns of expression but appear to function similarly. The C subunits are 40-kDa proteins that contain the binding sites for protein substrates and ATP. The R subunits are 49- to 51-kDa proteins that contain two cAMP binding sites. In addition, the R subunit dimer contains a region that interacts with cellular anchoring proteins that serve to localize PKA appropriately within the cell.

The binding of second messengers by PKA and the other second messenger-regulated kinases relieves an inhibitory constraint and thus activates the enzymes (Fig. 12.12). The C subunit has intrinsic protein kinase

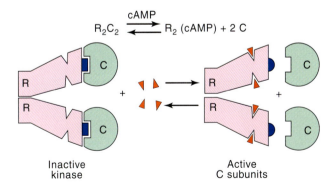

FIGURE 12.12 Activation of PKA by cyclic AMP. An autoinhibitory segment (blue) of the regulatory subunit (R) dimer interacts with the substrate binding domain of the catalytic (C) subunits of PKA, blocking access of substrates to their binding site. Binding of four molecules of cyclic AMP reduces the affinity of R for C, resulting in dissociation of constitutively active C subunits.

activity that remains inhibited as long as the C subunit is complexed with the R subunits in the tetrameric holoenzyme. As each R subunit binds two molecules of cAMP, its affinity for the C subunit is greatly reduced, and the C subunit dissociates as a free active kinase. Cyclic AMP therefore activates the C subunit by relieving it of its inhibitory R subunits. The steady-state level of cAMP determines the fraction of PKA that is in the dissociated or active form. In this way PKA decodes cAMP signals into the phosphorylation of proteins and the resultant change in various cellular processes.

PKA is a member of a large family of protein kinases that have in common a significant degree of homology in their catalytic domains and are likely derived from an ancestral gene (Hanks and Hunter, 1995) (Fig. 12.13). This homology extends to the three-dimensional crystal structure, based on X-ray crystal-

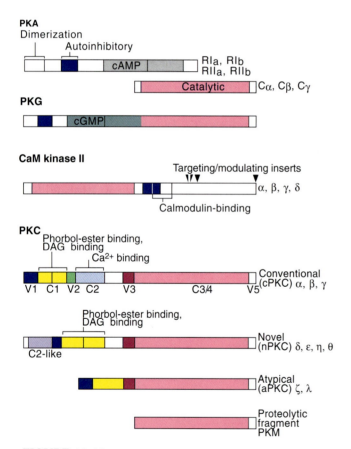

FIGURE 12.13 Domain structure of protein kinases. Protein kinases are encoded by proteins with recognizable structural sequences that encode specialized functional domains. Each of the kinases [PKA, PKG, CaM (Ca^{2+}-calmodulin-dependent) kinase II, and PKC] have homologous catalytic domains that are kept inactive by the presence of an autoinhibitory segment (blue segments). The regulatory domains contain sites for binding second messengers such as cAMP, cGMP, Ca^{2+}-calmodulin, DAG, and Ca^{2+}-phosphatidylserine. Alternative splicing creates additional diversity.

lography of PKA and a few other kinases. The catalytic domain comprises approximately 280 amino acids that may be in a subunit distinct from the regulatory domain, as in PKA, or in the same subunit, as in PKC and Ca^{2+}-calmodulin-dependent (CaM) kinases. The crystal structure of the C subunit complexed to a segment of protein kinase inhibitor (PKI), a selective high-affinity inhibitor of PKA, reveals that the C subunit is composed of two lobes (Knighton *et al.*, 1991). A small N-terminal lobe contains a highly conserved region that binds Mg^{2+}-ATP in a cleft between the two lobes. A larger C-terminal lobe contains the protein–substrate recognition sites and the appropriate amino acids for catalyzing the transfer of the phosphoryl moiety from ATP to the polypeptide chain of the substrate. Inhibition by PKI is diagnostic of PKA involvement; PKI contains an autoinhibitory sequence resembling PKA substrates and is positioned in the catalytic site like a substrate, thus blocking access for substrates.

How can protein kinases have a common structural homology yet exhibit phosphorylation target specificity? Although the C-terminal lobes of all kinases may use a similar scaffold for their peptide binding and catalytic sites, distinct amino acids are positioned on this scaffold to produce specificity in peptide binding. Dedicated protein kinases, such as myosin light chain kinase (MLCK), which phosphorylates only certain myosin light chains, have numerous sites of contact between the catalytic domain and phosphorylation site on the substrate and therefore have high specificity. In contrast, PKA has a smaller number of contact sites, thus enabling a much larger yet specific set of substrates to bind. Some substrates have been found to contain a distinct docking site for the kinase that regulates them that may serve to further increase the selectivity of phosphorylation (Smith *et al.*, 2000).

PKA phosphorylates Ser or Thr at specific sites in dozens of proteins. The sequences of amino acids at the phosphorylation sites are not identical, but a consensus sequence can be deduced from a comparison of these sequences (Kennelly and Krebs, 1991). PKA phosphorylates at sites with the consensus sequence Arg–Arg–X–Ser/Thr–Y, in which X can be one of many different amino acids and Y is a hydrophobic amino acid. Each kinase has a characteristic consensus sequence that forms the basis for distinct substrate specificities (Table 12.4). These consensus sequences are often used to identify putative sites of phosphorylation on newly cloned proteins, although many substrates are phosphorylated at "anomalous" sites, thus reducing the reliability of these predictions. Secondary and tertiary structures probably have a

TABLE 12.4 Consensus Phosphorylation Sites of
Some Protein Kinases[a]

Protein kinase	Consensus phosphorylation site
PKA	R-R/K-X-S*/T*
PKG	R/K$_{2-3}$-X-S*/T*
cPKC	(R/K$_{1-3}$,-X$_{2-0}$)-S*/T*-(X$_{2-0}$,-R/K$_{1-3}$)
CaM kinase II	R-X-X-S*/T*
MLCK (smooth muscle)	(K/R$_2$-X)-X$_{1-2}$-K/R$_3$-X$_{2-3}$-R-X$_2$-S*-N-V-F

[a]R, Arg; K, Lys; S*, phospho-Ser; T*, phospho-Thr; X, polar amino acid; N, Asp; V, Val; F, Phe.
Adapted from Kennelly and Krebs (1991).

role in substrate recognition, and the finding of anomalous phosphorylation sites may simply be due to our lack of knowledge of exactly how substrates are recognized (Walsh and Patten, 1994). The consensus sites of PKA, CaM kinase II, and PKC all include a basic residue on the substrate, and these kinases do share some target substrates.

A regulatory theme common to PKA, CaM kinase II, and PKC is that their second messengers activate them by displacing an autoinhibitory domain from the active site; that is, they relieve an inhibitory constraint rather than stabilize a conformation of the kinase that has higher activity (Kemp et al., 1994). Some of the contacts between the C and R subunits of PKA resemble those between the C subunit and its protein substrates. The R subunit blocks access of substrates by positioning a pseudosubstrate or autoinhibitory domain in the catalytic site. This segment of R resembles a substrate and binds to C as would a substrate or PKI (protein kinase inhibitor). Binding of cAMP to the R subunit near this autoinhibitory domain must disrupt its binding to the C subunit, thus leading to dissociation of an active C subunit (Su et al., 1995). CaM kinase II and PKC likewise have autoinhibitory segments that are near the second messenger binding sites and may be activated similarly (Kemp et al., 1994) (see Fig. 12.13).

Functional differences between Type I and Type II PKA (which have C subunits in common but have different R subunits) may arise from differential targeting in cells and from differences in regulation by autophosphorylation. RII, but not RI, is autophosphorylated by its C subunit when it is in the holoenzyme form. This potentiates cAMP action by reducing the rate of reassociation of RII and C after a stimulus. Only anchoring proteins for RII have been characterized thus far (Edwards and Scott, 2000).

Multifunctional CaM Kinase II Decodes Diverse Signals That Elevate Intracellular Ca^{2+}

Most of the effects of Ca^{2+} in neurons and other cell types are mediated by calmodulin, and many of the effects of Ca^{2+}-calmodulin are mediated by protein phosphorylation–dephosphorylation. In contrast with the cAMP system, both dedicated and multifunctional kinases are found in the Ca^{2+} signaling system (Schulman and Braun, 1999). Two kinases, MLCK and phosphorylase kinase, are each dedicated to the phosphorylation of a single substrate—myosin light chains and phosphorylase, respectively. The Ca^{2+} signaling system also contains a family of Ca^{2+}-calmodulin-dependent protein kinases with broad substrate specificity, including CaM kinases I, II, and IV; of these, CaM kinase II is the best characterized. CaM kinase II phosphorylates tyrosine hydroxylase, MAP-2, synapsin I, calcium channels, Ca^{2+}-ATPase, transcription factors, and glutamate receptors and thereby regulates synthesis of catecholamines, cytoskeletal function, synaptic release in response to high-frequency stimuli, calcium currents, calcium homeostasis, gene expression, and synaptic plasticity, respectively. The enzyme is activated by Ca^{2+} regardless of whether it is elevated by influx through Ca^{2+} channels or ligand-gated Ca^{2+} channels or is released from intracellular stores following stimulation of the PI signaling pathway. This kinase is found in every tissue but is particularly enriched in neurons, where it may account for as much as 2% of all hippocampal protein. This level is 50 times as high as the level of the kinase in other tissues; thus, CaM kinase II likely serves some special functions in such brain regions. It is found in the cytosol, in the nucleus, in association with cytoskeletal elements, and in postsynaptic thickening, termed the postsynaptic density, which is found in asymmetric synapses. It is a large multimeric enzyme, consisting of 12 subunits derived from four homologous genes (α, β, γ, and δ) that encode different isoforms of the kinase that range from 54 to 72 kDa per subunit. Multimers and heteromultimers of α- and β-CaM kinase II isoforms are found predominantly in brain, whereas the γ- and δ-CaM kinases are found throughout the body, including the brain.

The domain structure of CaM kinase II isoforms is shown in Fig. 12.13. Unlike those of PKA, the catalytic, regulatory, and targeting domains are all contained within a single polypeptide. The N-terminal half of each isoform contains the catalytic domain, which is highly homologous to the catalytic subunit of PKA and other Ser/Thr kinases. The middle region constitutes the regulatory domain, which contains an autoinhibitory domain with an overlapping calmodulin-binding sequence. The C-terminal end contains an association domain that allows 12–14 subunits (two rings of subunits) to assemble into a multimer (Kolodziej *et al.*, 2000, Hoelz *et al.*, 2003), as well as targets sequences that direct the kinase to distinct intracellular sites.

Regulation of the kinase by autophosphorylation is a critical feature of CaM kinase II. The basic three-dimensional conformation of the catalytic domain is likely to be similar to the structures determined for PKA and CaM kinase I. The kinase is inactive in the basal state because an autoinhibitory segment is positioned in the catalytic site, sterically blocking access to its substrates. Peptides corresponding to this region are useful inhibitors for functional studies and inhibit the kinase by competing for binding of both ATP and protein substrates. Elevation of Ca^{2+} generates a Ca^{2+}-calmodulin complex that wraps around the calmodulin binding domain of the kinase, which overlaps with part of the autoinhibitory domain. Thus, binding of Ca^{2+}-calmodulin displaces the autoinhibitory domain from the catalytic site and thus activates the kinase by enabling ATP and protein substrates to bind. Displacement of this domain also exposes a binding site for anchoring proteins to which the activated kinase can bind, perhaps positioning it for more selective phosphorylation (Bayer *et al.*, 2001). The site occupied by one particular amino acid in the autoinhibitory domain of all isoforms of this kinase, Thr-286 (in α-CaM kinase II), must be crucial in allowing the autoinhibitory domain to keep the active site shut before activation. If the kinase is activated, it can autophosphorylate this particular Thr residue. Phosphorylation disables the autoinhibitory segment by preventing it from reblocking the active site after calmodulin dissociates and thereby locks the kinase in a partially active state that is independent, or autonomous, of Ca^{2+}-calmodulin (Saitoh and Schwartz, 1985; Miller and Kennedy, 1986; Hanson *et al.*, 1989).

An additional dramatic effect of autophosphorylation is that it enhances the affinity of the bound calmodulin by 400-fold, which it achieves by reducing the rate of dissociation of calmodulin from the kinase after Ca^{2+} levels are reduced below threshold. In essence, autophosphorylation traps bound calmodulin for several seconds and keeps the kinase active for a while after Ca^{2+} levels decline to baseline. The consequence of calmodulin trapping and disruption of the autoinhibitory domain is to prolong the active state of the kinase, a potentiation that led to its description as a cognitive kinase (Schulman and Braun, 1999; Lisman, 1994). CaM kinase II responds to a large number of neurotransmitter receptors sub-

served by various signal transduction systems *in situ*, and stimulation of these pathways increases the autonomous activity of the enzyme. The level of autonomous activity (8–15% in brain) may reflect integration of multiple cellular inputs, and the autonomous activity may be adjusted either upward or downward.

CaM kinase II is targeted to distinct cellular compartments. Differences between the four genes encoding CaM kinase II and between the two or more isoforms that are encoded by each gene by apparent alternative splicing reside primarily in a variable region at the start of the association domain (see Fig. 12.13). In some isoforms, this region contains an additional sequence of 11 amino acids that targets those isoforms to the nucleus. One unusual isoform termed αKAP has a short hydrophobic segment in place of the catalytic domain and serves to target catalytically competent subunits that co-assemble with it to membranes. The major neuronal isoform, α-CaM kinase II, is largely cytosolic but is also found attached to postsynaptic densities and to synaptic vesicles and may therefore have several targeting sequences.

Protein Kinase C Is the Principal Target of the PI Signaling System

Protein kinase C (PKC) is a collective name for members of a relatively diverse family of protein kinases most closely associated with the PI signaling system. PKC is a multifunctional Ser/Thr kinase capable of modulating many cellular processes, including exocytosis and endocytosis of neurotransmitter vesicles, neuronal plasticity, gene expression, regulation of cell growth and cell cycle, ion channels, and receptors. A major breakthrough in understanding of the PI signaling pathway was the realization that diacylglycerol (DAG) and Ca^{2+}, two products of this pathway, function as second messengers to activate PKC (Takai *et al.*, 1979; Tanaka and Nishizuka 1994). The role of DAG in PI signaling was unclear until its link to PKC was established. Many PKC isoforms also require an acidic phospholipid such as phosphatidylserine for appropriate activation. The kinase is also of interest because it is the target of a class of tumor promoters called phorbol esters. They activate PKC by simulating the action of DAG, bypassing the normal receptor-based pathway and somehow inappropriately stimulating cell growth.

The PKC family of kinases is diverse in structure and regulatory properties (Tanaka and Nishizuka, 1994). Unlike PKA, PKC is a monomeric enzyme (78–90 kDa) with catalytic, regulatory, and targeting domains all on one polypeptide. Each isoform has a regulatory domain, with several subdomains, in its N-terminal half and a catalytic domain at the C terminal (Newton, 1995) (see Fig. 12.13). Only the first PKC isoforms to be characterized, now termed the conventional isoforms (or cPKC), have all of the domains. The domains are referred to as (1) V1, which contains the autoinhibitory or pseudosubstrate sequence present in all isoforms; (2) C1, a cysteine-rich domain that binds DAG and phorbol esters; (3) C2, a region necessary for Ca^{2+} sensitivity and for binding to phosphatidylserine and to anchoring proteins; (4) V3, a protease-sensitive hinge; (5) C3/4, the catalytic domain; and (6) V5, which may also mediate anchoring. Subsequent cloning revealed a larger and more diverse group of isoforms than that of cPKC (Fig. 12.13). One class of isoforms, termed novel PKCs (nPKC), lacks a true C2 domain and is therefore not Ca^{2+} sensitive. Another class is considered atypical (aPKC) because it lacks C2 and the first of two cysteine-rich domains that are necessary for DAG (or phorbol ester) sensitivity. This class is neither Ca^{2+} nor DAG sensitive.

Activation of PKC is best understood for the conventional isoforms. Generation of DAG resulting from stimulation of the PI signaling pathway increases the affinity of cPKC isoforms for Ca^{2+} and phosphatidylserine. Although triglyceride lipases also generate DAG, the sn-1,2-diacylglycerol isomer is derived only from PI turnover, and it is the only isomer effective in activating PKC. Cell stimulation results in the translocation of cPKC from a variety of sites to the membrane or cytoskeletal elements where it interacts with PS-Ca^{2+}-DAG at the membrane (Newton, 1995; Kraft and Anderson, 1983; Zhang et al., 1995; Ron et al., 1994). Binding of the second messengers to the regulatory domain disrupts the nearby autoinhibitory domain, leading to a reversible activation of PKC by deinhibition, as is found for PKA and CaM kinase II (Muramatsu et al., 1989).

Translocation is not restricted to the plasma membrane. Some PKC isoforms translocate to intracellular sites enriched with certain anchoring proteins for the activated form of PKC, termed RACK (receptors for activated C kinase) (Ron et al., 1994; Mochly-Rosen, 1994). Distinct PKC isoforms can translocate to the cytoskeleton, membrane, perinuclear area, and nucleus, probably by binding of the C2 and/or the V1 and V5 domains to distinct anchoring proteins. In the inactive state, a segment of PKC may occupy this RACK binding domain, thus preventing translocation until the activation of PKC exposes this domain. Activation may therefore consist of both displacement of the autoinhibitory segment to unblock the catalytic site and displacement of an "auto-anchor" site to

unblock the RACK binding site. Whether DAG is synthesized only in the plasma membrane and the PKC–PS–DAG complex subsequently diffuses to distant RACKs or whether some DAG is generated intracellularly is not known. The early phase of DAG derived from PI is followed by a longer phase in which DAG derived from PC is more prominent. After termination of the signal, DAG is recycled into phospholipids, and PKC is redistributed to its initial sites.

Prolonged activation of PKC can be produced by the addition of phorbol esters, which simulate activation by DAG but remain in the cell until they are washed out. In a matter of hours to days, such persistent activation by phorbol esters leads to a degradation of PKC. Either PKC may be more susceptible to proteolysis when activated or some of the compartments to which it translocates have a higher level of protease activity. This phenomenon is sometimes used experimentally to produce a PKC-depleted cell (at least for phorbol ester binding isoforms) and thereafter to test for a loss of putative PKC functions.

Spatial Localization Regulates Protein Kinases and Phosphatases

Protein kinases and protein phosphatases are often spatially positioned near their substrates or they translocate to their substrates on activation to improve speed and specificity in response to neurotransmitter stimulation. PKA is targeted to intracellular sites on the cytoskeleton, membrane, and Golgi through interactions between the RII subunit and specific anchoring proteins. MAP-2 is an anchoring protein for RII, which concentrates PKA near its substrate (including MAP-2) in dendrites. Anchoring proteins with no previously known function are referred to as A kinase anchoring proteins, or AKAPs (Edwards and Scott, 2000). One such anchor, AKAP79, is an anchor for PKA, for PKC, and for calcineurin, the Ca^{2+}-calmodulin-dependent phosphatase (Klauck et al., 1996). The three signaling molecules bind to different sites on this anchoring protein. Better coordination of the phosphorylation–dephosphorylation of the same or different substrates may be achieved by placing calcineurin, PKC, and PKA in the same compartment through AKAPs. Another example of a signaling complex is the protein termed yotiao, which binds to the NMDA-type glutamate receptor and serves as an anchor for both PKA and a phosphatase (PP-1) (Westphal et al., 1999).

The use of anchoring proteins has several consequences. First, it enhances the rate of phosphorylation when kinases or phosphatases are placed near some substrates. Specificity is enhanced when these enzymes are concentrated near proteins that are to be substrates and away from other proteins that are not to be substrates in a given cell. Second, it increases the signal-to-noise ratio for substrates that are not near anchoring proteins because phosphorylation–dephosphorylation would be reduced in the basal state. For example, PKA is anchored on the Golgi away from the nucleus in the basal state. Brief stimuli lead to transient dissociation of C subunits and provide for localized regulation of nearby substrates but little phosphorylation of nuclear proteins. Prolonged stimuli enable some C subunits to passively diffuse through nuclear pores and into nuclei, where they can participate in regulation of gene expression (Bacskai et al., 1993). Termination of the nuclear action of C subunits is aided by PKI, which binds and inhibits the C subunit in the nucleus and hastens its export out of the nucleus, where it can reassociate with R subunits (Wen et al., 1995). A high signal-to-noise ratio can be achieved with PKC by translocation toward substrates only after activation. Interaction with anchoring proteins should increase the rate and specificity of phosphorylation and may also prolong the active state of PKC by stabilizing it. Third, anchoring enables significant basal phosphorylation of substrates near anchoring proteins. Basal phosphorylation would be high if PKA were highly concentrated by an anchoring protein near high-affinity substrates. For example, if an anchoring protein concentrates PKA 50-fold relative to the rest of the cell, then the local concentration of free C subunits will be high at basal cAMP even though the percentage of dissociated C subunits is low. Such an arrangement provides for novel ways of blocking the pathway through disruption of anchoring. When a peptide similar to the site at which RII binds to the AKAP is introduced into a neuron, it disrupts the ability of an AKAP to concentrate PKA near the AMPA receptor and leads to decreased modulation of the AMPA receptor in the basal state (Rosenmund et al., 1994).

PKA, CaM Kinase II, and PKC Are Cognitive Kinases

The ability of three major Ser/Thr kinases (PKA, CaM kinase II, and PKC) in brain to initiate or maintain synaptic changes that underlie learning and memory may require that they themselves undergo some form of persistent change in activity. As mentioned earlier, they have been described as cognitive kinases because they are capable of sustaining their activated states after their second messengers return to basal level and because their target substrates modulate synaptic plasticity (Schwartz, 1993).

cAMP-Dependent Protein Kinase

As is discussed in greater detail in Chapter 18, a role for PKA as a cognitive kinase can be seen in long-term facilitation of the withdrawal reflexes in Aplysia and in long-term potentiation in the vertebrate hippocampus. In Aplysia, stimulation of the body with a strong stimulus such as an electric shock facilitates the withdrawal response to a light touch delivered to another part of the animal. This is an example of a simple form of learning called sensitization. A single shock produces a short-lasting memory, but repeated shocks (training) produce a memory that can last several days. The shock stimulates the release of serotonin, which increases cAMP and PKA activity in the sensory neurons. Features of the behavioral training can be simulated in a system in which a single sensory neuron is co-cultured with a single motor neuron and serotonin is applied to the bath to mimic the effects of sensitizing stimuli. A single exposure to serotonin (or cAMP) produces short-term facilitation and a short-term increase in the phosphorylation of more than a dozen PKA substrates in these cells. However, repeated or prolonged exposure to these agents leads to long-term facilitation and an enhanced state of phosphorylation of the same set of proteins. This phenomenon is due to a PKA that is persistently active despite the fact that cAMP is no longer elevated (Chain et al., 1999). A possible scheme for this phenomenon is shown in Fig. 12.14 (see also Fig. 18.19). The RII subunit is autophosphorylated on its autoinhibitory segment by the C subunit in the holoenzyme. Phospho-RII and C subunits dissociate on elevation of cAMP and reassociate when cAMP levels subside. However, the reassociation rate is greatly reduced by the presence of phosphate on the R subunit, thus prolonging phosphorylation of various target proteins by the C subunit. Furthermore, the R subunits are more susceptible than the C subunits to proteolytic degradation in their dissociated state. Thus, prolonged or repetitive stimulation leads to a preferential decrease in the inhibitory R subunits and thus a slight excess of C subunits that remain persistently active because of insufficient R subunits. The various targets of PKA can then be phosphorylated by this active C subunit long after cAMP levels return to basal or prestimulus levels. Prolonged activation of PKA enables the C subunit to enter the nucleus and induce gene expression, and one of these genes facilitates further proteolysis of R. Although short-term facilitation does not require transcription of new genes or protein synthe-

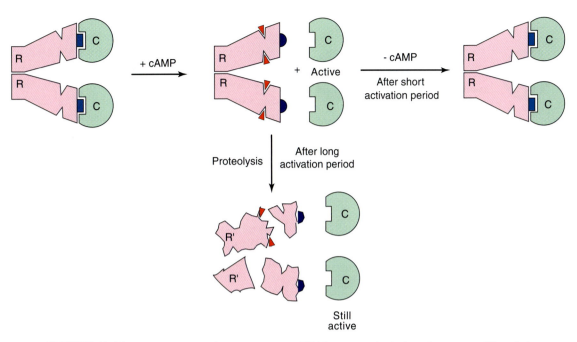

FIGURE 12.14 Long-term stimulation can convert PKA into a constitutively active enzyme. Dissociation of PKA R and C subunits is reversible with short-term elevation of cyclic AMP. More prolonged activation results in loss of R subunits to proteolysis, resulting in an insufficient amount of R subunits to associate with and inhibit all C subunits after the cAMP stimulus terminates.

sis, the long-term effects on phosphorylation and on facilitation do require transcription of new genes and protein synthesis (Kaang *et al.*, 1993). In this interesting process, a molecular memory of appropriate stimulation by serotonin is encoded by a persistence of PKA activity that is regenerative.

PKA may also function as a cognitive kinase in hippocampal long-term potentiation. This phenomenon is most clearly seen in the short- and long-term phases of long-term potentiation in CA3 neurons stimulated by means of the mossy fiber pathway, one of several sites of long-term potentiation in the hippocampus. Induction of both the early phase and the late phase of long-term potentiation at this site requires PKA (Huang *et al.*, 1994). The persistent phase of long-term potentiation requires both RNA synthesis and new protein synthesis, with the likelihood of PKA-mediated increases in gene expression. Although the induction of long-term potentiation requires a rise in Ca^{2+}, rather than of cAMP, one of the Ca^{2+}-stimulated adenylyl cyclases appears to convert some of the Ca^{2+} signal into a rise in cAMP. Mutant mice lacking one such cyclase isoform show a marked reduction in Ca^{2+}-sensitive cyclase and cAMP accumulation, a reduced long-term potentiation in hippocampal slices, and a weaker spatial memory in behavioral tests (Wu *et al.*, 1995).

Ca^{2+}-Calmodulin-Dependent Protein Kinase

CaM kinase II has features of a cognitive kinase because it has a molecular memory of its activation that is based on autophosphorylation and it phosphorylates proteins that modulate synaptic plasticity (Lisman, *et al.*, 2002; Schwartz, 1993; Hanson and Schulman, 1992). The biochemical properties of CaM kinase II suggest mechanisms by which appropriate stimulus frequencies can generate an autonomous enzyme (Fig. 12.15, see also Fig. 14.3 in Chapter 14). The critical site of phosphorylation is in the autoinhibitory segment within the ring of catalytic-regulatory domains in the holoenzyme. Each subunit can bind and be activated by calmodulin independently but requires neighboring subunits for autophosphorylation. Autophosphorylation takes place within each holoenzyme but requires the phosphorylation of one subunit by a proximate neighbor. Furthermore, calmodulin must be bound to the subunit that is to be phosphorylated, apparently to displace the autoinhibitory domain and expose the phosphorylation site to the active subunit. Individual stimuli may be too brief and available calmodulin may be limited, so a single stimulus may lead to binding and activation of only a few subunits per holoenzyme. Thus, each stimulus achieves only submaximal activation of the kinase. When two neighboring subunits are concurrently

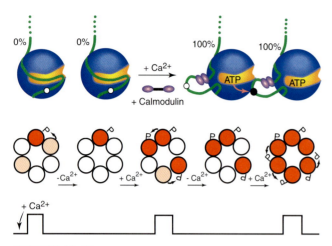

FIGURE 12.15 Frequency-dependent activation of CaM kinase II. Autophosphorylation occurs when both of two neighboring subunits in a holoenzyme are bound to calmodulin. At high frequency of stimulation (rapid Ca²⁺ spikes), the interspike interval is too short to allow significant dephosphorylation or dissociation of calmodulin, thereby increasing the probability of autophosphorylation with each successive spike. In a simplified CaM kinase with only six subunits, calmodulin-bound subunits are shown in pink, and autophosphorylated subunits with trapped calmodulin are shown in red. Adapted from Hanson and Schulman (1992).

bound to calmodulin during a stimulus, autophosphorylation is achieved. Autophosphorylation leads to a potentiation because the phosphorylated subunit traps its bound calmodulin for several seconds and thus remains active longer than the Ca²⁺ signal. Even after calmodulin dissociates, the autophorylated subunit remains partially active until dephosphorylated. At low stimulus frequency, the time between stimuli is sufficient for calmodulin to dissociate and the kinase to be dephosphorylated, and the same submaximal activation occurs with each stimulus. However, at higher frequencies, some subunits remain autophosphorylated and bound to calmodulin, so successive stimuli result in more calmodulin bound per holoenzyme, which makes autophosphorylation and subsequent calmodulin trapping more probable (Hanson *et al.*, 1994; De Koninck and Schulman, 1998). The enzyme is therefore able to decode the frequency of cellular stimulation; low-frequency stimulation leads to submaximal activation of the kinase at each stimulus, whereas higher frequencies exceed a threshold beyond which stimulation leads to recruitment of additional calmodulin and a higher level of activation and autonomy with each spike (De Koninck and Schulman, 1998). See Chapter 14 for additional discussion of the dynamics and effects of CaM kinase II autophosphorylation.

CaM kinase phosphorylates a number of substrates that affect synaptic strength. For example, phosphorylation of synapsin I, a synaptic vesicle protein (see

Chapter 8), by CaM kinase II reduces the attachment of synaptic vesicles to actin at nerve terminals. Phosphorylation of synapsin I by CaM kinase enables synaptic release to be maintained at high stimulus frequencies, perhaps by facilitating movement of vesicles toward release sites. Inhibition of CaM kinase II in hippocampal slices or just elimination of its autophosphorylation by an α-CaM kinase II mouse knock-in in which the critical Thr was replaced by Ala blocks the induction of long-term potentiation (Giese *et al.*, 1998). Although the kinase has normal activity *in vitro*, it shows no frequency-dependent activation and its activity cannot be prolonged without autophosphorylation. These mice are deficient in learning spatial navigational cues, one of the functions of the rodent hippocampus. The basis for its role is uncertain but may be the phosphorylation of AMPA receptors and their recruitment to the membrane, leading to a greater postsynaptic response (Hayashi *et al.*, 2000). The enzyme can therefore be appropriately described as a cognitive kinase with respect to its own molecular memory as well as its functional role in mediating aspects of synaptic plasticity.

Protein Kinase C

PKC also can be converted into a form that is independent, or autonomous, of its second messenger and can be described as a cognitive kinase. Before Ca²⁺ and DAG were known to have roles in the reversible activation of PKC, the PKC was identified as an inactive precursor that was activated *in vitro* by Ca²⁺-dependent proteolysis to a constitutively active fragment termed protein kinase M (PKM). Physiological activation of PKC may also convert it into a PKM-like species. Standard protocols lead to transient activation of both Ca²⁺-dependent and Ca²⁺-independent forms of PKC. However, during the persistent phase of long-term potentiation, some of the PKC remains active but is autonomous of Ca²⁺ and DAG. The activity appears to be a PKM or other modified form of an atypical PKC (Sacktor *et al.*, 1993). PKC has also been implicated in long-term potentiation, and its substrates include the NMDA and AMPA receptors.

Protein Tyrosine Kinases Take Part in Cell Growth and Differentiation

Protein kinases that phosphorylate tyrosine residues on key proteins participate in numerous cellular process and are usually associated with regulation of cell growth and differentiation. Signal transduction by protein tyrosine kinases often includes a cascade of kinases phosphorylating other kinases, eventually activating Ser/Thr kinases, which carry out the intended modification of a cellular process. There are two classes of protein tyrosine kinases. The first is a family of receptor tyrosine kinases that are activated by the binding of extracellular growth factors such as nerve growth factor, epidermal growth factor, insulin, and platelet-derived growth factor. The second family of protein tyrosine kinases such as c-Src are soluble kinases that also participate in regulation of cell growth, as well as in neuronal plasticity, but are indirectly activated by extracellular ligands.

Why have two sets of amino acids been chosen as targets for phosphorylation? First, the consequences of leaky, or "promiscuous," phosphorylation by a protein Ser/Thr kinase of an unintended target may affect metabolic activity or synaptic function but do not typically initiate irreversible and global functions such as cell growth and differentiation. The consequence of such inappropriate stimulation is seen in the effect of a variety of oncogenes that use altered forms of receptor tyrosine kinases or intermediates in their cascades to subvert normal cell growth. The cellular concentrations of protein Ser/Thr kinases and their targets are much higher than those of protein tyrosine kinases and their substrates. Inadvertent phosphorylation of targets that play critical roles in cell growth is less likely if these targets are regulated at tyrosine residues, which are not well recognized by the numerous protein Ser/Thr kinases. Second, introduction of a phosphotyrosine structure into a protein has a greater regulatory potential than does introduction of a phosphoserine or phosphothreonine. The three phosphorylated amino acids have in common an ability to produce conformational changes due to the extra charge or bulk of the phosphate. For example, in the activation of receptor tyrosine kinases,

autophosphorylation displaces an inhibitory domain. In addition, however, the phosphotyrosine and nearby amino acid sequences can be recognized by various signal transduction effectors, such as PLCγ, that contain structural domains that bind to the tyrosine-phosphorylated kinase. The receptor tyrosine kinase thus becomes a platform for concentrating various signaling molecules at specific phosphotyrosine sites in its sequence. These signaling molecules either are activated directly by binding or are activated after having been phosphorylated by the receptor tyrosine kinase. It is easier to bind with the necessary strict specificity to segments of protein around phosphotyrosines, because of the aromatic side chain in tyrosine, and this may be an additional reason for use of Tyr as phosphotransferase targets.

Protein Phosphatases Undo What Kinases Create

Protein phosphatases in neuronal signaling are categorized as either phosphoserine–phosphothreonine phosphatases (PSPs) or phosphotyrosine phosphatases (PTPs) (Hunter, 1995; Price and Mumby, 1999). The enzymes catalyze the hydrolysis of the ester bond of the phosphorylated amino acids to release inorganic phosphate and the unphosphorylated protein. Phosphatases control all of the cellular processes of protein kinases, including neurotransmission, neuronal excitability, gene expression, protein synthesis, neuronal plasticity, and cell growth. A limited number of multifunctional PSPs account for most of such phosphatase activity in cells (Hunter, 1995). They are categorized into six groups (1, 4, 5, 2A, 2B, and 2C) on the basis of their substrates, inhibitors,

TABLE 12.5   Categories of Protein Phosphatases

Phosphatase	Characteristic	Other inhibitors
PP-1	Sensitive to phospho-inhibitor 1, phospho-DARPP-32, and inhibitor 2; has targeting subunits	Weakly sensitive to okadaic acid
PP-4	Nuclear	Highly sensitive to okadaic acid
PP-5	Nuclear	Mildly sensitive to okadaic acid
PP-2A	Regulatory subunits Does not require divalent cation	Highly sensitive to okadaic acid
PP-2B (calcineurin)	Ca^{2+}/calmodulin-dependent CnB regulatory subunit	FK506, cyclosporin
PP-2C	Requires Mg^{2+}	EDTA
Receptor PTPs[a]	Plasma membrane	Vanadate, tyrphostin, erbstatin
Nonreceptor PTPs	Various cellular compartments	Vanadate, tyrphostin
Dual-specificity PTPs	Nuclear (e.g., cdc25A/B/C and VH family)	Vanadate

[a]Protein phosphotyrosine phosphatases.
Adapted from Hunter (1995).

and divalent cation requirements (Table 12.5). An additional historical distinction is that phosphatase 1 (PP-1) preferentially dephosphorylates the β subunit of phosphorylase kinase, whereas protein phosphatase 2A preferentially dephosphorylates its α subunit. Of these PSPs, only protein phosphatase 2B (PP-2B, or calcineurin) directly responds to a second messenger; it responds to increases in cellular Ca^{2+}. PP-1, –4, –5, –2A, and calcineurin are structurally related and differ from PP-2C. Little is known about the basis of substrate specificity of these phosphatases; examination of the primary sequences of their dephosphorylation sites reveals no obvious consensus. The specificity of PP-1 and PP-2A is particularly broad, and each can remove phosphates that were transferred by any of the protein kinases discussed herein as well as many other kinases. The phosphotyrosine phosphatases constitute a distinct and larger class of phosphatases, including PTPs with dual specificity for both phosphotyrosines and phosphoserine–phosphothreonines. PTPs are either soluble enzymes or membrane proteins with variable extracellular domains that enable regulation by extracellular binding of either soluble or membrane-bound signals.

Structure and Regulation of PP-1 and Calcineurin

PP-1 and calcineurin are the best characterized phosphatases with respect to both structure and regulation. The domain structures of the catalytic subunits of PP-1 and calcineurin are depicted in Fig. 12.16. PP-1 is a protein of 35–38 kDa; most of the sequence forms the catalytic domain; its C terminal is the site of regulatory phosphorylation. The catalytic domains of PP-1, PP-2A, and calcineurin are highly homologous (Price and Mumby, 1999).

Although PP-1 and PP-2A are usually prepared as free catalytic subunits, they are normally complexed in cells with specific anchoring or targeting subunits (Price and Mumby, 1999; Hubbard and Cohen, 1993). For example, PP-1 is attached to glycogen particles in

liver, myofibrils in muscle, and unidentified targeting subunits in brain. Phosphorylation of the PP-1 targeting subunit in liver releases the catalytic subunit and results in reduced dephosphorylation of substrates near the targeting subunit because diffusion reduces the local concentration of PP-1. Targeting of PP-1 also modulates its regulation by natural inhibitors. As PP-1 dissociates from targeting subunits, it becomes susceptible to inhibition by inhibitor 2.

Inhibition of PP-1 by two other inhibitors, inhibitor 1 and its homolog DARPP-32 (dopamine and cAMP-regulated phosphoprotein, M_r 32,000), is conditional on the phosphorylation state of these inhibitors (Hemmings et al., 1984). Inhibitor 1 has a broader distribution in brain than DARPP-32, which is found largely in the medium spiny neurons in the neostriatum and in their terminals in the globus pallidus and substantia nigra. Both proteins inhibit only after they are phosphorylated by PKA or PKG. PKA also increases the susceptibility of PP-1 to inhibition by stimulating its release from targeting subunits. Because the substrates for PKA and PP-1 overlap to a great extent, the rate and extent of phosphorylation of such substrates are enhanced by the ability of PKA to catalyze their phosphorylation while blocking their dephosphorylation. Inhibitor 1, DARPP-32, and inhibitor 2 are all selective for PP-1. Highly selective inhibitors capable of penetrating the cell membrane are available for these phosphatases. Okadaic acid, a natural product of marine dinoflagellates, is a tumor promoter, but, unlike phorbol esters, it acts on PP-2A and PP-1, rather than on PKC. The steady-state level of phosphorylation of dozens of proteins is elevated when cells are treated with okadaic acid and its derivatives.

Protein Phosphatase 1. The X-ray structure of the catalytic subunit of PP-1 bound to the toxin microcystin, a cyclic peptide inhibitor, reveals PP-1 to be a compact ellipsoid with hydrophobic and acidic surfaces forming a cleft for binding substrates (Goldberg et al., 1995). PP-1 is a metalloenzyme requiring two metals in the active site that likely take part in electrostatic interactions with the phosphate on substrates that aid in catalyzing the hydrolytic reaction. The phosphate would be positioned at the intersection of two grooves on the surface of the enzyme where binding to amino acid residues on the substrate would occur. Such binding would be blocked when phosphoinhibitor 1 or microcrystin LR binds to this surface. The same general structure of the catalytic domain is seen in calcineurin.

Calcineurin (PP-2B). Calcineurin is a Ca^{2+}-calmodulin-dependent phosphatase that is highly

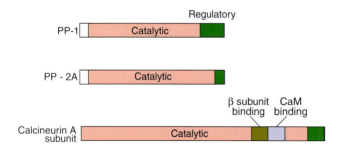

FIGURE 12.16 Domain structure of the catalytic subunits of some Ser/Thr phosphatases. The three major phosphoprotein phosphatases, PP–1, PP–2A, and calcineurin, have homologous catalytic domains but differ in their regulatory properties.

enriched in the brain. It is a heterodimer with a 60-kDa A subunit (CnA) that contains an N-terminal catalytic domain and a C-terminal regulatory domain that includes an autoinhibitory segment, a calmodulin-binding domain, and a binding site for the 19-kDa regulatory B subunit (CnB) (Rusnak and Mertz, 2000). CnB is a calmodulin-like Ca^{2+}-binding protein that binds to a hinge region of CnA. Regulation of calcineurin takes place in this region because it controls access of phosphoproteins to the catalytic site. Some activation of calcineurin is attained by binding of Ca^{2+} to CnB. Stronger activation is obtained by the binding of Ca^{2+}-camodulin.

The substrate specificity of calcineurin does not appear to be as broad as that of PP-1, and that of calcineurin and CaM kinase II has little overlap. Thus, a rise in Ca^{2+} does not lead to a futile cycle of phosphorylation and dephosphorylation by these Ca^{2+}-calmodulin-dependent enzymes. However, their Ca^{2+}-calmodulin sensitivity is quite different, and weak or low-frequency stimuli may selectively activate calcineurin, whereas strong or high-frequency stimuli activate CaM kinase II and calcineurin. This difference may play a role in bidirectional control of synaptic strength by low- and high-frequency stimulation (Lisman, 1994). (See also discussion of LTP and LTD in Chapter 18.)

Additional regulation may be accorded by interaction of this hinge region with cyclophilin and FKBP, proteins that bind the immunosuppressive agents cyclosporin and FK506, respectively. The FK506-binding protein, or FKBP, is highly abundant in the brain, and its distribution resembles that of calcineurin. Both FK506 and cyclosporin A are membrane permeant and are highly potent and selective inhibitors of calcineurin. They are referred to as immunophilins because their ability to block the essential role of calcineurin in lymphocyte activation makes them effective immunosuppressants. The X-ray structure of calcineurin complexed with FK506 reveals a ternary complex in which FK506 is bound at the interface between FKBP and the regulatory domain of CnA. (Griffith et al., 1995). Unlike calmodulin, CnB does not completely wrap around its target. CnB binds to one surface of an extended regulatory domain and FKBP–FK506 binds to the opposite surface. The FK506–FKBP complex is wedged between the regulatory domain and the catalytic site and likely inhibits calcineurin by making it difficult for phosphoproteins to have access to the catalytic site. It is unclear whether FKBP and calcineurin interact physiologically via a natural ligand that functions like FK506 to facilitate their interaction.

Protein Kinases, Protein Phosphatases, and Their Substrates Are Integrated Networks

Cross-talk between protein kinases and protein phosphatases is key to their ability to integrate inputs into neurons (Cohen, 1992). Such cross-talk is exemplified by the interaction of cyclic AMP and Ca^{2+} signals through PKA and calcineurin, respectively. The medium spiny neurons in the neostriatum receive cortical inputs from glutamatergic neurons that are excitatory and nigral inputs by dopaminergic neurons that inhibit them. A possible signal transduction scheme for this regulation is shown in Fig. 12.17. A more complex scheme is shown in Fig. 12.18 and described in Box 12.2. The key to the regulation is the bidirectional control of DARPP-32 phosphorylation (Greengard et al., 1998). Glutamate activates calcineurin by increasing intracellular Ca^{2+}, leading to the dephosphorylation and inactivation of phospho-DARPP-32. This releases inhibition of PP-1, which can then dephosphorylate a variety of substrates, including the Na^+, K^+-ATPase, and lead to membrane depolarization. This is countered by dopamine, which stimulates cAMP formation and activation of PKA, which then converts DARPP-32 into its phosphorylated (i.e., PP-1 inhibitory) state. Although PKA and calcineurin are acting in an antagonistic manner, they

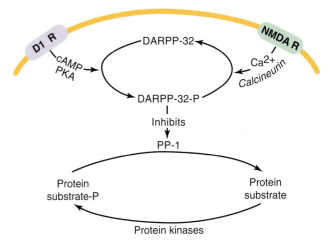

FIGURE 12.17 Cross-talk between kinases and phosphatases. The state of phosphorylation of protein substrates is regulated dynamically by protein kinases and phosphatases. In the striatum, for example, dopamine stimulates PKA, which converts DARPP–32 into an effective inhibitor of PP–1. This increases the steady-state level of phosphorylation of a hypothetical substrate subject to phosphorylation by a variety of protein kinases. This action can be countered by NMDA receptor stimulation by another stimulus that increases intracellular Ca^{2+} and activates calcineurin. PP–1 is deinhibited and dephosphorylates the phosphorylated substrate when calcineurin deactivates DARPP–32-P. Adapted from Greengard et al. (1998).

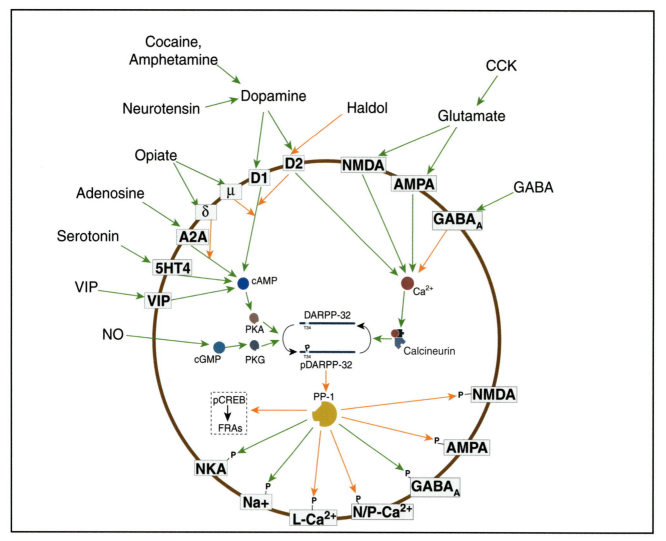

FIGURE 12.18 Signaling pathways in the neostriatum. Activation by dopamine of the D_1 subclass of dopamine receptors stimulates the phosphorylation of DARPP-32 at Thr-34. This is achieved through a pathway involving the activation of adenylyl cyclase, the formation of cAMP, and the activation of PKA. Activation by dopamine of the D_2 subclass of dopamine receptors causes the dephosphorylation of DARPP-32 through two synergistic mechanisms: D_2 receptor activation (i) prevents the D_1 receptor-induced increase in cyclic AMP formation and (ii) raises intracellular calcium, which activates a calcium-dependent protein phosphatase, namely, calcineurin, calcium/calmodulin-dependent protein phosphatase. Activated calcineurin dephosphorylates DARPP-32 at Thr-34. Glutamate acts as both a fast-acting and a slow-acting neurotransmitter. Activation by glutamate of AMPA receptors causes a rapid response through the influx of sodium ions, depolarization of the membrane, and firing of an action potential. Slow synaptic transmission, in response to glutamate, results in part from activation of the AMPA and NMDA subclasses of the glutamate receptor, which increases intracellular calcium and the activity of calcineurin, and causes the dephosphorylation of DARPP-32 on Thr-34. All other neurotransmitters that have been shown to act directly to alter the physiology of dopaminoceptive neurons also alter the phosphorylation state of DARPP-32 on Thr-34 through the indicated pathways. Neurotransmitters that act indirectly to affect the physiology of these dopaminoceptive neurons also regulate DARPP-32 phosphorylation; e.g., neurotensin, stimulating the release of dopamine, increases DARPP-32 phosphorylation; conversely, cholecystokinin (CCK), by stimulating the release of glutamate, decreases DARPP-32 phosphorylation. Anti-schizophrenic drugs and drugs of abuse, all of which affect the physiology of these neurons, also regulate the state of phosphorylation of DARPP-32 on Thr-34. For example, the anti-schizophrenic drug haloperidol (Haldol), which blocks the activation by dopamine of the D_2 subclass of dopamine receptors, increases DARPP-32 phosphorylation. Agonists for the μ and δ subclasses of opiate receptors block D_1 and A_{2A} receptor-mediated increases in cAMP, respectively, and block the resultant increases in DARPP-32 phosphorylation. Cocaine and amphetamine, by increasing extracellular dopamine levels, increase DARPP-32 phosphorylation. Marijuana, nicotine, alcohol, and LSD, all of which affect the physiology of the dopaminoceptive neurons, also regulate DARPP-32 phosphorylation. Finally, all drugs of abuse have greatly reduced biological effects in animals with targeted deletion of the DARPP-32 gene. 5HT4, 5-hydroxytryptophan (serotonin) receptor 4; NKA, Na^+, K^+-ATPase; VIP, vasoactive intestinal peptide; L- and N/P-Ca^{2+}, L-type and N/P-type calcium channels. From Greengard (1999).

BOX 12.2

INTERACTIONS OF SIGNAL TRANSDUCTION PATHWAYS IN THE BRAIN

An understanding of the signal transduction mechanisms by which neurotransmitters produce their effects on their target neurons and of the mechanisms by which coordination of various signal transduction pathways is achieved represents a major area of research in cellular neurobiology. The dopaminoceptive medium-sized spiny neurons, located in the neostriatum, have been studied in great detail with respect to these mechanisms. Figure 12.18 illustrates a portion of what is now known about interactions of signaling mechanisms in these neurons. Activation by dopamine of D_1 receptors increases cAMP, causing activation of PKA (cAMP-dependent protein kinase) and phosphorylation of DARPP-32 (dopamine + cAMP-regulated phosphoprotein, M_r 32,000) on threonine-34. Conversely, glutamate, acting on NMDA receptors, increases $[Ca^{2+}]_i$, leading to the activation of calcineurin (protein phosphatase 2B, PP-2B) and dephosphorylation of phosphothreonine-34–DARPP-32. Neurotensin, VIP, NO (nitric oxide), and some other neurotransmitters increase the phosphorylation of DARPP-32 through a variety of signaling mechanisms. Dopamine (acting on D_2 receptors), CCK, GABA, and some other neurotransmitters decrease the state of phosphorylation of DARPP-32 through a variety of other signaling mechanisms. CK1 (casein kinase I) and CK2 (casein kinase II)

phosphorylate DARPP-32 on residues other than threonine-34, causing it to undergo conformational changes. These changes result in phosphothreonine-34–DARPP-32 becoming a poorer substrate for calcineurin (in the case of CKI) or a better substrate for PKA (in the case of CKII). Antipsychotic drugs such as haloperidol (Haldol) increase the state of phosphorylation of DARPP-32 by blocking the dopamine-induced D_2 receptor-mediated activation of calcineurin.

The physiological consequences of phosphorylation of DARPP-32 on threonine-34 are profound. Thus, DARPP-32 in its threonine-34 phosphorylated, but not dephosphorylated, form acts as a potent inhibitor of PP-1 (protein phosphatase 1). PP-1 is a major serine–threonine protein phosphatase, which controls the state of phosphorylation of a variety of phosphoprotein substrates in the brain. These substrates include Na^+ channels; L-, N-, and P-type Ca^{2+} channels; the electrogenic ion pump Na^+, K^+-ATPase; the NR-1 subclass of glutamate receptors; and probably many more.

In summary, the DARPP-32/PP-1 cascade provides a mechanism by which a large number of neurotransmitters act in a complex, but coordinated, fashion to regulate the state of phosphorylation and activity of a variety of ion channels, ion pumps, and neurotransmitter receptors.

Adapted from Greengard *et al.* (1998).

---

are not doing it by phosphorylating and dephosphorylating the ATPase. By their actions upstream, at the level of DARPP-32, the regulation of numerous target enzymes (e.g., Ca^{2+} channels and Na^+ channels) in addition to the ATPase can be coordinated.

Studying Cellular Processes Controlled by Phosphorylation–Dephosphorylation Requires a Set of Criteria

Major goals of signal transduction research are to delineate pathways by which signals such as neurotransmitters transduce their signals, usually across the plasma membrane, and to determine how the transduced signal is transmitted to the ultimate cellular components that are to be modified by the extracellular signal. Many signaling components are yet to be discovered; in other circumstances, which of the

known pathways are used by a given physiological stimulus is unclear. For the transmission of the signal, the question is often: Which kinase(s) and which phosphatase(s) are responsible for the phosphorylation? Cellular and biochemical assays can often identify the entire signaling pathway, from stimulation of receptor to generation of a second messenger activation of a kinase or phosphatase, change in the phosphorylation state of the substrate, and an ultimate change in its functional state. Such investigations use a variety of pharmacological inhibitors or activators of the signaling molecules complemented by genetic approaches that use transfection of activated forms of the kinases or phosphatases in question, transgenic animals, and mice with individual signaling components knocked out. Clearly, these pathways are complex and even a complete knowledge of the components may not lead to an understanding of the function of the network. Consequently, as described in Chapter 14, cellular and biochemical assays and approaches need to be complemented with the techniques of mathematical analyses and modeling and computer simulation.

Summary

A cell's morphology is determined by protein constituents. Its function is regulated by the phosphorylation or dephosphorylation of the proteins. Phosphorylation modifies the function of regulatory proteins subsequent to their genetic expression. The activities of the protein kinases and protein phosphatases are typically regulated by second messengers and extracellular ligands. Kinases and phosphatases integrate and encode stimulation of a large group of cellular receptors. The number of possible effects is almost limitless and enables the tuning of cellular processes over a broad time scale. The kinases that regulate phosphorylation can exhibit conformational changes that potentiate the activity of the kinase. This may be one of the key elements in molecular memory and neuronal plasticity. Most of the effects of Ca^{2+} in cells are mediated by calmodulin, which in turn mediates changes in protein phosphorylation–dephosphorylation. The phosphoinositol signaling system is mediated through PKC, which modulates many cellular processes from exocytosis to gene expression. All three classes of enzymes discussed have been described as cognitive kinases because they are capable of sustaining their activated states after their second messenger stimuli have returned to basal levels. PKA has been implicated in learning and memory in *Aplysia* and in hippocampus, where it is involved in long-term potentiation. Protein phosphatases play an equally important role in neuronal signaling by dephosphorylating proteins. Cross-talk between protein kinases and protein phosphatases is key to their ability to integrate inputs into neurons. A major effort of signal transduction research is to delineate the pathways through which the neurotransmitters' signals across the plasma membrane are transmitted to the ultimate cellular components to be modified.

References

Allbritton, N. L., Meyer, T., and Stryer, L. (1992). Range of messenger action of calcium ion and inositol 1,4,5-trisphosphate. *Science* **258**, 1812–1815.

Bacskai, B. J., Hochner, B., Mahaut-Smith, M., Adams, S. R., Kaang, B. K., Kandel, E. R., and Tsien, R. Y. (1993). Spatially resolved dynamics of cAMP and protein kinase A subunits in *Aplysia* sensory neurons. *Science* **260**, 222–226.

Baranano, D. E., Ferris, C. D., and Snyder, S. H. (2001). Atypical neural messengers. *Trends Neurosci.* **24**, 99–106.

Bayer, K. -U., De Koninck, P., Leonard, A. S., Hell, J. W., and Schulman, H. (2001). Interaction with the NMDA receptor locks CaMKII in an active conformation. *Nature* **411**, 801–805.

Baylor, D. (1996). How photons start vision. *Proc. Natl. Acad. Sci. USA* **93**, 560–565.

Berridge, M. J. (1993). Inositol trisphosphate and calcium signalling. *Nature* **361**, 315–325.

Bootman, M. D., Berridge, M. J., and Lipp, P. (1997). Cooking with calcium: The recipes for composing global signals from elementary events. *Cell* **91**, 367–373.

Bourne, H. R., and Nicoll, R. (1993). Molecular machines integrate coincident synaptic signals. *Cell* **72**, 65–75.

Bourne, H. R., Sanders, D. A., and McCormick, F. (1990). The GTPase superfamily: A conserved switch for diverse cell functions. *Nature* **348**, 125–132.

Braun, A. P., and Schulman, H. (1995). The multifunctional calcium/calmodulin-dependent protein kinase: From form to function. *Annu. Rev. Physiol.* **57**, 417–445.

Carafoli, E., and Klee, C. (1999). *"Calcium as a Cellular Regulator"*. Oxford Univ. Press, New York.

Cassel, D., and Selinger, Z. (1978). Mechanism of adenylyl cyclase activation through the β-adrenergic receptor: Catecholamine-induced displacement of bound GDP by GTP. *Proc. Natl. Acad. Sci. USA.* **75**, 4155–4159.

Chain, D. G., Casadio, A., Schacher, S., Hegde, A. N., Valbrun, M., Yamamoto, N., Goldberg, A. L., Bartsch, D., Kandel, E. R., and Schwartz, J. H. (1999). Mechanisms for generating the autonomous cAMP-dependent protein kinase required for long-term facilitation in *Aplysia*. *Neuron* **22**, 147–156.

Clapham, D. E. (1995). Calcium signaling. *Cell* **80**, 259–268.

Clapham, D. E., and Neer, E. J. (1993). New roles for G-protein βγ-dimers in transmembrane signalling. *Nature* **365**, 403–406.

Cohen, P. (1992). Signal integration at the level of protein kinases, protein phosphatases and their substrates. *Trends Biochem. Sci.* **17**, 408–413.

Cohen, P., and Klee, C. B. (1988). Calmodulin. *In* "Molecular Aspects of Cellular Regulation," Vol. 5. Elsevier, Amsterdam.

Coleman, D. E., Berghuis, A. M., Lee, E., Linder, M. E., Gilman, A. G., and Sprang, S. R. (1994). Structures of active conformations of $G_{i\alpha 1}$ and the mechanism of GTP hydrolysis. *Science* **265**, 1405–1412.

Conklin, R. B., Farfel, Z., Lustig, K. D., Julius, D., and Bourne, H. R. (1993). Substitution of three amino acids switches receptor specificity of $G_{q\alpha}$ to that of $G_{i\alpha}$. *Nature* **369**, 274–276.

De Koninck, P., and Schulman, H. (1998). Sensitivity of Ca^{2+}/calmodulin-dependent protein kinase II to the frequency of Ca^{2+} oscillations. *Science* **279**, 227–230.

Delcour, A. H., and Tsien, R. W. (1993). Altered prevalence of gating modes in neurotransmitter inhibition of N-type calcium channels. *Science* **259**, 980–984.

Divecha, N., and Irvine, R. F. (1995). Phospholipid signaling. *Cell* **80**, 269–278.

Edwards, A. S., and Scott, J. D. (2000). A-kinase anchoring proteins: Protein kinase A and beyond. *Curr. Opin. Cell Biol.* **12**, 217–221.

Federman, A. D., Conklin, B. R., Schrader, K. A., Reed, R. R., and Bourne, H. R. (1992). Hormonal stimulation of adenylyl cyclase through G_i-protein βγ subunits. *Nature* **356**, 159–161.

Ferrell, J. E., Jr., and Machleder, G. M. (1998). The biochemical basis of an all-or-none switch in *Xenopus* oocytes. *Science* **280**, 895–898.

Furchgott, R. F., and Zawadzki, J. V. (1980). The obligatory role of the endothelial cells in the relaxation of arterial smooth muscle by acetylcholine. *Nature* **288**, 373–376.

Garbers, D. L., and Lowe, D. G. (1994). Guanylyl cyclase receptors. *J. Biol. Chem.* **269**, 30741–30744.

Giese, K. P., Fedorov, N. B., Filipkowski, R. K., and Silva, A. J. (1998). Autophosphorylation at Thr^{286} of the α calcium-calmodulin kinase II in LTP and learning. *Science* **279**, 870–873.

Goldberg, J., Huang, H., Kwon, Y., Greengard, P., Nairn, A. C., and Kuriyan, J. (1995). Three-dimensional structure of the catalytic subunit of protein serine/threonine phosphatase-1. *Nature* **376**, 745–753.

Greengard, P. (2001). The neurobiology of slow synaptic transmission. *Science* **294**, 1024–1030.

Greengard, P., Allen, P. B., and Nairn, A. C. (1999). Beyond the dopamine receptor: the DARPP-32/protein phosphatase-1 cascade. *Neuron* **23**, 435–447.

Greengard, P., Nairn, A. C., Girault, J. -A., Quimet, C. C., Snyder, G. L., Fisone, G., Allen, P. B., Fienberg, A., and Nishi, A. (1998). The DARPP–32/protein phosphatase–1 cascade: A model for signal integration. *Brain Res. Rev.* **26**, 274 284.

Griffith, J. P., Kim, J. L., Kim, E. E., Sintchak, M. D., Thomson, J. A., Fitzgibbon, M. J., Fleming, M. A., Caron, P. R., Hsiao, K., and Navia, M. A. (1995). X-ray structure of calcineurin inhibited by the immunophilin-immunosuppressant FKBP12-FK506 complex. *Cell* **82**, 507–522.

Hanks, S. K., and Hunter, T. (1995). Protein kinases 6. The eukaryotic protein kinase superfamily: Kinase (catalytic) domain structure and classification. *FASEB J.* **9**, 576–596.

Hanoune, J., and Defer, N. (2001). Regulation and role of adenylyl cyclase isoforms. *Annu. Rev. Pharmacol. Toxicol.* **41**, 145–174.

Hanson, P. I., and Schulman, H. (1992). Neuronal Ca^{2+}/calmodulin-dependent protein kinases. *Annu. Rev. Biochem.* **61**, 559–601.

Hanson, P. I., Kapiloff, M. S., Lou, L. L., Rosenfeld, M. G., and Schulman, H. (1989). Expression of a multifunctional Ca^{2+}/calmodulin-dependent protein kinase and mutational analysis of its autoregulation. *Neuron* **3**, 59–70.

Hanson, P. I., Meyer, T., Stryer, L., and Schulman, H. (1994). Dual role of calmodulin in autophosphorylation of multifunctional CaM kinase may underlie decoding of calcium signals. *Neuron* **12**, 943–956.

Hayashi, Y., Shi, S. H., Esteban, J. A., Piccini, A., Poncer, J. C., and Malinow, R. (2000). Driving AMPA receptors into synapses by LTP and CaMKII: Requirement for GluR1 and PDZ domain interaction. *Science* **287**, 2262–2267.

Hemmings, H. C., Jr., Greengard, P., Tung, H. Y., and Cohen, P. (1984). DARPP-32, a dopamine-regulated neuronal phosphoprotein, is a potent inhibitor of protein phosphatase–1. *Nature* **310**, 503–505.

Hille, B. (1992). G protein-coupled mechanisms and nervous signaling. *Neuron* **9**, 187–195.

Hoelz, A., Nairn, A. C., and Kuriyan, J. (2003). Crystal structure of tetradecameric assembly of Ca^{2+}/calmodulin-dependent kinase II. *Mol. Cell* **11**, 1241–1251.

Hokin, L. E., and Hokin, M. R. (1955). Effects of acetylcholine on the turnover of phosphoryl units in individual phospholipids of pancreas slices and brain cortex slices. *Biochim. Biophys. Acta.* **18**, 102–110.

Huang, C. -L., Slesinger, P. A., Casey, P. J., Jan, Y. N., and Jan, L. Y. (1995). Evidence that direct binding of G$_{\beta\gamma}$ to the GIRK1G protein-gated inwardly rectifying K$^+$ channel is important for channel activation. *Neuron* **15**, 1133–1143.

Huang, Y. Y., Li, X. C., and Kandel, E. R. (1994). cAMP contributes to mossy fiber LTP by initiating both a covalently mediated early phase and macromolecular synthesis-dependent late phase. *Cell* **79**, 69–79.

Hubbard, M. J., and Cohen, P. (1993). On target with a new mechanism for the regulation of protein phosphorylation. *Trends Biochem. Sci.* **18**, 172–177.

Hunter, T. (1995). Protein kinases and phosphatases: The yin and yang of protein phosphorylation and signaling. *Cell* **80**, 225–236.

Ikura, M., Clore, G. M., Fronenborn, A. M., Zhu, G., Klee, C. B., and Bax, A. (1992). Solution structure of a calmodulin—target peptide complex by multidimensional NMR. *Science* **256**, 632–638.

Kaang, B. K., Kandel, E. R., and Grant, S. G. (1993). Activation of cAMP-responsive genes by stimuli that produce long-term facilitation in *Aplysia* sensory neurons. *Neuron* **10**, 427–435.

Kemp, B. E., Faux, M. C., Means, A. R., House, C., Tiganis, T., Hu, S. -H., and Mitchelhill, K. I. (1994). Structural aspects: pseudosubstrate and substrate interactions. *In* "Protein Kinases" (J. R. Woodgett, Ed.), pp. 30–67. Oxford Univ. Press.

Kennelly, P. J., and Krebs, E. G. (1991). Consensus sequences as substrate specificity determinants for protein kinases and protein phosphatases. *J. Biol. Chem.* **266**, 15555–15558.

Klauck, T. M., Faux, M. C., Labudda, K., Langeberg, L. K., Jaken, S., and Scott, J. D. (1996). Coordination of three signaling enzymes by AKAP79, a mammalian scaffold protein. *Science* **271**, 1589–1592.

Kleuss, C., Scherübl, H., Hescheler, J., Schultz, G., and Wittig, B. (1992). Different β-subunits determine G-protein interaction with transmembrane receptors. *Nature* **358**, 424–426.

Kleuss, C., Scherübl, H., Hescheler, J., Schultz, G., and Wittig, B. (1993). Selectivity in signal transduction determined by γ subunits of heterotrimeric G proteins. *Science* **259**, 832–834.

Knighton, D. R., Zheng, J., Eyck, L. F. T., Xuong, N., Taylor, S. S., and Sowadski, J. M. (1991). Structure of a peptide inhibitor bound to the catalytic subunit of cyclic adenosine monophosphate-dependent protein kinase. *Science* **253**, 414–420.

Kolodziej, S. J., Hudmon, A., Waxham, M. N., and Stoops, J. K. (2000). Three-dimensional reconstructions of calcium/calmodulin-dependent (CaM) kinase IIα and truncated CaM kinase IIα reveal a unique organization for its structural core and functional domains. *J. Biol. Chem.* **275**, 14354–14359.

Kraft, A. S., and Anderson, W. B. (1983). Phorbol esters increase the amount of Ca^{2+}, phospholipid-dependent protein kinase associated with plasma membrane. *Nature* **301**, 621–623.

Krupinski, J., Coussen, F., Bakalyar, H. A., Tang, W. -J., Feinstein, P. G., Orth, K., Slaughter, C., Reed, R. R., and Gilman, A. G. (1989). Adenylyl cyclase amino acid sequence: possible channel- or transporter-like structure. *Science* **244**, 1558–1564.

Levin, L. R., Han, P. -L., Hwang, P. M., Feinstein, P. G., Davis, R. L., and Reed, R. R. (1992). The Drosophila learning and memory gene rutabaga encodes a Ca^{2+}/calmodulin-responsive adenylyl cyclase. *Cell* **68**, 479–489.

Lisman, J. (1994). The CaM kinase II hypothesis for the storage of synaptic memory. *Trends Neurosci* **17**, 406–412.

Lisman, J., Schulman, H., and Clive, H. (2002). The molecular basis of CMRII function in synaptic and behavioural memory. *Nat. Rev. Neurosci.* **71**, 175–190.

Logothetis, D. E., Kurachi, Y., Galper, J., Neer, E. J., and Clapham, D. E. (1987). The $\beta\gamma$ subunits of GTP-binding proteins activate the muscarinic K$^+$ channel in heart. *Nature* **325**, 321–326.

Miller, S. G., and Kennedy, M. B. (1986). Regulation of brain type II Ca^{2+}/calmodulin-dependent protein kinase by autophosphorylation: A Ca^{2+}-triggered switch. *Cell* **44**, 861–870.

Minta, A., and Tsien, R. Y. (1989). Fluorescent indicators for cytosolic calcium based on rhodamine and fluorescein chromophores. *J. Biol. Chem.* **264**, 8171–8178.

Mochly-Rosen, D. (1995). Localization of protein kinases by anchoring proteins: A theme in signal transduction. *Science* **268**, 247–251.

Muramatsu, M., Kaibuchi, K., and Arai, K. (1989). A protein kinase C cDNA without the regulatory domain is active after transfection *in vivo* in the absence of phorbol ester. *Mol. Cell Biol.* **9**, 831–836.

Nairn, A. C., Hemmings, H. C., Jr., and Greengard, P. (1985). Protein kinases in the brain. *Annu. Rev. Biochem.* **54**, 931–976.

Neer, E. J. (1995). Heterotrimeric G proteins: Organizers of transmembrane signals. *Cell* **80**, 249–257.

Newton, A. C. (1995). Protein kinase C: Structure, function, and regulation. *J. Biol. Chem.* **270**, 28495–28498.

Niemeyer, B. A., Suzuki, E., Scott, K., Jalink, K., and Zuker, C. S. (1996). The Drosophila light-activated conductance is composed of the two channels TRP and TRPL. *Cell* **85**, 651–659.

Nishizuka, Y. (1995). Protein kinase C and lipid signaling for sustained cellular responses. *FASEB J.* **9**, 484–496.

Northup, J. K., Smigel, M. D., Sternweis, P. C., and Gilman, A. G. (1983). The subunits of the stimulatory regulatory component of adenylyl cyclase: Resolution of the 45,000-dalton α subunit. *J. Biol. Chem.* **258**, 11369–11376.

Palczewski, K., Kumasaka, T., Hori, T., Behnke, C. A., Motoshima, H., Fox, B. A., Trong, I. L., Teller, D. C., Okada, T., Stenkamp, R. E., Yamamoto, M., and Miyano, M. (2000). Crystal structure of rhodopsin: A G protein-coupled receptor. *Science* **289**, 739–745.

Persechini, A., and Cronk, B. (1999). The relationship between the free concentrations of Ca²⁺ and Ca²⁺-calmodulin in intact cells. *J. Biol. Chem.* **274**, 6827–6830.

Price, N. E., and Mumby, M. C. (1999). Brain protein serine/threonine phosphatases. *Curr. Opin. Neurobiol.* **9**, 336–342.

Rameh, L. E., and Cantley, L. C. (1999). The role of phosphoinositide 3-kinase lipid products in cell function. *J. Biol. Chem.* **274**, 8347–8350.

Ron, D., Chen, C. H., Caldwell, J., Jamieson, L., Orr, E., and Mochly-Rosen, D. (1994). Cloning of an intracellular receptor for protein kinase C: A homolog of the beta subunit of G proteins. *Proc. Natl. Acad. Sci. USA* **91**, 839–843.

Rosenmund, C., Carr, D. W., Bergeson, S. E., Nilaver, G., Scott, J. D., and Westbrook, G. L. (1994). Anchoring of protein kinase A is required for modulation of AMPA/kainate receptors on hippocampal neurons. *Nature* **368**, 853–856.

Ross, E. M. (1989). Signal sorting and amplification through G protein-coupled receptors. *Neuron* **3**, 141–152.

Rusnak, F., and Mertz, P. (2000). Calcineurin: Form and function. *Physiol. Rev.* **80**, 1483–1521.

Sacktor, T. C., Osten, P., Valsamis, H., Jiang, X., Naik, M. U., and Sublette, E. (1993). Persistent activation of the ζ isoform of protein kinase C in the maintenance of long-term potentiation. *Proc. Natl. Acad. Sci. USA* **90**, 8342–8346.

Saitoh, T., and Schwartz, J. H. (1985). Phosphorylation-dependent subcellular translocation of a Ca²⁺/calmodulin-dependent protein kinase produces an autonomous enzyme in Aplysia neurons. *J. Cell Biol.* **100**, 835–842.

Schmidt, H. H., and Walter, U. (1994). NO at work. *Cell* **78**, 919–925.

Schulman, H., and Braun, A. (1999). Ca²⁺/calmodulin-dependent protein kinases. *In* "Calcium as a Cellular Regulator" (E. Carafoli and C. Klee, Eds.), pp. 311–343. Oxford Univ. Press, New York.

Schuman, E. M., and Madison, D. V. (1994). Locally distributed synaptic potentiation in the hippocampus. *Science* **263**, 532–536.

Schwartz, J. H. (1993). Cognitive kinases. *Proc. Natl. Acad. Sci. USA.* **90**, 8310–8313.

Scott, J. D. (1991). Cyclic nucleotide-dependent protein kinases. *Pharmacol. Ther.* **50**, 123–145.

Smith, J. A., Poteet-Smith, C. E., Lannigans, D. A., Freed, T. A., Zoltosk, A. J., and Sturgill, T. W. (2000). Creation of a stress-activated p90 ribosomal S6 kinase: The carboxyl-terminal tail of the MAPK-activated protein kinases dictates the signal transduction pathway in which they function. *J. Biol. Chem.* **275**, 31588–31593.

Sondek, J., Bohm, A., Lambright, D. G., Hamm, H. E., and Sigler, P. B. (1996). Crystal structure of a G protein βγ dimer at 2.1 Å resolution. *Nature* **379**, 369–374.

Sondek, J., Lambright, D. G., Noel, J. P., Hamm, H. E., and Sigler, P. B. (1994). GTPase mechanism of G proteins from the 1.7 Å crystal structure of transducin α-GDP-AlF₄⁻. *Nature* **372**, 276–279.

Streb, H., Irvine, R. F., Berridge, M. J., and Schulz, I. (1983). Release of Ca²⁺ from a nonmitochondrial intracellular store in pancreatic acinar cells by inositol-1,4,5-trisphosphate. *Nature* **306**, 67–69.

Stryer, L. (1991). Visual excitation and recovery. *J. Biol. Chem.* **266**, 10711–10714.

Stryer, L., and Bourne, H. R. (1986). G proteins: A family of signal transducers. *Ann. Rev. Cell Biol.* **2**, 391–419.

Su, Y., Dostmann, W. R., Herberg, F. W., Durick, K., Xuong, N. H., Ten Eyck, L., Taylor, S. S., and Varughese, K. I. (1995). Regulatory subunit of protein kinase A: Structure of deletion mutant with cAMP binding domains. *Science* **269**, 807–813.

Takai, Y., Kishimoto, A., Kikkawa, U., Mori, T., and Nishizuka, Y. (1979). Unsaturated diacylglycerol as a possible messenger for the activation of calcium-activated, phospholipid-dependent protein kinase system. *Biochem. Biophys. Res. Commun.* **91**, 1218–1224.

Tanaka, C., and Nishizuka, Y. (1994). The protein kinase C family for neuronal signaling. *Annu. Rev. Neurosci.* **17**, 551–567.

Tang, W. -J., and Gilman, A. G. (1991). Type-specific regulation of adenylyl cyclase by G protein βγ subunits. *Science* **254**, 1500–1503.

Taussig, R., and Gilman, A. G. (1995). Mammalian membrane-bound adenylyl cyclases. *J. Biol. Chem.* **270**, 1–4.

Tsien, R. W., and Tsien, R. Y. (1990). Calcium channels, stores, and oscillations. *Annu. Rev. Cell. Biol.* **6**, 715–760.

Venter, J. C., Adams, M. D., Myers, E. W., Li, P. W., Mural, R. J., Sutton, G. G., Smith, H. O., Yandell, M., Evans, C. A., Holt, R. A., Gocayne, J. D., Amanatides, P., Ballew, R. M., Huson, D. H., *et al.* (2001). The sequence of the human genome. *Science* **291**, 1304–1351.

Wall, M. A., Coleman, D. E., Lee, E., Iniguez-Lluhi, J. A., Posner, B. A., Gilman, A. G., and Sprang, S. R. (1996). The structure of the G protein heterotrimer Gᵢα₁β₁γ₂ *Cell* **83**, 1047–1058.

Walsh, D. A., and Patten, S. M. V. (1994). Multiple pathway signal transduction by the cAMP-dependent protein kinase. *FASEB J.* **8**, 1227–1236.

Wayman, G. A., Impey, S., Wu, Z., Kindsvogel, W., Prichard, L., and Storm, D. R. (1994). Synergistic activation of the type I adenylyl cyclase by Ca²⁺ and Gₛ-coupled receptors *in vivo*. *J. Biol. Chem.* **269**, 25400–25405.

Wen, W., Meinkoth, J. L., Tsien, R. Y., and Taylor, S. S. (1995). Identification of a signal for rapid export of proteins from the nucleus. *Cell* **82**, 463–473.

Westphal, R. S., Tavalin, S. J., Lin, J. W., Alto, N. M., Fraser, I. D., Langeberg, L. K., Sheng, M., and Scott, J. D. (1999). Regulation of NMDA receptors by an associated phosphatase–kinase signaling complex. *Science* **285**, 93–96.

Wickman, K. D., Iñiguez-Lluhi, J. A., Davenport, P. A., Taussig, R., Krapivinsky, G. B., Linder, M. E., Gilman, A. G., and Clapham, D. E. (1994). Recombinant G-protein βγ-subunits activate the muscarinic-gated atrial potassium channel. *Nature* **368**, 255–257.

Wu, Z. L., Thomas, S. A., Villacres, E. C., Xia, Z., Simmons, M. L., Chavkin, C., Palmiter, R. D., and Storm, D. R. (1995). Altered behavior and long-term potentiation in type I adenylyl cyclase mutant mice. *Proc. Natl. Acad. Sci. USA* **92**, 220–224.

Yue, L., Peng, J. -B., Hediger, M. A., and Clapham, D. E. (2001). CaT1 manifests the pore properties of the calcium-release-activated calcium channel. *Nature* **410**, 705–709.

Zhang, G., Kazanietz, M. G., Blumberg, P. M., and Hurley, J. H. (1995). Crystal structure of the cys2 activator-binding domain of protein kinase C delta in complex with phorbol ester. *Cell* **81**, 917–924.

Regulation of Neuronal Gene Expression and Protein Synthesis

James L. Roberts and James R. Lundblad

INTRACELLULAR SIGNALING AFFECTS NUCLEAR GENE EXPRESSION

The previous chapter describes how signaling systems regulate the function of cellular proteins already expressed; another critical level of control exerted by these systems is their ability to regulate the synthesis of cellular proteins by regulating the expression of specific genes. For all living cells, regulation of gene expression by intracellular signals is a fundamental mechanism of development, homeostasis, and adaptation to the environment. Protein phosphorylation and regulation of gene expression by intracellular signals are the most important mechanisms underlying the remarkable degree of plasticity exhibited by neurons. Alterations in gene expression underlie many forms of long-term changes in neural functioning, with a time course that ranges from hours to many years. Indeed, as discussed earlier and in Chapter 18, evidence now suggests that formation of long-term memories in many neural systems requires changes in gene expression and new protein synthesis. In addition to regulation of the levels of synthesis of new mRNAs from specific genes, changes in messenger RNA turnover and/or changes in the efficiency of synthesis of new proteins from a specific messenger RNA, both processes that result in altered levels of expression of a specific gene product, are observed.

Interactions of Specific DNA Sequences with Regulatory Proteins Control Both Basal and Signal-Regulated Transcription

The double helix of DNA is an ideal molecule for the storage of information. DNA is a stable, linear polymer, and because of the properties of complementarity between the strands of the double helix, can be readily replicated or serve as the template for the synthesis of RNA. DNA and RNA polymerases processing down its length can add a succession of nucleotides complementary to those in the template strand. However, its chemical simplicity and relatively rigid helical structure limit its functions in the cell to information storage and transfer. The information contained within DNA is therefore expressed through other molecules: RNA and proteins. The human genome contains approximately 30,000 to 40,000 genes that encode structural RNAs or protein-coding messenger RNAs (mRNAs) (International Human Genome Sequencing Consortium, 2001). Within genes, a fundamental distinction can be made between DNA sequences that code for RNAs—and, in the case of protein-coding genes, mRNAs that will eventually be translated—and DNA sequences that exert control functions. Certain control sequences determine the beginnings and ends of segments of DNA that can be transcribed into RNA. Other closely linked DNA sequences determine whether a potentially transcribed segment is actually transcribed in a particular cell and, if so, in what circumstances. In contrast to prokaryotic organisms, the genomic DNA of eukaryotes is packaged with histones into a nucleoprotein complex to form chromatin (Workman and Kingston, 1998). As a consequence of condensing the genome into the confines of the nucleus, chromatin plays an essential role in regulation of the dynamic range of gene expression, in part by limiting the accessibility of DNA to the transcriptional machinery. Regulated gene expression conferred by the nucleotide sequence of the DNA itself is called *cis*-regulation because the control regions are physically

Copyright 2004, Elsevier Science (USA).
All rights reserved.

linked on the DNA to the regions that can potentially be transcribed. The control sequences of genes influence the structure of chromatin by serving as high-affinity binding sites for regulatory proteins called transcription factors (or *trans*-acting factors because they may be encoded anywhere in the genome rather than on the same stretch of DNA that they regulate). These DNA-bound transcription factors then contact basal transcription factors either directly or indirectly through co-regulatory factors, or recruit enzyme complexes that modify the accessibility of the DNA in chromatin to the transcriptional machinery.

A multiprotein enzyme complex called RNA polymerase II carries out the transcription of protein-encoding genes into mRNA. This process is often divided into three steps: initiation of RNA synthesis, RNA chain elongation, and chain termination (Fig. 13.1) (Lee and Young, 2000). Although biologically significant regulation may occur at any of these steps, it is at the step of transcription initiation at which extracellular signals, such as neurotransmitters, hormones, drugs, and growth factors, exert their most significant control over the processes that gate the flow of information out of the genome.

The basic machinery for the synthesis of mRNA is largely common to all RNA polymerase II-transcribed genes; thus, additional layers of control permit the specificity that ensures appropriate temporal and conditional regulation of transcription. Controlling this process at the step of transcription initiation accomplishes this degree of specificity by (1) positioning RNA polymerase II at the correct start site of the gene to be transcribed and (2) controlling the efficiency of the initiation of RNA synthesis to produce the appropriate transcriptional rate for the circumstances of the cell (Tjian and Maniatis, 1994). The *cis*-regulatory elements that set the transcription start sites of genes are called *basal promoter elements* (Smale, 1994). Other *cis*-regulatory elements tether additional activator and repressor proteins to the DNA to regulate the overall transcriptional rate (Ptashne and Gann, 1997).

The Basal Promoter

The promoter regions of genes transcribed by RNA polymerase II contain distinct DNA sequences that function as basal promoter elements on which the basal transcription complex is assembled (Smale, 1994; Butler and Kadonaga, 2002). This complex is composed of a set of proteins, some of which recognize the specific DNA structural elements and some of which bind and position the RNA polymerase II. In contrast to other *cis*-regulatory sequences discussed in

this chapter that confer tissue-specific, neurotransmitter, or hormonal regulation, the basal promoter determines the site of transcription initiation and the direction of RNA synthesis (Smale, 2001). Although specific promoter architecture varies considerably, the core elements of the basal promoter are largely conserved among genes (Fig. 13.2) The basal promoter of most of these genes contains a sequence rich in the nucleotides adenine (A) and thymine (T) located between 25 and 30 bases upstream of the transcription start site, called the *TATA box* (Breathnach and Chambon, 1981). This sequence binds the transcription factor TATA-binding protein (TBP), a component of the TFIID multiprotein complex (Burley and Roeder, 1997). Another sequence located just upstream of the TATA box binds TFIIB in the complex with TBP and the TATA element (Lagrange *et al.*, 1998). Certain RNA polymerase II-transcribed promoters, most commonly those controlling "housekeeping genes," lack a TATA box; in such cases, cytosine–guanosine (CpG)-rich nucleotide sequences may stand in for the TATA box (Smale and Baltimore, 1989; Smale, 1997). For many promoters, the pyrimidine-rich Initiator sequence (Inr), overlapping the transcription initiation site (-3 to $+5$), serves as a recognition site for the TFIID associated factors $TAF_{II}250$ and $TAF_{II}150$ (Chalkley and Verrijzer, 1999) and RNA polymerase II (Weis and Reinberg, 1997). Other sequences sometimes present include the CCAAT box, located 70 to 80 bases upstream of the start site (McKnight and Kingsbury, 1982; Graves *et al.*, 1986), and a downstream promoter element (DPE) contacted by components of TFIID (Burke *et al.*, 1998; Butler and Kadonaga, 2002). The presence and combination of these elements determine the relative strength of the basal promoter. The DNA elements of the promoter sequentially recruit the basal transcription factors, beginning with the binding of TFIID to the TATA element (Burley and Roeder, 1997). These factors initiate events required for positioning the transcription initiation site, promoter melting, DNA unwinding, and the early events in the transition between transcription initiation and promoter clearance and elongation by the RNA polymerase complex (Orphanides *et al.*, 1996).

Sequence-Specific Transcription Factors

The basal transcription apparatus is not adequate to confer specificity of expression from a particular promoter or initiate more than low levels of transcription. To achieve the extremely broad dynamic range of regulated expression observed *in vivo*, this multiprotein assembly requires help from transcriptional

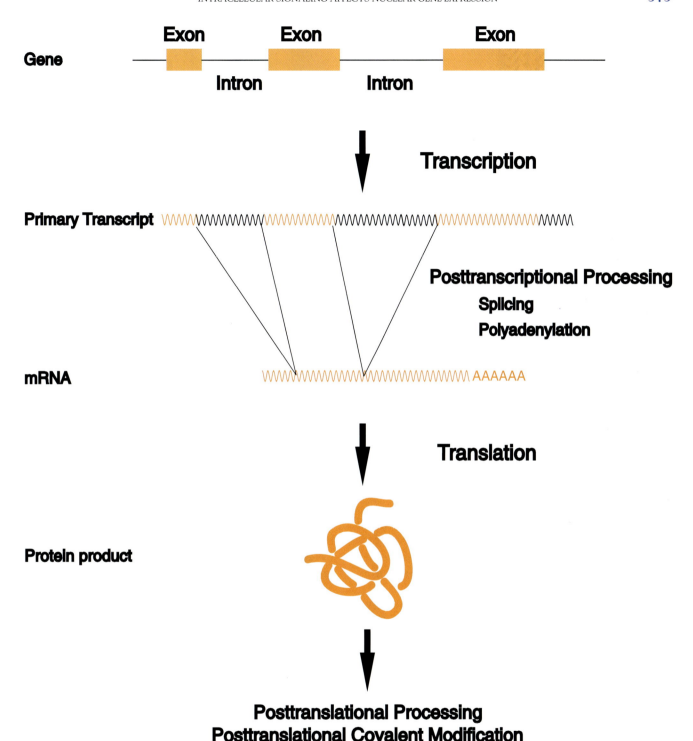

FIGURE 13.1 Expression of protein encoding genes. Transcription of protein encoding genes is mediated by RNA polymerase II. The primary RNA transcript undergoes posttranscriptional processing, including splicing of intron sequences and 3′ polyadenylation, to generate the mature messenger RNA (mRNA). Splicing, mRNA stability, and turnover are also important processes targeted by regulatory mechanisms. The function of the protein product of a gene may also be regulated by posttranslational proteolytic processing, or covalent modifications.

A

B

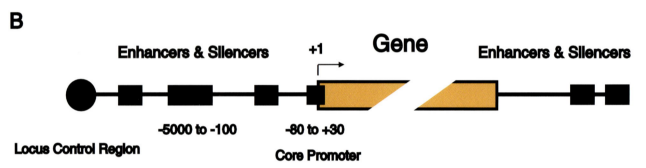

FIGURE 13.2   Promoter architecture. A Core promoter elements. Promoters may contain virtually any combination of these elements, although some promoters lack even the TATA sequence, the initiator (Inr), and downstream promoter elements (DPE). B Structure of the transcriptional regulatory region of a gene. Enhancer and silencer elements may be located at a distance flanking the structural gene sequence itself, or even within the transcribed portion of a gene. Other higher levels of regulation over larger domains of the genome are exerted by genetic control elements such as locus control regions.

activators and repressors, proteins that recognize and bind *cis*-regulatory elements found elsewhere within the gene. Because they are tethered to DNA by specific *cis*-regulatory recognition sequences, such proteins have been described as sequence-specific transcription factors (Pabo and Sauer, 1992).

Functional *cis*-regulatory elements are generally found within several hundred base pairs of the start site of the gene to which they are linked, but they can occasionally be found many thousands of nucleotides away, upstream or downstream of the start site, or even within the transcribed sequence (McKnight and Tjian, 1986; Guarente, 1988). They are generally composed of small modular DNA sequences, generally 7–12 bp in length and structured as inverted or direct repeats, each of which is a specific binding site for one or more transcription factors. Each gene has a particular combination of *cis*-regulatory elements, the nature, number, and spatial arrangement of which determine the gene's unique pattern of expression, including the cell types in which it is expressed, the times during development in which it is expressed, and the level at which it is expressed in adults both basally and in response to physiological signals. *Cis*-regulatory ele-

ments that activate expression of their target genes when appropriate transcription factors are bound are commonly termed *enhancers* (Fig. 13.2b). *Cis*-regulatory elements that repress expression of target genes are commonly termed *silencers*.

Sequence-specific transcription factors commonly comprise several physically distinct functional domains. In particular, such transcription factors frequently contain (1) a domain that recognizes and binds a specific nucleotide sequence (i.e., a *cis*-regulatory element), (2) transcription activation or repression domains that interact with either general transcription factors or transcriptional co-regulatory proteins, and (3) cooperativity domains that permit interactions with other transcription factors (Johnson and McKnight, 1989; Triezenberg, 1995).

Many transcription factors are active only as dimers or higher-order complexes with other proteins. Transcription factors frequently interact with DNA as dimers, using intrinsic dimerization domains or forming dimers in a cooperative manner on binding in a cooperative manner to direct or inverted DNA repeats. For some of these transcriptional regulatory proteins, oligomerization and DNA binding may be regulated by posttranslational modifications (e.g., phosphorylation) or by small molecule binding (e.g., the nuclear hormone receptors) (Beckett, 2001). In some cases heterodimerization may generate novel DNA recognition motifs from the contributions of the DNA binding determinants of each partner. Within active transcription factor dimers, whether homodimers or heterodimers, both partners commonly contribute jointly to both the DNA binding domain and the activation domain. Dimerization can be a mechanism of either positive or negative control of transcription since a partner may negatively influence the activation domain within a dimeric transcription factor or even prevent DNA binding of the complex. As is discussed in the next section, the second messenger responsive CREB, CREM, and ATF family of bZIP proteins may heterodimerize through a common dimerization domain, a coiled-coil motif termed the *leucine zipper*, potentially combining proteins with distinct activities (Vinson *et al.*, 2002). In this way, the ability of transcription factors to form heterodimers and other multimers increases the diversity of transcription factor complexes that can form in cells and, as a result, increases the types of specific regulatory information that can be exerted on gene expression at a particular *cis*-acting DNA element.

How do sequence-specific transcription factors binding at distant sites influence transcription from a promoter? Transcriptional activators and repressors are hypothesized to alter transcription at a distance by direct protein–protein interactions with the promoter-bound basal transcriptional machinery. DNA-bound transcriptional activator proteins may directly contact one or more proteins within the basal transcription complex such as TFIID, TFIIB, or the RNA polymerase itself. Frequently, however, they do not interact with the basal transcription apparatus directly but through the mediation of adapter proteins termed co-activators (Naar *et al.*, 2001). Repressors may inhibit transcription from a promoter by direct interference with activator function. For example, some repressors inhibit DNA binding of an activator by competing for a DNA binding site or by inactivating DNA binding through heterodimerization; for example, the repressor domain of CREMα may interfere with the CREB activation domain through heterodimerization; (Loriaux *et al.*, 1994). Others may interfere with activator function directly by physically masking a transactivation domain; for example, MDM2 masks the activity of the tumor suppressor and transcriptional activator p53 (Chene, 2003). In a manner analogous to activators, however, sequence-specific repressors may interfere with the basal transcriptional machinery directly or indirectly through recruitment of co-

repressor proteins (Burke and Baniahmad, 2000; Glass and Rosenfeld, 2000). In any of these models, transcription factors may communicate with the basal transcription apparatus, looping out intervening DNA and bringing linearly distant control regions into close proximity, providing a biochemical explanation for the distance- and orientation-independence of *cis*-acting elements (Fig. 13.3) (Ptashne and Gann, 1997; Blackwood and Kadonaga, 1998).

Co-activators, Co-repressors, and Chromatin Modification

The concept of transcriptional co-activators came from the observation that some activators contacted the TATA-binding protein (TBP)-associated components of the TFIID complex (Pugh and Tjian, 1990, 1992; Tanese *et al.*, 1991). Subsequently other factors have been found to mediate the effects of DNA-bound activators. For the most part, these proteins reside in large multifunctional complexes proposed to bridge activators to basal factors. For example, a variety of biochemically defined complexes functionally similar to the yeast Mediator complex have been found in metazoans; these complexes interact with the carboxyl-terminal domain of the largest subunit of RNA polymerase II to allow activator-dependent regulation from promoters *in vitro* (Malik and Roeder, 2000; Rachez and Freedman, 2001). The functions of the individual components of these large complexes remain unclear, but they may participate in preinitiation complex formation or the transition from initiation to transcription elongation (Naar *et al.*, 2001).

Chromatin modification has emerged as a key mechanism of transcriptional regulation. Other factors identified as co-activators for specific transcription factors include the general factors CBP/p300 and P/CAF and the nuclear hormone receptor co-activators SRC-1 and ACTR (Goodman and Smolik, 2000; Glass and Rosenfeld, 2000). These proteins harbor histone acetylase enzymatic activity (Roth *et al.*, 2001) linking DNA-bound transcriptional activators to a modification long known to be associated with transcriptionally active chromatin.

Chromatin exerts an important level of regulation on expression of genes; DNA assembled into chromatin is largely transcriptionally inactive in the absence of the binding of gene-specific activators (Workman and Kingston, 1998). The nucleosome, the basic repeating unit of chromatin, consists of DNA wrapped around an octameric complex (see Fig. 13.4a) composed of a dimer of a tetramer of core histone proteins, H2A, H2B, H3, and H4 (Kornberg and Lorch, 1999). In the core nucleosome, 146 bp of

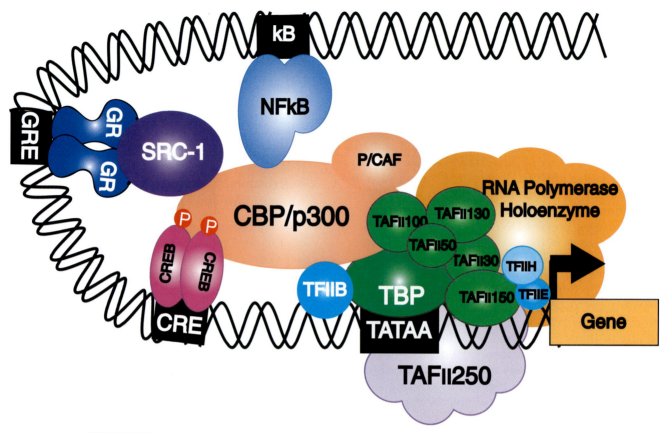

FIGURE 13.3 Transcriptional enhancer sequences act at a distance. DNA-bound activators (and repressors) may influence the activity of a distant promoter either directly by contacting the components of the basal transcription complex (such as TFIIB, TFIID, TFIIA, and even RNA polymerase) or indirectly through co-regulatory proteins (co-activators or co-repressors). The general co-activators CBP and p300 interact with a large number of DNA-bound transcriptional regulatory proteins including the phosphorylation-dependent transcriptional activator CREB, Fos/Jun, NF-κB, and the nuclear hormone receptors (e.g., the glucocorticoid receptor or GR). CBP and p300 also form complexes with other co-activators, including the nuclear hormone receptor co-activators (e.g., SRC-1) and PCAF, as well as components of the basal transcription complex, including TFIIB and TFIID, and the RNA polymerase holoenzyme.

DNA is wrapped 1.65 times around the octamer core. The core nucleosome with the linker histones H1 and H5 forms the 10-nm nucleosomal fiber or the "beads on a string" structure visible by electron microscopy. This chromatin fiber then assembles into higher-order structures to provide the degree of packaging required to condense nearly 2 m of mammalian chromosomal DNA into the nucleus.

The overall effect of this high degree of DNA condensation is to limit the accessibility of DNA to both DNA-binding regulatory proteins and the transcriptional apparatus itself. As part of the mechanism of increasing transcription at a promoter, activators increase accessibility of DNA to the transcription apparatus by recruiting co-activator complexes that alter the structure or remodel the chromatin encompassing the target gene (Naar *et al.*, 2001; Narlikar *et al.*, 2002) (Fig. 13.4). The activities of chromatin-directed co-activators may be divided into two broad

categories: (1) protein complexes with ATP-dependent chromatin remodeling activity that may "slide" or reposition nucleosomes on DNA (e.g., the SWI/SNF, ISWI, and Mi-2 families of complexes) and (2) proteins with histone acetyltransferase activity (e.g., the general co-activators CBP/p300 and P/CAF, the nuclear hormone receptor co-activators SRC-1 and ACTR, and the basal factor $TAF_{II}250$). Although the details of how these complexes work together remain unclear, both decondensation and histone acetylation are associated with activation of gene transcription. Countering this activity, co-repressor complexes contain proteins with deacetylase activity (Knoepfler and Eisenman, 1999).

Histone acetylation has long been known to be associated with active chromatin. Lysine residues within the N-terminal tails of the histones (H3, H4, and, to lesser extent, H2A and H2B) in particular are substrates of the acetyltransferase activity of

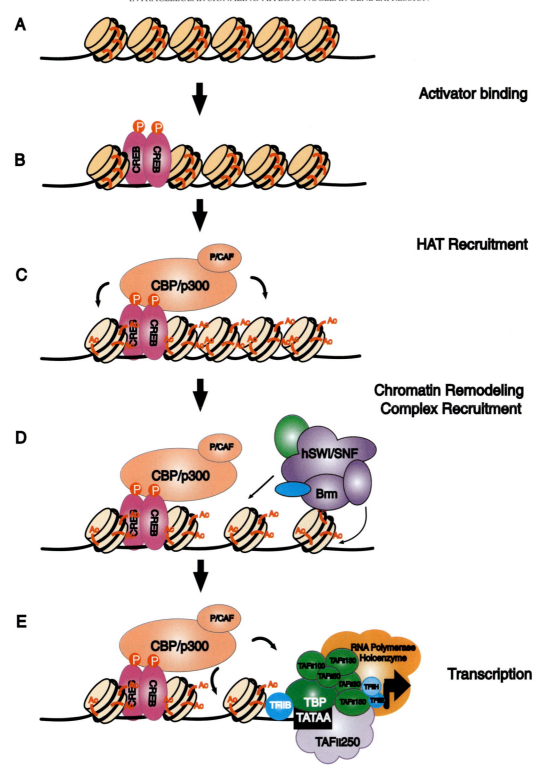

FIGURE 13.4 Transcriptional activators recruit co-activators with histone modifying and chromatin remodeling enzymatic activities. Histone acetylation is associated with activation of transcription from a promoter. A In the basal state, promoter-associated histones are relatively hypoacetylated. B A DNA-bound activator (e.g., protein kinase A-phosphorylated CREB) recruits co-activator complexes that harbor histone acetyltransferase activity (CBP/p300, PCAF). This leads to C acetylation of promoter-associated histones within N-terminal tails and D the recruitment of ATP-dependent chromatin remodeling complexes (e.g., hSWI/SNF). E Relief of chromatin repression permits access by components of the basal transcriptional machinery, which are also contacted by the general co-activators CBP/p300.

co-activators *in vitro* and *in vivo* (Roth *et al.*, 2001). The biochemical consequences of acetylation of these lysines and other posttranslational modifications of the N-terminal histone tails, including phosphorylation and methylation, on the structure of the nucleosome are unclear. The crystal structure of the nucleosome suggests that neutralization of the basic charge of lysines within the N-terminal tails by acetylation may decrease binding to DNA or alter intranucleosomal protein–protein interactions promoting chromatin decondensation (Luger *et al.*, 1997). An alternative mechanism is that modification of histone tails may serve to mark these specific regions of chromatin for decondensation by chromatin remodeling enzymes complexes by serving as docking sites for modification-specific protein interaction domains (Jenuwein and Allis, 2001; Turner, 2002). The bromodomain, found in many proteins implicated in transcriptional activation, including the histone acetylases CBP, p300, TAF$_{II}$250, P/CAF, and GCN5, is thought to target these proteins to acetylated histones in transcriptionally active chromatin, or euchromatin (see Fig. 13.4). The acetylated histones then attract bromodomain-containing components of the ATP-dependent SWI/SNF chromatin remodeling complex (Turner, 2002). In contrast, a class of nuclear transcriptional repressor proteins containing a module called the chromodomain are recruited to lysine 9-methylated histone H3 (MeK9 H3) in transcriptionally silenced or heterochromatin (Richards and Elgin, 2002). Repression of transcription by cytosine methylation at CpG sites also involves histone modification. Proteins containing 5-methylcytosine binding domains (MBD1, MBD2, MBD3, MeCP2) participate in methylation-dependent repression by recruiting histone deacetylases (Bird, 2002). Recruitment to this type of "marked" chromatin (either histones or DNA) is thought to stabilize either the active or inactive chromatin state. Thus, through a process of covalent histone modification, DNA-bound activators and repressors may influence chromatin remodeling.

Restriction of Expression of Neurally Expressed Genes to the Nervous System

Covalent modification of chromatin (DNA methylation, and histone acetylation and methylation) not only exerts important controls on the short-term regulation of gene expression within a cell or tissue type, but is also essential for establishing the stable patterns of gene expression that are heritable through cell division. The covalent modification of both histones and DNA underlies mitotically stable changes in gene function that are not explained by alterations in the sequence of DNA itself (Francis and Kingston, 2001). These *epigenetic* mechanisms determine the developmental patterns of genes that are either expressed or silenced within specific cell lineages, the process of genomic imprinting, and X-chromosome inactivation.

The processes of cellular memory maintain the programs of gene expression in the nervous system. Although the entire complement of regulatory mechanisms defining the patterns of neuronal gene expression remain elusive, a number of genes restricted to neuronal tissue including the type II sodium channel and SCG10 contain a regulatory element called the *neuron restrictive silencer element* (NRSE) (Mori *et al.*, 1990), also known as *repressor element 1* (RE-1)(Maue *et al.*, 1990). NRSE/RE-1 functions to silence target genes specifically in nonneuronal cell types. NRSE/RE-1 recruits a DNA-binding protein called repressor element 1 silencing transcription factor (REST) (Chong *et al.*, 1995), also known as neuron restrictive silencer factor (NRSF) (Mori *et al.*, 1992; Schoenherr and Anderson, 1995). The REST/NRSF DNA binding domain consists of eight GL1-Krüpple zinc fingers and at least two repression domains. The carboxyl-terminal repression domain recruits the transcriptional co-repressor CoREST (Andres *et al.*, 1999) whereas the amino-terminal repression domain associates with the mSin3/histone deacetylase complex (Holdener *et al.*, 2000) and the nuclear receptor co-repressor N-CoR (Jepsen *et al.*, 2000). CoREST recruitment to DNA-bound REST/NRSF leads to silencing over a broad chromosomal region encompassing even promoters not directly targeted by REST/NRSF, through propagation of both DNA methylation and histone H3 lysine 9 methylation and recruitment of heterochromatic protein 1 (HP1), leading to stable chromatin condensation and transcriptional silencing in nonneuronal tissue (Lunyak *et al.*, 2002) (Fig. 13.5).

A Significant Consequence of Intracellular Signaling Is the Regulation of Transcription

Intracellular signals play a major role in the regulation of gene expression. Many activator proteins can participate in the assembly of the mature transcription apparatus only after a signal-directed change in subcellular localization (e.g., from the cytoplasm to the nucleus) or a posttranslational modification, most commonly phosphorylation. Such alterations in location or conformation permit information obtained by the cell from its different signaling systems to regulate gene expression appropriate to the cells' status.

All diploid cells within an organism, starting with the fertilized one-cell embryo, contain a complete copy of the organism's genome. Differential expres-

A

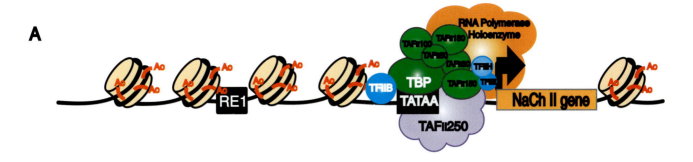

B

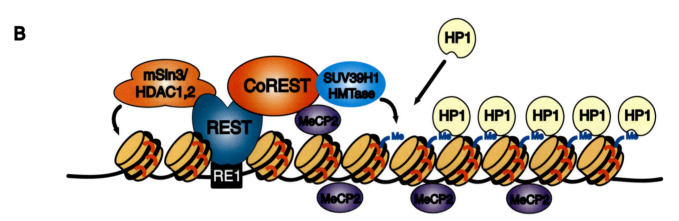

FIGURE 13.5 The transcriptional repressor REST/NRSF restricts the expression of neuronal genes to neural tissue. Neural tissue-specific expression of a number of genes is maintained by the transcriptional repressor REST (or NRSF). A In neuronal tissue, genes such as the voltage-gated sodium channel II gene are transcriptionally active. B In non-neuronal tissue, the transcriptional repressor REST/NRSF silences neuronal genes by recruitment of co-repressor complexes, including CoREST and Sin3-associated histone deacetylase complexes (HDACs 1 and 2). The co-repressor CoREST may also interact with other repressors, including MeCP2 (a 5-methylcytosine-directed DNA-binding protein) and the histone methyltransferase (HMTase) SUV39H1. Histone N-terminal tail methylation by HMTases at lysine 9 of histone H3 is associated with transcriptionally silenced chromatin or heterochromatin, recruiting heterochromatin-associated protein 1 (HP1). For some of these gene promoters, methylation at CpG sites is essential for CoREST recruitment and transcriptional repression. Through these mechanisms, DNA binding of REST/NRSF with CoREST recruitment may propagate a silenced state of chromatin to encompass adjacent clustered neuron-specific genes in nonneuronal tissue (modified from Lunyak *et al.*, 2002).

sion of this common genome is required for the formation of distinct cell types during development, of crucial importance in the differentiation of thousands of distinct types of neurons found in the brain (see Chapter 1). The mechanisms by which these differentiated cells form are highly dependent on intercellular signaling. Much work in this area has been done in *Drosophila* and *Xenopus*, organisms in which viable embryos can be well studied in isolation.

In certain cases, restriction of the expression of a gene to specific cell types depends on the presence of critical transcription factors only in those cell types. For example, the pituitary hormones—growth hormone and prolactin—are expressed only in pitu-

itary lactotrophs and somatotrophs because their required activator transcription factor, Pit 1, is expressed only in those two cell types in the mature organism (Nelson *et al.*, 1988). In other cases as discussed above, genes restricted to the nervous system contain *cis*-regulatory elements that bind transcriptional repressor proteins in nonneuronal tissues; the presence of repressor proteins locks expression of those genes in that cell type.

The sequential expression, during development, of hierarchies of activator and repressor proteins depends initially on the asymmetric distribution of critical signaling molecules within the embryo, leading to differential gene expression within embryonic cells. As cells gain individual identities during development, cell–cell interactions mediated by contact or by the elaboration of intercellular autocrine, paracrine, or longer-range signaling continue the process of specifying the complement of genes expressed in target cells. Genes that are silent during particular phases of development may become unavailable for subsequent activation because they become wrapped in inactive chromatin structures.

Transcriptional Regulation by Intracellular Signals

As discussed above, all protein-encoding genes contain cis-regulatory elements that permit the genes to which they are linked to be activated or repressed by physiological signals. Intracellular signals can activate transcription factors through a variety of different general mechanisms, but each requires a translocation step by which the signal is transmitted through the cytoplasm to the nucleus. Some transcription factors are themselves translocated to the nucleus. For example, the transcription factor NF-kB is retained in the cytoplasm by its binding protein IkB; this interaction masks the NF-kB nuclear localization signal. Signal-regulated phosphorylation of IkB by protein kinase C and other protein kinases leads to dissociation of NF-kB, permitting it to enter the nucleus. Other transcription factors must be directly phosphorylated or dephosphorylated to bind DNA. For example, in many cytokine-signaling pathways, plasma membrane receptor tyrosine phosphorylation of transcription factors known as signal transducers and activators of transcription (STATs) permits their multimerization, which in turn permits both nuclear translocation and construction of an effective DNA binding site within the multimer. Yet other transcription factors are already bound to their cognate cis-regulatory elements within the nucleus under basal conditions and become able to activate transcription after phosphorylation. The transcription factor CREB, for example, is constitutively bound to cAMP response elements (CREs) found within many genes. The critical nuclear translocation step in CREB activation involves not the transcription factor itself, but the catalytic subunit of protein kinase A, which, on entering the nucleus, can phosphorylate CREB. Phosphorylation of CREB converts it into its active state by permitting it to interact with the adapter protein CBP, which can then contact the basal transcription apparatus and acetylate histones leading to chromatin remodeling (see Fig. 13.4).

ROLE OF cAMP AND CA^{2+} IN THE ACTIVATION PATHWAYS OF TRANSCRIPTION

As described in Chapter 12, the cAMP second-messenger pathway is among the best characterized intracellular signaling pathways; a major feature of signaling by this pathway is the regulation of a large number of genes. Cyclic AMP response elements with the consensus nucleotide sequence of TGACGTCA have been identified in many genes expressed in the nervous system.

The consensus CRE sequence illustrates a common feature of many transcription factor binding sites; it is a palindrome or inverted repeat. Examination of the sequence TGACGTCA readily reveals that the sequences on the two complementary strands, which run in opposite directions, are identical. Many cis-regulatory elements are perfect or approximate palindromes because many transcription factors bind DNA as dimers, in which each member of the dimer recognizes one of the "half-sites." CREB binds to CREs as a homodimer, with a higher affinity for perfectly palindromic than for asymmetric CREs.

When bound to a CRE, CREB activates transcription when it is phosphorylated at Ser-133 (Mayr and Montminy, 2001). It does so, as described earlier, because phosphorylated CREB, but not unphosphorylated CREB, can recruit the adapter protein, CBP, into the transcription complex. CBP, in turn, interacts with the basal transcription complex and modifies histones to enhance the efficiency of transcription (Fig. 13.4).

The regulation of CREB activation by phosphorylation illustrates several general principles, including the requirement for nuclear translocation of protein kinases in cases where transcription factors are already found in the nucleus under basal conditions and the role of phosphorylation in regulating protein–protein interactions. An additional important principle illustrated by CREB is the convergence of signaling pathways. CREB is phosphorylated on Ser-133 by the free catalytic subunit of the cAMP-dependent protein kinase. However, CREB Ser-133 can also be phosphorylated by Ca^{2+}-calmodulin-dependent protein kinases types II and IV (Sheng et al., 1990) and by RSK2, a kinase activated in growth factor pathways including Ras and MAP kinase (Xing et al., 1996). When each individual signal is relatively weak, convergence may be a critical mechanism resulting in specificity of gene regulation, with some genes being activated only when multiple pathways are stimulated. Some genes that contain CREs are known to be induced in more than an additive fashion by the interaction of cAMP and Ca^{2+}, but how convergent phosphorylation on the same serine might produce synergy is not yet clear. Synergy is more readily understood in cases in which a particular protein is modified at two different sites, causing interacting conformational changes. In addition to Ser-133, CREB contains sites for phosphorylation by a variety of protein kinases, including glycogen synthase kinase 3 (GSK3), but the biological effects of phosphorylating these additional serines are not fully understood. These additional phosphorylation events may fine-tune the regulation of CREB-mediated transcription.

CREB illustrates yet another important principle of transcriptional regulation: CREB is a member of a family of related proteins. Many transcription factors are members of families; this permits complex forms of positive and negative regulation as discussed above. CREB is closely related to other proteins called activating transcription factors (ATFs) and CRE modulators (CREMs), which result from alternative splicing of a single CREM gene. All of these proteins bind CREs as dimers; many can dimerize with CREB itself. ATF-1 appears to be very similar to CREB in that it can be activated by both cAMP and Ca^{2+} pathways. Many of the other ATF proteins and CREM isoforms can activate transcription; however, certain CREMs may act to repress it. These CREM isoforms lack the glutamine-rich transcriptional activation domain of CREB–ATF family members that are activators of transcription. Thus CREB–CREM homodimers may bind DNA but fail to activate transcription. Like CREB, many of the ATF proteins are constitutively synthesized, but ATF 3 and certain CREM isoforms are inducible in response to environmental stimuli. The new synthesis of transcription factors is yet another mechanism of gene regulation.

The dimerization domain used by the CREB–ATF proteins and several other families of transcription factors is called a *leucine zipper*. This domain was first identified in transcription factor C/EBP and is also used by the AP-1 family of transcription factors. The so-called leucine zipper actually forms a coiled coil. The dimerization motif is an α helix in which every seventh residue is a leucine; based on the periodicity of α helices, the leucines line up along one face of the helix two turns apart. The aligned leucines of the two dimerization partners interact hydrophobically and stabilize the dimer. In CREB, C/EBP, and the AP-1 family of proteins, the leucine zipper is at the carboxy terminus of the protein. Just upstream of the leucine zipper is a region of highly basic amino acid residues that form the DNA binding domain. Dimerization by means of the leucine zipper domain juxtaposes the adjacent basic regions of each of the partners; these

juxtaposed basic regions then undergo a random coil-to-helix transition when they bind DNA in the major groove of DNA. This combination of motifs is why this superfamily of proteins is referred to as the basic leucine zipper proteins (bZIPs).

AP-1 Transcription Factors Are Derived Mainly from Cellular "Immediate-Early" Genes

Activator protein 1 (AP-1) is another family of bZIP transcription factors that play a central role in the regulation of neural gene expression by extracellular signals. The AP-1 family comprises multiple proteins that bind as heterodimers (and a few as homodimers) to the DNA sequence TGACTCA, the consensus AP-1 element that forms a palindrome flanking a central C or G. Although the AP-1 sequence differs from the CRE sequence by only a single base, this one-base difference strongly biases protein binding away from the CREB family of proteins. AP-1 sequences confer responsiveness to the protein kinase C pathway.

The AP-1 proteins generally bind DNA as heterodimers composed of one member each of two different families of related bZIP proteins, the Fos family and the Jun family. The known members of the Fos family are c-Fos, Fra-1 (Fos-related antigen-1), Fra-2, and FosB; there is also evidence for posttranslationally modified forms of FosB. The known members of the Jun family are c-Jun, JunB, and JunD. Heterodimers form between proteins of the Fos family and proteins of the Jun family by means of the leucine zipper. Unlike the Fos proteins, c-Jun and JunD, but not JunB, can form homodimers that bind to AP-1 sites, albeit with far lower affinity than Fos-Jun heterodimers. The potential complexity of transcriptional regulation is greater still because some AP-1 proteins can heterodimerize through the leucine zipper with members of the CREB–ATF family, for example, ATF2 with c-Jun. AP-1 proteins can also form higher-order complexes with unrelated families of transcription factors. In addition, AP-1 proteins can complex with and thus apparently inhibit the transcriptional activity of certain nuclear hormone receptors (discussed below).

Among the known Fos and Jun proteins, only JunD is expressed constitutively at high levels in many cell types. The other AP-1 proteins tend to be expressed at low or even undetectable levels under basal conditions but, with stimulation, may be induced to high levels of expression. Thus, unlike genes that are regulated by constitutively expressed transcription factors such as CREB, genes that are regulated by c-Fos–c-Jun heterodimers require new transcription and translation of their required regulatory factors.

Genes that are transcriptionally activated by synaptic activity, drugs, and growth factors have often been classified roughly into two groups. Genes, such as the *c-fos* gene itself, that are activated rapidly (within minutes), transiently, and without requiring new protein synthesis are often described as cellular *immediate-early genes* (IEGs). Genes that are induced or repressed more slowly (within hours) and are dependent on new protein synthesis have been described as *late-response genes*. The term IEG was initially applied to describe viral genes that are activated "immediately" on infection of eukaryotic cells by utilization of preexisting host cell transcription factors. Viral immediate-early genes generally encode transcription factors needed to activate viral "late" gene expression. This terminology has been extended to cellular (i.e., nonviral) genes with varying success. The terminology is problematic because many cellular genes are induced independent of protein synthesis, but with a time course intermediate between those of "classic" IEGs and late-response genes. In fact, some genes may be regulated with different time courses or requirements for protein synthesis in response to different intracellular signals. Moreover, many cellular genes regulated as IEGs encode proteins that are not transcription factors. Despite these caveats, the concept of IEG-encoded transcription factors in the nervous system is useful. Because of their rapid induction from low basal levels in response to neuronal depolarization (the critical signal being Ca^{2+} entry) and second-messenger and growth factor pathways, several IEGs have been used as cellular markers of neural activation, permitting novel approaches to functional neuroanatomy.

The protein products of those cellular IEGs that function as transcription factors bind to *cis*-regulatory elements contained within a subset of late-response genes to activate or repress them. IEGs such as *c-fos* have therefore been termed third messengers in signal transduction cascades, with neurotransmitters designated as intercellular first messengers and small intracellular molecules, such as cAMP and Ca^{2+}, as second messengers. However, IEGs are not always a necessary step in signal-regulated expression of genes having roles in the differentiated function of neurons. In fact, many such genes, including many genes encoding neuropeptides such as proenkephalin (Konradi *et al.*, 1993) and prodynorphin (Cole *et al.*, 1995) and some genes encoding neurotrophic factors, are activated in response to neuronal depolarization or cAMP by phosphorylation of the constitutively expressed transcription factor CREB rather than by IEG third messengers. In sum, neural genes that are regulated by extracellular signals are activated or repressed with varying time courses by reversible phosphorylation of constitutively synthesized transcription factors and by newly synthesized transcription factors, some of which are regulated as IEGs.

Activation of the c-fos Gene

The *c-fos* gene is activated rapidly by neurotransmitters or drugs that stimulate the cAMP pathway or Ca^{2+} elevation. Both pathways produce phosphorylation of transcription factor CREB (Sheng *et al.*, 1990). The *c-fos* gene contains three binding sites for CREB. The *c-fos* gene can also be induced by the Ras/MAP kinase pathway, which is activated by a number of growth factors. For example, neurotrophins, such as nerve growth factor (NGF), bind a family of receptor tyrosine kinases (Trks); NGF interacts with Trk A, which activates Ras. Ras then acts through a cascade of protein kinases including Raf and the cytoplasmic MAP kinase kinases (MAPKKs) MEK1/MEK2, which phosphorylate the MAP kinases ERK1/ERK2. These kinases translocate into the nucleus, where they can activate RSK2 to phosphorylate CREB, but they can also apparently directly phosphorylate other transcription factors such as the ternary complex factor Elk-1. Elk-1 binds along with the serum response factor (SRF) to the serum response element (SRE) within the *c-fos* gene and many other growth factor-inducible genes. Cross-talk between neurotransmitter and growth factor signaling pathways has been documented with increasing frequency and likely plays an important role in the precise tuning of neural plasticity to diverse environmental stimuli.

Regulation of c-Jun

Expression of most of the proteins of the Fos and Jun families that constitute transcription factor AP-1 and the binding of AP-1 to DNA is regulated by extracellular signals. However, both Fos and Jun family members are phosphoproteins themselves, and AP-1-mediated transcription requires not only the new synthesis of AP-1 but also the phosphorylation of proteins within AP-1 complexes. Phosphorylation of c-Jun within its N-terminal activation domain has been shown to markedly enhance its ability to activate transcription without affecting its ability to form dimers or bind DNA. Other phosphorylation sites within c-Jun regulate its ability to bind DNA.

Phosphorylation and activation of c-Jun can result from the action of Jun N-terminal kinase (JNK). JNK is a member of the mitogen-activated protein kinase (MAPK) family of protein kinases whose mammalian members include the ERKs, p38, and JNK. In addition to cellular stressors, the inflammatory cytokines interleukin-1β (IL-1β) and tumor necrosis factor α (TNFα) have been shown to activate both JNK and p38. JNK has also been shown to be activated by neurotransmitters, including glutamate. Thus, AP-1-mediated transcription within the nervous system requires multiple steps beginning with the activation of genes encoding AP-1 proteins.

Growth Factors and Cytokines as Modulators of Gene Expression in the Nervous System

With respect to function, the boundary between trophic, or growth, factors and cytokines in the nervous system has become increasingly arbitrary. However, cell signaling mechanisms offer a useful means of distinction. Growth factors, such as the neurotrophins (e.g., nerve growth factor, brain-derived neurotrophic factor, and neurotrophin 3), epidermal growth factor (EGF), and fibroblast growth factor (FGF), act through receptor protein tyrosine kinases, whereas the cytokines, such as leukemia inhibitory factor (LIF), ciliary neurotrophic factor (CNTF), and interleukin-6 (IL-6), act through nonreceptor protein tyrosine kinases (see Chapter 10).

LIF, CNTF, and IL-6 subserve a wide array of overlapping functions inside and outside the nervous system, including hematopoietic and immunological functions outside the nervous system and regulation of neuronal survival, differentiation, and, in certain circumstances, plasticity within the nervous system. These peptides have marked homologies of their tertiary structures rather than their primary sequences, which presumably permits them to interact with related receptor complexes that contain a common signal-transducing subunit, gp130.

The receptors for these cytokines consist of a signal-transducing β component, which includes gp130 and, in some cases, additional subunits. As is typical of cytokine receptors, the cytoplasmic tails of the signal-transducing β components of the IL-6, LIF, and CNTF receptors lack kinase domains. Rather, the cytoplasmic domains interact with nonreceptor protein tyrosine kinases (PTKs) of the Janus kinase (Jak) family, which include Jak1, Jak2, and Tyk2 (Fig. 13.6). Some cytokine receptors, such as the prolactin receptor, which interacts exclusively with Jak2, can interact only with a single Jak PTK. In contrast, the IL-6, LIF, and CNTF receptors can interact with multiple Jak PTKs, including Jak1, Jak2, and Tyk2. Presumably, the dimerization of receptors on ligand binding permits Jak-family PTKs to cross-phosphorylate each other.

Signal transduction to the nucleus includes tyrosine phosphorylation by the Jak PTKs of one or more of the STAT proteins mentioned earlier. The first STAT family members were identified as proteins binding to interferon-regulated genes but have subsequently been found to take part in the activity of multiple cytokines. On phosphorylation, STAT proteins form dimers through the association of SH2 domains, an important type of protein interaction domain described earlier. Dimerization is thought to trigger translocation to the nucleus, where STATS bind their cognate cytokine response elements. Different STATs become activated by different cytokine receptors, not because of differential use of Jak PTKs but because of specific coupling of certain STATs to certain receptors. Thus, for example, the IL-6 receptor preferentially activates STAT1 and STAT3; the CNTF receptor preferentially activates STAT3.

Signaling through the Jak/STAT systems is ultimately turned off by feedback from a STAT-induced gene product, suppressor of cytokine signaling (SOCS). SOCS binds to the activated Jak PTK and prevents it from phosphorylating the β component of the cytokine receptor and, thus, blocks subsequent STAT binding (see Fig. 13.6). Thus, like many other signaling cascades, this system has an inherent feedback mechanism that terminates the signal.

Many of these target genes integrate signals from a variety of signaling pathways. The *c-fos* gene, for example, contains an element called the SIS-inducible element (SIE), which binds STAT proteins; thus, *c-fos* gene expression can also be induced by cytokines. Cytokine response elements have now been identified within many neural genes, including vasoactive intestinal polypeptide and several other neuropeptide-encoding genes.

Nuclear Receptors Regulate Transcription Directly

The differentiation of many cell types in the brain is established by exposure to steroids and other small molecule hormones. For example, exposure to estrogen or testosterone during critical developmental periods results in sexually dimorphic development of certain brain nuclei. The small-molecule hormones, including the glucocorticoids, sex steroids, mineralocorticoids, retinoids, thyroid hormone, and vitamin D, are small lipid-soluble ligands that can diffuse across cell membranes. Unlike the other types of intercellular signals described above, these small-molecule hormones bind their receptor inside the cell and subsequently affect specific gene transcription in the nucleus. In some cases, such as the estrogen receptor or retinoic acid receptor, the unliganded receptor is resident in the nucleus, awaiting the diffusing

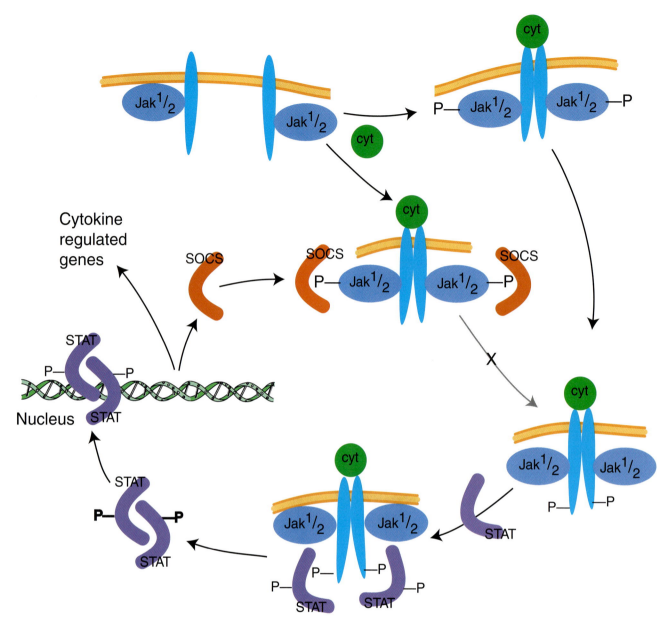

FIGURE 13.6 Cytokine regulation of gene transcription. Cytokines (Cyt) bind, causing dimerization and activation of Jak PTK (Jak 1/2) by phosphorylation, which subsequently phosphorylates the receptor. Phosphoreceptor recruits STAT proteins, which become phosphorylated by Jak PTK, which allows for dimerization and nuclear translocation. Phosphorylated STATs activate transcription of multiple genes, including SOCS, which feeds back to block cytokine receptor phosphorylation and turn off the signal.

molecule. In other systems, such as with the glucocorticoid receptor, the receptor is in the cytoplasm and, on hormone binding, undergoes conformational change and dissociation of associated proteins which unmasks a nuclear localization signal and DNA binding domain, similar to the activation of NFκB. A significant exception to this deals with the more recently accepted ability of small-molecule hormone receptors to affect signaling in the cytoplasm without cycling through the nucleus. This is described in detail in Chapter 10.

Like the other transcription factors described in this chapter, the steroid hormone receptors are modular in nature (see Fig. 13.7A). Each has a transcriptional activation domain at its amino terminus, a DNA binding domain, and a ligand binding domain at its carboxy terminus. The DNA binding domains recognize specific palindromic or direct repeat DNA

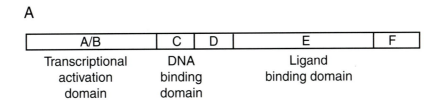

A

| A/B | C | D | E | F |

Transcriptional activation domain

DNA binding domain

Ligand binding domain

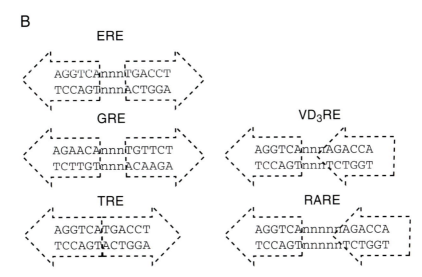

B

ERE

AGGTCAnnnTGACCT
TCCAGTnnnACTGGA

GRE

AGAACAnnnTGTTCT
TCTTGTnnnACAAGA

VD₃RE

AGGTCAnnnAGACCA
TCCAGTnnnTCTGGT

TRE

AGGTCATGACCT
TCCAGTACTGGA

RARE

AGGTCAnnnnnAGACCA
TCCAGTnnnnnTCTGGT

FIGURE 13.7 Nuclear hormone receptor structure and DNA binding sites. (A) Domain structure of the nuclear small molecule hormone receptor. Domain A/B contains the transcriptional activation structures for interacting with other proteins, C is the highly homologous DNA binding domain, D represents the structure involved in stabilizing the dimer form of the receptor, and E contains the ligand (hormone) binding domain. (B) The estrogen receptor regulatory element (ERE), glucocorticoid receptor element (GRE), and thyroid hormone receptor element (TRE) are constructed of palindromic sequences of six nucleotides separated by three or zero nucleotides. The vitamin D₃ receptor regulatory element (VD₃RE) and retinoic acid receptor regulatory element (RARE) contain imperfect repeats separated by three to five nucleotides. These represent the consensus sequences and there may be small variations in sequence that alter the affinity of the receptor for the DNA.

sequences, hormone response elements, within the regulatory regions of specific genes (Fig. 13.7B), binding in dimers. In the palidromic regulatory elements, the "spacer" DNA puts the same sequence (the complement from the opposite strand) one helical turn away and thus on the same side of the DNA surface, so the nuclear receptor homodimer can bind to an identical site in a head-to-head configuration.

As with the other transcription factors, once bound to the DNA, the nuclear hormone receptors interact with transcriptional co-activators and/or repressors to modulate levels of transcription appropriate for the physiological state of the animal. The same molecular mechanisms described above are used to modulate transcription levels.

Nuclear hormone receptors also elicit transcriptional actions via interactions with other nuclear tran-

scriptional machinery. This became apparent in earlier studies on specific gene promoters that contained no hormone receptor regulatory element in the DNA, yet were still transcriptionally sensitive to hormone action. For example, for some promoters, glucocorticoids modulated transcription, but required an AP-1 element in the promoter to mediate the action (Schüle et al., 1990, Jonat et al., 1990). At the same time, similar observations were made for estrogen (Gaub et al., 1990). Subsequent studies have shown that the small-molecule hormone receptor can affect the recruitment of co-activators/co-repressors without having to interact with the DNA (Kushner et al., 2000, Göttlicher et al., 1998). This type of action of nuclear hormone receptors has now been seen for other transcription factor systems such as NFκB and CREB, suggesting even greater complexity in transcriptional regulation.

It provides the basis for understanding the broad changes in gene expression that occur within the brain in association with major changes in hormone levels associated with chronic stress or developmental events such as puberty.

Gene Expression Is Also Regulated at the Posttranscriptional Level

The steady-state level of a specific mRNA is determined by its rate of synthesis (transcription) as well as its rate of degradation (mRNA turnover). Although modulation of the transcriptional rate remains the predominant mode for regulating gene expression in the nervous system, evidence continues to accumulate showing that regulation of mRNA turnover also plays an important role. Indeed, the coordination of transcription and mRNA turnover allows for exquisitely tight control over the level of an mRNA, allowing for rapid on/off signaling in some cases.

mRNA has several posttranscriptional modifications that help prevent it from endoribonuclease degradation: 7-Me-G capping at the 5′ end and polyadenylation at the 3′ end. During the lifetime of the mRNA, the poly(A) tail gradually shortens and finally the 5′ 7-Me-G cap is removed and the mRNA becomes subject to rapid nuclease degradation (see Fig. 13.8). Early biochemical experiments classified mRNAs as either short-lived, with a $t_{1/2}$ of about 30–60 min, or long-lived, with a $t_{1/2}$ of 24–36 h, but now we know there is a broad range of half-lives ranging from 5 min to many days. Further studies have identified multiple sequence elements within the mRNA that mediate the relative stability of the different mRNAs.

The first characterized RNA structure mediating mRNA stability was an AU-rich element (ARE), which is present in the 3′ untranslated region of the mRNA (see Fig. 13.8). These have been categorized into three classes, ARE I–III, based on sequence (Chen and Shyu, 1995), and all bind a variety of cytoplasmic proteins, which either confer stability or instability. Although the exact mechanism is still not fully understood, it appears that the proteins binding to the AREs in many cases cause the 3′ poly(A) tail to either shorten or stabilize, resulting in mRNA stability change. This mechanism is common with many cytokine, transcription factor, cell cycle regulator, or cell surface receptor mRNAs, all belonging to classes of neural genes that the organism would like to have under tight control. For example, the transcription factor c-Fos-encoding gene is rapidly induced on depolarization of a neuron, producing a rapid rise in c-fos mRNA. At the same time, however, there is also

an induction of proteins that interact with the c-Fos mRNA ARE, resulting in instability and rapid degradation of the newly made c-fos mRNA. The net result is a transcriptional signal that lasts for only an hour or so after a neuron depolarizes.

Although the exact mechanisms by which signaling in a neuron or glia is translated into alteration of stability of a specific mRNA are still being elucidated, we are beginning to see a pathway that has much in common with the regulation of transcription. The family of signal transduction activators of RNA (STAR) family of proteins is an excellent example (Lasko, 2003). These proteins bind mRNA through specific domains and can be subject to phosphorylation by activated tyrosine kinases. This modification changes how the STAR protein interacts with other proteins leading to alterations in mRNA stability and/or where the mRNA is located within a cell. For example, myelin basic protein (MBP) mRNA contains a STAR protein binding sequence in its 3′ untranslated region that enhances the stability of MBP mRNA. In the quaking mouse, where myelination is sharply reduced, the mutation was traced to the QK1 gene (Hardy et al., 1996), a STAR protein that interacts with this MBP mRNA element, resulting in destabilization and mislocation of MBP mRNA in the oligodendrocyte (Li et al., 2000).

The issue of where a specific mRNA is located and translated in a neuron is also important to neuronal function. Specific mRNAs can be found asymmetrically distributed in subcellular compartments where the proteins they encode are used. For example, several synaptic protein-encoding mRNAs are located at the neuromuscular junction (Chakkalakal and Jasmin, 2002). There are also extensive examples of dendritic localization of specific mRNAs and the mechanism of their translation is better understood (Kindler et al., 1997; Steward and Schuman, 2001). The presumption has been that there is local synthesis of proteins encoded by these dendritic mRNAs, allowing for rapid, local changes in proteins. Indeed, the various components of the translational machinery have been localized to the dendrites (Tiedge and Brosius, 1996) and protein synthesis has been shown to occur in isolated dendrites (Torre and Steward, 1992). The identification of localized synthesis of proteins is crucial to our understanding of how a neuron or glia can rapidly respond with changes in protein expression often at places far removed from the cell body.

Finally, the actual act of translation of an mRNA into a protein can also be regulated by modulation of proteins that bind the mRNA. In some cases, particularly in development, specific mRNAs are kept in an

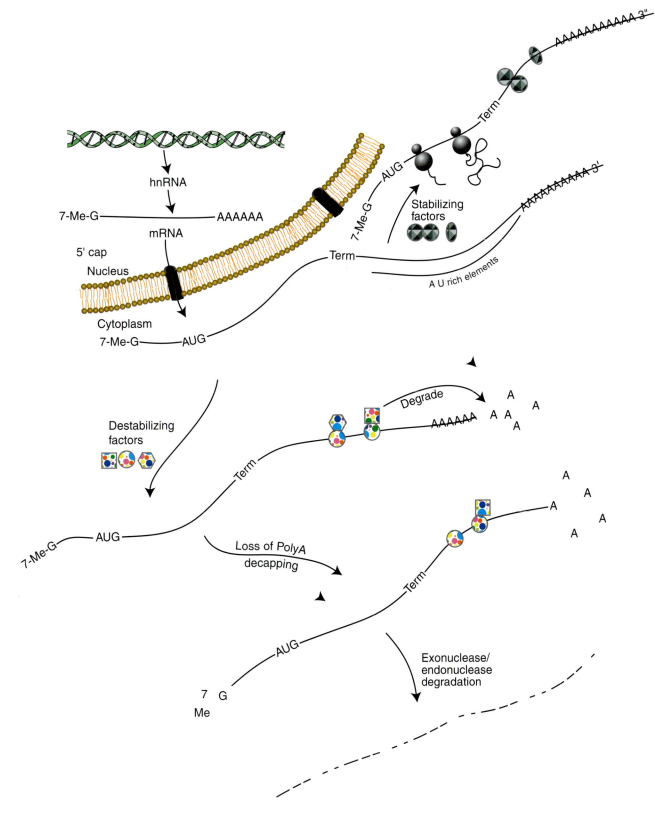

FIGURE 13.8   Schematic of regulation of mRNA turnover. After synthesis in the nucleus with posttranscriptional modification of both the 5′ and 3′ ends of the mRNA it is transported to the cytoplasm. There it can interact with destabilizing factors that lead to a loss of the poly (A) tail at the 3′ end, subsequent 7-Me-G decapping, and rapid degradation in the cytoplasm.

inactive state, awaiting the appropriate signals to begin translation. Again, proteins bind to specific mRNA sequences, effectively shutting down the mRNA, and cytoplasmic changes in polyadenylation of the mRNA appear to play an important role (Richter, 1999). Translation silencing is also important in avoiding translation until an mRNA has been transported to its subcellular site of translation (Richter and Lorenz, 2002; Wang *et al.*, 2002). Other mechanisms involve turning up or down the level of translation of particular mRNAs. Again, in a manner analogous to regulation of transcription, specific proteins, which bind to the mRNA, interact with translation initiation factors altering the level of translation. Thus, there are multiple mechanisms, in addition to transcriptional regulation, by which different neurons and glia can adjust the level of synthesis of specific proteins to respond to the different developmental and physiological states.

SUMMARY

The formation of long-term memories requires changes in gene expression and new protein synthesis (see also Chapter 18). Control sequences on DNA determine which segments of DNA can be transcribed into RNA. It is at transcription initiation that extracellular signals such as neurotransmitters, hormones, drugs, and growth factors exert their most significant control. The transcription itself is carried out by RNA polymerases. The transcription is modulated by transcription factors that recruit the polymerases to the DNA. For example, the critical nuclear translocation step in the activation of transcription factor CREB involves the catalytic subunit of PKA, which can phosphorylate CREB on entering the nucleus.

The genes that encode the transcription factors themselves may respond quickly or slowly. These genes have been coined "third messengers" in signal transduction cascades. Cross-talk between neurotransmitter and growth factor signaling pathways is likely to play an important role in the precise tuning of neuronal plasticity to diverse environmental stimuli.

The active, mature transcription complex is a remarkable architectural assembly of proteins assembled at the basal promoter—in most cases, at a sequence called the TATA box. In addition to RNA polymerase II, this complex includes a large number of associated general transcription factors, sequence-specific transcription factors, and intervening adapters. A wide variety of transcription factors bound to *cis*-regulatory elements elsewhere in the gene, but permitted to interact by the looping of DNA, join in the formation of the active transcription complex. This remarkable mechanism permits cells to exert exquisite control of the genes being transcribed in a variety of situations, for example, to govern appropriate entry or exit from the cell cycle, to maintain appropriate cellular identity, and to respond appropriately to extracellular signals.

Transcription can be regulated by many different extracellular signals modulated by a large array of signaling pathways (many including reversible phosphorylation) and a complex array of transcription factors. Most of these factors are members of families and regulate transcription only as multimers. Given this complexity, the potential for very precise regulation is clear, but the mechanisms by which such precision is achieved are not fully understood. In this chapter, regulation has been illustrated by only a few of the families of transcription factors. Those chosen appear to play important roles in the nervous system and illustrate many of the basic principles of gene regulation.

Gene expression is also regulated at the posttranscriptional level. The steady-state level of an mRNA is determined as much by its degradation as it is by its synthesis. As with transcriptional regulation, there are proteins that bind to specific sequences in the mRNA and vary the stability of the mRNA, its subcellular location within the cell, or the rate at which it is translated into protein.

References

Altshuller, Y. M., Frohman, M. A., Kraner, S. D., and Mandel, G. (1995). REST: A mammalian silencer protein that restricts sodium channel gene expression to neurons. *Cell* **80**, 949–957.

Andres, M. E., Burger, C., Peral-Rubio, M. J., Battaglioli, E., Anderson, M. E., Grimes, J., Dallman, J., Ballas, N., and Mandel, G. (1999). CoREST: A functional corepressor required for regulation of neural-specific gene expression. *Proc. Natl. Acad. Sci. USA* **96**, 9873–9878.

Beckett, D. (2001). Regulated assembly of transcription factors and control of transcription initiation. *J. Mol. Biol.* **314**(3), 335–352.

Bird, A. (2002). DNA methylation patterns and epigenetic memory. Genes Dev. **16**, 6–21.

Blackwood, E. M, and Kadonaga, J. T. (1998). Going the distance: A current view of enhancer action. *Science* **281**, 61–63.

Breathnach, R., and Chambon, P. (1981). Organization and expression of eucaryotic split genes coding for proteins. *Annu. Rev. Biochem.* **50**, 349–383.

Burke, L. J., and Baniahmad, A. (2000). Co-repressors 2000. *FASEB J.* **14**, 1876–1888.

Burke, T. W., Willy, P. J., Kutach, A. K., Butler, J. E., and Kadonaga, J. T. (1998). The DPE, a conserved downstream core promoter element that is functionally analogous to the TATA box. *Cold Spring Harbor Symp. Quant. Biol.* **63**, 75–82.

Burley, S. K., and Roeder, R. G. (1997). Biochemistry and structural biology of transcription factor IID (TFIID). *Annu. Rev. Biochem.* **65**, 769–799.

Butler, J. E., and Kadonaga, J. T. (2002). The RNA polymerase II core promoter: A key component in the regulation of gene expression. *Genes Dev.* **16**, 2583–2592.

Chakkalakal, J. V., Jasmin, B. J. (2002). Localizing synaptic mRNAs at the neuromuscular junction: It takes more than transcription. *BioEssays* **25**, 25–31.

Chalkley, G. E., and Verrijzer, C. P. (1999). DNA binding site selection by RNA polymerase II TAFs: A TAF(II)250–TAF(II)150 complex recognizes the initiator. *EMBO J.* **18**, 4835–4845.

Chen, C. Y., and Shyu, A. B. (1995). AU-rich elements: Characterization and importance in mRNA decay. *Trends Biochem. Sci.* **20**, 465–470.

Chene, P. (2003). Inhibiting the p53–MDM2 interaction: An important target for cancer therapy. *Nat. Rev. Cancer* **3**, 102–109.

Chong, J. A., Tapia-Ramirez, J., Kim, S., Toledo-Aral, J. J., Zheng, Y., Boutros, M. C., Altshuller, Y. M., Frohman, M. A., Kraner, S. D., and Mandel, G. (1995). REST: A mammalian silencer protein that restricts sodium channel gene expression to neurons. *Cell* **80**, 949–957.

Cole, R. L., Konradi, C., Douglass, J., and Hyman, S. E. (1995). Neuronal adaptation to amphetamine and dopamine: Molecular mechanisms of prodynorphin gene regulation in rat striatum. *Neuron* **14**, 813–823.

Francis, N. J., and Kingston, R. E. (2001). Mechanisms of transcriptional memory. *Nat. Rev. Mol. Cell. Biol.* **2**, 409–421.

Gaub, M. P., Bellard, M., Scheuer, I., Chambon, P., and Sassone, C. P. (1990). Activation of the ovalbumin gene by the estrogen receptor involves the fos–jun complex. *Cell* **63**, 1267–1276.

Glass, C. K., and Rosenfeld, M. G. (2000). The coregulator exchange in transcriptional functions of nuclear receptors. *Genes Dev.* **14**, 121–141.

Goodman, R. H., and Smolik, S. (2000). CBP/p300 in cell growth, transformation, and development. *Genes Dev.* **14**, 1553–1577.

Göttlicher, M., Heck, S., and Herrlich, P. (1998). Transcriptional cross-talk, the second mode of steroid hormone receptor action. *J. Mol. Med.* **76**, 480–489.

Graves, B. J., Johnson, P. F., and McKnight, S. L. (1986). Homologous recognition of a promoter domain common to the MSV LTR and the HSV tk gene. *Cell* **44**, 565–576.

Grimes, J. A., Nielsen, S. J., Battaglioli, E., Miska, E. A., Speh, J. C., Berry, D. L., Atouf, F., Holdener, B. C., Mandel, G., Kouzarides, T. (2000). The co-repressor mSin3A is a functional component of the REST-CoREST repressor complex. *J Biol Chem.* **275**, 9461–7.

Guarente, L. (1988). UASs and enhancers: common mechanism of transcriptional activation in yeast and mammals. *Cell* **52**, 303–305.

Hardy, R. J., Loushin, C. L., Friedrich, V. L., Chen, Q., Ebersole, T. A., Lazzarini, R. A., and Artzt, K. (1996). Neural cell type specific expression of QKI proteins is altered in the quaking viable mutant mouse. *J. Neurosci.* **16**, 7941–7949.

Holdener, B. C., Mandel, G., and Kouzarides, T. (2000). The co-repressor mSin3A is a functional component of the REST–CoREST repressor complex. *J. Biol. Chem.* **275**, 9461–9467.

International Human Genome Sequencing Consortium (2001). Initial sequencing and analysis of the human genome. *Nature* **409**, 860–921.

Jenuwein, T., and Allis, C. D. (2001). Translating the histone code. *Science* **293**, 1074–1080.

Jepsen, K., Hermanson, O., Onami, T. M., Gleiberman, A. S., Lunyak, V., McEvilly, R. J., Kurokawa, R., Kumar, V., Liu, F., Seto, E., Hedrick, S. M., Mandel, G., Glass, C. K., Rose, D. W., and Rosenfeld, M. G. (2000). Combinatorial roles of the nuclear receptor corepressor in transcription and development. *Cell* **102**, 753–763.

Johnson, P. F., and McKnight, S. L. (1989). Eukaryotic transcriptional regulatory proteins. *Annu. Rev. Biochem.* **58**, 799–839.

Jonat, C., Rahmsdorf, H. J., Park, K. K., Cato, A. C., Gebel, S., Ponta, H., and Herrlich, P. (1990). Antitumor promotion and antiinflammation: Down-modulation of AP-1 (Fos/Jun) activity by glucocorticoid hormone. *Cell* **62**, 1189–2104.

Kindler, S., Mohr, E., and Richter, D. (1997). Quo vadis: extrasomatic targeting of neuronal mRNAs in mammals. *Molecular Cellular Endocrinology* **128**, 7–10.

Knoepfler, P. S., and Eisenman, R. N. (1999). Sin meets NuRD and other tails of repression. *Cell* **99**, 447–450.

Konradi, C., Kobierski, L. A., Nguyen, T. V., Heckers, S. H., and Hyman, S. E. (1993). The cAMP-response-element-binding protein interacts, but Fos protein does not interact, with the proenkephalin enhancer in rat striatum. *Proc. Natl. Acad. Sci. USA* **90**, 7005–7009.

Kornberg, R. D., and Lorch, Y. (1999). Twenty-five years of the nucleosome, fundamental particle of the eukaryote chromosome. *Cell* **98**, 285–294.

Kushner, P. J., Agard, D. A., Greene, G. L., Scanlan, T. S., Shiau, A. K., Uht, R. M., Webb, P. (2000). Estrogen receptor pathways to AP-1. *J. Steroid Biochem. Mol. Bio.* **74**, 311–317.

Lagrange, T., Kapanidis, A. N., Tang, H., Reinberg, D., and Ebright, R. H. (1998). New core promoter element in RNA polymerase II-dependent transcription: Sequence-specific DNA binding by transcription factor IIB. *Genes Dev.* **12**, 34–44.

Lasko, P. (2003). Gene regulation at the RNA layer: RNA binding proteins in intercellular signaling networks. *Science's Stke* **179**, 1–8.

Lee, T. I., and Young, R. A. (2000). Transcription of eukaryotic protein-coding genes. *Annu. Rev. Genet.* **34**, 77–137.

Li, Z., Zhang, Y., Li, D., and Feng, Y. (2000). Destabilization and mislocalization of MBP mRNAs in quaking dysmyelination lacking the QKI RNA binding proteins. *J. Neurosci.* **20**, 4944–4953.

Loriaux, M. M., Brennan, R. G., and Goodman, R. H. (1994). Modulatory function of CREB.CREMα heterodimers depends upon CREMα phosphorylation. *J. Biol. Chem.* **269**, 28839–28843.

Luger, K., Mader, A. W., Richmond, R. K., Sargent, D. F., and Richmond, T. J. (1997). Crystal structure of the nucleosome core particle at 2.8 A resolution. *Nature* **389**, 251–260.

Lunyak, V. V., Burgess, R., Prefontaine, G. G., Nelson, C., Sze, S. H., Chenoweth, J., Schwartz, P., Pevzner, P. A., Glass, C., Mandel, G., and Rosenfeld, M. G. (2002). Corepressor-dependent silencing of chromosomal regions encoding neuronal genes. *Science* **298**, 1747–1752.

Malik, S., and Roeder, R. G. (2000). Transcriptional regulation through Mediator-like coactivators in yeast and metazoan cells. *Trends Biochem. Sci.* **25**, 277–283.

Maue, R. A., Kraner, S. D., Goodman, R. H., and Mandel, G. (1990). Neuron-specific expression of the rat brain type II sodium channel gene is directed by upstream regulatory elements. *Neuron* **4**, 223–231.

Mayr, B., and Montminy, M. (2001). Transcriptional regulation by the phosphorylation-dependent factor CREB. *Nat.Rev.Mol.Cell Biol.* **2**, 599–609.

McKnight, S., and Tjian, R. (1986). Transcriptional selectivity of viral genes in mammalian cells. *Cell* **46**, 795–805.

McKnight, S. L., Kingsbury, R. (1982). Transcriptional control signals of a eukaryotic protein-coding gene. *Science* **217**, 316–324.

Mori, N., Schoenherr, C., Vandenbergh, D. J., and Anderson, D. J. (1992). A common silencer element in the SCG10 and type II Na+ channel genes binds a factor present in nonneuronal cells but not in neuronal cells. *Neuron* **9**, 45–54.

Mori, N., Stein, R., Sigmund, O., Anderson, D. J. (1990). A cell type-preferred silencer element that controls the neural-specific expression of the SCG10 gene. *Neuron* **4**, 583–594.

Naar, A. M., Lemon, B. D., and Tjian, R. (2001). Transcriptional coactivator complexes. *Annu. Rev. Biochem.* **70**, 475–501.

Narlikar, G. J., Fan, H. Y., and Kingston, R. E. (2002). Cooperation between complexes that regulate chromatin structure and transcription. *Cell* **108**, 475–487.

Nelson, C., Albert, V. R., Elsholtz, H. P., Lu, L. I. W., Rosenfeld, M. G. (1988). Activation of cell-specific expression of rat growth hormone and prolactin genes by a common transcription factor. *Science* **239**, 1400–1405.

Orphanides, G., Lagrange, T., and Reinberg, D. (1996). The general transcription factors of RNA polymerase II. *Genes Dev.* **10**, 2657–2683.

Pabo, C. O., and Sauer, R. T. (1992). Transcription factors: Structural families and principles of DNA recognition. *Annu. Rev. Biochem.* **61**, 1053–1095.

Ptashne, M., and Gann, A. (1997). Transcriptional activation by recruitment. *Nature* **386**, 569–577.

Pugh, B. F., and Tjian, R. (1990). Mechanism of transcriptional activation by Sp1: Evidence for coactivators. *Cell* **61**, 1187–1197.

Pugh, B. F., and Tjian, R. (1992). Diverse transcriptional functions of the multisubunit eukaryotic TFIID complex. *J. Biol. Chem.* **267**, 679–682.

Rachez, C., and Freedman, L. P. (2001). Mediator complexes and transcription. *Curr. Opin. Cell. Biol.* **13**, 274–280.

Richards, E. J., and Elgin, S. C. (2002). Epigenetic codes for heterochromatin formation and silencing: Rounding up the usual suspects. *Cell* **108**, 489–500.

Richter, J. D. (1999). Cytoplasmic polyadenylation in development and beyond. *Microbio. Mol. Biol. Rev.* **63**, 446–456.

Richter, J. D., and Lorenz, L. J. (2002). Selective translation of mRNAs at synapses. *Curr. Opini. Neurobiol.* **12**, 300–304.

Roth, S. Y., Denu, J. M., and Allis, C. D. (2001). Histone acetyltransferases. *Annu. Rev. Biochem.* **70**, 81–120.

Schoenherr, C. J., and Anderson, D. J. (1995). The neuron-restrictive silencer factor (NRSF): A coordinate repressor of multiple neuron-specific genes. *Science* **267**, 1360–1363.

Schüle, R., Rangarajan, P., Kliewer, S., Ransone, L. J., Bolado, J., Yang, N., Verma, I. M., and Evans, R. M. (1990). Functional antagonism between oncoprotein c-Jun and the glucocorticoid receptor. *Cell* **62**, 1217–1226.

Sheng, M., McFadden, G., and Greenberg, M. E. (1990). Membrane depolarization and calcium induce c-fos transcription via phosphorylation of transcription factor CREB. *Neuron* **4**, 571–582.

Smale, S. T. (1994). Core promoter architecture for eukaryotic protein-coding genes. *In* "Transcription: Mechanisms and Regulation" (R. C., Conaway, and J. W. Conaway, Eds.), pp. 63–81. Raven Press, New York.

Smale, S. T. (1997). Transcription initiation from TATA-less promoters within eukaryotic protein-coding genes. *Biochim. Biophys. Acta* **1351**, 73–88.

Smale, S. T. (2001). Core promoters: Active contributors to combinatorial gene regulation. *Genes Dev.* **15**, 2503–2508.

Smale, S. T., and Baltimore, D. (1989). The "initiator" as a transcription control element. *Cell* **57**, 103–113.

Steward, O., and Schuman, E. M. (2001). Protein synthesis at synaptic sites on dendrites. *Annu. Rev. Neurosci.* **24**, 299–325.

Tanese, N., Pugh, B. F., and Tjian, R. (1991). Coactivators for a proline-rich activator purified from the multisubunit human TFIID complex. *Genes Dev.* **5**, 2212–2224.

Tiedge, H., and Brosius, J. (1996). Translational machinery in dendrites of hippocampal neurons in culture. *J. Neurosci.* **16**, 7171–7181.

Tjian, R., and Maniatis, T. (1994). Transcription activation: A complex puzzle with few easy pieces. *Cell* **77**, 5–8.

Torre, E. R., and Steward, O. (1992). Demonstration of local protein synthesis within dendrites using a new cell culture system which permits the isolation of living axons and dendrites from their cell bodies. *J. Neurosci.* **12**, 762–772.

Triezenberg, S. J. (1995). Structure and function of transcriptional activation domains. *Curr. Opin. Genet. Dev.* **5**, 190–196.

Turner, B. M. (2002). Cellular memory and the histone code. *Cell* **111**, 285–91.

Vernet, C., and Artzt, K. (1997). STAR, a gene family involved in signal transduction and activation of RNA. *Trends Genet.* **13**, 479–484.

Vinson, C., Myakishev, M., Acharya, A., Mir, A. A., Moll, J. R., and Bonovich, M. (2002). Classification of human B-ZIP proteins based on dimerization properties. *Mol. Cell. Biol.* **22**, 6321–6335.

Wang, H., Iacoangeli, A., Popp, S., Muslimov, I. A., Imataka, H., Sonenberg, N., Lomakin, I. B., and Tiedge, H. (2002). Dendritic BC1 RNA: Functional role in regulation of translation initiation. *J. Neurosci.* **22**(23), 10232–10241.

Weis, L., and Reinberg, D. (1997). Accurate positioning of RNA polymerase II on a natural TATA-less promoter is independent of TATA-binding-protein-associated factors and initiator-binding proteins. *Mol. Cell. Biol.* **17**, 2973–2984.

Workman, J. L., and Kingston, R. E. (1998). Alteration of nucleosome structure as a mechanism of transcriptional regulation. *Annu. Rev. Biochem.* **67**, 545–579.

Xing, J., Ginty, D. D., and Greenberg, M. E. (1996). Coupling of the RAS–MAPK pathway to gene activation by RSK2, a growth factor-regulated CRED kinase. *Science* 959–963.

Mathematical Modeling and Analysis of Intracellular Signaling Pathways

Paul D. Smolen, Douglas A. Baxter, and John H. Byrne

As discussed in Chapter 12, sequences of biochemical reactions termed intracellular signaling pathways transmit signals from the extracellular medium to cytoplasmic or nuclear targets. Ca^{2+} influx is one such signal. Binding of hormones, neurotransmitters, or growth factors to receptors is also a signal. Intracellular signaling pathways translate these signals into effects on the rates of specific cellular processes, such as the transcription of particular genes. In this way, extracellular signals can modulate long-term processes, such as neuronal growth and the formation of long-term memory. It is possible that, in addition to electrical activity, biochemical signals between neurons may be usefully regarded as a conduit of information within the nervous system (Katz and Clemens, 2001). This chapter provides an overview of concepts and techniques necessary to describe and analyze the operation of intracellular signaling pathways with concise mathematical language. By use of these techniques, it is possible to model series (either unbranched or branched) of enzyme reactions. Since such series are commonly termed *metabolic pathways* or *biochemical pathways*, we often use these terms as well as *signaling pathways*.

Because of recent technological advances, information on the operation of intracellular signaling pathways and relationships to long-term neuronal changes is rapidly accumulating. For example, the expression of large groups of genes subsequent to activation of signaling pathways by hormones, neurotransmitters, or drugs can be followed with microarrays ("DNA chips") of specific DNA sequences. Mathematical modeling can provide a conceptual framework to assemble these complex data into concise pictures of the structure and operation of signaling or metabolic pathways. Although all biologists

work with "word models" that describe for them how a system works, intuition fails to adequately deal with complex biochemical systems, and equations are needed to make "word models" rigorous and check them for consistency (Marder, 2000). Furthermore, once model equations are developed, computer simulations or algebraic calculations can display and predict the behavior of complex biochemical systems more accurately and completely than can intuition. This is particularly true if the individual biochemical reactions are nonlinear. A reaction is termed linear if its rate is directly proportional to the concentrations of reactants or of other components such as allosteric effectors of an enzyme. If the reaction rate varies in a more complex manner, the reaction is nonlinear. Computer simulations of equations describing nonlinear reactions commonly display unexpected and complex behaviors (see also Chapter 7, Box 7.4).

Models also help in understanding the effects of multiple feedback interactions within intracellular signaling pathways. Such interactions might include alteration of enzyme activities due to allosteric binding of reaction products further along in a pathway. Time delays, such as are required for macromolecular synthesis and intracellular transport, also add complexity to the behavior of biochemical systems. For example, time delays can be critical for sustaining oscillations in reaction rates and in metabolite concentrations. Such effects of time delays are often nonintuitive and can be appreciated only with a mathematical model.

A model is particularly valuable if it succeeds in reproducing, or predicting, behaviors in addition to those it was originally constructed to replicate (Mauk, 2000). For example, the Hodgkin–Huxley model of

Copyright 2004, Elsevier Science (USA). All rights reserved.

voltage-gated currents was constructed to reproduce the properties of those currents as observed in experiments with each current isolated (Hodgkin and Huxley, 1952). That the model would, in addition, succeed in simulating an action potential was not evident *a priori*, and was of great significance. The simulation helped to confirm that the action potential resulted from the voltage and time dependence of the known voltage-gated currents in the squid axon. The success of this model also helped to establish the utility of mathematical descriptions of neuronal properties. Chapters 5 and 7 discuss the Hodgkin–Huxley model in more detail.

Models are also valuable in establishing principles governing the behavior of biochemical systems. Modeling of signaling pathways and gene regulation has established a unifying principle that feedback interactions commonly underlie complex behaviors. Negative feedback, in which a biochemical species acts to inhibit its own production, often underlies oscillations in biochemical concentrations or gene expression rates. This principle will be illustrated by examples throughout the chapter.

Biochemical models can be constructed at several levels of detail. The most detailed is the stochastic level. Here, one keeps track of the numbers of each type of molecule over time. Such models can be very important if the average numbers of key macromolecules are low. For example, only tens or hundreds of molecules of specific transcription factors or mRNAs may be present. Then, random variations in molecule numbers due to the creation or destruction of individual molecules can have considerable consequences. For example, irreducible individual variability in laboratory animals may be due to random fluctuations in the numbers of key transcription factors and mRNAs during embryonic development (Gartner, 1990).

Lack of experimental data for molecule numbers and for their fluctuations, however, limits the applicability of stochastic models. More commonly, the approximation is made that biochemical concentrations can be written as continuous variables, not as discrete numbers of molecules. The continuous level of modeling uses differential equations for the rates of change of these biochemical concentration variables. At this level, familiar quantities from biochemical kinetics—rate constants, Michaelis constants, enzyme activities—govern the changes in concentrations. Most models discussed in this chapter use this continuous approximation. Computer simulations numerically integrate the differential equations to follow concentrations over time, and the simulations are compared with experimental data. Values of model parameters, such as binding constants and rate constants, are adjusted to improve agreement between simulation and experiment. Finally, if the experimental data are incomplete, a logical or boolean approach may be used. Here, genes or enzymes are simply regarded as switches that are either ON or OFF. For example, this approach has been used to model the changing activities of large numbers of genes during development or cell differentiation.

The choice of which level of modeling to use is determined mainly by the hypothesis to be tested. For example, if a specific enzyme is believed to be rate-limiting for the progress of a biochemical pathway, it is appropriate to use the continuous level of modeling. The activity of the enzyme will be a parameter determining the rate of the reaction. Then it is possible to explore, in a computer simulation, how changing the activity affects the concentrations both of the reaction product and of all chemical species "downstream" in the pathway.

Usually, it is desirable to construct the simplest model capable of capturing the features of the system most relevant to the hypothesis being tested. Such a model highlights the essential features without adding irrelevant details. For example, fluctuations in molecule numbers may not be important when average molecule numbers are large. A method based on separation of rapid and slow biochemical reactions is also often used to simplify models. It is often useful to reduce the number of equations by assuming that the fastest biochemical reactions in a pathway are always near equilibrium.

Another important distinction is between quantitative models, which describe the structure and behavior of specific intracellular biochemical pathways, and qualitative models, which generally have fewer equations and are constructed to embody biochemical elements common to many pathways. Such elements include feedback interactions, random fluctuations in molecule numbers, and formation of multiprotein complexes containing enzymes or transcription factors. This chapter relies primarily on qualitative models. Such models illustrate how typical biochemical elements can give rise to specific types of pathway dynamics, that is, pathway behavior over time. These behaviors include oscillations in reaction rates, or switching between steady states of reaction rates and macromolecule concentrations.

For further study, a selection of textbooks are listed at the beginning of the reference list. These provide a more comprehensive treatment of the ways in which models can be constructed and analyzed, and of the mathematics required, including differential equations.

METHODS FOR MODELING INTRACELLULAR SIGNALING PATHWAYS

Intracellular Transport of Signaling Molecules Can Be Modeled at Several Levels of Detail

Intracellular messages can be transmitted by ions such as Ca^{2+}, by small molecules such as cAMP, or by movement of macromolecules such as enzymes and transcription factors. Modeling intracellular transport of these species is of great importance for understanding the operation of intracellular signaling pathways.

Passive Diffusion Dominates for Ions and Small Molecules

Intracellular transport of ions and small molecules is generally diffusive. Therefore, modeling of diffusive transport—due to movement caused by random thermal collisions with other molecules—is discussed in some detail. Passive diffusion of macromolecules can be modeled by the same equations as for small molecules, but with much smaller diffusion coefficients. More detailed "electrodiffusion" models are used to consider movement of charged molecules due to electric potential gradients within neurons.

The discussion focuses on techniques for modeling the diffusion of a specific ion, Ca^{2+}. The intracellular movement of Ca^{2+} has been extensively studied experimentally and modeled in a variety of ways. The same equations and the same techniques for numerical simulations can be applied to diffusion of other small molecules and ions. The diffusion coefficient must be corrected for differences in molecular weight. For free diffusion in aqueous solution, the diffusion coefficient is inversely proportional to the cube root of molecular weight. However, diffusion *in vivo* is not free, because buffering, or binding of small molecules or ions to sites on macromolecules, is common throughout the cytoplasm. Buffering generally decreases diffusion coefficients below values in aqueous solution.

Concise Models of Ca^{2+} Exchange between Compartments Represent Ca^{2+} Release and Uptake Mechanisms

As discussed in Chapter 12, both free Ca^{2+} and Ca^{2+} bound to proteins such as calmodulin can activate numerous enzymes within intracellular signaling pathways. Therefore, many investigators have modeled aspects of the movement of Ca^{2+} from the plasma membrane to other regions. Two basic approaches are used. The simplest considers only Ca^{2+} exchange across membranes that separate intracellular compartments. The exchange can be driven by passive diffusion or by ion "pump" proteins that use ATP. Distinct membrane-bound compartments are each assigned a pool of Ca^{2+}. For example, cytoplasmic Ca^{2+}, Ca^{2+} in the endoplasmic reticulum, mitochondrial Ca^{2+}, and Ca^{2+} in the nucleus might each be considered to form a distinct "pool." Cytoplasmic Ca^{2+} could communicate with all the other pools. The time course of the amounts of Ca^{2+} in each pool is followed during computer simulations. In the second, more computationally intensive approach, the spaces within intracellular compartments are modeled as small volume elements, and the Ca^{2+} concentration in each element is followed separately. Diffusion of Ca^{2+} within compartments or across membranes that separate compartments can be modeled as fluxes between volume elements. This approach is required if it is necessary to model concentration gradients within the cytoplasm or other compartments.

Suppose a single pool describes cytoplasmic Ca^{2+}. A single ordinary differential equation is then written to describe the rate of change of cytoplasmic Ca^{2+} concentration, $[Ca^{2+}_{cyt}]$. The rate of change is assumed proportional to Ca^{2+} influx across the plasma membrane. It also depends on rates of exchange between cytoplasmic Ca^{2+} and Ca^{2+} in intracellular compartments such as the endoplasmic reticulum:

$$\frac{d[Ca^{2+}_{cyt}]}{dt} = \lambda I_{Ca} + k_{exch}\left([Ca^{2+}_{ER}] - [Ca^{2+}_{cyt}]\right) - k_{Ca}[Ca^{2+}_{cyt}]. \tag{14.1}$$

In Eq. 14.1, the first term on the right-hand side describes Ca^{2+} influx across the plasma membrane through an ion channel or set of channels. A scaling constant, here denoted by λ, is required to convert Ca^{2+} current to a rate of change of Ca^{2+} concentration. As discussed in Chapter 7, detailed expressions can be used to describe the voltage and time dependence of such ionic currents. The second term on the right-hand side of Eq. 14.1 describes exchange of Ca^{2+} between the endoplasmic reticulum (ER) and the cytoplasm. For illustrative purposes this term is very simple: it is merely proportional to the concentration difference between the two Ca^{2+} pools.

In models of specific cell types more detailed kinetic expressions might be needed to describe Ca^{2+} release from the ER. For example, special expressions are necessary to describe stimulation of ER Ca^{2+} release by the second messenger molecule inositol 1, 4, 5-triphosphate (IP_3). Recent models (Wagner *et al.*, 1998; Li and Rinzel, 1994) describe Ca^{2+} release through the IP_3 receptor on the ER as increasing steeply with the third or fourth power of the cytoplasmic IP_3 concentration. Ca^{2+} release through the IP_3 receptor also increases steeply with the cytoplasmic Ca^{2+} concentration. This is positive feedback, because an initial increase in cytoplasmic Ca^{2+} stimulates further release of Ca^{2+} from the ER and a further increase in cytoplasmic Ca^{2+}. Such positive feedback can help to generate Ca^{2+} waves and oscillations in a variety of cell types (Wagner *et al.*, 1998; Li and Rinzel, 1994; Nakamura *et al.*, 1999).

The last term in Eq. 14.1 describes Ca^{2+} extrusion across the membrane as a simple first-order process. However, a more detailed expression may be preferred if data concerning rates of extrusion are available to give estimates for the parameters in such an expression. For example, the expression

$$-k_{Ca} \frac{[Ca^{2+}_{cyt}]}{[Ca^{2+}_{cyt}] + K} \qquad (14.2)$$

describes Ca^{2+} extrusion across the plasma membrane by a saturable ion pump. In Eq. 14.2, the parameter K gives the value of $[Ca^{2+}]$ at which the pump rate is half-maximal. Saturable pumps driven by ATP hydrolysis are present in many cell types. A more complicated expression,

$$-k_{Ca} ([Ca^{2+}_{cyt}]^2 [Na^+_{external}]^3 -$$
$$[Ca^{2+}_{external}]^2 [Na^+_{cyt}]^3), \qquad (14.3)$$

is sometimes used to describe removal of Ca^{2+} by a $Na^+ - Ca^{2+}$ exchanger present in neurons and other excitable cells (De Schutter and Smolen, 1998). This exchanger brings Na^+ into the cytoplasm while concurrently removing Ca^{2+}. In Eq. 14.3 there are two terms. The first term describes Ca^{2+} removal and the second term is necessary because the exchanger can also operate in reverse, bringing in Ca^{2+}. The powers of Na^+ and Ca^{2+} concentrations reflect the mechanism of the exchanger protein, which undergoes repeated cycles of conformational change, bringing in approximately three Na^+ ions per cycle while extruding two Ca^{2+} ions.

To simulate changes in Ca^{2+} concentration as described by a differential equation such as Eq. 14.1, Euler's method is often used. A reasonable initial value for $[Ca^{2+}_{cyt}]$ is chosen (such as 50 nM, which is in the range expected if a neuron is hyperpolarized and there is little Ca^{2+} influx through ion channels). Then, at each small step in time, one calculates the rate of change of $[Ca^{2+}_{cyt}]$ in terms of the quantities on the right-hand side of Eq. 14.1. The rate of change is multiplied by the size of the time step and the result is added to $[Ca^{2+}_{cyt}]$. This procedure is repeated until a time length sufficient to display the behavior of

$[Ca^{2+}_{cyt}]$ during neuronal activity has been simulated. To determine whether the time step chosen is small enough, the entire simulation is typically repeated with a time step approximately one-third as long, and if the behavior of $[Ca^{2+}_{cyt}]$ does not vary significantly, the larger time step is considered adequate. This simple method of numerical integration can be applied to any differential equation discussed in this chapter. More complex integration methods also exist and these may, for many models, allow faster simulations. Textbooks on numerical analysis (Burden and Faires, 1993) can be consulted for details about these methods.

To Model Diffusion in Realistic Detail, Standard but Complex Equations Can be Used

To more accurately model the intracellular diffusion of Ca^{2+}, other ions, or small molecules, the intracellular space must be discretized or divided into small volume elements. The evolution of the Ca^{2+} concentration within each element must be simulated. Such discretization is the standard method for simulating the partial differential equation that describes diffusion within an extended region. A common implementation uses cubic or rectangular volume elements. The cytoplasm or other region of interest is approximated by a large set of small boxes in x, y, z space. The overall set is chosen to have a shape that is a reasonable representation of the region of interest. For example, an approximate sphere with a "boxy" boundary may be used to represent the nucleus. The flux of Ca^{2+} between two cubic boxes depends on the cube size and the Ca^{2+} diffusion coefficient, D_{Ca}. The dependence is

$$Flux_{1 \to 2} = D_{Ca} \frac{A}{L} \left([Ca^{2+}]_1 - [Ca^{2+}]_2\right). \qquad (14.4)$$

In Eq. 14.4, L is the length of a cube edge, and $A (= L^2)$ is the area of a cube face. $[Ca^{2+}]_1$ refers to the Ca^{2+} concentration in cube 1. A variant of Eq. 14.4 is used for rectangular boxes. Although Eq. 14.4 as written describes the diffusion of only free Ca^{2+}, the same type of equation can be used to describe the diffusion of Ca^{2+} bound to calmodulin or other species. Simulations using Eq. 14.4 can use Euler's numerical integration method, as used for Eq. 14.1. At each small time step, the flux of Ca^{2+} given by Eq. 14.4 is multiplied by the size of the time step, and the resulting quantity is added to the preceding Ca^{2+} concentration in volume element 2 while also being subtracted from the preceding Ca^{2+} concentration in volume element 1. At each time step this calculation must be repeated for every pair of adjacent volume elements in the system.

In one application (Simon and Llinas, 1985, see also Fig. 8.3), Eq. 14.4 was used to simulate dynamics of localized Ca^{2+} peaks near clusters of open Ca^{2+} channels. It was found that near the cytoplasmic mouth of a channel, free Ca^{2+} concentrations could reach values of tens of micromolar, or higher. These Ca^{2+} microdomains form and dissipate within microseconds of channel openings/closings. Enzymes or other proteins positioned near clusters of channels can serve as sensors for such elevations of concentration, and intracellular signaling pathways could thereby be activated. For example, pathway activation could lead to local synaptic potentiation or depression. The importance of Ca^{2+} microdomains for the release of neurotransmitter was discussed in Chapter 8.

Modeling diffusion with rectangular volume elements has the disadvantage of requiring very large amounts of computer time and memory because of the large number of elements required to discretize space in three dimensions. For example, if each axis is to be divided into 100 parts, a total of 1,000,000 volume elements are needed. A great reduction in time and memory requirements can be achieved if the biological system can be approximated by a model where, because of an inherent symmetry, diffusion in one or two dimensions can be neglected. Then only two or one dimensions, respectively, need to be divided into small volume elements.

For example, suppose diffusion within a spherically symmetric compartment, such as an ideal soma, is being modeled. If Ca^{2+} is assumed to enter and leave at equal rates over all regions of the outer surface, then Ca^{2+} concentration is independent of both angle coordinates within the sphere. Therefore, only the radial dimension needs to be discretized because diffusion is only along the radial direction from the surface to the center. The compartment can be divided radially into thin, concentric spherical shells (Fig. 14.1A). The concentration changes in each shell due to fluxes of Ca^{2+} between shells are simulated. An equation analogous to Eq. 14.4 is used, and Euler's integration method can be used. Computer algorithms for these simulations have been published. The algorithm in Blumenfeld *et al.* (1992) is particularly straightforward to implement.

As a second example in which symmetry reduces the number of dimensions that need to be discretized, consider a portion of a neuronal dendrite or axon, which can be modeled as a cylindrical compartment. Suppose Ca^{2+} is assumed to enter one end and leave at the other end, without significant influx from the sides. Then Ca^{2+} gradients across the width of the cylinder can be neglected, and Ca^{2+} diffusion occurs only along the length of the cylinder. Thus, only the dimension of length needs to be divided into volume elements. The cylinder can be divided into thin slabs, and the evolution of the concentration in each slab simulated (Fig. 14.1B). Also, consider a different cylindrical compartment for which Ca^{2+} flux through the side face, but not through the ends, is significant. If all regions of the face are assumed homogenous, then concentration gradients will develop only in the radial direction, from the outer surface to the center. Thus, the cylinder may be divided into a series of thin, concentric cylindrical shells. The algorithm in Blumenfeld *et al.* (1992) can readily be adapted to this case. As a last example of a useful symmetry, diffusion of neurotransmitter within a synaptic cleft has been modeled by describing the cleft as a thin disk-shaped region and neglecting the gradient of neurotransmitter across the cleft. If neurotransmitter is released at the center, only diffusion in the radial direction needs to be considered. The disk can be discretized into concentric annular rings (Fig. 14.1C) and diffusion between the rings is simulated. However, if neurotransmitter is released anywhere besides the center of the disk, diffusion in the angular direction cannot be neglected, and simulation requires discretization in both the radial and the angular dimensions.

These methods for modeling diffusion all assume that ionic movement is driven by concentration gradients to a much greater extent than by electrical potential gradients. This may not hold true within small structures such as dendritic spines. For example, influx of ions during electrical activity might cause large potential differences between the spine head and the dendritic shaft. The electrodiffusion modeling approach is more accurate for such situations. This approach uses more general and complex equations, with terms to account for ion movement due to diffusion and due to potential-driven drift. As an example, the Nernst–Planck equation and discretization into small volume elements have been used to model diffusion of several ionic species within a spine and the adjacent dendrite (Qian and Sejnowski, 1990). When constructing electrodiffusion models, it is necessary to carefully consider which charged ions or molecules are likely to play the largest roles in carrying charges and dynamically altering potential gradients. Both the concentrations and mobilities of common charged species need to be considered when making these assessments.

Buffering Must Be Considered in Models

Binding of ions or small molecules to macromolecules can greatly slow diffusion. For example, a Ca^{2+} ion spends much of its time bound, so that the

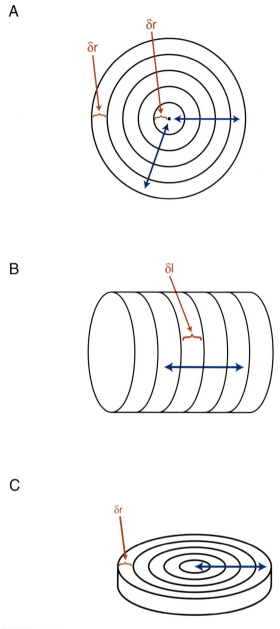

FIGURE 14.1 Symmetry can allow advantageous division of spatial regions into volume elements to model diffusion. (A) Spherically symmetric diffusion. A spherical region is divided into thin concentric cells of width δr. Diffusion of molecules between shells occurs in the radial direction (blue arrows). (B) Longitudinal diffusion in a cylinder (blue arrow). The cylinder is divided into thin slabs of width δl. (C) Radially symmetric diffusion in a disk-shaped region that approximates a synaptic cleft. The disk is divided into concentric annuli of width δr.

diffusion of Ca^{2+} ions is much slower in the cell than in water. Other ions, and small molecules, tend to have fewer binding sites than does Ca^{2+}, so that their diffusion is not slowed to such an extent. For example, there are relatively few intracellular binding sites for the second messenger IP_3. Therefore, diffu-

sion of IP_3 can be considerably more rapid than diffusion of Ca^{2+}, even though the molecular weight of a Ca^{2+} ion is much lower (Jaffri and Keizer, 1995).

For a given class of Ca^{2+} binding sites, denoted S, and present at a total concentration S_{tot}, consider equilibration between free and bound Ca^{2+}. By use of standard chemical kinetic notation, the dissociation and association rate constants for formation of S–Ca^{2+} complex are respectively denoted k_b and k_f. Equilibration is then described by the differential equation

$$\frac{d[Ca^{2+}]_{free}}{dt} = k_b[S-Ca^{2+}]$$
$$-k_f\left(S_{tot}-[S-Ca^{2+}]\right)[Ca^{2+}]_{free} \tag{14.5}$$

In Eq. 14.5, $[S-Ca^{2+}]$ denotes the concentration of S–Ca^{2+} complex, $[Ca^{2+}]_{free}$ denotes the concentration of unbound Ca^{2+}, and the expression $S_{tot} - [S-Ca^{2+}]$ gives the concentration of free S sites. An equilibrium dissociation constant, K, is defined as k_b/k_f. When $[Ca^{2+}]_{free}$ is at equilibrium and equal to K, half of the sites S are occupied by Ca^{2+}. This can be seen as follows. At equilibrium, the system is unchanging, so that

$$\frac{d[Ca^{2+}]_{free}}{dt} = 0.$$

Therefore, at equilibrium, the right-hand side of Eq. 14.5 can be set equal to zero. Then a rearrangement gives

$$K = k_b\Big/k_f = \frac{[S][Ca^{2+}]_{free}}{[S-Ca^{2+}]}. \tag{14.6}$$

If $[Ca^{2+}]_{free} = K$, Eq. 14.6 simplifies to

$$\frac{[S]}{[S-Ca^{2+}]} = 1;$$

i.e. 50% of the sites are occupied by Ca^{2+}.

If binding and unbinding of Ca^{2+} are fast processes (large k_f and k_b), then a rapid equilibrium approximation is justified. This assumes that the difference between the two terms on the right-hand side of Eq. 14.5 is always small in comparison with the absolute magnitudes of either term. Thus the terms can be considered equal for practical purposes. The right-hand side of Eq. 14.5 is therefore again set equal to zero. The equilibrium constant K is k_b/k_f. There are two conservation conditions that express the total concentration of buffer sites S_{tot} and the total Ca^{2+} concentration C_{tot} in terms of the individual biochemical species:

$$S_{\text{tot}} = [S] + [S - Ca^{2+}]$$

and

$$C_{\text{tot}} = [Ca^{2+}]_{\text{free}} + [S - Ca^{2+}].$$

Under these conditions, the concentration of free S can be expressed as

$$[S] = [Ca^{2+}]_{\text{free}} + S_{\text{tot}} - C_{\text{tot}}. \qquad (14.7)$$

Using Eq. 14.6 for the equilibrium constant K, and using Eq. 14.7, an equation can be written that expresses $[Ca^{2+}]_{\text{free}}$ in terms of constant parameters and the total calcium concentration:

$$\frac{[S][Ca^{2+}]_{\text{free}}}{[S - Ca^{2+}]} = \frac{\left([Ca^{2+}]_{\text{free}} + S_{\text{tot}} - C_{\text{tot}}\right)[Ca^{2+}]_{\text{free}}}{C_{\text{tot}} - [Ca^{2+}]_{\text{free}}} = K. \quad (14.8)$$

From Eq. 14.8, it is possible to solve for $[Ca^{2+}]_{\text{free}}$ in terms of the other quantities.

Euler's integration method, discussed above, can be used for simulations of the evolution of the concentrations of free and bound Ca^{2+}. At each time step of the simulation, the total Ca^{2+} concentration C_{tot} can first be updated by considering processes of Ca^{2+} influx into or removal from the cytoplasm. To generate a time course of free Ca^{2+}, Eq. 14.8 can then be used to update $[Ca^{2+}]_{\text{free}}$. Why use this somewhat cumbersome equation? Because if binding of Ca^{2+} to S is much faster than influx or removal of Ca^{2+} from the cytoplasm, then assuming binding to be always at equilibrium (Eq. 14.8) can allow much larger time steps to be taken in simulations, saving much computer time. The time steps only need to be small relative to Ca^{2+} influx or removal rates. Equation 14.8 is an example of a more general principle. Assuming that rapid biochemical processes are always at equilibrium can greatly simplify a model by reducing the number of differential equations.

When multiple classes of buffer sites (S_1, S_2, . . .) are present, each class will have its own values of the parameters k_f, k_b, and S_{tot}. An analog of Eq. 14.5 can be written. The right-hand side of the analog contains separate terms for each value of k_f and k_b for each class. However, the rapid-equilibrium approximation is no longer useful in simulations, because with multiple buffers, Eq. 14.8 becomes too complex to allow for savings in computer time.

Experimental data suggest that, to a reasonable approximation, total intracellular Ca^{2+} buffering can often be thought of as due to one or two "lumped" buffering species, or classes of buffer sites (Zhou and Neher, 1993). The effect of each species is described by different values of the equilibrium constant K and

the total site concentration S_{tot}. A useful experimental quantity that helps determine values of K and S_{tot} is the buffering capacity. This is the ratio of the change in total (free + bound) Ca^{2+} to the change in free Ca^{2+} when a small amount of Ca^{2+} is added to the cytoplasm. In an analysis of Ca^{2+} buffering in adrenal chromaffin cells (Zhou and Neher, 1993) it was found that buffering could be described by two "lumped" species: one immobile with a buffering capacity of ~30, and the other slowly diffusing and having a buffering capacity of ~10. In neurons, similar or greater values apply. Detailed calculations using these estimates of buffering capacity suggest that ~98–99% of cytosolic Ca^{2+} is bound in chromaffin cells (Zhou and Neher, 1993) and presumably in neurons.

Molecular probes have been developed that monitor the intracellular movement of Ca^{2+} and other ions or small molecules, including Mg^{2+}, Zn^{2+}, and cAMP. For example, dyes such as Fura-2 have a fluorescence emission spectrum that is altered by Ca^{2+} binding. Experimental data characterizing Ca^{2+} gradients and movement are obtained by monitoring changes in the fluorescence of such dyes (see Fig. 18.8). Data obtained with such probes are vital to refining mathematical models that include transport. Models to simulate such data need to consider the diffusion of all important species including dye with and without bound Ca^{2+}. Buffering due to introduced probes can have complex effects on the diffusion coefficients of ions and small molecules (Wagner and Keizer, 1994). Introduction of a small probe molecule that binds Ca^{2+} can speed up the net rate of Ca^{2+} diffusion, even though the free concentration of Ca^{2+} is decreased. This occurs because, after the probe is added, a significant fraction of total Ca^{2+} is bound to highly mobile probe molecules instead of poorly mobile macromolecules.

Models that incorporate the dynamics of IP_3—induced Ca^{2+} release, Fura–2 diffusion, and Ca^{2+} buffering have simulated experimental characterizations of intracellular Ca^{2+} waves following hormonal or electrical stimulation. These models—or any model—are most useful if they can predict the results of future experiments. If the predictions bear out, the model gains credibility as a valid physiological description. For example, one recent investigation (Fink *et al.*, 2000) quantitatively predicted alterations in Ca^{2+} waves due to microinjection of excess buffer or photorelease of intracellular IP_3.

Details of buffer and Ca^{2+} diffusion can be disregarded if Ca^{2+} dynamics are being modeled using separate intracellular "pools" (Eq. 14.1) with Ca^{2+} exchange across membranes separating pools. In this situation, it is common to regard the ratio of free Ca^{2+} to total Ca^{2+} as fixed within each pool. For example, if the overall buffering capacity of the cytoplasm is in the range of 100, a fixed ratio of 0.01 would be appropriate. In this case, only 1/100th of the Ca^{2+} entering or leaving the pool is reflected as a change in the concentration of free Ca^{2+}. The modification of Eq. 14.1 required is

$$\frac{d[Ca^{2+}{}_{cyt}]_{free}}{dt} = f\left(\lambda I_{Ca} + k_{exch}\left([Ca^{2+}{}_{ER}]_{free} - [Ca^{2+}{}_{cyt}]_{free}\right)\right.$$
$$\left. - k_{Ca}[Ca^{2+}{}_{cyt}]_{free}\right). \qquad (14.9)$$

In Eq. 14.9, f denotes the fixed ratio of free Ca^{2+} to total Ca^{2+}. In Eq. 14.9, only concentrations of free Ca^{2+} are used to determine total Ca^{2+} fluxes between compartments or out of the cell because bound Ca^{2+} is generally not transported across membranes.

From Eq. 14.9, it follows that the effect of a small ratio f is to greatly slow down the rate of change of Ca^{2+} concentration. As a result, $[Ca^{2+}{}_{cyt}]_{free}$ can become one of the slowest changing variables in models of electrical activity in neurons. With these models, oscillations in $[Ca^{2+}{}_{cyt}]_{free}$ with a period on the order of seconds can be simulated. These oscillations are often coupled to oscillations of the membrane potential, and sometimes bursts of action potentials are superimposed on the membrane potential oscillations (Amini *et al.*, 1999; Canavier *et al.*, 1993; Smolen and Keizer, 1992).

When modeling diffusion of ions other than Ca^{2+}, or of small molecules, including buffering is often less important because there are fewer binding sites than for Ca^{2+}. For modeling the diffusion of macromolecules, it is desirable to include buffering. However, there are usually insufficient data to determine the number and distribution of binding sites for specific macromolecules. With this limitation, a reasonable approach is to assign a plausible value for the macromolecular diffusion coefficient and then simulate experiments that measure macromolecular movement rates. A common experimental technique is fluorescence recovery after photobleaching (FRAP) (Kaufman and Jain, 1990). This technique uses strong brief illumination of an intracellular region with a laser. The laser eliminates emission from a fluorescently tagged macromolecule within the region by photobleaching the fluorophore. Afterward, the time course of fluorescence recovery is monitored as the macromolecule diffuses in from other regions. Fitting this time course with computer simulations provides an estimate for the diffusion coefficient.

Active Transport May Dominate over Diffusion for Macromolecules

For most macromolecules, intracellular movement is often via active transport requiring ATP. Active transport is especially important in cells with extended morphologies, such as neurons. In neurons, calculations indicate that passive diffusion of mRNAs and proteins from the soma into narrow dendrites would be much slower than the observed rates of movement (Sabry *et al.*, 1995). Active transport consists of directed movement along cytoskeletal elements of specific macromolecules or of small, membrane-bound vesicles containing macromolecules. The movement is mediated by motor proteins such as kinesin and dynein. Active transport has not been modeled as often as diffusion. For constraining models of active macromolecular transport, data specific to neurons need to be used because there is great variability in transport rates between neurons and other cell types. A reasonable approximation is to assume a constant drift of macromolecule, at a velocity that may vary between different regions of a neuron. If diffusive movement is also important, it should be modeled separately. As a first approximation, diffusion might be simply added to the motion due to active transport, providing an additional random motion. Diffusion might be ignored if data suggest active transport is much more important for particular macromolecules. As mentioned above, this appears to be the case for transport of many RNAs and proteins into extended neuronal processes.

Standard Equations Simplify Modeling of Enzymatic Reactions, Feedback Loops, and Allosteric Interactions

In cells, most biochemical reactions of interest are catalyzed by enzymes, and a variety of mathematical descriptions have been developed for these reactions. Many enzymatic reactions have complex kinetic mechanisms, and specialized equations are needed to describe their rates in detail (Segel, 1975). When a series or group of reactions is being modeled, however, it is more common to use simplified equations for the individual reaction rates. A few such forms have become standard.

To review these equations, it is first helpful to review the definition of reaction order. A zero-order reaction converts substrate into product at a fixed rate independent of substrate concentration. A first-order reaction converts substrate into product at a rate proportional to substrate concentration. A second-order reaction creates product at a rate proportional to either the product of two substrate concentrations or the square of a single substrate concentration. Higher-order reactions are similarly defined. In many cases, an enzyme transforms a single substrate, S, into a single product, P. The simplest assumption is that the rate of production of P is first-order with respect to S. This assumption can be useful when many reactions are being modeled, to minimize the complexity of the model and the computational time required for simulations (Bhalla and Iyengar, 1999). The next level of detail considers saturation of the enzymatic reaction rate at high concentrations of S. Then, the reaction no longer has a definite order, and a nonlinear equation must be used for the rate. The simplest such equation for the rate of production of P, $d[P]/dt$, is the standard Michaelis–Menten equation:

$$\frac{d[P]}{dt} = \frac{V_{\max}[S]}{[S]+K_m}. \tag{14.10}$$

This equation has been found *in vitro* to accurately describe the rates of many enzymatic reactions. $V_{\max}$ represents the maximal enzyme velocity when it is fully saturated with the substrate S, and K_m represents the concentration of S at which the velocity is half of its maximal value. Note that if the enzyme is saturated with substrate (high [S]), Eq. 14.10 reduces to a description of a zero-order reaction.

If two substrates, S_1 and S_2, are converted into a single product P, then the simplest assumption is that the rate of production of P is second-order, proportional to the product $[S_1][S_2]$. One can add more realism by including saturation effects. This can be done by using a product of Michaelis–Menten expressions. Suppose also a first-order degradation of P. Then a differential equation for the *in vivo* rate of change of the concentration of P is

$$\frac{d[P]}{dt} = k_{\max}\left(\frac{[S_1]}{[S_1]+K_1}\frac{[S_2]}{[S_2]+K_2}\right) - k_{\deg}[P]. \tag{14.11}$$

Here, the first term on the right-hand side is a product of Michaelis–Menten expressions for $[S_1]$ and $[S_2]$. The maximal velocities $V_{\max}$ have been combined into the parameter $k_{\max}$. The second term represents first-order degradation. In Eq. 14.11, K_1, K_2, $k_{\max}$, and $k_{\deg}$ are parameters to be estimated from experimental data.

A model of a biochemical reaction pathway is composed of a set of differential equations such as Eq. 14.11. There is an equation for the rate of change of the concentration of each reaction substrate and product. Given initial values for the concentrations of all biochemical species, numerical integration of the differential equations is done in the manner described previously for Eq. 14.1. The integration gives values of the concentrations for times subsequent to the initial time.

Allosteric Interactions between Enzymes and Small Molecules Can Alter Enzyme Activities and Mediate Feedback

Binding of small molecules can alter an enzyme's conformation and alter the rate of the reaction catalyzed by the enzyme. Often, such allosteric interactions are with effector molecules not involved in the reaction. But for some enzymes (e.g. pyruvate kinase, phosphofructokinase) substrates or products can serve as allosteric effectors. Allosteric interactions can therefore mediate feedback and feedforward interactions within a biochemical pathway, as well as cross-talk between pathways.

In a feedback interaction, a product of an enzymatic reaction affects the activity of another enzyme earlier in the pathway, whereas in a feedforward interaction, the affected enzyme is later in the pathway. With cross-talk, a metabolite from one pathway affects the activity of an enzyme in another pathway. Two types of feedback can be distinguished. If the product of a later reaction acts to speed up an earlier reaction, the feedback is positive; if the effect is to slow down the earlier reaction, the feedback is negative. Positive feedback tends to drive a biochemical pathway to a state of maximal activity determined by the saturated rate of the slowest individual reaction, whereas negative feedback tends to drive the pathway to a state of low activity. Graphically, such an interaction is often represented as in Fig. 14.2A, where the circled minus sign denotes negative feedback.

In models of enzyme regulation, allosteric interactions are commonly represented by Hill functions (Segel, 1975). These are saturable functions of the concentration of the effector molecule. With the concentration of effector denoted by [L], if [L] activates an enzyme, the enzyme activity is taken as proportional to the following increasing function of the nth power of [L]:

$$\frac{[L]^n}{[L]^n + K_H^n}. \tag{14.12a}$$

When the Hill function of Eq. 14.12a is plotted versus effector concentration [L], the graph has a characteristic sigmoid shape. The enzyme activity is near 0 for low values of [L]. In Eq. 14.12a, the parameter K_H has units of concentration. When [L] = K_H, the activity has a value of 0.5. Over a range of [L] centered about K_H, the activity increases rather steeply to near its maximal value of one. The parameter n is called the Hill coefficient. Greater values of n correspond to steeper sigmoids, that is, to a narrowing of the range of [L] over which the enzyme activity is significantly above 0 and also significantly below 1.

If L inhibits an enzyme, the enzyme activity is taken as proportional to a decreasing function of [L]:

$$\frac{K_H^n}{[L]^n + K_H^n}. \qquad (14.12b)$$

In Eq. 14.12b, the parameters n and K_H have the same meanings as in Eq. 14.12a. As an example of how these functions are used in rate equations for biochemical reactions, inclusion of an increasing Hill function in Eq. 14.10 gives

$$\frac{d[P]}{dt}\left(\frac{[L]^n}{[L]^n + K_H^n}\right)\frac{V_{max}[S]}{[S] + K_m}. \qquad (14.13)$$

Often, an enzyme has multiple binding sites for an allosteric effector, particularly if the enzyme is composed of multiple subunits. Greater values of the Hill coefficient often correspond to a larger number of binding sites for a given allosteric effector. Indeed, experimentally determined Hill coefficients are often taken as a rough indication of the number of binding sites. Instead of Hill functions, more complex expressions based on Monod–Wyman–Changeux allosteric theory, or similar theories, are sometimes used (Segel, 1975).

Figure 14.2B illustrates a relatively simple model (Cooper et al., 1995) that incorporates Hill functions, reaction rates directly proportional to substrate concentration, and first-order degradation. This model describes the relationship between cyclic AMP (cAMP) production and the rate of Ca²⁺ influx into a cell. The model can simulate persistent oscillations in [cAMP] if the parameter values fall within the proper range (legend to Fig. 14.2).

The model contains four biochemical species: cAMP, Ca²⁺, active adenylyl cyclase enzyme (denoted by AC), and active Ca²⁺ channels in the plasma membrane (denoted by Y). Some model species alter the production or degradation rates of others (Fig. 14.2B). The rate of cAMP production is assumed to be proportional to the amount of active adenylyl cyclase enzyme. cAMP is assumed to be degraded by a first-order process. Therefore, the differential equation describing the rate of change of cAMP concentration is

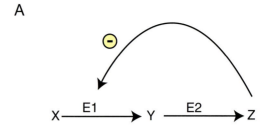

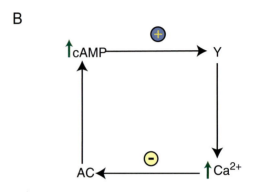

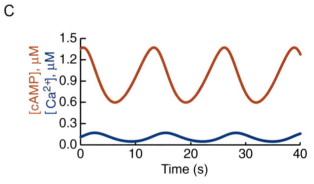

FIGURE 14.2   Feedback within two model signaling pathways. (A) Negative feedback within a simple, linear reaction scheme. Metabolite X is transformed into Y which is transformed into end product Z. Z inhibits the enzyme E1 that catalyzes the X → Y reaction. (B) A model that captures some relationships between cAMP production and Ca²⁺ influx. cAMP activates Ca²⁺ channels (denoted Y). Ca²⁺ influx is thereby increased. Ca²⁺ influx inhibits adenylyl cyclase (AC). The overall effect is a negative-feedback loop in which cAMP inhibits its own production. (C) Oscillations in cAMP and Ca²⁺ are sustained by the negative-feedback loop. For this simulation, the values of the parameters in Eqs. 14.14–14.17 are $k_1 = 1.0\,s^{-1}$, $k_2 = 0.5\,s^{-1}$, $k_3 = 1.0\,s^{-1}$, $k_4 = 0.5\,s^{-1}$, $k_5 = 0.45\,\mu M\,s^{-1}$, $k_6 = 1.0\,s^{-1}$, $k_7 = 0.11\,\mu M\,s^{-1}$, $k_8 = 2.0\,s^{-1}$, $K_1 = 2.0\,\mu M$, $K_2 = 0.2\,\mu M$. Initial values for the model variables at $t = 0$ are cAMP = 1.0 μM, AC = 0.5 μM, Y = 0.05 μM, and Ca²⁺ = 0.1 μM.

$$\frac{d[cAMP]}{dt} = k_1[AC] - k_2[cAMP]. \qquad (14.14)$$

The rate of Ca²⁺ influx is assumed to be proportional to the number of active Ca²⁺ channels in the plasma membrane. Ca^{2+} is extruded from the cell by a first-order process. Therefore, the rate of change of Ca^{2+} concentration can be written as,

$$\frac{d[Ca^{2+}]}{dt} = k_3 Y - k_4 [Ca^{2+}], \tag{14.15}$$

Ca^{2+} channels are assumed to be activated by the binding of cAMP. A Hill function, with a Hill coefficient of 3, characterizes activation by cAMP. Channels deactivate in a first-order fashion in the absence of cAMP. By use of Eq. 14.12a, the differential equation for the number of active channels, denoted Y, can be written as

$$\frac{dy}{dt} = \frac{k_5 [cAMP]^3}{[cAMP]^3 + K_1^3} - k_6 Y. \tag{14.16}$$

Ca^{2+} is assumed to inhibit adenylyl cyclase. Mathematically, this inhibition can be represented by an equation in which Ca^{2+} increases the rate at which the variable AC decreases. This effect of Ca^{2+} can be described by a Hill function. The resulting differential equation for AC is

$$\frac{dAC}{dt} = k_7 - AC \frac{k_8 [Ca^{2+}]^3}{[Ca^{2+}]^3 + K_2^3}. \tag{14.17}$$

As discussed in Chapter 12, the actual regulation of adenylyl cyclase is more complex than embodied in Eq. 14.17. Some neuronal isoforms of adenylyl cyclase are inhibited by elevations of Ca^{2+} such as occur during electrical activity, but others are stimulated (Chapters 12 and 18).

If Hill coefficients greater than 2 are chosen in Eqs. 14.6 and 14.17, the model can produce stable oscillations in [cAMP] and the other variables (Fig. 14.2C). The generation of oscillations by this model requires a negative-feedback loop. This operates as follows. An initial increase in the level of cAMP activates Ca^{2+} channels, allowing the influx of Ca^{2+}. Ca^{2+}, in turn, inhibits adenylyl cyclase. The inhibition acts to decrease the concentration of cAMP, closing the loop. The model helps to suggest a principle that negative-feedback loops favor the occurrence of oscillations in concentrations and reaction rates.

Positive and Negative Feedback Can Support Complex Dynamics of Biochemical Pathways

Feedback Can Sustain Oscillations or Multistability of Concentrations and Reaction Rates

One interesting behavior in biochemical pathways is persistent oscillations in the rates of enzymatic reac-

tions and in the concentrations of substrates and products. For example, oscillations have been observed in the metabolic flux through glycolysis and also in the rates of secretion of hormones such as insulin. The previous section discussed a model (Fig. 14.2) in which concentration oscillations were driven by a negative-feedback loop. More generally, oscillations in reaction rates and concentrations commonly rely on negative feedback to sustain oscillations (Gartner, 1990). Models of biochemical systems that rely on positive feedback tend to display another type of complex behavior. This behavior, termed multistability, is defined as the existence of multiple steady states for the concentrations of the chemical species. Each state is fixed in time and corresponds to a fixed metabolic flux through the pathway. Each state is stable in that small disturbances of the system die out with time so that the system, if originally in one of the states, returns to that state. Large disturbances of the system, however, can switch it between stable states.

Bistability is a specific type of multistability with two stable states. Bistability was discussed in Chapter 7 (Fig. 7.19) in the context of a model neuron that could exhibit a fixed, stable membrane potential but, following a brief current injection, would switch to a state of continuous electrical spiking. Ca^{2+}/calmodulin-activated protein kinase II (CAMKII) activity may provide an example of biochemical bistability relevant to learning and memory. As discussed in Chapter 12, CAMKII exists *in vivo* as a holoenzyme of multiple subunits (~12). CAMKII is activated by binding of Ca^{2+}-calmodulin complexes to the subunits. Each subunit of CAMKII can be phosphorylated on Thr-286. The enzymatic activity of a phosphorylated subunit is increased, and this activity persists when Ca^{2+}-calmodulin is not bound to the subunit. A holoenzyme can undergo autophosphorylation in which active subunits phosphorylate other subunits that have Ca^{2+}-calmodulin complexes bound (Hanson *et al.*, 1994; Zhabotinsky, 2000). Elevations of Ca^{2+} increase the fraction of calmodulin saturated with bound Ca^{2+} and thereby increase the frequency of CAMKII autophosphorylation reactions. Modeling has suggested that CAMKII might act as a bistable switch (Zhabotinsky, 2000; Lisman and Zhabotinsky, 2001). For a given CAMKII holoenzyme, if a brief electrical stimulus caused Ca^{2+} influx that phosphorylated a critical number of subunits on Thr-286, then these subunits might phosphorylate the remaining subunits. Thus, the holoenzyme would remain active after Ca^{2+} influx ceased. As mentioned in Chapters 12 and 18, prolonged activation of CAMKII may be important for long-term synaptic potentiation in response to correlated pre- and postsynaptic electrical activity (Lisman, 1994).

The most recent models (Zhabotinsky, 2000; Lisman and Zhabotinsky, 2001; Lisman, 1994) suggest that CAMKII could remain highly active and autophosphorylated only in the postsynaptic regions of dendrites. In other regions of the cytosol, the total concentration of CAMKII is much lower relative to phosphatases that counter the autophosphorylation reaction, so the autophosphorylated state of CAMKII could not be maintained after [Ca^{2+}] returned to its rest level. Thus, in the postsynaptic region, CAMKII might act as a bistable switch in which a brief influx of Ca^{2+} would lead to long-term activation of CAMKII via sustained autophosphorylation. Figure 14.3A shows schematically the autophosphorylation process, and Fig. 14.3B illustrates how bistable activity of the CAMKII holoenzyme can be represented as a plot of enzyme activity versus [Ca^{2+}].

Chapter 7 introduced the concept of a bifurcation diagram, in which the stable or unstable steady states of a variable in a model are plotted. The plot shows how the steady states change as a function of a model parameter. For example, in a model of neuronal excitability, a stable solution for membrane potential changes its value and perhaps loses stability if a conductance parameter or a background stimulus current is varied (see Figs. 7.18 and 7.20). Figure 14.3B is a bifurcation diagram showing how the activity of CAMKII varies as [Ca^{2+}] is varied. The solid (upper and lower) portions of the black curve give the values of CAMKII activity that are stable at any given [Ca^{2+}]. If [Ca^{2+}] lies within the blue region, there are two possible values of CAMKII activity. The higher activity corresponds to a higher degree of CAMKII autophosphorylation. The dashed portion of the black curve corresponds to an intermediate activity that is unstable—a small change in [Ca^{2+}] will cause CAMKII activity to move up or down to one of the two stable states. Prior to a stimulus that leads to an increase in Ca^{2+} influx, the point representing the state of CAMKII might lie in the lower steady state of the graph—low activity, at a low (resting) level of [Ca^{2+}]. This state is labeled 1. Electrical activity and consequent Ca^{2+} influx can move the Ca^{2+} concentration well to the right, from point 1 on the low steady state to point 2, where the lower steady state of CAMKII activity no longer exists. At point 2 only the state of high CAMKII activity is stable. If [Ca^{2+}] stays above the blue region for some critical time, Ca^{2+}-calmodulin—induced autophosphorylation will move the activity of CAMKII up to a high value, to point 3. Then, when Ca^{2+} concentration returns to a lower level, the state of CAMKII moves back to the left, to point 4. But the CAMKII activity still remains on the

upper portion of the curve—high activity, with sustained autophosphorylation.

An important caveat is regarding the significance of the bistability in Fig. 14.3. Long-lasting changes in CAMKII activity have not yet been shown to be important for long-term synaptic potentiation (LTP). Autophosphorylation of CAMKII does occur after electrical stimuli that produce LTP (Yamagata and Obata, 1998; Ouyang et al., 1997). When autophosphorylation is prevented by mutating Thr-286, LTP *induction* is severely impaired (Giese et al., 1998). However, even if CAMKII autophosphorylation and increased activity are long-lasting, the increased activity may not be required for LTP *maintenance*. Inhibitors of CAMKII applied subsequent to LTP induction are most often reported not to affect LTP maintenance (Chen et al., 2001).

Negative or positive feedback due to allosteric regulation of enzymes can generate oscillations in the rate of metabolic flux through a biochemical pathway. The best known example is oscillations in the rate of glycolysis. The enzyme phosphofructokinase (PFK) is largely responsible for glycolytic oscillations in a variety of experimental preparations. Sustained oscillations have been found in several systems, for example, in suspended yeast cells (Dano et al., 1999). Glycolytic oscillations may modulate the electrical activity of some excitable cells (Prentki et al., 1997). PFK is activated by one of its reaction products, adenosine diphosphate (ADP). The activation is cooperative, depending on the second or higher power of ADP concentration. Thus, positive feedback occurs, with production of ADP progressively activating PFK until PFK activity and ADP concentration plateau at high levels. No negative feedback is present. Usually, positive feedback alone cannot support oscillations. In glycolysis, however, the substrate of PFK (fructose 6-phosphate) is supplied at a limited rate. Therefore, cycles of substrate depletion can occur. When ADP concentration and PFK activity are high, the substrate of PFK is used faster than it is supplied. A fall in the substrate concentration leads to a drop in the production of ADP and a loss of PFK activity. PFK activity then remains low while substrate accumulates again. Eventually, substrate accumulation raises the rate of production of ADP, and enough ADP is formed to reach the threshold for positive feedback to become effective. Then the cycle can begin anew. Cooperativity of activation of PFK by ADP is essential for these oscillations. Several mathematical models have been developed that describe these oscillations (Goldbeter, 1996). With these models, bistability in PFK activity can also appear when different values of kinetic parameters are assumed. This result

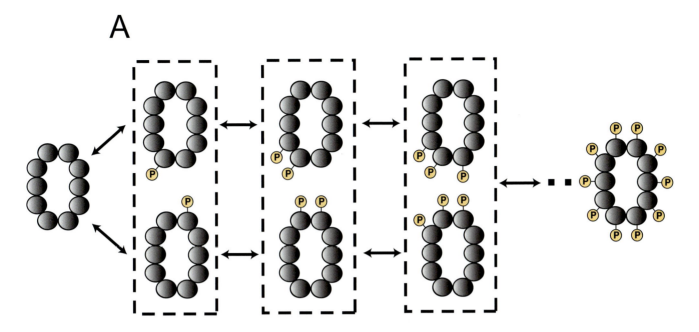

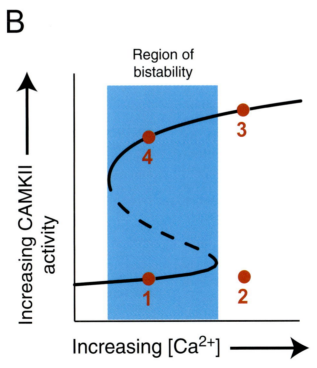

FIGURE 14.3 Autophosphorylation of CAMKII may lead to bistability in enzyme activity. (A) Schematic illustrating multiple phosphorylation states. If it is assumed that there are 10 subunits per holoenzyme, there are a large number of possible states, but if CAMKII activity depends only on the number of phosphorylated subunits and not their location within the holoenzyme, there are only 11 states of differing enzyme activity. Five of these states are shown, corresponding to 0, 1, 2, 3, and 10 phosphorylated subunits. Phosphate groups are orange. A calmodulin molecule with bound Ca^{2+} (not shown) must be bound to a subunit for phosphorylation to be possible. (B) Schematic illustrating bistability. The x axis represents intracellular Ca^{2+} concentration. For a range of Ca^{2+} concentrations, there is both a low and a high stable steady state for CAMKII activity (solid portions of curve, blue region of graph). The two stable states are separated by an unstable steady state (dashed portion of curve).

is as expected, since bistability is typical for models with positive feedback.

Several Mechanisms, Including Positive Feedback, Can Yield High Sensitivity of Response Magnitude to Signal Strength

The term ultrasensitivity refers to the following phenomenon. Sometimes a modest change in the concentration of substrate of an enzyme, or a modest change in the activity of an enzyme, greatly changes the net rate of the reaction catalyzed by that enzyme. For example, in a biochemical pathway, a relatively abrupt transition in metabolic flux from near zero to near maximum might be seen in response to a modest change in the concentration of the substrate of the first reaction in the pathway. Such switchlike behavior can be generated by several mechanisms. One mechanism relies on allosteric binding of a substrate to multiple sites on an enzyme in a cooperative fashion, such as that described by Eq. 14.12a. If the Hill coefficient of cooperativity is high (i.e., high n in Eq. 14.12a), then a small change in substrate level can yield a large change in enzyme activity. A second mechanism is zero-order ultrasensitivity (Goldbeter and Koshland, 1984). Here, two enzymes catalyze the same biochemical transformation, one in the forward direction and the other in the reverse direction. The enzyme pair could be a kinase and a phosphatase. Both enzymes must be nearly saturated with substrate, so that both the reverse and forward reactions are approximately zero-order. Then the rates of these reactions hardly change with the concentrations of their substrates. At steady state, the substrate concentrations are often such that the rates of both reactions are nearly equal. Then the net flux, or the difference between the rates of the forward and reverse reactions, is much smaller than the rates of the individual reactions. Now, suppose there is a small increase in the activity (V_{max}) of the enzyme catalyzing the forward reaction. This unbalances the rates so that the forward reaction is faster. Since this enzyme is saturated with substrate, a large drop in the concentration of substrate is required to bring the rates back into balance. Thus, a small change in enzyme activity results in a large change in the ratio of substrate to product.

A well-known signaling pathway in which zero-order ultrasensitivity may be important begins with hormonal or neurotransmitter receptors and continues through activation of several protein kinases to affect the activity of one or more isoforms of mitogen-activated protein kinase (MAPK) (Fig. 14.4A). In this pathway, MAPK is itself a substrate for a kinase termed mitogen-activated protein kinase kinase (MAPKK or MEK), and MAPKK is in turn a substrate

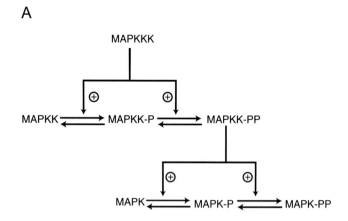

A

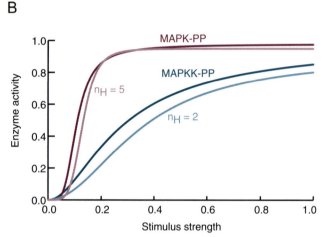

B

FIGURE 14.4 A mechanism for generating ultrasensitivity in the stimulus–response curve of a MAP kinase signaling cascade. (A) Kinetic scheme illustrating double phosphorylations of both MAPKK and MAPK. Phosphorylations are reversible. (B) Stimulus–response curves for the activities of MAPK (top, red curve) and MAPKK (blue curve) based on the kinetic scheme of (A). For both enzymes, the activity is assumed proportional to the fraction that is doubly phosphorylated. The stimulus is a constant level of MAPKKK activation. Also plotted, alongside the stimulus–response curves, are graphs of Hill functions generated using Eq. 14.12a. For the pink curve, the Hill coefficient n_H is 5; this curve is similar to the stimulus–response curve of MAPK. For the blue curve, the Hill coefficient n_H is 2; this curve is similar to the stimulus–response curve of MAPKK.

for MAP kinase kinase kinase (MAPKKK). Phosphorylation of MAPK by MAPKK is necessary for MAPK activation. Some isoforms of MAPK may be present in high enough concentrations to nearly saturate MAPKK. It has therefore been suggested that a small increase in the activity of MAPKK could cause, via zero-order ultrasensitivity, a large increase in phosphorylated and active MAPK. Modeling has suggested this occurs under physiological conditions (Ferrell, 1996; Huang and Ferrell, 1996).

In addition to zero-order ultrasensitivity, the MAPK pathway may exhibit ultrasensitivity as a result of a requirement for multiple phosphorylations of MAPK and MAPKK (Ferrell, 1996). MAPK must be doubly phosphorylated by MAPKK to be activated. In turn, to be activated, MAPKK must also be doubly phosphorylated by MAPKKK. The phosphorylations are reversible, and the phosphatase activities are often considered to be constant (Huang and Ferrell, 1996). Figure 14.4B illustrates stimulus–response curves for the doubly phosphorylated fractions of MAPK and MAPKK versus the strength of an input stimulus. These curves were calculated from a model (Huang and Ferrell, 1996) that uses the kinetic scheme of Fig. 14.4A. The stimulus corresponds to a fixed activation level of MAPKKK in Fig. 14.4A. The stimulus was assumed to be applied long enough that all reactions in Fig. 14.4A attained equilibrium. The rates of change of the amounts of active enzymes can therefore be set equal to zero. The resulting system of equations can be solved to yield the fractions of doubly phosphorylated MAPKK and MAPK as a function of input stimulus strength. Figure 14.4B shows that the steep curve for MAPK phosphorylation can be approximately reproduced by Eq. 14.12a with a high Hill coefficient (~5). To use this equation, effector concentration is replaced by input stimulus strength. The curve for MAPKK phosphorylation is less steep (Hill coefficient of ~2).

Cross-Talk between Neuronal Signaling Pathways May Influence Stimulus-Induced Synaptic Changes

Electrical activity or exposure to neurotransmitters or growth factors can simultaneously activate a variety of signaling pathways in neurons. Cross-talk between pathways can be mediated by a chemical species that is produced or consumed by reactions in two pathways. Cross-talk can also be mediated by a species produced in one pathway that interacts allosterically with an enzyme in another pathway. Modeling of pathways can help identify points of possible cross-talk and ranges of parameter values that would make this cross-talk significant. Experiments could then examine whether parameter values fall in these ranges and whether cross-talk causes observable effects on system behavior.

One recent study (Bhalla and Iyengar, 1999) modeled glutamate-activated signaling, which involved activation of four signaling pathways at dendritic synapses of hippocampal pyramidal neurons. These signaling pathways activate protein kinase A (PKA), protein kinase C (PKC), MAPK, and

CAMKII. Figure 14.5 illustrates points of cross-talk between these pathways. Experimental values for model parameters were sometimes not available for neurons, so values from other cell types had to be used in this model. Despite this disadvantage, the model is useful for suggesting points of cross-talk. One possibility is a positive-feedback loop that increases both MAPK and PKC activity. MAPK can phosphorylate and activate the enzyme phospholipase A_2, whose activity generates the second messenger arachidonic acid (AA). AA activates PKC. In turn, PKC can phosphorylate and activate a guanine nucleotide exchange factor, which then activates Ras GTPase. Ras can then further activate the MAPK signaling cascade, closing the positive-feedback loop. In simulations with the model, a brief glutamate exposure activated the positive-feedback loop. This caused the model to switch from a stable state with a low, basal activity of MAPK and PKC to a stable state with a high activity of both enzymes. In neurons, induction by glutamate of a prolonged activation of MAPK and PKC might correspond to induction of a large change in synaptic strength, dependent on a long-lasting phosphorylation by MAPK or PKC of transcription factors such as Ca^{2+}/cAMP—responsive element-binding protein (CREB) (see below and Chapter 13).

A second point of possible cross-talk is a gate involving cAMP and CAMKII. As discussed above, autophosphorylation of CAMKII after a brief influx of Ca^{2+} might convert CAMKII to a long-lasting, active state independent of Ca^{2+}. Such a state might be essential for long-term potentiation of synapses. However, Ca^{2+} influx also elevates the concentration of Ca^{2+}-calmodulin complex, which in turn may activate a phosphatase, protein phosphatase I (PP1) (Bhalla and Iyengar, 1999). PP1 can dephosphorylate CAMKII (see also Chapter 12). If cAMP levels are also elevated by the same stimulus that causes Ca^{2+} influx, the autophosphorylation of CAMKII might have a greater chance of becoming self-sustaining. This is because cAMP activates PKA, which in turn leads to inhibition of PP1, thus "opening a gate" to allow a high level of CAMKII phosphorylation. This high level could be self-sustaining after cAMP and PP1 activity return to basal levels, because once a critical number of subunits in a CAMKII holoenzyme have been phosphorylated, autophosphorylation of the remaining subunits may be rapid enough to overcome the effects of PP1. Then, the holoenzyme will remain phosphorylated and active until eventual degradation.

Simulations (Bhalla and Iyengar, 1999) suggested that this "gate" could be important. Glutamate exposure led to a long-lasting increase in CAMKII activity only if PP1 was inhibited by PKA. In that case, only

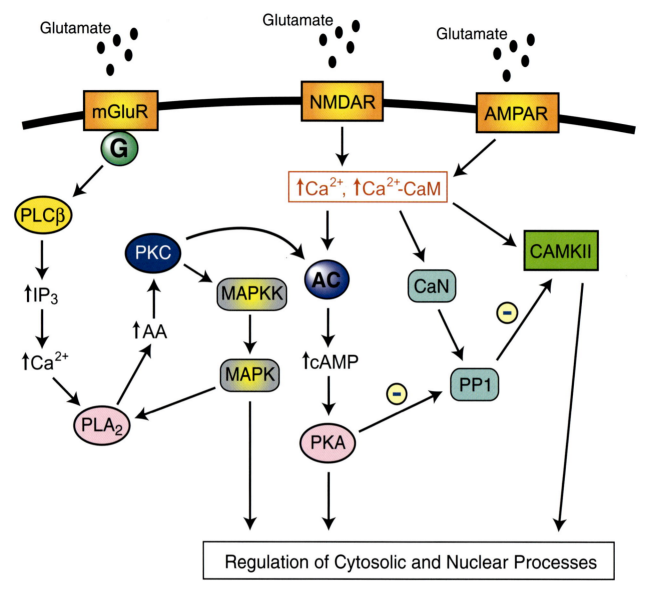

FIGURE 14.5   Aspects of a model that relates glutamate exposure at hippocampal synapses to long-term synaptic strengthening. Glutamate can act through metabotropic glutamate receptors (mGluR) to activate G proteins (G). Glutamate also acts through NMDA and AMPA receptors to increase intracellular levels of free Ca^{2+} and Ca^{2+} bound to calmodulin. These events lead to activation of phospholipase C (PLC), CAMKII, calcineurin (CaN), adenylyl cyclase (AC), and PKA. Two forms of cross-talk between these signaling pathways are illustrated. As discussed in the text, PKA activation leads to the inhibition of PP1. This inhibition relieves dephosphorylation of CAMKII by PP1, thus helping to sustain CAMKII activity. Also, MAPK activates phospholipase A_2 (PLA_2) and the resulting increase in arachidonic acid activates PKC. PKC in turn activates MAPKK, which further activates MAPK. As illustrated, MAPK, PKA, and CAMKII regulate gene expression and cytosolic components, such as the cytoskeleton, that are essential for long-term synaptic strengthening.

stimuli that increased both Ca^{2+} and cAMP past threshold levels were predicted to yield prolonged activation of CAMKII. Recent experiments (Brown *et al.*, 2000) have supported the importance of this dual threshold of Ca^{2+} and cAMP levels in determining the degree of synaptic modification in hippocampal neurons. Addition of a cAMP analog together with electrical stimulation was observed to yield strong activation of CAMKII. Neither stimulus alone could produce strong CAMKII activation. Furthermore, only the combination of electrical stimulation and cAMP analog could induce long-term synaptic potentiation.

The model of Fig. 14.5 also helps to order elements of signaling pathways in a set of cause–effect relationships. For example, Fig. 14.5 suggests that during the biochemical events following glutamate exposure, PKA activation lies "downstream" of both PKC and MAPK activation, but "upstream" of CAMKII activation.

The model of Fig. 14.5 can discriminate between stimulus patterns such as those known to produce long-term potentiation (LTP) and long-term depression (LTD) of synapses (Bhalla, 2002) (see also Chapter 18). To demonstrate discrimination between stimulus patterns, a compartmental model of a neuron was constructed, part of which is shown schematically in Fig. 14.6. The biochemical model (Fig. 14.5) was assumed to operate in the terminal (uppermost) compartment of the dendritic spine (Fig. 14.6). Stimulus patterns were modeled as pulses

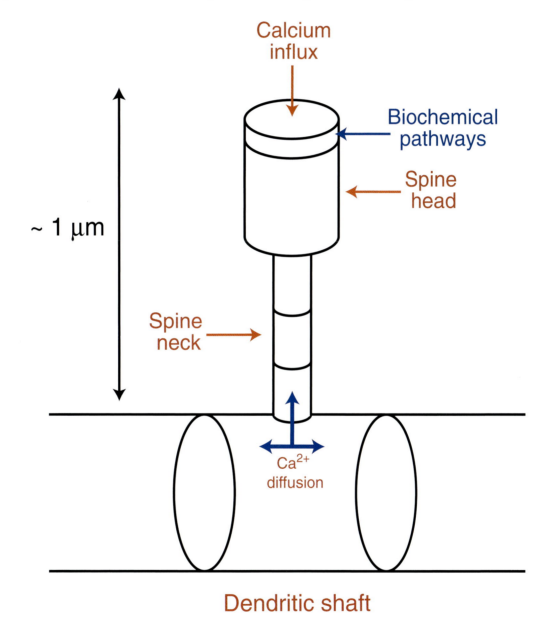

FIGURE 14.6 Part of a compartmental model of a hippocampal neuron. A dendritic spine and part of the dendritic shaft are shown. The spine head, the spine neck, and the dendritic shaft are each modeled as a series of cylindrical compartments. The compartments are necessary to model Ca^{2+} diffusion. Stimuli are modeled as pulses of Ca^{2+} influx into the spine. The components of the biochemical model of Fig. 14.5 are all assumed to exist in the terminal (uppermost) compartment of the spine. Ca^{2+} influx affects this model as discussed in the text, leading to activation of MAPK and other enzymes. MAPK activation can be prolonged via the positive feedback loop in which MAPK and PKC reciprocally activate each other (Fig. 14.5).

of Ca^{2+} influx into the spine, which activate enzymes such as MAPK (Fig. 14.5). The compartmental model provides a method for simulating Ca^{2+} diffusion between different portions of the spine (different compartments) and between the spine and the dendritic shaft. Within these small structures, Ca^{2+} diffusion plays an essential role in shaping Ca^{2+} concentration changes following stimulation. Therefore, Ca^{2+} diffusion must be included to model the activation of neuronal biochemical pathways. Diffusion between the cylindrical compartments is modeled with a variant of Eq. 14.4. A compartmental model can also simulate the spread of voltage changes between portions of a neuron (e.g. action potentials).

Activation of the MAPK–PKC feedback loop was simulated following eight different stimulus patterns. One pattern used was three tetanic stimuli, each of 100 pulses at a frequency of 100 Hz, with 60 s between tetani; this pattern has been used experimentally to produce LTP. Another pattern was 900 s of 1-Hz pulses; this pattern has been used experimentally to produce LTD. Even at low stimulus amplitude (small pulses of Ca^{2+} influx), long-lasting activation of the MAPK–PKC feedback loop resulted from the "LTD-producing" 1-Hz stimulus pattern. The "LTP-producing" tetanic stimulus pattern was also effective, but at a higher amplitude.

In an extensive series of simulations, different model parameters, such as the extent of Ca^{2+} buffering, were varied. The order of selectivity among stimulus patterns was defined by the stimulus amplitudes required to activate the MAPK–PKC feedback loop. The "LTD-producing" 1-Hz pattern was usually first in order (lowest amplitude). However, among the seven other patterns, numerous changes in selectivity order were observed when parameters were varied. Therefore, these simulations illustrate that the biochemical parameters of neurons may play a key role in determining which stimulus patterns are most effective for altering synapses.

Before proceeding to discuss the methods of flux control analysis, stochastic modeling, and genetic modeling, it is important to address two general topics related to modeling biochemical systems. First, it is necessary to analyze how sensitive modeling results are to parameter values and to consider how experimental data constrain those values. Second, it is useful to develop the simplest model that can qualitatively reproduce the dynamics of a biochemical pathway. Starting with a complex model including multiple enzymes and metabolite concentrations, how might the number of variables be reduced while preserving a behavior of interest, such as oscillations in concentrations?

GENERAL ISSUES IN THE MODELING OF BIOCHEMICAL SYSTEMS

Analyzing Sensitivity of Model Behavior to Parameter Changes

It is necessary to assess the sensitivity of models to modest variations in the values of parameters. Genetic and biochemical systems are commonly observed to be robust even to large changes in the values of parameters, such as genetic mutations that alter enzyme activities (Wagner, 2000; Little et al., 1999). The qualitative behavior of a gene network or biochemical pathway is often preserved due to evolved compensatory mechanisms. Therefore, models of these systems should most often not exhibit large variability in dynamics after small parameter changes. A standard method of assessing robustness to parameter changes is to define a set of sensitivities S_i, with the index i ranging over all model parameters p_i (Beck and Arnold, 1977; Frank, 1978). Let R denote the amplitude of the response of a signaling pathway to a fixed stimulus. For each p_i, a small change is made, and the resulting change in R is determined. The relative sensitivity S_i is then defined as the *relative*, or *fractional*, change in R divided by the *relative* change in p_i (Beck and Arnold, 1977, Frank, 1978)

$$S_i = \frac{\Delta R / R}{\Delta p_i / p_i}. \tag{14.18}$$

The relative sensitivity has the advantage that it is independent of the magnitudes and units of R and of the p_i.

R in Eq. 14.18 could also represent the time delay between a simulated stimulus and response, or the period of a simulated oscillation in gene transcription rate. In general, R is some quantity that has been measured and whose simulation is a major goal of model development.

The S_i in Eq. 14.18 are determined for small changes (10% or less) in the values of each individual parameter. For a model to be considered robust (i.e. not overly sensitive) the S_i should not be too large. These values should generally not exceed ~10, unless there is experimental evidence that a response is particularly sensitive to a specific parameter. In practice, determinations of S_i or similar measures are complemented by qualitative assessments of whether particular behaviors (e.g. oscillations in reaction rates) are preserved during modest parameter changes.

More detailed investigations vary multiple parameters simultaneously. The goal is to determine model behavior in regions of the multidimensional parameter space whose coordinate axes are the parameter values. For example, a region might support multiple stable solutions for metabolite concentrations and reaction rates, as does the blue region in Fig. 14.3B. Characterizing the location and size of such a region could predict that particular *in vivo* conditions would support multistability. However, a model of a biochemical pathway or of a network of co-regulated genes often has too many parameters to make a thorough characterization of the behavior in parameter space computationally feasible.

Bifurcation analysis (Wiggins, 1990; Rinzel, 1985) can help in determining and visualizing model behavior in parameter space. This technique allows determination of the curves or surfaces in parameter space at which model behavior undergoes significant changes. For example, with parameter values on one side of such a surface, the model variables may maintain a stable equilibrium. For parameter values on the other side of the surface, the equilibrium may disappear and persistent oscillations of model variables may appear. Software packages have been developed to determine and plot these curves or surfaces (Doedel, 1981).

Parameter Uncertainties Imply the Majority of Models are Qualitative, Not Quantitative, Descriptions

Models of biochemical pathways usually contain parameters whose values are not well constrained by experiment. It is obligatory for investigators to state which parameters in their models are poorly constrained, because parameter values might later be found to differ considerably from those used in a given model. Such differences could falsify the model because simulation of experimental results might no longer be possible.

Standard experiments can be done *in vitro* to estimate some parameter values. For example, enzyme Michaelis constants (e.g. the parameters K_1 and K_2 in Eq. 14.11) are often estimated with preparations containing small amounts of enzyme. Other parameters are harder to estimate. For example, it is difficult to estimate the amount of active enzyme per cell. Therefore the maximal velocities of enzymes *in vivo* are often poorly constrained.

Many parameters such as reaction rate constants, enzyme activities, and Michaelis constants depend on the activity coefficients of enzymes and reactants.

Activity coefficients are likely to be considerably different *in vivo* than *in vitro*. In cells, a high concentration of macromolecules creates a macromolecular crowding effect. Experiments with crowding by inert substances such as polyethylene glycol demonstrate that macromolecular crowding raises activity coefficients of all macromolecular species (Zimmerman and Minton, 1993). This increase in activities preferentially increases association rates and consequently increases levels of aggregates at the expense of monomers. For example, consider the association and dissociation rates, R_f and R_b, respectively, that govern the formation of an AB dimer from monomers of A and B:

$$R_f = k_f \lambda_A [A] \lambda_B [B], \qquad (14.19a)$$

$$R_b = k_b \lambda_{AB} [AB]. \qquad (14.19b)$$

In Eqs. 14.19a and 14.19b, k_f and k_b are standard rate constants. λ_A, λ_B, and λ_{AB}, are activity coefficients for A and B monomers and for the AB dimer. It is seen that the concentrations of A, B, and AB are each multiplied by the corresponding activity coefficients. The association rate R_f contains the product of two activity coefficients, so R_f tends to be enhanced more than R_b when activity coefficients are raised by macromolecular crowding. Therefore, formation of AB dimers is favored.

Increased association rates are expected to help stabilize signaling complexes containing multiple enzymes. This stabilization can enhance the efficacy of metabolite transfer between enzymes and, thereby, enhance the flux through a metabolic pathway (Rohwer *et al.*, 1998). The effect of macromolecular crowding on activity coefficients of small molecules is species-dependent: both charge and molecular weight are important.

Because of uncertainties in parameter values and activity coefficients, models of biochemical pathways should most commonly not be thought of as quantitative descriptions. Rather, they are most commonly qualitative descriptions. Simulations with such models are, as we have seen, very valuable for characterizing the types of dynamics likely to be supported by the biochemical architecture of a pathway. Also, simulations that adjust parameters to reproduce experimental results can help estimate the qualitative importance of interactions, such as feedback loops. Parameter adjustment could, for example, suggest that a dissociation constant for an allosteric effector is quite large and that, as a result, a feedback interaction mediated by that effector is weak. Further experiments are necessary to test such predictions.

Measurements of amounts of metabolites or second messengers after cell exposure to a hormone or neurotransmitter are particularly useful for estimating model parameter values. Groups of measurements are commonly taken at several time points following signaling pathway activation. Simulations of these experiments generally begin with reasonable "baseline" values for metabolite amounts. Then, the activity of key enzymes is temporarily increased to mimic the effect of the stimulus. An initial estimate is made for model parameter values. The time courses of metabolite amounts after the stimulus are simulated. These time courses are then compared with experiment. The simulation is repeated, adjusting model parameters, until optimal agreement is obtained between simulated and experimental data. For some parameters, such as Michaelis constants, *in vitro* estimates may exist. In this case, the *in vivo* value is usually kept close to the *in vitro* value, unless this assumption makes it impossible to simulate experimental data.

The adjustment of parameter values is commonly done by trial-and-error; however, it is often preferable to use a computer program that repeats the simulation with many different sets of parameter values. In this case, after each repetition, a measure of the "distance" between the simulated and experimental time courses is computed. A commonly used measure sums the squares of the differences between simulated and experimental metabolite concentrations. The sum is taken over all concentrations and also over all experimental time points. Some type of optimization routine (Press, 1994) is commonly included in such computer programs to adjust parameters in a direction suggested by the "distance" calculations for the previous few simulations. The goal is to find parameter values that minimize the distance measure.

Parameter adjustment continues until one or more value sets are found for which the shapes and amplitudes of the simulated metabolite time courses approximate the experimental time courses. If this approximation proves impossible, a new model based on new equations is needed. Following successful simulation, key control parameters can be defined as those parameters that are tightly constrained by the need to reproduce experimental data. Relatively small variations in these parameters cause significant changes in model behavior. Key control parameters are often candidate points for biological regulation of the signaling pathway. Experiments that determine values of these parameters constitute stringent tests that can falsify a model.

Separation of Fast and Slow Processes to Simplify Models

Experimental data generally determine the time scale of interest for a model. For example, experimental data concerning gene expression have a time scale of minutes or hours, whereas data concerning enzyme reaction rates may have a time scale of seconds or minutes. Models that simulate experimental data with a long time scale can be simplified, and the number of differential equations reduced, when processes or reactions with a fast time scale are assumed to be at equilibrium. Conversely, to simulate data with a fast time scale, the dynamics of slow processes can be neglected. A variable associated with a slow process, such as the concentration of a gene product, should simply be treated as a parameter. This separation into fast and slow time scales, and correspondingly fast and slow variables, is widely used for simplification of models. Chapter 7 has illustrated such a separation in a model of neuronal electrical activity. In that model, $[Ca^{2+}]$ was the slow variable, and the fast variables were membrane potential and a channel gating variable w (see Figs. 7.21 and 7.22).

As another example, if Ca^{2+} dynamics are being modeled, the number of differential equations can sometimes be reduced by assuming the rapid binding of Ca^{2+} to a buffer is always at equilibrium (see discussion of Eq. 14.8). Another example underlies the derivation of the standard Michaelis–Menten equation. Recall that the underlying reactions assumed in this derivation are

$$E + S \xrightarrow{k_1} ES, \quad ES \xrightarrow{k_1} E + S, \quad ES \xrightarrow{k_2} E + P.$$

Here, S is substrate, E is enzyme, and P is product. *In vitro* (although often not *in vivo*) the total concentration of enzyme, $E_{tot} = [E] + [ES]$, is usually much less than the concentration of substrate, $[S]$. In this case, relaxation of $[ES]$ to a quasi-steady-state value usually occurs on a fast time scale. By comparison, $[S]$ and $[P]$ change very slowly. It follows that $[ES]$ will always be very near its quasi-steady-state value. Therefore, $[S]$ and $[P]$ can be thought of as fixed parameters that determine what value $[ES]$ has at any time. A separate differential equation for $[ES]$ is therefore not necessary. One simply fixes $[P]$ and $[S]$ in the reaction scheme above and obtains the steady-state value for $[ES]$. To obtain $[ES]$, the rates of ES dissociation and formation are set equal:

$$(k_{-1} + k_2)[ES] = k_1 [E][S].$$

Next, one solves for [ES] and uses the conservation condition, $E_{\text{tot}} = [E] + [ES]$, to eliminate [E]:

$$[ES] = \left(\frac{k_1}{k_{-1} + k_2}\right)[E][S]$$

$$= \left(\frac{k_1}{k_{-1} + k_2}\right)(E_{\text{tot}} - [ES])[S].$$

Another rearrangement gives an expression for [ES] in terms of [S]:

$$[ES] = \frac{E_{\text{tot}}[S]}{K_m + [S]}, \quad with \; K_m = \left(\frac{k_{-1} + k_2}{k_1}\right). \quad (14.20)$$

This "steady-state approximation" allows the dynamics of the slow variables [S] and [P] to be described by a single differential equation:

$$\frac{d[P]}{dt} = -\frac{d[S]}{dt} = k_2[ES] = \frac{V_{\max}[S]}{K_m + [S]}, \quad (14.21)$$

with K_m as above and $V_{max.} = k_2 E_{\text{tot}}$

This is the standard Michaelis–Menten equation. In contrast, two differential equations are needed when the steady-state approximation is not used. One equation describes the rate of change of [P] and the other is for the rate of change of [ES].

The general method for model simplification is as follows. First, the equilibrium values of the "fast" variables ([ES] in the above example) are solved for as a function of the "slow" variables ([S] in the above example). The slow variables are treated as parameters in this calculation. Then, one changes perspective, using the equilibrium values of the fast variables in the differential equations for the slow variables. Since these equilibrium values are now functions of the slow variables, the fast variables, and their differential equations, are eliminated from the model. Only the differential equations for the slow variables are left. These differential equations together are termed the "slow subsystem."

A second simplified model also results from this method. This model consists of the differential equations for the fast variables alone, with the slow variables represented by parameters. As was mentioned in Chapter 7, these differential equations constitute the "fast subsystem." When the method is applied to the original Hodgkin–Huxley model with four differential equations (Hodgkin and Huxley, 1952), the two simplified models consist of two differential equations each. The qualitative behavior of the simplified models can be analyzed and represented in an intuitively appealing manner, similar to that in Chapter 7 (Figs. 7.21 and 7.22) [see also Rinzel (1985)]. For further examples and for discussion of simplification

by separation of spatial scales as well as time scales, see Ermentrout (2001).

SPECIFIC MODELING METHODS

Models Help to Analyze Metabolic Flux Regulation

One important goal of modeling biochemical pathways is to determine the parameters, such as amounts of particular enzymes, that control the rate at which substance flows from initial substrate to final product. An example would be the rate at which glucose is converted to pyruvate in glycolysis. This rate is termed the metabolic flux through the pathway. Flux regulation is necessary to keep metabolite concentrations within ranges appropriate to the state of the cell. Flux regulation of pathways such as glycolysis is also essential for balancing the production and use of ATP. Parameters to which metabolic fluxes are most sensitive are candidates for biological flux regulation.

The method of metabolic control analysis (MCA, also termed metabolic control theory) provides flux control coefficients to describe the importance of each enzyme in regulating the flux of metabolite through the pathway. These flux control coefficients provide a clear and concise representation of pathway regulation. The methodology of MCA is therefore briefly discussed. For a fuller treatment, Stephanopoulos *et al.* (1998) should be consulted. Limitations of MCA are as follows. The control coefficients are defined using the response of the pathway to small changes in enzyme activities. Large changes, such as may be caused by mutations in genes coding for enzymes, are not well handled. Also, MCA usually considers enzymes and metabolites to be at uniform concentrations throughout the cytoplasm.

In MCA, the sensitivities of the metabolic flux to changes in the activities of enzymes are expressed as flux control coefficients (FCCs). With the steady-state metabolic flux through a pathway denoted by J and the activity of the ith enzyme in the pathway denoted by E_i, an FCC, F^i, is defined as follows. An infinitesimal change in E_i, dE_i, yields an infinitesimal change in J, dJ. The *relative*, or *fractional*, change in J is then defined as dJ/J. This relative change is divided by the *relative* change in E_i:

$$F^i = \frac{dJ/J}{dE_i/E_i} = \frac{d\ln J}{d\ln E_i}. \quad (22)$$

To understand FCCs, it helps to examine the relationship between pathway flux and enzyme activity. Figure 14.7 illustrates that the flux J will saturate as the activity or concentration of any given enzyme is increased. The slope of the graph of J versus enzyme activity is related to the FCC. To obtain the FCC, the slope of the tangent line (the derivative dJ/dE_i) is multiplied by the enzyme activity and then divided by J. Figure 14.7 also illustrates why infinitesimal changes in the enzyme activity are used to define FCCs. For a large change in enzyme activity, the inferred slope of the flux-versus-activity relationship (the fraction $\Delta J/\Delta E_i$) can be considerably different than the actual slope (the derivative dJ/dE_i). Therefore, FCCs inferred from systems with large changes in enzyme activities can differ considerably from actual FCCs. An extension of MCA has been developed to approximate FCCs from responses to large changes (Small and Kacser, 1993).

A summation theorem states that all the FCCs—one for each enzyme in the pathway—must add up to 1. A consequence is that most FCCs in long pathways must be small, so that the sensitivity of pathway flux to alterations in the activity of any one enzyme is usually low. By contrast, in short pathways several

FCCs are often large, and the corresponding enzymes exert strong flux control.

One result of MCA has been to modify or overturn concepts of metabolic control that were estimated via intuition. For example, consider enzymes that are subject to feedback inhibition by their reaction products. These enzymes serve as control points of the metabolic flux through the pathway, because the feedback prevents the concentrations of their reaction products and of pathway end products from rising too high. Intuitively, one might expect that these enzymes are strongly rate-limiting, in the sense that small changes in their activity would have large effects on the pathway flux. However, MCA has demonstrated that this expectation is usually false (Stephanopoulos et al., 1998). Instead, enzymes whose activities are affected by allosteric interactions with their substrates or products tend to have low FCCs.

Some applications of MCA have helped increase understanding of neuronal metabolism. For example, in several regions of rat brain, the flux control coefficient of the enzyme nitric oxide synthase (NOS) for the in vivo synthesis of nitric oxide (NO) has been determined (Salter, 1996). Administering varying dosages of an NOS inhibitor and measuring effects on

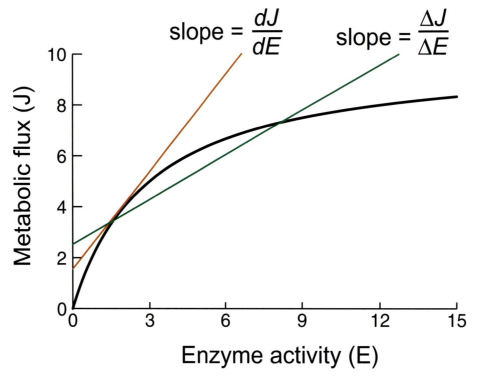

$$\text{slope} = \frac{dJ}{dE} \qquad \text{slope} = \frac{\Delta J}{\Delta E}$$

FIGURE 14.7   Metabolic flux versus activity of an arbitrary enzyme in a biochemical pathway. The flux saturates at a finite level as the enzyme activity is increased. An approximation of the rate of change of flux with enzyme activity is shown (slope of green line). This approximation is calculated using a large change ΔE in activity. For some values of E, the approximate rate of change can differ considerably from the true rate of change. This rate (slope of red line) is calculated using an infinitesimal activity change dE.

the NO synthesis rate determined the FCC of NOS. In all regions, this FCC was found to be close to 1. Because the sum of all FCCs in a biochemical pathway must equal 1 by the summation theorem, other processes essential for NO synthesis must have low FCCs, and NOS must be rate-limiting for NO production. Other processes, such as the transport of the NO precursor L-arginine from blood into neurons, must also not be rate-limiting for NO production. In other biochemical pathways, this method could be used to quantify the flux control of any enzyme for which a specific inhibitor is available.

If a detailed kinetic model can be constructed for a biochemical pathway, using differential equations such as Eq. 14.11 that describe changes in the levels of each substrate and product, then regulation of pathway flux can be assessed without MCA. However, because MCA flux control coefficients provide a concise representation of pathway regulation, their prediction may still be valuable. Such predictions might be compared with experimental values to test the model's validity. Also, biochemical data may not suffice to estimate many parameter values in a detailed model. It may then be better not to present such a model, because unjustified detail can obscure the state of knowledge of pathway regulation. It would be preferable to experimentally determine flux control coefficients to concisely characterize pathway regulation.

MCA has recently been extended to begin considering concentration gradients within cells. FCCs have quantified the control exerted by macromolecule diffusion coefficients on the flux through a simple model biochemical pathway (Kholodenko *et al.*, 2000); however, a problem is that no experimental method has been devised to estimate diffusion FCCs.

Special Modeling Techniques Are Required if Enzymes Are Organized in Macromolecular Complexes

To maximize the efficiency and specificity of reactions, many biochemical pathways are organized so that successive enzymes are positioned close to each other. This organization is mediated by multienzyme complexes or signaling complexes. These complexes are aggregates of macromolecules that contain groups of enzymes in combination with organizing "anchoring" proteins. Other proteins involved in signal transduction may be present, such as receptors for neurotransmitters or hormones. Further organization can be achieved by binding of the anchoring proteins to cytoskeletal elements. These signaling complexes

are often based on reversible protein–protein binding reactions (Teruel and Meyer, 2000) so that an individual complex may not be very long-lived. However, the lifetimes of such complexes are likely to be long compared with the time scale of enzyme reactions and of intracellular transport of metabolites.

Several instances of such organization are now known. A complex of glycolytic enzymes can be organized by binding to actin filaments (Fokina *et al.*, 1997; Kurganov, 1986). In neurons, signaling pathways using isoforms of MAPK appear to be organized into complexes. For example, scaffold proteins termed JIP proteins bind to particular isoforms of MAPK and MAPKK (Davis, 2000) and interact with kinases that activate MAPKK. As another example, PKA, calcineurin, and protein kinase C (PKC) bind to A-kinase anchoring proteins (AKAPS), which, in turn, bind cytoskeletal elements (Edwards and Scott, 2000). Finally, the NMDA receptor is in a large complex with more than 50 signaling proteins, kinases, and phosphatases (Husi *et al.*, 2000).

Two advantages of organizing signaling proteins into complexes are: (1) to favor a rapid passage of signals through an enzymatic cascade, and (2) to prevent unwanted cross-talk between signaling pathways. For example, if specific isoforms of MAPKK and MAPK are co-localized in complexes, then a given isoform of MAPKK will not be able to activate "improper" isoforms of MAPK. Also, a metabolite produced by an enzyme can be more efficiently passed on to the next enzyme in a pathway if both enzymes are co-localized.

Modeling signaling pathways organized into complexes may require equations specific to the system under study; however, some general comments can be made. Intermediate metabolites might not equilibrate with the cytoplasm *via* diffusion, but instead metabolic channeling may dominate. Metabolic channeling is defined as movement of a metabolite directly from one enzyme to the next within a multienzyme complex. In this situation, the rate of use of the intermediate metabolite by the next enzyme would not be a function of its bulk cytoplasmic concentration. Rather, it might be a direct function of the rate of production of the metabolite. If the metabolite was produced at a low rate, it might be immediately channeled to the next enzyme and used. If it was produced at a high rate, the channeling might saturate. Denoting the rate of metabolite production as $R_{\text{production}}$, this situation could be modeled with a Michaelis–Menten expression for the rate of usage of the metabolite, R_{usage}:

$$R_{\text{usage}} = \frac{V_{\text{max}} R_{\text{production}}}{R_{\text{production}} + K_{\text{S}}}. \tag{14.23}$$

Here, $V_{\max}$ and K_s correspond to the maximal velocity and Michaelis constant of the second enzyme in a standard Michaelis–Menten equation (Eq. 14.10). However, the units and therefore the numerical values of these parameters are different in Eq. 14.23 than in Eq. 14.10. Metabolic control analysis, discussed above, can be extended to assess flux regulation within a signaling pathway containing multi- enzyme complexes (Stephanopoulos *et al.*, 1998).

A recent model (Levchenko *et al.*, 2000) illustrates possible kinetic effects of a scaffold protein that is assumed to bind both MAPKK and MAPK in a complex, allowing interaction of the two enzymes. As shown in Fig. 14.8, this model assumes reversible binding of MAPKK and MAPK to the scaffold protein. Both MAPKK and MAPK must be doubly phosphorylated to be active. In solution, these phosphorylations occur sequentially. However, the model assumes that when either enzyme is bound to the scaffold protein, both phosphorylations occur simultaneously, hence the absence of singly phosphorylated MAPKK and MAPK in Fig. 14.8. Another assumption is that only inactive MAPK and MAPKK can bind to the scaffold. Thus, if MAPK or MAPKK is activated while bound to the scaffold, and then dissociates from

the scaffold, it cannot rebind until the phosphates are removed. At present, neither of these assumptions has been confirmed experimentally.

Both of these assumptions are, however, necessary to support the results of simulations (Levchenko *et al.*, 2000) with the model of Fig. 14.8. One result is that the presence of scaffold protein reduces the steepness of the dose–response curve for MAPK activation as a function of stimulus strength, when stimulus strength is represented by the activity of MAPKKK. This result depends on the assumption that the dual phosphorylations of MAPKK or MAPK in the scaffold are simultaneous. A second result is that for fixed stimulus strength, as the concentration of scaffold protein is increased, the percentage of activated MAPK at first increases, but then decreases. The increase is due to the assumption of simultaneous dual phosphorylations when MAPKK and MAPK are bound to the scaffold, because simultaneous phosphorylations are more efficient than the sequential phosphorylations that occur in solution. The decrease in MAPK activation when the level of scaffold protein is made very high does not depend on the above assumptions. Rather, it comes about because complexes of scaffold protein with only one inactive enzyme (MAPKK or

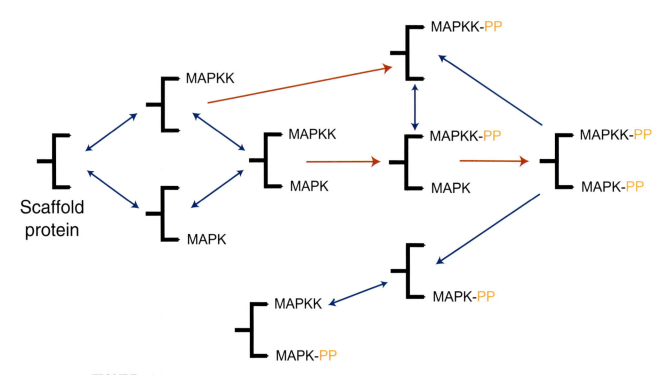

FIGURE 14.8 Model of a scaffold protein able to reversibly bind both MAPK and MAPKK. Nine biochemical species are possible: free scaffold, scaffold with unphosphorylated MAPK and/or MAPKK bound, and scaffold with doubly phosphorylated MAPK and/or MAPKK bound. Transitions between species can be due to kinase association or dissocation from the scaffold (blue arrows) or to phosphorylation of kinases (red arrows). Reverse transitions due to dephosphorylation are not modeled. Dissociation of phosphorylated kinases is irreversible (single-direction blue arrows). For clarity, only some possible dissociations are shown.

MAPK) account for an increasing proportion of total enzyme as scaffold concentration is increased.

Random Fluctuations in Molecule Numbers Influence the Dynamics of Biochemical Reactions

Rather than using chemical concentrations as variables, the stochastic level of modeling uses the actual numbers of each molecular species. In this way, one can include random variability in the times of synthesis and degradation of individual macromolecules or metabolites. For some macromolecules, such as specific mRNAs and transcription factors, average molecule numbers per cell can be modest. They may be on the order of a hundred or less. In this case, variability in the times of single-molecule synthesis or degradation lead to significant stochastic fluctuations in macromolecule numbers (McAdams and Arkin, 1999). These fluctuations are seen as random variations about a mean number of molecules, and their presence can considerably alter the dynamics of biochemical pathways. For example, stochastic fluctuations of the numbers of enzymes or metabolites in signaling pathways might cause large variability in the response of individual members of a cell population to a stimulus (Levchenko *et al.*, 2000; McAdams and Arkin, 1998). Only a fraction of cells might give an observable response, such as differentiation following application of a hormone. A rule of thumb is, if there are fewer than ~200 to 300 copies of any important macromolecule or metabolite, stochastic fluctuations may significantly affect dynamics.

Stochastic modeling has recently yielded qualitative insights into how fluctuations are expected to affect the dynamics of a biochemical pathway. For example, fluctuations are expected to most commonly decrease the steepness of a nonlinear stimulus–response relationship, such as that in Fig. 14.4B for MAPK activation (Berg *et al.*, 2000). However, in specific circumstances the steepness can increase (Paulsson *et al.*, 2000). Also, via a phenomenon known as stochastic resonance, fluctuations can increase the reliability or magnitude of responses to small signals (Stacey and Durand, 2001). These results are qualitative, and the effects of fluctuations within specific signaling pathways in neurons have not been studied. This is because of the difficulty in estimating average values for the absolute numbers of important macromolecules per cell. Such estimates are very difficult for cells with a complex morphology, such as neurons. These estimates are, however, essential prerequisites to quantitative modeling of fluctuations. Currently, data sufficient to allow detailed stochastic modeling

exist for only a few prokaryotic pathways (Arkin *et al.*, 1998).

To carry out computational simulations that include stochastic fluctuations, the equations of a biochemical model are first formulated according to the continuous, or differential-equation-based, approach. Concentrations are the variables. Differential equations such as Eqs. 14.11 and 14.14–14.17 determine how the variables evolve. Then, the variables are changed from concentrations to molecule numbers. This requires rescaling parameters such as maximal enzyme velocities and Michaelis constants to reflect the change in units; however, the form of the differential equations remains unchanged. The deterministic rate of each individual reaction is still computed from these equations. For example, in Eq. 14.11, the first term on the right-hand side gives the rate of the reaction that synthesizes P from the substrates S_1 and S_2.

These reaction rates have units of molecules per second. The reciprocals of these rates give the average time intervals between successive occurrences of each specific reaction. For a given reaction, such a time interval is denoted by T_{avg}. For simulating fluctuations in molecule numbers, one can proceed as follows. The time step is chosen small enough (much less than all the T_{avg} values) so that the probability of each reaction occurring during a step is only a few percent. At each time step, a random number is chosen for each reaction. Each number is chosen from a uniform distribution on the interval {0,1}. Whether a specific reaction occurs during the time step can be determined by whether its random number is greater or less than the reaction probability. That probability is the product of the reaction rate and the time step. If a reaction occurs, the appropriate molecule numbers are changed by 1 or –1.

Another accurate and rapid algorithm (Gillespie, 1977) uses the following result from statistical physics. If a particular biochemical reaction occurs at a time arbitrarily taken to be 0, the probability $P(t)$ that the *next* reaction of that type will occur within a small time interval Δt that is centered at $t > 0$ is

$$P(t) = \frac{\Delta t}{T_{avg}} \exp\left(\frac{-t}{T_{avg}}\right), \qquad (14.24)$$

with T_{avg} defined as in the last paragraph. For an implementation of Eq. 14.24 suited to systems with many molecular species, see Gibson and Bruck (2000).

Both of the above methods are limited in that only changes in the total number of molecules of each chemical species are simulated. Individual molecules are not represented. Thus, the fate of particular molecules cannot be traced, and therefore diffusion or spatial localization of molecules cannot be simulated. This limitation can be overcome by using an algorithm that represents each molecule as an individual object. At each time step this object is able to diffuse and also has probabilities of undergoing reactions. One software package that embodies this approach is StochSim, developed by Firth and Bray (2001) and colleagues. Such a package is particularly useful in simulating the dynamics of a signaling complex of associated proteins, containing enzymes and receptors specific to one or more intracellular signaling pathways. Conformational changes, covalent modifications, and ligand binding are assumed to switch the complex between a series of possible states, and the state trajectory can be followed (Morton-Firth et al., 1999). Such simulations are needed to help understand how the NMDA receptor complex, or other neuronal complexes, transmits signals.

Consider a pathway for which a model based on continuous concentration variables predicts oscillations or multiple stable steady states of metabolite concentrations. Large fluctuations could disrupt the oscillations or destabilize the steady states (Barkai and Leibler, 2000; Smolen et al., 1999). Therefore, predictions of behavior that neglect fluctuations may be false. Nevertheless, for many biochemical pathways, continuous models consisting of sets of differential equations (e.g. Eqs. 14.14–14.17) remain essential because of insufficient data to justify a stochastic model. Generally, much of the data used to construct a continuous model consist of large and relatively reproducible changes in pathway fluxes and metabolite concentrations following strong stimuli. Because those responses are reproducible, a continuous model may be expected to reliably predict responses to new stimuli similar in strength to those used in model construction.

Genes Can Be Organized into Networks That Are Activated by Signaling Pathways

Gene regulation is a common endpoint of biochemical signaling pathways. As discussed in Chapter 13, signaling pathways often activate proteins termed transcription factors (TFs) (Karin, 1994). The activation is often via phosphorylation of critical amino acid residues. Activated TFs regulate the transcription of genes by binding to nearby short segments of DNA. In Chapter 13, these segments were referred to as cis-regulatory elements. Another common terminology is response elements. If these elements activate transcription, they are commonly termed enhancers; if they repress transcription, they are commonly termed silencers. Many genes are regulated by multiple TFs.

Genes coding for TFs can be repressed or activated by TFs, including their own products.

Large clusters of genes are often regulated in concert by biochemical signaling pathways that activate specific TFs. For example, activation of MAPK can lead to activation of hundreds of genes and repression of many others (Roberts et al., 2000). Gene networks may be defined as gene clusters in which the expression of some members is regulated by the protein products of other members or by a common input such as a neurotransmitter binding to a G protein–coupled receptor. Thus, the expression of network genes varies in a coordinated manner. Gene networks mediate long-term processes such as development and memory formation. Understanding these processes therefore requires an understanding of the dynamics of gene networks.

One gene network often implicated in the control of synaptic plasticity and memory formation is the network based on transcriptional regulation by the family of TFs that includes CREB (Mayford and Kandel, 1999; De Cesare and Sassone–Corsi, 2000; Lonze and Ginty, 2002; Bartsch et al., 1998; Yin and Tully, 1986). Figure 14.9 illustrates aspects of this network, many of which are discussed in more detail in Chapter 13. Neurotransmitters, such as serotonin, bind to receptors and activate G proteins. Production of intracellular second messenger molecules, such as cAMP, is enhanced. Kinases, such as PKA and MAPK, are activated as a result. These kinases phosphorylate CREB and related TFs. A positive-feedback loop appears to exist in this network (Fig. 14.9). In this loop, phosphorylated CREB, when bound to CREs in the vicinity of the creb promoter, activates creb transcription (DeCesare and Sassone–Corsi, 2000; Walker et al., 1995). There is also a negative-feedback loop. CREB induces the gene for another TF of the CREM/CREB family (Chapter 13) termed inducible Ca^{2+}/cAMP-responsive early repressor (ICER) (Molina et al., 1993). ICER is a powerful transcriptional repressor. On binding to CREs, ICER represses icer transcription and also that of creb, closing the negative-feedback loop.

Experimental data essential for understanding the dynamics of gene networks are obtained by several methods. For example, sets of time courses of the mRNA levels expressed by large numbers of genes allow characterization of the response of cells or tissues to stimuli (Roberts et al., 2000) or characterization of genetic disease states (Mirnics et al., 2000). Current technology allows the expression time courses of ~10,000 genes to be simultaneously followed via quantification of their mRNAs. Cross-talk between signaling pathways can take the form of

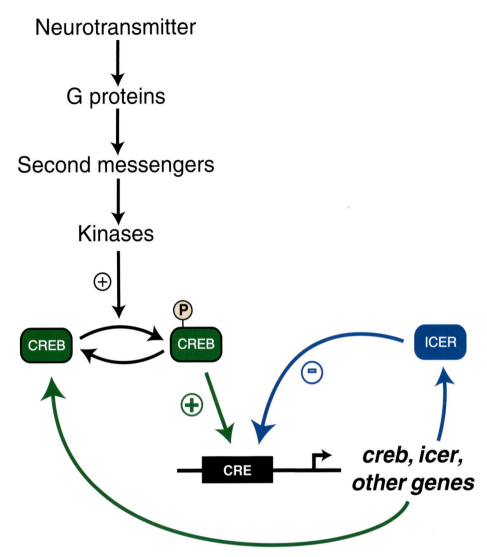

FIGURE 14.9 Signaling pathway involving transcriptional regulation by CREB. Neurotransmitters such as serotonin bind to receptors and act through G proteins to elevate levels of intracellular messengers (e.g. cAMP, Ca^{2+}). Kinases such as PKA are activated, resulting in phosphorylation of CREB and related TFs. Phosphorylated CREB stimulates the transcription of genes when bound to enhancer sequences termed Ca^{2+}/cAMP response elements (CREs). Possible feedback interactions among the genes coding for CREB and related TFs are shown. A positive-feedback loop (green arrows) may regulate CREB synthesis. In this loop, CREB binds to CREs near *creb* and activates *creb* transcription. The repressor ICER is an element of a negative-feedback loop (blue arrows). Transcription of *icer* is increased when the level of CREB increases, and ICER, in turn, can bind to CREs near the *creb* gene, repressing *creb* transcription.

convergent activation of groups of genes by several pathways, and can be identified from mRNA time courses. Understanding neural plasticity requires determining expression time courses for genes involved in synaptic modification. Further complexity exists because protein levels are often not well correlated with mRNA levels. Therefore, time courses of protein levels also need to be characterized (Ryu and Nam, 2000).

Understanding genetic regulation requires more than collecting large numbers of mRNA and protein time courses. A framework is required for expressing the biochemical architecture of genetic systems and for extracting genetic regulatory relationships from large sets of gene expression data. Modeling provides such a frame work (Smolen et. al., 2000). Models are also important for predicting the response of gene networks to novel pharmacological or chemical agents.

Modeling studies of gene networks have often used qualitative models with modest numbers of equations and parameters. These models assess the behaviors of networks with common biochemical elements (Smolen *et al.*, 2000). Such elements include formation of TF dimers or oligomers, and positive- and negative-feedback loops in which TFs activate or repress transcription of their own or each other's genes. These models give insight into the behaviors more complex gene networks exhibit, and are used below to illustrate the dynamics generated by typical regulatory schemes. Relatively few models have considered regulation of mRNA processing or translation (Cao and Parker, 2001). Therefore, the discussion below focuses on transcriptional regulation.

Methods Exist to Model Gene Networks at Very Different Levels of Detail

The Logical-Network Method Regards Genes as Simple ON–OFF Switches

In the *boolean*, or *logical-network*, *method*, modeling is rather coarse, but simulations can be carried out rapidly. The rate of transcription of each gene is assumed to be a boolean logical variable: either ON (a fixed transcription rate) or OFF. Control of gene expression is modeled by simple rules for which genes are activated or repressed by others. Relatively large time steps are taken in simulations. Figure 14.10 illustrates a simple model of this type.

In Fig. 14.10, three genes have their expression updated according to the following rules: (1) Gene 1 is ON only if Gene 2 was ON at the previous time step and Gene 3 was OFF, (2) Gene 2 is only ON if Gene 1 was OFF at the previous time step, (3) Gene 3 is only ON if Gene 1 was ON at the previous time step. Figure 14.10A shows schematically these interactions between genes with plus (activating) signs and genes with minus (suppressing) signs. The behavior of this model is rather simple, as illustrated in Fig. 14.10B. Irrespective of the initial state of the genes, the model rapidly enters and remains in a specific oscillation in which four network states occur in a fixed, repeating order. Boolean models of gene networks can be efficiently constructed from large sets of gene expression data (Wen *et al.*, 1998; Toh and Horimoto, 2002). Because it models regulation so coarsely, the logical-network method is suited only for large networks in which it is not feasible to study in detail the regulation of most of the genes.

The Continuous Method Is More Accurate but Computationally Intensive

With the continuous approach, differential equations are written to describe the rates of change of important mRNA and protein concentrations. These equations

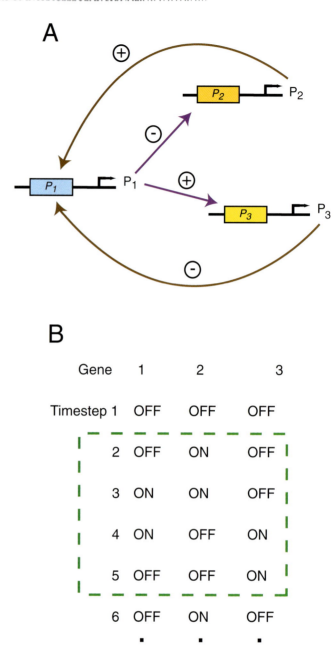

FIGURE 14.10 A simple logical-network gene model. (A) Interactions among three genes. The protein product of Gene 1 represses transcription of Gene 2 and activates transcription of Gene 3, the product of Gene 2 activates Gene 1, and the product of Gene 3 represses Gene 1. (B) When these interactions are implemented as the logical rules described in the text, then during a simulation, the activity state of the three genes changes at each time step. However, irrespective of the initial state, the model quickly enters a repeating cycle of four network activity states (outlined by the dashed box).

contain terms to represent processes such as macromolecular synthesis, degradation, and association of monomers into oligomers. Regulation of transcription is included in terms describing the synthesis of mRNA. As an example of a simple model that produces interesting behavior, consider the case of a single gene, *tf-a*, whose protein product is denoted TF–A. TF–A activates *tf-a* transcription, thereby forming a positive-feedback loop. It is assumed that two TF–A molecules must bind together, forming a dimer, which then binds to an enhancer sequence near the *tf-a* promoter (Fig. 14.11A). The corresponding differential equation for the rate of change of *tf-a* mRNA concentration is

$$\frac{d[tf-a \text{ mRNA}]}{dt} = \frac{k_{max}[TF-A]^2}{[TF-A]^2 + K_d}$$
$$-k_{degR}[tf-a \text{ mRNA}] + R_{bas}. \quad (14.25)$$

In Eq. 14.25, the first term on the right-hand side gives the transcription rate of *tf-a* as a saturable function of TF–A concentration. The second term accounts for first-order degradation of *tf-a* mRNA. The third term accounts for a low rate of transcription, R_{bas}, in the absence of activation of the *tf-a* gene by TF–A. At high [TF–A], the transcription rate approaches a maximal value k_{max}. TF–A dimers regulate transcription, and in Eq. 14.25, the square of TF–A concentration is used to approximate the concentration of TF–A dimers. The justification for this is as follows. From elementary kinetics, the rate of formation of TF–A dimers is proportional to the square of TF–A monomer concentration, and can be written as k_f [TF–A]$_{mon}$ [TF–A]$_{mon}$. The dissociation rate of dimers is proportional to dimer concentration, and can be written as k_b [TF–A]$_{dim}$. As discussed previously (following Eq. 14.6), a rapid-equilibrium approximation can be made by setting the association and dissociation rates equal to each other. Then, solving for dimer concentration yields a proportionality to the square of monomer concentration

$$[TF-A]_{dim} = \frac{k_f}{k_b} [TF-A]_{mon}[TF-A]_{mon}. \quad (14.26)$$

If most TF–A is monomeric, then Eq. 14.26 implies that [TF–A]$_{dim}$ is approximately proportional to the square of total [TF–A], which is the assumption of Eq. 14.25.

Along with Eq. 14.25, a second differential equation is needed for TF–A protein concentration. If TF–A protein was immediately translated from mRNA, a simple equation might be assumed, with two linear terms for TF–A synthesis and degradation:

$$\frac{d[TF-A]}{dt} = k_{2,f}[tf-a \text{ mRNA}] - k_{2,d}[TF-A]. \quad (14.27a)$$

If one wishes to take into account the time delay required for mRNA processing and transport, Eq. 14.27a can be modified to a *delay differential equation*:

$$\frac{d[TF-A]}{dt} = k_{2,f}[tf-a \text{ mRNA}](t-\tau)$$
$$-k_{2,d}[TF-A]. \quad (14.27b)$$

Here, the quantity $(t-\tau)$ in the first term on the right-hand side indicates that the concentration of *tf-a* mRNA at a given time is used to determine the rate of appearance of TF–A protein at a time τ min later. Delay differential equations are commonly used to represent slow kinetic processes. These equations exhibit a rich variety of dynamic behaviors (MacDonald, 1989).

The relatively simple model of Eqs. 14.25–14.27 can generate complex behavior (Smolen *et al.*, 1998, 1999). Figure 14.11B is a bifurcation diagram displaying the ways in which the dynamics of [TF–A] change as the parameter k_{max} is varied. With appropriate parameter values, the model is bistable. The curve traces the stable solutions of [TF–A] and of the *tf–a* transcription rate. Along this curve, the derivatives in Eqs. 14.25, 14.27a, and 14.27b are zero. At low values of k_{max} there is only one stable solution. Stimulus-induced phosphorylation of TF–A could make TF–A more effective at activating transcription, thereby increasing k_{max}. When k_{max} is raised into the brown region in Fig. 14.11B, two stable steady states exist (lower and upper portions of curve) with an unstable solution between them (middle, dashed portion of curve). For even higher values of k_{max} there is again a single stable solution.

The following discussion may help to further understand the two steady states. In the lower state, there is virtually no activation of the positive-feedback loop in which *tf–a* activates its own transcription, because [TF–A] is very low. A small degradation rate, proportional to [TF–A], is balanced by a small basal rate of *tf–a* transcription. If [TF–A] is increased, positive feedback occurs, and TF–A dimers activate *tf–a* transcription in a regenerative process. The rate of transcription and TF–A protein synthesis is then faster than the rate of TF–A degradation. At a high concentration of TF–A protein, the positive feedback saturates as the TF–RE enhancer becomes fully occupied with TF–A dimers. A maximal transcription rate is approached. But the TF–A degradation rate still increases with [TF–A] because it is directly proportional to [TF–A] in Eqs. 14.27a and 14.27b. Therefore, at a sufficiently high value of [TF–A], the degradation rate of TF–A must "catch up" with the synthesis rate.

A

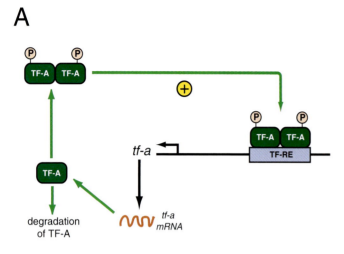

B

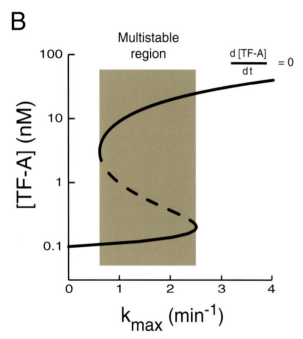

FIGURE 14.11 Positive autoregulation of a single gene. (A) Schematic of the model. Phosphorylated dimers of TF–A activate *tf-a* transcription when bound to a response element, termed a TF-RE, near the TF-A promoter. Degradation of TF-A protein is also indicated. (B) Bistability in the model of (A). For 0.6 min^{-1} < k_{max} < 2.5 min^{-1} (brown region in figure) two stable steady-state solutions of [TF–A] exist (lower and upper portions of the steady-state curve along which d[TF–A]$/dt = 0$). There is an unstable steady solution between (dashed). Outside the brown region there is only a single stable solution.

The second steady state is thereby created. If the model is initially at either stable solution, it will return to that solution after a small induced change in [TF–A]. However, a large induced change in [TF–A] can switch the system between steady states.

In comparing methods of modeling gene networks, the continuous method using differential equations is often preferred over the boolean method (model of Fig. 14.10). Greater physical accuracy is inherent in the use of continuous, rather than ON–OFF, gene expression rates. Furthermore, simulations using the boolean approach can erroneously predict behavior. Oscillations in gene expression seen with a boolean model may disappear when a more accurate, differential equation-based model is used (Bagley and Glass, 1996). Steady states may also disappear (Glass and Kauffman, 1973). Finally, methods of *bifurcation analysis* (Frank, 1978; Wiggins, 1990) can be applied to models based on differential equations. As discussed in Chapter 7, bifurcation analysis allows visualization of how the behavior of a model varies as a function of parameter values.

The continuous method, however, is more computationally intensive than the boolean method, because the former method requires much smaller time steps in simulations. For this reason, and because of the impracticality of thoroughly characterizing the expression kinetics of many genes, the boolean method is sometimes regarded as reasonable for large gene networks.

In both the boolean and continuous methods, intracellular transport of mRNA or protein can be modeled only in an *ad hoc* fashion. Time delays can be used, as in Eq. 14.27b, to represent the time between synthesis of a macromolecule in one intracellular compartment and its arrival and function in another compartment. For active transport of macromolecules, this is a reasonable approach (Smolen *et al.*, 1999). For diffusive transport, more accurate modeling would require discretizing the intracellular space into small volume elements, as was discussed for Ca^{2+} diffusion earlier in this chapter. It may be important to model transport, particularly in extended cells such as neurons. In neurons, delays of hours can be associated with transport of protein or mRNA from the soma to synapses targeted for long-term modification.

Simulations of macromolecular transport coupled with gene expression dynamics in neurons have not yet been reported. However, simulations of gene regulation in cells with simple shapes, such as a spherical nucleus in a spherical cell, have been carried out. These studies have found large differences in gene expression dynamics depending on whether transport of mRNA and protein from nucleus to cytoplasm and vice versa is assumed to be primarily active or primarily diffusive (Smolen *et al.*, 1999; Busenberg and Mahaffy, 1985). If active transport is modeled by a fixed time delay required for transit (Eq. 14.27b), then longer transit times tend to destabilize steady states of gene expression rates and favor oscillations. By contrast, in diffusion-dominated systems, slow diffusion tends to damp oscillations and yield steady states. These results suggest that dynamics of neuronal gene expression and synaptic modification may depend critically on the mode of macromolecular transport.

Gene Network Models Suggest That Feedback Loops and Protein Dimerization Can Generate Complex Dynamics

Transcription factors commonly bind to their target sequences as dimers. Dimerization of TFs steepens the relationship between the level of TF and the strength of the regulatory effect. Steepening this relationship favors complex dynamics, such as multiple stable gene expression rates and oscillations in gene expression rates. Complex dynamics are also favored if the regulated gene lies within a positive- or negative-feedback loop in which a gene activates or represses its own expression. In Keller (1995), four models are described that include feedback loops and dimerization of TFs. These models exhibit multiple stable gene expression rates. In each model, if dimerization of TFs does not occur, only a single steady state is obtained. However, dimerization of TFs is not an absolute requirement for multistability. Keller (1995) also gives conditions for multistability in two models with monomers. In one model, two TFs each repress the other's transcription, and in the second, a TF activates its transcription by binding to multiple enhancers.

Positive Feedback in Gene Networks Favors Multistability, and Negative Feedback Favors Oscillations

Positive feedback is usually essential for multistability of gene expression rates (Thomas *et al.*, 1995). For example, both positive feedback and TF oligomerization are essential for bistability in a model of neural tissue development in the *Drosophila* embryo (Kerszberg and Changeux, 1998). This model simulates the development of a group of undifferentiated cells into a structure of neurons that resembles an embryonic neural tube. Bistability in gene expression rates is essential for simulating the development of well-defined, smooth boundaries between neural and nonneural cells. In the simulation, proneural genes are ON within a spatially restricted region and OFF outside that region.

A key element of this model is a switch in which bistability of gene expression rates is created by competition between homodimers of two TFs. Each TF homodimer activates expression of its own gene while repressing the gene for the other TF. Figure 14.12 illustrates this competition. This scheme constitutes a positive-feedback loop. An increase in the level of one TF, by repressing the formation of the second TF, favors a further increase in the first TF. In each stable solution, the gene for one TF is strongly activated by its own product, which simultaneously represses the gene for the other TF. If TF-1 levels are high in a specific cell, that cell will differentiate into a neuron. If TF-2 levels are high, then TF-1 expression is repressed, and the cell will not become a neuron.

In the complete model, protein–protein interactions between receptors on the membranes of adjacent cells were also included so that the development of each cell would be influenced by the state of neighboring cells. For each cell, the state of the neighboring cells influences the transcription of the TFs in Fig. 14.12. Therefore, for each cell, the choice of whether to differentiate into a neuron (corresponding to a state with TF-1 expression high and TF-2 expression repressed) or not (corresponding to TF-2 expression high and TF-1 expression repressed) depends on the state of the neighboring cell. These interactions between cells are critical for the development of boundaries between regions of neural and nonneural tissue (Kerszberg and Changeux, 1998).

Multistability in gene networks might be important in several biological processes. Steady states of gene expression rates could constitute information that could be preserved through cycles of cell division (Keller, 1994). Multistability in gene networks may be

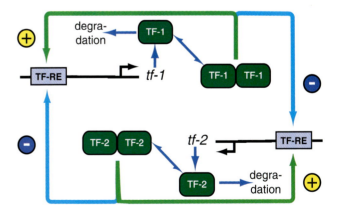

FIGURE 14.12   A simple gene regulatory scheme that generates bistability. The transcription factor TF–1 activates expression of its own gene and represses the TF–2 gene, whereas TF–2 activates its own gene and represses the TF–1 gene. Both TFs compete as homodimers for binding to response elements.

essential for cell differentiation (Thomas *et al.*, 1995; Laurent and Kellershohn, 1999). Because multistability is a manifestation of a positive-feedback loop, it has been suggested that genes involved in such loops are good candidates to regulate differentiation (Thomas *et al.*, 1995). However, none of these proposed roles for multistability have been experimentally verified.

Positive feedback alone usually cannot generate oscillations of rates of gene expression and therefore of macromolecule concentrations. In model gene networks, negative feedback is generally necessary to sustain oscillations (Smolen *et al.*, 2000). Many models describing oscillations also rely on dimerization of TFs (Goldbeter, 1996; Smolen *et al.*, 1998; Keller, 1995). A relatively simple model with negative feedback can be constructed by adding a second TF to the model of Fig. 14.11A. As diagrammed in Fig. 14.13A, this second TF, termed TF-R, represses transcription of both its own gene and that for the transcriptional activator, TF–A. In turn, TF–A activates the transcription of both *tf-a* and *tf-r*. The differential equations corresponding to this model consist of (1) Eqs. 14.25 and 14.27 modified to include repression of *tf-a* transcription by TF-R, and (2) two additional differential equations that describe the rates of change of [*tf-r* mRNA] and [TF-R] (Smolen *et al.*, 1999). The modified version of Eq. 25 is

$$\frac{d[tf-a\ \text{mRNA}]}{dt} = \left(\frac{k_{\max}[\text{TF}-\text{A}]^2}{[\text{TF}-\text{A}]^2 + K_d}\right)\left(\frac{K_R}{[\text{TF}-\text{R}]^2 + K_R}\right)$$
$$-k_{\text{degR}}[tf-a\ \text{mRNA}] + R_{\text{bas}} \quad (14.28)$$

The model of Fig. 14.13A can readily generate stable oscillations in the concentrations of both mRNA and protein species (Fig. 14.13B). An important synergistic application of modeling and experiment is construction of simple genetic systems that generate bistability or oscillations, combined with modeling based on equations similar to Eqs. 14.25–14.28. In one recent example (Gardner *et al.*, 2000) a bistable switch was constructed based on repression by two genes of each other's transcription. Modeling predicted perturbations that would flip the switch between states. In a second example (Elowitz and Leibler, 2000) a network of three genes was constructed. Gene 1 repressed gene 2, which repressed gene 3, which in turn repressed gene 1. This network constitutes a negative-feedback loop because expression of gene 1 ultimately leads to repression of gene 1. This system generated oscillations in transcription rates. Modeling predicted environmental conditions capable of sustaining oscillations.

A

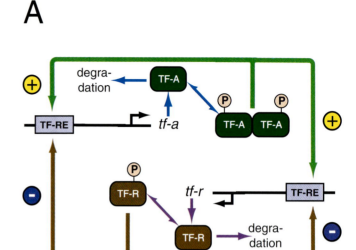

B

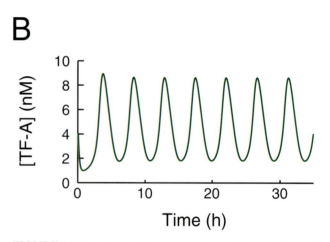

FIGURE 14.13   A simple two-gene network that sustains oscillations. (A) Schematic of the model. A second transcription factor, TF-R, is added to the model of Fig. 14.11A. TF–A activates transcription by binding to TF–RE response elements, which are present near the promoter regions of both *tf-a* and *tf-r*. TF-R represses transcription by competing with the TF–A dimer for binding to both TF–REs. (B) Sustained oscillations of [TF–A] and [TF–R] produced by the model of (A).

Simple, designed genetic systems similar to the bistable switch or the three-gene network might prove useful for gene therapy. For example, transfection with several genes might confer novel drug responsiveness on target cells. Correction of a polygenetic disorder, in which several genes contribute to the observed phenotype, might also require transfection with multiple genes. Several promising methods for gene delivery to the brain are being developed, including viral transfection and cerebral injection of neural stem cells (Hsieh *et al.*, 2002). Models are likely to be essential for predicting and understanding the responses of multiple transfected genes, or even a single gene, *in vivo*. Models have already helped to design gene vectors with maximal transfection efficiency (Varga *et al.*, 2001).

The positive- and negative-feedback loops within the gene network involving CREB and related TFs (Fig. 14.9) might support complex dynamics such as oscillations in transcription rates. Oscillations of CREB mRNA have been studied in endocrine cells (Walker *et al.*, 1995), and these feedback loops were suggested to be responsible.

Negative Feedback Underlies Circadian Rhythm Generation

A group of gene networks in which negative feedback appears essential for oscillations in gene expression is responsible for circadian rhythms, which have evolved in most organisms exposed to daylight (Gartner, 1990; Dunlap *et al.*, 1999; Reppert and Weaver, 2001). These behavioral rhythms are self-sustaining in constant darkness. In circadian oscillators that have been well characterized, the products of one or a few core genes repress their own transcription in a cyclic fashion. Following the onset of repression, there is a delay of several hours, after which the repressing proteins are degraded, relieving repression (Dunlap *et al.*, 1999). For example, in *Drosophila* neurons that drive the circadian behavior rhythm, a transcription factor termed PER represses transcription of its own gene. Over many hours, PER becomes multiply phosphorylated and then degrades. Degradation relieves autorepression of *per* by PER, and another surge of transcription initiates the next cycle. The multiple phosphorylations of PER prior to degradation introduce a time delay, which appears to be important for generating the ~24-h oscillation period. In addition to the negative-feedback loop based on *per* autorepression, positive regulation of *per* and other genes essential for oscillations is also important (Bae *et al.*, 2000; Cyran et al., 2003). Posttranscriptional regulation of protein synthesis may also be important (Roenneberg and Merrow, 1998). In mammals, several isoforms of PER with distinct functions exist, and the PER proteins in concert with others mediate both negative and positive regulation of core gene transcription (Shearman *et al.*, 2000). As in *Drosophila*, multiple phosphorylations of mammalian PER occur before degradation, and may be important for the 24-h period (Lee *et al.*, 2001).

Models based on differential equations have been developed to represent circadian rhythms and their responses to stimuli. The organism most commonly modeled is *Drosophila* (Goldbeter, 1996; Leloup and Goldbeter, 1998; Smolen *et al.*, 2000). Models for the generation of circadin rhythms in particular species are complex because of multiple genes and biochemical time delays. One recent *Drosophila* model incorporates negative and positive regulation of *per* and other genes, as well as stochastic fluctuations in the numbers of gene product proteins (Smolen *et al.*, 2001). A simpler model, presented here, captures essential processes common to many organisms. To formulate this model, a TF is assumed to repress transcription of its own gene. Newly synthesized TF protein must be multiply phosphorylated prior to degradation. A burst of *tf* transcription is ended by repression due to TF protein accumulation. Then, there is a period of approximately 1 day during which TF protein is phosphorylated and degraded. Degradation allows initiation of another burst of *tf* transcription. A time delay τ is included between transcription of *tf* mRNA and subsequent translation of protein. The model is shown schematically in Fig. 14.14A.

This model simulates circadian oscillations as illustrated in Fig. 14.14B, which displays time courses of mRNA and of total protein concentration. Circadian oscillations are simulated only if multiple (~8 or more) phosphorylations of the TF are assumed necessary prior to degradation. This observation is consistent with experiments in *Drosophila* and other organisms (Dunlap *et al.*, 1999). The necessity for multiple phosphorylations is not surprising. It follows from a general result that increasing the number of kinetic steps in a negative-feedback loop promotes oscillations (Griffith, 1968). Inclusion of the delay τ, with a value of 1–2 h, increases the amplitude and stability of oscillations.

The equations for this model are somewhat complex. P_{0-N} is used to denote a TF with 0, 1, . . ., N phosphorylations, and only P_N is degraded. N takes values of 8–10. k_{ph} is a first-order rate constant governing the rate of every phosphorylation reaction. $[P_{tot}]$ denotes the total concentration of protein, $[P_{tot}] = [P_0] + [P_1] + \ldots + [P_N]$. The differential equation for TF mRNA has a synthesis term containing repression by TF protein ($[P_{tot}]$ in the denominator) with a Hill coefficient of n. n is typically 2–4. First-order degradation of mRNA is assumed:

$$\frac{d[\text{mRNA}]}{dt} = v_R \frac{K_R^{\,n}}{\left[P_{\text{tot}}\right]^n + K_R^{\,n}} - k_d[\text{mRNA}].$$

A

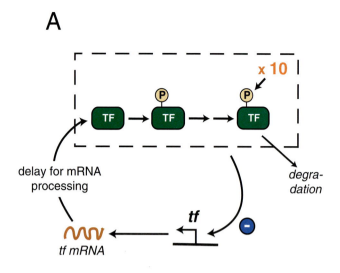

B

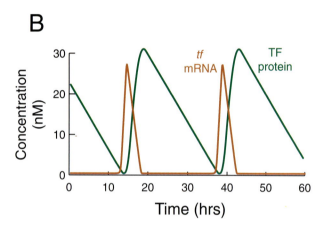

FIGURE 14.14 Generation of circadian rhythmicity by negative transcriptional feedback. (A) The kinetic scheme. A transcription factor (TF) can undergo multiple phosphorylation steps. Ten sequential phosphorylations are assumed. As indicated by the dashed box, all forms of TF protein are assumed capable of repressing *tf* transcription. A time delay is included between the appearance of *tf* mRNA and TF protein. Only fully phosphorylated TF can degrade. But it does so relatively rapidly, so that over the space of a day, virtually all TF protein becomes fully phosphorylated and then degrades. This relieves *tf* repression so that another "burst" of *tf* transcription can occur. (B) Circadian oscillations in the levels of *tf* mRNA and TF protein simulated by the model of (A).

Nonphosphorylated TF, P_0, is synthesized from mRNA after a delay τ (typically 1–2 h). P_0 disappears via phosphorylation to P_1:

$$\frac{d[P_0]}{dt} = k_P[\text{mRNA}](t - \tau) - k_{\text{ph}}[P_0].$$

A series of phosphorylation reactions ensue until the most highly phosphorylated form, P_N, is created. The

reactions are assumed sequential so that phosphorylation i cannot occur until phosphorylation i-1 has occurred:

$$\frac{d[P_i]}{dt} = -k_{\text{ph}}[P_i] + k_{\text{ph}}[P_{i-1}], \quad \text{for } i = 2, \ldots N - 1.$$

Finally, P_N is synthesized by phosphorylation and is degraded in a Michaelis–Menten enzymatic reaction:

$$\frac{d[P_N]}{dt} = k_{\text{ph}}[P_{N-1}] - \frac{v_P[P_N]}{K_P + [P_N]}.$$

Random Fluctuations in Molecule Numbers Can Strongly Influence Genetic Regulation

As discussed above (Eq. 14.24 and accompanying text), models using differential equations can be extended to include random variability in the times of individual macromolecular synthesis and degradation events. This variability gives rise to stochastic fluctuations in the numbers of individual mRNAs and proteins. Such fluctuations could strongly influence the dynamics of gene expression. Recent observations indicate that the timing of transcription of individual mRNAs is indeed quite random (Elowitz *et al.*, 2001; Zlokarnik *et al.*, 1998). Also, for important mRNA or protein species the average copy number present in the nucleus may be < 100, so that random fluctuations in copy number may be relatively large. In neurons, late long-term potentiation of synaptic connections lasting 24 h or longer is dependent on transcription and translation. Therefore, stochastic fluctuations in the numbers of mRNAs and proteins may well induce considerable synapse-to-synapse variability in the amount of potentiation.

An example of how fluctuations can destabilize steady states is shown in Fig. 14.15. The simulations for this illustration used a set of four differential equations describing reciprocal regulation of two genes. The regulatory interactions were those of Fig. 14.12. The equations were similar to Eqs. 14.27a and 14.28. Figure 14.15A shows that without fluctuations, this model exhibits multistability. Initially, the concentrations of both *tf-1* and *tf-2* mRNA and protein are in a low, stable steady state. As shown, if a temporary increase in the basal transcription rate of *tf-1* is large enough, a transition to a second, stable steady state occurs. Here, the concentrations of *tf-1* mRNA and protein are high, whereas those of *tf-2* mRNA and protein (not shown) are low. In the absence of the transient increase in *tf-1* transcription, the levels of *tf-1* and *tf-2* gene products would not undergo this transition.

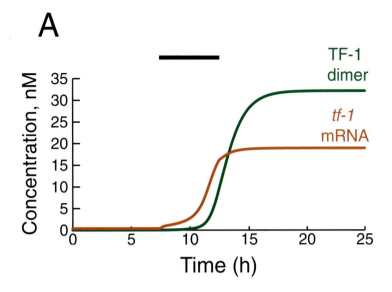

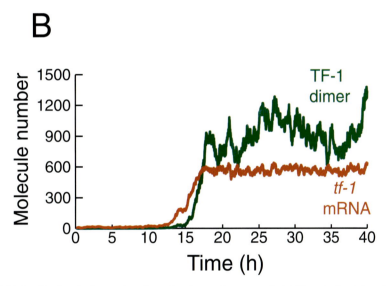

FIGURE 14.15 Stochastic fluctuations in molecule numbers may destabilize steady states of genetic regulatory systems. (A) Without fluctuations, a set of equations based on Fig. 14.12A exhibits multistability. Initial levels of *tf-1* (and *tf-2*) mRNA and protein are low (< 2 nM) and steady. At *t* = 7.5 h the rate of *tf-1* transcription is increased by a constant amount for 10 h (horizontal bar). The increase causes a transition to a stable state with high *tf-1* mRNA and protein levels. The high levels persist after the imposed increase in transcription rate ceases. (B) The initial state of (A) is spontaneously destabilized within ~ 20 h when stochastic fluctuations are incorporated. No induced increase in *tf-1* transcription is included in this simulation, which is otherwise the same as in (A) except for a scaling factor converting the units of the variables from concentrations to molecule numbers.

Stochastic fluctuations in molecule numbers were then added. The method based on Eq. 14.24 was used. In this case, when the numbers of each molecular species were initialized at values corresponding to the lower steady state of Fig. 14.15A, this state was no longer stable. After several hours, random fluctuations in molecule numbers would cross a threshold, initiating a positive-feedback loop in which either *tf-1* or *tf-2* would activate its own transcription. Figure 14.15B illustrates that the system would leave the initial state and settle in a new steady state, even without an extrinsic perturbation. In the new state, *tf-1* transcription was high and *tf-2* transcription low. These observations suggest that more generally, stochastic fluctuations can prevent gene networks from settling in states that are only weakly stable.

SUMMARY

Mathematical models are essential for synthesizing large amounts of data concerning intracellular signaling pathway dynamics and regulation of gene expression. Standard modeling techniques have been developed to represent common biochemical elements such as allosteric interactions, feedback interactions in biochemical pathways and gene networks, and intracellular transport of molecules. Modeling of biochemical signaling pathways and of gene networks has suggested a unifying principle—that positive- and negative-feedback interactions commonly underlie complex dynamics of signaling pathways and of gene expression. Such dynamics can include multiple stable states, or oscillations, of biochemical concentrations and reaction rates. Models can also help assess the importance of interactions between signaling pathways.

For extended cells such as neurons, it may be important that models of intracellular signaling include transport of macromolecules or metabolites. Inclusion of random fluctuations in the numbers of particular macromolecules is important if the molecule number per cell is small (several hundred or less). For example, understanding variability in stimulus responses may require consideration of fluctuations in critical mRNAs and proteins. However, sufficient data to constrain models of fluctuations are not available for most signaling pathways or gene networks.

A combination of theory (metabolic control analysis) and experiment has helped determine principles of metabolic flux control in biochemical pathways. For example, the metabolic flux through a pathway is least likely to be sensitive to changes in the activity of enzymes subject to feedback regulation. Use of metabolic control analysis to interpret experimental data can also help determine the rate-limiting reactions for the production of metabolites.

Regulation of gene expression can be modeled at a coarse level, in which genes are ON–OFF switches, or at a finer level, in which gene expression rates are continuous variables. For large gene networks, the coarse method may be the only one feasible, due to limited data. The finer level, using differential equations, is appropriate for networks that can be well characterized experimentally. In gene networks, feedback interactions can generate complex dynamics

such as multistability and oscillations. For more realism in models of small gene networks, random fluctuations in macromolecule numbers could be included; however, sufficient data to constrain such models are usually not available.

For all models, it is essential to determine the sensitivity of model behavior to parameters and to assess which parameters are not well constrained by data. Experiments that estimate parameters that are key control points of model behavior constitute stringent tests of a model. Uncertainties in parameters imply that biochemical models are most often qualitative rather than quantitative descriptions of *in vivo* systems. Finally, reduction of the number of model equations to the minimum necessary to simulate experimental data is desirable to help determine the biochemical elements most important for system behavior. Separation of variables according to whether they vary on a fast or slow time scale can help with this simplification.

In conclusion, it is evident that with the biochemical and gene expression data now becoming available, efficient characterization of coupled genetic and biochemical systems will require collaboration between biochemists, molecular biologists, and mathematical biologists. These collaborations will fit data into a mathematical framework and design experiments to test model predictions. This methodology is necessary to predict and analyze the responses of neurons and organisms to physiological stimuli, biologically active environmental contaminants, and novel pharmaceutical agents.

References

General

Bower, J., and Bolouri, H. (2001). "*Computational Modeling of Genetic and Biochemical Networks.*" MIT Press, Cambridge, MA.

Edelstein-Keshet, L. (1988). "*Mathematical Models in Biology.*" Random House, New York.

Goldbeter, A. (1996). "*Biochemical Oscillations and Cellular Rhythms.*" Cambridge Univ. Press, Cambridge.

Koch, C., and Segev, I. (1998). "*Methods in Neuronal Modeling: From Synapses to Networks,*" 2nd ed. MIT Press, Cambridge, MA.

Murray, J. D. (1989). "*Mathematical Biology.*" Springer-Verlag, New York.

Segel, I. H. (1975). "*Enzyme Kinetics.*" Wiley, New York.

Smolen, P., Baxter, D. A., and Byrne, J. H. (2000). Mathematical modeling of gene networks. *Neuron* **26**, 567–580.

Specific

Amini, B., Clark, J. W., and Canavier, C. C. (1999). Calcium dynamics underlying pacemaker-like and burst firing oscillations in midbrain dopaminergic neurons, a computational study. *J. Neurophysiol.* **82**, 2249–2261.

Arkin, A., Ross, J., and McAdams, H. (1998). Stochastic kinetic analysis of developmental pathway bifurcation in phage λ-infected *E. coli* cells. *Genetics* **149**, 1633–1648.

Bae, K., Lee, C., Hardin, P. E., and Edery, I. (2000). dCLOCK is present in limiting amounts and likely mediates daily interactions between the dCLOCK-CYC transcription factor and the PER–TIM complex. *J. Neurosci.* **20**, 1746–1753.

Bagley, R. J., and Glass, L. (1996). Counting and classifying attractors in high dimensional dynamical systems. *J. Theor. Biol.* **183**, 269–284.

Barkai, N., and Leibler, S. (2000). Circadian clocks limited by noise. *Nature* **403**, 267–268.

Bartsch, D., Casadio, A., Karl, K., Serodio, P., and Kandel, E. R. (1998). *CREB1* encodes a nuclear activator, a repressor, and a cytoplasmic modulator that form a regulatory unit critical for long-term facilitation. *Cell* **95**, 211–223.

Beck, J. V., and Arnold, K. J. (1977). "*Parameter Estimation In Engineering and Science,*" pp. 17–24, 481–487. Wiley, New York.

Berg, O. G., Paulsson, J., and Ehrenberg, M. (2000). Fluctuations and quality of control in biological cells: Zero-order ultrasensitivity reinvestigated. *Biophys. J.* **79**, 1228–1236.

Bhalla, U. S. (2002). Biochemical signaling networks decode temporal patterns of synaptic input. *J. Comp. Neurosci.* **13**, 49–62.

Bhalla, U. S., and Iyengar, R. (1999). Emergent properties of networks of biological signaling pathways. *Science* **283**, 381–386.

Blumenfeld, H., Zablow, L., and Sabatini, B. (1992). Evaluation of cellular mechanisms for modulation of Ca^{2+} transients using a mathematical model of fura-2 Ca^{2+} imaging in *Aplysia* sensory neurons. *Biophys. J.* **63**, 1146–1164.

Brown, G. P., Blitzer, R. D., Connor, J. H., Wong, T., Shenolikar, S., Iyengar, R., and Landau, E. M. (2000). Long-term potentiation induced by ϑ frequency stimulation is regulated by a protein phosphatase–1-operated gate. *J. Neurosci.* **20**, 7880–7887.

Burden, R. L., and Faires, J. D. (1993). "*Numerical Analysis.*" PWS-Kent, Boston.

Busenberg, S., and Mahaffy, J. M. (1985). Interaction of spatial diffusion and delays in models of genetic control by repression. *J. Math. Biol.* **22**, 313–333.

Canavier, C. C., Clark, J. W., and Byrne, J. H. (1993). Simulation of the bursting activity of neuron R15 in *Aplysia*: Role of ionic currents, calcium balance, and modulatory transmitters. *J. Neurophysiol.* **66**, 2107–2124.

Cao, D., and Parker, R. (2001). Computational modeling of eukaryotic mRNA turnover. *RNA* **7**, 1192–1212.

Chen, H. X., Otmakhov, N., Strack, S., Colbran, R. J., and Lisman, J. E. (2001). Is persistent activity of calcium/calmodulin-dependent kinase required for the maintenance of LTP? *J. Neurophysiol.* **85**, 1368–1376.

Cooper, D., Mons, N., and Karpen, J. (1995). Adenylyl cyclases and the interaction between calcium and cAMP signaling. *Nature* **374**, 421–424.

Cyran, S. A., Buchsbaum, A. M., Reddy, K. L., Lin, M. C., Glossop, N. R., and Hardin, P. E. (2003). *Vrille, Pdp1,* and *dClock* form a second feedback loop in the *Drosophila* circadian clock. *Cell* **112**, 329–341.

Dano, S., Sorensen, P. G., and Hynne, F. (1999). Sustained oscillations in living cells. *Nature* **402**, 320–322.

Davis, R. J. (2000). Signal transduction by the JNK group of MAP kinases. *Cell* **103**, 239–252.

De Cesare, D., and Sassone-Corsi, P. (2000). Transcriptional regulation by cyclic AMP-responsive factors. *Prog. Nucleic Acid Res. Mol. Biol.* **64**, 343–369.

De Schutter, E., and Smolen, P. (1998). Calcium dynamics in large neuronal models. *In "Methods in Neuronal Modeling: From Synapses to Networks."* (C. Koch and I. Segev, Eds.), 2nd ed. MIT Press, Cambridge, MA.

Doedel, E. (1981). AUTO: A program for the automatic bifurcation analysis of autonomous systems. *Congr. Num.* **30**, 265–284.

Dunlap, J. C., Loros, J. J., Liu, Y., and Crosthwaite, S. K. (1999). Eukaryotic circadian systems: Cycles in common. *Genes Cells* **4**, 1–10.

Edwards, A. S., and Scott, J. D. (2000). A-kinase anchoring proteins: Protein kinase A and beyond. *Curr. Opin. Cell Biol.* **12**, 217–221.

Elowitz, M. B., and Leibler, S. (2000). A synthetic oscillatory network of transcriptional regulators. *Nature* **403**, 335–338.

Elowitz, M. B., Levine, A. J., Siggia, E. D., and Swain, P. S. (2001). Stochastic gene expression in a single cell. *Science* **297**, 1183–1186.

Ermentrout, B. (2001). Simplifying and reducing complex models. *In "Computational Modeling of Genetic and Biochemical Networks"* (J. Bower and H. Bolouri, Eds.). MIT Press, Cambridge, MA

Ferrell, J. E. (1996). Tripping the switch fantastic: How a protein kinase cascade can convert graded inputs into switch-like outputs. *Trends Biochem. Sci.* **21**, 460–466.

Fink, C. C., Slepchenko, B., Moraru, I., Watras, J., Schaff, J. C., and Loew, L. M. (2000). An image-based model of calcium waves in differentiated neuroblastoma cells. *Biophys. J.* **79**, 163–183.

Firth, C. A., and Bray, D. (2001). Stochastic simulation of cell signaling pathways. *In "Computational Modeling of Genetic and Biochemical Networks."* (J. Bower and H. Bolouri, Eds.), MIT Press, Cambridge, MA.

Fokina, K. V., Dainyak, M. B., Nagradova, N. K., and Muronetz, V. I. (1997). A study on the complexes between human erythrocyte enzymes participating in the conversions of 1,3-diphosphoglycerate. *Arch. Biochem. Biophys.* **345**, 185–192.

Frank, P. M. (1978). "*Introduction to System Sensitivity Theory,*" pp. 9–10. Academic Press, New York.

Gardner, T. S., Cantor, C. R., and Collins, J. J. (2000). Construction of a genetic toggle switch in *Escherichia coli. Nature* **403**, 339–342.

Gartner, K. (1990). A third component causing variability beside environment and genotype: A reason for the limited success of a 30-year long effort to standardize laboratory animals? *Lab. Anim.* **24**, 71–77.

Gerhold, D., Rushmore, T., and Caskey, C. T. (1999). DNA chips: Promising toys have become powerful tools. *Trends Biochem. Sci.* **24**, 168–173.

Gibson, M. A., and Bruck, J. (2000). Efficient exact stochastic simulation of chemical systems with many species and many channels. *J. Phys. Chem.* A **104**, 1876–1889.

Giese, K. P., Fedorov, N. B., Filipkowski, R. K., and Silva, A. J. (1998). Autophosphorylation at Thr(286) of the alpha calcium-calmodulin kinase II in learning and memory. *Science* **279**, 870–873.

Gillespie, D. T. (1977). Exact stochastic simulation of coupled chemical reactions. *J. Phys. Chem.* **61**, 2340–2361.

Glass, L., and Kauffman, S. A. (1973). The logical analysis of continuous, non linear biochemical control networks. *J. Theor. Biol.* **39**, 103–129.

Goldbeter, A. (1996). "*Biochemical Oscillations and Cellular Rhythms.*" Cambridge Univ. Press, Cambridge.

Goldbeter, A., and Koshland, D. E. (1984). Ultrasensitivity in biochemical systems controlled by covalent modification: Interplay between zero-order and multistep effects. *J. Biol. Chem.* **259**, 14441–14447.

Griffith, J. S. (1968). Mathematics of cellular control processes. I. Negative feedback to one gene. *J. Theor. Biol.* **20**, 202–208.

Hanson, P. I., Meyer, T., Stryer, L., and Schulman, H. (1994). Dual role of calmodulin in autophosphorylation of multifunctional CaM kinase may underlie decoding of calcium signals. *Neuron* **12**, 943–956.

Hodgkin, A. L., and Huxley, A. F. (1952). A quantitative description of membrane current and its application to conduction and excitation in nerve. *J. Physiol. (London)* **117**, 500–544.

Holmes, W. R. (2000). Models of calmodulin trapping and CaM kinase II activation in a dendritic spine. *J. Comp. Neurosci.* **8**, 65–85.

Hsich, G., Sena-Esteves, M., and Breakefield, X. O. (2002). Critical issues in gene therapy for neurologic disease. *Hum. Gene Ther.* **13**, 579–604.

Huang, C. F., and Ferrell, J. E. (1996). Ultrasensitivity in the mitogen-activated protein kinase cascade. *Proc. Natl. Acad. Sci. USA* **93**, 10078–10083.

Husi, H., Ward, M. A., Choudhary, J. S., Blackstock, W. P., and Grant, S. G. (2000). Proteomic analysis of NMDA receptor–adhesion protein signaling complexes. *Nat. Neurosci.* **3**, 661–669.

Ihmels, J., Friedlander, G., Bergmann, S., Sarig, O., Ziv, Y., and Barkai, N. (2002). Revealing modular organization in the yeast transcriptional network. *Nat. Genet.* **31**, 370–377.

Jafri, M. S., and Keizer, J. (1995). On the roles of Ca^{2+} diffusion, Ca^{2+} buffers, and the endoplasmic reticulum in IP3-induced Ca^{2+} waves. *Biophys. J.* **69**, 2139–2153.

Karin, M. (1994). Signal transduction from the cell surface to the nucleus through the phosphorylation of transcription factors. *Curr. Opin. Cell Biol.* **6**, 415–424.

Katz, P. S., and Clemens, S. (2001). Biochemical networks in nervous systems: Expanding neuronal information capacity beyond voltage signals. *Trends Neurosci.* **24**, 18–25.

Kaufman, E. N., and Jain, R. K. (1990). Quantification of transport and binding parameters using fluorescence recovery after photobleaching, potential for *in vivo* applications. *Biophys. J.* **58**, 873–885.

Keller, A. (1994). Specifying epigenetic states with autoregulatory transcription factors. *J. Theor. Biol.* **170**, 175–181.

Keller, A. (1995). Model genetic circuits encoding autoregulatory transcription factors. *J. Theor. Biol.* **172**, 169–185.

Kerszberg, M., and Changeux, J. (1998). A simple molecular model of neurulation. *Bioessays* **20**, 758–770.

Kholodenko, B. N., Brown, G. C., and Hoek, J. B. (2000). Diffusion control of protein phosphorylation in signal transduction pathways. *Biochem. J.* **350**, 901–907.

Kurganov, B. I. (1986). The role of multienzyme complexes in integration of intracellular metabolism. *J. Theor. Biol.* **119**, 445–455.

Laurent, M., and Kellershohn, N. (1999). Multistability: A major means of differentiation and evolution in biological systems. *Trends Biochem. Sci.* **24**, 418–422.

Lee, C., Etchegaray, J., Cagampang, F., Loudon, A. S., and Reppert, S. M. (2001). Posttranslational mechanisms regulate the mammalian circadian clock. *Cell* **107**, 855–867.

Leloup, J. C., and Goldbeter, A. (1998). A model for circadian rhythms in *Drosophila* incorporating the formation of a complex between the PER and TIM proteins. *J. Biol. Rhythms* **13**, 70–87.

Levchenko, A., Bruck, J., and Sternberg, P. W. (2000). Scaffold proteins may biphasically affect the levels of mitogen-activated protein kinase signaling and reduce its threshold properties. *Proc. Natl. Acad. Sci. USA* **97**, 5818–5823.

Li, Y. X., and Rinzel, J. (1994). Equations for InsP$_3$ receptor-mediated [Ca]$_i$ oscillations derived from a detailed kinetic model: a Hodgkin–Huxley-like formalism. *J. Theor. Biol.* **166**, 461–473.

Lisman, J. (1994). The CaM kinase II hypothesis for the storage of synaptic memory. *Trends Neurosci.* **17**, 406–412.

Lisman, J. E., and Zhabotinsky, A. M. (2001). A model of synaptic memory: A CAMKII/PP1 switch that potentiates transmission by organizing an AMPA receptor anchoring assembly. *Neuron* **31**, 191–201.

Little, J. W., Shepley, D. P., and Wert, D. W. (1999). Robustness of a gene regulatory circuit. *EMBO J.* **18**, 4299–4307.

Lonze, B. E., and Ginty, D. D. (2002). Function and regulation of CREB family transcription factors in the nervous system. *Neuron* **35**, 605–623.

MacDonald, N. (1989). "*Biological Delay Systems: Linear Stability Theory*". Cambridge Univ. Press, New York.

Marder, E. (2000). Models identify hidden assumptions. *Nat. Neurosci. Suppl.* **3**, 1198.

Martin, S. J, Grimwood, P. D, and Morris, R. G. (2000). Synaptic plasticity and memory: An evaluation of the hypothesis. *Annu. Rev. Neurosci.* **23**, 649–711.

Mauk, M. D. (2000). The potential effectiveness of simulations verses phenomenological models. *Nat. Neurosci.* **3**, 649–651.

Mayford, M., and Kandel, E. R. (1999). Genetic approaches to memory storage. *Trends Genet.* **15**, 463–470.

McAdams, H., and Arkin, A. (1998). Simulation of prokaryotic genetic circuits. *Annu. Rev. Biophys. Biomol. Struct.* **27**, 199–224.

McAdams, H., and Arkin, A. (1999). It's a noisy business! Genetic regulation at the nanomolar scale. *Trends Genet.* **15**, 65–69.

Mirnics, K., Middleton, F. A., Marquez, A., Lewis, D. A., and Levitt, P. (2000). Molecular characterization of schizophrenia viewed by microarray analysis of gene expression in prefrontal cortex. *Neuron* **28**, 53–67.

Molina, C., Foulkes, N., Lalli, E., and Sassone-Corsi, P. (1993). Inducibility and negative autoregulation of CREM: An alternative promoter directs the expression of ICER, and early response repressor. *Cell* **75**, 875–886.

Morton-Firth, C. J., Shimizu, T. S., and Bray, D. (1999). A free-energy-based stochastic simulation of the Tar receptor complex. *J. Mol. Biol.* **286**, 1059–1074.

Nakamura, T., Barbara, J., Nakamura, K., and Ross, W. N. (1999). Synergistic release of Ca^{2+} from IP$_3$-sensitive stores evoked by synaptic activation of mGluRs paired with backpropagating action potentials. *Neuron* **24**, 727–737.

Ouyang, Y., Kantor, D., Harris, K. M., Schuman, E. M., and Kennedy, M. B. (1997). Visualization of the distribution of autophosphorylated calcium/calmodulin dependent protein kinase II after tetanic stimulation in the CA1 area of the hippocampus. *J. Neurosci.* **17**, 5416–5427.

Paulsson, J., Berg, O. G, and Ehrenberg, M. (2000). Stochastic focusing: Fluctuation-enhanced sensitivity of intracellular regulation. *Proc. Natl. Acad. Sci. USA* **97**, 7148–7153.

Prentki, M., Tornheim, K., and Corkey, B. E. (1997). Signal transduction mechanisms in nutrient-induced insulin secretion. *Diabetologia* **40** (Suppl. 2), S32–S34.

Press, W. H. (1994). "*Numerical Recipes in C: The Art of Scientific Computing*." Cambridge Univ. Press, Cambridge.

Qian, N., and Sejnowski, T. (1990). When is an inhibitory synapse effective? *Proc. Natl. Acad. Sci. USA* **87**, 8145–8149.

Ramoni, M. F., Sebastiani, P., and Kohane, I. S. (2002). Cluster analysis of gene expression dynamics. *Proc. Natl. Acad. Sci. USA* **99**, 9121–9126.

Reppert, S. M., and Weaver, D. R. (2001). Molecular analysis of mammalian circadian rhythms. *Annu. Rev. Physiol.* **63**, 647–676.

Rinzel, J. (1985). Excitation dynamics: Insights from simplified membrane models. *Fed. Proc.* **44**, 247–264.

Roberts, C. J., Nelson, B., Marton, M. J., Stoughton, R., Meyer, M. R., Bennett, H. A., He, Y. D., Dai, H., Walker, W. L., Hughes, T. R., Tyers, M., Boone, C., and Friend, S. H. (2000). Signaling and circuitry of multiple MAPK pathways revealed by a matrix of global gene expression profiles. *Science* **287**, 873–880.

Roenneberg, T., and Merrow, M. (1998). Molecular circadian oscillators: An alternative hypothesis. *J. Biol. Rhythms* **13**, 167–179.

Rohwer, J. M., Postma, P. W., Kholodenko, B. N., and Westerhoff, H. V. (1998). Implications of macromolecular crowding for signal transduction and metabolite channeling. *Proc. Natl. Acad. Sci. USA* **95**, 10547–10552.

Ryu, D. D., and Nam, D. H. (2000). Recent progress in biomolecular engineering. *Biotechnol. Prog.* **16**, 2–16.

Sabry, J., O'Connor, T., and Kirschner, M. W. (1995). Axonal transport of tubulin in Ti1 Pioneer neurons *in situ. Neuron* **14**, 1247–1256.

Salter, M. (1996). Determination of the flux control coefficient of nitric oxide synthase for nitric oxide synthesis in discrete brain regions *in vivo. J. Theor. Biol.* **182**, 449–452.

Segel, I. H. (1975). "*Enzyme Kinetics.*" Wiley, New York.

Shearman, L. P., Sriram, S., Weaver, D. R., Maywood, E. S., Chaves, I., Zheng, B., Kume, K., Lee, C. C., van der Horst, G., Hastings, M. H., and Reppert, S. M. (2000). Interacting molecular loops in the mammalian circadian clock. *Science* **288**, 1013–1019.

Simon, S., and Llinas, R. (1985). Compartmentalization of the submembrane calcium activity during calcium influx and its significance in transmitter release. *Biophys. J.* **48**, 485–498.

Small, J. R., and Kacser, H. (1993). Responses of metabolic systems to large changes in enzyme activities and effectors. *Eur. J. Biochem.* **213**, 613–640.

Smolen, P., and Keizer, J. (1992). Slow voltage inactivation of Ca²⁺ currents and bursting mechanisms for the mouse pancreatic beta-cell. *J. Membr. Biol.* **127**, 9–19.

Smolen, P., Baxter, D. A., and Byrne, J. H. (1998). Frequency selectivity, multistability, and oscillations emerge from models of genetic regulatory systems. *Am. J. Physiol.* **274**, C531–C542.

Smolen, P., Baxter, D. A., and Byrne, J. H. (1999). Effects of macromolecular transport and stochastic fluctuations on the dynamics of genetic regulatory systems. *Am. J. Physiol.* **277**, C777–C790.

Smolen, P., Baxter, D. A., and Byrne, J. H. (2000). Mathematical modeling of gene networks. *Neuron* **26**, 567–580.

Smolen, P., Baxter, D. A., and Byrne, J. H. (2001). Modeling circadian oscillations with interlocking positive and negative feedback loops. *J. Neurosci.* **21**, 6644–6656.

Stacey, W. C., and Durand, D. M. (2001). Synaptic noise improves detection of subthreshold signals in hippocampal CA1 neurons. *J. Neurophysiol.* **86**, 1104–1112.

Stephanopoulos, G. N., Aristidou, A., and Nielsen, J. (1998). "*Metabolic Engineering, Principles and Methodologies.*" Academic Press, San Diego.

Teruel, M. N., and Meyer, T. (2000). Translocation and reversible localization of signaling proteins: A dynamic future for signal transduction. *Cell* **103**, 181–184.

Thomas, R., Thieffry, D., and Kaufman, M. (1995). Dynamical behaviour of biological regulatory networks: I. Biological role of feedback loops and practical use of the concept of the loop-characteristic state. *Bull. Math. Biol.* **57**, 247–276.

Toh, H., and Horimoto, K. (2002). Inference of a genetic network by a combined approach of cluster analysis and graphical Gaussian modeling. *Bioinformatics* **18**, 287–297.

Varga, C. M., Hong, K., and Lauffenburger, D. A. (2001). Quantitative analysis of synthetic gene delivery vector design properties. *Mol. Ther.* **4**, 438–446.

Wagner, A. (2000). Robustness against mutations in genetic networks of yeast. *Nat. Genet.* **24**, 355–361.

Wagner, J., and Keizer, J. (1994). Effects of rapid buffers on Ca²⁺ diffusion and Ca²⁺ oscillations. *Biophys. J.* **67**, 447–456.

Wagner, J., Li, Y. X., Pearson, J., and Keizer, J. (1998). Simulation of the fertilization Ca²⁺ wave in *Xenopus laevis* eggs. *Biophys. J.* **75**, 2088–2097.

Walker, W., Fucci, L., and Habener, J. (1995). Expression of the gene encoding transcription factor CREB: Regulation by follicle-stimulating hormone-induced cAMP signaling in primary rat sertoli cells. *Endocrinology* **136**, 3534–3545.

Wen, X. L., Fuhrman, S., Michaels, G., Carr, D., Smith, S., Barker, J., and Somogyi, R. (1998). Large-scale temporal gene expression mapping of central nervous system development. *Proc. Natl. Acad. Sci. USA* **95**, 334–339.

Wiggins, S. (1990). "*Introduction to Applied Nonlinear Dynamical Systems and Chaos.*" Springer-Verlag, Heidelberg.

Yamagata, Y., and Obata, K. (1998). Dynamic regulation of the activated, autophosphorylated state of Ca²⁺/calmodulin-dependent protein kinase II by acute neuronal excitation *in vivo. J. Neurochem.* **71**, 427–439.

Yeung, M. K. S., Tegner, J., and Collins, J. J. (2002). Reverse engineering gene networks using singular value decomposition and robust regression. *Proc. Natl. Acad. Sci. USA* **99**, 6163–6168.

Yin, J., and Tully, T. (1996). CREB and the formation of long-term memory. *Curr. Opin. Neurobiol.* **6**, 264–267.

Zhabotinsky, A. M. (2000). Bistability in the Ca²⁺/calmodulin-dependent protein kinase–phosphatase system. *Biophys. J.* **79**, 2211–2221.

Zhou, Z., and Neher, E. (1993). Mobile and immobile calcium buffers in bovine adrenal chromaffin cells. *J. Physiol. (London)* **469**, 245–273.

Zimmerman, S., and Minton, A. P. (1993). Macromolecular crowding: biochemical, biophysical, and physiological consequences. *Annu. Rev. Biophys. Biomol. Struct.* **22**, 27–65.

Zlokarnik, G., Negulescu, P., Knapp, T. E., Mere, L., Burres, N., Feng, L., Whitney, M., Roemer, K., and Tsien, R. Y. (1998). Quantitation of transcription and clonal selection of single living cells with β-lactamase as a reporter. *Science* **279**, 84–88.

15

Cell–Cell Communication: An Overview Emphasizing Gap Junctions

David C. Spray, Eliana Scemes, Renato Rozental, and Rolf Dermietzel

Cells of the nervous system are anatomically and functionally specialized for the intercellular transmission of electrical and chemical signals and for the bulk transport of metabolites from one cell to another and of solute from cell interior to extracellular space. An enormous variety of different types of molecules are involved in distinct mechanisms of intercellular signaling. These mechanisms include release and recognition of hormones, growth factors, neurotransmitters and extracellular matrix molecules, cell–cell and cell–matrix complexes (such as adhesive and tight junctions), and both electrotonic and chemical synapses.

One unique feature of neurons is intercellular communication by way of synapses, which, until the 1950s, was viewed as chemically mediated. The properties of chemical synapses and the transmitters they release, as well as their receptors and transduction mechanisms, have been described in previous chapters. Chapters 16, 17, and 18 discuss further the features of chemical synapses. Chemical synapses are not the exclusive means by which cell-to-cell communication occurs, however. Indeed, the previous dogma was changed through the demonstrations by Harry Grundfest's group and by Edward Furshpan and David Potter that transmission between segments of cord giant fibers in crustaceans and at the giant cord–motor synapse of the crayfish was electrically mediated. Such transmission was initially termed *ephaptic* (not occurring at real synapses), a term that had been used more than a century before to describe electrical interactions between nerves placed in contact (for an extensive discussion of ephapses, see Grundfest, 1959). However, because the giant fiber segments forming the nerve cord did indeed meet the synaptic criterion of anatomical discontinuity

(Robertson, 1961), and the giant fiber–giant motor synapse exhibited the additional requirement of polarized transmission, these electrotonic synapses were a legitimate synaptic mechanism whereby nerve cells communicated with one another. Subsequently, neurons in many other arthropods, in mollusks and even in very primitive invertebrates have been shown to be electrically coupled (for reviews, see Bennett, 1977; Bennett *et al.*, 1981).

Furshpan and Potter's experiments (1959) showed that depolarizing current injected into the prefiber crosses the junctional membrane to depolarize the postfiber (Fig. 15.1). However, when the postfiber is depolarized, the electrotonic potential spread to the prefiber is small; conversely, hyperpolarization spreads well from postsynaptic to presynaptic cells, whereas antidromic depolarization is largely prevented. This junction thus exhibits rectification, with conductance in the orthodromic direction about 20 times as great as that in the opposite direction, so that action potential propagation toward the periphery is vastly favored over centripetal conduction.

Electrical transmission in vertebrate brain was demonstrated in 1959 by Michael Bennett, Stanley Crain, and Harry Grundfest, who showed that the synchronous activity of the supramedullary neurons of the puffer, a teleost fish whose ovaries are notable for production of the Na^+ channel blocker tetrodotoxin, was due to electrical coupling (Bennett *et al.*, 1959). This demonstration was soon followed by numerous examples of coupling between neurons in the medullary and spinal electromotor relay nuclei of gymnotid fish and between neurons in the supramedullary and oculomotor nuclei of teleost fish, clearly establishing that electrical synaptic connections are commonly found in synchronously active

Copyright 2004, Elsevier Science (USA).
All rights reserved.

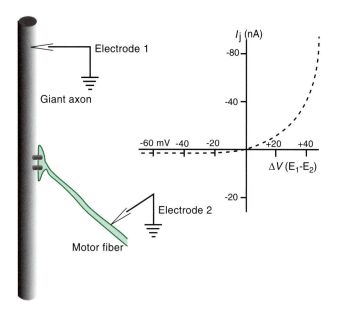

FIGURE 15.1   Anatomy and function of the rectifying electrotonic synapse between the giant axon and motor fiber mediating the escape tail flip in crayfish. Impulses normally flow from the head (not illustrated) down the giant axon (Electrode 1) in the nerve and across the electrotonic synapse to the giant motor fiber (Electrode 2) innervating the tail. The graph illustrates the current (I)–voltage (V) relationship of this synapse when Electrode 1 polarizes the giant fiber and Electrode 2 records the current. The I–V relationship strongly rectifies (steeper slope in upper right quadrant), so that positive (depolarizing) voltages spread well to the postsynaptic fiber, whereas negative (hyperpolarizing) voltages are strongly attenuated.

TABLE 15.1   Sites at Which Cx36 Is a Major Component of Neuronal Gap Junctions

Nuclear complex of the inferior olive

Cerebellar nuclei

Molecular layer of the cerebellar cortex

Granular layer of the cerebellar cortex

Olfactory bulb

Reticular thalamic nucleus

Interneurons of the hippocampus and cortex

Amacrine type AII cells of the retina

Motor neurons of the spinal cord

Pineal gland

Striatum

neuronal populations (Bennett *et al.*, 1967a,b; for recent reviews see Bennett, 1996, 1997). Although electrical transmission in mammalian cardiac and smooth muscles became well established in the 1960s (see Spray *et al.*, 2001, for review), the idea that electrical synapses were present in mammalian brain finally was demonstrated unequivocally by Rodolfo Llinas, Henri Korn, Don Faber, and colleagues in the early 1970s (see Llinas, 1985). Electrotonic synapses between mammalian neurons have now been described virtually throughout the brain, including the sensory neocortex, olfactory bulb, cerebellar cortex, inferior olive, mesencephalic fifth nucleus, vestibular system, hypothalamus, dentate gyrus, pyramidal cells of the hippocampus, GABAergic network of the neocortex and the hippocampal formation (see Bennett, 1997; Llinas, 1985; Dudek *et al.*, 1982; MacVicar and Dudek, 1982; Peinado *et al.*, 1993; Welsh and Llinas, 1997), and the retina (between photoreceptors and horizontal, bipolar, amacrine, and even ganglion cells) (see Penn *et al.*, 1994; Vaney, 1996; Weiler, 1996) (Table 15.1). Depending on their sites and developmental time courses of expression, different functions may be fulfilled by these electrotonic

synapses related to the particular requirements of the local networks.

Both electrical (electrotonic) and chemical synapses are mediated by specialized structures, with families of genes encoding the protein components of each of these two types of synapses. However, even though gap junction proteins interact with other membrane and cytoplasmic proteins to form a signaling complex (the "Nexus": Spray *et al.*, 1999; Duffy *et al.*, 2002), the components of chemical synapses are still much more diverse than those of electrical synapses. Not only do various gene families encode channel-forming proteins in chemical postsynaptic membranes (Chapter 11), but numerous other proteins take part in various aspects of the presynaptic release of vesicles containing neurotransmitters, in the binding of vesicles to the presynaptic release site (Chapter 8), and in holding the postsynaptic receptors in arrays proximal to sites of neurotransmitter release.

When compared with the multiple steps involved in chemical neurotransmission, transmission at electrical synapses is much simpler. Driven by their electrochemical gradients, ions flow from a stimulated presynaptic cell through gap junction channels that connect the cytoplasmic compartment of one cell with that of another, thereby changing the resting potential of the electrically coupled postsynaptic cell (Fig. 15.2). An aggregate of junctional channels is referred to as a gap junction, the site of the electrical synapse (Fig. 15.3). Under most conditions, the electrical equivalent current elements describing gap junction channels are purely resistive (although ionic driving force does dissipate under steady-state conditions owing to limited cell volume, which acts as a capacitive element).

Electrical synapses are formed of channels that in aggregates (seen ultrastructurally as junctional plaques) are termed *gap junctions*. In fact, the term *gap junction* is a misnomer, because it implies a separation

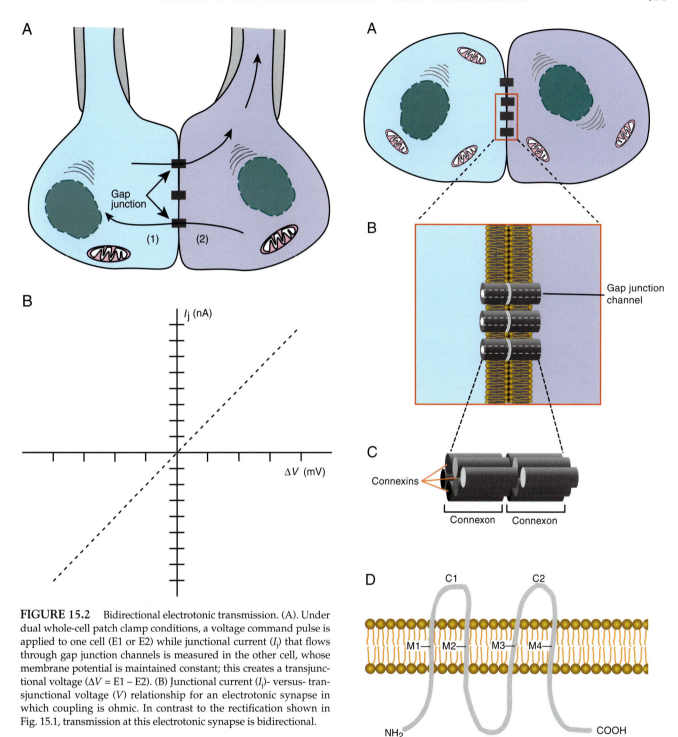

FIGURE 15.2 Bidirectional electrotonic transmission. (A). Under dual whole-cell patch clamp conditions, a voltage command pulse is applied to one cell (E1 or E2) while junctional current (I_j) that flows through gap junction channels is measured in the other cell, whose membrane potential is maintained constant; this creates a transjunctional voltage ($\Delta V = E1 - E2$). (B) Junctional current (I_j)- versus- transjunctional voltage (V) relationship for an electrotonic synapse in which coupling is ohmic. In contrast to the rectification shown in Fig. 15.1, transmission at this electrotonic synapse is bidirectional.

FIGURE 15.3 Diagrammatic structure of electrotonic synapses. (A) Two apposed cells linked by a junctional plaque composed of numerous gap junction channels. (B). Three gap junction channels in a junctional plaque. Each cell provides a hemichannel or connexon that is linked to its partner by protein domains extending across the extracellular gap. (C) The gap junction channel is formed by two hemichannels (connexons); each connexon is formed by six identical subunits of proteins called connexins surrounding the central pore through which ions flow. (D) Membrane topology of a connexin protein showing the two extracellular loops (C1, C2), four intramembranous segments (M1, M2, M3, M4), and cytosolic carboxyl and amino terminals (labeled COOH and NH$_2$ in the figure, the carboxyl and amino terminals are referred to as CT and NT).

at this type of junction; the term was initially coined by Revel and Karnovsky (1967), who described paracrystalline arrays in certain plasma membrane domains when infused with the electron-dense alkali metal lanthanum and viewed with transmission electron microscopy. At identical positions, cross sections through these domains revealed a lucent space of about 20 nm, which led to the term *gap junctions*. In contrast to tight junctions, these do not show a spacing between adjoining cell membranes. Later it turned out that the space is bridged by electron-dense particles, which apparently represent the channel-forming proteins. At sites of gap-junctional contact, appositional membranes of contiguous cells tend to be planar; thus, gap junctions viewed in grazing thin section or in freeze-fracture replicas generally appear as aggregates of particles forming plaquelike structures (schematized in Figs. 15.3A, B). Gap junction plaques can be quite large (as much as several micrometers across in liver, ventricular myocytes, and astroglia) or quite small (probably as small as a single channel or only a few gap junction channels between neurons).

Each gap junction channel is made of two mirror-image symmetric components (one contributed by each cell) called **connexons** or **hemichannels**, and each connexon in turn comprises six homologous subunits, the **connexin molecules** (Fig. 15.3C). Most connexin proteins, of which there are approximately 20 in both rodents and humans (see Table 15.2), are believed to share a common membrane topology and are encoded by a gene family with a generally common gene structure: a single intron separating a small upstream exon (Ex1) from an exon (Ex2) containing the entire connexin coding sequence (see Willecke *et al.*, 2002). Three exceptions to this simple gene structure have been described thus far: (a) the most dominant neuronal connexin (Cx36) and its fish ortholog

TABLE 15.2　Summary of Currently Known Mouse and Human Connexin Genes

Mouse connexin	Human connexin	Tissue distribution	Phenotype of connexin null mice	Human hereditary disease
—	hCx25	?	?	?
mCx26	hCx26	Breast, skin, cochlea, liver, placenta	Lethal on ED11	Sensorineural hearing loss, palmoplantar hyperkeratosis
mCx29	hCx30.2	Myelinating cells	?	?
mCx30	hCx30	Skin, brain, cochlea	Hearing impairment	Nonsyndromic hearing loss, hydrotic ectodermal dysplasia
mCx30.2	hCx31.9	Vascular smooth muscle	?	?
mCx30.3	hCx30.3	Skin	?	Erythrokeratodermia variabilis (EKV)
mCx31	hCx31	Skin, placenta	Transient placental dysmorphogenesis	Hearing impairment, erythrokeratodermia variabilis (EKV)
mCx31.1	hCx31.1	Skin	?	?
mCx32	hCx32	Liver, Schwann cells, oligodendrocytes	Decreased glycogen degradation, increased liver carcinogenesis	CMTX (hereditary peripheral neuropathy)
mCx33	—	Testis	?	—
mCx36	hCx36	Neurons	Visual deficits	?
mCx37	hCx37	Endothelium, ovary	Female sterility	?
mCx39	hCx40.1	?	?	?
mCx40	hCx40	Heart, endothelium	Atrial arrhythmias	?
mCx43	hCx43	Many cell types	Heart malformation	Visceroatrial heterotaxia
mCx45	hCx45	Heart, smooth muscle, neurons	Lethal on ED 10.5	?
mCx46	hCx46	Lens	Zonular nuclear cataract	Congenital cataract
mCx47	hCx47	Brain, spinal cord	?	?
mCx50	hCx50	Lens	Microphthalmia, zonular pulverulent	Zonular pulverulent cataract
—	hCx59	?	—	?
mCx57	hCx62	ovaries	?	?

Modifed from Willecke *et al.* (2002).

(Cx35) exhibit a reading frame that is localized on Ex1 and Ex2 interrupted by an intron; (b) for Cx32, which is prominent in myelinating cells of the nervous system, use of alternate promotors has been documented yielding at least three alternatively spliced 5′ untranslated Ex1 sequences; (c) analysis of Cx45 transcripts indicates that the 5′ untranslated region is composed either of two exons, Ex1 and Ex 2, or only of Ex1, whereas the complete reading frame is located on an additional exon (Ex3). Such differences in gene structure and splice patterns could imply functional consequences with respect to mRNA stability and translational control in the case of Cx32 and Cx45, while a hypothetical alternatively spliced Ex1 in Cx36 might yield heterogenous isoforms of Cx36. From the sequencing of the mouse and human genomes we expect that the connexin family in humans consists of 20 members whereas the mouse genome carries 19 connexin genes (see Willecke *et al.*, 2002, for review). A summary of the mouse and human connexin families is given in Table 15.2.

Connexin family members are quite homologous, with about 50% sequence identity at the amino acid level, and display a diverse pattern of tissue distribution (Dermietzel and Spray, 1993, 1998). Each connexin molecule crosses the membrane four times (segments M1, M2, M3, and M4 in Fig. 15.3D) and both its amino and carboxy termini are on the cytoplasmic side of the membrane. Extracellular loops (C1 and C2) are structurally conserved, with three cysteine residues in each loop identically positioned in all connexins (Fig. 15.3D) (see Bennett *et al.*, 1991). Presumably, this conservative motif accounts for the high-affinity interactions between many (but not all) different connexin molecules when connexons are paired at their extracellular extremities (White *et al.*, 1995; Elfgang *et al.*, 1995). The part of the connexin molecule between segments M2 and M3 forms a loop or hinge region in the cytoplasm whose length (long or short) can be used as a criterion to separate the different connexins into three subfamilies (α or group II, β or group I, and γ or group III). The cytoplasmic domains show the most divergent amino acid sequences, with the C terminus having the largest variability in length. Interestingly, the neuronal connexin Cx36 has the shortest C terminus and one of the longest cytoplasmic loops (100 amino acids). The third transmembrane domain of connexins is the most amphipathic (with charged amino acid residues at every third or fourth position in a predominantly hydrophobic sequence), and six of these M3 regions may provide the hydrophilic face lining the lumen of the channel (Bennett *et al.*, 1991), although other domains may also contribute (Zhou *et al.*, 1997).

CHEMICAL AND ELECTRICAL SYNAPSES DIFFER IN FUNCTIONAL CHARACTERISTICS

The major functional characteristics of chemical transmission that distinguish chemical from electrotonic synapses (Bennett, 1997) include transmission that is usually only in one direction; a delay (0.2–0.5 ms) between the pre- and postsynaptic signals necessary for transmitter release, intercellular diffusion, ligand binding, and postsynaptic channel opening; inhibitory action; and fatigue or facilitation in response to repeated stimuli. However, none of these criteria are absolute.

Directionality

Although chemical synaptic transmission is generally unidirectional and electrotonic transmission is generally bidirectional, electronic transmission strongly rectifies in some cases, such as the rectifying junction in crayfish (Fig. 15.1) (Furshpan and Potter, 1959) and in hatchetfish pectoral fin motor neurons (Auerbach and Bennett, 1969). Conversely, in the nerve net of the syphozoan jellyfish *Cyanea*, chemical transmission is bidirectional: neurotransmitter vesicles are present on both pre- and postsynaptic sides of apposed membranes, and transmission can be evoked by stimulation of either post- or presynaptic elements (Anderson, 1985; Anderson and Grunert, 1988). Nevertheless, the bidirectionality of most electrotonic synapses is an ideal property for the synchronization of activity, providing a mechanism by which depolarization (or hyperpolarization) can be effectively and reciprocally spread throughout the neural network (see below).

Speed

The importance of transmission speed in "escape" or "startle" responses is obvious, with shorter reaction times exerting evolutionary pressure to select for more successful survival. Electrotonic transmission characterizes stereotypic escape behaviors of invertebrates and lower vertebrates, which are mediated by the first gap junctions to be identified with structural and electrophysiological techniques (the crayfish nerve cord and the Mauthner cells of goldfish brain: Robertson, 1961, 1963; Furshpan and Potter, 1959; Furukawa, 1966) (Fig. 15.4) (see Korn and Faber, 1975, 1996). Rapid predator evasion by teleost fish is mediated by a neuronal circuit involving both chemical and electrotonic synapses. Mauthner fibers are

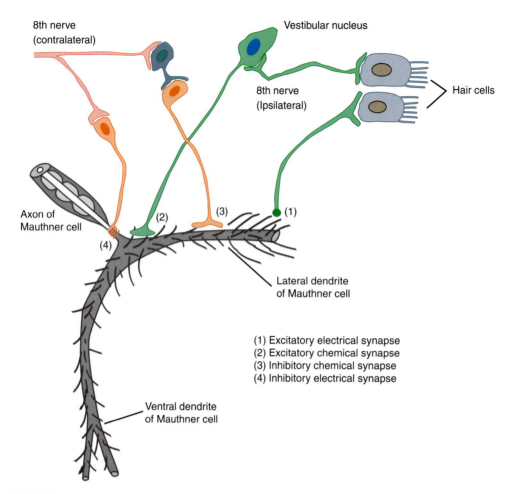

FIGURE 15.4 Mauthner cell synapses involved in the escape response in teleost fishes. From the hair cells that sense the approach of predators, the ipsilateral eighth nerve forms direct electrical excitatory synapses onto the lateral dendrite of the Mauthner cell (1) and indirect (through the vestibular nucleus) excitatory chemical synapses onto the cell body (2). The contralateral eighth nerve makes two inhibitory synapses to the Mauthner cell, one chemical on the lateral dendrite (3) and one electrical onto the cell body (4), both of which serve to prevent inappropriate "escape" toward the predator (see Korn and Faber, 1996).

involved in the control of the large pectoral adductor muscles, whose activation causes the fish to jump upward. These fibers also control the axial musculature that mediates a tail flip to both sides.

Synaptic Inhibition

Chemical synapses can be either excitatory or inhibitory, depending on whether the net flow of ions resulting from ligand-gated opening of postsynaptic channels is depolarizing or hyperpolarizing (see Chapter 16). Thus, if the channels opened by neurotransmitter are primarily K^+ selective, the postsynaptic cell will hyperpolarize owing to K^+ efflux, whereas if channels are nonselective to cations or selective primarily to Na^+ the cell will depolarize owing to Na^+ influx.

Electrotonic synapses are usually excitatory, with the depolarizing phase of the action potential giving rise to brief electrotonic postsynaptic potentials (electrotonic PSPs or "fast prepotentials") if coupling strength is low or to synchronized activity if coupling strength is high. Electrotonic synapses can also inhibit, however, through electrotonic spread of hyperpolarizing chemical PSPs as well as transmission of the hyperpolarizing phase of the action potential. Such inhibitory action of contralateral eighth nerve electrical synapse plays an important role in the escape behavior of teleost fish (Furukawa and Furshpan, 1993) (see Fig. 15.4), so that motor neuron activation propels the animal away from the predator. In addition, the interplay between chemical and electrotonic PSPs can provide flexibility in the output from coupled neurons.

Fatigue, Facilitation, and Modulation

Plasticity is the most fascinating aspect of chemical neurotransmission, allowing the establishment and storage of memory traces (see Chapter 18). Both pre- and postsynaptic structures appear to be endowed with mechanisms by which the amplitude of the postsynaptic event can be strengthened or depressed, depending on coincidence with the activity of other synapses or with the activity pattern within the same terminal.

Although the structural simplicity of electrotonic synapses limits the degree of plasticity of their transmission, the strength of coupling can be altered by changing either junctional conductance itself, conductance of an adjacent nonjunctional membrane, or the shape or duration of the presynaptic impulse. Junctional conductance depends on the number and type of gap junction proteins expressed between cells, as well as the probability of these channels being open (gating, the subject of the next section). Because the turnover of gap junction proteins is extraordinarily rapid (half-lives from synthesis to degradation may be 3 h or less for the connexins thus far assessed), transcriptional and posttranscriptional controls can in principle provide flexibility in coupling strength over this time frame. Moreover, because channels formed of different connexins exhibit different unitary conductances, sensitivities to gating stimuli, phosphorylation properties, and affinities to one another, modulation of the relative expression levels of different connexins may be an important factor in such processes as neuronal differentiation.

Because chemical synapses are generally near electrical synapses in the vertebrate nervous system, shunting of currents due to conductance changes by chemical synaptic inputs may modulate coupling strength, as in the mammalian inferior olivary nucleus and in invertebrate neural circuits (Sotelo et al., 1968; Llinas, 1985; Spray et al., 1981) (see Fig. 15.5). Under conditions where the circuitry includes coupling to an inhibitory interneuron, the sign of coupling may actually reverse enabling activity in one group of motor neurons either to excite or to inhibit another, depending on the degree of activity in the populations (Spira et al., 1976) (see Fig. 15.5).

Another way in which the interplay between chemical and electrotonic synapses may convey plasticity is in cases in which coupling synchronizes chemical synaptic inputs. In the establishment of retinal–tectal connections, for example, graded strength of ganglion cell coupling across the retina may lead to additional synchronous inputs by adjacent ganglion cells onto postsynaptic cells, thus specifically strengthening chemical synaptic connections onto appropriate targets in the optic tectum (Katz, 1993; Shatz, 1994; Kandler and Katz, 1995).

Plasticity at electrotonic synapses also arises from changes in form of the presynaptic impulse because of the time required to charge the postsynaptic membrane. Such effects may be especially significant when the postsynaptic time constant is long and the initially brief presynaptic impulse becomes prolonged by tetanic stimulation or other manipulations (see Pereda and Faber, 1996).

Finally, increases in coupling strength between neurons (Pereda et al., 1998) and among astrocytes (De Pina-Benabou et al., 2001) appear to be transduced through activation of calmodulin-dependent protein kinase (CaM kinase). The underlying mechanism is believed to involve phosphorylation of the connexins, although this remains to be demonstrated. Nevertheless, this long-term increase in coupling (LINC: De Pina-Benabou et al., 2001) may outlast the initial stimulus for an hour or longer.

Electrical Synapses Do Some Things That Chemical Synapses Cannot

Electrotonic synapses exert their postsynaptic effects through the passage of ionic current between cells. Although this current is carried predominantly by K^+ ions, owing to the high intracellular abundance and junctional permeability of this ion, the gap junctions that underlie electrotonic synapses are not nearly as selective as K^+ channels. Gap junction channels are permeant to molecules whose molecular mass is as great as ~1000 Da, including the second messenger molecules Ca^{2+}, 3′-5′-cyclic adenosine monophosphate (cAMP), and inositol 1,4,5-trisphosphate (IP_3) (see Saez et al., 1989). Diffusion of these molecules between coupled cells may therefore result in co-regulation of levels of these second messengers in pre- and postsynaptic elements. Among the expected consequences of such exchange at synaptic regions having both chemical and electrotonic components is modulation of presynaptic neurotransmitter release by changes in levels of second messenger molecules in postsynaptic cells (Pereda et al., 1995).

Gap junctions between glial cells provide a pathway for long range metabolite delivery (Giaume et al., 1997) and second messenger signaling throughout the brain (see Charles, 1998; Giaume and Venance, 1998; Dermietzel and Spray, 1998). Propagation of Ca^{2+} waves through astrocyte gap junctions in response to glutamate application or mechanical stimulation has been observed both in culture and in

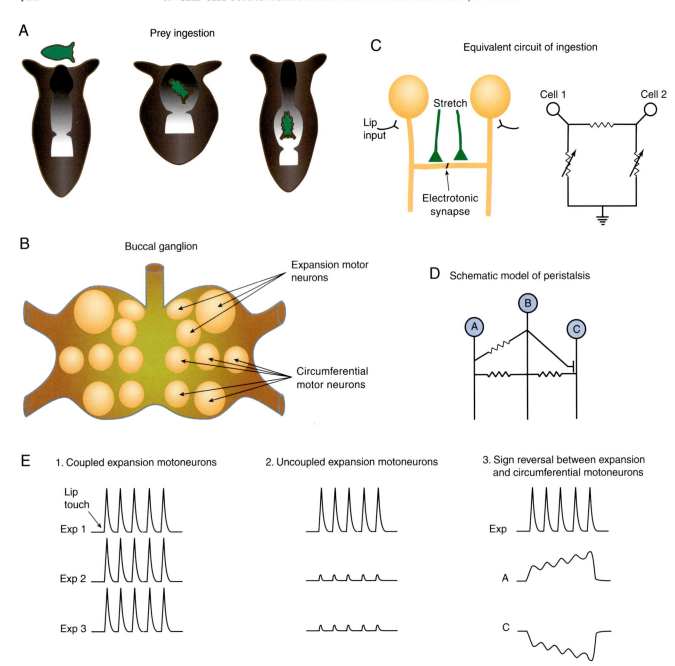

FIGURE 15.5   Neural circuitry responsible for prey ingestion and swallowing in the mollusk *Navanax*. (A) Prey is engulfed by rapid expansion of the buccal cavity, owing to simultaneous contraction of radial musculature innervated by coupled expansion motor neurons. Prey is then swallowed by regional pharyngeal expansion, together with peristaltic pharyngeal constriction by circumferential muscular bands innervated by weakly coupled circumferential motor neurons. (B) Buccal ganglion of *Navanax*, illustrating the positions of expansion and circumferential neurons mediating ingestion and swallowing. (C) Schematic diagram and equivalent electrical circuit describing behavior of expansion motor neurons. Initial pharyngeal expansion is due to activation of the coupled expansion motor neurons from olfactory and mechanoreceptors on the animal's lips. Thereafter, chemical synaptic input from pharyngeal stretch receptors provides a nonjunctional conductance increase, shunting the electrotonic current and uncoupling the cells to allow regional expansion. (D) Schematic diagram of circumferential motor neuron activity during peristalsis. (E) Normally coupled expansion motoneurons (1) are uncoupled (2), and the coupling of expansion motor neurons to interneurons with inhibitory contact onto circumferential motor neurons reverses the sign of the coupling (3) so that circumferential motor neurons are active when expansion motor neurons are inhibited (see Spira and Bennett, 1972; Spray *et al.*, 1980, Spira *et al.*, 1976, 1980).

hypothalamic slices (see Charles 1998; Enkvist and McCarthy, 1992; Cornell-Bell *et al.*, 1991) (Fig. 15.6). In many cases, this propagation is partially or completely blocked by gap junction inhibitors, indicating that conducting gap junction channels are required (see Sanderson, 1995; Scemes *et al.*, 1998; Charles, 1998), and the Ca^{2+} waves most probably consist of intracellular release of Ca^{2+} that is triggered by Ca^{2+}, cyclic ADP ribose (CADPR), and/or IP_3 diffusing across the junctional membrane.

With respect to the various functions that have been proposed for gap junctions between astrocytes (dissipation of K^+ ions, control of cell proliferation: Kuffler and Nicholls, 1966; Naus *et al.*, 1996; Kimelberg and Kettenmann, 1990; Bender *et al.*, 1993), the spread of intercellular Ca^{2+} waves has been hypothesized to provide a mechanism by which cooperative cell activity is coordinated (Sanderson, 1996) and, more recently, as a form by which astrocytes

modulate synaptic transmission (Kang *et al.*, 1998; Araque *et al.*, 1998a,b, 1999; Parpura *et al.*, 1994).

Besides the transmission through gap junctions, calcium waves can also spread between cells through a paracrine route, in which extracellular adenosine triphosphate (ATP) or glutamate released from injured or excited cells activates purinergic P2 or glutaminergic receptors, leading to propagated changes in intracellular Ca^{2+} over long distances and even across regions of cell separation (Enkvist and McCarthy, 1992; Hassinger *et al.*, 1996). It now appears most likely that Ca^{2+} waves among astrocytes depend on both extracellular and intercellular (gap junction-dependent) routes, with a finely tuned balance between the two mechanisms (Scemes *et al.*, 2000).

Summary

Cells within the nervous system are responsive to blood brain-permeant hormones and nutrients and interact with other neural cells through long-range and local signaling mechanisms. True synapses between neurons are of two types, electrical and chemical, and both consist of membrane specializations between discrete cells that can be unidirectional and functionally flexible. Gap junctions (the anatomical substrate of electrical synapses) are also found between glia, providing an intimate pathway for potentially long-range exchange of signaling molecules, nutrients, and osmolytes.

BIOPHYSICAL AND PHARMACOLOGICAL PROPERTIES OF GAP JUNCTIONS IN THE NERVOUS SYSTEM

It now appears that of the 20 or more connexins in the rodent genome at least 11 (Cx26, Cx29, Cx32, Cx33, Cx36, Cx37, Cx40, Cx43, Cx45, Cx47, Cx49) are expressed to various degrees in the developing and adult nervous system (see Table 15.2). Functional studies in diverse cell types and in various exogenous expression systems have revealed that gap junction channels formed by different connexins are regulated differently, both at the single-channel level (gating controls such as voltage sensitivity and variations in unitary conductance: Spray, 1994) and at the level of synthesis (expression, altered to different extents for different connexins by various hormones, extracellular matrix, and cell cycle conditions) (see Bennett *et al.*, 1991). Similarly to members of other ion channel and transport families, gap junction channels have an

A   WT Cx43 astrocytes

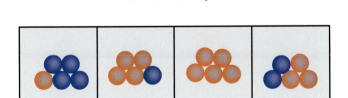

B   Cx43 KO astrocytes

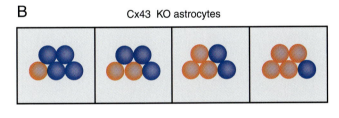

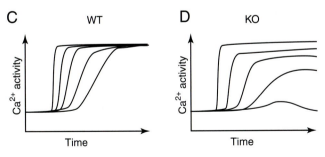

FIGURE 15.6 Intercellular Ca^{2+} waves between astrocytes from (A) wild-type and (B) Cx43-deficient mice. In astrocytes from wild-type mice, mechanical, electrical, or pharmacological stimulation can induce a rise in intracellular Ca^{2+} that spreads from cell to cell (high Ca^{2+} activity is represented by orange color) with a conduction velocity of about 10–20 μm/s [illustrated in frames at 2-s intervals in (A) and plotted schematically in (C)]. In astrocytes from Cx43-deficient (KO) mice, spread is slower and to fewer cells [illustrated in frames at 2-s intervals in (B) and plotted in (D)].

array of diverse properties. Some gap junction channels are more sensitive to certain gating stimuli than others; some display a degree of ionic selectivity; and whereas some pair promiscuously with other connexins (to form heterotypic channels), others are quite selective in their interaction (homotypic channels). Such differences are likely to be important from the standpoint of the physiological roles of gap junctions in different cell types, as well as in the establishment of communication compartments within the nervous system.

Gap Junctions Are Gated by Transjunctional Voltage

Voltage-clamp studies on cell pairs have revealed that like other ionic channels in the cell membrane (e.g., Chapter 16), individual gap junction channels open and close in discrete steplike transitions between states (Fig.15.7A). The average time that mammalian gap junction channels reside in the fully open state is sensitive to voltage across the membrane. Different from other voltage-dependent channels, gap junctions are sensitive primarily to the voltage difference between the coupled cells (transjunctional voltage) but much less to the cells' absolute membrane potentials (Spray *et al.*, 1981). When either cell of a pair is depolarized or hyperpolarized (i.e., when the transjunctional potential is increased), the time during which the channel is in the fully open state is reduced; however, for channels formed of most connexins, high transjunctional voltage induces a residual conductance. This residual conductance is believed to be due to conformational changes in gap junction channels, in which transjunctional voltage shifts most junction channels from a mainstate to a substate behavior

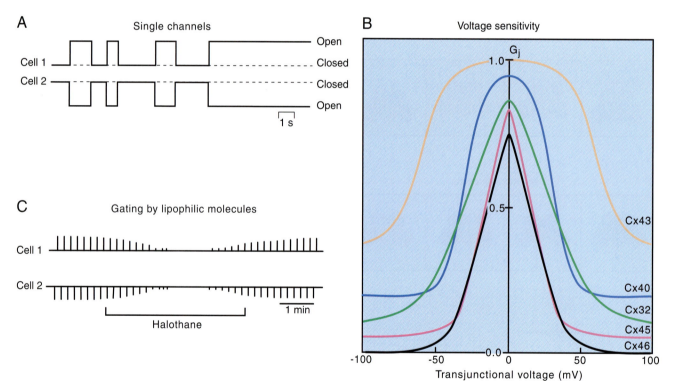

FIGURE 15.7 Gating of gap junction channels. (A). When each cell of a coupled pair is voltage clamped and a transjunctional voltage maintained, current flows between the cells through gap junction channels. When few channels are open, single channel currents can be resolved as simultaneous steplike transitions of opposite sign recorded in each cell's current trace. The probability of gap junction channels being open is determined by the transjunctional voltage (B) and by various other factors (C). (B). Voltage sensitivity of junctional conductance (G_j) is a connexin-specific property. All mammalian gap junction channels are maximally open at $V_j = 0$ mV and close when one cell is either depolarized or hyperpolarized relative to its partner. Note that Cx43 gap junction channels are not very sensitive to V_j below 50 mV, whereas Cx32 channels are more voltage sensitive and Cx46 channels close in response to even small imposed V_j values. (C). Sensitivity of junctional conductance to the volatile anesthetic halothane. In this illustration, hyperpolarizing pulses are applied to cell 2, eliciting outward currents in cell 1. Application of halothane at anesthetic doses reduces junctional current to zero; when halothane is removed, passage of junctional currents is quickly restored.

(Moreno *et al.*, 1994a). After subtraction of the residual conductance from the total junctional conductance, the voltage-sensitive decline in open time for potentials of either polarity is well fit by two Boltzmann equations, with parameters that are generally symmetric about the zero transjunctional voltage axis. Each connexin has a distinct voltage sensitivity (Fig.15.7B). For example, the predominant channel in astrocyte Cx43 and neuronal Cx36 channels (see Srinivas *et al.*, 1999) are insensitive to transjunctional voltages below 50 mV, whereas open time of Cx45 channels is dramatically reduced by transjunctional voltages as low as 30 mV. Differences in voltage sensitivity may reflect functional advantages for expression of certain connexins in certain tissues. For example, a highly voltage-sensitive gap junction channel would be of limited utility in cardiac tissues, where action potential plateaus can create large and long-lasting transjunctional voltage gradients, but could be useful in generating compartment boundaries during embryonic morphogenesis.

Junctional Channels Are Also Gated by Other Factors

Very high (micromolar to millimolar) intracellular Ca^{2+} levels may close some gap junction channels, either directly or through cell acidification or the generation of phospholipases and consequent elevations in phospholipid metabolites such as arachidonic acid (see Spray *et al.*, 2001). Moderate acidification to levels shown to be reached during ischemic episodes blocks gap junction channels (see Spray and Scemes, 1998). Some connexins are markedly more sensitive to acidification than others (e.g., Cx43 is more sensitive than Cx36), providing the possibility that under ischemic conditions coupling within certain cellular compartments might be selectively affected.

Experimentally the most effective way to block gap junction channels is to treat cells with lipophilic agents such as heptanol, octanol, halothane (illustrated in Fig.15.7C), or arachidonic, glyccyrhetinic, or oleic acids or carbenoxolone (see Spray *et al.*, 2002). Although the mechanism of action of these compounds is unknown, it is possible that they disrupt gap junction function by intercalating into the membrane surrounding the connexon or within hydrophobic connexin domains. Because structurally similar blocking compounds are generated through lipid peroxidation and by phospholipases activated under ischemic conditions, this mechanism may be the ultimate cause of the observed cell uncoupling in acute pathological conditions.

Gap junction channels are also blocked by certain agents that are known to affect other channel types. Such compounds include chloride channel blockers (e.g., flufenamic acid) and K^+ channel blockers (e.g., quinine and derivatives) (Spray *et al.*, 2002). At least for quinine, sensitivity of different connexins varies greatly, offering the opportunity for selective blockade of certain gap junction types, while sparing others (Srinivas *et al.*, 2001).

Most connexins (with the exception of Cx26) are phosphoproteins and are thus sensitive to the action of protein kinases and phosphatases (Saez *et al.*, 1993). Serine phosphorylation of specific residues within the carboxyl tail of gap junction proteins by cAMP-dependent protein kinase and protein kinase C has been demonstrated for Cx32 and Cx43. Phosphorylation of connexin molecules affects the conductance and open time of the channels. For Cx43, serine phosphorylation is correlated with decreased unitary conductance (Moreno *et al.*, 1994b); for Cx32, it appears to increase the time that the channel is open (Saez *et al.*, 1993). In contrast, phosphorylation of a specific tyrosine residue uncouples cells that are connected by Cx43 (Swenson *et al.*, 1990).

Gap Junction Channels Formed of Different Connexins Show Differences in Unitary Conductances and Ionic Selectivities

Gap junction channels allow the intercellular passage of molecules ranging widely in size. The highly charged anionic fluorescent dye Lucifer Yellow, designed by Walter Stewart to detect gap junctional communication (Stewart, 1981), has been used extensively as a diagnostic test for the presence of functional gap junctions between cells. The number of cells that are fluorescent after Lucifer Yellow has been injected into one cell gives a so-called dye-coupling measurement used to quantify coupling strength. A comparison of the extent of dye coupling by using dyes with different charges indicates the ionic selectivity of gap junction channels (e.g., Veenstra *et al.*, 1995).

Interestingly, recent studies in retina and central neurons have shown that the spread of injected neurobiotin (a positively charged molecule that is smaller in size than Lucifer Yellow) through coupled cellular networks is much more extensive than that of Lucifer Yellow (e.g., Vaney, 1996; Peinado *et al.*, 1993). Similar studies using dyes of different diameter and charge allow discrimination among gap junction channels with different permeabilities (see Mills and Massey, 2000). Such results indicate both that some gap junctions are more permeable to cations than anions (cationic selectivity) and that the channels may differ in pore size. In fact, gap junction channels have a range of ionic selectivities (Spray, 1994; Veenstra *et al.*, 1995). Perfusion of coupled cells with similar-sized compounds has shown that Cx32 favors the permeation of anions over cations, in a ratio of at least 2:1, whereas Cx43 is about equally anion and cation permeant and Cx45 is more permeant to cations than anions (P_{anion}:P_{cation} ~1:5).

Connexons Have Selective Affinities for One Another

Because gap junctions are formed by the pairing of two connexons, the unusual term *selective affinity* can be employed to describe the functional relevance of heterologous channels formed by connexons made of different connexin proteins. As mentioned earlier, the two extracellular loops (C1 and C2 in Fig. 15.3D) of each connexin join the connexons tightly together to form a gap junction channel. As would be expected from the high degree of amino acid homology among these extracellular loops, a large number of functional heterotypic channels have been observed (Elfgang *et al.*, 1995; White *et al.*, 1995). However, connexons formed of some connexins pair avidly with others (e.g., Cx45 with Cx43), whereas other heterologous pairings (e.g., Cx32 with Cx43) likely produce nonfunctional channels. In addition, Cx33 appears to form functional channels neither with itself nor with other connexins (Chang *et al.*, 1996). Such differential selective affinity may play a role in separating cells, including those of the nervous system, into functionally discrete compartments (Dermietzel and Spray, 1993).

Summary

Gap junction channels make up electrical synapses between neurons and form direct pathways for diffusional exchange of metabolites and ions among glia. Gap junction channels formed by different connexins have different unitary conductances and selective permeabilities and are differentially sensitive to transjunctional voltage and intracellular pH. Several types of agents provide the possibility of reducing junctional conductance experimentally, although none of these treatments is totally without effects on other channels. Gap junction connexons formed of one connexin type may or may not pair with hemichannels formed of another connexin, providing the possibility for the existence of compartmental boundaries created by connexin expression patterns in different cell types.

ROLE OF GAP JUNCTIONS IN FUNCTIONS OF NERVOUS TISSUE

Cells constituting the nervous system are specialized for different roles (such as impulse transmission in the case of neurons or metabolic support as in astrocytes) and form functional compartments within the tissue. The strength of intercellular communication within each compartment and the types of connexins expressed vary according to cell type; communication extends across some of these compartmental boundaries as well. Between neurons, chemical transmission is generally the major mechanism by which cells interact. Gap junctions between glial cells are common, providing direct pathways for intercellular communication of second messengers, metabolites, and ions. Moreover, gap junctions between some populations of neurons have been identified, as has chemical responsiveness in glia.

Gap Junctions Connect Neuronal Cell Populations

Most of the knowledge regarding structural features and functional properties of electrical synapses has been obtained from invertebrates and lower vertebrates; the extent and functional significance of electrical connections between neurons in higher vertebrates, including humans, remained elusive due to the technical difficulties in directly assessing their existence. However, improved methods for electrophysiological recording from visualized neurons in slice preparations of the neocortex and the hippocampus recently allowed recordings of electrical coupling directly from pairs of connected cells. In particular, GABAergic interneurons of the neocortex and the hippocampus have been a primary focus of these studies (Gibson *et al.*, 1999; Galarreta and Hestrin, 1999; Beierlein *et al.*, 2000), as a number of electron microscopical studies have revealed that gap junctions are frequent between these neurons. Paired recordings from GABAergic interneurons provide strong evidence that electrical synapses are fundamental features of local inhibitory circuits and suggest that this form of coupling defines functionally diverse networks of GABAergic interneurons throughout the brain (for review see Galaretta and Hestrin, 2001).

Electrical coupling between GABAergic interneurons has been shown in layers L2 to L6 in the neocortex and in the hippocampal formation (Galarreta and Hestrin, 2001; Gibson *et al.*, 1999). Two functional distinct classes of inhibitory interneurons have been carefully examined and their physiological properties characterized (Gibson *et al.*, 1999; Amitai *et al.*, 2002): so-called fast-spiking (FS) parvalbumin-positive interneurons that exhibit a high-frequency discharge of spikes without spike frequency adaptation, and low-threshold spiking somatostatin-positive cells (LTS) that show the tendency to fire from the rebound when depolarized from more negative potentials with clear frequency adaptation. Both distinct classes of interneurons formed homotypic electrical synapses (i.e., electrical synapses between cells of the same class) and did not interconnect or did so only rarely with cells of the other class. Combined electrophysiological and immunocytochemical studies in the rat somatosensory cortex indicate that these GABAergic interneurons form large electrically coupled networks, where each neuron contacts tens of other neurons (Amitai *et al.*, 2002). The estimated number of cells electrically connected to a single neuron seems to be between 20 and 50. The large number of electrical synapses implies that GABAergic interneurons participate in a large, continuous electrical syncytium.

In the context of the above-described general features of gap junctions, the physiological properties of electrical synapses between GABAergic interneuons can be defined as follows: (a) If a current step is injected into the soma of one interconnected cell the paired neuron responds with a voltage change. (b) Electrical coupling between these cells has been found to be reciprocal (i.e., current flows bidirectionally, unlike the rectification in the crayfish electrical synapse described above). (c) The coupling coefficient (i.e., the ratio of the voltage change in the noninjected cell to that in the injected cell) is in the range of 2.6 to 10%. (d) No voltage dependency of interneuronal electrical coupling has been detected so far. (e) The total junctional conductance (G_j) varies from 235 to 2100 pS, which would represent about 16–140 channels if formed of Cx36 (Srinivas *et al.*, 1999). It is important to note that these data characterize the electrical synapses of neocortical inhibitory interneurons, which constitute the best studied contacts to date. It seems most likely that electrical synapses between other classes of neurons will exhibit different electrophysiological profiles, as a variable number of connexins may be expressed in neurons (Dermietzel *et al.*, 2000).

In terms of network behavior, some features of interneuronal electrical synapses as described between GABAergic cells seem to be of prime importance. First, electrical synapses can act as a low-pass filter. When subthreshold sine wave currents of different frequencies are injected into one cell the efficacy of electrical transmission decreases with increasing frequency. Thus, signaling through electrical synapses favors the propagation of slow potentials (Galarreta and Hestrin, 1999; see also Spira and Bennett, 1972). The importance of this feature becomes apparent if one considers the divergence of electrical inputs flowing into the interneuronal network. Because different types of electrically coupled networks are suggested to exist, one may speculate that the filter characteristics also differ, thereby facilitating network-specific propagation of oscillations. Second, electrical synapses participate in inducing or enhancing the synchronization of coupled cells. Synchronization could occur as fluctuations in subthreshold membrane potentials or in a temporally precise coordination of action potentials. In the latter case the time interval between pre- and postsynaptic depolarization can be very short (< 1 ms), leading to submillisecond spike coordination. Such coupling in inhibitory networks could coordinate electrical impulses converging from different sources, thus leading to strong inhibitory postsynaptic potentials (IPSPs) on their targets. In addition, the synchronizing effect of inhibitory interneurons can produce nearly simultaneous IPSPs in a large number of cells, thereby coordinating their electrical activity. In this sense electrical coupling of inhibitory interneurons seems to play an important role in the detection and promotion of synchronous activity in the neocortex. Different types of oscillatory events in the brain have been attributed to interneuronal gap junction coupling, including γ oscillations (40–70 Hz) (Hormuzdi *et al.*, 2001) and high-frequency oscillations (>200 Hz) (Draguhn *et al.*, 1998; Traub *et al.*, 2002; Schmitz *et al.*, 2001). As coherent firing among neurons may be related to sensory information, memory consolidation, and behavioral states, electrical coupling is suggested to provide a prominent mechanism in the coordination of higher brain function.

The synchronization of activity of presynaptic sensory neurons by such electrotonic coupling may also be important in optimizing reflex activity, as in the stretch reflex involving the mesencephalic fifth nucleus (Hinricksen and Larramendi, 1968). Moreover, even weak electrotonic interaction may be important in generating rhythmic levels of increased or decreased excitability of neuronal populations, as in the inferior olivary nucleus (Llinas, 1985), or in strengthening synaptic connections during development (see below). The high speed of electrotonic synaptic transmission may optimize the performance of certain behaviors, as in "escape" or "startle" responses discussed earlier in this chapter, and rapid electrotonic transmission in Deiter's neurons could be critical for quick adjustments in balance in mammals (Korn *et al.*, 1973; Sotelo and Palay, 1970). Finally, the interplay between chemical and electrical synaptic events can enhance contrast detection, as in sensory cortex and olfactory bulb (Rall *et al.*, 1966) and in the cerebellum, where coupling between stellate and basket cell populations synchronizes the inhibition of Purkinje cells (Welsh and Llinas, 1997). Such sharpening of contrast provided by electrical and chemical synapses is especially prominent in the retina, where horizontal cell coupling provides surround effects (see Vaney, 1996; Weiler, 1996).

Among neurons the predominant connexin type has appeared to be Cx36 as evidenced by immunolabeling and *in situ* hybridization (Belluardo *et al.*, 2000; Rash *et al.*, 2000; Meier *et al.*, 2002a). Certain nuclei and neuronal cell populations have been found to express this type of connexin (see Table 15.1). Interestingly, Cx36 expression seems to be low or even absent in hippocampal pyramidal cells (Condorelli *et al.*, 2000; Hormuzdi *et al.*, 2001), which have been shown to be electrically coupled, raising the possibility that the gap junctions there are too small to be detected by immunocytochemistry or that additional connexin types might contribute to neuronal gap junctions. In fact, Cx26, Cx32, Cx43, Cx45, and Cx47 have been suggested to be present in diverse neuronal populations *in vivo* at certain developmental stages or *in vitro* under certain conditions (Dermietzel and Spray, 1998; Nagy and Dermietzel, 2000).

Expression of Connexins is Developmentally Regulated in the Brain

Numerous studies have demonstrated that the incidence of interneuronal coupling (and, by inference, the expression of gap junction proteins) dramatically decreases during the processes of brain embryogenesis and neuronal maturation (see Rozental *et al.*, 2000; Walton and Navarrete, 1991; Peinado *et al.*, 1993; Lo Turco and Kriegstein, 1991). In addition, during both early and late stages of neural differentiation, expression shifts from one type of connexin to another. For instance, Cx26 is highly expressed in fetal compared with adult brain, whereas Cx32 shows developmentally regulated increases in abundance that reach maximal levels only after birth (Dermietzel *et al.*, 1989). Because these changes in coupling strength and patterns of connexin expression coincide with the progressive differentiation and commitment of cells and of cell groups to the neuronal lineage, the early presence of gap junctions may hypothetically enable signaling molecules to diffuse among the cells, retarding neuronal differentiation.

Tissue culture preparations from embryonic neural tissue have allowed the manipulation of individual

cells and evaluation of changes in junctional distribution and expression during maturation (see Mehler *et al.*, 1993). Such studies have clarified the relationships between sequential changes in phenotypes of neural cells, revealing that the extent of coupling mediated by Cx43 (which is abundant in neural precursor populations) declines, whereas other gap junction proteins appear. For instance, Cx43 continues to be expressed in cells destined for the astrocytic lineage, but is progressively reduced in cells acquiring neuronal phenotypes (Rozental *et al.*, 2000).

Even after neuronal circuits are formed, however, coupling among neurons persists to various degrees in different brain regions. In regions such as the neocortex, coupling is gradually restricted during early neonatal life, as the necessity for synchrony in the establishment of projections onto adjacent cells declines (Lo Turco and Kreigstein, 1991; Peinado *et al.*, 1993; Kandler and Katz, 1995). In this way, the individual identity of neurons within functional subcircuits is optimized. In this context, the postnatal expression pattern of Cx36 is of interest. Using transgenic mice with a reporter gene under the control of the Cx36 promoter, Deans *et al.* (2001) showed that in early postanatal stages (P7) neocortical expression is prevalent in principal neurons and in interneurons. As brain maturation proceeds, Cx36 expresssion becomes progressively confined to subclasses of interneurons. Identical findings were obtained by *in situ* hybridization in the hippocampus, revealing a considerable reduction of Cx36 expression in hippocampal pyramidal cells (Hormuzdi *et al.*, 2001).

Interestingly, the persistent expression of gap junctions during pre- and postnatal development may have deleterious consequences. For example, the incidence of seizure activity that is more common in young children than later in development correlates well with the decline in coupling that is observed among neurons (Cepeda *et al.*, 1993).

Gap Junctions Connect Glial Cells

The brain can be envisioned as being separated into volumes that are filled with intracellular, extracellular, and cerebrospinal fluids. The intracellular volume of the brain is primarily glial, consisting of communication compartments formed by distinct cell types (ependymal cells, ependymoglial cells, astrocytes, oligodendrocytes, leptomeningeal cells; see also Chapter 1). Cells within each compartment that are well coupled to one another (homologous coupling) can be considered a *functional syncytium*. In addition to this communication within a compartment, certain cell types can interact with one another across compartmental boundaries (heterologous coupling). Within each glial compartment, the role of gap junction channels is most likely to allow the exchange of second messenger molecules and to permit the sharing of cellular metabolites and other regulatory molecules (see Giaume and McCarthy, 1996; Dermietzel and Spray, 1993).

Each specific cell type in the brain expresses a specific set of connexins and the expression patterns thus coincide with tissue compartmentalization (see Dermietzel and Spray, 1998). In adult brains the predominant connexin is Cx43, which is abundant in astrocytes and is also expressed in leptomeninges, endothelial cells, and ependyma. The second type of macroglia, the oligodendrocytes (and their peripheral counterparts, the Schwann cells), appear to express a different most abundant gap junction protein, Cx32, although to a lower extent *in situ* than the level of Cx43 expression exhibited by astrocytes. Besides these two proteins, Cx26 is found at modest levels in the adult brain, where it is confined to leptomeninges, ependyma, pinealocytes, and, to some extent, astrocytes (Nagy *et al.*, 2001).

Astrocytes

Astrocytes express mainly Cx43 and are well coupled *in vivo* and under culture conditions (Dermietzel *et al.*, 1991; Giaume *et al.*, 1991). However, the strength of coupling and degree of Cx43 expression between astrocytes vary depending on brain regions, being higher in the hypothalamus than in the striatum (Batter *et al.*, 1992; Lee *et al.*, 1994). Differences may also be due to the astrocytic cell type. The second major connexin that is expressed in astrocytes is Cx30 (Kunzelmann *et al.*, 1999; Nagy *et al.*, 1999). It is of some interest that this connexin occurs late (>P20) during postnatal differentiation, with a distinct locally resticted expression pattern being highest in type 2 astrocytes with low abundance in white matter and a considerable accumulation in perivascular astroglia. It has been suggested that the late expression of Cx30 is related to the postnatal maturation of the blood–brain barrier (BBB) (Chapter 1), which necessitates specific commitments of the astroglia to BBB functions. Recent studies using genetically engineered Cx43 knockout mice (Reaume *et al.*, 1995) reveal that astrocytes also express minor amounts of other connexins, including Cx26, Cx45, Cx46, and Cx40 (Dermietzel *et al.*, 2000). The significance of this minor expression of other connexins is unknown.

Oligodendrocytes and Schwann Cells

Myelinating cells of both the CNS (the oligodendrocytes) and the periphery (the Schwann cells) (see Chapter 1) express gap junction channels that may have a role in the coordination of the processes and myelination and demyelination, as well as in the metabolic maintenance of the inner, most adaxonal, portions of the cells.

In Schwann cells, Cx32 and Cx29 immunoreactivity has been described, and is most prominent in the Schmidt–Lanterman incisures and the paranodal parts of the nodes of Ranvier (Bergoffen *et al.*, 1993; Scherer *et al.*, 1998; Altevogt *et al.*, 2002). Additional sites of Cx32 expression occur between the plasma membranes of the internodal portion where it faces the first layer of the compact myelin (unpublished studies by Carola Meier, Rolf Dermietzel and John Rash). The reflexive junctional contacts in myelinating cells may promote nutrient exchange between the soma and the cytoplasmic processes that would otherwise be isolated by the myelin lamellae.

Additional studies indicate that expression of another gap junction protein, Cx46, is induced in Schwann cells in response to cell injury (Chandross *et al.*, 1996). During the bursts of Schwann cell proliferation in regenerating nerve, there is upregulation of Cx46, which is believed to form gap junction channels between dividing Schwann cells and to play a role in remyelination.

Cx32 provides intercellular coupling between oligodendrocytes. Unexpectedly, this coupling is weaker, with respect to the spread of Lucifer Yellow, than that observed between the less anion-selective Cx43 of astrocytes. This difference is due in part to the higher junctional conductances that are measured electrophysiologically in astrocytes, and it may be due to the expression of additional populations of Lucifer Yellow-impermeant gap junction channels between the oligodendrocytes (i.e., Cx45) (Dermietzel *et al.*, 1997). Finally, Cx29 is another member of the connexin family with expression sites in oligodendrocytes and Schwann cells (Altevogt *et al.*, 2002).

Leptomeningeal Cells

In culture, leptomeningeal cells are even more strongly coupled than are astrocytes, they do not show prominent voltage dependence, and they display at least two types of junctional single-channel currents, which is consistent with their expression of two junctional proteins, Cx43 and Cx26 (Spray *et al.*, 1991). Junctional conductance between leptomeningeal cells is doubled within 5 min of the addition of membrane-permeant cAMP or the adenyl cyclase activator forskolin and is strongly inhibited by addition of the kinase C-activating phorbol esters. Junctional responses to both types of stimuli are accompanied by rapid changes from flattened to rounder morphology, but the degree to which these changes contribute to the effects of these agents on junctional conductance is unknown. Leptomeningeal cell coupling presumably plays a role in volume regulation in the meningeal cell layer and in maintaining ionic balance between brain and nonneural tissue.

Ependymoglia

A striking feature of ependymoglial cells is their cilia, which project into the cerebrospinal fluid (Del Bigio, 1994) (Fig. 15.8). The cilia beat in a coordinated manner, tending to sweep experimentally applied foreign particles in the same direction as the bulk flow of cerebrospinal fluid (CSF). Such coordinated activity of ciliated cells in mammalian trachea and molluscan gills is attributed to the presence of gap junctions between the cells (see Sanderson, 1996). Freeze-fracture studies reveal abundant gap junctions and orthogonal arrays of particles between neighboring ependymal cells and occasionally between ependymal cells and astrocytes. Immunohistochemical studies of gap junction proteins reveal the presence of Cx26 and Cx43 in mature ependymal cells (Dermietzel *et al.*, 1989). The gap junctions allow transfer of tracer substances and electrical current between these cells, which probably serves to integrate cell function, not only between ependymal cells but also possibly between ependymoglia and astrocytes.

The Astrocytic Compartment Is a Functional Syncytium

The astrocyte syncytium extends to the circulation in the CNS capillaries and to the cerebrospinal fluid. Astrocytes buffer the accumulation of extracellular K$^+$ resulting from neuronal activity, and gap junctions in this system most likely provide a direct pathway to K$^+$ disposal into the perivascular compartments (Kuffler and Nicholls, 1966). An equally plausible notion is that coupling between astrocytes creates a large volume that could act as a buffer sink (Ransom, 1996).

To better comprehend the role of the astrocytic syncytium, let us briefly consider the blood-brain barrier (BBB) (see also Chapter 1). The way in which the brain maintains the constancy of its environment depends in part on isolation of brain tissue by the BBB, the cells of which can exclude certain molecules because they are joined by tight junctions of high electrical resistance (on the order of thousands of ohms per square centimeter). This exclusion ability prevents the transcellular movement of molecules and even ions; however, active ion exchange processes within these specialized cells are primarily responsible for maintenance of the brain microenvironment.

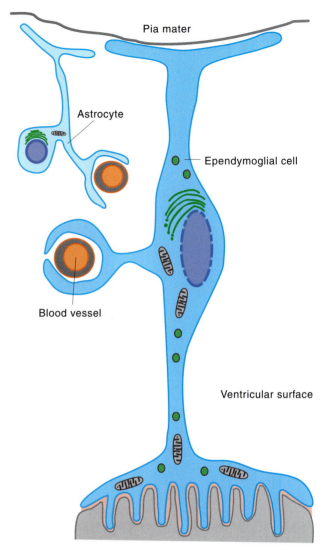

FIGURE 15.8 Ependymoglial cell. The ependyma is a single-layered, ciliated epithelium that covers the surface of the ventricular system of the brain and central canal of the spinal cord. Ependymoglial cells are polarized, having luminal, lateral, and basal sides. Highly specialized ependymal cells cover the choroid plexus and the circumventricular organs. In certain regions of the CNS, a substantial but unknown number of ependymal cells maintain basal processes of varying lengths. When these basal processes establish direct contact with the pia mater, such ependymoglial cells resemble in form the radial glia of the developing CNS. The pial end feet of these cells together with the end feet of central astroglial cells participate in the formation of the blood–brain barrier. The ependymoglial basal processes, in passing through neural tissue, usually extend appendages or side-arms that may end in contact with neuronal cell bodies, dendritic, axonal, and synaptic processes, and the walls of the blood vessels.

The stability of the brain water content is determined by the coupled transport of water and electrolytes across the BBB, by the pressure-dependent bulk flow between the interstitial space and the CSF, by exchange between the interstitial space and the intercellular compartment, and by the ways in which the balance between these exchanges is affected by extracellular osmotic perturbation (Van Harreveld, 1966). More is known about the contribution of extracellular compartments than about that of intracellular compartments with respect to adjustments regulating brain volume (= water content) *in vivo*; however, *in vitro* studies reveal that the cells themselves possess volume regulatory mechanisms enabling them to respond to conditions that produce osmotic imbalance (Kimelberg, 1991). Ultimately, maintenance of brain volume depends on coordinated regulation of intracellular as well as extracellular compartments. The increase in gap junctional coupling and in the velocity of calcium waves spreading between astrocytes exposed to hyposmotic shocks (Scemes *et al.*, 1998) suggests that the astroglial syncytium greatly contributes to such a role.

A variety of pathological situations, such as brain tumors, bacterial meningitis, traumatic brain edema, and chronically relapsing inflammatory diseases such as multiple sclerosis, disrupt the BBB (Chapter 1). The cytotoxic effects of edema include intracellular swelling of neurons, glia, and endothelial cells, with a concomitant reduction of brain extracellular space (see Kimelberg, 1995). Cytotoxic edema develops in hypoxia from asphyxia or global cerebral ischemia after cardiac arrest due to failure of the ATP-dependent Na$^+$–K$^+$ pump, allowing Na$^+$, and therefore water, to accumulate within cells (Rapoport, 1979). Another cause of cytotoxic edema is water intoxication, a consequence of acute systemic hyposmolality due to excessive ingestion of water or administration of intravenous fluids. Acute hyponatremia, induced, for example, by inappropriate secretion of antidiuretic hormone or atrial natriuretic hormone, also can cause swelling and brain edema.

The extracellular space constitutes about 27% of brain volume (Van Harreveld, 1966; Rapoport, 1979). Under numerous physiological and pathological conditions, this extracellular volume is reduced. For example, under anoxia and membrane depolarization elicited by elevation of extracellular K$^+$ concentration, or by repeated electrical pulses, the observed cell swelling (Fig. 15.9) may substantially reduce the size of the extracellular compartment, severely affecting neuronal and glial functions (see Kimelberg, 1995). Both high K$^+$ and glutamate elicit cortical spreading depression (the tie-in with gap junctions is discussed in more detail later in this chapter). This phenomenon consists of a wavelike progression of neuronal hyperexcitability across the cortical surface, followed by a period of unresponsiveness. During spreading

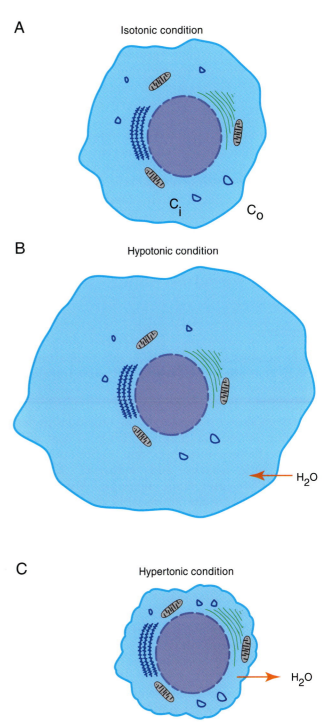

FIGURE 15.9 Cell volume changes. (A) Under conditions in which extra- and intracellular osmotic concentrations are the same ($C_o = C_i$), the osmotic pressure in both compartments is the same ($\pi_o = \pi_i$). Such an isotonic condition is characterized by $\Delta\pi = 0$ and by a steady-state cell volume given by $V = (A/S_o)(1-l/\rho)$, where A is concentration of the impermeant solute, S_o is that of the permeant solute, l is the partition coefficient, and ρ represents activity of the metabolic pump. (B) Under hypotonic *conditions* ($C_o < C_i$), water chemical potential increases in the external compartment, leading to a positive osmotic pressure gradient ($\Delta\pi > 0$); the influx of water (arrow) causes cell swelling. (C) Under hypertonic conditions ($C_o > C_i$), the osmotic pressure gradient is negative ($\Delta\pi < 0$), leading to an efflux of water from the cell and consequent cell shrinkage.

depression, the extracellular compartment is substantially reduced (to half its volume) as revealed by electron microscopy, ionic activity, and electrical impedance measurements (Van Harreveld and Khattab, 1967). For example, in response to repetitive stimulation of the cat motor sensory cortex, the extracellular compartment shrinks by about 50% in regions in which the stimulus-induced release of K^+ is maximal, whereas within the honeybee retina, extracellular space can be reduced by at least one-third during light stimulation (Dietzel et al., 1980; Orkand et al., 1984). Interestingly, activity-dependent shrinkage coincides temporally with the differentiation of glial cells during development, indicating that movements of electrolyte and water into glial cells may be responsible for shrinkage of the extracellular compartment (Bourke and Nelson, 1972).

Summary

The nervous system is divided into compartments of cells that are specialized for different functions. Communication within compartments is strong for leptomeningeal cells, astroglia, and ependymal cells and is weaker for neurons and oligodendroglia. Coupling among glial cells helps the brain cope with volume changes resulting from hyposmotic and hyperosmotic shocks and following K^+ and glutamate release from neurons. Myelinating Schwann cells are not appreciably coupled to one another, although Cx32 gap junction proteins appear to form reflexive junctional channels between cytoplasmic processes at internodes and Schmidt–Lantermann incisures; coupling between proliferating Schwann cells due to expression of another gap junction protein after injury may promote remyelination. Neuronal precursor cells are well coupled, and decline in connexin expression is correlated with neuronal differentiation. Coupling between cells of different compartments in the nervous system appears to be quite limited. Although astroglia may be connected to oligodendrocytes, creating a "panglial syncytium" (Rash et al., 1997; see also Mugnaini, 1986), connections between glia and neurons do not appear to be present in the adult brain. Separation of neural communication compartments is generally predicted on the basis of whether the types of connexin molecules expressed by each cellular population can functionally pair with the other cells' connexins. Nevertheless, the role that matrix and cell adhesion molecules play in establishing and maintaining such compartmental boundaries remains to be determined.

GAP JUNCTION-RELATED NEUROPATHOLOGIES

Most of the understanding of the physiology and molecular biology of gap junctions in the brain has been obtained from studies of neural cells in culture or of channels formed by the gap junction proteins expressed in exogenous systems such as Xenopus oocytes and transfected mammalian cells. However, alterations in gap junction expression, distribution, and function have been found in a number of naturally occurring or induced somatic disease states: thus, a peripheral neuropathy (X-linked Charcot–Marie–Tooth disease) is a gap junction genetic disease affecting specifically Schwann cells in the peripheral nervous system, and hereditary nonsyndromic deafness is related to Cx26 mutations. In this section, we consider various pathological conditions in the brain that may result from aberrant gap junction expression or function (see also Spray and Dermietzel, 1995). The focus of this section is on the ways in which these alterations in junctional communication affect nervous system function and the underlying mechanisms of these dysfunctional states (see Table 15.2).

Neuronal Hyper- and Hypoexcitability Underlie Epilepsy and Spreading Depression and May Involve Changes in Cell Coupling

Abnormal electrical activity is the hallmark of nervous system dysfunction. Paroxysmal neuronal excitability, as occurs in epilepsy, might arise from strengthened electrical coupling between neurons, thereby synchronizing activity patterns. Because astrocytes located at epileptic foci do not express inwardly rectifying K^+ currents (Bordey and Sontheimer, 1998), neuronal hyperexcitability resulting from the lack of this K^+ current (Janigro et al., 1997) might be a consequence of glial inability to buffer extracellular K^+. Alternatively, reduced coupling between glial cells could be involved in such enhanced neuronal hyperexcitability due to the impaired ability of uncoupled glia to efficiently remove excess extracellular K^+; in these circumstances, neurons would be constantly depolarized.

That electrical coupling is strengthened between hyperexcitable neurons and that electrotonic connections may contribute to the positive feedback that synchronizes neuronal firing in tissues prone to epileptogenesis are proposals made almost two decades ago (Taylor and Dudek, 1982; Dudek et al., 1986). Recent studies show that spontaneous synchronized activity of neurons (so-called "field bursting") continues to occur and is even enhanced under these conditions (Jefferys, 1995). Moreover, this synchrony (and the occurrence of electrotonic PSPs) is blocked when *in vitro* preparations are treated with gap junction-uncoupling agents such as weak acids (which lower intracellular pH), halothane, and octanol (Perez-Velazquez *et al.*, 1994).

A comparison of Cx43 mRNA levels in cerebral cortical samples from epileptic cortices gives further evidence that coupling strength may increase in hyperexcitable brain regions; Cx43 expression is markedly increased compared with nonepileptic controls (Naus *et al.*, 1991; but see Elisevich *et al.*, 1997). Because Cx43 in the brain is localized predominantly in astrocytes and possibly also in some neurons, this finding indicates that glial or neuronal gap junction expression, or both, may be altered in epileptogenic foci. Furthermore, some anti-epileptic drugs block gap junction channels, and preliminary evidence supports the notion that astrocytes cultured from hyperexcitable regions of human brain are more strongly coupled than those in normal regions, a property that persists for weeks in tissue culture (Lee *et al.*, 1995).

Spreading depression (SD), which may be involved in seizure discharges, migraine headaches, and cerebral ischemia, consists of a number of different reactions to local stimulation of the gray matter including slow negative potential, changes in electrical impedance, and the existence of a refractory period in a range of minutes. The increase in [K]$_o$ and variations in volume of the tissue during the reaction are the most common events observed during spreading depression. These reactions have been observed in the hippocampus, olfactory bulb, striatum, spinal cord, superior colliculus, and cerebellum (do Carmo and Somjen, 1994; Leao, 1987; McLachlan and Girvin, 1994; Herreras *et al.*, 1994; Streit *et al.*, 1995). Using genetic engineering strategies the group of Klaus Willecke succeeded in producing conditional knockout animals that lack astrocytic Cx43 expression. The resulting mice show a significant increase in spreading depression (20%), indicating a major disturbance of spatial potassium buffering, which may lead to changes in neuronal excitability (Willecke *et al.*, 2002).

Retinal tissue has been widely used to probe the phenomenon of spreading depression (see Martins-Ferreira, 1993). In this tissue, the SD reaction can be repeatedly observed over a period of several hours by visual observations of an enlarging milky circle starting from the point of focal stimulation (illustrated in Fig. 15.10). Because the rate of propagation of this wave of depression (20–50 μm/s) is so similar to that measured for Ca^{2+} waves among glial cells both in tissue culture (see Fig. 15.6) and in brain slice prepara-

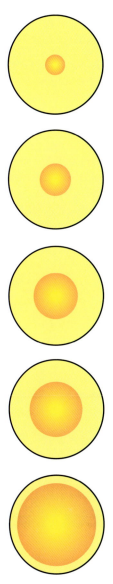

FIGURE 15.10 Chick eyecup model of spreading depression. Mechanical, electrical, or pharmacological focal stimulation of the chick eyecup results in a radially propagating change in opacity with a conduction velocity of about 10–50 μm/s. As this change sweeps across the eyecup, illustrated by an orange wavefront, the tissue is refractory to further stimulation, a phenomenon termed spreading depression.

tions, and because the phenomenon is blocked by gap junction inhibitors (Martins-Ferreira and Ribeiro, 1995; Nedergaard *et al.*, 1995), it has been suggested that spreading depression results from the propagation of Ca^{2+} waves through gap junctions between the cells. Whether the aura associated with migraine headaches is due to gap junction-mediated spreading depression and whether gap junction blockade might reduce damage caused by ischemic insults (Rawanduzy *et al.*, 1997) are intriguing ideas that are currently being investigated.

CMTX: Connexin32 Mutations and Peripheral Neuropathy

Charcot–Marie–Tooth (CMT) disease affects conduction in both motor and sensory axons. It is the most common hereditary peripheral neuropathy, occurring in the population with an incidence of about 1 in 2000 individuals. Three genetic loci and alterations in their encoded gene products are now believed to be responsible for the three major demyelinating forms of this disease. The most prevalent of these disorders, CMT1A, is mapped to duplications or missense mutations in human chromosome 17 (17 p11.2–12) and involves the myelin protein

PMP22; CMT1B maps to human chromosome 1 and appears to involve missense mutations in the myelin constituent P0, whereas the CMTX form is linked to mutations within the Cx32 locus of chromosome X (Bergoffen et al., 1993; Ouvrier, 1996). Reports on the X-linked form of the disease thus far have identified a total of more than 300 families in which sequencing of the coding region of the Cx32 gap junction gene revealed more than 200 different mutations (Scherer et al., 1998; Bone et al., 1997; Rabionet et al., 2002). Most of these mutations involve single base substitutions leading to single amino acid alterations; others result in formation of premature stop codons at various points in the sequence, produce frame shifts

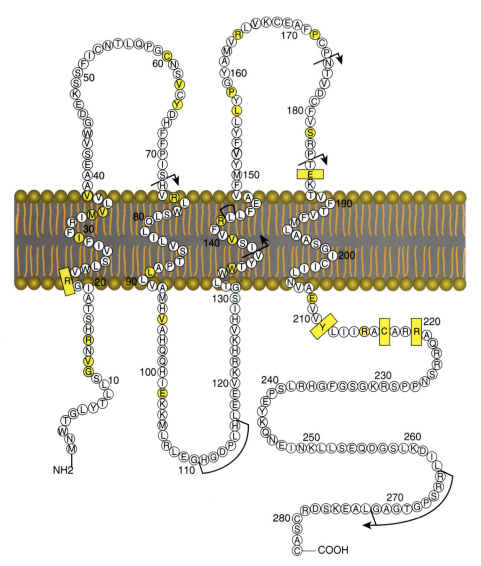

FIGURE 15.11 Examples of the diversity of coding region mutations in human Cx32 associated with the X-linked form of Charcot–Marie–Tooth disease. Each ball in the chain represents an amino acid residue indicated by a letter identifying the amino acid present in the wild-type sequence. Different symbols represent different types of mutations: Missense mutations are indicated by yellow circles, nonsense mutations by yellow bars, frameshift mutations by arrows, and deletions by brackets. Modified from Scherer et al. (1998).

leading to premature translational termination, or result in the elimination of one or more amino acid residues (Fig. 15.11). Most of these mutations occur in regions of the molecule believed to line the pore (M3, the third membrane-spanning domain of Cx32), to mate connexon subunits across extracellular space (C1, C2, the two extracellular loops), or to terminate the molecule proximal to its phosphorylation sites. Coding region mutations are absent in some CMTX families, in which cases defects of expression may arise from derangement of Cx32 gene regulatory elements.

The phenotype resulting from mutations of Cx32 is limited largely to peripheral nerves, which is surprising because Cx32 is a major component of gap junctions of diverse tissues including liver, exocrine pancreas, and epithelium of the gastrointestinal tract. Moreover, although Cx32 appears to be a major component of gap junctions in oligodendrocytes and in at least some neurons in the CNS, impairment of brain function in CMTX patients is uncommon. Because most tissues express other connexins in addition to Cx32, these other connexins may compensate functionally. Alternatively, the mutations of the Cx32 gene may result in specific loss of function in the peripheral myelin sheath because of this tissue's unique structural organization or metabolic requirements, or both.

As noted above, the role ascribed to Cx32 gap junction channels in myelinating Schwann cells has been to form reflexive gap junctions between cytoplasmic loops at nodes of Ranvier and Schmitt–Lantermann incisures (Bergoffen et al., 1993; Scherer et al., 1998; Bone et al., 1997). These channels would thus short-circuit otherwise very long and curcuitous pathways for nutrient and metabolite exchange between the nucleus and the Schwann cell cytoplasm nearest the axon. The coding region mutations in Cx32 seen in CMTX pedigrees could interfere with Schwann cell function through altered expression or trafficking to the membrane (Deschenes et al., 1997), through alterations in channel permeability (Oh et al., 1997; Omori et al., 1996; Bruzzone et al., 1994) or by negative-dominant effects on other connexins (Bruzzone et al., 1994; Omori et al., 1996). Conduction slowing and demyelination seen progressively in CMTX patients are also observed in Cx32 null mice (Anzini et al., 1997), strongly indicating that the loss of functional Cx32 channels is responsible for the disease.

Mutations in the Cx26 gene cause the most frequent nonsyndromic sensorineural hearing defect in humans (Kelsell et al., 1997). Mutations in the Cx31 (Liu et al., 2000) and Cx30 (Grifa et al., 1999) genes also lead to hearing defects. It has been suggested that these gap junction channels are involved in the buffering of K^+ and glutamate secreted from activated hair cells by neighboring supporting cells. A defect in this siphoning effect of supporting cells is considered to be responsible for an extracellular accumulation of both substances and eventually toxic events leading to destruction of the hair cells with the consequence of hearing impairment.

It has generally been assumed that the connexin mutations associated with these hereditary disorders cause dysfunction or loss of coupling between the cells, either because of deficits in conductance or gating of the mutant channels or as a result of disrupted intracellular trafficking (see VanSlyke et al., 2000). However, both disrupted trafficking and other effects on intercellular communication could conceivably arise from altered binding of connexins to other cytoplasmic proteins caused by the mutations, independent of channel function per se. Although gap junctions do not have the elaborate substructure of chemical synapses, cytoplasmic connexin domains do bind other proteins, forming a macromolecular complex termed the *Nexus* (Spray et al., 1999).

Gap Junctions May Play Roles in Tumorigenesis and Provide a Therapeutic Strategy

Along with the first description of gap junction-mediated intercellular coupling came the hypothesis that intercellular communication may affect the growth rate of embryonic cells (Furshpan and Potter, 1968), and that reduced gap junction expression or function could lead to the loss of growth control in tumors (Loewenstein, 1979). Although most studies of this correlation between coupling strength (or junctional expression) and growth rate have been performed on cells derived from tumors (such as hepatomas) of the digestive system, glial tumors offer an opportunity to extend this analysis to cells of the brain. Most nervous system tumors are of astrocytic origin (astrocytomas) and can be graded according to their severity and growth characteristics from the least invasive (type I) to the most malignant (type IV). Tumors developing from oligodendrocytes (oligodendrogliomas), although less common, are quite severe. These two glial populations express different connexins (Cx43 in astrocytes, Cx32 in oligodendrocytes), raising the question of whether the expression of either the type or the abundance of connexins correlates with severity of malignancy.

Thus, the escape of glial tumor cells from growth control regulation does not appear to result from the absence of gap junction communication. Moreover, most recent studies on tumor cells transfected with connexin cDNA sequences indicate that growth arrest is not restored in culture (Eghbali *et al.*, 1991; Elshami *et al.*, 1996; Bond *et al.*, 1994). However, growth of these cells as tumors in nude mice is retarded compared with communication-deficient cell lines, and both spontaneous and carcinogen-induced hepatomas are reported to be more common in Cx32 null mice than wild-type siblings (Temme *et al.*, 1997). Ironically, the presence of gap junctions between tumor cells is responsible for a phenomenon known as "bystander cell killing," which is the basis for a therapeutic strategy currently undergoing clinical trials (Fig. 15.12) (see Shinoura *et al.*, 1996). Cells transduced with viral thymidine kinase are rendered sensitive to the chemotherapeutic drug ganciclovir, which is metabolized by viral thymidine kinase to a phosphorylated low-molecular-weight cytotoxin that intercalates into DNA, causing apoptotic cell death. The phosphorylated acyclovir is presumably gap junction permeant, thereby killing cells that are coupled to the viral thymidine kinase transfectants (Fig. 15.12), both in culture and in tumors *in vivo* (Dilber *et al.*, 1997; Elshami *et al.*, 1996; Fick *et al.*, 1995; Mesnil *et al.*, 1996; Vrionis *et al.*, 1997).

Brain Lesions Affect Intercellular Coupling

In cardiac muscle, both acute and chronic infarctions lead first to loss of gap junctions and, after a period of days, to reexpression and reorganization of these channels (see Saffitz, 1997). Glial gap junction expression may be similarly responsive to such insult. For example, lesion of the facial nerve results in rapidly increased Cx43 expression by astrocytes in the ipsilateral facial nucleus (Laskawi *et al.*, 1997). Furthermore, traumatic injury to the spinal cord leads to both loss of Cx43 immunoreactivity in the center of the spinal cord and increased immunostaining in outer regions (Theriault *et al.*, 1997). Finally, subsequent to a local stab wound of the visual cortex, regional loss of dye coupling between astrocytes has been observed to persist for as long as 4 weeks after wounding (Spray and Dermietzel, 1995). By use of mouse models with ablated connexin genes, differences in the response to either traumatic or ischemic brain trauma were described. Occlusion of the middle cerebral artery elicits an increase in Cx36 and Cx32 proteins accompanied by an unchanged expression of Cx43 in astrocytes (Oguro *et al.*, 2001). Mice lacking the Cx32 gene (Y/−) showed a higher rate of neuronal survival as compared with wild-type animals, indicating that the persistence of Cx32 expression provides

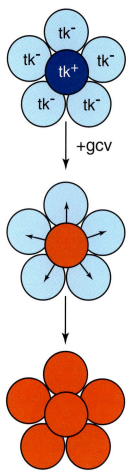

The bystander effect

FIGURE 15.12   Gap junction-mediated cell killing through the bystander effect. Cells transfected with herpes thymidine kinase (tk⁺ dark blue) die apoptotically (turn red in this figure) when they are exposed to the anticancer drug ganciclovir (gcv). When tk⁺ cells are co-cultured with untransfected (tk⁻) (light blue) cells to which they make gap junctional contact, the tk⁻ "bystanders" are also killed (turning red). This bystander effect also occurs in tumors, allowing the possibility of gap junction-mediated tumor therapy.

protective factors that may rescue the neurons from apoptotic death. Contrary to these data, persistence of Cx43 expression in astrocytes is considered to be deleterious to neurons as indicated by *in vitro* studies after exploiting metabolic arrest (Lin *et al.*, 1998) and in a mechanical trauma model since blocking of junctional coupling and ablation of Cx43 expression (Frantseva *et al.*, 2002) increase the rate of neuronal survival. A feasible explanation is that persistent junctional coupling offers some disadvantage to adjacent cells by transmitting trauma-induced stress factors, similar to "bystander cell killing."

Protozoan Infections Alter Coupling between CNS Cell Populations

Protozoan parasites associated with brain infection in immunosuppressed persons include *Trypanosoma cruzi*, which causes American trypanosomiasis or Chagas' disease, the prevalent cause of cardiomyopathy in South America, and *Toxoplasma gondii*, which causes toxoplasmosis, an opportunistic infection contributing to death from HIV infection. *T. cruzi* infection leads to disorganization of gap junctions and severe reductions in intercellular communication in cultured cardiac myocytes (Campos de Carvalho *et al.*, 1993). In astrocytes, both *T. gondii* and *T. cruzi* infections are associated with the loss of coupling and reduce expression of Cx43-containing junctional plaques (Campos de Carvalho *et al.*, 1998). In leptomeningeal cells, both protozoans reduce the coupling and the organized expression of both Cx26 and Cx43. Thus, the effects on junctional disruption are generalized with respect to both type of host cell and type of connexin interconnecting the cell populations. Because the changes in connexin distribution occur in parasitized cells without changes in connexin abundance (at either the protein or mRNA level), a trafficking disorder appears responsible for the loss of coupling between parasitized cells. Whatever the mechanism, parasite-induced loss of intercellular communication would be expected to affect astrocyte functions dramatically, disturbing K^+ balance, dissipation of cell volume changes, and propagation of Ca^{2+} waves between these glial cells.

Summary

Gap junctions in nervous tissue synchronize neuronal activity, provide pathways for second messenger and metabolite exchange, and may modulate cell growth, differentiation, and organization. Abnormal gap junction expression or function is associated with both genetic and somatic disease states, presumably contributing to the pathology through loss of the important intercellular pathway provided by these channels. As listed in Table 15.2, mice in which connexins are deleted by molecular genetic manipulation provide model systems in which the roles of specific gap junction types can be explored in detail and in which the contribution of functionally ablated gap junctions to different pathological situations can be directly assessed.

References

Altevogt, B. M., Kleopa, K. A., Postma, F. R., Scherer, S. S., and Paul, D. L. (2002). Connexin29 is uniquely distributed within myeli-nating glial cells of the central and peripheral nervous system. *J. Neurosci.* **22**, 6458–6470.

Amitai, Y., Gibson, J. R., Beierlein, M., Patrick, S. L., Ho, A. M., Connors, B. W., and Golomb, D. (2002). The spatial dimensions of electrically coupled networks of interneurons in the neocortex. *J. Neurosci.* **22**, 4142–4152.

Anderson, P. A. (1985). Physiology of a bidirectional, excitatory, chemical synapse. *J. Neurophysiol.* **53**, 821–835.

Anderson, P. A., and Grunert, U. (1988). Three-dimensional structure of bidirectional, excitatory chemical synapses in the jellyfish *Cyanea capillata*. Synapse **2**, 606–613.

Anzini, P., Neuberg, D. H., Schachner, M., Nelles, E., Willecke, K., Zielasek, J., Toyka, K. V., Suter, U., and Martini, R. (1997). Structural abnormalities and deficient maintenance of peripheral nerve myelin in mice lacking the gap junction protein connexin32. *J. Neurosci.* **17**, 4545–4551.

Araque, A., Parpura, V., Sanzgiri, R. P., and Haydon, P. G. (1998a). Glutamate-dependent astrocyte modulation of synaptic transmission between cultured hippocampal astrocytes. *Eur. J. Neurosci.* **10**, 2129–2142.

Araque, A., Parpura, V., Sanzgiri, R. P., and Haydon, P. G. (1999). Tripartite synapses: Glia, the unacknowledged partner. *Trends Neurosci.* **22**, 208–213.

Araque, A., Sanzgiri, R. P., Parpura, V., and Haydon, P. G. (1998b). Calcium elevation in astrocytes causes an NMDA receptor-dependent increase in the frequency of miniature synaptic currents in cultured hippocampal neurons. *J. Neurosci.* **18**, 6822–6829.

Auerbach, A. A., and Bennett, M. V. L. (1969). A rectifying synapse in the central nervous system of a vertebrate. *J. Gen. Physiol.* **53**, 211–237.

Batter, D. K., Corpina, R. A., Roy, C., Spray, D. C., Hertzberg, E. L., and Kessler, J. A (1992). Heterogeneity in gap junction expression in astrocytes cultured from different brain regions. *Glia* **6**, 213–221.

Beierlein, M., Gibson, J. R., and Connors, B. W. (2000). A network of electrically coupled interneurons drives synchronized inhibition in neocortex. *Nat. Neurosci.* **3**, 904–910.

Belluardo, N., Mud, G., Trovato-Salinaro, A., LeGurun, S., Charollais, A., Serre-Beinier, V., Amato, G., Haefliger, J. A., Meda, P., and Condorelli, D. F. (2000). Expression of connexin36 in the adult and developing rat brain. *Brain Res.* **865**, 121–138.

Bender, A. S., Neary, J. T., and Norenberg, M. D. (1993). Role of phosphoinositide hydrolysis in astrocyte volume regulation. *Neurochemistry* **6**, 1506–1514.

Bennett, M. V. L. (1977). Electrical transmission: A functional analysis and comparison to chemical transmission. *In* "Handbook of Physiology: The Nervous System I," pp. 357–416. American Physiological Society, Washington.

Bennett, M. V. L. (1996). Gap junctions as electrical synapses. *In* "Gap Junctions in the Nervous System" (D. C. Spray and R. Dermietzel, Eds.), pp. 61–74. R. G. Landes, New York.

Bennett, M. V. L. (1997). Gap junctions as electrical synapses. *J. Neurocytol.* **26**, 349–366.

Bennett, M. V. L., Barrio, L., Bargiello, T. A., Spray, D. C., Hertzberg, E. L., and Saez, J. C. (1991). Gap junctions: New tools, new answers, new questions. *Neuron* **6**, 305–320.

Bennett, M. V. L., Crain, S. M., and Grundfest, H. (1959). Electrophysiology of supramedullary neurons in *Spheroides maculatus*. I. Orthodromic and antidromic responses. *J. Gen. Physiol.* **43**, 159–188.

Bennett, M. V. L., Nakajima, Y., and Pappas, G. D. (1967a). Physiology amd ultrastructure of electrotonic junctions. I. Supramedullary neurons. *J. Neurophysiol.* **30**, 161–179.

Bennett, M. V. L., Nakajima, Y., and Pappas, G. D. (1967b). Physiology and ultrastructure of electrotonic junctions. III. Giant electromotor neurons of *Malapterurus electricus*. *J. Neurophysiol.* **30**, 209–235.

Bennett, M. V. L., Spray, D. C., and Harris, A. L. (1981). Electrical coupling in development. *Am. Zool.* **21**, 413–427.

Bergoffen, J., Scherer, S. S., Wang, S., Oronzi-Scott, Bone, L. J., Paul, D. L., Chen, K., Lensch, M. W., Chance, P., and Fischbeck, K. (1993). Connexin mutations in X-linked Charcot–Marie–Tooth disease. *Science* **262**, 2039–2042.

Bond, S. L., Bechberger, J. F., Khoo, N. K., and Naus, C. C. (1994). Transfection of C6 glioma cells with connexin32: The effects of expression of a nonendogenous gap junction protein. *Cell Growth Differ.* **5**, 179–186.

Bone, L. J., Deschenes, S. M., Balice-Gordon, R. J., Fishbeck, K. H., and Scherer, S. S. (1997). Connexin32 and x-linked Charcot–Marie–Tooth disease. *Neurobiol. Dis.* **4**, 221–230.

Bordey, A., and Sontheimer, H. (1998). Properties of human glial cells associated with epileptic seizure foci. *Epilepsy Res.* **32**, 286–303.

Bourke, R., and Nelson, K. (1972). Further studies on the K^+-dependent swelling of primate cerebral cortex in vivo: The enzymatic basis of the K^+-dependent transport of chloride. *J. Neurochem.* **19**, 663–685.

Bruzzone, R., White, T. W., Scherer, S. S., Fishbeck, K. H., and Paul, D. L. (1994). Null mutations of connexin32 in patients with X-linked Charcot–Marie–Tooth disease. *Neuron* **13**, 1253–1260.

Campos de Carvalho, A. C., Roy, C., Hertzberg, E. L., Tanowitz, H. B., Kessler, J. A., Weiss, L. M., Wittner, M., Gao, Y., and Spray, D. C. (1998). Gap junction disappearance in astrocytes and leptomeningeal cells as a consequence of protozoan infection. *Brain Res.* **790**, 304–314.

Campos de Carvalho, A. C., Tanowitz, H. B., Wittner, M., Dermietzel, R., Roy, C., Hertzberg, E. L., and Spray, D.C. (1993). Gap junction distribution is altered between cardiac myocytes infected with *Trypanosoma cruzi*. *Circ. Res.* **70**, 733–742.

Cepeda, C., Walsh, J. P., Peacock, W., Buchwald, N. A., and Levine, M. S. (1993). Dye-coupling in human neocortical tissue resected from children with intractable epilepsy. *Cereb. Cortex.* **3**, 95–107.

Chandross K. J., Kessler, J. A., Cohen, R. I., Simburger, E., Spray, D. C., Bieri, P., and Dermietzel, R. (1996). Altered connexin expression after peripheral nerve injury. *Mol. Cell. Neurosci.* **7**, 501–518.

Chang, M., Werner, R., and Dahl, G. (1996). A role for an inhibitory connexin in testis? *Dev. Biol.* **175**, 50–56.

Charles, A. (1998). Intercellular calcium waves in glia. *Glia* **24**: 39–49.

Christ, G. J., Moreno, A. P., Melman, A., and Spray, D. C. (1992). Gap junction-mediated intercellular diffusion of Ca^{2+} in cultured human smooth muscle cells. *Am. J. Physiol.* **263**, C373–C383.

Churchill, G., and Louis, C. (1998). Roles of Ca^{2+} inositol triphosphate and cyclic ADP-ribose in mediating intercellular Ca^{2+} signaling in sheep lens cells. *J. Cell Sci.* **111**, 1217–1223.

Condorelli, D. F., Belluardo, N., Trovato-Salinaro, A., and Mudo, G. (2000). Expression of C x 36 in mammalian neurons. *Brain Res. Brain Res Rev.* **32**, 72–85.

Cornell-Bell, A. H., Finkbeiner, S. M., Cooper, M. S., and Smith, S. J. (1991). Glutamate induces calcium waves in cultured astrocytes: Long-range glial signaling. *Science* **247**, 470–473.

Deans, M. R., Gibson, J. R., Sellitto, C., Connors, B. W., and Paul, D. L. (2001) Synchronous activity of inhibitory networks in neocortex requires electrical synapses containing connexin36. *Neuron* **31**, 477–485.

Del Bigio, M. R. (1994). The ependyma: A protective barrier between brain and cerebrospinal fluid. *Glia* **14**, 1–13.

De Pina-Benabou, M. H., Srinivas, M., Spray, D. C., and Scemes, E. (2001). Calmodulin kinase pathway mediates the K^+-induced increase in gap junctional communication between mouse spinal cord astrocytes. *J. Neurosci.* **21**, 6635–6643.

Dermietzel, R., Farooq, M., Kessler, J. A., Althaus, H., Hertzberg, E. L., and Spray, D. C. (1997). Oligodendrocytes express gap junction proteins connexin32 and connexin45. *Glia* **20**, 101–114.

Dermietzel, R., Hertzberg, E. L., Kessler, J. A., and Spray, D. C. (1991). Gap junctions between cultured astrocytes: Immunocytochemical, molecular, and electrophysiological analysis. *J. Neurosci.* **11**, 1421–1432.

Dermietzel, R., Kremer, M., Paputsoglu, G., Stang, A., Skerrett, I. M., Gomes, D., Srinivas, M., Janssen-Bienhold, U., Weiler, R., Nicholson, B. J., Bruzzone, R., and Spray, D. C. (2000). Molecular and functional diversity of neural connexins in the retina. *J. Neurosci.* **20**, 831–8343.

Dermietzel, R., and Spray, D. C. (1993). Gap junctions in the brain: Where, what type, how many and why. *Trends Neurosci.* **16**, 186–192.

Dermietzel, R., and Spray, D. C. (1998). From glue ("Nervenkitt") to glia: A prologue. *Glia* **24**, 1–7.

Dermietzel, R., Traub, O., Hwang, T. K., Beyer, E., Bennett, M. V., Spray, D. C., and Willecke, K. (1989). Differential expression of three gap junction proteins in developing and mature brain tissues. *Proc. Natl. Acad. Sci. USA* **86**, 10148–10152.

Deschenes, S. M., Walcott, J. L., Wexler, T. L., Scherer, S. S., and Fischbeck, K. H. (1997). Altered trafficking of mutant connexin32. *J. Neurosci.* **17**, 9077–9084.

Dietzel, L., Heinemann, V., Hofmeier, G., and Lux, H. (1980). Transient changes in the size of the extracellular space in the sensorimotor cortex of cells in relation to stimulus-induced changes in potassium concentrations. *Exp. Brain Res.* **40**, 432–439.

Dilber, M. S., Abedi, M. R., Christensson, B., Bjorkstrand, B., Kidder, G. M., Naus, C. C., Gahrton, G., and Smith, C. I. (1997). Gap junctions promote the bystander effect of herpes simplex virus thymidine kinase *in vivo*. *Cancer Res.* **57**, 1523–1528.

do Carmo, R. J., and Somjen, G. G. (1994). Spreading depression of Leao: 50 years since a seminal discovery. *J. Neurophysiol.* **72**, 1–2.

Draguhn, A., Traub, R. D., Schmitz, D., and Jeffrey, J. G. (1998). Electrical coupling underlies high-frequency oscillations in the hippocampus in vitro. *Nature* **394**, 132–133.

Dudek, F. E., Andrew, R. D., MacVicar, B. A., and Hatton, G. I. (1982). Intracellular electrophysiology of mammalian peptidergic neurons in rat hypothalamic slices. *Fed. Proc.* **41**, 2953–2958.

Dudek, E. E., Snow, R. W., and Taylor, C. P. (1986). Role of electrical interactions in synchronication of epileptiform bursts. *Adv. Neuro.* **44**, 593–617.

Duffy, H. S., Delmar, M., and Spray, D. C. (2002). Formation of the gap junction nexus: Binding partners for connexins. *J. Physiol. Paris* **96**: 243–249.

Eghbali, B., Kessler, J. A., Reid, L. M., Roy, C., and Spray, D. C. (1991). Involvement of gap junctions in tumorigenesis: Transfection of tumor cells with connexin 32 cDNA retards growth *in vivo*. *Proc. Natl. Acad. Sci. USA* **88**, 10701–10705.

Elfgang, C., Eckert, R., Lichtenberg-Frate, H., Butterweck, A., Traub, O., Klein, R. A., Hlser, D. F., and Willecke, K. (1995). Specific permeability and selective formation of gap junction channels in connexin-transfected HeLa cells. *J. Cell Biol.* **129**, 805–817.

Elisevich, K., Rempel, S. A., Smith, B. J., and Edvardsen, K. (1997). Hippocampal connexin 43 expression in human complex partial seizure disorder. *Exp. Neurol.* **45**, 154–164.

Elshami, A. A., Saavedra, A., Zhang, H., Kucharczuk, J. C., Spray, D. C., Fishman, G. I., Amin, K. M., Kaiser, L. R., and Albelda, S.,

S. M. (1996). Gap junctions play a role in the "bystander effect" of the herpes simplex virus thymidine kinase/ganciclovir system *in vitro. Gene Ther.* **3**, 85–92.

Enkvist, M. O., and McCarthy, K. D. (1992). Activation of protein kinase C blocks astroglial gap junction communication and inhibits the spread of calcium waves. *Neurochemistry* **59**, 519–526.

Feigenspan, A., Teubner, B., Willecke, K., and Weiler, R. (2001). Expression of neuronal connexin36 in AII amacrine cells of the mammalian retina. *J. Neurosci.* **21**, 230–239.

Fick, J., Barker, F. G., 2nd, Dazin, P., Westphale, E. M., Beyer, E. C., and Israel, M. A. (1995). The extent of heterocellular communication mediated by gap junctions is predictive of bystander tumor cytotoxicity in vitro. *Proc. Natl. Acad. Sci. USA* **92**, 11071–11075.

Frantseva, M. V., Kokarovtseva, L., Naus, C. G., Carlen, P. L., MacFabe, D., and Perez Velazquez, J. L. (2002). Specific gap junctions enhance the neuronal vulnerability to brain traumatic injury. *J. Neurosci.* **22**, 644–653.

Furshpan, E. J., and Potter, D. D. (1959). Transmission at the giant motor synapses of the crayfish. *J. Physiol. London* **145**, 289–325.

Furshpan, E. J., and Potter, D. D. (1968). Low-resistance junctions between cells in embryos and tissue culture. *In* "Current Topics in Developmental Biology" (A. A. Moscona and A. Monroy, Eds.), Vol. 3, pp. 95–127. Academic Press, New York.

Furukawa, T. (1966). Synaptic interaction at the mauthner cell of goldfish. *Prog. Br. Res.* **21**, 44–77.

Furukawa, T., and Furshpan, E. J. (1993). Two inhibitory mechanisms in the Mauthner neurons of goldfish. *J. Neurophysiol.* **26**, 140–176.

Galarreta, M., and Hestrin, S. (1999). A network of fast-spiking cells in the neocortex connected by electrical synapses. *Nature* **402**, 72–75.

Galarreta, M., and Hestrin, S. (2001). Electrical synapses between GABA-releasing interneurons. *Nat. Rev. Neurosci.* **2**, 425–433.

Giaume, C., Fromaget, C., el Aoumari, A., Cordier, J., Glowinski, J., and Gros, D. (1991). Gap junctions in cultured astrocytes: Single-channel currents and characterization of channel-forming protein. *Neuron* **6**, 133–143.

Giaume, C., and McCarthy, K. D. (1996). Control of junctional communication in astrocytic networks. *Trends Neurosci.* **19**, 319–325.

Giaume, C., Tabernero, A., and Medina, J. M. (1997). Metabolic trafficking through astrocytic gap junctions. *Glia* **21**, 114–123.

Giaume, C., and Venance, L. (1998). Intercellular calcium signaling and gap junction communication in astrocytes. *Glia* **24**, 50–64.

Gibson, J. R., Beierlein, M., and Connors, B. W. (1999). Two networks of electrically coupled inhibitory neurons in neocortex. *Nature* **402**, 75–79.

Grifa, A., Wagner, C. A., D'Ambrosio, L., Melchionda, S., Bernardi, F., Lopez-Bigas, N., Rabionet, R., Arbones, M., Monica, M. D., Estivill, X., Lang, F., and Gasparini, P. (1999). Mutations in GJB6 cause nonsyndromic autosomal dominant deafness at DFNA3 locus. *Nat. Genet.* **23**, 16–18.

Grundfest, H. (1959). Synaptic and ephaptic transmission. *In* "Handbook of Physiology. Neurophysiology" (J. Field, Ed.), Sect. 1, Vol. 1, pp. 147–197. *Am. Physiol. Soc.*, Washington DC.

Hassinger, T. D., Guthrie, P. B., Atkinson, P. B., Bennett, M. V., and Kater, S. B. (1996). An extracellular signaling component in propagation of astrocytic calcium waves. *Proc. Natl. Acad. Sci. USA* **93**, 13268–13273.

Herreras, O., Largo, C., Ibarz, J. M., Somjen, G. G., and Martin del Rio, R. (1994). Role of neuronal synchronizing mechanisms in the propagation of spreading depression in the in vivo hippocampus. *J. Neurosci.* **14**(11, Pt. 2), 7087–7098.

Hinricksen, C. F. L., and Larramendi, M. H. (1968). Synapses and cluster formation of the mouse mesencephalic fifth nucleus. *Brain Res.* **7**, 296–299.

Hormuzdi, S. G., Pais, I., LeBeau, F. E., Towers, S. K., Rozov, A., Buhl, E. H., Whittington, M. A., and Monyer, H. (2001). Impaired electrical signaling disrupts gamma frequency oscillations in connexin36-deficient mice. *Neuron* **31**, 487–495.

Janigro, D., Gasporini, S., D'Ambrosio, R., McKhann, G., and DiFrancesco, D. (1997). Reduction of K⁺ uptake in glia prevents long-term depression maintenance and causes epileptiform activity. *J. Neurosci.* **17**, 2813–1824.

Jefferys, J. G. R. (1995). Nonsynaptic modulation of neuronal activity in the brain: electric currents and extracellular ions (review). *Physiol. Rev.* **75**, 689–723.

Kandler, K., and Katz, L. C. (1995). Neuronal coupling and uncoupling in the developing nervous system. *Curr. Opin. Neurobiol.* **5**, 98–105.

Kang, J., Jiang, L., Goldman, A. S., and Nedergaard, M. (1998). Astrocyte-mediated potentiation of inhibitory synaptic transmission. *Nat. Neurosci.* **1**, 683–692.

Katz, L. C. (1993). Coordinate activity in retinal and cortical development. *Curr. Opin. Neurobiol.* **3**, 93–99.

Kelsell, D. P., Dunlop, J., Stevens, H. P., Lench, N. J., Liang, J. N., Parry, G., Mueller, R. F., and Leigh, I. M. (1997). Connexin26 mutations in hereditary non-syndromic sensorineural deafness. *Nature* **387**, 80–83.

Kimelberg, H. K. (1991). Swelling and volume control in brain astroglial cell. *In* "Comparative Environment Physiology" (R. Gilles, E. K. Hoffman, and L. Boles, Eds.), Vol. 9, pp. 81–117. Springer-Verlag, New York.

Kimelberg, H. K. (1995). Brain edema. *In* "Neuroglia" (H. Kettenmann and B. R. Ransom, Eds.), pp. 919–935. Oxford Univ. Press, London/New York.

Kimelberg, H. K., and Kettenmann, H. (1990). Swelling-induced changes in electrophysiological properties of cultured astrocytes and oligodendrocytes. I. Effects on membrane potential, input impedance and cell–cell coupling. *Brain Res.* **529**, 255–261.

Korn, H., and Faber, D. S. (1975). An electrically mediated inhibition in goldfish medulla. *J. Neurophysiol.* **38**, 452–471.

Korn, H., and Faber, D. S. (1996). Escape behavior: Brainstem and spinal cord circuitry and function. *Curr. Opin. Neurobiol.* **6**, 826–832.

Korn, H., Sotelo, F. and Crepel, F. (1973). Electronic coupling between neurons in the rat lateral vestibular nucleus. *Exp. Brain Res.* **16**, 255–275.

Kuffler, S. W., and Nicholls, J. G. (1966). The physiology of neuroglial cells. *Ergeb Physiol.* **57**, 1–90.

Kunzelmann, P., Schroder, W., Traub, O., Steinhauser, C., and Dermietzel, R. (1999). Late onset and increasing expression of the gap junction protein connexin30 in adult murine brain and long-term cultured astrocytes. *Glia* **25**, 111–119.

Laskawi, R., Rohlmann, A., Landgrebe, M., and Wolff, J. R. (1997). Rapid astroglial reactions in the motor cortex of adult rats following peripheral facial nerve lesions. *Eur. Arch. Otorhinolaryngol.* **254**, 81–85.

Leao, A. A. P. (1987). Spreading depression. *In* "Encyclopedia of Neuroscience" (G. Adelman, Ed.), pp. 1137–1138. Birkhauser, Stuttgart.

Lee, S. H., Kim, W. T., Cornell-Bell, A. H., and Sontheimer, H. (1994). Astrocytes exhibit regional specificity in gap-junction coupling. *Glia* **11**, 315–325.

Lee, S. H., Magge, S., Spencer, D. D., Sontheimer, H., and Cornell-Bell, A. H. H. (1995). Human epileptic astrocytes exhibit increased gap junction coupling. *Glia* **15**, 195–202.

Lin, J. H., Weigel, H., Cotrina, M. L., Liu, S., Bueno, E., Hansen, A. J., Goldman, S. and Nedergaard, M. (1998). Gap junction-mediated propagation and simplification of cell injury. *Nat. Neurosci.* **1**, 494–500.

Liu, X. Z., Xia, X. J., Xu, L. R., Pandya, A., Liang, C. Y., Blanton, S. H., Brown, S. D., Steel, K. P., and Nance, W. E. (2000). Mutations in connexin31 underlie recessive as well as dominant non-syndromic hearing loss. *Hum. Mol. Genet.* **9**, 63–67.

Llinas, R. (1985). Electrotonic transmission in the mammalian central nervous system. *In* "Gap Junctions" (M. V. L. Bennett and D. C. Spray, Eds.), pp. 337–353. Cold Spring Harbor Laboratory Press, Cold Spring Harbor, NY.

Loewenstein, W. R. (1979). Junctional intercellular communication and the control of growth. *Biochim. Biophys. Acta* **560**, 1–65.

Lo Turco, J. J., and Kriegstein, A. R. (1991). Clusters of coupled neuroblasts in embryonic neocortex. *Science* **252**, 563–566.

MacVicar, B. A., and Dudek, F. E. (1982). Electrotonic coupling between granule cells of rat dentate gyrus: Physiological and anatomical evidence. *J. Neurophysiol.* **47**, 579–592.

Martins-Ferreira, H. (1993). Propagation of spreading depression in isolated retina. *In* "Migraine: Basic Mechanisms and Treatment" (A. Lehmenkuller, K. H. Grotemeyer, and F. Tegmeier Urban of Schwartzenberg, Munich), pp. 533–546.

Martins-Ferreira, H., and Ribeiro, L. J. (1995). Biphasic effects of gap junctional uncoupling agents on the propagation of retinal spreading depression. *Braz. J. Med. Biol. Res.* **28**, 991–994.

McLachlan, R. S., and Girvin, J. P. (1994). Spreading depression of Leao in rodent and human cortex. *Brain Res.* **666**, 133–136.

Mehler, M. F., Rozental, R., Doudgherty, M., Spray, D. C., and Kessler, J. A. (1993). Cytokine regulation of neuronal differentiation in immortalized hippocampal progenitor cells. *Nature* **362**, 62–65.

Meier, C., Dermietzel, R., Yasumura, T., Davidson, U., and Rash, J. 2002a. Cx32-containing gap junctions at the site of myelin-compaction: A possible pathway for intracellular transport of water and ions. In preparation.

Meier, C., Petrasch-Parwez, E., Habbes, H. W., Teubner, B., Guldenagel, M., Degen, J., Sohl, G., Willecke, K., and Dermietzel, R. (2002b). Immunohistochemical detection of the neuronal connexin36 in the mouse central nervous system in comparison to connexin36-deficient tissues. *Histochem. Cell. Biol.* **117**, 461–471.

Mesnil, M., Piccoli, C., Tiraby, G., Willecke, K., and Yamasaki, H. (1996). Bystander killing of cancer cells by herpes simplex virus thymidine kinase gene is mediated by connexins. *Proc. Natl. Acad. Sci. USA* **93**, 1831–1835.

Mills, S. L., and Massey, S. C. (2000). A series of biotinylated tracers distinguishes three types of gap junction in retina. *J. Neurosci.* **20**, 8629–8636.

Moreno, A. P., Rook, M. B., Fishman, G. I., and Spray, D. C. (1994a). Gap junction channels: Distinct voltage-sensitive and -insensitive conductance states. *Biophys. J.* **67**, 113–119.

Moreno, A. P., Saez, J. C., Fishman, G. I., and Spray, D. C. (1994b). Human connexin43 gap junction channels: Regulation of unitary conductances by phosphorylation. *Circ. Res.* **74**, 1050–1057.

Mugnaini, E. (1986). Cell junctions of astrocytes, ependyma, and related cells in the mammalian central nervous system, with emphasis, on the hypothesis of a generalized functional syncytium of supporting cells. *In* "Astrocytes" (S. Fedoroff and A. Vernadakis, Eds.), Vol.I, pp. 329–371. Academic Press, New York.

Nagy, J. I., and Dermietzel, R. (2000). Gap junctions and connexin in the mammalian central nervous system. *Adv. Mol. Cell Biol.* **30**, 323–396.

Nagy, J. I., Li, X., Rempel, H., Stelmack, G., Patel, D., Staines, W. A., Yasumura, T., and Rash, J. E. (2001). Connexin26 in adult rodent central nervous system: Demonstration at astrocytic gap junctions and colocalization with connexin30 and connexin43. *J. Comp. Neurol.* **441**, 302–323.

Nagy, J. I., Patel, D., Ochalski, P. A., and Stelmack, G. L. (1999). Connexin30 in rodent, cat and human brain: Selective expression in gray matter astrocytes, co-localization with connexin43 at gap junctions and late developmental appearance. *Neuroscience* **88**, 447–468.

Naus, C. C. G., Becherger, J. F., and Bond, S. L. (1996). Effect of gap junctional communication on glioma cell function. *In* "Gap Junctions in the Nervous System" (D. C. Spray and R. Dermietzel, Eds.), pp. 193–202. R. G. Landes, Dallas, TX.

Naus, C. C., Bechberger, J. F., and Paul, D. L. (1991). Gap junction gene expression in human seizure disorder. *Exp. Neurol.* **111**, 198–203.

Nedergaard, M., Cooper, A. J. L., and Goldman, S. A. (1995). Gap junctions are required for the propagation of spreading depression. *J. Neurobiol.* **28**, 433–444.

Oguro, K., Jover, T., Tanaka, H., Lin, Y., Kojima, T., Oguro, N., Grooms, S. Y., Bennett, M. V., and Zukin, R. S. (2001). Global ischemia-induced increases in the gap junctional proteins connexin 32 (Cx32) and Cx36 in hippocampus and enhanced vulnerability of C×32 knock-out mice. *J. Neurosci.* **21**, 7534–7542.

Oh, S., Ri, Y., Bennett, M. V., Trexler, E. B., Verselis, V. K., and Bargiello, T. A. (1997). Changes in permeability caused by connexin 32 mutations underlie X-linked Charcot–Marie–Tooth disease. *Neuron* **19**, 927–938.

Omori, Y., Mesnil, M., and Yamasaki, H. (1996). Connexin32 mutations from X-linked Charcot–Marie–Tooth disease patients: Functional defects and dominant negative effects. *Mol. Biol. Cell* **7**, 907–916.

Orkand, R., Dietzel, I., and Coles, J. (1984). Light-induced changes in extracellular volume in the retina of the drone, *Apis mellifera. Neurosci. Lett.* **45**, 273–278.

Ouvier, R. (1996). Correlation between the histopathologic, genotypic and phenotypic features of hereditary peripheral neuropathies in childhood. *J. Child. Neurol.* **11**, 133–146.

Parpura, V., Basarsky, T. A., Liu, F., Jeftinijas, K., Jeftinija, S., and Haydon, P. G. (1994). Glutamate-mediated astrocyte neuron signaling. *Nature* **369**, 744–747.

Peinado, A., Yuste, R., and Katz, L. C. (1993). Extensive dye coupling between rat neocortical neurons during the period of circuit formation. *Neuron* **10**, 103–114.

Penn, A. A., Wong, R. O., and Shatz, C. J. (1994). Neuronal coupling in the developing mammalian retina. *J. Neurosci.* **14**, 3805–3815.

Pereda, A. E., Bell, T. D., Chang, B. H., Czernik, A. J., Nairn, A. C., Soderling, T. R., and Faber, D. S. (1998). Ca^{2+}/calmodulin-dependent kinase II mediates simultaneous enhancement of gap-junctional conductance and glutamatergic transmission. *Proc. Natl. Acad. Sci. USA* **95**, 13272–13277.

Pereda, A. E., Bell, T. D., and Faber, D. S. (1995). Retrograde synaptic communication via gap junctions coupling auditory afferents to the Mauthner cell. *J. Neurosci.* **15**, 5943–5955.

Pereda, A. E., and Faber, D. S. (1996). Activity-dependent short-term enhancement of intercellular coupling. *J. Neurosci.* **16**, 983–992.

Perez-Velazquez, J. L., Valiante, T. A., and Carlen, P. L. (1994). Modulation of gap junctional mechanisms during calcium-free induced field burst activity: A possible role for electrotonic coupling in epileptogenesis. *J Neurosci.* **14**, 4308–4317.

Rabionet, R., Lopez-Bigas, N., Lourdes Arbones, M., and Estivill, X. (2002). Connexin mutations in hearing loss, dermatological and neurological disorders. *Trends Mol. Med.* **5**, 205–210.

Rall, W., Shepherd, G. M., Reese, T. S., and Brightman, M. W. (1966). Dendrodendritic synaptic pathway for inhibition in the olfactory bulb. *Exp. Neurol.* **14**, 44–56.

Ransom, B. R. (1996). Do glial gap junctions play a role in extracellular homeostasis? *In* "Gap Jucntions in the Nervous System" (D. C. Spray and R. Dermietzel, Eds.), pp. 161–173. R. G. Landes, Dallas, TX.

Rapoport, S. I. (1979). Role of cerebrovascular permeability, brain compliance and brain hydraulic conductivity in vasogenic brain edema. *In* "Neural Trauma" (A. J. Popp, R. S. Bourke, L. R. Nelson, and H. K. Kimelberg, Eds.), pp. 51–62. Raven Press, New York.

Rash, J. E., Duffy, H. S., Dudek, F. E., Bilhartz, B. L., Whalen, L. R., and Yasumura, T. (1997). Grid-mapped freeze-fracture analysis of gap junctions in gray and white matter of adult rat central nervous system, with evidence for a "pan-glial syncytium" that is not coupled to neurons. *J. Comp. Neurol.* **388**, 1–28.

Rash, J. E., Staines, W. A., Yasumura, T., Patel, D., Furman, C. S., Stelmack, G. L., and Nagy, J. I. (2000). Immunogold evidence that neuronal gap junctions in adult rat brain and spinal cord contain connexin-36 but not connexin-32 or connexin-43. *Proc Natl Acad Sci USA.* **97(13)**, 7573–8.

Rawanduzy, A., Hansen, A., Hansen, T. W., and Nedergaard, M. E. (1997). Effective reduction of infarct volume by gap junction blockade in a rodent model of stroke. *J. Neurosurg.* **87**, 916–920.

Reaume, A. G., deSousa, P. A., Kulkarni, S., Langille, B. L., Zhu, D., Davies, T. C., Juneja, S. C., Kidder, G. M., and Rossant, J. (1995). Cardiac malformation in neonatal mice lacking connexin43. *Science* **267**, 1831–1834.

Revel, J. P., and Karnovsky, M. J. (1967). Hexagonal array of subunits in intercellular junctions of the mouse heart and liver. *J. Cell Biol.* **33**, C7–C12.

Robertson, J. D. (1961). Ultrastructure of excitable membranes and the crayfish median-giant synapse. *Ann. NY Acad Sci.* **94**, 339–389.

Robertson, J. D. (1963). The occurrence of a subunit pattern in the unit membranes of club endings in Mauthner cell synapses in goldfish brains. *J. Cell Biol.* **19**, 201–221.

Rozental, R., and Spray, D. C. (1996). Temporal expression of gap junctions during neuronal ontogeny. *In* "Gap Junctions in the Nervous System" (D. C. Spray and R. Dermietzel, Eds.), pp. 261–273. R. G. Landes (Medical Intelligence Publishing Co.), Dallas, TX.

Rozental, R., Srinivas, M., Gokhan, S., Urban, M., Dermietzel, R., Kessler, J. A., Spray, D. C., and Mehler, M. F. (2000). Temporal expression of neuronal connexins during hippocampal ontogeny. *Brain Res. Brain Res. Rev.* **32**, 57–71.

Saez, J. C., Berthoud, V. M., Moreno, A. P., and Spray, D. C. (1993). Gap junctions: Multiplicity of controls in differentiated and undifferentiated cells and possible functional implications. *Adv. Second Messenger Phosphoprotein Res.* **27**, 163–198

Saez, J. C., Connor, J. A., Spray, D. C., and Bennett, M. V. (1989). Hepatocyte gap junctions are permeable to the second messenger, inositol, 1,4,5-trisphosphate, and to calcium ions. *Proc. Natl. Acad. Sci. USA* **86**, 2708–2712.

Saffitz, J. E. (1997). Gap junctions: Functional effects of molecular structure and tissue distribution. *Adv. Exp. Med. Biol.* **430**, 291–301.

Sanderson, M. J. (1995). Intercellular calcium waves mediated by inositol trisphosphate. *In* "Calcium Waves, Gradients and Oscillations" (G. R. Bock and K. Akrill, Eds.), pp. 175–194. Wiley, Chichester. UK,

Sanderson, M. J. (1996). Intercellular waves of communication. *News Physiol. Sci.* **11**, 262–269.

Scemes, E., Dermietzel R., and Spray, D. C. (1998). Calcium waves between astrocytes from Cx43 knockout mice. *Glia* **24**, 65–73.

Scemes, E., Suadicani, S. O., and Spray, D. C. (2000). Intercellular communication in spinal cord astrocytes: Fine tuning between gap junctions and P2 nucleotide receptors in calcium wave propagation. *J. Neurosci.* **20**, 1435–1445.

Scherer, S. S., Xu, Y. T., Fischbeck, K., Willecke, K., and Bone, L. (1998). Connexin32-null mice develop demyelinating peripheral neuropathy. *Glia* **24**, 8–20.

Schmitz, D., Schuchmann, S., Fisahn, A., Draguhn, A., Buhl, E. H., Petrasch-Parwez, E., Dermietzel, R., Heinemann, U., and Traub, R. D. (2001). Axo-axonal coupling, a novel mechanism for ultrafast neuronal communication. *Neuron* **31**, 831–840.

Shatz, C. J. (1994). Viktor Hamburger Award review: Role for spontaneous neural activity in the patterning of connections between retina and LGN during visual system development. *J. Dev. Neurosci.* **12**, 531–546.

Shinoura, N., Chen, Ll, Wani, M. A., Kim, Y. G., Larson, J. J., Warnick, R. E., Simon, M., Menon, A. G., Bi, W. L., and Stambrook, P. J. (1996). Protein and messenger RNA expression of connexin43 in astrocytomas: Implications in brain tumor gene therapy. *J. Neurosurg.* **84**, 839–845.

Sotelo, C., Gotow, T., and Wassef, M. (1968). Localization of glutamic-acid-decarboxylase-immunoreactive axon terminals in the inferior olive of the rat, with special emphasis on anatomical relations between GABAergic synapses and dendrodendritic gap junctions. *J. Comp. Neurol.* **252**, 32–50.

Sotelo, C., and Palay, S. L. (1970). The fine structure of the later vestibular nucleus in the rat. II. Synaptic organization. *Brain Res.* **18**, 93–115.

Spira, M. E., and Bennett, M. V. L. (1972). Synaptic control of electrotonic coupling between neurons. *Brain Res.* **37**, 294–300.

Spira, M. E., Spray, D. C., and Bennett, M. V. L. (1976). Electrotonic coupling: Effective sign reversal by inhibitory neurons. *Science* **194**, 1065–1067.

Spira, M. E., Spray, D. C., and Bennett, M. V. (1980). Synaptic organization of expansion motoneurons of *Navanax inermis*. *Brain Res.* **195**, 241–269.

Spray, D. C. (1994). Physiological and pharmacological regulation of gap junction channels. *In* ed. "Molecular Mechanisms of Epithelial Cell Junctions: From Development to Disease" (S. Citi, EO pp. 195–215. R. G. Landes/Medical Intelligence Austin, TX.

Spray, D. C., and Dermietzel, R. (1995). X-linked dominant Charcot–Marie–Tooth disease and their potential gap-junction diseases of the nervous system. *Trends Neurosci.* **18**, 256–262.

Spray, D. C., Duffy, H. S., and Scemes, E. (1999). Gap junctions in glia: Types, roles and plasticity. *Adv. Exp. Med. Biol.* 468:339–359.

Spray, D. C., Harris, A. L., and Bennett, M. V. L. (1981). Equilibrium properties of a voltage-dependent junctional conductance. *J. Gen. Physiol.* **77**, 77–93.

Spray, D. C., Moreno, A. P., Kessler, J. A., and Dermietzel, R. (1991). Characterization of gap junctions between cultured leptomeningeal cells. *Brain Res.* **568**, 1–14.

Spray, D. C., Rook, M., Moreno, A. P., Saez, J. C., Christ, G., Campos de Carvalho, A. C., and Fishman, G. I. (1994). Cardiovascular gap junctions: Gating properties, function and dysfunction. *In* "Ion Channels in the Cardiovascular System: Function and Dysfunction" (P. M. Spooner, A. M. Brown, W. A. Catterall, G. J. Kaczorowski, and H. C. Strauss Eds.), p. 185–217. Futura, Mt. Kisco, NY.

Spray, D. C., Rozental, R., and Srinivas, M. (2002). Prospects for rational development of pharmacological gap junction channel blockers. *Curr. Drug Targets* **3**, 455–464.

Spray, D. C., and Scemes, E. (1998). Effects of pH (and Ca) on gap junction channels. *In* "pH and Brain Function" (K. Kaila and B. R. Ransom, Eds.). Academic Press, New York.

Spray, D. C., Spira, M. E., and Bennett, M. V. (1980). Peripheral fields and branching patterns of buccal mechanosensory neurons in the opisthobranch mollusc, *Navanax inermis. Brain Res.* **182**, 253–270.

Spray, D. C., Suadicani, S., Srinivas, M., Gutstein, D. E., and Fishman, G. I. (2001). Gap junctions in the cardiovascular system. *In* "Handbook of Physiology," Sect. 2: "The Cardiovascular System," Vol 1: "The Heart: Gap Junctions in the Cardiovascular System," pp. 169–212. Oxford Univ. Press, New York.

Srinivas, M., Hopperstad, M. G., and Spray, D. C. (2001). Quinine blocks specific gap junction channel subtypes. *Proc. Natl. Acad. Sci. USA* **98**, 10942–10947.

Srinivas, M., Rozental, R., Kojima, T., Dermietzel, R., Mehler, M., Condorelli, D. F., Kessler, J. A., and Spray, D. C. (1999). Functional properties of channels formed by the neuronal gap junction protein connexin36. *J. Neurosci.* **19**, 9848–9855.

Stewart, W. W. (1981). Lucifer dyes: Highly fluorescent dyes for biological tracing. *Nature* **292**, 17–21.

Streit, D. S., Ferreira Filho, C. R., and Martins-Ferreira, H. (1995). Spreading depression in isolated spinal cord. *J. Neurophysiol.* **74**, 888–890.

Swenson, K. I., Piwnica-Worms, H., McNamee, H., and Paul, D. L. (1990). Tyrosine phosphorylation of the gap junction protein connexin43 is required for the pp 60 v-src-induced inhibition of communication. *Cell Regul.* **1**, 989–1002.

Taylor, C. P., and Dudek, F. E. (1982). Synchronous neural afterdischarges in rat hippocampal slices without active chemical synapses. *Science* **218**, 810–812.

Temme, A., Buchmann, A., Gabriel, H. D., Nelles, E., Schwarz, M., and Willecke, K. (1997). High incidence of spontaneous and chemically induced liver tumors in mice deficient for connexin32. *Curr. Biol.* **7**, 713–716.

Theriault, E., Frankenstein, U. N., Hertzberg, E. L., and Nagy, J. I. (1997). Connexin43 and astrocytic gap junctions in the rat spinal cord after acute compression injury. *J. Comp. Neurol.* **382**, 199–214.

Traub, R. D., Draguhn, A., Whittington, M. A., Baldeweg, T., Bibbig, A., Buhl, E. H., and Schmitz, D. (2002). Axonal gap junctions between principal neurons: A novel source of network oscillation, and perhaps epileptogenesis. *Rev. Neurosci.* **13**, 1–30.

Teubner, B., Degen, J., Sohl, G., Guldenagel, M., Bukauskas, F. F., Trexler, E. B., Verselis, V. K., DeZeeuw, C. I., Lee, C. G., Kozak, C. A., Petrasch-Parwez, E., Dermietzel, R., and Willecke K. (2000). Functional expression fo the murine connexin36 gene coding for a neuron-specific gap junctional protein. *J. Membr. Biol.* **176**, 249–262.

Vaney, D. I. (1996). Cell coupling in the retina. *In* "Gap Junctions in the Nervous System" (D. C. Spray and R. Dermietzel, Eds.), pp. 79–102. R. G. Landes, Georgeotwn, TX.

Van Harreveld, A. (1966). "Brain Tissue Electrolytes," pp. 95–126. Butterworth, London.

Van Harreveld, A., and Khattab, F. (1967). Changes in cortical extracellular space during spreading depression investigated with the electron microscope. *J. Neurophysiol.* **30**, 911–929.

VanSlyke, J. K., Deschenes, S. M., and Musil, L. S. (2000). Intracellular transport, assembly, and degradation of wild-type and disease-linked mutant gap junction proteins. *Mol. Biol. Cell* **11**, 1933–1946.

Veenstra, R. D., Wang, H. Z., Beblo, D. A., Chilton, M. G., Harris, A. L., Beyer, E. C., and Brink, P. R. (1995). Selectivity of connexin-specific gap junctions does not correlate with channel conductance. *Circ. Res.* **77**, 1156–1165.

Veenstra, R. D., Wang, H. Z., Beyer, E. C., and Brink, P. R. (1994). Selective dye and ionic permeability of gap junction channels formed by connexin45. *Circ. Res.* **75**, 483–490.

Vrionis, F. D., Wu, J. K., Qi, P., Waltzman, M., Cherington, V., and Spray, D. C. (1997). The bystander effect exerted by tumor cells expressing the herpes simplex virus thymidine kinase (HSVtk) gene is dependent on connexin expression and cell communication. *Gene Ther.* **6**, 577–585.

Walton, K. D., and Navarrete, R. (1991). Postnatal changes in motoneurone electrotonic coupling studied in the in vitro rat lumbar spinal cord. *J. Physiol.* (*London*) **433**, 283–305.

Weiler, R. (1996). The modulation of gap junction premeability in the retina. *In* "Gap Junctions in the Nervous System" (D. C. Spray and R. Dermietzel, Eds.), pp. 103–122. R. G. Landes, Georgetown, TX.

Welsh, J. P., and Llinas, R. (1997). Some organizing princinples for the control of movement based on olivocerebellar physiology. *Prog. Brain. Res.* **114**, 449–461.

White, T. W., Paul, D. L., Goodenough, D. A., and Bruzzone, R. (1995). Functional analysis of selective interactions among rodent connexins. *Mol. Biol. Cell* **6**, 459–470.

Willecke, K., Eiberger, J., Degen, J., Eckardt, D., Romualdi, A., Guldenagel, M., Deutsch, U., and Sohl, G. (2002). Structural and functional diversity of connexin genes in the mouse and human genome. *Biol. Chem.* **383**, 725–737.

Zhou, X. W., Pfahnl, A., Werner, R., Hudder, A., Llanes, A., Luebke, A., and Dahl, G. 1997. Identification of a pore lining segment in gap junction hemichannels. *Biophys. J.* **72**, 1946–1953.

CHAPTER

16

Postsynaptic Potentials and Synaptic Integration

John H. Byrne

The study of synaptic transmission in the central nervous system provides an opportunity to learn more about the diversity and richness of mechanisms underlying this process and to learn the ways in which some of the fundamental signaling properties of the nervous system, such as action potentials and synaptic potentials, work together to process information and generate behavior.

Postsynaptic potentials (PSPs) in the CNS can be divided into two broad classes on the basis of mechanisms and, generally, duration of these potentials. One class is based on the *direct* binding of a transmitter molecule(s) with a receptor–channel complex; these receptors are *ionotropic*. The structure of these receptors is discussed in detail in Chapter 11. The resulting PSPs are generally short-lasting and hence are sometimes called fast PSPs; they have also been referred to as "classic" because they were the first synaptic potentials to be recorded in the CNS (Eccles, 1964; Spencer, 1977). The duration of a typical fast PSP is about 20 ms.

The other class of PSPs is based on the *indirect* effect of transmitter molecule(s) binding with a receptor. The receptors that produce these PSPs are *metabotropic*. As discussed in Chapter 11, the receptors activate G proteins and are therefore also called G protein coupled receptors (GPCR). They affect the channel either directly or through additional steps in which the level of a second messenger is altered. The responses mediated by GPCRs can be long-lasting and are therefore called slow PSPs. The mechanisms for fast PSPs mediated by ionotropic receptors are considered first.

IONOTROPIC RECEPTORS: MEDIATORS OF FAST EXCITATORY AND INHIBITORY SYNAPTIC POTENTIALS

The Stretch Reflex Is Useful to Examine the Properties and Functional Consequences of Ionotropic PSPs

The stretch reflex, one of the simpler behaviors mediated by the central nervous system, is a useful example with which to examine the properties and functional consequences of ionotropic PSPs. The tap of a neurologist's hammer to a ligament elicits a reflex extension of the leg, as illustrated in Fig. 16.1. The brief stretch of the ligament is transmitted to the extensor muscle and is detected by specific receptors in the muscle and ligament. Action potentials initiated in the stretch receptors are propagated to the spinal cord by afferent fibers. The receptors are specialized regions of sensory neurons with somata located in the dorsal root ganglia just outside the spinal column. The axons of the afferents enter the spinal cord and make excitatory synaptic connections with at least two types of postsynaptic neurons. First, a synaptic connection is made to the extensor motor neuron. As the result of its synaptic activation, the motor neuron fires action potentials that propagate out of the spinal cord and ultimately invade the terminal regions of the motor axon at neuromuscular junctions. There, acetylcholine (ACh) is released, nicotinic ACh receptors are activated, an end-plate potential (EPP) is produced, an action potential is initiated in the muscle cell, and

Copyright 2004, Elsevier Science (USA).
All rights reserved.

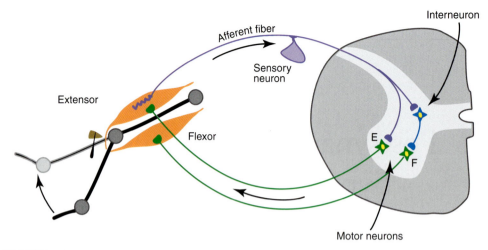

FIGURE 16.1 Features of the vertebrate stretch reflex. Stretch of an extensor muscle leads to the initiation of action potentials in the afferent terminals of specialized stretch receptors. The action potentials propagate to the spinal cord through afferent fibers (sensory neurons). The afferents make excitatory connections with extensor motor neurons (E). Action potentials initiated in the extensor motor neuron propagate to the periphery and lead to the activation and subsequent contraction of the extensor muscle. The afferent fibers also activate interneurons that inhibit the flexor motor neurons (F).

the muscle cell is contracted, producing the reflex extension of the leg. Second, a synaptic connection is made to another group of neurons called interneurons (nerve cells interposed between one type of neuron and another). The particular interneurons activated by the afferents are inhibitory interneurons, because activation of these interneurons leads to the release of a chemical transmitter substance that inhibits the flexor motor neuron. This inhibition tends to prevent an uncoordinated (improper) movement (i.e., flexion) from occurring. The reflex system illustrated in Fig. 16.1 is also known as the monosynaptic stretch reflex because this reflex is mediated by a single ("mono") excitatory synapse in the central nervous system.

Figure 16.2 illustrates procedures that can be used to experimentally examine some of the components of synaptic transmission in the reflex pathway for the stretch reflex. Intracellular recordings are made from one of the sensory neurons, the extensor and flexor motor neurons, and an inhibitory interneuron. Normally, the sensory neuron is activated by stretch to the muscle, but this step can be bypassed by simply injecting a pulse of depolarizing current of sufficient magnitude into the sensory neuron to elicit an action potential. The action potential in the sensory neuron leads to a potential change in the motor neuron known as an excitatory postsynaptic potential (EPSP) (Fig. 16.2).

Mechanisms responsible for fast EPSPs mediated by ionotropic receptors in the CNS are fairly well known. Moreover, the ionic mechanisms for EPSPs in the CNS are essentially identical to the ionic mech-

anisms at the skeletal neuromuscular junction. Specifically, the transmitter substance released from the presynaptic terminal (Chapters 8–10) diffuses across the synaptic cleft, binds to specific receptor sites on the postsynaptic membrane (Chapter 11), and leads to a simultaneous increase in permeability to Na^+ and K^+, which makes the membrane potential move *toward* a value of about 0 mV. However, the processes of synaptic transmission at the sensory neuron–motor neuron synapse and the motor neuron–skeletal muscle synapse differ in two fundamental ways: (1) in the transmitter used and (2) in the amplitude of the PSP. The transmitter substance at the neuromuscular junction is ACh, whereas that released by the sensory neurons is an amino acid, probably glutamate (see Chapter 9). Indeed, glutamate is the most common transmitter that mediates excitatory actions in the CNS. The amplitude of the postsynaptic potential at the neuromuscular junction is about 50 mV; consequently, each PSP depolarizes the postsynaptic cell beyond threshold, so there is a one-to-one relationship between an action potential in the spinal motor neuron and an action potential in the skeletal muscle cell. Indeed, the EPP must depolarize the muscle cell by only about 30 mV to initiate an action potential, allowing a safety factor of about 20 mV. In contrast, the EPSP in a spinal motor neuron produced by an action potential in an afferent fiber has an amplitude of only about 1 mV. The mechanisms by which these small PSPs can trigger an action potential in the postsynaptic neuron are discussed in a later section of this chapter and in Chapter 17.

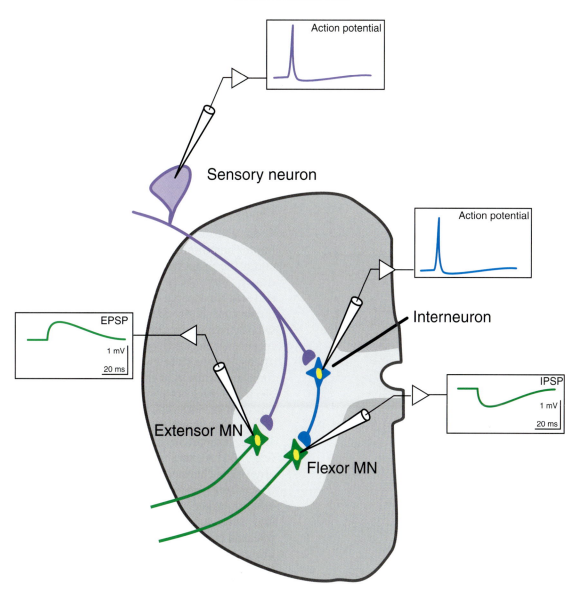

FIGURE 16.2 Excitatory (EPSP) and inhibitory (IPSP) postsynaptic potentials in spinal motor neurons. Idealized intracellular recordings from a sensory neuron, interneuron, and extensor and flexor motor neurons (MNs). An action potential in the sensory neuron produces a depolarizing response (an EPSP) in the extensor motor neuron. An action potential in the interneuron produces a hyperpolarizing response (an IPSP) in the flexor motor neuron.

Macroscopic Properties of PSPs Are Determined by the Nature of the Gating and Ion Permeation Properties of Single Channels

Patch-Clamp Techniques

Patch-clamp techniques (Hamill *et al.*, 1981) with which current flowing through single isolated receptors can be measured directly can be sources of insight into both the ionic mechanisms and the molecular properties of PSPs mediated by ionotropic receptors. This approach was pioneered by Erwin Neher and Bert Sakmann in the 1970s and led to their being awarded the Nobel Prize in Physiology or Medicine in 1991.

Figure 16.3 illustrates an idealized experimental arrangement of an "outside-out" patch recording of a single ionotropic receptor. The patch pipette contains a solution with an ionic composition similar to that of the cytoplasm, whereas the solution exposed to the outer surface of the membrane has a composition similar to that of normal extracellular fluid. The electrical potential across the patch, and hence the transmembrane potential (V_m), is controlled by the patch-clamp amplifier. The extracellular (outside)

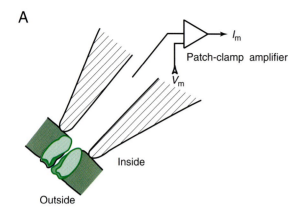

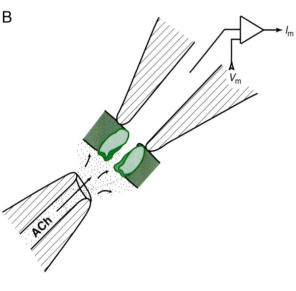

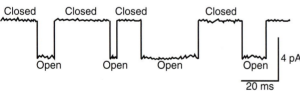

FIGURE 16.3 Single-channel recording of ionotropic receptors and their properties. (A) Experimental arrangement for studying properties of ionotropic receptors. (B) Idealized single-channel currents in response to application of ACh.

fluid is considered "ground." Transmitter can be delivered by applying pressure to a miniature pipette filled with an agonist (in this case, ACh), and the current (I_m) flowing across the patch of membrane is measured by the patch-clamp amplifier (Fig. 16.3B). Pressure in the pipette that contains ACh can be continuous, allowing a constant stream of ACh to contact the membrane, or can be applied as a short pulse to

allow a precisely timed and discrete amount of ACh to contact the membrane. The types of recordings obtained from such an experiment are illustrated in the traces in Fig. 16.3. In the absence of ACh, no current flows through the channel (Fig. 16.3A). When ACh is continuously applied, current flows across the membrane (through the channel), but remarkably, the current does not flow continuously; instead, small steplike changes in current are observed (Fig. 16.3B). These changes represent the probabilistic (random) opening and closing of the channel.

Channel Openings and Closings

As a result of the type of patch-recording techniques heretofore described, three general conclusions about the properties of ligand-gated channels can be drawn. First, ACh, as well as other transmitters that activate ionotropic receptors, causes the opening of individual ionic channels (for a channel to open, usually two molecules of transmitter must bind to the receptor). Second, when a ligand-gated channel opens, it does so in an all-or-none fashion. Increasing the concentration of transmitter in the ejection microelectrode does not increase the permeability (conductance) of the channel; it increases its probability (P) of being open. Third, the ionic current flowing through a single channel in its open state is extremely small (e.g., 10^{-12} A); as a result, the current flowing through any single channel makes only a small contribution to the normal postsynaptic potential. Physiologically, when a larger region of the postsynaptic membrane, and thus more than one channel, is exposed to released transmitter, the net conductance of the membrane increases owing to the increased probability that a larger population of channels will be open at the same time. The normal PSP, measured with standard intracellular recording techniques (e.g., Fig. 16.2), is then proportional to the sum of the currents that flow through these many individual open channels. The properties of voltage-sensitive channels (see Chapter 5) are similar in that they, too, open in all-or-none fashion, and, as a result, the net effect on the cell is due to the summation of the currents flowing through many individual open ion channels. The two types of channels differ, however, in that one is opened by a chemical agent, whereas the other is opened by changes in membrane potential.

Statistical Analysis of Channel Gating and the Kinetics of the PSP

The experiment illustrated in Fig. 16.3B was performed with continuous exposure to ACh. Under such conditions, the channels open and close repeatedly. When ACh is applied by a brief pressure pulse to more accurately mimic the transient release from the presynaptic terminal, the transmitter diffuses away before it can cause a second opening of the channel. A set of data similar to that shown in Fig. 16.4A would be obtained if an ensemble of these openings were collected and aligned with the start of each opening. Each individual trace represents the response to each successive "puff" of ACh. Note that, among the responses, the duration of the opening of the channel varies considerably—from very short (less than 1 ms) to more than 5 ms. Moreover, channel openings are independent events. The duration of any one channel opening does not have any relationship to the duration of a previous opening. Figure 16.4B illustrates a plot that is obtained by adding 1000 of these individual responses. Such an addition roughly

simulates the conditions under which transmitter released from a presynaptic terminal leads to the near-simultaneous activation of many single channels in the postsynaptic membrane. (Note that the addition of 1000 channels would produce a synaptic current equal to about 4 nA.) This simulation is valid given the assumption that the statistical properties of a single channel over time are the same as the statistical properties of the ensemble at one instant of time (i.e., an ergotic process). The ensemble average can be fit with an exponential function with a decay time constant of 2.7 ms. An additional observation (explored in the next section) is that the value of the time constant is equal to the mean duration of the channel openings. The curve in Fig. 16.4B is an indication of the probability that a channel will remain open for various times, with a high probability for short times and a low probability for long times.

The ensemble average of the single-channel currents (Fig. 16.4B) roughly accounts for the time course of the EPSP. However, note that the time course of the aggregate synaptic *current* can be somewhat faster than that of the excitatory postsynaptic *potential* in Fig. 16.2. This difference is due to the charging of the membrane capacitance by a rapidly changing synaptic current. Because the single-channel currents were recorded with the membrane voltage-clamped, the capacitive current [$I_c = C_{m^*} (dV/dt)$] is zero. In contrast, for the recording of the postsynaptic potential in Fig. 16.2, the membrane was not voltage-clamped, and therefore as the voltage changes (dV/dt), some of the synaptic current charges the membrane capacitance (see Eq. 16.17).

Analytical expressions that describe the shape of the ensemble average of the open lifetimes and the mean open lifetime can be derived by considering that single-channel opening and closing is a stochastic process (Colquhoun and Hawkes, 1977, 1981, 1982; Colquhoun and Sakmann, 1981; Johnston and Wu, 1995; Sakmann, 1992). Relations are formalized to describe the likelihood (probability) of a channel being in a certain state. Consider the following two-state reaction scheme:

$$C \underset{\alpha}{\overset{\beta}{\rightleftharpoons}} O$$

In this scheme, α represents the rate constant for channel closing, and β, the rate constant for channel opening. The scheme can be simplified further if we consider a case in which the channel has been opened by the agonist and the agonist is removed instantaneously. A channel so opened (at time 0) will then close after a certain random time (Fig. 16.4). We first formulate an analytical expression that describes the probability that the channel is open (o) at some time

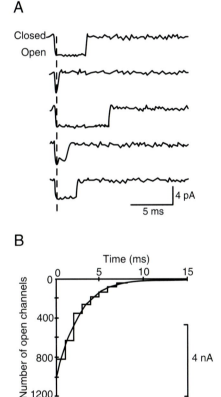

FIGURE 16.4 Determination of the shape of the postsynaptic response from the single-channel currents. (A) Each trace represents the response of a single channel to a repetitively applied puff of transmitter. The traces are aligned with the beginning of the channel opening (dashed line). (B) Addition of 1000 of the individual responses. If a current equal to 4 pA were generated by the opening of a single channel, then a 4-nA current would be generated by 1000 channels opening at the same time. The data are fitted with an exponential function having a time constant equal to $1/\alpha$ [see Eq. (9)]. Reprinted with permission from Sakmann (1992). Copyright 1992 American Association for the Advancement of Science.

(i.e., time t), given that it was open at time 0. This expression is referred to as $P_{o/o}(t)$. To formulate an analytical expression for $P_{o/o}(t)$, first consider the probability that a channel will be *closed* (c) at time $t + \Delta t$, given that it was open at time t, in the limit that Δt is so small that we can ignore multiple events such as an opening followed by a closing. This term, which is referred to as $P_{c/o}(\Delta t)$, will equal $\alpha \Delta t$ (the product of the reverse rate constant and the time interval). Therefore, the probability [$P_{c/o}(\Delta t)$] that a channel will be *open* at time $t + \Delta t$, given that it was open at time t, will equal $1 - \alpha \Delta t$ (i.e., 1 minus the probability that it will be closed at $t + \Delta t$). Finally, the probability that the channel will be open at time t *and* will be open at time $t + \Delta t$ can be described by

$$P_{o/o}(t + \Delta t) = P_{o/o}(t) \, P_{o/o}(\Delta t). \tag{16.1}$$

By substituting and factoring, we obtain

$$P_{o/o}(t + \Delta t) = P_{o/o}(t) \, (1 - \alpha \Delta t), \tag{16.2}$$

$$P_{o/o}(t + \Delta t) = P_{o/o}(t) - \alpha \Delta t P_{o/o}(t), \tag{16.3}$$

$$(P_{o/o}(t + \Delta t) - P_{o/o}(t))/\Delta t = -\alpha P_{o/o}(t). \tag{16.4}$$

Note that as $\Delta t \to 0$, the left-hand term defines the derivative. Thus,

$$\frac{dP_{o/o}(t)}{dt} = -\alpha P_{o/o}(t). \tag{16.5}$$

This differential equation is satisfied by an exponential function. Consequently,

$$P_{o/o}(t) = e^{-\alpha t}. \tag{16.6}$$

We can now determine the probability [$P_{c/o}(t)$] that the channel is closed at time t, given that it was open at time 0. This will simply be $1 - P_{o/o}(t)$. Therefore,

$$P_{c/o}(t) = 1 - e^{-\alpha t}. \tag{16.7}$$

The function $P_{c/o}(t)$ represents the cumulative distribution function (or simply the distribution function) for the channel (i.e., the probability that a channel will be closed by time t). This quantity is called the cumulative distribution because it is equal to the sum, or integral, over the probabilities that the channel closes at each of the preceding times. Distribution functions satisfy the relationship

$$0 \le P(t) \le 1. \tag{16.8}$$

Note that, for Eq. (16.7), at $t = 0$ the probability of a channel being closed is 0 and, at $t = \infty$, the probability

of a channel being closed is 1. To obtain an equation for the probability that a channel closing occurs in exactly some period $t + \Delta t$ as Δt approaches 0, we need to determine the probability density function [$p(t)$], which is defined as the first derivative of the cumulative distribution function (Papoulis, 1965). Thus, the probability density function is

$$p(t) = \alpha e^{-\alpha t}. \tag{16.9}$$

Note that the distribution of open lifetimes illustrated in Fig. 16.4B corresponds well to that predicted by Eq. 16.9. With an analytical expression for the probability density function in hand, we can now determine another important property of channels—the mean open lifetime. The mean open lifetime can be obtained by taking the average of the probability density function (i.e., the expected value). Operationally, we multiply t and $p(t)$ and integrate between time 0 and ∞. Thus,

$$\text{mean open time} = \int_0^\infty t \, \alpha \, e^{-\alpha t} dt = \frac{1}{\alpha}. \tag{16.10}$$

Note that the mean open time is the time constant of the cumulative distribution function (Eq. 16.7) and the probability density function (Eq. 16.9) of the channel.

Gating Properties of Ligand-Gated Channels

Although statistical analysis can be a valuable source of insight into the statistical nature of the gating process and the molecular determinants of the macroscopic postsynaptic potential, the description in the preceding section is a simplification of the actual processes. Specifically, a more complete description must include the kinetics of receptor binding and unbinding and the determinants of the channel opening, as well as the fact that channels display rapid transitions between open and closed states during a single agonist receptor occupancy. Thus, the open states illustrated in Figs. 16.3B and 16.4A represent the period of a *burst* of extremely rapid openings and closings. If the bursts of rapid channel openings and closings are thought of, and behave functionally, as a single continuous channel closure, the formalism developed in the preceding section is a reasonable approximation for many ligand-gated channels. Nevertheless, a more complex reaction scheme is necessary to quantitatively explain the available data. Such a scheme would include the states

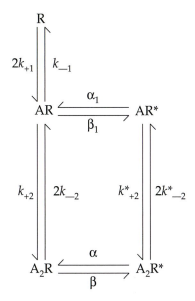

where R represents the receptor, A the agonist, and the α's, β's, and k's the forward and reverse rate constants for the various reactions. A_2R^* represents a channel opened as a result of the binding of two agonist molecules. The asterisk indicates an open channel (Colquhoun and Sakmann, 1981; Sakmann, 1992). (Note that the lower part of the reaction scheme is equivalent to that developed earlier, i.e., $C \overset{\beta}{\underset{\alpha}{\rightleftharpoons}} O$.) With the use of probability theory, equations describing the transitions between the states can be determined. The approach is identical to that used in the simplified two-state scheme. However, the mathematics and analytical expressions are more complex because of the interactions among transitions and the multiple dimensionality of the variables (Colquhoun and Hawkes, 1977, 1981, 1982). For some receptors, additional states must be represented. For example, as described in Chapter 11, some ligand-gated channels exhibit a process of desensitization in which continued exposure to a ligand results in channel closure.

The Null (Reversal) Potential and Slope of I–V Relationships

What ions are responsible for the synaptic current that produces the EPSP? Early studies of the ionic mechanisms underlying the EPSP at the skeletal neuromuscular junction yielded important information. Specifically, voltage-clamp and ion substitution experiments indicated that the binding of transmitter to receptors on the postsynaptic membrane led to a simultaneous increase in Na^+ and K^+ permeability that depolarized the cell toward a value of about 0 mV (Takeuchi and Takeuchi, 1960; Fatt and Katz,

1951). These findings are applicable to the EPSP in a spinal motor neuron produced by an action potential in an afferent fiber and have been confirmed and extended at the single-channel level.

Figure 16.5 illustrates the type of experiment in which the analysis of single-channel currents can be a source of insight into the ionic mechanisms of EPSPs. Transmitter is delivered to the patch while the membrane potential is systematically varied (Fig. 16.5A). In the upper trace, the patch potential is –40 mV. The ejection of transmitter produces a sequence of channel openings and closings, the amplitudes of which are constant for each opening (i.e., about 4 pA). Now consider the case in which the transmitter is applied when the potential across the patch is –20. mV. The frequency of the responses, as well as the mean open

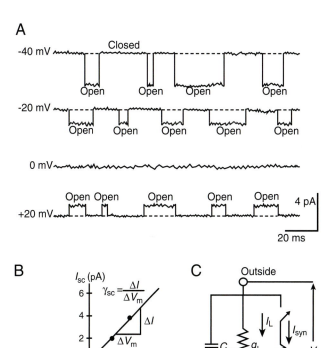

FIGURE 16.5 Voltage dependence of the current flowing through single channels. (A) Idealized recording of an ionotropic receptor in the continuous presence of agonist. (B) I–V relationship of the channel in (A). (C) Equivalent electrical circuit of a membrane containing that channel. Abbreviations: γ_{sc} single-channel conductance; I_L, leakage current; I_{sc}, single-channel current; g_L, leakage conductance; g_{syn} macroscopic synaptic conductance; E_L, leakage battery; E_r reversal potential.

lifetimes, is about the same as when the potential was at −40 mV, but now the amplitude of the single-channel currents is decreased uniformly. Even more interestingly, when the patch is artificially depolarized to a value of about 0 mV, an identical puff of transmitter produces no current in the patch. If the patch potential is depolarized to a value of about +20 mV and the puff again delivered, openings are again observed, but the flow of current through the channel is reversed in sign; a series of upward deflections indicates outward single-channel currents. In summary, there are downward deflections (inward currents) when the membrane potential is at −40 mV, no deflections (currents) when the membrane is at 0 mV, and upward deflections (outward currents) when the membrane potential is moved to +20 mV.

The simple explanation for these results is that no matter what the membrane potential, the effect of the transmitter binding with receptors is to produce a permeability change that tends to move the membrane potential toward 0 mV. If the membrane potential is more negative than 0 mV, an inward current is recorded. If the membrane potential is more positive than 0 mV, an outward current is recorded. If the membrane potential is at 0 mV, there is no deflection because the membrane potential is already at 0 mV. At 0 mV, the channels are opening and closing as they always do in response to the agonist, but there is no net movement of ions through them. This 0-mV level is known as the synaptic null potential or reversal potential, because it is the potential at which the sign of the synaptic current reverses. The fact that the experimentally determined reversal potential equals the calculated value obtained by using the Goldman–Hodgkin–Katz (GHK) equation (Chapter 5) provides strong support for the theory that the EPSP is due to the opening of channels that have equal permeabilities to Na^+ and K^+. Ion substitution experiments also confirm this theory. Thus, when the concentration of Na^+ or K^+ in the extracellular fluid is altered, the value of the reversal potential shifts in a way predicted by the GHK equation. (Some other cations, such as Ca^{2+}, also permeate these channels, but their permeability is low compared with that of Na^+ and K^+.)

Different families of ionotropic receptors have different reversal potentials because each has unique ion selectivity. In addition, it should now be clear that the sign of the synaptic action (excitatory or inhibitory) depends on the value of the reversal potential relative to the resting potential. If the reversal potential of an ionotropic receptor channel is more positive than the resting potential, opening of that channel will lead to a depolarization (i.e., an EPSP). In contrast, if the reversal potential of an ionotropic receptor channel is more negative than the resting potential, opening of that channel will lead to a hyperpolarization, that is, an inhibitory postsynaptic potential (IPSP), which is the topic of a later section in this chapter.

Plotting the average peak value of the single-channel currents (I_{sc}) versus the membrane potential (transpatch potential) at which they are recorded (Fig. 16.5B) can be a source of quantitative insight into the properties of the ionotropic receptor channel. Note that the current–voltage (I–V) relationship is linear; it has a slope, the value of which is the single-channel conductance, and an intercept at 0 mV. This linear relationship can be put in the form of Ohm's law ($I = G * \Delta V$). Thus,

$$I_{sc} = \gamma_{sc} * (V_m - E_r), \tag{16.11}$$

where γ_{sc} is the single-channel conductance and E_r is the reversal potential (here, 0 mV).

Summation of Single-Channel Currents

We now know that the sign of a synaptic action can be predicted by knowledge of the relationship between the resting potential (V_m) and the reversal potential (E_r), but how can the precise amplitude be determined? The answer to this question lies in understanding the relationship between the synaptic conductance and the extra synaptic conductances. These interactions can be rather complex (see Chapter 17), but some initial understanding can be obtained by analyzing an electrical equivalent circuit for these two major conductance branches. We first need to move from a consideration of single-channel conductances and currents to that of macroscopic conductances and currents. The postsynaptic membrane contains thousands of any one type of ionotropic receptor, and each of these receptors could be activated by transmitter released by a single action potential in a presynaptic neuron. Because conductances in parallel add, the total conductance change produced by their simultaneous activation would be

$$g_{syn} = \gamma_{sc} * P * N, \tag{16.12}$$

where γ_{sc}, as before, is the single-channel conductance, P is the probability of opening of a single channel (controlled by the ligand), and N is the total number of ligand-gated channels in the postsynaptic membrane. The macroscopic postsynaptic current produced by the transmitter released by a single presynaptic action potential can then be described by

$$I_{syn} = g_{syn} * (V_m - E_r). \tag{16.13}$$

Equation 16.13 can be represented physically by a voltage (V_m) measured across a circuit consisting of a resistor (g_{syn}) in series with a battery (E_r). An equivalent circuit of a membrane containing such a conductance is illustrated in Fig. 16.5C. Also included in this circuit is a membrane capacitance (C_m), a resistor representing the leakage conductance (g_L), and a battery (E_L) representing the leakage potential. (Voltage-dependent Na^+, Ca^{2+}, and K^+ channels that contribute to the generation of the action potential have been omitted for simplification.)

The simple circuit allows the simulation and further analysis of the genesis of the PSP. Closure of the switch simulates the opening of the channels by transmitter released from some presynaptic neuron (i.e., a change in P of Eq. 16.12 from 0 to 1). When the switch is open (i.e., no agonist is present and the ligand-gated channels are closed), the membrane potential (V_m) is equal to the value of the leakage battery (E_L). Closure of the switch (i.e., the agonist opens the channels) tends to polarize the membrane potential toward the value of the battery (E_r) in series with the synaptic conductance. Although the effect of the channel openings is to depolarize the postsynaptic cell *toward* E_r (0 mV), this value is never achieved, because the ligand-gated receptors are only a small fraction of the ion channels in the membrane. Other channels (such as the leakage channels, which are not affected by the transmitters) tend to hold the membrane potential at E_L and prevent the membrane potential from reaching the 0-mV level. In terms of the equivalent electrical circuit (Fig. 16.5C), g_L is much greater than g_{syn}.

An analytical expression that can be a source of insight into the production of an EPSP by the engagement of a synaptic conductance can be derived by examining the current flowing in each of the two conductance branches of the circuit in Fig. 16.5C. As previously shown (Eq. 16.13), the current flowing in the branch representing the synaptic conductance is equal to

$$I_{syn} = g_{syn} * (V_m - E_r).$$

Similarly, the current flowing through the leakage conductance is equal to

$$I_L = g_L * (V_m - E_L). \qquad (16.14)$$

By conservation of current, the two currents must be equal and opposite. Therefore,

$$g_{syn} * (V_m - E_r) = -g_L * (V_m - E_L).$$

Rearranging and solving for V_m, we obtain

$$V_m = \frac{g_{syn}E_r + g_L E_L}{g_{syn} + g_L}. \qquad (16.15)$$

Note that when the synaptic channels are closed (i.e., switch open), g_{syn} is 0 and

$$V_m = E_L.$$

Now consider the case of the ligand-gated channels being opened by release of transmitter from a presynaptic neuron (i.e., switch closed) and a neuron with $g_L = 10$ nS, $E_L = -60$ mV, $g_{syn} = 0.2$ nS, and $E_r = 0$ mV. Then

$$V_m = \frac{(0.2 \times 10^{-9} * 0) + (10 \times 10^{-9} * -60)}{10.2 \times 10^{-9}}$$
$$= -59 \text{ mV}.$$

Thus, as a result of the closure of the switch, the membrane potential has changed from its initial value of -60 mV to a new value of -59 mV; that is, an EPSP of 1 mV has been generated.

The preceding analysis ignored the membrane capacitance (C_m), the charging of which makes the synaptic potential slower than the synaptic current. Thus, a more complete analytical description of the postsynaptic factors underlying the generation of a PSP must account for the fact that some of the synaptic current will flow into the capacitative branch of the circuit. Again, by conservation of current, the sum of the currents in the three branches must equal 0. Therefore,

$$0 = C_m \frac{dV_m}{dt} + I_L + I_{syn}, \qquad (16.16)$$

$$0 = C_m \frac{dV_m}{dt} + g_L * (V_m - E_L) + g_{syn}(t) * (V_m - E_r), \qquad (16.17)$$

where $C_m (dV_m/dt)$ is the capacitative current.

By solving for V_m and integrating the differential equation, we can determine the magnitude and time course of a PSP. An accurate description of the kinetics of the PSP requires that the simple switch closure (all-or-none engagement of the synaptic conductance) be replaced with an expression [$g_{syn}(t)$] that describes the dynamics of the change in synaptic conductance with time. Equation 16.9, which describes the dynamics of channel closure, could be used as an approximation of these effects, but a more accurate simulation requires an expression that also describes the kinetics of channel opening (which in Eq. 16.9, is assumed to be instantaneous) (Magleby and Stevens, 1972).

Nonlinear I–V Relationships of Some Ionotropic Receptors

For many PSPs mediated by ionotropic receptors, the current–voltage relationship of the synaptic current is linear or approximately linear (Fig. 16.5B).

Such ohmic relations are typical of nicotinic ACh channels and non-NMDA (*N*-methyl-D-aspartate) glutamate channels (as well as many receptors mediating IPSPs). The linear *I–V* relationship is indicative of a channel whose conductance is not affected by the potential across the membrane. Such linearity should be contrasted with the steep voltage dependency of the conductance of channels underlying the initiation and repolarization of action potentials (Chapters 5 and 7).

NMDA glutamate channels are a class of ionotropic receptors that have nonlinear current–voltage relationships. At negative potentials, the channel conductance is low even when glutamate is bound to the receptor. As the membrane is depolarized, the conductance increases and the current flowing through the channel increases, resulting in the *I–V* relationship illustrated in Fig. 16.6A. This nonlinearity is represented by an arrow through the resistor representing this synaptic conductance in the equivalent circuit of Fig. 16.6B. The nonlinear *I–V* relationship of the NMDA receptor can be explained by a voltage-dependent block of the channel by Mg^{2+} (Fig. 16.7). At normal values of the resting potential, the pore of the channel is blocked by Mg^{2+}. Thus, even when glutamate binds to the receptor (Fig. 16.7B), the blocked channel prevents ionic flow (and an EPSP). The block can be relieved by depolarization, which presumably displaces the Mg^{2+} from the pore (Fig. 16.7B). When the pore is unblocked, cations (i.e., Na^+, K^+, and Ca^{2+}) can readily flow through the channel, and this flux is manifested in the linear part of the *I–V* relationship (Fig. 16.6A). Non-NMDA channels (Fig. 16.7A) are not blocked by Mg^{2+} and have linear *I–V* relationships (Fig. 16.5B).

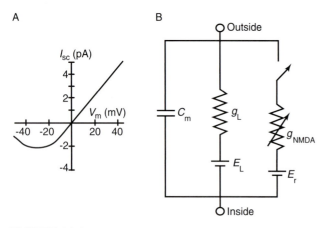

FIGURE 16.6 (A) *I–V* relationship of the NMDA receptor. (B) Equivalent electrical circuit of a membrane containing NMDA receptors.

Inhibitory Postsynaptic Potentials Decrease the Probability of Cell Firing

Some synaptic events *decrease* the probability of generating action potentials in the postsynaptic cell. Potentials associated with these actions are called inhibitory postsynaptic potentials. Consider the inhibitory interneuron illustrated in Fig. 16.2. Normally, this interneuron is activated by summating EPSPs from converging afferent fibers. These EPSPs summate in space and time such that the membrane potential of the interneuron reaches threshold and fires an action potential. This step can be bypassed by artificially depolarizing the interneuron to initiate an action potential. The consequences of that action potential from the point of view of the flexor motor neuron are illustrated in Fig. 16.2. The action potential in the interneuron produces a transient increase in the membrane potential of the motor neuron. This transient hyperpolarization (the IPSP) looks very much like the EPSP, but it is reversed in sign.

What are the ionic mechanisms for these fast IPSPs and what is the transmitter substance? Because the membrane potential of the flexor motor neuron is about –65 mV, one might expect an increase in the conductance to some ion (or ions) with an equilibrium potential (reversal potential) more negative than –65 mV. One possibility is K^+. Indeed, the K^+ equilibrium potential in spinal motor neurons is about –80 mV; thus, a transmitter substance that produced a selective increase in K^+ conductance would lead to an IPSP. The K^+ conductance increase would move the membrane potential from –65 mV toward the K^+ equilibrium potential of –80 mV. Although an increase in K^+ conductance mediates IPSPs at some inhibitory synapses (see below), it does not at the synapse between the inhibitory interneuron and the spinal motor neuron. At this particular synapse, the IPSP seems to be due to a selective increase in Cl^- conductance. The equilibrium potential for Cl^- in spinal motor neurons is about – 70 mV. Thus, the transmitter substance released by the inhibitory neuron diffuses across the cleft and interacts with receptor sites on the postsynaptic membrane. These receptors are normally closed, but when opened they become selectively permeable to Cl^-. As a result of the increase in Cl^- conductance, the membrane potential moves from a resting value of –65 mV toward the Cl^- equilibrium potential of –70 mV.

As in the sensory neuron–spinal motor neuron synapse, the transmitter substance released by the inhibitory interneuron in the spinal cord is an amino acid, but in this case the transmitter is glycine. The toxin strychnine is a potent antagonist of glycine

A

Non-NMDA

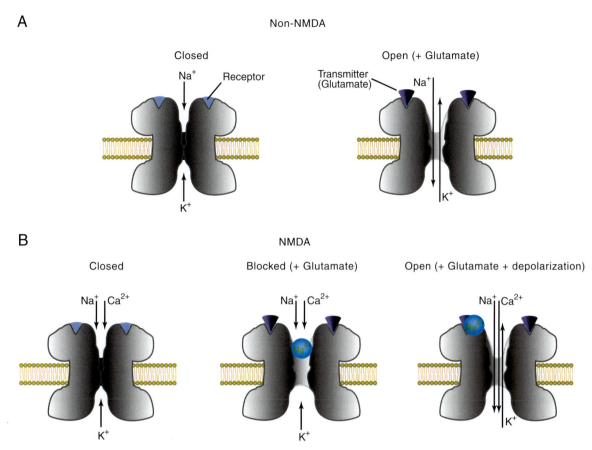

FIGURE 16.7 Features of non-NMDA and NMDA glutamate receptors. (A) Non-NMDA receptors: (left) in the absence of agonist, the channel is closed; (right) glutamate binding leads to channel opening and an increase in Na^+ and K^+ permeability. (B) NMDA receptors: (left) in the absence of agonist, the channel is closed; (middle) the presence of agonist leads to a conformational change and channel opening, but no ionic flux occurs, because the pore of the channel is blocked by Mg^{2+}; (right) in the presence of depolarization, the Mg^{2+} block is removed and the agonist-induced opening of the channel leads to changes in ion flux (including Ca^{2+} influx into the cell).

receptors. Although glycine was originally thought to be localized to the spinal cord, it is also found in other regions of the nervous system. The most common transmitter associated with inhibitory actions in many areas of the brain is γ-aminobutyric acid (GABA) (see Chapter 9).

GABA receptors are divided into three major classes: $GABA_A$, $GABA_B$, and $GABA_C$ (Billington *et al.,* 2001; Bormann, 1988, Bormann and Feigenspan, 1995; Bowery, 1993; Cherubini and Conti, 2001; Gage, 2001; Moss and Smart, 2001). As discussed in Chapter 11, $GABA_A$ receptors are ionotropic receptors, and, like glycine receptors, binding of transmitter leads to an increased conductance to Cl^-, which produces an IPSP. $GABA_A$ receptors are blocked by bicuculline and picrotoxin. A particularly striking aspect of $GABA_A$ receptors is their modulation by anxiolytic benzodiazepines. Figure 16.8 illustrates the response of a

neuron to GABA before and after treatment with diazepam (Bormann, 1988). In the presence of diazepam, the response is greatly potentiated. In contrast to $GABA_A$ receptors that are pore-forming channels, $GABA_B$ receptors are G protein-coupled (see also Chapter 11). $GABA_B$ receptors can be coupled to a variety of different effector mechanisms in different neurons. These mechanisms include decreases in Ca^{2+} conductance, increases in K^+ conductance, and modulation of voltage-dependent A-type K^+ current. In hippocampal pyramidal neurons, the $GABA_B$-mediated IPSP is due to an increase in K^+ conductance (see Fig. 16.10). Baclofen is a potent agonist of $GABA_B$ receptors, whereas phaclofen is a selective antagonist. $GABA_C$ receptors are pharmacologically distinct from $GABA_A$ and $GABA_B$ receptors and are found predominantly in the vertebrate retina. $GABA_C$ receptors, like $GABA_A$ receptors, are Cl^--selective pores.

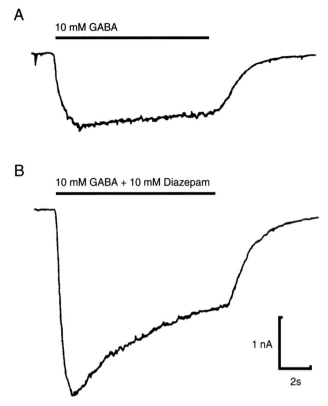

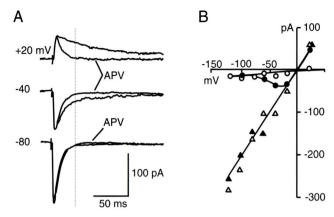

FIGURE 16.9 Dual-component glutamatergic EPSP. (A) The excitatory postsynaptic current was recorded before and during the application of APV at the indicated membrane potentials. (B) Peak current–voltage relationships are shown before (solid triangles) and during (open triangles) the application of APV. The current–voltage relationships measured 25 ms after the peak of the EPSC [dotted line in (A)] before (solid circles) and during (open circles) application of APV are also shown. Reprinted with permission from Hestrin *et al.* (1990).

FIGURE 16.8 Potentiation of GABA responses by benzodiazepine ligands. (A) Brief application (bar) of GABA leads to an inward Cl⁻ curent in a voltage-clamped spinal neuron. (B) In the presence of diazepam the response is significantly enhanced. From Bormann (1988).

Ionotropic receptors that lead to the generation of IPSPs and ionotropic receptors that lead to the generation of EPSPs have biophysical features in common. Indeed, the analyses of the preceding section are generally applicable. A quantitative understanding of the effects of the opening of glycine or GABA receptors can be obtained by using the electrical equivalent circuit of Fig. 16.5C and Eq. 15, with the values of g_{syn} and E_r appropriate for the respective ionotropic receptor. Interactions between excitatory and inhibitory conductances can be modeled by adding additional branches to the equivalent circuit (see Fig. 16.15D and Chapter 17).

Some PSPs Have More Than One Component

The transmitter released from a presynaptic terminal diffuses across the synaptic cleft, where it binds to ionotropic receptors. In many cases, the postsynaptic receptors are homogeneous. In other cases, the same transmitter activates more than one type of receptor. A major example of this type of heterogeneous postsynaptic action is the simultaneous activation by glu-

tamate of NMDA and non-NMDA receptors on the same postsynaptic cell. Figure 16.9 illustrates such a dual-component glutamatergic EPSP in the CA1 region of the hippocampus. The cell is voltage-clamped at various fixed holding potentials, and the macroscopic synaptic currents produced by activation of the presynaptic neurons are recorded. The experiment is performed in the presence and absence of the agent 2-amino-5-phosphonovalerate (APV), which is a specific blocker of NMDA receptors. When the cell is held at a potential of +20 or –40 mV, APV leads to a dramatic reduction of the late, but not the early, phase of the excitatory postsynaptic current (EPSC). In contrast, when the potential is held at –80 mV, the EPSC is unaffected by APV. These results indicate that the PSP consists of two components: (1) an early non-NMDA component and (2) a late NMDA component. In addition, the results indicate that the conductance of the non-NMDA component is linear, whereas the conductance of the NMDA component is nonlinear. The *I–V* relationships of the early (peak) and late (at approximately 25 ms) components of the EPSC are plotted in Fig. 16.9 (Hestrin *et al.*, 1990). Note the similarity in form of these plots of macroscopic currents to the plots of single channel currents in Figs. 16.5B and 16.6A.

Dual-component IPSPs are also observed in the CNS, but here the transmitter (GABA) that mediates the inhibitory actions may be released from different neurons that converge on a common postsynaptic neuron. Stimulation of afferent pathways to the hippocampus results in an IPSP in a pyramidal neuron,

A

B

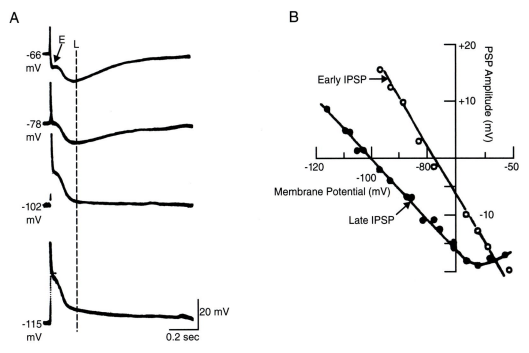

FIGURE 16.10 Dual-component IPSP. (A) Intracellular recordings from a pyramidal cell in the CA3 region of the rat hippocampus in response to activation of mossy fiber afferents. With the membrane potential of the cell at the resting potential, afferent stimulation produces an early (E) and late (L) IPSP. With increased hyperpolarizing produced by injecting constant current into the cell, the early component reverses first. At more negative levels of the membrane potential, the late component also reverses. This indicates the ionic conductance underlying the two phases is distinct. (B) Plots of the change in amplitude of the early (measured at 25 ms) and late (measured at 200 ms, dashed line) response as a function of the membrane potential. The reversal potentials of the early and late components are consistent with a GABA$_A$-mediated chloride conductance and a GABA$_B$-mediated potassium conductance, respectively. From Thalmann (1988).

which has a fast initial inhibitory phase followed by a slower inhibitory phase (Fig. 16.10). Application of the GABA$_A$ antagonists blocks the early inhibitory phase, whereas the GABA$_B$ receptor antagonist phaclofen blocks the late inhibitory phase (not shown). The early and late IPSPs can also be distinguished based on their ionic mechanisms. Hyperpolarizing the membrane potential to –78 mV nulls the early response, but at this value of membrane potential the late response is still hyperpolarizing (Figs. 16.10A and B). Hyperpolarizing the membrane potential to values more negative than –78 mV reverses the sign of the early response, but the slow response does not reverse until the membrane is made more negative than about –100 mV (Thalmann, 1988). The reversal potentials are consistent with a fast Cl⁻-mediated IPSP, mediated by fast opening of GABA$_A$ receptors, and a slower K⁺-mediated IPSP, mediated by the G-protein GABA$_B$ receptors.

Dual-component PSPs need not be strictly inhibitory or excitatory. For example, a presynaptic cholinergic neuron in the mollusk *Aplysia* produces a diphasic excitatory–inhibitory (E–I) response in its postsynaptic fol-

lower cell. The response can be simulated by local discrete application of ACh to the postsynaptic cell (Fig. 16.11) (Blankenship *et al.*, 1971). The ionic mechanisms underlying this synaptic action were investigated in ion substitution experiments, which revealed that the dual response is due to an early Na⁺-dependent component followed by a slower Cl⁻-dependent component. Molecular mechanisms underlying such slow synaptic potentials are discussed next.

Summary

Synaptic potentials mediated by ionotropic receptors are the fundamental means by which information is rapidly transmitted between neurons. Transmitters cause channels to open in an all-or-none fashion, and the currents through these individual channels summate to produce the macroscopic postsynaptic potential. The sign of the postsynaptic potential is determined by the relationship between the membrane potential of the postsynaptic neuron and the ion selectivity of the ionotropic receptor.

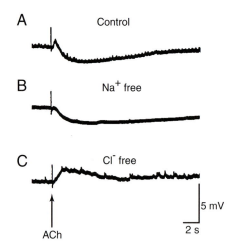

FIGURE 16.11 Dual-component cholinergic excitatory–inhibitory response. (A) Control in normal saline. Ejection of ACh produces a rapid depolarization followed by a slower hyperpolarization. (B) In Na⁺-free saline, ACh produces a purely hyperpolarizing response, indicating that the depolarizing component in normal saline includes an increase in g_{Na} (C) In Cl⁻-free saline, ACh produces a purely depolarizing response, indicating that the hyperpolarizing component in normal saline includes an increase in g_{Cl}. Reprinted with permission from Blankenship *et al.* (1971).

METABOTROPIC RECEPTORS: MEDIATORS OF SLOW SYNAPTIC POTENTIALS

A common feature of the types of synaptic actions heretofore described is the direct binding of the transmitter with the receptor–channel complex. An entirely separate class of synaptic actions has as its basis the indirect coupling of the receptor with the channel. So far, two types of coupling mechanisms have been identified: (1) coupling of the receptor and channel through an intermediate regulatory protein, such as a G protein; and (2) coupling through a diffusible second messenger system. Because the coupling through a diffusible second messenger system is the most common mechanism, it is the focus of this section.

A comparison of the features of direct, fast ionotropic-mediated and indirect, slow metabotropic-mediated synaptic potentials is shown in Fig. 16.12. Slow synaptic potentials are not observed at every postsynaptic neuron, but Fig. 16.12A illustrates an idealized case in which a postsynaptic neuron receives two inputs, one of which produces a conventional fast EPSP and the other of which produces a slow EPSP. An action potential in neuron 1 leads to an EPSP in the postsynaptic cell with a duration of about 30 ms (Fig. 16.12B). This type of potential might be produced in a spinal motor neuron by an action

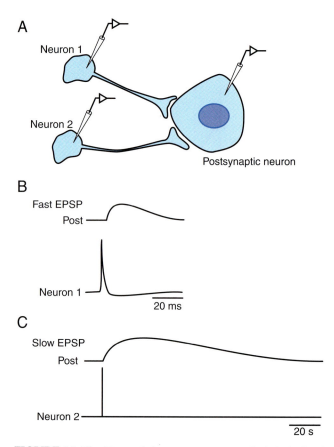

FIGURE 16.12 Fast and slow synaptic potentials. (A) Idealized experiment in which two neurons (1 and 2) make synaptic connections with a common postsynaptic follower cell (Post). (B) An action potential in neuron 1 leads to a conventional fast EPSP with a duration of about 30 ms. (C) An action potential in neuron 2 also produces an EPSP in the postsynaptic cell, but the duration of this slow EPSP is more than three orders of magnitude greater than that of the EPSP produced by neuron 1. Note the change in the calibration bar.

potential in an afferent fiber. Neuron 2 also produces a postsynaptic potential (Fig. 16.12C), but its duration (note the calibration bar) is more than three orders of magnitude greater than that of the EPSP produced by neuron 1.

How can a change in the postsynaptic potential of a neuron persist for many minutes as a result of a single action potential in the presynaptic neuron? Possibilities include a prolonged presence of the transmitter due to continuous release, slow degradation, or slow reuptake of the transmitter, but the mechanism here involves a transmitter-induced change in the metabolism of the postsynaptic cell. Figure 16.13 compares the general mechanisms for fast and slow synaptic potentials. Fast synaptic potentials are produced when a transmitter substance binds to a channel and produces a conformational change in the channel, causing it to become permeable to one or more ions (both Na^+ and K^+ in Fig. 16.13A). The increase in permeability leads to a depolarization associated with the EPSP. The duration of the synaptic event critically depends on the amount of time during which the transmitter substance remains bound to the receptors. Acetylcholine, glutamate, and glycine remain bound only for a very short period. These transmitters are removed by diffusion, enzymatic breakdown, or reuptake into the presynaptic cell. Therefore, the duration of the synaptic potential is directly related to the lifetimes of the opened channels, and these lifetimes are relatively short (see Fig. 16.4B).

One mechanism for a slow synaptic potential is shown in Fig. 16.13B. In contrast with the fast PSP for which the receptors are actually part of the ion channel complex, the channels that produce the slow synaptic potentials are not directly coupled to the transmitter receptors. Rather, the receptors are physically separated and exert their actions indirectly through changes in metabolism of specific second messenger systems. Figure 16.13B illustrates one type of response in *Aplysia*, for which the cAMP–protein kinase A system is the mediator, but other slow PSPs use other second messenger–kinase systems (e.g., the protein kinase C system). In the cAMP-dependent slow synaptic responses in *Aplysia*, transmitter binding to membrane receptors activates G proteins and stimulates an increase in the synthesis of cAMP. Cyclic AMP then leads to the activation of cAMP-dependent protein kinase (protein kinase A, PKA), which phosphorylates a channel protein or protein associated with the channel (Siegelbaum *et al.*, 1982). A conformational change in the channel is produced, leading to a change in ionic conductance. Thus, in contrast with a direct conformational change produced by the binding of a transmitter to the receptor–channel complex, in this case, a conformational change is produced by protein phosphorylation. Indeed, phosphorylation-dependent channel regulation is a fairly general feature of slow PSPs. However, channel regulation by second messengers is not exclusively produced by phosphorylation. In one family of ion channels, the channels are gated or regulated directly by cyclic nucleotides. These cyclic nucleotide-gated channels require cAMP or cGMP to open but have other features in common with members of the superfamily of voltage-gated ion channels. (Kaupp, 1995; Zimmermann, 1995).

Another interesting feature of slow synaptic responses is that they are sometimes associated with decreases rather than increases in membrane conductance. For example, the particular channel illustrated in Fig. 16.13B is selectively permeable to K^+ and is normally open. As a result of the activation of the second messenger, the channel closes and becomes less permeable to K^+. The resultant depolarization may seem paradoxical, but recall that the membrane potential is due to a balance between the resting K^+ and Na^+ permeability. The K^+ permeability tends to move the membrane potential toward the K^+ equilibrium potential (–80 mV), whereas the Na^+ permeability tends to move the membrane potential toward the Na^+ equilibrium potential (+55 mV). Normally, the K^+ permeability predominates, and the resting membrane potential is close to, but not equal to, the K^+ equilibrium potential. If K^+ permeability is decreased because some of the channels close, the membrane potential will be biased toward the Na^+ equilibrium potential and the cell will depolarize.

At least one reason for the long duration of slow PSPs is that second messenger systems are slow (from seconds to minutes). Take the cAMP cascade as an example. Cyclic AMP takes some time to be synthesized, but, more importantly, after synthesis, cAMP levels can remain elevated for a relatively long period (minutes). The duration of the elevation of cAMP depends on the actions of cAMP phosphodiesterase, which breaks down cAMP. However, duration of an effect could outlast the duration of the change in the second messenger because of persistent phosphorylation of the substrate protein(s). Phosphate groups are removed from the substrate proteins by protein phosphatases. Thus, the net duration of a response initiated by a metabotropic receptor depends on the actions of not only the synthetic and phosphorylation processes but also the degradative and dephosphorylation processes.

The activation of a second messenger by a transmitter can have a localized effect on the membrane potential through phosphorylation of membrane channels near the site of a metabotropic receptor. The effects can be more widespread and even longer-lasting than depicted in Fig. 16.13B. For example, second messengers and protein kinases can diffuse and affect more distant membrane channels. Moreover, a long-term effect can be induced in the cell by altering gene expression. For example, protein kinase A can diffuse to the nucleus, where it can activate proteins that regulate gene expression. Detailed descriptions of second messengers and their actions are given in Chapters 12 , 13, 14 and 18.

Summary

In contrast to the rapid responses mediated by ionotropic receptors, responses mediated by metabotropic receptors (i.e., GPCRs) are generally relatively slow to develop and persistent. These properties arise

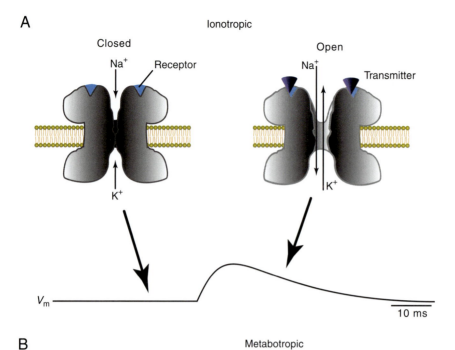

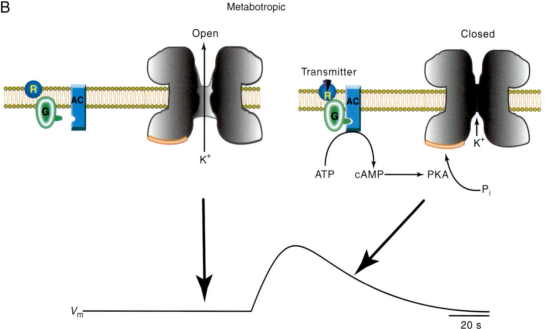

FIGURE 16.13   Ionotropic and metabotropic receptors and mechanisms of fast and slow EPSPs. (A, left) Fast EPSPs are produced by the binding of transmitter to specialized receptors that are directly associated with an ion channel (i.e., a ligand-gated channel). When the receptors are unbound, the channel is closed. (A, right) Binding of the transmitter to the receptor produces a conformational change in the channel protein such that the channel opens. In this example, the channel opening is associated with a selective increase in the permeability to Na^+ and K^+. The increase in permeability results in the EPSP shown in the trace. (B, left) Unlike fast EPSPs that are due to the binding of a transmitter with a receptor–channel complex, slow EPSPs are due to the activation of receptors (metabotropic) that are not directly coupled to the channel. Rather, the coupling takes place through the activation of one of several second-messenger cascades, in this example, the cAMP cascade. A channel that has a selective permeability to K^+ is normally open. (B, right) Binding of the transmitter to the receptor (R) leads to the activation of a G protein (G) and adenylyl cyclase (AC). The synthesis of cAMP is increased, cAMP-dependent protein kinase (protein kinase A, PKA) is activated, and a channel protein is phosphorylated. The phosphorylation leads to closing of the channel and the subsequent depolarization associated with the slow EPSP shown in the trace. The response decays owing to both the breakdown of cAMP by cAMP-dependent phosphodiesterase and the removal of phosphate from channel proteins by protein phosphatases (not shown).

because metabotropic responses can involve the activation of second messenger systems. By producing slow changes in the resting potential, metabotropic receptors provide long-term modulation of the effectiveness of responses generated by ionotropic receptors. Moreover, these receptors, through the engagement of second-messenger systems, provide a vehicle by which a presynaptic cell can not only alter the membrane potential but also produce widespread changes in the biochemical state of a postsynaptic cell.

INTEGRATION OF SYNAPTIC POTENTIALS

The small amplitude of the EPSP in spinal motor neurons (and other cells in the CNS) poses an interesting question. Specifically, how can an EPSP with an amplitude of only 1 mV drive the membrane potential of the motor neuron (i.e., the postsynaptic neuron) to threshold and fire the spike in the motor neuron that is necessary to produce the contraction of the muscle? The answer to this question lies in the principles of temporal and spatial summation.

When the ligament is stretched (Fig. 16.1), many stretch receptors are activated. Indeed, the greater the stretch, the greater the probability of activating a larger number of the stretch receptors; this process is referred to as recruitment. However, recruitment is not the complete story. The principle of frequency coding in the nervous system specifies that the greater the intensity of a stimulus, the greater the number of action potentials per unit time (frequency) elicited in a sensory neuron. This principle applies to stretch receptors as well. Thus, the greater the stretch, the greater the number of action potentials elicited in the stretch receptor in a given interval and therefore the greater the number of EPSPs produced in the motor neuron from that train of action potentials in the sensory cell. Consequently, the effects of activating multiple stretch receptors add together (spatial summation), as do the effects of multiple EPSPs elicited by activation of a single stretch receptor (temporal summation). Both of these processes act in concert to depolarize the motor neuron sufficiently to elicit one or more action potentials, which then propagate to the periphery and produce the reflex.

Temporal Summation Allows Integration of Successive PSPs

Temporal summation can be illustrated by firing action potentials in a presynaptic neuron and moni-

toring the resultant EPSPs. For example, in Figs. 16.14A and 16.14B, a single action potential in sensory neuron 1 produces a 1-mV EPSP in the motor neuron. Two action potentials in quick succession produce two EPSPs, but note that the second EPSP occurs during the falling phase of the first, and the depolarization associated with the second EPSP adds to the depolarization produced by the first. Thus, two action potentials produce a summated potential that is about 2 mV in amplitude. Three action potentials in quick succession would produce a summated potential of about 3 mV. In principle, 30 action potentials in quick succession would produce a potential of about 30 mV

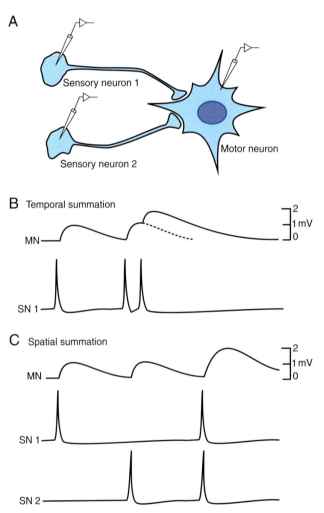

FIGURE 16.14 Temporal and spatial summation. (A) Intracellular recordings are made from two idealized sensory neurons (SN1 and SN2) and a motor neuron (MN). (B) Temporal summation: A single action potential in SN1 produces a 1-mV EPSP in the MN. Two action potentials in quick succession produce a dual-component EPSP, the amplitude of which is approximately 2 mV. (C) Spatial summation: Alternative firing of single action potentials in SN1 and SN2 produce 1-mV EPSPs in the MN. Simultaneous action potentials in SN1 and SN2 produce a summated EPSP, the amplitude of which is about 2 mV.

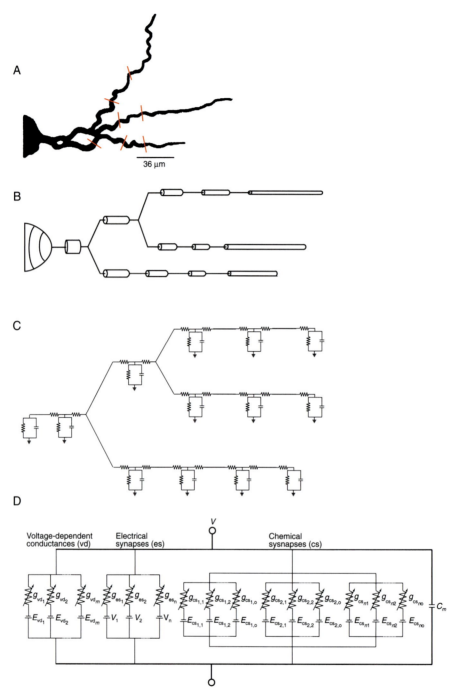

FIGURE 16.15   Modeling the integrative properties of a neuron. (A) Partial geometry of a neuron in the CNS revealing the cell body and pattern of dendritic branching. (B) The neuron modeled as a sphere connected to a series of cylinders, each of which represents the specific electrical properties of a dendritic segment. (C) Segments linked with resistors representing the intracellular resistance between segments, with each segment represented by the parallel combination of the membrane capacitance and the total membrane conductance. Reprinted with permission from Koch and Segev (1989). Copyright 1989 MIT Press. (D) Electrical circuit equivalent of the membrane of a segment of a neuron. In (D), the segment has a membrane potential V and a membrane capacitance C_m. Currents arise from three sources: (1) m voltage-dependent conductances ($g_{vd_1}-g_{vd_m}$), (2) n conductances due to electrical synapses ($g_{es_1}-g_{es_n}$), and (3) n times o time-dependent conductances due to chemical synapses with each of the n presynaptic neurons ($g_{cs_{1,1}}-g_{cs_{n,o}}$). E_{vd} and E_{cs} are constants and represent the values of the equilibrium potential for currents due to voltage-dependent conductances and chemical synapses, respectively. V_1-V_n represent the value of the membrane potential of the coupled cells. Reprinted with permission from Ziv *et al.* (1994).

and easily drive the cell to threshold. This summation is strictly a passive property of the cell. No special ionic conductance mechanisms are necessary. Specifically, the postsynaptic conductance change [g_{syn} in Eq. (16.13)] produced by the second of two successive action potentials adds to that produced by the first. In addition, the postsynaptic membrane has a capacitance and can store charge. Thus, the membrane temporarily stores the charge of the first EPSP, and the charge from the second EPSP is added to that of the first. However, the "time window" for this process of temporal summation very much depends on the duration of the postsynaptic potential, and temporal summation is possible only if the presynaptic action potentials (and hence postsynaptic potentials) are close in time to each other. The time frame depends on the duration of changes in the synaptic conductance and the time constant (Chapter 4). Temporal summation, however, is rarely observed to be linear as in the preceding examples, even when the postsynaptic conductance change (g_{syn} in Eq. 16.13) produced by the second of two successive action potentials is identical to that produced by the first (i.e., no presynaptic facilitation or depression), the synaptic current is slightly less because the first PSP reduces the driving force ($V_m - E_r$) for the second. Interested readers should try some numerical examples.

Spatial Summation Allows Integration of PSPs from Different Parts of a Neuron

Spatial summation (Fig. 16.14C) requires a consideration of more than one input to a postsynaptic neuron. An action potential in sensory neuron 1 produces a 1-mV EPSP, just as it did in Fig. 16.14B. Similarly, an action potential in a second sensory neuron by itself also produces a 1-mV EPSP. Now, consider the consequences of action potentials elicited simultaneously in sensory neurons 1 and 2. The net EPSP is equal to the summation of the amplitudes of the individual EPSPs. Here, the EPSP from sensory neuron 1 is 1 mV, the EPSP from sensory neuron 2 is 1 mV, and the summated EPSP is approximately 2 mV (Fig. 16.14C). Thus, spatial summation is a mechanism by which synaptic potentials generated at different sites can summate. Spatial summation in nerve cells is influenced by the space constant—the ability of a potential change produced in one region of a cell to spread passively to other regions of a cell (see Chapter 4).

Summary

Whether a neuron fires in response to synaptic input depends, at least in part, on how many action potentials are produced in any one presynaptic excitatory pathway and on how many individual convergent excitatory input pathways are activated. The summation of EPSPs in time and space is only part of the process, however. The final behavior of the cell is also due to the summation of inhibitory synaptic inputs in time and space, as well as to the properties of the voltage-dependent currents (Fig. 16.15) in the soma and along the dendrites (Koch and Segev 1989; Ziv et al., 1994). For example, voltage-dependent conductances such as the A-type K^+ conductance have a low threshold for activation and can thus oppose the effectiveness of an EPSP to trigger a spike. Low-threshold Na^+ and Ca^{2+} channels can boost an EPSP. Finally, we need to consider that the spatial distribution of the various voltage-dependent channels, ligand-gated receptors, and metabotropic receptors is not uniform. Thus, each segment of the neuronal membrane can perform selective integrative functions. Clearly, this system has an enormous capacity for the local processing of information and for performing logical operations. The flow of information in dendrites and the local processing of neuronal signals are discussed in Chapter 17.

References

General

Burke, R. E., and Rudomin, P. (1977). Spatial neurons and synapses. In "Handbook of Physiology" (E. R. Kandel, Ed.), Sect. 1, Vol. 1, Part 2, pp. 877–944. American Physiological Society, Bethesda, MD.

Byrne, J. H., and Schultz, S. G. (1994). "An Introduction to Membrane Transport and Bioelectricity", 2nd ed. Raven Press, New York.

Cowan, W. M., Sudhof, T. C., and Stevens, C. F. (Eds.) (2001). "Synapses". Johns Hopkins Univ. Press, Bethesda, MD.

Hille, B. (Ed.) (2001). "Ion Channels of Excitable Membranes," 3rd ed. Sinauer Associates, Sunderland, MA.

Shepherd, G. M. (Ed.) (2001). "The Synaptic Organization of the Brain," 4th ed., Oxford Univ. Press, New York.

Cited

Billinton, A., Ige, A. O., Bolam, J. P., White, J. H., Marshall, F. H., and Emson, P. C. (2001). Advances in the molecular understanding of GABA$_B$ receptors. Trends Neurosci. **24**, 277–282.

Blankenship, J. E., Wachtel, H., and Kandel, E. R. (1971). Ionic mechanisms of excitatory, inhibitory and dual synaptic actions mediated by an identified interneuron in abdominal ganglion of Aplysia. J. Neurophysiol. **34**, 76–92.

Bormann, J. (1988). Electrophysiology of GABA$_A$ and GABA$_B$ receptor subtypes. Trends Neurosci. **11**, 112–116.

Bormann, J., and Feigenspan, A. (1995). GABA$_C$ receptors. Trends Neurosci. **18**, 515–519.

Bowery, N. G. (1993). GABA$_B$ receptor pharmacology. Annu. Rev. Pharmacol. Toxicol. **33**, 109–147.

Cherubini, E., and Conti, F. (2001). Generating diversity at GABAergic synapses. Trends Neurosci. **24**, 155–162.

Colquhoun, D., and Hawkes, A. G. (1977). Relaxation and fluctuations of membrane currents that flow through drug-operated channels. *Proc. R. Soc. London Ser. B* **199**, 231–262.

Colquhoun, D., and Hawkes, A. G. (1981). On the stochastic properties of single ion channels. *Proc. R. Soc. London Ser. B* **211**, 205–235.

Colquhoun, D., and Hawkes, A. G. (1982). On the stochastic properties of bursts of single ion channel openings and of clusters of bursts. *Proc. R. Soc. London Ser. B* **300**, 1–59.

Colquhoun, D., and Sakmann, B. (1981). Fluctuations in the microsecond time range of the current through single acetylcholine receptor ion channels. *Nature (London)* **294**, 464–466.

Eccles, J. C. (1964). "The Physiology of Synapses." Springer-Verlag, New York.

Fatt, P., and Katz, B. (1951). An analysis of the end-plate potential recorded with an intra-cellular electrode. *J. Physiol. (London)* **115**, 320–370.

Gage, P. W. (2001). Activation and modulation of neuronal K$^+$ channels by GABA. *Trends Neurosci.* **15**, 46–51.

Hamill, O. P., Marty, A., Neher, E., Sakmann, B., and Sigworth, J. (1981). Improved patch-clamp techniques for high-resolution current recording from cells and cell-free membrane patches. *Pflügers Arch.* **391**, 85–100.

Hestrin, S., Nicoll, R. A., Perkel, D. J., and Sah, P. (1990). Analysis of excitatory synaptic action in pyramidal cells using whole-cell recording from rat hippocampal slices. *J. Physiol. (London)* **422**, 203–225.

Johnston, D., and Wu, S. M.-S. (1995). "Foundations of Cellular Neurophysiology". MIT Press, Cambridge, MA.

Kaupp, U. B. (1995). Family of cyclic nucleotide gated ion channels. *Curr. Opin. Neurobiol.* **5**, 434–442.

Koch, C., and Segev, I. (1989). "*Methods in Neuronal Modeling.*" MIT Press, Cambridge, MA.

Magleby, K. L., and Stevens, C. F. (1972). A quantitative description of end-plate currents. *J. Physiol. (London)* **223**, 173–197.

Moss, S. J., and Smart, T. G. (2001). Constructing inhibitory synapses. *Nat. Rev. Neurosci.* **2**, 240–250.

Papoulis, A. (1965). "*Probability, Random Variables, and Stochastic Processes*". McGraw–Hill, New York.

Sakmann, B. (1992). Elementary steps in synaptic transmission revealed by currents through single ion channels. *Science* **256**, 503–512.

Siegelbaum, S. A., Camardo, J. S., and Kandel, E. R. (1982). Serotonin and cyclic AMP close single K$^+$ channels in *Aplysia* sensory neurones. *Nature (London)* **299**, 413–417.

Spencer, W. A. (1977). The physiology of supraspinal neurons in mammals. *In* "Handbook of Physiology" (E. R. Kandel, Ed.), Sect. 1, pp. 969–1022, Vol. 1, Part 2. American Physiological Society, Bethesda, MD.

Takeuchi, A., and Takeuchi, N. (1960). On the permeability of end-plate membrane during the action of transmitter. *J. Physiol. (London)* **154**, 52–67.

Thalmann, R. H. (1988). Evidence that guanosine triphosphate (GTP)-binding proteins control a synaptic response in brain: Effect of pertussis toxin and GTPγS on the late inhibitory postsynaptic potential of hippocampal CA3 neurons. *J. Neurosci.* **8**, 4589–4602.

Zimmermann, A. L. (1995). Cyclic nucleotide gated channels. *Curr. Opin. Neurobiol.* **5**, 296–303.

Ziv, I., Baxter, D. A., and Byrne, J.H. (1994). Simulator for neural networks and action potentials: Description and application. *J. Neurophysiol.* **71**, 294–308.

Information Processing in Complex Dendrites

Gordon M. Shepherd

One of the hallmarks of neurons is the variety of their dendrites. The branching patterns are dazzling and the range of size is astounding, from the large trees of cortical pyramidal neurons to the tiny retinal bipolar cell, which would fit comfortably within the cell body of a pyramidal neuron (see Fig. 17.1)! The

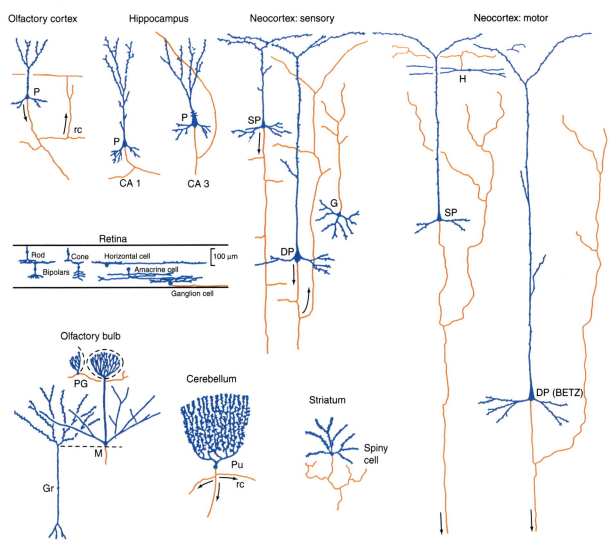

FIGURE 17.1 Varieties of neurons and dendritic trees. Abbreviations: CA1, CA3, hippocampal regions; P, pyramidal cell; rc, recurrent collateral; SP, superficial pyramidal cell; DP, deep pyramidal cell; G, granule (stellate) cell; PG, periglomerular cell; M, mitral cell; Gr, granule cell (olfactory); Pu, Purkinje cell. From Shepherd (1991).

Copyright 2004, Elsevier Science (USA).
All rights reserved.

functions of these dendritic trees have drawn increasing interest in recent years (Segev *et al.*, 1995; Yuste and Tank, 1996; Wilson, 1998; Stuart *et al.*, 1999; Segev and London, 2000; Matus and Shepherd, 2000; Stern and Marx, 2000). The fundamental questions asked in this chapter include: What are the principles of information processing in complex dendritic trees, and how are they adapted for the specific operational tasks of a particular type of dendrite?

STRATEGIES FOR STUDYING COMPLEX DENDRITES

As was discussed in Chapter 4, the neuron processes information through five basic types of activity: intrinsic, reception, integration, encoding, and output. As also discussed in Chapter 4, understanding the ways in which these activities are integrated within the neuron starts with the rules of passive current spread. We now ask how active, voltage-gated channels are involved in *complex information processing*, particularly within branching dendritic trees.

Many of the principles were first worked out in the dendrites of neurons that lack axons or the ability to generate action potentials. There are many examples in invertebrate ganglia. In vertebrates, they include the retinal amacrine cell and the olfactory granule cell. Studies of these neurons are covered in Shepherd (1991). In summary, a dendritic tree by itself is capable of performing many of the basic functions required for information processing, such as generation of intrinsic activity, input–output functions for feature extraction, parallel processing, signal-to-noise enhancement, and oscillatory activity. These cells demonstrate that there is no one thing that dendrites do; they do whatever is required to process information within their particular neuron or neuronal circuit.

Information in dendrites can take many forms. There are actions of neuropeptides on membrane receptors and internal cytoplasmic or nuclear receptors, actions of second and third messengers within the neuron, movement of substances within the dendrites by diffusion or by active transport, and changes occurring during development. All of these types of cellular traffic and information flow in dendrites are coming under direct study (Stuart *et al.*, 1999; Matus and Shepherd, 2000). The student should review these subjects in earlier chapters. Here, we focus on information processing involving electrical signaling mechanisms by synapses and voltage-gated channels, and consider how this takes place in neurons with axons.

Among cells with axons, long-axon (output) cells tend to be larger than short-axon (local) cells and have therefore been more accessible to experimental analysis. Indeed, virtually everything that we know about the functional relationships between dendrites and axons has been obtained from studies of long-axon cells. Much of what we think we understand about those relationships in short-axon cells is only by inference.

As in the analysis of the passive properties of neurons, there are a number of sites on the web that support the computational analysis of complex neurons and their active dendrites. They are included in the list in Box 4.2 of Chapter 4 (see also Chapter 7).

An Axon Places Constraints on Dendritic Processing

We immediately recognize that the presence of an axon places critical constraints on dendritic processing (Fig. 17.2).

The first principle is: *if a neuron has an axon, it has only one*. This near-universal single-axon rule is remarkable and still little understood. It results from developmental mechanisms that provide for differentiation of a single axon from among early undifferentiated processes; these mechanisms are currently being analyzed in neuronal cultures (Craig and Banker, 1994). The principle, which can be regarded virtually as a law for neurons, means that for dendritic integration to lead to output from the neuron to distant targets, all of the activity within the dendrites must eventually be funneled into the origin of the axon in the single axon hillock. Therefore, in these cells the flow of information in dendrites has an overall orientation, just as surmised by the classic neuroanatomists. We thus have a principle of *global output*:

> To transfer information between regions, the information distributed at different sites within a dendritic tree of an output neuron must be encoded for global output at a single site at the origin of the axon.

A related principle, and virtually another universal rule, is that the main function of the axon in long-axon cells is to support the generation of action potentials in the axon hillock–initial segment region. By definition, action potentials have thresholds for generation; thus, the principle of *frequency encoding of global output* in an axonal neuron is:

> The results of dendritic integration affect the output through the axon only by initiating or modulating action potential generation in the axon hillock–initial segment. Global output from

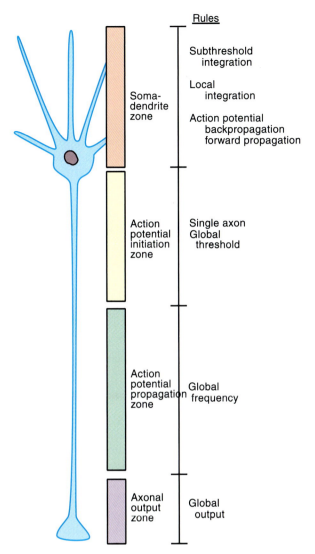

Rules

Soma-dendrite zone — Subthreshold integration / Local integration / Action potential backpropagation forward propagation

Action potential initiation zone — Single axon Global threshold

Action potential propagation zone — Global frequency

Axonal output zone — Global output

FIGURE 17.2 The presence of a single axon forces several organizational rules onto a neuron. See text for details.

dendritic integration is therefore encoded in impulse frequency in a single axon.

A further consequence of the spatial separation of dendrites and axon is the presence of *subthreshold dendritic activity*:

A considerable amount of subthreshold activity, including local active potentials, can affect the integrative states of the dendrites and their local outputs but not necessarily directly or immediately affect the global output of the neuron.

We turn now to the functional properties that allow dendritic trees to process information within these constraints.

Dendrodendritic Interactions between Axonal Cells

First recognize that with axonal cells, as with anaxonal cells, output can be through the dendrites (see principle of subthreshold dendritic activity above). This is against the common wisdom, which assumes that if a neuron has an axon, all the output goes through the axon. There are many examples in invertebrates.

Neurite–Neurite Synapses in the Lobster Stomatogastric Ganglion

One of the first examples in invertebrates was in the stomatogastric ganglion of the lobster (Selverston *et al.*, 1976). Neurons were recorded intracellularly and stained with Procion yellow. Serial electron micrographic reconstructions showed the synaptic relationships between stained varicosities in the processes and their neighbors (the processes are equivalent to dendrites, but are often referred to as neurites in the invertebrate literature). In many cases, a varicosity could be seen to be not only presynaptic to a neighboring varicosity, but also postsynaptic to that same process. It was concluded that synaptic inputs and outputs are distributed over the entire neuritic arborization. Polarization was not from one part of the tree to another. The "bifunctional" varicosities appeared to act as local input–output units, similar to the manner in which granule cell spines appear to operate (see below). Similar organization has been found in other types of stomatogastric neurons (Fig. 17.3A).

Sets of these local input–output units, distributed throughout the neuritic tree, participate in the generation and coordination of oscillatory activity involved in controlling the rhythmic movements of the stomach. In a current model of this oscillatory circuit these interactions are mutually inhibitory (see Fig. 17.3B).

Summary

A cell with an axon can have local outputs through its dendrites as well as its axon, which may be involved in specific functions such as oscillatory circuits.

Passive Dendritic Trees Can Perform Complex Computations

Another principle that carries over from axonless nonspiking cells is the ability of axonal cells to carry out complex computations in dendritic trees with passive properties. This is exemplified by neurons that are motion detectors.

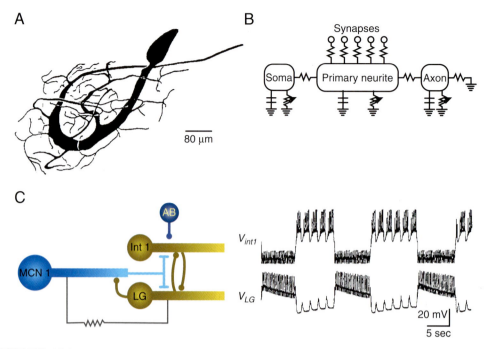

FIGURE 17.3 Local synaptic input–output sites are widely found within the neuropil of invertebrate ganglia. (A) Neurite–neurite interactons in the gastric mill ganglion of the lobster. (B) Compartment model of stomatogastric neuron. (C) Model of rhythm generating circuit of the gastric mill of the lobster, involving neurite-neurite interactions. (A) and (B) from Golowasch and Marder (1992). (C) from Manor *et al.* (1999).

Motion detection is a fundamental operation carried out by the nervous systems of most species; it is essential for detecting prey and predator alike. In invertebrates, motion detection has been studied especially in the brain of the blowfly. In the lobula plate of the third optic neuropil are tangential cells (LPTCs) that respond to preferential direction (PD) of motion with increased depolarization due to sequential responses across their dendritic fields. This response has been modeled by Reichardt and his colleagues by a series of elementary motion detectors (EMDs) in the dendrites. A compartmental model (Single and Borst, 1998) reproduces the experimental results and theoretical predictions by showing how local modulations at each EMD are smoothed by integration in the dendritic tree to give a smoothed high-fidelity global output at the axon (see Fig. 17.4A). In the model, spatial integration is largely independent of specific electrotonic properties but depends critically on the geometry and orientation of the dendritic tree.

In the vertebrates, motion detection is built into the visual pathway at various stages in different species, principally the retina, midbrain (optic tectum), and cerebral cortex. Recent studies in the optic tectum have revealed cells with splayed uniplanar dendritic trees and specialized distal appendages that appear highly homologous across reptiles, birds, and

mammals (Fig. 17.4B) (Luksch *et al.*, 1998). These are presumed to mediate motion detection. Physiological studies are needed to test the hypothesis that these cells perform operations through their dendritic fields similar to those of the LPTC cells in the insect. To the extent that this is borne out, it will support a principle of *motion detection through spatially distributed dendritic computations* that is conserved across vertebrates and invertebrates. This kind of directional selectivity of dendritic processing was predicted by Rall (1964) from his studies of dendritic electrotonus (see Chapter 4).

Distal Dendrites Can Be Closely Linked to Axonal Output

An obvious problem for a neuron with an axon is that the distal branches of dendritic trees are a long distance from site of axon origin at or near the cell body. As mentioned earlier, the common perception is that these distal dendrites are too distant from the site of axonal origin and impulse generation to have more than a slow and weak background modulation of impulse output.

This perception is disproved by many kinds of neurons in which specific inputs are located preferentially on their distal dendrites. Such is true of the mitral and tufted cells in the olfactory bulb, where the

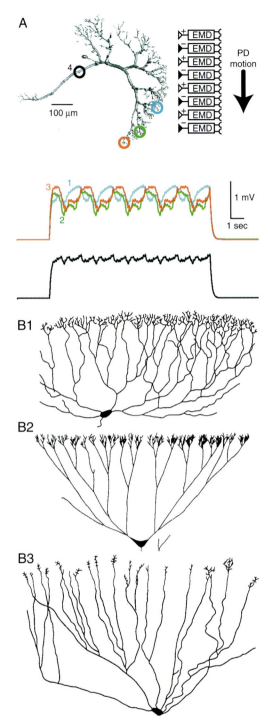

input from the olfactory nerves ends on the most distal dendritic branches in the glomeruli; in rat mitral cells this may be 400–500 μm or more from the cell body, in turtle, 600–700 μm. The same applies to their targets, the pyramidal neurons of the olfactory cortex, where the input terminates on the spines of the most distal dendrites in layer I. In many other neurons, a given type of input terminates over much or all of the dendritic tree; such is the case, for example, for climbing fiber and parallel fiber inputs to the cerebellar Purkinje cells. The relative significance of the more distal inputs in these cells is not so apparent. All of these examples are shown in Fig. 17.1.

How do distal dendrites effectively control axonal output? We consider several important properties.

Large Synaptic Conductances

Of key importance is the amplitude of the conductance generated by the synapse itself (Fig. 17.5A). In motor neurons, the conductances of the most distal excitatory synapses may be many times the amplitude of proximal synapses (Redman and Walmsley, 1983). This would account for the fact that the peak unitary synaptic response recorded at the soma varies in time course according to synaptic location but has a constant amplitude of approximately 100 μV (Fig. 17.5A). Recent studies have provided evidence for a similar increase in synaptic conductance in the distal dendrites of cortical pyramidal neurons (Magee, 2000).

High Specific Membrane Resistance

A second key property is the specific membrane resistance (R_m) of the dendritic membrane. Traditionally, the argument was that if R_m is relatively low, the characteristic length of the dendrites will be relatively short, the electrotonic length will be correspondingly long, and synaptic potentials will therefore decrement sharply in spreading toward the axon hillock. However, as discussed in Chapter 4, intracellular recordings indicated that R_m is sufficiently high that the electrotonic lengths of most dendrites are in the range of 1–2 (Johnston and Wu, 1995) and recent patch recordings suggest much higher R_m values, indicating electrotonic lengths less than 1. Thus, a relatively high R_m seems adequate for close electrotonic linkage, at least in the steady state (Fig. 17.5B).

Low K+ Conductances

An important factor controlling the effective membrane resistance is K+ conductances. Chapter 4 discussed how a K+ channel, I_h, can affect the summation of EPSPs in striatal spiny cells. There is increasing evidence for control of dendritic input conductance by

FIGURE 17.4 Dendritic systems as motion detectors. (A) A computational model of a motion detector neuron in the visual system of the fly, consisting of elementary motion detector (EMD) units in its dendritic tree activated by the preferential direction (PD) of motion. The local modulations of the individual EMDs (1–3) are integrated in the dendritic tree to give the smooth global output in the axon (bottom trace) (Single and Borst, 1998). (B) The dendritic trees of neurons in the optic tectum of lizard (B1), chick (B2), and gray squirrel (B3). The architecture of the dendritic branching patterns and distal specialization for reception of retinal inputs is highly homologous. (References in Luksch *et al.*, 1998.)

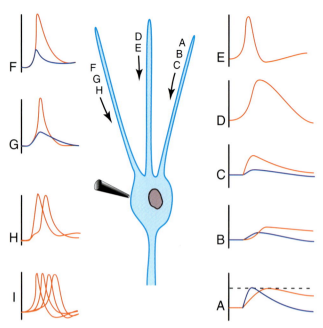

FIGURE 17.5 Mechanisms by which the synaptic responses in distal dendrites can have an enhanced effect in controlling impulse output from the axon hillock–initial segment region. Schematized neuron with patch electrode on the soma. Blue traces show responses with passive dendritic membrane; red traces show responses with dendritic boosting mechanisms. (A) Larger distal synaptic conductances (the response is of the same amplitude but slower compared with a soma synaptic conductance because of the electrotonic delay in the intervening dendrites). Spine stem diameter = 0.1 μm. (B) Higher membrane resistance (the response is slowed compared with the response to a soma input by the larger time constant). Voltage-gated channels (VGCs) may increase an EPSP amplitude as seen at the soma (C). Within the dendrites, VGCs can give rise to slow action potentials (D) or to forwardpropagating full action potentials (E). Dendritic VGCs can set up fast prepotentials as recorded at the soma (F); they can also function as coincidence detectors (G) which give rise to "pseudosaltatory conduction" toward the soma through individual sites (H) or clusters (I). (see text).

different types of K^+ currents (Midtgaard *et al.*, 1993; Magee, 1999). When dendritic K^+ conductances are turned off, R_m increases and dendritic coupling to the soma is enhanced. These conductances also control backpropagating action potentials, as we discuss below.

Voltage-Gated Depolarizing Conductances

For transient responses, the electrotonic linkage becomes weaker because of the filtering effect of the capacitance of the membrane, and it is made worse by a higher R_m, which increases the membrane time constant and thereby slows the spread of a passive potential (see Chapter 4). This disadvantage can be overcome by depolarizing voltage-gated conductances, either Na^+ or Ca^{2+}, or both. Box 17.1 discusses the variety of mechanisms by which these voltage-gated conductances can operate.

Summary

These examples illustrate an important principle of *distal dendritic processing*:

Distal dendrites can mediate relatively rapid, specific information processing, even at the weakest levels of detection, in addition to slower modulation of overall neuronal activity. Spread of potentials to the site of global output from the axon is enhanced by multiple passive and active mechanisms.

Depolarizing and Hyperpolarizing Dendritic Conductances Interact Dynamically

We see that depolarizing conductances increase the excitability of the distal dendrites and the effectiveness of distal synapses, whereas K^+ conductances reduce the excitability and control the temporal characteristics of the dendritic activity. This balance is thus crucial to the functions of dendrites. Figure 17.7 summarizes recent data, showing how these conductances vary along the extents of the dendrites of mitral cells, hippocampal and neocortical pyramidal neurons, and Purkinje cells.

The significance of a particular density of channel needs to be judged in relation to the electrotonic properties discussed in Chapter 4. For instance, a given conductance has more effect on membrane potential in smaller distal branches because of the higher input resistance (see Fig. 4.9). We discuss the significance of these conductance interactions for the firing properties of these different cell types below. Dendritic conductances can be crucial in setting the intrinsic excitability state of the neuron. In the motor neuron, for example, the neuron can alternate between bistable states dependent on activation of dendritic metabotropic glutamate receptors (Svirskie *et al.*, 2001).

Summary

The combination of conductances at different levels of the dendritic tree involves a delicate balance between depolarizing and hyperpolarizing actions acting over different time periods. The combination of conductances at different levels of the dendritic tree is characteristic for different morphological types of neurons.

The Axon Hillock–Initial Segment Encodes Global Output

In cells with long axons, activity in the dendrites eventually leads to activation and modulation of

BOX 17.1

VOLTAGE-GATED COMPUTATIONS IN DENDRITES

Active sites within branches or spines may act as *coincidence detectors* of simultaneous synaptic responses. Through such mechanisms, simple logic gates are set up, which can perform the basic logic operations of AND (e.g., Fig. 17.6), OR, and AND-NOT. Other types of computation in dendrites include linearization of synaptic interactions and basic types of arithmetic processing: addition, subtraction, multiplication, and division (Koch, 1999).

There may be a sequence of coincidence detection as an active response spreads from site to site through the branching tree. This means that *the effectiveness of a distal EPSP may depend not on spreading all the way to the soma, but rather on spreading to the nearest site containing voltage-gated Na⁺ or Ca²⁺ channels, for coincidence detection and conduction to the next local site*, and so on. Sequential spread of local active potentials from site to site provides for *pseudosaltatory conduction* through the dendritic tree (Fig. 17.5H). This may occur between individual sites or multiple sites forming clusters (Fig. 17.5I). There is experimental and/or theoretical evidence for all of these mechanisms, some of which is considered below.

How does forward propagation of active responses in dendrites fit with the classic model of action potential initiation at the axonal initial segment? We see that there are cells in which the initiation site actually shifts between initial segment and distal dendrite depending on the strengths and locations of synaptic excitation and inhibition. *The site of action potential initiation thus can vary depending on the dynamic state of the neuron.* It is also important to recognize that, because of the filtering effect of the dendritic cable properties, dendritic action potentials that spread to the soma may be indistinguishable at the soma from EPSPs (Fig. 17.6B; see Chapter 4). Thus, dendritic EPSPs that trigger the action potential at the initial segment, as in the classic model, may actually include significant contributions from active dendritic depolarization.

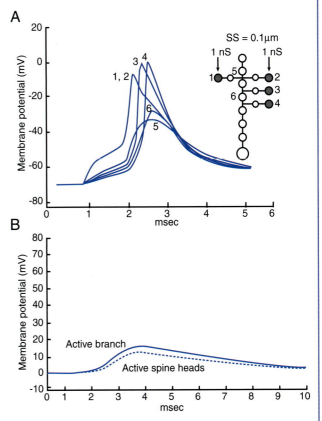

FIGURE 17.6 Logic operations are inherent in coincidence detection by active dendritic sites. The example is an AND operation performed by two dendritic spines with Hodgkin–Huxley–type active kinetics, with intervening passive dendritic membrane. (A) Simultaneous synaptic input of 1-nS conductance to spines 1 AND 2 gives rise to action potentials within both spines, which spread passively to activate action potentials in spines 3 and 4. Decreasing the synaptic conductance or increasing the spine stem diameter causes the coincidence mechanism to fail. Sequential coincidence detection by active spines can thus bring boosted synaptic responses close to the soma. (B) Recording of the boosted spine responses at the soma shows their similarity to the slow time course of classic EPSPs, due to the electrotonic properties of the intervening dendritic membrane. See text. (A) from Shepherd and Brayton (1979). (B) from Woolf *et al.* (1991).

action potential output in the axon. A key question is the precise site of origin of this action potential. This question was one of the first to be addressed in the rise of modern neuroscience; the historical background is summarized in Box 17.2.

These studies established the classic model: the lowest threshold site for action potential generation is in the *axonal initial segment*.

Further testing had to await the development of methods for recording directly from dendrites in tissue slices. In CA1 hippocampal pyramidal neurons, weak synaptic potentials elicited action potentials near the cell body (Richardson *et al.*, 1987), but this site shifted to the proximal dendrites with stronger synaptic excitation (Turner *et al.*, 1991). This confirmed the suggestion of Fuortes and his col-

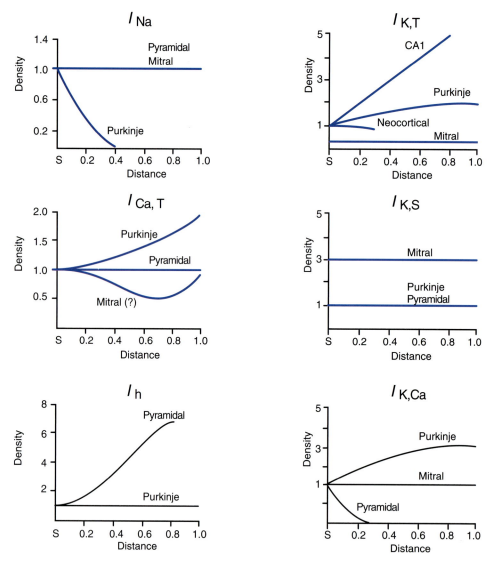

FIGURE 17.7 Graphs of the distribution of different types of conductances at different levels of the dendritic trees in different types of neurons. S: normalized extent of dendritic tree. From Magee (1999, p. 149).

leagues that the site can shift under different stimulus conditions, and was consistent with the stretch receptor, where larger receptor potentials shift the initiation site closer to the cell body.

Definitive analysis was achieved by Stuart and Sakmann (1994) using dual patch recordings from cortical pyramidal neurons under infrared differential contrast microscopy. As shown in Fig. 17.10, with depolarization of the distal dendrites by injected current or excitatory synaptic inputs, a large amplitude depolarization is produced in the dendrites which spreads to the soma. Despite its lower amplitude, the soma depolarization is the first to initiate the action potential. Subsequent studies with triple patch electrodes have shown that the action potential actually arises first in the initial segment and first node (as

we saw in Chapter 4, Fig. 4.15) This approach has provided the breakthrough for subsequent analyses of dendritic properties and their coupling to the axon, as is discussed below.

Retrograde Impulse Spread into Dendrites Can Have Several Functions

In addition to identifying the preferential site for action potential initiation in the axonal initial segment, the experiments of Stuart and Sackmann (1994) showed clearly that the action potential does not merely spread passively back into the dendrites but actively backpropagates. Note that we distinguish between passive "spread" and active "propagation" of the action potential.

BOX 17.2

CLASSIC STUDIES OF THE ACTION POTENTIAL INITIATION SITE

Fuortes and colleagues (1957) were the first to deduce, redundant that the EPSP spreads from the dendrites through the soma to initiate the action potential in the region of the axon hillock and the initial axon segment.

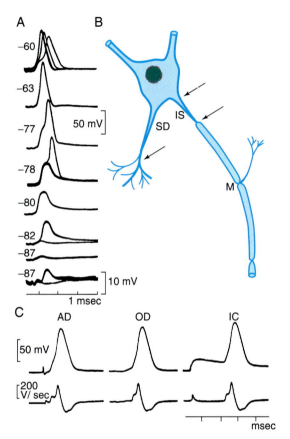

FIGURE 17.8 Classic evidence for the site of action potential initiation. Intracellular recordings were from the cell body of the motor neuron of an anesthetized cat. (A) Differential blockade of an antidromic impulse by adjusting the membrane potential by holding currents. The recordings reveal the sequence of impulse invasion in the myelinated axon (recordings at –87 mV, two amplifications), initial segment of the axon (first component of the impulse beginning at –82 mV), and soma–dendritic region (large component beginning at –78 mV). (B) Sites of the three regions of impulse generation (M, myelinated axon; IS, initial segment; SD, soma and dendrites); arrows show probable sites of impulse blockade in (A). (C) Comparison of intracellular recordings of impulses generated antidromically (AD), synaptically (orthodromically, OD), and by direct current injection (IC). Lower traces indicate electrical differentiation of these recordings, showing the separation of the impulse into the same two components and indicating that sequence of impulse generation from the initial segment into the soma–dendritic region is the same in all cases. From Eccles (1957).

They suggested that the action potential has two components: (1) an A component that is normally associated with the axon hillock and initial segment, and (2) a B component that is normally associated with retrograde invasion of the cell body. The site of action potential initiation can shift under different membrane potentials, so they preferred the noncommittal terms "A" and "B" for the two components as recorded from the cell body. In contrast, Eccles (1957) referred to the initial component as the initial segment (IS) component and to the second component as the soma dendritic (SD) component (Fig. 17.8).

Apart from the motor neuron, the best early model for intracellular analysis of neuronal mechanisms was the crayfish stretch receptor, described by Eyzaguirre and Kuffler (1955). Intracellular recordings from the cell body showed that stretch causes a depolarizing receptor potential equivalent to an EPSP, which spreads through the cell to initiate an action potential. It was first assumed that this action potential arose at or near the cell body.

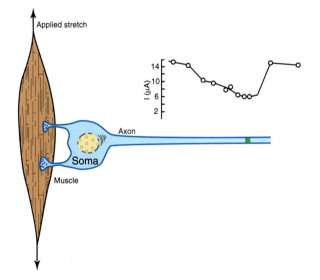

FIGURE 17.9 Classic demonstration of the site of impulse initiation in the stretch receptor cell of the crayfish. Moderate stretch of the receptor muscle generated a receptor potential that spread from the dendrites across the cell body to initiate the action potential in the axon. The excitability curve (shown above the axon) was obtained by passing current between extracellular electrodes along the axon and finding the current (I) intensity needed to evoke an action potential response; this shows the trigger zone (site with lowest threshold) to be several hundred micrometers out on the axon (green region in axon). From Ringham (1971).

BOX 17.2 (continued)

Edwards and Ottoson (1958), working in Kuffler's laboratory, tested this postulate by recording the local extracellular current to locate precisely the site of inward current associated with action potential initiation. Surprisingly, this site turned out to be far out on the axon, some 200 μm from the cell body (Fig. 17.9). This result showed that potentials generated in the distal dendrites can spread all the way through the dendrites and soma well out into the initial segment of the axon to initiate impulses. It further showed that the action potential recorded at the cell body is the backward spreading impulse from the initiation site. Edwards and Ottoson's study was important in establishing the basic model of impulse initiation in the axonal initial segment.

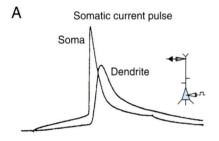

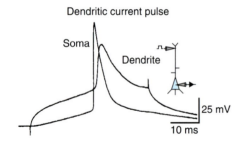

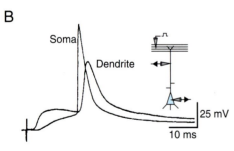

FIGURE 17.10   Direct demonstration of impulse initiation zone and backpropagation into dendrites, using dual patch recordings from soma and dendrites of a layer V pyramidal neuron in a slice preparation of the rat neocortex. (A) Depolarizing current injection in either the soma or the dendrites elicits an impulse first in the soma. (B) The same result is obtained with synaptic activation of layer I input to distal dendrites. Note the close similarity of these results to the earlier findings in the motor neuron (Fig. 17.8). From Stuart and Sakmann (1994).

What is the function of the backpropagating action potential? Experimental evidence shows that it can have a variety of functions.

Dendrodendritic Inhibition

A clear function for a backpropagating action potential was first suggested for the olfactory mitral cell, where the mitral-to-granule dendrodendritic synapses are triggered by the action potential spreading from the soma into the secondary dendrites (Fig. 17.11). Because of the delay in activating the reciprocal inhibitory synapses from the granule cells, self-inhibition of the mitral cell occurs in the wake of the passing impulse; the two do not collide. The mechanism functions similarly with both active backpropagation and passive electrotonic spread into the dendrites, as tested in computer simulations (Rall and Shepherd, 1968). The functions of the dendrodendritic inhibition include center–surround antagonism mediating the abstraction of molecular determinants underlying the discrimination of different odor molecules, storing of olfactory memories at the reciprocal synapses, and generation of oscillating activity in mitral and granule cell populations.

Boosting Synaptic Responses

In several types of pyramidal neurons, active dendritic properties appear to boost action potential invasion, so that summation with EPSPs occurs that makes them effective in spreading to the soma.

Resetting the Membrane Potential

A possible function is that the Na^+ and K^+ conductance increases associated with active propagation wipe out the existing membrane potential, resetting the membrane potential for new inputs.

Synaptic Plasticity

The action potential in the dendritic branches presumably depolarizes the spines (because of the favorable impedance matching, as discussed in Chapter 4), which means that the impulse depolarization would summate with the synaptic depolarization of the spines. This process would enable the spines to function as coincidence detectors and implement Hebb-like changes in synaptic plasticity (as discussed in detail in Chapter 18). This postulate has been tested by electrophysiological recordings (Spruston et al., 1995) and Ca²⁺ imaging (Yuste et al., 1994 and 1995). Activity-dependent changes in dendritic synaptic potency are not seen with passive retrograde depolarization but appear to require actively propagating retrograde impulses (Spruston et al., 1995).

Frequency Dependence

Trains of action potentials generated at the soma–axon hillock can invade the dendrites to varying extents. The proximal dendrites appear to be invaded throughout a high-frequency burst, whereas the distal dendrites appear to be invaded mainly by the early action potentials (Spruston et al., 1995; Yuste et al., 1994; Regehr et al., 1989; Callaway and Ross, 1995). Activation of Ca²⁺ -activated K⁺ conductances by the early impulses may effectively switch off the distal dendritic compartment.

Retrograde Actions at Synapses

The retrograde action potential may contribute to the activation of neurotransmitter release from the dendrites. Dynorphin released by synaptically stimulated dentate granule cells can affect the presynaptic terminals (Simmons et al., 1995). In the cerebral cortex there is evidence that GABAergic interneuronal dendrites act back on axonal terminals of pyramidal cells, and glutamatergic pyramidal cell dendrites act back on axonal terminals of the interneurons (Fig. 17.12). The combined effects of the axonal and dendritic compartments of both neuronal types regulate the normal excitability of pyramidal neurons, and may be a factor in the development of cortical hyperexcitability and epilepsy (Zilberter, 2000).

Conditional Axonal Output

Because of the long distance between distal dendrites and initial axonal segment, we may hypothesize that the coupling between the two is not automatic. Indeed, conditional coupling dependent on synaptic inputs and intrinsic activity states at intervening dendritic sites appears to be fundamental to the relationship between local dendritic inputs and global axonal output (Spruston, 2000).

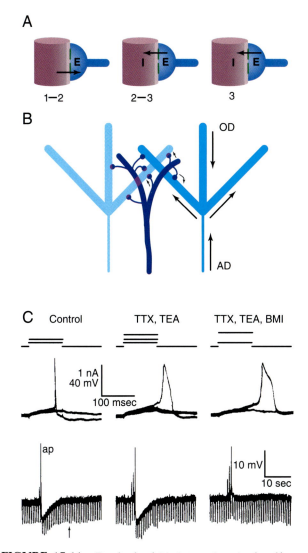

FIGURE 17.11 Dendrodendritic interactions in the olfactory bulb. (A) Depolarization by an action potential in period 1–2 activates excitatory output from mitral cell dendrite, setting up an EPSP (E) in a granule cell spine. During period 2–3, the granule spine EPSP activates a reciprocal inhibitory synapse, setting up an IPSP (I) in the mitral cell dendrite which lasts into period 3. (B) Either orthodromic (OD) or antidromic (AD) activation of the mitral cell sets up a backspreading/backpropagating impulse into the secondary dendrites, activating both feedback and lateral inhibition of the mitral cells through the dendrodendritic pathway. (C) Experimental demonstration of the dendrodendritic pathway. (Left) In an intracellular recording from a mitral cell in an isolated turtle olfactory bulb, injected depolarizing current elicits an action potential [fast trace and action potential (AP) in slow trace below] followed by a long-lasting hyperpolarizing IPSP; transient downward deflections used to measure input resistance are reduced during the IPSP, indicating an increase in membrane conductance during the IPSP. (Middle) Depolarizing current elicits a lower-amplitude and slower action potential when the preparation is bathed in TTX (which blocks the Na⁺ component of the impulse) and TEA (which blocks K⁺ conductances that would shunt the remaining Ca²⁺ component). The IPSP persists (bottom trace), activated by the Ca²⁺-dependent action potential. (Right) Addition of bicuculline (BMI) to the bath blocks the IPSP (bottom trace), presumably by blocking the granule-to-mitral reciprocal synapse. (A) and (B) from Rall and Shepherd (1968). (C) adapted from Jahr and Nicoll (1982).

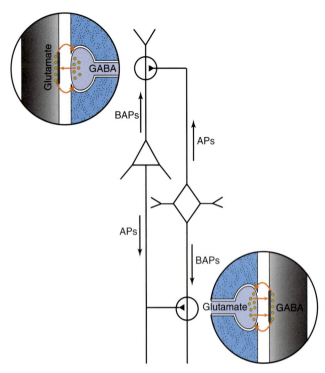

FIGURE 17.12 Pyramidal neurons and interneurons in the cerebral cortex interact through axodendritic and dendroaxonic contacts. APs, action potentials in axons; BAPs, backpropagating action potentials in dendrites. See text. From Zilberter (2000).

Examples of How Voltage-Gated Channels Take Part in Dendritic Integration

It is commonly believed that active dendrites are a modern concept, but in fact this idea is as old as Cajal. Box 17.3 gives a short history of this idea and the experimental evidence that has been obtained over the years.

Detailed analysis of active dendritic properties began with the computational studies of olfactory mitral cells and experimental studies of cerebellar Purkinje cells. Since then studies of active dendritic properties have proliferated, particularly since the introduction of the patch recording method. Several types of neurons have provided important models for the possible functional roles of active dendritic properties.

Purkinje Cells

The cerebellar Purkinje cell has the most elaborate dendritic tree in the nervous system, with more than 100,000 dendritic spines receiving synaptic inputs from parallel fibers and mossy fibers. The basic distribution of active properties in the Purkinje cell was indicated by the pioneering experiments of Llinas and Sugimori (1980) in tissue slices (Fig. 17.14). The action

potential in the cell body and axon hillock is due mainly to fast Na^+ and delayed K^+ channels; there is also a Ca^{2+} component. The action potential correspondingly has a large amplitude in the cell body and decreases by electrotonic decay in the dendrites. In contrast, the recordings in the dendrites are dominated by slower "spike" potentials that are Ca^{2+} dependent, owing to a P-type Ca^{2+} conductance (see Fig. 17.14). These spikes are generated from a plateau potential due to a persistent Na_p current (see Chapter 5).

There are two distinct operating modes of the Purkinje cell in relation to its distinctive inputs. Climbing fibers mediate strong depolarizing EPSPs throughout most of the dendrites that appear to give rise to synchronous Ca^{2+} dendritic action potentials throughout the dendritic tree, which then spread to the soma to elicit the bursting "complex spike" in the axon hillock. In contrast, parallel fibers are active in small groups, giving rise to smaller populations of individual EPSPs possibly targeted to particular dendritic regions (compartments). In this mode, *subthreshold amplification* through active dendritic properties may enhance the effect of a particular set of input fibers in controlling or modulating the frequency of Purkinje cell action potential output in the axon hillock. The Purkinje cell is subjected to local inhibitory control by stellate cell synapses targeted to specific dendritic compartments, and *global inhibitory control* of axonal output by basket cell synapses on the axonal initial segment.

Pyramidal Neurons

Active properties of the apical dendrite of hippocampal pyramidal neurons have been amply documented by patch recordings (Magee and Johnston, 1995). In contrast to the Purkinje cell, both fast Na^+ and Ca^{2+} conductances have been shown throughout the dendritic tree of the pyramidal neuron by electrophysiological and dye-imaging methods (see Fig. 17.7). Activation of the low-threshold Na^+ channels is believed to play an important role in triggering the higher-threshold Ca^{2+} channels. Similar results have been obtained in studies of pyramidal neurons of the cerebral cortex.

The output pattern of a neuron depends on its dendritic properties and their interaction with the soma. This is exemplified by the generation of a burst response in a pyramidal neuron. EPSPs spread through the dendrite, activating fast Na^+ and then high-threshold (HT) Ca^{2+} channels that give a subthreshold boost to the EPSP. The enhanced EPSP spreads to the soma–axon hillock, triggering a Na^+ action potential. This propagates into the axon and also backpropagates into the dendrites, eliciting a

BOX 17.3

CLASSIC STUDIES OF ACTIVE DENDRITIC PROPERTIES

The first intracellular recordings of active properties of dendrites were obtained in 1958 by Eccles and collaborators from motor neurons undergoing chromatolytic degeneration after amputation of their axons (Eccles *et al.*, 1958). Small spikes could be seen riding on EPSPs, which were thought to be due to impulse "booster" sites in the dendrites. Similar activity was seen in the first intracellular recordings from hippocampal pyramidal neurons (Spencer and Kandel, 1961). These "fast prepotentials" appeared to intervene between the EPSP in the dendrite and the impulse initiation in the soma–axon hillock region (Fig. 17.13). These active sites were suggested to be at branch points in the apical dendrite, where they would serve to boost the EPSPs generated by more distal dendritic inputs. This boosting property has provided an important model for the possible significance of active dendritic properties. Active properties of dendrites were the subject of increasing study from the 1950s on, with extracellular [see Fatt (1947) and Anderson (1960)] and intracellular recordings. The use of dual patch recordings finally enabled direct recordings from dendrites and comparisons with soma recordings, as discussed in the text.

A

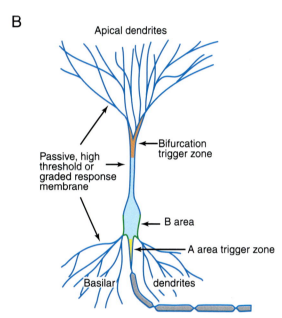

B

FIGURE 17.13   Early evidence for active dendritic properties in normal adult neurons. (A) Intracellular recordings from the soma of a hippocampal pyramidal neuron in an anesthetized cat; the large spontaneous action potentials are preceded by a small "fast prepotential" (small arrows), which occasionally occurs in isolation (large arrow). (B) Conceptual schema of how a "trigger zone" at bifurcating dendritic branches could give rise to the fast prepotential and boost the distal dendritic response. From Spencer and Kandel (1961).

---

slower all-or-nothing Ca^{2+} action potential. This large-amplitude, slow depolarization then spreads through the dendrites and back to the soma, triggering a train of action potentials that form a burst response.

This sequence of events is contained in a two-compartment model representing the soma and dendritic compartments (Fig. 17.15). The sequence emphasizes not only the importance of the interplay between the different types of channels, but also the critical role of the compartmentation of the neuron into dendritic and somatic compartments so that they can interact in controlling the intensity and time course of the impulse output.

Does the specific form of the input–output transformation depend on a specific distribution of active channels in the dendritic tree? Na^+ and Ca^{2+} channels are distributed widely in pyramidal neuron dendrites. In computational simulations, grouping channels in different distributions may have little effect on the input–output functions of a neuron (Mainen and

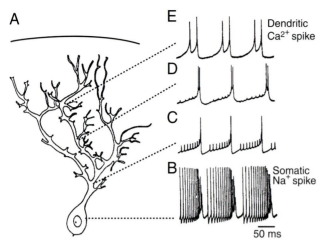

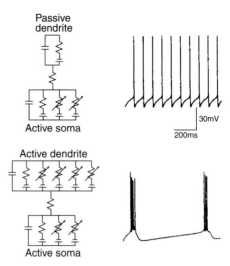

FIGURE 17.14   Classic demonstration of the difference between soma and dendritic action potentials. (A) Drawing of a Purkinje cell in the cerebellar slice. (B) Intracellular recordings from the soma, showing fast Na⁺ spikes. (C–E) Intracellular recordings from progressively more distant dendritic sites; the fast soma spikes become small, owing to electrotonic decrement, and are replaced by large-amplitude dendritic Ca²⁺ spikes. Spread of these spikes to the soma causes an inactivating burst that interrupts the soma discharge. Adapted from Llinas and Sugimori (1980).

FIGURE 17.15   Generation of a burst response by interactions between soma and dendrites. From Pinsky and Rinzel (1994).

Sejnowski, 1995). However, there is evidence that subthreshold amplification by voltage-gated channels may tend to occur in the more proximal dendrites of some neurons (Yuste and Denk, 1995). In addition, the dendritic trees of some neurons are clearly divided into different anatomical and functional subdivisions, as discussed in the next section.

Medium Spiny Cells

A third instructive example of the role of active dendritic properties is found in the medium spiny cell of the neostriatum (Figs. 17.16A,B). The passive electrotonic properties of this cell are described in Chapter 4 (Fig. 4.12). Inputs to a given neuron from the cortex are widely distributed, meaning that a given neuron must summate a significant number of synaptic inputs before generating an impulse response. The responsiveness of the cell is controlled by its cable properties; individual responses in the spines are filtered out by the large capacitance of the many dendritic spines, so individual EPSPs recorded at the soma are small.

With synchronous specific inputs, the larger summated EPSPs depolarize the dendritic membrane strongly. The dendritic membrane contains inwardly rectifying channels (I$_h$) (Fig. 17.16C), which reduce their conductance on depolarization and thereby increase the effective membrane resistance and shorten the electrotonic length of the dendritic tree. The large depolarization also activates HT Ca²⁺ chan-

nels, which contribute to the large-amplitude, slow depolarizations. These combined effects change the neuron from a state in which it is insensitive to small noisy inputs into a state in which it gives a large response to a specific input and is maximally sensitive to additional inputs. Through this voltage-gated mechanism a neuron can enhance the effectiveness of distal dendritic inputs, not by boosting inward Na⁺ and K⁺ currents, but by reducing outward shunting K⁺ currents. This exemplifies the principle of dynamic control over dendritic properties mentioned earlier (see above).

Multiple Impulse Initiation Sites Are under Dynamic Control

Can the active properties of dendrites give rise to full dendritic action potentials that propagate toward the cell body and precede the action potential in the soma–axon hillock–initial segment region? Evidence for this began with extracellular recordings of a "population spike" that appears to propagate along the apical dendrites toward the cell body in hippocampal pyramidal cells (Anderson, 1960), supported by the recording of "fast prepotentials" (see above) and by current source density calculations in cortical pyramidal neurons (Herreras, 1990). However, because of the indirect nature of this evidence, it is possible (Stuart *et al.*, 1997) that these active properties of distal dendrites can boost dendritic synaptic responses but may be too slow to lead to action potential initiation and forward propagation.

Evidence on this question has been obtained from the olfactory mitral cell, whose excitatory inputs are

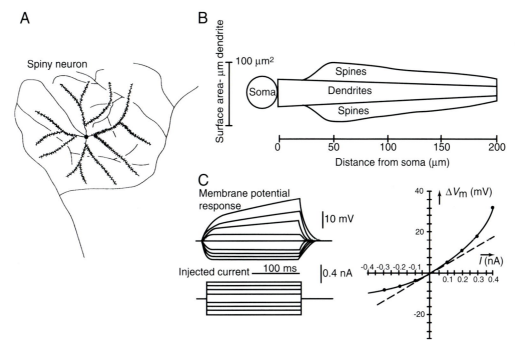

FIGURE 17.16 Dendritic spines and dendritic membrane properties interact to control neuronal excitability. (A) Diagram of a medium spiny neuron in the caudate nucleus. (B) Plot of surface areas of different compartments, showing large increase in surface area due to spines. (C) Intracellular patch-clamp analysis of medium spiny neuron, showing inward rectification of the membrane that controls the response of the dendrites to excitatory synaptic inputs (cf. Chapter 4, Fig. 4.12). From Wilson (1998).

on its distal dendritic tuft (see Fig. 17.17A). At weak levels of electrical shocks to the olfactory nerves, the site of action potential initiation is at or near the soma, as in the classic model (trace labeled 17 μA in Fig. 17.17B). This shows that despite its long length, the primary dendrite is not an impediment to the transfer of the EPSP carrying specific sensory infor-

mation from the distal dendrite to the soma and initial axonal segment. It adds another nail to the coffin of the common misconception that in neurons with axons specific excitatory inputs must be targeted near the axon hillock and that distal dendrites can mediate only slow background modulation of that site.

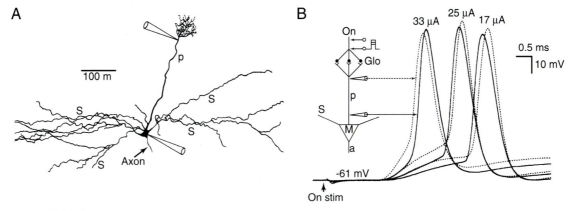

FIGURE 17.17 Shift of action potential initiation site between soma and distal dendrite. (A) A mitral cell in a slice preparation from the rat olfactory bulb stained with biocytin, showing placement of dual patch recording electrodes, one on the soma and one at the distal end of the primary dendrite 300 μm from the soma near the distal dendritic tuft in the glomerulus. (B) In another cell, an electrode site near the soma is paired with a distal dendritic site. With weak shocks to the olfactory nerves (17 μA) the soma action potential arises first; as the shocks are strengthened to 33 μA, the action potential initiation shifts to the distal dendrite recording site. Abbreviations: M, mitral cell; p, primary dendrite; s, secondary dendrite; a, axon; On, olfactory nerves; Glo, glomerulus. From Chen *et al.* (1997).

As the level of distal excitatory input is increased, the dual patch recordings show clearly that the action potential initiation site is not fixed; instead, the site shifts gradually from the soma to the distal dendrite (see 25- and 33-μA traces in Fig. 17.17B). Thus, the site of impulse initiation is not fixed in the mitral cell, but varies with the intensity of distal excitatory input. The action potential is due to tetrodotoxin-sensitive Na⁺ channels distributed along the extent of the primary dendrite. The way that passive potential spread along the dendrite controls the site of action potential initiation in these experiments has been discussed in Chapter 4 (Fig. 4.15). The site can also be shifted to the distal dendrites by synaptic inhibition applied to the soma through the dendrodendritic synapses.

Summary

These are only a few examples of the range of operations carried out by complex dendrites. These dendritic operations are embedded in the circuits that control behavior. Many further examples could be mentioned; for instance, the way that motoneuron intrinsic properties are involved in the activation patterns of motor units controlling the limbs (Gorassini *et al.*, 1999). Thus, for each neuron, the dendritic tree constitutes an expanded unit essential to the circuits underlying behavior.

Dendritic Spines Are Multifunctional Microintegrative Units

The very small size of dendritic spines has made it difficult to study them directly. However, examples have already been given of spines with complex information processing capacities, such as granule cell spines in the olfactory bulb and spines of medium spiny neurons in the striatum. In cortical neurons, spines have been implicated in cognitive functions from observations of dramatic changes in spine morphology in relation to different types of mental retardation and different hormonal exposures. Activity-dependent changes in spine morphology could be a mechanism contributing to learning and memory [summarized in Harris and Kater (1994), Shepherd (1996), Yuste and Denk (1995); see also Chapter 18].

Computational models have been very useful in testing these hypotheses, as well as suggesting other possible functions, such as the dynamic changes of electrotonic structure in medium spiny cells of the basal ganglia (see above). With the development of more powerful light microscopical methods, such as two-photon laser confocal microscopy, it has become possible to image Ca²⁺ fluxes in individual spines in relation to synaptic inputs and neuronal activity

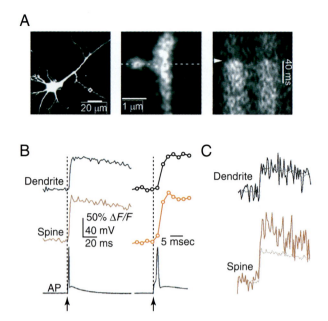

FIGURE 17.18 Calcium transients can be imaged in single dendritic spines in a rat hippocampal slice. (A) Fluo-4, a calcium sensitive dye, injected into a neuron enables an individual spine to be imaged under two-photon microscopy. (B) An action potential (AP) induces an increase in Ca²⁺ in the dendrite and a larger increase in the spine (averaged responses). Fluctuation analysis indicated that spines likely contain up to 20 voltage-sensitive Ca channels; single-channel openings could be detected, which had a high (0.5) probability of opening following a single action potential. From Sabatini and Svoboda (2000).

(Fig. 17.18). The evidence for active properties of dendrites has suggested that the spines may also have active properties. Thus, spines may be devices for nonlinear thresholding operations, either through voltage-gated ion channels or through voltage-dependent synaptic properties such as *N*-methyl-D-aspartate (NMDA) receptors. On the other hand, spines may function as compartments to isolate changes at the synapse, such as excess Ca²⁺, that would be harmful to the rest of the neuron (Volfovsky *et al.*, 1999).

The range of functions that have been hypothesized for spines is partly a reflection of how little direct evidence we have of specific properties of spines. It also indicates that the answer to the question "What is the function of the dendritic spine?" is unlikely to be only one function but rather a range of functions that is tuned in a given neuron to the specific operations of that neuron. The spine is increasingly regarded as a microcompartment that integrates a range of functions (Segev and London, 2000; Harris and Kater, 1994; Shepherd, 1996; Yuste and Denk, 1995). A spiny dendritic tree is thus covered with a large population of microintegrative units. As previously discussed, the effect of any given one of these units on the action potential output of the neuron should therefore not be assessed with regard only to the far-off cell body and axon hillock, but rather with regard first to its effect on its neighboring microintegrative units.

SUMMARY: THE DENDRITIC TREE AS A COMPLEX INFORMATION PROCESSING SYSTEM

Dendrites are the primary information processing substrate of the neuron. They allow the neuron wide flexibility in carrying out the operations needed for processing information in the spatial and temporal domains within nervous centers. The main constraints on these operations are the rules of passive electrotonic spread (Chapter 4) and the rules of nonlinear thresholding at multiple sites within the complex geometry of dendritic trees discussed in this chapter. Cells with and without axons and action potentials demonstrate many specific types of information processing that are possible in dendrites, such as motion detection, oscillatory activity, lateral inhibition, and

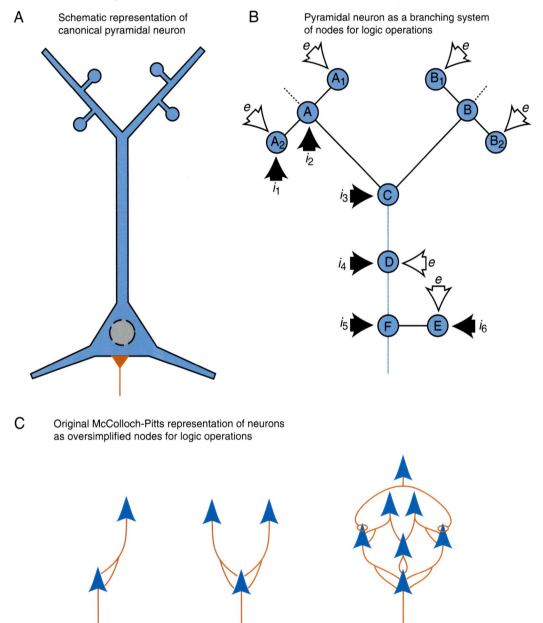

FIGURE 17.19   The dendritic tree as a computational system. Dendritic tree of a cortical pyramidal neuron (A) represented as a system of computational nodes (B). These new concepts illustrated in (B) contrast with (C), which depicts the earlier concept of McCulloch and Pitts (1943) in which the dendritic tree is ignored and the entire neuron is reduced to a single computational node. Abbreviations: e, excitatory synapse; i, inhibitory synapse. From Shepherd (1994).

network control of sensory processing and motor control. These types and more are possible for cells with axons, which in addition operate within constraints that govern local versus global outputs and subthreshold versus suprathreshold activities.

Spines add a dimension of local computation to dendritic function that is especially relevant to mechanisms for learning and memory (see Chapter 18). Although spines seem to distance synaptic responses from directly affecting axonal output, in fact many cells demonstrate that distal spine inputs carry specific information.

The key to understanding how all parts of the dendritic tree, including its distal branches and spines, can participate in mediating specific types of information processing is to recognize the tree as a complex system of active nodes. From this perspective, if a spine can affect its neighbor and that spine its neighbor, a dendritic tree becomes a cascade of decision points, with multiple cascades operating over multiple overlapping time scales. As illustrated in Fig. 17.19, a neocortical pyramidal neuron can be reduced to a canonical form (A) and then represented as a system of computational nodes. In this system, each node receiving excitatory (e) input is subjected to gating by inhibitory input at that site (i 1–6), and to further modulation and gating by inhibition between that site and the site of global output from the soma and axon hillock (site i 5). Thus, far from being a single node, as in the classic concept of McCulloch and Pitts (1943) and the classic neural network models, the complex neuron is a system of nodes in itself, within which *the dendrites constitute a kind of neural microchip for complex computations.* The neuron as a single node, so feeble in its information processing capacities, is replaced by the neuron as a complex multinodal system. The range of operations of which this complex system is capable continues to expand (see Shepherd, 1994). Exploring the information processing capacities of the brain at the level of real dendritic systems, by both experimental and theoretical methods, thus presents one of the most exciting challenges for neuroscientists in the future.

References

Anderson, P. (1960). Interhippocampal impulses: II. Apical dendritic activation of CA1 neurons. *Acta Physiol. Scand.* **48**, 178–208.

Callaway, J. C., and Ross, W. N. (1995). Frequency-dependent propagation of sodium action potentials in dendrites of hippocampal CA1 pyramidal neurons. *J. Neurophysiol.* **74**, 1395–1403.

Chen, W. R., Midtgaard, J., and Shepherd, G. M. (1997). Forward and backward propagation of dendritic impulses and their synaptic control in mitral cells. *Science* **278**, 463–467.

Craig, A. M., and Banker, G. (1994). Neuronal polarity. *Annu. Rev. Neurosci.* **17**, 267–310.

Eccles, J. C. (1957). "The Physiology of Nerve Cells." Johns Hopkins Univ. Press, Baltimore.

Eccles, J. C., Libet, B., and Young, R. R. (1958). The behaviour of chromatolysed motoneurons studied by intracellular recording. *J. Physiol. (London)* **143**, 11–40.

Edwards, C., and Ottoson, D. (1958). The site of impulse initiation in a nerve cell of a crustacean stretch receptor. *J. Physiol. (London)* **143**, 138–148.

Eyzaguirre, C., and Kuffler, S. W. (1955). Processes of excitation in the dendrites and in the soma of single isolated sensory nerve cells of the lobster and crayfish. *J. Gen. Physiol.* **39**, 87–119.

Fatt, P. (1957). Sequence of events in synaptic activation of a motoneurone. *J. Neurophysiol.* **20**, 61–80.

Fuortes, M. G. F., Frank, K., and Becker, M. C. (1957). Steps in the production of motor neuron spikes. *J. Gen. Physiol.* **40**, 735–752.

Golowasch, J., and Marder, E. (1992). Ionic currents of the lateral pyloric neuron of the stomatogastric ganglion of the crab. *J. Neurophysiol.* **67**, 318–331.

Gorassini, M., Bennett, D.J., Kiehn, O., Eken, T., and Hultborn, H. (1999). Activation patterns of hindlimb motor units in the awake rat and their relation to motoneuron intrinsic properties. *J. Neurophysiol.* **82**, 709–717.

Harris, K. M., and Kater, S. B. (1994). Dendritic spines: Cellular specializations imparting both stability and flexibility to synaptic function. *Annu. Rev. Neurosci.* **17**, 341–371.

Herreras, O. (1990). Propagating dendritic action potential mediates synaptic transmission in CA1 pyramidal cells *in situ*. *J. Neurophysiol.* **64**, 1429–1441.

Jahr, C. E., and Nicoll, R. A. (1982). An intracellular analysis of dendrodendritic inhibition in the turtle *in vitro* olfactory bulb. *J. Physiol. (London)* **326**, 213–234.

Johnston, D. A., and Wu, S. M.-S. (1995). "Foundations of Cellular Neurophysiology." MIT Press, Cambridge, MA.

Koch, C. (1999). "Biophysics of Computation: Information Processing in Single Neurons." Oxford Univ. Press, New York.

Llinas, R., and Sugimori, M. (1980). Electrophysiological properties of *in vitro* Purkinje cell dendrites in mammalian cerebellar slices. *J. Physiol. (London)* **305**, 197–213.

Luksch, H., Cox, K., and Karten, H.J. (1998). Bottlebrush dendritic endings and large dendritic fields: motion-detecting neurons in the tectofugal pathway. *J. Comp. Neurol.* **396**, 399–414.

Magee, J. C. (1999). Voltage-gated ion channels in dendrites. *In* "Dendrites" (G. Stuart, N. Spruston, and M. Hausser, Eds.), pp. 139–160. Oxford Univ. Press, New York.

Magee, J. C. (2000). Dendritic integration of excitatory synaptic input. *Nat. Neurosci.* **1**, 181–190.

Magee, J. C., and Johnston, D. (1995). Characterization of single voltage-gated Na$^+$ and Ca^{2+} channels in apical dendrites of rat CA1 pyramidal neurons. *J. Physiol. (London)* **487**, 67–90.

Mainen, Z. F., and Sejnowski, T. J.(1995). Influence of dendritic structure on firing pattern in model neocortical neurons. *Nature* **382**, 363–365.

Manor, Y., Nadim, F., Epstein, S., Ritt, J., Marder, E., and Kopell, N. (1999). Network oscillations generated by balancing graded asymmetric reciprocal inhibition in passive neurons. *J. Neurosci.* **19**, 2765–2779.

Matus, A., and Shepherd, G.M. (2000). The millennium of the dendrite? *Neuron*, **27**, 431–434.

McCulloch, W. S., and Pitts, W. H. (1943). A logical calculus of the ideas immanent in nervous activity. *Bull. Math. Biophys.* **5**, 115–133.

Midtgaard, J., Lasser-Ross, N., and Ross, W. N. (1993). Spatial distribution of Ca^{2+} influx in turtle Purkinje cell dendrites *in vitro*: Role of a transient outward current. *J. Neurophysiol.* **70**, 2455–2469.

Pinsky, P. F., and Rinzel, J. (1994) Intrinsic and network rhythmogenesis in a reduced Traub model for CA3 neurons. *J. Comput. Neurosci.* **1**, 39–60.

Poirazi, P., and Mel, B. W. (2001). Impact of active dendrites and structural plasticity on the memory capacity of neural tissue. *Neuron.* **29**, 779–796

Rall, W. (1964). Theoretical significance of dendritic trees for neuronal input–output relations. *In* "Neural Theory and Modelling" (R.F. Reiss, Ed.), pp. 73–97. Stanford Univ. Press, Stanford, CA.

Rall, W., and Shepherd, G. M. (1968). Theoretical reconstruction of field potentials and dendrodendritic synaptic interactions in olfactory bulb. *J. Neurophysiol.* **31**, 884–915.

Redman, S. J., and Walmsley, B. (1983). Amplitude fluctuations in synaptic potentials evoked in cat spinal motoneurons at identified group in synapses. *J. Physiol. (London)* **343**, 135–145.

Regehr, W. G., Connor, J. A., and Tank, D. W. (1989). Optical imaging of calcium accumulation in hippocampal pyramidal cells during synaptic activation. *Nature (London)*, 533–536.

Richardson, T. L., Turner, R. W., and Miller, J. J. (1987). Action-potential discharge in hippocampal CA1 pyramidal neurons. *J. Neurophysiol.* **58**, 98–996.

Ringham, G. L. (1971). Origin of nerve impulse in slowly adapting stretch receptor of crayfish. *J. Neurophysiol.* **33**, 773–786.

Sabatini, B. L., and Svoboda, K. (2000). Analysis of calcium channels in single spines using optical fluctuation analysis. *Nature* **408**, 589–593.

Segev I., and London, M. (2000). Untangling dendrites with quantitative models. *Science* **290**, 744–750.

Segev, I., Rinzel, J., and Shepherd, G.M. (Eds.) (1995). "The Theoretical Foundation of Dendritic Function. Selected Papers of Wilfrid Rall." MIT Press, Cambridge, MA.

Selverston, A. I., Russell, D. F., and Miller, J. P. (1976). The stomatogastric nervous system: Structure and function of a small neural network. *Prog. Neurobiol.* **37**, 215–289.

Shepherd, G. M. (1991). "Foundations of the Neuron Doctrine." Oxford Univ. Press, New York.

Shepherd, G. M. (1994). "Neurobiology," 3rd ed. Oxford Univ. Press, New York.

Shepherd, G. M. (1996). The dendritic spine: A multifunctional integrative unit. *J. Neurophysiol.* **75**, 2197–2210.

Shepherd, G. M., and R. K. Brayton (1979). Computer simulation of a dendrodendritic synaptic circuit for self- and lateral inhibition in the olfactory bulb. *Brain Res.* **175**, 377–382.

Simmons, M. L., Terman, G. W., Gibbs, S. M., and Chavkin, C. (1995). L -type calcium channels mediate dynorphin neuropeptide release from dendrites but not axons of hippocampal granule cells. *Neuron* **14**, 1265–1272.

Single, S., and Borst, A. (1998). Dendritic integration and its role in computing image velocity. *Science* **281**, 1848–1850.

Spencer, W. A., and Kandel, E. R. (1961). Electrophysiology of hippocampal neurons: IV. Fast potentials. *J. Neurophysiol.*, 272–285.

Spruston, N. (2000). Distant synapses raise their voices. *Nat. Neurosci.* **3**, 849–851.

Spruston, N., Schiller, Y., Stuart, G., and Sakmann, B. (1995). Activity-dependent action potential invasion and calcium influx into hippocampal CA1 dendrites. *Science* **268**, 297–300.

Stern, P., and Marx, J. (2000). Beautiful, complex and diverse specialists. *Science* **290**, 735.

Stuart, G. J., and Sakmann, B. (1994). Active propagation of somatic action potentials into neocortical pyramidal cell dendrites. *Nature (London)* **367**, 6–72.

Stuart, G., Spruston, N., and Hausser, M. (1999). "Dendrites". Oxford Univ. Press, New York.

Stuart, G., Spruston, N., Sakmann, B., and Hausser, M. (1997). Action potential initiation and backpropagation in neurons of the mammalian central nervous system. *Trends Neurosci.* **20**, 125–131.

Svirskie, G., Gutman, A., and Hounsgaard, J. (2001). Electrotonic structure of motoneurons in the spinal cord of the turtle: inferences for the mechanisms of bistability. *J. Neurophysiol.* **85**, 391–398.

Turner, R. W., Meyers, E. R., Richardson, D. L., and Barker, J. L. (1991). The site for initiation of action potential discharge over the somatosensory axis of rat hippocampal CA1 pyramidal neurons. *J. Neurosci.* **11**, 2270–2280.

Volfovsky, N., Parnas, H., Segal M., and Korkotian, E. (1999). Geometry of dendritic spines affects calcium dynamics in hippocampal neurons: theory and experiments. *J. Neurophysiol.* **82**, 450–462.

Wilson, C. (1998). Basal ganglia. *In* "The Synaptic Organization of the Brain" (G. Shepherd, Ed.), 4th ed., pp. 329–375. Oxford Univ. Press, New York.

Woolf, T. B., Shepherd, G. M., and Greer, C. A. (1991). Local information processing in dendritic trees: Subsets of spines in granule cells of the mammalian olfactory bulb. *J. Neurosci.* **11**, 1837–1854.

Yuste, R., and Denk, W. (1995). Dendritic spines as basic functional units of neuronal integration in dendrites. *Nature (London)* **375**, 682–684.

Yuste, R., and Tank, D. (1996). Dendritic integration in mammalian neurons, a century after Cajál. *Neuron* **13**, 23–43.

Yuste, R., Gutnick, M. J., Saar, D., Delaney, K. D., and Tank, D. W. (1994). Calcium accumulations in dendrites from neocortical neurons: An apical band and evidence for functional compartments. *Neuron* **13**, 23–43.

Zilberter, Y. (2000). Dendritic release of glutamate suppresses synaptic inhibition of pyramidal neurons in rat neocortex. *J. Physiol.* **528**, 489–496.

18

Learning and Memory: Basic Mechanisms

Thomas H. Brown, John H. Byrne, Kevin S. LaBar, Joseph E. LeDoux, Derick H. Lindquist,
Richard F. Thompson, and Timothy J. Teyler

Previous chapters in this book described the various components of nerve cells and their biophysical and biochemical properties. This chapter describes the ways in which these components and properties of the nervous system are used to mediate two of its most important functions: learning and memory. Possible subcellular modifications that underlie learning and memory are discussed in the first half of this chapter. The latter half of the chapter is an extension of a discussion from Chapter 16 on ways in which specific neural circuits can generate behavior and ways in which learning can change these behaviors and circuits.

Within a decade we expect to have a reasonably detailed cellular and molecular theory of simple forms or aspects of learning and memory. The field is just now experiencing great synergism from the fusion of two research traditions. The venerable, "top-down" approach started with the behavioral facts and laws, then identified the critical circuits, and is currently attempting to localize the neuronal mechanisms responsible for changes in the modified circuits. The more recent "bottom-up" approach began by exploring neuronal modifications that seemed like promising candidate mechanisms for supporting plasticity in circuits that control the behavior(s) of interest. Progress in this largely *in vitro* strategy has been closely tied to the development of suitable technology for analyzing synaptic structure and function.

LONG-TERM SYNAPTIC POTENTIATION AND DEPRESSION

The "bottom-up" approach is exemplified by studies on the cellular neurobiology of long-term synaptic potentiation (LTP) and depression (LTD). These activity-dependent neuronal changes have been traced for hours, days, and even months. Much of the recent research in this field has focused on the relationship of LTP and LTD to associative learning and memory.

The discovery that some forms of both LTP and LTD are "associative" strongly impacted subsequent theoretical and experimental research on the neurobiology of learning and memory. Precisely these types of modifications had been identified and explored theoretically as "associative memory elements" (Levy and Steward, 1979) by several influential thinkers (Albus, 1971; Anderson, 1985; Hebb, 1949; Hopfield, 1982, 1984; Klopf, 1982; Levy, 1985; Marr, 1969; Sejnowski, 1977a,b; Stent, 1973; von der Malsburg, 1973; Willshaw and von der Malsburg, 1976).

Leading memory researchers immediately realized that contiguity-driven, associative, synaptic modifications furnish an ideal foundation for the ~2000-year old theory that spatiotemporal contiguity is a major driving force in memory formation. First advanced by Aristotle (c. 350 BC), the idea was later revived and extended by the British Associationists (see Brown, 1818/1999), and then given progressively more refined neurophysiological interpretations by William James (1980), Ivan Pavlov (1927, 1932), Donald Hebb (1949), David Marr (1969), and James Albus (1971). Versions of this idea continue to figure prominently in postmodern theories of animal learning and memory (Brown and Lindquist, 2003; Donegan and Wagner, 1987; Ito, 1989; LeDoux, 2000; Mauk and Donegan, 1997; Medina *et al.*, 2002; Teyler, 2000; Thompson *et al.*, 1997; Wagner and Brandon, 2001) (see Classical Conditioning in Vertebrates).

For almost two decades LTP and LTD have remained the leading candidate synaptic mechanisms

Copyright 2004, Elsevier Science (USA).
All rights reserved.

for representing associative learning and memory (Brown *et al.*, 1988, 1990; Brown and Lindquist, 2003; Brown and Zador, 1990). This section describes the properties of LTP and LTD, discusses how these synaptic phenomena are analyzed, reviews what is known or suspected regarding the underlying biochemistry and molecular biology, and concludes with remarks about linkages between these synaptic modifications and representations of experience. We begin with LTP because it was discovered first and because the overall chronology of discovery has been driven more strongly by research on potentiation than depression.

The Defining Property of LTP Is That It Can Be *Rapidly* Induced by *Brief* Activity and Is Extremely *Persistent* on a Neural Time Scale

LTP Is Not the Only Activity-Dependent Increase in Synaptic Strength

Synaptic stimulation can induce many forms of activity-dependent neuronal plasticity. Three increases in synaptic strength—as measured by the size of the excitatory postsynaptic potential (EPSP) or current (EPSC)—are *facilitation* (see Fig. 18.5), *augmentation*, and *posttetanic potentiation* (PTP; see Figs. 18.1, 18.3, and 18.4 and Chapter 8). Facilitation usually lasts tens to hundreds of milliseconds, augmentation persists for seconds, and PTP is complete within a thousand seconds. Plasticity across this time scale is a normal concomitant of ordinary synaptic communication.

Synapses Exhibiting LTP can be Persistently Strengthened by Just Seconds of Stimulation

The durations of facilitation, augmentation, and PTP span four orders of magnitude. Facilitation was first carefully analyzed in crustacean neuromuscular synapses by Nobelist Bernard Katz (reviewed in Katz and Miledi, 1968). PTP was discovered a decade earlier by T. P. Feng (1939) in his pioneering studies of the cat neuromuscular junction and was later explored by Nobelist Sir John Eccles (Curtis and Eccles, 1960) in synapses of the spinal cord. For a third of a century, PTP remained the most persistent known activity-dependent form of synaptic plasticity, but memory researchers were seeking experimental evidence for a much more enduring modification. The first interesting results of this experimental search appeared in 1973, when Timothy Bliss and Terje Lomo demonstrated what we now call LTP in the anesthetized rabbit. In this classic paper, Bliss and Lomo (1973) reported that brief, high-frequency stimulation of the perforant pathway input to the dentate gyrus of the hippocampus produced a long-lasting enhancement of the extracellularly recorded field potential.

The term *plasticity*, which we and others use often in discussions of LTP and LTD, was first introduced as a hypothetical construct by William James (1890). As he defined the term, plasticity is the set of experience-dependent changes in neuronal pathways that support acquired habits. Our use of this term is essentially unchanged. The chief difference is that at the present time we can actually observe and study the kinds of neuronal modifications that James postulated to exist.

Changes in Transmission Have Been Recorded Weeks or Months after the Critical Synaptic Stimulation

A brief bout of synaptic activity can cause neuronal changes that persist for hours, days, weeks, or months. The actual duration depends in part on the experimental methods used. The changes can be traced for longer durations using *in vivo* methods, although the latter currently lack the resolution offered by *in vitro* approaches. *In vitro* studies of LTP are typically limited to about 6 h of continuous recording. The various forms of use-dependent synaptic plasticity span and help link together the time scale on which neurophysiologists routinely measured events ($\sim 10^{-5}$ to $\sim 10^{+3}$ s, counting PTP but not LTP) and the time scale over which psychologists analyze associative memory ($\sim 10^{+3}$ to $\sim 10^{+9}$ s, not counting working memory). The latency interval from $\sim 10^{+3}$ s to $\sim 10^{+5}$ s is critically important for understanding the relationship between synaptic plasticity and long-term associative memory. This interval begins after the decay of both PTP and working memory. We assume that working memory is supported by a form of neuronal plasticity that is less enduring and more rapidly reversible than LTP. The possibility of understanding causal connections across these time scales is responsible for some of the current excitement in modern neuroscience and neurophilosophy (Churchland and Sejnowski, 1992).

Repetitive, high-frequency synaptic stimulation (seconds or less) commonly induces both PTP and LTP, although either can occur without the other. In the example shown in Fig. 18.1, a single-microelectrode "switch clamp" was used to measure the mossy fiber (MF) synaptic currents in CA3 pyramidal cells before and after LTP induction (Barrionuevo *et al.*, 1986). Unitary synaptic responses (produced by stimulation of a single presynaptic axon) are considerably smaller (e.g., Fig. 18.5) than this compound response (produced by stimulation of several axons and therefore several synaptic inputs). Although the largest single-

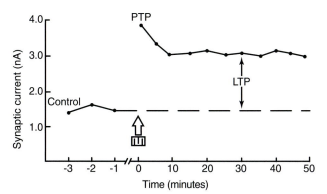

FIGURE 18.1 Mossy fiber EPSC amplitudes plotted over time, before and after the induction of LTP. Brief tetanic stimulation was applied at the time indicated (striped bar and arrow). Note the change in the time scale at the time of stimulation. Each data point is the average of five EPSCs obtained from a holding potential of –90 mV. The tetanic stimulation induced posttetanic potentiation (PTP) and long-term potentiation (LTP). Adapted, with permission, from G. Barrionuero, S. Kelso, D. Johnston, and T.H. Brown (1986). Conductance mechanism responsible for long-term potentiation monosynaptic and isolated excitatory synaptic input to the hippocampus. *J. Neurophysiol.* **55**, 540–550.

quantal events can be clearly resolved in intracellular recordings from CA3 pyramidal cells (Brown *et al.*, 1979; Johnston and Brown, 1984), the noise level of microelectrode recordings is too high to detect the majority of either spontaneous or evoked single-quantal responses. For this purpose, whole–cell recordings are required (Fig. 18.6) (Moyer & Brown, 2002; Xiang *et al.*, 1994).

The Unique Feature of LTP Is That It Can be Rapidly Induced and Is Extremely Persistent

The defining feature of LTP is not only the persistence of the synaptic modification but also the fact that it can be rapidly induced by brief bouts of activity. We define LTP as a rapid and persistent use-dependent increase in the postsynaptic conductance change, where "rapid" and "persistent" are understood to be relative to other known use-dependent synaptic changes. This high ratio of expression to induction duration is one feature of LTP that makes it so appealing as a candidate synaptic substrate for what we term "rapid learning." Fear conditioning (described later in this chapter), which can occur in a single trial, is an example of rapid learning.

The "Classic Properties" of LTP Include Cooperativity, Associativity, and Input Specificity

There are certain "classic properties" associated with LTP that are not part of the definition. These

properties are characteristic of LTP in certain extensively studied synapses in the hippocampal formation. Three properties that historically have been associated with these hippocampal synapses are "cooperativity," "associativity", and "input specificity" (Brown *et al.*, 1990). These classic properties are not orthogonal but rather are different manifestations of the same underlying Hebbian algorithm. "Associative LTP," the name given to this phenomenon (Barrionuevo and Brown, 1983), is most commonly studied in the Schaffer collateral/commissural (Sch/com) synaptic input to pyramidal neurons of hippocampal region CA1 (see Fig. 18.2 for the circuitry and terminology). It is important to recognize that other types of LTP, some of which are discussed below, have rather different signatures while sharing the same phenomenological definition. In other words, the classic properties of LTP are *not* universal.

Cooperativity Refers to the Fact That the Probability of Inducing LTP Increases with the Number of Stimulated Afferents

The number of stimulated afferents can be varied by changing the intensity of extracellular stimulation. Smaller or briefer currents (weak stimulations) activate fewer afferents than larger or longer currents (strong stimulations). Weak, high-frequency stimulation often does not induce LTP, whereas strong stimulation at the same frequency and for the same duration produces LTP more reliably (McNaughton *et al.*, 1978). This finding was termed *cooperativity* because it was thought that axons cooperated in triggering the induction of LTP. From the standpoint of the postsynaptic cell, the recruitment of additional afferents results in a greater postsynaptic depolarization. Postsynaptic depolarization (and the resulting Ca^{2+} influx) is a key factor in inducing associative LTP. The probability of inducing LTP (or the magnitude of the change) is a monotonically increasing function of postsynaptic depolarization.

Associativity is a "Pairing-Dependent" Increase in Synaptic Strength

Associativity, a hallmark of conditioning, involves learning that is driven by the spatiotemporal contiguity or covariation between or among stimuli. Levy and Steward (1979) furnished the first neurophysiological evidence for an "associative" neuronal modification based on *in vivo* extracellular recordings from the rat dentate gyrus. As soon as suitable *in vitro* methodology became available, Barrionuevo and Brown (1983) looked for and found "associative LTP" in the Schaffer collateral/commissural inputs to hippocampal CA1 pyramids. The *in vitro* brain slice

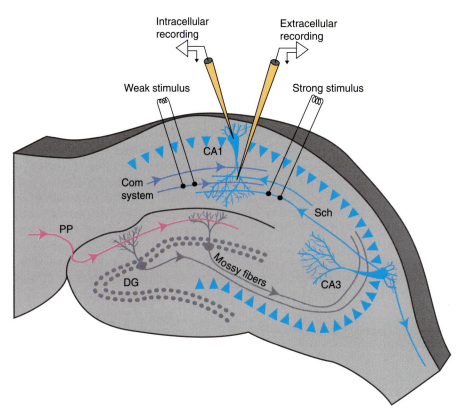

FIGURE 18.2   Schematic of a transverse hippocampal brain slice preparation from the rat. Two extracellular stimulating electrodes are used to activate two nonoverlapping inputs to pyramidal neurons of the CA1 region of the hippocampus. Both inputs consisted of axons of the Schaffer collateral/commissural (Sch/com) system. By suitably adjusting the current intensity delivered to the stimulating electrodes, different numbers of Sch/com axons can be activated. In this way, one stimulating electrode was made to produce a weak postsynaptic response and the other to produce a strong postsynaptic response. Sometimes three or more stimulating electrodes are used. Also illustrated is an extracellular recording electrode placed in the stratum radiatum (the projection zone of the Sch/com inputs) and an intracellular recording electrode in the stratum pyramidale (the cell body layer). Also indicated is the mossy fiber projection from the granule cells of the dentate gyrus (DG) to the pyramidal neurons of the CA3 region. Adapted from Barrionuevo and Brown (1983).

preparation (see Fig. 18.2; discussed later at length) helped elucidate the range of possible underlying mechanisms by enabling intracellular recordings to be combined with pharmacological manipulations.

By using brain slices it was possible to eliminate the possibility that the stimulation caused depression of inhibitory synapses versus enhancement of excitatory synapses as well as some less interesting alternatives. Note that depression of synaptic inhibition is LTD, not LTP. In the absence of pharmacological disinhibition, stimulation of the Sch/com fibers commonly elicits concomitant feedforward and/or feedback inhibition, so that the usual postsynaptic response is an excitatory–inhibitory conductance sequence (Griffith *et al.*, 1986). Appropriate changes in either inhibitory or excitatory synaptic conductances can therefore increase net excitation.

Partly to highlight similarities and differences between associative LTP and aspects of pavlovian conditioning, Brown and colleagues (Barrionuevo *et al.*, 1986; Brown and Zador, 1990; Kelso and Brown, 1986) examined interactions between weak (W, small number of stimulated afferents) and strong (S, large number of stimulated afferents) afferent inputs (Fig. 18.3). The W and S inputs were viewed as *in vitro* analogs to neuronal pathways conveying the conditioned stimulus (CS)- and unconditioned stimulus (US)-related information, respectively (see section on Associative Learning). This analogy was meant to be direct—exactly as captured in their recent models of Pavlovian fear conditioning (Brown and Faulkner, 1999; Faulkner *et al.*, 1997; McGann and Brown, 2000; Tieu *et al.*, 1999). The results were quite striking. Tetanic (high-frequency) stimulation of the W input

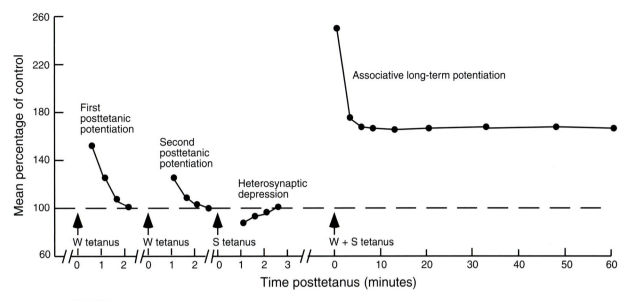

FIGURE 18.3 Plot of associative LTP in extracellular experiments (W, weak input; S, strong input). Experiment entailed recording population EPSPs in response to stimulation of the weak input, which was tested at 0.2 Hz throughout the experimental protocol. Data points represent either single response measurements (when the W-response amplitude was changing rapidly after a tetanus) or mean of five responses (when the W-response amplitudes had relaxed to a relatively constant value). The time scale changes after concurrent tetanic stimulation of both inputs (W + S tetanus). Modified from Barrioneuvo and Brown (1983).

by itself failed to produce LTP in that pathway unless this stimulation was contiguous with tetanic stimulation of the S input (Fig. 18.3). LTP was induced in W inputs only when their activity was accompanied by concurrent activity in the S input.

A simplified understanding of this associative synaptic change can be gained by simply taking into account the membrane potential of the postsynaptic cell. As already stated, postsynaptic depolarization (and perhaps also the elicitation of backward-propagating action potentials) is the key to understanding associative LTP. Repetitive stimulation of the W input alone fails to depolarize the postsynaptic cell sufficiently to induce LTP. However, the S input *is* able to depolarize the postsynaptic neuron sufficiently to allow LTP induction in the W input. Any synapses on the postsynaptic neuron that are active at the time of the critical depolarization are eligible for associative LTP induction. LTP is associatively induced in the W input by pairing stimulation of this input with concurrent stimulation of the S input.

Synapse Specificity Implies That LTP Is Restricted to the Synapses That Receive the Critical Stimulation

Synapse specificity, also termed "input specificity" (Lynch *et al.*, 1977), is another property of this form of LTP. The "spatiotemporal specificity" of associative LTP induction was examined in detail in the CA1 region by using two sets of weak inputs (W1 and W2) activated by two independent stimulating electrodes

(Kelso and Brown, 1986). Both W1 and W2 received the same tetanic stimulations, but only one of them was stimulated at the same time as the S input. LTP was induced only in the W input that was stimulated at the same time as the S input (Fig. 18.4).

The high degree of spatial specificity was emphasized by the fact that these W inputs were both Sch/com synapses that were anatomically intermingled on the apical dendrites of the CA1 pyramidal neurons. Temporal specificity was evident from the fact that LTP was not induced in the W input if its tetanic stimulation terminated a fraction of a second before the onset of the S stimulation. From the standpoint of network function, this implies that the same cell can participate in many different networks by virtue of potentiating different subsets of nearby synapses. As noted earlier, associative LTP is the obvious candidate synaptic substrate for the 2000-year-old idea (Aristotle, ca. 350 BC) that spatiotemporal contiguity drives learning and memory.

Development of the *in Vitro* Brain Slice Preparation Has Been Critical for Understanding LTP

Tooling up to Study LTP Mechanisms in Vitro Required about a Decade

In retrospect, it may seem odd that rigorous cellular neurophysiological analysis of LTP did not get

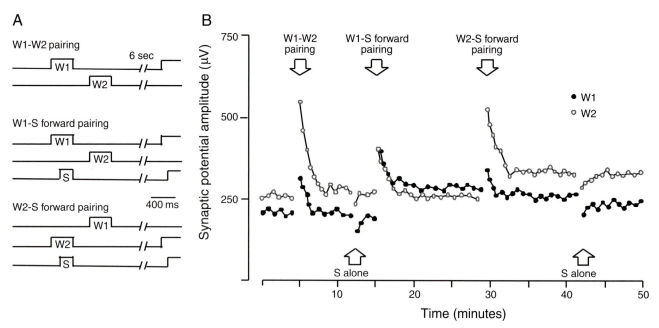

FIGURE 18.4 Differential conditioning of associative synaptic enhancement. Three bipolar stimulating electrodes—S, strong stimulation; W1, weak stimulation 1; W2, weak stimulation 2—were placed in stratum radiatum to stimulate Sch/com fibers projecting to area CA1. (A) Timing relations for three stimulation patterns. Square pulses indicate the onset and duration of 100-Hz afferent stimulation through the specified electrode. The first pattern was used as a control to determine the effects of stimulation of the weak synaptic inputs alone. In the second pattern, W1 was forward-paired with S, and in the third pattern, W2 was forward-paired with S. (B) Results from a slice in which the three stimulation patterns were applied sequentially. W1 (•), W2 (○). Modified, with permission, from S.R. Kelso and T.H. Brown (1986). Differential conditioning of associative synaptic enhancement in hippocampal brain slices. *Science* **232**, 85–87. Copyright 1986 American Association for the Advancement of Science.

underway until about a decade after its discovery. During the early 1980s, a general appreciation developed for the numerous experimental advantages of brain slices (Fig. 18.2) in studies of synaptic physiology (Dingledine, 1984). Among the most important advantages of brain slices are the exquisite degree of control the experimenter has over the chemical and physical environment of the tissue, coupled with easy access to the neurons of the slice (Moyer and Brown, 2002). Although not well suited for the study of networks of neurons (the network is not preserved), the brain slice has been the preparation of choice for studying synaptic function at identifiable classes of synapses. Partly for these reasons, most of the research that has been done on LTP thus far has used acute brain slices.

During the early development of the brain slice preparation, reasonable skepticism existed about the health of "chopped brain." Like any system in a novel environment, the brain slice may not be entirely normal. Caution must therefore be used when interpreting results from slices. One dramatic manifestation of "chopping" is the observation of a higher density of dendritic spines on hippocampal neurons

in brain slices than seen *in vivo*, a change thought to be due to the isolation of the *in vitro* neurons from their normally active synaptic inputs (Kirov and Harris, 1999). Slice health, which has historically been much more of a concern among cellular neurophysiologists than neurochemists, continues to be an important topic. Useful optical methods are now available (discussed below) for monitoring the health of the tissue as well as any particular neuron. Methods of slice preparation have become much more refined and precise partly because of the rapid optical feedback regarding neuronal and tissue health.

The Hippocampal Brain Slice Was Particularly Appealing for Architectural Reasons

The hippocampal brain slice proved to be particularly useful because much of the intrinsic circuitry remains intact in a transverse slice. A schematic illustration of a rat hippocampal brain slice preparation is shown in Fig. 18.2. Illustrated here is the classic "trisynaptic circuit": entorhinal cortex to dentate, dentate to region CA3, and region CA3 to region CA1. Because of its "lamellar" organization, much of the trisynaptic circuit remains intact in a transverse hippocampal brain slice. Although it is now known that the circuitry within a hippocampal brain slice is more complex than suggested by the trisynaptic circuit (Claiborne *et al.*, 1993; Xiang and Brown, 1998), the architecture of parts of the hippocampus—in particular, region CA1—continues to be the most convenient for many neurophysiological studies.

The Schaffer Collateral/Commissural Synapses onto Pyramids of Region CA1 Have Been Favorites

Within the hippocampus proper, the most commonly studied synapse is the Schaffer collateral/commissural input to pyramidal neurons of the CA1 region (Fig. 18.2). These afferents arise from ipsilateral CA3 cells and from the ipsilateral and contralateral hippocampus. The Sch/com-to-CA1 input is by far the most extensively studied synapse in the mammalian brain. Part of the interest is due to its relatively simple circuitry and its lamellar organization. The spatial distribution of excitatory synaptic current sinks and sources is such that useful interpretations can be made of extracellular recordings. The latter are much easier to perform and can be preferable to study in some instances. The CA1 subfield of the rat hippocampus is sufficiently large that Brown and co-workers were able to use as many as three independently controlled stimulating electrodes plus two recording electrodes, which was essential for their early studies of associative LTP and its underlying Hebbian mechanism (Barrionuevo and Brown, 1983; Kelso and Brown, 1986; Kelso *et al.*, 1986).

The Mossy Fiber Synapses onto Pyramidal Cells of Region CA3 are Large and Close to the Soma

Relatively few laboratories have studied the mossy fiber input from the granule cells of the dentate gyrus onto the pyramidal neurons of region CA3 (Fig. 18.2). In the CA3 region (as in the amygdala and perirhinal cortex, see Brown and Lindquist, 2003) the spatial distribution of current sinks and sources does not lend itself to secure interpretations of the relationship between synaptic currents and extracellular field potentials. Progress in understanding architecturally complex brain regions such as these requires higher-resolution stimulation and recording methods. The first voltage-clamp study of MF LTP (see Fig. 18.1) used microelectrodes in concert with a "switch clamp" (Barrionuevo *et al.*, 1986). The MF synapses seemed ideal for voltage-clamp studies because of their proximity to the pyramidal cell body (Brown and Johnston, 1983; Johnston and Brown, 1984). The MF synapses were also appealing because their unusually large size raised the possibility of perform-

ing physiological experiments on the presynaptic "expansions." CA3 pyramids were the first neurons in the brain from which single-quantal events were detected (Brown *et al.*, 1979). The reason why they were first discovered and studied in these neurons is that the amplitude of the largest quantal events—perhaps those from the MF synapses—is considerably greater than is typical of other classes of hippocampal projection neurons. The large size of individual quanta in these cells is favorable for quantal analysis (Greenwood *et al.*, 1999; Johnston and Brown, 1984; Xiang *et al.*, 1994).

Examples of unitary MF EPSCs recorded in rat brain slices before and after LTP induction are illustrated in Fig. 18.5. The upper waveforms are superimposed EPSCs before (Fig. 18.5A) and after (Fig. 18.5B) LTP induction, and the corresponding lower waveforms are their averages. LTP is indicated by the fact that the average responses to both the first and the second pair of stimulations were greater after LTP induction (Fig. 18.5B, bottom traces) than before (Fig. 18.5A, bottom traces). LTP was induced by stimulating the MF synapses with a short train of stimuli (5–10 stimulations) at 100 Hz. In rat brain slices, MF LTP can last for an hour or more with little if any decrement. Since brain slices have a limited life span, studies of the full time course of hippocampal LTP typically require *in vivo* studies. Of course, some degree of experimental control and access is relinquished in such studies, and the measurements are typically limited to extracellular recordings.

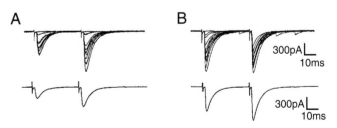

FIGURE 18.5 LTP in the mossy fiber synapses of the rat hippocampus. Whole-cell recordings were made from a CA3 pyramidal cell in which the soma was maintained under voltage clamp at −80 mV during minimum stimulation of the dentate gyrus. Both before and after LTP induction, the synapses were tested using a "paired-pulse" paradigm to reveal "paired-pulse facilitation" (PPF). (A) Upper traces are 15 superimposed records of mossy fiber EPSCs before LTP induction; the lower trace is their average. Note that the amplitude of the response increases in the second of the paired stimulations, demonstrating PPF. (B) Upper traces are 15 superimposed records of EPSCs after LTP induction; the lower trace is their average (same cell and conditions). Note that LTP induction increased the average EPSC amplitude in response to both the first and second pulses. Adapted from Xiang *et al.* (1994).

Optical Advances Have Been Critical for Developing the Brain Slice Preparation

The early brain slice experiments were performed while viewing the tissue with an ordinary dissecting microscope. This allowed discrimination of hippocampal subfields (dentate, region CA1, and region CA3; see Fig. 18.2) as well as the layers in each of these regions (stratum pyramidale, oriens, radiatum, and moleculare). Although individual neurons cannot be visualized using these optics, these gross landmarks were sufficient for extracellular field-potential recordings as well as many types of intracellular recordings using sharp microelectrodes. During the late 1980s, several hippocampal brain slice laboratories began contemplating changing from sharp intracellular electrodes to whole-cell recordings using patch pipets.

This transition brought with it the need to be able to visualize the cells before and during the recordings. The first key discovery was the finding by Brian McVicar in 1984 that the use of infrared (IR) illumination, applied to an ordinary compound microscope, enabled individual hippocampal neurons to be easily visualized. Image contrast was improved with the addition of differential interference contrast (DIC) optics and digital/analog video-enhanced microscopy (Keenan *et al.*, 1988; Moyer and Brown, 1998, 2002). Using these optics, the cell type of interest can be selected and one can know in advance whether the selected neuron is healthy. The health of the slice as a whole or of some portion of the slice, such as a particular cortical layer, is easily discerned from the frequency of normal-looking cells (Moyer and Brown, 1998, 2002). The ability to visualize the soma and dendrites has allowed multiple patch-clamp recordings from a single neuron (Magee and Johnston, 1997).

Development of New Brain Slice Preparations

Most brain regions do not offer the relatively simple architecture of the hippocampus. To develop new slice preparations, it is necessary to visualize both the individual cells and the local cytoarchitecture (gross tissue landmarks). The ability to visualize such landmarks, as well as the individual cells, is critical for developing and understanding previously unexplored brain slices (see Fig. 18.6). With IR-DIC optics, the gross architecture of the living tissue is at least as distinctive as the images on plates from a rat brain atlas. Matching slices with plates is therefore straightforward. Such matching is critical in communicating the *XYZ* coordinates of a recording or stimulation site, and the rapid visual feedback regarding cell health is essential for optimizing new procedures or verifying the adequacy of old ones (Moyer and Brown, 1998, 2002). We would not have attempted to study the neurophysiology of amygdalar and perirhinal brain slices without the benefit of modern visualization techniques (Moyer and Brown, 1998, 2002).

Confocal Laser Microscopy Combined with Whole-Cell Recordings

Confocal laser scanning microscopy (CLSM) allows optical slicing through tissue. By eliminating out-of-focus images, CLSM affords greater spatial resolution in living tissue and allows visualization of living structures as small as dendritic spines (Fig. 18.7). A CA3 pyramidal neuron is pictured with CLSM at low power in Fig. 18.7A and higher magnification in Fig. 18.7B. The spines on the dendrites are clearly evident. The spines shown in Figs. 18.7A–C are so-called "thorny excrescences," unusually large spines that serve as the postsynaptic structures for the MF synaptic "boutons" or "expansions." The thorns are located on dendritic processes near the cell soma. More typically, distally located spines are smaller (Fig. 18.7D).

A number of fluorescent dyes have been developed for monitoring intracellular events, such as calcium (Ca^{2+}) dynamics. Interest in Ca^{2+} dynamics stems partly from the fact that Ca^{2+} is thought to be critical for triggering LTP induction, as described below. Patch clamping (using IR-DIC optics) was combined with CLSM to image $[Ca^{2+}]_i$ dynamics in the dendritic thorns as well as other subcellular compartments (Jaffe and Brown, 1997) (Fig. 18.8). Figure 18.8A shows a thin optical section through a CA3 pyramidal cell that is filled with a Ca^{2+}-sensitive dye. Figure 18.8C shows the Ca^{2+} transients, produced by a depolarizing current step, in the dendritic shaft (labeled D) and the thorn (labeled T). It appears that the Ca^{2+} signals are larger and peak sooner in the thorns than in the shaft, suggesting that there might be voltage-dependent Ca^{2+} channels (VDCCs) in the thorns themselves. To obtain higher temporal resolution of Ca^{2+} dynamics, a "line scan" was performed through the dendritic shaft and thorn (Fig. 18.8B). With this higher temporal resolution, it appeared that the change in the thorns was at least as rapid as that in the shaft and it seemed to decay faster. One implication is that VDCCs may be located directly on the dendritic spine itself (Jaffe and Brown, 1997).

Brown and co-workers (1994) had previously raised this possibility based on studies of the smaller and more typical spines on these pyramidal neurons. These Ca^{2+}- imaging experiments used a single laser gun as the photon source. More recent experiments that have employed "dual-photon" or "multiphoton" imaging systems, which offer even better resolution (Mainen *et al.*, 1999; Shi *et al.*, 1999) of changes in

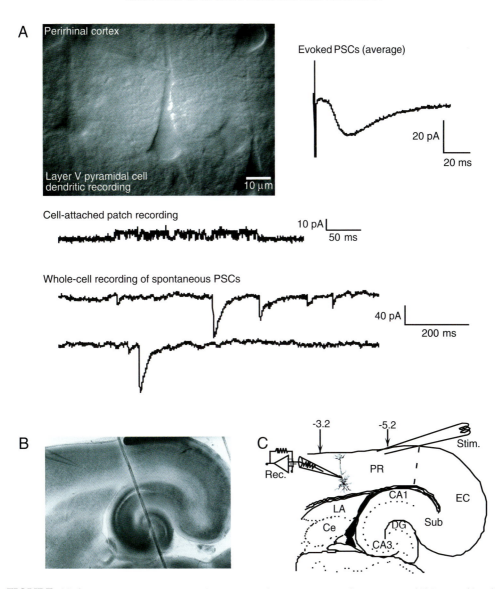

FIGURE 18.6 Dendritic recording without tissue clearing in nonsurface neurons. (A) Image file of a pyramidal cell in layer V of perirhinal cortex showing the location of the patch electrode on the apical dendrite. Data shown were collected before (cell-attached patch) and after (spontaneous and evoked PSCs) achieving the whole-cell configuration using standard electrode and recording solutions. This cell had a resting potential of –78 mV and an input resistance of 360 MΩ, and the recording remained stable for 2.5 h. (B) Photograph of the brain slice from which the cell in (A) was recorded. Such low-power photographs, taken after completion of the experiment, give convenient feedback regarding both the cell layer and the brain region from which the recording was made. Note the location of the recording electrode within layer V and the bipolar concentric stimulation electrode in layer I of perirhinal cortex. (C) Schematic diagram of the relative location of the recording site with respect to neighboring regions. CA1 and CA3, area CA1 and area CA3 of the hippocampus; Ce, central nucleus of the amygdala; DG, dentate gyrus; EC, entorhinal cortex; LA, lateral nucleus of the amygdala; PR, perirhinal cortex; Rec., recording electrode; Stim, stimulating electrode; Sub, subiculum. Numbers (relative to bregma) and arrows indicate the region within perirhinal cortex from which recordings were made. Adapted from Moyer and Brown (1998). Copyright 1998, with permission from Elsevier Science.

$[Ca^{2+}]_i$. Confirmation of L-type VDCCs in the postsynaptic membrane of excitatory synapses in the hippocampus has been achieved in an immunocytochemistry study, combined with electron microscopy

(Hell *et al.*, 1996). A dilemma associated with the existence of VDCCs on the spines is that postsynaptic activity alone might be expected to induce LTP—if Ca^{2+} influx into the spine is a sufficient condition

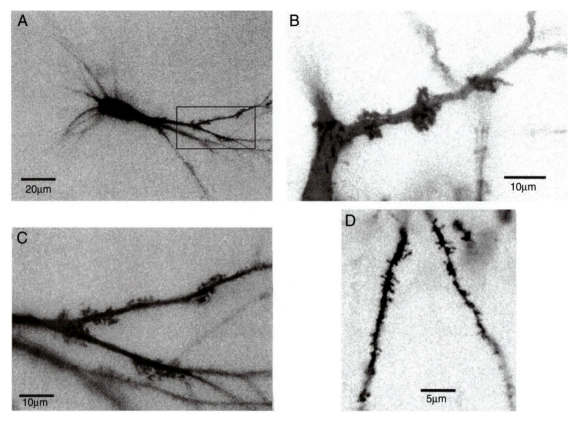

FIGURE 18.7 High-spatial-resolution imaging of thorns. Images were taken on an inverted confocal system with use of a 63× oil immersion objective (1.25 numerical aperture). (A) CA3 pyramidal neuron filled with the Ca²⁺-sensitive dye Ca²⁺ green. Image was signal averaged with photon counting. Thorns can clearly be seen on proximal apical dendrites of this neuron (boxed region). (B) Thorns visualized with higher magnification (with use of electronic zoom 3.1×) from boxed region in (A). (C) Thorns on different CA3 pyramidal neuron labeled with Fast DiI with use of oil drop method of Hosakawa *et al.* (1992). (D), basal dendrites from same CA3 region as in (C). Note difference in spine morphology and distribution between simple spines on basal dendrites and thorns in (B) and (C). Adapted, with permission, from D. B. Jaffe and T. H. Brown (1997). Calcium dynamics in thorny excrescences of CA3 pyramidal neurons. *J. Neurophysiol.* **78**, 10–18.

(Zador *et al.*, 1990). However, we know that postsynaptic activity alone does not have this effect (Kelso *et al.*, 1986). A global increase in the dendritic [Ca²⁺]ᵢ does not seem to be a sufficient condition for inducing LTP. One way around this apparent dilemma is to assume extreme spatial compartmentalization of the earliest Ca²⁺-dependent biochemical reactions that can ultimately have an enduring effect on synaptic strength. An alternative or additional possibility is that there is another second messenger whose level can be used to reflect presynaptic activity.

A Hebbian Mechanism Explains the Classic Properties of Long-Term Potentiation

How can these classic properties of LTP in the hippocampal formation be explained? In the late 1940s, the Canadian psychologist Donald Hebb (1949) advanced an idea regarding the *conditions* that cause

synapses to change. His thinking proved to be influential and informed later experiments that probed the mechanisms behind LTP. According to Hebb's now-famous postulate:

> When an axon of cell A is near enough to excite a cell B and repeatedly or persistently takes part in firing it, some growth process or metabolic change takes place in one or both cells such that A's efficiency, as one of the cells firing B, is increased. (Hebb, 1949, p. 62)

In short, coincident activity in two synaptically coupled neurons would cause increases in the synaptic strength between them. Hebb's postulate could be thought of as the synthesis of William James' (1890) "law of neural habit" and Eugenio Tanzi's (1893) synaptic hypothesis for memory (Brown *et al.*, 1990). Note that when Hebb published his thoughts the synaptic hypothesis for learning and memory had

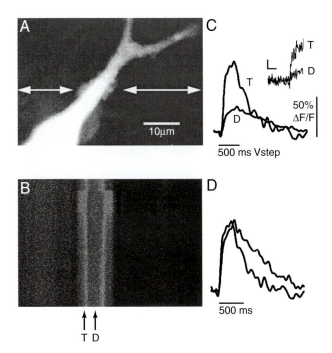

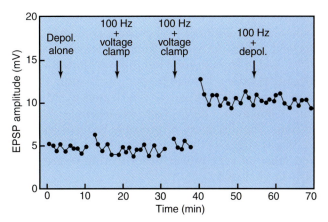

FIGURE 18.9 Demonstration of hebbian synapses. Intracellular recordings were made from a pyramidal cell in area CA1 of the hippocampus, with stimulation applied to the Sch/com inputs. Excitatory postsynaptic potential (EPSP) amplitudes are plotted as a function of time of occurrence (arrows) of three manipulations: an outward current step alone (Depol. alone) or synaptic stimulation trains delivered while applying either a voltage clamp (100 Hz + voltage clamp) or an outward current step (100 Hz + depol.). Only presynaptic stimulation combined with postsynaptic depolarization resulted in lasting potentiation. Each point is the average of five consecutive EPSP amplitudes. Adapted from Kelso *et al.* (1986).

FIGURE 18.8 High-time-resolution measurement of Ca^{2+} signals in thorns. (A) Full frame image (768 × 512 pixels) of a CA3 pyramidal neuron visualized with seminaphthorhodafluor 1(SNARF-1) emission. Horizontal arrows: position of scanned line. (B) Fluorescence line scan of cell in (A) in response to a 500-ms depolarizing voltage step (V_{step}, −80–0 mV). Thorn and dendritic time courses follow down vertical axis. (C) changes in fluorescence in thorns and dendritic shaft. Larger $\Delta F/F$ signals are observed in thorns (T) versus dendrite (D). Inset: Unfiltered data from first 100 ms of voltage step. Bar: 25% $\Delta F/F$, 50 ms. (D) Normalized time course from (C). Decay of Ca^{2+} in thorn and dendrite, in this example, was best fit by single exponentials. Decay time constants: thorn, 330 ms; dendrite, 670 ms. Adapted, with permission, from D. B. Jaffe and T. H. Brown (1997). Calcium dynamics in thorny excrescences of CA3 pyramidal neurons. *J. Neurophysiol.* **78**, 10–18.

been discussed for several decades already (Cajal, 1911; Konorski, 1948; Tanzi, 1893; Woodworth, 1921; Zeigler, 1900). There are also numerous contemporary interpretations of Hebb's postulate, most of which can be captured by the mnemonic *Cells that fire together, wire together*, a concept that also applies to developmental changes in neural connectivity.

Could the classic properties of LTP all be consequences of synapses that obey a Hebbian rule? A definitive test of this hypothesis was performed by Brown and co-workers (Kelso *et al.*, 1986). If the synapses were hebbian, then it should be possible to induce LTP under experimental conditions in which direct depolarization of the postsynaptic neuron via the recording microelectrode substitutes for the usual S input (see Fig. 18.9). In contrast, if the synapses were nonhebbian, and the critical role of the S input were instead to release an "LTP factor," as some had suggested (Goddard, 1982; Hopkins and Johnston,

1984), then pairing presynaptic stimulation of a W input with direct depolarization of the postsynaptic cell should fail to induce LTP.

The results of Kelso *et al.* (1986) are shown in Fig. 18.9, which plots the mean amplitude of the EPSPs evoked in CA1 pyramidal neurons as a function of time and the various experimental manipulations. Single extracellular stimuli were delivered to the Sch/com afferent input, producing stable EPSPs. Depolarization of the postsynaptic cell alone (in the absence of presynaptic tetanic stimulation) was without effect (Depol. alone). Nor were two successive tetanic stimuli (100 Hz) to the W input delivered while the postsynaptic soma was maintained at −80 mV under voltage-clamp conditions (100 Hz + voltage clamp). However, in agreement with the hebbian hypothesis, LTP was induced when the same presynaptic tetanic stimulation was given while the postsynaptic cell was depolarized under current-clamp conditions (100 Hz + Depol.).

From these studies and others it is clear that Hebb-type synapses do exist throughout the mammalian brain. Interestingly, hebbian synapses have also been reported in invertebrates. David Glanzman and co-workers (Lin and Glanzman, 1994; Murphy and Glanzman, 1997) discovered an associative synaptic enhancement in *Aplysia* that seems similar to the classic form of LTP in hippocampus in that it is triggered by an elevation in postsynaptic $[Ca^{2+}]_i$ that depends on *N*-methly-D-aspartate (NMDA)-type glutamate receptors. The NMDA receptor (NMDAR) antagonist D-2-amino-5-phosphonopentenoic acid (termed D-AP5, AP5, or APV) prevented an analogue of Pavlovian conditioning but did not block the serotonin-dependent heterosynaptic facilitation that is more commonly studied in this organism (Murphy and Glanzman, 1997) (see Invertebrate Studies section). The discovery of probable hebbian learning in *Aplysia* suggests that this form of neural computation must be very powerful for it to have arisen so early in phylogeny.

Mechanisms of LTP Induction, Expression, and Maintenance

Brown and co-workers suggested a conceptual distinction between LTP induction, expression, and maintenance (Baxter *et al.*, 1985; Briggs *et al.*, 1985a; Brown *et al.*, 1988). LTP *induction* refers to the early steps that trigger the modification process. LTP *expression* is the proximal cause of the altered synaptic efficacy. The processes that control LTP induction and expression could theoretically be localized on different sides of the synaptic cleft. LTP *maintenance* is a subset of the more general neurobiological problem of preserving function in the face of molecular and structural change.

There Is More Than One Biophysical Mechanism or Molecular Pathway for LTP Induction

A variety of experimental protocols can induce LTP. A persistent error has been to assume that the known or most common LTP mechanisms or conditions for inducing LTP are universal and complete. Three sets of variables seem to segregate different varieties of LTP: the conditions that reliably induce the synaptic change; the mechanisms responsible for its expression; and the resulting adaptive computations. Results from the laboratories of Timothy Teyler (Cavus and Teyler, 1996) and Daniel Johnston (Johnston *et al.*, 1992) suggest that even the same class of hippocampal synapses can undergo different forms of LTP, depending on the exact conditions used to induce the modification.

Three kinds of glutamate receptors are implicated.
To understand LTP induction, it is first necessary to consider glutamate and the receptors that it activates. Most of the synapses in which LTP has been studied use glutamate as the neurotransmitter, although notable exceptions exist (Briggs et al., 1985b; Brown and McAfee, 1982). In very broad terms, glutamate receptors (GluRs) can be subdivided into two general categories: ionotropic receptors and metabotropic

receptors (see Chapters 11 and 16 for additional details). The ionotropic receptors can in turn be divided further into two subpopulations: those that respond optimally to NMDA and those that respond to kainic acid (KA) or α-amino-3-methylsoxazole-4-propionic acid (AMPA). Metabotropic glutamate receptors (mGluRs) produce their effects by interacting with the family of trimeric G proteins that, when activated, initiate a variety of intracellular cascades.

Calcium ions trigger LTP induction.
General agreement exists that one aspect of LTP induction depends on $[Ca^{2+}]_i$ in some key compartment of the pre- and/or postsynaptic cells (Bliss and Collingridge, 1993; Johnston et al., 1992; Magee and Johnston, 1997). The exact role of Ca^{2+} in the induction process depends on the form of LTP and the particular cells and synapses involved. In the CA1 region of the hippocampus, LTP induction at the Sch/com synapse seems to depend critically on postsynaptic $[Ca^{2+}]_i$. This is also true for the MF synapses (Williams and Johnston, 1989).

It is worth pointing out here that LTP experiments purportedly all performed on the MF synapses have not always returned consistent results. The disparate findings appear to reflect the fact that the MF synapses are difficult to study in isolation, partly because so few exist (50 out of 20,000) and also because the circuitry of the dentate/CA3 region is so complex (Claiborne et al., 1993; Xiang and Brown, 1998; Xiang et al., 1994). Some of the published results on "MF LTP" clearly involved different synapses, often di- or polysynaptic inputs to the CA3 pyramids (Xiang and Brown, 1998).

The current view (Bliss and Collingridge, 1993; Teyler et al., 1994) is that multiple biochemical pathways can modulate or control $[Ca^{2+}]_i$ in critical subcellular compartment(s) (see Fig. 18.10). Three extensively studied pathways have been implicated in this aspect of LTP induction: Ca^{2+} influx through ionotropic GluRs, especially the NMDAR; Ca^{2+} influx through voltage-dependent Ca^{2+} channels (VDCCs); and Ca^{2+} release from intracellular stores via mGluR activation (Bliss and Collingridge, 1993; Magee and Johnston, 1997; Teyler et al., 1994). These three routes are described in detail below.

NMDAR-Dependent LTP

Recall that the classic (associative) form of LTP has properties that can be explained in terms of a Hebb-type rule. Considerable evidence points to a role for the NMDAR in associative LTP in the Sch/com synapses (Bliss and Collingridge, 1993). Numerous pharmacological studies have shown that competitive

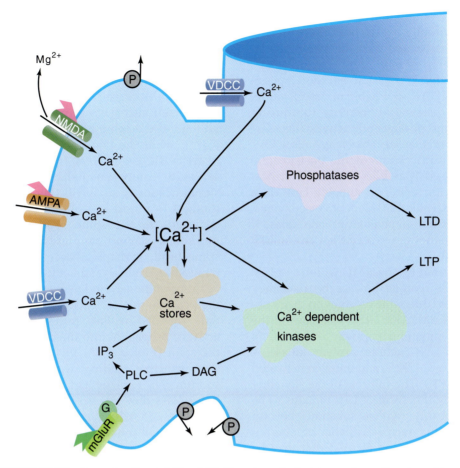

FIGURE 18.10   Events leading to LTP or LTD. The schematic depicts a postsynaptic spine with various sources of Ca^{2+}. The NMDA receptor–channel complex admits Ca^{2+} only after depolarization removes the Mg^{2+} block. Ca^{2+} may also enter through ligand-gated AMPA receptor channel or voltage-dependent Ca^{2+} channels (VDCC), which may be located on the spine head or dendritic shaft. Also, certain subtypes of metabotropic glutamate receptors (mGluRs) are coupled positively to phospholipase C (PLC), which cleaves membrane phospholipids into inositol trisphosphate (IP_3) and diacylglycerol (DAG). Increased levels of IP_3 lead to the release of intracellular Ca^{2+} stores, whereas increases in DAG activate Ca^{2+}-dependent enzymes. Ca^{2+} pumps (P) located on the spine head, neck, and dendritic shaft, are hypothesized to help isolate $[Ca^{2+}]_i$ changes in the spine head from those in the dendritic shaft.

antagonists of the NMDA receptor site (e.g., APV) or NMDA ion channel blockers (such as the non-competitive antagonist (+)-5-methyl-10,11-dihydro-5H–dibenzo[*a,d*]cyclohepten-5,10-imine; MK-801) can prevent the induction of associative LTP in these synapses.

The NMDAR has two properties that immediately suggest the nature of its role in LTP induction at hebbian synapses (Bliss and Collingridge, 1993; Brown *et al.*, 1990). First, NMDARs are permeable to Ca^{2+}, in addition to Na^+ and K^+. This property is significant given that postsynaptic $[Ca^{2+}]_i$ plays a critical role in inducing this form of LTP. Second, the channel permeability is a function of both pre- and postsynaptic factors. Channel opening requires the neurotransmitter glutamate (or some related agonist)

to bind to the NMDAR site. Glutamate binding normally signals presynaptic activity. At the usual resting membrane potential, the NMDA channels are blocked by Mg^{2+}, but this channel block is relieved by sufficient depolarization of the postsynaptic membrane. The NMDAR-mediated conductance is both ligand- and voltage-dependent, allowing Ca^{2+} entry only when presynaptic release is combined with sufficient postsynaptic depolarization.

Before proceeding further, recall some essential differences between the properties of the AMPA receptor (AMPAR) and the NMDAR (Chapters 11 and 16). The AMPAR does not exhibit a voltage-dependent Mg^{2+} block and has a much lower Ca^{2+} permeability. The AMPAR-mediated conductance is essentially voltage-independent and is permeable mainly to Na^+ and K^+, resulting in an EPSP on glutamate binding. Released glutamate can potentially act on both the AMPARs and the NMDARs associated with the membrane on the dendritic spine (Fig. 18.10).

With this knowledge, it is easy to envision a possible role for the NMDAR in a hebbian modification. Nearly concurrent presynaptic activity (producing glutamate release) and postsynaptic activity (depolarizing the neuron to relieve Mg^{2+} blockage) allow Ca^{2+} influx into the dendritic spine of the postsynaptic neuron. The increased $[Ca^{2+}]_i$ level in some critical region of the dendritic spine, presumably very close to the NMDAR, activates Ca^{2+}-dependent enzymes, such as Ca^{2+}/calmodulin-dependent protein kinase II (CaMKII), which are thought to play a key role in LTP induction (Fig. 18.10).

The kinases activated by the NMDAR form of LTP are termed serine/threonine kinases because they phosphorylate serine/threonine residues on their substrates (such as the AMPAR). Other enzymes in the second messenger signaling pathway activated by NMDA-dependent LTP include protein kinase A (PKA), protein kinase C (PKC), and mitogen-activated protein kinase (MAPK). The full role of the signaling pathway initiated by this form of LTP remains to be discovered, but probably involves substrates beyond the AMPAR, including factors capable of altering gene expression (transcription factors) that may be important for LTP maintenance.

In qualitative terms, it is easy to see how these molecular events could help account for the properties of cooperativity, associativity, and synapse specificity. Active synapses release glutamate, which can bind to the NMDAR, causing Ca^{2+} influx into dendritic spines on the postsynaptic cell. This Ca^{2+} influx acts locally, resulting in *input-specific LTP*. However, the Ca^{2+} influx occurs only when the synaptic input is strong enough to depolarize the postsynaptic membrane sufficiently to relieve the Mg^{2+} block, giving rise to *cooperativity*. The depolarization itself is mediated in large part by the ligand-gated AMPARs, which are also co-localized on the dendritic spine (Fig. 18.10). Note that activity in a weak input by itself would not depolarize the postsynaptic cell sufficiently to relieve the Mg^{2+} block unless this activity were properly timed in relationship to activity in a strong input to the same cell. The combined depolarization of the two inputs gives rise to *associativity* and synapse specificity.

Computational models of NMDAR-dependent LTP. Efforts to create biophysical models of hebbian (associative) LTP quickly led to the realization that properties of the NMDAR alone are insufficient to account

for the classic properties of LTP. For example, the voltage dependence of the NMDAR-mediated conductance is insufficient to cause an adequate rise in $[Ca^{2+}]_i$, a problem that can be rectified in part by assuming cooperative Ca^{2+} binding to the relevant kinase (CaM kinase). Characteristics of the dendritic spine itself also seem to play a key role. The small volume of the spine head results in amplification of $[Ca^{2+}]_i$, while the thin (~0.1 μm) spine neck severely restricts diffusion. Thus, the geometry of the spine serves to compartmentalize and amplify the Ca^{2+} signal (Zador *et al.*, 1990) (Fig. 18.10). The spine is thought to pump and bind Ca^{2+}, isolating the Ca^{2+} transients to the spine head. The relevant Ca^{2+}-dependent enzymes are presumed to be located in the spine head, perhaps adjacent to the Ca^{2+} channels themselves (Holmes and Levy, 1990; Zador *et al.*, 1990).

Computational models were also used to explore the possible role of "backpropagating" action potentials (BPAPs)—spikes initiated in the soma that actively propagate into the dendrites, as well as out the axon (Zador *et al.*, 1990) (see also Chapter 17). The principal finding was that backpropagating sodium (Na^+) spikes had little effect on LTP induction unless they in turn triggered Ca^{2+} spikes in the dendrites, in which case they had a pronounced effect. Daniel Johnston and co-workers have shown that there are backpropagating Na^+ spikes, that they can trigger Ca^{2+} spikes, that they critically participate in the induction of some forms of LTP, and that synaptic inputs can control the initiation and spread of these spikes (Magee and Johnston, 1997). Additional research of this kind is necessary to elucidate more precisely the nature of the synaptic control of dendritic Ca^{2+} spikes. This research does not negate the importance of NMDARs in understanding the role of Ca^{2+} in hebbian synaptic plasticity, but it clearly indicates that additional requirements or conditions are important. A dual requirement for (or synergistic interaction between) NMDAR-mediated Ca^{2+} currents and BPAPs furnishes an excellent biophysical platform for hebbian computations (Brown *et al.*, 1990; Magee and Johnston, 1997; Zador *et al.*, 1990).

In their original demonstration of the hebbian nature of associative LTP, Brown and co-workers (Kelso *et al.*, 1986) found that pharmacologically blocking Na^+ spikes (with internal QX–222) did not prevent associative LTP induction but that this treatment left intact Ca^{2+} spikes, which were clearly evident in the published recordings. The finding that Na^+ spikes were not necessary, at least under these artificial experimental conditions, led to discussion as to whether these synapses ought to be termed "hebbian" based on what Donald Hebb (1949) actually wrote. Hebb did not make a distinction between Na+ versus Ca2+ spikes or whether they occurred in the soma versus the dendrites. Instead, he referred only to "firing" the cell. Perhaps "BPAP elicitation" should be substituted for "firing," at least in some synapses under certain conditions.

Based on the history of this idea, combined with contemporary usage in the computational community, Brown and co-workers (1990) use the term *hebbian* or some variant in reference to synapses that change strength as a function of the contiguity or covariation between some measure of the pre- and postsynaptic "activity." From a computational perspective, associative LTP is quite different from the nonhebbian, "heterosynaptic" changes that have been most commonly studied in *Aplysia* (Carew *et al.* 1984) (see Invertebrate Studies).

Biochemical mechanism for NMDAR-dependent LTP. Sanes and Lichtman (1999) listed more than 100 different molecules that have been implicated in LTP induction and expression. Here, we consider only a few of these that are attracting considerable interest. It is known that glutamate receptors and signaling molecules that modulate synaptic transmission are closely associated with the postsynaptic density (PSD) of hippocampal neurons. Ca2+/calmodulin-dependent protein kinase II (CaMKII), a major constituent of the PSD (see Chapter 12), is a large multimeric enzyme that consists of 10–12 subunits. In its basal state it is inactive due to an autoinhibitory domain that blocks substrate binding. It has several interesting biochemical properties that make it particularly well suited to be a transducer of Ca2+ signals.

CaMKII can autophosphorylate, generating a constitutively active kinase that remains active long after $[Ca^{2+}]_i$ has returned to baseline levels. Phosphorylation disables the autoinhibitory domain by not allowing it to block the active site even after CaM dissociates. Autophosphorylation, which first occurs on Thr286, allows a transient elevation in Ca2+ concentration to be transduced into prolonged kinase activity that persists until the appropriate protein phosphatase dephosphorylates Thr286. Alcino Silva and colleagues explored the potential importance of this site in transgenic mice with a Thr286 point mutation, reporting that brain slices from these mice were deficient with respect to LTP (Giese *et al.*, 1998).

Although CaMKII can phosphorylate numerous protein substrates, one particularly important to LTP is the AMPAR, which appears to be altered during LTP. The AMPAR is a heteromeric complex composed of various combinations of subunits, GluR1 to GluR4,

each complex consisting of four or five such subunits. In the CA1 region of the hippocampus, the AMPAR is composed of GluR1 and GluR2. When in its active state, CaMKII phosphorylates Ser831 on GluR1 with no analogous site on the GluR2 subunit. Interestingly, a recently developed knockout mouse lacking GluR1 shows an LTP deficit in region CA1 (Zamanillo *et al.*, 1999). Phosphorylation of the AMPAR by CaMKII increases the AMPAR-mediated synaptic current. The GluR1 subunit can adopt multiple conductance states ranging from 9 to 28 pS (Derkach, *et al.*, 1999). Phosphorylation of the AMPAR is suggested to increase the single-channel conductance from its lower basal conductance state to a higher conductance state. This effect on the AMPAR current by CaMKII requires 15–30 min for maximal effect (Davies *et al.*, 1989).

Several other protein kinases, including extracellular signal-regulated kinase/mitogen-activated protein kinase (ERK/MAPK), PKC, PKA, and the SRC family of tyrosine kinases, have also been suggested to contribute to LTP. PKC can phosphorylate Ser831 on GluR1, whereas PKA can phosphorylate the adjacent Ser845. PKC has been suggested to play a role analogous to CaMKII, primarily because increasing PKC activity can enhance synaptic transmission (Carroll *et al.*, 1998). Phosphorylation of Ser845 appears to occur under basal conditions, which suggests that PKA may not play a primary role in LTP induction. However, PKA could indirectly support LTP by boosting CaMKII activity. By decreasing competing protein phosphorylation activity by means of phosphorylation of inhibitor 1, an endogenous protein phosphatase inhibitor, PKA can help sustain CaMKII activity (Blitzer *et al.*, 1998). A recently developed mouse model genetically inhibited calcineurin, a protein phosphatase, resulting in increased LTP both *in vitro* and *in vivo* (Malleret *et al.*, 2001). This facilitation was PKA-dependent, lending support to the importance of a balance between protein kinases and phosphatases in the intact brain. The facilitation of LTP was accompanied by an improvement in learning in hippocampus-dependent tasks and the improvement was reversed by suppression of transgene expression.

NMDAR function is also altered during LTP due to phosphorylation by tyrosine kinases. The NMDAR is a heteromeric complex consisting of the NR1 subunit in combination with the NR2A–D subunits. The NR2A and NR2B subunits are subject to phosphorylation of the intracellular C-terminal tyrosine residues, thereby relieving a basal zinc inhibition of the NMDAR. Phosphorylation of NR2A or NR2B thus potentiates the current through the NMDAR complex.

Consequently, induction of LTP can activate the SRC family of kinases, resulting in phosphorylation and enhancement of the NMDAR-mediated current. The increase in Ca^{2+} influx can then trigger autophosphorylation of CaMKII, which in due course leads to potentiation of the AMPAR-mediated current, as discussed above.

Neurons also contain many different protein phosphatases, including protein phosphatases 1, 2A, and 2B (PP1, PP2A, and PP2B or calcineurin). There are significant levels of PP1 and PP2A within the PSD, both of which are effective at dephosphorylating CaMKII, although PP1 is generally thought to be primarily responsible. PP1 normally is suppressed from dephosphorylating CaMKII during LTP via an inhibitor protein called inhibitor 1 (I1). I1 only becomes active once phophorylated by PKA. A plausible scenario is that in the initial stages of LTP induction, PKA activation causes it to phosphorylate I1, which in turn inhibits PP1, with the end result of prolonging CaMKII activity.

Hippocampal neurons contain a variety of voltage-dependent ion channels, some of which are expressed nonuniformly across the neuron. A transient K^+ channel whose density increases from the soma outward to the distal dendrites has a dampening effect on the BPAP. Activation of this K^+ current attenuates the BPAP amplitude, decreases the amplitude of EPSPs, and increases the threshold for dendritic spike initiation. Reducing this K^+ current therefore can be expected to have a significant effect on LTP induction. Johnston and co-workers have already identified two mechanisms that can control this K^+ current. First, activation of several different kinases (PKA, PKC, MAPK) can reduce this current (Hoffman and Johnston, 1998). Second, the current can be partly inactivated by trains of summating EPSPs. If a BPAP occurs within 15–20 ms of the onset of an EPSP train, its amplitude is larger in the dendrites. BPAP enlargement due to kinase activation and/or prior EPSP trains could enhance LTP induction through the increased dendritic depolarization (which should facilitate removal of the Mg^{2+} block in the NMDARs) and through enhanced Ca^{2+} influx (through VDCCs), resulting in an increase in dendritic $[Ca^{2+}]_i$ in or near the head of the dendritic spine (Johnston et al., 2000).

NMDAR-Independent LTP

The preceding account of the "classic" (hebbian, NMDAR-dependent, associative) form of LTP is sometimes incorrectly assumed to apply universally (Brown and Lindquist, 2003). It is important to recall that this account is applicable only to certain synapses under some conditions, and even then it may be only one piece of the story (Johnston et al., 1992; Magee and Johnston, 1997; Nicoll and Malenka, 1995; Teyler, 2000; Teyler et al., 1994). In many synapses, LTP induction does not involve NMDAR activation (Baxter et al., 1985; Briggs et al., 1985b; Brown and McAfee, 1982; Johnston et al., 1992; Teyler et al., 1994). Even within the hippocampus, some synapses exhibit NMDAR-independent forms of LTP. One example is the MF synaptic input to CA3 pyramidal cells. In the MF synapses, LTP is induced by a mechanism that does not involve NMDARs (Johnston et al., 1992). NMDAR-independent forms of LTP have also been reported at thalamic (Castro-Alamancos and Calcagnotto, 1999) and amygdala (Bauer et al., 2002; Brown and Lindquist, 2003; Chapman and Bellavance, 1992; Tsvetkov et al., 2002; Weisskopf Bauer and LeDoux et al., 1999) synapses.

Voltage-dependent calcium channel-dependent LTP. The Sch/com inputs to CA1 pyramidal neurons, which are known to exhibit the classic hebbian form of LTP that relies on the NMDAR, can, under the appropriate conditions, also exhibit an NMDAR-independent type of LTP. Grover and Teyler (1995) demonstrated that LTP can in fact be induced in these synapses in the presence of the competitive NMDAR antagonist APV. They were careful to evaluate the possibility that high-frequency tetanic release of glutamate might competitively unblock the APV by using a relatively high concentration of APV (200 mM) to saturate the APV binding site (Grover and Teyler, 1995). They discovered that the onset of this NMDAR-independent form of LTP was relatively slow (20–30 min), exhibited input specificity, and was prevented by nifedipine, an L-type Ca^{2+} channel blocker (Fig. 18.11). A distinction is therefore sometimes made between "nmdaLTP" and "vdccLTP."

Using a combination of electrophysiological and $[Ca^{2+}]_i$ imaging methods, Magee and Johnston (1997) further examined vdccLTP in the Sch/com synaptic input to CA1. Backpropagating dendritic spikes paired with stimulation of a weak Sch/com synaptic input to CA1 neurons produced LTP in the presence of an NMDAR blocker (100 mM DL -APV). Not surprisingly, stimulation of the weak synaptic input by itself failed to induce LTP in the presence of APV. The backpropagating spikes produced a strong Ca^{2+} signal in the dendrites, and the LTP produced by pairing was blocked by the Ca^{2+} channel blockers nimodipine and Ni^{2+}. vdccLTP is not restricted to the CA1 region of the hippocampus (Johnston et al., 1992).

Investigation of the second messenger systems activated during vdccLTP has suggested a reliance on a tyrosine kinase signaling pathway (Cavus and Teyler, 1996). Thus, nmdaLTP and vdccLTP appear to use different signaling pathways within the same cell, perhaps to mediate different cellular responses. Some hints about the different cellular effects of these two forms of LTP emanate from the suggestion that neurotrophin tyrosine kinase receptors (trkB) are upregulated following vdccLTP induction and that bathing a hippocampal slice in brain-derived neurotrophic factor (BDNF) mimics LTP (Kang and Schuman, 1995). Other work has shown that LTP in hippocampal slices is strongly inhibited following application of the BDNF scavenger TrkB-IgG (Figurov et al., 1996). Xu et al. (2000) recently reported that they were unable to induce LTP via tetanic stimulation in mutant mice lacking trkB in CA1 pyramidal neurons. They suggest the trkB reduction affected presynaptic function because postsynaptic glutamate receptors are not affected by trkB reduction, indicative perhaps that BDNF acts through trkB presynaptically, not postsynaptically, to modulate LTP. A direct connection between these latter studies and vdccLTP remains to be demonstrated.

Mnemonic functions of NMDAR-independent LTP. The possibility exists that both NMDAR-dependent and NMDAR-independent forms of LTP may co-occur in the same brain region, among different classes of synaptic inputs onto the same postsynaptic neuron, and even within the same class of synaptic inputs to the same postsynaptic neuron (Fig. 18.11). The question of the cellular and network significance of multiple forms of synaptic enhancement remains to be elucidated. One suspects that different forms of LTP may have different computational implications at the circuit level.

Timothy Teyler and colleagues hypothesized that nmdaLTP and vdccLTP play different roles in the encoding of experience. This suggestion was prompted by three observations. First, these two forms of LTP engage distinct signal transduction cascades. Second, the induction and decay of vdccLTP occurs on a slower time scale. Third, vdccLTP does not depotentiate readily or much. Combining these observations, Teyler and colleagues (Cavus and Teyler, 1996; Teyler, 2000) suggested that vdccLTP may be more important in long-term memory, whereas the more labile nmdaLTP could support short- or intermediate-term memory (Morgan et al., 1993, Teyler, 2000). Similar ideas have recently been put forward for memory formation in the amygdala (Bauer et al., 2002; Blair et al., 2001).

The Role of Metabotropic Glutamate Receptors in LTP Induction

Another of the more recently investigated LTP induction mechanisms involves the metabotropic

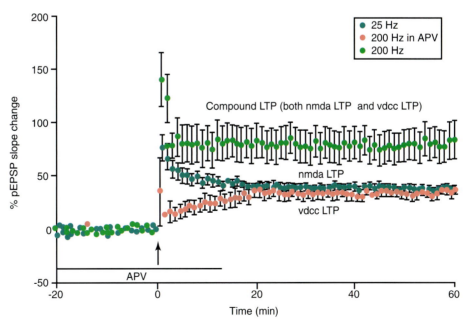

FIGURE 18.11 Composite graph of multiple forms of LTP. Superimposition of 25-Hz LTP (blue circles) and 200-Hz in APV LTP (red circles) reveal that the two forms achieve stable potentiation at similar magnitude (60 min posttetanus). A 200-Hz tetanus in standard medium (green circles) results in a compound LTP twice as large as the NMDA- and VDCC-mediated components. pEPSP, population EPSP. Adapted, with permission, from Cavus and Teyler (1996). Two forms of long-term potentiation in area CA 1 activate different signal transduction pathways. *J. Neurophysiol.* **76**, 3038–3047.

glutamate receptor, a receptor that is not coupled to an ion channel (Bashir *et al.*, 1993) (Chapter 11). Like the NMDAR, this receptor is found on the postsynaptic cell, but unlike the NMDAR, it is also found presynaptically. The role of mGluRs in LTP has been studied most extensively in the Sch/com input to hippocampal region CA1. Application of the mGluR antagonist (+)-α-methyl-4-carboxy-phenylglycine (MCPG) was reported to block LTP induction in synapses that had not previously received high-frequency stimulation, but not in synapses that had prior exposure to this stimulation (Bortolotto *et al.*, 1994). One interpretation of these and other experiments is that the mGluR acts as a "molecular switch" that must be activated as a prerequisite to LTP induction (Bortolotto *et al.*, 1994). This interesting idea has generated further experiments in other laboratories, but a consistent pattern has not yet emerged (Selig *et al.*, 1995).

Clearly, it is important to know more about the factors that affect the spatial distribution and relative proportions of the various mGluR subtypes and their ultimate roles in neuronal function. Class I mGluR subtypes (mGluR 1 and mGluR 5) are coupled to phospholipase C (PLC), which enzymatically breaks down membrane phospholipids to form diacylglycerol (DAG) and inositol 1,4,5-trisphosphate (IP$_3$) (see Fig. 18.10 and Chapter 12). DAG modulates channel activity through PKC, while IP$_3$ mobilizes release of Ca^{2+} from intracellular stores (Fig. 18.10), a process that does not raise intracellular Ca^{2+} concentrations as quickly as the opening of VDCCs. In contrast, class II mGluR subtypes (mGluR 2 and mGluR 3) are coupled to G protein-mediated inhibition of adenylyl cyclase, an action that causes a depression of the second messenger cAMP (not shown in Fig. 18.10). Perhaps more selective manipulations—at the level of mGluR subtypes—will yield a consistent and rational pattern in terms of LTP induction.

Numerous Mechanisms Have Been Implicated in LTP Expression

Up to this point, we have focused on early events in the causal chain that trigger the synaptic modification process. Now we shift to the subsequent physical changes that are responsible for the synaptic enhancement after it has been induced. Most of this research has looked for direct or indirect evidence of either increased transmitter release or increased sensitivity to released transmitter. Evidence for increased release has been used to suggest a presynaptic modification, whereas increased sensitivity has been taken as evidence for a postsynaptic change.

Presynaptic Changes in LTP Expression

The expression of LTP could theoretically result from either presynaptic and/or postsynaptic modifications. One obvious presynaptic change that could be responsible for LTP expression is an increase in neurotransmitter release. Although this seems like a straightforward hypothesis, convincing experimental data have been difficult to obtain. Both direct and indirect approaches have been developed for monitoring transmitter release. The following section discusses several of these different approaches.

Supporting evidence from direct measurement of released neurotransmitter. Some of the earliest evidence for a presynaptic mechanism in LTP expression came from attempts to measure released glutamate in the hippocampus. The early findings by Bliss and co-workers reported an increase in what was presumed to be released glutamate after LTP induction (Dolphin *et al.*, 1982). Their results were not universally accepted, however, partly because so little was known at that time about glutamatergic synapses and the measurement of synaptically released glutamate.

An excellent assay did exist, however, for acetylcholine (ACh) release. Brown and colleagues therefore searched for and found LTP in cholinergic synapses, selecting the *in vitro* rat superior cervical sympathetic ganglion (SCG) preparation for this work (Briggs *et al.*, 1985a; Brown and McAfee, 1982). The SCG preparation was the gold standard for studies of mammalian cholinergic neuropharmacology. Brown and McAfee decided to study the nicotinic component of the postsynaptic response. Nicotinic LTP in these synapses was as large and persistent as LTP reported in any *in vitro* preparation.

The SCG preparation was so stable that LTP could routinely be studied in two separate sets of synaptic inputs to the ganglion for several hours. The separate inputs could be stimulated separately or together similar to the experiments that were being done at the same time in hippocampal brain slices (Barrionuevo and Brown, 1983; Kelso and Brown, 1986). In the SCG preparation, the induction of nicotinic LTP was associated with an increase in ACh release (Briggs *et al.*, 1985b). LTP in the SCG was shown to differ from LTP in CA1 in that there is no requirement for postsynaptic depolarization, postsynaptic firing was not required, and ACh binding to the nicotinic AChR was not necessary.

Mixed evidence from quantal analysis of neurotransmitter release. Quantal analysis (see Chapter 8) is a classic method for exploring plasticity in simple synaptic systems, such as the vertebrate and crustacean neuromuscular junctions. The idea behind this method is that the mean postsynaptic response amplitude, $\bar{R}$, is the product of two quantal parameters, m and q. The mean quantal content, m, is the average number of quanta released per presynaptic action potential. The mean quantal size, q, is the average postsynaptic response to the release of each quantal packet of neurotransmitter. One goal of quantal analysis is to determine how these two quantal parameters change in a way that can account for an observed change in $\bar{R}$.

Classic methods of quantal analysis (Katz, 1969) have been used in an attempt to determine whether LTP results from an increase in m or q or both. In certain simple systems, such as the crayfish claw-opener neuromuscular junction (NMJ), experience with quantal analysis can generate confidence in the method. Brown and colleagues set out to determine whether the quantal basis of LTP could be studied in these crustacean neuromuscular synapses. The magnitude of LTP in these synapses was large and it could be studied for several hours (Baxter et al., 1985). The three classic methods of quantal analysis (coefficient of variation, proportion of failures, and "direct") were in excellent agreement with each other both before and after LTP induction.

All three quantal methods agreed that LTP reflects an increase in the quantal parameter m. One could be confident that q did not change much relative to the change in m (Baxter et al., 1985). The simplest interpretation of this result is that LTP in these synapses is due to a presynaptic change that increases the average number of nerve impulse-evoked quantal packets of neurotransmitter. Any change in the sensitivity or responsiveness of the muscle to the release of each quantum of transmitter was undetectably small.

Applications of quantal analysis to LTP expression in synapses of the mammalian central nervous system have been less successful, as judged by consistency of inferences among different laboratories. At least part of the reason is that quantal analysis is sensitive to some of the underlying assumptions, and these assumptions tend not to be easily satisfied except in a few model systems. In the case of the Sch/com synaptic inputs to CA1 pyramidal neurons (Fig. 18.2), different laboratories have often obtained seemingly contradictory results (see Larkman and Jack, 1995) in regard to determining which quantal parameter changes and also with respect to the biological interpretation of apparent changes in quantal parameters.

The MF synaptic input to CA3 pyramidal cells (Fig. 18.2) offers several potential advantages for quantal studies. First, the quantal size appears to be larger, perhaps in part because the MF synapses are

closer to the cell soma (the usual recording site). We have long known that it is possible to record unusually large, spontaneous, single-quantal events in hippocampal CA3 pyramidal neurons (Brown et al., 1979; Johnston and Brown, 1984). The large size of these tetrodotoxin-resistant, spontaneous synaptic potentials ("minis") is not unrelated to the fact that CA3 pyramids were the first identifiable brain cells in which single-quantal events were detected. Quantal size is significant because the ratio of the mean quantal size to the recording noise level is critical for accurate parameter estimation (Greenwood et al., 1999). Second, because the MF synapses are closer to the cell soma, voltage-clamp experiments record more faithfully the actual EPSCs (Brown and Johnston, 1983; Johnston and Brown, 1983, 1984).

Three independent lines of evidence implicated a presynaptic mechanism for MF LTP expression (Xiang et al., 1994). Two of the three classic methods of quantal analysis (coefficient of variation and proportion of failures) were used to assess changes in m and/or q. Both methods gave the correct answers when applied to test cases in which the answer was known in advance. When applied to the LTP data, both methods indicated that m increased and there was no evidence of a change in q, consistent with a presynaptic modification that increases transmitter release. Recall that the same conclusion was reached in studies of the claw-opener neuromuscular synapses. The third method looked for an interaction between synaptic facilitation, which is known to involve a presynaptic modification, and LTP. The rationale was that if facilitation and LTP both result from a presynaptic modification, then the presence of one of them could alter the expression of the other. There was in fact a significant interaction between facilitation and LTP in some but not all of the synapses. The proximity of the MF synapses to the somatic recording site should have minimized space-clamp errors, which can influence the outcome of an interaction test.

We conclude this section with some general comments regarding the three important steps in a quantal analysis. The first step is model selection. Simpler quantal models—those with fewer free parameters—are more likely to yield unique solutions and/or small errors of estimate. However, the realism of the simplest models can sometimes be questioned. If the model is inappropriately simple, then its estimated parameters may be badly biased and even meaningless (Brown et al., 1976). The second step is parameter estimation, given a model. Unless the model is very simple, the experimental data may not be able to restrict the outcome to a unique set of

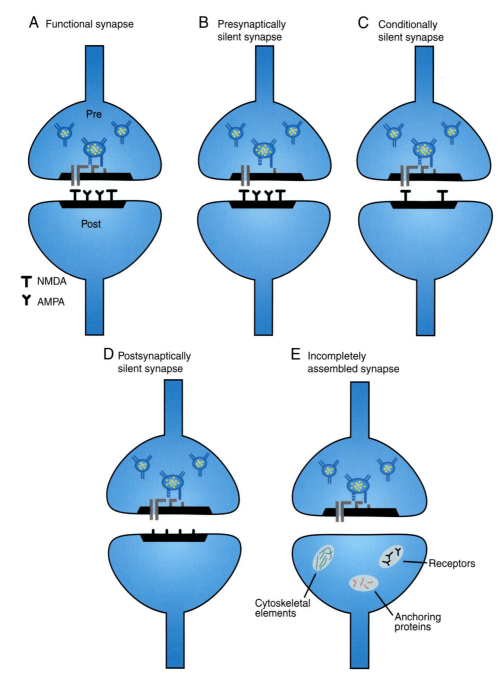

FIGURE 18.12   Possible types of silent synapse. (A) Complete synapse, possessing all of the necessary pre- and postsynaptic components. (B) Synapse silent due to presynaptic molecular deficiency. (C) Conditionally silent mammalian synapse (NMDA receptors only). (D) Synapse silent due to postsynaptic deficiency (receptors nonfunctional). (E) Synapse in transition (postsynaptic components being assembled). Adapted from Atwood and Wojtowicz (1999).

quantal parameters. Even when using a simple model, the quantal signal-to-noise ratio and other factors determine the accuracy of the parameter estimates and confidence intervals need to be established (Greenwood *et al.*, 1999). The third step is to furnish a biological interpretation to the estimated quantal parameters or changes in their values (Brown *et al.*,

1976). The probability of *not* making an error at any of these three steps is not large, which may explain some published discrepancies.

Pharmacological analysis yields mixed inferences. Pharmacological analyses of LTP have used many different compounds to examine their effects on LTP induction and expression. One such study used MK-801, a noncompetitive NMDAR antagonist, to examine the probability of neurotransmitter release (*p*) at hippocampal CA1 synapses (Kullmann *et al.*, 1996). The analysis assumes a binomial quantal model of transmitter release, where the mean quantal content, *m*, is the product of two parameters: *n* and *p*. One interpretation is that *p* is the quantal release probability associated with each of *n* release sites (Brown *et al.*, 1976). Because MK-801 is an open-channel blocker (Jahr, 1992), only those receptors under synapses that release transmitter can be blocked, thereby preventing their participation in subsequent transmission.

If *p* is large, then a large fraction (*p*) of the synapses or release sites will become inactivated with each trial, and EPSC amplitudes will decay rapidly across successive tests. However, if *p* is small, then only a small number of synapses will be eliminated with each trial and transmission will decay much more slowly across trials. If LTP expression results from a presynaptic modification that increases *p*, then LTP induction should increase the rate of decline of successive NMDAR-mediated EPSCs produced by MK-801. Kullman and co-workers (1996) reported that MK-801 *does* in fact increase the rate of decline of NMDAR-mediated signals following high-frequency tetanization. The authors interpret this result as suggesting that LTP in hippocampal CA1 neurons is expressed presynaptically via an increase in *p* (Kullmann *et al.*, 1996). This result is inconclusive, however, as others have failed to find an increase in the rate of decline (e.g., Manabe and Nicoll, 1994). Since these two groups used different LTP induction paradigms, it is possible that they were studying different forms of LTP.

Electron microscopy hints at presynaptic factors.
"Unsilencing" existing synapses is another possibility for a presynaptic modification (Brown *et al.*, 1989). Based on his pioneering ultrastructural work on the crustacean neuromuscular junction, Harold Atwood has long recognized the possibility that "presynaptically silent synapses" could be unsilenced (Fig. 18.12). One reason was that the number of ultrastructurally identified neuromuscular synapses exceeded physiological estimates of the number of release sites. Jahromi and Atwood's (1974) ultrastructural studies showed numerous synapses lacking presynaptic dense bodies (active zones) as well as docked synaptic vesicles. Subsequent studies in *Drosophila* mutants have shown that the disruption or absence of several molecules involved in presynaptic vesicle docking or protein fusion impedes synaptic transmission

(Littleton *et al.*, 1993; Umbach *et al.*, 1994). These studies emphasize the importance of molecular and ultrastructural conditions in the presynaptic terminal that can control the functionality of synapses (Fig. 18.12B).

Postsynaptic Changes in LTP Expression

Postsynaptic changes in sensitivity to released neurotransmitter could theoretically result from changes in the number or properties of postsynaptic receptors or from changes in the axial resistance of the dendritic spines. Interest in the latter possibility has diminished for reasons discussed elsewhere (Brown and Zador, 1990; Brown *et al.*, 1992). In what follows we therefore focus on changes in receptor properties and numbers.

Increased GluR conductance and/or increased numbers of GluRs. The synaptic conductance increase associated with LTP (Barrionuevo *et al.*, 1986) could result from an increase in the current through existing GluRs or from the addition of new GluRs. Both appear to occur in the hippocampus. Single AMPAR channel conductance increases can be quickly expressed via CaMKII/PKC phosphorylation of the GluR1 subunit (Benke *et al.*, 1998). The involvement of CaMKII is particularly significant because this kinase autophosphorylates, causing its action to outlast the Ca^{2+} transient that initiated the process (Kennedy, 1989).

Rapid insertion of the AMPAR subunit GluR1 into dendritic spines (signifying perhaps more AMPARs) following tetanic stimulation has been demonstrated in cultured hippocampal neurons (Shi *et al.*, 1999). Tagging the GluR1 subunit with green fluorescent protein (GFP) allowed Shi *et al.* (1999) to visualize the GluR1–GFP protein using two-photon laser scanning microscopy. Whereas most of the GluR1–GFP protein was initially located intracellularly, tetanic stimulation resulted in the subsequent detection of GluR1–GFP in dendritic spines and clustered in the dendritic shaft.

"Conditionally silent synapses" can be unsilenced. The proposal that insertion, or unveiling, of new glutamate receptors may underlie LTP was first put forth by Lynch and Baudry (1984). Their proposal has been experimentally verified in recent years (Liao *et al.*, 1995, 2001; Poncer and Malinow, 2001; Shi *et al.*, 1999). Specifically, excitatory synapses have been described that contain NMDA receptor responses in the absence of functional AMPA receptors, making them postsynaptically silent at resting membrane potentials. Activation of these synapses results in the rapid recruitment of AMPA receptors within minutes.

Following the terminology of Atwood and Wojtowicz (1999), these synapses lacking functional AMPARs could be termed "conditionally silent synapses" (Fig. 18.12C) because they express a physiological response only if the membrane is depolarized (the synapses are silent only when the transmembrane potential is near the resting potential).

Roger Nicoll, Robert Malenka, and colleagues recently reported that a drug that interferes with membrane fusion reduced LTP in CA1 (Lledo et al., 1998). One interpretation is that blocking membrane fusion prevents the insertion of endogenous AMPARs. Recall that Shi et al. (1999) actually tracked the movement of fluorescent protein-tagged AMPAR subunits (GluR1–GFP), documenting an increase in fluorescence in dendritic spines within 15 min of an LTP-inducing tetanus. Even spines that displayed some fluorescence prior to the LTP induction protocol showed an increase in fluorescence following the tetanus. These changes in the distribution of GFP–GluR1 were prevented by the NMDAR antagonist APV, suggesting that NMDAR activation is required for the delivery of AMPARs into dendritic spines. Their results demonstrate that AMPARs can be rapidly inserted into the postsynaptic membrane in response to synaptic stimulation, supportive of the hypothesis (Beggs et al., 1999; Brown et al., 1989) that "silent synapses" can be "unsilenced."

Neuroanatomical/immunohistochemical suggestions of a postsynaptic change. Further evidence for silent synapses comes from several immunohistochemical studies. One such study reported (in Sch/com synapses) that the ratio of AMPARs to NMDARs is a linear function of the PSD diameter (Takumi et al., 1999). Their results suggest that AMPAR number drops to approximately zero at a PSD diameter of ~180 nm, indicating that the total number of AMPARs in a synapse is proportional to the available PSD area (the area unoccupied by NMDARs). AMPA and NMDA receptors are co-localized at approximately 75% of Sch/com synapses. The silent synapses (those containing no AMPARs) are suggested to be among the smallest synapses with respect to PSD diameter (Takumi et al., 1999).

Interestingly, animals trained in trace eyeblink conditioning, a hippocampus-dependent task (Moyer et al., 1990) (see section on Eyeblink Conditioning) have a larger PSD area in hippocampal nonperforated axospinous synapses than animals receiving trace eyeblink pseudoconditoning (Geinisman et al., 2000). Comparison of conditioned and pseudoconditioned animals showed that the proportion of nonperforated synapses with PSDs in the smallest size category (PSD area $< 20\ mm^2 \times 10^3$) decreased in the conditioned animals. Conversely, the proportions of nonperforated synapses that had PSDs belonging to larger size categories were increased in conditioned animals (Geinisman et al., 2000). Only the smallest nonperforated PSDs, presumably those lacking AMPARs, were increased in size in the conditioned animals.

A relatively consistent observation of electron microscopy studies is that the total number of synapses is not changed by hippocampal LTP. However, further morphological synaptic changes have been observed to occur in response to LTP induction. Changes in the proportion of perforated synapses (those displaying a disruption in the PSD) to nonperforated synapses have been found under multiple conditions. Such changes were first reported by Greenough and colleagues (Greenough et al., 1978) who found an increase in perforated synapses in the cortex of rats raised in a complex environment. Other studies have found that within the dentate gyrus of rats, a 40% increase in perforated to nonperforated synapses is observed 1 h after LTP induction (Geinisman et al., 1991) (but see Fiala et al., 2002). Furthermore, the proportion of perforated axospinous synapses expressing AMPARs is at least twice as large as that of nonperforated synapses (Desmond and Weinberg, 1998). One interesting hypothesis is that the increase in the ratio of perforated to nonperforated synapses is a morphological manifestation of silent synapses changing into functional synapses containing AMPARs.

The pattern of AMPAR and NMDAR expression in Sch/com synapses differs substantially from that observed in the MF synapses. Compared with Sch/com synapses, the mossy fibers are more uniformly labeled, with no observable synapses lacking AMPARs (Nusser et al., 1998). In addition, the MF synapses are less immunoreactive for NMDARs (Fritschy et al., 1998). The fact that MF synapses do not show NMDA-dependent LTP is consistent with these immunolabeling studies. Available evidence is consistent with the inference that MF LTP expression results from a presynaptic modification (Tong et al., 1996; Xiang et al., 1994).

LTP Maintenance

Regardless of the ultimate nature and locus of the modification that gives rise to LTP expression, the more general problem remains as to how a synaptic change can endure over long periods in the face of constant molecular turnover. The maintenance problem is important because most memory models require a persistent synaptic modification.

Gene Expression and Protein Synthesis

Long-lasting forms of synaptic plasticity are associated with modifications in gene expression and protein synthesis. For instance, synaptic stimulation that induces LTP also promotes the expression of the immediate early gene activity-regulated cytoskeleton-associated protein (*Arc*) (Steward and Worley, 2001a). Newly synthesized *Arc* mRNA is targeted to synapses that have recently undergone specific forms of synaptic activity where it is locally translated (Steward and Worley, 2001b). Kelly and Deadwyler (2002) have also recently shown an experience-dependent upregulation of *Arc* mRNA in several temporal lobe structures following acquisition of an operant lever-pressing task. The targeting of *Arc* mRNA to specific synapses is interrupted by application of NMDA receptor antagonists (MK-801 or APV), suggesting its dependence on the NMDA receptor (Steward and Worley, 2001b). In addition to mRNA, an accumulation of *Arc* protein is observed in activated synapses. Using an *Arc*-specific antibody, Steward and co-workers (Steward *et al.*, 1998) uncovered a band of newly synthesized *Arc* protein in the same dendritic laminae in which *Arc* mRNA was concentrated.

A large and old literature has consistently pointed to the importance of protein synthesis in certain aspects of learning and/or memory (Matthies, 1989). Application of protein synthesis inhibitors can interfere with the formation and retention of memories. One *in vivo* study suggested that a late stage of LTP, which can last for weeks, depends on protein synthesis (Krug *et al.*, 1984). Brain slice experiments report that an early stage of LTP, which lasts for about 2 h and is not dependent on protein synthesis, is followed by a later stage of LTP, which does depend on protein synthesis (Nguyen *et al.*, 1994). Recent work on reconsolidation casts a whole new light on the role of protein synthesis in various forms of learning, memory, and synaptic plasticity (Kida *et al.*, 2002; Nader *et al.*, 2000).

The Problem of Targeting Newly Synthesized Proteins

The involvement of protein synthesis in LTP immediately raises the question of how the synthesized proteins are targeted to just those synapses whose enhanced efficacy is to be maintained. This targeting is required because LTP can be remarkably "input specific" (Kelso and Brown, 1986). If neural activity ultimately affects gene expression in the nucleus, then the proteins produced in the soma could in principle be transported to synapses at any location on the dendritic arbor. One obvious solution is that a synapse-specific molecular marker or tag—set by neuronal activity during tetanization—can sequester plasticity-related proteins being transported within the dendrites. In search of evidence consistent with this idea, Frey and Morris (1998) stimulated pathways S1 and S2, respectively, 30 min before and 30 min after application of the protein synthesis inhibitor anisomycin. The rationale was that stimulation of S1 would trigger the synthesis and dendritic transport of plasticity-related proteins that could subsequently be captured by the marker set by stimulation of S2. Accordingly, they reported that stimulation of S2 induced persistent LTP in that pathway, even in the presence of anisomycin. The idea of an activity-dependent synaptic marker is not unappealing, but it begs the question of how the molecular marker is maintained and controlled.

Long-Term Depression May Solve Two Problems

Some of the mechanisms responsible for inducing LTP may also play a role in triggering another synaptic phenomenon termed *long-term depression*, a persistent reduction in EPSC magnitude following appropriate stimulation. In the following sections, a distinction is made between "homosynaptic" (or "telencephalic") LTD and "cerebellar LTD" (Fig. 18.13). The latter is studied at the parallel-fiber synapses onto Purkinje cells of the cerebellar cortex, whereas the former has been studied most extensively in excitatory cortical synapses, mainly in the hippocampus. There is no evidence that cerebellar-type LTD can occur in the telencephalon.

LTP Saturation in Excitatory Synapses

Donald Hebb did not offer a "learning rule" or computational algorithm for synaptic modifications; that is, he did not specify quantitatively the synaptic activity–modification relationship. The simplest possible "learning rule" or modification algorithm that is consistent with Hebb's postulate is termed (Brown *et al.*, 1990) a "product rule," according to which

$$\Delta W_{BA} = \varepsilon\, V_B\, V_A, \qquad (1)$$

where V_B represents the firing rate of the postsynaptic neuron, V_A is the firing rate of the presynaptic neuron, ε is the learning rate constant, and ΔW_{BA} is the change in synaptic strength between the pre- and postsynaptic neurons.

This version of the Hebb rule is graphed in Figure 18.14 (straight line that starts at the origin). The abscissa is the level of postsynaptic activity and the slope of the line represents the level of presynaptic

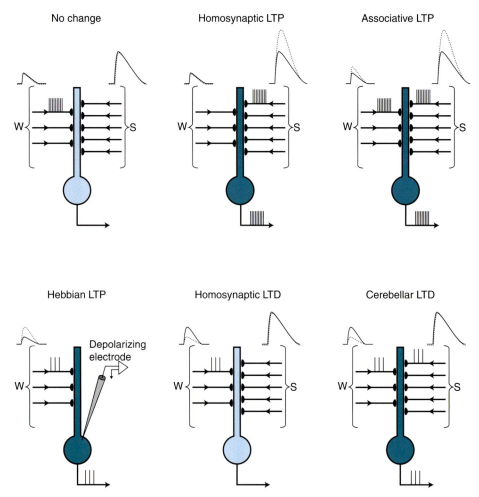

FIGURE 18.13   Various types of synaptic activity–modification relationships. Schematic neurons are shown receiving nonoverlapping inputs that are either weak (W, few axons) or strong (S, many axons). Action potentials on axons are denoted by short vertical lines that are closely spaced (high frequency) or widely spaced (low frequency). Activity in synapses or cells is indicated by black shading. Postsynaptic potentials before inducing stimulation are given by solid lines; those after are given by dotted lines. No change is produced when a single weak input is stimulated at high frequency. Homosynaptic LTP is induced in the synapses of a strong input when it receives high-frequency stimulation, but the unstimulated synapses of the weak input are unchanged. Associative LTP occurs when a weak input (that would not potentiate if stimulated by itself) is stimulated concurrently with a strong input, resulting in potentiation of both inputs. Hebbian LTP occurs when a weak input (that would not potentiate if stimulated by itself) is stimulated at the same time depolarization is induced in the postsynaptic cell through injection of an intracellular current. Homosynaptic ("telencephalic") LTD is induced only in synapses receiving low-frequency (1–5 Hz) stimulation; unstimulated synapses are unchanged. Cerebellar LTD occurs when parallel fibers (represented here by the W input) are stimulated simultaneously with climbing fibers (represented here by the S input). Parallel fibers undergo LTD, but climbing fibers do not. Adapted from Brown *et al.* (1990).

activity. One obvious limitation of this formulation of Hebb's postulate is that it almost guarantees saturation because fortuitous coincidences of pre- and postsynaptic activity will cause the synaptic strengths to increase. Over time, therefore, the synapses tend to strengthen to their maximum values, which causes information that is stored or represented by the matrix of synaptic strengths (see Figs 18.14 and 18.36) to be degraded.

One way to solve the saturation problem inherent in Eq. 1 is to add a passive decay term. However, such a term implies forgetting or the loss of all information simply as a function of time. Within the computational community, Hebb's postulate has always been interpreted to imply that synaptic strengthening is a function of the correlation or contingency between pre- and postsynaptic activity. Fortuitous coincidences were not envisioned to produce learning.

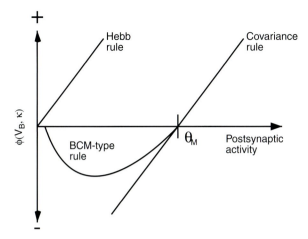

FIGURE 18.14 Graphical comparison of three "learning rules." The synaptic modification is a function of Φ, the quantity on the ordinate. Φ is a linear function (in the case of Hebb- and covariance-type rules) or nonlinear function (in the case of a BCM-type rule) of postsynaptic activity. The actual synaptic modification, ΔW_{AB}, is the product of three terms: V_A, the level of presynaptic activity; Φ, already mentioned; and K, a learning rate constant. Both the covariance and BCM rules are variants of the Hebb rule and postulate a threshold, Θ_M, above which there is LTP (positive change in synaptic strength) and below which there is LTD (negative change in synaptic strength). Adapted, with permission, from Brown and Chattarji (1994).

Terrence Sejnowski proposed making changes in synaptic strength depending on the covariance between pre- and postsynaptic activity (Sejnowski, 1977a, b). In this case,

$$\Delta W_{BA} = \varepsilon(V_B - \langle V_B\rangle)(V_A - \langle V_A\rangle), \quad (2)$$

where $\langle V_B\rangle$ and $\langle V_A\rangle$ are, respectively, the expected (or average) firing rates of the postsynaptic and presynaptic neurons and the other terms have their usual meanings. Note that if either the pre- or postsynaptic cells are firing *more slowly* than their average values, then the corresponding term in Eq. 2 will be *negative*, causing the synaptic strength to *decrease*. Equation 2 addresses the problem of saturation through fortuitous coincidences because change is driven by covariance rather than coincidence. Taking a time average of the change in synaptic weight in Eq. 2 gives

$$\langle\Delta W_{BA}\rangle = \varepsilon(\langle V_B V_A\rangle - \langle V_B\rangle\langle V_A\rangle), \quad (3)$$

where the first term on the right-hand side has the same form as the product rule (Eq. 1) and the second term is a modification threshold (see discussion below) whose value (Θ) is a function of the product of the time-averaged pre- and postsynaptic activity levels (see Fig. 18.14 and associated text). Θ prevents persistent changes in synaptic strength when the

covariance between pre- and postsynaptic activity levels is zero. Note that Θ is determined by both pre- and postsynaptic activity levels, unlike the BCM rule (Fig. 18.14), where it is controlled exclusively by postsynaptic activity.

The covariance learning rule is the straight line that crosses the abscissa at Θ. As before, the abscissa could represent the level of postsynaptic activity while the slope of the line reflects the level of presynaptic activity (see Fig. 18.14). Θ is the level of postsynaptic activity, or $[Ca^{2+}]_i$, that causes no change in synaptic strength. LTD or LTP can be induced when postsynaptic activity is, respectively, less than or greater than Θ, which is why it is termed the "synaptic modification threshold." In principle, Θ could be either a constant or a variable. In the latter case, Θ could be some function of previous activity, the current synaptic strength, the $[Ca^{2+}]_i$, or a number of other physiological factors discussed below.

According to the covariance rule, synaptic strength will *increase* if pre- and postsynaptic firing are positively correlated, *decrease* if they are negatively correlated, and remain *unchanged* if they are uncorrelated (Fig. 18.14). The covariance rule is one of several modern "fixes" for the saturation problem in a way that captures the spirit of Hebb's postulate. Implicit in Hebb's language and thinking is the idea that synaptic strengthening should reflect *correlated* pre- and postsynaptic activity. This property is obviously built into Eq. 2. By allowing bidirectional synaptic change (LTP or LTD), Eq. 2 captures the general sense of two useful mnemonics for Hebb-type synapses. According to the first of these, *"cells that fire together, wire together,"* which is close to Hebb's original claim. The second mnemonic asserts that *"cells that are out of sync, weaken their link."* This addition was not addressed by Hebb, who considered only synaptic strengthening. However, it is a necessary addition if synaptic modifications are to be driven by correlation or contingency, which seems implicit in Hebb's thinking and certainly that of subsequent generations of computational neuroscientists.

Synaptic Encoding in Inhibitory Neurons

In cerebellar Purkinje cells (PCs), whose output is inhibitory, LTD in the afferent parallel-fiber (PF) synapses causes disinhibition (decreased inhibition) in the deep nuclei (Fig. 18.13). Associative LTD in PF-to-PC synapses is therefore functionally similar to associative LTP in excitatory neurons. Cerebellar LTD (Fig. 18.13) has been termed *antihebbian* because the synaptic modification is in the opposite direction from what Hebb proposed (Brown *et al.*, 1990).

Mechanisms Underlying Induction and Expression of Long-Term Depression

There has been much less research on LTD than LTP, but most of the suggested mechanisms are ones that have already arisen in conjunction with LTP.

Is LTD the Reversal of LTP or a Separate Process?

An ongoing problem is whether the depression represents a process that reverses the effects of LTP (that is, depotentiation) or, alternatively, a separate, counterprocess that is superimposed on potentiation. Although "depotentiation" and "LTD" (terms for the two, possibly different, processes) can often be induced using the same protocol, they are defined differently. Depotentiation is defined as depression of a potentiated synapse; LTD is defined as depression of a nonpotentiated synapse. Stated differently, depotentiation is the undoing of a process, whereas LTD is the superimposition of a process.

A number of contradictory results have been reported regarding the induction of LTD and depotentiation in the hippocampus. Using hippocampal slice preparations, Dudek and Bear (1993) and Mulkey and Malenka (1992) induced stable LTD using low-frequency stimulation (LFS). Subsequently, Bashir et al., (1993) failed to induce LTD using a similar LFS induction protocol. In the anesthetized rat, some have reported successful induction of LTD (Heynen et al., 1996; Thiels et al., 1994) and depotentiation (Heynen et al., 1996); whereas others (Staubli and Scafidi, 1997) have been unable to induce LTD using similar stimulation patterns. In freely moving rats, Manahan-Vaughan (1997) reported reliable LTD in hippocampal area CA1, whereas others (Staubli and Lynch, 1990) have reported depotentiation but no LTD, or neither LTD nor depotentiation (Errington et al., 1995).

The preceding studies point to the importance of procedural details. Experimental outcomes have been suggested to depend on some combination of the following: the stimulation protocol, the age of the animal, the prior synaptic history of the recorded input, and the particular stock of rat. At present, it is unclear whether LTD and depotentiation are separable and independent processes or instead represent two procedures that engage the same mechanism. In what follows, we use the term LTD in reference to either of these hypothesized processes.

Biochemistry of LTD in the Telencephalon

LTD induction depends on calcium influx. In the hippocampus, brief, high-frequency stimulation (four trains of 10 shocks at 100 Hz) can induce classic LTP, whereas low-frequency stimulation over longer periods (900 shocks at 1 Hz for 15 min) can induce LTD. In studies of the Sch/com input to CA1, both the LTP and LTD stimulation procedures are Ca^{2+}-dependent processes that can be blocked by injecting Ca^{2+} chelators into the postsynaptic cell (Mulkey and Malenka, 1992). The fact that both modifications are triggered by Ca^{2+} entry naturally raises the question of how and where in the messenger signaling pathway the signals diverge.

Free calcium controls NMDA-dependent LTD. A presumption in most models is that more Ca^{2+} influx occurs during an LTP-inducing tetanus than during LTD-inducing, low-frequency stimulation (Teyler et al., 1994). One computational molecular model explicitly incorporates this Ca^{2+}-dependent and bidirectional control of synaptic strength (Lisman, 1989). In this model, high $[Ca^{2+}]_i$ (> 5 μM) activates a protein kinase that phosphorylates a protein leading to LTP induction, whereas low $[Ca^{2+}]_i$ (< 5 μM) activates a protein phosphatase that dephosphorylates this protein and causes LTD (Fig. 18.10).

NMDA-dependent LTD is regulated by kinases and phosphatases. It has been proposed that a balance between the activity of CaMKII and protein phosphotase 1 (PP1) influences synaptic strength by controlling the phosphorylation state of some unidentified phosphoproteins. Small rises in $[Ca^{2+}]_i$ favor activation of PP1, whereas larger rises are necessary for increasing CaMKII activity. Because PP1 is not directly influenced by $[Ca^{2+}]_i$, a Ca^{2+}-dependent protein phosphatase cascade has been invoked to translate the Ca^{2+} signal into an increase in PP1 activity. The cascade activates the Ca^{2+}/CaM-dependent phosphatase calcineurin, which then dephosphorylates inhibitor 1, a phosphoprotein. In its phosphorylated state, inhibitor 1 is a potent inhibitor of PP1.

This proposal suggests that activation of calcineurin causes an increase in PP1 activity through disinhibition. When pharmacological inhibitors of PP1 or calcineurin were loaded directly into CA1 cells, the generation of LTD was prevented (Mulkey et al., 1994). Furthermore, the phosphorylated form of inhibitor 1, but not the unphosphorylated form, blocked LTD (Mulkey et al., 1994). Critical to this interpretation is the fact that calcineurin has a higher affinity for Ca^{2+}/CaM than does CaMKII and, therefore, would be preferentially activated by small rises in $[Ca^{2+}]_i$.

Analogous to the role that AMPAR phosphorylation is believed to play in LTP induction, it is also thought that AMPAR dephosphorylation is a component of LTD. Using phosphorylation site-specific antibodies, the induction of LTD produced a persistent dephosphorylation of the GluR1 subunit at Ser[845], the PKA substrate, but not at Ser[831], a substrate of PKC and CaMKII (Lee *et al.*, 1998). These results suggest that PKA may play a role in the postsynaptic modulation of synaptic transmission and that AMPAR phosphorylation or dephosphorylation may contribute to the expression of LTP and LTD, respectively.

Interestingly, the Sch/com input to region CA1 may exhibit both NMDAR-dependent (Christie *et al.*, 1996; Dudek and Bear, 1992) and NMDAR-independent (Christie *et al.*, 1996) forms of LTD, with the latter possibly mediated by Ca^{2+} entry through VDCCs. One question that remains to be answered is why synapses possess so many different forms of synaptic plasticity.

AMPAR Endocytosis May Be a Component of LTD

It is possible that AMPAR endocytosis may contribute to LTD. In cultured hippocampal neurons, NMDAR-dependent LTD was accompanied by a decrease in the number of synaptic surface AMPARs, with no significant effect on the distribution of NMDARs (Carroll *et al.*, 1999). NMDARs are thought to be critically involved in AMPAR internalization by allowing an influx of Ca^{2+} that activates the phosphatase calcineurin (Beattie *et al.*, 2000). AMPAR endocytosis can occur even when the AMPARs have been blocked, emphasizing the importance of the NMDARs (Beattie *et al.*, 2000).

Biochemistry of LTD in the Cerebellar Cortex

Long-term depression has been demonstrated to occur at the synapses between PFs and PCs in the cerebellum. Similar to the hippocampal LTP/LTD discussed above, it is believed that cerebellar LTD is triggered by an influx of Ca^{2+} into PCs. However, PC Ca^{2+} entry occurs via both ionotropic and metabotropic GluRs and VDCCs. In contrast to many neurons in the brain, adult PCs do not contain functional postsynaptic NMDARs. Fast excitatory transmission at PF–PC synapses depends on AMPARs.

The PF–PC synapses contain mGluR1 receptors that are coupled to phospholipase C, whose activation leads to production of IP_3 and DAG. PCs also have both ryanodine- and IP_3-sensitive intracellular Ca^{2+} stores. In cultured PCs, Ca^{2+} release from ryanodine (Kohda *et al.*, 1995) or IP_3 (Kasono and Hirano, 1995) stores, or both, appears to be required for LTD induction. However, there is still some uncertainty as to whether these intracellular stores are required under normal physiological conditions in the intact brain.

PKC is abundantly expressed in PCs, and its activation is thought to be modulated both by VDCC-dependent Ca^{2+} entry and DAG produced by mGluR activation. Accordingly, it is now believed that PKC activation is required for LTD induction in PCs. In addition to PKC, Ca^{2+} entry can also induce the formation of nitric oxide (NO) from arginine by activating calmodulin-dependent NO synthase. Since NO is highly diffusible, it probably targets soluble guanylyl cyclase within its own cell and surrounding cellular elements. This transduction cascade then activates cGMP-dependent protein kinases in the PCs. Production of cGMP activates cGMP-dependent protein kinase (PKG), a potent inhibitor of protein phosphatases.

It has long been thought that LTD induction might lead to a long-term desensitization of the postsynaptic ionotropic GluRs, presumably the AMPARs (Ito *et al.*, 1982). Levenes *et al.* (1998) note two independent lines of evidence that support a modification of the AMPAR conductance during LTD. First, the nootropic compound aniracetam, which reduces desensitization of AMPARs and/or attenuates the ion channel closing rate, has a larger potentiating effect on PF-mediated EPSCs during expression of LTD than under control conditions (Hemart *et al.*, 1994). Second, perfusing the acute cerebellar slice with agents known to induce LTD, such as 8-bromo-cGMP, dibutyryl-cGMP, and the phosphatase inhibitor calyculin A, induces a persistent (30 min or longer) phosphorylation of the PC AMPAR subunits (Nakazawa *et al.*, 1995).

More recent work has suggested two alternative hypotheses. David Linden (2001) examined the EPSC shape in synaptically connected granule cell–PC pairs (in culture), finding no changes following LTD induction—contrary to predictions of the desensitization hypothesis. He proposed that cerebellar LTD is caused by a reduction in the number of functional AMPARs. Another hypothesis proposes that cerebellar LTD is mediated by *presynaptic* NMDARs (Casado *et al.*, 2002). The authors recorded EPSCs in PF-to-PC synapses, reporting that bath application of NMDA and glycine caused LTD and that APV reversibly blocked it. They propose that NMDAR-mediated Ca^{2+} entry into the presynaptic nerve terminals activates NO synthase, which (through a cascade reaction) decreases postsynaptic sensitivity to glutamate.

Synapses Are in a Continuous State of Morphological Change

The brain is thought to reorganize continuously as a function of experience, hormones, aging, and neuropathology. An emerging view is one of constant change in the presynaptic terminals as well as the dendritic spines on which they are located. Dendritic spines are morphological specializations present on many mammalian neuronal dendrites that are specialized for receiving synaptic inputs (Chapter 17) (Sheppard, 1996). Spines have several different shapes, some of which may relate to different developmental or functional roles (see Harris *et al.*, 1992; Parnass *et al.*, 2000). Thin spines are characterized by a bulbous head attached to the dendrite by a long, thin neck. Mushroom-shaped spines are similar to thin spines, but have larger heads. Stubby spines do not have a clear neck, but are closely connected to the shaft of the dendrite. Developing neurons also contain protrusions termed *filopodia* which are typically longer, thinner processes than spines. Filopodia do not end in bulbous heads and, because of their longer lengths, are generally thought to be structurally different from dendritic spines (Fiala *et al.*, 1998).

Rapid spine motility and structural plasticity, in periods as short as seconds, have been observed in dissociated cultures and acute brain slices from the hippocampus, neocortex, and cerebellum (Dunaevsky *et al.*, 1999; Fischer *et al.*, 1998; Kaech *et al.*, 1999). Parnass *et al.* (2000) conducted an analysis of a population of spines over a period of 2–4 h, using the four categories of dendritic protrusions outlined above. Considerable morphological conversion among the four categories was found. Especially high levels of conversion were found among the filopodia, although interestingly, the conversion was not unidirectional. In other words, the filopodia can convert into one of the three types of spines, but they can also originate from the spines themselves (Parnass *et al.*, 2000).

Spine density has also been found to vary as a function of hormonal concentrations (Woolley and McEwen, 1992). As levels of the ovarian steroid estradiol fluctuate during a female rat's normal 5 day estrous cycle, there is a concomitant fluctuation in the density of synapses in the CA1 region of the hippocampus. Specifically, low levels of estradiol are correlated with low synaptic density, whereas high levels of estradiol are correlated with a higher density of synapses. The changes in synaptic density can occur in relatively short periods, inasmuch that in only 24 h (from proestrus to estrus) the authors noted a reduction in spine density of almost one-third.

The persistence of memory across time is quite remarkable considering that the storage medium is in a state of continuous structural change and turnover in molecular composition. Memory continuity in the face of constant change in the memory circuits is one of the interesting puzzles that remains to be solved.

Metaplasticity Is a Change in the Synaptic Modification Process Itself

Synaptic plasticity is modulated and shaped by the pattern of ongoing extrinsic activity, concomitant hormone circulation, and prior patterns of synaptic activity. "Plasticity of synaptic plasticity," sometimes termed *metaplasticity*, is a quantitative change in the synaptic modification rule. The so-called "BCM rule," described next, has been used to characterize metaplasticity.

Origin and Properties of the BCM Learning Rule

Recall that the covariance rule (Eq. 2) prevents LTP saturation as a function of fortuitous coincidences between pre- and postsynaptic activity. However, if these activities are strongly correlated over time, saturation *can* occur and probably will in the absence of some type of negative feedback mechanism. The BCM learning rule (see Fig. 18.14), originally designed to help understand the development of orientation specificity and binocular interactions in the visual cortex (Bienenstock *et al.*, 1982), is sometimes used in circuit simulations to help maintain synaptic strengths within bounds. In the BCM algorithm, synapses are strengthened when the postsynaptic activity exceeds a given threshold Θ and weakened when the postsynaptic activity falls below Θ.

Two features of the Bienenstock–Cooper–Munro (BCM) rule are notable (Fig. 18.14). First, synaptic modification varies as a nonlinear function (Φ) of postsynaptic activity. Presynaptic stimulation causes LTP or LTD, respectively, when postsynaptic activity is greater than or less than Θ. Second, Θ can be made to "slide" along the abscissa as a function, for example, of the previous history of postsynaptic activity. Shifting Θ to the left or right favors LTP or LTD, respectively. The sliding threshold can reduce the probability of extreme deviations from the midpoint of the LTP/LTD range (see Fig. 18.14).

Changes in the Synaptic Modification Threshold Associated with Experience

Cortical development depends critically on experience (Fregnac *et al.*, 1988; Hirsch and Spinelli, 1970; Pettigrew, 1974; Shatz and Stryker, 1978; Wiesel and Hubel, 1965). In the visual system, sensory deprivation prevents many of the maturational changes in cortical function that normally occur with age. During normal development, the proportion of synapses with detectable AMPAR responses increases, while the decay kinetics of NMDAR responses become faster (Crair and Malenka, 1995). The change in NMDAR kinetics is believed to result from a developmental switch in the subunit composition of the NMDAR (Monyer *et al.*, 1994). Specifically, early in development the NMDAR is composed of the NR1 and NR2B subunits; as the animal matures the NR2B subunits are replaced or supplemented with NR2A subunits.

This developmental change in subunit expression appears to be partly experience-dependent. The developmental decay in NMDAR kinetics in visual cortical neurons is delayed when animals are deprived of vision, suggesting that the proportion of NR2A and NR2B subunits expressed are different in dark-reared and light-reared animals (Carmignoto and Vicini, 1992). Building on this finding, Mark Bear and colleagues (Quinlan *et al.*, 1999) demonstrated that dark-reared rats have low levels of NR2A-containing NMDARs in the visual cortex. Remarkably, they found that only 1 to 2 h of visual experience can alter the expression pattern of NR2A-containing NMDARs. This experience-dependent change in synaptic composition is believed to be controlled by NMDAR activation.

In functional terms, this change in subunit composition is likely to have a significant impact on the properties of synaptic plasticity. Within the context of the BCM rule, shortening NMDAR currents would be expected to shift Θ to the right, making LTD more likely to occur with synaptic activation than LTP (see Fig. 18.14). Thus, as the animal matures, the plasticity is shifted in the favor of LTD. That is, conditions of pre- and postsynaptic activity that cause LTP in a developing animal could cause LTD in the adult.

Changes in Synaptic Modification Threshold Caused by Hormones

Stress impairs learning or memory on tasks that are known to depend on hippocampal function (Diamond *et al.*, 1996). Interestingly, LTP induction is impaired in hippocampal brain slices taken from stressed rats (Foy *et al.*, 1987). Although the stress response involves the release of many different neurochemicals, the effects of glucocorticoids correlate well with stress effects in the hippocampus. Corticosterone is the primary glucocorticoid synthesized in response to stress and the hippocampus contains abundant receptors for this hormone. Both high-affinity (Type I) mineralocorticoid and lower-affinity (Type II) glucocorticoid receptors are found within the hippocampus.

Type I and Type II receptor activation respectively increases and decreases the probability or magnitude of LTP (Pavlides *et al.*, 1995). During nonstressful conditions, corticosterone may act primarily on the Type I receptors, thereby enhancing the magnitude of LTP. However, in times of high stress, corticosterone levels may rise enough to activate Type II receptors, in which case LTP may be attenuated while the induction of LTD is increased. Stress effects on LTP/LTD induction have also been characterized in terms of shifting Θ (Kim and Yoon, 1998). Accordingly, small influxes of Ca^{2+}, as would be encouraged by basal levels of glucocorticoids, shift Θ to the left, allowing for easier induction of LTP; whereas larger influxes of Ca^{2+}, as could occur with higher levels of stress, shift Θ to the right, facilitating LTD (Fig. 18.14).

Neuropharmacological Linkages between LTP and Information Storage

The most important and difficult remaining challenge is to link particular forms of LTP to specific aspects of learning and memory (Barnes, 1995; Brown *et al.*, 1988). A common approach is to examine the effect on a learning or memory task of drugs that antagonize LTP induction. The most commonly used drugs antagonize the NMDAR and VDCC channels, which normally allow Ca^{2+} influx leading to LTP induction. Timothy Teyler and co-workers recently reported that antagonists of NMDARs and VDCCs produced deficits in acquisition and retention, respectively, suggesting different roles for the two forms of LTP (Borroni *et al.*, 2000). According to Teyler's hypothesis, mentioned earlier, nmdaLTP tends to subserve short-term storage, whereas vdccLTP tends to support long-term storage (Borroni *et al.*, 2000; Teyler, 2000). The relationship between LTP and learning and memory is clearly more complex than many have assumed (Brown and Lindquist, 2003; Teyler, 2000).

Genetically Engineered Mice Seek Linkages between Synaptic Modifications and Learning

Genetic approaches have proven to be more specific, in both molecular and spatial terms, than pharmacological methods. Two lines of work that have been particularly promising include genetically engineered transgenic and knockout mice. The first generation of knockouts were temporally nonspecific and could not be confined to a specific part of the brain. However, techniques have recently been developed to impose selective regional and/or temporal restrictions to the genetic deletions. One such technique uses the Cre/loxP system in which any region (i.e., one or more of a particular gene's exons) flanked by two loxP sites (a "floxed gene") is deleted by active Cre, enabling deletion of specific portions of a gene. Gene deletion only occurs when the Cre gene-associator promoter is active; in this way, only a distinct region of the brain, at a particular developmental period, is inactivated. One group using this method (Tsien *et al.*, 1996) developed a mutant knockout mouse in which the deletion of the NMDAR1 gene appeared to be restricted to the CA1 pyramidal cells of the hippocampus. Furthermore, this mutation was expressed only after the third postnatal week, thereby reducing influences on early development.

Tests of these mice revealed that they have LTP deficits in hippocampal region CA1, but none were found in the dentate gyrus, where the NMDAR was expressed normally. The behavioral effects of this manipulation are also interesting and include impaired performance on tests of spatial memory, as well as altered firing properties of CA1 pyramidal cells during navigational behaviors. The firing rate of a CA1 pyramidal cell commonly increases when the mouse visits certain spatial locations, which constitute the "place field" for that neuron. Cells coding for overlapping place fields also tend to show correlated firing. However, in knockout mice the place fields were spatially more diffuse than those seen in the controls, and the firing patterns of neurons from overlapping place fields did not correlate with each other as highly. These results are consistent with the hypothesis that the NMDAR participates in some aspect of the formation of place fields through a hebbian mechanism, and deficits in spatial memory could result, in part, from untuned and/or improperly associated place fields.

Transgenic mice have also been created that either express novel proteins or overexpress proteins normally present (usually enzymes or structural proteins). One such study overexpressed the NR2B subunit of the NMDAR (Tang *et al.*, 1999). The effect of this mutation was longer NMDAR-mediated EPSPs and enhanced LTP expression, which is consistent with an enhancement in NMDAR activity and Ca^{2+} influx. Interestingly, the transgenic mice also displayed superior ability to master a battery of behavioral tasks, which strongly suggests, but does not prove, a link between LTP and learning and memory.

Inducible Transgenic and Knockout Mice

A more recent and elegant example of genetic engineering is the use of inducible knockouts and transgenics. These techniques allow not only for regional restriction of gene expression, but also for control over the timing of gene expression. One method of engineering inducible knockouts uses the Cre/loxP recombination system described above. Normally, any region flanked by two loxP sites is deleted by active Cre. The floxed gene of interest can be deleted, under experimental control, by fusing Cre to the ligand binding domain (LBD) of a specific ligand. This enables Cre to be turned on and off by injecting the LBD at a particular time. In the absence of the ligand, the LBD/Cre protein is inactivated. Ligand binding, however, enables the LBD/Cre protein to become active and promotes loxP-guided gene deletion. After deletion of the gene, the remaining protein must be degraded naturally before the mutant can be studied, which usually takes a few days to a few weeks.

In addition to gene deletion, it is also possible to study the functional impact of the expression of mutant genes in mice. When an inducible transgenic mouse has a mutated gene added to its genome, the resulting transgenic animal will have the normal complement of genes in addition to the added mutant gene. One technique used in such studies involves the creation of two lines of transgenically altered mice. One line carries the transgene of interest attached to the promoter *tet-O*. The second line of mice carries a hybrid transcriptional regulator called tetracycline transactivator (tTA). When the two lines of mice are crossed, some of the offspring will carry both transgenes. In these mice, the tTA binds to the *tet-O* promoter and activates the mutated transgene of interest. The elegance of this technique is seen on presentation of a drug that binds to and inactivates the transcription factor tTA, thereby switching off the expression of the mutated gene. One commonly used drug, mixed into the drinking water, is doxycycline, a tetracycline analogue with high permeability through the blood–brain barrier. Removal of doxycycline from the drinking water restores expression of the mutant gene. Mice can also be generated that express reverse tTA (rtTA), such that the transgene is inactivated until doxycycline is administered to the animal.

The use of inducible transgenic mice has two primary advantages. First, any effects caused by the mutated protein can be reversed through inactivation of the transgene. Second, the induction and expression of the mutant protein can be quite rapid. Despite these advantages, it must be kept in mind that the "original" protein is still expressed and is in competition with the mutant protein.

These more recent genetic approaches allow the researcher a greater degree of control as to where and when the deleted or mutated protein is expressed. This, in turn, allows observed behavioral effects to be more confidently matched to the genetic modification. Although mutant studies do not directly connect LTP to learning and memory, they are impressive because they combine evidence and ideas from molecular, cellular, systems, and behavioral levels into a consistent picture. This general approach holds tremendous promise for testing hypotheses about the functional role of various forms of LTP in the brain.

Summary

The discovery of LTP accomplished the remarkable goal of furnishing a causal connection between the time scale on which neurophysiological events had traditionally been measured and the time scale over which pertinent forms of learning and memory are studied. The complexity and varieties of LTP and LTD are only just now becoming widely appreciated. The finding that some forms of LTP and LTD are "associative" has enormously important historical and contemporary implications. The idea that spatiotemporal contiguity drives memory formation has a ~2000-year-long history that continues to be reflected in postmodern theories of animal learning and memory. The discovery that synapses can actually operate as contiguity-driven "associative memory elements" helps us understand these persistently influential ideas about learning and memory at the cellular and molecular levels. The rapid growth of relevant technology during the last decade should guarantee progress toward the ultimate goal of understanding the connection between specific forms of LTP and/or LTD and particular forms or aspects of learning and memory.

PARADIGMS HAVE BEEN DEVELOPED TO STUDY ASSOCIATIVE AND NONASSOCIATIVE LEARNING

Associative Learning

Associative learning is a broad category that includes much of our daily learning activities: learning to be afraid, learning to talk, learning a foreign language, learning to play the piano. In essence, associative learning involves the formation of associations among stimuli and/or responses. It is usually subdivided into classical conditioning and instrumental conditioning. Classical (or Pavlovian) conditioning is induced by a procedure in which a generally neutral stimulus, termed a conditioned stimulus (CS), is paired with a stimulus that generally elicits a response, termed an unconditioned stimulus (US). Two examples of unconditioned stimuli are food that elicits salivation and a shock to the foot that elicits limb withdrawal.

Instrumental (or operant) conditioning is a process by which an organism learns to associate consequences with its own behavior. In an operant conditioning paradigm, the delivery of a reinforcing stimulus is contingent on the expression of a designated behavior. The probability that this behavior will actually be expressed is then altered. This chapter focuses on classical conditioning because it is mechanistically the best understood type of associative conditioning.

An astute observation by Ivan Pavlov, a Russian physiologist who had been studying digestion in dogs, led to his discovery of classical conditioning in a celebrated case of serendipity. He first noticed that the mere sight of the food dish caused dogs to salivate. He continued the experiments to see if dogs would also salivate in response to a bell rung at feeding time. Pavlov trained dogs to stand in a harness and, after the sound of a bell, fed them meat powder (Fig. 18.15) (Rachlin, 1991). He then recorded the salivary responses of the dogs. At first, the bell by itself did not elicit any response, but the meat powder elicited reflex salivation, which was termed the unconditioned response. He noted that after a few pairings of the bell and meat powder the dogs began to salivate when the bell rang, before they received the meat powder. This response is termed the conditioned response. This type of conditioning came to be called reward or appetitive classical conditioning. If the bell or another stimulus was followed by an unpleasant event, such as an electric shock, then a variety of autonomic responses became conditioned. This type of conditioning is often termed *aversive* or *fear conditioning*. Skeletal muscle movements appropriate to deal with the US (e.g., leg flexion after a shock delivered to a paw) are also learned in aversive classical conditioning.

Traditionally, classical or Pavlovian conditioning is an operation that pairs one stimulus, the conditioned stimulus, with a second stimulus, the unconditioned stimulus, as noted earlier. The US reliably elicits a response termed the unconditioned response (UR). Repeated pairings of the CS and US result in the CS eliciting a response, which is defined as the conditioned response (CR).

Conditioning procedures in which the CS and US overlap in time are called delay conditioning, whereas in trace conditioning a short interval is interposed between the CS and the US. The CR often is similar to the UR in such procedures (e.g., in Pavlov's experiment both were salivation). Although the traditional view of Pavlovian conditioning emphasized the contiguity of the CS and US, a more general and contemporary view of Pavlovian conditioning emphasizes the informational relationship between the CS and the US. In other words, the information that the CS provides about the occurrence of the US is the critical feature for learning (Rescorla, 1988).

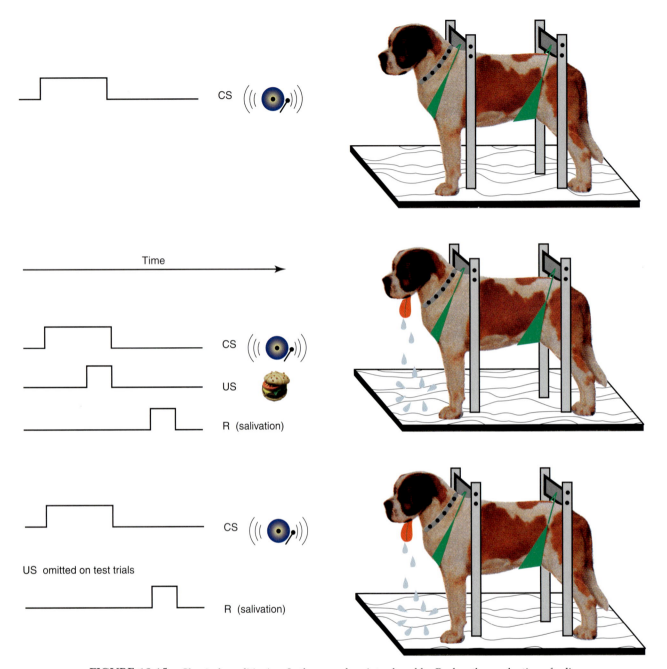

FIGURE 18.15 Classical conditioning. In the procedure introduced by Pavlov, the production of saliva is monitored continuously. Presentation of meat powder reliably leads to salivation, whereas some "neutral" stimulus such as a bell initially does not. With repeated pairings of the bell and meat powder, the animal learns that the bell predicts the food and salivates in response to the bell alone. Modified from Rachlin (1991).

Nonassociative Learning

Three examples of nonassociative learning have received the most experimental attention: habituation, dishabituation, and sensitization. Habituation is defined as a reduction in the response to a stimulus that is repeatedly delivered. Dishabituation refers to the restoration or recovery of a habituated response due to the presentation of another, typically strong, stimulus to the animal. Sensitization is an enhancement or augmentation of a response produced by the presentation of a strong stimulus. The following sections introduce the *Aplysia* and focus on the neural and molecular mechanisms of sensitization.

INVERTEBRATE STUDIES: KEY INSIGHTS FROM *Aplysia* INTO BASIC MECHANISMS OF LEARNING

Since the mid–1960s, the marine mollusk *Aplysia* has proven to be an extremely useful model system for gaining insights into the neural and molecular mechanisms of simple forms of memory. Indeed, the pioneering discoveries of Eric Kandel using this animal were recognized by his receipt of the Nobel Prize in Physiology or Medicine in 2000. A number of characteristics make *Aplysia* well suited for examination of the molecular, cellular, morphological, and network mechanisms underlying neuronal modifications (plasticity) and learning and memory. The animal has a relatively simple nervous system with large, individually identifiable neurons that are accessible for detailed anatomical, biophysical, biochemical, and molecular studies. Neurons and neural

circuits that mediate many behaviors in *Aplysia* have been identified precisely. In several cases, these behaviors have been shown to be modified by learning. Moreover, specific loci within neural circuits at which modifications occur during learning have been identified, and aspects of the cellular mechanisms underlying these modifications have been analyzed and modeled (Byrne and Kandel, 1996; Byrne *et al.*, 1993; Hawkins *et al.*, 1993).

The Siphon–Gill and Tail–Siphon Withdrawal Reflexes of *Aplysia*

Within the mantle cavity of *Aplysia* lies the respiratory organ of the animal, the gill, and protruding from the mantle cavity is the siphon (Fig. 18.16). The siphon–gill withdrawal reflex is elicited when a tactile or electrical stimulus is delivered to the siphon; the stimulus causes withdrawal of the siphon and gill

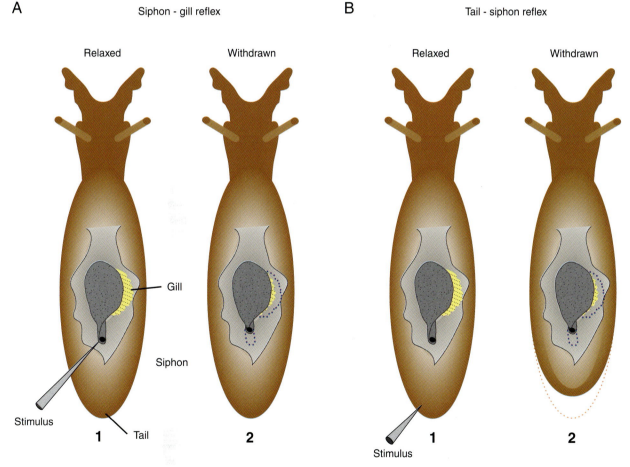

A Siphon - gill reflex   **B** Tail - siphon reflex

Relaxed   Withdrawn   Relaxed   Withdrawn

Gill
Siphon
Stimulus   1   Tail   2   1   2
Stimulus

FIGURE 18.16 Siphon–gill and tail–siphon withdrawal reflexes of *Aplysia*. (A) Siphon-gill withdrawal. Dorsal view of *Aplysia*: (1) Relaxed position. (2) A stimulus (e.g., a water jet, brief touch, or weak electric shock) applied to the siphon causes the siphon and the gill to withdraw into the mantle cavity. (B) Tail–siphon withdrawal reflex: (1) Relaxed position. (2) A stimulus applied to the tail elicits a reflex withdrawal of the tail, the siphon, and the gill.

(Fig. 18.16A). A second behavior that has been examined extensively is the tail–siphon withdrawal reflex. Tactile or electrical stimulation of the tail elicits a coordinated set of defensive responses composed of a reflex withdrawal of the tail and the siphon (Fig. 18.16B).

These two defensive reflexes in *Aplysia* can exhibit three forms of nonassociative learning: habituation, dishabituation, and sensitization. A single sensitizing stimulus, such as a brief several-second-duration electric shock, can produce a reflex enhancement that lasts minutes (short-term sensitization), whereas prolonged training (e.g., multiple stimuli over an hour or longer) produces an enhancement that lasts from days to weeks (long-term sensitization). *Aplysia* also exhibit several forms of associative learning, including classical conditioning (see below) and operant conditioning (Botzer *et al.*, 1998; Brembs *et al.*, 2002; Cook and Carew, 1989).

A prerequisite for successful analysis of the neural and molecular bases of these different forms of learning is an understanding of the neural circuit that controls the behavior. The afferent limb of the siphon–gill withdrawal reflex consists of a population of approximately 24 sensory neurons with somata in the abdominal ganglion. The siphon sensory neurons (SN) monosynaptically excite a population of approximately 13 gill and siphon motor neurons (MN) that are also located in the abdominal ganglion (Fig. 18.17A). Activation of the gill and siphon motor neurons leads to contraction of the gill and siphon. Excitatory, inhibitory, and modulatory interneurons (IN) in the withdrawal circuit have also been identified, although only excitatory interneurons are illustrated in Figure 18.17. The afferent limb of the tail–siphon withdrawal reflex consists of a bilaterally

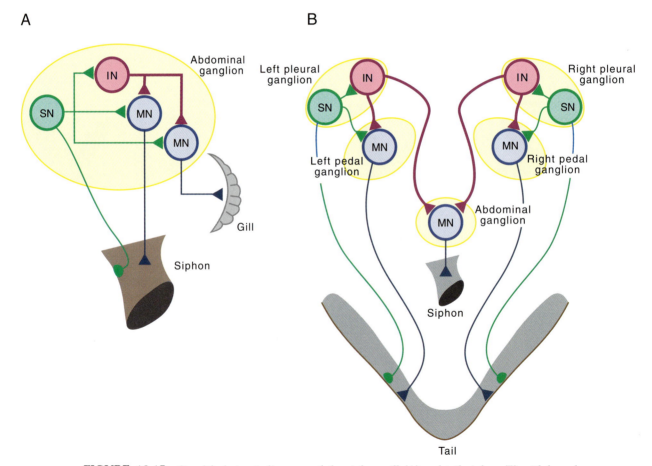

FIGURE 18.17 Simplified circuit diagrams of the siphon–gill (A) and tail–siphon (B) withdrawal reflexes. Stimuli activate the afferent terminals of mechanoreceptor sensory neurons (SN), the somata of which are located in central ganglia (abdominal, pedal, and pleural). The sensory neurons make excitatory synaptic connections (triangles) with interneurons (IN) and motor neurons (MN). The excitatory interneurons provide a parallel pathway for excitation of the motor neurons. Action potentials elicited in the motor neurons, triggered by the combined input from the SNs and INs, propagate out peripheral nerves to activate muscle cells and produce the subsequent reflex withdrawal of the organs. Modulatory neurons (not shown here but see Fig. 18.18A1), such as those containing serotonin (5-HT), regulate the properties of the circuit elements and, consequently, the strength of the behavioral responses.

symmetrical cluster of approximately 200 sensory neurons located in the left and right pleural ganglia. These sensory neurons make monosynaptic excitatory connections with at least three motor neurons in the adjacent pedal ganglion, which produce withdrawal of the tail (Fig. 18.17B). In addition, the tail sensory neurons form synapses with various identified excitatory and inhibitory interneurons. Some of these interneurons activate motor neurons in the abdominal ganglion, which control reflex withdrawal of the siphon. Moreover, several additional neurons modulate the tail–siphon withdrawal reflex (Cleary *et al.*, 1995) (see Fig. 18.18A1).

The sensory neurons for both the siphon–gill and tail–siphon withdrawal reflexes are similar and appear to be important, although probably not exclusive (e.g., Cleary *et al.*, 1998), sites of plasticity in their respective neural circuits. Changes in their membrane properties and the strength of their synaptic connections (synaptic efficacy) are associated with sensitization.

Multiple Cellular Processes and Short- and Long-Term Sensitization in *Aplysia*

Short-Term Sensitization

Short-term (minutes) sensitization is induced when a single brief train of shocks to the body wall results in the release of modulatory transmitters, such as serotonin (5-HT), from a separate class of interneurons referred to as facilitatory neurons (Fig. 18.18A1). These facilitatory neurons regulate the properties of the sensory neurons and the strength of their connections with postsynaptic interneurons and motor neurons through a process called heterosynaptic facilitation (Byrne and Kandel, 1996) (Figs. 18.18A2 and A3). The molecular mechanisms contributing to heterosynaptic facilitation are illustrated in Fig. 18.18B. The first step is the binding of 5-HT to one class of receptors on the outer surface of the membrane of the sensory neurons. This leads to the activation of adenylyl cyclase, which in turn leads to an elevation of the intracellular level of the second messenger cyclic AMP (cAMP) in sensory neurons. When cAMP binds to the regulatory subunit of cAMP-dependent protein kinase (protein kinase A, or PKA), the catalytic subunit is released and can now add phosphate groups to specific substrate proteins, thereby altering their functional properties. One consequence of this protein phosphorylation is an alteration in the properties of membrane channels. Specifically, the increased levels of cAMP lead to a decrease in the serotonin-sensitive potassium current [S-K$^+$ current ($I_{K,S}$)], a component of the calcium-activated K$^+$ current ($I_{K,Ca}$) and the delayed K$^+$ current ($I_{K,V}$). (See Chapters 5 and 6 for more information on these channel types.) These changes in membrane currents lead to depolarization of the membrane potential, enhanced excitability, and an increase in the duration of the action potential (i.e., spike broadening). Reflections of enhanced excitability include an increase in the number of action potentials elicited in a sensory neuron by a fixed extrinsic current injected into the cell or by a fixed stimulus to the skin.

Cyclic AMP also appears to activate a facilitatory process that is independent of membrane potential and spike duration. This process is represented in Fig. 18.18B (large arrow) as the translocation or mobilization of transmitter vesicles from a reserve pool to a releasable pool. The translocation makes more transmitter-containing vesicles available for release with subsequent action potentials in the sensory neuron. The overall effect is a short-term cAMP-dependent enhancement of transmitter release.

Serotonin also acts through another class of receptors to increase the level of the second messenger DAG. DAG activates PKC, which, like PKA, contributes to facilitation that is independent of spike duration (e.g., mobilization of vesicles). In addition, PKC regulates a nifedipine-sensitive Ca^{2+} channel ($I_{Ca, Nif}$) and the delayed K$^+$ channel ($I_{K,V}$). Thus, the delayed K$^+$ channel ($I_{K,V}$), is dually regulated by PKC and PKA. The modulation of $I_{K,V}$ contributes importantly to the increase in duration of the action potential (Fig. 18.18A3). Because of its small magnitude, the modulation of $I_{Ca, Nif}$ appears to play a minor role in the facilitatory process.

Prolonged treatments of 5-HT (1.5 h) activate MAPK (Martin *et al.*, 1997). This pathway was originally suggested to be important only for the induction of long-term processes (see below). However, recent work indicates that a brief (5 min) application of 5-HT results in the phosphorylation of synapsin, possibly through a MAPK-dependent process (Angers *et al.*, 2002). Synapsin is a synaptic vesicle-associated protein that tethers synaptic vesicles to cytoskeletal elements and thus helps to control the reserve pool of vesicles in synaptic terminals (see Chapter 8). Phosphorylation of synapsin would allow vesicles in the reserve pool to migrate to the releasable pool and thus contribute to enhanced transmitter release. Of general significance is the observation that a single modulatory transmitter (i.e., 5-HT) activates at least two (PKA, PKC) and possibly a third (MAPK) kinase systems. The involvement of multiple second messenger systems in synaptic plasticity also appears to be a theme emerging from mammalian studies. For example, the induction of LTP in the CA1 area of the hippocampus (see section on LTP/LTD) appears to

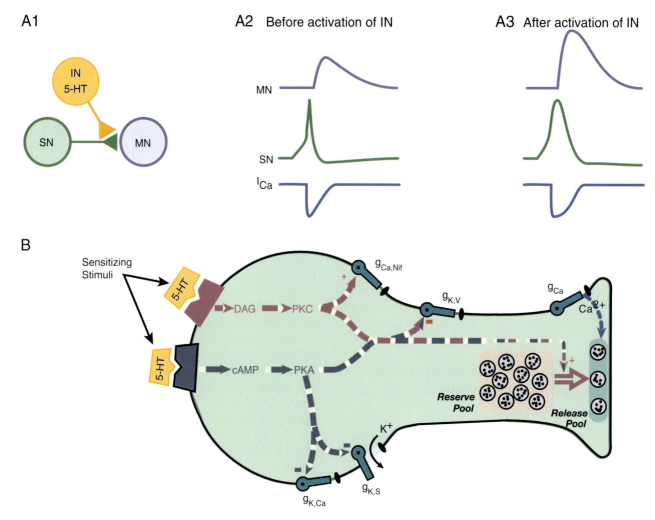

FIGURE 18.18 Model of short-term heterosynaptic facilitation of the sensorimotor connection that contributes to short- and long-term sensitization in *Aplysia*. (A1) Sensitizing stimuli activate facilitatory interneurons (IN) that release modulatory transmitters, one of which is 5-HT. The modulator leads to an alteration of the properties of the sensory neuron (SN). (A2, A3) An action potential in SN after the sensitizing stimulus results in greater transmitter release and hence a larger postsynaptic potential in the motor neuron (MN, A3) than an action potential before the sensitizing stimulus (A2). For short-term sensitization, the enhancement of transmitter release is due, at least in part, to broadening of the action potential and an enhanced flow of Ca^{2+} (I_{Ca}) into the sensory neuron. (B) Model of a sensory neuron that depicts the multiple processes for short-term facilitation that contribute to short-term sensitization. 5-HT released from facilitatory neurons binds to at least two distinct classes of receptors on the outer surface of the membrane and leads to the transient activation of two intracellular second messengers, DAG and cAMP, and their respective kinases (PKC and PKA). These two kinases affect multiple cellular processes, the combined effects of which lead to enhanced transmitter release when subsequent action potentials are fired in the sensory neuron (see text for additional details). Modified from Byrne and Kandel (1996).

involve MAPK, PKC, CaMKII, and tyrosine kinase (reviewed in Dineley *et al.*, 2001).

The consequences of activating these multiple second messenger systems and modulating these various cellular processes are expressed when test stimuli elicit action potentials in the sensory neuron at various times after the presentation of the sensitizing stimuli (Fig. 18.18A3). More transmitter is available for release as a result of the mobilization process, and each action potential is broader, allowing a larger influx of Ca^{2+} to trigger release of the available transmitter. The combined effects of mobilization and spike broadening lead to facilitation of transmitter release from the sensory neuron and, consequently, to a larger postsynaptic potential in the motor neuron. Larger postsynaptic potentials lead to enhanced activation of interneurons and motor neurons and, thus, to an enhanced behavioral response. The maintenance of short-term sensitization is dependent on the persistence of the PKA-, PKC-, and possibly MAPK-induced phosphorylations of the various substrate proteins.

Long-Term Sensitization

Sensitization also exists in a long-term form, which persists for at least 24 h. Whereas short-term sensitization can be produced by a single brief stimulus, the induction of long-term sensitization requires a more extensive training period of an hour or longer.

A substantial amount of data indicates that both short- and long-term sensitization share some common cellular pathways during their induction. For example, both forms activate the cAMP/PKA cascade (Fig. 18.19). However, in the long-term form, unlike the short-term form, activation of the cAMP/PKA cascade induces gene transcription and new protein synthesis (Byrne *et al.*, 1993; Hawkins *et al.*, 1993). Repeated training leads to a translocation of PKA to the nucleus where it phosphorylates the transcriptional activator CREB1 (cAMP-responsive element-binding protein). CREB1 binds to a regulatory region of genes known as CRE (cAMP-responsive

element). Next, this bound and phosphorylated form of CREB1 leads to increased transcription. Serotonin also leads to the activation of MAPK either via cAMP (Martin *et al.*, 1995) or through another pathway (Dyer *et al.*, 2003) which phosphorylates the transcriptional repressor CREB2. Phosphorylation of CREB2 by MAPK leads to a derepression of CREB2 and therefore promotes CREB1-mediated transcriptional activation (Bartsch *et al.*, 1995). The combined effects of activation of CREB1 and derepression of CREB2 lead to changes in the synthesis of specific proteins. So far, more than 10 gene products that are regulated by sensitization training have been identified, and others are likely to be found in the future. These results indicate that there is not a single memory gene or protein, but that multiple genes are regulated, and they act in a coordinated way to alter neuronal properties and synaptic strength. In the following section, three regulated proteins of particular significance are discussed.

The downregulation of a homologue of a neuronal cell adhesion molecule (NCAM), ApCAM, plays a key role in long-term facilitation. This downregulation has two components. First, the synthesis of ApCAM is reduced. Second, preexisting ApCAM is internalized via increased endocytosis. The internalization and

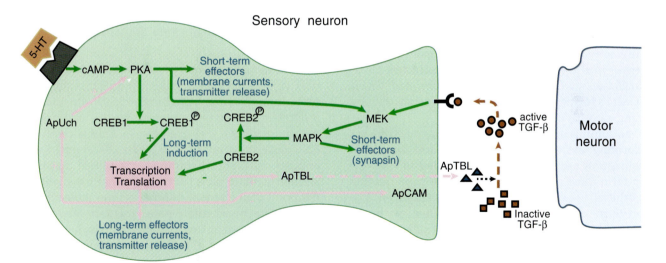

FIGURE 18.19 Simplified scheme of the mechanisms in sensory neurons that contribute to long-term sensitization and some aspects of short-term sensitization. Sensitization training leads to cAMP-dependent regulation of short-term effectors (see Fig. 18.18 for details) and phosphorylation of CREB1. cAMP also activates MAPK, which regulates the repressor CREB2. The combined effects of activation of CREB1 and derepression of CREB2 lead to regulation of the synthesis of at least 10 proteins, only 3 of which (ApTBL, ApCAM, ApUch) are shown. Two of these proteins (ApTBL and ApUch) appear to be components of positive feedback cycles. ApTBL is believed to activate latent forms of TGF-β, which can then bind to receptors on the sensory neuron. TGF-β activates MAPK, which can have both acute and long-term effects. One of its acute effects is the regulation of transmitter release. MAPK may also act by initiating a second round of gene regulation by affecting CREB2-dependent pathways. Increased synthesis of ApUch leads to enhanced degradation of the regulatory subunit of PKA, leading to enhanced activation of the catalytic subunit and increased phosplorylation of CREB1. The third protein, ApCAM, is downregulated. Downregulation of ApCAM is involved in regulating growth processes associated with long-term facilitation.

degradation of ApCAM allow for the restructuring of the axon arbor (Bailey *et al.*, 1992). The sensory neuron can now form additional connections with the same postsynaptic target or make new connections with other cells. Another protein whose synthesis is regulated is *Aplysia* tolloid/BMP-like protein (ApTBL–1). Tolloid and the related molecule BMP–1 appear to function as secreted Zn^{2+} proteases. In some preparations, they activate members of the transforming growth factor β (TGF-β) family. Indeed, in sensory neurons, TGF-β mimics the effects of 5-HT in that it produces long-term increases in synaptic strength of the sensory neurons (Zhang *et al.*, 1997). Interestingly, TGF-β activates MAPK in the sensory neurons and induces its translocation to the nucleus. Thus, TGF-β could be part of an *extracellular* positive feedback loop possibly leading to another round of protein synthesis (Fig. 18.19) to further consolidate the memory (Zhang *et al.*, 1997). A third important protein, *Aplysia* ubiquitin hydrolase (ApUch), appears to be involved in an *intracellular* positive feedback loop. During induction of long-term facilitation, ApUch levels in sensory neurons are increased, possibly via CREB phosphorylation and a consequent increase in ApUch transcription. The increased levels of ApUch increase the rate of degradation, via the ubiquitin–proteosome pathway, of proteins including the regulatory subunit of PKA (Chain *et al.*, 1999). The catalytic subunit of PKA, when freed from the regulatory subunit, is highly active. Thus, increased ApUch will lead to an increase in PKA activity and more protracted phosphorylation of CREB1. This phosphorylated CREB may act to further prolong ApUch expression, thereby closing a positive feedback loop.

One simplifying hypothesis is that the mechanisms underlying the expression of short- and long-term sensitization are the same, but extended in time for long-term sensitization. Some evidence supports this hypothesis. For example, long-term sensitization, like short-term sensitization, is associated with an enhancement of sensorimotor connections. In addition, K⁺ currents and the excitability of sensory neurons are modified by long-term sensitization (Cleary *et al.*, 1998). Based on the model for short-term sensitization, one would expect action potential duration to be affected by long-term sensitization. Surprisingly, this hypothesis has never been rigorously examined. Although some of the expression mechanisms are common, the expression of long-term sensitization has been associated with unique mechanisms. Structural changes such as neurite outgrowth and active zone remodeling have been correlated with long-term sensitization, but not with short-term sensitization (Bailey and Kandel, 1993; Wainwright *et al.*,

2002). Another recently identified correlate of long-term sensitization, and a correlate of procedures that mimic sensitization training, is an increase in high-affinity glutamate uptake (Levenson *et al.*, 2000). A change in glutamate uptake could potentially exert a significant effect on synaptic efficacy by regulating the amount of transmitter available for release, the rate of clearance from the cleft, and thereby the duration of the EPSP and the degree of receptor desensitization. Moreover, long-term sensitization has been correlated with changes in the postsynaptic cell (i.e., the motor neuron, Cleary *et al.*, 1998). Thus, as with other examples of memory, multiple sites of plasticity exist even within this simple reflex system.

Persistent phosphorylation also contributes to intermediate-term facilitation (see below) and may contribute to the induction and maintenance of long-term facilitation as well. For example, one way that new protein synthesis regulates synaptic strength is by reducing levels of the regulatory subunit of PKA, resulting in persistent phosphorylation of target proteins (Chain *et al.*, 1999). An interesting hypothesis is that TGF-β could also be part of the maintenance mechanism. Its late activation by ApTBL-1 could feed back through an extracellular loop to reactivate MAPK in the sensory neurons and therefore engage some acute MAPK-dependent effectors (Fig. 18.19).

Other Temporal Domains for the Memory of Sensitization

Operationally, memory has frequently been divided into two temporal domains: short-term and long-term. It has become increasingly clear from studies of a number of memory systems that this distinction is overly restrictive. For example, in *Aplysia* Carew and his colleagues (Sutton *et al.*, 2001, 2002) and Kandel and his colleagues (Ghirardi, *et al.*, 1995) discovered an intermediate phase of memory that has distinctive temporal characteristics and a unique molecular signature. The intermediate-phase memory for sensitization is expressed at times approximately 30 min to 3 h after the beginning of training. It declines completely prior to the onset of long-term memory. Like long-term sensitization, its induction requires protein synthesis, but unlike long-term memory it does not require mRNA synthesis. The expression of the intermediate-phase memory requires the persistent activation of PKA.

In addition to intermediate-phase memory, it is likely that *Aplysia* has different phases of long-term memory. For example, at 24 h after sensitization training, there is increased synthesis of a number of proteins, some of which are different from those whose synthesis is increased during and immediately after training. Yet blocking protein synthesis between 12 and 24 h after training does not block long-term facilitation at 24 h. These results suggest that the memory for sensitization that persists for times longer than 24 h may be dependent on the synthesis of proteins occurring at 24 h and may have a different molecular signature than the 24-h memory.

Mechanisms Underlying Associative Learning of Withdrawal Reflexes in *Aplysia*

The withdrawal reflexes of *Aplysia* are subject to classical conditioning (Byrne *et al.*, 1993; Hawkins *et al.*, 1993). The short-term classical conditioning observed at the behavioral level reflects, at least in part, a cellular mechanism called activity-dependent neuromodulation. A diagram of the general scheme is presented in Fig. 18.20. The US pathway is activated by a shock to the animal, which elicits a withdrawal response (the UR). When a CS is consistently paired with the US, the animal develops an enhanced withdrawal response (CR) to the CS. Activity-dependent neuromodulation is proposed as the mechanism for this pairing-specific effect. The US activates both a motor neuron (UR) and a modulatory system. The modulatory system delivers the neurotransmitter serotonin (5-HT) to all the sensory neurons (parts of the various CS pathways), which leads to a nonspecific enhancement of transmitter release from the sensory neurons. This nonspecific enhancement contributes to short-term sensitization (see prior discussion). Sensory neurons whose activity is temporally contiguous with the US-mediated reinforcement are additionally modulated. Spiking in a sensory neuron during the presence of 5-HT leads to changes in that cell relative to other sensory neurons whose activity was not paired with the US. Thus, a subsequent CS will lead to an enhanced activation of the reflex (Fig. 18.20B).

Figure 18.21 illustrates a more detailed model of the proposed cellular mechanisms responsible for this

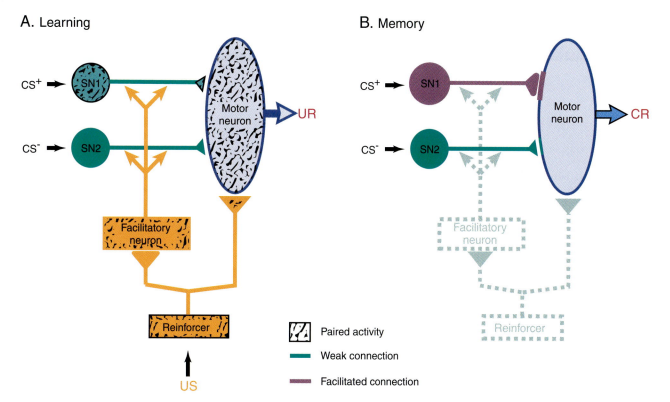

FIGURE 18.20   Model of classical conditioning of a withdrawal reflex in *Aplysia*. (A) Activity in a sensory neuron (SN1) along the CS+ (paired) pathway is coincident with activity in neurons along the reinforcement pathway (US). However, activity in the sensory neuron (SN2) along the CS– (unpaired) pathway is not coincident with activity in neurons along the US pathway. The US directly activates the motor neuron, producing the UR. The US also activates a modulatory system in the form of the facilitatory neuron, resulting in the delivery of a neuromodulatory transmitter to the two sensory neurons. The pairing of activity in SN1 with the delivery of the neuromodulator yields the associative modifications. (B) After the paired activity in (A), the synapse from SN1 to the motor neuron is selectively enhanced. Thus, it is more likely to activate the motor neuron and produce the conditioned response (CR) in the absence of US input. Modified from Lechner and Byrne (1998).

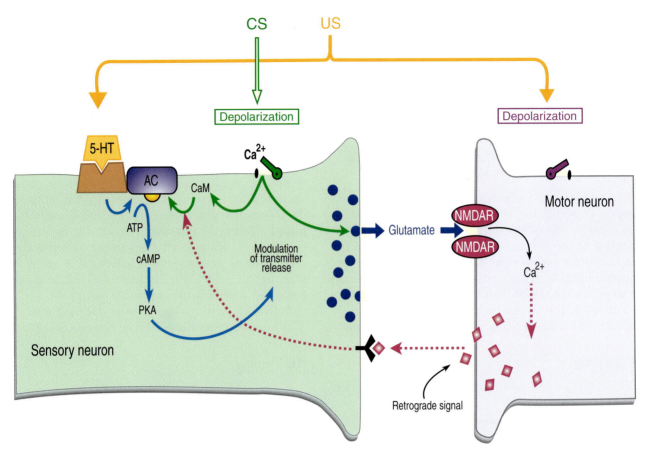

FIGURE 18.21   Model of associative facilitation at the *Aplysia* sensorimotor synapse. This model has both a presynaptic and a postsynaptic detector for the coincidence of the CS and the US. Furthermore, a putative retrograde signal allows for the integration of these two detection systems at the presynaptic level. The CS leads to activity in the sensory neuron, yielding presynaptic calcium influx, which enhances the US-induced cAMP cascade. The CS also induces glutamate release, which results in postsynaptic calcium influx through NMDA receptors if paired with the US-induced depolarization of the postsynaptic neuron. The postsynaptic calcium influx putatively induces a retrograde signal, which further enhances the presynaptic cAMP cascade. The end result of the cAMP cascade is to modulate transmitter release and enhance the strength of the synapse. Modified from Lechner and Byrne (1998).

example of classical conditioning. The modulator (US) acts by increasing the activity of adenylyl cyclase (AC), which in turn increases the levels of cAMP in the sensory neuron. Spiking in the sensory neurons (CS) leads to increased levels of intracellular calcium, which greatly enhances the action of the modulator to increase the cAMP cascade. This system determines CS–US contiguity by a method of coincidence detection at the presynaptic terminal. Now, consider the postsynaptic side of the synapse. The postsynaptic region contains NMDA type receptors (see section on LTP/LTD and Chapters 9 and 11). These receptors need concurrent delivery of glutamate and depolarization to allow calcium to enter. The glutamate is provided by the activated sensory neuron (CS), and the depolarization is provided by the US (Bao *et al.*, 1998; Murphy and Glanzman, 1997) (for review see

Lechner and Byrne, 1998). Thus, the postsynaptic neuron provides another example of coincidence detection. This aspect of plasticity at the sensorimotor synapse is similar to the phenomenon of LTP in the CA1 region of hippocampus (see section on LTP/LTD). The increase in intracellular calcium putatively causes a retrograde signal to be released from the postsynaptic to the presynaptic terminal, ultimately acting to further enhance the cAMP cascade in the sensory neuron. The overall amplification of the cAMP cascade acts to raise the level of PKA, which in turn leads to the modulation of transmitter release. These activity-dependent changes enhance synaptic efficacy between the specific sensory neuron of the CS pathway and the motor neuron. Thus, the sensory neuron along the CS pathway is better able to activate the motor neuron and produce the CR.

Operant Conditioning of Feeding Behavior in *Aplysia*

Although this section has focused on the use of withdrawal reflexes of *Aplysia* as a model system to study sensitization and classical conditioning, the study of feeding behavior of *Aplysia* has recently begun to reveal the mechanisms underlying operant conditioning. Feeding behavior in *Aplysia* exhibits several features that make it amenable to the study of learning. For example, the behavior occurs in an all-or-nothing manner and is therefore easily quantified, and the neural circuitry underlying the generation of the behavior, termed the central pattern generator (CPG), is well characterized to the extent that many of the key individual neurons responsible for the generation of feeding movements have been identified. Like the study of classical conditioning of the defensive withdrawal reflex, researchers are exploiting the advantages of *Aplysia* and are identifying loci of plasticity and changes in membrane properties in the key neurons of the CPG that occur during operant conditioning.

In an operant conditioning paradigm, the delivery of a reinforcing stimulus is contingent on the expression of a designated behavior. In the case of feeding behavior, the operant behavior is ingestive or biting movements and the reinforcement is a stimulus to the esophageal nerve (En), which based on previous work is a neural pathway enriched in dopamine processes and which signals the presence of food in the mouth (Kabotyanski, *et al.*, 1998; Lechner *et al.*, 2000). A training protocol was developed in which, over a 10-min training period, the En was stimulated each time the animal performed a spontaneous biting movement (contingent reinforcement) (Fig. 18.22) (Brembs *et al.*, 2002). Control animals received the same number of stimulations over the 10-min period; however, they were explicitly unpaired. Learning was assessed by measuring the number of spontaneous bites during a test period 1 and 24 h following training. The group of animals that had received paired training showed a significantly larger number of spontaneous bites both 1 and 24 h after training compared with control animals.

The development of behavioral protocols for operant conditioning allowed for further investigations into the mechanisms underlying the learning. Neural correlates of operant conditioning were identified by removing the buccal ganglia, where the CPG underlying feeding behavior is located, from recently trained animals and studying the change in cellular properties of key cells that are essential for producing the neural activity underlying biting move-

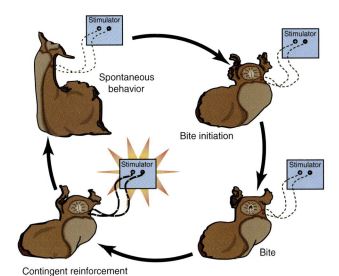

FIGURE 18.22 Operant conditioning of feeding behavior. Throughout the experiment, the animal was observed and all bites were recorded. In the contingent reinforcement group a bite was immediately followed by a brief electric stimulation of the esophageal nerve. A control group received the same sequence of stimulations as the contingent group, but the stimulation was uncorrelated with the animal's behavior. Experimental sessions consisted of a 5-min pretest, 10 min of training, and a final test period. In each period, the number of bites was recorded. The final test period was either 1 or 24 h after training. The group of animals that had received contingent reinforcement showed a significantly larger number of spontaneous bites both 1 and 24 h after training compared with control animals. Modified from Brembs *et al.* (2002).

ments. The CPG in the buccal ganglia produces activity termed *buccal motor programs* (BMPs) which underlie the various types of feeding movements. These BMPs have been well characterized (Morton and Chiel, 1993a,b; Church and Lloyd, 1994), and one cell that has proven to be essential in the generation of BMPs underlying biting movements (ingestive-like BMPs) is cell B51 (Nargeot *et al.*, 1999b, c). Therefore, changes in the cellular properties of B51 were measured in ganglia taken from trained and untrained animals. This analysis revealed that the input resistance was significantly higher and the burst threshold was significantly lower in B51 of ganglia from trained animals compared with untrained animals. These types of changes would serve to increase the probability that B51 would become active and would therefore facilitate the generation of the neural activity underlying biting movements in the trained animals.

Because the ganglia of *Aplysia* can be removed and studied for extended periods in isolation, *in vitro* and single-cell analogues of operant conditioning can be developed and used not only to further probe the network, cellular, and membrane properties supporting training, but also to validate the identified mechanisms underlying learning at multiple levels of analysis. An *in vitro* analogue of operant conditioning has been developed in which buccal ganglia from naïve animals are removed and placed in a chamber (Nargeot *et al.*, 1997, 1999b,c). Over a 10-min training period, every ingestion-like BMP that occurred was reinforced by stimulation of the En (paired or contingent reinforcement). Controls consisted of ganglia that received explicitly unpaired (noncontingent) delivery of the same number of En stimulations. "Learning" was assessed by measuring the number of ingestion-like BMPs generated 1 h following training in both groups. The ganglia from the contingent reinforcement group generated significantly more ingestion-like BMPs than the control group. In addition, similar to what was seen in the neural correlates in *in vivo* operant conditioning, in ganglia that had received contingent reinforcement, the burst threshold of cell B51 was significantly lower and the input resistance was significantly higher compared with control ganglia (Nargeot *et al.*, 1999b).

Taken together, the studies described above suggested that changes in the properties of B51 are correlated with learning-induced changes that occur during operant conditioning. However, they did not indicate whether the cellular mechanisms underlying this associative plasticity were intrinsic to B51 or whether they were due to extrinsic factors such as plasticity that may occur in other cells within the CPG that exert their influence on B51. This issue was addressed in a series of experiments in which a single-cell analogue of operant conditioning that consisted of cell B51 isolated in culture was developed (Brembs *et al.*, 2002). In this procedure, an individual B51 cell was removed from a naïve ganglion and maintained in culture. Because a series of previous experiments had provided strong evidence that dopamine (DA) was a critical neurotransmitter in the reinforcement pathway (Kabotyanski *et al.*, 1998; Nargeot *et al.*, 1999a), reinforcement was mimicked by the application of a brief "puff" of DA onto the cell. In addition, previous experiments had revealed that B51 exhibits a plateau potential during each ingestion-like BMP (Nargeot *et al.*, 1999b, c). Therefore, DA reinforcement was made contingent on a plateau potential that was elicited in B51 by injection of a brief depolarizing current pulse. Training consisted of delivering seven depolarizations, which elicited plateau potentials, over a 10-min interval. Each plateau potential was immediately followed by a DA puff onto the cell. Controls consisted of cells that received the DA puff 40 s after the plateau potential. Training produced a significant increase in input resistance and significant decrease in burst threshold

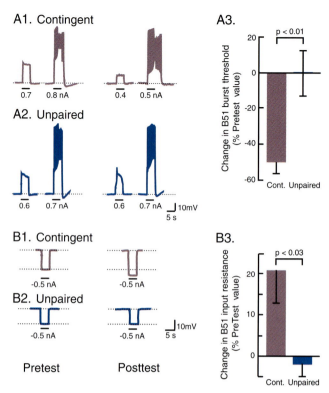

FIGURE 18.23 Changes in B51 produced by single-cell *in vitro* analogue of operant conditioning. Contingent-dependent changes in burst threshold and input resistance in cultured B51. (A) Burst threshold. (A1, A2) Intracellular recordings from a pair of contingently reinforced and unpaired neurons. Depolarizing current pulses were injected into B51 before (Pretest) and after (Posttest) training. In this example, contingent reinforcement led to a decrease in burst threshold from 0.8 to 0.5 nA (A1), whereas it remained at 0.7 nA in the corresponding unpaired cell (A2). (A3) Summary data. The contingently reinforced cells had significantly decreased burst thresholds. (B) Input resistance. (B1, B2) Intracellular recordings from a pair of contingently reinforced and unpaired control neurons. Hyperpolarizing current pulses were injected into B51 before (Pretest) and after (Posttest) training. In this example, contingent reinforcement led to an increased deflection of the B51 membrane potential in response to the current pulse (B1), whereas the deflection remained constant in the corresponding unpaired cell (B2). (B3) Summary data. The contingently reinforced cells had significantly increased input resistances. Modified from Brembs *et al.* (2002).

(Fig. 18.23), similar to the changes observed in the neural correlates (Brembs *et al.*, 2002) and *in vitro* analogues (Nargeot *et al.*, 1999b) of operant conditioning. These data suggest B51 is an important locus of plasticity in operant conditioning of feeding behavior and that intrinsic cellwide plasticity may be one important mechanism underlying this type of learning. The continued analysis will provide insights into the molecular mechanisms of operant conditioning as well as into the mechanistic relationship between operant and classical conditioning.

BOX 18.1

SOME INVERTEBRATES THAT ARE USEFUL FOR PROVIDING INSIGHTS INTO THE MECHANISMS UNDERLYING LEARNING

Gastropod Mollusks

Aplysia

See text.

Hermissenda

The gastropod mollusk *Hermissenda* exhibits associative learning of light-elicited locomotion and foot length (CRs). The conditioning procedure consists of pairing visual stimuli (light, unconditioned stimulus, CS) with vestibular stimuli (high-speed rotation, US). After conditioning, the CS suppresses normal light-elicited locomotion and elicits foot shortening. The associative memory can be retained from days to weeks depending on the number of conditioning trials administered during initial acquisition. The type A and B photoreceptors mediating the CS pathway have been identified as critical sites of plasticity for associative learning. Cellular correlates of conditioning have identified changes in membrane properties and in several K$^+$ conductances of the photoreceptors, and have identified several second messenger systems that mediate the plasticity supporting learning (Crow, 2003).

Pleurobranchaea

The opisthobranch *Pleurobranchaea* is a voracious marine carnivore. When exposed to food, the animal exhibits a characteristic bite–strike response. After pairing of a food stimulus (CS) with a strong electric shock to the oral veil (US), the CS, instead of eliciting a bite–strike response, elicits a withdrawal and suppression of feeding responses (CR). The CR is acquired within a few trials and is retained for up to 4 weeks. Neural correlates of associative learning have been analyzed by examining responses of various identified neurons in the circuit to chemosensory inputs in animals that have been conditioned. One correlate is an enhanced inhibition of command neurons for feeding (Gillette, 1992).

Tritonia diomedea

To escape a noxious stimulus, the opisthobranch *Tritonia diomedea* initiates stereotypical rhythmic swimming. This response exhibits both habituation and sensitization and involves changes in many different components of swim behavior in each case (Frost *et al.*, 1996). The neural circuit consists of sensory neurons, pre-central pattern-generating (CPG) neurons, and motor neurons. Habituation appears to involve plasticity at multiple loci, including decrement at the first afferent synapse. Sensitization appears to involve enhanced excitability and synaptic strength in one of the CPG interneurons (Frost, 2003).

Pond Snail (*Lymnaea stagnalis*)

The pulmonate *Lymnaea stagnalis* exhibits fairly rapid nonaversive conditioning of feeding behavior. A neutral chemical or mechanical stimulus (CS) applied to the lips is paired with a strong stimulant of feeding such as sucrose (US) (Kemenes and Benjamin, 1994). Greater levels of rasping, a component of the feeding behavior, can be produced by a single trial, and this response can persist for at least 19 days. The circuit consists of a network of three types of CPG neurons, 10 types of motor neurons, and a variety of modulatory interneurons. An analog of the behavioral response occurs in the isolated central nervous system. The enhancement of the feeding motor program appears to be due to facilitation of the motor neurons, the modulatory neurons, and presumably the CPG neurons, resulting in increased activation of the CPG cells by mechanosensory inputs from the lips (Benjamin *et al.*, 2000).

Land Snail (*Helix*)

Land snails can be conditioned to avoid food using procedures similar to those used with *Pleurobranchaea*. A food stimulus such as a piece of carrot (CS) is paired with an electric shock to the dorsal surface of the snail (US). After 5–15 pairings, the carrot, instead of eliciting a feeding response, elicits withdrawal and suppression of feeding responses. The transmitter serotonin appears to have a critical role in learning. Animals injected with a toxin that destroys serotonergic neurons exhibit normal responses to the food and the shocks alone, but are incapable of learning. *Helix* also exhibit habituation and sensitization of avoidance responses elicited by tactile stimuli (Balaban *et al.*, 1994).

Limax

The pulmonate *Limax* is an herbivore that locomotes toward desirable food odors. This behavior makes it well suited to food-avoidance conditioning. The slug's normal attraction to a preferred food odor (CS) is significantly reduced when the preferred odor is paired with a bitter taste (US). In addition to this example of classical conditioning, food avoidance in *Limax* exhibits higher-order

BOX 18.1 *(continues)*

features of classical conditioning, such as blocking and second-order conditioning. An analog of taste-aversion learning occurs in the isolated central nervous system, facilitating subsequent cellular analyses of learning in *Limax*. The procerebral lobe in the cerebral ganglion processes olfactory information and is a likely site for the plasticity. The procerebral lobe contains several types of neuropeptides, and FMRFamide as well as SCPB have been shown to affect activity of the rhythmic motor network inderlying the *Limax* feeding system, making these neuropeptides attractive candidates for modulating plasticity within the network. In addition, NO synthase is present in the procerebral lobe and NO has been shown to affect oscillations within the procerebral lobe. Thus, NO, acting through the cGMP signaling pathway, may modulate plasticity via long-term changes in cells within the procerebral lobe (Gelperin, 2003).

Arthropods

Cockroach (Periplaneta americana) and Locust (Schistocerca gregaria)

Learned modifications of leg positioning in the cockroach and locust may be useful in the cellular analysis of operant conditioning. When the animal is suspended over a dish containing a fluid, initially, it makes many movements, including those that cause the leg to come in contact with the liquid surface. When contact with the fluid is paired with an electric shock, the insect rapidly learns to hold its foot away from the fluid. Neural correlates of the conditioning have been observed in somata of the leg motor neurons. These correlates include changes in intrinsic firing rate and membrane conductance (Eisenstein and Carlson, 1994).

Crayfish (Procambarus clarkii)

The crayfish tail-flip response exhibits habituation and sensitization. A key component of the circuit is a pair of large neurons called the lateral giants (LGs), which run the length of the animal's nerve cord. The LGs are the decision and command cells for the tail flip. Learning is related to changes in the strength of synaptic input driving the LGs (Edwards *et al.*, 1999; Krasne, 1992; Sahley and Crow, 1998).

Honeybee (Apis mellifera)

Honeybees, like other insects, are superb at learning. For example, sensitization of the antenna reflex of *Apis mellifera* is produced as a result of presenting gustatory stimuli to the antennae. Classical conditioning of feeding behavior can be produced by pairing a visual or olfactory CS with sugar solution (US) to the antennae. The small size of bee neurons is an obstacle in pursuing detailed cellular analyses of these behavioral modifications. Nevertheless, regions of the brain necessary for associative learning have been identified. In particular, intracellular recordings have revealed that one identified cell, the ventral unpaired median (VUM) neuron that is putatively octopaminergic, mediates reinforcement during olfactory conditioning and represents the neural correlate of the US. The learning has been dissected into several phases of memory, including short-term, midterm, and long-term, and the cellular and molecular mechanisms supporting the various phases are beginning to be unraveled. For example, numerous studies have revealed that, as in other species, the molecular mechanisms underlying memory formation in the honeybee involve upregulation of the cAMP pathway and activation of PKA, resulting in CREB-mediated transcription of downstream genes (Menzel, 2001).

Drosophila

Because the neural circuitry in the fruit fly is both complex and inaccessible, the fly might seem to be an unpromising subject for studying the neural basis of learning. However, the ease with which genetic studies are performed compensates for the difficulty in performing electrophysiological studies (DeZazzo and Tully, 1995). A frequently used protocol employs a two-stage differential odor–shock avoidance procedure, which is performed on large groups of animals simultaneously rather than on individual animals. Animals learn to avoid odors paired (CS+) with shock but not odors explicitly unpaired (CS–). This learning is typically retained for 4–6 h, but retention for 24 h to 1 week can be produced by a spaced training procedure. Several mutants deficient in learning have been identified. Analysis of the affected genes has revealed elements of the cAMP signaling pathway as key in learning and memory. It is now known that the formation of long-term memory requires activation of the cAMP signaling pathway, which, in turn, activates members of the CREB transcription family. These proteins appear to be the key molecular switch that enables expression of genes necessary for the formation of long-term memories. Further, the expression pattern of these genes, as well as other mutational analyses, has revealed that brain structures called mushroom bodies are crucial sites for olfactory learning. These transcription factors are also important for long-term memory in *Aplysia* and in vertebrates (Dudai and Tully, 2003; Waddel and Quinn, 2001).

BOX 18.1 *(continues)*

Annelids

Leech

Defensive reflexes in the leech (*Hirudo medicinalis*) exhibit habituation, dishabituation, sensitization, and classical conditioning. For example, the shortening response is enhanced following pairing of a light touch to the head (CS) with electric shock to the tail (US). The identified S neurons appear critical for sensitization, as their ablation disrupts sensitization. Interestingly, ablation of the S cells only partly disrupts dishabituation, indicating that separate processes contribute to dishabituation and sensitization (Sahley, 1995). Separate processes also contribute to dishabituation and sensitization in *Aplysia*. The transmitter serotonin (5-HT) appears to mediate at least part of the reinforcing effects of sensitizing stimuli and the US by mediating a 5-HT-dependent increase in cAMP. Serotonin appears to play similar roles in *Aplysia*, *Helix*, *Hermissenda*, and *Tritonia* (Sahley and Crow, 1998).

Nematoda

Caenorhabditis elegans

C. elegans is a valuable model system for cellular and molecular studies of learning. Its principal advantages are threefold. First, its nervous system is extremely simple. It has a total of 302 neurons, the anatomical connectivity of which has been described at the electron microscopy level. Second, the developmental lineage of each neuron is completely specified. Third, its entire genome has been sequenced, making it highly amenable to a number of genetic and molecular manipulations. *C. elegans* responds to a vibratory stimulus applied to the medium in which they locomote by swimming backward. This reaction, known as the tap withdrawal reflex, exhibits habituation, dishabituation, sensitization, long-term (24 h) retention of habituation training, and context conditioning. Laser ablation studies have been used to elucidate the neural circuitry supporting the tap withdrawal reflex and to identify likely sites of plasticity within the network. Analysis of several *C. elegans* mutants has revealed that synapses at the locus of plasticity in the network may be glutamatergic (Rose and Rankin, 2001; Sahley and Crow, 1998; Steidl and Rankin, 2003).

Summary

Certain invertebrates display an enormous capacity for learning and offer particular experimental advantages for analyzing the cellular and molecular mechanisms of learning (see Box 18.1). For example, behaviors in *Aplysia* are mediated by relatively simple neural circuits, which can be analyzed with conventional electroanatomical approaches. Once the circuit is specified, the neural locus for the particular example of learning can be found, and biophysical, biochemical, and molecular approaches can then be used to identify mechanisms underlying the change. The relatively large size of some of these cells allows these analyses to take place at the level of individually identified neurons. Individual neurons can be surgically removed and assayed for changes in the levels of second messengers, protein phosphorylation, and RNA and protein synthesis. Moreover, peptides and nucleotides can be injected into individual neurons. This chapter has focused exclusively on *Aplysia*, but many other invertebrates have proven to be valuable model systems for the cellular and molecular analysis of learning and memory (for review see Byrne, 1987). Each has its own unique advantages. For example, *Aplysia* is excellent for applying cell biological approaches to the analysis of learning and memory mechanisms. Other invertebrate model systems such as *Drosophila* and *Caenorhabditis elegans* are not well suited for cell biological approaches because of their small neurons, but offer tremendous advantages for obtaining insights into mechanisms of learning and memory through the application of genetic approaches. [See Box 18.1 and Carew (2000) for a review of several selected invertebrate model systems that have contributed significantly to the understanding of memory mechanisms.]

CLASSICAL CONDITIONING IN VERTEBRATES: DISCRETE RESPONSES AND FEAR AS MODELS OF ASSOCIATIVE LEARNING

When animals, including humans, are faced with an aversive or threatening situation, at least two complementary processes of learning occur. Learned fear or arousal develops very rapidly, often in one trial. Subsequently, the organism learns to make the most adaptive behavioral motor responses to deal with the situation. These observations led to so-called two-process theories of learning: development of an initial learned fear or arousal, followed by slower learning of discrete behavioral responses (Rescorla & Solomon, 1967). As the latter form of learning develops, fear subsides. It is now believed that, at least in mammals, a third process typically takes place in which declarative memory for the events and their relationships develops. This section focuses on the learning of discrete responses, using eyeblink conditioning as the model system, and on conditioned fear.

Eyeblink Is a Model System for Studying the Conditioning of Discrete Behavioral Responses in Vertebrates

A vast amount of research has used Pavlovian conditioning of the eyeblink response in humans and other mammals in various types of investigations (Gormezano *et al.*, 1983). The eyeblink response exhibits all the basic laws and properties of Pavlovian conditioning equally in humans and other mammals. The basic procedure is to present a neutral CS, such as a tone or a light, followed approximately a quarter of a second later (while the CS is still present) by a puff of air to the eye or by a periorbital (around the eye) shock (US). This is an example of what is known as a delay procedure. Initially, there is no response to the CS and a reflex eyeblink to the US. After a number of such trials, the eyelid begins to close in response to the CS before the US occurs. In a well-trained subject, the eyelid closure (CR) becomes very precisely timed so that it is maximally closed about the time that the airpuff or shock (US) onset occurs (see Fig. 18.24). This very adaptive timing of the eyeblink CR develops over the range of CS–US onset intervals in which learning occurs, which is about 100 ms to 1 s. Thus, the conditioned eyeblink response is a very precisely timed and elementary learned motor skill. The same is true of other discrete behavioral responses learned to deal with aversive stimuli (e.g., the forelimb or hindlimb flexion response and the head turn).

Classical conditioning consists of two basic procedures: delay (see above) and trace. Pavlov (1927) was the first to describe trace classical conditioning. He stressed that the organism must maintain a "trace" of the CS in the brain in order for the CS and the US to become associated. In eyeblink conditioning in animals, a typical trace interval between CS offset and US onset is 500 ms. The trace eyeblink procedure is much more difficult to learn than the delay procedure.

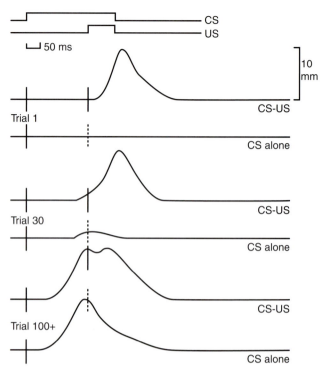

FIGURE 18.24   Adaptive nature of classical conditioning. This example shows the development of the conditioned eyeblink response over the trials of training. The CS is typically a "neutral" light or tone; the US here is a puff of air to the cornea. The eyelid closure response is indicated by upward movement of the tracing. The first marker is tone CS onset; the second is airpuff US onset. In Trial 1 the eyeblink does not move to the CS but closes (blinks) following onset of the US. The conditioned response (CR) is any measurable degree of eyelid closure prior to the onset of the US. Note that after learning, the CR peaks at the onset of the US, i.e., maximum eyelid closure at airpuff onset. If the CS–US onset interval were longer (e.g., 500 ms), the CR would now peak at the onset of the US, 500 ms after CS onset. The conditioned response is adaptive. For this type of learning, a period (ISI) of about 250 ms between CS onset and US onset (shown here) yields the best learning. This best learning time varies widely depending on the type of response (e.g., for fear learning, several seconds is best).

Two brain systems, the hippocampus and the cerebellum, become massively engaged in eyeblink conditioning (Thompson and Kim, 1996). The amygdala also plays a role when the US is sufficiently aversive to elicit learned fear. In an experimental design using a click CS and glabellar (forehead) tap US in restrained cats, a very short latency (< 20 ms) eyeblink muscle EMG (electrical response recorded from muscles around the eye) CR involving the motor cortex develops (Woody *et al.*, 1974; Woody and Yarowsky, 1972). However, bilateral removal of the motor cortex does not appear to affect either learning or expression of the standard longer-latency adaptive delay or trace CRs (Ivkovich and Thompson, 1997). This short-latency EMG response may be a component of the startle response elicited by a sudden acoustic stimulus (Davis, 1984).

Hippocampus and Classic Conditioning

In eyeblink conditioning, neuronal unit activity in hippocampal fields CA1 and CA3 increases very rapidly in paired (tone CS–corneal airpuff US) training trials, shifts forward in time as learning develops, and forms a predictive "temporal model" of the learned behavioral response both within trials and during repeated trials of training (Berger *et al.*, 1976; Berger and Thompson, 1978) (Fig. 18.25). To summa-

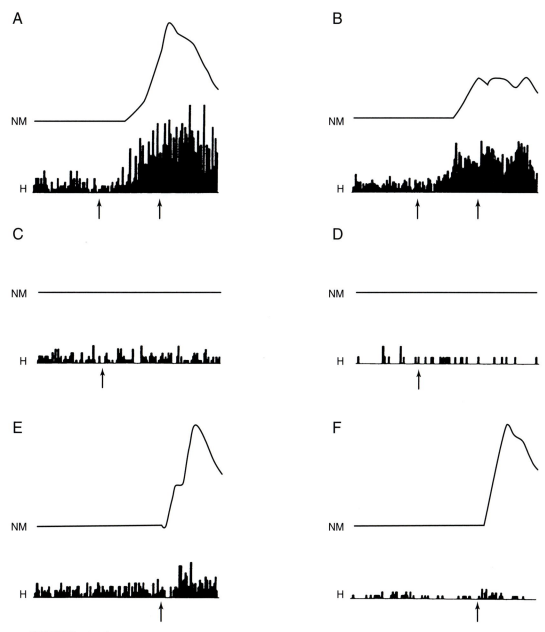

FIGURE 18.25   Engagement of hippocampal neurons in eyeblink conditioning. Responses of identified pyramidal neurons during paired (A, B) and unpaired (C–F) presentations of tone and corneal airpuff. The upper traces show the averaged nictitating membrane (NM, a component of the eyeblink) response for all trials during which a given cell was recorded. The bottom traces show the response of the recorded neuron in the form of a peristimulus time histogram. The total length of both NM responses and histograms was 750 ms. Arrows occurring early in the trial period indicate tone onset; arrows occurring late in the trial period indicate airpuff onset. H, hippocampus. In this particular figure, (A) and (B) show examples of responses of two pyramidal neurons recorded from two different animals during delay conditioning. The results in (C) and (E) show the response of a pyramidal neuron recorded from an animal given unpaired tone-alone (C) and airpuff-alone (E) presentations. (D, F) Same for a different pyramidal cell recorded from a different control animal. From Berger *et al.* (1986).

rize a large body of research on this topic, the growth of the hippocampal unit response, under normal conditions, invariably and strongly predicts subsequent behavioral learning (Berger *et al.*, 1986). This increase in neuronal activity in the hippocampus becomes significant by the second or third trial of training, long before behavioral signs of learning develop. Neurons in the hippocampus become engaged in many other types of learning as well (Isaacson and Pribram, 1986).

In eyeblink conditioning, many neurons identified as pyramidal neurons in fields CA1 and CA3 show learning-related increases in discharge frequency during the trial period (Berger *et al.*, 1983) (Fig. 18.25). Typically, a given neuron models only some limited period of the trial, although some pyramidal neurons model the entire learned behavioral response, as shown in Figs. 18.25A and B. Thus, the pyramidal neuron representation of the behavioral learned response is distributed over both space and time in the hippocampus (Thompson, 1990).

The results just described were obtained using the basic delay procedure, in which hippocampal lesions do not impair simple response acquisition in rabbits. Similarly, humans with hippocampal–temporal lobe anterograde amnesia are able to learn simple acquisition of the eyeblink conditioned response, but cannot recall the learning experience (Weiskrantz and Warrington, 1979). The involvement of the hippocampus depends on the difficulty of the task. For example, such amnesic humans are massively impaired on conditional discriminations in eyeblink conditioning (e.g., blink to tone only if preceded by light) but not on simple discriminations (Daum *et al.*, 1991). Bilateral hippocampal lesions in rabbits markedly impair subsequent acquisition of the trace CR (Moyer *et al.*, 1990; Solomon *et al.*, 1986). Interestingly, when the hippocampal lesion is made immediately after trace learning in rabbits, the CR is abolished; when the lesion is made a month after training, the CR is not impaired (Kim, *et al.*, 1995) (Fig. 18.26).

These results are striking in light of reports of declarative memory deficit following damage to the temporal lobe of the hippocampal system in humans and monkeys. These deficits have two key temporal characteristics: (1) profound and permanent anterograde amnesia and (2) profound but clearly time-limited retrograde amnesia. Subjects have great difficulty learning new declarative tasks and/or information (anterograde amnesia) and have substantial memory loss for events for some period preceding brain damage (retrograde amnesia), but relatively intact memory for earlier events (Zola-Morgan and Squire, 1990).

One of the hallmarks of declarative memory in humans is awareness. In the case of eyeblink conditioning, awareness refers to the ability to describe the experience and the contingencies between the CS and the US. Patients with temporal lobe–hippocampal amnesia can learn the delay eyeblink procedure at a normal rate, as noted above, but are unable to acquire the trace eyeblink procedure (Clark and Squire, 1998; McGlinchey-Berroth *et al.*, 1997). For normal humans, awareness of the situation is unrelated to successful learning in the delay paradigm (procedural learning) but is a prerequisite for successful trace conditioning. Trace conditioning is temporal lobe–hippocampus-dependent because, as in other tasks of declarative memory, conscious knowledge must be acquired across training sessions. Trace conditioning may thus provide a simple means for studying awareness in nonhuman animals in the context of current ideas about multiple memory systems in the brain (Clark and Squire, 1998). Therefore, even in simple memory tasks such as eyeblink conditioning, hippocampus-dependent "declarative" memory processes can develop. Consistent with this is the striking observation that extreme deficits in ability to learn eyeblink conditioning occur in Alzheimer's patients; indeed, marked deficits in normal elderly humans may be diagnostic of the subsequent development of Alzheimer's disease (Woodruff-Pak, *et al.*, 1990).

What are the mechanisms of the changes in the hippocampus? The process of LTP is widely considered to be the most likely mechanism of memory storage in the hippocampal system (see section on LTP/LTD). In the case of classical conditioning, a number of parallels exist between the properties of LTP and the properties of the learning-induced increase in neuronal activity in the hippocampus (Berger *et al.*, 1986). Both LTP and the learning-induced increase in hippocampal neuron activity are associated with pyramidal neurons, and both begin to develop after very brief periods (e.g., 100 Hz for 1 s for LTP; one to three trials of training in eyeblink conditioning). Also, both approach a limit asymptotically over a period of many minutes, show the same magnitude of increase, and are developed only with very specific parameters of stimulation. Furthermore, there is a persistent increase in the extracellularly recorded monosynaptic population spike in the dentate gyrus in response to stimulation of the perforant path as a result of eyeblink conditioning, just as occurs when LTP is induced by tetanus of the perforant path (Weisz *et al.*, 1984).

There are strikingly parallel and persistent increases in AMPA receptor binding on hippocampal membranes in both eyeblink conditioning (well-trained animals) and *in vivo* expression of LTP by stimulation of the perforant path projection to the hippocampal dentate gyrus. The pattern of increased binding is similar in both (Baudry *et al.*, 1993; Maren *et al.*, 1993; Tocco *et al.*, 1992). NMDA receptors play a critical role in induction of LTP (at least in the dentate gyrus and CA1) and also appear to be involved in acquisition of the trace eyeblink CR (Thompson *et al.*, 1992). Mechanisms of LTP are discussed at length in the section above.

In addition to alterations in synaptic strength, as in LTP, there are also changes in the intrinsic excitability of pyramidal neurons in the hippocampus following eyeblink conditioning (Disterhoft *et al.*, 1986) (see also Box 18.2). Specifically, the calcium-dependent slow afterhyperpolarization (AHP), mediated by a voltage-gated potassium conductance, is markedly reduced following learning but not in control animals given unpaired CS and US trials (deJonge *et al.*, 1990). These observations were first made in the delay paradigm, where pyramidal neurons in the hippocampus markedly increase their responsiveness as a result of learning (Fig. 18.25).

A similar decrease in the AHP in hippocampal pyramidal neurons is seen following trace eyeblink conditioning (Moyer *et al.*, 1996). Importantly, this trace learning-induced decrease in the AHP is time-limited, being maximum 1 to 24 h after training and returning toward baseline after a week, consistent with the time-dependent effect of hippocampal lesions on the memory for trace conditioning (see Fig. 18.26) (Thompson *et al.*, 1996).

The Cerebellar System and Classic Conditioning of Discrete Responses

Since publication of the classic papers of Marr (1969) and Albus (1971), the cerebellum has been favored as a structure for modeling neuronal learning. Figure 18.27 is a highly simplified diagram of a current qualitative working model of the role of the cerebellum in basic classic (delay) conditioning of eyeblink and other discrete responses (Thompson, 1986; Thompson and Krupa, 1994). Laterality is not addressed in Fig. 18.27; the critical region of the cerebellum is ipsilateral to the trained eye (or limb), whereas the critical regions of the pontine nuclei, red nucleus, and inferior olive are contralateral. For a more realistic representation, see Fig. 18.30. In this section, the data refer to the basic delay eyeblink CR, unless otherwise noted (Lavond *et al.*, 1993; Thompson and Krupa, 1994; Yeo, 1991).

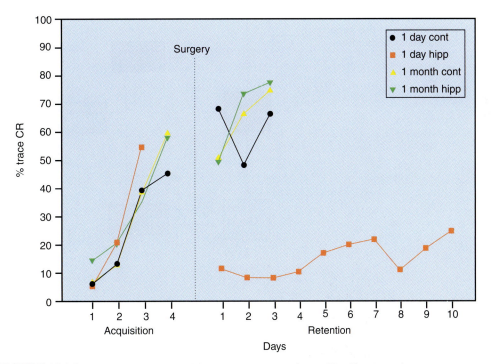

FIGURE 18.26 Effects of hippocampal lesions on retention of trace CRs. Shown are the mean percentage of CRs during initial training and following postoperative training: 1 day cont, controls given cortical or sham lesions 1 day after training; 1 month cont, controls given lesions 1 month after training; 1 day hipp, bilateral hippocampal lesions made 1 day after training; 1 month hipp, hippocampal lesions made 1 month after training. Only the hippocampal lesions made immediately after training abolished the trace CR. From Kim *et al.* (1995).

In brief, the reflex eyeblink response pathways activated by corneal airpuff (or periorbital shock) include the trigeminal nucleus, direct projections to the relevant motor nuclei (mostly the seventh and accessory sixth), and indirect projections to the motor nuclei via the brainstem reticular formation (Fig. 18.27). Analysis of response latencies rules out any direct role of the cerebellum in the reflex response. The tone (and light) CS pathways project to the cerebellum as mossy fibers, mostly relaying through the pontine nuclei. The mossy fibers, in turn, activate granule cells, which project to Purkinje cells via parallel fibers. The US pathway projects from the trigeminal nucleus to the inferior olive and then as climbing fibers to the cerebellum. The CS-activated mossy fiber–parallel fiber pathway and the US-activated climbing fiber pathway converge and make synaptic connections on Purkinje

neurons in the cerebellar cortex (parallel fiber–climbing fiber) and on neurons in the interpositus nucleus (mossy fiber–climbing fiber). The CR pathway projects from the interpositus nucleus of the cerebellum via the superior cerebellar peduncle to the red nucleus, and from there to the premotor and motor nuclei (mostly seventh and accessory sixth), controlling the eyeblink response.

This circuitry has been identified using a number of methods, including lesion studies, electrophysiological recordings, electrical microstimulation, and anatomical characterization of projection pathways. For example, in animals, neurons in the cerebellar cortex and interpositus nucleus respond to the CS and US before training and develop amplitude–time course models (e.g., Fig. 18.28) of the learned behavioral response. These models precede and predict the

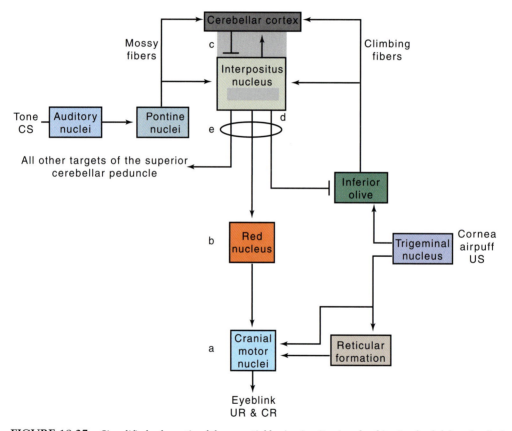

FIGURE 18.27 Simplified schematic of the essential brain circuitry involved in standard delay classical conditioning of discrete responses (e.g., eyeblink response). Shadowed boxes represent areas that have been reversibly inactivated during training. (a) Inactivation of motor nuclei including facial (seventh) and accessory sixth. (b) Inactivation of magnocellular red nucleus. (c) Inactivation of dorsal aspect of the anterior interpositus and overlying cerebellar cortex. (d) Inactivation of ventral anterior interpositus nucleus and associated white matter. (e) Complete inactivation of the superior cerebellar peduncle (scp), essentially all output from the cerebellar hemisphere. Inactivation of each of these regions in trained rabbits abolishes performance of the CR. Significantly, inactivation of the motor nuclei (a), the red nucleus (b), the superior cerebellar peduncle (e), and the output of the interpositus nucleus (d) during training do not prevent learning at all, but inactivation of a localized region of the anterior interpositus nucleus and overlying cortex (c) during training completely prevents learning. From Thompson and Krupa (1994).

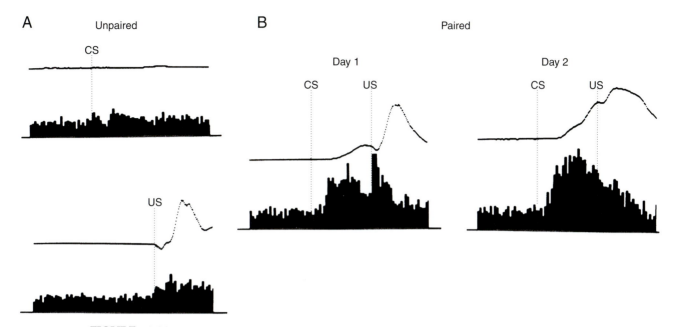

FIGURE 18.28   Engagement of neurons in the cerebellar interpositus nucleus in eyeblink conditioning. Histograms of a unit cluster recording from the anterior interpositus nucleus over the course of training are shown. The eyeblink response (here nictitating membrane extension) is shown on the tracing above each histogram. (A) Results of a day of unpaired CS and US presentations. There is some activity to the US. However, when paired training (B) is given (Days 1 and 2), as behavioral learning develops (eyelid closure prior to US onset) there is a massive increase in neuronal discharges in the CS period that precedes and correlates with performance of the conditioned eyeblink response. Total trace duration, 750 ms, CS–US onset interval, 250 ms. Each trace and histogram is the average or cumulation of 1 day of training (120 trials). From McCormick and Thompson (1984).

occurrence and form of the CR within trials and over the trials of training (McCormick and Thompson, 1984) (see Fig. 18.28). By inference from PET analysis, the same process is thought to occur in humans (Logan and Grafton, 1995) (see Fig. 18.29). In animals, appropriate lesions of the anterior interpositus nucleus completely and permanently prevent learn-

ing. Similar lesions inflicted after training permanently abolish the CR but are without effect on the UR (McCormick and Thompson, 1984; Steinmetz *et al.*, 1992; Yeo *et al.*, 1985a,b). In the same way, in humans, cerebellar lesions can completely prevent learning of the CR, but again are without effect on the UR (Daum *et al.*, 1993). Interestingly, when the interpositus lesion

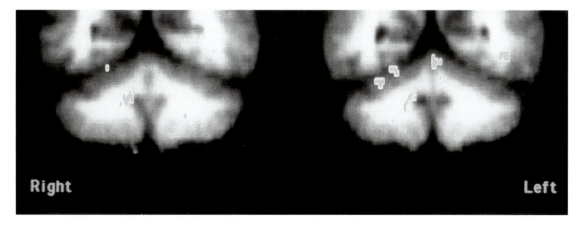

FIGURE 18.29   Functional localization of brain cerebellar activity (PET scan) in human eyeblink conditioning. Regions showing bright yellow are the regions where activation correlated significantly with degree of learning. The cerebellar anterior interpositus nucleus and several regions of cerebellar cortex show highly significant increases in activation with learning. From Logan and Grafton (1995).

(in rabbits) is incomplete, resulting in a marked impairment in the CR but not complete abolition, the attenuated CR does not recover with further training (Clark *et al.*, 1984; Welsh and Harvey, 1989).

Appropriate lesions of the pontine nuclei (the CS pathway) can selectively abolish the CR to one modality of CS, and stimulation of the pontine nuclei serves as a supernormal CS, yielding learning faster than peripheral CSs (Steinmetz *et al.*, 1986, 1987). Finally, lesions of the appropriate region of the inferior olive completely prevent learning if they are made before training. Lesions made at the same location after training result in extinction and abolition of the CR (McCormick, *et al.*, 1985; Voneida *et al.*, 1990; Yeo *et al.*, 1986). Electrical microstimulation of this same region elicits discrete movements, and the exact movements elicited can be trained to occur in response to any neutral stimulus (Mauk *et al.*, 1986). Note that the inferior olive–climbing fiber system is the only system in the brain other than reflex afferents where this type of training can take place (Thompson, 1989).

These results constitute a verification of the theories initially developed in the classic papers of Marr (1969) and Albus (1971) and elaborated on by Eccles (1977), Ito (1984), and Thach *et al.* (1992). These theories proposed that the cerebellum was a neuronal learning system in which there was a convergence of mossy fibers–parallel fibers that conveyed information about stimuli and movement contexts (CSs) and the climbing fibers that conveyed information about specific movement errors and aversive events (USs). This convergence of fibers might occur on Purkinje neurons in the cerebellar cortex to alter the synaptic efficacy of the parallel fiber synapses on Purkinje dendrites. A similar convergence of mossy and climbing fibers exists on neurons in the interpositus nucleus (see Figs. 18.27 and 18.30).

The Cerebellum: The Locus of the Long-Term Memory Trace

The results described so far demonstrate that the cerebellum is necessary for learning, retention, and expression of classical conditioning of the eyeblink and other discrete responses. The next and more critical issue concerns the locus of the memory traces. Considerable evidence has accrued pointing to the cerebellum as the location where long-term memory traces for this type of learning are formed and stored.

Reversible inactivation has proven to be a powerful tool in the localization of sites of memory storage in systems where the essential circuitry is known, such as eyeblink conditioning (Figs. 18.27 and 18.30). In brief, if inactivation of a structure abolishes the learned response, the structure is considered to be part of the circuitry necessary for expression of the learned response. If the structure is inactivated during training and the animal immediately shows complete learning when the inactivation is subsequently removed, then the structure is not involved in acquiring the learned response but lies on the efferent path from the memory trace. However, if the animal shows no evidence of learning following inactivation training, then either the memory trace is normally located in the structure or the structure is a necessary afferent to the trace. Reversible inactivation can be produced by local cooling using a cold probe or by infusion of a drug. A variety of drugs can produce reversible inactivation, including muscimol, a GABA agonist that inactivates only neuron somata and not axons, and tetrodotoxin (TTX), a sodium channel blocker that blocks both neuron somata and axons.

Several parts of the circuit shown in Fig. 18.27 have been reversibly inactivated during training in naïve animals (Clark and Lavond, 1993; Hardiman *et al.*, 1996; Krupa *et al.*, 1993; Lavond *et al.*, 1993; Thompson and Kim, 1996; Thompson and Krupa, 1994). In brief, inactivation of the motor nuclei (Fig. 18.27a), red nucleus (Fig. 18.27b), superior cerebellar peduncle (Figs. 18.27d and e), or a localized region of the anterior interpositus nucleus and overlying cerebellar cortex (Fig. 18.27c) prevents expression of the CR (only motor nucleus inactivation also prevents expression of the UR). When tested without inactivation after training with inactivation of motor nuclei, the red nucleus, or the superior cerebellar peduncle, the CR is found to be fully learned as soon as the inactivation has ceased (Figs. 18.27a,b, d, and e). However, localized cerebellar inactivation (Fig. 18.27c) completely prevents learning; animals must learn from scratch as if completely untrained. These results argue strongly for cerebellar localization of the memory trace. This hypothesis is supported by the observation that inhibition of protein synthesis in the cerebellar interpositus nucleus appears to prevent long-term retention of the conditioned eyeblink response (Bracha and Bloedel, 1996).

Experimentally, it has proven extremely difficult to determine the relative roles of the cerebellar cortex and interpositus nucleus in eyeblink conditioning using the lesion method (Lavond *et al.*, 1987; Yeo *et al.*, 1985a,b). These difficulties were overcome by making use of the mutant Purkinje cell degeneration (*pcd*) mouse strain (Chen *et al.*, 1996). In this mutant, Purkinje neurons (and all other neurons studied) are normal throughout pre- and perinatal development. Approximately 2–4 weeks after birth, the Purkinje neurons in the cerebellar cortex degenerate and disappear. For about 2 months after this

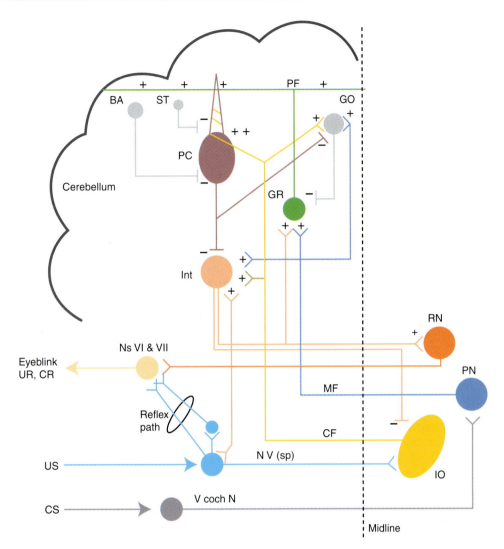

FIGURE 18.30   Commonly accepted eyeblink conditioning circuit based on experimental findings and anatomy of the cerebellum and the brainstem. The conditioned stimulus (CS) pathway consists of excitatory (+) mossy fiber (MF) projections primarily from the pontine nuclei (PN) to the interpositus nucleus (Int) and to the cerebellar cortex. In the cortex, the mossy fibers form synapses with granule cells (GR), which in turn send excitatory parallel fibers to the Purkinje cells (PC). Purkinje cells are the exclusive output neurons from the cortex and they send inhibitory (–) fibers to deep nuclei such as the interpositus. The unconditioned stimulus (US) pathway consists of excitatory climbing fiber (CF) projections from the inferior olive to the interpositus nucleus and to the Purkinje cells in the cerebellar cortex. Within the cerebellar cortex, Golgi (GO), stellate (ST), and basket (BA) cells exert inhibitory actions on their respective target neurons. The efferent conditioned response (CR) pathway projects from the interpositus nucleus to the red nucleus (RN) and via the descending rubral pathway to act ultimately on the motor neurons generating the eyeblink response. V Coch N, ventral cochlear nucleus; N V (sp), spinal fifth cranial nucleus; N VI, sixth and accessory sixth cranial nuclei; N VII, seventh cranial nucleus; UR, unconditioned response. Note that the reflex pathways do not involve the cerebellar circuitry. From Kim and Thompson (1997).

time, other neuronal structures appear relatively normal. Thus, during the period of young adulthood, the animals have a complete, selective functional decortication of the cerebellum (Goldowitz and Eisenman, 1992).

The *pcd* mice learned very slowly, poorly, and at a much lower level than wild-type controls but did show substantial and significant learning and showed extinction with subsequent training to the CS alone. Thus, the cerebellar cortex plays a critically important role in normal learning (of discrete behavioral responses), but some degree of learning is possible without the cerebellar cortex. Appropriate lesions of the interpositus nucleus in *pcd* mice completely prevent eyeblink conditioning (Chen *et al.*, 1999), demonstrating that the residual learning in nonlesioned *pcd* mice requires the interpositus.

Considerable evidence exists that the cerebellar cortex plays a key role in adaptive timing of the learned response (McCormick and Thompson, 1984; Perrett *et al.*, 1993). It also plays a role in the normal rate of learning. The Purkinje neurons in the cerebellar cortex were counted (in representative sections) in a group of rabbits aged 3–50 months. The correlation between trials to criterion and number of Purkinje neurons was −0.79 (Woodruff-Pak *et al.*, 2000). A consistent finding in normal humans is a marked and virtually linear increase in difficulty of learning with increasing age (Solomon *et al.*, 1989; Woodruff-Pak *et al.*, 1996; Woodruff-Pak and Thompson, 1988). It seems clear from the above that processes of neuronal plasticity known as "memory traces" develop in both the cerebellar cortex and the interpositus nucleus as a result of classical conditioning of the eyeblink and other discrete responses.

Putative Mechanisms of Memory Storage in the Cerebellum

Classic theories of the cerebellum as a learning machine (see earlier) proposed that conjoint activation of Purkinje neurons by parallel fibers and climbing fibers would lead to alterations in synaptic strength of the parallel fiber synapses. Ito (1982) discovered that such conjoint activation leads to a long-lasting depression in the efficacy of parallel fiber synapses to Purkinje neuron dendrites. This process is known as cerebellar LTD (see section on LTP/LTD). Ito and associates showed that such a process plays a key role in adaptation of the vestibulo-ocular reflex (VOR) (duLac *et al.*, 1995; Ito, 1984, 1989, 1993).

In eyeblink conditioning, many of the Purkinje neurons that exhibit learning-related changes show decreases in simple spike responses in the CS period (Thompson, 1990) that are consistent with LTD. As described in more detail in the LTP/LTD section above, the current view at the molecular level is that LTD is caused by a persistent decrease in AMPA receptor function at parallel fiber synapses on Purkinje neuron dendrites (Ito, 1993; Linden and Connor, 1995; Linden *et al.*, 1991). This decrease in AMPA receptor function is, in turn, the result of glutamate activation of AMPA and metabotropic receptors on Purkinje neuron dendrites, together with increased levels of intracellular calcium, which is normally caused by climbing fiber activation.

Classical conditioning studies using "gene knockout" mice have strengthened the argument that LTD is a key mechanism of memory storage in the cerebellar cortex (Kim *et al.*, 1996) (see section on LTP/LTD).

Thus, mice that lack metabotropic glutamate receptors (mGluR1) critical for LTD show marked impairments in cerebellar cortical LTD as expected, but also in eyeblink conditioning (Aiba *et al.*, 1994). They also show generalized motor impairments, that is, some degree of ataxia, as do the *pcd* mice mentioned previously.

Current studies present evidence supporting the view that LTD is more important for learning (e.g., eyeblink conditioning) than for motor coordination. Thus, the PKCγ knockout mutant mouse maintains to adulthood the perinatal condition of more than one climbing fiber per Purkinje neuron. In contrast, wild-type adults have only one climbing fiber per Purkinje neuron. The mutant mouse exhibits normal LTD but impaired motor coordination, which is due primarily to the multiple climbing fiber innervation of Purkinje neurons. In striking contrast, these animals learn the conditioned eyeblink response more rapidly than do wild-type controls (Chen *et al.*, 1995). This finding is consistent with the view that the climbing fiber system is the reinforcing or teaching pathway.

On the other hand, a quite different mutant, the glial fibrillary acidic protein (GFAP) knockout mouse (Shibuki *et al.*, 1996), shows marked deficiency in cerebellar cortical LTD and in eyeblink conditioning. The performance of such mutants is very similar to that of *pcd* mice. Unlike PKCγ mutants, these animals do not show any obvious impairments in motor coordination or general motor behavior. GFAP, which is expressed following neuronal injury, is not present in neurons, but only in glial cells. It is normally present in the cerebellum in substantial amounts in the Bergmann glia that surround the parallel fiber and the synapses between the climbing fibers and the Purkinje neuron dendrites. Although Bergmann glia appear morphologically normal in GFAP knockout mice, they have no GFAP. This observation illustrates the point that an abnormality limited to glial cells markedly impairs a form of synaptic plasticity (LTD) and a form of basic associative learning and memory, suggesting that the glia plays a key role in processes of learning and memory.

The memories formed in classical conditioning of discrete responses are long-term, relatively permanent memories. As with other forms of long-term memories (see section on LTP/LTD), these memories appear to require protein synthesis. Thus, infusion of anisomycin and protein kinase inhibitor H7 into the interpositus markedly impairs learning, and infusion of actinomycin D completely prevents learning (Bracha *et al.*, 1998; Chen and Steinmetz, 2000; Gomi *et al.*, 1999). However, when learning occurs in normal circumstances, infusion of these substances in the interpositus does not impair performance of the learned response. Thus, it is apparent that protein synthesis is necessary for formation of the memory trace but not for expression of the memory once it is formed. One study indicates that eyeblink conditioning (young adult rabbit) induces expression of a cell division cycle-type protein kinase in the interpositus nucleus (Gomi *et al.*, 1999). Since neurons do not divide postnatally, the finding suggests an interesting alternative role for this class of enzymes.

This section has focused on the essential role of the cerebellum in the classical conditioning of discrete behavioral responses, which is a basic form of associative learning and memory. To date, this is perhaps the clearest and most decisive example of evidence for the localization of a memory trace to a particular brain region (the cerebellum) in mammals. The cerebellum has also been pinpointed as the location where complex, multijoint movements are learned and stored (Thach *et al.*, 1992).

In addition, a growing body of evidence suggests that the cerebellum is critically involved in many other forms of learning and memory, including cardiovascular conditioning (Supple and Leaton, 1990), discrete response instrumental avoidance learning (Steinmetz *et al.*, 1993), maze learning (Pelligrino and Altman, 1979), spatial learning and memory (Goodlett *et al.*, 1992; Lalonde and Botez, 1990), and adaptive timing (Keele and Ivry, 1990). There is even a growing literature implicating the cerebellum in complex cognitive processes (Schmahmann, 1997).

The type of learning exemplified by the cerebellar circuitry underlying classical conditioning of discrete responses has been termed *supervised learning* (Knudsen, 1994). Information from one network of neurons acts as an instructive signal (US) to influence the pattern of connectivity in another network (e.g., the CS); other examples include adaptation of the VOR and calibration of the auditory space map in the barn owl. In eyeblink conditioning, the neutral CSs (e.g., tone or light) only weakly influence the activity of neurons in the cerebellum and do not yield the behavioral response. The strong connections established between networks of neurons as a result of training are not functionally coupled prior to learning. That is, diffuse cerebellar mossy and/or parallel fibers activated by the CS develop sufficient strength of their synaptic connections to successfully signal the specific circuit initially formed by the very localized climbing fiber projections to the cerebellum activated by the corneal airpuff US. In summary, weak and ineffective anatomical connections become powerful and effective through learning. However, note that the connections do indeed exist before training. This may

be considered a general principle in all aspects of learning and memory.

Fear Conditioning Is a Model System for Investigating the Neural Substrates of Emotional Memory

Significant progress has been made in identifying the neural substrates of emotional memory processing. Many of the advances have involved studies of classical fear conditioning. This model system has been used extensively in animal studies and, more recently, in human experiments as well.

What is Fear Conditioning?

Classical fear conditioning, also known as aversive classical conditioning or Pavlovian defensive conditioning, consists of repeated temporal pairings of an emotionally neutral stimulus (CS) with an aversive unconditioned stimulus (US) (Fig. 18.31). The US elicits a multitude of physiological and behavioral unconditioned responses, and over one or more conditioning trials, conditioned responses develop in reaction to the CS itself. In a typical experiment, the subject is presented with a tone followed by a brief electric shock. After several tone–shock pairings, the tone becomes aversive to the subject and begins to elicit a set of responses (the CRs) characteristic of a state of fear: freezing; autonomic responses (such as changes in heart rate, blood pressure, skin conductance, or pupillary dilation); endocrine responses (such as the release of so-called stress hormones, especially ACTH, adrenal steroids, and adrenal catecholamines); and potentiation of somatic reflexes (like eyeblink and startle) (see Choi *et al.*, 2001; Davis, 1998; Fendt and Fanselow, 1999; Kapp *et al.*, 1992; LeDoux, 2000). The CRs thus form a set of observable indices that can be used to gauge emotional learning and memory as the organism acquires the association between the CS and the US. This emotional stimulus learning becomes extinguished if the subject receives subsequent CS-alone presentations, as shown by a decrease in the expression of the CRs over trials (Fig. 18.31).

As discussed in the previous section, memory researchers often distinguish between explicit or declarative (conscious) and implicit or nondeclarative (unconscious) memory. Fear conditioning falls into the latter category; it occurs normally in humans with a loss of explicit memory due to brain damage (Bechara *et al.*, 1995). It has been found that brain damage disrupting fear conditioning fails to interfere with explicit memory of the conditioning situation (Bechara *et al.*, 1995; LaBar *et al.*, 1995). Thus, in an

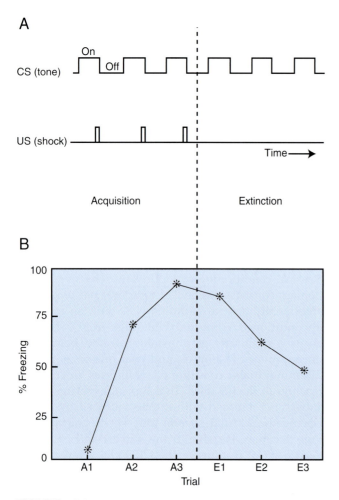

A

CS (tone)   On
             Off

US (shock)

                                    Time →

         Acquisition   Extinction

B

FIGURE 18.31   Fear conditioning paradigm: (A) Typical parametric arrangement of stimuli during the acquisition and extinction phases of a simple delay conditioning task. CS, conditioned stimulus; US, unconditioned stimulus. (B) Hypothetical (but realistic) acquisition and extinction learning curves for defense (freezing) responses conditioned to the CS. Note the rapid increase of freezing during the CS during acquisition trials (A1–A3) and the decline of freezing during extinction trials (E1–E3).

emotionally arousing situation in normal life, one brain system will form implicit (unconscious) emotional memories (expressed as conditioned fear responses) and another brain system will form explicit (conscious) memories about the experience. These have been called emotional memories and memories about emotions, respectively (LeDoux, 1996). The section below focuses on the neural circuits involved in the implicit formation of emotional memories. The circuits involved in the formation of conscious memories about emotions are the same circuits involved in any other kind of explicit memory. It should also be noted that the two kinds of memories can interact. In other words, conscious explicit memories about an emotional situation can be modulated

(enhanced or diminished) by the concurrent activation of an implicitly functioning emotional system (Cahill and McGaugh, 1998).

Specific Neural Circuits within the Amygdala Are Involved in Fear Conditioning

Across various paradigms, species, and response measures, the amygdala has consistently emerged as a brain structure essential to the acquisition and expression of conditioned fear (see Davis, 1998; Fendt and Fanselow, 1999; Kapp et al., 1992; LeDoux, 2000; Lee et al., 2001). Pretraining lesions of the amygdala prevent the development of CRs, whereas posttraining lesions of the amygdala disrupt the expression of CRs that have already been learned, even after extensive overtraining (Maren, 1998).

In combination with lesion data, neuroanatomical tract-tracing studies have begun to elucidate the afferent and efferent connections of the amygdala to sensory and motor areas involved in transmitting information about the CS and US and generating emotional responses (for reviews see LeDoux, 2000; McDonald, 1998; Pitkanen et al., 1997). Most of the work on CS pathways has involved auditory stimuli, although visual CS pathways have also been described. Information regarding an auditory CS reaches the amygdala by way of two neural routes: a direct thalamo-amygdala pathway from the auditory thalamus (medial portion of the medial geniculate nucleus and posterior intralaminar nucleus) to the lateral nucleus of the amygdala, and an indirect thalamo-cortico-amygdala pathway linking the auditory thalamus with the lateral nucleus of the amygdala by way of connections within the auditory cortex (Fig. 18.32) (LeDoux et al., 1991). The direct thalamo-amygdala pathway is more rapid but provides a relatively crude representation of the incoming sensory stimulus compared with the thalamo-cortico-amygdala pathway (LeDoux, 1986). The direct pathway is thought to function in two ways: first as a quick route for simple stimulus features to evoke defensive emotional responses; and second as a method of priming the amygdala to set up appropriate emotional responses to incoming stimuli that are more highly processed by the indirect pathway. Lesion studies have shown that either of these routes in isolation is sufficient to mediate responding in simple conditioning protocols involving one CS (Romanski and LeDoux, 1992). However, in situations involving more complex stimulus processing it is likely that the auditory cortex projection to the amygdala will be involved (Jarrell et al., 1987), but the exact conditions requiring the auditory cortex are poorly understood (Armony et al., 1997).

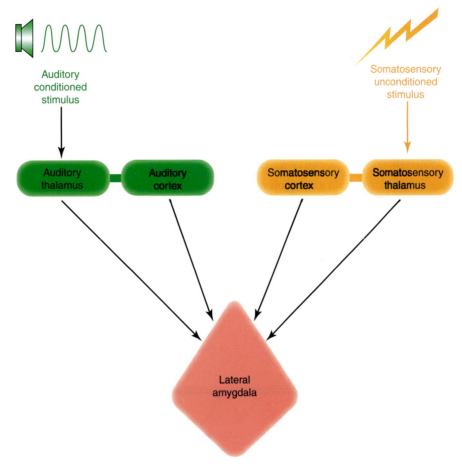

FIGURE 18.32 CS and US transmission pathways in fear conditioning. The auditory conditioned stimulus and somatosensory unconditioned stimulus are transmitted through brainstem to the thalamic stations in each pathway (the line is broken to reflect that the pathways include stations in the brainstem that are not shown). The thalamic regions give rise to connections to cortical receiving areas. Components of both the thalamic and cortical processing regions in each pathway then connect with the lateral nucleus of the amygdala. Brain lesion studies suggest that the CS and US can reach the amygdala through either thalamic or cortical areas.

The two incoming CS pathways converge in the lateral nucleus of the amygdala (LeDoux *et al.*, 1991; Li *et al.*, 1996), which functions as the sensory gateway to the amygdala (Amaral *et al.*, 1992; Pitkanen *et al.*, 1997). Indeed, damage to the lateral amygdala (Campeau and Davis, 1995; LeDoux *et al.*, 1990) or inactivation of this region and adjacent areas (Muller *et al.*, 1997; Wilensky *et al.*, 1999) prior to conditioning prevents fear learning from taking place. In addition, cells in the lateral nucleus that receive incoming acoustic CS information also respond to somatosensory signals elicited by footshock. Although the exact sources of US inputs to the amygdala are still not fully understood, it appears that stimulation of the posterior intralaminar nucleus can serve as an effective US in fear conditioning (Cruikshank *et al.*, 1992). However, some evidence exists that the US pathway, like the CS pathway,

involves both direct thalamo-amygdala and indirect thalamo-cortico-amygdala circuits that converge in the lateral amygdala (Shi and Davis, 1999). Therefore, it appears that the lateral nucleus of the amygdala is a likely region of integration, both within CS and US systems and between them. At the same time, plasticity in the acoustic thalamus and auditory cortex may also play some role in auditory fear conditioning (Weinberger, 1995).

Incoming sensory information reaching the lateral nucleus is then projected by intra-amygdala circuitry to several targets within the amygdala, including the basal, accessory basal, medial, and central nuclei (Pare *et al.*, 1995; Pitkanen *et al.*, 1997). However, for simple fear conditioning, only the connection to the central nucleus is required, as damage to other amygdala regions has little effect (Nader *et al.*, 2001). The central nucleus, in turn, mediates emotional responses through efferent connections with motor and auto-nomic centers (see Choi *et al.*, 2001; Davis, 1998; Fendt and Fanselow, 1999; Kapp *et al.*, 1992; LeDoux, 2000) (Fig. 18.33). Interestingly, lesions of the target struc-tures or transection of fibers projecting to these target areas selectively disrupt CRs expressed in specific response modalities, leaving other CR measures intact. For example, lesions of the central gray region disrupt freezing to the CS, but increases in blood pres-sure elicited by the same stimulus remain evident. These results, along with others, suggest that brain-stem projections of the central amygdala are involved in the expression of conditioned responses in differ-ent modalities. Therefore, along the neural routes involved in fear conditioning, the amygdala appears to function as the key station where emotional significance is assessed. This assessment takes place independently of the manner in which the stimulus enters the brain and independently of the way the responses are expressed.

Contributions of Other Brain Systems

Whereas the roles of the amygdala and its projec-tions are well established as being critical for the regulation of conditioned fear associations, the contri-butions of other brain regions are only now receiving increased attention. Hippocampal lesions have little effect on emotional responses conditioned to an explicit CS in simple conditioning, but interfere with fear conditioning to contextual or background stimuli (Kim & Fanselow, 1992; Maren *et al.*, 1997; Philips and LeDoux, 1992, 1995), and higher-order conditioning processes, such as blocking (Rickert *et al.*, 1978) and trace conditioning (Clark and Squire, 1998; Huerta *et al.*, 2000; McEchron *et al.*, 1998). In these situations, the hippocampus may be exerting its influence through

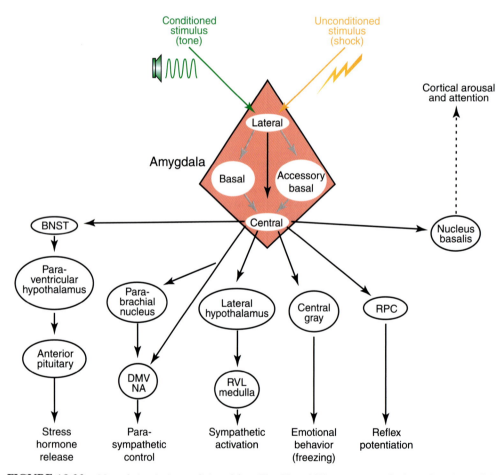

FIGURE 18.33   Neural circuits in conditioned fear. The CS and US converge in the lateral nucleus of the amygdala (for CS and US pathways, see Fig. 18.32). Through intra-amygdala circuitry, the output of the lateral nucleus is transmitted to the central nucleus, which serves to activate various effector systems involved in the expression of emotional responses. BNST, bed nucleus of the stria terminalis; DMV, dorsal motor nucleus of the vagus; NA, nucleus ambiguous; RPC, nucleus reticularis pontis caudalis; RVL medulla, rostral ventrolateral nuclei of the medulla.

its anatomical interactions with the amygdala (LeDoux, 2000; Maren and Fanselow, 1996). These findings are consistent with current theories regarding hippocampal involvement in complex stimulus processing (Eichenbaum, 1999; O'Keefe, 1999; Rudy and O'Reilly, 1999). Although the role of the hippocampus in contextual processing is not universally accepted (Gisquet-Verrier et al., 1999; McNish et al., 1997), it seems that when animals are conditioned and tested in ways that minimize the availability of non-contextual processing strategies, the hippocampal role in context is clearer (Frankland et al., 1998; Maren et al., 1997).

Some evidence suggests that lesions of the dorsal medial prefrontal cortex (anterior cingulate) result in a potentiation of fear responses (Morgan et al., 1993; Vouimba et al., 2000), whereas ventral medial prefrontal cortex can selectively interfere with the extinction of conditioned fear responses (Morgan et al., 1993). Although some studies have failed to find effects of such lesions on extinction (Gerwitz et al., 1997), other studies have confirmed that damage to the prefrontal region alters extinction (Quirk et al., 2000). Physiological studies also show that the amygdala and medial prefrontal cortex are interrelated (Garcia et al., 1999). These observations, when extended to humans, are potentially important for understanding pathological fear and anxiety, conditions in which fear regulation is disrupted.

Cellular Mechanisms of Fear Conditioning

The fact that CS and US pathways converge in the lateral amygdala suggests that this sensory interface of the amygdala may be the site of plasticity in the circuit. Indeed, single-unit recording studies have shown that some cells in the lateral amygdala have short-latency responses to acoustic stimuli (Bordi and LeDoux, 1992) and that the firing rate of these cells increases dramatically during training (Collins and Pare, 2000; Maren, 2000; Quirk et al., 1995, 1997; Repa et al., 2001) (Fig. 18.33). Recent studies have shown that such changes predict the acquisition of conditioned fear behavior and that the cellular responses in at least some of the cells persist during extinction until the behavioral response has been eliminated by presentations of the CS without the US (Repa et al., 2001). Although conditioned neural responses have also been observed in the basal (Maren et al., 1991) and central (Pascoe and Kapp, 1985) nuclei, the latency of these is longer than that in the lateral nucleus, suggesting the changes in the lateral amygdala may play a primary or driving role in amygdala plasticity.

Some have suggested that the amygdala is not involved in plasticity during fear learning, but only in the modulation of memory in other areas (Cahill and McGaugh, 1998; Cahill et al., 1999). Typically, the role of the amygdala in modulation is shown by infusing a drug into this brain region immediately after training. With this design, learning occurs in the absence of the drug, and the drug affects only posttraining storage processes. For example, infusion of the GABA agonist muscimol into the amygdala after training on several tasks that depend on other brain regions for storage impairs memory for the task. However, inactivation of the amygdala immediately after fear conditioning has no effect on memory for this task (Wilensky et al., 1999, 2000). The amygdala may indeed modulate memories formed in other areas, but does not seem to modulate the plasticity of its own neurons (Fanselow and LeDoux, 1999).

That plasticity occurs in the amygdala is indisputable. For example, a number of studies have shown in vivo (Clugnet and LeDoux, 1990; Rogan and LeDoux, 1995) and in vitro (Chapman et al., 1990; Huang and Kandel, 1998; Weisskopf et al., 1999) that LTP occurs in the amygdala, including pathways that transmit auditory CS information to the lateral amygdala. Furthermore, induction of LTP in the CS pathways alters the processing of auditory stimuli (Rogan and LeDoux, 1995), which is similar to the way that fear conditioning alters the processing of auditory stimuli in the lateral amygdala (Rogan et al., 1997). In addition, fear conditioning in vivo alters the responses of amygdala cells to afferent stimulation in vitro (McKernan and Shinnick-Gallagher, 1997). These various findings have strengthened the hypothesis that meaningful plasticity occurs in the lateral amygdala during fear conditioning, and also highlight the potential of fear conditioning as an approach to understanding the relationship of LTP to memory formation.

Molecular Basis of Fear Conditioning

Following the discovery that long-term potentiation in the hippocampus depends on NMDA receptors (see section on LTP/LTD), researchers pursued the possibility that fear conditioning involves an NMDA-dependent form of synaptic plasticity in the amygdala (Fig. 18.34). Several studies have shown that conditioning is disrupted by a blockade of NMDA receptors in the lateral (and basal) amygdala prior to fear conditioning (Gerwitz and Davis, 1997; Lee et al., 1998; Maren et al., 1996; Miserendino et al., 1990; Walker and Davis, 2000). Other recent studies have specifically implicated the NR2B subunit of the NMDA receptor in fear conditioning (Rodrigues et al., 2001; Tang et al., 1999). In these later studies, manipulation of NR2B impaired both short-term memory (STM) and long-term memory (LTM) of fear

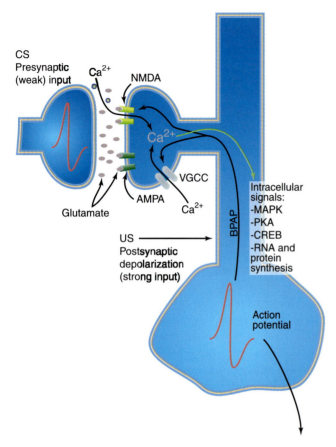

FIGURE 18.34 Model of synaptic plasticity during fear conditioning. The CS provides weak glutamate-mediated activation of presynatpic inputs to lateral amygdala cells. The glutamate activates AMPA receptors but cannot activate NMDA receptors in the absence of strong, depolarizing inputs to the cell. If the weak presynaptic input from the CS arrives at about the same time that the postsynaptic cell is also depolarized by a strong input, such as by the US, postsynaptic depolarization occurs, allowing calcium to enter through NMDA receptors and voltage-gated calcium channels (VGCCs). This combined calcium signal then activates a variety of signaling pathways, including MAPK, PKA, and CREB, and initiates RNA and protein synthesis. The proteins then act at the CS input synapses to strengthen and stabilize the connection.

conditioning, suggesting that NMDA receptors play an essential role in the initial synaptic plasticity underlying fear learning (Rodrigues *et al.*, 2001; Tang *et al.*, 1999; Walker and Davis, 2000).

Other studies have found that LTP in the amygdala is NMDA-receptor dependent, which is consistent with the behavioral findings (Bauer *et al.*, 2002; Gean *et al.*, 1993; Huang and Kandel, 1998). However, under some conditions, associative and synapse-specific LTP in the amygdala requires calcium influx in the postsynaptic cell through L-type voltage-gated calcium channels (VGCCs; termed VDCCs in LTP/LTD section above) rather than NMDA receptors (Bauer *et al.*, 2002; Weisskopf *et al.*, 1999). Thus, as in the hip-

pocampus (Magee and Johnston, 1997), both NMDA receptors and L-type VGCCs are involved in aspects of synaptic plasticity in the amygdala (see section on LTP/LTD), suggesting that fear conditioning may require some combination of these mechanisms at the cellular level (Blair *et al.*, 2001). This hypothesis is supported by new studies showing that blockade of L-type VGCCs in the lateral amygdala disrupts fear conditioning.

Some progress has also been made in pursuing the molecular basis of fear conditioning in the amygdala. Fear memory consolidation (LTM, but not STM), has been shown to require the synthesis of RNA and new proteins in the lateral (and basal) amygdala, as well as the activity of several kinases, including MAPK and PKA (Bailey *et al.*, 1999; Schafe *et al.*, 1999, 2000; Schafe and LeDoux, 2000). These kinases have also been implicated in LTP in the lateral amygdala (Huang *et al.*, 2000; Huang and Kandel, 1998; Schafe *et al.*, 2000). Both MAPK and PKA phosphorylate the transcription factor CREB. Mice deficient in CREB have impaired fear conditioning (Bourtchouladze *et al.*, 1994), and upregulation of CREB in the amygdala enhances fear conditioning (Josselyn *et al.*, 2001). Although much remains to be done in this area, the basic findings are consistent with the general roles of CREB, PKA, and MAPK discovered in studies as diverse as synaptic facilitation in *Aplysia* (see section on Invertebrate Studies), conditioning in *Drosophila* (Yin *et al.*, 1994, 1995), hippocampal LTP (English and Sweatt, 1997; Frey *et al.*, 1993; Huang *et al.*, 1994; Impey *et al.*, 1996), and hippocampus-dependent spatial learning (Abel *et al.*, 1997; Blum *et al.*, 1999; Bourtchouladze *et al.*, 1994; Guzowski and McGaugh, 1997). The conservation of these molecular contributions across species and training conditions suggests that the uniqueness of memory is not due to the underlying molecules so much as to the circuit in which these molecules act. It is important to note that the recent discovery that protein synthesis in the lateral amygdala is also required for the stabilization of memories after retrieval (so-called reconsolidation) is stimulating new research and raising questions about classic notions of consolidation (Nader *et al.*, 2000).

Human Fear Conditioning and Anxiety

Fear conditioning has been readily demonstrated in human subjects, and many experimental preparations for measuring fear in animals can be adapted for use with human populations. The brain structures that mediate aspects of fear learning in animals appear to perform similar functions in humans. For instance, patients with lesions of the amygdala fail to acquire conditioned fear reactions (Bechara *et al.*, 1995; LaBar *et al.*, 1995; Phelps *et al.*, 1998), and fear conditioning activates the amygdala during the early phases of acquisition and extinction training in normal subjects (Buchel *et al.*, 1998; LaBar *et al.*, 1998) (Fig. 18.35). In contrast, patients with hippocampal lesions show intact learning during simple conditioning tasks, but fail to associate their learning with

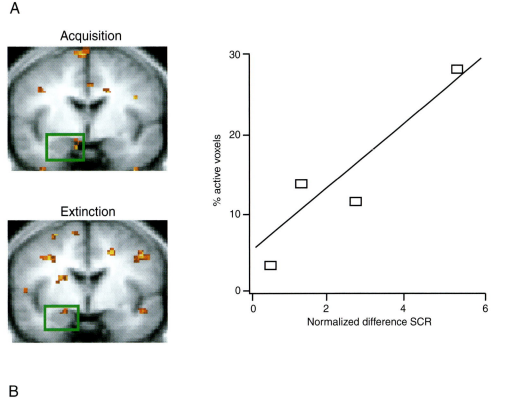

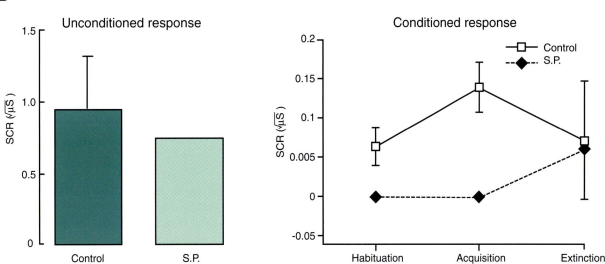

FIGURE 18.35   Role of the human amygdala in conditioned fear. (A) Conditioning-related hemodynamic changes in the amygdala of normal human subjects during functional magnetic resonance imaging. The amygdala (green box) was activated during the early phases of both acquisition and extinction training, and the extent of amygdala activity during acquisition correlated with skin conductance responses (SCRs) in individual subjects. Adapted from LaBar *et al.* (1998). (B) Impaired acquisition of fear conditioning in patient S.P., who has bilateral amygdala lesions. Despite showing intact declarative memory of the CS–US contingency and normal SCRs to the US, S.P. does not acquire conditioned responses to the CS. μS, microsiemens. Adapted from Phelps *et al.* (1998).

contextual cues in the environment (reviewed in LaBar and Disterhoft, 1998; O'Connor *et al.*, 1999).

Selective amygdala damage in humans also leads to deficits in the recognition of facial expressions of fear (Adolphs *et al.*, 1994), and fearful facial expressions activate the amygdala in normal subjects, even when presented subliminally (Breiter *et al.*, 1996; Morris *et al.*, 1996; Whalen *et al.*, 1998). Stimulation of the amygdala in epileptic patients undergoing surgical treatment typically evokes feelings of fear and anxiety (Gloor, 1992). Thus, in humans as well as other species, the amygdala appears to be centrally involved in regulating mechanisms of fear.

There is a marked similarity in the clinical symptoms of anxiety in humans and measures of conditioned fear in animals, and drugs with anxiolytic effects in animal models of fear have been applied to treat human anxiety (Davis, 1992). Some researchers have thus proposed using fear conditioning as a model for studying human affective disorders, such as posttraumatic stress disorder, phobia, and panic (Charney *et al.*, 1993; Wolpe and Rowan, 1988). For example, one clinical marker for posttraumatic stress disorder is an increase in startle. These patients exhibit increased fear-potentiated startle and conditioned autonomic responses relative to control subjects (Orr *et al.*, 2000), and have exaggerated amygdala activation to fearful faces (Rausch *et al.*, 2000).

Finally, fear conditioning using phobic stimuli (e.g., snakes and spiders) is more resistant to extinction than fear conditioning using nonphobic stimuli (Marks and Tobena, 1990). Behavioral therapies based on classical conditioning procedures have been relatively successful in the treatment of patients with phobic disorders. These therapies have evolved to incorporate contemporary theories of conditioning, instrumental learning, and cognition.

HOW DOES A CHANGE IN SYNAPTIC STRENGTH STORE A COMPLEX MEMORY?

The relationship between the synaptic changes and the behavior of conditioned reflexes can be straightforward, because the locus for the plastic change is part of the mediating circuit. Thus, the change in the strength of the sensorimotor synapse in *Aplysia* can be related to the memory for sensitization (e.g., Fig. 18.18A). However, the idea that an increase in synaptic strength leads to an enhanced behavioral response, and a decrease in synaptic strength leads to a decreased behavioral response, can be misleading. For example, a decrease in synaptic strength in a postsynaptic neuron that exerts an inhibitory action can be translated into an enhanced behavioral response. Indeed, in the parallel fiber-to-Purkinje cell connection in the cerebellum, such a disinhibition is precisely the mechanism that has been proposed to mediate classical conditioning of the eyeblink reflex. Nevertheless, for relatively simple reflex systems in which the circuit is well understood, it is possible to directly relate a change in synaptic strength to learning. However, in most other examples of memory, it is considerably less clear how the synaptic changes are induced and, once induced, how the information is retrieved. This is especially true in memory systems that involve the storage of information for patterns, facts, and events. Neurobiologists have turned to

---

BOX 18.2

MEMORY BEYOND THE SYNAPSE

The search for the biological basis of learning and memory has led many of the 20th century's leading neuroscientists to direct their efforts to investigating the synapse. The focus of most recent work on LTP (and indeed the bulk of this chapter) has been to elucidate the mechanisms underlying changes in synaptic strength. Although changes in synaptic strength are certainly ubiquitous, they are not the exclusive means for the expression of neuronal plasticity associated with learning and memory. Both short-term and long-term sensitization and classical conditioning in *Aplysia* are associated with an enhancement of excitability of the sensory neurons in addition to changes in synaptic strength. Changes in excitability of sensory neurons in the mollusk *Hermissenda* are produced by classical conditioning. In vertebrates, classical conditioning of eyeblink reflexes produces changes in excitability of cortical neurons. Eyeblink conditioning also produces changes in the spike afterpotential of hippocampal pyramidal neurons. Finally, as described in their original report on LTP, Bliss and Lomo (1973) found that the expression of LTP was also associated with enhanced excitability.

artificial neural circuits to gain insights into these issues.

A simple network that can store and "recognize" patterns is illustrated in Fig. 18.36. The network is artificial, but nevertheless is inspired by actual circuitry in the CA3 region of the hippocampus. In this example, six different input projections make synaptic connections with the dendrites of each of six postsynaptic neurons (Fig. 18.36A). The postsynaptic neurons serve as the output of the network. Input projections can carry multiple types of patterned information, and these patterns can be complex. To

simplify the present discussion, consider that the particular input pathway in Fig. 18.36A carries information regarding the pattern of neural activity induced by a single brief flash of a spatial pattern of light. For example, activity in the top pathway (line *a*) might represent light falling on the temporal region of the retina, whereas activity in the pathway on the bottom (line *f*) might represent light falling on the nasal region of the retina. Thus, the spatial pattern of an image falling on the retina could be reconstructed from the pattern of neuronal activity over the *n* (in this case 6) input projections to the network.

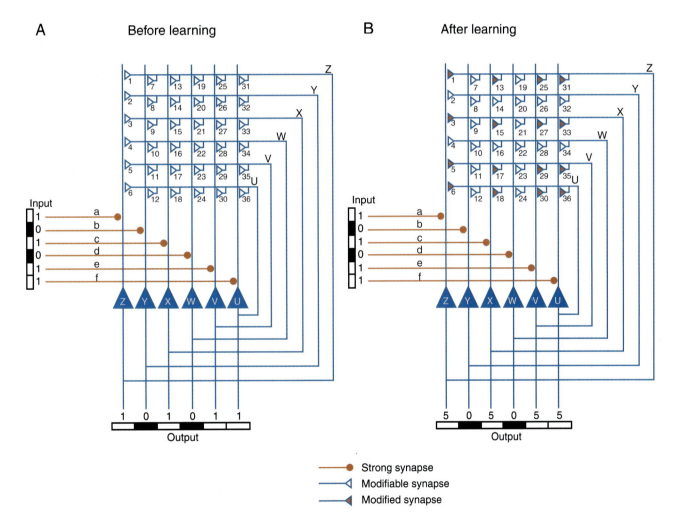

FIGURE 18.36 Autoassociation network for recognition memory. The artificial circuit consists of six input pathways that make strong connections to each of six output neurons. The output neurons have axon collaterals that make synaptic connections with each of the output cells. (A) A pattern represented by activity in the input lines or axons (*a, b, c, d, e, f*) is presented to the network. A *1* represents an active axon (e.g., a spike) whereas a *0* represents an inactive axon. The input pathways make strong synapses (filled circles) with the postsynaptic output cells. Thus, the output cells (*u, v, w, x, y, z*) generate a pattern that is a replica of the input pattern. The collateral synapses were initially weak and do not contribute to the output. Nevertheless, the activity in the collaterals that occurred in conjunction (assume minimal delays within the circuit) with the input pattern led to a strengthening of a subset of the 36 synapses. (B) A second presentation of the input produces an output pattern that is an amplified, but an otherwise intact, replica of the input. An incomplete input pattern can be used as a cue to retrieve the complete pattern.

Three aspects of the circuit endow it with the ability to store and retrieve patterns. First, each of the input lines makes a sufficiently strong connection with its corresponding postsynaptic cell to activate it reliably. Second, each output cell (z to u) sends an axon collateral that makes an excitatory connection with itself as well as the other five output cells. This pattern of synaptic connectivity leads to a network of 36 synapses (a total of 42, including the 6 input synapses). Third, each of the 36 synaptic connections are modifiable through an LTP-like mechanism (see above). Specifically, the strength of a particular synaptic connection is initially weak, but it will increase if the presynaptic *and* postsynaptic neurons are active at the same time. The circuit configuration with the embedded synaptic "learning rule" leads to an autoassociation or autocorrelation matrix. The autoassociation is derived from the fact that the output is fed back to the input where it associates with itself.

Now consider the consequences of presenting the patterned input to the network of Fig. 18.36A. The input pattern will activate the six postsynaptic cells in such a way as to produce an output pattern that will be a replica of the input pattern. In addition, the pattern will induce changes in the synaptic strength of the active synapses in the network. For example, synapse 3 will be strengthened because the postsynaptic cell, cell z, and the presynaptic cell, cell x, will be active at the same time. Note also that synapses 1, 5, and 6 will be strengthened as well. This occurs because these input pathways to cell z are also active. Thus, all synapses that are active at the same time as cell z will be strengthened. When the pattern is presented again as in Fig. 18.36B, the output of the cell will be governed not only by the input, but also by the feedback connections, a subset of which were strengthened (Fig. 18.36B, filled synapses) by the initial presentation of the stimulus. Thus, for output cell z a component of its activity will be derived from input a, but components will also come from synapses 1, 3, 5, and 6. If each of the initially strong and newly modified synapses is assumed to contribute equally to the firing of output cell z, the activity would be five times greater than the activity produced by input a before the learning. After learning, the output is an amplified version of the input but the basic features of the pattern are preserved.

Note that the "memory" for the pattern does not reside in any one synapse or in any one cell. Rather, it is *distributed* throughout the network at multiple sites. The properties of these types of autoassociation networks have been examined by James Anderson, Teuvo Kohonen, David Marr, Edmond Rolls, David Wilshaw, and their colleagues (see Kohonen, 1989)

and found to exhibit a number of phenomena that would be desirable for a biological recognition memory system. For example, such networks exhibit pattern completion. If a partial input pattern is presented, the autoassociation network can complete the pattern in the sense that it can produce an output that is approximately what is expected for the full input pattern. Thus, any part of the stored pattern can be used as a cue to retrieve the complete pattern. For the example of Fig. 18.36, the input pattern was {101011}. This pattern led to an output pattern of {505055}. If the input pattern was degraded to {101000}, the output pattern would be {303022}. Some change in the strength of firing occurs but the basic pattern is preserved. Autoassociation networks also exhibit a phenomenon known as graceful degradation, which means that the network can still function even if some of the input connections or postsynaptic cells are lost. This property arises from the distributed representation of the memory within the circuit.

SUMMARY

The search for the biological basis of learning and memory has occupied the research of many of the 20th century's leading neuroscientists. This work has ranged from studies on the behavior of whole organisms down to the molecular changes occurring at individual synapses. Within the past two decades, great strides have been made in increasing our understanding of learning and memory. For example, the neural and molecular mechanisms of short- and long-term sensitization in *Aplysia* have been elucidated. The search for the engram underlying classical eyeblink conditioning has been continually narrowed and refined. The anatomical and physiological properties subserving Pavlovian fear conditioning have also been elucidated, at least in general terms. Finally, although much work remains to be done, neuroscientists have garnered a greater understanding of the subcellular changes underlying synaptic plasticity, particularly within the hippocampal formation and cerebellum.

In the near future, a major experimental question to be answered is the extent to which the mechanisms for learning are common both within any one animal and between different species. Although many common features are emerging, there seem to be some differences. Thus, it will be important to understand the extent to which specific mechanisms are used selectively for one type of learning and not another. Irrespective of the particular example of learning and memory that is analyzed, whether it be simple or complex, it will be important to pay attention to three major details: the details of the circuit interactions, the details of the learning rule, and the details of the intrinsic biophysical properties of the neurons within the circuit.

References

Suggested Readings

Aggleton, J. P. (2000). *The Amygdala: A Functional Analysis*, 2nd ed. Oxford Univ. Press, New York.

Bear, M. F., and Malenka, R. C. (1994). Synaptic plasticity: LTP and LTD. *Curr. Opin. Neurobiol.* **4**, 389–399.

Bliss, T. V. P., and Collingridge, G. L. (1993) A synaptic model of memory: Long-term potentiation in the hippocampus. *Nature (London)* **361**, 31–39.

Byrne, J. H. (1987). Cellular analysis of associative learning. *Physiol. Rev.* **67**, 329–439.

Carew, T. J. (2000). *"Behavioral Neurobiology"*. Sinauer Associates, Sunderland, MA.

Dineley, K. T., Weeber, E. J., Atkins, C, Adams, J. P., Anderson, A. E., and Sweatt, J. D. (2001) Leitmotifs in the biochemistry of LTP induction: Amplification, integration and coordination. *J. Neurochem.* **77**, 961–971.

Fendt, M., and Fanselow, M. S. (1999). The neuroanatomical and neurochemical basis of conditioned fear. *Neurosci. Biobehav. Rev.* **23**, 743–760.

Ito, M. (1989) Long-term depression. *Annu. Rev. Neurosci.* **12**, 85–102.

Kandel, E. R. (2001). The molecular biology of memory storage: A dialogue between genes and synapses. *Science* **294**, 1030–1038

Malenka, R. C., and Nicoll, R. A. (1999). Long-term potentiation—a decade of progress? *Science* **285**, 1870–18744.

Martin, S. J., Grimwood, P. D., and Morris, R. G. (2000). Synaptic plasticity and memory: An evaluation of the hypothesis. *Annu. Rev. Neurosci.* **23**, 649–711.

Squire, L. R., and Kandel, E. R. (1999) *"Memory: From Mind to Molecules"*. Freeman, New York.

Cited

Abel, T., Nguyen, P. V., Barad, M., Deuel, T. A., Kandel, E. R., and Bourtchouladze, R. (1997). Genetic demonstration of a role for PKA in the late phase of LTP and in hippocampus-based long-term memory. *Cell* **88**, 615–626.

Adolphs, R., Tranel, D., Damasio, H., and Damasio, A. R. (1994). Impaired recognition of emotion in facial expressions following bilateral damage to the human amygdala. *Nature*, **372**, 669–672.

Aiba, A., Kano, M., Chen, C., Stanton, M. E., Fox, G. D., Herrup, K., Zwingman, T. A., and Tonegawa, S. (1994). Deficient cerebellar long-term depression and impaired motor learning in mGluR1 mutant mice. *Cell* **79**, 377–388.

Albus, J. S. (1971). A theory of cerebellar functions. *Math. Biosci.* **10**, 25–61.

Amaral, D. G., Price, J. L., Pitkanen, A., and Carmichael, S. T. (1992). Anatomical organization of the primate amygdaloid complex. *In The Amygdala: Neurobiological aspects of emotion, memory, and mental dysfunction* J. Aggleton, (Ed.), pp. 1–66. Wiley–Liss, New York.

Anderson, J. A. (1985). What Hebb synapses build. *In "Synaptic Modification, Neuron Selectivity, and Nervous System Organization"*, (W. B. Levy, J. A. Anderson, and S. Lehmkuhle Eds.), pp. 153–173. Erlbaum, Hillsdale, NJ.

Angers, A., Fioravante, D., Chin, J., Cleary, L. J., Bean, A. J., and Byrne, J. H. (2002). Serotonin stimulates phosphorylation of *Aplysia* synapsin and alters its subcellular distribution in sensory neurons. *J. Neurosci.* **22**, 5412–5422.

Armony, J. L., Servan-Schreiber, D., Romanski, L. M., Cohen, J. D., and LeDoux, J. E. (1997). Stimulus generalization of fear responses: Effects of auditory cortex lesions in a computational model and in rats. *Cerebral Cortex* **7**(2), 157–165.

Atwood, H., L., and Wojtowicz, J. M. (1999). Silent synapses in neural plasticity: Current evidence. *Learning & Memory* **6**(6), 542–571.

Bailey, C. H., Chen, M., Keller, F., and Kandel, E. R. (1992). Serotonin-mediated endocytosis of apCAM: An early step of learning-related synaptic growth in *Aplysia*. *Science* **256**, 645–649.

Bailey, D. J., Kim, J. J., Sun, W., Thompson, R. F., and Helmstetter, F. J. (1999). Acquisition of fear conditioning in rats requires the synthesis of mRNA in the amygdala. *Behav. Neurosci.* **113**, 276–282.

Bailey, H. C., and Kandel, E. R. (1993). Structural changes accompanying memory storage. *Ann. Rev. Physiol.* **55**, 397–426.

Balaban, P. M., Maksimova, O. A., and Bravarenko, H. I. (1994). Behavioral plasticity in a snail and its neural mechanisms. *Neuros. Behav. Physiol.* **24**, 97–104.

Bao, J. X., Kandel, E. R., and Hawkins, R. D. (1998). Involvement of presynaptic and postsynaptic mechanisms in a cellular analog of classical conditioning at *Aplysia* sensory–motor neuron synapses in isolated cell culture. *J. Neurosci.* **18**, 458–466.

Barnes, C. A. (1995). Involvement of LTP in memory: Are we "Searching under the street light?" *Neuron* **15**, 751–754.

Barrionuevo, G., and Brown, T. H. (1983). Associative long-term potentiation in hippocampal slices. *Proc. Natl. Acad. Sci. USA* **80**, 7347–7351.

Barrionuevo, G., Kelso, S., Johnston, D., and Brown, T. H. (1986). Conductance mechanism responsible for long-term potentiation in monosynaptic and isolated excitatory synaptic inputs to the hippocampus. *J. Neurophysiol.* **55**, 540–550.

Bartsch, D., Ghirardi, M., Skehel, P. A., Karl, K. A., Herder, S. P., Chen, M., Bailey, C. H., and Kandel, E. R. (1995). *Aplysia* CREB2 represses long-term facilitation: Relief of repression converts transient facilitation into long-term functional and structural change. *Cell* **83**, 979–992.

Bashir, Z. I., Bortolotto, Z. A., Davies, C. H., Beretta, N., Irving, A. J., Seal, A. J., Henley, J. M., Jane, D. E., Watkins, J. C., and Collingridge, G. L. (1993). Induction of LTP in hippocampus needs synaptic activation of glutamate metabotropic receptors. *Nature* **363**, 347–350.

Baudry, M., Davis, J. L., and Thompson, R. F. (Eds.) (1993). *"Synaptic Plasticity: Molecular and Functional Aspects,"* MIT Press, Cambridge, MA.

Bauer, E. P., Schafe, G. E., and LeDoux, J. E. (2002). NMDA receptors and L-type voltage-gated calcium channels contribute to long-term potentiation and different components of fear memory formation in the lateral amygdala. *J. Neurosci.* **22**, 5239–5249.

Baxter, D. A., Bittner, G. D., and Brown, T. H. (1985). Quantal mechanisms of long-term potentiation. *Proc. Natl. Acad. Sci. USA* **82**, 5978–5982.

Beattie, E. C., Carroll, R. C., Yu, X., Morishita, W., Yasuda, H., von Zastrow, M., and Malenka, R. C. (2000). Regulation of AMPA receptor endocytosis by a signaling mechanism shared with LTD. *Nat. Neurosci.* **3**, 1291–1300.

Bechara, A., Tranel, D., Damasio, H., Adolphs, R., Rockland, C., and Damasio, A. R. (1995). Double dissociation of conditioning and declarative knowledge relative to the amygdala and hippocampus in humans. *Science* **269**, 1115–1118.

Beggs, J. M., Brown, T. H., Byrne, J. H., Crow, T., LeDoux, J. E., LeBar, K., and Thompson, R. F. (1999). Learning and memory: Basic mechanisms and systems. *In "Fundamentals of Neuroscience"* (M. J. Zigmond, F. E. Bloom, S. C. Landis, J. L. Roberts, and L. S. Squire, Eds.), pp. 1411–1454). Academic Press San Diego.

Benjamin, P. R., Staras, K., and Kemenes, G. (2000). A systems approach to the cellular analysis of associative learning in the pond snail *Lymnaea. Learn. Memory* **7**, 124–131.

Benke, T. A., Luthi, A., Issac, J. R., and Collingridge, G. L. (1998). Modulation of AMPA receptor unitary conductance by synaptic activity. *Nature* **393**, 793–797.

Berger, T. W., Alger, B. E., and Thompson, R. F. (1976). Neuronal substrate of classical conditioning in the hippocampus. *Science* **192**, 483–485.

Berger, T. W., Berry, S. D., and Thompson, R. F. (1986). Role of the hippocampus in classical conditioning of aversive and appetitive behaviors. *In "The Hippocampus"* (R. L. Isaacson and K. H. Pribram, Eds.), (pp. 203–239). Plenum, New York.

Berger, T. W., Rinaldi, P. C., Weisz, D. J., and Thompson, R. F. (1983). Single-unit analysis of different hippocampal cell types during classical conditioning of the rabbit nictitating membrane response. *J. Neurophysiol.* **50**, 1197–1219.

Berger, T. W., and Thompson, R. F. (1978). Identification of pyramidal cells as the critical elements in hippocampal neuronal plasticity during learning. *Proc. Natl. Acad. Sci. USA* **75**, 1572–1576.

Bienenstock, E. I., Cooper, L. N., and Munro, P. W. (1982). Theory for the development of neuron selectivity: Orientation specificity and binocular interaction in the visual cortex. *J. Neurosci.* **2**, 32–48.

Blair, H. T., Schafe, G. E., Bauer, E. P., Rodrigues, S. M., and LeDoux, J. E. (2001). Synaptic plasticity in the lateral amygdala: A cellular hypothesis of fear conditioning. *Learn. Memory* **8**, 229–242.

Bliss, T. V. P., and Collingridge, G. L. (1993). A synaptic model of memory: Long-term potentiation in the hippocampus. *Nature* **361**, 31–39.

Bliss, T. V. P., and Lomo, T. (1973). Long-lasting potentiation of synaptic transmission in the dentate area of the anaesthetized rabbit following stimulation of the perforant path. *J. Physiol.* **232**, 331–356.

Blitzer, R. D., Connor, J. H., Brown, G. P., Wong, T., Shenolikar, S., Iyengar, R., and Landau, E. M. (1998). Gating of CaMKII by cAMP-regulated protein phosphatase activity during LTP. *Science* **280**, 1940–1942.

Blum, S. A., Moore, N., Adams, F., and Dash, P. K. (1999). A mitogen-activated protein kinase cascade in the CA1/CA2 subfield of the dorsal hippocampus is essential for long-term spatial memory. *J. Neurosci.* **19**, 3535–3544.

Bordi, F., and LeDoux, J. (1992). Sensory tuning beyond the sensory system: An initial analysis of auditory response properties of neurons in the lateral amygdaloid nucleus and overlying areas of striatum. *J. Neurosci.* **12**, 2493–2503.

Borroni, A., Fichtenholtz, H., Woodside, B., and Teyler, T. J. (2000). Role of vdccLTP and nmdaLTP in spatial memory. *J. Neurosci.* **20**, 9272–9276.

Bortolotto, Z. A., Bashir, Z. I., Davies, C. H., and Collingridge, G. L. (1994). A molecular switch activated by metabotropic glutamate receptors regulates induction of long-term potentiation. *Nature* **368**, 740–743.

Botzer, D., Markovich, S., and Susswein, A. J. (1998). Multiple memory processes following training that a food is inedible in *Aplysia. Learn. Memory* **5**, 204–219.

Bourtchouladze, R., Frenguelli, B., Blendy, D., Cioffi, D., Schultz, G., and Silva, A. J. (1994). Deficient long-term memory in mice with a targeted mutation of the cAMP-responsive element-binding protein. *Cell* **79**, 59–68.

Bracha, V., and Bloedel, J. R. (1996). In J. R. Bloedel, T. J. Ebner, and S. P. Wise (Eds.), *The acquisition of motor behavior in vertebrates* (pp. 175–204). Cambridge, MA: MIT Press.

Bracha, V., Irwin, K. B., Webster, M. L., Wunderlich, D. A., Stachowiak, M. K., and Bloedel, J. R. (1998). Microinjections of anisomycin into the intermediate cerebellum during learning affect the acquisition of classically conditioned responses in the rabbit. *Brain Research* **788**, 169–178.

Breiter, H. C., Etcoff, N. L., Whalen, P. J., Kennedy, W. A., Rauch, S. L., Buchner, R. L., Strauss, M. M., Hyman, S. E., and Rosen, B. R. (1996). Response and habituation of the human amygdala during visual processing of facial expression. *Neuron* **17**, 875–877.

Brembs, B., Lorenzetti, F. D., Reyes, F. D., Baxter, D., and Byrne, J. H. (2002). Operant reward learning in Aplysia: neuronal correlates and mechanisms. *Science* **296**, 1706–1709.

Briggs, C. A., Brown, T. H., and McAfee, D. A. (1985a). Neurophysiology and pharmacology of long-term potentiation in the rat sympathetic ganglion. *Journal of Physiology* **359**, 503–521.

Briggs, C. A., McAfee, D. A., and McCaman, R. E. (1985b). Long-term potentiation of synaptic acetylcholine release in the superior cervical ganglion of the rat. *Journal of Physiology* **363**, 181–190.

Brown, T. (1818, 1999). *Inquiry into the relation of cause and effect* (4th ed.): Scholars Facsimiles and Reprint.

Brown, T. H., Chapman, P., Kairiss, E. W., and Keenan, C. L. (1988). Long-term potentiation. *Science*, **242**, 724–728.

Brown, T. H., and Chattarji, S. (1994). Hebbian synaptic plasticity: Evolution of the contemporary concept. *In "Models of Neural Networks II"* (E. Domany, J. L. Van Hemmen, and K. Schulten, Eds.), pp. 287–314. Springer-verlag, New York.

Brown, T. H., and Faulkner, B. (1999). Hebbian synapses. *In "Elsevier's Encyclopedia of Neuroscience"* (G. Adelman and B. H. Smith, Eds.), pp. 865–868. Elsevier Science, Amsterdam.

Brown, T. H., Ganong, A. H., Kairiss, E. W., and Keenan, C. L. (1990). Hebbian synapses: Biophysical mechanisms and algorithms. *Annu. Rev. Neurosci.* **13**, 475–512.

Brown, T. H., Ganong, A. H., Kairiss, E. W., Keenan, C. L., and Kelso, S. R. (1989). Long-term potentiation in two synaptic systems of the hippocampal brain slice. *In "Neural Models of Plasticity: Experimental and Theoretical Approaches"* (J. H. Byrne and W. O. Berry Eds.), pp. 266–306. Academic Press, San Diego.

Brown, T. H., and Jaffe, D. B. (1994). Calcium imaging in hippocampal neurons using confocal microscopy. *Annals of the New York Academy of Science*, **747**, 313–24.

Brown, T. H., and Johnston, D. (1983). Voltage-clamp analysis of mossy-fiber synaptic input to hippocampal neurons. *J. Physiol.* **50**, 487–507.

Brown, T. H., and Lindquist, D. H. (2003). Long-term potentiation: Amygdala. *In "Learning and Memory"* (J. H. Byrne, Ed.), 2nd ed., pp. 342–343. Macmillan, New York.

Brown, T. H., and McAfee, D. (1982). Long-term synaptic potentiation in superior cervical ganglion. *Science* **215**, 1411–1413.

Brown, T. H., Perkel, D. H., and Feldman, M. W. (1976). Evoked neurotransmitter release: Statistical effects of nonuniformity and nonstationarity. *Proc. Nat. Acad. Sci.* **73**, 2913–2917.

Brown, T. H., Wong, R. K., and Prince, D. A. (1979). Spontaneous miniature synaptic potentials in hippocampal neurons. *Brain Res.* **177**, 94–99.

Brown, T. H., and Zador, A. M. (1990). The hippocampus. *In* *"Synaptic Organization of the Brain"* (G. Shepherd, Ed.), (pp. 346–388): Oxford Univ. Press, New York.

Brown, T. H., Zador, A. M., Mainen, Z. F., and Claiborne, B. J. (1992). Hebbian computations in hippocampal dendrites and spines. *In* *"Neuron Computation"* (T. McKenna, J. Davis, and S. F. Zornetzer Eds.), pp. 81–116. Academic Press, San Diego.

Buchel, C., Morris, J., Dolan, R. J., and Friston, K. J. (1998). Brain systems mediating aversive conditioning: An event-related fMRI study. *Neuron* **20**, 947–957.

Byrne, J. H. (1987). Cellular analysis of associative learning. *Physiol. Rev.* **67**, 329–439.

Byrne, J. H., and Kandel, E. R. (1996). Presynaptic facilitation revisited: State- and time-dependence. *J. Neurosci.* **16**, 425–435.

Byrne, J. H., Zwartjes, R., Homayouni, R., Critz, S., and Eskin, A. (1993). Roles of second messenger pathways in neuronal plasticity and in learning and memory: Insights gained from *Aplysia*. *Adv. Second Messenger Phosphoprotein Res.* **27**, 47–108.

Cahill, L., and McGaugh, J. L. (1998). Mechanisms of emotional arousal and lasting declarative memory. *Trends Neurosci.* **21**, 294–299.

Cahill, L., Weinberger, N. M., Roozendaal, B., and McGaugh, J. L. (1999). Is the amygdala a locus of "conditioned fear"? Some questions and caveats. *Neuron* **23**, 227–278.

Cajal, S. R. y. (1911). *"Histologie du systeme nerveux"*, Vol. 2. Maline, Paris.

Campeau, S., and Davis, M. (1995). Involvement of subcortical and cortical afferents to the lateral nucleus of the amygdala in fear conditioning measured with fear-potentiated startle in rats trained concurrently with auditory and visual conditioned stimuli. *J. Neurosci.* **15**, 2312–2327.

Carew, T. J. (2000). *"Behavioral Neurobiology"*. Sinauer Associates, Sunderland, MA.

Carew, T. J., Hawkins, R. D., Abrams, T. W., and Kandel, E. R. (1984). A test of Hebb's postulate at identified synapses which mediate classical conditioning in *Aplysia*. *J. Neurosci.* **4**, 217–224.

Carmignoto, G., and Vicini, S. (1992). Activity-dependent decrease in NMDA receptor responses during development of the visual cortex. *Science* **258**, 1007–1011.

Carroll, R. C., Beattie, E. C., Xia, H., Luscher, C., Altschuler, Y., Nicoll, R. A., Malenka, R. C., and von Zastrow, M. (1999). Dynamin-dependent endocytosis of ionotropic glutamate receptors. *Proc. Natl. Acad. Sci.* **96**, 14112–14117.

Carroll, R. C., Nicoll, R. A., and Malenka, R. C. (1998). Effects of PKA and PKC on miniature excitatory postsynaptic currents in CA1 pyramidal cells. *J. Neurophysiol.* **80**, 2797–2800.

Casado, M., Isope, P., and Ascher, P. (2002). Involvement of presynaptic N-methyl-D-aspartate receptors in cerebellar long-term depression. *Neuron* **33**, 123–130.

Castro-Alamancos, M., and Calcagnotto, M. E. (1999). Presynaptic long-term potentiation in corticothalamic synapses. *J. Neurosci.* **19**, 9090–9097.

Cavus, I., and Teyler, T. J. (1996). Two forms of long-term potentiation in area CA1 activate different signal transduction pathways. *J. Neurophysiol.* **76**, 3038–3047.

Chain, D. G., Casadio, A., Schacher, S., Hegde, A. N., Valbrun, M., Yamamoto, N., Goldberg, A. L., Bartsch, D., Kandel, E. R., and Schwartz, J. H. (1999). Mechanisms for generating the autonomous cAMP-dependent protein kinase required for long-term facilitation in *Aplysia*. *Neuron* **22**, 147–156.

Chapman, P. F., and Bellavance, L. L. (1992). Induction of long-term potentiation in the basolateral amygdala does not depend on NMDA receptor activation. *Synapse* **11**, 310–318.

Chapman, P. F., Kairiss, E. W., Keenan, C. L., and Brown, T. H. (1990). Long-term synaptic potentiation in the amygdala. *Synapse* **6**, 271–278.

Charney, D. S., Deutch, A. V., Krystal, J. H., Southwick, A. M., and Davis, M. (1993). Psychobiologic mechanisms of posttraumatic stress disorder. *Arch. Gen. Psychiatry* **50**, 294–305.

Chen, C., Masanobu, K., Abeliovich, A., Chen, L., Bao, S., Kim, J. J., Hashimoto, K., Thompson, R. F., and Tonegawa, S. (1995). Impaired motor coordination correlates with persistent multiple climbing fiber innervation in PKCy mutant mice. *Cell* **83**, 1233–1242.

Chen, G., and Steinmetz, J. E. (2000). Microinfusion of protein kinase inhibitor H7 into the cerebellum impairs the acquisition but not the retention of classical eyeblink conditioning in rabbits. *Brain Res.* **856**, 193–201.

Chen, L., Bao, S., Lockard, J. M., Kim, J. J., and Thompson, R. F. (1996). Impaired classical eyeblink conditioning in cerebellar lesioned and Purkinje cell degeneration (pcd) mutant mice. *J. Neurosci.* **16**, 2829–2838.

Chen, L., Bao, S., and Thompson, R. F. (1999). Bilateral lesions of the interpositus nucleus completely prevent eyeblink conditioning in Purkinje cell-degeneration mutant mice. *Behav. Neurosci.* **113**, 204–216.

Choi, J.-S., Lindquist, D. H., and Brown, T. H. (2001). Amygdala lesions prevent conditioned enhancement of the rat eyeblink reflex. *Behav. Neurosci.* **115**, 764–775.

Church, P., and Lloyd, P. (1994). Activity of multiple identified motor neurons recorded intracellulary during evoked feeding like motor programs in *Aplysia*. *J. Neurophysiol.* **72**, 1794–1809.

Churchland, P. S., and Sejnowski, T. J. (1992). *The computational brain*. Cambridge, MA: The MIT Press.

Christie, B. R., Magee, J. C., and Johnston, D. (1996). The role of dendritic action potentials and Ca²⁺ influx in the induction of homosynaptic long-term depression in hippocampal CA1 pyramidal neurons. *Learn. Memory* **3**, 160–169.

Claiborne, B. J., Xiang, Z., and Brown, T. H. (1993). Hippocampal circuitry complicates analysis of long-term potentiation in mossy fiber synapses. *Hippocampus* **3**, 115–122.

Clark, G. A., McCormick, D. A., Lavond, D. G., and Thompson, R. F. (1984). Effect of lesions of cerebellar nuclei on conditioned behavioral and hippocampal neuronal responses. *Brain Res.* **291**, 125–136.

Clark, R. E., and Lavond, D. G. (1993). Reversible lesions of the red nucleus during acquisition and retention of a classically conditioned behavior in rabbits. *Behav. Neurosci.* **107**, 264–270.

Clark, R. E., and Squire, L. R. (1998). Classical conditioning and brain systems: The role of awareness. *Science* **280**, 77–81.

Cleary, L. J., Byrne, J. H., and Frost, W. N. (1995). Role of interneurons in defensive withdrawal reflexes in *Aplysia*. *Learn. Memory* **2**, 133–151.

Cleary, L. J., Lee, W. L., and Byrne, J. H. (1998). Cellular correlates of long-term sensitization in *Aplysia*. *J. Neurosci.* **18**, 5988–5998.

Clugnet, M. C., and LeDoux, J. E. (1990). Synaptic plasticity in fear conditioning circuits: Induction of LTP in the lateral nucleus of the amygdala by stimulation of the medial geniculate body. *J. Neurosci.* **10**, 2818–2824.

Collins, D. R., and Pare, D. (2000). Differential fear conditioning induces reciprocal changes in the sensory responses of lateral amygdala neurons to the CS(+) and CS(–). *Learn. Memory* **7**, 97–103.

Cook, D. G., and Carew, T. J. (1989). Operant conditioning of head-waving in *Aplysia*. III. Cellular analysis of possible reinforcement pathways. *J. Neurosci.* **9**, 3115–3122.

Crair, M. C., and Malenka, R. C. (1995). A critical period for long-term potentiation at thalamocortical synapses. *Nature* **375**, 325–328.

Crow, T. (2003). Invertebrate learning: Associative learning in *Hermissenda*. *In* "*Learning and Memory*" (J. H. Byrne, Ed.), 2nd ed., pp. 277–281. Macmillan, New York.

Cruikshank, S. J., Edeline, J. M., and Weinberger, N. M. (1992). Stimulation at a site of auditory–somatosensory convergence in the medial geniculate nucleus is an effective unconditioned stimulus for fear conditioning. *Behav. Neurosci.* **106**, 471–483.

Curtis, D. R., and Eccles, J. C. (1960). Synaptic action during and after repetitive stimulation. *J. Physiol.* **150**, 374–398.

Daum, I., Channon, S., Polkey, C. E., and Gray, J. A. (1991). Classical conditioning after temporal lobe lesions in man: Impairment in conditional discrimination. *Behav. Neurosci.* **105**, 396–408.

Daum, I., Schugens, M. M., Ackerman, H., Lutzenberger, W., Dichgans, J., and Birbaumer, N. (1993). Classical conditioning after cerebellar lesions in human. *Behav. Neurosci.* **107**, 748–756.

Davies, S. N., Lester, R. A., Reymann, K. G., and Collingridge, G. L. (1989). Temporally distinct pre- and post synaptic mechanisms maintain long-term potentiation. *Nature* **338**, 500–503.

Davis, M. (1984). Mammalian startle response. *In* "*Neural Mechanisms of Startle Behavior*" (R. C. Eaton, Ed.), pp. 287–351. Plenum, New York.

Davis, M. (1992). The role of the amygdala in fear and anxiety. *Annu. Rev. Neurosci.* **15**, 353–375.

Davis, M. (1998). Are different parts of the extended amygdala involved in fear versus anxiety? *Biol. Psychiatry* **44**, 1239–1247.

de Jonge, M. C., Black, J., Deyo, R. A., and Disterhoft, J. F. (1990). Learning-induced afterhyperpolarization reductions in hippocampus are specific for cell type and potassium conductance. *Exp. Brain Res.* **80**, 456–462.

Derkach, V., Barria, A., and Soderling, T. R. (1999). Ca^{2+}/calmodulin-kinase II enhances channel conductance of alpha-amino-3-hydroxy-5-methyl-4-isoxazolepropionate type glutamate receptors. *Proc. Natl. Acad. Sci. USA* **96**, 3269–3274.

Desmond, N. L., and Weinberg, R. J. (1998). Enhanced expression of AMPA receptor at perforated axospinous synapses. *NeuroReport* **9**, 857–860.

DeZazzo, J., and Tully, T. (1995). Dissection of memory formation: From behavioral pharmacology to molecular genetics. *Trends Neurosci.* **18**, 212–218.

Diamond, D. M., Fleshner, M., Ingersoll, N., and Rose, G. M. (1996). Psychological stress impairs spatial working memory: Relevance to electrophysiological studies of hippocampal function. *Behav. Neurosci.* **110**, 661–72.

Dineley, K. T., Weeber, E. J., Atkins, C., Adams, J. P., Anderson, A. E., and Sweatt, J. D. (2001). Leitmotifs in the biochemistry of LTP induction: amplification, integration and coordination. *J. Neurochem.* **77**, 961–971.

Dingledine, R. (Ed.). (1984). Brain Slices. Plenum, New York.

Disterhoft, J. F., Coulter, D. A., and Alkon, D. L. (1986). Conditioning-specific membrane changes of rabbit hippocampal neurons measured *in vitro*. *Proc. Natl. Acad. Sci.* **83**, 2733–2737.

Dolphin, A. C., Errington, M. L., and Bliss, T. V. (1982). Long-term potentiation of the perforant path *in vivo* is associated with increased glutamate release. *Nature* **297**, 496–498.

Donegan, N. H., and Wagner, A. R. (1987). Conditioned diminution and facilitation of the UR: A sometimes opponent-process interpretation. *In* Classical Conditioning, (I. Gormezano and W. F. Prokasy, Eds.), 3rd ed., pp. 339–369. Lawrence Erlbaum Associates, Hillsdale, NJ.

Dudai, Y., and Tully, T. (2003). Invertebrate learning: Neurogenetics of memory in *Drosophila*. *In* "*Learning and Memory*" (J. H. Byrne, Ed.), 2nd ed., pp. 292–296. Macmillan, New York.

Dudek, S. M., and Bear, M. F. (1992). Homosynaptic long-term depression in area CA1 of hippocampus and effects of *N*-methyl-D-aspartate receptor blockade. *Proc. Natl. Acad. Sci.* **89**, 4363–4367.

Dudek, S. M., and Bear, M. F. (1993). Bidirectional long-term modification of synaptic effectiveness in the adult and immature hippocampus. *J. Neurosci.* **13**, 2910–2918.

duLac, S., Raymond, J. L., Sejnowski, T. J., and Lisberger, S. G. (1995). Learning and memory in the vestibulo-ocular reflex. *Annu. Rev. Neurosci.* **18**, 409–441.

Dunaevsky, A., Tashiro, A., Majewska, A., Mason, C. A., and Yuste, R. (1999). Developmental regulation of spine motility in mammalian CNS. *Proc. Natl. Acad. Sci.* **96**, 13438–13443.

Dyer, J. R., Manseau, F., Castellucci, V. F., Sossin, W. S. (2003). Serotonin persistently activates the extracellular signal-related kinase in sensory neurons of Aplysia independently of cAMP or protein kinase C. *Neuroscience* **116**, 13–17.

Eccles, J. C. (1977). An instruction-selection theory of learning in the cerebellar cortex. *Brain Res.* **127**, 327–352.

Edwards, D. H., Heitler, W. J., and Krasne, F. B. (1999). Fifty years of a command neuron: The neurobiology of escape behavior in the crayfish. *Trends Neurosci.* **22**, 153–161.

Eichenbaum, H. (1999). Conscious awareness, memory and the hippocampus. *Nat. Neurosci.* **2**, 775–776.

Eisenstein, E., and Carlson, A. (1994). Leg position learning in the cockroach nerve cord using an analog technique. *Physiol. Behav.* **56**, 687–691.

English, J. D., and Sweatt, J. D. (1997). A requirement for the mitogen-activated protein kinase cascade in hippocampal long-term potentiation. *J. Biol. Chem.* **272**, 19103–19106.

Errington, M. L., Bliss, T. V. P., Richter-Levin, G., Yenk, K., Doyere, V., and Laroche, S. (1995). Stimulation at 1–5 Hz does not produce long-term depression or depotentiation in the hippocampus of the adult rat *in vivo*. *J. Neurophysiol.* **74**, 1793–1799.

Fanselow, M. S., and LeDoux, J. E. (1999). Why we think plasticity underlying Pavlovian fear conditioning occurs in the basolateral amygdala. *Neuron* **23**, 2239–2242.

Faulkner, B., Tieu, K. H., and Brown, T. H. (1997). Mechanism for temporal encoding in fear conditioning: Delay lines in perirhinal cortex and lateral amygdala. *In* "*Computational Neuroscience*" (J. Bower, Ed.), pp. 641–645. Plenum, New York: Publishing Corp.

Fendt, M., and Fanselow, M. S. (1999). The neuroanatomical and neurochemical basis of conditioned fear. *Neurosci. Biobehav. Rev.* **23**, 743–760.

Feng, T. P., Tsung-Han, L., and Ting, Y.-C. (1939). Repetitive discharges and inhibitory after-effect in post-tetanically facilitated responses of cat muscles to single nerve volleys. *Chin. J. Physiol.* **14**, 55–80.

Fiala, J. C., Allwardt, B., and Harris, K. M. (2002). Dendritic spines do not split during hippocampal LTP or maturation. *Nat. Neurosci.* **5**, 297–298.

Fiala, J. C., Feinberg, M., Popov, V., and Harris, K. M. (1998). Synaptogenesis via dendritic filopodia in developing hippocampal area CA1. *J. Neurosci.* **18**, 8900–8911.

Figurov, A., Pozzo-Miller, L., Olafsson, P., Wang, T., and Lu, B. (1996). Regulation of synaptic responses to high-frequency stimulation and LTP by neurotrophins in the hippocampus. *Nature* **381**, 706–709.

Fischer, M., Kaech, S., Knutti, D., and Matus, A. (1998). Rapid actin-based plasticity in dendritic spines. *Neuron* **20**, 847–854.

Foy, M. R., Stanton, M. E., Levine, S., and Thompson, R. F. (1987). Behavioral stress impairs long-term potentiation in rodent hippocampus. *Behav. Neural Biol.* **48**, 138–149.

Frankland, P. W., Cestari, V., Filipkowski, R. K., McDonald, R. J., and Silva, A. J. (1998). The dorsal hippocampus is essential for context discrimination but not for contextual conditioning. *Behav. Neurosci.* **112**, 863–874.

Fregnac, Y., Shulz, D., Thorpe, S., and Bienenstock, E. (1988). A cellular analogue of visual cortical plasticity. *Nature* **333**, 367–370.

Frey, U., Huang, Y. Y., and Kandel, E. R. (1993). Effects of cAMP stimulate a late stage of LTP in hippocampal CA1 neurons. *Science* **260**, 1661–1664.

Frey, U., and Morris, R. G. M. (1998). Synaptic tagging: Implications for late maintenance of hippocampal long-term potentiation. *Trends Neurosci.* **21**, 181–188.

Fritschy, J.-M., Weinmann, O., Wenzel, A., and Benke, D. (1998). Synapse-specific localization of NMDA and GABA_A receptor subunits revealed by antigen-retrieval immunohistochemistry. *J. Comp. Neurol.* **390**, 194–210.

Frost, W. N. (2003). Invertebrate learning: Habituation and sensitization in *Tritonia. In "Learning and Memory"* (J. H. Byrne, Ed.), 2nd ed., pp. 291–292. Macmillan, New York.

Frost, W. N., Brown, G. D., and Getting, P. A. (1996). Parametric features of habituation of swim cycle in the marine mollusc *Tritonia diomedea. Neurobiol. Learn. Memory* **65**, 125–135.

Garcia, R., Vouimba, R. M., Baudry, M., and Thompson, R. F. (1999). The amygdala modulates prefrontal cortex activity relative to conditioned fear. *Nature* **402**, 294–296.

Gean, P. W., Chang, F. C., Huang, C. C., Lin, J. H., and Way, L. J. (1993). Long-term enhancement of EPSP and NMDA receptor mediated synaptic transmission in the amygdala. *Brain Res. Bull.* **31**, 7–11.

Geinisman, Y., deToledo-Morrell, L., and Morrell, F. (1991). Induction of long-term potentiation is associated with an increase in the number of axospinous synapses with segmented postsynaptic densities. *Brain Res.* **566**, 77–88.

Geinisman, Y., Disterhoft, J. F., Gundersen, H. J. G., McEchron, M. D., Persina, I. S., Power, J. M., Van der Zee, E. A., and West, M. J. (2000). Remodeling of hippocampal synapses following hippocampus-dependent associative learning. *J. Comp. Neurol.* **417**, 49–59.

Gelperin, A. (2003). Invertebrate learning: Associative learning in *Limax. In "Learning and Memory"* (J. H. Byrne, Ed.), 2nd ed., pp. 281–287. Macmillan, New York.

Gerwitz, J. C., and Davis, M. (1997). Second-order fear conditioning prevented by blocking NMDA receptors in amygdala. *Nature* **388**, 471–474.

Gerwitz, J. C., Falls, W. A., and Davis, M. (1997). Normal conditioned inhibition and extinction of freezing and fear-potentiated startle following electrolytic lesions of medial prefrontal cortex in rats. *Behav. Neurosci.* **111**, 712–726.

Ghirardi, M., Montarolo, P. G., and Kandel, E. R. (1995). A novel intermediate stage in the transition between short- and long-term facilitation in the sensory to motor neuron synapse of *Aplysia. Neuron* **14**, 413–420.

Giese, K. P., Fedorov, N. B., Filipkowski, R. K., and Silva, A. J. (1998). Autophosphorylation at the Thr[286] of the alpha calcium-calmodulin kinase II in LTP and learning. *Science* **279**, 870–873.

Gillette, R. (1992). Invertebrate learning: Associative learning in *Pleurobranchaea. In "Encyclopedia of Learning and Memory"* (L. R. Squire, Ed.), pp. 302–305. Macmillan, New York.

Gisquet-Verrier, P., Dutrieux, G., Richer, P., and Doyere, V. (1999). Effects of lesions to the hippocampus on contextual fear: Evidence for a disruption of freezing and avoidance behavior but not context conditioning. *Behavi. Neurosci.* **113**, 507–522.

Gloor, P. (1992). Role of the amygdala in temporal lobe epilepsy. *In "The Amygdala: Neurobiological Aspects of Emotion, Memory, and Memory Dysfunction"* (J. P. Aggleton, Ed.), pp. 505–522 Wiley–Liss, New York.

Goddard, G. (1982). Hippocampal long-term potentiation: Mechanisms and implications for memory. *Neurosci. Res. Program Bull.* **20**, 676–680.

Goldowitz, D., and Eisenman, L. M. (1992). Genetic mutations affecting murine cerebellar structure and function. *In "Genetically Defined Animal Models of Neurobehavioral Dysfunctions"* (P. Driscoll, Ed.), pp. 66–88. Birkhauser. Boston.

Gomi, H., Sun, W., Finch, C. E., Itohara, S., Yoshimi, K., and Thompson, R. F. (1999). Learning induces a CDC2-related protein kinase, KKIAMRE. *J. Neurosci.* **19**, 9530–9537.

Goodlett, C. R., Hamre, K. M., and West, J. R. (1992). Dissociation of spatial navigation and visual guidance performance in Purkinje cell degeneration (*pcd*) mutant mice. *Behav. Brain Res.* **47**, 129–44.

Gormezano, I., Kehoe, E. J., and Marshall-Goodell, B. S. (1983). Twenty years of classical conditioning research with the rabbit. *In "Progress in Physiological Psychology"* (J. M. Sprague and A. N. Epstein, Eds.), (pp. 197–275). Academic Press, New York.

Greenough, W. T., Rest, R. W., and DeVoogd, T. J. (1978). Subsynaptic plastic perforations: Changes with age and experience in the rat. *Science* **202**, 1096–1098.

Greenwood, A. C., Landaw, E. M., and Brown, T. H. (1999). Testing the fit of a quantal model of neurotransmission. *Biophys. J.* **76**, 1847–1855.

Griffith, W. H., Brown, T. H., and Johnston, D. (1986). Voltage-clamp analysis of synaptic inhibition during long-term potentiation in hippocampus. *J. Neurophysiol.* **55**, 767–775.

Grover, L. M., and Teyler, T. J. (1995). Different mechanisms may be required for maintenance of NMDA receptor-dependent and independent forms of long-term potentiation. *Synapse* **19**, 121–133.

Guzowski, J. F., and McGaugh, J. L. (1997). Antisense oligodeoxynucleotide-mediated disruption of hippocampal cAMP response element binding protein levels impairs consolidation of memory for water maze training. *Proc. Natl. Acad. Sci. USA* **94**, 2693–2698.

Hardiman, M. J., Ramnani, N., and Yeo, C. H. (1996). Reversible inactivations of the cerebellum with muscimol prevent the acquisition and extinction of conditioned nictitating membrane responses in the rabbit. *Exp. Brain Res.* **110**, 235–247.

Harris, K. M., Jensen, F. E., and Tsao, B. (1992). Three-dimensional structure of dendritic spines and synapses in rat hippocampus (CA1) at postnatal day 15 and adult ages: Implications for the maturation of synaptic physiology and long-term potentiation. *J. Neurosci.* **12**, 2685–2705.

Hawkins, R. D., Kandel, E. R., and Siegelbaum, S. (1993). Learning to modulate transmitter release: Themes and variations in synaptic plasticity. *Annu. Rev. Neurosci.* **16**, 625–665.

Hebb, D. O. (1949). The Organization of Behavior. Wiley–Interscience, New York.

Hell, J. W., Westenbrook, R. E., Breeze, L. J., Wang, K. K. W., Chavkin, C., and Catterall, W. A. (1996). *N*-Methyl-D-aspartate receptor-induced proteolytic conversion of postsynaptic class C L-type channels in hippocampal neurons. *Proc. Natl. Acad. Sci. USA* **93**, 3362–3367.

Hemart, H., Daniel, H., Jaillard, D., and Crepel, F. (1994). Properties of glutamate receptors are modified during long-term depression in rat Purkinje cells. *Neurosci. Res.* **19**, 213–221.

Heynen, A. J., Abraham, W. C., and Bear, M. F. (1996). Bidirectional modification of CA1 synapses in the adult hippocampus *in vivo. Nature* **381**, 163–166.

Hirsch, H. V., and Spinelli, D. N. (1970). Visual experience modifies distribution of horizontally and vertically oriented receptive fields in cats. *Science* **168**, 869–871.

Hoffman, D. A., and Johnston, D. (1998). Downregulation of transient K$^+$ channels in dendrites of hippocampal CA1 pyramidal neurons by activation of PKA and PKC. *J. Neurosci.* **18**, 3521–3528.

Holmes, W. R., and Levy, W. B. (1990). Insights into associative long-term potentiation from computational models of NMDA receptor-mediated calcium influx and intracellular calcium changes. *J. Neurophysiol.* **63**, 1148–1168.

Hopfield, J. J. (1982). Neural networks and physical systems with emergent collective computational abilities. *Proc. Natl. Acad. Sci. USA* **79**, 2554–2558.

Hopfield, J. J. (1984). Neurons with graded response have collective computational properties like those of two-state neurons. *Proc. Natl. Acad. Sci.* **81**, 3088–3092.

Hopkins, W. F., and Johnston, D. (1984). Frequency-dependent noradrenergic modulation of long-term potentiation in the hippocampus. *Science* **19**, 350–352.

Hosakawa, T., Bliss, T. V. P., and Fine, A. (1992). Persistance of individual dendritic spines in living brain slices. *Neuroreport*, **3**, 477–80.

Huang, Y.-Y., and Kandel, E. R. (1998). Postsynaptic induction and PKA-dependent expression of LTP in the lateral amygdala. *Neuron* **21**, 169–178.

Huang, Y. Y., Li, X. C., and Kandel, E. R. (1994). cAMP contributes to mossy fiber LTP by initiating both a covalently mediated early phase and macromolecular synthesis-dependent late phase. *Cell* **79**, 69–79.

Huang, Y. Y., Martin, K. C., and Kandel, E. R. (2000). Both protein kinase A and mitogen-activated protein kinase are required in the amygdala for the macromolecular synthesis-dependent late phase of long-term potentiation. *J. Neurosci.* **20**, 6317–6325.

Huerta, P. T., Sun, L. D., Wilson, M. A., and Tonegawa, S. (2000). Formation of temporal memory requires NMDA receptors within CA1 pyramidal neurons. *Neuron* **25**, 473–480.

Impey, S., Mark, M., Villacres, E. C., Poser, C., Chavkin, C., and Storm, D. R. (1996). Induction of CRE-mediated gene expression by stimuli that generate long-lasting LTP in area CA1 of the hippocampus. *Neuron* **16**, 973–982.

Isaacson, R. L., and Pribram, K. H. (Eds.) (1986). *"The Hippocampus"*, Vol. 4. Plenum, New York.

Ito, M. (1984). *"The Cerebellum and Neural Control"*. Appleton–Century–Crofts, New York.

Ito, M. (1989). Long-term depression. *Annu. Rev. Neurosci.* **12**, 85–102.

Ito, M. (1993). Cerebellar mechanisms of long-term depression. *In "Synaptic Plasticity: Molecular and Functional Aspects"* (M. Baudry, J. L. Davis, and R. F. Thompson, Eds pp. 117–46). MIT Press, Cambridge, MA.

Ito, M., Sukurai, M., and Tongroach, P. (1982). Climbing fibre induced depression of both mossy fibre responsiveness and glutamate sensitivity of cerebellar Purkinje cells. *J. Physiol.* **324**, 113–134.

Ivkovich, D., and Thompson, R. F. (1997). Motor cortex lesions do not affect learning or performance of the eyeblink response in rabbits. *Behavi. Neurosci.* **111**, 727–738.

Jaffe, D. B., and Brown, T. H. (1997). Calcium dynamics in thorny excrescences of CA3 pyramidal neurons. *J. Neurophysiol.* **78**, 10–18.

Jahr, C. E. (1992). High probability opening of NMDA receptor channels by L-glutamate. *Science* **255**, 470–472.

Jahromi, S. S., and Atwood, H. L. (1974). Three-dimensional ultrastructure of the crayfish neuromuscular apparatus. *J. Cell Biol.* **63**, 599–613.

James, W. (1890). *"Psychology"*. Harvard Univ. Press, Cambridge, MA.

Jarrell, T. W., Gentile, C. G., Romanski, L. M., McCabe, P. M., and Schneiderman, N. (1987). Involvement of cortical and thalamic auditory regions in retention of differential bradycardiac conditioning to acoustic conditioned stimuli in rabbits. *Brain Res.* **412**, 285–294.

Johnston, D., and Brown, T. H. (1983). Interpretation of voltage-clamp measurements of hippocampal neurons. *J. Neurophysiol.* **50**, 464–486.

Johnston, D., and Brown, T. H. (1984). Biophysics and microphysiology of synaptic transmission in hippocampus. *In "Brain Slices"* (R. Dingledine, Ed.), pp. 51–86: Plenum, New York.

Johnston, D., Hoffman, D. A., Magee, J. C., Poolos, N. P., Watanabe, S., Colbert, C. M., and Migliore, M. (2000). Dendritic potassium channels in hippocampal pyramidal neurons. *J. Physiol.* **525**, 75–81.

Johnston, D., Williams, D., Jaffe, D., and Gray, R. (1992). NMDA-receptor-independent long-term potentiation. *Annu. Rev. Physiol.* **54**, 489–505.

Josselyn, S. A., Shi, C., Carlezon, W. A., Jr., Neve, R. L., Nestler, E. J., and Davis, M. (2001). Long-term memory is facilitated by cAMP response element-binding protein overexpression in the amygdala. *J. Neurosci.* **21**, 2402–2412.

Kabotyanski, E., Baxter, D., and Byrne, J. (1998). Identification and characterization of catecholaminergic neuron B65, which initiates and modifies patterned activity in the buccal ganglia of *Aplysia*. *J. Neurophysiol.* **79**, 605–621.

Kaech, S., Brinkhas, H., and Matus, A. (1999). Volatile anesthetics block actin-based motility in dendritic spines. *Proc. Natl. Acad. Sci. USA* **96**, 10433–10437.

Kang, H., and Schuman, E. (1995). Long-lasting neurotrophin-induced enhancement of synaptic transmission in the adult hippocampus. *Science* **267**, 1658–1662.

Kapp, B. S., Whalen, P. J., Supple, W. F., and Pascoe, J. P. (1992). Amygdaloid contributions to conditioned arousal and sensory information processing. *In "The Amygdala: Neurobiological Aspects of Emotion, Memory, and Mental Dysfunction"* (J. Aggleton, Ed.), pp. 229–254. Wiley–Liss, New York.

Kasono, K., and Hirano, T. (1995). Involvement of inositol trisphosphate in cerebellar long-term depression. *NeuroReport* **6**, 569–572.

Katz, B. (1969). *"The Release of Neural Transmitter Substances"*. Liverpool Univ. Press, Liverpool.

Katz, B., and Miledi, R. (1968). The role of calcium in neuromuscular facilitation. *J. Physiol* **195**, 481–492.

Keele, S. W., and Ivry, R. B. (1990). Does the cerebellum provide a common computation for diverse tasks: A timing hypothesis. *In "The Development and Neural bases of Higher Cognitive Functions"* (A. Diamond, Ed.), pp. 179–211. Academic Press, New York.

Keenan, C. L., Chapman, P. F., Chang, V., and Brown, T. H. (1988). Videomicroscopy of acute brain slices from hippocampus and amygdala. *Brain Res. Bull.* **21**, 373–383.

Kelly, M. P., and Deadwyler, S. A. (2002). Acquisition of a novel behavior induces higher levels of Arc mRNA than does overtrained performance. *Neuroscience* **110**, 617–626.

Kelso, S. R., and Brown, T. H. (1986). Differential conditioning of associative synaptic enhancement in hippocampal brain slices. *Science* **232**, 85–87.

Kelso, S. R., Ganong, A. H., and Brown, T. H. (1986). Hebbian synapses in hippocampus. *Proc. Natl. Acad. Sci.* **83**, 5326–5330.

Kemenes, G., and Benjamin, P. R. (1994). Training in a novel environment improves the appetitive learning performance of the snail, *Lymnaea stagnalis*. *Behav Neurosci.* **61**, 139–149.

Kennedy, M. B. (1989). Regulation of synaptic transmission in the central nervous system: Long-term potentiation. *Cell* **59**, 777–787.

Kida, S., Josselyn, S. A., de Ortiz, S. P., Kogan, J. H., Chevere, I., Masushige, S., and Silva, A. J. (2002). CREB required for the stability of new and reactivated fear memories. *Nat. Neurosci.* **5**, 348–355.

Kim, J. J., Chen, L., Bao, S., Sun, W., and Thompson, R. F. (1996). Genetic dissections of the cerebellar circuitry involved in classical eyeblink conditioning. *In "Gene Targeting and New Developments in Neurobiology"*. (S. Nakanishi, A. J. Silva, S.

Aizawa, and M. Katsuki, Eds.), (pp. 3–15. Japan Scientific Societies Press, Tokyo.

Kim, J. J., Clark, R. E., and Thompson, R. F. (1995). Hippocampectomy impairs the memory of recently, but not remotely, acquired trace eyeblink conditioned responses. *Behav. Neurosci.* **109**, 195–203.

Kim, J. J., and Fanselow, M. S. (1992). Modality-specific retrograde amnesia of fear. *Science* **256**, 675–677.

Kim, J. J., and Thompson, R. F. (1997). Cerebellar circuits and synaptic mechanisms involved in classical eyeblink conditioning. *Trends Neurosci.* **20**, 177–181.

Kim, J. J., and Yoon, K. S. (1998). Stress: metaplastic effects in the hippocampus. *Trends Neurosci.* **21**, 505–509.

Kirov, S. A., and Harris, K. A. (1999). Dendrites are more spiny on mature hippocampal neurons when synapses are activated. *Nat. Neurosci.* **2**, 878–883.

Klopf, A. H. (1982). *"The Hedonistic Neuron: A Theory of Memory, Learning, and Intelligence"*. Hemisphere, Washington, DC.

Knudsen, E. I. (1994). Supervised learning in the brain. *J. Neurosci.* **14**, 3985–3997.

Kohda, K., Inoue, T., and Mikoshiba, K. (1995). Ca^{2+} release from Ca^{2+} stores, particularly from ryanodine-sensitive Ca^{2+} stores, is required for the induction of LTD in cultured cerebellar Purkinje cells. *J. Neurophysiol.* **74**, 2184–2188.

Kohonen, T. (1989). *"Self-Organization and Associative Memory"*, 3rd ed. Springer-Verlag, Heidelberg.

Konorski, J. (1948). *"Conditioned Reflexes and Neuronal Organization"*. Cambridge Univ. Press, London.

Krasne, F. B. (1992). Invertebrate learning: Nonassociative learning in crayfish. *In "Encyclopedia of Learning and Memory"* (L. R. Squire, Ed.), pp. 310–311. Macmillan, New York.

Krug, M., Loessner, B., and Otto, T. (1984). Anisomycin blocks the late phase of long-term potentiation in the dentate gyrus of freely moving rats. *Brain Res. Bull.* **13**, 39–42.

Krupa, D. J., Thompson, J. K., and Thompson, R. F. (1993). Localization of a memory trace in the mammalian brain. *Science* **260**, 989–991.

Kullmann, D. M., Erdemli, G., and Asztely, F. (1996). LTP of AMPA and NMDA receptor-mediated signals: Evidence for presynaptic expression and extrasynaptic glutamate spillover. *Neuron* **17**, 461–474.

LaBar, K. S., and Disterhoft, J. F. (1998). Conditioning, awareness, and the hippocampus. *Hippocampus* **8**, 620–626.

LaBar, K. S., Gatenby, J. C., Gore, J. C., LeDoux, J. E., and Phelps, E. A. (1998). Human amygdala activation during conditioned fear acquisition and extinction: A mixed-trial fMRI study. *Neuron* **20**, 937–945.

LaBar, K. S., LeDoux, J. E., Spencer, D. D., and Phelps, E. A. (1995). Impaired fear conditioning following unilateral temporal lobectomy in humans. *J. Neurosci.* **15**, 6846–6855.

Lalonde, R., and Botez, M. I. (1990). The cerebellum and learning processes in animals. *Brain Res. Rev.* **15**, 325–332.

Larkman, A. U., and Jack, J. J. (1995). Synaptic plasticity: Hippocampal LTP. *Curr. Opin. Neurobiol.* **5**, 324–334.

Lavond, D. G., Kim, J. J., and Thompson, R. F. (1993). Mammalian brain substrates of aversive classical conditioning. *Annu. Rev. Psychol.* **44**, 317–342.

Lavond, D. G., Steinmetz, J. E., Yokaitis, M. H., and Thompson, R. F. (1987). Reacquisition of classical conditioning after removal of cerebellar cortex. *Exp. Brain Res.* **67**, 569–593.

Lechner, H. A., and Byrne, J. H. (1998). New perspectives on classical conditioning: A synthesis of Hebbian and non-Hebbian mechanisms. *Neuron* **20**, 355–358.

Lechner, H., Baxter, D. and Byrne, J. (2000). Classical conditioning of feeding in *Aplysia*. I. Behavioral analysis. *J. Neurosci.* **20**, 3369–3376.

LeDoux, J. E. (1986). Sensory systems and emotion. *Integrative Psychiatry* **4**, 237–248.

LeDoux, J. E. (1996). *"The Emotional Brain: The Mysterious Underpinnings of Emotional Life"*. Simon and Schuster, New York.

LeDoux, J. E. (2000). Emotion circuits in the brain. *Annu. Rev. Neurosci.* **23**, 155–184.

LeDoux, J. E., Cicchetti, P., Xagoraris, A., and Romanski, L. M. (1990). The lateral amygdaloid nucleus: Sensory interface of the amygdala in fear conditioning. *J. Neurosci.* **10**, 1062–1069.

LeDoux, J. E., Farb, C. R., and Romanski, L. M. (1991). Overlapping projections to the amygdala and striatum from auditory processing areas of the thalamus and cortex. *Neurosci. Lett.* **134**, 139–144.

Lee, H. J., Choi, J.-S., Brown, T. H., and Kim, J. J. (2001). Amygdalar *N*-methyl-D-aspartate (NMDA) receptors are critical for the expression of multiple conditioned fear responses. *J. Neurosci.* **21**, 4116–4124.

Lee, H.-K., Kameyama, K., Huganir, R. L., and Bear, M. F. (1998). NMDA induces long-term synaptic depression and dephosphorylation of the GluR1 subunit of AMPA receptors in hippocampus. *Neuron* **21**, 1151–1162.

Levenes, C., Daniel, H., and Crepel, F. (1998). Long-term depression of synaptic transmission in the cerebellum: Cellular and molecular mechanisms revisited. *Prog. Neurobiol.* **55**, 79–91.

Levenson, J., Endo, S., Kategaya, L. S., Fernandez, R. I., Brabham, D. G., Chin, J., Byrne, J. H., and Eskin, A. (2000). Long-term regulation of neuronal high-affinity glutamate and glutamine uptake in *Aplysia*. *Proc. Natl. Acad. Sci. USA* **97**, 12858–12863.

Levy, W. B. (1985). Associative changes at the synapse: LTP in the hippocampus. *In "Synaptic Modification, Neuron Selectivity, and Nervous System Organization"* (W. B. Levy, J. A. Anderson, and S. Lehmkuhle, Eds.), pp. 5–33. Erlbaum, Hillsdale, NJ.

Levy, W. B., and Steward, O. (1979). Synapses as associative memory elements in the hippocampal formation. *Brain Res.* **175**, 233–245.

Li, X. F., Armony, J. L., and LeDoux, J. E. (1996). $GABA_A$ and $GABA_B$ receptors differentially regulate synaptic transmission in the auditory thalamo-amygdala pathway: An *in vivo* microiontophoretic study and a model. *Synapse* **24**, 115–124.

Liao, D., Hessler, N. A., and Malinow, R. (1995). Activation of postsynaptically silent synapses during pairing-induced LTP in CA1 region of the hippocampal slice. *Nature* **375**, 400–404.

Liao, D., Scannevin, R. H., and Huganir, R. (2001). Activation of silent synapses by rapid activity-dependent synaptic recruitment of AMPA receptors. *J. Neurosci.* **21**, 6008–6017.

Lin, X. Y., and Glanzman, D. L. (1994). Hebbian induction of long-term potentiation of *Aplysia* sensorimotor synapses: Partial requirement for activation of an NMDA-related receptor. *Proc. R. Soc. London Ser. B*, **255**, 215–221.

Linden, D. J. (2001). The expression of cerebellar LTD in culture is not associated with changes in AMPA-receptor kinetics, agonist affinity, or unitary conductance. *Proc. Natl. Acad. Sci. U.* **98**, 14066–14071.

Linden, D. J., and Connor, J. A. (1995). Long-term synaptic depression. *Annu. Rev. Neurosci.* **18**, 319–35.

Linden, D. J., Dickinson, M. H., Smeyne, M., and Connor, J. A. (1991). A long-term depression of AMPA currents in cultured cerebellar Purkinje neurons. *Neuron* **7**, 81–89.

Lisman, J. (1989). A mechanism for the Hebb and anti-Hebb processes underlying learning and memory. *Proc. Natl. Acad. Sci. USA* **86**, 9574–9578.

Littleton, J. T., Stern, M., Schulze, K., Perin, M., and Bellen, H. J. (1993). Mutational analysis of *Drosophila synaptotagmin* demonstrates its essential role in Ca^{2+} activated neurotransmitter release. *Cell* **74**, 1125–1134.

Lledo, P. M., Zhang, X., Sudhof, T. C., Malenka, R. C., and Nicoll, R. A. (1998). Postsynaptic membrane fusion and long-term potentiation. *Science* **279**, 399–403.

Logan, C. G., and Grafton, S. T. (1995). Functional anatomy of human eyeblink conditioning determined with regional cerebral glucose metabolism and positron-emission tomography. *Proc. Natl. Acad. Sci. USA* **92**, 7500–7504.

Lynch, G., and Baudry, M. (1984). The biochemistry of memory: A new and specific hypothesis. *Science* **224**, 1057–1063.

Lynch, G. S., Dunwiddie, T., and Gribkoff, V. (1977). Heterosynaptic depression: A postsynaptic correlate of long-term potentiation. *Nature* **266**, 737–739.

Magee, J., and Johnston, D. (1997). A synaptically controlled, associative signal for hebbian plasticity in hippocampal neurons. *Science* **275**, 209–213.

Mainen, Z. F., Maletic-Savatic, M., Shi, S. H., Hayashi, Y., Malinow, R., and Svoboda, K. (1999). Two-photon imaging in living brain slices. *Methods* **18**, 231–239.

Malleret, G., Haditsch, U., Genoux, D., Jones, M. W., Bliss, V. P., Vanhoose, A. M., Weitlauf, C., Kandel, E. R., Winder, D. G., and Mansuy, I. M. (2001). Inducible and reversible enhancement of learning, memory and long-term potentiation by genetic inhibition of calcineurin. *Cell* **104**, 675–686.

Manabe, T., and Nicoll, R. A. (1994). Long-term potentiation: Evidence against an increase in transmitter release probability in the CA1 region of the hippocampus. *Science* **265**, 1888–1892.

Manahan-Vaughan, D. (1997). Group 1 and 2 metabotropic glutamate receptors play differential roles in hippocampal long-term depression and long-term potentiation in freely moving rats. *J. Neurosci.* **17**, 3303–3311.

Maren, S. (1998). Overtraining does not mitigate contextual fear conditioning deficits produced by neurotoxic lesions of the basolateral amygdala. *J. Neurosci.* **18**, 3088–3097.

Maren, S. (2000). Auditory fear conditioning increases CS-elicited spike firing in lateral amygdala neurons even after extensive overtraining. *Eur. J. Neurosci.* **12**, 4047–4054.

Maren, S., Aharonov, G., and Fanselow, M. (1997). Neurotoxic lesions of the dorsal hippocampus and Pavlovian fear conditioning in rats. *Behav. Brain Res.* **88**, 261–274.

Maren, S., Aharonov, G., Stote, D. L., and Fanselow, M. S. (1996). N-Methyl-D-aspartate receptors in the basolateral amygdala are required for both acquisition and expression of conditional fear in rats. *Behav. Neurosci.* **110**, 1365–1374.

Maren, S., and Fanselow, M. S. (1996). The amygdala and fear conditioning: Has the nut been cracked? *Neuron* **16**, 237–240.

Maren, S., Poremba, A., and Gabriel, M. (1991). Basolateral amygdaloid multi-unit neuronal correlates of discriminative avoidance learning in rabbits. *Brain Res.* **549**, 311–316.

Maren, S., Tocco, G., Standley, S., Baudry, M., and Thompson, R. F. (1993). Postsynaptic factors in the expression of long-term potentiation (LTP): Increased glutamate receptor binding following LTP induction *in vivo*. *Proc. Natl. Acad. Sci. USA* **90**, 9654–9658.

Marks, I., and Tobena, A. (1990). Learning and unlearning fear: A clinical and evolutionary perspective. *Neurosci. Biobehavi. Rev.* **14**, 365–384.

Marr, D. (1969). A theory of cerebellar cortex. *J. Physiol.* **202**, 437–470.

Martin, K. C., Michael, D., Rose, J. C., Barad, M., Casadio, A., Zhu, H., and Kandel, E. R. (1997). MAP kinase translocates into the nucleus of the presynaptic cell and is required for long-term facilitation in *Aplysia*. *Neuron* **18**, 899–912.

Matthies, H. (1989). In search of cellular mechanisms of memory. *Prog. Neurobiol.* **32**, 277–349.

Mauk, M. D., and Donegan, N. H. (1997). A model of pavlovian eyelid conditioning based on the synaptic organization of the cerebellum. *Learn. Memory* **4**, 130–158.

Mauk, M. D., Steinmetz, J. E., and Thompson, R. F. (1986). Classical conditioning using stimulation of the inferior olive as the unconditioned stimulus. *Proc. Natl. Acad. Sci.* **83**, 5349–5353.

McCormick, D. A., Steinmetz, J. E., and Thompson, R. F. (1985). Lesions of the inferior olivary complex cause extinction of the classically conditioned eyeblink response. *Brain Res.* **359**, 120–130.

McCormick, D. A., and Thompson, R. F. (1984). Cerebellum: Essential involvement in the classically conditioned eyelid response. *Science* **223**, 296–299.

McDonald, A. J. (1998). Cortical pathways to the mammalian amygdala. *Prog. Neurobiol.* **55**, 257–332.

McEchron, M. D., Bouwmeester, H., Tseng, W., Weiss, C., and Disterhoft, J. F. (1998). Hippocampectomy disrupts auditory trace fear conditioning and contextual fear conditioning in rats. *Hippocampus* **8**, 638–646.

McGann, J., and Brown, T. H. (2000). Fear conditioning model predicts key temporal aspects of conditioned response production. *Psychobiology* **28**, 303–313.

McGann, J. P., McCreeless, M. P., Moyer, J. R., Jr., and Brown, T. H. (Submitted). Pavlovian interstimulus interval function emerges from interaction between random synaptic amplitude fluctuations and intrinsic postsynaptic biophysics.

McGlinchey-Berroth, R., Carrillo, M. C., Gabrieli, J. D., Brawn, C. M., and Disterhoft, J. F. (1997). Impaired trace eyeblink conditioning in bilateral, medial-temporal lobe amnesia. *Behav. Neurosci.* **111**, 873–882.

McKernan, M. G., and Shinnick-Gallagher, P. (1997). Fear conditioning induces a lasting potentiation of synaptic currents *in vitro*. *Nature* **390**, 607–611.

McNaughton, B. L., Douglass, R. M., and Goddard, G. V. (1978). Synaptic enhancements in fascia dentata: Cooperativity among coactive efferents. *Brain Res.* **157**, 277–293.

McNish, K. A., Gerwitz, J. C., and Davis, M. (1997). Evidence of contextual fear after lesions of the hippocampus: A disruption of freezing but not fear-potentiated startle. *J. Neurosci.* **17**, 9353–9360.

Medina, J. F., Repa, J., Mauk, M. D., and LeDoux, J. E. (2002). Parallels between cerebellum- and amygdala-dependent conditioning. *Nat. Rev. Neurosci.* **3**, 122–131.

Menzel, R. (2001). Searching for the memory trace in a Mini-brain: The honeybee. *Learn. Memory* **8**, 53–62.

Miserendino, M. J. D., Sananes, C. B., Melia, K. R., and Davis, M. (1990). Blocking of acquisition but not expression of conditioned fear-potentiated startle by NMDA antagonists in the amygdala. *Nature* **345**, 716–718.

Monyer, H., Burnashev, N., Laurie, D. J., Sakmann, B., and Seeburg, P. H. (1994). Developmental and regional expression in the rat brain and functional properties of four NMDA receptors. *Neuron* **12**, 529–540.

Morgan, M. A., Romanski, L. M., and LeDoux, J. E. (1993). Extinction of emotional learning: Contribution of medial prefrontal cortex. *Neurosci. Lett.* **163**, 109–113.

Morgan, S. L., Coussens, C. M., and Teyler, T. J. (2001). Depotentiation of vdccLTP requires NMDAR activation. *Neurobiol. Learn. Memory* **76**, 229–238.

Morris, J. S., Frith, C. D., Perret, D. I., Rowland, D., Young, A. W., Calder, A. J., and Dolan, R. J. (1996). A differential neural response in the human amygdala to fearful and happy facial expressions. *Nature* **383**, 812–815.

Morton, D. W. and Chiel, H. J. (1993a). *In vivo* buccal nerve activity that distinguishes ingestion from rejection can be used to predict behavioral transitions in *Aplysia. J. Comp. Physiol. A* **172**, 17–32.

Morton, D., and Chiel, H. (1993b). The timing of activity in motor neurons that produce radula movements distinguishes ingestion from rejection in *Aplysia. J. Comp. Physiol. A* **173**, 519–536.

Moyer, J. R., Jr., and Brown, T. H. (1998). Methods for whole-cell recording from visually preselected neurons of perirhinal cortex in brain slices from young and aging rats. *J. Neurosci. Methods* **86**, 35–54.

Moyer, J. R., Jr., and Brown, T. H. (2002). Patch-clamp techniques applied to brain slices. In "*Advanced Techniques for Patch-Clamp Analysis*" (A. Walz, A., Boulton, and G. B. Baker, Eds.), pp. 135–194. Humana Press, Totowa, NJ.

Moyer, J. R., Jr., Deyo, R. A., and Disterhoft, J. F. (1990). Hippocampectomy disrupts trace eye-blink conditioning in rabbits. *Behav. Neurosci.* **104**, 243–252.

Moyer, J. R., Thompson, L. T., and Disterhoft, J. F. (1996). Trace eye-blink conditioning increases CA1 excitability in a transient and learning-specific manner. *J. Neurosci.* **16**, 5536–5546.

Mulkey, R. M., Endo, S., Shenolikar, S., and Malenka, R. C. (1994). Involvement of a calcineurin/inhibitor-1 phosphatase cascade in hippocampal long-term depression. *Nature* **369**, 486–488.

Mulkey, R. M., and Malenka, R. C. (1992). Mechanisms underlying induction of homosynaptic long-term depression in area CA1 of the hippocampus. *Neuron* **9**, 967–975.

Muller, J., Corodimas, K. P., Fridel, Z., and LeDoux, J. E. (1997). Functional inactivation of the lateral and basal nuclei of the amygdala by muscimol infusion prevents fear conditioning to an explicit conditioned stimulus and to contextual stimuli. *Behav. Neurosci.* **111**, 863–891.

Murphy, G. G., and Glanzman, D. L. (1997). Mediation of classical conditioning in *Aplysia californica* by long-term potentiation of sensorimotor synapses. *Science* **278**, 467–471.

Nader, K., Majidishad, P., Amorapanth, P., and LeDoux, J. E. (2001). Damage to the lateral and central, but not other, amygdaloid nuclei prevents the acquisition of auditory fear conditioning. *Learn. Memory* **8**, 156–163.

Nader, K., Schafe, G. E., and LeDoux, J. E. (2000). Fear memories require protein synthesis in the amygdala for reconsolidation after retrieval. *Nature* **406**, 722–726.

Nakazawa, K., Mikawa, S., Hashikawa, T., and Ito, M. (1995). Transient and persistent phosphorylation of AMPA-type glutamate receptor subunits in cerebellar Purkinje cells. *Neuron* **15**, 697–709.

Nargeot, R., Baxter, D., and Byrne, J. (1997). Contingent-dependent enhancement of rhythmic motor patterns: An *in vitro* analog of operant conditioning. *J. Neurosci.* **17**, 8093–8105.

Nargeot, R., Baxter, D., and Byrne, J. (1999a). Dopaminergic synapses mediate neuronal changes in an analogue of operant conditioning. *J. Neurophysiol.* **81**, 1983–1987.

Nargeot, R., Baxter, D., and Byrne, J. (1999b). *In vitro* analog of operant conditioning in *Aplysia*. I. Contingent reinforcement modifies the functional dynamics of an identified neuron. *J. Neurosci.* **15**, 2247–2260.

Nargeot, R., Baxter, D., and Byrne, J. (1999c). *In vitro* analog of operant conditioning in *Aplysia*. II. Modifications of the functional dynamics of an identified neuron contribute to motor pattern selection. *J. Neurosci.* **19**, 2261–2272.

Nguyen, P. V., Abel, T., and Kandel, E. R. (1994). Requirement of a critical period of transcription for induction of a late phase of LTP. *Science* **265**, 1104–1107.

Nicoll, R. A., and Malenka, R. C. (1995). Contrasting properties of two forms of long-term potentiation in the hippocampus. *Nature* **377**, 115–118.

Nusser, Z., Lujan, R., Laube, G., Roberts, J. D. B., Molnar, E., and Somogyi, P. (1998). Cell type and pathway dependence of synaptic AMPA receptor number and variability in the hippocampus. *Neuron* **21**, 545–559.

O'Connor, K. J., LaBar, K. S., and Phelps, E. A. (1999). Impaired contextual fear conditioning in amnesics. *J. Cogn. Neurosci. Suppl.* **19**,

O'Keefe, J. (1999). Do hippocampal pyramidal cells signal nonspatial as well as spatial information? *Hippocampus* **9**, 352–364.

Orr, S. P., Metzger, L. J., Lasko, N. B., Macklin, M. L., Peri, T., and Pitman, R. K. (2000). De novo conditioning in trauma-exposed individuals with and without posttraumatic stress disorder. *J. Abnormal Psychol.* **109**, 290–298.

Pare, D., Smith, Y., and Pare, J. F. (1995). Intra-amygdaloid projections of the basolateral and basomedial nuclei in the cat: *Phaseolus vulgaris*-leucoagglutinin anterograde tracing at the light and electron microscope level. *Neuroscience* **69**, 567–583.

Parnass, Z., Tashiro, A., and Yuste, R. (2000). Analysis of spine morphological plasticity in developing hippocampal pyramidal neurons. *Hippocampus* **10**, 561–568.

Pascoe, J. P., and Kapp, B. S. (1985). Electrophysiological characteristics of amygdaloid central nucleus neurons in the awake rabbit. *Brain Res. Bull.* **14**, 331–338.

Pavlides, C., Watanabe, Y., Magarinos, A. M., and McEwen, B. S. (1995). Opposing roles of type I and type II adrenal steroid receptors in hippocampal long-term potentiation. *Neuroscience* **68**, 387–394.

Pavlov, I. P. (1927). "*Conditioned Reflexes*" (G.V., Anrep, Tran.). Oxford Univ. Press, London.

Pavlov, I. P. (1932). The reply of a physiologist to psychologists. *Psychol. Rev.* **39**, 91–127.

Pelligrino, L. J., and Altman, J. (1979). Effects of differenatial interference with postnatal cerebellar neurogenesis on motor performance, activity level and maze learning of rats: A developmental study. *J. Comp. Physiol. Psychol.* **93**, 1–33.

Perrett, S. P., Ruiz, B. P., and Mauk, M. D. (1993). Cerebellar cortex lesions disrupt learning-dependent timing of conditioned eyelid responses. *J. Neurosci.* **13**, 1708–1718.

Pettigrew, J. D. (1974). The effect of visual experience on the development of stimulus specificity by kitten cortical neurons. *J. Physiol.* **237**, 49–74.

Phelps, E. A., LaBar, K. S., Andersen, A. K., O'Connor, K. J., and Fulbright, R. K. (1998). Specifying the contributions of the human amygdala to emotional memory: A case study. *NeuroCase* **4**, 527–540.

Philips, R. G., and LeDoux, J. E. (1992). Differential contribution of amygdala and hippocampus to cued and contextual fear conditioning. *Behav. Neurosci.* **106**, 274–285.

Phillips, R. G., and LeDoux, J. E. (1995). Lesions of the fornix but not the entorhinal or perirhinal cortex interfere with contextual fear conditioning. *J. Neurosci.* **15**, 5308–5315.

Pitkanen, A., Savander, V., and LeDoux, J. E. (1997). Organization of intra-amygdaloid circuitries in the rat: An emerging framework for understanding functions of the amygdala. *Trends Neurosci.* **20**, 517–523.

Poncer, J. C., and Malinow, R. (2001). Postsynaptic conversion of silent synapses during LTP affects synaptic gain and transmission dynamics. *Nat. Neurosci.* **4**, 989–996.

Quinlan, E. M., Benjamin, D. P., Huganir, R. L., and Bear, M. F. (1999). Rapid, experience-dependent expression of synaptic NMDA receptors in visual cortex *in vivo*. *Nat. Neurosci.* **2**, 352–357.

Quirk, G. J., Armony, J. L., and LeDoux, J. E. (1997). Fear conditioning enhances different temporal components of tone-evoked spike trains in auditory cortex and lateral amygdala. *Neuron* **19**, 613–624.

Quirk, G. J., Repa, J. C., and LeDoux, J. E. (1995). Fear conditioning enhances short-latency auditory responses of lateral amygdala neurons: Parallel recordings in the freely behaving rat. *Neuron* **15**, 1029–1039.

Quirk, G. J., Russo, G. K., Barron, J. L., and Lebron, K. (2000). The role of ventromedial prefrontal cortex in the recovery of extinguished fear. *J. Neurosci.* **20**, 6225–6231.

Rachlin, H. (1991). *"Introduction to Modern Behaviorism,"* 3rd ed. Freeman, New York.

Rausch, S. L., Whalen, P. J., Shin, L. M., McInerney, S. C., Macklin, M. L., Lasko, N. B., Orr, S. P., and Pitman, R. K. (2000). Exaggerated amygdala response to masked facial stimuli in posttraumatic stress disorder: A functional MRI study. *Biol. Psychiatry* **47**, 769–776.

Repa, J. C., Muller, J., Apergis, J., Desrochers, T. M., Zhou, Y., and LeDoux, J. E. (2001). Two different lateral amygdala cell populations contribute to the initiation and storage of memory. *Nat. Neurosci.* **4**, 724–731.

Rescorla, R. A. (1988). Behavioral studies of pavlovian conditioning. *Annu. Rev. Neurosci.* **11**, 329–352.

Rescorla, R. A., and Solomon, R. L. (1967). Two process learning theory: Relationships between pavlovian conditioning and instrumental learning. *Psychol. Rev.* **55**, 151–182.

Rickert, E. J., Bennett, T. L., Lane, P. and French, J. (1978). Hippocampectomy and the attenuation of blocking. *Behav. Biol.* **22**, 147–160.

Rodrigues, S. M., Schafe, G. E., and LeDoux, J. E. (2001). Intraamygdala blockade of the NR2B subunit of the NMDA receptor disrupts the acquisition but not the expression of fear conditioning. *J. Neurosci.* **21**, 6889–6896.

Rogan, M. T., and LeDoux, J. E. (1995). LTP is accompanied by commensurate enhancement of auditory-evoked responses in a fear conditioning circuit. *Neuron* **15**, 127–136.

Rogan, M. T., Staubli, U. V., and LeDoux, J. E. (1997). Fear conditioning induces associative long-term potentiation in the amygdala. *Nature* **390**, 604–607.

Romanski, L. M., and LeDoux, J. E. (1992). Equipotentiality of thalamo-amygdala and thalamo-cortico-amygdala circuits in auditory fear conditioning. *J. Neurosci.* **12**, 4501–4509.

Rose, J. K., and Rankin, C. H. (2001). Analysis of habituation in *Caenorhabditis elegans*. *Learn. Memory* **8**, 63–69.

Rudy, J. W., and O'Reilly, R. C. (1999). Contextual fear conditioning, conjunctive representations, pattern completion, and the hippocampus. *Behav. Neurosci.* **113**, 867–880.

Sahley, C., and Crow, T. (1998). Invertebrate learning: Current perspectives. *In "Neurobiology of Learing and Memory"* (J. Martinez and R. Lesner, Eds.), pp. 177–199. Academic Press, San Diego.

Sahley, C. L. (1995). What we have learned from the study of learning in the leech. *J. Neurobiol.* **27**, 434–445.

Sanes, J. R., and Lichtman, J. W. (1999). Can molecules explain long-term potentiation? *Nat. Neurosci.* **2**, 597–604.

Schafe, G. E., Atkins, C. M., Swank, M. W., Bauer, E. P., Sweatt, J. D., and LeDoux, J. E. (2000). Activation of ERK/MAP kinase in the amygdala is required for memory consolidation of pavlovian fear conditioning. *J. Neurosci.* **20**, 8177–8187.

Schafe, G. E., and LeDoux, J. E. (2000). Memory consolidation of auditory pavlovian fear conditioning requires protein synthesis and protein kinase A in the amygdala. *J. Neurosci.* **20**, RC96.

Schafe, G. E., Nadel, N. V., Sullivan, G. M., Harris, A., and LeDoux, J. E. (1999). Memory consolidation for contextual and auditory fear conditioning is dependent on protein synthesis, PKA, and MAP kinase. *Learn. Memory* **6**, 97–110.

Schmahmann, J. D. (Ed.). (1997). *"International Review of Neurobiology"*, Vol. 41. Academic Press, San Diego.

Sejnowski, T. J. (1977a). Statistical constraints on synaptic plasticity. *J. Theor. Biol.* **69**, 385–38.

Sejnowski, T. J. (1977b). Storing covariance with nonlinearly interacting neurons. *J. Math. Biol.* **4**, 303–321.

Selig, D. K., Lee, H. K., Bear, M. F., and Malenka, R. C. (1995). Reexamination of the effects of MCPG on hippocampal LTP, LTD, and depotentiation. *J. Neurophysiol.* **74**, 1075–1082.

Shatz, C. J., and Stryker, M. P. (1978). Ocular dominance in layer IV of the cat's visual cortex and the effects of monocular deprivation. *J. Physiol.* **281**, 267–283.

Sheppard, G. M. (1996). The dendritic spine: A multifunctional integrated unit. *J. Neurophysiol.* **75**, 2197–2210.

Shi, C., and Davis, M. (1999). Pain pathways involved in fear conditioning measured with fear-potentiated startle: Lesion studies. *J. Neurosci.* **19**, 420–430.

Shi, S.-H., Hayashi, Y., Petralia, R. S., Zaman, S. H., Wenthold, R. J., and Malinow, R. (1999). Rapid spine delivery and redistribution of AMPA receptors after synaptic NMDA receptor activation. *Science* **284**, 1811–1816.

Shibuki, K., Gomi, H., Chen, C., Bao, S., Kim, J. J., Wakatsuki, H., Fujisaki, T., Fujimoto, K., Katoh, A., Ikeda, T., Chen, C., Thompson, R. F., and Itohara, S. (1996). Deficient cerebellar long-term depression, impaired eyeblink conditioning and normal motor coordination in GFAP mutant mice. *Neuron* **16**, 587–599.

Solomon, P. R., Pomerleau, D., Bennett, L., James, J., and Morse, D. L. (1989). Acquisition of the classically conditioned eyeblink responses in humans over the life span. *Psychol. Aging* **4**, 34–41.

Solomon, P. R., Vander Schaaf, E. R., Thompson, R. F., and Weisz, D. J. (1986). Hippocampus and trace conditioning of the rabbit's classically conditioned nictitating membrane response. *Behav. Neurosci.* **100**, 729–744.

Staubli, U., and Lynch, G. (1990). Stable depression of potentiated synaptic responses in the hippocampus with 1–5 Hz stimulation. *Brain Res.* **513**, 113–118.

Staubli, U., and Scafidi, J. (1997). Studies on long-term depression in area CA1 of the anesthetized and freely moving rat. *J. Neurosci.* **17**, 4820–4828.

Steidl, S., and Rankin, C. H. (2003). Invertebrate learning: *C. elegans*. *In "Learning and Memory"* (J. H. Byrne, Ed.), 2nd ed., pp. 287–291. Macmillan, New York.

Steinmetz, J. E., Lavond, D. G., Ivkovich, D., Logan, C. G., and Thompson, R. F. (1992). Disruption of classical eyelid conditioning after cerebellar lesions: Damage to a memory trace system or a simple performance deficit? *J. Neurosci.* **12**, 4403–4426.

Steinmetz, J. E., Logan, C. G., Rosen, D. J., Thompson, J. K., Lavond, D. G., and Thompson, R. F. (1987). Initial localization of the acoustic conditioned CS projection system to the cerebellum essential for classical eyelid conditioning. *Proc. Natl. Acad. Sci.* **84**, 3531–3535.

Steinmetz, J. E., Logue, S. F., and Miller, D. P. (1993). Using signaled barpressing tasks to study the neural substrates of appetitive and aversive learning in rats: Behavioral manipulations and cerebellar lesions. *Behav. Neurosci.* **107**, 941–954.

Steinmetz, J. E., Rosen, D. J., Chapman, P. F., Lavond, D. G., and Thompson, R. F. (1986). Classical conditioning of the rabbit eyelid response with a mossy-fiber stimulation CS. I. Pontine nuclei and middle cerebellar peduncle stimulation. *Behav. Neurosci.* **100**, 878–887.

Stent, G. S. (1973). A physiological mechanism for Hebb's postulate of learning. *Proc. Natl. Acad. Sci. USA* **70**, 997–1001.

Steward, O., Wallace, C. S., Lyford, G. L., and Worley, P. F. (1998). Synaptic activation causes the mRNA for the IEG Arc to localize selectively near activated postsynaptic sites on dendrites. *Neuron* **21**, 741–751.

Steward, O., and Worley, P. F. (2001a). A cellular mechanism for targeting newly synthesized mRNAs to synaptic sites on dendrites. *Proc. Nat. Acad. Sci. USA* **98**, 7062–7068.

Steward, O., and Worley, P. F. (2001b). Selective targeting of newly synthesized Arc mRNA to activated synapses requires NMDA receptor activation. *Neuron* **30**, 227–240.

Supple, W. F., Jr., and Leaton, R. N. (1990). Lesions of the cerebellar vermis and cerebellar hemispheres: Effects on heart rate conditioning in rats. *Behav. Neurosci.* **104**, 934–947.

Sutton, M. A., Ide, J., Masters, S. E., and Carew, T. J. (2002). Interaction between amount and pattern of training in the induction of intermediate- and long-term memory for sensitization in *Aplysia*. *Learn. Memory* **9**, 29–40.

Sutton, M. A., Masters, S. E., Bagnall, M. W., and Carew, T. J. (2001). Molecular mechanisms underlying a unique intermediate phase of memory in *Aplysia*. *Neuron* **31**, 143–154.

Takumi, Y., Ramirez-Leon, V., Laake, P., Rinvik, E., and Ottersen, O. P. (1999). Different modes of expression of AMPA and NMDA receptors in hippocampal synapses. *Nat. Neurosci.* **2**, 618–624.

Tang, Y.-P., Shimizu, E., Dube, G. R., Rampon, C., Kerchner, G. A., Zhuo, M., Liu, G., and Tsien, Z. (1999). Genetic enhancement of learning and memory in mice. *Nature* **401**, 63–69.

Tanzi, E. (1893). I fatti e la induzioni nell'odierna istologia del sistema nervoso. *Riv. Sper. Freniatr. Med. Leg. Alienazioni Met. Soc. Ital. Psichiatria* **19**, 419–472.

Teyler, T. J. (2000). Forms of associative synaptic plasticity. In *"Model Systems and the Neurobiology of Associative Learning"* (J. E. Steinmetz, M. Gluck, and P. Solomon, Eds.), pp. 23–40. Erlbaum, Hillsdale, NJ.

Teyler, T. J., Cavus, I., Coussens, C., Discenna, P., Grover, L., Lee, Y. P., and Little, Z. (1994). Multideterminant role of calcium in hippocampal synaptic plasticity. *Hippocampus* **4**, 623–634.

Thach, W. T., Goodkin, H. G., and Keating, J. G. (1992). The cerebellum and the adaptive coordination of movement. *Annu. Rev. Neurosci.* **15**, 403–442.

Thiels, E., Barrioneuvo, G., and Berger, T. W. (1994). Excitatory stimulation during postsynaptic inhibition induces long-term depression in hippocampus *in vivo*. *J. Neurophysiol.* **72**, 3009–3016.

Thompson, L. T., Deyo, R. A., and Disterhoft, J. F. (1992). Hippocampus-dependent learning facilitated by a monoclonal antibody or D-cycloserine. *Nature* **359**, 838–841.

Thompson, L. T., Moyer, J. R., and Disterhoft, J. F. (1996). Transient changes in excitability of rabbit CA3 neurons with a time course appropriate to support memory consolidation. *J. Neurophysiol.* **76**, 1836–1849.

Thompson, R. F. (1986). The neurobiology of learning and memory. *Science* **233**, 941–947.

Thompson, R. F. (1989). Role of inferior olive in classical conditioning. In *"The Olivecerebellar System in Motor Control"* (P. Strata, Ed.), pp. 347–362. Springer-Verlag, New York.

Thompson, R. F. (1990). Neural mechanisms of classical conditioning in mammals. *Philos. Trans. R. Soc. London Ser. B* **329**, 161–170.

Thompson, R. F., Bao, S., Chen, L., Cipriano, B. D., Grethe, J. S., Kim, J. J., Thompson, J. K., Tracy, J. A., Weninger, M. S., and Krupa, D. J. (1997). Associative learning. *Int. Rev. Neurosci.* **41**, 151–189.

Thompson, R. F., and Kim, J. J. (1996). Memory systems in the brain and localization of a memory. *Proc. Nat. Acad. Sci. USA* **93**, 13438–13444.

Thompson, R. F., and Krupa, D. J. (1994). Organization of memory traces in the mammallian brain. *Annu. Rev. Neurosci.* **17**, 519–549.

Tieu, K. H., Keidel, A. L., McGann, J. P., Faulkner, B., and Brown, T. H. (1999). Perirhinal-amygdala circuit model of temporal encoding in fear conditioning. *Psychobiology* **27**, 1–25.

Tocco, G., Maren, S., Shors, T. J., Baudry, M., and Thompson, R. F. (1992). Long-term potentiation is associated with increased 3H-AMPA binding n the rat hippocampus. *Brain Res.* **573**, 228–234.

Tong, G., Malenka, R. C., and Nicoll, R. A. (1996). Long-term potentiation in cultures of single hippocampal granule cells: A presynaptic form of plasticity. *Neuron* **16**, 1147–1157.

Tsien, J. Z., Huerta, P. T., and Tonegawa, S. (1996). The essential role of hippocampal CA1 NMDA receptor-dependent synaptic plasticity in spatial memory. *Cell* **87**, 1327–1338.

Tsvetkov, E., Carlezon, W. A., Benes, F. M., Kandel, E. R., and Bolshakov, V. Y. (2002). Fear conditioning occludes LTP-induced presynaptic enhancement of synaptic transmission in the cortical pathway to the lateral amygdala. *Neuron* **34**, 289–300.

Umbach, J. A., Zinsmaier, K. E., Eberle, K. K., Buchner, E., Benzer, S., and Gundersen, C. B. (1994). Presynaptic dysfunction in *Drosophila csp* mutants. *Neuron* **13**, 899–907.

von der Malsburg, C. (1973). Self-organization of orientation sensitive cells in the striate cortex. *Kybernetik* **14**, 85–100.

Voneida, T., Christie, D., Bogdanski, R., and Chopko, B. (1990). Changes in instrumentally and classically conditioned limb-flexion responses following inferior olivary lesions and olivo-cerebellar tractotomy in the cat. *J. Neurosci.* **10**, 3583–3593.

Vouimba, R. M., Garcia, R., Baudry, M., and Thompson, R. F. (2000). Potentiation of conditioned freezing following dorsomedial prefrontal cortex lesions does not interfere with fear reduction in mice. *Behav. Neurosci.* **114**, 720–724.

Waddel, S., and Quinn, W. G. (2001). Flies, genes and learning. *Annu. Rev. Neurosci.* **24**, 1283–1309.

Wagner, A. R., and Brandon, S. E. (2001). A componential theory of Pavlovian conditioning. In *"Handbook of Contemporary Learning Theories"* (R. R. Mowrer and S. B. Klein, Eds.), pp. 23–64. Erlbaum, Mahwah, NJ.

Wainwright, M. L., Zhang, H., Byrne, J. H., and Cleary, L. J. (2002). Localized neuronal outgrowth induced by long-term sensitization training in aplysia. *J. Neurosci.* **22**, 4132–4141.

Walker, D. L., and Davis, M. (2000). Involvement of NMDA receptors within the amygdala in short- versus long-term memory for fear conditioning as assessed with fear-potentiated startle. *Behav. Neurosci.* **114**, 1019–1033.

Weinberger, N. M. (1995). Retuning the brain by fear conditioning. In *"The Cognitive Neurosciences"* (M. S. Gazzaniga, Ed.), pp. 1071–1090. MIT Press, Cambridge, MA.

Weiskrantz, L., and Warrington, E. K. (1979). Conditioning in amnesic patients. *Neuropsychologia* **17**, 187–194.

Weisskopf, M. G., Bauer, E. P., and LeDoux, J. E. (1999). L-Type voltage gated calcium channels mediate NMDA-independent associative long-term potentiation at thalamic input synapses to the amygdala. *J. Neurosci.* **19**, 10512–10519.

Weisz, D. J., Clark, G. A., and Thompson, R. F. (1984). Increased activity of dentate granule cells during nictitating membrane response conditioning in rabbits. *Behav. Brain Res.* **12**, 145–154.

Welsh, J. P., and Harvey, J. A. (1989). Cerebellar lesions and the nictitating membrane reflex: Performance deficits of the conditioned and unconditioned response. *J. Neurosci.* **9**, 299–311.

Whalen, P. J., Rauch, S. L., Etcoff, N. L., McInerney, S. C., Lee, M. B., and Jenike, M. A. (1998). Masked presentation of emotional facial expressions modulate amygdala activity without explicit knowledge. *J. Neurosci.* **18**, 411–418.

Wiesel, T. N., and Hubel, D. H. (1965). Comparison of the effects of unilateral and bilateral eye closure on cortical unit responses in kittens. *J. Neurophysiol.* **28**, 1029–1040.

Wilensky, A. E., Schafe, G. E., and LeDoux, J. E. (1999). Functional inactivation of the amygdala before but not after auditory fear conditioning prevents memory formation. *J. Neurosci.* **19**(RC48), 1–5.

Wilensky, A. E., Schafe, G. E., and LeDoux, J. E. (2000). The amygdala modulates memory consolidation of fear-motivated inhibitory avoidance learning but not classical fear conditioning. *J. Neurosci.* **20**, 7059–7066.

Williams, S., and Johnston, D. (1989). Long-term potentiation of hippocampal mossy fiber synapses is blocked by postsynaptic injection of calcium chelators. *Neuron* **3**, 583–588.

Willshaw, D. J., and von der Malsburg, C. (1976). How patterned neural connections can be set up by self-organization. *Proc. R. Soc. London Sec B* **194**, 431–445.

Wolpe, J., and Rowan, V. C. (1988). Panic disorder: A product of classical conditioning. *Behav. Res. Ther.* **26**, 441–450.

Woodruff-Pak, D. S., Finkbiner, R. G., and Sasse, D. K. (1990). Eyeblink conditioning discriminates Alzheimer's patients from non-demented aged. *NeuroReport* **1**, 45–48.

Woodruff-Pak, D. S., Goldenberg, G., Downey-Lamb, M. M., Boyko, O. B., and Lemieux, S. K. (2000). Cerebellar volume in humans related to magnitude of classical conditioning. *NeuroReport* **11**, 609–615.

Woodruff-Pak, D. S., Romano, S., and Papka, M. (1996). Training to criterion in eyeblink classical conditioning in Alzheimer's disease, Down's syndrome with Alzheimer's disease, and healthy elderly. *Behav. Neurosci.* **110**, 22–29.

Woodruff-Pak, D. S., and Thompson, R. F. (1988). Classical conditioning of the eyeblink response in the delay paradigm in adults aged 18–83 years. *Psychol. Aging* **3**, 219–229.

Woodworth, R. S. (1921). *"Psychology: A Study of Mental Life"*. Holt, New York.

Woody, C. D., Yarowsky, P., Owens, J., Black-Cleworth, P., and Crow, T. (1974). Effect lesions of cortical motor areas on acquisition of conditioned eye blink in the cat. *J. Neurophysiol.* **37**, 385–394.

Woody, C. D., and Yarowsky, P. J. (1972). Conditioned eyeblink using electrical stimulation of coronal-precruciate cortex of cats. *J. Physiol.* **35**, 242–252.

Woolley, C. S., and McEwen, B. S. (1992). Estradiol mediates fluctuation in hippocampal synapse density during the estrous cycle in the adult rat. *J. Neurosci.* **12**, 2549–2554.

Xiang, Z., and Brown, T. H. (1998). Complex synaptic current waveforms evoked in hippocampal pyramidal neurons by extracellular stimulation of dentate gyrus. *J. Neurophysiol.* **79**, 2475–2484.

Xiang, Z., Greenwood, A. C., Kairiss, E. W., and Brown, T. H. (1994). Quantal mechanisms of long-term potentiation in hippocampal mossy-fiber synapses. *J. Neurophysiol* **71**, 2552–2556.

Xu, B., Gottschalk, W., Chow, A., Wilson, R. I., Schnell, E., Zang, K., Wang, D., Nicoll, R. A., Lu, B., and Reichardt, L. F. (2000). The role of brain-derived neurotrophic factor receptors in the mature hippocampus: Modulation of long-term potentiation through a presynaptic mechanism involving TrkB. *J. Neurosci.* **20**, 6888–6897.

Yeo, C. H. (1991). Cerebellum and classical conditioning of motor responses. *Ann. NY Acad. Sc* **627**, 292–304.

Yeo, C. H., Hardiman, M. J., and Glickstein, M. (1985a). Classical conditioning of the nictitating membrane response of the rabbit. I. Lesions of the cerebellar nuclei. *Exp. Brain Res.* **60**, 87–98.

Yeo, C. H., Hardiman, M. J., and Glickstein, M. (1985b). Classical conditioning of the nictitating membrane response of the rabbit. II. Lesions of the cerebellar cortex. *Exp. Brain Res.* **60**, 99–113.

Yeo, C. H., Hardiman, M. J., and Glickstein, M. (1986). Classical conditioning of the nictitating membrane response of the rabbit. IV. Lesions of the inferior olive. *Exp. Brain Res.* **63**, 81–92.

Yin, J. C., Del Vecchio, M., Zhou, H., and Tully, T. (1995). CREB as a memory modulator: Induced expression of a dCREB2 activator isoform enhances long-term memory in *Drosophila. Cell* **81**, 107–115.

Yin, J. C., Wallach, J. S., Del Vecchio, M., Wilder, E. L., Zhou, H., Quinn, W. G., and Tully, T. (1994). Induction of a dominant negative CREB transgene specifically blocks long-term memory in *Drosophila. Cell* **79**, 49–58.

Zador, A., Koch, C., and Brown, T. H. (1990). Biophysical model of a hebbian synapse. *Proc. Nat. Acad. Sci. USA* **87**, 6718–6722.

Zamanillo, D., Sprengel, R., Hvalby, O., Jensen, V., Burnashev, N., Rozov, A., Kaiser, K. M., Koster, H. J., Borchardt, T., Worley, P., Lubke, J., Frotscher, M., Kelly, P. H., Sommer, B., Andersen, P., Seeburg, P. H., and Sakmann, B. (1999). Importance of AMPA receptors for hippocampal synaptic plasticity but not for spatial learning. *Science* **284**, 1805–1811.

Zeigler, H. E. (1900). Theoretisches zur tierpsychologie und vergleichender neurophysiologie. *Biol. Central* **20**.

Zhang, F., Endo, S., Cleary, L. J., Eskin, A., and Byrne, J. H. (1997). Role of transforming growth factor-beta in long-term facilitation in *Aplysia. Science* **275**, 1318–1320.

Zola-Morgan, S. M., and Squire, L. R. (1990). The primate hippocampal formation: Evidence for a time-limited role in memory storage. *Science* **250**, 288–290.

Index

A

AADC, *see* L-Aromatic amino acid decarboxylase

Acetylcholine (ACh)
autoreceptor regulation of storage and release, 271, 273
biosynthesis, 271
history of study, 270–271
inactivation
acetylcholinesterase, 273
reuptake with choline transporter, 274
nontransmitter roles, 275–276
receptors, *see* Muscarinic acetylcholine receptor; Nicotinic acetylcholine receptor
vesicular cholinergic transporter, 271

Acetylcholinesterase (AChE)
acetylcholine degradation, 273
inhibitor applications, 273

ACh, *see* Acetylcholine

AChE, *see* Acetylcholinesterase

Actin, *see* Microfilament

Action potential
activity patterns in central nervous system neurons, 129, 131
ATP consumption in propagation, 76
backpropagation, 156
bursts, 129, 135–136
conductance of sodium and potassium in generation, 121–125
coupling to synaptic vesicle fusion, 219
dendrite effects, *see* Dendrite
initiation studies, 487–488
neuronal ionic current types and functions, 131–132
recording, 122–124, 129
refractory period prevention of reverberation, 125
voltage-clamp studies, 122–124

Action potential propagation
ATP consumption, 76
dendrite backpropagation
functions
conditional axonal output, 489
dendrodendritic inhibition, 488
frequency response, 489
membrane potential resetting, 488
neurotransmitter release, 489
synaptic plasticity, 488–489

synaptic response boosting, 488
overview, 156, 486, 488
myelination effects on speed, 125

Adenylyl cyclase
G protein-coupled receptor signaling
coincidence detection, 344–345
coupling to receptor, 344
inhibitory control, 344
isoforms and differential regulation, 343–344

Adrenergic receptors
classification, 327
pharmacology, 327–328

Afterhyperpolarization, 115, 125

γ-Aminobutyric acid (GABA)
autoreceptor regulation of release, 267–268
biosynthesis, 265, 267
discovery, 265
inactivation
enzymatic degradation, 268
reuptake, 268
inhibitory interneurons, 6, 8
receptors, *see* GABA$_A$; GABA$_B$
transporters, 267–268

Amino-3-hydroxy-5-methylisoxazoleproprionic acid (AMPA) receptors
assembly and subunit diversity, 313–314
long-term depression role, 524–525
long-term potentiation induction, 513–514
RNA splicing and editing of subunits, 314–315
structure, 315

AMPA receptors, *see* Amino-3-hydroxy-5-methylisoxazoleproprionic acid receptors

Amygdala, fear conditioning role, 554–556

Aplysia, 531

AP-1, gene expression regulation, 381–382

L-Aromatic amino acid decarboxylase (AADC)
catecholamine biosynthesis, 255
serotonin biosynthesis, 262

Aspartate, excitatory neurotransmission, 268

Astrocyte
astrocyte–neuron metabolic unit, 87
astrogliosis, 20

blood–brain barrier, 18, 27
calcium signaling, 288
calcium waves, 20
development, 18
energy consumption, 77, 79–80
function, 18–22
gap junctions
calcium flux, 437, 439
connexin expression, 446
syncytium, 446–448
glutamate metabolism, 84–87
glutamate stimulation of glucose uptake, 80–83
glycogen metabolism, 83–84
hepatic encephalopathy role, 85
ion channels, 19
markers, 18–19
neuropathology
brain lesion effects on intercellular coupling, 452
Charcot–Marie–Tooth disease, 450–451
connexin mutation in disease, 434
epilepsy, 448–449
glial tumors, 451–452
protozoan infection effects on intercellular coupling, 453
types, 17–18

ATP, neuron processes in consumption, 76

Augmentation
calcium modulation, 236–237
definition, 235
features, 500

Axoaxonic cell, *see* Chandelier cell

Axon
basket cells, 6
conduction, 35
definition, 33
dendrite interactions, *see* Dendrite
diameters, 33, 35
differentiation, 33
electrotonic spread, *see* Cable theory
electrotonus, history of study, 92
hillock, 2, 33
initial segment
encoding of global neuron output, 484–486
fascicles, 33, 35
microtubules, 50
morphology, 36

Copyright 2004, Elsevier Science (USA).
All rights reserved.

Axon (*continued*)
  presynaptic terminals, 33
  single-axon rule, 480
  spiny stellate cells, 6
Axonal transport
  ATP consumption, 76
  fast, 59, 62
  regulation, 62–63
  slow, 59

B

Basket cell
  axons, 6
  markers, 7
BCM rule, *see* Bienstock–Cooper–Munro
  rule
BDNF, *see* Brain-derived neurotrophic
  factor
Bienstock–Cooper–Munro (BCM) rule,
  526
Bipolar neuron
  cortical distribution, 8
  markers, 8–9
Bistability, 187–189, 402
Blood–brain barrier
  astrocytes, 27
  basement membrane, 26–27
  disruption and edema, 27
  endothelial cell characteristics, 26
  function, 25–26
  glucose transporters, 26
  pericytes, 27
  zonula occludens, 26
Blood supply
  brain vasculature, 22, 24–25
  imaging of blood flow in brain, 71–73
  neuronal activity coupling with blood
    flow, 70–71
Boltzmann equation, 166
Brain-derived neurotrophic factor (BDNF)
  neurotransmitter functions, 293–294
  storage and release, 292–293
  synthesis, 291–292

C

Cable theory
  assumptions, 93–96
  dynamic changes in neuron electrotonic
    structure, 106–109
  electrotonic length, 98
  electrotonic spread
    dendrites
      boundary conditions of
        termination and branching
        effects, 103–104
      synaptic potential modulation by
        electrotonic spread, 104–105
    dependence on characteristic length,
      94, 96
    dependence on diameter, 96–97
    impulse propagation in unmyelinated
      axons, 100–101

propagation comparison, 102
transient signal dependence on
  membrane capacitance, 98–100
history of study, 92
myelination effects on conduction speed,
  101–102
Calcineurin
  regulation, 365
  structure, 364–365
Calcium/calmodulin-dependent protein
  kinase II
  autophosphorylation, 258
  cognitive kinase activity, 361–362
  consensus target sequence, 357
  gene expression regulation, 380
  isoforms, 357–358
  long-term depression role, 524
  long-term potentiation induction,
    513–514
  modeling, 402, 404
  subcellular targeting, 358
  substrates, 357
Calcium channels
  distribution in neurons, 156
  ion selectivity, 152–154
  long-term potentiation induction by
    voltage-dependent channels, 515
  synaptic vesicle exocytosis role, 206
Calcium currents
  high-threshold currents in neurons,
    134–135
  low-threshold currents in neurons,
    135–136
  signal transduction, 134
  types and functions, 131, 134
Calcium flux
  fluorescent dyes for study, 348, 398
  gap junction mediation, 437, 439
  long-term depression role, 524
  long-term potentiation induction, 510
  modeling
    allosteric interactions and Hill
      functions, 400–401
    buffering considerations, 396–398
    discretization of cellular space, 395
    electrodiffusion modeling, 396
    Euler's method, 394–395
    experimental validation, 398
    ordinary differential equations, 394
    passive diffusion, 393
    positive feedback, 394
    symmetry reduction of dimension
      number, 395–396
  signal transduction, 348
  subdomains, 348
Calmodulin
  ion channel modulation, 148
  N-methyl-D-aspartate receptor binding,
    318
  protein–protein interactions, 349
  signal transduction, 348–349
Carbon monoxide (CO)
  mechanism of neuronal effects, 291
  neurotransmission, 291
  synthesis, 291

Catecholamines
  autoreceptor regulation of synthesis and
    release, 257–258
  biosynthesis
    L-aromatic amino acid decarboxylase,
      255
    dopamine β-hydroxylase, 255–256
    overview, 250, 253
    phenylethanolamine
      N-methyltransferase, 256
    tyrosine hydroxylase, 253–255
  functions, 250
  inactivation, 258–259
  neuronal transporters
    dopamine, 259, 261
    norepinephrine, 259, 261
  receptors, *see* Adrenergic receptors;
    Dopamine receptors
  release from vesicles, 257
  storage in vesicles, 256
  structure, 250
  vesicular monoamine transporters,
    256–257
Catechol-*O*-methyltransferase (COMT)
  catecholamine inactivation, 258–259
  inhibitors in neuropsychiatric disorder
    management, 260
Cerebellum
  eyeblink conditioning role, 544, 547–553
  long-term memory traces, 550–552
  memory storage mechanisms, 552–553
Chandelier cell
  cortical distribution, 7, 8
  morphology, 7
Charcot–Marie–Tooth disease (CMT)
  gap junctions in neuropathology, 450–451
  peripheral myelin protein-22 mutations,
    17
ChAT, *see* Choline acetyltransferase
Chloride channels
  inhibitory receptor association, 311–312
  structure, 143–144
Choline acetyltransferase (ChAT),
  acetylcholine synthesis, 271
Chromatin, modification in gene
  expression regulation, 375–376, 378
Classical conditioning
  eyeblink conditioning
    cerebellum role, 544, 547–553
    hippocampus role, 544–547
    stimuli, 544
  fear conditioning
    amygdala role, 554–556
    cellular mechanisms, 557
    hippocampus role, 557–558
    human fear conditioning and anxiety,
      558–560
    medial prefrontal cortex role, 557
    molecular basis, 557–558
    stimuli, 553
Clutch cell
  cortical distribution, 8
  function, 8
CMT, *see* Charcot–Marie–Tooth disease
CO, *see* Carbon monoxide

Coat protein (COP), roles in protein transport, 40
CLSM, *see* Confocal laser scanning microscopy
Cochlear hair cell, features, 10–11
Compartmental models
    history of study, 92
    signal transduction modeling, 407–408
    two-compartment model and signal spread, 99–100
COMT, *see* Catechol-*O*-methyltransferase
Cone, features, 10–11
Confocal laser scanning microscopy (CLSM), combination with whole-cell recording, 506–508
Connexins, *see* Gap junction
COP, *see* Coat protein
CREB, *see* Cyclic AMP response element-binding protein
Cyclic AMP-dependent protein kinase, *see* Protein kinase A
Cyclic AMP response element-binding protein (CREB)
    gene expression regulation, 380–381
    long-term sensitization role in *Aplysia*, 535–536
    modeling of gene network regulation, 416–418
Cytoskeleton
    axonal transport, 59
    interaction of cytoskeletal systems, 54–55
    intermediate filaments, 52–54
    microfilaments, 50–52
    microtubules, 48–50
    motors, 49, 55–58
    protein types, 47

D

DAG, *see* Diacylglycerol
DBH, *see* Dopamine β-hydroxylase
Dendrite
    action potential backpropagation functions
        conditional axonal output, 489
        dendrodendritic inhibition, 488
        frequency response, 489
        membrane potential resetting, 488
        neurotransmitter release, 489
        synaptic plasticity, 488–489
        synaptic response boosting, 488
        overview, 156, 486, 488
    arborization, 1–2, 35
    axonal output control by distal dendrite mechanisms
        membrane resistance, 483
        overview, 482–484
        potassium conductance, 483–484
        synaptic conductance, 483
        voltage-gated depolarizing conductances, 484
    axon effects on processing, 480–481
    axon initial segment encoding of global output, 484–486

complex computations of passive dendritic trees, 481–482
dendrodendritic interactions between axonal cells, 481
depolarizing and hyperpolarizing conductance interactions, 484
electrotonics, *see also* Cable theory
    active conductances and relation to cable properties, 109
    compartmentalization of electrotonic and biochemical properties, 109–110
    nonlinear interactions of synaptic conductances, 108–109
    properties, 110
ion channel distribution, 157
morphology, 1–2, 5, 35–36, 479–480
Nissl substance, 35
plasticity, 35
spines
    microintegrative function, 494–495
    plasticity, 526
    structure, 2–3
subthreshold dendritic activity, 481
voltage-gated computations, 485
voltage-gated channels in dendritic integration
    medium spiny cells, 492
    Purkinje cells, 490
    pyramidal neurons, 490–492
    multiple impulse initiation site dynamic control, 492–494
    information processing overview, 479–480, 495–496
Depression, catecholamine-degrading enzyme inhibitors in treatment, 260
Diacylglycerol (DAG)
    signal transduction, 336, 345–346
    sources, 345–346
DNase I, actin binding, 52
Dopamine, *see* Catecholamines
Dopamine β-hydroxylase (DBH), catecholamine biosynthesis, 255–256
Dopamine receptors
    classification, 328
    pharmacology, 328
Double bouquet cell
    classes, 8
    cortical distribution, 8
Dynamin, synaptic vesicle recycling, 219
Dynein
    ATPase inhibitor sensitivity, 55
    microtubule association, 57
Dystrophin, 51

E

Electrotonic spread, *see* Cable theory
Enteric plexus
    neurons, 11–12
    neurotransmitters, 12
Ependymoglial cell, gap junctions, 446
Epilepsy, gap junctions in neuropathology, 448–449

Epinephrine, *see* Catecholamines
Excitable membrane modeling, *see* Membrane excitability models
Eyeblink conditioning, *see* Classical conditioning

F

Facilitation, *see* Synaptic facilitation
Fear conditioning, *see* Classical conditioning
FitzHugh–Nagumo model, 192
Fluorescence recovery after photobleaching (FRAP), 398
Flux control coefficient (FCC), 412–413
fMRI, *see* Functional magnetic resonance imaging
Fos, gene expression regulation, 382
FRAP, *see* Fluorescence recovery after photobleaching
Functional magnetic resonance imaging (fMRI), principles, 71, 73

G

GABA, *see* γ-Aminobutyric acid
GABA$_A$
    chloride channel, 311
    glycine receptor homology, 312
    pharmacology, 311–312, 469
    structure, 310–311
GABA$_B$
    pharmacology, 469
    types and functions, 330
GAD, *see* Glutamic acid decarboxylase
Gap junction
    astrocyte syncytium, 446–448
    calcium flux mediation, 437, 439
    chemical transmission comparison
        directionality, 435
        plasticity, 437
        speed, 435–436
        synaptic inhibition, 436
    connexins
        developmental regulation of expression, 444
        gap junction composition effects on properties, 441–442
        homology, 435
        mutation in disease, 434
        structure, 435
        types, 434–435, 439
    definition, 432–433
    gating
        ligands, 441
        voltage, 440–441
    glial cell connectivity
        astrocytes, 445
        ependymoglial cells, 446
        leptomeningeal cells, 445
        oligodendrocytes, 445
        Schwann cells, 445
    history of study, 431–432
    neuronal connectivity, 442–444
    permeability, 437

Gap junction (*continued*)
  phosphorylation, 441
  selective affinity of connexons, 442
  structure, 433–435
Gelsolin, 51–52
Gene expression regulation
  chromatin modification, 375–376, 378
  cis versus *trans*-regulating factors,
    371–372
  co-activators, 375–376, 378
  co-repressors, 375–376, 378
  cytokines in regulation, 383
  growth factors in regulation, 383
  neuron-specific expression, 378
  nuclear receptor mechanisms, 383–386
  posttranscriptional regulation, 386, 388
  promoters
    binding proteins, 372
    structure, 372, 374
  signal-regulated transcription
    AP-1, 381–382
    calcium-calmodulin-dependent protein
      kinase, 380
    CREB, 380–381
    Fos, 382
    Jun, 382–383
    mitogen-activated protein kinase,
      382–393
    nuclear factor-κB, 380
    overview, 371–372, 378–380
  transcription factors
    dimerization, 374–375
    domains, 374–375
  transcription overview, 372
Glial cells, *see* Neuroglia
Glucose
  brain cell-specific uptake and
    metabolism, 79–80
  brain utilization rate, 67–68
  compartmentalization of uptake, 82–83
  glutamate stimulation of uptake in
    astrocytes, 80–83
  imaging of metabolism in brain, 71–73,
    80
  metabolism, 67–68, 73–75
Glucose transporters (GLUTs)
  blood–brain barrier, 26
  types in brain, 78–79
Glutamate
  astrocyte metabolism, 84–87
  biosynthesis, 2688–269
  excitatory neurotransmission, 268
  inactivation, 270
  release regulation, 270
  reuptake, 84
  stimulation of glucose uptake in
    astrocytes, 80–83
  vesicular transporter, 269–270
Glutamate receptors
  assembly and subunit diversity, 313–314,
    317–318
  classification, 312
  evolutionary relationships, 312–313
  history of study, 313

ionotropic receptors, *see* Amino-3-
    hydroxy-5-methylisoxazole-
    proprionic acid receptors; Kainate
    receptors; *N*-Methyl-D-aspartate
    receptors
  metabotropic receptors
    long-term potentiation induction,
      515–516
    overview, 329–330
  RNA splicing and editing of subunits,
    314–315, 317
  structure, 315
Glutamic acid decarboxylase (GAD),
    γ-aminobutyric acid biosynthesis,
    267
GLUTs, *see* Glucose transporters
Glycine receptor
  chloride channel, 312
  GABA$_A$ homology, 312
Glycogen
  astrocyte stores, 83
  metabolism coupling with neuronal
    activity, 83
  neurotransmitter regulation of
    metabolism in astrocytes, 83–84
Glycolysis, 67–68, 73
Goldman–Hodgkin–Katz equation, 120
Golgi apparatus
  cis-Golgi network, 39
  coat protein roles in protein transport,
    40
  endocytosis and membrane cycling,
    41–42
  trans-Golgi network, 39, 41–43
  vesicle fusion, 40
GPCR, *see* G protein-coupled receptor
G protein
  GTPase activity, 337–338, 343
  ion channel modulation, 148–149
  subunit functions in coupled receptor
    activation
    α-subunit, 338, 341
    βγ-subunits, 341
G protein-coupled receptor (GPCR), *see also*
    specific receptors
  activation, 318–319, 321–322
  desensitization
    downregulation, 325–326
    overview, 323
    phosphorylative modulation, 324–325
    sequestration, 325–326
  disulfide bonds, 326
  G protein coupling
    activation transduction sites, 321–322
    neurotransmitter affinity effects,
      322–323
    sites, 321
    specificity and potency of activation,
      323
  glycosylation, 326
  ionotropic receptor interactions, 326
  metabotropic glutamate receptors,
    329–330
  neurotransmitter binding

conformational change and G protein
    activation, 320–321
  site localization, 320
  oligomerization, 322
  peptide receptors, 330–331
  signal transduction
    adenylyl cyclase fine-tuning of cyclic
      AMP, 343–345
    advantages in neurotransmission,
      352–353
    calcium, 348
    complexity, 339–340
    experimental manipulation, 341–342
    G protein cycle, 337–338
    guanylyl cyclases, 349–351
    ion channel modulation, 351–352
    overview, 335–337
    phosphodiesterases, 351
    phospholipids, 345–348
    requirements, 335
    response specificity, 342–343
    structure, 319–320, 322, 326, 331
Guanylate synthase
  nitric oxide modulation, 289–290,
    349–351
  types, 351

H

Hair cell, features, 10–11
Hebbian rule, long-term potentiation,
    508–510, 521
Hepatic encephalopathy, metabolic
    features, 85
Hill function, allosteric interaction
    modeling, 400–401
Hindmarsh–Rose model, 193
Hippocampus
  brain slice studies in learning and
    memory advantages, 504–505
  Schaeffer collateral/commissural
    synapses, 505
  eyeblink conditioning role, 544–547
  fear conditioning role, 557–558
Hodgkin–Huxley model
  action potential simulation, 391–392
  complexity of equations, 179
  components of membrane current,
    162–165
  computation of equation solutions,
    174–175
  equivalent electrical circuit, 161–162
  instantaneous current–voltage
    relationship, 164–165
  limitations, 177
  parameter modification in equations, 172
  patch-clamped membranes, 178
  potassium conductance characterization,
    172–173
  simplification, 411
  simulations, 173, 175–177
  sodium conductance characterization,
    169–172

time and voltage dependency of ion conductances, 165–169
total membrane current, 173
two-dimensional reduction, 180, 182
voltage-clamp studies, 122–124, 161, 179
Hyperpolarization
definition, 115
dendrite depolarizing and hyperpolarizing conductance interactions, 484
ion flux, 118–119
ionic current activation in rhythmic activity, 137
membrane potential ion distribution, 118–119
Inositol trisphosphate (IP3)
calcium mobilization, 347
signal transduction, 336, 345, 347
termination of signal, 347–348

I

Intermediate filament
glutamate region, 53–54
interaction of cytoskeletal systems, 54–55
neurofilament triplet proteins, 53–54
packing, 53
pathology, 54
structure, 52
types, 52–54
Interneuron, *see* Basket cell; Bipolar neuron; Chandelier cell; Clutch cell; Double bouquet cell; Spiny stellate cell
Ion channels, *see also specific channels*
calcium modulation, 147–148
classification, 141, 143–144
distribution in neurons, 154–157
functional overlap, 144
G protein modulation, 351–352
inactivation mechanisms, 126, 129
ion binding sites, 150–152
ion selectivity, 141, 149–154
membrane potential maintenance, 120–121
metabolic modulation, 149
multiple ions in a pore, 150–152
mutations in disease, 127–128
neurotransmitter modulation, 148–149
noise in quantal analysis, 226
second messenger integration of signaling, 149
voltage gating
functions, 144
inactivation types
C-type, 147
N-type, 146
P-type, 146–147
S4 segment as voltage sensor, 144–146
Ionotropic receptors, types, 299–300
IP3, *see* Inositol trisphosphate

J

Jellyfish, ion channel studies, 135
Jun, gene expression regulation, 382–383

K

Kainate receptors
assembly and subunit diversity, 313–314
functions, 315–316
RNA splicing and editing of subunits, 314–315
structure, 315
Katz model, *see* Quantal analysis
Ketone bodies, brain utilization for energy, 68–69
Kinesin
discovery, 55
mechanism of action, 55–56
structure, 55

L

Learning and memory
Aplysia studies
activity-dependent neuromodulation, 537–538
advantages, 531
intermediate-term sensitization, 536–537
long-term sensitization, 535
operant conditioning of feeding behavior, 539–540
short-term sensitization, 533–535
siphon–gill withdrawal reflexes, 531–533
tail–siphon withdrawal reflexes, 532–533
arthropod species for study, 542
associative learning, 529
Caenorhabditis elegans studies, 543
classical conditioning studies, *see* Classical conditioning
gastropod mollusk species for study, 541–542
leech studies, 542
mechanisms, *see* Long-term depression; Long-term potentiation
nonassociative learning, 530
prospects for study, 562–563
synaptic strength change and complex memory storage, 560–562
Leptomeningeal cell, gap junctions, 445
Long-term depression (LTD)
definition, 521
depotentiation, 524
functions
long-term potentiation saturation problem at excitatory synapses, 521–523
synaptic encoding in inhibitory neurons, 523
mechanisms

amino-3-hydroxy-5-methylisoxazole-proprionic acid receptor, 524–525
calcium flux, 524
cerebellar cortex, 525
N-methyl-D-aspartate receptor, 524–525
telecephalon, 524–525
overview, 499–500
Long-term potentiation (LTP)
associativity, 501–503
Bienstock–Cooper–Munro rule, 526
brain slice studies
confocal laser scanning microscopy with whole-cell recording, 506–508
hippocampus
advantages, 504–505
Schaeffer collateral/commissural synapses, 505
mossy fibers in CA3, 505
novel preparations, 506
optical studies, 506
origins, 503–504
brain-derived neurotrophic factor role, 294
cooperativity, 501
definition, 235
expression
postsynaptic changes
glutamate receptor responses, 519–520
unsilencing of synapses, 519–520
presynaptic changes
electron microscopy studies, 519
neurotransmitter release, 516–518
pharmacological analysis, 518–519
Hebbian mechanism of synaptic plasticity, 508–510
history of study, 499–500, 503–504
induction mechanisms
calcium, 510
glutamate receptors, 510
metabotropic glutamate receptors, 515–516
N-methyl-D-aspartate receptor induction and modeling, 510–514
N-methyl-D-aspartate receptor-independent induction, 514–515
voltage-dependent calcium channels, 515
maintenance, gene expression and protein synthesis, 520–521
neuropharmacological linkages with information storage, 527
persistence, 500–501
rapid induction, 500–501
synapse specificity, 503
transgenic mouse studies, 527–528
LTD, *see* Long-term depression
LTP, *see* Long-term potentiation
Lysosome
functions, 43
hereditary diseases, 43
protein trafficking, 43–44

M

mAChR, Muscarinic acetylcholine receptor
Mannose, brain utilization for energy, 69
MAO, *see* Monoamine oxidase
MAPK, *see* Mitogen-activated protein kinase
MCA, *see* Metabolic control analysis
Medium-sized spiny cell
  morphology, 9
  voltage-gated channels in dendritic integration, 492
Membrane excitability models, *see also* Hodgkin–Huxley model
  FitzHugh–Nagumo model, 192
  geometric analysis
    action potential trajectory in the phase plane, 182, 185–187
    advantages, 179–180
    bifurcation, 187–188
    bistability, 187–189
    bursting activity, dynamical underpinnings, 189–191, 193–194
    nonlinear dynamical systems, fixed points and bifurcations, 183–185
    two-dimensional reduction of Hodgkin–Huxley model, 180, 182
  Hindmarsh–Rose model, 193
  Morris–Lecar model, 192–193
Membrane potential
  ion distribution
    concentrations and equilibrium potentials in neuron systems, 117
    differential distribution, 116
    hyperpolarization versus depolarization, 118–119
    ion pump maintenance, 120–121
    passive distribution, 116–118
  Nernst equation, 118
  resting membrane potential
    Goldman–Hodgkin–Katz equation, 120
    ion contributions, 119
    neuron type specificity, 120
Memory, *see* Learning and memory
MET, *see* Morpho-electrtonic transform
Metabolic control analysis (MCA), 412–413, 426
N-Methyl-D-aspartate (NMDA) receptors
  assembly and subunit diversity, 317–318
  calmodulin binding, 318
  gating, 316
  long-term depression role, 524–525
  long-term potentiation induction induction and modeling, 510–514
  pharmacology, 316
  RNA splicing of subunits, 317
  structure, 315
  subunit homology with other glutamate receptors, 316
Michaelis–Menten equation, 399, 411
Microfilament
  actin genes, 50–51
  capping proteins, 51
  functions, 50
  interacting proteins, 51–52

interaction of cytoskeletal systems, 54–55
structure, 47
Microglia
  development, 21
  immune response mediation, 21
  morphology, 21
  reactive microglia, 21–22
Microtubule
  associated proteins, 49–50, 57
  axons, 50
  dynamics, 50
  functions, 48
  interaction of cytoskeletal systems, 54–55
  motor proteins, 49, 55–58
  structure, 47–48
  tubulin modifications, 48
Mitochondria, protein targeting, 44–45
Mitogen-activated protein kinase (MAPK)
  gene expression regulation, 382–393
  modeling of cascade, 404–405, 414
Monoamine oxidase (MAO)
  catecholamine inactivation, 258–259
  inhibitors in neuropsychiatric disorder management, 260
Monod–Wyman–Changeaux allosteric theory, 400
Morpho-electrotonic transform (MET), 106 also
Morris–Lecar model, 192–193
Muscarinic acetylcholine receptor (mACHR)
  functions, 326–327
  types, 327
Myelin sheath
  action potential propagation speed effects, 125
  composition, 15
  internodes, 14
  myelination program in development, 13
  nerve conduction facilitation, 13, 101–102
  P_0 function, 15–16
  peripheral myelin protein-22 mutations in Charcot–Marie–Tooth disease, 17
Myosin
  ATPase inhibitor sensitivity, 55
  structure
    myosin I, 58
    myosin II, 57–58
  types, 57–58

N

nAChR, *see* Nicotinic acetylcholine receptor
Nernst equation, 118
Nerve growth factor (NGF)
  neurotransmitter functions, 293–294
  storage and release, 292–293
  synthesis, 292
Neuroglia, *see also* Astrocyte; Microglia; Myelin sheath; Oligodendrocyte; Schwann cell
  cytoskeleton, *see* Cytoskeleton
  definition, 13

energy consumption, 77
gap junctions
  astrocytes, 445
  ependymoglial cells, 446
  leptomeningeal cells, 445
  oligodendrocytes, 445
  Schwann cells, 445
  tumors, 451–452
myelin synthesis, 13–19
Neuromuscular junction
  active zone, 218
  end plates, 198
  neurotransmitter release, 199–200
  quantal analysis, *see* Quantal analysis
Neuron
  astrocyte–neuron metabolic unit, 87
  cytoskeleton, *see* Cytoskeleton
  functions, 91
  information processing theory, 91–92
  interneurons, *see* Interneuron
  modeling, *see also* Cable theory; Compartmental models
    relating passive and active potentials, 111
    Web site resources, 93
  morphology, 1–2
  protein synthesis, *see* Protein synthesis
  pyramidal cell, *see* Pyramidal neuron
  spiny stellate cell, *see* Spiny stellate cell
  subcortical neurons
    medium-sized spiny cells, 9
    Purkinje cells, 10
    spinal motor neurons, 10
    substabtia nigra dopaminergic neurons, 9
Neuropeptides
  coexistence with classic neurotransmitters, 286–287
  gene, 285
  inactivation, 286
  receptor, 286
  storage and release, 286
  synthesis, 285–286
Neurotransmitter, *see also specific transmitters*
  classic neurotransmitters, 250
  definitive criteria, 246, 248, 295
  history of study, 245–246
  nontransmitter roles, 275–276
  peptide transmitters, *see also* Neuropeptides
    experimental approaches for study, 283, 285
    inactivation, 283–284
    precursor processing, 282
    synthesis and storage, 281, 283
  rationale for multiple transmitters
    afferent convergence on a common neuron, 274, 279–280
    colocalization of neurotransmitters, 274–275, 280
    fast versus slow responses of target neurons, 275, 281
    release from different neuron processes, 275, 280

synaptic specializations versus nonjunctional appositions between neurons, 275, 280–281
receptors, *see specific receptors*
steps in neurotransmission
biosynthesis, 248
inactivation and termination of action, 249
peptides versus classic neurotransmitters, 281, 283
receptor binding, 248–249
release, 248
storage, 248
Neurotransmitter release, *see also* Quantal analysis; Synaptic vesicle
calcium in synapse exocytosis
binding features
high-affinity binding for slow transmitter secretion, 207–208
low-affinity binding for triggering, 206–207
microdomains in release, 204, 206
mobilization of vesicles to docking sites, 207
slow and fast transmitter corelease from same neuron terminal, 208
triggering, 203–204
catecholamines, 257–258
long-term potentiation expression, 516–518
quantal release, 197–199
rapidity, 209
synaptic membrane proteins in fusion
docking and priming of vesicles, 217–218
genetic screening, 212, 214
identification from purified vesicles, 211–212
NSF, 217
Rab3, 217
SNAP-25, 214, 216–217
SNARE complex, 214–217
synapsin, 218
synaptobrevin, 214–215, 217
synaptotagmin, 219
syntaxin, 214
topology, 209–210
types and functions, 211
Neurotrophins
neurotransmitter functions, 293–294
storage and release, 292–293
synthesis, 292
types, 291
NF-κB, *see* Nuclear factor-kB
NGF, *see* Nerve growth factor
Nicotinic acetylcholine receptor (nAChR)
assembly, 307
desensitization
history of study, 300–301
phosphorylation, 307–308
structure
acetylcholine binding sites, 305
architecture, 301
conformational change on opening, 305–306

ion selectivity and current flow relationships, 303
ionotropic receptor homology, 309–310
membrane-spanning segments, 301
neuromuscular junction versus *Torpedo* receptors, 306
neuronal subunit types and diversity, 309–310
Xenopus oocyte expression studies, 308
Nissl substance, 31–32, 35
Nitric oxide (NO)
guanylate synthase modulation, 289–290, 349–35
mechanism of neuronal effects, 289–290
modeling of synthesis, 413
neuronal activity coupling with blood flow, 70–71
neurotransmission, 287, 289
regulation of levels, 289
stability and inactivation, 289
synthesis, 289, 350
NMDA receptors, *see* N-Methyl-D-aspartate receptors
NO, *see* Nitric oxide
Norepinephrine, *see* Catecholamines
NSF, 40, 217
Nuclear factor-κB (NF-κB), gene expression regulation, 380

O

Oligodendrocyte
gap junctions, 445
myelin synthesis, 13–14

P

P_0, function, 15–16
Parkinson's disease (PD), catecholamine-degrading enzyme inhibitors in treatment, 260
Patch-clamp, 178, 461–462
PD, *see* Parkinson's disease
PDEs, *see* Phosphodiesterases
Pentose phosphate pathway, 74–75
Perikaryon
cytoskeleton, 2
Nissl substance, 31–32
nuclei, 31
translational cytoplasm, 31
Peripheral myelin protein-22 (PMP22), mutations in Charcot–Marie–Tooth disease, 17
Peroxisome, protein targeting, 45–46
PFK, *see* Phosphofructokinase
Phenylethanolamine N-methyltransferase (PNMT), catecholamine biosynthesis, 256
Phenylketonuria (PKU), features, 252–253
Phosphodiesterases (PDEs)
cyclic GMP phosphodiesterase, 351
functions, 343
Phosphofructokinase (PFK), modeling, 404

Phosphoinositols, signal transduction, 347
Phospholipase C (PLC), G protein coupling, 345–346
Phospholipase D (PLD), signal transduction, 345–346
PKA, *see* Protein kinase A
PKC, *see* Protein kinase C
PKU, *see* Phenylketonuria
PLC, *see* Phospholipase C
PLD, *see* Phospholipase D
PMP22, *see* Peripheral myelin protein-22
PNMT, *see* Phenylethanolamine N-methyltransferase
POMC, *see* Proopiomelanocortin
Positron emission tomography (PET), principles, 71–72, 82
Postsynaptic potentials (PSPs)
classification, 459
excitatory versus inhibitory postsynaptic potentials, 197
integration
recruitment, 475
spatial summation, 476
temporal summation, 475–476
ionotropic postsynaptic potentials
dual-component potentials, 470–471
gating and kinetic analysis, 462–465
inhibitory postsynaptic potentials, 469–469
nonlinear current–voltage relationships, 467–468
null potential and slope of current–voltage relationships, 465–466
patch-clamp studies, 461–462
stretch reflex, 459–460
summation of single-channel currents, 466–467
metabotropic postsynaptic potentials, 472–473
Posttetanic potentiation (PTP)
definition, 235
features, 500
Potassium channels
ATP-sensitive channels, 149
calcium activation, 148
distribution in neurons, 154, 156
inwardly rectifying, 143
two-pore channels, 143
voltage gating, S4 segment as voltage sensor, 145–146
Potassium conductance, dendrites, 483–484
Potassium currents
Hodgkin–Huxley model, 172–173
types and functions, 131–132
voltage sensitivity and kinetics, 132–134
PP-1, *see* Protein phosphatase-1
Promoter
binding proteins, 372
structure, 372, 374
Proopiomelanocortin (POMC), processing, 282
Protein kinase A (PKA)
cognitive kinase activity, 360–361

Protein kinase A (PKA) (*continued*)
  consensus target sequence, 356–357
  history of study, 355
  regulation, 355–356
  signal transduction, 336, 368
  spatial localization in regulation, 359–360
  structure, 355–357
  types, 357
Protein kinase C (PKC)
  activation, 359
  cognitive kinase activity, 362
  isoforms, 358–359
  signal transduction, 358, 368
  spatial localization in regulation, 359–360
  translocation, 359
Protein kinases, *see also specific kinases*
  catalytic reaction, 353
  cognitive kinases, 360–362
  criteria for study, 367–368
  functional overview, 353–355
  long-term depression role, 524–525
  long-term potentiation induction,
      513–514
  regulation, 354
  signal transduction network integration,
      365, 367
  spatial localization in regulation, 359–360
  structural homology, 355–356
  substrate specificity, 354
  tyrosine kinases in cell growth and
      differentiation, 362–363
Protein phosphatase-1 (PP-1)
  regulation, 364
  structure, 364
Protein phosphatases, *see also specific*
    phosphatases
  criteria for study, 367–368
  functional overview, 353–355
  regulation, 354
  signal transduction network integration,
      365, 367
  spatial localization in regulation, 359–360
  substrate specificity, 354
  types, 354, 363–364
Protein synthesis
  cotranslational modifications, 39
  cytoplasmic protein
      compartmentalization, 46–47
  Golgi transport
      cis-Golgi network, 39
      coat protein roles in protein transport,
          40
      endocytosis and membrane cycling,
          41–42
      trans-Golgi network, 39, 41–43
      vesicle fusion, 40
  integral membrane proteins, 37–38
  peripheral membrane protein synthesis
      and targeting, 37
  rough endoplasmic reticulum, 37–39, 47
  secretory protein pathway, 37–38
  signal sequences, 38–39
  transcriptional regulation, *see* Gene
      expression regulation

Pseudounipolar sensory neuron, dorsal
    root ganglia, 35, 36
PSP, *see* Postsynaptic potentials
PTP, *see* Posttetanic potentiation
Purinergic receptors, types and functions,
    312, 328–329
Purkinje cell
  morphology, 10
  voltage-gated channels in dendritic
      integration, 490
Pyramidal neuron
  excitatory inputs, 5
  excitatory outputs, 5
  morphology, 3–5
  neocortical layer features, 4–5
  voltage-gated channels in dendritic
      integration, 490–492

Q

Quantal analysis
  binomial models, 223, 229–230
  central nervous system synapses versus
      neuromuscular junction
      nonuniform release probabilities, 227
      one-quantum release, 226
      receptor sampling of different quantal
          contents, 227
      spontaneous miniature postsynaptic
          signals, 227–229
  coefficient of variation, 232–233
  long-term potentiation expression,
      516–518
  Poisson model, 223, 229
  quantal parameter estimation from
      evoked and spontaneous signals
      confidence intervals, 232
      model discrimination, 230
      noise deconvolution, 230
      overview, 227–228
      spontaneous miniature postsynaptic
          signals, 228–229
      Q, 229
  quantal parameter estimation from
      experimental manipulation of
      release probability, 233–235
  rationale, 221
  standard Katz model
      limiting considerations
      ion channel noise, 226
      postsynaptic summation and signal
          distortion, 226
      quantal uniformity, 224–226
      rapid and synchronous transmitter
          release, 226
      stationarity, 226
      uniform and independent release
          probabilities, 226
  overview, 221–222
  quantal parameters
      n, 223–224, 227
      p, 224, 227
      Q, 222, 227, 229

R

Rab
  Rab3, 217
  SNARE regulation, 40
Retina, photoreceptors, 10–11
RNA polymerase II, 372
Rod, features, 10–11

S

Saltatory conduction, 14, 125
Schizophrenia, catecholamine-degrading
    enzyme inhibitors in treatment, 260
Schwann cell
  gap junctions, 445
  myelin synthesis, 15
Serotonin
  autoreceptor regulation of synthesis and
      release, 264
  biosynthesis
      alternative tryptophan metabolic
          pathways, 262, 264
      L-aromatic amino acid decarboxylase,
          262
      overview, 262
      tryptophan hydroxylase, 262
  cell distribution, 261
  discovery, 261
  inactivation
      enzymatic degradation, 265
      reuptake, 264–265
  receptors
      classification, 329
      5-HT$_3$ structure, 310
  release from vesicles, 264
  short-term sensitization role in *Aplysia*,
      533, 537
  storage in vesicles, 264
  vesicular monoamine transporters, 264
shiverer mouse, 16–17
Signal transduction
  G protein-coupled receptors
      requirements, 335
      adenylyl cyclase fine-tuning of cyclic
          AMP, 343–345
      advantages in neurotransmission,
          352–353
      calcium, 348
      complexity, 339–340
      experimental manipulation, 341–342
      G protein cycle, 337–338
      guanylyl cyclases, 349–351
      ion channel modulation, 351–352
      overview, 335–337
      phosphodiesterases, 351
      phospholipids, 345–348
      response specificity, 342–343
  gene expression regulation
      AP-1, 381–382
      calcium-calmodulin-dependent
          protein kinase, 380
      CREB, 380–381
      Fos, 382

Jun, 382–383
  mitogen-activated protein kinase,
    382–393
  nuclear factor-κB, 380
  overview, 371–372, 378–380
modeling
  allosteric interactions and Hill
    functions, 400–401
  bifurcation analysis, 402, 409
  bistability, 402
  cross-talk between pathways, 405–408
  enzymatic reactions, 399–400
  feedback interactions, 392, 400–401,
    403–404
  gene network modeling
    continuous method, 419–421
    feedback loops, 421–425
    logical-network method, 418–419
    overview, 416–418
    stochastic fluctuations, 425)
  intracellular transport of signaling
    molecules
    active transport considerations, 399
    buffering considerations, 396–398
    discretization of cellular space, 395
    electrodiffusion modeling, 396
    Euler's method, 394–395
    experimental validation, 398
    ordinary differential equations, 394
    passive diffusion, 393
    positive feedback, 394
    symmetry reduction of dimension
      number, 395–396
  key control parameters, 410
  levels of detail, 392
  linear versus nonlinear systems, 391
  metabolic control analysis, 412–413, 426
  multienzyme complexes, 413–414
  persistent oscillations in pathways,
    401–402
  quantitative versus qualitative models,
    392, 409–410
  random variability and stochastic
    fluctuations, 414–416, 425
  sensitivity of model behavior to
    parameter changes, 409
  separation of fast and slow processes,
    411
  ultrasensitivity, 404–405
SNAP, 40–41
SNAP-25, 214, 216–217, 220
SNARE complex, 40–41, 214–217, 220
Sodium channels
  ion selectivity, 152–154
  structure, 126
  toxins in electrophysiology studies, 126
  voltage gating, S4 segment as voltage
    sensor, 144–145
Sodium currents
  Hodgkin–Huxley model, 169–172
  transient versus persistent, 132
  types and functions, 131

Sodium/potassium-ATPase
  activation and glucose utilization, 82
  astrocytes, 81
Spectrin, 51
Spinal motor neuron
  pathology, 10
  types, 10
Spiny stellate cell
  axons, 6
  cortical distribution, 5–6
  dendrites, 5
  functions, 12–13
Steroid hormones
  neurosteroid metabolites, 294–295
  neurotransmission, 295
  synthesis in brain, 294–295
Substantia nigra, dopaminergic neurons, 9
Synapse
  asymmetric versus symmetric, 1
  chemical synapse organization, 197
  discovery, 245
  fusion, see Neurotransmitter release
  plasticity
    experience effects, 526–527
    Hebbian mechanism, 508–510
    hormonal modification, 526–527
    learning and memory, 525–527
    metaplasticity, 526–527
    short-term synaptic plasticity, 235–238
Synapsin, synaptic vesicle targeting, 63, 218
Synaptic depression
  definition, 235
  mechanisms, 235–236
Synaptic facilitation
  calcium modulation, 236–237
  definition, 235
  features, 500
Synaptic potential
  generation and ion channel distribution,
    157
  integration, see Postsynaptic potentials
Synaptic vesicle, see also Neurotransmitter
    release
  action potential coupling to fusion, 219
  calcium in exocytosis
    binding features
      high-affinity binding for slow
        transmitter secretion, 207–208
      low-affinity binding for triggering,
        206–207
    microdomains in release, 204, 206
    mobilization of vesicles to docking
      sites, 207
    slow and fast transmitter corelease
      from same neuron terminal, 208
  membrane proteins in fusion
    docking and priming of vesicles,
      217–218
    genetic screening, 212, 214
    identification from purified vesicles,
      211–212
    NSF, 217

Rab3, 217
    SNAP-25, 214, 216–217
    SNARE complex, 214–217
    synapsin, 218
    synaptobrevin, 214–215, 217
    synaptotagmin, 219
    syntaxin, 214
    topology, 209–210
    types and functions, 211
  neurotransmitter packaging, 219–220
  quantal analysis, see Quantal analysis
  quantal release of neurotransmitters,
    197–199
  recycling
    endocytosis, 220
    histological tracers in study, 203
    steps, 201–203
    trafficking cycle, 208–209
    transmitter cycle, 200, 202
    vesicle membrane cycle, 200
Synaptobrevin, 214–215, 217, 220
Synaptotagmin, 219
Syntaxin, 214

T

TCA cycle, see Tricarboxylic acid cycle
TH, see Tyrosine hydroxylase
Transcription, see Gene expression
    regulation
Tricarboxylic acid (TCA) cycle, 67–68,
    73–74
Tryptophan hydroxylase, serotonin
    biosynthesis, 262
Tubulin, see Microtubule
Tyrosine hydroxylase (TH), catecholamine
    biosynthesis, 253–255

V

VAChT, see Vesicular cholinergic
    transporter
Vesicular cholinergic transporter (VAChT),
    271
Vesicular monoamine transporters
    (VMATs), 256–257, 264
VMATs, see Vesicular monoamine
    transporters
Voltage-clamp, see also Hodgkin–Huxley
    model
  action potential studies, 122–124
  principles, 122

W

Wernicke–Korsakoff syndrome, features, 76

Z

Zonula occludens, 26